Advances in Modal Logic

Volume 11

Advances in Modal Logic
Volume 11

Edited by

Lev Beklemishev

Stéphane Demri

and

András Máté

ISBN 978-1-84890-201-5

College Publications
Scientific Director: Dov Gabbay
Managing Director: Jane Spurr

http://www.collegepublications.co.uk

Printed by Lightning Source, Milton Keynes, UK

Contents

Preface

Advances in Modal Logic (AiML) is an initiative founded in 1995 and aimed at presenting an up-to-date picture of the state of the art in modal logic and its many applications. It consists of a conference series together with volumes based on the conferences. The conference is the main international forum at which research on all aspects of modal logic is presented. The first one was held in 1996 in Berlin, Germany, and since then it has been organised biennially, with meetings in 1998 in Uppsala, Sweden; in 2000 in Leipzig, Germany (jointly with ICTL-2000); in 2002 in Toulouse, France; in 2004 in Manchester, UK; in 2006 in Noosa, Australia; in 2008 in Nancy, France; in 2010 in Moscow, Russia; in 2012 in Copenhagen, Denmark and in 2014 in Groningen, The Netherlands. Information about AiML and related events, including conference proceedings, is available at the website `www.aiml.net`.

The eleventh conference in the AiML series was organised by András Máté (Eötvös University, Budapest) with the assistance of Tamás Bitai, Réka Markovich, Péter Mekis, Attila Molnár, Edi Pavlovic and Gergely Székely. It was held on from August 30th to September 2nd, 2016 at the Central European University (CEU) in Budapest, Hungary. The conference web page can be found at `http://phil.elte.hu/aiml2016/`.

This volume contains invited and contributed papers from the conference. The conference included invited lectures from the following people:

- Guram Bezhanishvili (New Mexico State University),
- Sonja Smets (ILLC, Universiteit van Amsterdam),
- Yde Venema (ILLC, Universiteit van Amsterdam),

and a special CEU talk by Hanoch ben Yami (CEU, Budapest).

A special session in memoriam of Professor Alexander V. Chagrov has been organized by Valentin Shehtman and Michael Zakharyaschev and an invited talk was given by

- Valentin Goranko (Stockholm University).

The Programme Committee received 53 regular paper submissions. Of these, 30 were selected for this volume by a reviewing process where every paper received three independent expert reviews. The volume includes papers on general problems in proof theory and algorithmic properties of modal logics, on systems for temporal and epistemic reasoning, on related kinds of logics - public announcement, dynamic, paraconsistent, substructural, intuitionistic, and on related topics in algebraic logic. In addition, there were 35 submissions for short presentations at the conference, and 29 were accepted for presentation.

Here are the members of the Programme Committee for the conference:

- Natasha Alechina (University of Nottingham, UK)

- Carlos Areces (FaMAF, Universidad Nacional de Córdoba, Argentina)
- Philippe Balbiani (IRIT, Toulouse, France)
- Alexandru Baltag (ILLC, University of Amsterdam, The Netherlands)
- Lev Beklemishev (co-chair, Steklov Mathematical Institute of RAS, Moscow M.V. Lomonosov State University National Research University Higher School of Economics, Russia)
- Thomas Bolander (Technical University of Denmark)
- Torben Braüner (Roskilde University, Denmark)
- Serenella Cerrito (Laboratoire IBISC, Evry France)
- Stéphane Demri (co-chair, LSV, CNRS, ENS Cachan, France)
- David Fernández-Duque (International Centre for Mathematics and Computer Science, Toulouse University, France)
- Melvin Fitting (Graduate Center, CUNY, USA)
- David Gabelaia (TSU Razmadze Mathematical Institute, Tbilisi, Georgia)
- Silvio Ghilardi (Università degli Studi di Milano, Italy)
- Valentin Goranko (Stockholm University, Sweden)
- Rajeev Goré (The Australian National University, Canberra, Australia)
- Andreas Herzig (IRIT, Toulouse, France)
- Rosalie Iemhoff (Utrecht University, The Netherlands)
- Agi Kurucz (King's College London, UK)
- Roman Kuznets (TU Wien, Austria)
- Martin Lange (University of Kassel, Germany)
- Carsten Lutz (Universität Bremen, Germany)
- András Máté (Local Organizing Committee Chair, Eötvös University, Budapest, Hungary)
- Angelo Montanari (University of Udine, Italy)
- Larry Moss (Indiana University, USA)
- Sara Negri (University of Helsinki, Finland)
- Sergei Odintsov (Sobolev Institute of Mathematics, Novosibirsk, Russia)
- Hiroakira Ono (Japan Advanced Insitute of Science and Technology, Japan)
- Valeria de Paiva (Nuance, USA)
- Mark Reynolds (The University of Western Australia)
- Ilya Shapirovsky (Institute for the Information Transmission Problems, Russia)
- Renate Schmidt (The University of Manchester, UK)
- Valentin Shehtman (Institute for the Information Transmission Problems, Russia)
- Viorica Sofronie-Stokkermans (Universität Koblenz-Landau, Germany)

- Thomas Studer (Universität Bern, Switzerland)
- Rineke Verbrugge (Groningen, The Netherlands)
- Heinrich Wansing (Ruhr University Bochum, Germany)
- Michael Zakharyaschev (Birkbeck College London, UK)

The Steering Committee of AiML for 2014–2016 consisted of:

- Lev Beklemishev (Steklov Mathematical Institute)
- Stéphane Demri (LSV, CNRS, ENS Cachan)
- Silvio Ghilardi (Università degli Studi di Milano)
- Valentin Goranko (Stockholm University)
- Rajeev Goré (Australian National University)
- Agi Kurucz (King's College London)
- András Máté (Eötvös University, Budapest) (local organizer AiML 2016)
- Larry Moss (Indiana University)
- Valentin Shehtman (Moscow State University)

Many other people assisted with the reviewing process, including: Juan P. Aguilera, Ryuta Arisaka, Felix Bou, Florian Bruse, Zoé Christoff, Marta Cialdea Mayer, Wojciech Dzik, Harley Eades Iii, Raul Fervari, Camillo Fiorentini, Hector Freytes, Christopher Hampson, Jens Ulrik Hansen, Lloyd Humberstone, Joost Joosten, Daniel Kernberger, Milad Ketab Ghale Haji Ali, Stanislav Kikot, Ioannis Kokkinis, Andrey Kudinov, Bjoern Lellmann, Tadeusz Litak, Johannes Marti, Szabolcs Mikulas, Pierluigi Minari, Hitoshi Omor, Fedor Pakhomov, Fabio Papacchini, Tudor Protopopescu, Revantha Ramanayake, Greg Restall, Mikhail Rybakov, Vladimir Rybakov, David Rydeheard, Joshua Sack, Katsuhiko Sano, Lutz Schröder, Dmitrij Skvortsov, John Stell, Lutz Strassburger, Nobu-Yuki Suzuki, Bruno Teheux, Hans van Ditmarsch, Fernando R. Velázquez-Quesada, Bartosz Wcisło, Alberto Zanardo, Shengyang Zhong, Evgeny Zolin. We apologise to anyone whose name was inadvertently left off this list.

We deeply thank the organisers of the conference for their hard and dedicated work. We thank the members of the Programme Committee and all other reviewers for the time, professional effort and the expertise that they invested in ensuring the high scientific standards of the conference and its proceedings. We thank the authors for their excellent contributions. We also thank Agi Kurucz for assistance with formatting the proceedings, and Jane Spurr for bringing this volume to publication.

We would like to thank Central European University for hosting the conference and for providing all the technical conditions for free. This is so much appreciated! Moreover, we would like to thank WECO travels for all the organizational work.

June 17th, 2016
Lev Beklemishev, Stéphane Demri and András Máté

Verification Logic: An Arithmetical Interpretation for Negative Introspection

Juan P. Aguilera [1]

Institut für diskrete Mathematik und Geometrie, Vienna University of Technology, Austria.

David Fernández-Duque [2]

Centre International de Mathématiques et d'Informatique, University of Toulouse, France;
Department of Mathematics, Instituto Tecnológico Autónomo de México, Mexico.

Abstract

We introduce *verification logic*, a variant of Artemov's logic of proofs with new terms of the form $¡\varphi!$ satisfying the axiom schema $\varphi \to ¡\varphi!{:}\varphi$. The intention is for $¡\varphi!$ to denote a proof of φ in Peano arithmetic, whenever such a proof exists. By a suitable restriction of the domain of $¡\cdot!$, we obtain the verification logic VS5, which realizes the axioms of Lewis' system S5. Our main result is that VS5 is sound and complete for its arithmetical interpretation.

Keywords: verification logic, logic of proofs, negative introspection.

1 Introduction

Over twenty years ago, Artemov introduced the *logic of proofs* (LP) to give a provability interpretation of the modal logic S4 [1]. As opposed to the provability logic GL, which uses a single modal operator $\Box$ to represent provability [4], LP uses *proof terms,* meant to denote derivations in Peano arithmetic. If t is a proof term and φ a formula, $t{:}\varphi$ is interpreted as "t is a proof of φ." Complex terms may be built from simpler ones using a handful of operations; in particular, to every term t there corresponds a 'proof-checking' term $!t$, allowing us to realize the modal axiom 4—or *positive introspection,* $\Box\varphi \to \Box\Box\varphi$—by the LP axiom $t{:}\varphi \to !t{:}t{:}\varphi$ (see Section 7 for a definition of 'realization'). To be precise, Artemov proved that LP enjoyed two essential properties:

[1] Email: aguilera@logic.at. Partially supported by FWF grants P-26976-N25, I-1897-N25, I-2671-N35, and W1255-N23.

[2] Email: david.fernandez@irit.fr. Partially supported by ANR-11-LABX-0040-CIMI within the program ANR-11-IDEX-0002-02.

(i) every theorem of S4 can be realized by a theorem of LP, and

(ii) LP is sound and complete for its arithmetical interpretation.

Since then, LP has inspired a family of logics called *justification logics,* which use simlar constructions for 'justification terms' but are not necessarily motivated by mathematical proof (see [2] for an overview). In particular, for applications in epistemic or doxastic logics, one may wish to work with a justification logic realizing the axiom 5—or *negative introspection,* $\neg\Box\varphi \to \Box\neg\Box\varphi$— by some term refuting $t{:}\varphi$. Such a logic may be obtained by extending LP with 'proof-refuting' terms of the form $?t$, satisfying $\neg t{:}\varphi \to ?t{:}\neg t{:}\varphi$. Rubstova proved that the resulting logic, now known as JS5, indeed realizes all theorems of S5 [12]. Her proof was non-constructive, but constructive proofs of this result have since been found; see, for example, [6]. Moreover, there are relational semantics for LP due to Fitting [5], which may also be used to interpret JS5 as well as other justification logics realizing many well-known modal logics [11].

Unfortunately, terms of the form $?t$ satisfying $\neg t{:}\varphi \to ?t{:}\neg t{:}\varphi$ cannot be interpreted using standard PA proofs, since any standard PA derivation represented by t proves at most finitely many formulae, and therefore there are infinitely many formulae φ for which $t{:}\varphi$ fails. But this means that $?t$ would have to be a proof with *infinitely many conclusions* (one for every formula φ not proven by t), contrary to the nature of standard PA proofs (although Kuznets and Studer consider derivations having infinitely many conclusions in [10]).

In order to circumvent this issue, Artemov et. al. [3] instead consider a logic LP(S5), where the term $?t$ satisfies $t{:}(\psi \to \neg s{:}\varphi) \to (\psi \to ?t{:}\neg s{:}\varphi)$. This allows $?t$ to prove finitely many instances of $\neg s{:}\varphi$ (only those already appearing in t), which in turns makes it possible for us to realize any finite collection of instances of negative introspection—one introduces proof constants c such that $c{:}(\neg t{:}\varphi \to \neg t{:}\varphi)$, whence $\neg t{:}\varphi \to ?c{:}\neg t{:}\varphi$ holds. Alternately, Artemov et. al. propose a variant where one is allowed to introduce new proof constants c satisfying $\neg t{:}\varphi \to c{:}\neg t{:}\varphi$. However, either version of LP(S5) has the drawback that the constant specification may have to be extended each time we want a suitable term satisfying a new instance of negative introspection.

In this work, we use a different approach to justification logics, introducing a new family of systems where proof-checkers are replaced with *fact-verifiers* of the form $¡\varphi!$, similar to the *update terms* in [9]. We call such systems *verification logics,* and while we define a few natural examples, we focus our attention on a particular verification logic which we denote VS5. As we show, VS5 realizes both axioms 4 and 5, while enjoying a natural arithmetical interpretation similar to that of LP. It has the advantage that all instances of negative introspection are uniformly realized, while maintaining the finiteness of proofs.

Our main result is that VS5 is complete for this arithmetical interpretation. Our completeness proof follows the basic structure of Artemov's in [1], but a few additional subtleties arise when dealing with the 'negative introspection verifiers.'

Layout. The paper is structured as follows. In Section 2, we introduce proof-checking terms, the language $L_{¡!}$, and the logic VL, along with its natural sublogics, which include VT, VS4 and VS5. In Section 4, we introduce arithmetical interpretations. The arithmetical completeness proof is divided between Section 5, which provides a constructive Lindenbaum lemma, and Section 6, which constructs suitable proof predicates using the fixed point theorem. Finally, in section 7, we present some comments on the possible realizability of S5 by VS5.

2 The logics

In this section we introduce verification logics, and show that some natural verification logics enjoy the internalization property.

Definition 2.1 We define the *terms* and *formulae* of the *full language* $L_{¡!}$ of verification logic by simultaneous recursion as follows:

$$\begin{array}{l} t :: x \mid t \cdot s \mid \ t + s \ \mid ¡\varphi! \\ \varphi :: p \mid \neg\varphi \mid \varphi \to \psi \mid t{:}\varphi, \end{array} \tag{1}$$

where each of x and p is an element of a countably-infinite set of proof or propositional variables. We assume that the two sets of variables are disjoint.

An *expression* is either a term or a formula. We define the *subexpressions* $\mathrm{se}(\epsilon)$ of an expression ϵ by $\mathrm{se}(\epsilon) = \{\epsilon\}$ if ϵ is a variable, $\mathrm{se}(\epsilon) = \{\epsilon\} \cup \mathrm{se}(\varphi)$ if $\epsilon = ¡\varphi!$ or $\epsilon = \neg\varphi$, $\mathrm{se}(\epsilon) = \{\epsilon\} \cup \mathrm{se}(\eta) \cup \mathrm{se}(\rho)$ if $\epsilon = \eta + \rho$, $\eta \cdot \rho$, $\eta \to \rho$ or $\eta{:}\rho$. The *subformulae* are the subexpressions that are formulae, denoted $\mathrm{sf}(\epsilon)$, and the *subterms* are the subexpressions that are terms, denoted $\mathrm{st}(\epsilon)$. If $\eta \in \mathrm{se}(\epsilon)$ we may say that η *occurs* in ϵ, and if Γ is a set of expressions, we say that η *occurs* in Γ if η is a subexpression of some $\epsilon \in \Gamma$.

Next, let us introduce the axioms of full verification logic.

Definition 2.2 The logic VL is defined as the logic generated by modus ponens and the following sets of axioms:

(i) all propositional tautologies,

(ii) $t{:}\varphi \to \varphi$,

(iii) $t{:}(\varphi \to \psi) \to (s{:}\varphi \to s \cdot t{:}\psi)$,

(iv) $t{:}\varphi \to (t + s){:}\varphi$ and $s{:}\varphi \to (t + s){:}\varphi$,

(v) $\varphi \to ¡\varphi!{:}\varphi$.

We will not work with full verification logic, but with one of its natural sublogics, as defined below.

Definition 2.3 A *natural sublanguage* is one obtained by restricting the domain of $¡\cdot!$ to specific choices of φ in (1). In particular, we define:

$L_{¡\mathsf{T}!}$, obtained by restricting the domain of $¡\cdot!$ to all axioms (including tautologies);

$L_{¡4!}$, obtained by restricting the domain of $¡\cdot!$ to axioms and formulae of the form $t{:}\varphi$; and

$L_{¡5!}$, obtained by restricting the domain of $¡\cdot!$ to tautologies, formulae of the form $t{:}\varphi$, and formulae of the form $\neg t{:}\varphi$.

If L is a natural restriction of $L_{¡!}$, the *natural restriction of* VL *associated to* L is the logic $\mathsf{V}(L)$ obtained by restricting the axioms to L and closing under modus ponens. Any logic obtained in this way is a *natural verification logic.* We denote by VS4, VS5 the natural restrictions associated to $L_{¡4!}$ and $L_{¡5!}$, respectively.

If $\lambda = \mathsf{V}(L)$ is a natural verification logic, $\Gamma \subset L$ and $\varphi \in L$, we write $\Gamma \vdash_\lambda \varphi$ if φ belongs to the smallest set containing all axioms of λ, all formulae of Γ, and closed under modus ponens. When $\Gamma = \varnothing$, we write $\vdash_\lambda \varphi$. We say Γ is *consistent (over λ)* if $\Gamma \nvdash_\lambda \varphi \wedge \neg\varphi$ for any formula φ.

Observe that $L_{¡5!}$ is not a proper extension of the other two languages since it does not include formulae of the forms $t{:}\varphi \to \varphi$ or $\varphi \to ¡\varphi!{:}\varphi$. However, there is good reason for defining it in this way:

Lemma 2.4 *For any axiom φ of* VS5 *of the forms* (ii)–(v), *there exists an $L_{¡5!}$-term s such that $\vdash_{\mathsf{VS5}} s{:}\varphi$.*

Proof. If $\varphi = t{:}\psi \to \psi$ is an instance of (ii), set

$$s = ¡\psi \to \varphi! \cdot ¡t! + ¡\neg t{:}\psi \to \varphi! \cdot ¡\neg t{:}\psi!.$$

From the assumption that $t{:}\psi$, it is easy to derive $¡\psi \to \varphi! \cdot ¡t!{:}\varphi$, while from the assumption that $\neg t{:}\psi$, one obtains $¡\neg t{:}\psi \to \varphi! \cdot ¡\neg t{:}\psi!{:}\varphi$; in either case, we obtain $s{:}\varphi$, and this reasoning may be performed within VS5 using the tautology $t{:}\psi \vee \neg t{:}\psi$.

If $\varphi = \psi \to ¡\psi!{:}\psi$ is an instance of (v), then there are three cases to consider; if ψ is a tautology, set $s = ¡\,(¡\psi!{:}\psi \to \varphi)\,! \cdot ¡\,(¡\psi!{:}\psi)\,!$. If $\psi = t{:}\theta$, we set $s = ¡\,(¡t{:}\theta!{:}t{:}\theta \to \varphi)\,! \cdot ¡(¡t{:}\theta!{:}t{:}\theta)! + ¡\neg t{:}\theta \to \varphi! \cdot ¡\neg t{:}\theta!$. Finally, if $\psi = \neg t{:}\theta$, set $s = ¡\,(¡\neg t{:}\theta!{:}\neg t{:}\theta \to \varphi)\,! \cdot ¡(¡\neg t{:}\theta!{:}\neg t{:}\theta)! + ¡t{:}\theta \to \varphi! \cdot ¡t{:}\theta!$.

In all cases, similar reasoning shows that $s{:}\varphi$ is derivable in VS5. The axioms (iii) and (iv) may be treated similarly and are left to the reader. □

Henceforth we will write $¡\varphi! = s$ for the term s given by Lemma 2.4; we will use this notation in the following proof. Observe that verification logics are only a minor variation of justification logics, and as such we may expect them to share many of their basic properties, such as the following familiar 'lifting lemma'. Below, if $\boldsymbol{t} = (t_i)_{i<n}$ is a tuple of terms and $\Gamma = (\gamma_i)_{i<n}$ a tuple of formulas, then $\boldsymbol{t}{:}\Gamma$ denotes the tuple $(t_i{:}\gamma_i)_{i<n}$.

Lemma 2.5 *Let $\lambda \in \{\mathsf{VT}, \mathsf{VS4}, \mathsf{VS5}\}$ and Γ, Δ, Ξ be tuples of formulae such that $\Gamma = \Delta = \varnothing$ if $\lambda = \mathsf{VT}$ and $\Delta = \varnothing$ if $\lambda = \mathsf{VS4}$. Let $\boldsymbol{t}$, $\boldsymbol{s}$ be tuples of terms and $\boldsymbol{x}$ of variables with the same length as Γ, Δ, Ξ, respectively. Then, if $\boldsymbol{t}{:}\Gamma, \neg\boldsymbol{s}{:}\Delta, \Xi \vdash_{\mathsf{VS5}} \varphi$, there is a term $u(\boldsymbol{x}, \boldsymbol{y}, \boldsymbol{z})$ such that*

$$\boldsymbol{t}{:}\Gamma, \neg\boldsymbol{s}{:}\Delta, \boldsymbol{x}{:}\Xi \vdash_{\mathsf{VS5}} u(\boldsymbol{x}, \boldsymbol{t}, \boldsymbol{s}){:}\varphi.$$

Proof. We proceed by induction on the derivation of φ. There are several base cases.

If φ is an axiom, we set $u = ¡\varphi!$, where in the case of $\lambda = \mathsf{VS5}$ this $¡\varphi!$ is given by Lemma 2.4. If $\varphi \in \boldsymbol{t}{:}\Gamma$ (so that $\lambda \neq \mathsf{VT}$) or $\varphi \in \neg\boldsymbol{s}{:}\Delta$ (so that $\lambda = \mathsf{VS5}$), set $u = ¡\varphi!$. Similarly, if $\varphi \in \Xi$, say $\varphi = \xi_i$, set $u = x_i$.

Otherwise, we obtain φ by modus ponens, say from formulae ψ and $\psi \to \varphi$. By the induction hypothesis, there are terms v, w such that $\boldsymbol{t}{:}\Gamma, \neg\boldsymbol{s}{:}\Delta, \boldsymbol{x}{:}\Xi \vdash_{\mathsf{VS5}} v(\boldsymbol{x}, \boldsymbol{t}, \boldsymbol{s}){:}\psi$ and $\boldsymbol{t}{:}\Gamma, \neg\boldsymbol{s}{:}\Delta, \boldsymbol{x}{:}\Xi \vdash_{\mathsf{VS5}} w(\boldsymbol{x}, \boldsymbol{t}, \boldsymbol{s}){:}(\psi \to \varphi)$, and we may set $u = w \cdot v$. □

As an immediate consequence, we obtain the internalization theorem for these three verification logics, which is an explicit version of the necessitation rule:

Corollary 2.6 *Let $\lambda \in \{\mathsf{VT}, \mathsf{VS4}, \mathsf{VS5}\}$ and suppose that $\vdash_\lambda \varphi$. Then, there is a λ-term t such that $\vdash_\lambda t{:}\varphi$.*

3 Theories of arithmetic

In this section we review some basic notions of first-order arithmetic and settle some notation and terminology. The material presented here is treated in detail, for example, in [8].

3.1 Conventions of syntax

We will consider arithmetical theories in languages extending that of first-order arithmetic with exponential, which includes the symbols $\mathsf{0}, \mathsf{1}, x+y, x\cdot y, \mathsf{2}^x$ and $=$, representing the standard constants, operations and relations on the natural numbers, along with the Booleans $\neg, \to$ and the quantifiers $\forall, \exists$; the language generated by these symbols will be denoted L_{PA}. We assume that L_{PA} has a countably infinite set of first-order variables $x, y, z, \ldots$. We may define $x \leq y$ by $\exists z (y = x + z)$ and $x < y$ by $x + \mathsf{1} \leq y$. We define inductively $\mathsf{2}^x_0 = \mathsf{2}^x$ and $\mathsf{2}^x_{l+1} = \mathsf{2}^{\mathsf{2}^x_l}$.

As is customary, we use Δ_0 to denote the set of all formulae where all quantifiers are *bounded*, that is, of the form $\forall x < t\ \varphi$ or $\exists x < t\ \varphi$. We will use pseudo-terms to simplify notation, where an expression $\varphi(t(\boldsymbol{x}))$ should be understood as a shorthand for $\exists y < s(\boldsymbol{x})\ \big(\psi(\boldsymbol{x}, y) \wedge \varphi(y)\big)$, with ψ a Δ_0 formula defining the graph of the intended interpretation of t and s a standard term bounding the values of $t(\boldsymbol{x})$. The domain of the functions defined by these pseudo-terms may be a proper subset of $\mathbb{N}$. Functions definable by pseudo-terms of this form are *elementary*.

We assume that every finite sequence $(s_1, \ldots, s_n)$ may be represented by a natural number $\boldsymbol{s}$, with the following properties:

(i) there is a Δ_0 formula $\mathtt{seq}(x)$ such that $\mathtt{seq}(\boldsymbol{s})$ is true if and only if $\boldsymbol{s}$ codes a sequence;

(ii) there is a pseudo-term $|x|$ such that $|\boldsymbol{s}|$ returns the length of $\boldsymbol{s}$, and we assume that $|\boldsymbol{s}| \leq \boldsymbol{s}$ for all sequences $\boldsymbol{s}$;

(iii) there is a pseudo-term $(x)_y$ such that $(\boldsymbol{s})_i$ returns the ith element of s, also with the assumption that $(\boldsymbol{s})_i \leq \boldsymbol{s}$, and

(iv) there is a term $B(x,y)$ such that whenever $\boldsymbol{s}$ is a sequence of length at most N and with each s_i bounded by M, it follows that $\boldsymbol{s} < B(N,M)$.

Finite sets may also be represented using sequences (by ordering them arbitrarily), in which case we say that x belongs to $\boldsymbol{s}$ if $x = (\boldsymbol{s})_i$ for some i. Similarly, we can represent some functions using sequences. Call a function *small* if its domain is finite. Small functions can be coded by pairs $f = (\boldsymbol{d}, \boldsymbol{r})$, where $\boldsymbol{d}$, $\boldsymbol{r}$ are sequences of the same length, and we write $f(x) = y$ if there is some i such that $(\boldsymbol{d})_i = x$ and $(\boldsymbol{r})_i = y$. We call $\boldsymbol{d}$ the *domain* of f and denote it by $\text{dom}(f)$.

3.2 Peano Arithmetic

For the sake of concreteness, we will be working in Peano arithmetic. PA is formed by adding the axiom schema of successor induction (equation (2) below for any formula φ of L_{PA}) to Robinson Arithmetic, e.g., as axiomatized by:

(i) $\forall x\ (x = x)$

(ii) $\forall x \forall y\ (x \neq y \vee \alpha \vee {\sim}\alpha[x/y])$

(iii) $\forall x \forall y\ (x \neq y \vee y = x)$

(iv) $\forall x \forall y \forall z\ (x \neq y \vee y \neq z \vee x = z)$

(v) $\forall x\ (0 \neq x + 1)$

(vi) $\forall x\ (x = 0 \vee \exists y\ x = y + 1)$

(vii) $\forall x\ (x + 0 = x)$

(viii) $\forall x \forall y\ (x + (y+1) = (x+y)+1)$

(ix) $\forall x\ (x \times 0 = 0)$

(x) $\forall x \forall y\ (x \times (y+1) = (x \times y) + y)$

(xi) $2^0 = 1$

(xii) $\forall x\ (2^{x+1} = 2^x + 2^x)$

(xiii) $\forall x \forall y\ (x + 1 \neq y + 1 \vee x = y)$

(in (ii) above, α is any atomic formula). The axiom schema of successor induction is given by

$$\varphi(0) \wedge \forall x\ (\varphi(x) \to \varphi(x+1)) \ \to\ \forall x\ \varphi(x). \tag{2}$$

We remark, however, that our proof is fairly general and would readily work for many theories stronger (or weaker) than PA, or even theories in other languages, such as that of set theory.

3.3 Gödel numberings

Fix some Gödel numbering $\ulcorner \cdot \urcorner : L_{\text{PA}} \cup L_{\text{¡5!}} \to \mathbb{N} \setminus \{0\}$, such that the set of codes from each language is elementary. Note that 0 is not the Gödel number of any expression. For a natural number n, define a term $\bar{n}$ recursively by $\bar{0} = 0$ and $\overline{n+1} = (\bar{n}) + 1$. As in the case of sequences, we will assume that if N, M are such that the expression ϵ has at most N symbols, each with code bounded by M, then $\ulcorner \epsilon \urcorner < B(N,M)$; that if η is a proper subexpression of ϵ, then $\ulcorner \eta \urcorner < \ulcorner \epsilon \urcorner$; and that if ϵ has n symbols, then $n < \ulcorner \epsilon \urcorner$.[3]

[3] These conditions are achieved, for example, if ϵ is coded by putting all of its symbols in a sequence and using our previous assumptions on the coding of sequences.

If $\Gamma = (\epsilon_i)_{i<n}$ is a sequence of expressions, we define $\ulcorner\Gamma\urcorner$ to be the code of the sequence $(\ulcorner\epsilon_i\urcorner)_{i<n}$. We will sometimes abuse notation by identifying expressions with their Gödel numbers and sets or sequences with their codes.

4 Arithmetical interpretations

In this section we will define the arithmetical interpretation of verification logic. It requires the notion of a *normal proof system*:

Definition 4.1 A *normal proof system* is a triple $(\pi, \mathsf{m}, \mathsf{a})$ such that:

(i) $\pi = \pi(x, y)$ is a Δ_0 formula of L_{PA} and, for every $\varphi \in L_{\mathrm{PA}}$, $\mathrm{PA} \vdash \varphi$ if and only if $\exists x\, \pi(x, \ulcorner\varphi\urcorner)$ holds.

(ii) Say that y is a *π-conclusion* of x if $\pi(x, y)$ holds. Then, the set $[\![x]\!]_\pi$ of all π-conclusions of x is finite for all x and the function $x \mapsto [\![x]\!]_\pi$ is computable.

(iii) m is a computable function such that, for all n, k and formulae φ, ψ, if $\pi(n, \ulcorner\varphi \to \psi\urcorner)$ and $\pi(k, \ulcorner\varphi\urcorner)$, then $\pi(\mathsf{m}(n, k), \ulcorner\psi\urcorner)$.

(iv) a is a computable function such that, for all n, k and any formula φ, if $\pi(n, \ulcorner\varphi\urcorner)$ or $\pi(k, \ulcorner\varphi\urcorner)$, then $\pi(\mathsf{a}(n, k), \ulcorner\varphi\urcorner)$.

We assume that a canonical normal proof system $(\mathtt{Proof}, \mathsf{m}_{\mathrm{PA}}, \mathsf{a}_{\mathrm{PA}})$ is given, and $[\![x]\!] = [\![x]\!]_{\mathtt{Proof}}$. The existence of normal proof systems is well known; for example, we may use a multi-conclusion variant of Gödel's proof predicate. Observe that it is not necessary to work with Δ_0 proof predicates (Δ_1 is sufficient), but we follow [7] and work within bounded arithmetic when possible.

Definition 4.2 A *potential arithmetical interpretation* is a tuple $\mathfrak{S} = (f, \pi, \mathsf{m}, \mathsf{a}, \mathsf{v})$ such that f maps propositional variables to formulas of L_{PA} and term variables to natural numbers, $(\pi, \mathsf{m}, \mathsf{a})$ is a normal proof system, and $\mathsf{v}\colon L_{\mathrm{PA}} \to \mathbb{N}$ is a computable function. We define a function $\cdot^{\mathfrak{S}}\colon L_{¡5!} \to L_{\mathrm{PA}}$ by letting

(i) $v^{\mathfrak{S}} = f(v)$ for any variable v,

(ii) $\cdot^{\mathfrak{S}}$ commute with Booleans,

(iii) $(t{:}\varphi)^{\mathfrak{S}} = \pi(\overline{t^{\mathfrak{S}}}, \overline{\ulcorner\varphi^{\mathfrak{S}}\urcorner})$,

(iv) $(t \cdot s)^{\mathfrak{S}} = \mathsf{m}(t^{\mathfrak{S}}, s^{\mathfrak{S}})$,

(v) $(t + s)^{\mathfrak{S}} = \mathsf{a}(t^{\mathfrak{S}}, s^{\mathfrak{S}})$, and

(vi) $¡\varphi!^{\mathfrak{S}} = \mathsf{v}(\varphi^{\mathfrak{S}})$.

We say that $\mathfrak{S}$ is an *arithmetical interpretation of* VS5 if whenever $\varphi \in L_{¡5!}$, then $\varphi^{\mathfrak{S}} \to \pi(\mathsf{v}(\varphi^{\mathfrak{S}}), \ulcorner\varphi^{\mathfrak{S}}\urcorner)$ holds. The arithmetical interpretation $\mathfrak{S}$ is *robust* if, whenever $\varphi \in \Delta_0$, then $\varphi^{\mathfrak{S}} \to \pi(\mathsf{v}(\varphi^{\mathfrak{S}}), \ulcorner\varphi^{\mathfrak{S}}\urcorner)$ also holds.

We remark that robustness is not necessary for our proofs to go through, but it is a desirable property since such interpretations automatically satisfy all expressions $\varphi \to ¡\varphi!{:}\varphi$ whenever $\varphi \in \Delta_0$, thus internalizing the completeness of Peano arithmetic for Δ_0 formulas [8].

Proposition 4.3 (Arithmetical soundness) *If* VS5 $\vdash \varphi$, *then* $\mathrm{PA} \vdash \varphi^{\mathfrak{S}}$ *for any arithmetical interpretation* $\mathfrak{S}$.

Proof. By a straightforward induction. Clearly, modus ponens preserves validity. We check the axiom $t{:}\varphi \to \varphi$, which translates as $\pi(\ulcorner t^{\mathfrak{S}}\urcorner, \ulcorner \varphi^{\mathfrak{S}}\urcorner) \to \varphi^{\mathfrak{S}}$. Either $\pi(\ulcorner t^{\mathfrak{S}}\urcorner, \ulcorner \varphi^{\mathfrak{S}}\urcorner)$ is true, whence $\mathrm{PA} \vdash \varphi^{\mathfrak{S}}$ (by the soundness of a normal proof predicate, i.e., condition (i)) and so $\mathrm{PA} \vdash \pi(\ulcorner t^{\mathfrak{S}}\urcorner, \ulcorner \varphi^{\mathfrak{S}}\urcorner) \to \varphi^{\mathfrak{S}}$, or $\pi(\ulcorner t^{\mathfrak{S}}\urcorner, \ulcorner \varphi^{\mathfrak{S}}\urcorner)$ is false, whence it is refutable in PA (as it is Δ_0), whereby again $\mathrm{PA} \vdash \pi(\ulcorner t^{\mathfrak{S}}\urcorner, \ulcorner \varphi^{\mathfrak{S}}\urcorner) \to \varphi^{\mathfrak{S}}$. The rest of the axioms are proved similarly. □

Our main objective is now to prove that VS5 is also complete. We will in fact prove a slightly strengthened version of completeness:

Theorem 4.4 (Arithmetical completeness) *If φ is consistent with* VS5, *then there is a robust arithmetical interpretation $\mathfrak{S}^*$ such that* $\mathrm{PA} \vdash \varphi^{\mathfrak{S}^*}$.

We defer the proof of this result to Section 6. Its general structure is similar to that of the arithmetical completeness proof in [1], although some additional care must be taken to deal with the 'negative information' conveyed by proof terms of the form $¡\neg t{:}\varphi!$. Among the technical difficulties is that many of the steps must be done constructively, including a version of the Lindenbaum lemma.

5 A constructive Lindenbaum's lemma

As in many familiar completeness proofs, the first step is to expand the consistent set $\{\varphi\}$ to a larger set that can be dealt with more easily. We do this by first expanding the set "downwards" and then "upwards." It will be convenient to introduce the notation $\sim\varphi$ defined by $\sim\varphi = \neg\varphi$ if φ does not begin with a negation, and $\sim\varphi = \psi$ if $\varphi = \neg\psi$.

Definition 5.1 A set of formulae Γ is *saturated* if:

(i) whenever ψ occurs in Γ, either $\psi \in \Gamma$ or $\sim\psi \in \Gamma$,

(ii) whenever $\neg(s+t){:}\varphi \in \Gamma$, then $\neg s{:}\varphi \in \Gamma$ and $\neg t{:}\varphi \in \Gamma$, and

(iii) whenever $\neg(s \cdot t){:}\varphi \in \Gamma$ and $\psi \to \varphi$ occurs in Γ, then either $\neg s{:}(\psi \to \varphi) \in \Gamma$, or $\neg t{:}\psi \in \Gamma$.

Lemma 5.2 *For any consistent formula φ there exists a finite, consistent, saturated set that contains φ.*

Proof. Choose a subformula of φ and either add it or its negation so as to maintain consistency, as well as the required formulae for $\neg(s \cdot t){:}\varphi$ and $\neg(s + t){:}\varphi$. Rinse and repeat. Note that either the formulae or the terms that we add are simpler than those previously appearing and thus the process terminates. □

As an immediate consequence of the definition, we note that saturated sets of formulae have some basic closure properties.

Lemma 5.3 *Suppose Γ is a saturated set of formulae.*

(i) $\psi \to \varphi \in \Gamma$ *implies that* $\sim\psi \in \Gamma$ *or* $\varphi \in \Gamma$ *and* $\neg(\psi \to \varphi) \in \Gamma$ *implies that* $\psi, \sim\varphi \in \Gamma$;

(ii) $\neg¡\varphi!{:}\varphi \in \Gamma$ *implies that* $\neg\varphi \in \Gamma$ *for φ of the form* $t{:}\psi$ *or* $\neg t{:}\psi$, *and*

(iii) $t{:}\varphi \in \Gamma$ *implies* $\varphi \in \Gamma$.

These properties are easily verified and left to the reader. Once we have included a formula φ in a saturated, consistent set Γ, the next step would be to use a suitable variant of Lindenbaum's lemma to extend Γ to a maximal-consistent set of formulae. This amounts to selecting which instances of $t{:}\varphi$ must be true, as propositional variables not appearing in Γ may all be assigned the value 'false.' In fact, instances of $t{:}\varphi$ may also be assumed false, unless they are forced to be true by axiom (v) or are sufficiently witnessed by Γ. We make this notion precise below. Recall that an expression ϵ *occurs* in Γ if it is a subexpression of a formula in Γ.

Definition 5.4 Let Γ be a set of formulae A *verification instance of* $L_{¡5!}$ (or simply: a *verification*) is any formula of the form $t{:}\varphi \in L_{¡5!}$. We say that $t{:}\varphi$ is Γ-*balanced* (or simply 'balanced') if φ occurs in either t or Γ. The set of Γ-balanced verifications will be denoted by $[\Gamma]$.

Remark. If $\Gamma \subseteq L_{¡5!}$ is a (code for a) finite set of formulae and $t{:}\varphi \in \Gamma$, then $t{:}\varphi \in [\Gamma]$. Moreover, $¡\varphi!{:}\varphi \in L_{¡5!}$ is always Γ-balanced, regardless of Γ or φ. Observe that if $t{:}\varphi$ is Γ-balanced then $\ulcorner\varphi\urcorner < \max(\ulcorner t\urcorner, \ulcorner\Gamma\urcorner)$ by our convention on Gödel numbering.

For balanced verifications, the truth value of $t{:}\varphi$ will be decided recursively, according to an order determined by the Gödel number of each formula:

Definition 5.5 Fix a (code for a) set of formulae Γ. If $t{:}\varphi$, $s{:}\psi$ are verifications, we write $t{:}\varphi \lhd s{:}\psi$ whenever $(\ulcorner t\urcorner, \ulcorner\varphi\urcorner) <_{\mathtt{lex}} (\ulcorner s\urcorner, \ulcorner\psi\urcorner)$; i.e., either $\ulcorner t\urcorner < \ulcorner s\urcorner$ or $t = s$ and $\ulcorner\varphi\urcorner < \ulcorner\psi\urcorner$. For any $\tau \in [\Gamma]$, we define $\downarrow\tau = \{\sigma \in [\Gamma] : \sigma \lhd \tau\}$.

In words, $\downarrow\tau$ is the set of balanced predecessors of τ. It will be convenient to give a simple bound on the size of this set:

Lemma 5.6 *Let* Γ *be saturated and* $t{:}\varphi \in [\Gamma]$. *Then,* $|\downarrow(t{:}\varphi)| < \ulcorner t\urcorner\cdot(\ulcorner t\urcorner + \ulcorner\Gamma\urcorner)$.

Proof. This follows from a straightforward counting argument: if $s{:}\psi \in \downarrow(t{:}\varphi)$, there are at most $\ulcorner t\urcorner$ choices for s, and $\ulcorner s\urcorner + \ulcorner\Gamma\urcorner \leq \ulcorner t\urcorner + \ulcorner\Gamma\urcorner$ choices for ψ, since $s{:}\psi$ must be balanced and, by our conventions, any subexpression ϵ of any formula in Γ must satisfy $\ulcorner\epsilon\urcorner < \ulcorner\Gamma\urcorner$. The inequality is strict because $t{:}\varphi$ is excluded. □

An immediate consequence of Lemma 5.6 is that $\lhd \restriction [\Gamma]$ is a well-order of order type ω.

Definition 5.7 Let Γ be any set of $L_{¡5!}$-formulae. We define a set $\widetilde{\Gamma} \subseteq [\Gamma]$ by recursion on $\lhd$ as follows.

Fix $\tau \in [\Gamma]$ and assume, inductively, that $\widetilde{\Gamma} \cap \downarrow\tau$ has been defined. Then, $\tau \in \widetilde{\Gamma}$ if and only if one of the following holds:

(i) $\tau \in \Gamma$,

(ii) $\tau = ¡\varphi!{:}\varphi$ and φ is a tautology;

(iii) $\tau = (s \cdot t){:}\varphi$, and there is a formula ψ such that both $s{:}(\psi \to \varphi)$ and $t{:}\psi$ belong to $\widetilde{\Gamma} \cap \downarrow \tau$;

(iv) $\tau = (t + s){:}\varphi$ and either $t{:}\varphi$ or $s{:}\varphi$ belongs to $\widetilde{\Gamma} \cap \downarrow \tau$;

(v) $\tau = ¡t{:}\varphi!{:}(t{:}\varphi)$ and $t{:}\varphi \in \widetilde{\Gamma} \cap \downarrow \tau$, or

(vi) $\tau = ¡\neg t{:}\varphi!{:}(\neg t{:}\varphi)$ and $t{:}\varphi \notin \widetilde{\Gamma} \cap \downarrow \tau$.

Although we have only closed $\widetilde{\Gamma}$ under restricted versions of the clauses for the term operations, $\widetilde{\Gamma}$ is actualy closed under the unrestricted versions, as we show in the following lemma.

Lemma 5.8 *Given finite $\Gamma \subseteq L_{¡5!}$ and arbitrary terms t, s and arbitrary formulae φ, ψ,*

(i) *if φ is a tautology then $¡\varphi!{:}\varphi \in \widetilde{\Gamma}$;*

(ii) *if both $s{:}(\psi \to \varphi)$ and $t{:}\psi$ belong to $\widetilde{\Gamma}$ then $(s \cdot t){:}\varphi \in \widetilde{\Gamma}$;*

(iii) *if either $t{:}\varphi$ or $s{:}\varphi$ belongs to $\widetilde{\Gamma}$ then $(s + t){:}\varphi \in \widetilde{\Gamma}$;*

(iv) *if $t{:}\varphi \in \widetilde{\Gamma}$ then $¡t{:}\varphi!{:}(t{:}\varphi) \in \widetilde{\Gamma}$;*

(v) *if $t{:}\varphi \notin \widetilde{\Gamma}$ then $¡\neg t{:}\varphi!{:}\neg(t{:}\varphi) \in \widetilde{\Gamma}$.*

Proof. (i) As observed above, $¡\varphi!{:}\varphi$ is always balanced. If φ is a tautology, it follows that $¡\varphi!{:}\varphi \in [\Gamma]$ and thus $¡\varphi!{:}\varphi \in \widetilde{\Gamma}$ by definition.

(ii) If both $s{:}(\psi \to \varphi)$ and $t{:}\psi$ belong to $\widetilde{\Gamma}$ then by definition, $s{:}(\psi \to \varphi), t{:}\psi \in [\Gamma]$. Moreover, $\ulcorner s \urcorner, \ulcorner t \urcorner < \ulcorner s \cdot t \urcorner$, whence $s{:}(\psi \to \varphi), t{:}\psi \lhd (s \cdot t){:}\varphi$. Meanwhile, φ occurs in $\psi \to \varphi$, which occurs in s, which occurs in $s \cdot t$, hence $(s \cdot t){:}\varphi$ is balanced. It follows by definition that $(s \cdot t){:}\varphi \in \widetilde{\Gamma}$.

(iii) If $s{:}\varphi \in \widetilde{\Gamma}$, then it is balanced. Reasoning as above, $(s + t){:}\varphi$ is also balanced and $s{:}\varphi \lhd (s + t){:}\varphi$. It follows that $(s + t){:}\varphi \in \widetilde{\Gamma}$; the case for $t{:}\varphi \in \widetilde{\Gamma}$ is symmetric.

(iv) As before, $¡t{:}\varphi!{:}t{:}\varphi$ is balanced regardless of t, φ, and since $\ulcorner t \urcorner < \ulcorner t{:}\varphi \urcorner < \ulcorner ¡t{:}\varphi! \urcorner$, we have that $t{:}\varphi \lhd ¡t{:}\varphi!{:}t{:}\varphi$. Moreover, if $t{:}\varphi \in \widetilde{\Gamma}$, then it must be balanced, so $t{:}\varphi \in \downarrow (¡t{:}\varphi!{:}t{:}\varphi)$, which by definition implies that $¡t{:}\varphi!{:}t{:}\varphi \in \widetilde{\Gamma}$.

(v) Assume that $t{:}\varphi \notin \widetilde{\Gamma}$. Then, $t{:}\varphi \notin \widetilde{\Gamma} \cap \downarrow (¡\neg t{:}\varphi!{:}\neg t{:}\varphi)$, and we know that $¡\neg t{:}\varphi!{:}\neg t{:}\varphi \in [\Gamma]$, so by definition $¡\neg t{:}\varphi!{:}\neg t{:}\varphi \in \widetilde{\Gamma}$, as desired. □

Of course, $\widetilde{\Gamma}$ is not actually a maximal-consistent set, but the set of its consequences is.

Definition 5.9 Let Γ be a saturated, consistent set of $L_{¡5!}$-formulae. We define $\Gamma \mathrel{|\!\!\sim} \varphi$ by induction on φ as follows:

(i) For any propositional variable p, $\Gamma \mathrel{|\!\!\sim} p$ if and only if $p \in \Gamma$.

(ii) If $t{:}\varphi$ is a verification, then $\Gamma \mathrel{|\!\!\sim} t{:}\varphi$ if and only if $t{:}\varphi \in \widetilde{\Gamma}$.

(iii) $\Gamma \mathrel{|\!\!\sim} \neg\varphi$ if and only if $\Gamma \not\mathrel{|\!\!\sim} \varphi$, and

(iv) $\Gamma \mathrel{|\!\!\sim} \psi \to \varphi$ if and only if either $\Gamma \not\mathrel{|\!\!\sim} \psi$ or $\Gamma \mathrel{|\!\!\sim} \varphi$.

The following lemma shows that $\widetilde{\Gamma}$ is not that far away from Γ.

Lemma 5.10 *Let Γ be a consistent set of formulae.*

(i) *If $t{:}\varphi \in \widetilde{\Gamma}$ and t occurs in Γ, then φ also occurs in Γ, and*

(ii) *if $\neg t{:}\varphi \in \Gamma$ then $t{:}\varphi \notin \widetilde{\Gamma}$.*

Lemma 5.10 is proved by induction. We omit the details.

Lemma 5.11 *Let Γ be any finite, consistent, saturated set of formulae.*

(i) *If φ is a tautology, then $\Gamma \mathrel{|\!\!\sim} \varphi$.*

(ii) *If $\varphi \in \Gamma$ then $\Gamma \mathrel{|\!\!\sim} \varphi$.*

(iii) *Whenever $t{:}\varphi \in \widetilde{\Gamma}$, it follows that $\Gamma \mathrel{|\!\!\sim} \varphi$.*

Proof. We only sketch the proof. For the first item: proceed by a simple induction using clauses (iii) and (iv) of the definition, treating expressions of the form $t{:}\varphi$ as separate propositional variables. The second item follows by an easy induction on the length of φ using the fact that Γ is consistent and saturated.

For the last item: if $t{:}\varphi \in \Gamma$, then $\varphi \in \Gamma$ by Lemma 5.3(iii), and thus by the previous item, $\Gamma \mathrel{|\!\!\sim} \varphi$. Hence we may assume that $t{:}\varphi \notin \Gamma$. The result then follows by induction on t. □

Our goal in the remainder of this section is to prove the next lemma. It will be crucial for defining the arithmetical interpretations needed for the proof of Theorem 4.4, and implies that membership in $\widetilde{\Gamma}$ is elementary.

Lemma 5.12 *There is a Δ_0 formula $\mathtt{Comp}_\Gamma(x)$ such that, for all verifications τ, $\tau \in \widetilde{\Gamma}$ if and only if $\mathtt{Comp}_\Gamma(\ulcorner\tau\urcorner)$ holds.*

Towards a proof of Lemma 5.12, we define some auxiliary notions.

Definition 5.13 Given any finite set $\Gamma \subseteq L_{¡5!}$, we say that a sequence $(\sigma_0, \ldots, \sigma_n)$ of Γ-balanced verifications is an *initial $\lhd$-segment* if, for all verifications τ and all $j \leq n$, $\tau \in \downarrow\sigma_j$ if and only if $\tau = \sigma_i$ for some $i \leq j$. We say that $(\sigma_0, \ldots, \sigma_n)$ *contains* τ if $\tau = \sigma_i$ for some $i \leq n$.

A sequence $(x_0, \ldots, x_n)$ *codes an initial $\lhd$-segment* if there is an initial $\lhd$-segment $(\sigma_0, \ldots, \sigma_n)$ such that $x_i = \ulcorner\sigma_i\urcorner$ for all $i \leq n$.

An initial $\lhd$-segment is simply an initial segment of the well-ordering $\lhd \restriction [\Gamma]$, so that it is uniquely determined by its last element.

Lemma 5.14 *There is an elementary function $K(x)$ such that if $\tau \in [\Gamma]$, then there is some $y < K(\ulcorner\tau\urcorner)$ coding an initial $\lhd$-segment containing τ.*

Proof. For each x that does not code some $\tau \in [\Gamma]$, we set $K(x) = 0$. If x does code such a τ, we will give a value for $K(x)$ such that for some $y < K(x)$, y codes the initial $\lhd$-segment whose last element is τ.

By definition, if $t{:}\varphi \lhd s{:}\psi$, then either $\ulcorner t\urcorner < \ulcorner s\urcorner$ or $t = s$ and $\ulcorner\varphi\urcorner < \ulcorner\psi\urcorner$. By the remark following Definition 5.4, if $\tau = s{:}\psi$ is Γ-balanced then $\ulcorner\psi\urcorner < \max(\ulcorner s\urcorner, \ulcorner\Gamma\urcorner)$. By our conventions on the Gödel numbering, $\ulcorner t{:}\varphi\urcorner$ is elementarily bounded in $\ulcorner t \urcorner + \ulcorner \varphi \urcorner$, for any verification $t{:}\varphi$. Hence, the Gödel numbers of τ and all of its $\lhd \restriction [\Gamma]$-predecessors are bounded by some M that is elementary in $\ulcorner s \urcorner + \ulcorner \Gamma \urcorner$. It is also clear that the cardinality of the set of $\lhd \restriction [\Gamma]$-predecessors of τ is bounded by M, so that we may set, for example, $K(\ulcorner \tau \urcorner) = B\,(M, M)$. □

Below, recall that a function on the natural numbers is *small* if its domain is finite.

Definition 5.15 Fix a finite set Γ of $L_{¡5!}$-formulae. We say that a small function e is an *initial evaluation* if $\mathrm{dom}(e)$ is an initial $\lhd$-segment, e takes values in $\{0, 1\}$, and for all $\tau \in \mathrm{dom}(e)$ we have that $e(\tau) = 1$ if and only if one of the following occurs:

(i) $\tau \in \Gamma$,

(ii) $\tau = ¡\varphi!{:}\varphi$ and φ is a tautology;

(iii) $\tau = (s \cdot t){:}\varphi$, and $e(s{:}(\psi \to \varphi)) = e(t{:}\psi) = 1$;

(iv) $\tau = (t + s){:}\varphi$ and either $e(t{:}\varphi) = 1$ or $e(s{:}\varphi) = 1$;

(v) $\tau = ¡t{:}\varphi!{:}(t{:}\varphi)$ and $e(t{:}\varphi) = 1$, or

(vi) $\tau = ¡\neg t{:}\varphi!{:}(\neg t{:}\varphi)$ and $e(t{:}\varphi) \neq 1$ (i.e., it is 0 or undefined).

Given any verification τ, define $E(\tau) = 1$ if and only if there exists an initial evaluation e which assigns 1 to τ, and set $E(\tau) = 0$ otherwise.

Observe that if τ is not Γ-balanced, then $E(\tau) = 0$. Otherwise, the value of $E(\tau)$ may be computed in any of several equivalent ways:

Lemma 5.16 *Let* $\tau \in [\Gamma]$, $b \in \{0, 1\}$. *Then, there is an elementary function* $C(x)$ *such that the following are equivalent:*

(i) $E(\tau) = b$,

(ii) $e(\tau) = b$ *for every initial evaluation* e,

(iii) $e(\tau) = b$ *for some initial evaluation* e,

(iv) $e(\tau) = b$ *for some initial evaluation* $e = (\boldsymbol{d}, \boldsymbol{r})$ *such that* $\ulcorner \boldsymbol{d} \urcorner, \ulcorner \boldsymbol{r} \urcorner < C(\ulcorner \tau \urcorner)$.

Proof. The lemma is proven by checking that if e, e' are two initial evaluations, then $e(\tau) = e'(\tau)$ whenever the two are defined; this is straightforward and we omit the details. Thus to evaluate $E(\tau)$, it suffices to consider *any* such initial evaluation. The function C bounding the smallest witness can be constructed using the funcion K from Lemma 5.14 and the function B from Section 3.1. □

The function $E(\cdot)$ is useful because it gives us an elementary method to determine whether $\tau \in \widetilde{\Gamma}$.

Lemma 5.17 *For any finite set of formulas* $\Gamma \subseteq L$ *and any verification* τ, $E(\tau) = 1$ *if and only if* $\tau \in \widetilde{\Gamma}$.

Proof. If τ is not Γ-balanced, then it is not included in any initial $\lhd$-segment, so $E(\tau) = 0$; on the other hand, $\tau \notin \widetilde{\Gamma}$, so the equivalence holds trivially.

Otherwise, we proceed by induction on τ along $\lhd \upharpoonright [\Gamma]$. In view of Lemma 5.16, it suffices to consider an arbitrary initial evaluation e such that $e(\tau)$ is defined, and prove that $e(\tau) = 1$ if and only if $\tau \in \widetilde{\Gamma}$. If $\tau \in \Gamma$, then $e(\tau) = 1$ and $\tau \in \widetilde{\Gamma}$, so we may assume otherwise. We must then consider several cases depending on τ; we work out only a few as examples.

If, for example, $\tau = (s \cdot t){:}\varphi$ and $e(\tau) = 1$, then $e(s{:}(\psi \to \varphi)) = e(t{:}\psi) = 1$ for some ψ, which by the induction hypothesis means that $s{:}(\psi \to \varphi), t{:}\psi \in \widetilde{\Gamma}$, so that by Lemma 5.8, $\tau \in \widetilde{\Gamma}$. Conversely, if $\tau \in \widetilde{\Gamma}$, then by definition $s{:}(\psi \to \varphi), t{:}\psi \in \widetilde{\Gamma} \cap \downarrow\tau$ for some ψ. But since the domain of e is an initial $\lhd$-segment and $e(\tau)$ is defined, we must also have that $s{:}(\psi \to \varphi), t{:}\psi \in \mathrm{dom}(e)$ and thus by the induction hypothesis, $e(s{:}(\psi \to \varphi)) = e(t{:}\psi) = 1$, which means that $e(\tau) = 1$.

Next, we consider the case when τ is of the form $¡\neg t{:}\varphi!{:}\neg t{:}\varphi$. If $t{:}\varphi$ is not Γ-balanced, then $e(t{:}\varphi)$ is undefined, whence $e(\tau) = 1$; similarly, $t{:}\varphi \notin \widetilde{\Gamma}$, so $\tau \in \widetilde{\Gamma}$. If, instead, $t{:}\varphi$ is Γ-balanced, then again we have that $t{:}\varphi \in \mathrm{dom}(e)$ and thus $e(\tau) = 1 \Leftrightarrow e(t{:}\varphi) = 0 \overset{\mathrm{IH}}{\Leftrightarrow} t{:}\varphi \notin \widetilde{\Gamma} \Leftrightarrow \tau \in \widetilde{\Gamma}$.

Each of the remaining cases is similar to one of the above and is left to the reader. □

Proof of Lemma 5.12. Let $C(\cdot)$ be the bound given by Lemma 5.16. We use $C(\cdot)$ to define $\mathtt{Comp}_\Gamma(x)$ by a natural translation into L_{PA} of:

x codes a verification τ and there is a number $y < C(x)$ such that y codes an initial evaluation e with $e(\tau) = 1$.

The only thing that needs to be verified is that the property 'y codes an initial evaluation' is Δ_0. This is straightforward from Definitions 5.13 and 5.15 and our conventions on coding of sequences, as all quantifiers involved are bounded by x and $\ulcorner\Gamma\urcorner$. □

6 Fixed-point proof predicates

The construction of $\widetilde{\Gamma}$ allows us to constructively extend any saturated, consistent set Γ to a maximal-consistent set of formulas $\{\varphi \in L : \Gamma \mathrel{|\!\sim} \varphi\}$. Next, we construct an arithmetical interpretation for L tailored specifically for this extended set. This construction relies on a fixed-point argument that we detail in this subsection, very similar to that of [1]. For this completeness proof, it is sufficient to consider simple propositional assignments, in the following sense:

Definition 6.1 A propositional assignment f is *simple* if f is elementary and, for every variable p, either $f(p) = (\ulcorner p\urcorner = \ulcorner p\urcorner)$ or $f(p) = (\ulcorner p\urcorner = 0)$.

Moreover, the arithmetical interpretations resulting from the proof will coincide with the function f given below, for a particular choice of π:

Definition 6.2 Given a formula $\pi = \pi(x, y)$ and a propositional assignment f, we extend f to an auxiliary function $f_\pi \colon L_{¡5!} \to L_{\mathrm{PA}}$ by letting

- $f_\pi(p) = f(p)$ for any propositional variable p,
- f_π commute with Booleans, and
- $f_\pi(t{:}\varphi) = \pi(\overline{\ulcorner t \urcorner}, \overline{\ulcorner f_\pi(\varphi) \urcorner})$.

Lemma 6.3 *If f is a simple propositional assignment and $\pi(x, y)$ is Δ_0, then $f_\pi(\varphi)$ is Δ_0 for all $\varphi \in L_{\text{¡5!}}$. Moreover, if π contains quantifiers and each of x and y appears free in π at least once, then f_π is injective.*

Proof. We prove by induction on $\ulcorner\varphi\urcorner + \ulcorner\psi\urcorner$ that if $f_\pi(\varphi) = f_\pi(\psi)$, then $\varphi = \psi$. If $\varphi = p$ is a propositional variable, then $f_\pi(\varphi)$ does not contain quantifiers or Booleans, and hence neither does $f_\pi(\psi)$. It follows that ψ must also be a propositional variable, which by the injectivity of the Gödel numbering yields $\psi = p$ as well.

Next, consider the case where neither φ nor ψ is of the form $t{:}\theta$. If $\varphi = \varphi_0 \to \varphi_1$, then we must also have that ψ is of the form $\psi_0 \to \psi_1$, since otherwise the outermost connective of $f_\pi(\psi)$ could not be an implication. Then, by the induction hypothesis, $\varphi_0 = \psi_0$ and $\varphi_1 = \psi_1$, so $\varphi = \psi$. The case where $\varphi = \neg\varphi_0$ is analogous.

Finally, suppose that one of φ, ψ is of the form $t{:}\theta$ (say, φ). Then, since $f_\pi(\varphi) = \pi(\overline{\ulcorner t \urcorner}, \overline{\ulcorner f_\pi(\theta) \urcorner})$ contains quantifiers (by our assumption on π), so does $f_\pi(\psi)$, which implies that ψ contains some occurrence of a subformula $\psi' = s{:}\gamma$. But then, if we let $\#\xi$ denote the total number of logical symbols in ξ (Booleans and quantifiers), we see that $\#f_\pi(\varphi) = \#f_\pi(\psi) \geq \#f_\pi(\psi') = \#f_\pi(\varphi)$, where the last equality holds because both formulas are instances of π. Thus $\#f_\pi(\psi) = \#f_\pi(\psi')$, which imples that $\psi = \psi'$. It follows by the injectivity of the Gödel numbering that $t = s$ and $f_\pi(\theta) = f_\pi(\gamma)$, and by the induction hypothesis that $\gamma = \theta$, so that $\varphi = \psi$, as needed. □

Note that if π does not contain quantifiers then f_π may fail to be injective, but proof predicates may always be assumed to contain quantifiers (if they don't, dummy quantifiers can always be introduced).

Lemma 6.4 *Given a simple propositional assignment f, there are elementary functions $f^+ : \mathbb{N}^2 \to \mathbb{N}$ and $f^- : \mathbb{N}^2 \to \mathbb{N}$ such that, whenever $\pi = \pi(x, y)$ is a predicate that contains quantifiers and has both variables free, and $\varphi \in L_{\text{¡5!}}$, then*

- $f^+(\ulcorner\varphi\urcorner, \ulcorner\pi\urcorner) = \ulcorner f_\pi(\varphi)\urcorner$,
- $f^-(\ulcorner f_\pi(\varphi)\urcorner, \ulcorner\pi\urcorner) = \ulcorner\varphi\urcorner$, *and*
- $f^-(\ulcorner\psi\urcorner, \ulcorner\pi\urcorner) = 0$ *for any $\psi \in L_{\mathrm{PA}}$ not lying in the range of f_π.*

Proof. By our assumptions on Gödel numbers, we have that, if n is the number of symbols in $\varphi \in L_{\text{¡5!}}$ and the greatest Gödel numbers of a symbol appearing in φ is m, then $n, m \leq \ulcorner\varphi\urcorner$. Meanwhile, for any number k, the length of $\overline{k}$ is $2k+1$, so the number of symbols in $f_\pi(\varphi)$ is bounded by $(2(2\ulcorner\varphi\urcorner + 1) + 3)\ulcorner\varphi\urcorner \cdot \ulcorner\pi\urcorner$. Similarly, $\ulcorner v\urcorner < \ulcorner\pi\urcorner$ if v is any variable appearing in π. If we let c be a bound for the Gödel codes of all constants and logical symbols of L_{PA}, we thus see that every symbol in $f_\pi(\varphi)$ is bounded by $\ulcorner\pi\urcorner + c$. It follows that

$$\ulcorner f_\pi(\varphi)\urcorner < B'(\ulcorner\varphi\urcorner, \ulcorner\pi\urcorner) := B((2(2\ulcorner\varphi\urcorner + 1) + 3)\ulcorner\varphi\urcorner \cdot \ulcorner\pi\urcorner, \ulcorner\pi\urcorner + c),$$

whenever $\varphi \in L_{¡5!}$ and $\pi(x, y) \in L_{\mathrm{PA}}$. Moreover, $f_\pi(\varphi)$ can be defined recursively on the complexity of φ, and thus we may set $z = f^+(x, y)$ if $z < B'(x, y)$ and there are φ, π such that $x = \ulcorner\varphi\urcorner$, $y = \ulcorner\pi\urcorner$ and $z = \ulcorner f_\pi(\varphi)\urcorner$, and $f^+(x, y) = 0$ if there is no such z. The function f^+ thus defined is clearly elementary.

To define f^-, note that $\ulcorner\varphi\urcorner$ can also be bounded elementarily by some elementary function $B''(\ulcorner f_\pi(\varphi)\urcorner, \ulcorner\pi\urcorner)$ (the actual function may be computed analogously to B' and is inessential). Thus we may set $f^-(x, y) = z$ if $x = f^+(z, x)$ with $z < B''(x, y)$, and otherwise, set $f^-(x, y) = 0$. □

Now that we have studied simple propositional assignments in general, let us relate them to our previous results on saturated sets of formulae. Given a saturated set Γ, we will define a propositional assignment designed to 'agree' with Γ.

Definition 6.5 Fix a set of formulae $\Gamma \subseteq L_{¡5!}$. Define a propositional assignment f^Γ given by $f^\Gamma(p) = (\ulcorner p\urcorner = \ulcorner p\urcorner)$ if $p \in \widetilde{\Gamma}$, and $f^\Gamma(p) = (\ulcorner p\urcorner = 0)$ if not.

Clearly, if Γ is finite, then f^Γ is simple. Next, we tailor our proof predicates so that they, too, match well with Γ (and f^Γ):

Definition 6.6 Given a formula π and a finite set of formulae $\Gamma \subseteq L_{¡5!}$, define a new formula

$$\begin{aligned}\texttt{NewProof}_\pi(x, y) = &\texttt{Proof}(x, y) \\ &\vee \exists t \, \exists\varphi \, (x = \ulcorner t\urcorner \wedge \texttt{Comp}_\Gamma(\ulcorner t{:}\varphi\urcorner) \wedge y = (f^\Gamma)^+(\ulcorner\varphi\urcorner, \ulcorner\pi\urcorner)).\end{aligned} \quad (3)$$

Lemma 6.7 *If Γ is any finite set of $L_{¡5!}$ formulae, there is a Δ_0 formula π^Γ containing at least one quantifier, with free variables $\{x, y\}$, and such that*

$$\mathrm{PA} \vdash \forall x \forall y \left(\pi^\Gamma(x, y) \leftrightarrow \texttt{NewProof}_{\pi^\Gamma}(x, y)\right). \quad (4)$$

Proof. Apply the usual fixed-point theorem. □

The reason why we want the formula π^Γ in Lemma 6.7 to contain at least one quantifier is so that it satisfy the hypotheses of Lemma 6.3. Lemma 6.8 below shows that the valuations given by Lemma 6.10 behave just like we would like them to.

Lemma 6.8 *If Γ is a finite, consistent, saturated set of $L_{¡5!}$-formulae, then for every formula φ, $\Gamma \mathrel{|\!\!\sim} \varphi$ if and only if* $\mathrm{PA} \vdash f^\Gamma_{\pi^\Gamma}(\varphi)$.

Proof. In view of Lemma 6.3, $f^\Gamma_{\pi^\Gamma}(\varphi)$ is always Δ_0, so $\mathrm{PA} \vdash f^\Gamma_{\pi^\Gamma}(\varphi)$ is equivalent to $f^\Gamma_{\pi^\Gamma}(\varphi)$. Thus we will only prove that $\Gamma \mathrel{|\!\!\sim} \varphi$ is equivalent to $f^\Gamma_{\pi^\Gamma}(\varphi)$. We proceed by induction on the complexity of φ, considering several cases.

If $\varphi = p$ for some propositional variable p, then PA $\vdash f^{\Gamma}_{\pi^{\Gamma}}(\varphi)$ exactly when $f^{\Gamma}_{\pi^{\Gamma}}(\varphi) = \ulcorner p \urcorner = \ulcorner p \urcorner$, which is the case if and only if $p \in \Gamma$. But this is equivalent, by definition, to $\Gamma \mid\!\sim \varphi$.

If $\varphi = \psi \to \theta$, we have that $f^{\Gamma}_{\pi^{\Gamma}}(\varphi) = f^{\Gamma}_{\pi^{\Gamma}}(\psi) \to f^{\Gamma}_{\pi^{\Gamma}}(\theta)$. By definition, $\Gamma \mid\!\sim \varphi$ if and only if either $\Gamma \not\mid\!\sim \psi$ or $\Gamma \mid\!\sim \theta$. But this is equivalent, by induction hypothesis, to having that either $\neg f^{\Gamma}_{\pi^{\Gamma}}(\psi)$ or $f^{\Gamma}_{\pi^{\Gamma}}(\theta)$ holds, i.e., that $f^{\Gamma}_{\pi^{\Gamma}}(\psi) \to f^{\Gamma}_{\pi^{\Gamma}}(\theta)$ holds. The case where $\varphi = \neg\psi$ for some formula ψ is similar.

Finally, consider $\varphi = t{:}\psi$. Suppose that $\Gamma \mid\!\sim \varphi$, so that $t{:}\psi \in \widetilde{\Gamma}$ by definition. This is equivalent, by Lemma 5.12, to $\mathtt{Comp}_{\Gamma}(\ulcorner t{:}\psi \urcorner)$ holding, which by (3) yields $\mathtt{NewProof}_{\pi^{\Gamma}}(\ulcorner t \urcorner, \ulcorner \psi \urcorner)$, as well as $\pi^{\Gamma}(\ulcorner t \urcorner, \ulcorner f^{\Gamma}_{\pi^{\Gamma}}(\psi) \urcorner)$ by (4). But the latter is just $f^{\Gamma}_{\pi^{\Gamma}}(\varphi)$, and since this is a Δ_0 formula it follows that PA $\vdash f^{\Gamma}_{\pi^{\Gamma}}(\varphi)$, as needed.

Conversely, suppose that $\Gamma \not\mid\!\sim \varphi$, so that $t{:}\psi \notin \widetilde{\Gamma}$, i.e., $\mathtt{Comp}_{\Gamma}(\ulcorner t{:}\psi \urcorner)$ is false. Since by assumption, $\mathtt{Proof}(\ulcorner t \urcorner, k)$ is false for all k, it follows that $\mathtt{NewProof}_{\pi^{\Gamma}}(\ulcorner t \urcorner, \ulcorner \psi \urcorner)$ is false and hence so is $\pi^{\Gamma}(\ulcorner t \urcorner, \ulcorner f^{\Gamma}_{\pi^{\Gamma}}(\psi) \urcorner)$, i.e., $\neg\pi^{\Gamma}(\ulcorner t \urcorner, \ulcorner f^{\Gamma}_{\pi^{\Gamma}}(\psi) \urcorner)$ is true. But this formula is Δ_0, whence PA $\vdash \neg\pi^{\Gamma}(\ulcorner t \urcorner, \ulcorner f^{\Gamma}_{\pi^{\Gamma}}(\psi) \urcorner) = f^{\Gamma}_{\pi^{\Gamma}}(\varphi)$.

Since we have considered all cases, the lemma follows. □

The following lemma shows that the predicate π^{Γ} is extensionally correct.

Lemma 6.9 *Let Γ be any finite, consistent, and saturated set of $L_{¡5!}$-formulae. Then, for any $\varphi \in L_{\mathrm{PA}}$, PA $\vdash \varphi$ if and only if $\exists x\, \pi^{\Gamma}(x, \ulcorner \varphi \urcorner)$ holds.*

Proof. One direction is obvious from (3), since $\mathtt{Proof}(k, \ulcorner \varphi \urcorner)$ implies that $\mathtt{NewProof}_{\pi^{\Gamma}}(k, \ulcorner \varphi \urcorner)$, and thus that $\pi^{\Gamma}(k, \ulcorner \varphi \urcorner)$.

For the other, we note that if $\pi^{\Gamma}(k, \ulcorner \varphi \urcorner)$ holds, either we already have that $\mathtt{Proof}(k, \ulcorner \varphi \urcorner)$, or else $k = \ulcorner t \urcorner$ for some term t and $\varphi = f^{\Gamma}_{\pi^{\Gamma}}(\theta)$ for some $\theta \in L_{¡5!}$ such that $t{:}\theta \in \widetilde{\Gamma}$, which by definition means that $\Gamma \mid\!\sim t{:}\theta$. By Lemma 5.11, $\Gamma \mid\!\sim \theta$, and thus PA $\vdash \varphi$ by Lemma 6.8. □

The following is the main technical lemma:

Lemma 6.10 *Given a finite, consistent, saturated set of formulae Γ, there are elementary functions $\mathsf{m}^*, \mathsf{a}^*, \mathsf{v}^*$ such that $\mathfrak{S}^* = (f^{\Gamma}, \pi^{\Gamma}, \mathsf{m}^*, \mathsf{a}^*, \mathsf{v}^*)$ is a robust arithmetical interpretation and satisfies $\varphi^{\mathfrak{S}} = f^{\Gamma}_{\pi^{\Gamma}}(\varphi)$ for all $\varphi \in L_{¡5!}$.*

Theorem 4.4 readily follows from Lemma 6.10:

Proof of Theorem 4.4. By Lemma 5.2, if φ is consistent, then it can be included in a finite, consistent, saturated set Γ. By Lemma 6.8, if $\Gamma \mid\!\sim \gamma$, then PA $\vdash f^{\Gamma}_{\pi^{\Gamma}}(\gamma)$; in particular, PA $\vdash f^{\Gamma}_{\pi^{\Gamma}}(\varphi)$. By Lemma 6.10, there is a robust arithmetical interpretation $\mathfrak{S}^*$ which satisfies $\psi^{\mathfrak{S}^*} = f^{\Gamma}_{\pi^{\Gamma}}(\psi)$ for all $\psi \in L_{¡5!}$. It follows that PA $\vdash \varphi^{\mathfrak{S}^*}$, as needed. □

Hence, it remains to prove Lemma 6.10. The proof uses the following fact:

Lemma 6.11 *If φ is Δ_0, then there is an elementary function $b(\cdot)$ such that then PA $\vdash \varphi(x) \wedge x < u \to \exists z \leq b(u)\; \pi_f(z, \ulcorner \varphi(x) \urcorner)$.*

Proof. If φ is Δ_0, then PA proves that for all $x \in \mathbb{N}$ such that $\varphi(x)$ holds, there is some $z \leq 2_l^u$ with $\texttt{Proof}(z, \ulcorner\varphi(x)\urcorner)$, where l is some constant that depends only on φ (see, for example, [8]). But then, once we have $\texttt{Proof}(z, \ulcorner\varphi(x)\urcorner)$, we also have $\pi_f(z, \ulcorner\varphi(x)\urcorner)$ using Lemma 6.7 and the definition of $\texttt{NewProof}$, so we may take $b(u) = 2_l^u$. □

Proof of Lemma 6.10. Let a_{PA} and m_{PA} be the computable functions introduced immediately after Definition 4.1, and $\mu x.\varphi(x)$ denote the least x satisfying φ, if it exists. We define $\mathsf{a}^*(n,k) = \ulcorner t + s\urcorner$ if both $n = \ulcorner t\urcorner$ and $k = \ulcorner s\urcorner$ for some proof terms s and t. If neither n nor k code proof terms, set $\mathsf{a}^*(n,k) = \mathsf{a}_{\mathrm{PA}}(n,k)$. Otherwise if, say, only n codes a proof term, define $\mathsf{a}^*(n,k) = \mathsf{a}_{\mathrm{PA}}(y,k)$, where $y = \mu x.\forall\varphi \in [\![n]\!]_{\pi^\Gamma}\, \texttt{Proof}(x, \ulcorner f^\Gamma_{\pi^\Gamma}(\varphi)\urcorner)$. As we show, the existence of such a y follows from the fact that by Lemma 6.9, PA $\vdash f^\Gamma_{\pi^\Gamma}(\varphi)$ for each $\varphi \in [\![n]\!]_{\pi^\Gamma}$. Indeed, this implies that for each $\varphi_i \in [\![n]\!]_{\pi^\Gamma} =: \{\varphi_0, \ldots, \varphi_n\}$, we can find some z_i such that $\texttt{Proof}(z_i, \ulcorner f^\Gamma_{\pi^\Gamma}(\varphi_i)\urcorner)$. Define $x_0 = z_0$ and $x_{i+1} = \mathsf{a}_{\mathrm{PA}}(x_i, z_{i+1})$. Hence, $x = x_n$ witnesses that y above is well-defined. If only k codes a proof term, define $\mathsf{a}^*(n,k)$ symmetrically.

Analogously, set $\mathsf{m}^*(n,k) = \ulcorner t \cdot s\urcorner$ if both $n = \ulcorner t\urcorner$, and $k = \ulcorner s\urcorner$ for some proof terms s, t; $\mathsf{m}^*(n,k) = \mathsf{m}_{\mathrm{PA}}(n,k)$ if n and k do not code proof terms; $\mathsf{m}^*(n,k) = \mathsf{m}_{\mathrm{PA}}(y,k)$ if only n codes a proof term, where $y = \mu x.\forall\varphi \in [\![n]\!]_{\pi^\Gamma}\, \texttt{Proof}(x, \ulcorner f^\Gamma_{\pi^\Gamma}(\varphi)\urcorner)$. The existence of such a y is proved as before. The remaining case is defined symmetrically.

We want to show that a^* and m^* are computable. From Lemma 5.12 and Lemma 6.7 follows that π^Γ is computable. By assumption, the set of codes of proof terms is elementary. Thus, it suffices to show that $x \mapsto [\![x]\!]_{\pi^\Gamma}$ is computable.

Claim 1 *For all x, $[\![x]\!]_{\pi^\Gamma}$ is finite, and the function $x \mapsto [\![x]\!]_{\pi^\Gamma}$ is computable.*

Proof. By Lemma 6.7, $y \in [\![x]\!]_{\pi^\Gamma}$ if and only if either $\texttt{Proof}(x,y)$ or x and y respectively code expressions t and $f^\Gamma_{\pi^\Gamma}(\varphi)$ such that $t{:}\varphi \in \widetilde{\Gamma}$. By our conventions on Gödel numbering, at most one of these alternatives occurs, and which one does—if any—is determined by whether x codes a proof term. If it does not, then the desired result is immediate, as $[\![\cdot]\!]_{\texttt{Proof}}$ is finite-valued and computable by assumption. Otherwise, $x = \ulcorner t\urcorner$ for some proof term t and $\pi^\Gamma(x,y)$ holds. Then y codes some formula $f^\Gamma_{\pi^\Gamma}(\varphi)$ with $\varphi \in \widetilde{\Gamma}$. Recovering φ from y is elementary by Lemma 6.4, and so too is deciding whether $\varphi \in \widetilde{\Gamma}$, by Lemma 5.12. The expression $t{:}\varphi$ is Γ-balanced, whence φ either occurs in t or in Γ. Hence, the Gödel number of any π^Γ-conclusion of x is bounded by $x + \ulcorner\Gamma\urcorner$, and we can computably enumerate all of them. □

Claim 2 *$(\pi^\Gamma, \mathsf{m}^*, \mathsf{a}^*)$ is a normal proof system.*

Proof. Note that the formula π^Γ is Δ_0 by Lemma 6.7, and each $x \in \mathbb{N}$ has only finitely-many π^Γ-conclusions by Claim 1. Moreover, a^* and m^* satisfy the required conditions—we prove this for m^*. Clearly, whenever k and n do not code proof terms, then $\pi^\Gamma(\mathsf{m}^*(n,k), \ulcorner\psi\urcorner)$ is true if so too are $\pi^\Gamma(n, \ulcorner\varphi \to \psi\urcorner)$ and $\pi^\Gamma(k, \ulcorner\varphi\urcorner)$, as π^Γ is essentially the predicate $\texttt{Proof}$ in this case. Meanwhile,

$\mathsf{m}^*(n,k) = \mathsf{m}_{\mathrm{PA}}(n,k)$ in this case, which is computable by assumption.

If only k codes a proof term s and $\varphi = f^{\Gamma}_{\pi^\Gamma}(\theta)$ for some $\theta \in L_{¡5!}$, then by (3), $\pi^\Gamma(n, \ulcorner\varphi \to \psi\urcorner)$ and $\pi^\Gamma(k, \ulcorner\varphi\urcorner)$ yield $\mathtt{Proof}(n, \ulcorner\varphi \to \psi\urcorner)$ and $\mathtt{Comp}_\Gamma(\ulcorner s{:}\theta\urcorner)$. In this case, $\mathsf{m}^*(n,k)$ is defined as $\mathsf{m}_{\mathrm{PA}}(n,y)$, where y is least such that $\forall\phi \in [\![k]\!]_{\pi^\Gamma}\ \mathtt{Proof}(y, \ulcorner f^{\Gamma}_{\pi^\Gamma}(\phi)\urcorner)$. Moreover, y can be computed from k, since $[\![\cdot]\!]_{\pi^\Gamma}$ and $\mathtt{Proof}$ are computable, and hence so is $\mathsf{m}_{\mathrm{PA}}(n,k)$. The case where only n codes a proof term is symmetric, and if n and k respectively code proof terms t and s, so that, say, $\varphi = f^{\Gamma}_{\pi^\Gamma}(\theta)$ and $\psi = f^{\Gamma}_{\pi^\Gamma}(\chi)$, then we obtain $\mathtt{Comp}_\Gamma(\ulcorner s{:}\theta\urcorner)$, as well as $\mathtt{Comp}_\Gamma(\ulcorner t{:}\theta \to \chi\urcorner)$, which together imply $\mathtt{Comp}_\Gamma(\ulcorner t\cdot s{:}\chi\urcorner)$ and thus $\pi^\Gamma(\mathsf{m}^*(n,k), \ulcorner\psi\urcorner)$. Clearly, the map $t, s \mapsto \ulcorner t\cdot s\urcorner$ is computable.

A similar argument shows that a^* also has the required properties, and thus $(\pi^\Gamma, \mathsf{m}^*, \mathsf{a}^*)$ is a normal proof system, as claimed. □

To define v^*, we need an auxiliary function. Let $b_t(x)$ be an elementary function such that whenever x codes a tautology, then $\mathtt{Proof}(k,z)$ holds for some $k < b_t(x)$; such a function can easily be computed for any reasonable proof system and is typically exponential. Whenever x codes a Δ_0 formula ϕ, define $b_x(\cdot)$ to be the function obtained by applying Lemma 6.11 to ϕ, so that $\phi(k)$ and $k < u$ imply $\pi^\Gamma(z, \ulcorner\phi(k)\urcorner)$ for some $z < b_x(u)$. Set $\mathsf{v}_{\mathrm{PA}}(x) = \mu y < b_x(x) + b_t(x).\,\pi^\Gamma(y,x)$ if such a y exists, and $\mathsf{v}_{\mathrm{PA}}(x) = 0$ otherwise. The function b_x is computable, whence so too is v_{PA}.

If $\pi^\Gamma(k,\varphi)$ holds, then, as it is a true Δ_0 sentence and $k < \ulcorner\pi^\Gamma(k,\varphi)\urcorner$, it follows that $\mathsf{v}_{\mathrm{PA}}(\ulcorner\pi^\Gamma(k,\varphi)\urcorner)$ is nonzero (here, we take $\phi(\cdot) = \pi^\Gamma(\cdot,\varphi)$). Hence,

$$\pi^\Gamma(k,\varphi) \to \pi^\Gamma(\mathsf{v}_{\mathrm{PA}}(\ulcorner\pi^\Gamma(k,\varphi)\urcorner), \ulcorner\pi^\Gamma(k,\varphi)\urcorner)$$

is valid. More generally, for any Δ_0-formula ϕ— in particular, for $\phi = \neg\pi^\Gamma(\cdot,\varphi)$— we have

$$\phi \to \pi^\Gamma(\mathsf{v}_{\mathrm{PA}}(\ulcorner\phi\urcorner), \ulcorner\phi\urcorner). \tag{5}$$

We also have $\varphi \to \pi^\Gamma(\mathsf{v}_{\mathrm{PA}}(\ulcorner\varphi\urcorner), \varphi)$, whenever φ is a tautology. Given this, define $\mathsf{v}^*(x) = \ulcorner ¡\varphi!\urcorner$ if $x = \ulcorner f^{\Gamma}_{\pi^\Gamma}(\varphi)\urcorner$ for some φ in the domain of ¡·!, and $\mathsf{v}^*(x) = \mathsf{v}_{\mathrm{PA}}(x)$ if no such φ exists. We may now define $\mathfrak{S}^* = (f^\Gamma, \pi^\Gamma, \mathsf{m}^*, \mathsf{a}^*, \mathsf{v}^*)$.

Claim 3 *If t is any term of $L_{¡5!}$, then $t^{\mathfrak{S}^*} = \ulcorner t\urcorner$, and if φ is any formula, $\varphi^{\mathfrak{S}^*} = f^{\Gamma}_{\pi^\Gamma}(\varphi)$.*

Proof. We prove both claims simultaneously by induction on any expression ϵ of $L_{¡5!}$. Consider first the case where ϵ is a formula. We have that $p^{\mathfrak{S}^*} = f^\Gamma(p) = f^{\Gamma}_{\pi^\Gamma}(p)$ is true for atomic p and, by definition, $f^{\Gamma}_{\pi^\Gamma}$ and $\cdot^{\mathfrak{S}^*}$ both commute with Booleans. For a formula $t{:}\theta$, we use the identity $t^{\mathfrak{S}^*} = \ulcorner t\urcorner$ to see that $f^{\Gamma}_{\pi^\Gamma}(t{:}\theta) = \pi^\Gamma(\overline{\ulcorner t\urcorner}, \overline{\ulcorner f^{\Gamma}_{\pi^\Gamma}(\theta)\urcorner}) \overset{\mathrm{IH}}{=} \pi^\Gamma(\overline{t^{\mathfrak{S}^*}}, \overline{\ulcorner\theta^{\mathfrak{S}}\urcorner}) = (t{:}\theta)^{\mathfrak{S}}$.

Now assume that ϵ is a term. We must consider several cases, depending on the form of ϵ, and use the definitions of a^*, m^*, and v^* to show that $\epsilon^{\mathfrak{S}^*} = \ulcorner\epsilon\urcorner$. We only consider two cases as examples. If $\epsilon = ¡\varphi!$, $¡\varphi!^{\mathfrak{S}^*} = \mathsf{v}^*(\varphi^{\mathfrak{S}^*}) \overset{\mathrm{IH}}{=}$

$\mathsf{v}^*(f^{\Gamma}_{\ulcorner\pi\urcorner}(\varphi)) = \ulcorner¡\varphi!\urcorner$; similarly, $(t\cdot s)^{\mathfrak{S}^*} = \mathsf{m}^*(t^{\mathfrak{S}^*}, s^{\mathfrak{S}^*}) \overset{\text{IH}}{=} \mathsf{m}^*(\ulcorner t\urcorner, \ulcorner s\urcorner) = \ulcorner t\cdot s\urcorner$. Considering $\epsilon = t+s$ and $\epsilon = x$ concludes the proof. □

Finally, note that $x = \ulcorner f^{\Gamma}_{\ulcorner\pi\urcorner}(\varphi)\urcorner$ for some φ in the domain of ¡ · ! iff x codes a formula in the range of $f^{\Gamma}_{\ulcorner\pi\urcorner}$, iff $f^{-}(x, \ulcorner\pi^{\Gamma}\urcorner) > 0$, and the function f^{-} is elementary by Lemma 6.4. Therefore, v^* is computable. It follows that $\mathfrak{S}^*$ is an arithmetical interpretation, and by (5), it is robust.

This finishes the proof of Lemma 6.10. □

7 An afterword on realizability

Recall that S5 is the normal modal logic generated by positive and negative introspections, and the reflection axiom T: $\Box p \to p$. The purpose of introducing VS5 is to obtain a justification logic which combines two properties:

(i) it realizes S5, and

(ii) it is sound and complete for its arithmetical interpretation.

In this article we have proven the second point and we leave the first for future work. However, in this section we will say a few words about it.

Let us use $(\cdot)^{\Box}$ to denote the "forgetful projection," which recursively replaces instances of $t{:}\varphi$ by $\Box(\varphi)^{\Box}$, commutes with Booleans and fixes propositional variables. Then, the following can be easily verified by induction on the length of a derivation:

Theorem 7.1 *For $\varphi \in L_{¡5!}$, if φ is derivable in* VS5 *then $(\varphi)^{\Box}$ is derivable in* S5.

A more interesting property would be the converse of this result; if we are given φ in the modal language such that S5 $\vdash \varphi$, can we find a formula φ^r in $L_{¡5!}$ such that $(\varphi^r)^{\Box} = \varphi$ and VS5 $\vdash \varphi$? Such a φ^r is a *realization* of φ. Let us see that all axioms of the respective modal logics have realizations.

Theorem 7.2 *If φ is an axiom of* S5, *then there is $\varphi^r \in L_{¡5!}$ such that $(\varphi^r)^{\Box} = \varphi$ and* VS5 $\vdash \varphi$.

Proof. Let x, y be arbitrary term variables.

- $\Box(p \to q) \to (\Box p \to \Box q)$ is realized by $x{:}(p \to q) \to (y{:}p \to (x\cdot y){:}q)$;
- $\Box p \to p$ is realized by $x{:}p \to p$;
- $\Box p \to \Box\Box p$ is realized by $x{:}p \to ¡x{:}p!{:}(x{:}p)$, and
- $\neg\Box p \to \Box\neg\Box p$ is realized by $\neg x{:}p \to ¡\neg x{:}p!{:}(\neg x{:}p)$.

□

Thus we see that it is relatively straightforward to realize axioms of S5. One would then expect to be able to 'cobble together' such realizations so as to realize more complex theorems, and while doing so is not trivial, it is possible for JS5. We believe this to also be the case for VS5, but leave it for future work. Thus we conclude our discussion on realizations with the following conjecture:

Conjecture 7.3 *If* $\mathsf{S5} \vdash \varphi$*, then there is* $\varphi^r \in L_{¡5!}$ *such that* $(\varphi^r)^\Box = \varphi$ *and* $\mathsf{VS5} \vdash \varphi^r$.

Acknowledgements. This work was inspired by discussions with Sergei Artemov and Melvin Fitting at the Second International Wormshop in Mexico City, 2014. Specifically, the issues with modeling negative introspection were brought to our attention by Fitting, and both later provided helpful comments for preparing this paper. We are also indebted to Lev Beklemishev, Eric Pacuit and Fernando Velázquez-Quesada for kindly answering our questions about justification logics.

References

[1] Artemov, S. N., *Explicit provability and constructive semantics*, Bulletin of Symbolic Logic **7** (2001), pp. 1–36.

[2] Artemov, S. N. and M. Fitting, *Justification logic*, in: E. N. Zalta, editor, *The Stanford Encyclopedia of Philosophy*, 2012, fall 2012 edition .
URL
`http://plato.stanford.edu/archives/fall2012/entries/logic-justification/`

[3] Artemov, S. N., E. Kazakov and D. Shapiro, *On logic of knowledge with justifications*, Technical Report CFIS **99-12** (1999).

[4] Boolos, G. S., "The Logic of Provability," Cambridge University Press, Cambridge, 1993.

[5] Fitting, M., *The logic of proofs, semantically*, Annals of Pure and Applied Logic **132** (2004), pp. 1–25.

[6] Fitting, M., *The realization theorem for* S5*: a simple, constructive proof*, in: *Games, Norms and Reasons*, Synthese Library **353**, Springer Science+Business Media B.V., 2011 pp. 61–76.

[7] Goris, E., *Feasible operations on proofs: the logic of proofs for bounded arithmetic*, Theory of Computing Systems **43** (2008), p. 185203.

[8] Hájek, P. and P. Pudlák, "Metamathematics of First Order Arithmetic," Springer-Verlag, Berlin, Heidelberg, New York, 1993.

[9] Kuznets, R. and T. Studer, *Update as evidence: Belief expansion*, in: S. N. Artemov and A. Nerode, editors, *Logical Foundations of Computer Science: International Symposium, LFCS 2013, San Diego, CA, USA, January 6-8, 2013. Proceedings*, Springer Berlin Heidelberg, Berlin, Heidelberg, 2013 pp. 266–279.

[10] Kuznets, R. and T. Studer, *Weak arithmetical interpretations for the logic of proofs*, Logic Journal of IGPL (2016).

[11] Pacuit, E., *A note on some explicit modal logics*, ILLC tech report (2006).

[12] Rubtsova, N., *On realization of* S5*-modality by evidence terms*, J. Log. Comput. **16** (2006), pp. 671–684.

A Paraconsistent View on B and S5

Arnon Avron [1]

Tel Aviv University, Ramat Aviv, Israel

Anna Zamansky

Haifa University, Haifa, Israel

Abstract

Paraconsistent logics are logics that in contrast to classical and intuitionistic logic, do not trivialize inconsistent theories. In this paper we show that the famous modal logics **B** and **S5**, can be viewed as paraconsistent logics with several particularly useful properties.

Keywords: KTB, B, modal logic, paraconsistent logic

1 Introduction

One of most counter-intuitive properties of classical logic (as well as of its most famous rival, intuitionistic logic) is the fact that it allows the inference of any proposition from a single pair of contradicting statements. This principle (known as the principle of explosion, 'ex falso sequitur quodlibet') has repeatedly been attacked on philosophical ground, as well as because of practical reasons: in its presence every inconsistent theory or knowledge base is totally trivial, and so useless. Accordingly, over the last decades a lot of work and efforts have been devoted to develop alternatives to classical logic that do not have this drawback. Such alternatives are known as *paraconsistent* logics.

In this paper we embark on a search for a paraconsistent logic which has particularly important properties. The most important of them is what is known as the *replacement* property, which basically means that equivalence of formulas implies their congruence. We show that the minimal paraconsistent logic which satisfies our criteria is in fact the famous Brouwerian modal logic **B** (also known as KTB [13,22]). This logic, in turn, is also shown to be a member of the well-studied family of paraconsistent logics known as C-systems ([16,18,19]). We further show that **B** is very robust paraconsistent logic in the sense that almost any axiom which has been used in the context of C-systems is either a theorem of **B**, or its addition to **B** leads to a logic which is no longer

[1] Both authors were supported by the Israel Science Foundation under grant agreement 817/15.

paraconsistent. There is exactly one (rather notable) exception, and the result of adding this exception to **B** is another famous modal logic: **S5**.

2 Paraconsistent Logics

We assume that all propositional languages share the same set $\{P_1, P_2, \ldots\}$ of atomic formulas, and use p, q, r to vary over this set. The set of well-formed formulas of a propositional language $\mathcal{L}$ is denoted by $\mathcal{W}(\mathcal{L})$, and φ, ψ, σ will vary over its elements.

Definition 2.1 A (Tarskian) *consequence relation* (tcr) for a language $\mathcal{L}$ is a binary relation $\vdash$ between sets of $\mathcal{L}$-formulas and $\mathcal{L}$-formulas, satisfying the following three conditions:

Reflexivity: if $\psi \in T$ then $T \vdash \psi$.
Monotonicity: if $T \vdash \psi$ and $T \subseteq T'$ then $T' \vdash \psi$.
Transitivity: if $T \vdash \psi$ and $T, \psi \vdash \varphi$ then $T \vdash \varphi$.

Definition 2.2 A *propositional logic* is a pair $\mathbf{L} = \langle \mathcal{L}, \vdash \rangle$, where $\vdash$ is a tcr for $\mathcal{L}$ which satisfies the following two conditions:

Structurality: if $T \vdash \varphi$ then $\sigma(T) \vdash \sigma(\varphi)$ for any substitution σ in $\mathcal{L}$.
Non-triviality: $p \not\vdash q$ for any distinct propositional variables p, q.

Various general notions of paraconsistent logics have been considered (see, e.g., [14,2,3,1]). In this paper we focus on a particular class of paraconsistent logics, which extend the positive fragment of classical logic as follows.

Notation: $\mathcal{L}_{CL^+} = \{\wedge, \vee, \supset\}$, $\mathcal{L}^{\mathsf{F}}_{CL} = \{\wedge, \vee, \supset, \mathsf{F}\}$ and $\mathcal{L}_{CL} = \{\wedge, \vee, \supset, \neg\}$.

Definition 2.3 $\mathbf{IL}^+$ is the minimal logic $\mathbf{L}$ in $\mathcal{L}_{CL^+}$ such that:

- $\mathcal{T} \vdash_{\mathbf{L}} A \supset B$ iff $\mathcal{T}, A \vdash_{\mathbf{L}} B$
- $\mathcal{T} \vdash_{\mathbf{L}} A \wedge B$ iff $\mathcal{T} \vdash_{\mathbf{L}} A$ and $\mathcal{T} \vdash_{\mathbf{L}} B$
- $\mathcal{T}, A \vee B \vdash_{\mathbf{L}} C$ iff $\mathcal{T}, A \vdash_{\mathbf{L}} C$ and $\mathcal{T}, B \vdash_{\mathbf{L}} C$

$\mathbf{CL}^+$, the $\mathcal{L}_{CL^+}$-fragment of classical logic, is obtained by extending $\mathbf{IL}^+$ with the axiom $A \vee (A \supset B)$. $\mathbf{CL}^{\mathsf{F}}$, the full classical logic (in $\mathcal{L}^{\mathsf{F}}_{CL}$) is obtained by extending $\mathbf{CL}^+$ with the axiom $\mathsf{F} \supset \psi$ (making F a bottom element.) [2]

Definition 2.4 A propositional logic $\mathbf{L} = \langle \mathcal{L}, \vdash_{\mathbf{L}} \rangle$ is $\neg$-*classical* if $\mathcal{L}_{CL} \subseteq \mathcal{L}$, the $\mathcal{L}_{CL^+}$-fragment of $\mathbf{L}$ is $\mathbf{CL}^+$, and $\mathbf{L}$ satisfies the three conditions concerning $\vee, \wedge$ and $\supset$ that were used in Definition 2.3 to characterize $\mathbf{IL}^+$.

---

[2] Another natural alternative for obtaining classical logic is to use the language $\mathcal{L}_{CL}$ rather than $\mathcal{L}^{\mathsf{F}}_{CL}$, and extend $\mathbf{CL}^+$ with the axioms [t] $\neg\psi \vee \psi$ and [$\neg \supset$] $(\neg\psi \supset (\psi \supset \varphi))$. To avoid confusions with the paraconsistent negation which is added to $\mathcal{L}_{CL^+}$ below, we use in this paper the approach with F.

Definition 2.5 A $\neg$-classical logic is *paraconsistent (with respect to $\neg$)* if $\neg$ satisfies the following conditions: (i) $\nvdash_{\mathbf{L}} (p \wedge \neg p) \supset q$, (ii) $\nvdash_{\mathbf{L}} p \supset \neg p$, and (iii) $\nvdash_{\mathbf{L}} \neg p \supset p$.

Remark 2.6 Most of the earlier definitions of paraconsistent logics do not explicitly require conditions (ii) and (iii). However, they have been required in the literature for negation in general (cf. [25,26,29,4,2]. In the latter two papers a connective which satisfies these two conditions is called *weak negation*). In [3,1] paraconsistent logics were defined using an even more restrictive condition (called $\neg$-*containment in classical logic*) on a connective $\neg$ to be counted as a negation in a logic **L**.That condition implies conditions (ii) and (iii) above, but they do suffice for the purposes of this paper.

The above definition mentions only negative properties of negation. Below is a list of positive properties that negation has in classical logic that might be desirable also in the context of paraconsistent logics:

Definition 2.7 Let $\mathbf{L} = \langle \mathcal{L}, \vdash_{\mathbf{L}} \rangle$ be a propositional logic for a language $\mathcal{L}$ with a unary connective $\neg$.

- $\neg$ is *complete* (for **L**) if it satisfies the following version of the *law of excluded middle*: (LEM) $\mathcal{T} \vdash_{\mathbf{L}} \varphi$ whenever $\mathcal{T}, \psi \vdash_{\mathbf{L}} \varphi$ and $\mathcal{T}, \neg\psi \vdash_{\mathbf{L}} \varphi$.
- $\neg$ is *right-involutive* (for **L**) if $\varphi \vdash_{\mathbf{L}} \neg\neg\varphi$ every formula φ (equivalently: for atomic φ), and is *left-involutive* (for **L**) if $\neg\neg\varphi \vdash_{\mathbf{L}} \varphi$ for every formula φ (equivalently: for atomic φ). $\neg$ is *involutive* if it is both right- and left-involutive.
- $\neg$ is *contrapositive* (in **L**) if $\neg\varphi \vdash_{\mathbf{L}} \neg\psi$ whenever $\psi \vdash_{\mathbf{L}} \varphi$.

Remark 2.8 It is easy to verify that if **L** is $\neg$-classical then:

- $\neg$ is complete for **L** iff $\vdash_{\mathbf{L}} \neg\varphi \vee \varphi$ for every φ.
- $\neg$ is right-involutive for **L** iff $\vdash_{\mathbf{L}} \varphi \supset \neg\neg\varphi$ for every φ.
- $\neg$ is left-involutive for **L** iff $\vdash_{\mathbf{L}} \neg\neg\varphi \supset \varphi$ for every φ.
- $\neg$ is contrapositive for **L** iff $\vdash_{\mathbf{L}} \neg\varphi \supset \neg\psi$ whenever $\vdash_{\mathbf{L}} \psi \supset \varphi$.

The next proposition shows that $\neg$-classical paraconsistent logics cannot enjoy *all* of the above properties of negation at the same time:

Proposition 2.9 *A $\neg$-classical logic in which $\neg$ is complete, right-involutive, and contrapositive cannot be paraconsistent.*

Completeness (or the law of excluded middle) is a very basic and natural property of negation, which is particularly important to retain in paraconsistent logics (which reject the other basic principle which characterizes classical negation - see Footnote 2). The minimal extension of $\mathbf{CL}^{+}$ which has complete negation is the paraconsistent logic **CLuN**, introduced by Batens under the name of PI in [6] and further studied in [7,9]; a Hilbert-style system for it is given in Fig. 1.

Demanding just completeness of our paraconsistent negation is obviously not sufficient, though, and we would like to preserve as much of the main classical properties of negation as possible. Proposition 2.9 means that we cannot have a complete paraconsistent negation which is both contrapositive and right-involutive. So we should choose between these two properties. On the other hand the demand of being left-involutive causes no problem, and so we have no reason not to impose it. Doing this leads to the paraconsistent logic known in the literature as $\mathbf{C}_{min}$ (used by the authors in [15,14] as the basis [3] for their taxonomy of C-systems), which is the minimal extension of $\mathbf{CL}^+$ which has both a complete and left-involutive negation. Fig. 1 contains a Hilbert-style system for this logic as well.

Inference Rule: [MP] $\dfrac{\psi \quad \psi \supset \varphi}{\varphi}$

Axioms of ***HCL***$^+$**:**

[$\supset$1]	$\psi \supset (\varphi \supset \psi)$
[$\supset$2]	$(\psi \supset (\varphi \supset \tau)) \supset ((\psi \supset \varphi) \supset (\psi \supset \tau))$
[$\wedge\supset$]	$\psi \wedge \varphi \supset \psi, \ \psi \wedge \varphi \supset \varphi$
[$\supset\wedge$]	$\psi \supset (\varphi \supset \psi \wedge \varphi)$
[$\supset\vee$]	$\psi \supset \psi \vee \varphi, \ \varphi \supset \psi \vee \varphi$
[$\vee\supset$]	$(\psi \supset \tau) \supset ((\varphi \supset \tau) \supset (\psi \vee \varphi \supset \tau))$
[$\supset$3]	$((\psi \supset \varphi) \supset \psi) \supset \psi$

Axioms of ***HCLuN*****:** The axioms of *HCL*$^+$ and:

[t]	$\neg\psi \vee \psi$

Axioms of ***HC***$_{min}$**:** The axioms of *HCLuN* and:

[c]	$\neg\neg\psi \supset \psi$

Fig. 1. The proof systems *HCL*$^+$, *HCLuN* and *HC*$_{min}$

Returning to the choice between having a negation which is involutive and a negation which is contrapositive, we note that almost no paraconsistent logic studied in the literature has a contrapositive negation [4]. In this paper we investigate what happens if we do follow this choice. As we show below, doing this is more challenging than securing the other properties, as it cannot be achieved by just adding axioms, and so another, more sophisticated way is needed.

---

[3] In [5] it is argued that the logic **BK**, introduced in the next subsection, is more appropriate as the basic C-system.

[4] An early notable exception is the system CC_ω studied in [38], and extended to a S5-like system in [21]. More recent related works are mentioned in Remark 4.3.

2.1 The family of C-systems

One of the oldest and best known approaches to paraconsistency is da Costa's approach ([16,18,19]), which seeks to allow the use of classical logic whenever it is safe to do so, but behaves completely differently when contradictions are involved. This approach has led to the introduction of the family of *Logics of Formal (In)consistency (LFIs)* ([14,15]). This family is based on the idea that the notion of consistency can be expressed in the language of the logic itself. In most of the LFIs studied in the literature this is done via a *consistency operator*. The expected "classical" behavior of a "consistent" formula ψ for which $\circ\psi$ holds, is expressed via the following conditions:

Definition 2.10 Let **L** be a logic for $\mathcal{L}$. A (primitive or defined) connective $\circ$ of **L** is a *consistency operator* with respect to $\neg$ if the following conditions are satisfied:

- **(b)** $\vdash_{\mathbf{L}} (\circ\psi \wedge \neg\psi \wedge \psi) \supset \varphi$ for every $\psi, \varphi \in \mathcal{W}(\mathcal{L})$.
- $(\mathbf{n}_1)$ $\nvdash_{\mathbf{L}} (\circ p \wedge \neg p) \supset q$
- $(\mathbf{n}_2)$ $\nvdash_{\mathbf{L}} (\circ p \wedge p) \supset q$

We say that $\circ$ is a *strong consistency operator* with respect to $\neg$ if it is a consistency operator which satisfies also **(k)** $\circ\psi \vee (\neg\psi \wedge \psi)$ for every $\psi \in \mathcal{W}(\mathcal{L})$.

Proposition 2.11 *Let* **L** *be a* $\neg$*-classical paraconsistent logic.*

- *If* $\circ$ *is a consistency operator with respect to* $\neg$*, then* $\otimes\psi =_{Df} (\neg\psi \wedge \psi) \supset \circ\psi$ *is a strong consistency operator with respect to* $\neg$*.*
- *For every* φ*,* $\otimes_{\varphi}\psi =_{Df} (\neg\psi \wedge \psi) \supset (\circ\varphi \wedge \neg\varphi \wedge \varphi)$ *is a strong consistency operator.*
- *If* $\circ$ *is a consistency operator while* $\otimes$ *is a strong consistency operator, then* $\vdash_{\mathbf{L}} \circ\psi \supset \otimes\psi$*.*
- *A strong consistency operator for* **L** *is unique up to equivalence.*

Definition 2.12 *Let* **L** *be a* $\neg$*-classical logic.* **L** *is a* C-system *if it is paraconsistent (w.r.t.* $\neg$*) and has a strong consistency operator* $\circ$ *(w.r.t.* $\neg$*).*

Theorem 2.13 *A* $\neg$*-classical paraconsistent logic is a C-system iff it is an extension (perhaps by definitions) of* $\mathbf{CL}^{\mathsf{F}}$ *(see Definition 2.3).*

The following is what we believe best deserves to be called the basic C-system (the argument is given in [5], see also Footnote 2):

Definition 2.14 The logic **BK** is obtained by extending $\mathbf{CL}^{+}$ with the axioms **(b)** and **(k)**.

Extensions of **BK** with various subsets[5] of the following axioms constitute the main C-systems studied in the literature ($\sharp \in \{\vee, \wedge, \supset\}$):

[5] There are several subsets the addition of which to **BK** results in the loss of paraconsistency, for their full list see [5].

(c)	$\neg\neg\varphi \supset \varphi$	**(e)**	$\varphi \supset \neg\neg\varphi$
$(\mathbf{n}^{\mathbf{l}}_{\wedge})$	$\neg(\varphi \wedge \psi) \supset (\neg\varphi \vee \neg\psi)$	$(\mathbf{n}^{\mathbf{r}}_{\wedge})$	$(\neg\varphi \vee \neg\psi) \supset \neg(\varphi \wedge \psi)$
$(\mathbf{n}^{\mathbf{l}}_{\vee})$	$\neg(\varphi \vee \psi) \supset (\neg\varphi \wedge \neg\psi)$	$(\mathbf{n}^{\mathbf{r}}_{\vee})$	$(\neg\varphi \wedge \neg\psi) \supset \neg(\varphi \vee \psi)$
$(\mathbf{n}^{\mathbf{l}}_{\supset})$	$\neg(\varphi \supset \psi) \supset (\varphi \wedge \neg\psi)$	$(\mathbf{n}^{\mathbf{r}}_{\supset})$	$(\varphi \wedge \neg\psi) \supset \neg(\varphi \supset \psi)$
$(\mathbf{o}^{\mathbf{1}}_{\sharp})$	$\circ\varphi \supset \circ(\varphi\sharp\psi)$	$(\mathbf{o}^{\mathbf{2}}_{\sharp})$	$\circ\psi \supset \circ(\varphi\sharp\psi)$
$(\mathbf{a}_{\sharp})$	$(\circ\varphi \wedge \circ\psi) \supset \circ(\varphi\sharp\psi)$	$(\mathbf{a}_{\neg})$	$\circ\varphi \supset \circ\neg\varphi$
(l)	$\neg(\varphi \wedge \neg\varphi) \supset \circ\varphi$	**(d)**	$\neg(\neg\varphi \wedge \varphi) \supset \circ\varphi$
$(\mathbf{i_1})$	$\neg\circ\varphi \supset \varphi$	$(\mathbf{i_2})$	$\neg\circ\varphi \supset \neg\varphi$

In what follows we will explore which of the axioms above is already derivable in the logic **NB** studied below, and which can be added to it while preserving its paraconsistency.

3 The Logic NB

3.1 Motivation:

Recall that we set out to find a paraconsistent logic which has a negation which is complete, left-involutive and contrapositive (note that by Proposition 2.9, we cannot also demand right-involutiveness). As we show below, having a contrapositive negation ensures the desirable property (which both classical and intuitionistic logics enjoy) of substitution of equivalents, also known as the replacement property (or self-extensionality [41]):

Definition 3.1 Let $\mathbf{L} = \langle \mathcal{L}, \vdash_{\mathbf{L}} \rangle$ be a logic.

- Formulas $\psi, \varphi \in \mathcal{W}(\mathcal{L})$ are *equivalent* in **L**, denoted by $\psi \dashv\vdash_{\mathbf{L}} \varphi$, if $\psi \vdash_{\mathbf{L}} \varphi$ and $\varphi \vdash_{\mathbf{L}} \psi$.
- Formulas $\psi, \varphi \in \mathcal{W}(\mathcal{L})$ are *congruent* (or *indistinguishable*) in **L**, denoted by $\psi \equiv_{\mathbf{L}} \varphi$, if for every formula σ and atom p it holds that $\sigma[\psi/p] \dashv\vdash_{\mathbf{L}} \sigma[\varphi/p]$.
- **L** has the *replacement property* if any two formulas which are equivalent in **L** are congruent in it.

The majority of paraconsistent logics considered in the literature do not have the replacement property. Can we construct ($\neg$-classical) paraconsistent logics which do enjoy the replacement property and have a reasonable (at least complete) negation? The next proposition shows that as long as we want $\neg$ to be complete, this goal cannot be achieved for extensions of **CLuN** by the usual way of adding axioms that force the *strong replacement condition*, where a $\neg$-classical logic **L** satisfies this condition if $\varphi \supset \psi, \psi \supset \varphi \vdash_{\mathbf{L}} \sigma[\psi/p] \supset \sigma[\varphi/p]$ for every atom p and formulas φ, ψ, σ.

Proposition 3.2 *Let* **CAR**[6] *be the logic which is obtained from* **CLuN** *by adding to it the following schema as an axiom:*

$$(\psi \supset \varphi) \wedge (\varphi \supset \psi) \supset (\neg\psi \supset \neg\varphi)$$

[6] It is easy to show that our **CAR** is equivalent to the logic that is called **CAR** in [17]. In Chapter 3 of [35] the same logic (with yet another axiomatization) is called **Le**.

Then **CAR** *is not paraconsistent.*

The above proposition entails that in order to develop paraconsistent extensions of **CLuN** that enjoy the replacement property, the inference of $\neg\varphi \supset \neg\psi$ from $\varphi \supset \psi$ and $\psi \supset \varphi$ should be forced only in the case where the premises are theorems of the logic. This can be done by including this rule in the corresponding proof systems not as a rule of derivation, but just as a *rule of proof*, that is: a rule that is used only to define the set of axioms of the system, but not its consequence relation. To make $\neg$ also contrapositive, it would be better to adopt as a rule of proof the inference of $\neg\varphi \supset \neg\psi$ from $\psi \supset \varphi$ alone. The next proposition implies that as long as we use the language $\mathcal{L}_{CL}$, it would also suffice for forcing the replacement property.

Proposition 3.3 *Let* **L** *be a* $\neg$*-classical logic in* $\mathcal{L}_{CL}$ *which extends* $\mathbb{L}^+$*, in which* $\vdash_{\mathbf{L}} \neg\varphi \supset \neg\psi$ *whenever* $\vdash_{\mathbf{L}} \psi \supset \varphi$*. Then* **L** *has the replacement property.*

The above considerations lead to the following definition of the logic **NB**:

Definition 3.4 $Th(NB)$ is the minimal set S of formulas in $\mathcal{L}_{CL}$, such that:

(i) S includes all axioms of HC_{min}.

(ii) S is closed under [MP] and the following rule:

$$[\text{CP}]\ \frac{\vdash \psi \supset \varphi}{\vdash \neg\varphi \supset \neg\psi}$$

HNB is the Hilbert-type system whose set of axioms is $Th(NB)$ and has [MP] for $\supset$ as its sole rule of inference.

NB is the logic in $\mathcal{L}_{CL}$ which is induced by HNB.

Obviously, $\vdash_{\mathbf{NB}} \varphi$ iff $\varphi \in Th(NB)$. Note again that [CP] is *not* a rule of inference of HNB, but only a *rule of proof*, i.e., it is used only for defining its set of axioms. This is similar to the role that the necessitation rule (from ψ infer $\Box\psi$) usually has in Hilbert-type systems in modal logics.[7]

The following lemma will be useful in the sequel:

Lemma 3.5

(i) *If* $\vdash_{\mathbf{NB}} \varphi$ *then for every* ψ*,* $\neg\varphi \vdash_{\mathbf{NB}} \psi$*.*

(ii) $\vdash_{\mathbf{NB}} \neg(\varphi \wedge \psi) \supset (\neg\varphi \vee \neg\psi)$

(iii) $\vdash_{\mathbf{NB}} \neg\neg\neg\varphi \equiv \neg\varphi$ *(that is,* $\vdash_{\mathbf{NB}} \neg\neg\neg\varphi \supset \neg\varphi$ *and* $\vdash_{\mathbf{NB}} \neg\varphi \supset \neg\neg\neg\varphi$*).*

Remark 3.6 The first item of Lemma 3.5 implies that if we take F to be an abbreviation of $\neg(P_1 \supset P_1)$ (say), then for every φ, $\vdash_{\mathbf{NB}} \mathsf{F} \supset \varphi$. Hence we

[7] More precisely: whether necessitation is taken in modal logics as a rule of proof or a rule of derivation depends on the intended consequence relation. If the *local* one of preserving truth in worlds is used, then the rule can be taken only as a rule of proof. In contrast, if the *global* one of preserving validity in frames (that is, truth in all worlds of a frame) is used, then the rule should be taken as a rule of derivation.

may assume that the language of **NB** is an extension of $\mathcal{L}^{\mathsf{F}}_{CL}$, and that every instance of a classical tautology in $\mathcal{L}^{\mathsf{F}}_{CL}$ is in $Th(NB)$.

Next we define a Gentzen-style system for **NB**.

Notation: $\neg S = \{\neg\varphi \mid \varphi \in S\}$.

Definition 3.7 The system GNB is obtained from the Gentzen-style system LK for **CL** ([20]) by replacing its left introduction rule for negation ($[\neg\Rightarrow]$) by the rule:

$$[\neg\Rightarrow]_B \quad \frac{\Gamma, \neg\Delta \Rightarrow \psi}{\neg\psi \Rightarrow \neg\Gamma, \Delta}$$

Theorem 3.8 *$\mathcal{T} \vdash_{GNB} \varphi$ (i.e., there is a finite $\Gamma \subseteq \mathcal{T}$ such that $\Gamma \Rightarrow \varphi$ is derivable in GNB) iff $\mathcal{T} \vdash_{\mathbf{NB}} \varphi$.*

Proposition 3.9 *GNB does not admit cut-elimination.*

Proof. Obviously, $\vdash_{GNB} \neg(p\vee q), \neg(p\vee q) \to r \Rightarrow r$. By applying $[\neg\Rightarrow]_B$ we get from this that $\vdash_{GNB} \neg r \Rightarrow \neg(\neg(p\vee q) \to r), p\vee q$. Since also $\vdash_{GNB} p\vee q \Rightarrow p, q$, an application of the Cut rule yields that $\vdash_{GNB} \neg r \Rightarrow \neg(\neg(p \vee q) \to r), p, q$. On the other hand a straightforward (though tedious) search reveals that this sequent has no cut-free proof in GNB. □

In the next subsection we will see that GNB does admit a weaker version of cut-elimination and this suffices for making it a decidable system which has the crucial subformula property.

3.2 Kripke-style Semantics for NB

For providing adequate semantics for **NB**, we use the following framework of Kripke frames for modal logics.

Definition 3.10 A triple $\langle W, R, \nu\rangle$ is called a **NB**-*frame* for $\mathcal{L}_{CL}$ [8], if W is a nonempty (finite) set (of "worlds"), R is a reflexive and symmetric relation on W, and $\nu : W \times \mathcal{W}(\mathcal{L}_{CL}) \to \{t, f\}$ satisfies the following conditions:

- $\nu(w, \psi \wedge \varphi) = t$ iff $\nu(w, \psi) = t$ and $\nu(w, \varphi) = t$.
- $\nu(w, \psi \vee \varphi) = t$ iff $\nu(w, \psi) = t$ or $\nu(w, \varphi) = t$.
- $\nu(w, \psi \supset \varphi) = t$ iff $\nu(w, \psi) = f$ or $\nu(w, \varphi) = t$.
- $\nu(w, \neg\psi) = t$ iff there exists $w' \in W$ such that wRw', and $\nu(w', \psi) = f$.

Definition 3.11 Let $\langle W, R, \nu\rangle$ be a **NB**-frame.

- A formula φ is *true* in a world $w \in W$ ($w \Vdash \varphi$) if $\nu(w, \varphi) = t$.
- A sequent $s = \Gamma \Rightarrow \Delta$ is *true* in a world $w \in W$ ($w \Vdash s$) if $\nu(w, \varphi) = f$ for some $\varphi \in \Gamma$, or $\nu(w, \varphi) = t$ for some $\varphi \in \Delta$. Equivalently, $w \Vdash s$ if $w \Vdash I(s)$,

[8] In the literature on modal logics one usually means by a "frame" just the pair $\langle W, R\rangle$, while we find it convenient to follow [34], and use this technical term a little bit differently, so that the valuation ν is a part of it.

where $I(s)$ is the usual interpretation of s (as defined, e.g., in the proof of Theorem 3.8).

- A formula φ is *valid* in $\langle W, R, \nu \rangle$ ($\langle W, R, \nu \rangle \models \varphi$) if it is true in every world $w \in W$.
- A sequent s is *valid* in $\langle W, R, \nu \rangle$ ($\langle W, R, \nu \rangle \models s$) if it is true in every world $w \in W$.

Definition 3.12

- Let $\mathcal{T} \cup \{\varphi\}$ be a set of formulas in $\mathcal{L}_{CL}$. φ *semantically follows in* **NB** from $\mathcal{T}$ if for every **NB**-frame $\langle W, R, \nu \rangle$ and every $w \in W$: if $w \Vdash \psi$ for every $\psi \in \mathcal{T}$ then $w \Vdash \varphi$.
- Let $S \cup \{s\}$ be a set of sequents in $\mathcal{L}_{CL}$. s *semantically follows in* **NB** from S if for every **NB**-frame $\mathcal{W}$, if $\mathcal{W} \models s'$ for every $s' \in S$, then $\mathcal{W} \models s$. s is **NB**-*valid* if s semantically follows in **NB** from $\emptyset$ (that is, s is valid in every **NB**-frame).

Proposition 3.13

(i) *If φ is a theorem of* **NB** *(that is, $\varphi \in Th(NB)$), then φ is valid in every* **NB**-*frame.*

(ii) *If $\mathcal{T} \vdash_{\mathbf{NB}} \varphi$ then φ semantically follows in* **NB** *from $\mathcal{T}$.*

(iii) *Let $S \cup \{s\}$ be a set of sequents. If $S \vdash_{GNB} s$ then s semantically follows in* **NB** *from S. In particular: if $\vdash_{GNB} s$ then s is* **NB**-*valid.*

Now we turn to prove the completeness of **NB** for its possible-worlds semantics, as well as the analyticity of GNB. The latter property is defined as follows:

Definition 3.14 Let G be a Gentzen-type system in a language $\mathcal{L}$.

- Let $\mathcal{F}$ be a set of formulas in $\mathcal{L}$. A proof in G is called $\mathcal{F}$-*analytic* if every formula which occurs in it belongs to $\mathcal{F}$.
- Let $S \cup \{s\}$ be a set of sequents in $\mathcal{L}$. A proof in G of s from S is called *analytic* if it is $\mathcal{F}$-analytic, where $\mathcal{F}$ is the set of subformulas of formulas in $S \cup \{s\}$.
- G has the *(strong) subformula property* if whenever $\vdash_{\mathsf{G}} s$ ($S \vdash_{\mathsf{G}} s$), there is an analytic proof of s (from S).

Theorem 3.15 *Let $S \cup \{s\}$ be a finite set of sequents in $\mathcal{L}_{CL}$. If s semantically follows in* **NB** *from S then s has an analytic proof in GNB from S.*

Proof. Suppose s does not have an analytic proof in GNB from S. We construct a **NB**-frame in which the elements of S are valid, but s is not.

Denote by $\mathcal{F}$ the set of subformulas of formulas in $S \cup \{s\}$. Call a sequent $\Gamma \Rightarrow \Delta$ $\mathcal{F}$-*maximal* if the following conditions (i) $\Gamma \cup \Delta = \mathcal{F}$, and (ii) $\Gamma \Rightarrow \Delta$ has no $\mathcal{F}$-analytic proof from S.

Lemma 1. Suppose $\Gamma \cup \Delta \subseteq \mathcal{F}$, and $\Gamma \Rightarrow \Delta$ has no $\mathcal{F}$-analytic proof from

S. Then $\Gamma \Rightarrow \Delta$ can be extended to an $\mathcal{F}$-maximal sequent $\Gamma' \Rightarrow \Delta'$ (that is, $\Gamma \subseteq \Gamma'$ and $\Delta \subseteq \Delta'$).

Proof of Lemma 1. Let $\Gamma' \Rightarrow \Delta'$ be a maximal extension of $\Gamma \Rightarrow \Delta$ that consists of formulas in $\mathcal{F}$, and has no $\mathcal{F}$-analytic proof from S. (Such $\Gamma' \Rightarrow \Delta'$ exists, because $\mathcal{F}$ is finite.) To show that $\Gamma' \Rightarrow \Delta'$ is $\mathcal{F}$-maximal, assume for contradiction that there is $\varphi \in \mathcal{F}$ such that $\varphi \notin \Gamma' \cup \Delta'$. Then the maximality of $\Gamma' \Rightarrow \Delta'$ implies that both $\Gamma' \Rightarrow \Delta', \varphi$ and $\varphi, \Gamma' \Rightarrow \Delta'$ have $\mathcal{F}$-analytic proofs from S. But then we can get an $\mathcal{F}$-analytic proof from S using these two proofs together with an application of a cut on φ to their conclusions. (Note that since $\varphi \in \mathcal{F}$, the resulting proof of $\Gamma' \Rightarrow \Delta'$ is still $\mathcal{F}$-analytic.) This contradicts our assumption about $\Gamma' \Rightarrow \Delta'$.

Since s has no $\mathcal{F}$-analytic proof from S, it follows from Lemma 1 that s can be extended to an $\mathcal{F}$-maximal sequent $\Gamma^* \Rightarrow \Delta^*$.

Let W be the set of all $\mathcal{F}$-maximal sequents. Since $\mathcal{F}$ is finite, so is W. Since $(\Gamma^* \Rightarrow \Delta^*) \in W$, W is also nonempty. Define a relation R on W as follows: $(\Gamma_1 \Rightarrow \Delta_1)R(\Gamma_2 \Rightarrow \Delta_2)$ iff for every formula φ, if $\neg\varphi \in \Delta_1$ then $\varphi \in \Gamma_2$, and if $\neg\varphi \in \Delta_2$ then $\varphi \in \Gamma_1$. Obviously, R is symmetric. That it is also reflexive follows from the fact that if $\{\neg\varphi, \varphi\} \subseteq \Delta$ then $\Gamma \Rightarrow \Delta$ has a cut-free proof in GNB (since it can be derived from the axiom $\varphi \Rightarrow \varphi$ using $[\Rightarrow\neg]$ and weakenings). This fact and the $\mathcal{F}$-maximality of a sequent $\Gamma \Rightarrow \Delta$ in W imply that if $\neg\varphi \in \Delta$ then $\varphi \in \Gamma$, and so $(\Gamma \Rightarrow \Delta)R(\Gamma \Rightarrow \Delta)$. Next, let $\mathcal{W}$ be the **NB**-frame $\langle W, R, \nu\rangle$ in which W and R are as above, and ν is obtained by letting $\nu(\Gamma \Rightarrow \Delta, p) = t$ iff $p \in \Gamma$ (p atomic).

Lemma 2. Let $\Gamma \Rightarrow \Delta \in W$ and $\varphi \in \mathcal{F}$. Then $\nu(\Gamma \Rightarrow \Delta, \varphi) = t$ if $\varphi \in \Gamma$, and $\nu(\Gamma \Rightarrow \Delta, \varphi) = f$ if $\varphi \in \Delta$.

Proof of Lemma 2. By induction on the complexity of φ.

Since $\Gamma^* \Rightarrow \Delta^*$ is an extension of s, it follows from Lemma 2 that if $s = \Gamma \Rightarrow \Delta$, then $\nu(\Gamma^* \Rightarrow \Delta^*, \varphi) = t$ for every $\varphi \in \Gamma$, while $\nu(\Gamma^* \Rightarrow \Delta^*, \varphi) = f$ for every $\varphi \in \Delta$. It follows that $\Gamma^* \Rightarrow \Delta^* \not\Vdash s$, and so $\mathcal{W} \not\Vdash s$.

Finally, let $s' = (\Gamma' \Rightarrow \Delta') \in S$, and let $w = (\Gamma \Rightarrow \Delta) \in W$. It is impossible that w is an extension of s', because w has no $\mathcal{F}$-analytic proof from S. It follows that either $\varphi \in \Delta$ for some $\varphi \in \Gamma'$, or $\varphi \in \Gamma$ for some $\varphi \in \Delta'$. By Lemma 2 this implies that either $\nu(w, \varphi) = f$ for some $\varphi \in \Gamma'$, or $\nu(w, \varphi) = t$ for some $\varphi \in \Delta'$. Hence $w \Vdash s'$ for every $w \in W$, and so $\mathcal{W} \Vdash s'$ for every $s' \in S$. □

Corollary 3.16 *GNB has the subformula property: if $S \vdash_{GNB} s$ then s has an analytic proof in GNB from S. In particular: if $\vdash_{GNB} s$ then s has an analytic proof in GNB.*[9]

Corollary 3.17 *If Γ is finite then $\Gamma \vdash_{\mathbf{NB}} \varphi$ iff φ semantically follows in* **NB** *from Γ.*

---

[9] This implies that if $\vdash_{GNB} s$ then s has a proof in GNB in which all cuts are analytic (that is, the cut formulas are subformulas of s).

The above can be strengthened to full completeness (we leave the proof to the reader):

Theorem 3.18 *For every theory $\mathcal{T}$, $\mathcal{T} \vdash_{\mathbf{NB}} \varphi$ iff φ semantically follows in* **NB** *from $\mathcal{T}$.*

Remark 3.19 *GNB* is a version of the Gentzen-type system for **B** given in [39] (and described in [40]). Unlike our proof, the analyticity of the assumptions-free fragment of that system is proved in [39] by syntactic[10] means.

3.3 Basic Properties of NB

From Proposition 2.9 and Note 2.8 it follows that a logic which has a complete, left-involutive and contrapositive negation should contain the logic **NB**. The next proposition shows that **NB** is in fact the minimal logic of this type:

Proposition 3.20 **NB** *is the minimal extension of* $\mathbf{CL}^+$ *in* $\mathcal{L}_{CL}$ *in which* $\neg$ *is complete, contrapositive, and left-involutive.*

Proposition 3.21 **NB** *is paraconsistent and has the replacement property.*

Theorem 3.22 **NB** *is decidable.*

Proof. Given a formula φ (or a finite set of formulas $\Gamma \cup \{\varphi\}$), the number of sequents which consist only of subformulas of φ (or $\Gamma \Rightarrow \varphi$) is finite. Hence it easily follows[11] from Theorem 3.16 that it is decidable whether $\vdash_{GNB} \varphi$ (or $\vdash_{GNB} \Gamma \Rightarrow \varphi$) or not. □

3.4 NB as a C-system

From Note 3.6 it follows that we may assume (as we do from this point on) that the language of **NB** includes F, and that **NB** is an extension of $\mathbf{CL}^{\mathsf{F}}$. Therefore Theorem 2.13 implies that **NB** (and any of its paraconsistent extensions) is a *C*-system. From Proposition 2.11, the replacement property of **NB**, and the first item of Lemma 3.5 it further follows that (any paraconsistent extension in $\mathcal{L}_{CL}$ of) **NB** has a *unique* (up to congruence) strong consistency operator $\circ$, which can be defined as $\circ\varphi =_{def} \varphi \wedge \neg\varphi \supset \mathsf{F}$, or as $\circ\varphi =_{def} (\varphi \wedge \neg\varphi) \supset \neg(\varphi \supset \varphi)$.

We now check which of the schemas listed in subsection 2.1 is valid in **NB**, and which can be added to it without losing its paraconsistency.

Proposition 3.23 *The following schemas from the list given in subsection 2.1 are provable in* **NB** *(in addition to* (**b**), (**k**), (**t**) *and* (**c**)*):* $(\mathbf{n}^{\mathbf{l}}_{\wedge})$, $(\mathbf{n}^{\mathbf{r}}_{\wedge})$, $(\mathbf{n}^{\mathbf{l}}_{\vee})$, $(\mathbf{n}^{\mathbf{1,2}}_{\supset})$, $(\mathbf{a}_{\neg})$, $(\mathbf{a}_{\wedge})$, *and* $(\mathbf{a}_{\vee})$.

Remark 3.24 The fact that $(\mathbf{a}_{\neg})$, $(\mathbf{a}_{\wedge})$, and $(\mathbf{a}_{\vee})$ are all valid in **NB** means that **NB** is almost perfectly adequate to serve as a C-system according to da

---

[10] A semantic proof appeared in [23] (Example 5.54), as a particular instance of a general method for proving analyticity.

[11] Instead of using the Gentzen-type system *GNB*, one can use the semantics of **NB** in order to provide a decision procedure for it. This is due to the fact that from the proof of Theorem 3.15 it follows that a sequent s is **NB**-valid iff it is valid in every **NB**-frame in which the number of worlds is at most 2^n, where n is the number of subformulas of s.

Costa's ideas. The only principle that it misses (as we show is the next theorem) is $(\mathbf{a}_{\supset})$. This is the price it pays for being contrapositive and for having the replacement property. However, this is not a high price, since the language of $\{\neg, \wedge, \vee\}$ suffices for classical reasoning (since its set of primitive connectives is functionally complete for two-valued matrices).

It is also remarkable that $\neg(\varphi \wedge \psi)$ is equivalent (and so congruent) in **NB** to $\neg\varphi \vee \neg\psi$. However, the next theorem shows that the other De Morgan rules are only partially valid in **NB**. Another important fact that is shown in the next theorem is that with one exception (to be dealt with in the sequel), all the schemas from the list in Subsection 2.1 that are not already derivable in **NB** cannot even be added to it without losing its paraconsistency. This shows that **NB** is rather robust as a paraconsistent logic.

Theorem 3.25 *Let* **L** *be obtained by adding to HNB as an axiom any element of the set* $\{(\mathbf{e}), (\mathbf{n}^{\mathbf{l},\mathbf{1}}_{\supset}), (\mathbf{n}^{\mathbf{r}}_{\vee}), (\mathbf{n}^{\mathbf{r}}_{\supset}), (\mathbf{a}_{\supset}), (\mathbf{i_1}), (\mathbf{l}), (\mathbf{d})\}$, *or any axiom of the form* $(\mathbf{o}^{\mathbf{i}}_{\#})$ $(\sharp \in \{\wedge, \vee, \supset\}, i \in \{1, 2\})$. *Then* **L** *is not -paraconsistent.*

Proof. We show for the cases of $(\mathbf{e})$ and $(\mathbf{n}^{\mathbf{l},\mathbf{1}}_{\supset})$, leaving the rest of the cases to the reader.

- Suppose $(\mathbf{e})$ is valid in **L**. Then from Proposition 2.9 it follows that $\neg\varphi \supset (\varphi \supset \neg\neg\psi)$ for every φ, ψ. Since $(\mathbf{c})$ is valid in **NB**, this implies that $\vdash_{\mathbf{L}} \neg\varphi \supset (\varphi \supset \psi)$.
- Suppose $(\mathbf{n}^{\mathbf{l},\mathbf{1}}_{\supset})$ is valid in **L**. Then $\vdash_{\mathbf{L}} \neg(\varphi \supset \psi) \supset \varphi$. By applying [CP] we get that $\vdash_{\mathbf{L}} \neg\varphi \supset \neg\neg(\varphi \supset \psi)$. Hence $\vdash_{\mathbf{L}} \neg\varphi \supset (\varphi \supset \psi)$.

□

Remark 3.26 The exception which has not been dealt with in Proposition 3.23 and Theorem 3.25 is the schema $(\mathbf{i_2})$. Now it is not difficult to show that $(\mathbf{i_2})$ is not provable in **NB**. In the sequel we show that it can nevertheless be added to **NB** without losing its paraconsistency, and that this addition leads to another interesting logic.

3.5 NB is the Modal Logic B

The notion of '**NB**-frame' is very similar to the notion of a Kripke frame used in the study of modal logics. Indeed, **NB** is actually (equivalent to) the famous modal logic which is usually called **B** or **KTB** (see e.g. [13]). The language of **B** is usually taken to be $\{\wedge, \vee, \supset, \mathsf{F}, \Box\}$ (or $\{\wedge, \vee, \supset, \neg, \Box\}$, where $\neg$ denotes the *classical* negation). Its semantics is given by Kripke frames in which the accessibility relation R is again reflexive and symmetric, where the notion of a 'Kripke frame' is defined like in Defn 3.10, except that instead of the clause there for $\neg$ we have the following clause for $\Box$:

- $\nu(w, \Box\psi) = t$ iff $\nu(w', \psi) = t$ for every $w' \in W$ such that wRw'.

Now it is easy to see that with respect to Kripke frames, the language of our **NB** and the language of the modal logic **B** are equivalent in their expressive power. $\Box$ is definable in the former by $\Box\varphi =_{def} \sim\neg\varphi$, where $\sim\psi =_{def} \psi \supset \mathsf{F}$. On the other hand $\neg$ is definable in the language of **B** by $\neg\varphi =_{def} \sim\Box\varphi$. It follows that the paraconsistent logic **NB** (whose language is just $\mathcal{L}_{CL}$) and the modal logic **B** are practically identical.

It is worth noting that the presentation of the modal **B** in the form **NB** is more concise (and in our opinion also clearer) than the usual one in two ways. First, **NB** really has only two basic connectives: $\supset$ and $\neg$. (F can be defined as $\neg(\varphi \supset \varphi)$, where φ is arbitrary, and $\vee$ and $\wedge$ can of course be defined in terms of $\supset$ and F.) The standard presentation of **B** needs three connectives: $\supset$, F, and $\Box$. Second, the standard Hilbert-type proof system for **B** is more complicated than HNB. It is obtained from (the full) HCL by the addition of one rule of proof and three axioms. The rule is the necessitation rule (if $\vdash \varphi$ then $\vdash \Box\varphi$). The three axioms are: **(K)** $\Box(\varphi \supset \psi) \supset (\Box\varphi \supset \Box\psi)$, **(T)** $\Box\varphi \supset \varphi$ and **(B)** $\varphi \supset \Box\Diamond\varphi$, where $\Diamond\varphi =_{def} \sim\Box\sim\varphi$. In contrast, HNB is obtained from HCL^+ by the addition of one rule of proof (which admittedly is somewhat more complex than the necessitation rule), and just two extremely simple and natural axioms.

4 The Logic NS5

Following Note 3.26, in this section we investigate the system that is obtained from **NB** by the addition of ($\mathbf{i_2}$). We start by presenting two schemas which are equivalent to ($\mathbf{i_2}$) over **NB**.

Lemma 4.1 *The logics which are obtained by extending* **NB** *with one of the following schemas are identical.*

(i) ($\mathbf{i_2}$) *(that is: $\neg\circ\varphi \supset \neg\varphi$).*

(ii) $\circ(\neg\varphi)$ *(that is: $\neg\varphi \wedge \neg\neg\varphi \supset \mathsf{F}$).*

(iii) $\neg\neg\varphi \supset (\neg\varphi \supset \psi)$.

Definition 4.2 Let $HNS5$ be the Hilbert-type system which is obtained from HNB by adding to it as an axiom schema one of the three schemas which were proved equivalent in Lemma 4.1. **NS5** is the logic induced by $HNS5$.

Remark 4.3 Béziau ([10,11,12] and Batens ([8]) have introduced systems equivalent to **NS5**. (**NS5** was called $\mathbb{Z}$ by Béziau, and **A** by Batens.) Further study of this system was done in [37]. The Hilbert-type system $HNS5$ presented above is an improved version of the Hilbert-type system for $\mathbb{Z}$ presented in that paper. The same simplified axiomatization of HNS5 was also independently discovered by Omori and Waragai and presented in [36]. Actually, our axiomatization HNB of the modal logic **KTB** is implicitly given there as well.[12] At this point it is worth noting also that the realization on which the present paper is based (that the same method that was applied to **S5** can be applied to other modal logics in order to produce interesting paraconsistent logics) was first pursued independently in [28,27,30] and in [31,32,33]. Another investigation of paraconsistent logics from a modal viewpoint, studying also

[12] We are grateful to an anonymous referee for bringing this paper and these facts to our attention.

analytic and cut-free sequent calculi for such logics, is presented in the current volume ([24]).

Remark 4.4 The following observation leads to a simpler version of $HNS5$:

$$\neg\neg\varphi \supset (\neg\varphi \supset \varphi), \neg\varphi \vee \varphi \vdash_{\mathbf{IL}^+} \neg\neg\varphi \supset \varphi$$

It follows that in order to axiomatize **NS5**, it suffices to add to HCL^+ the schemas $\neg\varphi \vee \varphi$, $\neg\neg\varphi \supset (\neg\varphi \supset \psi)$, and the rule [CP] (or to add to **CLuN** the schema $\neg\neg\varphi \supset (\neg\varphi \supset \psi)$, and the rule [CP]).

Like in the case of **NB**, we provide a Gentzen-type system and Kripke-style semantics for it, leaving all proofs in the section to the reader.

Definition 4.5 [$GNS5$] The system $GNS5$ is the system which is obtained from LK by replacing its rule $[\neg\Rightarrow]$ by the rule:

$$[\neg\Rightarrow]_5 \quad \frac{\neg\Gamma \Rightarrow \psi, \neg\Delta}{\neg\Gamma, \neg\psi \Rightarrow \neg\Delta}$$

Theorem 4.6 *$\mathcal{T} \vdash_{GNS5} \varphi$ iff $\mathcal{T} \vdash_{\mathbf{NS5}} \varphi$.*

Definition 4.7

- An **NB**-frame $\langle W, R, \nu\rangle$ is called a **NS5**-*frame* for $\mathcal{L}_{CL}$ if R is transitive (in addition to its being reflexive and symmetric).
- The notions of truth (in worlds) and validity in **NS5**-frames (of formulas and sequents) are defined like in Definition 3.11.
- Semantic consequence in **NS5** is defined like in Definition 3.12, using **NS5**-frames instead of **NB**-frames.

Proposition 4.8

(i) *If φ is a theorem of* **NS5** *then φ is valid in every* **NS5**-*frame.*

(ii) *If $\mathcal{T} \vdash_{\mathbf{NS5}} \varphi$ then φ semantically follows in* **NS5** *from $\mathcal{T}$.*

(iii) *Let $S \cup \{s\}$ be a set of sequents. If $S \vdash_{GNS5} s$ then s semantically follows in* **NS5** *from S. In particular: if $\vdash_{GNS5} s$ then s is* **NS5**-*valid.*

Theorem 4.9 *Let $S \cup \{s\}$ be a finite set of sequents in $\mathcal{L}_{CL}$. If s semantically follows in* **NS5** *from S then s has an analytic proof in $GNS5$ from S.*

Theorem 4.9 has for **NS5** the same important corollaries as Theorem 3.15 has for **NB**.

Theorem 4.10 *$GNS5$ has the subformula property: if $S \vdash_{GNS} s$ then s has an analytic proof in $GNS5$ from S. In particular: if $\vdash_{GNS5} s$ then s has an analytic proof in $GNS5$.*

Theorem 4.11 *For finite Γ, $\Gamma \vdash_{\mathbf{NS5}} \varphi$ iff φ semantically follows in* **NS5** *from Γ.*

Theorem 4.12 **NS5** *is decidable.*

Proposition 4.13 **NS5** *is a* $\neg$*-classical paraconsistent logic with a complete, contrapositive, and left-involutive negation. It is also a C-system in which all the schemas listed in Proposition 3.23 are valid, as well as* ($\mathbf{i_2}$) *and* $\circ\neg\varphi$.

Theorem 4.14 **NS5** *is equivalent to the famous modal logic* **S5** *(also known as* **KT5** *or* **KT45***).*

Remark 4.15 The modal logic **S5** is the logic induced by the class of Kripke frames in which the accessibility relation is an equivalence relation. Theorem 4.14 follows from this characterization of **S5**. Note that the standard Hilbert-type system for **S5** is obtained from that of **B** by replacing the axiom (B) by the axiom **(5)** $\diamond\varphi \supset \Box\diamond\varphi$. It is worth noting also that in **NS5** $\Box\varphi$ (that is: $\sim\neg\varphi$) is equivalent to $\neg\neg\varphi$. (This can easily be shown by using the semantics, or by using $GNS5$.)

References

[1] Arieli, O., A. Avron and A. Zamansky, *Ideal paraconsistent logics*, Studia Logica **99** (2011), pp. 31–60.

[2] Arieli, O., A. Avron and A. Zamansky, *Maximal and premaximal paraconsistency in the framework of three-valued semantics*, Studia Logica **97** (2011), pp. 31–60.

[3] Arieli, O., A. Avron and A. Zamansky, *What is an ideal logic for reasoning with inconsistency?*, in: *Proc. of 22nd Int. Joint Conf. on Artificial Intelligence* (IJCAI 2011), 2011, pp. 706–711.

[4] Avron, A., O. Arieli and A. Zamansky, *On strong maximality of paraconsistent finite-valued logics*, in: *Proc. of Logic in Computer Science (LICS), 2010 25th Annual IEEE Symposium on*, IEEE, 2010, pp. 304–313.

[5] Avron, A., B. Konikowska and A. Zamansky, *Cut-free sequent calculi for C-systems with generalized finite-valued semantics*, Journal of Logic and Computation **23** (2013), pp. 517–540.

[6] Batens, D., *Paraconsistent extensional propositional logics*, Logique et Analyse Louvain **23** (1980), pp. 195–234.

[7] Batens, D., *Inconsistency-adaptive logics*, in: E. Orlowska, editor, *Logic at Work*, Physica Verlag, 1998 pp. 445–472.

[8] Batens, D., *On some remarkable relations between paraconsistent logics, modal logics, and ambiguity logics*, in: W. A. Carnielli, M. E. Coniglio and I. D'Ottaviano, editors, *Paraconsistency: The Logical Way to the Inconsistent*, number 228 in Lecture Notes in Pure and Applied Mathematics, Marcel Dekker, 2002 pp. 275–293.

[9] Batens, D., K. de Clercq and N. Kurtonina, *Embedding and interpolation for some paralogics. the propositional case.*, Reports on Mathematical Logic **33** (1999), pp. 29–44.

[10] Béziau, J. Y., *S5 is a paraconsistent logic and so is first-order classical logic*, Logical Investigations **8** (2002), pp. 301–309.

[11] Béziau, J. Y., *Paraconsistent logic from a modal viewpoint*, Journal of Applied Logic **3** (2005), pp. 6–14.

[12] Béziau, J. Y., *The paraconsistent logic Z. a possible solution to Jaśkowski's problem*, Logic and Logical Philosophy **15** (2006), pp. 99–111.

[13] Bull, R. and K. Segerberg, *Basic modal logic*, in: *Handbook of philosophical logic*, Springer, 1984 pp. 1–88.

[14] Carnielli, W., M. E. Coniglio and J. Marcos, *Logics of Formal Inconsistency*, in: *Handbook of philosophical logic*, Springer, 2007 pp. 1–93.

[15] Carnielli, W. A. and J. Marcos, *A taxonomy of C-systems*, in: W. A. Carnielli, M. E. Coniglio and I. D'Ottaviano, editors, *Paraconsistency: The Logical Way to the Inconsistent*, number 228 in Lecture Notes in Pure and Applied Mathematics, Marcel Dekker, 2002 pp. 1–94.

[16] da Costa, N. C. A., *On the theory of inconsistent formal systems*, Notre Dame Journal of Formal Logic **15** (1974), pp. 497–510.

[17] da Costa, N. C. A. and J. Y. Béziau, *Carnot's logic*, Bul. of the Section of Logic **22** (1993), pp. 98–105.

[18] da Costa, N. C. A., J. Y. Béziau and O. A. S. Bueno, *Aspects of paraconsistent logic*, Bul. of the IGPL **3** (1995), pp. 597–614.

[19] D'Ottaviano, I., *On the development of paraconsistent logic and da Costa's work*, Journal of Non-classical Logic **7** (1990), pp. 89–152.

[20] Gentzen, G., *Investigations into logical deduction* (1934), in German. An English translation appears in 'The Collected Works of Gerhard Gentzen', edited by M. E. Szabo, North-Holland, 1969.

[21] Gordienko, A. B., *A paraconsistent extension of sylvans logic*, Algebra and Logic **46** (2007), pp. 289–296.

[22] Hughes, G. E. and M. J. Cresswell, "A new introduction to modal logic," Psychology Press, 1996.

[23] Lahav, O. and A. Avron, *A unified semantic framework for fully structural propositional sequent systems*, ACM Trans. Comput. Logic **14** (2013), pp. 27:1–27:33.

[24] Lahav, O., J. Marcos and Y. Zohar, *It ain't necessarily so: Basic sequent systems for negative modalities*, this volume.

[25] Lenzen, W., *Necessary conditions for negation operators*, in: H. Wansing, editor, *Negation in Focus*, Walter de Gruyter, 1996 pp. 37–58.

[26] Lenzen, W., *Necessary conditions for negation operators (with particular applications to paraconsistent negation)*, in: *Reasoning with Actual and Potential Contradictions*, Springer, 1998 pp. 211–239.

[27] Marcos, J., *Modality and paraconsistency*, in: M. Bilkova and L. Behounek, editors, *The Logica Yearbook 2004*, Filosofia, 2005 pp. 213–222.

[28] Marcos, J., *Nearly every normal modal logic is paranormal*, Logique et Analyse **48** (2005), pp. 279–300.

[29] Marcos, J., *On negation: Pure local rules*, Journal of Applied Logic **3** (2005), pp. 185–219.

[30] Marcos, J., *Negative modalities, consistency and determinedness*, Electronic Notes in Theoretical Computer Science **300** (2014), pp. 21–45.

[31] Mruczek-Nasieniewska, K. and M. Nasieniewski, *Syntactical and semantical characterization of a class of paraconsistent logics*, Bul. of the Section of Logic **34** (2005), pp. 229–248.

[32] Mruczek-Nasieniewska, K. and M. Nasieniewski, *Paraconsistent logics obtained by J.-Y. Béziaus method by means of some non-normal modal logics*, Bull. Sect. Log **37** (2008), pp. 185–196.

[33] Mruczek-Nasieniewska, K. and M. Nasieniewski, *Béziau's logics obtained by means of quasi-regular logics*, Bul. of the Section of Logic **38** (2009), pp. 189–203.

[34] Nerode, A. and R. A. Shore, "Logic for Applications," Springer, 1997.

[35] Odintsov, S. P., "Constructive Negations and Paraconsistency," Trends in Logic **26**, Springer, 2008.

[36] Omori, H. and T. Waragai, *Negative modalities in the light of paraconsistency*, in: *The Road to Universal Logic*, Springer, 2015 pp. 539–555.

[37] Osorio, M., J. L. Carballido and C. Zepeda, *Revisiting* $\mathbb{Z}$, Notre Dame Journal of Formal Logic **55** (2014), pp. 129–155.

[38] Sylvan, R., *Variations on da Costa C-systems and dual-intuitionistic logics I. analyses of* C_ω *and* CC_ω, Studia Logica **49** (1990), pp. 47–65.

[39] Takano, M., *Subformula property as a substitute for cut-elimination in modal propositional logics*, Mathematica Japonica **37** (1992), pp. 1129–1145.

[40] Wansing, H., *Sequent systems for modal logics*, in: *Handbook of Philosophical Logic*, Springer, 2002 pp. 61–145.

[41] Wójcicki, R., "Theory of Logical Calculi: Basic Theory of Consequence Operations," Kluwer Academic Publishers, 1988.

Algebraic Semantics of Refinement Modal Logic

Zeinab Bakhtiari [1] Hans van Ditmarsch [2]

LORIA, CNRS — Université de Lorraine, France

Sabine Frittella [3]

Delft University of Technology, The Netherlands

Abstract

We develop algebraic semantics of refinement modal logic using duality theory. Refinement modal logic has quantifiers that are interpreted using a refinement relation. A refinement relation is like a bisimulation, except that from the three relational requirements only 'atoms' and 'back' have to be satisfied. We study the dual notion of refinement on algebras and present algebraic semantics of refinement modal logic. To this end, we first present the algebraic semantics of action model logic quantifier, and we then introduce an algebraic model based on the semantics of the refinement quantifier in terms of the refinement relation. Then we show that refinement modal logic is sound and complete with respect to this algebraic semantics.

Keywords: Refinement modal logic, arbitrary action model logic, dynamic epistemic logic, algebraic semantics.

1 Introduction

In *modal logic* we attempt to formalize propositions about *possibility* and *necessity*. In epistemic modal logics the modal operator is interpreted as knowledge or belief [19], initially for a single knowing agent but later for a set of agents, including their higher-order knowledge (i.e., what they know about each other) [11]. The knowledge of agents is encoded in a relational structure known as a *Kripke model* or *relational structure*, consisting of a domain of worlds, a binary accessibility relation for each agent, and a valuation of atomic propositions over the worlds. Informative updates can be formalized as yet another modal operator, a dynamic modality, that is interpreted as a relation between such Kripke models. A well-known form of informative updates are action models [5], wherein the updates themselves also take the shape of a relational structure.

The Kripke model resulting from executing an action model in an initial Kripke model can also can be seen as a *refinement* of that initial model. A

[1] bakhtiarizeinab@gmail.com

[2] hans.van-ditmarsch@loria.fr

[3] s.s.a.frittella@tudelft.nl

refinement relation is like a bisimulation relation, except that from the three relational requirements only 'atoms' and 'back' need to be satisfied. This therefore results in structural loss. From the perspective of knowledge change, this implies that in the refined model agents know more, namely they are less uncertain between different worlds. In [7] refinement modal logic (RML) is introduced, wherein modal logic is augmented with a new operator $\exists$ (and with its dual $\forall$, they are interdefinable as usual), which quantifies over all refinements of a given pointed model. In this logic the expression $\exists\varphi$ stands for "there is a refinement after which φ." In other words, $\exists\varphi$ is true in a Kripke model M with point s (we write M_s for such a pair) if there is a pointed model $M'_{s'}$ such that $(M_s, M'_{s'})$ are a pair in the refinement relation, (we also say that $M'_{s'}$ is a refinement of M_s), and such that φ is true in $M'_{s'}$. The logic is equally expressive as basic modal logic. A well-known result is that action model execution results in a refinement. We can similarly (although not trivially) augment the logic of knowledge with refinement quantifiers, and also the multi-agent logic of knowledge.

A different form of quantification is over action models. This has been investigated in [18]. This logic is called arbitrary action model logic. It is an extension of action model logic with an action model quantifier such that $\exists\varphi$ stands for "there is an action model such that after its execution φ (is true)." Given such an expression $\exists\varphi$, in [18] Hales presents a method for synthesizing a multi-pointed action model α_T after which φ is true (in the sense that $\exists\varphi$ is logically equivalent to $\langle\alpha_\mathsf{T}\rangle\varphi$), and he also proved that the action model quantifier is equivalent to the refinement quantifier.

In this paper we develop an algebraic semantics of refinement modal logic. Already from close to the inception of dynamic epistemic logics, there has been a strong current to model such logics in algebraic or coalgebraic settings [3,4]. More recently, in [21,22] an algebraic semantics was proposed for public announcement logic and action model logic. This methodology has further been productively used in [9] for a probabilistic dynamic epistemic logic and in [2] for epistemic updates on bilattices.

In [21,22], product updates are dually characterized through a construction that transforms the complex algebra associated with a given Kripke model into the complex algebra associated with the model updated by means of an action model. Given a Kripke model M and an action model α, the result of executing that action model can be seen as a submodel of a so-called intermediate model that contains copies of M indexed by the domain of α. In this way, action model logic can be endowed with an algebraic semantics that is dual (and equivalent) to the relational one, via a Jónsson-Tarski-type duality [6]. In particular, this holds for the multi-pointed action model α_T such that $\exists\varphi$ is equivalent to $\langle\alpha_\mathsf{T}\rangle\varphi$, according [18] mentioned above.

We use this result to define the algebraic semantics of RML. Indeed, we can dually characterize the algebraic notion of refinement relation as a lax-morphism (named *refinement morphism*) between the complex algebras associated with a given initial Kripke model and a 'resulting' Kripke model that is in the refinement relation with the initial model. Then, via the Jónsson-Tarski duality, we associate that resulting Kripke model to a boolean algebra with operators (BAO). Given the set of all refinements of the initial Kripke model, we then take the product of all corresponding BAOs in order to define a unique algebra and the required refinement morphism. The motivation behind our approach is to capture the non-constructive notion of refinement. Whereas arbitrary action model logic approaches the notion of refinement with brute force by having a witnessing action model that enforces the same post-condition φ bound by the quantifier, refinement modal logic only needs the existence of such an epistemic action (and thus the possibility of synthesizing it) but not the actual construction.

Structure of the paper. In Section 2, we introduce modal logic, refinement modal logic, action model logic, and arbitrary action model logic. In Section 3, we introduce relevant algebraic terminology. In Section 4, we present the methodology to define the algebraic semantics of dynamic epistemic logics. Finally, in Section 5, we present the algebraic semantics of refinement modal logic. Section 6 describes our results in view of prior works and concludes. The appendix contains the proofs of the results in Section 5.

2 Logical preliminaries

In this section, we succinctly introduce modal logic, action model logic [5], arbitrary action model logic [18], and refinement modal logic [23,7]. As all these logics are equally expressive, we can present them all as fragments of one logical language. Throughout the paper, we assume a non-empty, countable set of propositional atoms AtProp. We present here the single-agent version of these logics. All results in this section generalize to the multi-agent setting.

Models. A *(Kripke) frame* is a pair $\mathcal{F} = (S, R)$ where S is the *domain* consisting of *worlds* (or *states*), and $R \subseteq S \times S$ is a binary *accessibility relation.* Given $s \in S$, a pair (F, s), written as F_s, is a *pointed frame*, and a pair $(\mathcal{F}, T)$ with $T \subseteq S$ is a *multi-pointed frame* denoted $\mathcal{F}_T$. A *Kripke model* is a triple $M = (S, R, V)$ where (S, R) is a frame and where $V : \mathsf{AtProp} \to \mathcal{P}(S)$ is a *valuation* assigning to each propositional variable $p \in \mathsf{AtProp}$ the subset of the domain where the proposition p is true. Given a logical language $\mathcal{L}$, an *action model* over $\mathcal{L}$ is a triple $\alpha = (\mathsf{S}, \mathsf{R}, \mathsf{Pre})$ where (S, R) is a frame and where $\mathsf{Pre} : \mathsf{S} \to \mathcal{L}$ is a *precondition function.* The elements of the domain of an action model are called *actions*, or *action points.* Similarly to frames, we also define (multi-)pointed action models: given $\mathsf{s} \in \mathsf{S}$, a pair (α, u), written as α_{u}, is a *pointed action model*, and a pair (α, T) with $\mathsf{T} \subseteq \mathsf{S}$ is a *multi-pointed action model* and denoted as α_{T}. A (multi-)pointed action model is also called an *epistemic action.* The class of all action models *with finite domains* is $\mathcal{AM}$.

Let $M = (S, R, V)$ and $M' = (S', R', V')$ be given Kripke models. A non-empty relation $\mathfrak{R} \subseteq S \times S'$ is a *bisimulation* between M and M' if for all $(s, s') \in \mathfrak{R}$:

atoms $s \in V(p)$ iff $s' \in V'(p)$, for all $p \in \mathsf{AtProp}$;

forth $\forall t \in S$, if $R(s,t)$, $\exists t' \in S'$ such that $R'(s',t')$ and $(t,t') \in \mathfrak{R}$;

back $\forall t' \in S'$, if $R'(s',t')$, $\exists t \in S$ such that $R(s,t)$ and $(t,t') \in \mathfrak{R}$.

We write $M \simeq M'$ (M and M' *are bisimilar*) iff there is a bisimulation between M and M', and we write $M_s \simeq M'_{s'}$ (M_s and $M'_{s'}$ are bisimilar) iff this bisimulation links s and s'. A relation $\mathfrak{R}$ that satisfies **atoms**, **back** is called a *refinement*. We say that $M'_{s'}$ *refines* M_s and we write $M_s \succeq M'_{s'}$.

Languages. The language $\mathcal{L}_{\Box\forall\otimes\overline{\forall}}$ is inductively defined as:

$$\varphi ::= p \mid \neg\varphi \mid (\varphi \wedge \varphi) \mid \Box\varphi \mid \forall\varphi \mid [\alpha_{\mathsf{u}}]\varphi \mid \overline{\forall}\varphi$$

where $p \in \mathsf{AtProp}$, $(\mathsf{S}, \mathsf{R}, \mathsf{Pre}) = \alpha \in \mathcal{AM}$ and $\mathsf{u} \in \mathsf{S}$.

We assume the usual abbreviations for propositional logical connectives, and also $\Diamond\varphi ::= \neg\Box\neg\varphi$, $[\alpha_{\mathsf{T}}]\varphi ::= \bigwedge_{\mathsf{u}\in\mathsf{T}}[\alpha_{\mathsf{u}}]\varphi$, $\langle\alpha_{\mathsf{u}}\rangle\varphi ::= \neg[\alpha_{\mathsf{u}}]\neg\varphi$ $\langle\alpha_{\mathsf{T}}\rangle\varphi ::= \neg[\alpha_{\mathsf{T}}]\neg\varphi$, $\exists\varphi ::= \neg\forall\neg\varphi$, and $\overline{\exists}\varphi ::= \neg\overline{\forall}\neg\varphi$.

The following fragments of the language will occur in the paper (with the obvious restrictions): $\mathcal{L}_{\Box}$ of modal logic (K); $\mathcal{L}_{\Box\otimes}$ of *action model logic* (AML); $\mathcal{L}_{\Box\otimes\overline{\forall}}$ of *arbitrary action model logic* (AAML); $\mathcal{L}_{\Box\forall}$ of *refinement modal logic* (RML).

Semantics. Let $M = (S, R, V)$ be a Kripke model, $s \in S$, $(\mathsf{S}, \mathsf{R}, \mathsf{Pre}) = \alpha \in \mathcal{AM}$ be an action model (note that it is finite), and $\mathsf{u} \in \mathsf{S}$. The interpretation of $\varphi \in \mathcal{L}_{\Box\forall\otimes\overline{\forall}}$ is defined inductively by

$$\begin{array}{lll}
M_s \models p & \text{iff} & s \in V(p) \\
M_s \models \varphi \wedge \psi & \text{iff} & M_s \models \varphi \text{ and } M_s \models \psi \\
M_s \models \neg\varphi & \text{iff} & M_s \not\models \varphi \\
M_s \models \Box\varphi & \text{iff} & \text{for all } t \in R(s) : M_t \models \varphi \\
M_s \models \forall\varphi & \text{iff} & \text{for all } M'_{s'} : M_s \succeq M'_{s'} \text{ implies } M'_{s'} \models \varphi \\
M_s \models [\alpha_{\mathsf{u}}]\varphi & \text{iff} & M_s \models \mathsf{Pre}(\mathsf{u}) \text{ implies } (M \otimes \alpha)_{(s,\mathsf{u})} \models \varphi \\
M_s \models \overline{\forall}\varphi & \text{iff} & \text{for all } \alpha_{\mathsf{u}} \in \mathcal{AM} : M_s \models [\alpha_{\mathsf{u}}]\varphi
\end{array}$$

where $M \otimes \alpha = (S^\alpha, R^\alpha, V^\alpha)$ is the *product update* defined as

$$\begin{array}{rcl}
S^\alpha & = & \{(s,\mathsf{u}) \in S \times \mathsf{S} \mid M_s \models \mathsf{Pre}(\mathsf{u})\} \\
(s,\mathsf{u})R^\alpha(s',\mathsf{u}') & \text{iff} & sRs' \text{ and } \mathsf{uRu}' \\
V^\alpha(p) & = & (V(p) \times \mathsf{S}) \cap S^\alpha
\end{array}$$

We say that $M \otimes \alpha$ is the model resulting from applying the epistemic action α on the model M. The *extension map* $[\![\cdot]\!]_M : \mathcal{L}_{\Box\forall\otimes\overline{\forall}} \to \mathcal{P}(S)$ for a $\varphi \in \mathcal{L}_{\Box\forall\otimes\overline{\forall}}$ is $[\![\varphi]\!]_M := \{s \in S \mid M_s \models \varphi\}$. A formula φ is *valid on* M, notation $M \models \varphi$, if for all $s \in S$, $M_s \models \varphi$. A formula φ is *valid*, if for all M, $M \models \varphi$. Instead of $(M \otimes \alpha)_{(s,\mathsf{u})}$ we may write $M_s \otimes \alpha_{\mathsf{u}}$.

Axiomatization AML. The axiomatization of AML consists of the rules and axioms of K [18, Definition IV.1] along with the following axioms and the rule of necessitation for dynamic box modalities:

$$\begin{array}{rl}
\textbf{AP} & [\alpha_{\mathsf{u}}]p \leftrightarrow (\mathsf{Pre}(\mathsf{u}) \rightarrow p) \text{ for all } p \in \mathsf{AtProp} \\
\textbf{AN} & [\alpha_{\mathsf{u}}]\neg\varphi \leftrightarrow (\mathsf{Pre}(\mathsf{u})\imath\neg[\alpha_{\mathsf{u}}]\varphi) \\
\textbf{AC} & [\alpha_{\mathsf{u}}](\varphi \wedge \psi) \leftrightarrow ([\alpha_{\mathsf{u}}]\varphi \wedge [\alpha_{\mathsf{u}}]\psi) \\
\textbf{AK} & [\alpha_{\mathsf{u}}]\Box\varphi \leftrightarrow (\mathsf{Pre}(\mathsf{u}) \rightarrow \bigwedge\{\Box[\alpha_{\mathsf{u}'}]\varphi \mid \mathsf{uRu}'\}) \\
\textbf{AU} & [\alpha_{\mathsf{T}}]\varphi \leftrightarrow \bigwedge_{\mathsf{u}\in\mathsf{T}}[\alpha_{\mathsf{u}}]\varphi \\
\textbf{NecA} & \text{From } \varphi \text{ infer } [\alpha_{\mathsf{u}}]\varphi
\end{array}$$

Axiomatization RML**.** The axiomatization of RML consists of the rules and axioms of K and all substitution instances of the axioms and rules

$$\begin{array}{rlrl}
\textbf{R} & \forall(\varphi \rightarrow \psi) \rightarrow \forall\varphi \rightarrow \forall\psi & \textbf{MP} & \text{From } \varphi \rightarrow \psi \text{ and } \varphi \text{ infer } \psi \\
\textbf{RProp} & \forall p \leftrightarrow p \text{ and } \forall\neg p \leftrightarrow \neg p & \textbf{NecK} & \text{From } \varphi \text{ infer } \Box\varphi \\
\textbf{RK} & \exists\nabla\Phi \leftrightarrow \bigwedge\Diamond\exists\Phi & \textbf{NecR} & \text{From } \varphi \text{ infer } \forall\varphi
\end{array}$$

where for any finite set Φ of $\mathcal{L}_{\forall\Box}$ formulas we define by abbreviation $\nabla\Phi$ as $\Box\bigvee_{\varphi\in\Phi}\varphi \wedge \bigwedge_{\varphi\in\Phi}\Diamond\varphi$ and $\bigwedge\Diamond\exists\Phi$ as $\bigwedge_{\varphi\in\Phi}\Diamond\exists\varphi$, where $\bigvee_{\varphi\in\emptyset}\varphi := \bot$ and $\bigwedge_{\varphi\in\emptyset}\varphi := \top$.

Axiomatization AAML**.** The axiomatization of AAML is a substitution schema consisting of the rules and axioms of the logics RML and AML.

Some results about K**,** RML**,** AAML**,** AML**.**

- [23, Prop. 4&5]: The result of executing an epistemic action in a pointed model is a refinement of that model. Dually, for every refinement of a *finite* pointed model there is an epistemic action such that its execution results in a model bisimilar to that refinement.
- [18, Theorem V.3]: Let $\varphi \in \mathcal{L}_{\Box\forall\otimes\overline{\forall}}$. Then $\models \forall\varphi \leftrightarrow \overline{\forall}\varphi$.
- [5,7,18]: The logics K, RML, AAML, AML are all equally expressive.

Given a logic L, $\vdash \varphi$ means that φ is a theorem of the logic L, namely one can derive φ from the axioms and rules of L. In the Section 5, we use extensively the following theorem and lemma.

Theorem 2.1 *[18, Theorem V.3]: Let $\varphi \in \mathcal{L}_{\Box\otimes\forall}$. Then there exists a multi-pointed action model $\alpha^{\varphi}_{\mathsf{T}}$ such that $\vdash [\alpha^{\varphi}_{\mathsf{T}}]\varphi$ and $\vdash \langle\alpha^{\varphi}_{\mathsf{T}}\rangle\varphi \leftrightarrow \exists\varphi$.*

In [18] an algorithm is given to compute $\alpha^{\varphi}_{\mathsf{T}}$ from φ. In this construction a normal form is used that is called *cover disjunctive form* [20].

Lemma 2.2 *Let M_s be a pointed Kripke model and $\varphi \in \mathcal{L}_{\Box\otimes\forall}$. Then there exists a multi-pointed action model $\alpha^{\varphi}_{\mathsf{T}} = ((\mathsf{S},\mathsf{R},\mathsf{Pre}),\mathsf{T})$ such that $M_s \models \langle\alpha^{\varphi}_{\mathsf{T}}\rangle\varphi$ iff $M_s \models \exists\varphi$.*

Proof. Suppose that $M_s \models \exists\varphi$. Then by [18, Theorem V.3], there exists a multi-pointed action model $\alpha_{\mathsf{T}} = ((\mathsf{S},\mathsf{R},\mathsf{Pre}),\mathsf{T})$ such that $M_s \models \langle\alpha_{\mathsf{T}}\rangle\varphi$, which means $M_s \models \bigvee_{\mathsf{u}\in\mathsf{T}}\langle\alpha_{\mathsf{u}}\rangle\varphi$ with $\mathsf{T} \subseteq \mathsf{S}$. Therefore $M_s \models \langle\alpha_{\mathsf{S}}\rangle\varphi$.

For the other direction, suppose that there is a multi-pointed action model $\alpha_{\mathsf{T}} = ((\mathsf{S}, \mathsf{R}, \mathsf{Pre}), \mathsf{T})$ such that $M_s \models \langle \alpha_{\mathsf{T}} \rangle \varphi$. Then there exists $\mathsf{u} \in \mathsf{T}$ such that $M_s \models \langle \alpha_{\mathsf{u}} \rangle \varphi$, i.e. $M_s \otimes \alpha_{\mathsf{u}} \models \varphi$. □

3 Preliminaries on Algebras

In this section, we introduce relevant definitions and results on boolean algebras.

Boolean algebras. A *Boolean algebra* (**BA**) $\mathbb{A} = \langle A, \vee, \wedge, \neg, \bot, \top \rangle$ is an algebra with two binary operations $\vee$ (called 'join' or 'or') and $\wedge$ (called 'meet' or 'and'), one unary operation $\neg$ (called 'not' or 'complement'), and two nullary operations $\bot$ and $\top$ (called 'bottom' and 'top') which satisfy the following equations:

$a \wedge b = b \wedge a$	$a \vee b = b \vee a$	(commutativity)
$a \vee (b \vee c) = (a \vee b) \vee c$	$a \wedge (b \wedge c) = (a \wedge b) \wedge c$	(associativity)
$a \wedge (a \vee b) = a$	$a \vee (a \wedge b) = a$	(absorption)
$a \wedge \top = a$	$a \vee \bot = a$	(identity)
$a \wedge (b \vee c) = (a \wedge b) \vee (a \wedge c)$	$a \vee (b \wedge c) = (a \vee b) \wedge (a \vee c)$	(distributivity)
$a \wedge \neg a = \bot$	$a \vee \neg a = \top$	(complementation)

Let X be a set and $\mathcal{P}(X)$ be the set of all the subsets of X. Denote with $\cup$, $\cap$ and $(-)^c$ the operations union, intersection and complement on $\mathcal{P}(X)$, respectively. Then $(\mathcal{P}(X), \cup, \cap, (-)^c, \emptyset, X)$ forms a **BA**.

Underlying poset of a boolean algebra. A **BA** $\mathbb{A} = (A, \vee, \wedge, \neg, \bot, \top)$ can also be seen as a poset (partially ordered set) $(A, \leq)$ where the order $\leq$ is defined as follows: $x \leq y$ iff $x \wedge y = x$ iff $x \vee y = y$, for any $x, y \in A$. We call $(A, \leq)$ the underlying poset of $\mathbb{A}$. Let $(A, \leq)$ be a poset, $a \in A$ and $S \subseteq A$, a is an *upper bound* (resp. *lower bound*) of S, if $s \leq a$ (resp. $a \leq s$) for every $s \in S$. The element $a \in A$ is the *least upper bound* of S if it is an upper bound of S and if $a \leq s$ for every upper bound s of S. The element $a \in \mathbb{A}$ is the *greatest lower bound* of S if it is a lower bound of S and if $s \leq a$ for every lower bound s of S. If they exist, the *least upper bound* of S is denoted by $\bigvee S$ and the greatest lower bound of S by $\bigwedge S$. For any **BA** $\mathbb{A} = \langle A, \vee, \wedge, \neg, \bot, \top \rangle$, $\bigvee S$ and $\bigwedge S$ of a finite subset $S \subseteq A$ always exist and are unique, however they may not exist if S is infinite.

Complete boolean algebras. A **BA** $\mathbb{A}$ is *complete* if $\bigvee S$ and $\bigwedge S$ exist for every $S \subseteq A$. The **BA** $(\mathcal{P}(X), \cup, \cap, (-)^c, \emptyset, X)$ is complete. The underlying order is given by the inclusion $\subseteq$, and for $S \subseteq \mathcal{P}(X)$, $\bigvee S$ and $\bigwedge S$ are respectively given by the union and the intersection. Formally, if I is an index set and $X_i \subseteq X$ for all $i \in I$, then $\bigvee_{i \in I} X_i = \bigcup_{i \in I} X_i$ and $\bigwedge_{i \in I} X_i = \bigcap_{i \in I} X_i$.

Boolean algebras with operators.

A *boolean algebra with operators* (**BAO**) is a structure $(\mathbb{A}, \{\Diamond_i\}_{i \in I})$ such that $\mathbb{A}$ is a **BA**, I is a non-empty finite set and $\Diamond_i : \mathbb{A} \to \mathbb{A}$ for every $i \in I$. A *normal* boolean algebra with operators is a **BAO** $(\mathbb{A}, \{\Diamond_i\}_{i\in I})$ such that the unary operations $\{\Diamond_i\}_{i\in I}$ on $\mathbb{A}$ satisfy $\Diamond_i\bot = \bot$ and $\Diamond_i(a \vee b) = \Diamond_i a \vee \Diamond_i b$, for all $a, b \in A$. The following definitions for boolean algebras are given for only one operator $\Diamond$. This can be done without loss of generality.

Congruence. A *congruence* θ on a **BAO** $\mathbb{A}$ is an equivalence relation which satisfies this compatibility property: for all $a, a', b, b' \in A$, if $a\theta b$ and $a'\theta b'$ then:

$$(\neg a)\theta(\neg b), \quad (a \vee a')\theta(b \vee b'), \quad (a \wedge a')\theta(b \wedge b') \quad \textit{and} \quad (\Diamond a)\theta(\Diamond b).$$

We denote by A/θ the set of the equivalence classes defined by the congruence θ, namely $A/\theta = \{[a]_\theta \mid a \in A\}$ with $[a]_\theta = \{b \in a \mid a\theta b\}$. There is a natural way to define the operations $\vee', \wedge', \neg'$ on the set A/θ of equivalence classes of $\mathbb{A}$ over θ. Namely, for all $a, b \in A$, we define

$$[a]_\theta \vee' [b]_\theta := [a \vee b]_\theta, \quad [a]_\theta \wedge' [b]_\theta := [a \wedge b]_\theta, \quad \textit{and} \quad \neg'[a]_\theta := [\neg a]_\theta.$$

It can be shown that $\mathbb{A}/\theta := \langle A/\theta, \vee', \wedge', \neg', [\bot]_\theta, [\top]_\theta\rangle$ is a **BA**. We call it the *quotient algebra* of $\mathbb{A}$ modulo θ.

Complex algebras. Let $\mathcal{F} := (S, R)$ be a Kripke frame. The *complex algebra* of $\mathcal{F}$, denoted $\mathcal{F}^+$, is the power set algebra $(\mathcal{P}(S), \cup, \cap, (-)^c, \emptyset, S)$ enriched with the operator $\Diamond_R : \mathcal{P}S \to \mathcal{P}S$ defined as $\Diamond_R(X) := \{s \in S \mid sRt$ for some $t \in X\} = R^{-1}[X]$ for every $X \in \mathcal{P}S$. We note that $\mathcal{F}^+$ is a normal **BAO**.

Complex algebras are the concrete **BAO**s that algebraize relational semantics [6, Theorem 5.25]. By means of complex algebras one can construct a **BAO** from a frame. For the other direction we need to construct the *ultrafilter frame* [6, Definition 5.34]. By transforming this frame in a complex algebra we get the *Jònson-Tarski Theorem* underlying the algebraization of modal logic:

> *Every* **BAO** *can be embedded in the complex algebra of its ultrafilter frame [6, Theorem 5.43].*

Adjunction. A map $f : (A, \leq_A) \to (B, \leq_B)$ between two posets is *monotone* if $a \leq_A b$ implies $f(a) \leq_B f(b)$ for all $a, b \in A$. A pair (f, g) of monotone maps $f : \mathbb{A} \to \mathbb{B}$ and $g : \mathbb{B} \to \mathbb{A}$ between two posets forms an *adjunction* (denoted $f \dashv g$) between $\mathbb{A}$ and $\mathbb{B}$ if $f(a) \leq_B b$ is equivalent to $a \leq_A g(b)$, for all $a \in A$ and $b \in B$. If $f \dashv g$, then g is a *right adjoint* and f a *left adjoint.*

Let $(\mathbb{A}, \Diamond)$ be a complex algebra of some Kripke frame, and $\Box := \neg\Diamond\neg$, then $\Diamond$ is a left adjoint and $\Box$ is a right adjoint. Moreover, there exist $\blacklozenge$ and $\blacksquare$ such that $\Diamond \dashv \blacksquare$ and $\blacklozenge \dashv \Box$.

Algebraic models.

An *algebraic model* is a tuple $\mathcal{A} = (\mathbb{A}, V)$ such that $\mathbb{A}$ is a normal **BAO** and $V : \mathsf{AtProp} \to \mathbb{A}$. Let $M := (\mathcal{F}, V)$ with $V : \mathsf{AtProp} \to \mathcal{P}S$ be a Kripke model, the *algebraic model associated with* M is the tuple $\mathcal{A} = (\mathcal{F}^+, V)$ where $\mathcal{F}^+$ is the complex algebra of $\mathcal{F}$. Notice that the valuation V (resp. the extension map $[\![\cdot]\!]_M$) sends atomic propositions (resp. formulas) to elements in $\mathbb{A}$.

We will rely on the duality between Kripke frames and normal boolean algebras with operators to define the algebraic semantics of arbitrary action model logic and refinement modal logic.

4 Algebraic semantics of Arbitrary Action Model Logic

In this section, first we present the methodology to define epistemic updates on algebras (Section 4.1), and then we give the algebraic semantics of Action Model Logic (Section 4.2). Sections 4.1 and 4.2 report on results introduced in [21,22].

4.1 Epistemic update on algebras

We first describe the methodology to define the epistemic update on normal boolean algebras with operators, and then the mathematical steps to compute the updated algebra.

Methodology. Let $M = (S, R, V)$ be a Kripke model and $\alpha = (\mathsf{S}, \mathsf{R}, \mathsf{Pre})$ be an action model over $\mathcal{L}_{\Box\otimes}$. The product update $M \otimes \alpha$ defined in Section 2 can be built in an algebraic way in two steps as follows.

<u>STEP 1.</u> We define the following intermediate model

$$\coprod_{\alpha} M = (\coprod_{\mathsf{S}} S, R \times \mathsf{R}, \coprod_{\alpha} V)$$

where

(i) $\coprod_{\mathsf{S}} S \simeq S \times \mathsf{S}$ is the $|\mathsf{S}|$-fold coproduct of S, which is set-isomorphic to cartesian product of $S \times \mathsf{S}$,

(ii) $R \times \mathsf{R}$ is the binary relation on $\coprod_{\mathsf{S}} S$ defined as

$$(s, \mathsf{u})(R \times \mathsf{R})(s', \mathsf{u}') \quad \text{iff} \quad sRs' \text{ and } \mathsf{u}\mathsf{R}\mathsf{u}',$$

(iii) $\coprod_{\alpha} V : \mathsf{AtProp} \to \mathcal{P}(\coprod_{\mathsf{S}} S)$ such that for every $p \in \mathsf{AtProp}$

$$\coprod_{\alpha} V(p) = \coprod_{\alpha} (V(p)) = V(p) \times \mathsf{S}.$$

<u>STEP 2.</u> $M \otimes \alpha$ is the submodel of $\coprod_{\alpha} M$ that contains the tuples $(s, u) \in \coprod_{\mathsf{S}} S$ such that $M_s \models \mathsf{Pre}(u)$.

This two-step-account of the product update construction can be seen as a pseudo-coproduct, as illustrated by the following diagram

$$M \hookrightarrow \coprod_{\alpha} M \hookleftarrow M \otimes \alpha.$$

This perspective makes it possible to use the duality between products and coproducts in category theory (cf. [10,1]): coproducts can be dually characterized as products, and subobjects as quotients. Using this result, the update of M

with the action model α, regarded as a "subobject after coproduct" concatenation, can be dually characterized on its algebraic counterpart $(\mathbb{A}, V)$ by means of a "quotient after product" concatenation, as illustrated in the following diagram:

$$\mathbb{A} \twoheadleftarrow \prod_{\alpha} \mathbb{A} \twoheadrightarrow \mathbb{A}^{\alpha}.$$

Indeed, the pseudo-coproduct $\coprod_{\alpha} M$ is dually characterized as a *pseudo-product* $\prod_{\alpha} \mathbb{A}$ and an appropriate *quotient* of $\prod_{\alpha} \mathbb{A}$ is then taken to dually characterize the submodel step. This construction we now define.

Product on Sets. Recall that, in the category of sets, the *product* is the Cartesian product. Namely, given a family of sets $(X_i)_{i\in I}$, as $\prod_{i\in I} X_i := \{(x_i)_{i\in I} \mid \forall i \in I, x_i \in X_i\}$ with the canonical projections $\pi_j : \prod_{i\in I} X_i \to X_j$ defined as $\pi_j((x_i)_{i\in I}) := x_j$.

Dual characterization of the intermediate structure.

Definition 4.1 [Action model on algebras] For every algebra $\mathbb{A}$, we define an *action model over* $\mathbb{A}$ as a tuple $a = (\mathsf{S}, \mathsf{R}, \mathsf{Pre}_a)$ such that S is a finite nonempty set, $\mathsf{R} \subseteq \mathsf{S} \times \mathsf{S}$ and $\mathsf{Pre}_a : \mathsf{S} \to \mathbb{A}$. As for Kripke models, one can define pointed action models (a, u) over $\mathbb{A}$ with $\mathsf{u} \in a$ denoted a_{u}.

Clearly, for every Kripke model $M = (S, R, V)$, each action model $\alpha = (\mathsf{S}, \mathsf{R}, \mathsf{Pre})$ over $\mathcal{L}_{\Box\otimes}$ induces a corresponding action model a over the complex algebra $\mathbb{A}$ of the underlying frame (S, R) of M, via the valuation $V : \mathsf{AtProp} \to \mathbb{A}$, namely, a is defined as $a = (\mathsf{S}, \mathsf{R}, \mathsf{Pre}_a)$, with $\mathsf{Pre}_a = V \circ \mathsf{Pre}_a$.

For every **BA** $\mathbb{A}$ and every action model $a = (\mathsf{S}, \mathsf{R}, \mathsf{Pre}_a)$ over $\mathbb{A}$, let $\prod_a \mathbb{A}$ be the $|\mathsf{S}|$-fold product of $\mathbb{A}$, which is set-isomorphic to the collection $\mathbb{A}^{\mathsf{S}}$ of the set maps $f : \mathsf{S} \to \mathbb{A}$. The set $\mathbb{A}^{\mathsf{S}}$ can be canonically endowed with the same algebraic structure as $\mathbb{A}$ by pointwise lifting the operations on $\mathbb{A}$[4]; as such, it satisfies the same equations as $\mathbb{A}$.

Definition 4.2 Let $(\mathbb{A}, \Diamond)$ be a normal **BAO**, $\Box := \neg\Diamond\neg$ and $a = (\mathsf{S}, \mathsf{R}, \mathsf{Pre}_a)$ be an action model over $\mathbb{A}$, we define the operations $\Diamond^{\prod_a \mathbb{A}}$ and $\Box^{\prod_a \mathbb{A}}$ on the product $\prod_a \mathbb{A}$ as follows: for every $f : \mathsf{S} \to \mathbb{A}$,

$$\begin{aligned} \Diamond^{\prod_a \mathbb{A}} f : \mathsf{S} &\to \mathbb{A} \\ \mathsf{u} &\mapsto \bigvee \{\Diamond^{\mathbb{A}} f(\mathsf{u}') \mid \mathsf{uRu}'\} \end{aligned} \qquad \begin{aligned} \Box^{\prod_a \mathbb{A}} f : \mathsf{S} &\to \mathbb{A} \\ \mathsf{u} &\mapsto \bigwedge \{\Box^{\mathbb{A}} f(\mathsf{u}') \mid \mathsf{uRu}'\}. \end{aligned}$$

The operators $\Diamond^{\prod_a \mathbb{A}}$ and $\Box^{\prod_a \mathbb{A}}$ are normal modal operators such that $\Box^{\prod_a \mathbb{A}} = \neg\Diamond^{\prod_a \mathbb{A}}\neg$, and the product algebra $(\prod_a \mathbb{A}, \Diamond^{\prod_a \mathbb{A}})$ is a normal **BAO** [21, Proposition 3.2]. Also, if $\mathbb{A}$ is the complex algebra of the underlying frame of the Kripke model $(\mathcal{F}, V)$ and if the action model a over $\mathbb{A}$ is derived from the action model $\alpha = (\mathsf{S}, \mathsf{R}, \mathsf{Pre})$ over $\mathcal{L}_{\Box\otimes}$, then $(\prod_a \mathbb{A}, \Diamond^{\prod_a \mathbb{A}})$ is isomorphic to

---

[4] For all $f, g : \mathsf{S} \to \mathbb{A}$, the maps $(f \wedge^{\prod_a \mathbb{A}} g), \neg^{\prod_a \mathbb{A}} f \bot^{\prod_a \mathbb{A}} : \mathsf{S} \to \mathbb{A}$ are respectively defined as follows: $(f \wedge^{\prod_a \mathbb{A}} g)(\mathsf{u}) := f(\mathsf{u}) \wedge^{\mathbb{A}} g(\mathsf{u})$, $(\neg^{\prod_a \mathbb{A}} f)(\mathsf{u}) := \neg^{\mathbb{A}} f(\mathsf{u})$ and $\bot^{\prod_a \mathbb{A}}(\mathsf{u}) := \bot$.

the complex algebra of the underlying frame $\coprod_a \mathcal{F}$ of the intermediate model $\coprod_a M$ [21, Proposition 3.1].

Quotient of the intermediate structure. Let $\mathbb{A}$ be a normal **BAO** and $a = (\mathsf{S}, \mathsf{R}, \mathsf{Pre}_a)$ be an action model over $\mathbb{A}$. The equivalence relation $\equiv_a$ on $\prod_a \mathbb{A}$ is defined as follows: for all $f, g \in \mathbb{A}^{\mathsf{S}}$,

$$f \equiv_a g \text{ iff } f \wedge \mathsf{Pre}_a = g \wedge \mathsf{Pre}_a.$$

For any $f \in \mathbb{A}^{\mathsf{S}}$, we denote by $[f]_a$ its equivalence class. The subscript will be dropped whenever it causes no confusion. Let the quotient algebra $\mathbb{A}^{\mathsf{S}}/{\equiv_a}$ be denoted by $\mathbb{A}^a$. This quotient is compatible with the boolean operations, however it is not compatible with the modal operators, indeed $f \equiv g$ does not imply that $\Diamond f \equiv \Diamond g$. So we need to choose a definition for the modalities on $\mathbb{A}^a$: let for every $f \in \mathbb{A}^{\mathsf{S}}$,

$$\Diamond^a[f] := [\Diamond^{\prod_a \mathbb{A}}(f \wedge \mathsf{Pre}_a)] \qquad \text{and} \qquad \Box^a[f] := [\Box^{\prod_a \mathbb{A}}(f \rightarrow \mathsf{Pre}_a)].$$

The operators $\Diamond^a$ and $\Box^a$ are normal modal operators such that $\Box^a = \neg\Diamond^a\neg$ and $(\mathbb{A}^a, \Diamond^a)$ is a normal **BAO**. Moreover, if $(\mathbb{A}, V) = (\mathcal{F}^+, V)$ for some Kripke model $M = (\mathcal{F}, V)$, and if the action model a over $\mathbb{A}$ is derived from some action model $\alpha = (\mathsf{S}, \mathsf{R}, \mathsf{Pre})$ over $\mathcal{L}_{\Box\otimes}$, then $\mathbb{A}^a \cong_{\mathbf{BAO}} \mathcal{F}^{\alpha+}$, in which $\mathcal{F}^\alpha$ is the underlying frame of the updated model $M \otimes \alpha$.

Definition 4.3 Let $(\mathbb{A}, \Diamond)$ be a normal **BAO** and $a = (\mathsf{S}, \mathsf{R}, \mathsf{Pre}_a)$ be an action model over $\mathbb{A}$. The update of $\mathbb{A}$ with a is $\mathbb{A}^a := (\mathbb{A}^{\mathsf{S}}/\equiv_a, \Diamond^a)$, where $(\mathbb{A}^{\mathsf{S}}/\equiv_a)$ is the quotient algebra and $\Diamond^a$ is the normal modality, as above.

4.2 Algebraic semantics of action model logic

In this section we report on the algebraic semantics of action model logic proposed by [21,22]. We recall that that an *algebraic model* is a tuple $\mathcal{A} = (\mathbb{A}, V)$ such that $\mathbb{A}$ is a normal **BAO** and $V : \mathsf{AtProp} \rightarrow \mathbb{A}$.

Definition 4.4 Let $\mathcal{A} = (\mathbb{A}, V)$ be an algebraic model, $\alpha = (\mathsf{S}, \mathsf{R}, \mathsf{Pre}_\alpha)$ an action model over $\mathcal{L}_{\Box\otimes}$ and $a = (\mathsf{S}, \mathsf{R}, \mathsf{Pre}_a)$ the action model induced by α via V.
The *intermediate algebraic model* $\prod_\alpha \mathcal{A}$ is defined as

$$\prod_\alpha \mathcal{A} := (\prod_a \mathbb{A}, \prod_a V)$$

where, for any $p \in \mathsf{AtProp}$, the map $(\prod_a V)(p) : \mathsf{S} \rightarrow \prod_a \mathbb{A}$ is such that $((\prod_a V)(p))\,(\mathsf{u}) = V(p)$.
The *updated algebraic model* $\mathcal{A}^\alpha$ is defined as

$$\mathcal{A}^\alpha := (\mathbb{A}^a, V^a)$$

where $V^a : \mathsf{AtProp} \rightarrow \mathbb{A}^a$ is the map such that $V^a(p) = [\prod_a V(p)]_a$.

Let $\mathcal{A} = (\mathbb{A}, V)$ be an algebraic model, $\alpha = (\mathsf{S}, \mathsf{R}, \mathsf{Pre}_\alpha)$ be an action model over $\mathcal{L}_{\Box\otimes}$, and a the action model induced by α via V. Let $\pi_\mathsf{u} : \prod_a \mathbb{A} \to \mathbb{A}$ be the projection on the u-indexed coordinate that maps every $f \in \prod_a \mathbb{A}$ to $f(\mathsf{u})$, and let $i' : \mathbb{A}^a \to \prod_a \mathbb{A}$ be defined as $i'([f]) = f \wedge \mathsf{Pre}_a$ for all $[f] \in \mathbb{A}^a$. Then:

Definition 4.5 For every algebraic model $\mathcal{A} = (\mathbb{A}, V)$, its extension map $[\![.]\!]_\mathcal{A} : \mathcal{L}_{\Box\otimes} \to \mathbb{A}$ is defined recursively as

$$
\begin{aligned}
[\![p]\!]_\mathcal{A} &:= v(p) \\
[\![\bot]\!]_\mathcal{A} &:= \bot^\mathbb{A} \\
[\![\circ\varphi]\!]_\mathcal{A} &:= \circ^\mathbb{A}[\![\varphi]\!]_\mathcal{A} && \text{for } \circ \in \{\neg, \Diamond, \Box\} \\
[\![\varphi \bullet \psi]\!]_\mathcal{A} &:= [\![\varphi]\!]_\mathcal{A} \bullet^\mathbb{A} [\![\psi]\!]_\mathcal{A} && \text{for } \bullet \in \{\vee, \wedge, \rightarrow\} \\
[\![\langle\alpha_\mathsf{u}\rangle\varphi]\!]_\mathcal{A} &:= [\![\mathsf{Pre}(\mathsf{u})]\!]_\mathcal{A} \wedge^\mathbb{A} \pi_\mathsf{u} \circ i'([\![\varphi]\!]_{\mathcal{A}^\alpha}) \\
[\![[\alpha_\mathsf{u}]\varphi]\!]_\mathcal{A} &:= [\![\mathsf{Pre}(\mathsf{u})]\!]_\mathcal{A} \rightarrow^\mathbb{A} \pi_\mathsf{u} \circ i'([\![\varphi]\!]_{\mathcal{A}^\alpha})
\end{aligned}
$$

5 Algebraic semantics of refinement modal logic

In this section, we present our main result, namely an algebraic semantics for RML. First we introduce the notion of *refinement morphism*. It is the analogue on normal boolean algebras with operators of the notion of refinement between Kripke models. Then we define *refinement algebra*. This is used in the definition of the *algebraic semantics of refinement modal logic*. Finally we prove that RML is sound and complete w.r.t. this semantics.

Throughout this section we adopt the following notational conventions.

- $\mathbb{A}$ is a complete normal **BAO**;
- $\mathcal{A} = (\mathbb{A}, V)$ is an algebraic model;
- $\alpha^\varphi_\mathsf{S} = (\mathsf{S}, \mathsf{R}, \mathsf{Pre}^\varphi)$ denotes the multi-pointed action model obtained by using Hales algorithm on the formula φ (cf. Lemma 2.2 and [18, Lemmas V.I and V.II]);
- $a^\varphi = (\mathsf{S}, \mathsf{R}, \mathsf{Pre}_a)$ denotes the action model over $\mathbb{A}$ induced by $\alpha^\varphi_\mathsf{S}$ via V;
- $\mathcal{A}^\varphi$ denotes the algebraic model $\mathcal{A}^{\alpha^\varphi_\mathsf{S}}$ (Definition 4.4);
- $\mathbb{A}^\varphi$ denotes the **BAO** underlying the algebraic model $\mathcal{A}^\varphi$;
- $[\![.]\!]_\mathcal{A} : \varphi \in \mathcal{L}_{\Box\forall\otimes\overline{\forall}} \to \mathbb{A}$ is the *extension map* of V on $\mathcal{A}$ such that $[\![\varphi]\!]_\mathcal{A}$ follows Definition 4.5 for all logical connectives except quantifiers and such that

$$
\begin{aligned}
[\![\overline{\exists}\varphi]\!]_\mathcal{A} = [\![\exists\varphi]\!]_\mathcal{A} &:= \bigvee_{\mathsf{u}\in\mathsf{S}}[\![\langle\alpha^\varphi_\mathsf{u}\rangle\varphi]\!]_\mathcal{A} = \bigvee_{\mathsf{u}\in\mathsf{S}}([\![\mathsf{Pre}^\varphi(\mathsf{u})]\!]_\mathcal{A} \wedge \pi_\mathsf{u} \circ i'([\![\varphi]\!]_{\mathcal{A}^\varphi})) \\
[\![\forall\varphi]\!]_\mathcal{A} &:= [\![\neg\exists\neg\varphi]\!]_\mathcal{A} \\
[\![\overline{\forall}\varphi]\!]_\mathcal{A} &:= [\![\neg\overline{\exists}\neg\varphi]\!]_\mathcal{A}
\end{aligned}
$$

Refinement morphisms and their adjoints. The dual notion of refinement on algebras is the *refinement morphism*. We prove that refinement morphisms are right adjoints.

Definition 5.1 Let $\mathbb{A}$ and $\mathbb{A}'$ be two normal **BAO**s. A map $f : \mathbb{A} \to \mathbb{A}'$ is a *refinement morphism* if it is monotone, preserves $\bot$ and $\vee$, and satisfies the inequality $\blacklozenge^{\mathbb{A}'} \circ f^{\varphi} \leq f^{\varphi} \circ \blacklozenge^{\mathbb{A}}$ where $\blacklozenge \dashv \Box$ (cf. page 44).

The inequality $\blacklozenge^{\mathbb{A}'} \circ f^{\varphi} \leq f^{\varphi} \circ \blacklozenge^{\mathbb{A}}$ is the dual notion on algebras of the back condition in the refinement relation (cf. page 41).

Definition 5.2 For any algebraic model $\mathcal{A} = (\mathbb{A}, V)$ and any formula $\varphi \in \mathcal{L}_{\otimes\forall}$, we define the maps f^{φ} and g^{φ} as follows:

$$f^{\varphi} : \mathbb{A} \to \mathbb{A}^{\varphi} \qquad\qquad g^{\varphi} : \mathbb{A}^{\varphi} \to \mathbb{A}$$
$$b \mapsto [f_b] \qquad\qquad [h] \mapsto \bigvee_{\mathsf{u}\in\mathsf{S}} (h(\mathsf{u}) \wedge \mathsf{Pre}_a(\mathsf{u}))$$

where $f_b : \mathsf{S} \to \mathbb{A}$ is the map such that $f_b(\mathsf{u}) := b \wedge \mathsf{Pre}_a(\mathsf{u})$ and $a = (\mathsf{S}, \mathsf{R}, \mathsf{Pre}_a)$ is the action model induced by $\alpha^{\varphi}_{\mathsf{S}}$ via V.

Lemma 5.3 *For any algebraic model $\mathcal{A}$ and any formula $\varphi \in \mathcal{L}_{\otimes\forall}$,*

(i) *the map f^{φ} is a refinement morphism,*

(ii) *the map g^{φ} is monotone and preserves arbitrary joins,*

(iii) $g^{\varphi} \dashv f^{\varphi}$.

Lemma 5.4 *Let $M = (S, R, V)$ and $M' = (S', R', V')$ be Kripke models, and let, respectively, $\mathcal{A}$ and $\mathcal{A}'$ be their algebraic models.*

(There exists a refinement morphism $f : \mathbb{A} \to \mathbb{A}'$) iff $M_s \succeq M'_{s'}$.

Lemma 5.5 *For any algebraic model $\mathcal{A}$ and any formula $\varphi \in \mathcal{L}_{\otimes\forall}$,*

$$[\![\exists\varphi]\!]_{\mathcal{A}} = g^{\varphi}([\![\varphi]\!]_{\mathcal{A}^{\varphi}}). \tag{1}$$

Refinement. We aim at proposing an algebraic semantics for the refinement modality $\exists$, i.e. for any algebraic model $\mathcal{A} = (\mathbb{A}, V)$, we want to find a normal **BAO** $\mathfrak{A}_{\mathcal{A}}$ and a map $G : \mathfrak{A}_{\mathcal{A}} \to \mathbb{A}$ such that for any $\varphi \in \mathcal{L}_{\Box\forall}$,

$$[\![\exists\varphi]\!]_{\mathcal{A}} = G([\![\varphi]\!]_{\mathfrak{A}_{\mathcal{A}}}).$$

To do so, we introduce a normal **BAO** $\mathfrak{A}_{\mathcal{A}}$ such that $\mathbb{A}^{\varphi}$ is a subalgebra of $\mathfrak{A}_{\mathcal{A}}$ for any $\varphi \in \mathcal{L}_{\Box\forall}$. Hence, for every algebraic model $\mathcal{A} = (\mathbb{A}, V)$, we define the following algebraic structure:

$$\mathfrak{A}_{\mathcal{A}} := \prod_{\varphi\in\mathcal{L}_{\Box\forall}} \mathbb{A}^{\varphi}.$$

Elements of $\mathfrak{A}_{\mathcal{A}}$ are tuples $(b^{\varphi})_{\varphi\in\mathcal{L}_{\Box\forall}}$ where $b^{\varphi} \in \mathbb{A}^{\varphi}$. When there is no risk of confusion, we write $(b^{\varphi})_{\varphi}$ instead of $(b^{\varphi})_{\varphi\in\mathcal{L}_{\Box\forall}}$ and $\mathfrak{A}$ instead of $\mathfrak{A}_{\mathcal{A}}$.

Recall that the product of any family $\{\mathbb{A}_i\}_{i\in I}$ of normal **BAO**s, where I may be an uncountable set, is a normal **BAO** [8, Section 7]. The operations on the algebra

$$\mathfrak{A} = \left(\prod_{\varphi \in \mathcal{L}_{\Box\forall}} \mathbb{A}^{\varphi}, \vee^{\mathfrak{A}}, \wedge^{\mathfrak{A}}, \neg^{\mathfrak{A}}, \bot^{\mathfrak{A}}, \top^{\mathfrak{A}}, \Diamond^{\mathfrak{A}} \right),$$

are the following, where $\Box^{\mathfrak{A}} := \neg^{\mathfrak{A}} \Diamond^{\mathfrak{A}} \neg^{\mathfrak{A}}$. For all $(b^{\varphi})_{\varphi}, (c^{\varphi})_{\varphi} \in \mathfrak{A}$,

Constants
$\bot^{\mathfrak{A}} = (\bot^{\varphi})_{\varphi}$, $\top^{\mathfrak{A}} = (\top^{\varphi})_{\varphi}$

Join and meet
$(b^{\varphi})_{\varphi} \vee^{\mathfrak{A}} (c^{\varphi})_{\varphi} = (b^{\varphi} \vee c^{\varphi})_{\varphi}$
$(b^{\varphi})_{\varphi} \wedge^{\mathfrak{A}} (c^{\varphi})_{\varphi} = (b^{\varphi} \wedge c^{\varphi})_{\varphi}$

Negation
$\neg^{\mathfrak{A}} (b^{\varphi})_{\varphi} = (\neg b^{\varphi})_{\varphi}$

Modal operators
$\Diamond^{\mathfrak{A}} (b^{\varphi})_{\varphi} = (\Diamond^{\mathbb{A}^{\varphi}} b^{\varphi})_{\varphi}$
$\Box^{\mathfrak{A}} (b^{\varphi})_{\varphi} = (\Box^{\mathbb{A}^{\varphi}} b^{\varphi})_{\varphi}$

One can easily verify that $\mathfrak{A}$ is a normal boolean algebra. We call $\mathfrak{A}$ the *refinement algebra* of the algebraic model $\mathcal{A} = (\mathbb{A}, V)$. One can also define the modal operators $\blacklozenge^{\mathfrak{A}}$ and $\blacksquare^{\mathfrak{A}}$ as $\blacklozenge^{\mathfrak{A}}(b^{\varphi})_{\varphi} = \left(\blacklozenge^{\mathbb{A}^{\varphi}} b^{\varphi}\right)_{\varphi}$ and $\blacksquare^{\mathfrak{A}}(b^{\varphi})_{\varphi} = \left(\blacksquare^{\mathbb{A}^{\varphi}} b^{\varphi}\right)_{\varphi}$, for any $(b^{\varphi})_{\varphi} \in \mathfrak{A}$ such that $\blacklozenge^{\mathfrak{A}} \dashv \Box^{\mathfrak{A}}$ and $\Diamond^{\mathfrak{A}} \dashv \blacksquare^{\mathfrak{A}}$.

Definition 5.6 For every algebraic model $\mathcal{A} = (\mathbb{A}, V)$, the maps $F_{\mathcal{A}}$ and $G_{\mathcal{A}}$ are defined as

$$\begin{aligned} F_{\mathcal{A}} : \mathbb{A} &\to \mathfrak{A}_{\mathcal{A}} \\ a &\mapsto \prod_{\varphi \in \mathcal{L}_{\Box\forall}} (f^{\varphi}(a)) \end{aligned} \qquad \begin{aligned} G_{\mathcal{A}} : \quad \mathfrak{A}_{\mathcal{A}} &\to \mathbb{A} \\ ([h]^{\varphi})_{\varphi} &\mapsto \bigvee_{\varphi} g^{\varphi}([h]^{\varphi}) \end{aligned}$$

where $f^{\varphi} : \mathbb{A} \to \mathbb{A}^{\varphi}$ and $g^{\varphi} : \mathbb{A}^{\varphi} \to \mathbb{A}$ are the maps of Def. 5.2.

Lemma 5.7 *$\mathcal{A} = (\mathbb{A}, V)$ be an algebraic model and $\mathfrak{A}_{\mathcal{A}}$ its refinement algebra. Then*

(i) *the map $F_{\mathcal{A}}$ is a refinement morphism,*

(ii) *the map $G_{\mathcal{A}}$ is monotone and preserves $\bot$, $\top$ and finite joins,*

(iii) $G_{\mathcal{A}} \dashv F_{\mathcal{A}}$.

Unless confusion results we write F for $F_{\mathcal{A}}$ and G for $G_{\mathcal{A}}$.

Algebraic semantics of refinement modal logic. We now conclude with the algebraic semantics of refinement modal logic and the corresponding completeness result.

Definition 5.8 Let $\mathcal{A} = (\mathbb{A}, V)$ be an algebraic model and $\mathfrak{A}$ its refinement algebra. Let $\mathcal{A}'$ be the algebraic model $(\mathfrak{A}, \mathcal{V})$ with $\mathcal{V} : \mathsf{AtProp} \to \mathfrak{A}$ and $\mathcal{V}(p) = (F \circ V)(p)$. The extension map $[\![.]\!]'_{\mathcal{A}} : \mathcal{L}_{\Box\forall} \to \mathbb{A}$ is defined as follows.

$$\begin{aligned} [\![p]\!]'_{\mathcal{A}} &:= V(p) \\ [\![\bot]\!]'_{\mathcal{A}} &:= \bot^{\mathbb{A}} \\ [\![\circ\varphi]\!]'_{\mathcal{A}} &:= \circ^{\mathbb{A}} [\![\varphi]\!]'_{\mathcal{A}} && \text{for } \circ \in \{\neg, \Diamond, \Box\} \\ [\![\varphi \bullet \psi]\!]'_{\mathcal{A}} &:= [\![\varphi]\!]'_{\mathcal{A}} \bullet^{\mathbb{A}} [\![\psi]\!]'_{\mathcal{A}} && \text{for } \bullet \in \{\vee, \wedge, \to\} \\ [\![\exists\varphi]\!]'_{\mathcal{A}} &:= G([\![\varphi]\!]'_{\mathcal{A}'}) \end{aligned}$$

Theorem 5.9 *The axiomatization of* RML *(cf. page 42) is sound and complete with respect to the algebraic semantics defined above.*

6 Conclusion and further research

We have proposed an algebraic semantics for refinement modal logic. Using action model synthesis and the algebraic characterization of epistemic updates, we have introduced the abstract notion of refinement on normal boolean algebras, and showed the soundness and completeness of refinement modal logic with respect to this algebraic semantics.

Our methodology builds on and further develops recent work [21,22] applying duality theory to dynamic epistemic logic. As part of this research program, proof systems for intuitionistic AML have been introduced [17,15], and gave rise to the novel methodology of multi-type display calculi [14], which has been applied not only to AML [13], but also to propositional dynamic logic [12] and inquisitive logic [16]. A natural direction is to pursue this research program also on refinement modal logic. We plan to weaken the classical propositional modal logical base to a non-classical propositional modal logical base, and to develop multi-type calculi for such non-classical modal logics with refinement quantiers, for example refinement intuitionistic (modal) logic.

An other step to take would be to generalize the algebraic semantics of RML to the multi-agent framework. In this framework, the refinement modality $\exists$ is indexed by a agent, hence we have modalities $\{\exists_i\}_{i \in \mathsf{Ag}}$ where Ag is the set of agents. The only difficulty to generalize our result is to prove, algebraically, the soundness of the additional axioms:

$$\exists_i \nabla_j \Phi \leftrightarrow \nabla_j \{\exists_i \varphi\}_{\varphi \in \Phi} \qquad \text{where } i \neq j$$
$$\exists_i \bigwedge_{j \in J} \nabla_j \Phi^j \leftrightarrow \bigwedge_{j \in J} \exists_i \nabla_j \Phi^j \qquad \text{where } J \subseteq \mathsf{Ag}.$$

Indeed, as the reader can see in the appendix, the soundness proofs can be quite involved. This step is however very useful to propose a good proof system for the multi-agent refinement modal logic.

Appendix

A Proofs Section 5

This section contains the proof of lemmas and theorems from section 5.

Proof of Lemma 5.3. (i). That f^φ is monotone and preserves $\bot$ and $\vee$ follows from [21, Fact 11.4]. That $(\blacklozenge^{\mathbb{A}^\varphi} \circ f^\varphi) \leq (f^\varphi \circ \blacklozenge^{\mathbb{A}})$ follows from the following chain of inequalities. For every $b \in \mathbb{A}$ and every $\mathsf{u} \in \mathsf{S}$,

$$
\begin{aligned}
&(\blacklozenge^{\mathbb{A}^\varphi} \circ f^\varphi(b))(\mathsf{u}) = \blacklozenge^{\mathbb{A}^\varphi}([f_b])(\mathsf{u}) = [\blacklozenge^{\Pi_\alpha \mathbb{A}}(f_b \wedge \mathsf{Pre}_a)](\mathsf{u}) \\
&= \left(\blacklozenge^{\Pi_\alpha \mathbb{A}}(f_b \wedge \mathsf{Pre}_a) \wedge \mathsf{Pre}_a\right)(\mathsf{u}) = \left(\blacklozenge^{\Pi_\alpha \mathbb{A}}(f_b \wedge \mathsf{Pre}_a)(\mathsf{u})\right) \wedge \mathsf{Pre}_a(\mathsf{u}) \\
&= \bigvee\{\blacklozenge^{\mathbb{A}}(f_b \wedge \mathsf{Pre}_a)(\mathsf{t}) \mid \mathsf{uRt}\} \wedge \mathsf{Pre}_a(\mathsf{u}) \\
&\leq \bigvee\{\blacklozenge^{\mathbb{A}} f_b(\mathsf{t}) \wedge \blacklozenge^{\mathbb{A}}\mathsf{Pre}_a(\mathsf{t}) \mid \mathsf{tRu}\} \wedge \mathsf{Pre}_a(\mathsf{u}) \\
&= \bigvee\{(\blacklozenge^{\mathbb{A}}(b \wedge \mathsf{Pre}_a(\mathsf{t})) \wedge \blacklozenge^{\mathbb{A}}\mathsf{Pre}_a(\mathsf{t})) \mid \mathsf{tRu}\} \wedge \mathsf{Pre}_a(\mathsf{u}) \\
&\leq \bigvee\{\blacklozenge^{\mathbb{A}} b \wedge \blacklozenge\mathsf{Pre}_a(\mathsf{t}) \wedge \blacklozenge^{\mathbb{A}}\mathsf{Pre}_a(\mathsf{t}) \mid \mathsf{tRu}\} \wedge \mathsf{Pre}_a(\mathsf{u}) \\
&\leq \bigvee\{\blacklozenge^{\mathbb{A}} b \wedge \blacklozenge^{\mathbb{A}}\mathsf{Pre}_a(\mathsf{t}) \mid \mathsf{tRu}\} \wedge \mathsf{Pre}_a(\mathsf{u}) \leq \bigvee\{\blacklozenge^{\mathbb{A}} b \mid \mathsf{tRu}\} \wedge \mathsf{Pre}_a(\mathsf{u}) \\
&\leq \blacklozenge^{\mathbb{A}} b \wedge \mathsf{Pre}_a(\mathsf{u}) = \blacklozenge^{\mathbb{A}} b \wedge \mathsf{Pre}_a(\mathsf{u}) \wedge \mathsf{Pre}_a(\mathsf{u}) = (f_{\blacklozenge^{\mathbb{A}} b} \wedge \mathsf{Pre}_a)(\mathsf{u}) = [f_{\blacklozenge^{\mathbb{A}} b}](\mathsf{u}) \\
&= f^\varphi(\blacklozenge^{\mathbb{A}} b)(\mathsf{u}) = (f^\varphi \circ \blacklozenge^{\mathbb{A}} b)(\mathsf{u}).
\end{aligned}
$$

(ii) Let $[h], [k] \in \mathbb{A}^\varphi$, assume that $[h] \leq [k]$. Hence, $h(\mathsf{u}) \wedge \mathsf{Pre}_a(\mathsf{u}) \leq k(\mathsf{u}) \wedge \mathsf{Pre}_a(\mathsf{u})$ for every $\mathsf{u} \in \mathsf{S}$. Then, $\bigvee_{\mathsf{u} \in \mathsf{S}} (h(\mathsf{u}) \wedge \mathsf{Pre}_a(\mathsf{u})) \leq \bigvee_{\mathsf{u} \in \mathsf{S}} (k(\mathsf{u}) \wedge \mathsf{Pre}_a(\mathsf{u}))$, which proves that $g^\varphi([h]) \leq g^\varphi([k])$.

Let $[h_i] \in \mathbb{A}^\varphi$, where $i \in I$ for an index set I. Then we have that

$$
\begin{aligned}
&g^\varphi(\bigvee_{i \in I}[h_i]) = g^\varphi([\bigvee_{i \in I} h_i]) = \bigvee_{\mathsf{u} \in \mathsf{S}} \left(\bigvee_{i \in I} h_i(\mathsf{u}) \wedge \mathsf{Pre}_a(\mathsf{u})\right) \\
&= \bigvee_{\mathsf{u} \in \mathsf{S}} \left(\bigvee_{i \in I} (h_i(\mathsf{u}) \wedge \mathsf{Pre}_a(\mathsf{u})\right) = \bigvee_{i \in I} \left(\bigvee_{\mathsf{u} \in \mathsf{S}} (h_i(\mathsf{u}) \wedge \mathsf{Pre}_a(\mathsf{u})\right) = \bigvee_{i \in I} (g^\varphi(h_i))
\end{aligned}
$$

(S is finite)

(iii) Let $[h] \in \mathbb{A}^\varphi$ and $b \in \mathbb{A}$, then we need to show that $[h] \leq f^\varphi(b)$ iff $g^\varphi([h]) \leq b$. By definition of f^φ, $[h] \leq f^\varphi(b)$ iff $[h] \leq [f_b]$. It follows from [21, Fact 9.2] that $[h] \leq [f_b]$ iff $h \wedge \mathsf{Pre}_a \leq f_b \wedge \mathsf{Pre}_a$. From this we obtain: for every $\mathsf{u} \in \mathsf{S}$,

$$
\begin{aligned}
h(\mathsf{u}) \wedge \mathsf{Pre}_a(\mathsf{u}) \leq f_b(\mathsf{u}) \wedge \mathsf{Pre}_a(\mathsf{u}) \quad &\text{iff} \quad h(\mathsf{u}) \wedge \mathsf{Pre}_a(\mathsf{u}) \leq (b \wedge \mathsf{Pre}_a(\mathsf{u})) \wedge \mathsf{Pre}_a(\mathsf{u}) \\
&\text{iff} \quad h(\mathsf{u}) \wedge \mathsf{Pre}_a(\mathsf{u}) \leq b \wedge \mathsf{Pre}_a(\mathsf{u}) \leq b \\
&\text{iff} \quad \bigvee_{\mathsf{u} \in \mathsf{S}} h(\mathsf{u}) \wedge \mathsf{Pre}_a(\mathsf{u}) \leq b \\
&\text{iff} \quad g^\varphi([h]) \leq b
\end{aligned}
$$

Proof of Lemma 5.4. Assume that $f : \mathbb{A} \to \mathbb{A}'$ is a refinement morphism. Define the relation $\mathfrak{R} = \{(s, s') \in S \times S' \mid s' \in f(\{s\})\}$. It is easy to see that $\mathfrak{R}$ is a refinement.

For the other direction, assume that $M_s \succeq M'_{s'}$. Hence the is a refinement relation $\mathfrak{R}$ from M_s to $M'_{s'}$. Define $f : \mathbb{A} \to \mathbb{A}'$ such that $f(X) = \mathfrak{R}[X]$, for every $X \subseteq S$. It is easy to see that f is a refinement morphism.

Proof of Lemma 5.7. (i) We show that F is a refinement morphism. Since for each $\varphi \in \mathcal{L}_{\Box\forall}$, f^φ is monotone, and preserves $\bot^{\mathbb{A}^\varphi}$ and $\vee^{\mathbb{A}^\varphi}$, it follows that $\prod_{\varphi\in\mathcal{L}_{\Box\forall}} f^\varphi$ also satisfies those conditions. It remains to prove that $\blacklozenge^{\mathfrak{A}} \circ F \leq F \circ \blacklozenge^{\mathbb{A}}$. Now note that $\blacklozenge^{\mathbb{A}^\varphi}(f^\varphi) \leq f^\varphi(\blacklozenge^{\mathbb{A}})$ for every $\varphi \in \mathcal{L}_{\Box\forall}$. Let $b \in \mathbb{A}$, then we have that

$$\blacklozenge^{\mathfrak{A}} \circ F(b) = \blacklozenge^{\mathfrak{A}}\big(f^\varphi(b)\big)_{\varphi\in\mathcal{L}_{\Box\forall}} = \big(\blacklozenge^{\mathbb{A}^\varphi}(f^\varphi)(b)\big)_{\varphi\in\mathcal{L}_{\Box\forall}} \leq$$
$$\big(f^\varphi(\blacklozenge^{\mathbb{A}}(b))\big)_{\varphi\in\mathcal{L}_{\Box\forall}} = \prod_{\varphi\in\mathcal{L}_{\Box\forall}} (f^\varphi(\blacklozenge^{\mathbb{A}}(b))) = (\prod_{\varphi\in\mathcal{L}_{\Box\forall}} f^\varphi)(\blacklozenge^{\mathbb{A}}(b)) = F \circ \blacklozenge^{\mathbb{A}}(b).$$

(ii) It is easy to see that G is monotone and preserves $\bot$. Also, G preserves $\top$, since the action model constructed for $\top$ is defined as follows: $\alpha^\top = (\mathsf{S}^\top, \mathsf{R}^\top, \mathsf{Pre}^\top)$, where $\mathsf{S}^\top = \{skip\}$, $\mathsf{R}^\top = \{(skip, skip)\}$, $\mathsf{Pre}^\top(skip) = \top$. Then, we have $g^\top(\llbracket\top\rrbracket_{\mathbb{A}^\top}) = \llbracket\mathsf{Pre}(skip)\rrbracket_{\mathbb{A}} = \top^{\mathbb{A}}$. We proceed to show that G preserves binary joins and then by induction we can easily prove that it preserves finite joins. Let $([h]^\varphi)_{\varphi\in\mathcal{L}_{\Box\forall}}, ([k]^\varphi)_{\varphi\in\mathcal{L}_{\Box\forall}} \in \mathfrak{A}$. Then

$$G\Big(\big(([h]^\varphi)_\varphi \vee ([k]^\varphi)_\varphi\big)\Big) = \bigvee_{\varphi\in\mathcal{L}_{\Box\forall}} g^\varphi\Big(\big(([h]^\varphi)_\varphi \vee ([k]^\varphi)_{\varphi\in\mathcal{L}_{\Box\forall}}\big)\Big)$$
$$= \bigvee_{\varphi\in\mathcal{L}_{\Box\forall}} g^\varphi([h]^\varphi)_\varphi \vee g^\varphi([k]^\varphi)_\varphi$$
$$= \bigvee_{\varphi\in\mathcal{L}_{\Box\forall}} g^\varphi([h]^\varphi)_\varphi \vee \bigvee_{\varphi\in\mathcal{L}_{\Box\forall}} g^\varphi([k]^\varphi)_\varphi = G(([h]^\varphi)_\varphi \vee G(([k]^\varphi)_\varphi).$$

(iii) $F \dashv G$ follows from the fact that $f^\varphi \dashv g^\varphi$, for each $\varphi \in \mathcal{L}_{\Box\forall}$.

Lemma A.1 *For any formula $\varphi \in \mathcal{L}_{\forall\Box}$, we have: $G(\llbracket\Diamond\varphi\rrbracket'_{\mathcal{A}'}) \leq \Diamond G(\llbracket\varphi\rrbracket'_{\mathcal{A}'})$.*

Proof. Fix an algebraic model $\mathcal{A}$ and a formula $\varphi \in \mathcal{L}_{\forall\Box}$. We want to prove $G(\llbracket\Diamond\varphi\rrbracket'_{\mathcal{A}'}) \leq \Diamond G(\llbracket\varphi\rrbracket'_{\mathcal{A}'})$.

$$G(\llbracket\Diamond\varphi\rrbracket'_{\mathcal{A}'}) = \bigvee_\gamma g^\gamma(\llbracket\Diamond\varphi\rrbracket'_{\mathcal{A}^\gamma}) = \bigvee_\gamma (\llbracket\langle\alpha^\gamma_{\mathsf{S}^\gamma}\rangle\Diamond\varphi\rrbracket'_{\mathcal{A}}) \quad \text{(definition of } G \text{ and } g^\gamma)$$
$$= \bigvee_\gamma \bigvee_{\mathsf{u}\in\mathsf{S}^\gamma} \left(\llbracket\mathsf{Pre}^\gamma(\mathsf{u})\rrbracket'_{\mathcal{A}} \wedge \llbracket \bigvee_{\mathsf{v}\in\mathsf{R}^\gamma(\mathsf{u})} \Diamond\langle\alpha^\gamma_{\mathsf{v}}\rangle\varphi\rrbracket'_{\mathcal{A}} \right) \quad \text{(A.1)}$$

$$\begin{aligned}
&\leq \bigvee_{\gamma} \bigvee_{\mathsf{u}\in\mathsf{S}^\gamma} [\![\bigvee_{\mathsf{v}\in\mathsf{R}^\gamma(\mathsf{u})} \Diamond\langle \alpha^\gamma_{\mathsf{v}}\rangle\varphi]\!]'_{\mathcal{A}} && (a\wedge b\leq a)\\
&\leq \bigvee_{\gamma} \bigvee_{\mathsf{u}\in\mathsf{S}^\gamma} \Diamond[\![\bigvee_{\mathsf{v}\in\mathsf{R}^\gamma(\mathsf{u})} \langle \alpha^\gamma_{\mathsf{v}}\rangle\varphi]\!]'_{\mathcal{A}} && (\Diamond(\varphi\vee\psi)=\Diamond\varphi\vee\Diamond\psi)\\
&\leq \bigvee_{\gamma} \bigvee_{\mathsf{u}\in\mathsf{S}^\gamma} \Diamond[\![\bigvee_{\mathsf{u}\in\mathsf{S}^\gamma} \langle \alpha^\gamma_{\mathsf{u}}\rangle\varphi]\!]'_{\mathcal{A}} && (\mathsf{R}^\gamma(\mathsf{u})\subseteq\mathsf{S}^\gamma)\\
&= \bigvee_{\gamma} \bigvee_{\mathsf{u}\in\mathsf{S}^\gamma} \Diamond g^\gamma([\![\varphi]\!]'_{\mathcal{A}^\gamma}) && (\text{definition of } g^\gamma)\\
&= \bigvee_{\gamma} \Diamond g^\gamma([\![\varphi]\!]'_{\mathcal{A}^\gamma}) = \Diamond \bigvee_{\gamma} g^\gamma([\![\varphi]\!]'_{\mathcal{A}^\gamma})\\
&= \Diamond G([\![\varphi]\!]'_{\mathcal{A}'}) && (\text{definition of } G)
\end{aligned}$$

□

Equivalence (A.1) follows from $\langle\alpha_u\rangle\varphi \leftrightarrow \mathsf{Pre}(\mathsf{u}) \wedge \bigvee_{\mathsf{v}\in\mathsf{R}(\mathsf{u})}\langle\alpha_v\rangle\varphi$ [21, Page 14].

Proof of Theorem 5.9.

Soundness. The definition of $[\![\cdot]\!]'_{\mathcal{A}}$ for $\mathcal{L}_\Box$ is identical to the algebraic semantics proposed in [21]. Hence the axioms and rules of K are sound w.r.t. this semantics.

(i) Axiom **RProp**. We need to prove $[\![\exists p]\!]'_{\mathcal{A}} = [\![p]\!]'_{\mathcal{A}}$ and $[\![\exists\neg p]\!]'_{\mathcal{A}} = [\![\neg p]\!]'_{\mathcal{A}}$. For every $p \in \mathsf{AtProp}$, $g^\varphi([\![p]\!]'_{\mathcal{A}^\varphi}) \leq [\![p]\!]'_{\mathcal{A}}$, so that $\bigvee_{\varphi\in\mathcal{L}_{\Box\forall}} g^\varphi([\![p]\!]'_{\mathcal{A}^\varphi}) \leq [\![p]\!]'_{\mathcal{A}}$. So, $G([\![p]\!]'_{\mathcal{A}'}) \leq [\![p]\!]'_{\mathcal{A}}$. For the other direction, according to the construction of multi-pointed action model $\alpha^p_{\mathsf{S}^p}$ for the atomic proposition p [18, Lemma V.2], $[\![\langle\alpha^p_{\mathsf{S}^p}\rangle p]\!]'_{\mathcal{A}} = [\![p]\!]'_{\mathcal{A}}$. This implies that $[\![p]\!]'_{\mathcal{A}} = g^p([\![p]\!]'_{\mathcal{A}^p})$ and $[\![p]\!]'_{\mathcal{A}} \leq \bigvee_{\varphi\in\mathcal{L}_{\Box\forall}} g^\varphi([\![p]\!]'_{\mathcal{A}^\varphi}) = G([\![p]\!]'_{\mathcal{A}'})$ as required. The other equality can be proved in a similar way.

(ii) Axiom **R**. We need to show $[\![\forall(\varphi\to\psi)]\!]' \leq [\![\forall\varphi\to\forall\psi]\!]'$.
First, observe that $[\![\forall(\varphi\to\psi)]\!]'_{\mathcal{A}} = [\![\neg\exists\neg(\varphi\to\psi)]\!]'_{\mathcal{A}} = \neg G([\![\neg(\varphi\to\psi)]\!]'_{\mathcal{A}'})$ and $[\![\forall\varphi\to\forall\psi]\!]'_{\mathcal{A}} = [\![\exists\neg\varphi\vee\neg\exists\neg\psi]\!]'_{\mathcal{A}} = G([\![\neg\varphi]\!]'_{\mathcal{A}'}) \vee \neg G([\![\neg\psi]\!]'_{\mathcal{A}'})$.
Hence, it is enough to show that

$$\neg G([\![\neg(\varphi\to\psi)]\!]'_{\mathcal{A}'}) \leq G([\![\neg\varphi]\!]'_{\mathcal{A}'}) \vee \neg G([\![\neg\psi]\!]'_{\mathcal{A}'})$$

First note that $\neg\varphi\vee\neg\psi \leftrightarrow \neg\varphi\vee\neg(\varphi\to\psi)$. So, $[\![\neg\varphi\vee\neg\psi]\!]'_{\mathcal{A}'} = [\![\neg\varphi\vee\neg(\varphi\to\psi)]\!]'_{\mathcal{A}'}$. Then, $G([\![\neg\varphi\vee\neg\psi]\!]'_{\mathcal{A}'}) = G([\![\neg\varphi\vee\neg(\varphi\to\psi)]\!]'_{\mathcal{A}'}))$. Since G preserves $\vee$, we get

$$G([\![\neg\varphi]\!]'_{\mathcal{A}'}) \vee G([\![\neg\psi]\!]'_{\mathcal{A}'}) = G([\![\neg\varphi]\!]'_{\mathcal{A}'}) \vee G([\![\neg(\varphi\to\psi)]\!]'_{\mathcal{A}'})).$$

By applying negation and DeMorgan laws, we get

$$\neg G([\![\neg\varphi]\!]'_{\mathcal{A}'}) \wedge \neg G([\![\neg\psi]\!]'_{\mathcal{A}'}) = \neg G([\![\neg\varphi]\!]'_{\mathcal{A}'}) \wedge \neg G([\![\neg(\varphi\to\psi)]\!]'_{\mathcal{A}'})$$

The equality above implies

$$\neg G([\![\neg\varphi]\!]'_{\mathcal{A}'}) \wedge \neg G([\![\neg(\varphi\to\psi)]\!]'_{\mathcal{A}'}) \leq \neg G([\![\neg\varphi]\!]'_{\mathcal{A}'}) \wedge \neg G([\![\neg\psi]\!]'_{\mathcal{A}'}).$$

It is easy to see that in any **BA**, $a\wedge b \leq a\wedge c$ implies that $b \leq \neg a\vee c$, which implies $\neg G([\![\neg(\varphi\to\psi)]\!]'_{\mathcal{A}'}) \leq G([\![\neg\varphi]\!]'_{\mathcal{A}'}) \vee \neg G([\![\neg\psi]\!]'_{\mathcal{A}'})$, as required.

(iii) Axiom **RK**. We need to show that $[\![\exists\nabla\Phi]\!]'_{\mathcal{A}} = [\![\bigwedge\Diamond\exists\Phi]\!]'_{\mathcal{A}}$ for every algebraic model $\mathcal{A}$. Fix an algebraic model $\mathcal{A}$.

Proof of $[\![\exists\nabla\Phi]\!]'_{\mathcal{A}} \leq [\![\bigwedge\Diamond\exists\Phi]\!]'_{\mathcal{A}}$.

$$
\begin{aligned}
[\![\exists\nabla\Phi]\!]'_{\mathcal{A}} &= G\left([\![\Box\left(\bigvee_{\varphi\in\Phi}\varphi\right) \wedge \bigwedge_{\varphi\in\Phi}\Diamond\varphi]\!]'_{\mathcal{A}'}\right) && \text{(with } G : \mathfrak{A}_{\mathcal{A}} \to \mathbb{A}\text{)}\\
&\leq G\left([\![\bigwedge_{\varphi\in\Phi}\Diamond\varphi]\!]'_{\mathcal{A}'}\right) \leq \bigwedge_{\varphi\in\Phi} G([\![\Diamond\varphi]\!]'_{\mathcal{A}'}) && \text{(monotonicity of } G\text{)}\\
&\leq \bigwedge_{\varphi\in\Phi}\Diamond G([\![\varphi]\!]'_{\mathcal{A}'}) = [\![\bigwedge\Diamond\exists\Phi]\!]'_{\mathcal{A}}. && \text{(Lemma A.1)}
\end{aligned}
$$

Proof of $[\![\bigwedge\Diamond\exists\Phi]\!]'_{\mathcal{A}} \leq [\![\exists\nabla\Phi]\!]'_{\mathcal{A}}$.

Let Φ be a finite set of formulas. We first show that $\bigwedge\Diamond G([\![\Phi]\!]'_{\mathcal{A}'}) \leq g^{\nabla\Phi}([\![\nabla\Phi]\!]'_{\mathcal{A}^{\nabla\Phi}})$. In order to show this we use the inductive structure of formulas $\nabla\Phi$ that may contain action models [18, Lemma V.2] and the algebraic semantics of AML. Let $\Phi \subseteq \mathcal{L}_{\Box\forall}$ be a set of formulas, and for each $\varphi \in \Phi$, $\alpha^{\varphi} = (\mathsf{S}^{\varphi}, \mathsf{R}^{\varphi}, \mathsf{Pre}^{\varphi})$, $\alpha^{\nabla\Phi} = (\mathsf{S}^{\nabla\Phi}, \mathsf{R}^{\nabla\Phi}, \mathsf{Pre}^{\nabla\Phi})$ be action models for the formulas $\varphi \in \Phi$ and $\nabla\Phi$, respectively. Note that $\mathsf{S}^{\nabla\Phi} = \{\mathsf{u}^\star\} \cup \bigcup_{\varphi\in\Phi}\mathsf{S}^{\varphi}$, $\mathsf{R}^{\nabla\Phi} = \{(\mathsf{u}^\star, \mathsf{u}) \mid \varphi \in \Phi, \mathsf{u} \in \mathsf{S}^{\varphi}\} \cup \bigcup\mathsf{R}^{\varphi}$ and $\mathsf{Pre}^{\nabla\Phi} = \{(\mathsf{u}^\star, \Diamond\exists\varphi \wedge \Diamond\exists\psi)\} \cup \bigcup\mathsf{Pre}^{\varphi}$. Then,

$$
\begin{aligned}
g^{\nabla\Phi}([\![\nabla\Phi]\!]'_{\mathcal{A}^{\nabla\Phi}}) &= \bigvee_{\mathsf{u}\in\mathsf{S}^{\nabla\Phi}} [\![\langle\alpha^{\nabla\Phi}\rangle\nabla\Phi]\!]'_{\mathcal{A}}\\
&= [\![\langle\alpha^{\nabla\Phi}_{\mathsf{u}^\star}\rangle\nabla\Phi]\!]'_{\mathcal{A}} \vee \bigvee_{\varphi\in\Phi}\bigvee_{\mathsf{u}\in\mathsf{S}^{\varphi}} [\![\langle\alpha^{\nabla\Phi}_{\mathsf{u}}\rangle\nabla\Phi]\!]'_{\mathcal{A}}.
\end{aligned}
$$

Also,

$$
\begin{aligned}
[\![\langle\alpha^{\nabla\Phi}_{\mathsf{u}^\star}\rangle\nabla\Phi]\!]'_{\mathcal{A}} &= [\![\langle\alpha^{\nabla\Phi}_{\mathsf{u}^\star}\rangle(\Box(\bigvee\Phi) \wedge \bigwedge\Diamond\Phi)]\!]'_{\mathcal{A}}\\
&= [\![\langle\alpha^{\nabla\Phi}_{\mathsf{u}^\star}\rangle\Box(\bigvee\Phi)]\!]'_{\mathcal{A}} \wedge [\![\langle\alpha^{\nabla\Phi}_{\mathsf{u}^\star}\rangle\bigwedge\Diamond\Phi]\!]'_{\mathcal{A}}
\end{aligned}
$$

Moreover,

$$
[\![\langle\alpha^{\nabla\Phi}_{\mathsf{u}^\star}\rangle\Box(\bigvee\Phi)]\!]'_{\mathcal{A}} = [\![\mathsf{Pre}^{\nabla\Phi}(\mathsf{u}^\star)]\!]'_{\mathcal{A}} \wedge [\![\bigwedge_{\mathsf{u}\in\mathsf{R}^{\nabla\Phi}(\mathsf{u}^\star)}\Box[\alpha^{\nabla\Phi}_{\mathsf{u}}](\bigvee\Phi)]\!]'_{\mathcal{A}}
$$

$$
\begin{aligned}
&= [\![\mathsf{Pre}^{\nabla\Phi}(\mathsf{u}^\star)]\!]'_{\mathcal{A}} \wedge \Box^{\mathbb{A}}\bigwedge_{\mathsf{u}\in\mathsf{R}^{\nabla\Phi}(\mathsf{u}^\star)} [\![[\alpha^{\nabla\Phi}_{\mathsf{u}}](\bigvee\Phi)]\!]'_{\mathcal{A}}\\
&= [\![\mathsf{Pre}^{\nabla\Phi}(\mathsf{u}^\star)]\!]'_{\mathcal{A}} \wedge \Box^{\mathbb{A}}\left(\bigwedge_{\varphi\in\Phi}\bigwedge_{\mathsf{u}\in\mathsf{S}^{\varphi}} [\![[\alpha^{\nabla\Phi}_{\mathsf{u}}](\bigvee\Phi)]\!]'_{\mathcal{A}}\right)\\
&= [\![\mathsf{Pre}^{\nabla\Phi}(\mathsf{u}^\star)]\!]'_{\mathcal{A}} \wedge \Box^{\mathbb{A}}\left(\bigwedge_{\varphi\in\Phi}\bigwedge_{\mathsf{u}\in\mathsf{S}^{\varphi}} [\![[\alpha^{\varphi}_{\mathsf{u}}](\bigvee\Phi)]\!]'_{\mathcal{A}}\right)\\
&= [\![\mathsf{Pre}^{\nabla\Phi}(\mathsf{u}^\star)]\!]'_{\mathcal{A}} \wedge \Box^{\mathbb{A}}\left(\bigwedge_{\varphi\in\Phi} [\![[\alpha^{\varphi}_{\mathsf{S}^{\varphi}}](\bigvee\Phi)]\!]'_{\mathcal{A}}\right)
\end{aligned}
$$

It follows from the definition of the structure of action models $\alpha^{\nabla\Phi}$ and α^{φ} that $\vdash [\alpha^{\varphi}]\varphi$, for every $\varphi \in \Phi$. So, we have $\vdash [\alpha^{\varphi}](\bigvee\Phi)$ which means that $[\![[\alpha^{\varphi}](\bigvee\Phi)]\!]'_{\mathcal{A}} = \top$. Therefore,

$$[\![\langle\alpha^{\nabla\Phi}_{\mathsf{u}^\star}\rangle\Box(\bigvee\Phi)]\!]'_{\mathcal{A}} = [\![\mathsf{Pre}^{\nabla\Phi}(\mathsf{u}^\star)]\!]'_{\mathcal{A}} = [\![\bigwedge\Diamond\exists\Phi]\!]'_{\mathcal{A}}. \qquad (A.2)$$

For every pointed action model $\alpha_{\mathsf{u}} = (\mathsf{S}, \mathsf{R}, \mathsf{Pre})$ over $\mathbb{A}$,

$$[\![\langle\alpha_{\mathsf{u}}\rangle\gamma]\!]'_{\mathcal{A}} = [\![\neg[\alpha_{\mathsf{u}}]\neg\gamma]\!]'_{\mathcal{A}} = [\![\neg(\mathsf{Pre}(\mathsf{u}) \to \neg[\alpha_{\mathsf{u}}]\gamma)]\!]'_{\mathcal{A}} = [\![\mathsf{Pre}(\mathsf{u}) \wedge [\alpha_{\mathsf{u}}]\gamma]\!]'_{\mathcal{A}}$$

So we have $[\![\langle\alpha^{\nabla\Phi}_{\mathsf{u}^\star}\rangle\Diamond\varphi]\!]'_{\mathcal{A}} = [\![\mathsf{Pre}^{\nabla\Phi}(\mathsf{u}^\star)]\!]'_{\mathcal{A}} \wedge [\![[\alpha^{\nabla\Phi}_{\mathsf{u}^\star}]\Diamond\varphi]\!]'_{\mathcal{A}}$ Since $\vdash \bigwedge[\alpha^{\nabla\Phi}_{\mathsf{u}^\star}]\Diamond\Phi$ [18, proof of Lemma V.2], we can deduce that for every $\varphi \in \Phi$, $\vdash [\alpha^{\nabla\Phi}_{\mathsf{u}^\star}]\Diamond\varphi$. We then get that for every $\varphi \in \Phi$,

$$[\![\langle\alpha^{\nabla\Phi}_{\mathsf{u}^\star}\rangle\Diamond\varphi]\!]'_{\mathcal{A}} = [\![\mathsf{Pre}^{\nabla\Phi}(\mathsf{u}^\star)]\!]'_{\mathcal{A}}$$

which together with (A.2) yields that

$$[\![\langle\alpha^{\nabla\Phi}_{\mathsf{u}^\star}\rangle\nabla\Phi]\!]_{\mathcal{A}} = [\![\mathsf{Pre}^{\nabla\Phi}(\mathsf{u}^\star)]\!]'_{\mathcal{A}} = [\![\Diamond\exists\varphi]\!]'_{\mathcal{A}} \wedge [\![\Diamond\exists\psi]\!]'_{\mathcal{A}}$$

To complete the proof, we have

$$\begin{aligned} G([\![\nabla\Phi]\!]'_{\mathcal{A}'}) &= \bigvee_{\gamma\in\mathcal{L}_{\forall\Box}} g^{\gamma}([\![\nabla\Phi]\!]'_{\mathcal{A}^{\gamma}}) \geq g^{\nabla\Phi}([\![\nabla\Phi]\!]'_{\mathcal{A}^{\nabla\Phi}}) \\ &\geq [\![\langle\alpha^{\nabla\Phi}_{\mathsf{u}^\star}\rangle\nabla\Phi]\!]'_{\mathcal{A}} = \bigwedge[\![\Diamond\exists\Phi]\!]'_{\mathcal{A}} = \bigwedge\Diamond G([\![\Phi]\!]'_{\mathcal{A}'}) \end{aligned}$$

Completeness. RML contains all the axioms of K. The algebraic semantics is complete w.r.t. K. RML is equivalent to K, hence the algebraic semantics is complete w.r.t. RML.

Acknowledgements

Zeinab Bakhtiari and Hans van Ditmarsch gratefully acknowledge support from European Research Council grant EPS 313360, and Sabine Frittella gratefully acknowledges support from the NWO Aspasia grant 015.008.054, and from a Delft Technology Fellowship awarded in 2013. Hans van Ditmarsch is also affiliated to IMSc, Chennai, India and to Zhejiang University, China.

References

[1] Awodey, S., "Category Theory," Oxford Logic Guides, Ebsco Publishing, 2006.

[2] Bakhtiarinoodeh, Z. and U. Rivieccio, *Epistemic updates on bilattices*, in: *Proc. of LORI*, 2015, pp. 426–428.

[3] Baltag, A., *A coalgebraic semantics for epistemic programs*, Electr. Notes Theor. Comput. Sci. **82** (2003), pp. 17–38.

[4] Baltag, A., B. Coecke and M. Sadrzadeh, *Algebra and sequent calculus for epistemic actions*, Electr. Notes Theor. Comput. Sci. **126** (2005), pp. 27–52.

[5] Baltag, A., L. S. Moss and S. Solecki, *The logic of public announcements, common knowledge and private suspicious*, Technical Report SEN-R9922 (1999).

[6] Blackburn, P., M. de Rijke and Y. Venema, "Modal logic," Theoretical Computer Science **53**, Cambridge University Press, 2001.

[7] Bozzelli, L., H. P. van Ditmarsch, T. French, J. Hales and S. Pinchinat, *Refinement modal logic*, Inf. Comput. **239** (2014), pp. 303–339.

[8] Burris, S. and H. Sankappanavar, "A course in universal algebra," Graduate texts in mathematics, Springer-Verlag, 1981.

[9] Conradie, W., S. Frittella, A. Palmigiano and A. Tzimoulis, *Probabilistic epistemic updates on algebras*, in: *Proc. of LORI*, 2015, pp. 64–76.

[10] Davey, B. A. and H. A. Priestley, "Lattices and Order," Cambridge Univerity Press, 2002.

[11] Fagin, R., J. Halpern, Y. Moses and M. Vardi, "Reasoning about Knowledge," MIT Press, Cambridge MA, 1995.

[12] Frittella, S., G. Greco, A. Kurz and A. Palmigiano, *Multi-type display calculus for propositional dynamic logic* (2014), (forthcoming).

[13] Frittella, S., G. Greco, A. Kurz, A. Palmigiano and V. Sikimić, *A multi-type display calculus for dynamic epistemic logic* (2014), (Forthcoming).

[14] Frittella, S., G. Greco, A. Kurz, A. Palmigiano and V. Sikimić, *Multi-type sequent calculi*, in: M. Z. Andrzej Indrzejczak, Janusz Kaczmarek, editor, *Trends in Logic XIII* (2014), pp. 81–93.

[15] Frittella, S., G. Greco, A. Kurz, A. Palmigiano and V. Sikimić, *A proof-theoretic semantic analysis of dynamic epistemic logic* (2014), (forthcoming).

[16] Frittella, S., G. Greco, A. Palmigiano and F. Yang, *Structural multi-type sequent calculus for inquisitive logic*, submitted **arXiv:1604.00936** (2016).

[17] Greco, G., A. Kurz and A. Palmigiano, *Dynamic epistemic logic displayed*, in: H. Huang, D. Grossi and O. Roy, editors, *Proceedings of the 4th International Workshop on Logic, Rationality and Interaction (LORI-4)*, LNCS **8196**, 2013.

[18] Hales, J., *Arbitrary action model logic and action model synthesis*, in: *Proc. of LICS*, 2013, pp. 253–262.

[19] Hintikka, J., "Knowledge and Belief," Cornell University Press, Ithaca, NY, 1962.

[20] Janin, D. and I. Walukiewicz, *Automata for the modal mu-calculus and related results*, in: *Proc. of 20th MFCS*, LNCS 969 (1995), pp. 552–562.

[21] Kurz, A. and A. Palmigiano, *Epistemic updates on algebras*, Logical Methods in Computer Science (2013).

[22] Ma, M., A. Palmigiano and M. Sadrzadeh, *Algebraic semantics and model completeness for intuitionistic public announcement logic*, Ann. Pure Appl. Logic **165** (2014), pp. 963–995.

[23] van Ditmarsch, H. and T. French, *Simulation and information: Quantifying over epistemic events*, in: *Proc. of KRAMAS*, 2008, pp. 51–65.

Before announcement

Philippe Balbiani [1]

Institut de recherche en informatique de Toulouse
Toulouse University

Hans van Ditmarsch [2]

Laboratoire lorrain de recherche en informatique et ses applications
CNRS — Université de Lorraine — Inria

Andreas Herzig[1]

Institut de recherche en informatique de Toulouse
Toulouse University

Abstract

We axiomatize the mono-agent logic of knowledge with public announcements and converse public announcements. A special variant of our logic is determined by the model of maximal ignorance wherein the agent considers all valuations of atomic formulas possible.

Keywords: Public announcement logic, subset space logic, complete axiomatization.

1 Introduction

Public Announcement Logic [15] models the effect of publicly observable events on information states, $[\phi]^+\psi$ standing for "after ϕ's announcement, ψ is true". Now, suppose you want to go in the other direction, $[\phi]^-\psi$ standing for "before ϕ's announcement, ψ was true". Just as $[\phi]^+$ has a diamond-version noted $\langle\phi\rangle^+$, $[\phi]^-$ also has a diamond-version which we note $\langle\phi\rangle^-$. Surely the relation between public announcement and converse public announcement resembles the behaviour of future and past constructs in temporal logic. Indeed, one expects the validity of $\psi \to [\phi]^+\langle\phi\rangle^-\psi$ and $\psi \to [\phi]^-\langle\phi\rangle^+\psi$. However, unlike the announcement operation, the converse announcement operation is not deterministic: different states of information may lead to the same outcome state

[1] IRIT, Toulouse University, 118 route de Narbonne, 31062 Toulouse Cedex 9, FRANCE; {Philippe.Balbiani,Andreas.Herzig}@irit.fr.

[2] LORIA, CNRS — Université de Lorraine — Inria, Campus scientifique, BP 239, 54506 Vandœuvre-lès-Nancy Cedex, FRANCE; hans.van-ditmarsch@loria.fr. Hans van Ditmarsch is also affiliated to IMSc, Chennai, India and to Zhejiang University, China.

of information.
Dynamic epistemic logics with constructs to denote what was the case before the executions of actions have been investigated in [1,18] where they are interpreted over history-based structures. See also [11,20]. In such structures, what agents currently know are mere snapshots forming parts of a larger structure that also contains possible prior and posterior states of information. These states can then be accessed in some way. In [18], this access is realized with the history-based structures of [14]. An actual state is accompanied by a list of prior actions: going back in the past means removing the last event from that history. This changed perspective then makes it possible to check what was true before now. In [16,17], this history-based approach has also been generalized to cover non-public events.
In dynamic epistemic logics, fresh atomic formulas can be used to store the values of formulas. This is useful, as the denotation of atomic formulas is constant throughout action execution but the denotation of formulas is not. This feature is used in the satisfiability preserving transformations demonstrating complexity arguments in dynamic epistemic logic [12]. It is used as well for the purpose of keeping denotations constant, for instance in [8] where the authors model the announcement "I knew that you know the number pair" in the Sum-and-Product riddle. Similarly, this technique is used to model "Do not turn on the light if you have turned it on already in the past" in the "One hundred prisoners and a light bulb" riddle [6].
As discussed above, the converse announcement operation is not deterministic and has similarities with quantification: in a given model, a converse announcement can be interpreted as one of the multifarious expansions of that model satisfying ϕ. About quantification, two analogues come to mind: Arbitrary Public Announcement Logic [2] and Refinement Modal Logic [4]. In Arbitrary Public Announcement Logic, $\blacksquare^+\phi$ stands for "ϕ is true after any arbitrary announcement". This quantifies over all modally definable restrictions of the actual information state whereas in converse announcements we quantify over all expansions containing the actual information state. In Refinement Modal Logic, $\blacksquare^+\phi$ means that ϕ is true in any structure that is a modal refinement of the actual information state.
A less likely place to look for reasoning about the past is in the setting of Subset Space Logic [5,13]. Apart from the epistemic modalities $\Box$ and $\Diamond$, which behave like $S5$-modalities, we now also have the so-called effort modalities $\blacksquare^+$ and $\blacklozenge^+$, which behave like $S4$-modalities. A typical schema in Subset Space Logic is $\blacklozenge^+\Box\phi$ which stands for "after some effort, the agent knows that ϕ". This logic is interpreted on models consisting of a domain plus a set of "enabled" subsets of the domain. A formula is true after some effort if there is an enabled subset of the current set that satisfies it. In Subset Space Logics, converse $\blacksquare^-$ and $\blacklozenge^-$ effort modalities were also investigated [9]. In Subset Space Logics with converse, we can now formalize statements like "before some effort, the agent knew that ϕ" by $\blacklozenge^-\Box\phi$.
An appealing semantics for converse announcement of ϕ is that of truth in all

models of which the current model is the ϕ restriction. We did not manage to axiomatize that logic. Instead, we chose a setting similar to that of Subset Space Logic with converse: given a state-subset pair in a model, a formula ψ is true before announcement of ϕ if ψ is true for all state-subset pairs in that model whose subset component contains the corresponding component in the given pair. This semantics comes at the price of losing some validities of the previous semantics, such as $\Box p \rightarrow \langle p \rangle^- \Diamond \neg p$. However, we can recover some of our desiderata in the largest subset space model: the model consisting of all valuations and all subsets of that. One possible logic of "before announcement" is then the special case of the theory of that model.

The section-by-section breakdown of the paper is as follows. In Sections 2 and 3, we present the syntax and the semantics of a mono-agent logic of knowledge with public announcements and converse public announcements. The aim of Section 4 is to demonstrate that the constructs $[\cdot]^+$ and $[\cdot]^-$ cannot be eliminated from our language. In Section 5, we compare our mono-agent logic to subset space logic. In Section 6, we give an axiomatization and in Sections 7 and 8, we prove its completeness. The purpose of Section 9 is to analyse public announcements and converse public announcements in the largest subset space model. Easy proofs have been omitted whereas some others can be found in the Annex.

2 Syntax

Let VAR be a countable set of atomic formulas (with typical members p, q, etc). The formulas are inductively defined as follows:

- $\phi, \psi ::= p \mid \bot \mid \neg\phi \mid (\phi \vee \psi) \mid \Box\phi \mid [\phi]^+\psi \mid [\phi]^-\psi$.

We define the other Boolean constructs as usual. The formulas $\Diamond\phi$, $\langle\phi\rangle^+\psi$ and $\langle\phi\rangle^-\psi$ are obtained as the following abbreviations: $\Diamond\phi$ is $\neg\Box\neg\phi$, $\langle\phi\rangle^+\psi$ is $\neg[\phi]^+\neg\psi$ and $\langle\phi\rangle^-\psi$ is $\neg[\phi]^-\neg\psi$. For the collection of boxes, we propose the following readings:

- $\Box\phi$: "the agent considers it necessary according to her knowledge that ϕ",
- $[\phi]^+\psi$: "every execution of the announcement ϕ that comes from the present situation leads to a situation bearing ψ",
- $[\phi]^-\psi$: "every execution of the announcement ϕ that leads to the present situation comes from a situation bearing ψ".

For the collection of diamonds, we propose the following readings:

- $\Diamond\phi$: "the agent considers it possible according to her knowledge that ϕ",
- $\langle\phi\rangle^+\psi$: "some execution of the announcement ϕ coming from the present situation leads to a situation bearing ψ",
- $\langle\phi\rangle^-\psi$: "some execution of the announcement ϕ leading to the present situation comes from a situation bearing ψ".

The key point to note about the constructs $[\cdot]^+$ and $[\cdot]^-$ is that they allow to make modalities out of formulas. Our language can be used to reason about the knowledge of some agent after and before announcements are executed. Let ϕ^0 and ϕ^1 respectively denote the formulas $\neg\phi$ and ϕ. We adopt the usual rules for omission of the parentheses. Let $(p_1, p_2, \ldots)$ be a non-repeating enumeration of VAR. Let $k \in \mathbb{N}$. A k-formula is a formula whose atomic formulas form a sublist of $(p_1, \ldots, p_k)$. A k-world is a formula of the form $p_1^{a_1} \wedge \ldots \wedge p_k^{a_k}$ where $a_1, \ldots, a_k \in \{0, 1\}$ and a k-step is a non-empty set of k-worlds. Obviously, there exist exactly 2^k k-worlds and there exist exactly $2^{2^k} - 1$ k-steps. A pair (α, A) consisting of a k-world α and a k-step A such that $\alpha \in A$ is called a k-tip. For all sets Γ of formulas, let

- $\Box\Gamma = \{\phi : \Box\phi \in \Gamma\}$,
- $[\phi]^+\Gamma = \{\psi : [\phi]^+\psi \in \Gamma\}$,
- $[\phi]^-\Gamma = \{\psi : [\phi]^-\psi \in \Gamma\}$.

The degree of a formula ϕ, in symbols $deg(\phi)$, is inductively defined as follows:

- $deg(p) = 3$,
- $deg(\bot) = 3$,
- $deg(\neg\phi) = deg(\phi)$,
- $deg(\phi \vee \psi) = \max\{deg(\phi), deg(\psi)\}$,
- $deg(\Box\phi) = deg(\phi)$,
- $deg([\phi]^+\psi) = deg(\phi) + deg(\psi)$,
- $deg([\phi]^-\psi) = deg(\phi) + deg(\psi)$.

Let the size of a formula ϕ, in symbols $size(\phi)$, be the number of occurrences of symbols it contains. For all finite sets Γ of formulas and for all formulas ψ, the formulas $\bigvee\Gamma$, $\nabla\Gamma$ and $\nabla_\psi\Gamma$ are defined by the following abbreviations:

- $\bigvee\Gamma ::= \bigvee\{\phi : \phi \in \Gamma\}$,
- $\nabla\Gamma ::= \Box\bigvee\Gamma \wedge \bigwedge\{\Diamond\phi : \phi \in \Gamma\}$,
- $\nabla_\psi\Gamma ::= \psi \wedge \Box(\psi \rightarrow \bigvee\Gamma) \wedge \bigwedge\{\Diamond(\psi \wedge \phi) : \phi \in \Gamma\}$.

3 Semantics

A model is a triple of the form $\mathcal{M} = (W, X, V)$ where W is a non-empty set (with typical members x, y, etc), X is a non-empty set of non-empty subsets of W (with typical members S, T, etc) and V is a function associating to each $p \in VAR$ a subset $V(p)$ of W. Elements of W will be called worlds. Each of them is an epistemic alternative to the real world: due to her lack of knowledge, the agent is not able to distinguish between the real world and its epistemic alternatives. Elements of X will be called steps. Each of them contains the real world together with its epistemic alternatives at some moment of their history. We shall say that a world-step pair (x, S) is a tip iff $x \in S$. Each tip determines the real world and the current restriction of the model containing the real world together with its epistemic alternatives. To flesh this out a little, the universal relation in the step component of tips should be interpreted as a contemporaneity relation between moments. As for the function V, it assigns to all atomic formulas, the set of all $\mathcal{M}$-worlds in which it holds. It will be called valuation of $\mathcal{M}$. The satisfiability of a formula ϕ in a model $\mathcal{M} = (W, X, V)$ at tip (x, S), in symbols $\mathcal{M}, (x, S) \models \phi$, is inductively defined as follows:

- $\mathcal{M}, (x, S) \models p$ iff $x \in V(p)$,
- $\mathcal{M}, (x, S) \not\models \bot$,
- $\mathcal{M}, (x, S) \models \neg\phi$ iff $\mathcal{M}, (x, S) \not\models \phi$,
- $\mathcal{M}, (x, S) \models \phi \vee \psi$ iff $\mathcal{M}, (x, S) \models \phi$ or $\mathcal{M}, (x, S) \models \psi$,
- $\mathcal{M}, (x, S) \models \Box\phi$ iff for all $y \in S$, $\mathcal{M}, (y, S) \models \phi$,
- $\mathcal{M}, (x, S) \models [\phi]^+\psi$ iff for all $T \in X$, if $x \in T$ and $T = \{z \in S : \mathcal{M}, (z, S) \models \phi\}$ then $\mathcal{M}, (x, T) \models \psi$,
- $\mathcal{M}, (x, S) \models [\phi]^-\psi$ iff for all $T \in X$, if $x \in T$ and $S = \{z \in T : \mathcal{M}, (z, T) \models \phi\}$ then $\mathcal{M}, (x, T) \models \psi$.

Let us reflect upon these truth conditions. First, the satisfiability of an atomic formula does not depend on the step components of tips: it only depends on their world components. This indicates that the concept of validity that we will define at the end of this section will give rise to a non-normal set of valid formulas. Second, the Boolean constructs are classically interpreted. Third, the $\Box$ construct behaves like a universal modality in the step component of tips. This makes it similar to the epistemic construct in subset space logic. Fourth, the $[\cdot]^+$ construct behaves like an announcement modality: if a tip satisfies ϕ then $[\phi]^+$ further restricts the model to those epistemic alternatives in the step component of the tip satisfying ϕ. As a result, the announcement modality $[\cdot]^+$ is always deterministic. The reader may have understood from the definition that a true announcement is executable in a tip iff the above-mentioned restriction of the model is itself a step. With the concept of freedom that we will define at the end of this section, we will concentrate on the class of all models in which true announcements are always executable. Fifth, the $[\cdot]^-$ construct behaves like the converse of the announcement modality $[\cdot]^+$. Seeing that the above-mentioned restriction of the model to those epistemic alternatives in the step components of differents tips satisfying an announced formula may produce the same result, there is no reason to expect the $[\cdot]^-$ construct to be deterministic. In any case, obviously,

- $\mathcal{M}, (x, S) \models \Diamond\phi$ iff there exists $y \in S$ such that $\mathcal{M}, (y, S) \models \phi$,
- $\mathcal{M}, (x, S) \models \langle\phi\rangle^+\psi$ iff there exists $T \in X$ such that $x \in T$, $T = \{z \in S : \mathcal{M}, (z, S) \models \phi\}$ and $\mathcal{M}, (x, T) \models \psi$,
- $\mathcal{M}, (x, S) \models \langle\phi\rangle^-\psi$ iff there exists $T \in X$ such that $x \in T$, $S = \{z \in T : \mathcal{M}, (z, T) \models \phi\}$ and $\mathcal{M}, (x, T) \models \psi$.

Let $R_\Box^\mathcal{M}$ be the binary relation between tips such that $(x, S)\ R_\Box^\mathcal{M}\ (y, T)$ iff $S = T$. Obviously, $R_\Box^\mathcal{M}$ is an equivalence relation between tips such that: $\mathcal{M}, (x, S) \models \Box\phi$ iff for all tips (y, T), if $(x, S)\ R_\Box^\mathcal{M}\ (y, T)$ then $\mathcal{M}, (y, T) \models \phi$. Let $R_{[\phi]^+}^\mathcal{M}$ and $R_{[\phi]^-}^\mathcal{M}$ be the binary relations between tips such that (i) $(x, S)\ R_{[\phi]^+}^\mathcal{M}\ (y, T)$ iff $x = y$ and $T = \{z \in S : \mathcal{M}, (z, S) \models \phi\}$ and

(ii) $(x, S)\ R^{\mathcal{M}}_{[\phi]^-}\ (y, T)$ iff $x = y$ and $S = \{z \in T : \mathcal{M}, (z, T) \models \phi\}$. Obviously, $R^{\mathcal{M}}_{[\phi]^+}$ and $R^{\mathcal{M}}_{[\phi]^-}$ are mutually converse relations between tips such that: (i) $\mathcal{M}, (x, S) \models [\phi]^+\psi$ iff for all tips (y, T), if $(x, S)\ R^{\mathcal{M}}_{[\phi]^+}\ (y, T)$ then $\mathcal{M}, (y, T) \models \phi$, (ii) $\mathcal{M}, (x, S) \models [\phi]^-\psi$ iff for all tips (y, T), if $(x, S)\ R^{\mathcal{M}}_{[\phi]^-}\ (y, T)$ then $\mathcal{M}, (y, T) \models \phi$. Moreover, the binary relation $R^{\mathcal{M}}_{[\phi]^+}$ is always deterministic. Remark that $R^{\mathcal{M}}_{[\top]^+}$ and $R^{\mathcal{M}}_{[\top]^-}$ are equal to the identity relation between tips. Let $\equiv^{\mathcal{M}}$ be the transitive closure of $\bigcup\{R^{\mathcal{M}}_{[\phi]^+} : \phi$ is a formula$\} \cup \bigcup\{R^{\mathcal{M}}_{[\phi]^-} : \phi$ is a formula$\}$. Obviously, $\equiv^{\mathcal{M}}$ is an equivalence relation between tips. Let $\equiv^{\mathcal{M}}_{\Box}$ be the transitive closure of $R^{\mathcal{M}}_{\Box} \cup \equiv^{\mathcal{M}}$. Obviously, $\equiv^{\mathcal{M}}_{\Box}$ is an equivalence relation between tips. Moreover, $\equiv^{\mathcal{M}}_{\Box}$ is coarser than $R^{\mathcal{M}}_{\Box}$ and $\equiv^{\mathcal{M}}$. We shall say that a formula ϕ is globally true in a model $\mathcal{M}$, in symbols $\mathcal{M} \models \phi$, if ϕ is satisfied at all tips in $\mathcal{M}$. There are two ways for the announcement of ϕ to fail in a model $\mathcal{M} = (W, X, V)$ at tip (x, S). One is for ϕ to be false at (x, S) (which matches the traditional semantics of announcements). The other is for the set of worlds in S for which ϕ is true not to be in X. A model $\mathcal{M} = (W, X, V)$ is said to be free if for all formulas ϕ and for all tips (x, S), if $\mathcal{M}, (x, S) \models \phi$ then there exists $T \in X$ such that $x \in T$ and $T = \{z \in S : \mathcal{M}, (z, S) \models \phi\}$. Obviously, in free models, true announcements are executable. Moreover, any model $\mathcal{M} = (W, X, V)$ in which X is closed under non-empty subsets is free. However, note that, in a free model $\mathcal{M} = (W, X, V)$, X is not necessarily closed under subsets.

Lemma 3.1 *Let $\mathcal{M}$ be a model. $\mathcal{M}$ is free iff for all formulas ϕ, $\mathcal{M} \models \phi \to \langle\phi\rangle^+\top$.*

A formula ϕ is said to be valid, in symbols $\models \phi$, if ϕ is globally true in all free models. In Sections 6–8, we will give a complete axiomatization of the set of all valid formulas. In the meantime, it is well worth noting some interesting properties.

Lemma 3.2 *The following formulas are valid:*

- $[\phi]^+p \leftrightarrow (\phi \to p)$,
- $[\phi]^+\bot \leftrightarrow \neg\phi$,
- $[\phi]^+\neg\psi \leftrightarrow (\phi \to \neg[\phi]^+\psi)$,
- $[\phi]^+(\psi \vee \chi) \leftrightarrow [\phi]^+\psi \vee [\phi]^+\chi$,
- $[\phi]^+\Box\psi \leftrightarrow (\phi \to \Box[\phi]^+\psi)$.

Lemma 3.3 *Let ϕ be a formula. If ϕ is $\{[\cdot]^+, [\cdot]^-\}$-free then $\models \phi$ iff $\phi \in S5$.*

Lemma 3.4 *Let ϕ be a formula. If ϕ is $[\cdot]^-$-free then $\models \phi$ iff $\phi \in PAL$.*

4 Expressivity

We tackle the problem of the definability of $[\cdot]^+$ and $[\cdot]^-$ in the class of all free models.

Proposition 4.1 (i) *$[\cdot]^+$ cannot be eliminated from the language in the class of all free models.*

(ii) $[\cdot]^-$ *cannot be eliminated from the language in the class of all free models.*

Proof. (1) Suppose $[\cdot]^+$ can be eliminated from the language in the class of all free models. Hence, there exists a formula $\phi(p,q)$ in $\Box$ and $[\cdot]^-$ such that $(*)$ for all free models $\mathcal{M} = (W, X, V)$ and for all tips (x, S), $\mathcal{M}, (x, S) \models \langle p\rangle^+\langle q\rangle^-\Diamond(p \wedge \neg q)$ iff $\mathcal{M}, (x, S) \models \phi(p,q)$. Let $\mathcal{M} = (W, X, V)$ and $\mathcal{M}' = (W', X', V')$ be the models such that $W = W' = \{x, y, z\}$, $X = \{\{x\}, \{y\}, \{z\}, \{x, y\}, \{y, z\}\}$, $X' = \{\{x\}, \{y\}, \{z\}, \{x, y\}\}$, $V(p) = V'(p) = \{y, z\}$ and $V(q) = V'(q) = \{y\}$. Obviously, $\mathcal{M}$ and $\mathcal{M}'$ are free. Moreover, $\mathcal{M}, (y, \{x, y\}) \models \langle p\rangle^+\langle q\rangle^-\Diamond(p \wedge \neg q)$ and $\mathcal{M}', (y, \{x, y\}) \not\models \langle p\rangle^+\langle q\rangle^-\Diamond(p \wedge \neg q)$. By $(*)$, $\mathcal{M}, (y, \{x, y\}) \models \phi(p,q)$ and $\mathcal{M}', (y, \{x, y\}) \not\models \phi(p,q)$. Nevertheless, a proof by induction, based on the function $size(\cdot)$ defined in Section 2, would lead to the conclusion that for all formulas $\psi(p,q)$ in $\Box$ and $[\cdot]^-$, $\mathcal{M}, (y, \{x, y\}) \models \psi(p,q)$ iff $\mathcal{M}', (y, \{x, y\}) \models \psi(p,q)$.
(2) Suppose $[\cdot]^-$ can be eliminated from the language in the class of all free models. Hence, there exists a formula $\phi(p,q)$ in $\Box$ and $[\cdot]^+$ such that $(*)$ for all free models $\mathcal{M} = (W, X, V)$ and for all tips (x, S), $\mathcal{M}, (x, S) \models \langle p\rangle^-\Diamond q$ iff $\mathcal{M}, (x, S) \models \phi(p,q)$. Let $\mathcal{M} = (W, X, V)$ and $\mathcal{M}' = (W', X', V')$ be the models such that $W = W' = \{x, y\}$, $X = \{\{x\}, \{y\}, \{x, y\}\}$, $X' = \{\{x\}, \{y\}\}$, $V(p) = V'(p) = \{x\}$ and $V(q) = V'(q) = \{y\}$. Obviously, $\mathcal{M}$ and $\mathcal{M}'$ are free. Moreover, $\mathcal{M}, (x, \{x\}) \models \langle p\rangle^-\Diamond q$ and $\mathcal{M}', (x, \{x\}) \not\models \langle p\rangle^-\Diamond q$. By $(*)$, $\mathcal{M}, (x, \{x\}) \models \phi(p,q)$ and $\mathcal{M}', (x, \{x\}) \not\models \phi(p,q)$. Nevertheless, a proof by induction, based on the function $size(\cdot)$ defined in Section 2, would lead to the conclusion that for all formulas $\psi(p,q)$ in $\Box$ and $[\cdot]^+$, $\mathcal{M}, (x, \{x\}) \models \psi(p,q)$ iff $\mathcal{M}', (x, \{x\}) \models \psi(p,q)$. □

Proposition 4.1 implies that the constructs $[\cdot]^+$ and $[\cdot]^-$ cannot be eliminated from our language.

5 Relationships with subset space logic

Let the language be extended with the constructs $\blacksquare^+$ and $\blacksquare^-$ with diamond-versions $\blacklozenge^+$ and $\blacklozenge^-$ and let the truth-conditions of formulas $\blacksquare^+\phi$ and $\blacksquare^-\phi$ in model $\mathcal{M} = (W, X, V)$ at tip (x, S) be defined as follows:

- $\mathcal{M}, (x, S) \models \blacksquare^+\phi$ iff for all $T \in X$, if $x \in T$ and $T \subseteq S$ then $\mathcal{M}, (x, T) \models \phi$,
- $\mathcal{M}, (x, S) \models \blacksquare^-\psi$ iff for all $T \in X$, if $x \in T$ and $S \subseteq T$ then $\mathcal{M}, (x, T) \models \phi$.

Obviously, $\blacksquare^+$ is the so-called effort modality of Subset Space Logic [5,13] and $\blacksquare^-$ is the converse effort modality introduced by Heinemann [9].

Proposition 5.1 $\blacksquare^+$ *and* $\blacksquare^-$ *cannot be both eliminated from the language in the class of all models.*

Proof. Suppose $\blacksquare^+$ and $\blacksquare^-$ can be both eliminated from the language in the class of all models. Hence, there exists a formula $\phi(p,q)$ in $\Box$, $[\cdot]^+$ and $[\cdot]^-$ such that $(*)$ for all models $\mathcal{M} = (W, X, V)$ and for all tips (x, S), $\mathcal{M}, (x, S) \models \blacklozenge^-\Diamond(p \wedge \blacklozenge^+\blacklozenge^-\Diamond q)$ iff $\mathcal{M}, (x, S) \models \phi(p,q)$. Let $\mathcal{M} = (W, X, V)$ and $\mathcal{M}' = (W', X', V')$ be the models such that $W = W' = \{x, y, z, t\}$, $X =$

$\{\{x\},\{z\},\{z,t\},\{x,y,z\}\}$, $X' = \{\{x\},\{z,t\},\{x,y,z\}\}$, $V(p) = V'(p) = \{y,z\}$ and $V(q) = V'(q) = \{t\}$. Obviously, $\mathcal{M},(x,\{x\}) \models \blacklozenge^-\Diamond(p \wedge \blacklozenge^+\blacklozenge^-\Diamond q)$ and $\mathcal{M}',(x,\{x\}) \not\models \blacklozenge^-\Diamond(p \wedge \blacklozenge^+\blacklozenge^-\Diamond q)$. By $(*)$, $\mathcal{M},(x,\{x\}) \models \phi(p,q)$ and $\mathcal{M}',(x,\{x\}) \not\models \phi(p,q)$. Nevertheless, a proof by induction, based on the function $size(\cdot)$ defined in Section 2, would lead to the conclusion that for all formulas $\psi(p,q)$ in $\Box$, $[\cdot]^+$ and $[\cdot]^-$, $\mathcal{M},(x,\{x\}) \models \psi(p,q)$ iff $\mathcal{M}',(x,\{x\}) \models \psi(p,q)$.$\Box$

Proposition 5.1 implies that the constructs $\blacksquare^+$ and $\blacksquare^-$ of subset space logics cannot be both defined in our language.

6 Axiomatization

Let $PAL^{\pm}$ be the least set of formulas containing the following axioms and closed under the following inference rules:

(A_1) all instances of CPL,

(A_2) $\Box(\phi \rightarrow \psi) \rightarrow (\Box\phi \rightarrow \Box\psi)$,

(A_3) $\Box\phi \rightarrow \phi$,

(A_4) $\Diamond\phi \rightarrow \Box\Diamond\phi$,

(A_5) $\Box\phi \rightarrow \Box\Box\phi$,

(A_6) $[\phi]^+(\psi \rightarrow \chi) \rightarrow ([\phi]^+\psi \rightarrow [\phi]^+\chi)$,

(A_7) $[\phi]^-(\psi \rightarrow \chi) \rightarrow ([\phi]^-\psi \rightarrow [\phi]^-\chi)$,

(A_8) $\psi \rightarrow [\phi]^+\langle\phi\rangle^-\psi$,

(A_9) $\psi \rightarrow [\phi]^-\langle\phi\rangle^+\psi$,

(A_{10}) $\langle\phi\rangle^+\psi \rightarrow [\phi]^+\psi$,

(A_{11}) $\neg\phi \rightarrow [\phi]^+\bot$,

(A_{12}) $[\phi]^+\bot \rightarrow \neg\phi$,

(A_{13}) $[\top]^+\phi \rightarrow \phi$,

(A_{14}) $p \rightarrow [\phi]^+p$,

(A_{15}) $\neg p \rightarrow [\phi]^+\neg p$,

(A_{16}) $\langle\phi\rangle^+\Box\psi \rightarrow \Box[\phi]^+\psi$,

(A_{17}) $\Box[\phi]^+\psi \rightarrow [\phi]^+\Box\psi$,

(R_1) $\frac{\phi,\ \phi\rightarrow\psi}{\psi}$,

(R_2) $\frac{\phi}{\Box\phi}$,

(R_3) $\frac{\psi}{[\phi]^+\psi}$,

(R_4) $\frac{\psi}{[\phi]^-\psi}$.

We briefly explain the importance of the above axioms and inference rules:

- (A_1) and (R_1) are all we need to prove Lindenbaum Lemma,
- (A_2) and (R_2) are all we need to prove the $\Diamond$-Lemma,
- (A_3)–(A_5) are all we need to prove that $\Box$ gives rise to an equivalence relation between maximal consistent sets of formulas,
- (A_6), (A_7), (R_3) and (R_4) are all we need to prove the $\langle\phi\rangle^{\pm}$-Lemma,
- (A_8) and (A_9) are all we need to prove that $[\phi]^+$ and $[\phi]^-$ give rise to mutually converse relations between maximal consistent sets of formulas,
- (A_{10}) means that announcements are deterministic,
- (A_{11}) and (A_{12}) mean that announcements are executable iff they are true,
- (A_{13}) means, together with (A_{10}), that announcing $\top$ has no effect at all,
- (A_{14}) and (A_{15}) mean that announcements have no effect on the valuation,
- (A_{16}) and (A_{17}) relate what becomes known after an announcement to what was known before it.

As the reader can see, (A_7)–(A_9) are the only axioms explicitly concerning the $[\cdot]^-$ construct. About axioms (A_{10})–(A_{17}), seeing that, apparently, they are less innocent than axioms (A_7)–(A_9), from now on, we will indicate their use.

Lemma 6.1 *The following formulas are in $PAL^{\pm}$:*

- $\phi \wedge [\phi]^+\psi \rightarrow \langle\phi\rangle^+\psi$,
- $p \rightarrow [\phi]^-p$,
- $[\phi]^+p \rightarrow (\phi \rightarrow p)$,
- $[\phi]^+\neg\psi \rightarrow (\phi \rightarrow \neg[\phi]^+\psi)$,
- $[\phi]^+(\psi \vee \chi) \rightarrow [\phi]^+\psi \vee [\phi]^+\chi$.

Proposition 6.2 (Soundness) *Let ϕ be a formula. If $\phi \in PAL^{\pm}$ then $\models \phi$.*

Proof. It suffices to verify that axioms (A_1)–(A_{17}) are valid and inference rules $(R1)$–$(R4)$ are validity-preserving. □

7 Canonical model

A set Γ of formulas is said to be consistent iff for all $n \in \mathbb{N}$ and for all $\phi_1, \ldots, \phi_n \in \Gamma$, $\neg(\phi_1 \wedge \ldots \wedge \phi_n) \notin PAL^{\pm}$. We shall say that a set Γ of formulas is maximal iff for all formulas ϕ, $\phi \in \Gamma$ or $\neg\phi \in \Gamma$. Let U_c be the set of all maximal consistent sets of formulas (with typical members Γ, Δ, etc).

Lemma 7.1 (Lindenbaum Lemma) *Let Γ be a set of formulas. If Γ is consistent then there exists a maximal consistent set Δ of formulas such that $\Gamma \subseteq \Delta$.*

Let $R_\Box$ be the binary relation on U_c such that $\Gamma\ R_\Box\ \Delta$ iff $\Box\Gamma \subseteq \Delta$.

Lemma 7.2 ($\Diamond$-Lemma) *Let ϕ be a formula. Let $\Gamma \in U_c$. If $\Diamond\phi \in \Gamma$ then there exists $\Delta \in U_c$ such that $\Gamma\ R_\Box\ \Delta$ and $\phi \in \Delta$.*

Lemma 7.3 *$R_\Box$ is an equivalence relation on U_c.*

For all formulas ϕ, let $R_{[\phi]^+}$ and $R_{[\phi]^-}$ be the binary relations on U_c such that (i) $\Gamma\ R_{[\phi]^+}\ \Delta$ iff $[\phi]^+\Gamma \subseteq \Delta$ and (ii) $\Gamma\ R_{[\phi]^-}\ \Delta$ iff $[\phi]^-\Gamma \subseteq \Delta$.

Lemma 7.4 ($\langle\phi\rangle^{\pm}$-Lemma) *Let ϕ be a formula. Let $\Gamma \in U_c$.*

- *If $\langle\phi\rangle^+\psi \in \Gamma$ then there exists $\Delta \in U_c$ such that $\Gamma\ R_{[\phi]^+}\ \Delta$ and $\psi \in \Delta$,*
- *if $\langle\phi\rangle^-\psi \in \Gamma$ then there exists $\Delta \in U_c$ such that $\Gamma\ R_{[\phi]^-}\ \Delta$ and $\psi \in \Delta$.*

Lemma 7.5 *Let ϕ be a formula. $R_{[\phi]^+}$ and $R_{[\phi]^-}$ are mutually converse on U_c.*

Lemma 7.6 *Let ϕ be a formula. Let $\Gamma, \Delta \in U_c$. If $\Gamma\ R_{[\phi]^+}\ \Delta$ then $\phi \in \Gamma$ and $[\phi]^+\Gamma = \Delta$.*

Lemma 7.7 *$R_{[\top]^+}$ and $R_{[\top]^-}$ are equal to the identity relation on U_c.*

Let $\equiv$ be the transitive closure of $\bigcup\{R_{[\phi]^+} : \phi$ is a formula$\} \cup \bigcup\{R_{[\phi]^-} : \phi$ is a formula$\}$.

Lemma 7.8 *$\equiv$ is an equivalence relation on U_c.*

The equivalence class of $\Gamma \in U_c$ modulo $\equiv$ will be simply noted $|\Gamma|$. Let $\equiv_\Box$ be the transitive closure of $R_\Box \cup \equiv$.

Lemma 7.9 *$\equiv_\Box$ is an equivalence relation on U_c. Moreover, $\equiv_\Box$ is coarser than $R_\Box$ and $\equiv$.*

Proposition 7.10 *Let $\Gamma, \Delta, \Lambda, \Theta \in U_c$. Let ϕ be a formula. If $\Box\Gamma \subseteq \Delta$, $[\phi]^+\Gamma \subseteq \Lambda$ and $[\phi]^+\Delta \subseteq \Theta$ then $\Box\Lambda \subseteq \Theta$.*

Proof. Suppose $\Box\Gamma \subseteq \Delta$, $[\phi]^+\Gamma \subseteq \Lambda$ and $[\phi]^+\Delta \subseteq \Theta$. Suppose $\Box\Lambda \not\subseteq \Theta$. Let ψ be a formula such that $\psi \in \Box\Lambda$ and $\psi \notin \Theta$. Hence, $\Box\psi \in \Lambda$. Since $[\phi]^+\Gamma \subseteq \Lambda$, therefore $\langle\phi\rangle^+\Box\psi \in \Gamma$. Using (A_{16}), $\Box[\phi]^+\psi \in \Gamma$. Since $\Box\Gamma \subseteq \Delta$, therefore $[\phi]^+\psi \in \Delta$. Since $[\phi]^+\Delta \subseteq \Theta$, therefore $\psi \in \Theta$: a contradiction. Thus, $\Box\Lambda \subseteq \Theta$. □

Proposition 7.11 *Let $\Gamma, \Delta, \Lambda \in U_c$. Let ϕ be a formula. If $[\phi]^+\Gamma \subseteq \Delta$ and $\Box\Delta \subseteq \Lambda$ then there exists $\Theta \in U_c$ such that $\Box\Gamma \subseteq \Theta$ and $[\phi]^+\Theta \subseteq \Lambda$.*

Proof. Suppose $[\phi]^+\Gamma \subseteq \Delta$ and $\Box\Delta \subseteq \Lambda$. Suppose $\Box\Gamma \cup \{\langle\phi\rangle^+\varphi' : \varphi' \in \Lambda\}$ is not consistent. Consequently, there exist $\varphi_1, \ldots, \varphi_m \in \Box\Gamma$ and there exist $\varphi'_1, \ldots, \varphi'_n \in \Lambda$ such that $\neg(\varphi_1 \wedge \ldots \wedge \varphi_m \wedge \langle\phi\rangle^+\varphi'_1 \wedge \ldots \wedge \langle\phi\rangle^+\varphi'_n) \in PAL^\pm$. Hence, $\varphi_1 \wedge \ldots \wedge \varphi_m \to [\phi]^+\neg(\varphi'_1 \wedge \ldots \wedge \varphi'_n) \in PAL^\pm$. Thus, $\Box(\varphi_1 \wedge \ldots \wedge \varphi_m) \to \Box[\phi]^+\neg(\varphi'_1 \wedge \ldots \wedge \varphi'_n) \in PAL^\pm$. Since $\varphi_1, \ldots, \varphi_m \in \Box\Gamma$, therefore $\Box(\varphi_1 \wedge \ldots \wedge \varphi_m) \in \Gamma$. Since $\Box(\varphi_1 \wedge \ldots \wedge \varphi_m) \to \Box[\phi]^+\neg(\varphi'_1 \wedge \ldots \wedge \varphi'_n) \in PAL^\pm$, therefore $\Box[\phi]^+\neg(\varphi'_1 \wedge \ldots \wedge \varphi'_n) \in \Gamma$. Using (A_{17}), $[\phi]^+\Box\neg(\varphi'_1 \wedge \ldots \wedge \varphi'_n) \in \Gamma$. Since $[\phi]^+\Gamma \subseteq \Delta$, therefore $\Box\neg(\varphi'_1 \wedge \ldots \wedge \varphi'_n) \in \Delta$. Since $\Box\Delta \subseteq \Lambda$, therefore $\neg(\varphi'_1 \wedge \ldots \wedge \varphi'_n) \in \Lambda$. Consequently, $\varphi'_1 \notin \Lambda$ or … or $\varphi'_n \notin \Lambda$: a contradiction. Hence, $\Box\Gamma \cup \{\langle\phi\rangle^+\varphi' : \varphi' \in \Lambda\}$ is consistent. Let $\Theta \in U_c$ be such that $\Box\Gamma \cup \{\langle\phi\rangle^+\varphi' : \varphi' \in \Lambda\} \subseteq \Theta$. Thus, $\Box\Gamma \subseteq \Theta$ and $[\phi]^+\Theta \subseteq \Lambda$.□

For all $\Gamma_0 \in U_c$, let $\mathcal{M}_{\Gamma_0} = (W_{\Gamma_0}, X_{\Gamma_0}, V_{\Gamma_0})$ be the model such that $W_{\Gamma_0} = \{|\Gamma| : \Gamma_0 \equiv_\Box \Gamma\}$, $X_{\Gamma_0} = \{S_\Box(\Gamma) : \Gamma_0 \equiv_\Box \Gamma\}$ where $S_\Box(\Gamma) = \{|\Delta| : \Gamma\ R_\Box\ \Delta\}$ and $V_{\Gamma_0}(p) = \{|\Gamma| : \Gamma_0 \equiv_\Box \Gamma \text{ and } p \in \Gamma\}$. For all $\Gamma_0 \in U_c$, $\mathcal{M}_{\Gamma_0}$ will be called Γ_0-canonical model. Each maximal consistent set of formulas equivalent with Γ_0 modulo $\equiv_\Box$ should be seen as a moment in the history of a world. If two of them are different but equivalent modulo $\equiv$, this means that they correspond to different moments in the history of the same world. For this reason, $\mathcal{M}_{\Gamma_0}$-worlds are equivalence classes modulo $\equiv$ of maximal consistent sets of formulas equivalent with Γ_0 modulo $\equiv_\Box$. As for $\mathcal{M}_{\Gamma_0}$-steps, each of them is determined by a moment in the history of a world and consists of the set of all $\mathcal{M}_{\Gamma_0}$-worlds that are equivalent with this moment modulo $R_\Box$. The thing is that one should understand $R_\Box$ as an equivalence relation of contemporaneity between moments. Concerning the $\mathcal{M}_{\Gamma_0}$-valuation, as expected, it associates to each atomic formula the set of all $\mathcal{M}_{\Gamma_0}$-worlds that contain a moment containing the atomic formula.

8 Truth Lemma

For an arbitrary $\Gamma_0 \in U_c$, let P be the set of all formulas ϕ such that for all $\Gamma, \Delta \in U_c$, if $\Gamma_0 \equiv_\Box \Gamma$, $\Gamma_0 \equiv_\Box \Delta$ and $|\Gamma| \in S_\Box(\Delta)$ then the 3 following conditions C_1–C_3 are equivalent:

(C_1) $\mathcal{M}_{\Gamma_0}, (|\Gamma|, S_\Box(\Delta)) \models \phi$,

(C_2) there exists $\Lambda \in U_c$ such that $\Gamma \equiv \Lambda$, $\Delta\ R_\Box\ \Lambda$ and $\phi \in \Lambda$,

(C_3) for all $\Lambda' \in U_c$, if $\Gamma \equiv \Lambda'$ and $\Delta\ R_\Box\ \Lambda'$ then $\phi \in \Lambda'$.

Proposition 8.1 *Let* $\psi \in P$. *Let* $\Delta \in U_c$. *Let* $T = \{|\Pi| \in S_\Box(\Delta) : \mathcal{M}_{\Gamma_0}, (|\Pi|, S_\Box(\Delta)) \models \psi\}$. *Let* $\Sigma, \Lambda \in U_c$. *If* $\Delta\ R_\Box\ \Sigma$, $\psi \in \Sigma$ *and* $[\psi]^+\Sigma = \Lambda$ *then* $T = S_\Box(\Lambda)$.

Proof. Suppose $\Delta\ R_\Box\ \Sigma$, $\psi \in \Sigma$ and $[\psi]^+\Sigma = \Lambda$. Let $\Pi \in U_c$.
Suppose $|\Pi| \in T$. Hence, $|\Pi| \in S_\Box(\Delta)$ and $\mathcal{M}_{\Gamma_0}, (|\Pi|, S_\Box(\Delta)) \models \psi$. Let $\Pi' \in U_c$ be such that $\Pi \equiv \Pi'$, $\Delta\ R_\Box\ \Pi'$ and $\psi \in \Pi'$. Such $\Pi' \in U_c$ exists because $\psi \in P$. Suppose $\Box\Lambda \not\subseteq [\psi]^+\Pi'$. Let φ be a formula such that $\varphi \in \Box\Lambda$ and $\varphi \notin [\psi]^+\Pi'$. Thus, $\Box\varphi \in \Lambda$ and $[\psi]^+\varphi \notin \Pi'$. Since $[\psi]^+\Sigma = \Lambda$, therefore $\Box\varphi \in [\psi]^+\Sigma$. Consequently, $[\psi]^+\Box\varphi \in \Sigma$. By Lemma 6.1, since $\psi \in \Sigma$, therefore $\langle\psi\rangle^+\Box\varphi \in \Sigma$. Using ($A_{16}$), $\Box[\psi]^+\varphi \in \Sigma$. Since $\Delta\ R_\Box\ \Sigma$ and $\Delta\ R_\Box\ \Pi'$, therefore $\Sigma\ R_\Box\ \Pi'$. Since $\Box[\psi]^+\varphi \in \Sigma$, therefore $[\psi]^+\varphi \in \Pi'$: a contradiction. Hence, $\Box\Lambda \subseteq [\psi]^+\Pi'$. Thus, $\Lambda\ R_\Box\ [\psi]^+\Pi'$. Since $\Pi \equiv \Pi'$, therefore $\Pi \equiv [\psi]^+\Pi'$. Since $\Lambda\ R_\Box\ [\psi]^+\Pi'$, therefore $|\Pi| \in S_\Box(\Lambda)$.
Suppose $|\Pi| \in S_\Box(\Lambda)$. Let $\Pi' \in U_c$ be such that $\Pi \equiv \Pi'$ and $\Lambda\ R_\Box\ \Pi'$. By Proposition 7.11, since $[\psi]^+\Sigma = \Lambda$, let $\Theta' \in U_c$ be such that $\Sigma\ R_\Box\ \Theta'$ and $\Theta'\ R_{[\psi]^+}\ \Pi'$. Since $\Delta\ R_\Box\ \Sigma$, therefore $\Delta\ R_\Box\ \Theta'$. Since $\Pi \equiv \Pi'$ and $\Theta'\ R_{[\psi]^+}\ \Pi'$, therefore $\Pi \equiv \Theta'$. Since $\Delta\ R_\Box\ \Theta'$, therefore $|\Pi| \in S_\Box(\Delta)$. By Lemma 7.6, since $\Theta'\ R_{[\psi]^+}\ \Pi'$, therefore $\psi \in \Theta'$. Since $\Pi \equiv \Theta'$, $\Delta\ R_\Box\ \Theta'$ and $\psi \in P$, therefore $\mathcal{M}_{\Gamma_0}, (|\Pi|, S_\Box(\Delta)) \models \psi$. Consequently, $|\Pi| \in T$. $\Box$

Proposition 8.2 (Truth Lemma) *For all formulas* ϕ, $\phi \in P$.

Proof. The proof is done by induction, based on the function $size(\cdot)$ defined in Section 2. Let ϕ be a formula such that for all formulas ψ, if $size(\psi) < size(\phi)$ then $\psi \in P$. We demonstrate $\phi \in P$. Let $\Gamma, \Delta \in U_c$ be such that $\Gamma_0 \equiv_\Box \Gamma$, $\Gamma_0 \equiv_\Box \Delta$ and $|\Gamma| \in S_\Box(\Delta)$. We demonstrate the 3 above conditions C_1–C_3 are equivalent. Let $\Theta \in U_c$ be such that $\Gamma \equiv \Theta$ and $\Delta\ R_\Box\ \Theta$. We have to consider the following 7 cases: $\phi = p$, $\phi = \bot$, $\phi = \neg\psi$, $\phi = \psi \vee \chi$, $\phi = \Box\psi$, $\phi = [\psi]^+\chi$ and $\phi = [\psi]^-\chi$. For the sake of brevity, we only present the most difficult of them, the case $\phi = [\psi]^-\chi$. The cases $\phi = \Box\psi$ and $\phi = [\psi]^+\chi$ are presented in the Annex.
Case $\phi = [\psi]^-\chi$. Since $size(\psi) < size(\phi)$ and $size(\chi) < size(\phi)$, therefore $\psi \in P$ and $\chi \in P$.
($C_1 \Rightarrow C_2$). Suppose $\mathcal{M}_{\Gamma_0}, (|\Gamma|, S_\Box(\Delta)) \models [\psi]^-\chi$. Suppose $[\psi]^-\chi \notin \Theta$. By Lemma 7.4, let $\Lambda \in U_c$ be such that $\Theta\ R_{[\psi]^-}\ \Lambda$ and $\chi \notin \Lambda$. By Lemma 7.6, $\psi \in \Lambda$ and $[\psi]^+\Lambda = \Theta$. Since $\Gamma \equiv \Theta$ and $\Theta\ R_{[\psi]^-}\ \Lambda$, therefore $\Gamma \equiv \Lambda$. Let $T = \{|\Pi| \in S_\Box(\Lambda) : \mathcal{M}_{\Gamma_0}, (|\Pi|, S_\Box(\Lambda)) \models \psi\}$. By Proposition 8.1, since $\psi \in \Lambda$ and $[\psi]^+\Lambda = \Theta$, therefore $T = S_\Box(\Theta)$. Since $\Delta\ R_\Box\ \Theta$, therefore $T = S_\Box(\Delta)$. Since $\Gamma \equiv \Lambda$, therefore $|\Gamma| \in S_\Box(\Lambda)$. Since $\mathcal{M}_{\Gamma_0}, (|\Gamma|, S_\Box(\Delta)) \models [\psi]^-\chi$ and $T = S_\Box(\Delta)$, therefore $\mathcal{M}_{\Gamma_0}, (|\Gamma|, S_\Box(\Lambda)) \models \chi$. Since $\Gamma \equiv \Lambda$ and $\chi \in P$, therefore $\chi \in \Lambda$: a contradiction. Thus, $[\psi]^-\chi \in \Theta$.

$(C_2 \Rightarrow C_3)$. Suppose $\Lambda \in U_c$ is such that $\Gamma \equiv \Lambda$, $\Delta\ R_\Box\ \Lambda$ and $[\psi]^-\chi \in \Lambda$. Let $\Lambda' \in U_c$ be such that $\Gamma \equiv \Lambda'$ and $\Delta\ R_\Box\ \Lambda'$. Suppose $[\psi]^-\chi \notin \Lambda'$. By Lemma 7.4, let $\Lambda'' \in U_c$ be such that $\Lambda'\ R_{[\psi]^-}\ \Lambda''$ and $\chi \notin \Lambda''$. By Lemma 7.6, $\psi \in \Lambda''$ and $[\psi]^+\Lambda'' = \Lambda'$. Since $\Delta\ R_\Box\ \Lambda$ and $\Delta\ R_\Box\ \Lambda'$, therefore $\Lambda'\ R_\Box\ \Lambda$. Since $\Lambda''\ R_{[\psi]^+}\ \Lambda'$, therefore by Proposition 7.11, let $\Lambda''' \in U_c$ be such that $\Lambda''\ R_\Box\ \Lambda'''$ and $\Lambda'''\ R_{[\psi]^+}\ \Lambda$. Since $[\psi]^-\chi \in \Lambda$, therefore $\chi \in \Lambda'''$. Since $\Gamma \equiv \Lambda$, $\Gamma \equiv \Lambda'$, $\Lambda'\ R_{[\psi]^-}\ \Lambda''$ and $\Lambda'''\ R_{[\psi]^+}\ \Lambda$, therefore $\Lambda''' \equiv \Lambda''$. Since $\Lambda''\ R_\Box\ \Lambda'''$, $\chi \in \Lambda'''$ and $\chi \in P$, therefore $\chi \in \Lambda''$: a contradiction. Hence, $[\psi]^-\chi \in \Lambda'$.

$(C_3 \Rightarrow C_1)$. Suppose for all $\Lambda' \in U_c$, if $\Gamma \equiv \Lambda'$ and $\Delta\ R_\Box\ \Lambda'$ then $[\psi]^-\chi \in \Lambda'$. Suppose $\mathcal{M}_{\Gamma_0}, (|\Gamma|, S_\Box(\Delta)) \not\models [\psi]^-\chi$. Hence, there exists $T \in X_{\Gamma_0}$ such that $|\Gamma| \in T$, $S_\Box(\Delta) = \{|\Pi| \in T : \mathcal{M}_{\Gamma_0}, (|\Pi|, T \models \psi\}$ and $\mathcal{M}_{\Gamma_0}, (|\Gamma|, T) \not\models \chi$. Let $\Delta' \in U_c$ be such that $\Gamma_0 \equiv_\Box \Delta'$ and $T = S_\Box(\Delta')$. Since $\Gamma \equiv \Theta$, $\Delta\ R_\Box\ \Theta$ and for all $\Lambda' \in U_c$, if $\Gamma \equiv \Lambda'$ and $\Delta\ R_\Box\ \Lambda'$ then $[\psi]^-\chi \in \Lambda'$, therefore $[\psi]^-\chi \in \Theta$. Since $|\Gamma| \in T$, $S_\Box(\Delta) = \{|\Pi| \in T : \mathcal{M}_{\Gamma_0}, (|\Pi|, T \models \psi\}$, $\mathcal{M}_{\Gamma_0}, (|\Gamma|, T) \not\models \chi$ and $T = S_\Box(\Delta')$, therefore $|\Gamma| \in S_\Box(\Delta')$, $S_\Box(\Delta) = \{|\Pi| \in S_\Box(\Delta') : \mathcal{M}_{\Gamma_0}, (|\Pi|, S_\Box(\Delta') \models \psi\}$ and $\mathcal{M}_{\Gamma_0}, (|\Gamma|, S_\Box(\Delta')) \not\models \chi$. Let $\Theta' \in U_c$ be such that $\Gamma \equiv \Theta'$, $\Delta'\ R_\Box\ \Theta'$ and $\chi \notin \Theta'$. Such $\Theta' \in U_c$ exists because $\chi \in P$. Since $|\Gamma| \in S_\Box(\Delta)$ and $S_\Box(\Delta) = \{|\Pi| \in S_\Box(\Delta') : \mathcal{M}_{\Gamma_0}, (|\Pi|, S_\Box(\Delta') \models \psi\}$, therefore $\mathcal{M}_{\Gamma_0}, (|\Gamma|, S_\Box(\Delta') \models \psi$. Since $\Gamma \equiv \Theta'$, $\Delta'\ R_\Box\ \Theta'$ and $\psi \in P$, therefore $\psi \in \Theta'$. Let $R^\psi_\Box(\Delta')$ be the set of all $\Pi' \in R_\Box(\Delta')$ such that $\psi \in \Pi'$. Since $\Delta'\ R_\Box\ \Theta'$ and $\psi \in \Theta'$, therefore $\Theta' \in R^\psi_\Box(\Delta')$. Since $\Gamma \equiv \Theta$ and $\Gamma \equiv \Theta'$, therefore $\Theta \equiv \Theta'$. Let Tri be the set of all triples of the form (d, m, φ) where $d \in \mathbb{N}$, $m \in \mathbb{N}$ and φ is a formula. Let Q be the set of all $(d, m, \varphi) \in Tri$ such that for all formulas $\varphi_1, \ldots, \varphi_m$, if $(deg(\varphi_1) \cdot \ldots \cdot deg(\varphi_m)) + deg(\varphi) \leq d$ then for all $s_1, \ldots, s_m \in \{+, -\}$, for all $\Pi' \in R^\psi_\Box(\Delta')$ and for all $\Pi \in R_\Box(\Delta)$, if $\Pi' \equiv \Pi$ then the 2 following conditions hold:

(D_1) for all $\Pi'_1, \ldots, \Pi'_m \in U_c$, if $[\psi]^+\Pi'\ R_{[\varphi_1]^{s_1}}\ \Pi'_1$, $\Pi'_1\ R_{[\varphi_2]^{s_2}}\ \Pi'_2$, $\ldots$, $\Pi'_{m-1}\ R_{[\varphi_m]^{s_m}}\ \Pi'_m$ then there exist $\Pi_1, \ldots, \Pi_m \in U_c$ such that $\Pi\ R_{[\varphi_1]^{s_1}}\ \Pi_1$, $\Pi_1\ R_{[\varphi_2]^{s_2}}\ \Pi_2$, $\ldots$, $\Pi_{m-1}\ R_{[\varphi_m]^{s_m}}\ \Pi_m$ and if $\varphi \in \Pi'_m$ then $\varphi \in \Pi_m$,

(D_2) for all $\Pi_1, \ldots, \Pi_m \in U_c$, if $\Pi\ R_{[\varphi_1]^{s_1}}\ \Pi_1$, $\Pi_1\ R_{[\varphi_2]^{s_2}}\ \Pi_2$, $\ldots$, $\Pi_{m-1}\ R_{[\varphi_m]^{s_m}}\ \Pi_m$ then there exist $\Pi'_1, \ldots, \Pi'_m \in U_c$ such that $[\psi]^+\Pi'\ R_{[\varphi_1]^{s_1}}\ \Pi'_1$, $\Pi'_1\ R_{[\varphi_2]^{s_2}}\ \Pi'_2$, $\ldots$, $\Pi'_{m-1}\ R_{[\varphi_m]^{s_m}}\ \Pi'_m$ and if $\varphi \in \Pi_m$ then $\varphi \in \Pi'_m$.

In the above definition, we use the product $deg(\varphi_1) \cdot \ldots \cdot deg(\varphi_m)$ of the degrees of the formulas $\varphi_1, \ldots, \varphi_m$. Since m may be equal to 0, we will consider that in this case, such product is equal to 2. The following claims illustrate the interest to consider the set Tri and its subset Q.

Claim (a):

(i) For all $\Pi' \in R^\psi_\Box(\Delta')$, there exists $\Pi \in R_\Box(\Delta)$ such that $\Pi' \equiv \Pi$,

(ii) for all $\Pi \in R_\Box(\Delta)$, there exists $\Pi' \in R^\psi_\Box(\Delta')$ such that $\Pi' \equiv \Pi$.

Claim (b)**:** If $Q = Tri$ then $[\psi]^+\Theta' \subseteq \Theta$.

Claim (c)**:** $Q = Tri$.

Claim (a) clearly shows the tight relationships between $R^{\psi}_{\Box}(\Delta')$ and $R_{\Box}(\Delta)$. It is only used in the proof of Claim (c). Now, by Claims (b) and (c), $[\psi]^+\Theta' \subseteq \Theta$. Hence, $[\psi]^-\Theta \subseteq \Theta'$. Since $[\psi]^-\chi \in \Theta$, therefore $\chi \in \Theta'$: a contradiction.□

Lemma 8.3 *For all* $\Gamma_0 \in U_c$, $\mathcal{M}_{\Gamma_0}$ *is free.*

Proposition 8.4 (Completeness) *Let* ϕ *be a formula. If* $\models \phi$ *then* $\phi \in PAL^{\pm}$.

Proof. Suppose $\models \phi$ and $\phi \not\in PAL^{\pm}$. Let $\Gamma_0 \in U_c$ be such that $\phi \not\in \Gamma_0$. By Proposition 8.2, $\mathcal{M}_{\Gamma_0}, (|\Gamma_0|, S_{\Box}(\Gamma_0)) \not\models \phi$. Moreover, by Lemma 8.3, $\mathcal{M}_{\Gamma_0}$ is free. Thus, $\not\models \phi$: a contradiction. □

9 Maximal ignorance

The model of maximal ignorance is the triple $\mathcal{M}_0 = (W_0, X_0, V_0)$ where $W_0 = 2^{VAR}$, $X_0 = 2^{2^{VAR}} \setminus \{\emptyset\}$ and V_0 is the function associating to each $p \in VAR$ the subset $V_0(p)$ of W_0 defined as follows: $x \in V_0(p)$ iff $p \in x$. In $\mathcal{M}_0$, each subset x of VAR represents an epistemic alternative for the real world and each non-empty set S of subsets of VAR contains the real world together with its epistemic alternatives at some moment of their history. Moreover, each world-step pair (x, S) such that $x \in S$ determines the real world x and the current restriction of the model containing the real world together with its epistemic alternatives.

Lemma 9.1 $\mathcal{M}_0$ *is free.*

From now on in this section, we will say that a formula ϕ is 0-valid, in symbols $\models_0 \phi$, if ϕ is globally true in $\mathcal{M}_0$. In this section, we investigate the set of all 0-valid formulas. This set is of special interest as we recover some of the original intuitions for the logic of "what is true before an announcement", for example the validity of the formula $\Box p \to \langle p\rangle^- \Diamond\neg p$ mentioned in the introduction.

Proposition 9.2 *Let* ϕ *be a formula. If* ϕ *is* $\{[\cdot]^+, [\cdot]^-\}$*-free then* $\models_0 \phi$ *iff* $\phi \in S5$.

Proof. By [10, Pages 29 and 30]. □

Let $k \in \mathbb{N}$.

Lemma 9.3 *Let* A *and* B *be k-steps. If* $A \subseteq B$ *then the formulas* $\nabla B \to [\bigvee A]^+\nabla A$ *and* $\nabla A \to \langle\bigvee A\rangle^-\nabla B$ *are 0-valid.*

Proposition 9.4 *Let* ϕ *be a k-formula. If* ϕ *is* $\{[\cdot]^+, [\cdot]^-\}$*-free then there exists a family* $\{(\alpha_1, A_1), \ldots, (\alpha_m, A_m)\}$ *of k-tips such that* $\models_0 \phi \leftrightarrow \bigvee\{\alpha_i \wedge \nabla A_i : 1 \leq i \leq m\}$.

For all $\mathcal{M}_0$-worlds x, let $f_k(x)$ be the unique k-world V-agreeing with x. For all $\mathcal{M}_0$-steps S, let $F_k(S) = \{f_k(x) : x \in S\}$ be the unique k-step consisting of all k-worlds V-agreeing with an $\mathcal{M}_0$-world in S. Obviously, for all $\mathcal{M}_0$-tips (x, S), the pair $(f_k(x), F_k(S))$ is a k-tip. Moreover, $f_k(x)$ is a finite conjunction of literals over $p_1, \ldots, p_k$ and $F_k(S)$ is a non-empty finite set of finite conjunctions of literals over $p_1, \ldots, p_k$.

Lemma 9.5 *For all $\mathcal{M}_0$-tips (x, S) and for all k-tips (α, A), $\mathcal{M}_0, (x, S) \models \alpha \wedge \nabla A$ iff $f_k(x) = \alpha$ and $F_k(S) = A$.*

Lemma 9.6 *For all k-formulas ϕ and for all $\mathcal{M}_0$-tips $(x, S), (y, T)$, if $f_k(x) = f_k(y)$ and $F_k(S) = F_k(T)$ then $\mathcal{M}_0, (x, S) \models \phi$ iff $\mathcal{M}_0, (y, T) \models \phi$.*

Proposition 9.7 *Let ϕ be a k-formula. The formula $\phi \leftrightarrow \bigvee\{f_k(x) \wedge \nabla F_k(S) : x \in W_0 \;\&\; S \in X_0 \;\&\; x \in S \;\&\; \mathcal{M}_0, (x, S) \models \phi\}$ is 0-valid.*

Proof. Let (y, T) be an $\mathcal{M}_0$-tip.
Suppose $\mathcal{M}_0, (y, T) \models \phi$. By Lemma 9.5, $\mathcal{M}_0, (y, T) \models f_k(y) \wedge \nabla F_k(T)$. Since $\mathcal{M}_0, (y, T) \models \phi$, therefore $\mathcal{M}_0, (y, T) \models \bigvee\{f_k(x) \wedge \nabla F_k(S) : x \in W_0 \;\&\; S \in X_0 \;\&\; x \in S \;\&\; \mathcal{M}_0, (x, S) \models \phi\}$.
Suppose $\mathcal{M}_0, (y, T) \models \bigvee\{f_k(x) \wedge \nabla F_k(S) : x \in W_0 \;\&\; S \in X_0 \;\&\; x \in S \;\&\; \mathcal{M}_0, (x, S) \models \phi\}$. Let $x \in W_0$ and $S \in X_0$ be such that $x \in S$, $\mathcal{M}_0, (x, S) \models \phi$ and $\mathcal{M}_0, (y, T) \models f_k(x) \wedge \nabla F_k(S)$. Hence, by Lemma 9.5, $f_k(y) = f_k(x)$ and $F_k(T) = F_k(S)$. Since $\mathcal{M}_0, (x, S) \models \phi$, therefore by Lemma 9.6, $\mathcal{M}_0, (y, T) \models \phi$. □

Proposition 9.8 *Let ϕ be a k-formula. There exists a $\{[\cdot]^+, [\cdot]^-\}$-free formula ψ such that $\models_0 \phi \leftrightarrow \psi$.*

Proof. By Proposition 9.7. □

Proposition 9.8 says that the constructs $[\cdot]^+$ and $[\cdot]^-$ can be eliminated from the language as far as 0-validity is concerned. It does not say how, though. To be able to say how, it suffices to be able to determine in particular which $\{[\cdot]^+, [\cdot]^-\}$-free formulas are 0-equivalent to $\langle\phi\rangle^+\psi$ and $\langle\phi\rangle^-\psi$ when the formulas ϕ and ψ are already $\{[\cdot]^+, [\cdot]^-\}$-free.

Proposition 9.9 *Let $\{(\alpha_1, A_1), \ldots, (\alpha_m, A_m)\}$ be a family of k-tips and (β, B) be a k-tip. Let $\phi = \bigvee\{\alpha_i \wedge \nabla A_i : 1 \leq i \leq m\}$ and $\psi = \beta \wedge \nabla B$. If $B_\phi = \{\alpha_i : 1 \leq i \leq m \;\&\; A_i = B\}$ then the formulas $\langle\phi\rangle^+\psi \leftrightarrow \beta \wedge \nabla_\phi B$ and $\langle\phi\rangle^-\psi \leftrightarrow \beta \wedge \nabla B_\phi$ are 0-valid.*

By Propositions 9.4 and 9.9, one can easily design a procedure computing for any given input formula a 0-equivalent $\{[\cdot]^+, [\cdot]^-\}$-free formula. For instance, the formula $\langle p\rangle^-\top$ is 0-equivalent to the $\{[\cdot]^+, [\cdot]^-\}$-free formula $\Box p$.

10 Conclusion

There are several ways to continue this research.
Firstly, there are computability issues. Within the context of the model of maximal ignorance, using the fact that for each $k \in \mathbb{N}$, there exist exactly

2^k k-worlds and there exist exactly $2^{2^k} - 1$ k-steps, one readily sees that the validity problem is decidable, although its exact complexity is still unknown. Within the context of the class of all free models, the computability of the validity problem is still open. Other computability issues are related to the problem consisting of given a formula ϕ, either to determine if there exists a formula ψ such that $\langle\psi\rangle^{+}\phi$ is valid, or to determine if there exists a formula ψ such that $\langle\psi\rangle^{-}\phi$ is valid.
Secondly, there are multi-agent issues. There exist already many multi-agent variants of subset space logic. Within the context of our logic of knowledge with public announcements and converse public announcements, we did not manage to find its acceptable multi-agent variant.
Thirdly, there are introspection issues. In our setting, the agent is both positively and negatively introspective. Suppose the agent is non negatively introspective. This implies that we have to get rid of axiom (A_4). But this also implies that the binary relation $R_{\Box}$ defined in Section 7 is no more an equivalence relation on U_c. And we did not manage to completely axiomatize the corresponding logic of knowledge with public announcements and converse public announcements. Remark that subset space logics of a merely positively introspective agent do not seem to exist.
Fourthly, there is the issue of the extension with the constructs $\blacksquare^{+}$ and $\blacksquare^{-}$ considered in Section 5 and corresponding to the effort modality of Subset Space Logic [5,13] and the converse effort modality introduced by Heinemann [9]. This extension is of great interest as, in our setting, $\blacksquare^{+}$ is like an arbitrary announcement modality [2], and thus $\blacksquare^{-}$ an "arbitrary before the announcement" modality. How to completely axiomatize this extension is still open.
Fifthly, there are characterization issues. For example, the characterization of the set of all pairs (ϕ, ψ) of formulas such that $[\phi]^{+}\psi$ is valid and the characterization of the set of all pairs (ϕ, ψ) of formulas such that $[\phi]^{-}\psi$ is valid.

Acknowledgements

We would like to thank the referees for the feedback we have obtained from them. Philippe Balbiani and Hans van Ditmarsch were partially supported by the ERC project EPS 313360. This work is also supported by the "Agence nationale de la recherche" (contract ANR-11-BS02-011). Special thanks are due to the members of the DynRes project for their extensive remarks concerning a preliminary version of the present paper.

References

[1] Aucher, G., Herzig, A.: *From DEL to EDL: exploring the power of converse events.* In: *Proceedings of ECSQARU 2007.* Springer (2007) 199–209.

[2] Balbiani, P., Baltag, A., van Ditmarsch, H., Herzig, A., Hoshi, T., de Lima, T.: *"Knowable" as "known after an announcement".* The Review of Symbolic Logic **1** (2008) 305–334.

[3] Baltag, A., Moss, L.: *Logics for epistemic programs.* Synthese **139** (2004) 165–224.

[4] Bozzelli, L., van Ditmarsch, H., French, T., Hales, J., Pinchinat, S.: *Refinement modal logic.* Information and Computation **239** (2014) 303–339.

[5] Dabrowski, A., Moss, L., Parikh, R.: *Topological reasoning and the logic of knowledge.* Annals of Pure and Applied Logic **78** (1996) 73–110.

[6] Van Ditmarsch, H., van Eijck, J., Wu, W.: *One hundred prisoners and a lightbulb — logic and computation.* In: *Proceedings of KR 2010.* AAAI (2010) 90–100.

[7] Van Ditmarsch, H., van der Hoek, W., Kooi, B.: *Dynamic Epistemic Logic.* Springer (2007).

[8] Van Ditmarsch, H., Ruan, J., Verbrugge, R.: *Sum and product in dynamic epistemic logic.* Journal of Logic and Computation **18** (2007) 563–588.

[9] Heinemann, B.: *Including the past in topologic.* In: *Proceedings of LFCS 2007.* Springer (2007) 269–283.

[10] Van der Hoek, W., Meyer, J.-J.: *Epistemic Logic for AI and Computer Science.* Cambridge University Press (1995).

[11] Hoshi, T., Yap, A.: *Dynamic epistemic logic with branching temporal structures.* Synthese (2009) 259–281.

[12] Lutz, C.: *Complexity and succinctness of public announcement logic.* In: *Proceedings of AAMAS 2006.* ACM (2006) 137–143.

[13] Parikh, R., Moss, L., Steinsvold, C.: *Topology and epistemic logic.* In: *Handbook of Spatial Logics.* Springer (2007) 299–341.

[14] Parikh, R., Ramanujam, R.: *A knowledge based semantics of messages.* Journal of Logic, Language and Information **12** (2003) 453–467.

[15] Plaza, J.: *Logics of public communications.* Synthese **158** (2007) 165–179.

[16] Renne, B., Sack, J., Yap, A.: *Dynamic epistemic temporal logic.* In: *Proceedings of LORI 2009.* Springer (2009) 263–277.

[17] Renne, B., Sack, J., Yap, A.: *Logics of temporal-epistemic actions.* Synthese **193** 813–849.

[18] Sack, J.: *Temporal languages for epistemic programs.* Journal of Logic, Language and Information (2008) 183–216.

[19] Wáng, Y., Ågotnes, T.: *Subset space public announcement logic.* In: *Proceedings of ICLA 2013.* Springer (2013) 245–257.

[20] Yap, A.: *Dynamic epistemic logic and temporal modality.* In: *Dynamic Formal Epistemology.* Springer (2011) 33–50.

Annex

Proof of Proposition 8.2: Case $\phi = \Box\psi$. Since $size(\psi) < size(\phi)$, therefore $\psi \in P$.

$(C_1 \Rightarrow C_2)$. Suppose $\mathcal{M}_{\Gamma_0}, (|\Gamma|, S_\Box(\Delta)) \models \Box\psi$. Suppose $\Box\psi \notin \Theta$. By Lemma 7.2, let $\Lambda \in U_c$ be such that $\Theta\ R_\Box\ \Lambda$ and $\psi \notin \Lambda$. Since $\Delta\ R_\Box\ \Theta$, therefore $\Delta\ R_\Box\ \Lambda$. Hence, $|\Lambda| \in S_\Box(\Delta)$. Since $\mathcal{M}_{\Gamma_0}, (|\Gamma|, S_\Box(\Delta)) \models \Box\psi$, therefore $\mathcal{M}_{\Gamma_0}, (|\Lambda|, S_\Box(\Delta)) \models \psi$. Since $\psi \in P$ and $\Delta\ R_\Box\ \Lambda$, therefore $\psi \in \Lambda$: a contradiction. Thus, $\Box\psi \in \Theta$.

$(C_2 \Rightarrow C_3)$. Suppose $\Lambda \in U_c$ is such that $\Gamma \equiv \Lambda$, $\Delta\ R_\Box\ \Lambda$ and $\Box\psi \in \Lambda$. Let $\Lambda' \in U_c$ be such that $\Gamma \equiv \Lambda'$ and $\Delta\ R_\Box\ \Lambda'$. Suppose $\Box\psi \notin \Lambda'$. By Lemma 7.2, let $\Pi \in U_c$ be such that $\Lambda'\ R_\Box\ \Pi$ and $\psi \notin \Pi$. Since $\Delta\ R_\Box\ \Lambda$ and $\Delta\ R_\Box\ \Lambda'$, therefore $\Lambda\ R_\Box\ \Pi$. Since $\Box\psi \in \Lambda$, therefore $\psi \in \Pi$: a contradiction. Hence, $\Box\psi \in \Lambda'$.

$(C_3 \Rightarrow C_1)$. Suppose for all $\Lambda' \in U_c$, if $\Gamma \equiv \Lambda'$ and $\Delta\ R_\Box\ \Lambda'$ then $\Box\psi \in \Lambda'$. Since $\Gamma \equiv \Theta$ and $\Delta\ R_\Box\ \Theta$, therefore $\Box\psi \in \Theta$. Suppose $\mathcal{M}_{\Gamma_0}, (|\Gamma|, S_\Box(\Delta)) \not\models \Box\psi$. Let $|\Lambda| \in S_\Box(\Delta)$ be such that $\mathcal{M}_{\Gamma_0}, (|\Lambda|, S_\Box(\Delta)) \not\models \psi$. Let $\Pi \in U_c$ be such that $\Lambda \equiv \Pi$, $\Delta\ R_\Box\ \Pi$ and $\psi \notin \Pi$. Such $\Pi \in U_c$ exists because

$\psi \in P$. Since $\Delta\ R_\Box\ \Theta$, therefore $\Theta\ R_\Box\ \Pi$. Since $\Box\psi \in \Theta$, therefore $\psi \in \Pi$: a contradiction. Hence, $\mathcal{M}_{\Gamma_0}, (|\Gamma|, S_\Box(\Delta)) \models \Box\psi$.

Proof of Proposition 8.2: Case $\phi = [\psi]^+\chi$**.** Since $size(\psi) < size(\phi)$ and $size(\chi) < size(\phi)$, therefore $\psi \in P$ and $\chi \in P$.
($C_1 \Rightarrow C_2$). Suppose $\mathcal{M}_{\Gamma_0}, (|\Gamma|, S_\Box(\Delta)) \models [\psi]^+\chi$. Suppose $[\psi]^+\chi \notin \Theta$. By Lemma 7.4, let $\Lambda \in U_c$ be such that $\Theta\ R_{[\psi]^+}\ \Lambda$ and $\chi \notin \Lambda$. By Lemma 7.6, $\psi \in \Theta$ and $[\psi]^+\Theta = \Lambda$. Let $T = \{|\Pi| \in S_\Box(\Delta) : \mathcal{M}_{\Gamma_0}, (|\Pi|, S_\Box(\Delta)) \models \psi\}$. By Proposition 8.1, since $\Delta\ R_\Box\ \Theta$, $\psi \in \Theta$ and $[\psi]^+\Theta = \Lambda$, therefore $T = S_\Box(\Lambda)$. Hence, $T \in X_{\Gamma_0}$. Since $\Gamma \equiv \Theta$, $\Delta\ R_\Box\ \Theta$, $\psi \in \Theta$ and $\psi \in P$, therefore $\mathcal{M}_{\Gamma_0}, (|\Gamma|, S_\Box(\Delta)) \models \psi$. Thus, $|\Gamma| \in T$. Since $\mathcal{M}_{\Gamma_0}, (|\Gamma|, S_\Box(\Delta)) \models [\psi]^+\chi$, therefore $\mathcal{M}_{\Gamma_0}, (|\Gamma|, S_\Box(\Lambda)) \models \chi$. Since $\Gamma \equiv \Theta$ and $\Theta\ R_{[\psi]^+}\ \Lambda$, therefore $\Gamma \equiv \Lambda$. Since $\mathcal{M}_{\Gamma_0}, (|\Gamma|, S_\Box(\Lambda)) \models \chi$ and $\chi \in P$, therefore $\chi \in \Lambda$: a contradiction. Consequently, $[\psi]^+\chi \in \Theta$.
($C_2 \Rightarrow C_3$). Suppose $\Lambda \in U_c$ is such that $\Gamma \equiv \Lambda$, $\Delta\ R_\Box\ \Lambda$ and $[\psi]^+\chi \in \Lambda$. Let $\Lambda' \in U_c$ be such that $\Gamma \equiv \Lambda'$ and $\Delta\ R_\Box\ \Lambda'$. Suppose $[\psi]^+\chi \notin \Lambda'$. By Lemma 7.4, let $\Lambda'' \in U_c$ be such that $\Lambda'\ R_{[\psi]^+}\ \Lambda''$ and $\chi \notin \Lambda''$. By Lemma 7.6, $\psi \in \Lambda'$ and $[\psi]^+\Lambda' = \Lambda''$. Since $\Gamma \equiv \Lambda'$, $\Delta\ R_\Box\ \Lambda'$, $\psi \in P$, $\Gamma \equiv \Lambda$ and $\Delta\ R_\Box\ \Lambda$, therefore $\psi \in \Lambda$. By Lemma 6.1, since $[\psi]^+\chi \in \Lambda$, therefore $\langle\psi\rangle^+\chi \in \Lambda$. By Lemma 7.4, let $\Lambda''' \in U_c$ be such that $\Lambda\ R_{[\psi]^+}\ \Lambda'''$ and $\chi \in \Lambda'''$. By Lemma 7.6, $[\psi]^+\Lambda = \Lambda'''$. Since $\Delta\ R_\Box\ \Lambda$ and $\Delta\ R_\Box\ \Lambda'$, therefore $\Lambda\ R_\Box\ \Lambda'$. By Proposition 7.10, since $[\psi]^+\Lambda = \Lambda'''$ and $[\psi]^+\Lambda' = \Lambda''$, therefore $\Lambda'''\ R_\Box\ \Lambda''$. Since $\Gamma \equiv \Lambda$ and $[\psi]^+\Lambda = \Lambda'''$, therefore $\Gamma \equiv \Lambda'''$. Since $\Gamma \equiv \Lambda'$ and $[\psi]^+\Lambda' = \Lambda''$, therefore $\Gamma \equiv \Lambda''$. Since $\Gamma \equiv \Lambda'''$, $\chi \in \Lambda'''$, $\chi \in P$ and $\Lambda'''\ R_\Box\ \Lambda''$, therefore $\chi \in \Lambda''$: a contradiction. Hence, $[\psi]^+\chi \in \Lambda'$.
($C_3 \Rightarrow C_1$). Suppose for all $\Lambda' \in U_c$, if $\Gamma \equiv \Lambda'$ and $\Delta\ R_\Box\ \Lambda'$ then $[\psi]^+\chi \in \Lambda'$. Suppose $\mathcal{M}_{\Gamma_0}, (|\Gamma|, S_\Box(\Delta)) \not\models [\psi]^+\chi$. Hence, there exists $T \in X_{\Gamma_0}$ such that $|\Gamma| \in T$, $T = \{|\Pi| \in S_\Box(\Delta) : \mathcal{M}_{\Gamma_0}, (|\Pi|, S_\Box(\Delta)) \models \psi\}$ and $\mathcal{M}_{\Gamma_0}, (|\Gamma|, T) \not\models \chi$. Thus, $\mathcal{M}_{\Gamma_0}, (|\Gamma|, S_\Box(\Delta)) \models \psi$. Since $\Gamma \equiv \Theta$, $\Delta\ R_\Box\ \Theta$, $\psi \in P$ and for all $\Lambda' \in U_c$, if $\Gamma \equiv \Lambda'$ and $\Delta\ R_\Box\ \Lambda'$ then $[\psi]^+\chi \in \Lambda'$, therefore $\psi \in \Theta$ and $[\psi]^+\chi \in \Theta$. By Lemma 6.1, $\langle\psi\rangle^+\chi \in \Theta$. By Lemma 7.4, let $\Lambda \in U_c$ be such that $\Theta\ R_{[\psi]^+}\ \Lambda$ and $\chi \in \Lambda$. By Lemma 7.6, $[\psi]^+\Theta = \Lambda$. By Proposition 8.1, since $\Delta\ R_\Box\ \Theta$ and $\psi \in \Theta$, therefore $T = S_\Box(\Lambda)$. Since $\mathcal{M}_{\Gamma_0}, (|\Gamma|, T) \not\models \chi$, therefore $\mathcal{M}_{\Gamma_0}, (|\Gamma|, S_\Box(\Lambda)) \not\models \chi$. Since $\Gamma \equiv \Theta$ and $\Theta\ R_{[\psi]^+}\ \Lambda$, therefore $\Gamma \equiv \Lambda$. Since $\mathcal{M}_{\Gamma_0}, (|\Gamma|, S_\Box(\Lambda)) \not\models \chi$ and $\chi \in P$, therefore $\chi \notin \Lambda$: a contradiction. Consequently, $\mathcal{M}_{\Gamma_0}, (|\Gamma|, S_\Box(\Delta)) \models [\psi]^+\chi$.

Proof of Claim (a)**:** (i) Let $\Pi' \in R_\Box^\psi(\Delta')$. Hence, $\Pi' \in R_\Box(\Delta')$ and $\psi \in \Pi'$. Since $\psi \in P$, therefore $\mathcal{M}_{\Gamma_0}, (|\Pi'|, S_\Box(\Delta') \models \psi$. Since $S_\Box(\Delta) = \{|\Pi| \in S_\Box(\Delta') : \mathcal{M}_{\Gamma_0}, (|\Pi|, S_\Box(\Delta') \models \psi\}$, therefore $|\Pi'| \in S_\Box(\Delta)$. Let $\Pi \in U_c$ be such that $\Pi' \equiv \Pi$ and $\Delta\ R_\Box\ \Pi$. Thus, $\Pi \in R_\Box(\Delta)$ and $\Pi' \equiv \Pi$.
(ii) Let $\Pi \in R_\Box(\Delta)$. Since $S_\Box(\Delta) = \{|\Pi| \in S_\Box(\Delta') : \mathcal{M}_{\Gamma_0}, (|\Pi|, S_\Box(\Delta') \models \psi\}$, therefore $|\Pi| \in S_\Box(\Delta')$ and $\mathcal{M}_{\Gamma_0}, (|\Pi|, S_\Box(\Delta') \models \psi$. Let $\Pi' \in U_c$ be such that $\Pi \equiv \Pi'$, $\Delta'\ R_\Box\ \Pi'$ and $\psi \in \Pi'$. Such $\Pi' \in U_c$ exists because $\psi \in P$.

Hence, $\Pi' \in R^{\psi}_{\Box}(\Delta')$ and $\Pi' \equiv \Pi$.

Proof of Claim (b)**:** Suppose $Q = Tri$ and $[\psi]^{+}\Theta' \not\subseteq \Theta$. Hence, there exists a formula φ such that $\varphi \in [\psi]^{+}\Theta'$ and $\varphi \notin \Theta$. Since $Q = Tri$, therefore $(deg(\varphi)+1, 0, \varphi) \in Q$. By condition (D_1), since $\Theta' \in R^{\psi}_{\Box}(\Delta')$, $\Theta \in R_{\Box}(\Delta)$ and $\Theta' \equiv \Theta$, therefore if $\varphi \in [\psi]^{+}\Theta'$ then $\varphi \in \Theta$. Since $\varphi \in [\psi]^{+}\Theta'$, therefore $\varphi \in \Theta$: a contradiction.

Proof of Claim (c)**:** The proof is done by induction on (d, m, φ), using the well-founded partial order $\ll$ on Tri defined as follows:

- $(d, m, \varphi) \ll (d', m', \varphi')$ iff one of the 3 following conditions holds:

(i) $d < d'$,
(ii) $d = d'$ and $m < m'$,
(iii) $d = d'$, $m = m'$ and $size(\varphi) < size(\varphi')$.

Let $(d, m, \varphi) \in Tri$ be such that for all $(d', m', \varphi') \in Tri$, if $(d', m', \varphi') \ll (d, m, \varphi)$ then $(d', m', \varphi') \in Q$. We demonstrate $(d, m, \varphi) \in Q$. Let $\varphi_1, \ldots, \varphi_m$ be formulas such that $(deg(\varphi_1) \cdot \ldots \cdot deg(\varphi_m)) + deg(\varphi) \leq d$, $s_1, \ldots, s_m \in \{+, -\}$, $\Pi' \in R^{\psi}_{\Box}(\Delta')$ and $\Pi \in R_{\Box}(\Delta)$ be such that $\Pi' \equiv \Pi$. We demonstrate the 2 above conditions D_1 and D_2. Since $(deg(\varphi_1) \cdot \ldots \cdot deg(\varphi_m)) + deg(\varphi) \leq d$, therefore $d \geq 4$. We consider the following 2 cases.

Case $m = 0$**.**

(D_1). Suppose $\varphi \in [\psi]^{+}\Pi'$. We demonstrate $\varphi \in \Pi$.

Subcase $\varphi = p$**.** Since $p \in [\psi]^{+}\Pi'$, therefore $[\psi]^{+}p \in \Pi'$. By Lemma 6.1, $\psi \to p \in \Pi'$. Since $\psi \in \Pi'$, therefore $p \in \Pi'$. Since $\Pi' \equiv \Pi$, therefore using (A_{14}) and Lemma 6.1, $p \in \Pi$.

Subcase $\varphi = \bot$**.** Since $\bot \in [\psi]^{+}\Pi'$, therefore $[\psi]^{+}\bot \in \Pi'$. Hence, using (A_{12}), $\psi \notin \Pi'$: a contradiction.

Subcase $\varphi = \neg\varphi'$**.** Since $\neg\varphi' \in [\psi]^{+}\Pi'$, therefore $[\psi]^{+}\neg\varphi' \in \Pi'$. By Lemma 6.1, $\psi \to \neg[\psi]^{+}\varphi' \in \Pi'$. Since $\psi \in \Pi'$, therefore $\neg[\psi]^{+}\varphi' \in \Pi'$. Hence, $[\psi]^{+}\varphi' \notin \Pi'$. Thus, $\varphi' \notin [\psi]^{+}\Pi'$. Obviously, $(d, 0, \varphi') \ll (d, 0, \neg\varphi')$. Consequently, $(d, 0, \varphi') \in Q$. Since $\varphi' \notin [\psi]^{+}\Pi'$, therefore $\varphi' \notin \Pi$. Hence, $\neg\varphi' \in \Pi$.

Subcase $\varphi = \varphi' \vee \varphi''$**.** Since $\varphi' \vee \varphi'' \in [\psi]^{+}\Pi'$, therefore $[\psi]^{+}(\varphi' \vee \varphi'') \in \Pi'$. By Lemma 6.1, $[\psi]^{+}\varphi' \in \Pi'$ or $[\psi]^{+}\varphi'' \in \Pi'$. Obviously, $(d, 0, \varphi') \ll (d, 0, \varphi' \vee \varphi'')$ and $(d, 0, \varphi'') \ll (d, 0, \varphi' \vee \varphi'')$. Consequently, $(d, 0, \varphi') \in Q$ and $(d, 0, \varphi'') \in Q$. Since $[\psi]^{+}\varphi' \in \Pi'$ or $[\psi]^{+}\varphi'' \in \Pi'$, therefore $\varphi' \in \Pi$ or $\varphi'' \in \Pi$. Hence, $\varphi' \vee \varphi'' \in \Pi$.

Subcase $\varphi = \Box\varphi'$**.** Since $\Box\varphi' \in [\psi]^{+}\Pi'$, therefore $[\psi]^{+}\Box\varphi' \in \Pi'$. Suppose $\Box\varphi' \notin \Pi$. By Lemma 7.2, let $\Pi_1 \in U_c$ be such that $\Pi \; R_{\Box} \; \Pi_1$ and $\varphi' \notin \Pi_1$. Since $\Pi \in R_{\Box}(\Delta)$, therefore $\Pi_1 \in R_{\Box}(\Delta)$. By item (ii) of Claim (a), let $\Pi'_1 \in R^{\psi}_{\Box}(\Delta')$ be such that $\Pi'_1 \equiv \Pi_1$. Obviously, $(d, 0, \varphi') \ll (d, 0, \Box\varphi')$. Consequently, $(d, 0, \varphi') \in Q$. Since $\varphi' \notin \Pi_1$, therefore $\varphi' \notin [\psi]^{+}\Pi'_1$. Hence, $[\psi]^{+}\varphi' \notin \Pi'_1$. Since $\Pi' \in R^{\psi}_{\Box}(\Delta')$ and $\Pi'_1 \in R^{\psi}_{\Box}(\Delta')$, therefore $\Box[\psi]^{+}\varphi' \notin \Pi'$. Thus, using (A_{16}), $\langle\psi\rangle^{+}\Box\varphi' \notin \Pi'$. Since $\psi \in \Pi'$, therefore

by Lemma 6.1, $[\psi]^+\Box\varphi' \not\in \Pi'$: a contradiction.
Subcase $\varphi = [\varphi']^s\varphi''$. Since $[\varphi']^s\varphi'' \in [\psi]^+\Pi'$, therefore $[\psi]^+[\varphi']^s\varphi'' \in \Pi'$. Suppose $[\varphi']^s\varphi'' \not\in \Pi$. By Lemma 7.4, let $\Pi_1 \in U_c$ be such that $\Pi\ R_{[\varphi']^s}\ \Pi_1$ and $\varphi'' \not\in \Pi_1$. Hence, $\neg\varphi'' \in \Pi_1$. Obviously, $(d-1,1,\neg\varphi'') \ll (d,0,[\varphi']^s\varphi'')$. Consequently, $(d-1,1,\neg\varphi'') \in Q$. Since $2 + deg([\varphi']^s\varphi'') \leq d$, therefore $deg(\varphi') + deg(\neg\varphi'') \leq d-1$. Since $(d-1,1,\neg\varphi'') \in Q$, $\Pi\ R_{[\varphi']^s}\ \Pi_1$ and $\neg\varphi'' \in \Pi_1$, therefore let $\Pi'_1 \in U_c$ be such that $[\psi]^+\Pi'\ R_{[\varphi']^s}\ \Pi'_1$ and $\neg\varphi'' \in \Pi'_1$. Thus, $\langle\varphi'\rangle^s\neg\varphi'' \in [\psi]^+\Pi'$. Consequently, $[\psi]^+\langle\varphi'\rangle^s\neg\varphi'' \in \Pi'$. Since $[\psi]^+[\varphi']^s\varphi'' \in \Pi'$, therefore $[\psi]^+\bot \in \Pi'$. Hence, using (A_{12}), $\psi \not\in \Pi'$. Thus, $\Pi' \not\in R^{\psi}_{\Box}(\Delta')$: a contradiction.
(D_2). Suppose $\varphi \in \Pi$. We demonstrate $\varphi \in [\psi]^+\Pi'$.
Subcase $\varphi = p$. Since $p \in \Pi$ and $\Pi' \equiv \Pi$, therefore $[\psi]^+p \in \Pi'$. Hence, $p \in [\psi]^+\Pi'$.
Subcase $\varphi = \bot$. Obviously, $\bot \not\in \Pi$.
Subcase $\varphi = \neg\varphi'$. Since $\neg\varphi' \in \Pi$, therefore $\varphi' \not\in \Pi$. Since $(d,0,\varphi') \ll (d,0,\neg\varphi')$, therefore $(d,0,\varphi') \in Q$. Since $\varphi' \not\in \Pi$, therefore $\varphi' \not\in [\psi]^+\Pi'$. Hence, $[\psi]^+\varphi' \not\in \Pi'$. Thus, $\langle\psi\rangle^+\neg\varphi' \in \Pi'$. Consequently, using (A_{10}), $[\psi]^+\neg\varphi' \in \Pi'$.
Subcase $\varphi = \varphi' \vee \varphi''$. Since $\varphi' \vee \varphi'' \in \Pi$, therefore $\varphi' \in \Pi$ or $\varphi'' \in \Pi$. Without loss of generality, suppose $\varphi' \in \Pi$. Since $(d,0,\varphi') \ll (d,0,\varphi' \vee \varphi'')$, therefore $(d,0,\varphi') \in Q$. Since $\varphi' \in \Pi$, therefore $\varphi' \in [\psi]^+\Pi'$. Hence, $[\psi]^+\varphi' \in \Pi'$. Thus, $[\psi]^+(\varphi' \vee \varphi'') \in \Pi'$. Consequently, $\varphi' \vee \varphi'' \in [\psi]^+\Pi'$.
Subcase $\varphi = \Box\varphi'$. Suppose $\Box\varphi' \not\in [\psi]^+\Pi'$. Hence, $[\psi]^+\Box\varphi' \not\in \Pi'$. Thus, using (A_{17}), $\Box[\psi]^+\varphi' \not\in \Pi'$. By Lemma 7.2, let $\Pi'_1 \in U_c$ be such that $\Pi'\ R_{\Box}\ \Pi'_1$ and $[\psi]^+\varphi' \not\in \Pi'_1$. Consequently, $\varphi' \not\in [\psi]^+\Pi'_1$ and using (A_{11}), $\psi \in \Pi'_1$. Since $\Pi' \in R_{\Box}(\Delta')$ and $\Pi'\ R_{\Box}\ \Pi'_1$, therefore $\Pi'_1 \in R^{\psi}_{\Box}(\Delta')$. By item (i) of Claim (a), let $\Pi_1 \in R_{\Box}(\Delta)$ be such that $\Pi'_1 \equiv \Pi_1$. Obviously, $(d,0,\varphi') \ll (d,0,\Box\varphi')$. Consequently, $(d,0,\varphi') \in Q$. Since $\varphi' \not\in [\psi]^+\Pi'_1$, therefore $\varphi' \not\in \Pi_1$. Since $\Pi \in R_{\Box}(\Delta)$ and $\Pi_1 \in R_{\Box}(\Delta)$, therefore $\Pi\ R_{\Box}\ \Pi_1$. Since $\varphi' \not\in \Pi_1$, therefore $\Box\varphi' \not\in \Pi$: a contradiction.
Subcase $\varphi = [\varphi']^s\varphi''$. Suppose $[\varphi']^s\varphi'' \not\in [\psi]^+\Pi'$. By Lemma 7.4, let $\Pi'_1 \in U_c$ be such that $[\psi]^+\Pi'\ R_{[\varphi']^s}\ \Pi'_1$ and $\varphi'' \not\in \Pi'_1$. Hence, $\neg\varphi'' \in \Pi'_1$. Obviously, $(d-1,1,\neg\varphi'') \ll (d,0,[\varphi']^s\varphi'')$. Consequently, $(d-1,1,\neg\varphi'') \in Q$. Since $2 + deg([\varphi']^s\varphi'') \leq d$, therefore $deg(\varphi') + deg(\neg\varphi'') \leq d-1$. Since $(d-1,1,\neg\varphi'') \in Q$, $[\psi]^+\Pi'\ R_{[\varphi']^s}\ \Pi'_1$ and $\neg\varphi'' \in \Pi'_1$, therefore let $\Pi_1 \in U_c$ be such that $\Pi\ R_{[\varphi']^s}\ \Pi_1$ and $\neg\varphi'' \in \Pi_1$. Thus, $[\varphi']^s\varphi'' \not\in \Pi$: a contradiction.
Case $m \geq 1$.
(D_1). Let $\Pi'_1, \ldots, \Pi'_m \in U_c$ be such that $[\psi]^+\Pi'\ R_{[\varphi_1]^{s_1}}\ \Pi'_1$, $\Pi'_1\ R_{[\varphi_2]^{s_2}}\ \Pi'_2$, $\ldots$, $\Pi'_{m-1}\ R_{[\varphi_m]^{s_m}}\ \Pi'_m$. If $\varphi \in \Pi'_m$ then let $\varphi' = \varphi$ else let $\varphi' = \neg\varphi$. Obviously, $deg(\varphi') = deg(\varphi)$ and $\langle\varphi_m\rangle^{s_m}\varphi' \in \Pi'_{m-1}$. Moreover, $(d,m-1,\langle\varphi_m\rangle^{s_m}\varphi') \ll (d,m,\varphi)$. Hence, $(d,m-1,\langle\varphi_m\rangle^{s_m}\varphi') \in Q$. Since $(deg(\varphi_1)\cdot\ldots\cdot deg(\varphi_{m-1})) + deg(\langle\varphi_m\rangle^{s_m}\varphi') = (deg(\varphi_1)\cdot\ldots\cdot deg(\varphi_{m-1})) + deg(\varphi_m) + deg(\varphi')$, $(deg(\varphi_1)\cdot\ldots\cdot deg(\varphi_m)) + deg(\varphi) \leq d$ and $deg(\varphi') = deg(\varphi)$, therefore $(deg(\varphi_1)\cdot\ldots\cdot deg(\varphi_{m-1})) + deg(\langle\varphi_m\rangle^{s_m}\varphi') \leq d$. Since $[\psi]^+\Pi'\ R_{[\varphi_1]^{s_1}}\ \Pi'_1$, $\Pi'_1\ R_{[\varphi_2]^{s_2}}\ \Pi'_2$, $\ldots$, $\Pi'_{m-2}\ R_{[\varphi_{m-1}]^{s_{m-1}}}\ \Pi'_{m-1}$, $\langle\varphi_m\rangle^{s_m}\varphi' \in$

Π'_{m-1} and $(d, m-1, \langle\varphi_m\rangle^{s_m}\varphi') \in Q$, therefore let $\Pi_1, \ldots, \Pi_{m-1} \in U_c$ be such that $\Pi\ R_{[\varphi_1]^{s_1}}\ \Pi_1$, $\Pi_1\ R_{[\varphi_2]^{s_2}}\ \Pi_2$, ..., $\Pi_{m-2}\ R_{[\varphi_{m-1}]^{s_{m-1}}}\ \Pi_{m-1}$ and $\langle\varphi_m\rangle^{s_m}\varphi' \in \Pi_{m-1}$. Thus, let $\Pi_m \in U_c$ be such that $\Pi_{m-1}\ R_{[\varphi_m]^{s_m}}\ \Pi_m$ and $\varphi' \in \Pi_m$. Consequently, $\Pi_1, \ldots, \Pi_m \in U_c$ are such that $\Pi\ R_{[\varphi_1]^{s_1}}\ \Pi_1$, $\Pi_1\ R_{[\varphi_2]^{s_2}}\ \Pi_2$, ..., $\Pi_{m-1}\ R_{[\varphi_m]^{s_m}}\ \Pi_m$ and if $\varphi \in \Pi'_m$ then $\varphi \in \Pi_m$.
(D_2). Let $\Pi_1, \ldots, \Pi_m \in U_c$ be such that $\Pi\ R_{[\varphi_1]^{s_1}}\ \Pi_1$, $\Pi_1\ R_{[\varphi_2]^{s_2}}\ \Pi_2$, ..., $\Pi_{m-1}\ R_{[\varphi_m]^{s_m}}\ \Pi_m$. If $\varphi \in \Pi_m$ then let $\varphi' = \varphi$ else let $\varphi' = \neg\varphi$. Obviously, $deg(\varphi') = deg(\varphi)$ and $\langle\varphi_m\rangle^{s_m}\varphi' \in \Pi_{m-1}$. Moreover, $(d, m-1, \langle\varphi_m\rangle^{s_m}\varphi') \ll (d, m, \varphi)$. Hence, $(d, m-1, \langle\varphi_m\rangle^{s_m}\varphi') \in Q$. Since $(deg(\varphi_1) \cdot \ldots \cdot deg(\varphi_{m-1})) + deg(\langle\varphi_m\rangle^{s_m}\varphi') = (deg(\varphi_1) \cdot \ldots \cdot deg(\varphi_{m-1})) + deg(\varphi_m) + deg(\varphi')$, $(deg(\varphi_1) \cdot \ldots \cdot deg(\varphi_m)) + deg(\varphi) \leq d$ and $deg(\varphi') = deg(\varphi)$, therefore $(deg(\varphi_1) \cdot \ldots \cdot deg(\varphi_{m-1})) + deg(\langle\varphi_m\rangle^{s_m}\varphi') \leq d$. Since $\Pi\ R_{[\varphi_1]^{s_1}}\ \Pi_1$, $\Pi_1\ R_{[\varphi_2]^{s_2}}\ \Pi_2$, ..., $\Pi_{m-2}\ R_{[\varphi_{m-1}]^{s_{m-1}}}\ \Pi_{m-1}$, $\langle\varphi_m\rangle^{s_m}\varphi' \in \Pi_{m-1}$ and $(d, m-1, \langle\varphi_m\rangle^{s_m}\varphi') \in Q$, therefore let $\Pi'_1, \ldots, \Pi'_{m-1} \in U_c$ be such that $[\psi]^+\Pi'\ R_{[\varphi_1]^{s_1}}\ \Pi'_1$, $\Pi'_1\ R_{[\varphi_2]^{s_2}}\ \Pi'_2$, ..., $\Pi'_{m-2}\ R_{[\varphi_{m-1}]^{s_{m-1}}}\ \Pi'_{m-1}$ and $\langle\varphi_m\rangle^{s_m}\varphi' \in \Pi'_{m-1}$. Thus, let $\Pi'_m \in U_c$ be such that $\Pi'_{m-1}\ R_{[\varphi_m]^{s_m}}\ \Pi'_m$ and $\varphi' \in \Pi'_m$. Consequently, $\Pi'_1, \ldots, \Pi'_m \in U_c$ are such that $[\psi]^+\Pi'\ R_{[\varphi_1]^{s_1}}\ \Pi'_1$, $\Pi'_1\ R_{[\varphi_2]^{s_2}}\ \Pi'_2$, ..., $\Pi'_{m-1}\ R_{[\varphi_m]^{s_m}}\ \Pi'_m$ and if $\varphi \in \Pi_m$ then $\varphi \in \Pi'_m$.

Proof of Lemma 8.3: Let $\Gamma_0 \in U_c$. Using (A_{12}), by Proposition 8.2, for all formulas ϕ, $\mathcal{M}_{\Gamma_0} \models \phi \to \langle\phi\rangle^+\top$. By Lemma 3.1, $\mathcal{M}_{\Gamma_0}$ is free.

Axiomatizing the lexicographic products of modal logics with linear temporal logic

Philippe Balbiani [1]

Institut de recherche en informatique de Toulouse
Toulouse University

David Fernández-Duque [2]

Centre International de Mathématiques et d'Informatique, Toulouse University
Department of Mathematics, Instituto Tecnológico Autónomo de México

Abstract

Given modal logics λ_1, λ_2, their lexicographic product $\lambda_1 \rhd \lambda_2$ is a new logic whose frames are the Cartesian products of a λ_1-frame and a λ_2-frame, but with the new accessibility relations reminiscent of a lexicographic ordering. This article considers the lexicographic products of several modal logics with linear temporal logic (LTL) based on "next" and "always in the future". We provide axiomatizations for logics of the form $\lambda \rhd \mathsf{LTL}$ and define *cover-simple* classes of frames; we then prove that, under fairly general conditions, our axiomatizations are sound and complete whenever the class of λ-frames is cover-simple. Finally, we prove completeness for several concrete logics of the form $\lambda \rhd \mathsf{LTL}$.

Keywords: Modal logics. Linear temporal logic. Lexicographic product. Axiomatization/completeness.

1 Introduction

There are a great many applications of modal logic to computer science and artificial intelligence that require the use of propositional languages mixing different sorts of modal connectives. By just considering the logical aspects of multi-agent systems, there are, for example, the combination of dynamic logic with epistemic logic [9] or the combination of temporal logic with epistemic logic [10]. There exist many ways to mix together given normal modal logics λ_1 and λ_2 defined over disjoint sets of modal connectives; the appropriate way to do so depends on the application at hand.

[1] Address: Institut de recherche en informatique de Toulouse, Toulouse University, 118 route de Narbonne, 31062 Toulouse Cedex 9, FRANCE. Email: Philippe.Balbiani@irit.fr.

[2] Partially supported by ANR-11-LABX-0040-CIMI within the program ANR-11-IDEX-0002-02. Email: david.fernandez@irit.fr.

If λ_1 and λ_2 are axiomatically presented, their fusion simply consists of putting together their axiomatical presentations. If λ_1 and λ_2 are semantically defined by means of the classes $\mathcal{C}_1$ and $\mathcal{C}_2$ of frames, their fusion simply consists of the modal logic determined by the class of all frames $(W, R_1, \ldots, R_m, S_1, \ldots, S_n)$ such that $(W, R_1, \ldots, R_m)$ is a $\mathcal{C}_1$-frame and $(W, S_1, \ldots, S_n)$ is a $\mathcal{C}_2$-frame. In both cases, the question arises whether the fusion operation preserves properties like decidability, interpolation, etc. [14,15,21].

However, in temporal epistemic logics with no learning and perfect recall, the fusion operation is not the most appropriate way of combining modal logics. In such a setting we may consider a different way to mix logics, given by the asynchronous product operation. Given modal logics λ_1 and λ_2, their asynchronous product is the logic of the products $(W_1 \times W_2, R'_1, \ldots, R'_m, S'_1, \ldots, S'_n)$ of $\mathcal{C}_1$-frames $(W_1, R_1, \ldots, R_m)$ and $\mathcal{C}_2$-frames $(W_2, S_1, \ldots, S_n)$ where $(x_1, x_2)R'_i(y_1, y_2)$ iff $x_1 R_i y_1$ and $x_2 = y_2$, and $(x_1, x_2)S'_j(y_1, y_2)$ iff $x_1 = y_1$ and $x_2 S_j y_2$. See [11,12,15].

More recently, first within the context of qualitative temporal reasoning and then within the context of ordinary modal logics, the first author [2,3] has introduced a third way of mixing modal logics: the lexicographic way. Given modal logics λ_1 and λ_2, their lexicographic product is the logic of the products $(W_1 \times W_2, R'_1, \ldots, R'_m, S'_1, \ldots, S'_n)$ of $\mathcal{C}_1$-frames $(W_1, R_1, \ldots, R_m)$ and $\mathcal{C}_2$-frames $(W_2, S_1, \ldots, S_n)$ where $(x_1, x_2)R'_i(y_1, y_2)$ iff $x_1 R_i y_1$ and $x_2 = y_2$ and $(x_1, x_2)S'_j(y_1, y_2)$ iff $x_2 S_j y_2$.

It has appeared later that the operation of lexicographic products has strong similarities with the operation of ordered sum considered, for example, by Beklemishev [7] within the context of the provability logic GLP. See also Babenyshev and Rybakov [1] and Shapirovsky [19]. The similarity between lexicographic products and ordered sums consists of the fact that, in many situations, the lexicographic product of two Kripke complete modal logics is equal to their ordered sum [6].

Layout of the article. This article considers the lexicographic products of modal logics with linear temporal logic based on "next" and "always in the future". It provides complete axiomatizations of the sets of valid formulas they give rise to. The section-by-section breakdown of the paper is as follows. In Sections 2–4, we present the syntax, the semantics and a minimal axiomatization of our lexicographic products. The aim of Section 5 is to define a requirement allowing us to assert a general completeness theorem. In Section 6, we provide more specific requirements making it possible to apply this general completeness theorem. Finally, Section 7 contains the proof that many familiar modal logics satisfy these more specific requirements. Easy proofs will be omitted.

2 Preliminaries

Let us review a few preliminary notions that we will use throughout the text. We assume the reader feels at home with tools and techniques in modal logic (generated subframes, bounded morphisms, etc.). For more on this, see [8].

2.1 Syntax

Let $\mathbb{P}$ be a countable set of propositional variables (with typical members denoted p, q, etc.) and $\mathfrak{S}$ be a countable set of modalities (with typical members denoted a, b, etc.). The set $\mathcal{L}_{\mathfrak{S}}$ of all formulas (with typical members denoted φ, ψ, etc.) is inductively defined as follows:

$$\varphi, \psi ::= p \mid \bot \mid \neg\varphi \mid (\varphi \vee \psi) \mid [a]\varphi.$$

The Boolean connectives $\top$, $\wedge$, $\rightarrow$ and $\leftrightarrow$ are defined by the usual abbreviations. As usual, $\langle a \rangle$ is the modal connective defined by $\langle a \rangle\varphi ::= \neg[a]\neg\varphi$.

Let $\mathfrak{S}^{\circ,G} = \mathfrak{S} \cup \{\circ, G\}$ and $\mathcal{L}_{\mathfrak{S}}^{\circ,G}$ be the corresponding set of formulas. We will simply write $\circ\varphi$ ("at all next moments of time, φ") instead of $[\circ]\varphi$ and $\hat{\circ}\varphi$ ("at some next moment of time, φ") instead of $\langle\circ\rangle\varphi$. Similarly, we will simply write $G\varphi$ ("at all future moments of time, φ") instead of $[G]\varphi$ and $F\varphi$ ("at some future moment of time, φ") instead of $\langle G \rangle\varphi$. We adopt the standard rules for omission of parentheses.

For all formulas φ, let $SF(\varphi)$ be the set of all subformulas of φ and $SF^{\neg}(\varphi)$ be the closure of $SF(\varphi)$ under negations. In the sequel, we use $\varphi(p_1, \ldots, p_n)$ to denote a formula whose propositional variables form a subset of $\{p_1, \ldots, p_n\}$. For all sets Γ of $\mathcal{L}_{\mathfrak{S}}$-formulas, let $\neg\Gamma = \{\neg\varphi : \varphi \in \Gamma\}$. For all $I \subseteq \mathfrak{S}$, let Γ^I be the set of all $\mathcal{L}_{\mathfrak{S}}$-formulas in Γ of the form $[a]\varphi$ or of the form $\langle a \rangle\varphi$ for some $a \in I$; we will usually omit parentheses when elements of I are written extensionally, as in, e.g., $\Gamma^{\circ,G}$.

2.2 Semantics

A *$\mathfrak{S}$-frame* is a relational structure of the form $\mathcal{F} = (W, R)$ where W is a non-empty set of *states* (with typical members denoted s, t, etc.) and R is a function associating to each $a \in \mathfrak{S}$ a binary relation $R(a)$ on W. For all $a \in \mathfrak{S}$ and for all states s in W, let $R(a)(s) = \{t \in W : sR(a)t\}$. If $\mathfrak{S}'$ is a countable set of modalities containing $\mathfrak{S}$ and $\mathcal{F}' = (W', R')$ is a $\mathfrak{S}'$-frame then $\mathcal{F}'^{|\mathfrak{S}} = (W, R)$ is the $\mathfrak{S}$-frame defined as follows:

- $W = W'$,
- for all $a \in \mathfrak{S}$, $R(a) = R'(a)$.

A *model based on a $\mathfrak{S}$-frame* $\mathcal{F} = (W, R)$ is a relational structure — also called an *$\mathfrak{S}$-model* — of the form $\mathcal{M} = (W, R, V)$, where $V : \mathbb{P} \rightarrow 2^W$. The function V is called the *valuation of* $\mathcal{M}$. The relation "the $\mathcal{L}_{\mathfrak{S}}$-formula φ is true in the $\mathfrak{S}$-model $\mathcal{M}$ at state s" (in symbols $\mathcal{M}, s \models \varphi$) is inductively defined as follows:

- $\mathcal{M}, s \models p$ iff $s \in V(p)$,
- $\mathcal{M}, s \not\models \bot$,
- $\mathcal{M}s \models \neg\varphi$ iff $\mathcal{M}, s \not\models \varphi$,
- $\mathcal{M}, s \models \varphi \vee \psi$ iff either $\mathcal{M}, s \models \varphi$, or $\mathcal{M}, s \models \psi$,
- $\mathcal{M}, s \models [a]\varphi$ iff for all states t in $\mathcal{M}$, if $sR(a)t$ then $\mathcal{M}, t \models \varphi$.

We shall say that φ is globally true in $\mathcal{M}$ (in symbols $\mathcal{M} \models \varphi$) if $\mathcal{M}, s \models \varphi$ for every state s in $\mathcal{M}$.

Let $\mathcal{C}$ be a class of $\mathfrak{S}$-frames. We will denote by Mod($\mathcal{C}$) the class of all $\mathfrak{S}$-models based on some $\mathfrak{S}$-frame in $\mathcal{C}$. We shall say that a $\mathcal{L}_{\mathfrak{S}}$-formula φ is $\mathcal{C}$-satisfiable if there exists a $\mathfrak{S}$-frame $\mathcal{F} = (W, R)$ in $\mathcal{C}$, there exists a $\mathfrak{S}$-model $\mathcal{M} = (W, R, V)$ based on $\mathcal{F}$ and there exists $s \in W$ such that $\mathcal{M}, s \models \varphi$. Finally, we denote the class of elements of $\mathcal{C}$ with finite set of states by $\mathcal{C}^{\text{fin}}$.

2.3 Relative covers

It will be convenient to work with relative cover modalities as well as global covers; these are a variation of the cover modalities $\nabla_i \Gamma$ [18]. Let $a \in \mathfrak{S}$ and Φ, Σ be finite sets of $\mathcal{L}_{\mathfrak{S}}$-formulas such that $\Phi \subseteq \Sigma$. We define the formula $\binom{\Sigma}{\Phi}_a$ as follows:

- $\binom{\Sigma}{\Phi}_a = \bigwedge_{\varphi \in \Phi} \langle a \rangle \varphi \wedge \bigwedge_{\varphi \in \Sigma \setminus \Phi} [a] \neg \varphi$.

This expression states that every formula in Φ holds at some $R(a)$-successor, and if $\sigma \in \Sigma \setminus \Phi$, then σ does not hold at any $R(a)$-successor. To be precise:

Lemma 2.1 *Let $\mathcal{M} = (W, R, V)$ be an $\mathfrak{S}$-model, and Φ, Σ be finite sets of $\mathcal{L}_{\mathfrak{S}}$-formulas such that $\Phi \subseteq \Sigma$. Then, for all $a \in \mathfrak{S}$ and all $s \in W$, the two following conditions are equivalent:*

(i) $\mathcal{M}, s \models \binom{\Sigma}{\Phi}_a$,

(ii) *for all $\varphi \in \Phi$, there exists $t \in R(a)(s)$ such that $\mathcal{M}, t \models \varphi$, and for all $\varphi \in \Sigma \setminus \Phi$ and $t \in R(a)(s)$, $\mathcal{M}, t \not\models \varphi$.*

Given a $\mathfrak{S}$-model $\mathcal{M} = (W, R, V)$ and sets Φ, Σ of $\mathcal{L}_{\mathfrak{S}}$-formulas such that $\Phi \subseteq \Sigma$, we say Φ *covers* $\mathcal{M}$ *relative to* Σ if for all $\varphi \in \Phi$, there exists $s \in W$ such that $\mathcal{M}, s \models \varphi$ and for all $\varphi \in \Sigma \setminus \Phi$ and for all $s \in W$, $\mathcal{M}, s \not\models \varphi$; in other words, Φ is precisely the set of formulas from Σ that are satisfied on $\mathcal{M}$. If Φ covers $\mathcal{M}$ relative to Σ, we will write $\mathcal{M} \models \binom{\Sigma}{\Phi}_\forall$. We warn the reader that $\binom{\Sigma}{\Phi}_\forall$ is not actually a formula of $\mathcal{L}_{\mathfrak{S}}$. In fact, when $\mathfrak{S}$ includes a universal modality, we will denote it by '$[U]$' rather than '$[\forall]$'.

Finally, we may also consider relative covers based on definable modalities. Broadly construed, a definable modality is any formula $\psi(p)$, where p is a propositional variable. We may then define $\widehat{\psi}(p) = \neg\psi(\neg p)$, and as above set

- $\binom{\Sigma}{\Phi}_\psi = \bigwedge_{\varphi \in \Phi} \widehat{\psi}(\varphi) \wedge \bigwedge_{\varphi \in \Sigma \setminus \Phi} \psi(\neg \varphi)$.

2.4 Generated subframes

If $\mathcal{F}$ is an $\mathfrak{S}$-frame and s is a state of $\mathcal{F}$, we denote by $\mathcal{F}_s$ the subframe of $\mathcal{F}$ generated by s. Let $\leq_{\mathcal{F}}$ be the binary relation on W defined by $s \leq_{\mathcal{F}} t$ iff t is a state in $\mathcal{F}_s$. If $\mathcal{M}$ is an $\mathfrak{S}$-model, we define $\mathcal{M}_s$ analogously.

We want to give a syntactic characterization of validity in a generated submodel. To do this, for all sets Γ of $\mathcal{L}_{\mathfrak{S}}$-formulas and $I \subseteq \mathfrak{S}$, let $[I]^*\Gamma$ be the set of all $\mathcal{L}_{\mathfrak{S}}$-formulas of the form $[a_1] \ldots [a_n]\varphi$ such that $a_1, \ldots, a_n \in I$ and $\varphi \in \Gamma$ (where we allow $n = 0$). Then, the following is straighforward and we omit the proof:

Lemma 2.2 *Let $\mathcal{M} = (W, R, V)$ be a $\mathfrak{S}$-model and Γ be a set of $\mathcal{L}_{\mathfrak{S}}$-formulas. For all $s \in W$, the three following conditions are equivalent:*

(i) $\mathcal{M}_s \models \Gamma$,

(ii) $\mathcal{M}_s, s \models [\mathfrak{S}]^*\Gamma$,

(iii) $\mathcal{M}, s \models [\mathfrak{S}]^*\Gamma$.

With these preliminary notions in mind, we are ready to define lexicographic products of modal logics.

3 Lexicographic products

In this paper, we will be interested in $\mathfrak{S}^{\circ,G}$-frames, but of a specific kind: lexicographic products of $\mathfrak{S}$-frames with $(\mathbb{N}, +1, <)$. If (W, R) is a $\mathfrak{S}^{\circ,G}$-frame, we will often write S instead of $R(\circ)$, and $<$ instead of $R(G)$. The *lexicographic product of a $\mathfrak{S}$-frame $\mathcal{F} = (W, R)$ with* $(\mathbb{N}, +1, <)$ is the relational structure $\mathcal{F}' = (W', R', S', <')$ defined as follows:

- $W' = W \times \mathbb{N}$,
- R' is the function associating to each $a \in \mathfrak{S}$ the binary relation $R'(a)$ on W' defined by $(s, i)R'(a)(t, j)$ iff $sR(a)t$ and $i = j$,
- S' is the binary relation on W' defined by $(s, i)S'(t, j)$ iff $i + 1 = j$,
- $<'$ is the binary relation on W' defined by $(s, i) <' (t, j)$ iff $i < j$.

Lemma 3.1 *Let $\mathcal{F}' = (W', R', S', <')$ be the lexicographic product of a $\mathfrak{S}$-frame with $(\mathbb{N}, +1, <)$. Then, $<' = S'^{+}$ and*

(i) *S' is serial,*

(ii) $S'^{-1} \circ S' \circ S' \subseteq S'$,

(iii) $<' \circ <' \subseteq <'$,

(iv) $S' \subseteq <'$,

(v) *for all $a \in \mathfrak{S}$, $S' \circ R'(a) \subseteq S'$,*

(vi) *for all $a \in \mathfrak{S}$, $R'(a) \circ S' \subseteq S'$,*

(vii) *for all $a \in \mathfrak{S}$, $R'(a)^{-1} \circ S' \subseteq S'$,*

(viii) *for all $a \in \mathfrak{S}$, $<' \circ R'(a) \subseteq <'$,*

(ix) *for all $a \in \mathfrak{S}$, $R'(a) \circ <' \subseteq <'$,*

(x) *for all $a \in \mathfrak{S}$, $R'(a)^{-1} \circ <' \subseteq <'$.*

These conditions may be easily verified by the reader. Obviously, lexicographic products of $\mathfrak{S}$-frames with $(\mathbb{N}, +1, <)$ can be considered as $\mathfrak{S}^{\circ,G}$-frames. We shall say that a $\mathfrak{S}^{\circ,G}$-frame is *concrete* if it is isomorphic to the lexicographic product of a $\mathfrak{S}$-frame with $(\mathbb{N}, +1, <)$. For all classes $\mathcal{C}$ of $\mathfrak{S}$-frames, let $\mathcal{C}^{\rhd}$ be the class of concrete $\mathfrak{S}^{\circ,G}$-frames it corresponds to. In the next sections, for multifarious classes $\mathcal{C}$ of $\mathfrak{S}$-frames, we will consider the axiomatization/completeness of the sets of valid $\mathcal{L}_{\mathfrak{S}}^{\circ,G}$-formulas given rise to by $\mathcal{C}^{\rhd}$.

4 The basic logic

A *logic in the signature* $\mathfrak{S}$ is a set λ of $\mathcal{L}_{\mathfrak{S}}$-formulas containing all propositional tautologies and closed under modus ponens and substitution. The logic λ is said to be *normal* if it also contains $[a](p \to q) \to ([a]p \to [a]q)$ for each $a \in \mathfrak{S}$ and is closed under *necessitation*, $\frac{\varphi}{[a]\varphi}$. We write $\lambda \vdash \varphi$ instead of $\varphi \in \lambda$. A $\mathfrak{S}$-frame $\mathcal{F}$ is said to be a λ-*frame* when for all $\mathcal{L}_{\mathfrak{S}}$-formulas φ, if $\lambda \vdash \varphi$ then $\mathcal{F} \models \varphi$. We shall say that a $\mathcal{L}_{\mathfrak{S}}$-formula φ is λ-consistent if $\lambda \nvdash \neg\varphi$. A set Σ of $\mathcal{L}_{\mathfrak{S}}$-formulas is said to be λ-consistent if for all finite subsets Γ of Σ, the $\mathcal{L}_{\mathfrak{S}}$-formula $\bigwedge\Gamma$ is λ-consistent.

Now, let us define the minimal lexicographic logic:

Definition 4.1 Given a logic λ in the signature $\mathfrak{S}$, let $(\lambda \rhd \mathsf{LTL})_0$ be the least set of $\mathcal{L}_{\mathfrak{S}}^{\circ,G}$-formulas containing λ, closed under modus ponens, necessitation for all modalities, and all substitution instances of the induction rule $\frac{p \to \circ p}{p \to Gp}$ and of the following axioms:

(i) $\circ(p \to q) \to (\circ p \to \circ q)$,
(ii) $G(p \to q) \to (Gp \to Gq)$,
(iii) $\hat{\circ}\top$,
(iv) $\hat{\circ}\hat{\circ}p \to \circ\hat{\circ}p$,
(v) $Gp \to GGp$,
(vi) $Gp \to \circ p$,
(vii) $\circ p \to \circ[a]p$,
(viii) $\circ p \to [a]\circ p$,
(ix) $\langle a\rangle\circ p \to \circ p$,
(x) $Gp \to G[a]p$,
(xi) $Gp \to [a]Gp$,
(xii) $\langle a\rangle Gp \to Gp$.

The next result follows from the well-known completeness of LTL, but can also be verified directly. We omit the proofs.

Lemma 4.2 *Given any normal logic λ in the signature $\mathfrak{S}$, the following formulas are derivable in $(\lambda \rhd \mathsf{LTL})_0$:*

(i) $\circ p \to \hat{\circ}p$,
(ii) $\hat{\circ}\circ p \to \circ\circ p$,
(iii) $\circ Gp \leftrightarrow G\circ p$,
(iv) $Gp \leftrightarrow \circ p \wedge \circ Gp$.

Given a logic λ in the signature $\mathfrak{S}$, a *lexicographic λ-logic* is any logic Λ in the signature $\mathfrak{S}^{\circ,G}$ containing $(\lambda \rhd \mathsf{LTL})_0$. Below, we make use of the notations Γ^a and $[I]^*\Gamma$ introduced in Sections 2.1 and 2.4, respectively.

Lemma 4.3 *Let Λ be a lexicographic λ-logic. Let Φ, Σ be finite sets of $\mathcal{L}_{\mathfrak{S}}^{\circ,G}$-formulas such that $\Phi \subseteq \Sigma$. If λ is normal, then*

(i) *if Φ is Λ-consistent then $\Phi \cup [\Phi^{\circ,G}]^{\mathfrak{S}}$ is Λ-consistent,*

(ii) *if $\binom{\Sigma}{\Phi}_\circ$ is Λ-consistent then for every $\varphi \in \Phi$, $\{\varphi\} \cup [\mathfrak{S}]^*\neg(\Sigma \setminus \Phi)$ is Λ-consistent.*

Proof. (i) Suppose Φ is Λ-consistent. Using Axioms (viii), (ix), (xi) and (xii), the reader may easily obtain that $\Lambda \vdash \circ\varphi \to [a_1]\ldots[a_n]\circ\varphi$, $\Lambda \vdash \hat{\circ}\varphi \to [a_1]\ldots[a_n]\hat{\circ}\varphi$, $\Lambda \vdash G\varphi \to [a_1]\ldots[a_n]G\varphi$ and $\Lambda \vdash F\varphi \to [a_1]\ldots[a_n]F\varphi$ for any $a_1,\ldots,a_n \in \mathfrak{S}$. Hence, for all $\mathcal{L}_{\mathfrak{S}}^{\circ,G}$-formulas $\psi \in [\mathfrak{S}]^*\Phi^\circ \cup \Phi^G$, $\Lambda \vdash \bigwedge\Phi \to \psi$. Since Φ is Λ-consistent, therefore $\Phi \cup [\mathfrak{S}]^*\Phi^\circ \cup \Phi^G$ is Λ-consistent.

(ii) Suppose $\left(\begin{smallmatrix}\Sigma\\\Phi\end{smallmatrix}\right)_\circ$ is Λ-consistent. Let $\varphi \in \Phi$. Suppose $\{\varphi\} \cup [\mathfrak{S}]^*\neg(\Sigma \setminus \Phi)$ is Λ-inconsistent. Hence, let Γ be a finite subset of $[\mathfrak{S}]^*\neg(\Sigma \setminus \Phi)$ such that $\Lambda \vdash \neg(\varphi \wedge \bigwedge \Gamma)$. Using Axiom (vii), the reader may easily obtain that for all $\psi \in \Gamma$, $\Lambda \vdash \left(\begin{smallmatrix}\Sigma\\\Phi\end{smallmatrix}\right)_\circ \to \circ\psi$. Thus, $\Lambda \vdash \left(\begin{smallmatrix}\Sigma\\\Phi\end{smallmatrix}\right)_\circ \to \circ\bigwedge \Gamma$. Since $\varphi \in \Phi$, therefore $\Lambda \vdash \left(\begin{smallmatrix}\Sigma\\\Phi\end{smallmatrix}\right)_\circ \to \hat{\circ}\varphi$. Since $\Lambda \vdash \left(\begin{smallmatrix}\Sigma\\\Phi\end{smallmatrix}\right)_\circ \to \circ\bigwedge \Gamma$, therefore $\Lambda \vdash \left(\begin{smallmatrix}\Sigma\\\Phi\end{smallmatrix}\right)_\circ \to \hat{\circ}(\varphi \wedge \bigwedge \Gamma)$. Since $\Lambda \vdash \neg(\varphi \wedge \bigwedge \Gamma)$, by $\circ$-necessitation we see that $\left(\begin{smallmatrix}\Sigma\\\Phi\end{smallmatrix}\right)_\circ$ is Λ-inconsistent: a contradiction. □

Next we will show that, under fairly general conditions, logics extending $(\lambda \rhd \mathsf{LTL})_0$ are complete for their class of lexicographic products.

5 A general completeness theorem

To state our general completeness results, we will need a few preliminary notions. Let $\mathcal{C}$ be a class of $\mathfrak{S}$-frames. A *universal frame for* $\mathcal{C}$ is a $\mathfrak{S}$-frame $\mathcal{F} \in \mathcal{C}$ such that for all $\mathcal{L}_\mathfrak{S}$-formulas φ, if φ is satisfiable in $\mathcal{C}$ then φ is satisfiable in $\mathcal{F}$. We shall say that $\mathcal{C}$ is *simple* if $\mathcal{C}$ possesses a universal frame. A *cover-universal frame for* $\mathcal{C}$ is a $\mathfrak{S}$-frame $\mathcal{F} \in \mathcal{C}$ such that for all finite sets Φ, Σ of $\mathcal{L}_\mathfrak{S}$-formulas with $\Phi \subseteq \Sigma$, if there is a model $\mathcal{M} \in \mathrm{Mod}(\mathcal{C})$ such that $\mathcal{M} \models \left(\begin{smallmatrix}\Sigma\\\Phi\end{smallmatrix}\right)_\forall$, then there is a model $\mathcal{M}$ based on $\mathcal{F}$ such that $\mathcal{M} \models \left(\begin{smallmatrix}\Sigma\\\Phi\end{smallmatrix}\right)_\forall$. We shall say that $\mathcal{C}$ is *cover-simple* if $\mathcal{C}$ possesses a cover-universal frame. Observe that if $\mathcal{C}$ contains a single frame, then it is trivially cover-simple, but larger classes of frames may also be cover-simple.

Lemma 5.1 *Let $\mathcal{C}$ be a class of $\mathfrak{S}$-frames. If $\mathcal{C}$ is cover-simple then $\mathcal{C}$ is simple.*

Proof. Given a $\mathcal{L}_\mathfrak{S}$-formula φ, simply take $\Phi = \{\varphi\}$ and $\Sigma = \{\varphi\}$. □

In this article, we will consider logics λ in the signature $\mathfrak{S}$ such that the class of all $\mathfrak{S}$-frames for λ is cover-simple. As we will see, a large number of familiar logics have this property. Now, our goal is to prove that under certain general conditions, a given λ-logic Λ in the signature $\mathfrak{S}^{\circ,G}$ is complete with respect to a class of concrete $\mathfrak{S}^{\circ,G}$-frames. We will first focus on constructing the temporal part of a concrete $\mathfrak{S}^{\circ,G}$-frame and then building a lexicographic product on top of it; a similar strategy is used in [11] for establishing completeness of other products of modal logics.

Let x, y be sets of $\mathcal{L}_\mathfrak{S}^{\circ,G}$-formulas. We shall say that the couple (x, y) is *temporally adequate* if for all $\mathcal{L}_\mathfrak{S}^{\circ,G}$-formulas φ, the following conditions hold:

- if $\hat{\circ}\varphi \in x$ then $\varphi \in y$,
- if $\circ\varphi \in x$ then $\neg\varphi \notin y$,
- if $G\varphi \in x$ then $\neg\varphi \notin y$ and $G\varphi \in y$,
- if $\neg G\varphi \in x$ then $\neg\varphi \in y$ or $\neg G\varphi \in y$.

Thus, temporally adequate pairs are similar to bricks in mosaics [5]. Given a finite set Σ of $\mathcal{L}_\mathfrak{S}^{\circ,G}$-formulas closed under subformulas and single negations, let $\mathcal{T}_\Sigma^\Lambda = (T_\Sigma^\Lambda, S_\Sigma^\Lambda, <_\Sigma^\Lambda)$ be the relational structure defined as follows:

- T_Σ^Λ is the set of all $x \subseteq \Sigma$ such that $\left(\begin{smallmatrix}\Sigma\\x\end{smallmatrix}\right)_\circ$ is Λ-consistent,

- S^{Λ}_{Σ} is the binary relation on T^{Λ}_{Σ} defined by $xS^{\Lambda}_{\Sigma}y$ if and only if (x, y) is temporally adequate,
- $<^{\Lambda}_{\Sigma}$ is the transitive closure of S^{Λ}_{Σ}.

The next lemma lists the basic properties of $\mathcal{T}^{\Lambda}_{\Sigma}$. Below, we follow the convention that $\bigvee \varnothing = \bot$.

Lemma 5.2 *Let Λ be any logic in the signature $\mathfrak{S}^{\circ,G}$ and $\Sigma \subseteq \mathcal{L}^{\circ,G}_{\mathfrak{S}}$ be finite and closed under subformulas and single negations. Then:*

(i) $\Lambda \vdash \bigvee \left\{ \binom{\Sigma}{x}_{\circ} : x \in T^{\Lambda}_{\Sigma} \right\}$.

(ii) *For all $\Phi \subseteq \Sigma$,*

$$\Lambda \vdash \bigwedge \Phi \to \bigvee \left\{ \binom{\Sigma}{y}_{\circ} : (\Phi, y) \textit{ is temporally adequate} \right\}.$$

In particular, the latter set is non-empty whenever Φ is consistent.

(iii) *For all $x \in T^{\Lambda}_{\Sigma}$, $\Lambda \vdash \binom{\Sigma}{x}_{\circ} \to \bigvee \left\{ \circ\binom{\Sigma}{y}_{\circ} : xS^{\Lambda}_{\Sigma}y \right\}$. The latter set is always non-empty.*

(iv) *For all $\neg G\varphi \in x$, there exists $y \in T^{\Lambda}_{\Sigma}$ such that $x <^{\Lambda}_{\Sigma} y$ and $\neg\varphi \in y$.*

Proof. (i) First note that $\bigvee \left\{ \binom{\Sigma}{x}_{\circ} : x \subseteq \Sigma \right\}$ is a tautology. Hence, $\Lambda \vdash \bigvee \left\{ \binom{\Sigma}{x}_{\circ} : x \subseteq \Sigma \right\}$. Let $x \subseteq \Sigma$. If $x \notin T^{\Lambda}_{\Sigma}$ then $\Lambda \vdash \neg\binom{\Sigma}{x}_{\circ}$. Thus, $\Lambda \vdash \bigvee \left\{ \binom{\Sigma}{x}_{\circ} : x \in T^{\Lambda}_{\Sigma} \right\}$.

(ii) Suppose $\bigwedge \Phi$ is Λ-consistent. By item (i), $\Lambda \vdash \bigvee \left\{ \binom{\Sigma}{y}_{\circ} : y \in T^{\Lambda}_{\Sigma} \right\}$. Let Q be the set of all $y \in T^{\Lambda}_{\Sigma}$ such that $\bigwedge \Phi \wedge \binom{\Sigma}{y}_{\circ}$ is Λ-consistent. Since $\Lambda \vdash \bigvee \left\{ \binom{\Sigma}{y}_{\circ} : y \in T^{\Lambda}_{\Sigma} \right\}$, therefore $\Lambda \vdash \bigwedge \Phi \to \bigvee \left\{ \binom{\Sigma}{y}_{\circ} : y \in Q \right\}$. The reader can then check that if $y \in Q$, then (Φ, y) is temporally adequate; the proof is very similar to that of item (iii) below. It follows that if $\bigwedge \Phi$ is Λ-consistent, then $Q \neq \varnothing$.

(iii) Let $x \in T^{\Lambda}_{\Sigma}$. Using Lemma 5.2, necessitation, axiom (iv) of $(\lambda \rhd \mathsf{LTL})_0$ and formulas (i) and (ii) of Lemma 4.2, we see that $\Lambda \vdash \bigvee \left\{ \circ\binom{\Sigma}{y}_{\circ} : y \in T^{\Lambda}_{\Sigma} \right\}$. Let $y \in T^{\Lambda}_{\Sigma}$ be such that not $xS^{\Lambda}_{\Sigma}y$. Hence, we have to consider the following four cases:

- **Case "there is $\hat{\circ}\varphi \in x$ such that $\varphi \notin y$":** First, observe that $\Lambda \vdash \binom{\Sigma}{x}_{\circ} \to \hat{\circ}\hat{\circ}\varphi$. Hence, by Axiom (iv), we also have $\Lambda \vdash \binom{\Sigma}{x}_{\circ} \to \circ\hat{\circ}\varphi$. Since $\varphi \notin y$, therefore $\Lambda \vdash \binom{\Sigma}{y}_{\circ} \to \circ\neg\varphi$. Thus, using necessitation and formula (i) of Lemma 4.2, $\Lambda \vdash \circ\binom{\Sigma}{y}_{\circ} \to \hat{\circ}\circ\neg\varphi$. Since $\Lambda \vdash \binom{\Sigma}{x}_{\circ} \to \circ\hat{\circ}\varphi$, therefore $\binom{\Sigma}{x}_{\circ} \wedge \circ\binom{\Sigma}{y}_{\circ}$ is Λ-inconsistent.
- **Case "there is $\circ\varphi \in x$ such that $\neg\varphi \in y$":** First, observe that $\Lambda \vdash$

$\binom{\Sigma}{x}_{\circ} \rightarrow \hat{\circ}\circ\varphi$. Since $\neg\varphi \in y$, therefore $\Lambda \vdash \binom{\Sigma}{y}_{\circ} \rightarrow \hat{\circ}\neg\varphi$. Thus, using necessitation, $\Lambda \vdash \circ\binom{\Sigma}{y}_{\circ} \rightarrow \circ\hat{\circ}\neg\varphi$. Since $\Lambda \vdash \binom{\Sigma}{x}_{\circ} \rightarrow \hat{\circ}\circ\varphi$, therefore $\binom{\Sigma}{x}_{\circ} \wedge \circ\binom{\Sigma}{y}_{\circ}$ is Λ-inconsistent.

- **Case "there is $G\varphi \in x$ such that $\neg\varphi \in y$ or $G\varphi \notin y$":** Suppose $\neg\varphi \in y$. Since $G\varphi \in x$, therefore using Axiom (vi), $\Lambda \vdash \binom{\Sigma}{x}_{\circ} \rightarrow \hat{\circ}\circ\varphi$. Since $\neg\varphi \in y$, therefore we can proceed as in the second case in order to conclude that $\binom{\Sigma}{x}_{\circ} \wedge \circ\binom{\Sigma}{y}_{\circ}$ is Λ-inconsistent. Suppose $G\varphi \notin y$. Since $G\varphi \in x$, therefore using Axioms (v), (vi), and formula (i) of Lemma 4.2, $\Lambda \vdash \binom{\Sigma}{x}_{\circ} \rightarrow \circ\hat{\circ}G\varphi$. Since $G\varphi \notin y$, therefore $\Lambda \vdash \binom{\Sigma}{y}_{\circ} \rightarrow \circ\neg G\varphi$. Hence, we can proceed as in the first case in order to conclude that $\binom{\Sigma}{x}_{\circ} \wedge \circ\binom{\Sigma}{y}_{\circ}$ is Λ-inconsistent.
- **Case "there is $\neg G\varphi \in x$ such that $\neg\varphi \notin y$ and $\neg G\varphi \notin y$":** Hence, $\Lambda \vdash \binom{\Sigma}{x}_{\circ} \rightarrow \hat{\circ}\neg G\varphi$. Since $\neg\varphi \notin y$ and $\neg G\varphi \notin y$, therefore $\Lambda \vdash \binom{\Sigma}{y}_{\circ} \rightarrow \circ\varphi \wedge \circ G\varphi$. Thus, by formula (iv) in Lemma 4.2, $\Lambda \vdash \binom{\Sigma}{y}_{\circ} \rightarrow G\varphi$. Thus, using necessitation, $\Lambda \vdash \circ\binom{\Sigma}{y}_{\circ} \rightarrow \circ G\varphi$. Since $\Lambda \vdash \binom{\Sigma}{x}_{\circ} \rightarrow \hat{\circ}\neg G\varphi$, therefore $\binom{\Sigma}{x}_{\circ} \wedge \circ\binom{\Sigma}{y}_{\circ}$ is Λ-inconsistent.

Consequently, for all $y \in T_{\Sigma}^{\Lambda}$, if not $xS_{\Sigma}^{\Lambda}y$ then $\binom{\Sigma}{x}_{\circ} \wedge \circ\binom{\Sigma}{y}_{\circ}$ is Λ-inconsistent. Since $\Lambda \vdash \bigvee\left\{\circ\binom{\Sigma}{y}_{\circ} : y \in T_{\Sigma}^{\Lambda}\right\}$, therefore $\Lambda \vdash \binom{\Sigma}{x}_{\circ} \rightarrow \bigvee\left\{\circ\binom{\Sigma}{y}_{\circ} : xS_{\Sigma}^{\Lambda}y\right\}$.

(iv) Let $\neg G\varphi \in x$. Thus, $\Lambda \vdash \binom{\Sigma}{x}_{\circ} \rightarrow \hat{\circ}\neg G\varphi$. By (iii), $\Lambda \vdash \binom{\Sigma}{x}_{\circ} \rightarrow \bigvee_{xS_{\Sigma}^{\Lambda}y} \circ\binom{\Sigma}{y}_{\circ}$ and $\Lambda \vdash \bigvee_{x<_{\Sigma}^{\Lambda}y} \binom{\Sigma}{y}_{\circ} \rightarrow \bigvee_{x<_{\Sigma}^{\Lambda}y} \bigvee_{yS_{\Sigma}^{\Lambda}z} \circ\binom{\Sigma}{z}_{\circ}$. Hence, $\Lambda \vdash \binom{\Sigma}{x}_{\circ} \rightarrow \circ\bigvee_{x<_{\Sigma}^{\Lambda}y} \binom{\Sigma}{y}_{\circ}$ and $\Lambda \vdash \bigvee_{x<_{\Sigma}^{\Lambda}y} \binom{\Sigma}{y}_{\circ} \rightarrow \circ\bigvee_{x<_{\Sigma}^{\Lambda}z} \binom{\Sigma}{z}_{\circ}$ Consequently, using induction, $\Lambda \vdash \bigvee_{x<_{\Sigma}^{\Lambda}y} \binom{\Sigma}{y}_{\circ} \rightarrow G\bigvee_{x<_{\Sigma}^{\Lambda}z} \binom{\Sigma}{z}_{\circ}$. Since $\Lambda \vdash \binom{\Sigma}{x}_{\circ} \rightarrow \circ\bigvee_{x<_{\Sigma}^{\Lambda}y} \binom{\Sigma}{y}_{\circ}$, therefore $\Lambda \vdash \binom{\Sigma}{x}_{\circ} \rightarrow \circ G\bigvee_{x<_{\Sigma}^{\Lambda}y} \binom{\Sigma}{y}_{\circ}$ and, by formula (iv) in Lemma 4.2, $\Lambda \vdash \binom{\Sigma}{x}_{\circ} \rightarrow G\bigvee_{x<_{\Sigma}^{\Lambda}y} \binom{\Sigma}{y}_{\circ}$. Suppose there is no $y \in T_{\Sigma}^{\Lambda}$ such that $x <_{\Sigma}^{\Lambda} y$ and $\neg\varphi \in y$. Thus, $\Lambda \vdash \bigvee_{x<_{\Sigma}^{\Lambda}y} \binom{\Sigma}{y}_{\circ} \rightarrow \circ\varphi$. Consequently, using necessitation, $\Lambda \vdash G\bigvee_{x<_{\Sigma}^{\Lambda}y} \binom{\Sigma}{y}_{\circ} \rightarrow G\circ\varphi$. Since $\Lambda \vdash \binom{\Sigma}{x}_{\circ} \rightarrow G\bigvee_{x<_{\Sigma}^{\Lambda}y} \binom{\Sigma}{y}_{\circ}$, therefore $\Lambda \vdash \binom{\Sigma}{x}_{\circ} \rightarrow G\circ\varphi$. Hence, by formula (iii) in Lemma 4.2, $\Lambda \vdash \binom{\Sigma}{x}_{\circ} \rightarrow \circ G\varphi$. Since $\Lambda \vdash \binom{\Sigma}{x}_{\circ} \rightarrow \hat{\circ}\neg G\varphi$, therefore $\binom{\Sigma}{x}_{\circ}$ is Λ-inconsistent: a contradiction. □

Now, we are ready to build the temporal part of our models. A *good Σ-path* is an infinite sequence $(x_i)_{i\in\mathbb{N}}$ of subsets of Σ such that for all $i \in \mathbb{N}$,

- (x_i, x_{i+1}) is temporally adequate,
- for all $\neg G\varphi \in x_i$, there exists $j \in \mathbb{N}$, $i < j$, such that $\neg\varphi \in x_j$.

We can then use Lemma 5.2 to construct good paths in $\mathcal{T}_{\Sigma}^{\Lambda}$. Let us denote the set of positive integers by $\mathbb{N}^{\star}$.

Lemma 5.3 *Let $\Phi \subseteq \Sigma$. If Φ is Λ-consistent then there exists a good path $(x_i)_{i\in\mathbb{N}^{\star}}$ such that for all $i \in \mathbb{N}^{\star}$, $x_i \in T_{\Sigma}^{\Lambda}$ and (Φ, x_1) is temporally adequate.*

Proof. For each $i \in \mathbb{N}^\star$, we build a sequence $x_1, \ldots, x_{n_i}$ in T^Λ_Φ recursively on i. First, for $i = 1$, we merely take $n_1 = 1$ and x_1 to be any element of T^Λ_Φ such that $(\Phi, , x_1)$ is temporally adequate, which exists by Lemma 5.2(ii). Then, consider $i \in \mathbb{N}^\star$ and let $\psi_1, \ldots, \psi_k$ be a list of all formulas ψ such that $\neg G\psi \in x_{n_i}$. In the case that $k = 0$, just let $n_{i+1} = n_i + 1$ and choose $x_{n_{i+1}}$ arbitrary so that $x_{n_i} S^\Lambda_\Sigma x_{n_{i+1}}$. This will ensure that the construction does not get stuck at x_{n_i}. If the case that $k \geq 1$, we construct a path $x_{n_i+1}, \ldots, x_{m_j}$ for all $j = 1 \ldots k$ as follows. By 5.2(iv), let $y \in T^\Lambda_\Sigma$ be such that $x_{n_i} <^\Lambda_\Sigma y$ and $\neg\psi_1 \in y$. Since $<^\Lambda_\Phi$ is the transitive closure of S^Λ_Φ, therefore we have a path $x_{n_i}, \ldots, x_{n_i+m_1} = y$. Now, suppose we have constructed $x_{n_i}, \ldots, x_{n_i+m_j}$ and consider two cases. If $\neg\psi_{j+1} \in x_{n_i+r}$ for some r, then $m_{j+1} = 0$ and we do nothing. Otherwise, it follows that $\neg G\psi_{j+1} \in x_{n_i+m_j}$, and thus we can construct $x_{n_i+m_{j+1}}$ with $\neg\psi_{j+1} \in x_{n_i+m_{j+1}}$ by once again using 5.2(iv). It is straightforward to check that the path thus constructed is a good path. □

Before stating our general completeness result, we need a last technical assumption. In order to state it, it will be convenient to treat temporal formulas as if they were propositional variables. For each formula of $\mathcal{L}^{\circ,G}_{\mathfrak{S}}$ of the form $\psi = \circ\varphi$ or $\psi = G\varphi$, let $\underline{\psi}$ denote a fresh propositional variable. If Σ is a set of formulas, let $\mathbb{P}_\Sigma$ denote the set of propositional variables which include all variables $\underline{\psi}$ for $\psi \in \Sigma$. We extend the notation $\underline{\psi}$ to an arbitrary $\mathcal{L}^{\circ,G}_{\mathfrak{S}}$-formula ψ by replacing the outermost occurrences of its $\circ$-subformulas and G-subformulas belonging to Σ by the corresponding variables, and if Φ is a set of formulas set $\underline{\Phi} = \{\underline{\psi} : \psi \in \Phi\}$. A *$\Sigma$-valuation* on an $\mathfrak{S}$-frame $\mathcal{F} = (W, R)$ is a function $V : \mathbb{P}_\Sigma \mapsto 2^W$ such that for all $\varphi \in \Sigma$ of the form $\circ\psi$ or $G\psi$, $V(\underline{\varphi}) = W$ or $V(\underline{\varphi}) = \varnothing$.

If $\mathcal{C}$ is a class of $\mathfrak{S}$-frames, a *$(\mathcal{C}, \Sigma)$-moment* is a pair $\mathfrak{m} = (\mathcal{F}, V)$ where $\mathcal{F} \in \mathcal{C}$ and V is a Σ-valuation on $\mathcal{F}$. Given a $(\mathcal{C}, \Sigma)$-moment $\mathfrak{m} = (W, R, V)$, we define $\chi^\Sigma(\mathfrak{m})$ to be the set of all $\varphi \in \Sigma$ such that $V(\underline{\varphi}) \neq \varnothing$. A *$(\mathcal{C}, \Sigma)$-quasimodel* is a sequence $(\mathfrak{m}_i)_{i\in\mathbb{N}}$ of $(\mathcal{C}, \Sigma)$-moments such that $(\chi^\Sigma(\mathfrak{m}_i))_{i\in\mathbb{N}}$ is a good path. As one would expect, quasimodels are useful because they may be used to construct models.

Lemma 5.4 *Let $(\mathfrak{m}_i)_{i\in\mathbb{N}}$ be a $(\mathcal{C}, \Sigma)$-quasimodel. If $\mathcal{C}$ is cover-simple then there exists a frame $\mathcal{F} = (W, R)$ in $\mathcal{C}$ and a valuation V' on the lexicographic product $\mathcal{F}' = (W', R', S', <')$ of $\mathcal{F}$ with $(\mathbb{N}, +1, <)$ such that for all $\varphi \in \Sigma$ and for all $i \in \mathbb{N}$, $\varphi \in \chi^\Sigma(\mathfrak{m}_i)$ iff there exists $w \in W$ such that $(\mathcal{F}', V'), (w, i) \models \varphi$.*

Proof. Suppose $\mathcal{C}$ is cover-simple. Let $\mathcal{F} = (W, R)$ be a cover-universal frame for $\mathcal{C}$. Since $(\mathfrak{m}_i)_{i\in\mathbb{N}}$ is a $(\mathcal{C}, \Sigma)$-quasimodel, therefore for all $i \in \mathbb{N}$, $\mathfrak{m}_i$ is a $(\mathcal{C}, \Sigma)$-moment. Hence, for all $i \in \mathbb{N}$, $\mathfrak{m}_i$ is a pair $(\mathcal{F}_i, V_i)$ where $\mathcal{F}_i \in \mathcal{C}$ and V_i is a Σ-valuation on $\mathcal{F}_i$. Let $\Phi = \chi^\Sigma(\mathfrak{m}_i)$; note that by definition, for all $i \in \mathbb{N}$, $\mathfrak{m}_i \models \left(\frac{\underline{\Sigma}}{\underline{\Phi}}\right)_\forall$. Since $\mathcal{F}$ is a cover-universal frame for $\mathcal{C}$, therefore for all $i \in \mathbb{N}$, let V'_i be a valuation on $\mathcal{F}$ such that $(\mathcal{F}, V'_i) \models \left(\frac{\underline{\Sigma}}{\underline{\Phi}}\right)_\forall$. Let V' be the valuation on the lexicographic product $\mathcal{F}' = (W', R', S', <')$ of $\mathcal{F}$ with $(\mathbb{N}, +1, <)$ such that for all $p \in \mathbb{P}$, for all $w \in W$ and for all $i \in \mathbb{N}$, $(w, i) \in V'(p)$ iff $w \in V'_i(p)$. A routine induction would show that for any $\varphi \in \Sigma$, for all $w \in W$ and for all $i \in \mathbb{N}$, $(\mathcal{F}', V'), (w, i) \models \psi$ if and only if $(\mathcal{F}, V_i'), w \models \underline{\psi}$. Thus, for all $\varphi \in \Sigma$ and for all $i \in \mathbb{N}$, $\varphi \in \chi^{\Sigma}(\mathfrak{m}_i)$ iff there exists $w \in W$ such that $(\mathcal{F}', V'), (w, i) \models \varphi$.□

As a result, we may turn our attention to building quasimodels rather than concrete models. A lexicographic λ-logic Λ is said to be *moment-complete for* $\mathcal{C}$ if for all finite sets Σ of $\mathcal{L}_{\mathfrak{S}}^{\circ,G}$-formulas closed under subformulas and single negations and for all $\Phi \subseteq \Sigma$,

(C_1) if Φ is Λ-consistent then there is a $(\mathcal{C}, \Sigma)$-moment $\mathfrak{m}$ satisfying $\bigwedge \underline{\Phi}$,

(C_2) if $\binom{\Sigma}{\Phi}_\circ$ is Λ-consistent then there is $(\mathcal{C}, \Sigma)$-moment $\mathfrak{m}$ such that $\mathfrak{m} \models \binom{\Sigma}{\underline{\Phi}}_\forall$.

With this, we are ready to state our general completeness theorem from which completeness for many specific logics will follow later.

Theorem 5.5 *Let $\mathcal{C}$ be a cover-simple class of $\mathfrak{S}$-frames and Λ be a λ-logic in the signature $\mathfrak{S}^{\circ,G}$. If Λ is moment-complete for $\mathcal{C}$ then for all $\mathcal{L}_{\mathfrak{S}}^{\circ,G}$-formulas φ, if φ is Λ-consistent then φ is satisfiable in $\mathcal{C}^{\triangleright}$.*

Proof. Suppose Λ is moment-complete for $\mathcal{C}$. Let φ be a Λ-consistent $\mathcal{L}_{\mathfrak{S}}^{\circ,G}$-formula and $\Sigma = SF^{\neg}(\varphi)$. Since $\mathcal{C}$ is a cover-simple class of $\mathfrak{S}$-frames, let $\mathcal{F} = (W, R)$ be a cover-universal frame for $\mathcal{C}$. Since φ is Λ-consistent, let Φ_0 be a maximal Λ-consistent subset of Σ containing φ. By Lemma 5.3, let $(x_i)_{i \in \mathbb{N}^\star}$ be a good path such that for all $i \in \mathbb{N}^\star$, $x_i \in T_{\Sigma}^{\Lambda}$ and $(\Phi_0, , x_1)$ is temporally adequate.

Since Λ is moment-complete for $\mathcal{C}$, by (C_1), let $\mathfrak{m}_0$ be a $(\mathcal{C}, \Sigma)$-moment satisfying $\bigwedge \underline{\Phi}_0$. Note that, for all, $i \geq 1$, $\binom{\Sigma}{x_i}_\circ$ is Λ-consistent. Hence, for all $i \geq 1$, by (C_2), there is a $(\mathcal{C}, \Sigma)$-moment $\mathfrak{m}_i$ satisfying $\binom{\Sigma}{\underline{x_i}}_\forall$. Obviously, the sequence $(\mathfrak{m}_i)_{i \in \mathbb{N}}$ of $(\mathcal{C}, \Sigma)$-moments constitute a $(\mathcal{C}, \Sigma)$-quasimodel. Consequently, by Lemma 5.4, there exists a valuation V' on the lexicographic product $\mathcal{F}' = (W', R', S', <')$ of $\mathcal{F}$ with $(\mathbb{N}, +1, <)$ such that for all $\psi \in \Sigma$ and for all $i \in \mathbb{N}$, $\psi \in \chi^{\Sigma}(\mathfrak{m}_i)$ iff there exists $w \in W$ such that $(\mathcal{F}', V'), (w, i) \models \psi$. Since $\varphi \in \chi^{\Sigma}(\mathfrak{m}_0)$, therefore there exists $w \in W$ such that $(\mathcal{F}', V'), (w, 0) \models \varphi$. □

6 Completeness for special classes of frames

In this section, we consider classes of frames satisfying specific conditions which will allow us to apply Theorem 5.5. Many of the well-known modal logics satisfy at least one of the conditions we will give.

6.1 Local classes

A class $\mathcal{C}$ of $\mathfrak{S}$-frames is *local* if it is closed under generated subframes and disjoint unions. The idea behind local classes of $\mathfrak{S}$-frames is that we can build models by looking only at what individual states see. Below, recall that a logic λ is *strongly complete* for $\mathcal{C}$ if whenever Φ is a (possibly infinite) set of formulas that is λ-consistent, then there is a model $\mathcal{M} \in \mathcal{C}$ with a world w such that $\mathcal{M}, w \models \varphi$ for every $\varphi \in \Phi$.

Lemma 6.1 *Let λ be a logic in the signature $\mathfrak{S}$, $\mathcal{C}$ be a cover-simple local class of $\mathfrak{S}$-frames and Λ be a lexicographic λ-logic. If λ is strongly complete with respect to $\mathcal{C}$ then Λ is moment-complete.*

Proof. Suppose λ is strongly complete with respect to $\mathcal{C}$. Since $\mathcal{C}$ is a cover-simple class of $\mathfrak{S}$-frames, let $\mathcal{F} = (W, R)$ be a cover-universal frame for $\mathcal{C}$. Let Σ be a finite set of $\mathcal{L}_{\mathfrak{S}}^{\circ;G}$-formulas closed under subformulas and single negations and let $\Phi_0 \subseteq \Sigma$.

(i) Suppose Φ_0 is Λ-consistent. Let Φ be a maximal Λ-consistent subset of Σ containing Φ_0. Hence, by Lemma 4.3, $\Phi \cup [\mathfrak{S}]^*\Phi^\circ \cup \Phi^G$ is Λ-consistent and $\Phi \cup [\mathfrak{S}]^*\Phi^\circ \cup \Phi^G$ is λ-consistent. Since λ is strongly complete with respect to $\mathcal{C}$, let $\mathcal{M}$ be a $\mathcal{C}$-model and s be a state in $\mathcal{M}$ such that $\mathcal{M}, s \models \Phi \cup [\mathfrak{S}]^*\Phi^\circ \cup \Phi^G$. Thus, by Lemma 2.2, $\mathcal{M}_s, s \models \Phi^\circ \cup \Phi^G$. Consequently, $\mathcal{M}_s$ is a $(\mathcal{C}, V)$-moment.

(ii) Suppose $\binom{\Sigma}{\Phi}_{\circ}$ is Λ-consistent and let $\varphi \in \Phi$. Hence, by Lemma 4.3, $\{\varphi\} \cup [\mathfrak{S}]^*\neg(\Sigma \setminus \Phi)$ is Λ-consistent and $\{\varphi\} \cup [\mathfrak{S}]^*\neg(\Sigma \setminus \Phi)$ is λ-consistent. Since λ is strongly complete with respect to $\mathcal{C}$, let $\mathcal{M}^\varphi$ be a $\mathcal{C}$-model and s_φ be a state in $\mathcal{M}^\varphi$ be such that $\mathcal{M}^\varphi, s_\varphi \models \{\varphi\} \cup [\mathfrak{S}]^*\neg(\Sigma \setminus \Phi)$. Thus, $\mathcal{M}^\varphi_{s_\varphi}, s_\varphi \models \{\varphi\} \cup [\mathfrak{S}]^*\neg(\Sigma \setminus \Phi)$. Since $\mathcal{C}$ is a local class of $\mathfrak{S}$-frames, therefore $\mathcal{M}^\varphi_{s_\varphi}$ belongs to $Mod(\mathcal{C})$. Let $\mathcal{M}_\Phi$ be the disjoint union of $\{\mathcal{M}^\varphi_{s_\varphi} : \varphi \in \Phi\}$. Since $\mathcal{C}$ is a local class of $\mathfrak{S}$-frames, therefore $\mathcal{M}_\Phi$ belongs to $Mod(\mathcal{C})$. Since for all $\varphi \in \Phi$, $\mathcal{M}^\varphi_{s_\varphi}, s_\varphi \models \{\varphi\} \cup [\mathfrak{S}]^*\neg(\Sigma \setminus \Phi)$, therefore $\mathcal{M}_\Phi \models \binom{\Sigma}{\underline{\Phi}}_{\forall}$. Since $\mathcal{F}$ is a cover-universal frame for $\mathcal{C}$, there exists a valuation V on $\mathcal{F}$ such that $(\mathcal{F}, V) \models \binom{\Sigma}{\underline{\Phi}}_{\forall}$.

□

From this, we may use Theorem 5.5 to immediately obtain the following:

Corollary 6.2 *Let $\mathcal{C}$ be a cover-simple local class of $\mathfrak{S}$-frames and Λ be a lexicographic λ-logic. If λ is strongly complete with respect to $\mathcal{C}$ then Λ is complete for $\mathcal{C}^{\triangleright}$.*

We remark that even if λ is strongly complete, we cannot expect Λ to be; this is because the set of formulas $\{\neg Gp\} \cup \{\circ^n p : n \in \mathbb{N}\}$ is consistent, but unsatisfiable.

6.2 Linear classes

Recall that the notation $\leq_{\mathcal{F}}$ was introduced in Section 2.4. A cover-simple class $\mathcal{C}$ of $\mathfrak{S}$-frames is *linear* if there exists a $\mathcal{L}_{\mathfrak{S}}$-formula $\theta_{\mathcal{C}}(p)$ and a cover-universal frame $\mathcal{F}_{\mathcal{C}}$ for $\mathcal{C}$ such that $\leq_{\mathcal{F}_{\mathcal{C}}}$ is a total order and for all valuations V on $\mathcal{F}_{\mathcal{C}}$ and for all states $s \in \mathcal{F}_{\mathcal{C}}$, $(\mathcal{F}_{\mathcal{C}}, V), s \models \theta_{\mathcal{C}}(p)$ iff for all states $t \in \mathcal{F}_{\mathcal{C}}$, if $s \leq_{\mathcal{F}_{\mathcal{C}}} t$ then $(\mathcal{F}_{\mathcal{C}}, V), t \models p$; in other words, $\theta_{\mathcal{C}}(p)$ expresses that p is true in the generated submodel of s.

Lemma 6.3 *Suppose $\mathcal{C}$ is a linear class of $\mathfrak{S}$-frames and the $\mathfrak{S}$-logic λ is complete for $\mathcal{C}$. Then,*

(i) $\lambda \vdash \theta_{\mathcal{C}}(p) \to p$,

(ii) $\lambda \vdash \theta_{\mathcal{C}}(p) \to \theta_{\mathcal{C}}(\theta_{\mathcal{C}}(p))$,

(iii) $\lambda \vdash \widehat{\theta_{\mathcal{C}}}(p) \wedge \widehat{\theta_{\mathcal{C}}}(q) \to \widehat{\theta_{\mathcal{C}}}(p \wedge \widehat{\theta_{\mathcal{C}}}(q)) \vee \widehat{\theta_{\mathcal{C}}}(q \wedge \widehat{\theta_{\mathcal{C}}}(p))$.

Proof. It is a well-known fact that these formulas are valid on any total order $(T, \leq)$ when one interprets $\theta_{\mathcal{C}}(\cdot)$ in T as a modal connective with $\leq$ playing the role of its accessibility relation. Since $\leq_{\mathcal{F}_{\mathcal{C}}}$ is a total order, therefore these formulas are valid in $\mathcal{F}_{\mathcal{C}}$. Since $\mathcal{F}_{\mathcal{C}}$ is a cover-universal frame for $\mathcal{C}$ and λ is complete for $\mathcal{C}$, therefore these formulas are in λ. □

We will need a few extra axioms to axiomatize lexicographic logics over linear classes of frames.

Definition 6.4 Let $(\lambda \rhd \mathsf{LTL})_{\theta_{\mathcal{C}}}$ be the least lexicographic λ-logic containing the following formulas:

(LC$_1$) $\hat{\circ} p \wedge \hat{\circ} q \to \hat{\circ}(\widehat{\theta_{\mathcal{C}}}(p) \wedge \widehat{\theta_{\mathcal{C}}}(q))$

(LC$_2$) $\circ p \to \circ \theta_{\mathcal{C}}(p)$

(LC$_3$) $\circ p \to \theta_{\mathcal{C}}(\circ p)$

(LC$_3$) $\widehat{\theta_{\mathcal{C}}}(\circ p) \to \circ p$

(LC$_3$) $Gp \to \theta_{\mathcal{C}}(Gp)$

(LC$_3$) $\widehat{\theta_{\mathcal{C}}}(Gp) \to Gp$

Our goal for the remainder of the section is to show that, in many cases where $\mathcal{C}$ is a linear class of $\mathfrak{S}$-frames and the $\mathfrak{S}$-logic λ is complete for $\mathcal{C}$, $(\lambda \rhd \mathsf{LTL})_{\theta_{\mathcal{C}}}$ is complete with respect to $\mathcal{C}^{\rhd}$. We will obtain this using Theorem 5.5, and in the following lemmas, we establish that $(\lambda \rhd \mathsf{LTL})_{\theta_{\mathcal{C}}}$ has the required conditions to apply said theorem.

Lemma 6.5 *Let $\Phi \subseteq \Sigma$ be finite sets of $\mathcal{L}_{\mathfrak{S}}$-formulas. If $\mathcal{C}$ is closed under generated subframes and $\binom{\Sigma}{\Phi}_{\theta_{\mathcal{C}}}$ is λ-consistent then there exists a valuation V on $\mathcal{F}_{\mathcal{C}}$ such that $(\mathcal{F}_{\mathcal{C}}, V) \models \binom{\Sigma}{\Phi}_{\forall}$.*

Proof. Suppose $\mathcal{C}$ is closed under generated subframes and $\binom{\Sigma}{\Phi}_{\theta_{\mathcal{C}}}$ is λ-consistent. Since $\mathcal{F}_{\mathcal{C}}$ is a cover-universal frame for $\mathcal{C}$ and λ is complete for $\mathcal{C}$, there exists a valuation $V_{\mathcal{C}}$ on $\mathcal{F}_{\mathcal{C}}$ and there exists a state $s \in \mathcal{F}_{\mathcal{C}}$ such that $(\mathcal{F}_{\mathcal{C}}, V_{\mathcal{C}}), s \models \binom{\Sigma}{\Phi}_{\theta_{\mathcal{C}}}$. Let $\mathcal{F}_s$ be the subframe of $\mathcal{F}_{\mathcal{C}}$ generated from s. Obviously, $(\mathcal{F}_s, V_{\mathcal{C}}) \models \binom{\Sigma}{\Phi}_{\forall}$. Moreover, since $\mathcal{C}$ is closed under generated subframes, therefore $\mathcal{F}_s$ is in $\mathcal{C}$. Since $\mathcal{F}_{\mathcal{C}}$ is a cover-universal frame for $\mathcal{C}$ and $(\mathcal{F}_s, V_{\mathcal{C}}) \models \binom{\Sigma}{\Phi}_{\forall}$, therefore there exists a valuation V on $\mathcal{F}_{\mathcal{C}}$ such that $(\mathcal{F}_{\mathcal{C}}, V) \models \binom{\Sigma}{\Phi}_{\forall}$. □

Lemma 6.6 *Let Λ be a lexicographic λ-logic containing $(\lambda \rhd \mathsf{LTL})_{\theta_{\mathcal{C}}}$. Let $\Phi \subseteq \Sigma$ be finite sets of $\mathcal{L}_{\mathfrak{S}}^{\circ;G}$-formulas. Then:*

(i) *If Φ is consistent then so is $\Phi \cup \{\theta_{\mathcal{C}}(\psi) : \psi \in \Phi^{\circ} \cup \Phi^{G}\}$.*

(ii) *If $\binom{\Sigma}{\Phi}_{\circ}$ is Λ-consistent then $\binom{\Sigma}{\Phi}_{\theta_{\mathcal{C}}}$ is Λ-consistent.*

Proof. (i) Immediate using axioms (LC$_3$)-(LC$_6$).
(ii) Suppose $\binom{\Sigma}{\Phi}_{\circ}$ is Λ-consistent. Using Formulas (LC$_1$) and (LC$_2$), the reader may easily verify that $\Lambda \vdash \bigwedge_{\varphi \in \Phi} \hat{\circ} \varphi \to \hat{\circ} \bigwedge_{\varphi \in \Phi} \widehat{\theta_{\mathcal{C}}}(\varphi)$ and $\Lambda \vdash \bigwedge_{\varphi \in \Sigma \setminus \Phi} \circ \neg \varphi \to$

$\circ \bigwedge_{\varphi \in \Sigma \setminus \Phi} \theta_C(\neg\varphi)$. Hence, $\Lambda \vdash \binom{\Sigma}{\Phi}_\circ \to \hat{\circ}\binom{\Sigma}{\Phi}_{\theta_C}$. Since $\binom{\Sigma}{\Phi}_\circ$ is Λ-consistent, therefore $\hat{\circ}\binom{\Sigma}{\Phi}_{\theta_C}$ is Λ-consistent. Thus, $\binom{\Sigma}{\Phi}_{\theta_C}$ is Λ-consistent. □

Lemma 6.7 *If $\mathcal{C}$ is closed under generated subframes, λ is complete for $\mathcal{C}$, and Λ is a lexicographic λ-logic containing $(\lambda \triangleright \mathsf{LTL})_{\theta_C}$ then Λ is moment-complete.*

Proof. Suppose $\mathcal{C}$ is closed under generated subframes and Λ is a lexicographic λ-logic containing $(\lambda \triangleright \mathsf{LTL})_{\theta_C}$. Let Σ be a finite set of $\mathcal{L}_{\mathfrak{S}}^{\circ;G}$-formulas closed under subformulas and single negations and let $\Phi_0 \subseteq \Sigma$.

(i) Suppose $\Phi_0 \subseteq \Sigma$ is Λ-consistent; without loss of generality, we may assume it is maximal consistent. Then, by Lemma 6.6(i), $\Gamma = \Phi_0 \cup \{\theta_C(\psi) : \psi \in \Phi_0^\circ \cup \Phi_0^G\}$ is consistent, so that $\underline{\Gamma}$ is also λ-consistent. Thus there is a model $\mathcal{M}$ based on $\mathcal{F}_C$ and a state s of $\mathcal{M}$ such that $\mathcal{M}, s \models \bigwedge \underline{\Gamma}$, so that $\mathcal{M}_s, s \models \bigwedge \underline{\Gamma}$. But since $\mathcal{M}_s$ is generated by s, if $x\psi \in \Phi_0$ with $x \in \{\circ, G\}$, from $\mathcal{M}_s, s \models \theta_C(\underline{x\psi})$ we obtain $\mathcal{M}_s \models \underline{x\psi}$, and from $\mathcal{M}_s, s \models \theta_C(\neg \underline{x\psi})$ we obtain $\mathcal{M}_s \models \neg\underline{x\psi}$; since Φ_0 was maximal consistent one of these two occurs for each such $x\psi$ so $\mathcal{M}$ is a Σ-moment satisfying $\bigwedge \Phi_0$, as needed.

(ii) Suppose $\binom{\Sigma}{\Phi}_\circ$ is Λ-consistent. Hence, by Lemma 6.6(ii), $\binom{\Sigma}{\Phi}_{\theta_C}$ is Λ-consistent. By Lemma 6.5, since $\mathcal{C}$ is closed under generated subframes, therefore there exists a valuation V on $\mathcal{F}_C$ such that $(\mathcal{F}_C, V) \models \binom{\underline{\Sigma}}{\underline{\Phi}}_\forall$.

□

With this we obtain a general completeness result for linear classes of frames.

Corollary 6.8 *Let $\mathcal{C}$ be a linear class of $\mathfrak{S}$-frames and Λ be a lexicographic λ-logic containing $(\lambda \triangleright \mathsf{LTL})_{\theta_C}$. If $\mathcal{C}$ is closed under generated subframes and λ is complete for $\mathcal{C}$ then Λ is complete for $\mathcal{C}^\triangleright$.*

Proof. By Lemma 6.7 and Theorem 5.5. □

6.3 Global classes

A class $\mathcal{C}$ of $\mathfrak{S}$-frames is *global* if there exists a $\mathcal{L}_{\mathfrak{S}}$-formula $\mu_C(p)$ such that for all models $\mathcal{M}$ in Mod($\mathcal{C}$) and for all states s in $\mathcal{M}$, $\mathcal{M}, s \models \mu_C(p)$ iff for all states t in $\mathcal{M}$, $\mathcal{M}, t \models p$.

Our work on linear classes readily applies to global classes.

Lemma 6.9 *If $\mathcal{C}$ is cover-simple and global then $\mathcal{C}$ is linear.*

Proof. Suppose $\mathcal{C}$ is cover-simple and global. Let $\mathcal{F}$ be a cover-universal frame for $\mathcal{C}$ and $\mu_C(p)$ be a $\mathcal{L}_{\mathfrak{S}}$-formula such that for all models $\mathcal{M}$ in Mod($\mathcal{C}$) and for all states s in $\mathcal{M}$, $\mathcal{M}, s \models \mu_C(p)$ iff for all states t in $\mathcal{M}$, $\mathcal{M}, t \models p$. Obviously, the only generated subframe of $\mathcal{F}$ is $\mathcal{F}$ itself. Taking $\theta_C(p) = \mu_C(p)$, the reader may easily verify that $\mathcal{C}$ is linear. □

The main reason that global classes are particularly useful is that they allow us to define global covers directly within our language.

Lemma 6.10 *Let Φ, Σ be finite sets of $\mathcal{L}_{\mathfrak{S}}$-formulas such that $\Phi \subseteq \Sigma$. If $\mathcal{C}$ is global then $\mathcal{M} \models \binom{\Sigma}{\Phi}_{\forall}$ if and only if $\mathcal{M} \models \binom{\Sigma}{\Phi}_{\mu\mathcal{C}}$ for every model $\mathcal{M}$ based on any frame from $\mathcal{C}$.*

In view of this, the property of being simple becomes equivalent to that of being cover-simple over any global class of frames.

Lemma 6.11 *If $\mathcal{C}$ is simple and global then $\mathcal{C}$ is cover-simple.*

Proof. Suppose $\mathcal{C}$ is simple and global. Let $\mathcal{F}$ be a universal frame for $\mathcal{C}$. Let Φ, Σ be finite sets of $\mathcal{L}_{\mathfrak{S}}$-formulas such that $\Phi \subseteq \Sigma$. Let $\mathcal{M}$ be a model in $\mathrm{Mod}(\mathcal{C})$ such that $\mathcal{M} \models \binom{\Sigma}{\Phi}_{\forall}$. Hence, by Lemma 6.10, $\mathcal{M} \models \binom{\Sigma}{\Phi}_{\mu\mathcal{C}}$. Since $\mathcal{F}$ is a universal frame for $\mathcal{C}$, therefore let V be a valuation on $\mathcal{F}$ such that $(\mathcal{F}, V) \models \binom{\Sigma}{\Phi}_{\mu\mathcal{C}}$. Thus, by Lemma 6.10, $(\mathcal{F}, V) \models \binom{\Sigma}{\Phi}_{\forall}$. □

In fact, the above result applies to classes of frames that are *not* global, provided they can be made global by appropriately extending the signature.

Lemma 6.12 *Let $\mathfrak{S}'$ be a countable set of modalities containing $\mathfrak{S}$. Let $\mathcal{C}'$ be a simple and global class of $\mathfrak{S}'$-frames. If $\mathcal{C}$ is the class of all $\mathcal{F}'^{|\mathfrak{S}}$ where $\mathcal{F}'$ is in $\mathcal{C}'$ then $\mathcal{C}$ is cover-simple.*

Proof. Suppose $\mathcal{C}'$ is simple and global. Let $\mathcal{F}'$ be a universal frame for $\mathcal{C}'$ and $\mathcal{F} = \mathcal{F}'^{|\mathfrak{S}}$. Let Φ, Σ be finite sets of $\mathcal{L}_{\mathfrak{S}}$-formulas such that $\Phi \subseteq \Sigma$. Let $\mathcal{M}$ be a model in $\mathrm{Mod}(\mathcal{C})$ such that $\mathcal{M} \models \binom{\Sigma}{\Phi}_{\forall}$. Let $\mathcal{M}'$ be the corresponding model in $\mathrm{Mod}(\mathcal{C}')$. Since $\mathcal{M} \models \binom{\Sigma}{\Phi}_{\forall}$, therefore $\mathcal{M}' \models \binom{\Sigma}{\Phi}_{\forall}$. Since $\mathcal{C}'$ is global, therefore by Lemma 6.10, $\mathcal{M}' \models \binom{\Sigma}{\Phi}_{\mu\mathcal{C}}$. Since $\mathcal{F}'$ is a universal frame for $\mathcal{C}'$, therefore let V' be a valuation on $\mathcal{F}'$ such that $(\mathcal{F}', V') \models \binom{\Sigma}{\Phi}_{\mu\mathcal{C}}$. Thus, by Lemma 6.10, $(\mathcal{F}', V') \models \binom{\Sigma}{\Phi}_{\forall}$. Remark that V' is also a valuation on $\mathcal{F}$. Moreover, $(\mathcal{F}, V') \models \binom{\Sigma}{\Phi}_{\forall}$. It follows that $\mathcal{F}$ is a cover-universal frame for $\mathcal{C}$. Consequently, $\mathcal{C}$ is cover-simple. □

This will often mean that logics that are known to have a simple extension with a universal modality will be cover-simple. We will use this technique in the following section.

7 Completeness for specific logics

Now we proceed to show that many familiar classes of frames fall into one of the above frameworks, and use this to prove that several lexicographic logics are complete for their class of concrete frames.

7.1 Logics with tree-like models

We begin with unimodal logics containing the D axiom. Recall that KD is the least normal logic in a unimodal signature containing the formula $\Box p \rightarrow \Diamond p$. Then, T is the least extension of KD with the formula $\Box p \rightarrow p$, $\mathsf{KD4}$ is the least extension of KD with the formula $\Box p \rightarrow \Box\Box p$, $\mathsf{S4}$ is the least extension of KD with the formulas $\Box p \rightarrow p$ and $\Box p \rightarrow \Box\Box p$ and $\mathsf{S5}$ is the least extension of KD with the formulas $\Box p \rightarrow \Box\Box p$ and $p \rightarrow \Box\Diamond p$. Let $\mathcal{C}_{\mathsf{KD}}$ be the class of all serial frames, $\mathcal{C}_{\mathsf{T}}$ be the class of all reflexive frames, $\mathcal{C}_{\mathsf{KD4}}$ be the class of all serial, transitive frames, $\mathcal{C}_{\mathsf{S4}}$ be the class of all reflexive, transitive frames and $\mathcal{C}_{\mathsf{S5}}$ be the class of all serial, transitive, symmetric frames.

If our unimodal signature is extended by the universal modality U then for $\lambda \in \{\mathsf{KD}, \mathsf{T}, \mathsf{KD4}, \mathsf{S4}, \mathsf{S5}\}$, we let λU be the least normal logic in the extended signature containing the formulas $[U]p \to \Box p$, $[U]p \to p$, $[U]p \to [U][U]p$ and $p \to [U]\langle U \rangle p$. For $\lambda \in \{\mathsf{KD}, \mathsf{T}, \mathsf{KD4}, \mathsf{S4}, \mathsf{S5}\}$, let $\mathcal{C}_{\lambda U}$ be the class of all frames in $\mathcal{C}_\lambda$ extended by the universal accessibility relation. Obviously, $\mathcal{C}_{\lambda U}$ is global.

The following is well-known (see, e.g., [8]):

Theorem 7.1 *Let $\eta \in \{\mathsf{KD}, \mathsf{T}, \mathsf{KD4}, \mathsf{S4}, \mathsf{S5}\}$ and λ be either η or ηU. Then, the following are equivalent:*

(i) $\lambda \vdash \varphi$,

(ii) $\mathcal{C}_\lambda \models \varphi$,

(iii) $\mathcal{C}_\lambda^{\mathrm{fin}} \models \varphi$.

Let us show that all of these classes of frames are cover-simple. For this, define the *ω-forest* to be the set $\mathbb{N}^{[0,\omega)}$ of all sequences $\boldsymbol{a} = (a_0, \ldots, a_n)$ where $n \geq 0$ and each $a_i \in \mathbb{N}$ (note that the empty sequence is *not* allowed). Let R_{KD} be the binary relation on $\mathbb{N}^{[0,\omega)}$ such that $\boldsymbol{a} R_{\mathsf{KD}} \boldsymbol{b}$ iff there exists $n \geq 0$ such that $\boldsymbol{a} = (a_0, \ldots, a_n)$ and $\boldsymbol{b} = (a_0, \ldots, a_n, b)$ for some $a_0, \ldots, a_n, b \in \mathbb{N}$. Let R_{T} be the reflexive closure of R_{KD}, R_{KD4} be its transitive closure, R_{S4} be its reflexive, transitive closure and R_{S5} be its transitive, symmetric closure. For $\lambda \in \{\mathsf{KD}, \mathsf{T}, \mathsf{KD4}, \mathsf{S4}, \mathsf{S5}\}$, we let $\mathcal{F}_\lambda = (\mathbb{N}^{[0,\omega)}, R_\lambda)$ and $\mathcal{F}_{\lambda U}$ be the extension of $\mathcal{F}_\lambda$ by the universal accessibility relation.

Lemma 7.2 *Let λ be in $\{\mathsf{KD}, \mathsf{T}, \mathsf{KD4}, \mathsf{S4}, \mathsf{S5}\}$. Then,*

(i) *$\mathcal{F}_\lambda$ is a λ-frame,*

(ii) *$\mathcal{F}_{\lambda U}$ is a λU-frame.*

To show that $\mathcal{F}_\lambda$ is cover-universal, we will use well-known results on bounded morphisms, along with the following:

Lemma 7.3 *Let $\lambda \in \{\mathsf{KD}, \mathsf{T}, \mathsf{KD4}, \mathsf{S4}, \mathsf{S5}\}$ and $\mathcal{F}$ be any λU-frame. If $\mathcal{F}$ is finite then $\mathcal{F}$ is a bounded morphic image of $\mathcal{F}_{\lambda U}$.*

Proof. Suppose $\mathcal{F}$ is finite. Our aim is to construct a surjective bounded morphism $f \colon \mathbb{N}^{[0,\omega)} \to W$. We will define $f(\boldsymbol{a})$ by induction on the length of $\boldsymbol{a}$. For the base case, let $g \colon \mathbb{N} \to W$ be an arbitrary surjection. We define $f((a_0)) = g(a_0)$ for any $(a_0) \in \mathbb{N}^{[0,\omega)}$. Now, suppose that $f((a_0, \ldots, a_n)) = w$ is defined and let $h \colon \mathbb{N} \to R(w)$ be an arbitrary surjection. Then, for any $b \in \mathbb{N}$, set $f((a_0, \ldots, a_n, b)) = h(b)$. One can then check that for each of the listed λ, the map f thus defined is a bounded morphism. Surjectivity comes from the way we chose g. $\Box$

Lemma 7.4 *If $\lambda \in \{\mathsf{KD}, \mathsf{T}, \mathsf{KD4}, \mathsf{S4}, \mathsf{S5}\}$ then both $\mathcal{C}_\lambda$ and $\mathcal{C}_{\lambda U}$ are cover-simple.*

Proof. Let λ be in $\{\mathsf{KD}, \mathsf{T}, \mathsf{KD4}, \mathsf{S4}, \mathsf{S5}\}$. It is well-known that λU has the finite model property over $\mathrm{Mod}(\mathcal{C}_{\lambda U})$. Hence, by Lemma 7.3 and the fact that bounded morphic images preserve validity, we obtain that $\mathcal{F}_{\lambda U}$ is a universal frame for $\mathcal{C}_{\lambda U}$. Thus, $\mathcal{C}_{\lambda U}$ is simple. Since $\mathcal{C}_{\lambda U}$ is global, therefore by Lemmas 6.12 and 6.11, both $\mathcal{C}_{\lambda}$ and $\mathcal{C}_{\lambda U}$ are cover-simple. □

Corollary 7.5 *If $\lambda \in \{\mathsf{KD}, \mathsf{T}, \mathsf{KD4}, \mathsf{S4}, \mathsf{S5}\}$ then $(\lambda \rhd \mathsf{LTL})_0$ is complete for $\mathcal{C}^{\rhd}_{\lambda}$.*

Proof. Let λ be in $\{\mathsf{KD}, \mathsf{T}, \mathsf{KD4}, \mathsf{S4}, \mathsf{S5}\}$. By Lemma 7.4, $\mathcal{C}_{\lambda}$ is cover-simple. Since $\mathcal{C}_{\lambda}$ is closed under generated subframes and disjoint unions, therefore $\mathcal{C}_{\lambda}$ is local. In other respect, it is well-known that λ is strongly complete with respect to $\mathcal{C}_{\lambda}$. Since $\mathcal{C}_{\lambda}$ is cover-simple, therefore by Corollary 6.2, $(\lambda \rhd \mathsf{LTL})_0$ is complete for $\mathcal{C}^{\rhd}_{\lambda}$. □

7.2 Examples of linear logics

Next we turn to giving examples of logics with linear classes of frames. We begin by those that contain a universal modality. Let λ be in $\{\mathsf{KD}, \mathsf{T}, \mathsf{KD4}, \mathsf{S4}, \mathsf{S5}\}$. Obviously, $\leq_{\mathcal{F}_{\mathcal{C}_{\lambda U}}}$ is the universal relation on $\mathbb{N}^{[0,\omega)}$. Hence, $\leq_{\mathcal{F}_{\mathcal{C}_{\lambda U}}}$ is a total order on $\mathbb{N}^{[0,\omega)}$. Let $\theta_{\mathcal{C}_{\lambda U}}(p)$ be the formula $[U]p$.

Corollary 7.6 *If $\lambda \in \{\mathsf{KD}, \mathsf{T}, \mathsf{KD4}, \mathsf{S4}, \mathsf{S5}\}$ then $(\lambda \rhd \mathsf{LTL})_{\theta_{\mathcal{C}_{\lambda U}}}$ is complete for $\mathcal{C}^{\rhd}_{\lambda U}$.*

Proof. Let λ be in $\{\mathsf{KD}, \mathsf{T}, \mathsf{KD4}, \mathsf{S4}, \mathsf{S5}\}$. By Lemma 7.4, $\mathcal{C}_{\lambda U}$ is cover-simple. Since $\mathcal{C}_{\lambda U}$ is global, therefore by Lemma 6.9, $\mathcal{C}_{\lambda U}$ is linear. Since $\mathcal{C}_{\lambda U}$ is closed under generated subframes and λU is complete for $\mathcal{C}_{\lambda U}$, therefore by Corollary 6.8, $(\lambda \rhd \mathsf{LTL})_{\theta_{\mathcal{C}_{\lambda U}}}$ is complete for $\mathcal{C}^{\rhd}_{\lambda U}$. □

Finally, we conclude with examples of linear logics which are not global.

Lemma 7.7 *Let $\mathcal{C}$ be either:*

(i) *the class of all linear orders, with signature $\{\leq\}$,*

(ii) *the class of all linear orders without a maximal element, with signature $\{<\}$,*

(iii) *$\{\mathbb{N}\}$, with signature $\{S, <\}$ or $\{S, \leq\}$.*

Then, $\mathcal{C}$ is linear.

Proof. Note that in signatures with $<$, we may define $[\leq]p = p \wedge [<]p$. Now, in the first two cases, $[0, \infty) \cap \mathbb{Q}$ is readily checked to be a cover-universal frame, while in the third $\mathbb{N}$ is already the only allowed frame, hence it is cover-universal. Moreover, the class of all linear orders is closed under generated subframes, while $\mathbb{N}$ is isomorphic to its own generated subframes. □

Let $\mathsf{S4.3}$ be the ordinary normal modal logic determined by the class of all linear orders (reflexive, transitive and weakly-connected frames) and $\mathsf{KD4.3}$ be the ordinary normal modal logic determined by the class of all strict linear orders without a maximal element (serial, transitive and weakly-connected frames). See [13, Chapter 3].

Corollary 7.8 *Let* $\lambda \in \{\mathsf{S4.3}, \mathsf{KD4.3}, \mathsf{LTL}\}$ *and* $\mathcal{C}_\lambda$ *be the class of all linear orders, the class of all strict linear orders, or* $\{\mathbb{N}\}$*, respectively. Then,* $(\lambda \rhd \mathsf{LTL})_{\leq}$ *is complete for* $\mathcal{C}_\lambda \rhd \{\mathbb{N}\}$.

Proof. Immediate from Lemma 7.7 and Corollary 6.8. □

8 Conclusion

In this article, we have considered the lexicographic products of modal logics with linear temporal logic based on "next" and "always in the future". We have provided axiomatizations of the sets of valid formulas they give rise to. The proof of their completeness uses tools and techniques like universal frames, cover-universal frames, etc. Much remains to be done.

There is the issue of the axiomatization of the lexicographic products of modal logics with a linear temporal logic based on "until". We believe the tools and techniques that we have developed can be applied as well. Can they still be applied if one considers the lexicographic product of a linear temporal logic based on "until" with a linear temporal logic based on "until"? Considering a linear temporal logic based either on "always in the future" or on "until", this time interpreted over the class of all dense linear orders without endpoints, how to axiomatize its lexicographic products with modal logics? Is it possible to obtain in our lexicographic setting complete axiomatizations by following the line of reasoning suggested by [4]?

There is also the question of the complexity of the temporal logic characterized by the lexicographic products of modal logics with linear temporal logic. Is it possible to obtain in our lexicographic setting complexity results by following either of the lines of reasoning suggested by [5] or [20]?

Acknowledgements

Special acknowledgement is heartily granted to the anonymous referees for their feedback on the submitted version of our paper.

References

[1] Babenyshev, S., Rybakov, V. *Logics of Kripke meta-models.* Logic Journal of the IGPL **18** (2010) 823–836.

[2] Balbiani, P. Time representation and temporal reasoning from the perspective of non-standard analysis. In Brewka, G., Lang, J. (editors): Eleventh International Conference on Principles of Knowledge Representation and Reasoning. AAAI (2008) 695–704.

[3] Balbiani, P. Axiomatization and completeness of lexicographic products of modal logics. In Ghilardi, S., Sebastiani, R. (editors): Frontiers of Combining Systems. Springer (2009) 165–180.

[4] Balbiani, P. Axiomatizing the temporal logic defined over the class of all lexicographic products of dense linear orders without endpoints In Markey, N., Wijsen, J. (editors): Temporal Representation and Reasoning. IEEE (2010) 19–26.

[5] Balbiani, P., Mikulás, S. Decidability and complexity via mosaics of the temporal logic of the lexicographic products of unbounded dense linear orders. In Fontaine, P., Ringeissen, C., Schmidt, R. (editors): Frontiers of Combining Systems. Springer (2013) 151–164.

[6] Balbiani, P., Shapirovsky, I., Shehtman, V. Complete axiomatizations of lexicographic sums and products of modal logics. To appear.
[7] Beklemishev, L. *Kripke semantics for provability logic GLP*. Annals of Pure and Applied Logic **161** (2010) 756–774.
[8] Blackburn, P., de Rijke, M., Venema, Y. Modal Logic. Cambridge University Press (2001).
[9] Van Ditmarsch, H., van der Hoek, W., Kooi, B. Dynamic Epistemic Logic. Springer (2007).
[10] Fagin, R., Halpern, J., Moses, Y., Vardi, M. Reasoning About Knowledge. MIT Press (1995).
[11] Gabbay, D., Kurucz, A., Wolter, F., Zakharyaschev, M. Many-Dimensional Modal Logics: Theory and Applications. Elsevier (2003).
[12] Gabbay, D., Shehtman, V. Products of modal logics, part 1. Logic Journal of the IGPL **6** (1998) 73–146.
[13] Goldblatt, R. *Logics of Time and Computation*. CSLI (1992).
[14] Kracht, M., Wolter, F. Properties of independently axiomatizable bimodal logics. Journal of Symbolic Logic **56** (1991) 1469–1485.
[15] Kurucz, A. Combining modal logics. In Blackburn, P., van Benthem, J., Wolter, F. (editors): Handbook of Modal Logic. Elsevier (2007) 869–924.
[16] Litak, T., Wolter, F. All finitely axiomatizable tense logics of linear time flows are CoNP-complete. Studia Logica **81** (2005) 153–165.
[17] Marx, M., Mikulás, S., Reynolds, M. The mosaic method for temporal logics. In Dyckhoff, R. (editor): Automated Reasoning with Analytic Tableaux and Related Methods. Springer (2000) 324–340.
[18] Moss, L. S. Coalgebraic logic. Annals of Pure and Applied Logic **96** (1999) 277–317.
[19] Shapirovsky, I. *PSPACE-decidability of Japaridze's polymodal logic*. In Areces, C., Goldblatt, R. (editors): Advances in Modal Logic, Volume 7. College Publications (2008) 289–304.
[20] Sistla, A., Clarke, E.: *The complexity of propositional linear temporal logics*. Journal of the Association for Computing Machinery **32** (1985) 733–749.
[21] Wolter, F. Fusions of modal logics revisited. In Kracht, M., de Rijke, M., Wansing, H., Zakharyaschev, M. (editors): Advances in Modal Logic. CSLI Publications (1998) 361–379.

About intuitionistic public announcement logic

Philippe Balbiani [1]

Institut de recherche en informatique de Toulouse
Toulouse University

Didier Galmiche [2]

Laboratoire lorrain de recherche en informatique et ses applications
CNRS — Université de Lorraine — Inria

Abstract

Public announcement logic (PAL) is a logic for reasoning about the dynamic of knowledge in a multi-agent system in which public announcements are made. Syntactically, public announcements are modal formulas. Semantically, they correspond to restrictions of models. In [10], Ma *et al.* use the standard toolkit of duality theory in modal logic to define an algebraic semantics for a combination of IPL and PAL into intuitionistic public announcement logic ($IPAL$). In this paper, grounding our approach on relational semantics rather than on algebraic semantics, we give a sound and complete axiomatization of $IPAL$ and we consider a complete sequent calculus for the associated membership problem.

Keywords: Public announcement logic. Intuitionistic propositional logic. Axiomatization/completeness. Decidability/complexity. Sequent calculus.

1 Introduction

Public announcement logic (PAL) is a logic for reasoning about the dynamic of knowledge in a multi-agent system [16]. Syntactically, public announcements are modal formulas. Semantically, they correspond to restrictions of models. There exist multifarious variants of PAL: PAL with arbitrary public announcements [1], PAL with common knowledge [7], etc. In all these variants, the construct $(\cdot \rightarrow \cdot)$ is the one of classical propositional logic. In [10], Ma *et al.* introduce a variant of PAL in which this construct is the one of intuitionistic propositional logic (IPL). By using the standard toolkit of duality theory in modal logic, they define an algebraic semantics for a combination of IPL and PAL into intuitionistic public announcement logic ($IPAL$). In

[1] IRIT, Toulouse University, 118 route de Narbonne, 31062 Toulouse Cedex 9, FRANCE; Philippe.Balbiani@irit.fr.

[2] LORIA, CNRS — Université de Lorraine — Inria, Campus scientifique, BP 239, 54506 Vandœuvre-lès-Nancy Cedex, FRANCE; Didier.Galmiche@loria.fr.

this paper, grounding our approach on relational semantics rather than on algebraic semantics, we provide supplementary results about $IPAL$. Firstly, we give a sound and complete axiomatization of $IPAL$ and we prove its completeness. Secondly, we study the features that, according to Simpson [17], might be expected of any intuitionistic modal logic and we examine whether $IPAL$ possesses them. Thirdly, we propose an alternative semantics for $IPAL$, dealing with stacks of annoucements, following the approach developed in [2] for PAL, and then we derive from this semantics a new sequent calculus for $IPAL$ that is sound and complete. Fourthly, we define a translation of $IPAL$'s formulas into formulas of a multimodal logic in which the construct $(\cdot \rightarrow \cdot)$ is the one of classical propositional logic.

2 Syntax and semantics

Let VAR be a countable set of atomic formulas called variables (denoted p, q, etc). The set of all formulas is inductively defined as follows:

- $\phi ::= p \mid \bot \mid (\phi \vee \psi) \mid (\phi \wedge \psi) \mid (\phi \rightarrow \psi) \mid \Box\phi \mid \Diamond\phi \mid [\phi]\psi \mid \langle\phi\rangle\psi$.

$\bot$, $(\cdot \vee \cdot)$, $(\cdot \wedge \cdot)$ and $(\cdot \rightarrow \cdot)$ are the ordinary constructs of IPL, $\Box\cdot$ ("it is necessary that ...") and $\Diamond\cdot$ ("it is possible that ...") are the alethic constructs of modal logic and $[\cdot]\cdot$ ("if ... then, after announcing it, ...") and $\langle\cdot\rangle\cdot$ ("... and, after announcing it, ...") are the announcement constructs of PAL. The IPL constructs $\neg\cdot$ and $(\cdot \leftrightarrow \cdot)$ are defined as usual.

- $\neg\phi ::= (\phi \rightarrow \bot)$,
- $(\phi \leftrightarrow \psi) ::= ((\phi \rightarrow \psi) \wedge (\psi \rightarrow \phi))$.

We adopt the standard rules for omission of the parentheses. Note that, following the line of reasoning suggested by [17, Chapter 3], we have added the new alethic constructs $\Box\cdot$ and $\Diamond\cdot$ and the new announcement constructs $[\cdot]\cdot$ and $\langle\cdot\rangle\cdot$ to the ordinary language of IPL. As proved in Section 7 (see Propositions 7.4 and 7.5), the constructs $\Box\cdot$ and $\Diamond\cdot$ are independent in $IPAL$ but $[\cdot]\cdot$ and $\langle\cdot\rangle\cdot$ are interdefinable. For all formulas ϕ, let ϕ^* be the formula obtained by recursively eliminating the alethic constructs and the announcement constructs occurring in ϕ. For all sets x of formulas, let $\Box x = \{\phi\colon \Box\phi \in x\}$ and $\Diamond x = \{\Diamond\phi\colon \phi \in x\}$. Let the size of a formula ϕ (denoted $size(\phi)$) be the number of occurrences of symbols ϕ contains. The size of a finite sequence $(\phi_1, \ldots, \phi_n)$ of formulas (denoted $size(\phi_1, \ldots, \phi_n)$) is the nonnegative integer defined as follows:

- $size(\phi_1, \ldots, \phi_n) = size(\phi_1) + \ldots + size(\phi_n) + n$.

By ϵ, we will denote the empty sequence of formulas. Obviously, $size(\epsilon) = 0$. A frame is a tuple of the form $\mathcal{F} = (W, \leq, R)$ where W is a nonempty set (denoted x, y, etc), $\leq$ is a partial order on W and R is a binary relation on W. The frame $\mathcal{F} = (W, \leq, R)$ is said to be standard if

- $R^{-1} \circ \leq \subseteq \leq \circ R^{-1}$,
- $R \circ \leq \subseteq \leq \circ R$.

A valuation on a frame $\mathcal{F} = (W, \leq, R)$ is a function $V\colon\ VAR \mapsto 2^W$. The valuation V on the frame $\mathcal{F} = (W, \leq, R)$ is said to be upward closed if

- for all $p \in VAR$ and for all $x \in W$, if $x \in V(p)$ then for all $y \in W$, if $x \leq y$ then $y \in V(p)$.

A model is a tuple of the form $\mathcal{M} = (W, \leq, R, V)$ where $\mathcal{F} = (W, \leq, R)$ is a frame and V is a valuation on $\mathcal{F}$. We shall say that the model $\mathcal{M} = (W, \leq, R, V)$ is standard if the frame $\mathcal{F} = (W, \leq, R)$ is standard. The model $\mathcal{M} = (W, \leq, R, V)$ is said to be upward closed if the valuation V on the frame $\mathcal{F} = (W, \leq, R)$ is upward closed. The satisfiability relation between a model $\mathcal{M} = (W, \leq, R, V)$, an element $x \in W$ and a formula ϕ (denoted $\mathcal{M}, x \models \phi$) is inductively defined as follows:

- $\mathcal{M}, x \models p$ iff $x \in V(p)$,
- $\mathcal{M}, x \not\models \bot$,
- $\mathcal{M}, x \models \phi \vee \psi$ iff either $\mathcal{M}, x \models \phi$, or $\mathcal{M}, x \models \psi$,
- $\mathcal{M}, x \models \phi \wedge \psi$ iff $\mathcal{M}, x \models \phi$ and $\mathcal{M}, x \models \psi$,
- $\mathcal{M}, x \models \phi \rightarrow \psi$ iff for all $y \in W$, if $x \leq y$ and $\mathcal{M}, y \models \phi$ then $\mathcal{M}, y \models \psi$,
- $\mathcal{M}, x \models \Box\phi$ iff for all $y, z \in W$, if $x \leq y$ and yRz then $\mathcal{M}, z \models \phi$,
- $\mathcal{M}, x \models \Diamond\phi$ iff there exists $y \in W$ such that xRy and $\mathcal{M}, y \models \phi$,
- $\mathcal{M}, x \models [\phi]\psi$ iff for all $y \in W$, if $x \leq y$ and $\mathcal{M}, y \models \phi$ then $\mathcal{M}_{|\phi}, y \models \psi$,
- $\mathcal{M}, x \models \langle\phi\rangle\psi$ iff $\mathcal{M}, x \models \phi$ and $\mathcal{M}_{|\phi}, x \models \psi$.

In the above definition, $\mathcal{M}_{|\phi} = (W_{|\phi}, \leq_{|\phi}, R_{|\phi}, V_{|\phi})$ is the model such that $W_{|\phi} = \{x \in W\colon\ \mathcal{M}, x \models \phi\}$, $\leq_{|\phi} = \leq \cap (W_{|\phi} \times W_{|\phi})$, $R_{|\phi} = R \cap (W_{|\phi} \times W_{|\phi})$ and for all $p \in VAR$, $V_{|\phi}(p) = V(p) \cap W_{|\phi}$. Notice that the clauses concerning the modal constructs $\Box\cdot$ and $[\cdot]\cdot$ imitate the clauses for the quantifier $\forall$ in first-order intuitionistic logic whereas the clauses concerning $\Diamond\cdot$ and $\langle\cdot\rangle\cdot$ imitate the clauses for $\exists$. See [6, Lemma 5.3.2] for details. Obviously, in any model $\mathcal{M} = (W, \leq, R, V)$,

- $\mathcal{M}, x \models \neg\phi$ iff for all $y \in W$, if $x \leq y$ then $\mathcal{M}, y \not\models \phi$,
- $\mathcal{M}, x \models \phi \leftrightarrow \psi$ iff for all $y \in W$, if $x \leq y$ then $\mathcal{M}, y \models \phi$ iff $\mathcal{M}, y \models \psi$.

Note that if $\mathcal{M}$ is upward closed then $\mathcal{M}_{|\phi}$ is upward closed too. The next lemma states that the set of elements satisfying a formula in an upward closed standard model is upward closed too.

Lemma 2.1 *Let ϕ be a formula. For all upward closed standard models $\mathcal{M} = (W, \leq, R, V)$ and for all $x \in W$, if $\mathcal{M}, x \models \phi$ then $\mathcal{M}_{|\phi}$ is upward closed standard and for all $y \in W$, if $x \leq y$ then $\mathcal{M}, y \models \phi$.*

A formula ϕ is said to be globally satisfied in a model $\mathcal{M} = (W, \leq, R, V)$ (denoted $\mathcal{M} \models \phi$) if for all $x \in W$, $\mathcal{M}, x \models \phi$. The following Lemma will be used in Section 7.

Lemma 2.2 *Let ϕ be a formula. Let $\mathcal{M} = (W, \leq, R, V)$ be a model such that*

$\leq$ is the identity relation on W. If $\phi \in PAL$ then $\mathcal{M} \models \phi$.

There are several reasons for being interested in upward closed standard models. Following the usual paradigm for IPL saying that facts should persist in a model as we ascend its partial order, the fact that xRy in a model $\mathcal{M} = (W, \leq, R, V)$ should persist too. Hence, the condition of being standard. Similarly, the fact that $x \in V(p)$ in a model $\mathcal{M} = (W, \leq, R, V)$ should persist too. Thus, the condition of being upward closed.

3 Validities

We shall say that a formula ϕ is ucs-valid (denoted $\models_{ucs} \phi$) if for all upward closed standard models $\mathcal{M}$, $\mathcal{M} \models \phi$.

Proposition 3.1 *The following formulas are ucs-valid and the following inference rules are ucs-validity preserving:*

A1 *All instances of IPL*,

A2 $\Box(\phi \to \psi) \to (\Box\phi \to \Box\psi)$,

A3 $\Box(\phi \to \psi) \to (\Diamond\phi \to \Diamond\psi)$,

A4 $(\Diamond\phi \to \Box\psi) \to \Box(\phi \to \psi)$,

A5 $\Diamond(\phi \vee \psi) \to (\Diamond\phi \vee \Diamond\psi)$,

A6 $\neg\Diamond\bot$,

A7 $[\phi]p \leftrightarrow (\phi \to p)$,

A8 $[\phi]\bot \leftrightarrow \neg\phi$,

A9 $[\phi](\psi \vee \chi) \leftrightarrow (\phi \to ([\phi]\psi \vee [\phi]\chi))$,

A10 $[\phi](\psi \wedge \chi) \leftrightarrow ([\phi]\psi \wedge [\phi]\chi)$,

A11 $[\phi](\psi \to \chi) \leftrightarrow ([\phi]\psi \to [\phi]\chi)$,

A12 $[\phi]\Box\psi \leftrightarrow (\phi \to \Box[\phi]\psi)$,

A13 $[\phi]\Diamond\psi \leftrightarrow (\phi \to \Diamond\langle\phi\rangle\psi)$,

A14 $\langle\phi\rangle\psi \leftrightarrow (\phi \wedge [\phi]\psi)$,

R1 *from* ϕ *and* $\phi \to \psi$ *infer* ψ,

R2 *from* ϕ *infer* $\Box\phi$,

R3 *from* $\phi \leftrightarrow \psi$ *infer* $[\chi]\phi \leftrightarrow [\chi]\psi$.

Proof. When restricted to announcement-free formulas, the formulas $A1$–$A6$ and the inference rules $R1$ and $R2$ have been used by Fischer Servi [8] and Simpson [17, Chapter 3] who have considered the intuitionistic analogue IK of modal logic K. The formulas $A7$, $A8$, $A10$, $A12$ and $A13$ have been used by Ma *et al.* [10] as reduction axioms. Hence, leaving to the reader the proof of the proposition for the formulas $A9$ and $A11$ and the inference rule $R3$, we only prove the proposition for the formula $A14$.
Suppose $\not\models_{ucs} \langle\phi\rangle\psi \leftrightarrow (\phi \wedge [\phi]\psi)$. Let $\mathcal{M} = (W, \leq, R, V)$ be an upward closed standard model and $x \in W$ be such that $\mathcal{M}, x \not\models \langle\phi\rangle\psi \leftrightarrow (\phi \wedge [\phi]\psi)$. Hence, either $\mathcal{M}, x \not\models \langle\phi\rangle\psi \to (\phi \wedge [\phi]\psi)$, or $\mathcal{M}, x \not\models (\phi \wedge [\phi]\psi) \to \langle\phi\rangle\psi$. In the former case, let $y \in W$ be such that $x \leq y$, $\mathcal{M}, y \models \langle\phi\rangle\psi$ and $\mathcal{M}, y \not\models \phi \wedge [\phi]\psi$. Thus, $\mathcal{M}, y \models \phi$, $\mathcal{M}_{|\phi}, y \models \psi$ and $\mathcal{M}, y \not\models [\phi]\psi$. Let $z \in W$ be such that $y \leq z$, $\mathcal{M}, z \models \phi$ and $\mathcal{M}_{|\phi}, z \not\models \psi$. Since $\mathcal{M}, y \models \phi$, therefore $y \leq_{|\phi} z$. Since $\mathcal{M}_{|\phi}, y \models \psi$, therefore by Lemma 2.1, $\mathcal{M}_{|\phi}, z \models \psi$: a contradiction. In the latter case, let $y \in W$ be such that $x \leq y$, $\mathcal{M}, y \models \phi \wedge [\phi]\psi$ and $\mathcal{M}, y \not\models \langle\phi\rangle\psi$. Consequently, $\mathcal{M}, y \models \phi$, $\mathcal{M}, y \models [\phi]\psi$ and $\mathcal{M}_{|\phi}, y \not\models \psi$. Hence, $\mathcal{M}_{|\phi}, y \models \psi$: a contradiction. Thus, $\models_{ucs} \langle\phi\rangle\psi \leftrightarrow (\phi \wedge [\phi]\psi)$. $\Box$

Proposition 3.2 *The following formulas are ucs-valid and the following inference rule is ucs-validity preserving:*

$A15$ $[\phi](\psi \to \chi) \to ([\phi]\psi \to [\phi]\chi)$,

$A16$ $[\phi](\psi \to \chi) \to (\langle\phi\rangle\psi \to \langle\phi\rangle\chi)$,

$A17$ $(\langle\phi\rangle\psi \to [\phi]\chi) \to [\phi](\psi \to \chi)$,

$A18$ $\langle\phi\rangle(\psi \vee \chi) \to (\langle\phi\rangle\psi \vee \langle\phi\rangle\chi)$,

$R4$ *from* ϕ *infer* $[\psi]\phi$.

Proof. Left to the reader. □

Proposition 3.3 *The following formulas are ucs-valid:*

$A19$ $[\phi]\top \leftrightarrow \top$,

$A20$ $\langle\phi\rangle\bot \leftrightarrow \bot$,

$A21$ $\langle\phi\rangle\top \leftrightarrow \phi$,

$A22$ $[\phi](\psi\vee\chi) \leftrightarrow (\phi \to \langle\phi\rangle\psi\vee\langle\phi\rangle\chi)$,

$A23$ $\langle\phi\rangle(\psi \vee \chi) \leftrightarrow \langle\phi\rangle\psi \vee \langle\phi\rangle\chi$,

$A24$ $[\phi](\psi \to \chi) \leftrightarrow (\langle\phi\rangle\psi \to \langle\phi\rangle\chi)$,

$A25$ $\langle\phi\rangle(\psi \to \chi) \leftrightarrow \phi \wedge (\langle\phi\rangle\psi \to \langle\phi\rangle\chi)$,

$A26$ $\langle\phi\rangle\Diamond\psi \leftrightarrow \phi \wedge \Diamond\langle\phi\rangle\psi$,

$A27$ $\langle\phi\rangle p \leftrightarrow \phi \wedge p$,

$A28$ $\langle\phi\rangle(\psi \wedge \chi) \leftrightarrow \langle\phi\rangle\psi \wedge \langle\phi\rangle\chi$,

$A29$ $\langle\phi\rangle\Box\psi \leftrightarrow \phi \wedge \Box[\phi]\psi$.

Proof. Left to the reader. □

Note that the set of all ucs-valid formulas is not closed under the inference rule of uniform substitution. For example, the formula $[p]p$ is ucs-valid but its instance $[q \wedge \Diamond\neg q](q \wedge \Diamond\neg q)$ is not globally satisfied in the upward closed standard model $\mathcal{M} = (W, \leq, R, V)$ where $W = \{x, y\}$, $\leq = \{(x, x), (y, y)\}$, $R = \{(x, y)\}$ and $V(q) = \{x\}$. Hence, we should be very careful when applying to $IPAL$ tools and techniques designed for normal modal logic.

4 Axiomatization/completeness

Let $IPAL$ be the least set of formulas containing the formulas $A1$–$A14$ and closed under the inference rules $R1$–$R3$. The soundness of $IPAL$ relative to its relational semantics is straightforward, seeing that

Proposition 4.1 (Soundness) *Let* ϕ *be a formula. If* $\phi \in IPAL$ *then* $\models_{ucs} \phi$.

Proof. By Proposition 3.1. □

Without using the standard toolkit of duality theory in modal logic and the results in [10], the completeness of $IPAL$ relative to its relational semantics is more difficult to establish than its soundness and we defer proving that $IPAL$ is complete with respect to the class of all upward closed standard models till the end of this section. A useful result is the following

Proposition 4.2 *Let* ϕ *be a formula and* ψ *be an announcement-free formula such that* $\phi \leftrightarrow \psi \in IPAL$. *Let* χ *be an announcement-free formula. There exists an announcement-free formula* θ *such that* $[\phi]\chi \leftrightarrow \theta \in IPAL$. *Moreover, if* ψ *and* χ *are* $\Box$*-free (respectively,* $\Diamond$*-free) then* θ *is* $\Box$*-free (respectively,* $\Diamond$*-free).*

Proof. Let FOR be the set of all announcement-free formulas χ such that there exists an announcement-free formula θ such that $[\phi]\chi \leftrightarrow \theta \in IPAL$ and, moreover, if ψ and χ are $\Box$-free (respectively, $\Diamond$-free) then θ is $\Box$-free (respectively, $\Diamond$-free). Proposition 4.2 says that for all announcement-free formulas χ, $\chi \in FOR$. We will demonstrate it by an induction on χ based on the function $size(\cdot)$ defined in Section 2. Let χ be an announcement-free formula such that for all announcement-free formulas μ, if $size(\mu) < size(\chi)$ then $\mu \in FOR$. We demonstrate $\chi \in FOR$. We only consider the case $\chi = \Diamond\mu$.
Note that $size(\mu) < size(\chi)$. Hence, $\mu \in FOR$. Let θ be an announcement-free formula such that $[\phi]\mu \leftrightarrow \theta \in IPAL$. By $A13$, $[\phi]\Diamond\mu \leftrightarrow (\phi \rightarrow \Diamond\langle\phi\rangle\mu) \in IPAL$. Since $\phi \leftrightarrow \psi \in IPAL$, therefore $[\phi]\Diamond\mu \leftrightarrow (\psi \rightarrow \Diamond\langle\phi\rangle\mu) \in IPAL$. By $A14$, $\langle\phi\rangle\mu \leftrightarrow (\phi \wedge [\phi]\mu) \in IPAL$. Since $\phi \leftrightarrow \psi \in IPAL$ and $[\phi]\mu \leftrightarrow \theta \in IPAL$, therefore $\langle\phi\rangle\mu \leftrightarrow (\psi \wedge \theta) \in IPAL$. Thus, $\Diamond\langle\phi\rangle\mu \leftrightarrow \Diamond(\psi \wedge \theta) \in IPAL$. Since $[\phi]\Diamond\mu \leftrightarrow (\psi \rightarrow \Diamond\langle\phi\rangle\mu) \in IPAL$, therefore $[\phi]\Diamond\mu \leftrightarrow (\psi \rightarrow \Diamond(\psi\wedge\theta)) \in IPAL$.□

From Proposition 4.2, it follows that

Proposition 4.3 *For all formulas ϕ, there exists an announcement-free formula ψ such that $\phi \leftrightarrow \psi \in IPAL$. Moreover, if ϕ is $\Box$-free (respectively, $\Diamond$-free) then ψ is $\Box$-free (respectively, $\Diamond$-free).*

Proof. Let FOR be the set of all formulas ϕ such that there exists an announcement-free formula ψ such that $\phi \leftrightarrow \psi \in IPAL$ and, moreover, if ϕ is $\Box$-free (respectively, $\Diamond$-free) then ψ is $\Box$-free (respectively, $\Diamond$-free). Proposition 4.3 says that for all formulas ϕ, $\phi \in FOR$. We will demonstrate it by an induction on ϕ based on the function $size(\cdot)$ defined in Section 2. Let ϕ be a formula such that for all announcement-free formulas ψ, if $size(\psi) < size(\phi)$ then $\psi \in FOR$. We demonstrate $\phi \in FOR$. We only consider the case $\phi = [\psi]\chi$.
Note that $size(\psi) < size(\phi)$ and $size(\chi) < size(\phi)$. Hence, $\psi \in FOR$ and $\chi \in FOR$. Let θ be an announcement-free formula such that $\psi \leftrightarrow \theta \in IPAL$ and μ be an announcement-free formula such that $\chi \leftrightarrow \mu \in IPAL$. By $R3$, $[\psi]\chi \leftrightarrow [\psi]\mu \in IPAL$. Let ν be an announcement-free formula such that $[\psi]\mu \leftrightarrow \nu \in IPAL$. Such ν exists by Proposition 4.2 because $\psi \leftrightarrow \theta \in IPAL$. Since $[\psi]\chi \leftrightarrow [\psi]\mu \in IPAL$, therefore $[\psi]\chi \leftrightarrow \nu \in IPAL$. □

Now, we are ready for the proof of the completeness of $IPAL$ relative to its relational semantics.

Proposition 4.4 (Completeness) *Let ϕ be a formula. If $\models_{ucs} \phi$ then $\phi \in IPAL$.*

Proof. Suppose $\models_{ucs} \phi$ and $\phi \notin IPAL$. Let ψ be an announcement-free formula such that $\phi \leftrightarrow \psi \in IPAL$. Such formula exists by Proposition 4.3. Since $\phi \notin IPAL$, therefore $\psi \notin IPAL$. By the Canonical Model Construction described in [17, Chapter 3], $\not\models_{ucs} \psi$. Since $\phi \leftrightarrow \psi \in IPAL$, therefore by Proposition 4.1, $\models_{ucs} \phi \leftrightarrow \psi$. Since $\not\models_{ucs} \psi$, therefore $\not\models_{ucs} \phi$: a contradiction. Hence, if $\models_{ucs} \phi$ then $\phi \in IPAL$. □

In the definition of $IPAL$, we did not use the formulas $A15$–$A29$ and the inference rule $R4$ considered in Propositions 3.2 and 3.3. Why not? The reason is that neither the formulas $A15$–$A29$ nor the inference rule $R4$ are used in the proof of Propositions 4.2 and 4.3. Moreover,

Proposition 4.5 *The formulas $A15$–$A29$ are in $IPAL$ and the inference rule $R4$ is admissible in $IPAL$.*

Proof. By Propositions 3.2, 3.3, 4.1 and 4.4. □

5 Canonical model

Let L be an extension of $IPAL$, i.e. L is a set of formulas containing the formulas $A1$–$A14$ and closed under the inference rules $R1$–$R3$. For all sets x, y of formulas, y is said to be an L-consequence of x (denoted $x \vdash_L y$) if there exists nonnegative integers m, n and there exists formulas $\phi_1, \ldots, \phi_m, \psi_1, \ldots, \psi_n$ such that $\phi_1, \ldots, \phi_m \in x$, $\psi_1, \ldots, \psi_n \in y$ and $\phi_1 \wedge \ldots \wedge \phi_m \rightarrow \psi_1 \vee \ldots \psi_n \in L$. In this definition, if $m = 0$ then we will consider that $\phi_1 \wedge \ldots \wedge \phi_m$ is equal to $\top$ and if $n = 0$ then we will consider that $\psi_1 \vee \ldots \psi_n$ is equal to $\bot$. In the sequel, we will always assume that $\emptyset \not\vdash_L \emptyset$, i.e. we will always assume that $\top \rightarrow \bot \notin L$. We shall say that a set x of formulas is L-prime if the following conditions hold:

- for all formulas ϕ, if $x \vdash_L \{\phi\}$ then $\phi \in x$,
- $x \not\vdash_L \{\bot\}$,
- for all formulas ϕ, ψ, if $\phi \vee \psi \in x$ then either $\phi \in x$, or $\psi \in x$.

Lemma 5.1 (Prime Lemma) *For all sets x, y of formulas, if $x \not\vdash_L y$ then there exists an L-prime set x' of formulas such that $x \subseteq x'$ and $x' \not\vdash_L y$.*

Since $\emptyset \not\vdash_L \emptyset$, therefore the set of all L-prime sets of formulas is nonempty. L's Canonical Model is the tuple $\mathcal{M}_c = (W_c, \leq_c, R_c, V_c)$ where W_c is the set of all L-prime sets of formulas, $\leq_c$ is the partial order on W_c defined by $x \leq_c y$ iff $x \subseteq y$, R_c is the binary relation on W_c defined by xR_cy iff $\Box x \subseteq y$ and $\Diamond y \subseteq x$ and V_c: $VAR \mapsto 2^{W_c}$ is the function defined by $x \in V_c(p)$ iff $p \in x$.

Lemma 5.2 *The model $\mathcal{M}_c$ is upward closed standard.*

Lemma 5.3 (Restricted Truth Lemma) *Let ϕ be an announcement-free formula. For all L-prime sets x of formulas, the following conditions are equivalent: (i) $\mathcal{M}_c, x \models \phi$, (ii) $\phi \in x$.*

Lemma 5.4 (Truth Lemma) *Let ϕ be a formula. For all L-prime sets x of formulas, the following conditions are equivalent: (i) $\mathcal{M}_c, x \models \phi$, (ii) $\phi \in x$.*

In Section 7, we will consider an extension of $IPAL$ that contains all formulas of the form $\phi \vee \neg\phi$.

Proposition 5.5 *Let L be an extension of $IPAL$ that contains all formulas of the form $\phi \vee \neg\phi$. For all L-primes sets x, y of formulas, if $x \subseteq y$ then $x = y$.*

Proof. Let x, y be L-primes sets of formulas. Suppose $x \subseteq y$ and $x \neq y$. Hence, $y \not\subseteq x$. Let ψ be a formula such that $\psi \in y$ and $\psi \notin x$. Since L is an extension of $IPAL$ that contains all formulas of the form $\phi \vee \neg\phi$, therefore $\psi \vee \neg\psi \in x$. Thus, either $\psi \in x$, or $\neg\psi \in x$. Since $\psi \notin x$, therefore

$\neg\psi \in x$. Since $x \subseteq y$, therefore $\neg\psi \in y$. Since $\psi \in y$, therefore $y \vdash_L \{\bot\}$: a contradiction. Consequently, if $x \subseteq y$ then $x = y$. □

6 Relationship with Ma *et al.* [10]

A formula ϕ is said to be a-valid (denoted $\models_a \phi$) if for all algebraic models $\mathcal{M} = (A, 0_A, 1_A, +_A, \times_A, \Rightarrow_A, l_A, m_A, V)$ (called Fischer Servi models in [10]), $\mid \phi \mid_{\mathcal{M}} = 1_A$.

Proposition 6.1 *Let ϕ be a formula. If $\models_{ucs} \phi$ then $\models_a \phi$.*

Proof. Suppose $\models_{ucs} \phi$ and $\not\models_a \phi$. By Proposition 4.4, $\phi \in IPAL$. Since the formulas considered in Proposition 3.1 are a-valid and the inference rules considered in Proposition 3.1 are a-validity preserving, therefore $\models_a \phi$: a contradiction. □

Proposition 6.2 *Let ϕ be a formula. If $\models_a \phi$ then $\models_{ucs} \phi$.*

Proof. Suppose $\models_a \phi$ and $\not\models_{ucs} \phi$. By [10], ϕ is derivable from the axioms and the inference rules considered in [10, Section 4.1]. Obviously, these axioms are standard-valid and these inference rules are standard-validity preserving. Hence, $\models_{ucs} \phi$: a contradiction. □

Let $IPAL'$ be the least set of formulas containing the formulas $A1$–$A8$, $A10$, $A13$ and $A19$–$A29$ and closed under the inference rules $R1$ and $R2$. The deducibility relation between a finite set X of variables and a formula ϕ (denoted $X \rhd \phi$) is inductively defined as follows:

- $X \rhd p$ iff $p \in X$,
- $X \not\rhd \bot$,
- $X \rhd \phi \vee \psi$ iff either $X \rhd \phi$, or $X \rhd \psi$,
- $X \rhd \phi \wedge \psi$ iff $X \rhd \phi$ and $X \rhd \psi$,
- $X \rhd \phi \to \psi$ iff if $X \rhd \phi$ then $X \rhd \psi$,
- $X \rhd \Box\phi$ iff $X \rhd \phi$,
- $X \rhd \Diamond\phi$ iff $X \rhd \phi$,
- $X \rhd [\phi]\psi$ iff if $X \rhd \phi$ then $X \rhd \psi^*$,
- $X \rhd \langle\phi\rangle\psi$ iff $X \rhd \phi$ and $X \rhd \psi^*$.

Note that the axioms and the inference rules considered in [10, Section 4.1] do not explicitly contain the inference rule $R3$. Hence, they are those of $IPAL'$. We believe that this absence of the inference rule $R3$ is only a careless mistake, seeing that

Lemma 6.3 *Let X be a finite set of variables and ϕ be a formula. If $\phi \in IPAL'$ then $X \rhd \phi$.*

Lemma 6.4 *Let X be a finite set of variables. If $p \in X$, $q \notin X$ and $r \in X$ then $X \rhd \langle p\rangle\langle q\rangle r$ and $X \not\rhd \langle\langle p\rangle q\rangle r$.*

Proposition 6.5 (i) $\langle p\rangle\langle q\rangle r \to \langle\langle p\rangle q\rangle r \in IPAL$.

(ii) $\langle p\rangle\langle q\rangle r \to \langle\langle p\rangle q\rangle r \notin IPAL'$.

Proof. (i) It suffices to use the completeness of $IPAL$ (Proposition 4.4) and the fact that $\langle p\rangle\langle q\rangle r \to \langle\langle p\rangle q\rangle r$ is ucs-valid.
(ii) By Lemmas 6.3 and 6.4. □

7 Other properties of $IPAL$

In [17], Simpson discusses what it means to combine IPL and modal logic into intuitionistic modal logic (IML) and isolates features that might be expected of an IML. In the following proposition, we examine whether $IPAL$ complains with Simpson's requirements.

Proposition 7.1 (i) *$IPAL$ is conservative over IPL.*

(ii) *$IPAL$ contains all instances of IPL.*

(iii) *$IPAL$ is closed under modus ponens.*

(iv) *The addition of the formulas of the form $\phi \vee \neg\phi$ to $IPAL$ yields PAL.*

(v) *If $\phi \vee \psi \in IPAL$ then either $\phi \in IPAL$, or $\psi \in IPAL$.*

Proof. (i) Let ϕ be a modality-free formula. To prove that $\phi \in IPAL$ iff $\phi \in IPL$, it suffices to use the soundness/completeness of $IPAL$ (Propositions 4.1 and 4.4) and IPL (Theorem 2.43 in [5]).
(ii) By definition, $IPAL$ contains all instances of IPL.
(iii) By definition, $IPAL$ is closed under modus ponens.
(iv) Let $IPAL^+$ be the axiom system consisting of the addition of the formulas of the form $\phi \vee \neg\phi$ to $IPAL$. Suppose $IPAL^+$ does not yield PAL. Hence, $IPAL^+ \neq PAL$. Obviously, $IPAL^+ \subseteq PAL$. Since $IPAL^+ \neq PAL$, therefore $PAL \not\subseteq IPAL^+$. Let ψ be a formula such that $\psi \in PAL$ and $\psi \notin IPAL^+$. Let χ be an announcement-free formula such that $\psi \leftrightarrow \chi \in IPAL$. Such formula exists by Proposition 4.3. Thus, $\psi \leftrightarrow \chi \in IPAL^+$. Since $\psi \notin IPAL^+$, therefore $\chi \notin IPAL^+$. Let $\mathcal{M}_c = (W_c, \leq_c, R_c, V_c)$ be $IPAL^+$'s Canonical Model. Since $\chi \notin IPAL^+$, therefore by Lemmas 5.1 and 5.4, there exists $x \in W_c$ such that $\mathcal{M}_c, x \not\models \chi$. Since $IPAL^+$ is an extension of $IPAL$ that contains all formulas of the form $\phi \vee \neg\phi$, therefore by Proposition 5.5, $\leq_c$ is the identity relation on W_c. Since $\mathcal{M}_c, x \not\models \chi$, therefore by Lemma 2.2, $\chi \notin PAL$. Obviously, $IPAL \subseteq PAL$. Since $\psi \leftrightarrow \chi \in IPAL$, therefore $\psi \leftrightarrow \chi \in PAL$. Since $\chi \notin PAL$, therefore $\psi \notin PAL$: a contradiction. Consequently, $IPAL^+$ yields PAL.
(v) Suppose $\phi \vee \psi \in IPAL$, $\phi \notin IPAL$ and $\psi \notin IPAL$. By Propositions 4.1 and 4.4, $\models_{ucs} \phi \vee \psi$, $\not\models_{ucs} \phi$ and $\not\models_{ucs} \psi$. Let $\mathcal{M}_1 = (W_1, \leq_1, R_1, V_1)$ be an upward closed standard model such that $\mathcal{M}_1 \not\models \phi$ and $\mathcal{M}_2 = (W_2, \leq_2, R_2, V_2)$ be an upward closed standard model such that $\mathcal{M}_2 \not\models \psi$. Let x be a new element and $\mathcal{M} = (W, \leq, R, V)$ be the model where $W = W_1 \cup W_2 \cup \{x\}$, $\leq = \leq_1 \cup \leq_2 \cup (\{x\} \times W_1) \cup (\{x\} \times W_2)$, $R = R_1 \cup R_2$ and for all $p \in VAR$, $V(p) = V_1(p) \cup V_2(p)$. The reader may easily verify that $\mathcal{M}$ is upward closed standard. Moreover, $\mathcal{M}_1$ and $\mathcal{M}_2$ are generated submodel of $\mathcal{M}$. A result similar to Proposition 2.6 in [4] would lead to the conclusion that the global satisfiability relation is invariant under generated submodels. Since $\mathcal{M}_1 \not\models \phi$ and $\mathcal{M}_2 \not\models \psi$, therefore $\mathcal{M} \not\models \phi$ and $\mathcal{M} \not\models \psi$. Since $\models_{ucs} \phi \vee \psi$, therefore $\mathcal{M}, x \models \phi \vee \psi$. Hence, either $\mathcal{M}, x \models \phi$, or $\mathcal{M}, x \models \psi$. In the former case, let $y \in W_1$ be arbitrary. Thus, $x \leq y$. Since $\mathcal{M}, x \models \phi$, therefore by Lemma 2.1, $\mathcal{M}, y \models \phi$. Consequently, $\mathcal{M}_1, y \models \phi$. Since y was arbitrary, therefore $\mathcal{M}_1 \models \phi$:

a contradiction. In the latter case, let $y \in W_2$ be arbitrary. Hence, $x \leq y$. Since $\mathcal{M}, x \models \psi$, therefore by Lemma 2.1, $\mathcal{M}, y \models \psi$. Thus, $\mathcal{M}_2, y \models \psi$. Since y was arbitrary, therefore $\mathcal{M}_2 \models \psi$: a contradiction. Consequently, if $\phi \vee \psi \in IPAL$ then either $\phi \in IPAL$, or $\psi \in IPAL$. □

In [17, Chapter 3], Simpson proves the following

Proposition 7.2 (i) *For all announcement-free formulas ϕ, if $\phi \in IK$ then $\models_{ucs} \phi$.*

(ii) *For all announcement-free formulas ϕ, if $\models_{ucs} \phi$ then $\phi \in IK$.*

(iii) *There exists no $\Box\cdot$-free announcement-free formula ϕ such that $\Box p \leftrightarrow \phi \in IK$.*

(iv) *There exists no $\Diamond\cdot$-free announcement-free formula ϕ such that $\Diamond p \leftrightarrow \phi \in IK$.*

From the soundness/completeness of $IPAL$ (Propositions 4.1 and 4.4) and IK (Items 1 and 2 of Proposition 7.2), we obtain the following

Proposition 7.3 *$IPAL$ is conservative over IK.*

The following propositions characterize a main difference between, on one hand, the modal constructs $\Box\cdot$ and $\Diamond\cdot$ and, on the other hand, $[\cdot]\cdot$ and $\langle\cdot\rangle\cdot$.

Proposition 7.4 (i) *There exists no $\Box\cdot$-free formula ϕ such that $\Box p \leftrightarrow \phi \in IPAL$.*

(ii) *There exists no $\Diamond\cdot$-free formula ϕ such that $\Diamond p \leftrightarrow \phi \in IPAL$.*

Proof. (i) By Proposition 4.3, Item 3 of Proposition 7.2 and Proposition 7.3.
(ii) By Proposition 4.3, Item 4 of Proposition 7.2 and Proposition 7.3. □

Proposition 7.5 (i) $[\phi]\psi \leftrightarrow (\phi \to \langle\phi\rangle\psi) \in IPAL$.

(ii) $\langle\phi\rangle\psi \leftrightarrow (\phi \wedge [\phi]\psi) \in IPAL$.

Proof. (i) Suppose $[\phi]\psi \leftrightarrow (\phi \to \langle\phi\rangle\psi) \notin IPAL$. By Proposition 4.4, $\not\models_{ucs} [\phi]\psi \leftrightarrow (\phi \to \langle\phi\rangle\psi)$. Let $\mathcal{M} = (W, \leq, R, V)$ be an upward closed standard model and $x \in W$ be such that $\mathcal{M}, x \not\models [\phi]\psi \leftrightarrow (\phi \to \langle\phi\rangle\psi)$. Hence, either $\mathcal{M}, x \not\models [\phi]\psi \to (\phi \to \langle\phi\rangle\psi)$, or $\mathcal{M}, x \not\models (\phi \to \langle\phi\rangle\psi) \to [\phi]\psi$. In the former case, let $y \in W$ be such that $x \leq y$, $\mathcal{M}, y \models [\phi]\psi$ and $\mathcal{M}, y \not\models \phi \to \langle\phi\rangle\psi$. Let $z \in W$ be such that $y \leq z$, $\mathcal{M}, z \models \phi$ and $\mathcal{M}, z \not\models \langle\phi\rangle\psi$. Thus, $\mathcal{M}_{|\phi}, z \not\models \psi$. Since $y \leq z$ and $\mathcal{M}, z \models \phi$, therefore $\mathcal{M}, y \not\models [\phi]\psi$: a contradiction. In the latter case, let $y \in W$ be such that $x \leq y$, $\mathcal{M}, y \models \phi \to \langle\phi\rangle\psi$ and $\mathcal{M}, y \not\models [\phi]\psi$. Let $z \in W$ be such that $y \leq z$, $\mathcal{M}, z \models \phi$ and $\mathcal{M}_{|\phi}, z \not\models \psi$. Consequently, $\mathcal{M}, z \not\models \langle\phi\rangle\psi$. Since $y \leq z$ and $\mathcal{M}, z \models \phi$, therefore $\mathcal{M}, y \not\models \phi \to \langle\phi\rangle\psi$: a contradiction. Hence, $[\phi]\psi \leftrightarrow (\phi \to \langle\phi\rangle\psi) \in IPAL$.
(ii) By definition, $\langle\phi\rangle\psi \leftrightarrow (\phi \wedge [\phi]\psi) \in IPAL$. □

8 An alternative semantics

A proof-theoretical analysis of PAL has been proposed in [11] in terms of a sequent calculus following the approach of [13]. Unfortunately, this sequent calculus is not complete as it cannot prove the valid formula $[p \wedge p]q \leftrightarrow [p]q$. For details, see [2] where an alternative semantics for PAL and a sequent calculus with labels that were based on a specific management of a stack of announcements have been proposed. A similar alternative semantics for $IPAL$ can be proposed too. Its definition necessitates the satisfiability relation between a model $\mathcal{M} = (W, \leq, R, V)$, an element $x \in W$, a finite sequence $\varphi = (\phi_1, \ldots, \phi_n)$ of formulas and a formula ϕ (denoted $\mathcal{M}, x, (\varphi) \Vdash \phi$) inductively defined as follows:

- $\mathcal{M}, x, \epsilon \Vdash p$ iff $x \in V(p)$,
- $\mathcal{M}, x, (\varphi, \phi_{n+1}) \Vdash p$ iff $\mathcal{M}, x, (\varphi) \Vdash \phi_{n+1}$ and $\mathcal{M}, x, (\varphi) \Vdash p$,
- $\mathcal{M}, x, (\varphi) \nVdash \bot$,
- $\mathcal{M}, x, (\varphi) \Vdash \phi \vee \psi$ iff either $\mathcal{M}, x, (\varphi) \Vdash \phi$, or $\mathcal{M}, x, (\varphi) \Vdash \psi$,
- $\mathcal{M}, x, (\varphi) \Vdash \phi \wedge \psi$ iff $\mathcal{M}, x, (\varphi) \Vdash \phi$ and $\mathcal{M}, x, (\varphi) \Vdash \psi$,
- $\mathcal{M}, x, \epsilon \Vdash \phi \rightarrow \psi$ iff for all $y \in W$, if $x \leq y$ and $\mathcal{M}, y, \epsilon \Vdash \phi$ then $\mathcal{M}, y, \epsilon \Vdash \psi$,
- $\mathcal{M}, x, (\varphi, \phi_{n+1}) \Vdash \phi \rightarrow \psi$ iff for all $y \in W$, if $x \leq y$, $\mathcal{M}, y, (\varphi) \Vdash \phi_{n+1}$ and $\mathcal{M}, y, (\varphi, \phi_{n+1}) \Vdash \phi$ then $\mathcal{M}, y, (\varphi, \phi_{n+1}) \Vdash \psi$,
- $\mathcal{M}, x, \epsilon \Vdash \Box\phi$ iff for all $y, z \in W$, if $x \leq y$ and yRz then $\mathcal{M}, z, \epsilon \Vdash \phi$,
- $\mathcal{M}, x, (\varphi, \phi_{n+1}) \Vdash \Box\phi$ iff for all $y, z \in W$, if $x \leq y$, yRz, $\mathcal{M}, y, (\varphi) \Vdash \phi_{n+1}$ and $\mathcal{M}, z, (\varphi) \Vdash \phi_{n+1}$ then $\mathcal{M}, z, (\varphi, \phi_{n+1}) \Vdash \phi$,
- $\mathcal{M}, x, \epsilon \Vdash \Diamond\phi$ iff there exists $y \in W$ such that xRy and $\mathcal{M}, y, \epsilon \Vdash \phi$,
- $\mathcal{M}, x, (\varphi, \phi_{n+1}) \Vdash \Diamond\phi$ iff there exists $y \in W$ such that xRy, $\mathcal{M}, y, (\varphi) \Vdash \phi_{n+1}$ and $\mathcal{M}, y, (\varphi, \phi_{n+1}) \Vdash \phi$,
- $\mathcal{M}, x, \epsilon \Vdash [\phi]\psi$ iff for all $y \in W$, if $x \leq y$ and $\mathcal{M}, y, \epsilon \Vdash \phi$ then $\mathcal{M}, y, (\phi) \Vdash \psi$,
- $\mathcal{M}, x, (\varphi, \phi_{n+1}) \Vdash [\phi]\psi$ iff for all $y \in W$, if $x \leq y$, $\mathcal{M}, y, (\varphi) \Vdash \phi_{n+1}$ and $\mathcal{M}, y, (\varphi, \phi_{n+1}) \Vdash \phi$ then $\mathcal{M}, y, (\varphi, \phi_{n+1}, \phi) \Vdash \psi$,
- $\mathcal{M}, x, (\varphi) \Vdash \langle\phi\rangle\psi$ iff $\mathcal{M}, x, (\varphi) \Vdash \phi$ and $\mathcal{M}, x, (\varphi, \phi) \Vdash \psi$.

The reader may easily verify that the above definition of $\mathcal{M}, x, (\varphi) \Vdash \phi$ is correct decreasing on $size(\varphi, \phi)$. A similar stack-based semantics has been proposed by Balbiani *et al.* [2] within the context of PAL. The main difference with the semantics proposed by [11] lies in our interpretation of $\Box$-based formulas.

Lemma 8.1 *Let $(\phi_1, \ldots, \phi_n)$ be a sequence of formulas and ϕ be a formula. For all models $\mathcal{M} = (W, \leq, R, V)$ and for all $x \in W$, the following conditions are equivalent: (i) $\mathcal{M}, x \models [\phi_1] \ldots [\phi_n]\phi$, (ii) if $\mathcal{M}, x, \epsilon \Vdash \phi_1$, ..., $\mathcal{M}, x, (\phi_1, \ldots, \phi_{n-1}) \Vdash \phi_n$ then $\mathcal{M}, x, (\phi_1, \ldots, \phi_n) \Vdash \phi$.*

9 A labelled sequent calculus

Now, we present a sequent calculus for $IPAL$ that is derived from the stack-based semantics given in the previous section. We propose a labelled calculus

$$\frac{}{x(\epsilon):p,\ \Gamma\vdash\Delta,\ x(\epsilon):p}\,\text{ax} \qquad \frac{}{x(\epsilon):\bot,\ \Gamma\vdash\Delta}\,\text{L}\bot$$

$$\frac{x(\varphi):\phi,\ x(\varphi):p,\ \Gamma\vdash\Delta}{x(\varphi,\phi):p,\ \Gamma\vdash\Delta}\,\text{Lp} \qquad \frac{\Gamma\vdash\Delta,\ x(\varphi):\phi \quad \Gamma\vdash\Delta,\ x(\varphi):p}{\Gamma\vdash\Delta,\ x(\varphi,\phi):p}\,\text{Rp}$$

$$\frac{x(\varphi):\phi,\ x(\varphi):\psi,\ \Gamma\vdash\Delta}{x(\varphi):\phi\wedge\psi,\ \Gamma\vdash\Delta}\,\text{L}\wedge \qquad \frac{\Gamma\vdash\Delta,\ x(\varphi):\phi \quad \Gamma\vdash\Delta,\ x(\varphi):\psi}{\Gamma\vdash\Delta,\ x(\varphi):\phi\wedge\psi}\,\text{R}\wedge$$

$$\frac{x(\varphi):\phi,\ \Gamma\vdash\Delta \quad x(\varphi):\psi,\ \Gamma\vdash\Delta}{x(\varphi):\phi\vee\psi,\ \Gamma\vdash\Delta}\,\text{L}\vee$$

$$\frac{\Gamma\vdash\Delta,x(\varphi):\phi}{\Gamma\vdash\Delta,x(\varphi):\phi\vee\psi}\,\text{R}\vee^1 \qquad \frac{\Gamma\vdash\Delta,x(\varphi):\psi}{\Gamma\vdash\Delta,x(\varphi):\phi\vee\psi}\,\text{R}\vee^2$$

$$\frac{x\leq y,\Gamma\vdash y(\epsilon):\phi \quad x\leq y,\Gamma,y(\epsilon):\psi\vdash\Delta}{x(\epsilon):\phi\rightarrow\psi,\Gamma\vdash\Delta}\,\text{L}\!\rightarrow^{\epsilon} \qquad \frac{\Gamma,x\leq y,y(\epsilon):\phi\vdash\Delta,\ y(\epsilon):\psi}{\Gamma\vdash\Delta,\ x(\epsilon):\phi\rightarrow\psi}\,\text{R}\!\rightarrow^{\epsilon}$$

$$\frac{\Gamma,\ x\leq y,\ y(\varphi):\phi_{n+1},\vdash y(\varphi,\phi_{n+1}):\phi \quad \Gamma,\ x\leq y,\ y(\varphi,\phi_{n+1}):\psi\vdash\Delta}{\Gamma,\ x(\varphi,\phi_{n+1}):\phi\rightarrow\psi\vdash\Delta}\,\text{L}\!\rightarrow^{\varphi}$$

$$\frac{\Gamma,\ x\leq y,y(\varphi):\phi_{n+1},\ y(\varphi,\phi_{n+1}):\phi\vdash\Delta,\ y(\varphi,\phi_{n+1}):\psi}{\Gamma\vdash\Delta,\ x(\varphi,\phi_{n+1}):\phi\rightarrow\psi}\,\text{R}\!\rightarrow^{\varphi}$$

Fig. 1. Inference rules for IPAL - intuitionistic rules.

in which labels are defined for capturing the semantics inside the sequent calculus. This approach based on labels is a uniform approach for designing calculi in various logics like modal or intuitionistic logics [13,17] from Kripke-style semantics. We want to emphasize that starting from our stack-based semantics is central here because the similar semantics proposed for PAL allowed us to propose a new labelled calculus for PAL that corrected the deficiency about completeness of an existing labelled sequent calculus [11]. Therefore we propose a sound and complete calculus with sequents that are with multiconclusions, and with distinguished rules for dealing with empty and non-empty stacks of announcements. Let Var be a countable set of variables (denoted x, y, etc). The sequents are pairs of finite sets of expressions either of the form $x(\varphi):\phi$ read "state x satisfies ϕ with respect to the sequence (φ)", or of the form xRy read "state x is related to state y by means of R". The sequent $\Gamma\vdash\Delta$ means that the conjunction of the expressions in Γ implies the disjunction of the expressions in Δ. Provability is defined as usual: formula ϕ is provable iff the sequent $\vdash x(\epsilon):\phi$ is derivable from the inference rules of the calculus presented in Figures 1 and 2. Let $\mathcal{M}=(W,R,V)$ be a model and $f: Var\mapsto W$. Sequents are pairs of finite sets of expressions either of the form $x(\phi_1,\ldots,\phi_n):\phi$, or of the form xRy. We define the property "$\mathcal{M}$ and f satisfy the expression exp" (denoted $\mathcal{M},f\Vdash exp$) as follows:

- $\mathcal{M},f\Vdash x(\phi_1,\ldots,\phi_n):\phi$ iff $\mathcal{M},f(x),(\phi_1,\ldots,\phi_n)\Vdash\phi$,
- $\mathcal{M},f\Vdash xRy$ iff $f(x)Rf(y)$.

$$\frac{x(\epsilon):\Box\phi,\ x\leq y, yRz,\ z(\epsilon):\phi,\ \Gamma\vdash\Delta}{x(\epsilon):\Box\phi,\ x\leq y, yRz, \Gamma\vdash\Delta}\text{L}\Box^{\epsilon} \qquad \frac{x\leq y, yRz,\ \Gamma\vdash\Delta,\ z(\epsilon):\phi}{\Gamma\vdash\Delta,\ x(\epsilon):\Box\phi}\text{R}\Box^{\epsilon}$$

$$\frac{x(\varphi,\phi_{n+1}):\Box\psi,\ x\leq y, yRz,\ z(\varphi,\phi_{n+1}):\phi,\ \Gamma\vdash\Delta}{x(\varphi,\phi_{n+1}):\Box\psi,\ x\leq y, yRz,\ z(\varphi):\phi_{n+1},\ \Gamma\vdash\Delta}\text{L}\Box^{\varphi}$$

$$\frac{x\leq y, yRz,\ z(\varphi):\phi_{n+1},\ \Gamma\vdash\Delta,\ z(\varphi,\phi_{n+1}):\phi}{\Gamma\vdash\Delta,\ x(\varphi,\phi_{n+1}):\Box\phi}\text{R}\Box^{\varphi}$$

$$\frac{y(\epsilon):\phi,\ xRy,\ \Gamma\vdash\Delta}{x(\epsilon):\Diamond\phi,\ \Gamma\vdash\Delta}\text{L}\Diamond^{\epsilon} \qquad \frac{\Gamma\vdash\Delta,\ y(\epsilon):\phi,\ xRy}{\Gamma\vdash\Delta,\ x(\epsilon):\Diamond\phi}\text{R}\Diamond^{\epsilon}$$

$$\frac{y(\varphi):\phi,\ xRy,\ y(\varphi,\phi_{n+1}):\phi,\ \Gamma\vdash\Delta}{x(\varphi,\phi_{n+1}):\Diamond\phi,\ \Gamma\vdash\Delta}\text{L}\Diamond^{\varphi} \qquad \frac{\Gamma\vdash\Delta,\ y(\varphi):\phi_{n+1},\ xRy,\ y(\varphi,\phi_{n+1}):\phi}{\Gamma\vdash\Delta,\ x(\varphi,\phi_{n+1}):\Diamond\phi}\text{R}\Diamond^{\varphi}$$

$$\frac{\Gamma,\ x\leq y\vdash\Delta,\ y(\epsilon):\phi \quad y(\phi):\psi,\ \Gamma\vdash\Delta}{x(\epsilon):[\phi]\psi,\ \Gamma\vdash\Delta}\text{L}[]^{\epsilon} \qquad \frac{\Gamma,\ x\leq y,\ y(\epsilon):\phi\vdash\Delta,\ y(\phi):\psi}{\Gamma\vdash\Delta,\ x(\epsilon):[\phi]\psi}\text{R}[]^{\epsilon}$$

$$\frac{x\leq y, y(\varphi):\phi_{n+1},\ \Gamma\vdash\Delta,\ y(\varphi,\phi_{n+1}):\phi \quad y(\varphi,\phi_{n+1}):\psi,\ \Gamma\vdash\Delta}{x(\varphi,\phi_{n+1}):[\phi]\psi,\ \Gamma\vdash\Delta}\text{L}[]^{\varphi}$$

$$\frac{x\leq y, y(\varphi):\phi_{n+1}, y(\varphi,\phi_{n+1}):\phi,\ \Gamma\vdash\Delta,\ y(\varphi,\phi_{n+1},\phi):\psi}{\Gamma\vdash\Delta,\ x(\varphi,\phi_{n+1}):[\phi]\psi}\text{R}[]^{\varphi}$$

$$\frac{\Gamma,\ x(\varphi):\phi,\ x(\varphi,\phi):\psi\ \vdash\Delta}{\Gamma,\ x(\varphi):\langle\phi\rangle\psi,\ \vdash\Delta}\text{L}{<>} \qquad \frac{\Gamma,\ \vdash\Delta,\ x(\varphi):\phi \quad \Gamma,\ \vdash\Delta,\ x(\varphi,\phi):\psi}{\Gamma,\ \vdash x(\varphi):\langle\phi\rangle\psi}\text{R}{<>}$$

Fig. 2. Inference rules for IPAL - modal rules.

We say that a sequent $\Gamma\vdash\Delta$ is valid iff for all models $\mathcal{M}=(W,R,V)$ and for all $f : Var \mapsto W$, if $\mathcal{M}$ and f satisfy every expression in Γ, then $\mathcal{M}$ and f satisfy some expression in Δ.

Proposition 9.1 *Let ϕ be a formula. If ϕ is provable then ϕ is ucs-valid.*

Proof. It suffices to demonstrate that the inference rules considered in Figures 1 and 2 are validity preserving. □

Proposition 9.2 *Let ϕ be a formula. If ϕ is ucs-valid then ϕ is provable.*

Proof. By Proposition 4.4, it suffices to demonstrate that the formulas considered in Proposition 3.1 are provable and the inference rules considered in Proposition 3.1 are provability preserving. □

In Nomura et al. [14], a labelled sequent calculus has been recently given for $IPAL$. It is basically the same as the one for PAL [15] but with, in some rules, restrictions on labelled expressions on the right-hand side of sequents. As this calculus does not use an announcement stack discipline and has such restrictions, it cannot be directly and easily compared with our new calculus. In future work, we will try to compare them with respect, for instance, to proof-search issues and also to explore possible translations between these calculi.

10 Translation into $S4PAL$

By Gödel's Translation, any formula of the IPL's language can be translated into a formula of the $S4$'s language such that the resulting translation is in $S4$ iff the translated formula is in IPL. See [5, Chapter 3] for details. Within the context of $IPAL$, the translation of a formula ϕ (denoted $\tau(\phi)$) is the formula inductively defined as follows:

- $\tau(p) = \blacksquare p$,
- $\tau(\bot) = \bot$,
- $\tau(\phi \vee \psi) = \tau(\phi) \vee \tau(\psi)$,
- $\tau(\phi \wedge \psi) = \tau(\phi) \wedge \tau(\psi)$,
- $\tau(\phi \rightarrow \psi) = \blacksquare(\tau(\phi) \rightarrow \tau(\psi))$,
- $\tau(\Box\phi) = \blacksquare\Box\tau(\phi)$,
- $\tau(\Diamond\phi) = \Diamond\tau(\phi)$,
- $\tau([\phi]\psi) = \blacksquare[\tau(\phi)]\tau(\psi)$,
- $\tau(\langle\phi\rangle\psi) = \langle\tau(\phi)\rangle\tau(\psi)$.

The resulting translations belong to the $S4PAL$'s language, i.e. the set of all formulas inductively defined as follows:

- $\phi ::= p \mid \bot \mid \neg\phi \mid (\phi \vee \psi) \mid \blacksquare\phi \mid \Box\phi \mid [\phi]\psi$.

In the $S4PAL$'s language, the Boolean constructs $(\cdot \wedge \cdot)$ and $(\cdot \rightarrow \cdot)$, the modal constructs $\blacklozenge\cdot$ and $\Diamond\cdot$ and the announcement construct $\langle\cdot\rangle\cdot$ are defined as usual. Moreover, the standard rules for omission of the parentheses are adopted. The formulas of the $S4PAL$'s language are interpreted in models, their $\leq$ binary relations being used to interpret $\blacksquare$-based formulas and their R binary relations being used to interpret $\Box$-based formulas. More precisely, the satisfiability relation between a model $\mathcal{M} = (W, \leq, R, V)$, an element $x \in W$ and a formula ϕ in the $S4PAL$'s language (denoted $\mathcal{M}, x \models \phi$) is inductively defined as follows:

- $\mathcal{M}, x \models p$ iff $x \in V(p)$,
- $\mathcal{M}, x \not\models \bot$,
- $\mathcal{M}, x \models \phi \vee \psi$ iff either $\mathcal{M}, x \models \phi$, or $\mathcal{M}, x \models \psi$,
- $\mathcal{M}, x \models \blacksquare\phi$ iff for all $y \in W$, if $x \leq y$ then $\mathcal{M}, y \models \phi$,
- $\mathcal{M}, x \models \Box\phi$ iff for all $y \in W$, if xRy then $\mathcal{M}, y \models \phi$,
- $\mathcal{M}, x \models [\phi]\psi$ iff if $\mathcal{M}, x \models \phi$ then $\mathcal{M}_{|\phi}, x \models \psi$.

In the above definition, $\mathcal{M}_{|\phi} = (W_{|\phi}, \leq_{|\phi}, R_{|\phi}, V_{|\phi})$ is the model such that $W_{|\phi} = \{x \in W\colon \mathcal{M}, x \models \phi\}$, $\leq_{|\phi} = \leq \cap (W_{|\phi} \times W_{|\phi})$, $R_{|\phi} = R \cap (W_{|\phi} \times W_{|\phi})$ and for all $p \in VAR$, $V_{|\phi}(p) = V(p) \cap W_{|\phi}$. Note that if $\mathcal{M}$ is upward closed then $\mathcal{M}_{|\phi}$ is upward closed too. However, there exists a standard model $\mathcal{M} = (W, R, V)$, there exists $x \in W$ and there exists an announcement formula ϕ in the $S4PAL$'s language such that $\mathcal{M}, x \models \phi$ and $\mathcal{M}_{|\phi}$ is not standard. For example, in the standard model $\mathcal{M} = (W, \leq, R, V)$ where $W = \{x, y, z, t, u\}$, $\leq = \{(x, x), (x, t), (y, y), (y, z), (z, z), (t, t), (u, u)\}$, $R = \{(x, y), (t, z), (t, u)\}$ and $V(p) = \{y, z\}$, we have $\mathcal{M}, x \models \Box p$, $\mathcal{M}, y \models \Box p$, $\mathcal{M}, z \models \Box p$, $\mathcal{M}, t \not\models \Box p$ and $\mathcal{M}, u \models \Box p$. Hence, $\mathcal{M}_{|\Box p} = (W_{|\Box p}, \leq_{|\Box p}, R_{|\Box p}, V_{|\Box p})$ where $W_{|\Box p} =$

$\{x, y, z, u\}$, $\leq_{|\Box p}= \{(x,x),(y,y),(y,z),(z,z),(u,u)\}$, $R_{|\Box p} = \{(x,y)\}$ and $V_{|\Box p}(p) = \{y,z\}$ is not standard. Nevertheless, this never happens when the announcement formula ϕ is the resulting translation of a formula in the $IPAL$'s language.

Lemma 10.1 *Let ϕ be a formula in the $IPAL$'s language. For all standard models $\mathcal{M} = (W, \leq, R, V)$ and for all $x \in W$, if $\mathcal{M}, x \models \tau(\phi)$ then $\mathcal{M}_{|\tau(\phi)}$ is standard and for all $y \in W$, if $x \leq y$ then $\mathcal{M}, y \models \tau(\phi)$.*

Lemma 10.2 *Let ϕ be a formula in the $IPAL$'s language. The formula $\tau(\phi) \rightarrow \blacksquare\tau(\phi)$ is s-valid.*

A formula ϕ in the $S4PAL$'s language is said to be globally satisfied in a model $\mathcal{M} = (W, \leq, R, V)$ (denoted $\mathcal{M} \models \phi$) if for all $x \in W$, $\mathcal{M}, x \models \phi$. We shall say that a formula ϕ in the $S4PAL$'s language is s-valid (denoted $\models_s \phi$) if for all standard models $\mathcal{M}$, $\mathcal{M} \models \phi$.

Lemma 10.3 *Let ϕ be a formula in the $IPAL$'s language. For all upward closed standard models $\mathcal{M} = (W, \leq, R, V)$ and for all $x \in W$, the following conditions are equivalent: (i) $\mathcal{M}, x \models \phi$, (ii) $\mathcal{M}, x \models \tau(\phi)$.*

Proposition 10.4 *Let ϕ be a formula in the $IPAL$'s language. The following conditions are equivalent: (i) $\models_{ucs} \phi$, (ii) $\models_s \tau(\phi)$.*

Proof. (i)⇒(ii): By Proposition 4.4, it suffices to demonstrate that the resulting translations of the formulas $A1$–$A14$ are s-valid and that the resulting translations of the inference rules $(R1)$–$(R3)$ are s-validity preserving.
(ii)⇒(i): By Lemma 10.3. □

Obviously, for all formulas ϕ in the $IPAL$'s language, $size(\tau(\phi)) \leq 2 \times size(\phi)$. Nevertheless, seeing that the complexity of the membership problem in the set of all s-valid formulas in $S4PAL$'s language is unknown, Proposition 10.4 does not give us any upper bound on the complexity of the membership problem in the set of all ucs-valid formulas in $IPAL$'s language.

11 Conclusion

In this paper, firstly, we have given a sound and complete axiomatization of $IPAL$ and we have proved its completeness. Secondly, we have studied the features that might be expected of any intuitionistic modal logic and we have examined whether $IPAL$ possesses them. Thirdly, we have proposed an alternative semantics for $IPAL$ and we have designed a new sequent calculus for $IPAL$ that is sound and complete. Fourthly, we have defined a translation of $IPAL$'s formulas into formulas of a multimodal logic in which the construct $(\cdot \rightarrow \cdot)$ is the one of classical propositional logic. Much remains to be done: computability of the membership problem in the set of all ucs-valid formulas in $IPAL$'s language; multi-agent variants with or without positive introspection, negative introspection, common knowledge, distributed knowledge, etc; extension of our framework to intermediate logics.

Acknowledgements

This work is supported by the "Agence nationale de la recherche" (contract ANR-11-BS02-011). Special thanks are due to the members of the DynRes project for their extensive remarks concerning a preliminary version of the present paper. We also would like to thank the referees for the feedback we have obtained from them.

References

[1] Balbiani, P., Baltag, A., van Ditmarsch, H., Herzig, A., Hoshi, T., de Lima, T.: *'Knowable' as 'known after an announcement'.* The Review of Symbolic Logic **1** (2008) 305–334.

[2] Balbiani, P., Demange, V., Galmiche, D.: *A sequent calculus with labels for PAL.* Short paper presented during AiML 2014.

[3] Balbiani, P., van Ditmarsch, H., Herzig, A., de Lima, T.: *Tableaux for public announcement logic.* Journal of Logic and Computation **20** (2010) 55–76.

[4] Blackburn, P., de Rijke, M., Venema, Y.: *Modal Logic.* Cambridge University Press (2001).

[5] Chagrov, A., Zakharyaschev, M.: *Modal Logic.* Oxford University Press (1997).

[6] Van Dalen, D.: *Logic and Structure.* Springer (1983).

[7] Van Ditmarsch, H., van der Hoek, W., Kooi, B.: *Dynamic Epistemic Logic.* Springer (2007).

[8] Fischer Servi G.: *Axiomatizations for some intuitionistic modal logics.* Rendiconti del Seminario Matematico Università e Politecnico di Torino **42** (1984) 179–194.

[9] Lutz, C.: *Complexity and succinctness of public announcement logic.* In: Proceedings of the Fifth International Joint Conference on Autonomous Agents and Multiagent Systems. ACM (2007) 137–143.

[10] Ma, M., Palmigiano, A., Sadrzadeh, M.: *Algebraic semantics and model completeness for intuitionistic public announcement logic.* Annals of Pure and Applied Logic **165** (2014) 963–995.

[11] Maffezioli, P., Negri, S.: *A Gentzen-style analysis of public announcement logic.* In Arrazola, X., Ponte, M. (editors): Proceedings of the International Workshop on Logic and Philosophy of Knowledge, Communication and Action. University of the Basque Country Press (2010) 303–323.

[12] Maffezioli, P., Negri, S.: *A proof-theoretical perspective on public announcement logic.* Logic and Philosophy of Science **9** (2011) 49–59.

[13] Negri, S.: *Proof analysis in modal logic.* Journal of Philosophical Logic **34** (2005) 507–544.

[14] Nomura, S., Sano, K., Tojo, S.: *A labelled sequent calculus for intuitionistic public announcement logic.* In Davis, M., Fehnker, A., McIver, A., Voronkov, A. (editors): Logic for Programming, Artificial Intelligence, and Reasoning. Springer (2015) 187–202.

[15] Nomura, S., Sano, K., Tojo, S.: *Revising a sequent calculus for public announcement logic.* In Fenrong, L., Ono, H. (editors): Structural Analysis of Non-classical Logics. Springer (2016) 131-157.

[16] Plaza, J.: *Logics of public communications.* Synthese **158** (2007) 165–179.

[17] Simpson, A.: *The Proof Theory and Semantics of Intuitionistic Modal Logic.* PhD Thesis at the University of Edinburgh (1994).

Annex

Proof of Lemma 2.1: Let FOR be the set of all formulas ϕ such that for all upward closed standard models $\mathcal{M} = (W, \leq, R, V)$ and for all $x \in W$, if $\mathcal{M}, x \models \phi$ then $\mathcal{M}_{|\phi}$ is upward closed standard and for all $y \in W$, if $x \leq y$ then $\mathcal{M}, y \models \phi$. Lemma 2.1 says that for all formulas ϕ, $\phi \in FOR$.

We will demonstrate it by an induction on ϕ based on the function $size(\cdot)$ defined in Section 2. Let ϕ be a formula such that for all formulas ψ, if $size(\psi) < size(\phi)$ then $\psi \in FOR$. We demonstrate $\phi \in FOR$. We only consider the case $\phi = \Diamond\psi$. Note that $size(\psi) < size(\phi)$. Hence, $\psi \in FOR$. Let $\mathcal{M} = (W, \leq, R, V)$ be an upward closed standard model and $x \in W$ be such that $\mathcal{M}, x \models \Diamond\psi$.
Let $y, z, t \in W_{|\Diamond\psi}$ be such that $y \leq_{|\Diamond\psi} z$ and $yR_{|\Diamond\psi}t$. We demonstrate there exists $u \in W_{|\Diamond\psi}$ such that $zR_{|\Diamond\psi}u$ and $t \leq_{|\Diamond\psi} u$. Since $y \leq_{|\Diamond\psi} z$ and $yR_{|\Diamond\psi}t$, therefore $y \leq z$ and yRt. Let $u \in W$ be such that zRu and $t \leq u$. Such u exists because $\mathcal{M}$ is standard. Since $t \in W_{|\Diamond\psi}$, therefore $\mathcal{M}, t \models \Diamond\psi$. Hence, there exists $v \in W$ such that tRv and $\mathcal{M}, v \models \psi$. Let $w \in W$ be such that uRw and $v \leq w$. Such w exists because $\mathcal{M}$ is standard and $t \leq u$. Since $\mathcal{M}$ is upward closed standard, $\psi \in FOR$ and $\mathcal{M}, v \models \psi$, therefore $\mathcal{M}, w \models \psi$. Since uRw, therefore $\mathcal{M}, u \models \Diamond\psi$. Thus, $u \in W_{|\Diamond\psi}$. Since $z, t \in W_{|\Diamond\psi}$, zRu and $t \leq u$, therefore $zR_{|\Diamond\psi}u$ and $t \leq_{|\Diamond\psi} u$.
Let $y, z, t \in W_{|\Diamond\psi}$ be such that $yR_{|\Diamond\psi}z$ and $z \leq_{|\Diamond\psi} t$. We demonstrate there exists $u \in W_{|\Diamond\psi}$ such that $y \leq_{|\Diamond\psi} u$ and $uR_{|\Diamond\psi}t$. Since $yR_{|\Diamond\psi}z$ and $z \leq_{|\Diamond\psi} t$, therefore yRz and $z \leq t$. Let $u \in W$ be such that $y \leq u$ and uRt. Such u exists because $\mathcal{M}$ is standard. Since $y \in W_{|\Diamond\psi}$, therefore $\mathcal{M}, y \models \Diamond\psi$. Hence, there exists $v \in W$ such that yRv and $\mathcal{M}, v \models \psi$. Let $w \in W$ be such that uRw and $v \leq w$. Such w exists because $\mathcal{M}$ is standard and $y \leq u$. Since $\mathcal{M}$ is upward closed standard, $\psi \in FOR$ and $\mathcal{M}, v \models \psi$, therefore $\mathcal{M}, w \models \psi$. Since uRw, therefore $\mathcal{M}, u \models \Diamond\psi$. Thus, $u \in W_{|\Diamond\psi}$. Since $y, t \in W_{|\Diamond\psi}$, $y \leq u$ and uRt, therefore $y \leq_{|\Diamond\psi} u$ and $uR_{|\Diamond\psi}t$.
Let $y \in W$ be such that $x \leq y$. We demonstrate $\mathcal{M}, y \models \Diamond\psi$. Since $\mathcal{M}, x \models \Diamond\psi$, therefore there exists $z \in W$ such that xRz and $\mathcal{M}, z \models \psi$. Let $t \in W$ be such that yRt and $z \leq t$. Such t exists because $\mathcal{M}$ is standard and $x \leq y$. Since $\mathcal{M}$ is upward closed standard, $\psi \in FOR$ and $\mathcal{M}, z \models \psi$, therefore $\mathcal{M}, t \models \psi$. Since yRt, therefore $\mathcal{M}, y \models \Diamond\psi$.

Proof of Lemma 2.2: Suppose $\phi \in PAL$. Hence, ϕ is globally PAL-satisfied in $\mathcal{M}$. Since $\leq$ is the identity relation on W, therefore one can demonstrate by an induction on ψ based on the function $size(\cdot)$ defined in Section 2, that for all formulas ψ and for all $x \in W$, $\mathcal{M}, x \models \psi$ iff ψ is PAL-satisfied at x in $\mathcal{M}$. Since ϕ is globally PAL-satisfied in $\mathcal{M}$, therefore $\mathcal{M} \models \phi$.

Proof of Lemma 5.1: The proof is similar to the proof in [17, Chapter 3].

Proof of Lemma 5.2: The proof is similar to the proof in [17, Chapter 3].

Proof of Lemma 5.3: The proof is similar to the proof in [17, Chapter 3].

Proof of Lemma 5.4: By Proposition 4.3, let ψ be an announcement-free formula such that $\phi \leftrightarrow \psi \in IPAL$. Hence, the following conditions are equivalent: (i) $\mathcal{M}_c, x \models \phi$, (ii) $\mathcal{M}_c, x \models \psi$, (iii) $\psi \in x$, (iv) $\phi \in x$. The equivalence between (i) and (ii) follows from Proposition 4.1, Lemma 5.2 and the fact that $\phi \leftrightarrow \psi \in IPAL$. The equivalence between (ii) and (iii) follows from Lemma 5.3. The equivalence between (iii) and (iv) follows from the fact that L is an extension of $IPAL$ and $\phi \leftrightarrow \psi \in IPAL$.

Proof of Lemma 6.3: It suffices to demonstrate that the formulas $A1$–$A14$ are X-deducible and that the inference rules $(R1)$ and $(R2)$ are X-deducibility preserving.

Proof of Lemma 8.1: Let FOR^+ be the set of all nonempty sequences $(\phi_1, \ldots, \phi_n, \phi)$ of formulas such that for all models $\mathcal{M} = (W, \leq, R, V)$ and for all $x \in W$, $\mathcal{M}, x \models [\phi_1] \ldots [\phi_n]\phi$ iff if $\mathcal{M}, x, \epsilon \Vdash \phi_1$, ..., $\mathcal{M}, x, (\phi_1, \ldots, \phi_{n-1}) \Vdash \phi_n$ then $\mathcal{M}, x, (\phi_1, \ldots, \phi_n) \Vdash \phi$. Lemma 8.1 says that for all nonempty sequences $(\phi_1, \ldots, \phi_n, \phi)$ of formulas $(\phi_1, \ldots, \phi_n, \phi) \in FOR^+$. We will demonstrate it by an induction on $(\phi_1, \ldots, \phi_n, \phi)$ based on the function $size(\cdot)$ defined in Section 2. Let $(\phi_1, \ldots, \phi_n, \phi)$ be a nonempty sequence of formulas such that for all nonempty sequences $(\phi'_1, \ldots, \phi'_{n'}, \phi')$, if $size(\phi'_1, \ldots, \phi'_{n'}, \phi') < size(\phi_1, \ldots, \phi_n, \phi)$ then $(\phi'_1, \ldots, \phi'_{n'}, \phi') \in FOR^+$. We demonstrate $(\phi_1, \ldots, \phi_n, \phi) \in FOR^+$. We only consider the case $\phi = \Diamond\psi$. Note that for all $i = 1 \ldots n$, $size(\phi_1, \ldots, \phi_{i-1}, \phi_i) < size(\phi_1, \ldots, \phi_n, \phi)$. Moreover, $size(\phi_1, \ldots, \phi_n, \psi) < size(\phi_1, \ldots, \phi_n, \phi)$. Hence, for all $i = 1 \ldots n$, $(\phi_1, \ldots, \phi_{i-1}, \phi_i) \in FOR^+$. Moreover, $(\phi_1, \ldots, \phi_n, \psi) \in FOR^+$. Let $\mathcal{M} = (W, \leq, R, V)$ be a model and let $x \in W$. Leaving the case $n = 0$ to the reader, we assume that $n \geq 1$.

Suppose $\mathcal{M}, x \models [\phi_1] \ldots [\phi_n]\Diamond\psi$. Suppose $\mathcal{M}, x, \epsilon \Vdash \phi_1$, ..., $\mathcal{M}, x, (\phi_1, \ldots, \phi_{n-1}) \Vdash \phi_n$. Since for all $i = 1 \ldots n$, $(\phi_1, \ldots, \phi_{i-1}, \phi_i) \in FOR^+$, therefore for all $i = 1 \ldots n$, $\mathcal{M}, x \models [\phi_1] \ldots [\phi_{i-1}]\phi_i$. Let $y \in W$ be such that for all $i = 1 \ldots n$, $\mathcal{M}, y \models [\phi_1] \ldots [\phi_{i-1}]\phi_i$, xRy and $\mathcal{M}, y \models [\phi_1] \ldots [\phi_n]\psi$. Such y exists because $\mathcal{M}, x \models [\phi_1] \ldots [\phi_n]\Diamond\psi$ and for all $i = 1 \ldots n$, $\mathcal{M}, x \models [\phi_1] \ldots [\phi_{i-1}]\phi_i$. Since for all $i = 1 \ldots n$, $(\phi_1, \ldots, \phi_{i-1}, \phi_i) \in FOR^+$ and $(\phi_1, \ldots, \phi_n, \psi) \in FOR^+$, therefore for all $i = 1 \ldots n$, $\mathcal{M}, y, (\phi_1, \ldots, \phi_{i-1}) \Vdash \phi_i$ and $\mathcal{M}, y, (\phi_1, \ldots, \phi_n) \Vdash \psi$. Since xRy, therefore $\mathcal{M}, x, (\phi_1, \ldots, \phi_n) \Vdash \Diamond\psi$.

Suppose if $\mathcal{M}, x, \epsilon \Vdash \phi_1$, ..., $\mathcal{M}, x, (\phi_1, \ldots, \phi_{n-1}) \Vdash \phi_n$ then $\mathcal{M}, x, (\phi_1, \ldots, \phi_n) \Vdash \Diamond\psi$. Suppose $\mathcal{M}, x \not\models [\phi_1] \ldots [\phi_n]\Diamond\psi$. Hence, for all $i = 1 \ldots n$, $\mathcal{M}, x \models [\phi_1] \ldots [\phi_{i-1}]\phi_i$ and for all $y \in W$, if for all $i = 1 \ldots n$, $\mathcal{M}, y \models [\phi_1] \ldots [\phi_{i-1}]\phi_i$ and xRy then $\mathcal{M}, y \not\models [\phi_1] \ldots [\phi_n]\psi$. Since for all $i = 1 \ldots n$, $(\phi_1, \ldots, \phi_{i-1}, \phi_i) \in FOR^+$, therefore for all $i = 1 \ldots n$, $\mathcal{M}, x, (\phi_1, \ldots, \phi_{i-1}) \Vdash \phi_i$. Since if $\mathcal{M}, x, \epsilon \Vdash \phi_1$, ..., $\mathcal{M}, x, (\phi_1, \ldots, \phi_{n-1}) \Vdash \phi_n$ then $\mathcal{M}, x, (\phi_1, \ldots, \phi_n) \Vdash \Diamond\psi$, therefore $\mathcal{M}, x, (\phi_1, \ldots, \phi_n) \Vdash \Diamond\psi$. Let $y \in W$ be such that for all $i = 1 \ldots n$, $\mathcal{M}, y, (\phi_1, \ldots, \phi_{i-1}) \Vdash \phi_i$, xRy and

$\mathcal{M}, y, (\phi_1, \ldots, \phi_n) \Vdash \psi$. Since for all $i = 1 \ldots n$, $(\phi_1, \ldots, \phi_{i-1}, \phi_i) \in FOR^+$ and $(\phi_1, \ldots, \phi_n, \psi) \in FOR^+$, therefore for all $i = 1 \ldots n$, $\mathcal{M}, y \models [\phi_1] \ldots [\phi_{i-1}]\phi_i$ and $\mathcal{M}, y \models [\phi_1] \ldots [\phi_n]\psi$. Since xRy, therefore $\mathcal{M}, x \models [\phi_1] \ldots [\phi_n]\Diamond\psi$: a contradiction.

Proof of Lemma 10.1: Let FOR be the set of all formulas ϕ in the $IPAL$'s language such that for all standard models $\mathcal{M} = (W, \leq, R, V)$ and for all $x \in W$, if $\mathcal{M}, x \models \tau(\phi)$ then $\mathcal{M}_{|\tau(\phi)}$ is standard and for all $y \in W$, if $x \leq y$ then $\mathcal{M}, y \models \tau(\phi)$. Lemma 10.1 says that for all formulas ϕ in the $IPAL$'s language, $\phi \in FOR$. We will demonstrate it by an induction on ϕ based on the function $size(\cdot)$ defined in Section 2. Let ϕ be a formula such that for all formulas ψ, if $size(\psi) < size(\phi)$ then $\psi \in FOR$. We demonstrate $\phi \in FOR$. We only consider the case $\phi = \Diamond\psi$. Note that $size(\psi) < size(\phi)$. Hence, $\psi \in FOR$. Let $\mathcal{M} = (W, \leq, R, V)$ be a standard model and $x \in W$ be such that $\mathcal{M}, x \models \Diamond\tau(\psi)$.

Let $y, z, t \in W_{|\Diamond\tau(\psi)}$ be such that $y \leq_{|\Diamond\tau(\psi)} z$ and $yR_{|\Diamond\tau(\psi)}t$. We demonstrate there exists $u \in W_{|\Diamond\tau(\psi)}$ such that $zR_{|\Diamond\tau(\psi)}u$ and $t \leq_{|\Diamond\tau(\psi)} u$. Since $y \leq_{|\Diamond\tau(\psi)} z$ and $yR_{|\Diamond\tau(\psi)}t$, therefore $y \leq z$ and yRt. Let $u \in W$ be such that zRu and $t \leq u$. Such u exists because $\mathcal{M}$ is standard. Since $t \in W_{|\Diamond\tau(\psi)}$, therefore $\mathcal{M}, t \models \Diamond\tau(\psi)$. Hence, there exists $v \in W$ such that tRv and $\mathcal{M}, v \models \tau(\psi)$. Let $w \in W$ be such that uRw and $v \leq w$. Such w exists because $\mathcal{M}$ is standard and $t \leq u$. Since $\mathcal{M}$ is standard, $\psi \in FOR$ and $\mathcal{M}, v \models \tau(\psi)$, therefore $\mathcal{M}, w \models \tau(\psi)$. Since uRw, therefore $\mathcal{M}, u \models \Diamond\tau(\psi)$. Thus, $u \in W_{|\Diamond\tau(\psi)}$. Since $z, t \in W_{|\Diamond\tau(\psi)}$, zRu and $t \leq u$, therefore $zR_{|\Diamond\tau(\psi)}u$ and $t \leq_{|\Diamond\tau(\psi)} u$.

Let $y, z, t \in W_{|\Diamond\tau(\psi)}$ be such that $yR_{|\Diamond\tau(\psi)}z$ and $z \leq_{|\Diamond\tau(\psi)} t$. We demonstrate there exists $u \in W_{|\Diamond\tau(\psi)}$ such that $y \leq_{|\Diamond\tau(\psi)} u$ and $uR_{|\Diamond\tau(\psi)}t$. Since $yR_{|\Diamond\tau(\psi)}z$ and $z \leq_{|\Diamond\tau(\psi)} t$, therefore yRz and $z \leq t$. Let $u \in W$ be such that $y \leq u$ and uRt. Such u exists because $\mathcal{M}$ is standard. Since $y \in W_{|\Diamond\tau(\psi)}$, therefore $\mathcal{M}, y \models \Diamond\tau(\psi)$. Hence, there exists $v \in W$ such that yRv and $\mathcal{M}, v \models \tau(\psi)$. Let $w \in W$ be such that uRw and $v \leq w$. Such w exists because $\mathcal{M}$ is standard and $y \leq u$. Since $\mathcal{M}$ is standard, $\psi \in FOR$ and $\mathcal{M}, v \models \tau(\psi)$, therefore $\mathcal{M}, w \models \tau(\psi)$. Since uRw, therefore $\mathcal{M}, u \models \Diamond\tau(\psi)$. Thus, $u \in W_{|\Diamond\tau(\psi)}$. Since $y, t \in W_{|\Diamond\tau(\psi)}$, $y \leq u$ and uRt, therefore $y \leq_{|\Diamond\tau(\psi)} u$ and $uR_{|\Diamond\tau(\psi)}t$.

Let $y \in W$ be such that $x \leq y$. We demonstrate $\mathcal{M}, y \models \Diamond\tau(\psi)$. Since $\mathcal{M}, x \models \Diamond\tau(\psi)$, therefore there exists $z \in W$ such that xRz and $\mathcal{M}, z \models \tau(\psi)$. Let $t \in W$ be such that yRt and $z \leq t$. Such t exists because $\mathcal{M}$ is standard and $x \leq y$. Since $\mathcal{M}$ is standard, $\psi \in FOR$ and $\mathcal{M}, z \models \tau(\psi)$, therefore $\mathcal{M}, t \models \tau(\psi)$. Since yRt, therefore $\mathcal{M}, y \models \Diamond\tau(\psi)$.

Proof of Lemma 10.2: Let FOR be the set of all formulas ϕ in the $IPAL$'s language such that the formula $\tau(\phi) \rightarrow \blacksquare\tau(\phi)$ is s-valid. Lemma 10.2 says that for all formulas ϕ in the $IPAL$'s language, $\phi \in FOR$. We will demonstrate it by an induction on ϕ based on the function $size(\cdot)$ defined in Section 2. Let ϕ be a formula such that for all formulas ψ, if $size(\psi) < size(\phi)$ then $\psi \in FOR$.

We demonstrate $\phi \in FOR$. We only consider the case $\phi = \langle\psi\rangle\chi$. Note that $size(\psi) < size(\phi)$ and $size(\chi) < size(\phi)$. Hence, $\psi \in FOR$ and $\chi \in FOR$. Thus, the formulas $\tau(\psi) \rightarrow \blacksquare\tau(\psi)$ and $\tau(\chi) \rightarrow \blacksquare\tau(\chi)$ are s-valid. Let us consider the following formulas: (i) $\langle\tau(\psi)\rangle\tau(\chi)$, (ii) $\tau(\psi) \wedge [\tau(\psi)]\tau(\chi)$, (iii) $\blacksquare\tau(\psi) \wedge [\tau(\psi)]\blacksquare\tau(\chi)$, (iv) $\blacksquare\tau(\psi) \wedge (\tau(\psi) \rightarrow \blacksquare[\tau(\psi)]\tau(\chi))$, (v) $\blacksquare\tau(\psi) \wedge \blacksquare[\tau(\psi)]\tau(\chi)$, (vi) $\blacksquare(\tau(\psi) \wedge [\tau(\psi)]\tau(\chi))$, (vii) $\blacksquare\langle\tau(\psi)\rangle\tau(\chi)$. The s-validity of the formula (i)→(ii) follows from the definition of the satisfiability of formulas in the $S4PAL$'s language. The s-validity of the formula (ii)→(iii) follows from the s-validity of the formulas $\tau(\psi) \rightarrow \blacksquare\tau(\psi)$ and $\tau(\chi) \rightarrow \blacksquare\tau(\chi)$. The s-validity of the formulas (iii)→(iv), (iv)→(v), (v)→(vi) and (vi)→(vii) follows from the definition of the satisfiability of formulas in the $S4PAL$'s language.

Proof of Lemma 10.3: Let FOR be the set of all formulas ϕ in the $IPAL$'s language such that for all upward closed standard models $\mathcal{M} = (W, \leq, R, V)$ and for all $x \in W$, $\mathcal{M}, x \models \phi$ iff $\mathcal{M}, x \models \tau(\phi)$. Lemma 10.3 says that for all formulas ϕ in the $IPAL$'s language, $\phi \in FOR$. We will demonstrate it by an induction on ϕ based on the function $size(\cdot)$ defined in Section 2. Let ϕ be a formula such that for all formulas ψ, if $size(\psi) < size(\phi)$ then $\psi \in FOR$. We demonstrate $\phi \in FOR$. We only consider the case $\phi = \langle\psi\rangle\chi$. Note that $size(\psi) < size(\phi)$ and $size(\chi) < size(\phi)$. Hence, $\psi \in FOR$ and $\chi \in FOR$. Let $\mathcal{M} = (W, \leq, R, V)$ be an upward closed standard model and $x \in W$.
Suppose $\mathcal{M}, x \models \langle\psi\rangle\chi$. Hence, $\mathcal{M}, x \models \psi$ and $\mathcal{M}_{|\psi}, x \models \chi$. Since $\psi \in FOR$, therefore $\{y \in W\colon \mathcal{M}, y \models \psi\} = \{y \in W\colon \mathcal{M}, y \models \tau(\psi)\}$ and $\mathcal{M}_{|\psi} = \mathcal{M}_{|\tau(\psi)}$. Moreover, since $\chi \in FOR$, $\mathcal{M}, x \models \psi$ and $\mathcal{M}_{|\psi}, x \models \chi$, therefore $\mathcal{M}, x \models \tau(\psi)$ and $\mathcal{M}_{|\psi}, x \models \tau(\chi)$. Since $\mathcal{M}_{|\psi} = \mathcal{M}_{|\tau(\psi)}$, therefore $\mathcal{M}_{|\tau(\psi)}, x \models \tau(\chi)$. Thus, $\mathcal{M}, x \models \langle\tau(\psi)\rangle\tau(\chi)$.
Suppose $\mathcal{M}, x \models \langle\tau(\psi)\rangle\tau(\chi)$. Hence, $\mathcal{M}, x \models \tau(\psi)$ and $\mathcal{M}_{|\tau(\psi)}, x \models \tau(\chi)$. Since $\psi \in FOR$, therefore $\{y \in W\colon \mathcal{M}, y \models \psi\} = \{y \in W\colon \mathcal{M}, y \models \tau(\psi)\}$ and $\mathcal{M}_{|\psi} = \mathcal{M}_{|\tau(\psi)}$. Moreover, since $\chi \in FOR$, $\mathcal{M}, x \models \tau(\psi)$ and $\mathcal{M}_{|\tau(\psi)}, x \models \tau(\chi)$, therefore $\mathcal{M}, x \models \psi$ and $\mathcal{M}_{|\tau(\psi)}, x \models \chi$. Since $\mathcal{M}_{|\psi} = \mathcal{M}_{|\tau(\psi)}$, therefore $\mathcal{M}_{|\psi}, x \models \chi$. Thus, $\mathcal{M}, x \models \langle\psi\rangle\chi$.

Unification in modal logic Alt_1

Philippe Balbiani [1]

Institut de recherche en informatique de Toulouse
Toulouse University

Tinko Tinchev [2]

Department of Mathematical Logic and Applications
Sofia University

Abstract

Alt_1 is the least modal logic containing the formula $\Diamond x \to \Box x$. It is determined by the class of all deterministic frames. The unification problem in Alt_1 is to determine, given a formula $\phi(x_1, \ldots, x_\alpha)$, whether there exists formulas $\psi_1, \ldots, \psi_\alpha$ such that $\phi(\psi_1, \ldots, \psi_\alpha)$ is in Alt_1. In this paper, we show that the unification problem in Alt_1 is in $PSPACE$. We also show that there exists an Alt_1-unifiable formula that has no minimal complete set of unifiers. Finally, we study sub-Boolean variants of the unification problem in Alt_1.

Keywords: Modal logic Alt_1. Computability of unifiability. Unification type. Sub-Boolean variants.

1 Introduction

Modal logics are essential to the design of logical systems that capture elements of reasoning about knowledge, time, etc. There exists variants of these logics with one or several modalities, with or without the universal modality, etc. The logical problems addressed in their setting usually concern their axiomatizability, their decidability, etc. Other desirable properties which one should establish whenever possible concern, for example, the admissibility problem and the unifiability problem. About admissibility, an inference rule $\frac{\phi_1, \ldots, \phi_n}{\psi}$ is admissible in a modal logic L if for all instances $\frac{\phi'_1, \ldots, \phi'_n}{\psi'}$ of the inference rule, if $\phi'_1, \ldots, \phi'_n$ are in L then ψ' is in L too [18]. About unifiability, a formula ϕ is unifiable in a modal logic L if there exists an instance ϕ' of the formula such that ϕ' is in L [11]. When a modal logic L is axiomatically presented, its

[1] Address: Institut de recherche en informatique de Toulouse, Toulouse University, 118 route de Narbonne, 31062 Toulouse Cedex 9, FRANCE; balbiani@irit.fr.

[2] Address: Department of Mathematical Logic and Applications, Sofia University, Blvd James Bouchier 5, 1126 Sofia, Bulgaria; tinko@fmi.uni-sofia.bg.

admissible inference rules can be added to its axiomatical presentation without changing the set of its theorems. As a result, in order to improve the efficiency of automated theorem provers for modal logics, methods for deciding the admissibility of inference rules can be used [8]. The unifiability problem is easily reducible to the admissibility problem, seeing that the formula ϕ is unifiable in L iff the inference rule $\frac{\phi}{\perp}$ is non-admissible in L. In some cases, when L's unification type is finitary, the admissibility problem is reducible to the unifiability problem. Therefore, in order to improve the efficiency of automated theorem provers for modal logics, methods for deciding the unifiability of formulas can be used as well.
Results about unification have been already obtained in many modal logics. Rybakov [17] demonstrated that unification in $S4$ is decidable. Wolter and Zakharyaschev [19] showed that unification is undecidable for $K4$ or K extended with the universal modality. The notion of projectivity has been introduced by Ghilardi [13] to determine the unification type, finitary, of $S4$ and $K4$. Jeřábek [16] established the unification type, nullary, of K. Within the context of description logics, checking subsumption of concepts is not sufficient and new inference capabilities are required. One of them, unification of concept terms, has been introduced by Baader and Narendran [6] for $\mathcal{FL}_0$. Baader and Küsters [4] established the $EXPTIME$-completeness of unification in $\mathcal{FL}_{reg}$ whereas Baader and Morawska [5] established the $NPTIME$-completeness of unification in $\mathcal{EL}$. Much remains to be done, seeing that the computability of unifiability and the unification types are unknown in multifarious modal logics. In this paper, we consider the unification problem in Alt_1. Its section-by-section breakdown is organized as follows. Section 2 defines the syntax, Section 3 introduces the semantics and Section 4 presents unification. In Section 5, useful Lemmas are proved. They are used in Section 6 to prove the soundness/completeness of a nondeterministic algorithm solving unification in polynomial space. In Section 7, it is shown that there exists a unifiable formula that has no minimal complete set of unifiers. In Section 8, we study sub-Boolean variants of unification.

2 Syntax

Let AF be a countable set of atomic formulas (denoted x, y, etc). The set F of all formulas (denoted ϕ, ψ, etc) is inductively defined as follows:

- $\phi ::= x \mid \perp \mid \neg\phi \mid (\phi \vee \psi) \mid \Box\phi$.

We define the other Boolean constructs as usual. The formula $\Diamond\phi$ is obtained as an abbreviation:

- $\Diamond\phi ::= \neg\Box\neg\phi$.

The modal connective $\Box^k$ is inductively defined as follows for each $k \in \mathbb{N}$:

- $\Box^0\phi ::= \phi$,
- $\Box^{k+1}\phi ::= \Box\Box^k\phi$.

The modal connective $\Box^{<k}$ is inductively defined as follows for each $k \in \mathbb{N}$:

- $\Box^{<0}\phi ::= \top$,
- $\Box^{<k+1}\phi ::= \Box^{<k}\phi \wedge \Box^{k}\phi$.

We adopt the standard rules for omission of the parentheses. Let $deg(\phi)$ denote the degree of a formula ϕ and $var(\phi)$ its atom-set. We shall say that a formula ϕ is atom-free iff $var(\phi) = \emptyset$. Let AFF be the set of all atom-free formulas. In the sequel, we use $\phi(x_1, \ldots, x_\alpha)$ to denote a formula whose atomic formulas form a subset of $\{x_1, \ldots, x_\alpha\}$. A substitution is a function σ associating to each variable x a formula $\sigma(x)$. We shall say that a substitution σ is closed if for all variables x, $\sigma(x) \in AFF$. For all formulas $\phi(x_1, \ldots, x_\alpha)$, let $\sigma(\phi(x_1, \ldots, x_\alpha))$ be $\phi(\sigma(x_1), \ldots, \sigma(x_\alpha))$. The composition $\sigma \circ \tau$ of the substitutions σ and τ associates to each atomic formula x the formula $\tau(\sigma(x))$. Remark that for all substitutions σ, τ, if τ is closed then $\sigma \circ \tau$ is closed.

3 Semantics

Our modal language receives a relational semantics and a tuple semantics.

3.1 Relational semantics

A frame is a relational structure of the form $\mathcal{F} = (W, R)$ where W is a nonempty set of states (with typical members denoted s, t, etc) and R is a binary relation on W. A model based on a frame $\mathcal{F} = (W, R)$ is a relational structure of the form $\mathcal{M} = (W, R, V)$ where V is a function associating to each variable x a set $V(x)$ of states. We inductively define the truth of a formula ϕ in a model $\mathcal{M}$ at state s, in symbols $\mathcal{M}, s \models \phi$, as follows:

- $\mathcal{M}, s \models x$ iff $s \in V(x)$,
- $\mathcal{M}, s \not\models \bot$,
- $\mathcal{M}s \models \neg\phi$ iff $\mathcal{M}, s \not\models \phi$,
- $\mathcal{M}, s \models \phi \vee \psi$ iff either $\mathcal{M}, s \models \phi$, or $\mathcal{M}, s \models \psi$,
- $\mathcal{M}, s \models \Box\phi$ iff for all states $t \in W$, if sRt then $\mathcal{M}, t \models \phi$.

Obviously,

- $\mathcal{M}, s \models \Diamond\phi$ iff there exists a state $t \in W$ such that sRt and $\mathcal{M}, t \models \phi$,
- $\mathcal{M}, s \models \Box^{k}\phi$ iff for all states $t \in W$, if $sR^{k}t$ then $\mathcal{M}, t \models \phi$,
- $\mathcal{M}, s \models \Box^{<k}\phi$ iff for all states $t \in W$ and for all $i \in \mathbb{N}$, if $sR^{i}t$ and $i < k$ then $\mathcal{M}, t \models \phi$.

Let $\mathcal{C}$ be a class of frames. We shall say that a formula ϕ is $\mathcal{C}$-valid, in symbols $\mathcal{C} \models \phi$, if for all frames $\mathcal{F} = (W, R)$ in $\mathcal{C}$, for all models $\mathcal{M} = (W, R, V)$ based on $\mathcal{F}$ and for all states $s \in W$, $\mathcal{M}, s \models \phi$.

3.2 Tuple semantics

For all $n \in \mathbb{N}$, an n-valuation is an $(n+1)$-tuple $(U_0, \ldots, U_n)$ of subsets of AF. We inductively define the truth of a formula ϕ in an n-valuation $(U_0, \ldots, U_n)$, in symbols $(U_0, \ldots, U_n) \models \phi$, as follows:

- $(U_0, \ldots, U_n) \models x$ iff $x \in U_n$,
- $(U_0, \ldots, U_n) \not\models \bot$,
- $(U_0, \ldots, U_n) \models \neg\phi$ iff $(U_0, \ldots, U_n) \not\models \phi$,
- $(U_0, \ldots, U_n) \models \phi \vee \psi$ iff either $(U_0, \ldots, U_n) \models \phi$, or $(U_0, \ldots, U_n) \models \psi$,
- $(U_0, \ldots, U_n) \models \Box\phi$ iff if $n \geq 1$ then $(U_0, \ldots, U_{n-1}) \models \phi$.

Obviously,

- $(U_0, \ldots, U_n) \models \Diamond\phi$ iff $n \geq 1$ and $(U_0, \ldots, U_{n-1}) \models \phi$,
- $(U_0, \ldots, U_n) \models \Box^k\phi$ iff if $n \geq k$ then $(U_0, \ldots, U_{n-k}) \models \phi$,
- $(U_0, \ldots, U_n) \models \Box^{<k}\phi$ iff for all $i \in \mathbb{N}$, if $n \geq i$ and $i < k$ then $(U_0, \ldots, U_{n-i}) \models \phi$.

We shall say that a formula ϕ is n-tuple-valid, in symbols $\models_n \phi$, iff for all n-valuations $(U_0, \ldots, U_n)$, $(U_0, \ldots, U_n) \models \phi$.

3.3 Correspondence between the two semantics

In this paper, we will be only interested in the class $\mathcal{C}_{det}$ of all deterministic frames, i.e. frames $\mathcal{F} = (W, R)$ such that for all states $s, t, u \in W$, if sRt and sRu then $t = u$.

Proposition 3.1 *Let ϕ be a formula. The following conditions are equivalent:*

(i) $\mathcal{C}_{det} \models \phi$.

(ii) *For all $n \in \mathbb{N}$, $\models_n \phi$.*

When the conditions from Proposition 3.1 hold, we shall simply say that ϕ is valid, in symbols $\models \phi$.

4 Unification

We shall say that a formula $\phi(x_1, \ldots, x_\alpha)$ is unifiable iff there exists $\psi_1, \ldots, \psi_\alpha \in F$ such that $\models \phi(\psi_1, \ldots, \psi_\alpha)$. In that case, the substitution σ defined by $\sigma(x_1) = \psi_1$, ..., $\sigma(x_\alpha) = \psi_\alpha$ is called unifier of ϕ. For instance, the formula $\phi = \Box x \vee \Box y$ is unifiable. The substitution σ defined by $\sigma(x) = z$ and $\sigma(y) = \neg z$ is a unifier of ϕ. Remark that if a formula possesses a unifier then it possesses a closed unifier. This follows from the fact that for all unifiers σ of a formula ϕ and for all closed substitutions τ, $\sigma \circ \tau$ is a closed unifier of ϕ. The unification problem is the decision problem defined as follows:

- given a formula $\phi(x_1, \ldots, x_\alpha)$, determine whether $\phi(x_1, \ldots, x_\alpha)$ is unifiable.

We shall say that a substitution σ is equivalent to a substitution τ, in symbols $\sigma \simeq \tau$, if for all variables x, $\models \sigma(x) \leftrightarrow \tau(x)$. We shall say that a substitution σ is more general than a substitution τ, in symbols $\sigma \preceq \tau$, if there exists a substitution υ such that $\sigma \circ \upsilon \simeq \tau$. We shall say that a set Σ of unifiers of a unifiable formula ϕ is complete if for all unifiers σ of ϕ, there exists a unifier τ of ϕ in Σ such that $\tau \preceq \sigma$. An important question is the following: when a formula is unifiable, has it a minimal complete set of unifiers? When the answer is "yes", how large is this set? We shall say that a unifiable formula

- ϕ is unitary if there exists a minimal complete set of unifiers of ϕ with cardinality 1,
- ϕ is finitary if there exists a finite minimal complete set of unifiers of ϕ but there exists no with cardinality 1,
- ϕ is infinitary if there exists an infinite minimal complete set of unifiers of ϕ but there exists no finite one,
- ϕ is nullary if there exists no minimal complete set of unifiers of ϕ.

For instance, the formula x is unitary: the substitution σ defined by $\sigma(x) = \top$ constitutes a minimal complete set of unifiers of it. We do not know whether there exists finitary, or infinitary formulas. We will show in Section 7 that the formula $x \to \Box x$ is nullary.

5 Unification problem: lemmas

Let $\psi(x)$ be an arbitrary formula with at most one atomic formula.

Lemma 5.1 *For all $k \in \mathbb{N}$, the following conditions are equivalent:*

(i) *$\psi(x)$ is unifiable;*

(ii) *there exists $\phi \in AFF$ such that $\models \psi(\phi)$;*

(iii) *there exists $\phi \in AFF$ such that $\models \Box^k \bot \to \psi(\phi)$ and $\models \Diamond^k \top \to \psi(\phi)$.*

Remark that Lemma 5.1 still holds when one considers a formula $\psi(x_1, \ldots, x_\alpha)$ with more than one atomic formula. In this case, simply replace the "there exists ϕ ..." by "there exists $\phi_1, \ldots, \phi_\alpha$ ...". Concerning the remainder of this Section and Section 6, the same remark is on as well. Hence, without loss of generality, we will always consider in the remainder of this Section and in Section 6 that ψ is a formula with at most one atomic formula. In this case, for all $n \in \mathbb{N}$, an n-valuation is comparable to an $(n+1)$-tuple of bits. Let $k \in \mathbb{N}$ be such that $deg(\psi(x)) \leq k$. For all $\phi \in AFF$ and for all $n \in \mathbb{N}$, if $k \leq n$ then let $V_k(\phi, n, i) =$ "if $\models_{n-k+i} \phi$ then 1 else 0" for each $i \in \mathbb{N}$ such that $i \leq k$.

Lemma 5.2 *For all $\phi \in AFF$ and for all $n \in \mathbb{N}$, if $k \leq n$ then the following conditions are equivalent:*

(i) $\models_n \psi(\phi)$;

(ii) $(V_k(\phi, n, 0), \ldots, V_k(\phi, n, k)) \models \psi(x)$.

Lemma 5.3 *For all $\phi \in AFF$, the following conditions are equivalent:*

(i) $\models \Diamond^k \top \to \psi(\phi)$;

(ii) *for all $n \in \mathbb{N}$, if $k \leq n$ then $(V_k(\phi, n, 0), \ldots, V_k(\phi, n, k)) \models \psi(x)$.*

For all $\phi \in AFF$ and for all $n \in \mathbb{N}$, if $k \leq n$ then let $\boldsymbol{V}_k(\phi, n) = (V_k(\phi, n, 0), \ldots, V_k(\phi, n, k))$. For all $\phi \in AFF$, let $f_k(\phi) = \{\boldsymbol{V}_k(\phi, n) \colon n \in$

$\mathbb{N}$ is such that $k \leq n\}$. The atom-free formulas ϕ' and ϕ'' are said to be k-equivalent, in symbols $\phi' \equiv_k \phi''$, iff $f_k(\phi') = f_k(\phi'')$.

Proposition 5.4 *$\equiv_k$ is an equivalence relation on AFF possessing finitely many equivalence classes.*

Proof. By definitions of $\equiv_k$ and f_k, knowing that for all $\phi \in AFF$, $f_k(\phi)$ is a nonempty set of $(k+1)$-tuples of bits. □

Lemma 5.5 *For all $\phi', \phi'' \in AFF$, if $\phi' \equiv_k \phi''$ then the following conditions are equivalent:*

(i) $\models \Diamond^k \top \to \psi(\phi')$;

(ii) $\models \Diamond^k \top \to \psi(\phi'')$.

For all $\phi \in AFF$ and for all $n \in \mathbb{N}$, let $\boldsymbol{a}_k(\phi, n) = \boldsymbol{V}_k(\phi, n \cdot (k+1) + k)$. For all $\phi \in AFF$, let $g_k(\phi) = \{(\boldsymbol{a}_k(\phi, n), \boldsymbol{a}_k(\phi, n+1))\colon n \in \mathbb{N}\}$. We shall say that the atom-free formulas ϕ' and ϕ'' are k-congruent, in symbols $\phi' \cong_k \phi''$, iff $g_k(\phi') = g_k(\phi'')$.

Proposition 5.6 *$\cong_k$ is an equivalence relation on AFF possessing finitely many equivalence classes.*

Proof. By definitions of $\cong_k$ and g_k, knowing that for all $\phi \in AFF$, $g_k(\phi)$ is a nonempty set of pairs of $(k+1)$-tuples of bits. □

Proposition 5.7 *For all $\phi', \phi'' \in AFF$, if $\phi' \cong_k \phi''$ then $\phi' \equiv_k \phi''$.*

Proof. Let $\phi', \phi'' \in AFF$. Suppose $\phi' \cong_k \phi''$ and $\phi' \not\equiv_k \phi''$. Hence, $g_k(\phi') = g_k(\phi'')$ and $f_k(\phi') \neq f_k(\phi'')$. Thus, either there exists $n' \in \mathbb{N}$ such that $k \leq n'$ and $\boldsymbol{V}_k(\phi', n') \notin f_k(\phi'')$, or there exists $n'' \in \mathbb{N}$ such that $k \leq n''$ and $\boldsymbol{V}_k(\phi'', n'') \notin f_k(\phi')$. Without loss of generality, assume there exists $n' \in \mathbb{N}$ such that $k \leq n'$ and $\boldsymbol{V}_k(\phi', n') \notin f_k(\phi'')$. By the division algorithm, let $m, l \in \mathbb{N}$ be such that $n' = m \cdot (k+1) + l$ and $l < k+1$.
Case $m = 0$. Since $k \leq n'$, $n' = m \cdot (k+1) + l$ and $l < k+1$, therefore $n' = k$. Hence, $\boldsymbol{V}_k(\phi', n') = \boldsymbol{a}_k(\phi', 0)$. Since $g_k(\phi') = g_k(\phi'')$, therefore let $n'' \in \mathbb{N}$ be such that $(\boldsymbol{a}_k(\phi', 0), \boldsymbol{a}_k(\phi', 1)) = (\boldsymbol{a}_k(\phi'', n''), \boldsymbol{a}_k(\phi'', n''+1))$. Since $\boldsymbol{V}_k(\phi', n') = \boldsymbol{a}_k(\phi', 0)$, therefore $\boldsymbol{V}_k(\phi', n') = \boldsymbol{V}_k(\phi'', n'' \cdot (k+1) + k)$.
Case $m \neq 0$. Since $g_k(\phi') = g_k(\phi'')$, therefore let $n'' \in \mathbb{N}$ be such that $(\boldsymbol{a}_k(\phi', m-1), \boldsymbol{a}_k(\phi', m)) = (\boldsymbol{a}_k(\phi'', n''), \boldsymbol{a}_k(\phi'', n''+1))$. Hence, $V_k(\phi', (m-1) \cdot (k+1) + k, i) = V_k(\phi'', n'' \cdot (k+1) + k, i)$ and $V_k(\phi', m \cdot (k+1) + k, i) = V_k(\phi'', (n''+1) \cdot (k+1) + k, i)$ for each $i \in \mathbb{N}$ such that $i \leq k$. Since either $n' = m \cdot (k+1) + l$ and $i \leq k - (l+1)$ and $V_k(\phi', m \cdot (k+1) + l, i) = V_k(\phi', (m-1) \cdot (k+1) + k, i + (l+1))$, or $k - l \leq i$ and $V_k(\phi', m \cdot (k+1) + l, i) = V_k(\phi', m \cdot (k+1) + k, i - (k-l))$ for each $i \in \mathbb{N}$ such that $i \leq k$, therefore either $i \leq k - (l+1)$ and $V_k(\phi', n', i) = V_k(\phi'', n'' \cdot (k+1) + k, i + (l+1))$, or $k - l \leq i$ and $V_k(\phi', n', i) = V_k(\phi'', (n''+1) \cdot (k+1) + k, i - (k-l))$ for each $i \in \mathbb{N}$ such that $i \leq k$. Thus, $V_k(\phi', n', i) = V_k(\phi'', (n''+1) \cdot (k+1) + l, i)$ for each $i \in \mathbb{N}$ such that $i \leq k$. Consequently, $\boldsymbol{V}_k(\phi', n') = \boldsymbol{V}_k(\phi'', (n''+1) \cdot (k+1) + l)$.
In both cases, $\boldsymbol{V}_k(\phi', n') \in f_k(\phi'')$: a contradiction. □

Lemma 5.8 *For all $\phi', \phi'' \in AFF$, if $\phi' \cong_k \phi''$ then the following conditions are equivalent:*

(i) $\models \Diamond^k\top \to \psi(\phi')$;

(ii) $\models \Diamond^k\top \to \psi(\phi'')$.

We shall say that a nonempty set B of pairs of $(k+1)$-tuples of bits is modally definable iff there exists $\phi \in AFF$ such that $B = g_k(\phi)$. For all nonempty sets B of pairs of $(k+1)$-tuples of bits, let $\rhd_B$ be the domino relation on B defined as follows:

- $(b'_1, b''_1) \rhd_B (b'_2, b''_2)$ iff $b''_1 = b'_2$.

We shall say that a path in the directed graph $(B, \rhd_B)$ is weakly Hamiltonian iff it visits each vertex at least once. Let $\mathbf{1}_{k+1}$ be the $(k+1)$-tuple of 1 and $\mathbf{0}_{k+1}$ be the $(k+1)$-tuple of 0.

Proposition 5.9 *For all nonempty sets B of pairs of $(k+1)$-tuples of bits, the following conditions are equivalent:*

(i) *B is modally definable;*

(ii) *the directed graph $(B, \rhd_B)$ contains a weakly Hamiltonian path either ending with $(\mathbf{1}_{k+1}, \mathbf{1}_{k+1})$, or ending with $(\mathbf{0}_{k+1}, \mathbf{0}_{k+1})$.*

Proof. Let B be a nonempty set of pairs of $(k+1)$-tuples of bits.
If. Suppose the directed graph $(B, \rhd_B)$ contains a weakly Hamiltonian path either ending with $(\mathbf{1}_{k+1}, \mathbf{1}_{k+1})$, or ending with $(\mathbf{0}_{k+1}, \mathbf{0}_{k+1})$. Let $s \in \mathbb{N}$ and $(b'_0, b''_0), \ldots, (b'_s, b''_s) \in B$ be such that $((b'_0, b''_0), \ldots, (b'_s, b''_s))$ is a weakly Hamiltonian path either ending with $(\mathbf{1}_{k+1}, \mathbf{1}_{k+1})$, or ending with $(\mathbf{0}_{k+1}, \mathbf{0}_{k+1})$. Let $(\beta_0, \ldots, \beta_{s\cdot(k+1)+k})$ be the sequence of bits determined by the sequence $(b'_0, , \ldots, b'_s)$ of $(k+1)$-tuples of bits.
Case $(b'_s, b''_s) = (\mathbf{1}_{k+1}, \mathbf{1}_{k+1})$**.** Let $\phi = \bigvee\{\Diamond^i\Box\bot \colon i \in \mathbb{N}$ is such that $i < s\cdot(k+1)$ and $\beta_i = 1\} \vee \Diamond^{s\cdot(k+1)}\top$.
Case $(b'_s, b''_s) = (\mathbf{0}_{k+1}, \mathbf{0}_{k+1})$**.** Let $\phi = \bigvee\{\Diamond^i\Box\bot \colon i \in \mathbb{N}$ is such that $i < s\cdot(k+1)$ and $\beta_i = 1\}$.
In both cases, the reader may easily verify that for all $n \in \mathbb{N}$, if $n \leq s$ then $V_k(\phi, n\cdot(k+1)+k, i) = \beta_{n\cdot(k+1)+i}$ for each $i \in \mathbb{N}$ such that $i \leq k$. Hence, for all $n \in \mathbb{N}$, if $n \leq s$ then $\boldsymbol{V}_k(\phi, n\cdot(k+1)+k) = b'_n$. Thus, for all $n \in \mathbb{N}$, if $n \leq s$ then $(\boldsymbol{a}_k(\phi, n), \boldsymbol{a}_k(\phi, n+1)) = (b'_n, b''_n)$. Hence, $B = g_k(\phi)$.
Only if. Suppose B is modally definable. Let $\phi \in AFF$ be such that $B = g_k(\phi)$. Let $n_0 \in \mathbb{N}$ be such that either for all $n \in \mathbb{N}$, if $n_0 \leq n$ then $\boldsymbol{a}_k(\phi, n) = \mathbf{1}_{k+1}$, or for all $n \in \mathbb{N}$, if $n_0 \leq n$ then $\boldsymbol{a}_k(\phi, n) = \mathbf{0}_{k+1}$. Thus, $((\boldsymbol{a}_k(\phi, 0), \boldsymbol{a}_k(\phi, 1)), \ldots, (\boldsymbol{a}_k(\phi, n_0), \boldsymbol{a}_k(\phi, n_0+1)))$ is a weakly Hamiltonian path either ending with $(\mathbf{1}_{k+1}, \mathbf{1}_{k+1})$, or ending with $(\mathbf{0}_{k+1}, \mathbf{0}_{k+1})$. □

6 Unification problem: algorithm

As in Section 5, let $\psi(x)$ be an arbitrary formula with at most one atomic formula and $k \in \mathbb{N}$ be such that $deg(\psi(x)) \leq k$. We shall say that an infinite sequence $(\beta_0, \beta_1, \ldots)$ of bits respects $\psi(x)$ iff the following conditions hold:

- for all $i \in \mathbb{N}$, if $i \leq k$ then $(\beta_0, \ldots, \beta_i) \models \psi(x)$,
- for all $i \in \mathbb{N}$, $(\beta_{i+1}, \ldots, \beta_{i+k+1}) \models \psi(x)$.

Using the above results, $\psi(x)$ is unifiable iff there exists a modally definable set B of pairs of $(k+1)$-tuples of bits from which, by means of its domino relation, an infinite sequence of bits respecting $\psi(x)$ and either ending with 1s, or ending with 0s can be constructed. Hence, in order to determine whether $\psi(x)$ is unifiable, it suffices to consider the following procedure:

```
procedure UNI(ψ(x))
begin
k := deg(ψ(x))
guess a tuple (b(0), ..., b(k)) of bits of size k + 1
bool := ⊤
i := 0
while bool ∧ i ≤ k do
    begin
    bool := MC(b(0), ..., b(i), ψ(x))
    i := i + 1
    end
if ¬bool then reject
while (b(0), ..., b(k)) ≠ 0_{k+1} ∧ (b(0), ..., b(k)) ≠ 1_{k+1} do
    begin
    guess a tuple (b(k + 1), ..., b(2k + 1)) of bits of size k + 1
    bool := ⊤
    i := 0
    while bool ∧ i ≤ k do
        begin
        bool := MC(b(i + 1), ..., b(i + k + 1), ψ(x))
        i := i + 1
        end
    if ¬bool then reject
    (b(0), ..., b(k)) := (b(k + 1), ..., b(2k + 1))
    end
accept
end
```

The function $MC(\cdot)$ takes as input a tuple $(b(i), \ldots, b(i+j))$ of bits and a formula $\psi(x)$ and returns the Boolean value

- $MC(b(i), \ldots, b(i+j), \psi(x))$ = "if $(b(i), \ldots, b(i+j)) \models \psi(x)$ then $\top$ else $\bot$".

It can be implemented as a deterministic Turing machine working in polynomial time. The procedure $UNI(\cdot)$ takes as input a formula $\psi(x)$ and accepts it iff, when $k = deg(\psi(x))$, there exists a modally definable set B of pairs of $(k+1)$-tuples of bits from which, by means of its domino relation, an infinite sequence of bits respecting $\psi(x)$ and either ending with 1s, or ending with 0s can be constructed. By Proposition 5.9, the procedure $UNI(\cdot)$ accepts its input $\psi(x)$ iff $\psi(x)$ is unifiable. It can be implemented as a nondeterministic Turing machine working in polynomial space. Hence, the unification problem is in $NPSPACE$. Since $NPSPACE = PSPACE$, therefore

Proposition 6.1 *The unification problem is in $PSPACE$.*

Still, we do not know whether the unification problem is $PSPACE$-hard.

7 Unification type

Following the line of reasoning suggested by Jeřábek [16], we consider the formula $\phi(x) = x \to \Box x$. We also consider the substitution $\sigma_\top$ defined by $\sigma_\top(x) = \top$ and for all $k \in \mathbb{N}$, the substitution σ_k defined by $\sigma_k(x) = \Box^{<k} x \wedge \Box^k \bot$.

Lemma 7.1 • *$\sigma_\top$ is a unifier of $\phi(x)$,*

• *for all $k \in \mathbb{N}$, σ_k is a unifier of $\phi(x)$.*

Lemma 7.2 *Let $k, l \in \mathbb{N}$. If $k \leq l$ then $\sigma_l \preceq \sigma_k$.*

Lemma 7.3 *Let $k, l \in \mathbb{N}$. If $k < l$ then $\sigma_k \not\preceq \sigma_l$.*

Proposition 7.4 *Let σ be a substitution. The following conditions are equivalent:*

(i) $\sigma_\top \circ \sigma \simeq \sigma$.

(ii) $\sigma_\top \preceq \sigma$.

(iii) $\models \sigma(x)$.

Proof. (i⇒ii) By definition of $\simeq$ and $\preceq$.
(ii⇒iii) Suppose $\sigma_\top \preceq \sigma$. Let τ be a substitution such that $\sigma_\top \circ \tau \simeq \sigma$. Thus, $\models \tau(\sigma_\top(x)) \leftrightarrow \sigma(x)$. Hence, $\models \top \leftrightarrow \sigma(x)$. Consequently, $\models \sigma(x)$.
(iii⇒i) Suppose $\models \sigma(x)$. Hence, $\models \top \leftrightarrow \sigma(x)$. Thus, $\models \sigma(\sigma_\top(x)) \leftrightarrow \sigma(x)$. Consequently, $\sigma_\top \circ \sigma \simeq \sigma$. $\Box$

Proposition 7.5 *Let σ be a unifier of $\phi(x)$ and $k \in \mathbb{N}$. The following conditions are equivalent:*

(i) $\sigma_k \circ \sigma \simeq \sigma$.

(ii) $\sigma_k \preceq \sigma$.

(iii) $\models \sigma(x) \to \Box^k \bot$.

Proof. (i⇒ii) By definition of $\simeq$ and $\preceq$.
(ii⇒iii) Suppose $\sigma_k \preceq \sigma$. Let τ be a substitution such that $\sigma_k \circ \tau \simeq \sigma$. Thus, $\models \tau(\sigma_k(x)) \leftrightarrow \sigma(x)$. Hence, $\models \Box^{<k}\tau(x) \wedge \Box^k \bot \leftrightarrow \sigma(x)$. Consequently, $\models \sigma(x) \to \Box^k \bot$.
(iii⇒i) Suppose $\models \sigma(x) \to \Box^k \bot$. Obviously, $\models \Box^{<k}\sigma(x) \wedge \Box^k \bot \leftrightarrow \sigma(\sigma_k(x))$. Hence, $\models \sigma(\sigma_k(x)) \to \sigma(x)$. Since σ is a unifier of $\phi(x)$, therefore $\models \sigma(x) \to \Box\sigma(x)$. Thus, $\models \sigma(x) \to \Box^{<k}\sigma(x)$. Since $\models \sigma(x) \to \Box^k \bot$, therefore $\models \sigma(x) \to \Box^{<k}\sigma(x) \wedge \Box^k \bot$. Since $\models \Box^{<k}\sigma(x) \wedge \Box^k \bot \leftrightarrow \sigma(\sigma_k(x))$, therefore $\models \sigma(x) \to \sigma(\sigma_k(x))$. Since $\models \sigma(\sigma_k(x)) \to \sigma(x)$, therefore $\models \sigma(\sigma_k(x)) \leftrightarrow \sigma(x)$. Consequently, $\sigma_k \circ \sigma \simeq \sigma$. $\Box$

Proposition 7.6 *Let σ be a unifier of $\phi(x)$ and $k \in \mathbb{N}$ be such that $\deg(\sigma(x)) \leq k$. One of the following conditions holds:*

(i) $\models \sigma(x)$.

(ii) $\models \sigma(x) \to \Box^k \bot$.

Proof. Suppose $\not\models \sigma(x)$ and $\not\models \sigma(x) \to \Box^k \bot$. Let $m, n \in \mathbb{N}$, $(U_0, \ldots, U_m)$ be an m-valuation such that $(U_0, \ldots, U_m) \not\models \sigma(x)$ and $(V_0, \ldots, V_n)$ be an n-valuation such that $(V_0, \ldots, V_n) \not\models \sigma(x) \to \Box^k \bot$. Hence, $(V_0, \ldots, V_n) \models \sigma(x)$ and $(V_0, \ldots, V_n) \not\models \Box^k \bot$. Thus, $n \geq k$. Since $\deg(\sigma(x)) \leq k$, therefore $n \geq \deg(\sigma(x))$. Since $(V_0, \ldots, V_n) \models \sigma(x)$, therefore $(U_0, \ldots, U_m, V_0, \ldots, V_n) \models \sigma(x)$. Since σ is a unifier of $\phi(x)$, therefore $\models \sigma(x) \to \Box\sigma(x)$. Consequently, $\models \sigma(x) \to \Box^{n+1}\sigma(x)$. Since $(U_0, \ldots, U_m, V_0, \ldots, V_n) \models \sigma(x)$, therefore $(U_0, \ldots, U_m, V_0, \ldots, V_n) \models \Box^{n+1}\sigma(x)$. Hence, $(U_0, \ldots, U_m) \models \sigma(x)$: a contradiction. □

Proposition 7.7 *$\phi(x)$ is nullary.*

Proof. Let $\Sigma = \{\sigma_\top\} \cup \{\sigma_k : k \in \mathbb{N}\}$. By Lemma 7.1 and Propositions 7.4–7.6, Σ is a complete set of unifiers of $\phi(x)$. Suppose there exists a minimal complete set of unifiers of $\phi(x)$. Let Γ be a minimal complete set of unifiers of $\phi(x)$. Let $\gamma \in \Gamma$ be such that $\gamma \preceq \sigma_0$. Since Σ is a complete set of unifiers of $\phi(x)$, therefore let $\sigma \in \Sigma$ be such that $\sigma \preceq \gamma$. Now, we consider the following 2 cases.

Case $\sigma = \sigma_\top$. Since $\gamma \preceq \sigma_0$, therefore $\sigma \preceq \sigma_0$. Let υ be a substitution such that $\sigma \circ \upsilon \simeq \sigma_0$. Hence, $\models \upsilon(\sigma(x)) \leftrightarrow \sigma_0(x)$. Thus, $\models \top \leftrightarrow \bot$: a contradiction.

Case $\sigma = \sigma_k$ for some $k \in \mathbb{N}$. By Lemma 7.3, $\sigma \not\preceq \sigma_{k+1}$. Let $\gamma' \in \Gamma$ be such that $\gamma' \preceq \sigma_{k+1}$. By Lemma 7.2, since $\sigma \preceq \gamma$, therefore $\gamma' \preceq \gamma$. Since Γ is a minimal complete set of unifiers of $\phi(x)$, therefore $\gamma' = \gamma$. Since $\gamma' \preceq \sigma_{k+1}$ and $\sigma \preceq \gamma$, therefore $\sigma \preceq \sigma_{k+1}$: a contradiction. □

8 Sub-Boolean variants

In this section, we study sub-Boolean variants of the unification problem.

8.1 $\{\Box, \top, \wedge\}$-fragment

In the $\{\Box, \top, \wedge\}$-fragment, formulas are defined as follows:

- $\phi ::= x \mid \top \mid (\phi \wedge \psi) \mid \Box\phi$.

Lemma 8.1 *Let σ, τ, υ be substitutions such that for all variables x, $var(\sigma(x)) \cap var(\tau(x)) = \emptyset$ and $\models \upsilon(x) \leftrightarrow \sigma(x) \wedge \tau(x)$. Then $\upsilon \preceq \sigma$ and $\upsilon \preceq \tau$.*

Lemma 8.2 *Let ϕ be a formula and σ, τ, υ be substitutions. If for all variables x, $\models \upsilon(x) \leftrightarrow \sigma(x) \wedge \tau(x)$ then $\models \upsilon(\phi) \leftrightarrow \sigma(\phi) \wedge \tau(\phi)$.*

The $\Box$-integer-set of a variable x with respect to a formula ϕ, in symbols $is_\Box(x, \phi)$, is inductively defined as follows:

- $is_\Box(x, y) = \{0\}$ if $x = y$,

- $is_\Box(x, y) = \emptyset$ if $x \neq y$,
- $is_\Box(x, \top) = \emptyset$,
- $is_\Box(x, \phi \wedge \psi) = is_\Box(x, \phi) \cup is_\Box(x, \psi)$,
- $is_\Box(x, \Box\phi) = \{i + 1 : \ i \in is_\Box(x, \phi)\}$.

For instance, $is_\Box(x, y \wedge \Box\Box z) = \emptyset$, $is_\Box(y, y \wedge \Box\Box z) = \{0\}$ and $is_\Box(z, y \wedge \Box\Box z) = \{2\}$. Let ϕ be an arbitrary formula.

Lemma 8.3 *For all $x \in AF$, $is_\Box(x, \phi)$ is a finite set such that $is_\Box(x, \phi) \neq \emptyset$ iff $x \in var(\phi)$. Moreover, $\models \phi \leftrightarrow \bigwedge\{\Box^i x : \ x \in AF \ \& \ i \in is_\Box(x, \phi)\}$.*

Lemma 8.4 *Let σ be a substitution and x be a variable. If $\models \sigma(\phi) \leftrightarrow x$ then there exists a variable y such that $0 \in is_\Box(y, \phi)$ and $\models \sigma(y) \leftrightarrow x$.*

Lemma 8.5 *Let σ be a substitution, $i \geq 0$ and x, y be variables. If $i \in is_\Box(x, \phi)$ and $\models \sigma(x) \leftrightarrow y$ then $i \in is_\Box(y, \sigma(\phi))$.*

In the $\{\Box, \top, \wedge\}$-fragment, unification problems are finite sets of pairs of formulas. We shall say that a finite set $S = \{(\phi_1, \psi_1), \ldots, (\phi_n, \psi_n)\}$ of pairs of formulas is unifiable iff there exists a substitution σ such that $\models \sigma(\phi_1) \leftrightarrow \sigma(\psi_1)$, $\ldots$, $\models \sigma(\phi_n) \leftrightarrow \sigma(\psi_n)$. In that case, σ is called a unifier of S. Of course, now, substitutions are functions associating to each variable a formula in the $\{\Box, \top, \wedge\}$-fragment. Obviously, if a finite set of pairs of formulas possesses a unifier then it possesses a closed unifier. Moreover, by Lemma 8.3, every atom-free formula is equivalent to $\top$. As a result,

Proposition 8.6 *Every finite set of pairs of formulas possesses a unifier.*

The simplicity of unification problems in the $\{\Box, \top, \wedge\}$-fragment does not entail that every finite set of pairs of formulas possesses a minimal complete set of unifiers. Following the line of reasoning suggested by Baader [2], we consider the formulas $\phi(x, y, z) = \Box x \wedge \Box y$ and $\psi(x, y, z) = y \wedge \Box\Box z$. We also consider for all $k \in \mathbb{N}$, the substitution σ_k defined by $\sigma_k(x) = t_k$, $\sigma_k(y) = \Box\Box^{<k+1} t_k$ and $\sigma_k(z) = \Box^k t_k$. We will assume that for all $k, l \in \mathbb{N}$, $k \neq l$, the variables t_k and t_l are distinct.

Lemma 8.7 *For all $k \in \mathbb{N}$, σ_k is a unifier of $\{(\phi(x, y, z), \psi(x, y, z))\}$.*

For all $k \in \mathbb{N}$, we consider

- the substitution γ_k inductively defined as follows:
  - $\gamma_0 = \sigma_0$,
  - γ_{k+1} is the substitution defined by $\gamma_{k+1}(x) = \gamma_k(x) \wedge \sigma_{k+1}(x)$, $\gamma_{k+1}(y) = \gamma_k(y) \wedge \sigma_{k+1}(y)$ and $\gamma_{k+1}(z) = \gamma_k(z) \wedge \sigma_{k+1}(z)$.

Lemma 8.8 *For all $k \in \mathbb{N}$, γ_k is a unifier of $\{(\phi(x, y, z), \psi(x, y, z))\}$.*

Lemma 8.9 *Let $k, l \in \mathbb{N}$. If $k \leq l$ then $\gamma_l \preceq \sigma_k$.*

Lemma 8.10 *Let $k, l \in \mathbb{N}$. If $k \leq l$ then $\gamma_l \preceq \gamma_k$.*

Proposition 8.11 *Let σ be a unifier of $\{(\phi(x, y, z), \psi(x, y, z))\}$ and $k \in \mathbb{N}$. If $\sigma \preceq \sigma_k$ then there exists a variable u such that $k \in is_\Box(u, \sigma(z))$.*

Proof. Suppose $\sigma \preceq \sigma_k$. Let τ be a substitution such that $\sigma \circ \tau \simeq \sigma_k$. Thus, $\models \tau(\sigma(x)) \leftrightarrow \sigma_k(x)$. Hence, $\models \tau(\sigma(x)) \leftrightarrow t_k$. By Lemma 8.4, let u be a variable such that $0 \in is_\Box(u, \sigma(x))$ and $\models \tau(u) \leftrightarrow t_k$. Since σ is a unifier of $\{(\phi(x,y,z), \psi(x,y,z))\}$, therefore $\models \Box\sigma(x) \wedge \Box\sigma(y) \leftrightarrow \sigma(y) \wedge \Box\Box\sigma(z)$. Since $0 \in is_\Box(u, \sigma(x))$, therefore $1 \in is_\Box(u, \sigma(y))$.
Claim: Let $i \geq 1$. If $i \in is_\Box(u, \sigma(y))$ and $i - 1 \notin is_\Box(u, \sigma(z))$ then $i + 1 \in is_\Box(u, \sigma(y))$.
Proof of the Claim: Suppose $i \in is_\Box(u, \sigma(y))$ and $i - 1 \notin is_\Box(u, \sigma(z))$. Since $\models \Box\sigma(x) \wedge \Box\sigma(y) \leftrightarrow \sigma(y) \wedge \Box\Box\sigma(z)$, therefore $i + 1 \in is_\Box(u, \sigma(y))$.
By the above Claim, let $i \geq 1$ be such that $i - 1 \in is_\Box(u, \sigma(z))$. By Lemma 8.5, since $\models \tau(u) \leftrightarrow t_k$, therefore $i - 1 \in is_\Box(t_k, \tau(\sigma(z)))$. Since $\sigma \circ \tau \simeq \sigma_k$, therefore $\models \tau(\sigma(z)) \leftrightarrow \sigma_k(z)$. Hence, $\models \tau(\sigma(z)) \leftrightarrow \Box^k t_k$. Since $i - 1 \in is_\Box(t_k, \tau(\sigma(z)))$, therefore $i - 1 = k$. Since $i - 1 \in is_\Box(u, \sigma(z))$, therefore $k \in is_\Box(u, \sigma(z))$. □

Lemma 8.12 *Let σ be substitution. If σ is a unifier of $\{(\phi(x,y,z), \psi(x,y,z))\}$ then $\sigma \preceq \gamma_k$ for at most finitely many $k \in \mathbb{N}$.*

Proposition 8.13 *There exists no minimal complete set of unifiers of $\{(\phi(x,y,z), \psi(x,y,z))\}$.*

Proof. Let Δ be a minimal complete set of unifiers of $\{(\phi(x,y,z), \psi(x,y,z))\}$. Let $\delta \in \Delta$ be such that $\delta \preceq \sigma_0$. Hence, $\delta \preceq \gamma_0$. By Lemmas 8.10 and 8.12, let $k \in \mathbb{N}$ be such that $\delta \preceq \gamma_k$ and $\delta \not\preceq \gamma_{k+1}$. Without loss of generality, we can assume that $var(\delta(x)) \cap var(\gamma_{k+1}(x)) = \emptyset$, $var(\delta(y)) \cap var(\gamma_{k+1}(y)) = \emptyset$ and $var(\delta(z)) \cap var(\gamma_{k+1}(z)) = \emptyset$. Let ϵ be the substitution defined by $\epsilon(x) = \delta(x) \wedge \gamma_{k+1}(x)$, $\epsilon(y) = \delta(y) \wedge \gamma_{k+1}(y)$ and $\epsilon(z) = \delta(z) \wedge \gamma_{k+1}(z)$. By Lemmas 8.1, 8.2 and 8.8, ϵ is a unifier of $\{(\phi(x,y,z), \psi(x,y,z))\}$, $\epsilon \preceq \delta$ and $\epsilon \preceq \gamma_{k+1}$. Since Δ is a minimal complete set of unifiers of $\{(\phi(x,y,z), \psi(x,y,z))\}$, therefore let $\delta' \in \Delta$ be such that $\delta' \preceq \epsilon$. Since $\epsilon \preceq \delta$, therefore $\delta' \preceq \delta$. Since Δ is a minimal complete set of unifiers of $\{(\phi(x,y,z), \psi(x,y,z))\}$, therefore $\delta' = \delta$. Since $\epsilon \preceq \gamma_{k+1}$ and $\delta' \preceq \epsilon$, therefore $\delta \preceq \gamma_{k+1}$: a contradiction. □

8.2 $\{\Diamond, \top, \wedge\}$-fragment

In the $\{\Diamond, \top, \wedge\}$-fragment, formulas are defined as follows:

- $\phi ::= x \mid \top \mid (\phi \wedge \psi) \mid \Diamond\phi$.

The $\Diamond$-integer-set of a variable x with respect to a formula ϕ, in symbols $is_\Diamond(x, \phi)$, is inductively defined as has been defined the $\Box$-integer-set of x with respect to ϕ. Let ϕ be an arbitrary formula.

Lemma 8.14 *For all $x \in AF$, $is_\Diamond(x, \phi)$ is a finite set such that $is_\Diamond(x, \phi) \neq \emptyset$ iff $x \in var(\phi)$. Moreover, $\models \phi \leftrightarrow \bigwedge\{\Diamond^i x : x \in AF \ \&\ i \in is_\Diamond(x, \phi)\} \wedge \Diamond^{deg(\phi)}\top$.*

As before, if a finite set of pairs of formulas possesses a unifier then it possesses a closed unifier. Unlike the $\{\Box, \top, \wedge\}$-fragment, there exists non-unifiable finite sets of pairs of formulas. The truth is that many atom-free formulas are not equivalent to $\top$. Nevertheless, by Lemma 8.14, for all atom-free formulas ϕ, ϕ is equivalent to $\Diamond^{deg(\phi)}\top$.

Lemma 8.15 *Let ϕ be a formula. For all closed substitutions σ, $\models \sigma(\phi) \leftrightarrow \bigwedge\{\Diamond^i\sigma(x) : x \in var(\phi) \ \& \ i = \max is_\Diamond(x,\phi)\} \wedge \Diamond^{deg(\phi)}\top$.*

Lemma 8.16 *Let S be a finite set of pairs of formulas. Let ϕ, ψ, ϕ', ψ' be formulas such that $deg(\phi) = deg(\phi')$, $var(\phi) = var(\phi')$, $deg(\psi) = deg(\psi')$ and $var(\psi) = var(\psi')$. If for all $x \in AF$, $\max is_\Diamond(x,\phi) = \max is_\Diamond(x,\phi')$ and $\max is_\Diamond(x,\psi) = \max is_\Diamond(x,\psi')$ then $S \cup \{(\phi,\psi)\}$ possesses a unifier iff $S \cup \{(\phi',\psi')\}$ possesses a unifier.*

Let the normal formulas be defined as follows:

- $\phi ::= (x_1 \wedge \ldots \wedge x_\alpha) \mid \top \mid ((x_1 \wedge \ldots \wedge x_\alpha) \wedge \Diamond\phi) \mid \Diamond\phi$.

For example, the formula $\Diamond x \wedge \Diamond y$ is not normal and the formula $y \wedge \Diamond\Diamond z$ is normal. In the above definition of normal formulas, we use the conjunction $(x_1 \wedge \ldots \wedge x_\alpha)$ of the variables $x_1, \ldots, x_\alpha$. In such a situation, we will always consider that $\alpha \geq 1$. We shall say that a formula ϕ is minimalist if for all $x \in AF$, x occurs at most once in ϕ. For instance, the formula $\bigwedge\{\Diamond^i x : x \in var(\phi) \ \& \ i = \max is_\Diamond(x,\phi)\}$ is minimalist for each formula ϕ.

Lemma 8.17 *Let ϕ be a formula. There exists a normal formula ϕ' such that $\models \phi \leftrightarrow \phi'$. Moreover, if ϕ is minimalist then ϕ' is minimalist too. Finally, ϕ' can be easily computed from ϕ in polynomial time.*

For example, the non-normal formula $\Diamond x \wedge \Diamond y$ is equivalent to the normal formula $\Diamond(x \wedge y)$ and the non-normal formula $y \wedge \Diamond\top \wedge \Diamond\Diamond z$ is equivalent to the normal formula $y \wedge \Diamond\Diamond z$.

Lemma 8.18 *Let S be a finite set of pairs of formulas. There exists a finite set S' of pairs of minimalist normal formulas such that S possesses a unifier iff S' possesses a unifier. Moreover, S' can be easily computed from S in polynomial time.*

Let the thin formulas be defined as follows:

- $\phi ::= x \mid \top \mid (x \wedge \Diamond\phi) \mid \Diamond\phi$.

For example, the formula $\Diamond x \wedge \Diamond y$ is not thin and the formula $y \wedge \Diamond\Diamond z$ is thin. Remark that for all formulas ϕ, if ϕ is thin then ϕ is normal.

Proposition 8.19 *Let S be a finite set of pairs of minimalist normal formulas with variables $x_1, \ldots, x_\alpha$. Let $\preceq$ be a total order on $1, \ldots, \alpha$. Let S' be a finite set of pairs of thin minimalist formulas obtained from S and $\preceq$ by replacing each conjunct of the form $(x_{\beta_1} \wedge \ldots \wedge x_{\beta_n})$ in S by x_β where $\beta = \max_\preceq\{\beta_1, \ldots, \beta_n\}$. Suppose S' possesses a closed unifier σ such that*

- *for all $\beta = 1 \ldots \alpha$, there exists $k_\beta \in \mathbb{N}$ such that $\sigma(x_\beta) = \Diamond^{k_\beta}\top$,*
- *for all $\beta, \gamma = 1 \ldots \alpha$, if $\beta \preceq \gamma$ then $k_\beta \leq k_\gamma$.*

Then σ is also a unifier of S.

Proof. Let $(x_{\beta_1} \wedge \ldots \wedge x_{\beta_n})$ be a conjunct in S and $\beta = \max_\preceq\{\beta_1, \ldots, \beta_n\}$. Let $k_{\beta_1}, \ldots, k_{\beta_n} \in \mathbb{N}$ be such that $\sigma(x_{\beta_1}) = \Diamond^{k_{\beta_1}}\top$, ..., $\sigma(x_{\beta_n}) = \Diamond^{k_{\beta_n}}\top$. Since for all $\gamma, \delta = 1 \ldots \alpha$, if $\gamma \preceq \delta$ then $k_\gamma \leq k_\delta$ and $\beta = \max_\preceq\{\beta_1, \ldots, \beta_n\}$, therefore

$k_\beta = \max_{\preceq}\{k_{\beta_1}, \ldots, k_{\beta_n}\}$. Hence, $\models \Diamond^{k_{\beta_1}}\top \wedge \ldots \wedge \Diamond^{k_{\beta_n}}\top \leftrightarrow \Diamond^{k_\beta}\top$. Thus, $\models \sigma(x_{\beta_1} \wedge \ldots \wedge x_{\beta_n}) \leftrightarrow \sigma(x_\beta)$. Since σ is a unifier of S', therefore σ is a unifier of S. □

Proposition 8.20 *Let S be a finite set of pairs of minimalist normal formulas with variables $x_1, \ldots, x_\alpha$. Suppose S possesses a closed unifier σ such that*

- *for all $\beta = 1 \ldots \alpha$, there exists $k_\beta \in \mathbb{N}$ such that $\sigma(x_\beta) = \Diamond^{k_\beta}\top$.*

Let $\preceq$ be a total order on $1, \ldots, \alpha$ such that

- *for all $\beta, \gamma = 1 \ldots \alpha$, if $\beta \preceq \gamma$ then $k_\beta \leq k_\gamma$.*

Let S' be a finite set of pairs of thin minimalist formulas obtained from S and $\preceq$ by replacing each conjunct of the form $(x_{\beta_1} \wedge \ldots \wedge x_{\beta_n})$ in S by x_β where $\beta = \max_{\preceq}\{\beta_1, \ldots, \beta_n\}$. Then σ is also a unifier of S'.

Proof. Similar to the proof of Proposition 8.19. □

In Propositions 8.19 and 8.20, the finite set S' of pairs of thin minimalist formulas obtained from S and $\preceq$ is called a thin $\preceq$-subset of S. Using the above results, a given finite set S of pairs of minimalist normal formulas with variables $x_1, \ldots, x_\alpha$ is unifiable iff there exists a total order $\preceq$ on $1, \ldots, \alpha$ and a thin $\preceq$-subset of S possessing a unifier. Now, in order to determine whether a given finite set S of pairs of thin minimalist normal formulas is unifiable, it suffices to consider the following procedure:

```
procedure UNISET(S)
begin
recursively replace each pair of the form (◇φ, ◇ψ) in S by (φ, ψ)
bool := BC(S)
if bool ∧ var(S) ≠ ∅ then
   begin
   guess a subset AF(S) of var(S)
   for all x ∈ AF(S) do
          replace in S each occurrence of x by ⊤
   for all x ∈ var(S) \ AF(S) do
      replace in S each occurrence of x in S by ◇x
   transform S into an equivalent finite set of pairs of thin minimalist normal
      formulas
   UNISET(S)
   end
if ¬bool then reject
accept
end
```

The function $BC(\cdot)$ takes as input a finite set S of pairs of thin minimalist normal formulas and returns the Boolean value

- $BC(S)$ ="if neither S contains pairs of the form $(\Diamond\phi, \top)$, nor S contains pairs of the form $(\top, \Diamond\psi)$ then $\top$ else $\bot$".

It can be implemented as a deterministic Turing machine working in polynomial time. The procedure $UNISET(\cdot)$ takes as input a finite set of pairs of thin minimalist normal formulas and accepts it iff it is unifiable. It can be implemented as a nondeterministic Turing machine working in polynomial space. Hence, the unification problem is in $NPSPACE$. Since $NPSPACE = PSPACE$, therefore

Proposition 8.21 *The unification problem is in $PSPACE$.*

Still, we do not know whether the unification problem is $PSPACE$-hard.

9 Conclusion

Much remains to be done. For example, there is the related admissibility problem: given an inference rule $\frac{\psi_1(x_1,\ldots,x_n),\ldots,\psi_k(x_1,\ldots,x_n)}{\chi(x_1,\ldots,x_n)}$, determine whether for all formulas $\phi_1,\ldots,\phi_n$, if $\models \psi_1(\phi_1,\ldots,\phi_n)$, ..., $\models \psi_k(\phi_1,\ldots,\phi_n)$ then $\models \chi(\phi_1,\ldots,\phi_n)$. One may also consider the unification problem when the ordinary modal language is extended by a set AP of parameters (denoted p, q, etc). In this case, the unification problem is to determine, given a formula $\psi(p_1,\ldots,p_\alpha,x_1,\ldots,x_\beta)$, whether there exists formulas $\phi_1,\ldots,\phi_\beta$ such that $\models \psi(p_1,\ldots,p_\alpha,\phi_1,\ldots,\phi_\beta)$. For each $k \geq 2$, one may also consider the unification problem in Alt_k, the least normal logic containing the formula $\Diamond(x_1\wedge\neg x_2\wedge\ldots\wedge\neg x_{k-1}\wedge\neg x_k)\wedge\ldots\wedge\Diamond(\neg x_1\wedge\neg x_2\wedge\ldots\wedge\neg x_{k-1}\wedge x_k) \rightarrow \Box(x_1\vee x_2\vee\ldots\vee x_{k-1}\vee x_k)$. Its decidability is open. Finally, what becomes of these problems when the ordinary modal language is extended by the master modality, the universal modality or the difference modality?

Acknowledgements

Special acknowledgement is heartly granted to Joseph Boudou, Wojciech Dzik, Çiğdem Gencer, Silvio Ghilardi, Rosalie Iemhoff, Emil Jeřábek and Michael Rusinowitch. We also would like to thank the referees for the feedback we have obtained from them. Philippe Balbiani was partially supported by the Bulgarian National Science Fund (contract DID02/32/209) and Tinko Tinchev was partially supported by the Centre international de mathématiques et d'informatique (contract ANR-11-LABX-0040-CIMI within the programme ANR-11-IDEX-0002-02). Philippe Balbiani and Tinko Tinchev were partially supported by the programme RILA (contracts 34269VB and DRILA01/2/2015).

References

[1] Anantharaman, S., Narendran, P., Rusinowitch, M.: *Unification modulo ACUI plus distributivity axioms.* Journal of Automated Reasoning **33** (2004) 1–28.

[2] Baader, F.: *Unification in commutative theories.* Journal of Symbolic Computation **8** (1989) 479–497.

[3] Baader, F., Ghilardi, S.: *Unification in modal and description logics.* Logic Journal of the IGPL **19** (2011) 705–730.

[4] Baader, F., Küsters, R.: *Unification in a description logic with transitive closure of roles.* In Nieuwebhuis, R., Voronkov, A. (editors): *Logic for Programming and Automated Reasoning.* Springer (2001) 217–232.

[5] Baader, F., Morawska, B.: *Unification in the description logic $\mathcal{EL}$.* In Treinen, R. (editor): *Rewriting Techniques and Applications.* Springer (2009) 350–364.

[6] Baader, F., Narendran, P.: *Unification of concept terms in description logics.* Journal of Symbolic Computation **31** (2001) 277–305.

[7] Babenyshev, S., Rybakov, V.: *Unification in linear temporal logic LTL.* Annals of Pure and Applied Logic **162** (2011) 991–1000.

[8] Babenyshev, S., Rybakov, V., Schmidt, R., Tishkovsky, D.: *A tableau method for checking rule admissibility in $S4$.* Electronic Notes in Theoretical Computer Science **262** (2010) 17–32.

[9] Chagrov, A.: *Decidable modal logic with undecidable admissibility problem.* Algebra and Logic **31** (1992) 53–61.

[10] Dzik, W.: *Unitary unification of $S5$ modal logics and its extensions.* Bulletin of the Section of Logic **32** (2003) 19–26.

[11] Dzik, W.: *Unification Types in Logic.* Wydawnicto Uniwersytetu Slaskiego (2007).

[12] Gencer, Ç., de Jongh, D.: *Unifiability in extensions of $K4$.* Logic Journal of the IGPL 17 (2009) 159–172.

[13] Ghilardi, S.: *Best solving modal equations.* Annals of Pure and Applied Logic **102** (2000) 183–198.

[14] Iemhoff, R.: *On the admissible rules of intuitionistic propositional logic.* The Journal of Symbolic Logic **66** (2001) 281–294.

[15] Jeřábek, E.: *Complexity of admissible rules.* Archive for Mathematical Logic **46** (2007) 73–92.

[16] Jeřábek, E.: *Blending margins: the modal logic K has nullary unification type.* Journal of Logic and Computation **25** (2015) 1231–1240.

[17] Rybakov, V.: *A criterion for admissibility of rules in the model system $S4$ and the intuitionistic logic.* Algebra and Logic **23** (1984) 369–384.

[18] Rybakov, V.: *Admissibility of Logical Inference Rules.* Elsevier (1997).

[19] Wolter, F., Zakharyaschev, M.: *Undecidability of the unification and admissibility problems for modal and description logics.* ACM Transactions on Computational Logic **9** (2008) 25:1–25:20.

Annex

Proof of Proposition 3.1: (i)⇒(ii): It suffices to remark that for all $n \in \mathbb{N}$, every n-valuation can be considered as a model based on a deterministic frame. (ii)⇒(i): It suffices to remark that for all $n \in \mathbb{N}$, every generated submodel of a model based on a deterministic frame is n-bisimilar to a k-valuation for some $k \in \mathbb{N}$ such that $k \leq n$.

Proof of Lemma 5.2: By induction on $\psi(x)$.

Proof of Lemma 5.3: Let $\phi \in AFF$. The following conditions are equivalent: (1) $\models \Diamond^k\top \to \psi(\phi)$; (2) for all $n \in \mathbb{N}$, $\models_n \Diamond^k\top \to \psi(\phi)$; (3) for all $n \in \mathbb{N}$, if $\models_n \Diamond^k\top$ then $\models_n \psi(\phi)$; (4) for all $n \in \mathbb{N}$, if $k \leq n$ then $(V_k(\phi, n, 0), \ldots, V_k(\phi, n, k)) \models \psi(x)$. The reasons for these equivalences to hold are the following: the equivalence between (1) and (2) follows from the definition of $\models$, the equivalence between (2) and (3) follows from the fact that $\phi \in AFF$ and the equivalence between (3) and (4) follows from Lemma 5.2.

Proof of Lemma 5.5: By definitions of $\equiv_k$ and f_k and Lemma 5.3.

Proof of Lemma 5.8: By Lemma 5.5 and Proposition 5.7.

Proof of Lemma 7.2: Suppose $k \leq l$. Let υ be the substitution defined by $\upsilon(x) = x \wedge \Box^k\bot$. The reader may easily verify that $\models \upsilon(\sigma_l(x)) \leftrightarrow \sigma_k(x)$. Hence, $\sigma_l \preceq \sigma_k$.

Proof of Lemma 7.3: Suppose $k < l$ and $\sigma_k \preceq \sigma_l$. Let υ be a substitution such that $\sigma_k \circ \upsilon \simeq \sigma_l$. Hence, $\models \upsilon(\sigma_k(x)) \leftrightarrow \sigma_l(x)$. Thus, $\models \Box^{<l}x \wedge \Box^l\bot \rightarrow \Box^k\bot$. Consequently, $\models \Box^l\bot \rightarrow \Box^k\bot$. Hence, $l \leq k$: a contradiction.

Proof of Lemma 8.1: Let θ and μ be the substitutions defined by

- $\theta(x) = x$ if $x \in var(\sigma(x))$ and $\theta(x) = \top$ otherwise,
- $\mu(x) = x$ if $x \in var(\tau(x))$ and $\mu(x) = \top$ otherwise.

The reader may easily verify that for all variables x, $\models \theta(\upsilon(x)) \leftrightarrow \sigma(x)$ and $\models \mu(\upsilon(x)) \leftrightarrow \tau(x)$. Hence, $\upsilon \preceq \sigma$ and $\upsilon \preceq \tau$.

Proof of Lemma 8.2: By induction on ϕ.

Proof of Lemma 8.3: By induction on ϕ.

Proof of Lemma 8.4: By induction on ϕ.

Proof of Lemma 8.5: By induction on ϕ.

Proof of Lemma 8.8: By Lemmas 8.2 and 8.7.

Proof of Lemma 8.9: Suppose $k \leq l$. Let υ be the substitution defined by $\upsilon(t_i) = t_k$ if $i = k$ and $\upsilon(t_i) = \top$ otherwise. The reader may easily verify that $\models \upsilon(\gamma_l(x)) \leftrightarrow \sigma_k(x)$, $\models \upsilon(\gamma_l(y)) \leftrightarrow \sigma_k(y)$ and $\models \upsilon(\gamma_l(z)) \leftrightarrow \sigma_k(z)$. Hence, $\gamma_l \preceq \sigma_k$.

Proof of Lemma 8.10: Suppose $k \leq l$. Let υ be the substitution defined by $\upsilon(t_i) = t_i$ if $i \leq k$ and $\upsilon(t_i) = \top$ otherwise. The reader may easily verify that $\models \upsilon(\gamma_l(x)) \leftrightarrow \gamma_k(x)$, $\models \upsilon(\gamma_l(y)) \leftrightarrow \gamma_k(y)$ and $\models \upsilon(\gamma_l(z)) \leftrightarrow \gamma_k(z)$. Hence, $\gamma_l \preceq \gamma_k$.

Proof of Lemma 8.12: By Lemma 8.9 and Proposition 8.11.

Proof of Lemma 8.14: By induction on ϕ.

Proof of Lemma 8.15: The equivalence between $\sigma(\phi)$, $\bigwedge\{\Diamond^i\sigma(x)$:

$x \in AF$ & $i \in is_\diamond(x,\phi)\} \wedge \diamond^{deg(\phi)}\top$ and $\bigwedge\{\diamond^i\sigma(x) : x \in var(\phi)$ & $i \in is_\diamond(x,\phi)\} \wedge \diamond^{deg(\phi)}\top$ is a consequence of Lemma 8.14. The equivalence between $\bigwedge\{\diamond^i\sigma(x) : x \in var(\phi)$ & $i \in is_\diamond(x,\phi)\} \wedge \diamond^{deg(\phi)}\top$ and $\bigwedge\{\diamond^i\sigma(x) : x \in var(\phi)$ & $i = \max is_\diamond(x,\phi)\} \wedge \diamond^{deg(\phi)}\top$ is a consequence of the fact that for all $i, j \in \mathbb{N}$, if $i \leq j$ then $\models \diamond^i\top \wedge \diamond^j\top \leftrightarrow \diamond^j\top$.

Proof of Lemma 8.16: Suppose for all $x \in AF$, $\max is_\diamond(x,\phi) = \max is_\diamond(x,\phi')$ and $\max is_\diamond(x,\psi) = \max is_\diamond(x,\psi')$. Let σ be a closed substitution. By Lemma 8.15, $\models \sigma(\phi) \leftrightarrow \bigwedge\{\diamond^i\sigma(x) : x \in var(\phi)$ & $i = \max is_\diamond(x,\phi)\} \wedge \diamond^{deg(\phi)}\top$, $\models \sigma(\phi') \leftrightarrow \bigwedge\{\diamond^i\sigma(x) : x \in var(\phi')$ & $i = \max is_\diamond(x,\phi')\} \wedge \diamond^{deg(\phi')}\top$, $\models \sigma(\psi) \leftrightarrow \bigwedge\{\diamond^i\sigma(x) : x \in var(\psi)$ & $i = \max is_\diamond(x,\psi)\} \wedge \diamond^{deg(\psi)}\top$ and $\models \sigma(\psi') \leftrightarrow \bigwedge\{\diamond^i\sigma(x) : x \in var(\psi')$ & $i = \max is_\diamond(x,\psi')\} \wedge \diamond^{deg(\psi')}\top$. Since $deg(\phi) = deg(\phi')$, $var(\phi) = var(\phi')$, $deg(\psi) = deg(\psi')$, $var(\psi) = var(\psi')$ and for all $x \in AF$, $\max is_\diamond(x,\phi) = \max is_\diamond(x,\phi')$ and $\max is_\diamond(x,\psi) = \max is_\diamond(x,\psi')$, therefore $\models \sigma(\phi) \leftrightarrow \sigma(\phi')$ and $\models \sigma(\psi) \leftrightarrow \sigma(\psi')$. Hence, $S \cup \{(\phi,\psi)\}$ possesses a unifier iff $S \cup \{(\phi',\psi')\}$ possesses a unifier.

Proof of Lemma 8.18: Let S' be the finite set of pairs of minimalist formulas obtained by replacing each pair (ϕ,ψ) in S of formulas by the pair $(\bigwedge\{\diamond^i x : x \in var(\phi)$ & $i = \max is_\diamond(x,\phi)\} \wedge \diamond^{deg(\phi)}\top, \bigwedge\{\diamond^i x : x \in var(\psi)$ & $i = \max is_\diamond(x,\psi)\} \wedge \diamond^{deg(\psi)}\top)$ of minimalist formulas. By Lemma 8.16, S possesses a unifier iff S' possesses a unifier. By Lemma 8.17, S' can be easily transformed into an equivalent finite set of pairs of minimalist normal formulas.

To Know is to Know the Value of a Variable

Alexandru Baltag [1]

ILLC, University of Amsterdam

Abstract

We develop an epistemic logic that can express *knowledge of a dependency between variables* (or complex terms). An epistemic dependency formula $K_a^{t_1,\ldots,t_n} t$ says that agent a knows the value of term t conditional on being given the values of terms $t_1, ..., t_n$. We add dynamic operators $[!t_1, \ldots, t_n]\phi$, capturing the effect of publicly (and simultaneously) announcing the values of terms $t_1, \ldots, t_n$. We prove completeness, decidability and finite model property.

Keywords: knowing what, knowledge de re, dynamic epistemic logic.

1 Introduction

In this paper we build on the work of Plaza [14,15], and Wang and Fan [24,25] on formalizing the notion of 'knowledge de re' (knowledge of an object, "knowledge what") over Kripke models [2]. We understand this as *knowing the value of a variable.* Here, a variable is what in first-order modal logic is called a "non-rigid designator" x, taking different values (in some fixed domain D) at different possible worlds. If we denote by $w(x)$ the value of variable x at world w, and we denote by $\sim_a$ the epistemic accessibility relation of some agent a, then Plaza's semantics for 'knowledge de re' is given by putting [3]: $w \models K_a x$ iff $\forall v \sim_a w\ (v(x) = w(x))$. This is a natural analogue of the usual semantics of "knowledge that" in epistemic logic: an agent knows the value of x if that value is the same in all her epistemic alternatives. When the range D of possible values of x is finite, then this operator is obviously reduceable to the usual one, via a finite disjunction $\bigvee_{d \in D} K_a(x = d)$. But in general this is not possible. Plaza [15] had a very simple axiomatization of this operator (in combination with the usual epistemic operator $K_a\varphi$ for "knowing that"), and claimed its completeness, based on a reduction to standard epistemic logic. He also extended

---

[1] thealexandrubaltag@gmail.com

[2] In its turn, the work of Wang and Fan builds on previous research in Security on knowledge of keys and passwords, e.g. [9,11,23].

[3] We use the the same symbol K_a for "knowledge what" as the usual epistemic operator for "knowledge that", and we use variables $x, y, \ldots$ to denote the non-rigid designators. Plaza, and Wang and Fan, use a different notation Kv_a for "knowledge what", and denote the non-rigid designators by constants c. In our framework, doing this would be very confusing, since we also have *rigid* designators, which are naturally denoted by constants.

this logic with public announcement operators[4] $[!\varphi]\psi$ (of which he is the main originator [14]), and used the resulting logic to treat the classical "Sum and Product" puzzle. But he could not prove completeness of this extended logic. Wang and Fan solved this problem by introducing a *conditional version* of the above operator $K_a^\varphi x$ ("conditionally knowing what"), with the intuitive meaning that agent a could find the value of x *if given* the additional information that φ was the case. The introduction of such a conditional operator allows one to "pre-encode" the dynamics of $[!\varphi]\psi$, following a strategy pioneered by van Benthem [20], obtaining Reduction Axioms that allow us to reduce any dynamic formula to a static one. Their completeness proof was very complex (in the multi-agent case), going via a detour through first-order intentional logic. A more natural type of announcement in this context is the action $!x$ of *publicly announcing the value of* x. In a recent talk (Univ. of Amsterdam 2015), Wang stated as an open question the problem of finding a complete axiomatization for a logic that combines the operators for "knowledge that" $K\phi$, "knowledge of a value" Kx, propositional public announcements $[!\phi]\psi$ and public announcements of values $[!x]\psi$. This problem remained open until now, despite efforts in this direction by van Eijck, Gattinger and Wang.[5]

In this paper we solve this problem, by introducing another kind of conditional version of the above operator. An *epistemic dependency formula* $K_a^{x_1,\ldots,x_n} y$ says that an agent knows the value of some variable y *conditional* on being given the values of variables $x_1, \ldots, x_n$. The semantics is the obvious generalization of the above clause: if we use the abbreviation $w(\overrightarrow{x}) = v(\overrightarrow{x})$ for the conjunction $w(x_1) = v(x_1) \wedge \ldots w(x_n) = v(x_n)$, then we put

$$w \models K_a^{x_1,\ldots,x_n} y \text{ iff } \forall v \sim_a w \left(w(\overrightarrow{x}) = v(\overrightarrow{x}) \Rightarrow v(y) = w(y) \right).$$

In words: an agent knows y given $x_1, \ldots, x_n$ if the value of y is the same in all the epistemic alternatives that agree with the actual world on the values of $x_1, \ldots, x_n$. This operator has connections with Dependence Logic[6] and allows us to "pre-encode" the dynamics of the value-announcement operator $[!x]\varphi$.

Besides the epistemic dependency formulas and the dynamic public value-announcement operator, we introduce a number of other formal innovations, that are useful for both technical and conceptual purposes. One is that, in addition to variables, we also allow *constants* (i.e. rigid designators) c, whose value is the same in all possible worlds, as well as more complex terms t (built

[4] We use the dynamic-logic style notation for this operator that is by now standard in Dynamic Epistemic Logic, which we regard as natural: this is a dynamic modality, capturing weakest precondition of an action exactly as in *PDL*, except that the action is the one of publicly announcing φ. Plaza uses the more opaque notation $\varphi + \psi$.

[5] After submitting the AiML abstract, we became aware of an unpublished draft by van Eijck, Gattinger and Wang, containing work in progress on a *partial* solution to this problem. The logic axiomatized there has knowledge de re operators Kx and value announcements $[!x]\psi$, but it *cannot* express the usual "knowledge that" $K\phi$, *nor* the usual (propositional) public announcements $[!\phi]\psi$, so it is *not* a complete solution to the above problem.

[6] See Section 4.

from variables and constants using functions). Moreover, we have *relational atoms* $R(t_1, \ldots, t_n)$ expressing relationships between terms, and in particular an *equality predicate* $t = t'$, which captures *identity of values* and plays an essential role in our system. Although statements $K_a t$ *cannot* in general be reduced to "knowledge that" statements $K_a\varphi$, formulas of the form $x = c$ can be used to provide "local reductions": semantically, at each possible world w, $K_a x$ is equivalent to $K_a(x = c)$, where c is denotes the value $w(x)$ of variable x in world w. In our axiomatic system, this "local reduction" takes the shape of our "Knowledge De Re" axiom, whose relevant instance in this case is the validity

$$(x = c) \Rightarrow (K_a x \Leftrightarrow K_a(x = c)).$$

In words: *when the value of x is c, then knowing the value of x is the same as knowing that this value is c.* Our Knowledge De Re axiom generalizes this to epistemic dependency formulas. Combined with our "Existence of Value" Rule (saying that *variables always have a value*), this allows us to prove complex properties of epistemic dependence (e.g. the well-known Armstrong axioms [2]) in a simple way, from basic epistemic axioms. It also allows us to provide a rather simple completeness proof, based on a variation of the canonical model construction, in which constants act as 'witnesses' for the values of variables.

Another technical innovation is that we include a special type of variables $?_\varphi$, storing the *truth value of formula φ*. On the one hand, this introduces another layer of (interesting) technical complexity, since terms of the form $?_\varphi$ are even "more non-rigid" than the generic variables x, in that they can change their value while this value is being learnt. Indeed, while x keeps its value when that value is publicly announced, terms $?_\varphi$ corresponding to Moore sentences φ (such as "$x = 0$ but you don't know it") may change their values after being learnt. On the other hand, the use of such "fluctuating variables" allow us to simplify the syntax, by reducing the usual 'knowledge that' operator to 'knowledge what', via the equivalence $K_a\varphi \Leftrightarrow (\varphi \wedge K_a?\varphi)$. This is in the spirit of the Schaffer quote above: to "know that" φ is to know the answer to the question "what is the truth value of φ?" So, unlike Plaza, and Wang and Fan, we do not need two epistemic operators: there is only one kind of knowledge, namely knowing the value of a variable. Similarly, propositional announcements $[!\varphi]\psi)$ can be reduced to learning the value of variable $?_\varphi$.

So one could say, without exaggerating too much, that all knowledge "is", or can at least be *represented as*, knowledge of the value of a variable. Hence, this paper's title: itself a paraphrase of Quine's famous *dictum.*[7] This epistemic-modeling variant seems less problematic than the original, ontological version! And, in fact, our formalism suggests that the unary knowledge operator is just a special case. The more general version of our motto is: *to know is to know the dependence between (values of) variables.* This fits well with the popular view of knowledge-acquisition as a process of *learning correlations* (with the goal of eventually *tracking causal relationships* in the actual world).

[7] "To be is, purely and simply, to be the value of a variable" [16].

2 The Logic of Epistemic Dependency

In the following, we assume given a finite set $\mathcal{A}$ of "agents", and four countable sets of symbols: a set P of *propositional atoms*; a set Var of *variables*; a set C of *constants*, among which there are two distinguished constants 0 and 1 (with $0 \neq 1$); a set $\mathcal{F}$ of *functional symbols* and a set $\mathcal{R}$ of *relational symbols*, together with an *arity map* $ar : \mathcal{F} \cup \mathcal{R} \to N^*$, associating to all symbols $f \in F, R \in Rel$ natural numbers $ar(f), ar(R) \in N^*$. $\mathcal{R}$ includes an *equality* symbol $=$, with $ar(=) = 2$. Intuitively, the difference between variables and constants is that constants are rigid designators, while variables are non-rigid: so a variable can take different values in different possible worlds, while a constant denotes the same objects in all the worlds of a given model. We use letters $p, q, \ldots$ to denote atoms in P, letters $x, y, \ldots$ to denote variables in Var, and letters $c, d, \ldots$ to denote constants in C. We denote by $\overrightarrow{x}$ finite strings $\overrightarrow{x} = (x_1, \ldots, x_k) \in Var^*$ of variables (of any length $k \geq 0$), and similarly use $\overrightarrow{c}$ to denote finite strings $\overrightarrow{c} = (c_1, \ldots, c_k) \in C^*$ of objects. We denote by λ the empty string.

Syntax. The *Logic of Epistemic Dependency* (*LED*) has a twofold syntax, consisting of a set $\mathcal{L} = \mathcal{L}(P, Var, C, F, ar)$ of propositional *formulas* φ and a set $\mathcal{T} = \mathcal{T}(P, Var, C, \mathcal{F}, \mathcal{R}, ar)$ of *terms* t, defined by double recursion:

$$\varphi ::= p \mid R(\overrightarrow{t}) \mid \varphi \to \varphi \mid K_a^{\overrightarrow{t}} t$$

$$t ::= x \mid c \mid ?_\varphi \mid f(\overrightarrow{t})$$

where $a \in \mathcal{A}$ are agents, $x \in Var$ are variables, $c \in C$ are constants, $t \in \mathcal{T}$ are terms, $\overrightarrow{t}$ are finite tuples of terms and $f \in \mathcal{F}, R \in \mathcal{R}$ are symbols of arity equal to the length of $\overrightarrow{t}$. We abbreviate $= (t, t')$ as $t = t'$.

Semantics. A *model (for $\mathcal{L}$ and $\mathcal{T}$)* is a structure

$$M = (W, D, [0], [1], \sim_a, \| \bullet \|, \bullet(\bullet), \mathbf{f}, \mathbf{R})_{a \in \mathcal{A}, f \in \mathcal{F}, R \in \mathcal{R}}$$

where: W is a set of *possible worlds*; D is a set of *objects*, containing at least two designated objects $[0] \neq [1]$; $\sim_a \subseteq W \times W$ are equivalence relations, called *epistemic indistinguishability* relations; $\| \bullet \|$ is a *valuation function* mapping each atomic sentence $p \in P$ to a set $\|p\| \subseteq W$ of possible worlds; $\bullet(\bullet) : W \times (Var \cup C) \to D$ is a map associating to each world $w \in W$ and each variable or constant $\alpha \in Var \cup C$ some object $w(\alpha) \in D$, called *the value of α at world w*, and satisfying the requirement that *the value of each constant is the same in all the worlds*: i.e., $w(c) = w'(c)$ for all $c \in C$ and all $w, w' \in W$; and for all symbols $f \in \mathcal{F}, R \in \mathcal{R}$ of arity $ar(f) = n$, we are given n-ary maps $\mathbf{f} : D^n \to D$ and n-ary relations $\mathbf{R} \subseteq D^n$, with the *standard interpretation of equality* $=$ as the *diagonal* $\{(d, d) : d \in D\}$ of D.

For the semantics, we simultaneously define an *extended valuation* (the "truth map") $\|\varphi\|_M$ for all formulas φ, and an *extended value map* $w(t)_M$ for all terms t and all worlds $w \in W$. We will use the notation $w(\overrightarrow{t}) := (w(t_1), \ldots, w(t_k))$ for the string of values corresponding to any given string of terms $\overrightarrow{t} = (t_1, \ldots, t_k) \in \mathcal{T}^*$. The truth map is given for propositional atoms $p \in P$ by the valuation $\|p\|$, and extended to other formulas by recursively putting: $\|R(\overrightarrow{t})\| = \{w \in W | w(\overrightarrow{t}) \in \mathbf{R}\}$; $\|\varphi \to \psi\| = (W \setminus \|\varphi\|) \cup \|\psi\|$; $\|K_a^{\overrightarrow{t}} t'\| = \{w \in W | \forall v \in W (w \sim_a v \wedge w(\overrightarrow{t}) = v(\overrightarrow{t}) \Rightarrow w(t') = v(t'))\}$. The extended value map $w(t')$ is given by the value map $w(\alpha)$ for variables and constants $\alpha \in Var \cup C$, and extended to other terms by recursion: $w(?_\varphi) = [1]$ iff $w \in \|\varphi\|$; $w(?_\varphi) = 0$ iff $w \notin \|\varphi\|$; $w(f(\overrightarrow{t})) = \mathbf{f}(w(\overrightarrow{t}))$.

Abbreviations. We put $\top := (1 = 1)$; $\bot := (1 = 0)$: $\neg\varphi := \varphi \to \bot$; $\varphi \vee \psi := \neg\varphi \to \psi$; $\varphi \wedge \psi := \neg(\neg\varphi \vee \neg\psi)$; $\varphi \leftrightarrow \psi := (\varphi \to \psi) \wedge (\psi \to \varphi)$; $K_a^{\overrightarrow{t}} \varphi := \varphi \wedge K_a^{\overrightarrow{t}} ?_\varphi$; $\langle K_a^{\overrightarrow{t}} \rangle \varphi := \neg K_a^{\overrightarrow{t}} \neg\varphi$; $K_a \varphi := K_a^\lambda \varphi$ (where λ is the empty string); $\langle K_a \rangle \varphi := \neg K_a \neg \varphi$; $K_a^\varphi \psi := K_a(\varphi \to \psi)$. We also put $K_a^{\overrightarrow{t}} \overrightarrow{t'} := \bigwedge_{1 \leq i \leq k} K_a^{\overrightarrow{t}} t'_i$, and $(\overrightarrow{t} = \overrightarrow{t'}) := \bigwedge_{1 \leq i \leq k} t_i = t'_i$, where k is the length of $\overrightarrow{t'}$.

Ground Terms A *ground term* is a term that contains no variables and no propositional formulas (hence, no ?); in other words, ground terms are built only from constants $c, d, \ldots \in C$ by recursively applying function symbols $f, g, \ldots$. Let us denote by $\mathcal{T}^0$ the set of all ground terms.

Propositional Substitution: For atoms $p \in P$ and formulas θ, the *substitution of p with θ* is an operation mapping every formula $\varphi \in \mathcal{L}$ into a new formula $\varphi[p/\theta] \in \mathcal{L}$, and similarly mapping every tuple of term $\overrightarrow{t} \in \mathcal{T}$ into a new tuple $\overrightarrow{t}[p/\theta]$, obtained by uniformly substituting p with θ as usual.[8]

Variable Substitution: For variables $x \in Var$ and terms $t \in \mathcal{T}$, the *substitution of x with t* is an operation mapping every formula $\varphi \in \mathcal{L}$ into a new formula $\varphi[x/t] \in \mathcal{L}$, and mapping every term $t' \in \mathcal{T}$ into a new term $t'[x/t]$, obtained by uniformly substituting x with t in the usual way.[9]

Example 1. Alice and Bob have each a natural number written on their foreheads. It is common knowledge that Alice's number x_a is the immediate successor of Bob's number x_b. Both are blindfolded, so nobody can see the numbers. The *model* has: $Var = \{x_a, x_b\}$, $D = C = N$ is the set of natural numbers; $\mathcal{F} = \{+, \times\}$ and $\mathcal{R} = \{=, >\}$ contain the usual operations and relations on N; the set W of worlds consists of all functions $w : Var \to N$, satisfying the given constraint $w(x_a) = w(x_b) + 1$; the epistemic relations are given by the universal relations: $\sim_a = \sim_b = W \times W$. Note that the sentence $\neg K_a x_a \wedge \neg K_b x_b \wedge K_a(x_a > x_b) \wedge K_b(x_a > x_b) \wedge K_a^{x_b} x_a \wedge K_b^{x_a} x_b$ is true in all worlds. So *nobody knows his/her number, but both know that Alice's number is larger, and both could come to know the numbers if given only the other's number.*

---

[8] More precisely: $p[p/\theta] := \theta$; $q[p/\theta] := q$; $(R(t_1, \ldots, t_n))[p/\theta] := R(t_1[p/\theta], \ldots, t_n[p/\theta])$; $(\varphi \to \psi)[p/\theta] := \varphi[p/\theta] \to \psi[p/\theta]$; $K_a^{t_1, \ldots, t_n} t[p/\theta] := K_a^{t_1[p/\theta], \ldots, t_n[p/\theta]} t[p/\theta]$; $\lambda[p/\theta] := \lambda$; $c[p/\theta] := c$; $x[p/\theta] := x$; $?_\varphi[p/\theta] := ?_{\varphi[p/\theta]}$; $f(t_1, \ldots, t_n)[p/\theta] := f(t_1[p/\theta], \ldots, t_n[p/\theta])$.

[9] I.e., $\lambda[x/t] := \lambda$; $c[x/t] := c$; $x[x/t] := t$; $y[x/t] := y$; $?_\varphi[x/t] := ?_{\varphi[x/t]}$; $f(t_1, \ldots, t_n)[x/t] := f(t_1[x/t], \ldots, t_n[x/t])$; $p[x/t] := p$; $(R(t_1, \ldots, t_n))[x/t] := R(t_1[x/t], \ldots, t_n[x/t])$; $(\varphi \to \psi)[x/t] := \varphi[x/t] \to \psi[x/t]$; $(K_a^{t_1, \ldots, t_n} t')[x/t] := K_a^{t_1[x/t], \ldots, t_n[x/t]} t'[x/t]$.

Proof system. The proof system LED consists of the following:

RULES:

- *Propositional Substitution*: From φ, infer $\varphi[p/\theta]$.
- *Variable Substitution*: From φ, infer $\varphi[x/t]$.
- *Modus Ponens Rule*: From φ and $\varphi \to \psi$, infer ψ.
- *Necessitation*: From φ, infer $K_a\varphi$.
- *Existence-of-Value Rule* (EVR):

$$\text{From } x = c \to \varphi \text{ , infer } \varphi, \text{ provided that } c \text{ does not occur in } \varphi.$$

AXIOMS:

- All the classical propositional tautologies.
- All the $S5$ axioms for K_a.
- *Knowledge De Re*:

$$(\overrightarrow{x} = \overrightarrow{c} \wedge y = d) \to \left(K_a^{\overrightarrow{x}} y \leftrightarrow K_a^{\overrightarrow{x}=\overrightarrow{c}} y = d)\right)$$

- *Equality Axioms*:

$$x = x$$

$$x = y \to y = x$$

$$(x = y \wedge y = z) \to x = z$$

$$\overrightarrow{x} = \overrightarrow{y} \to f(\overrightarrow{x}) = f(\overrightarrow{y})$$

$$(x = y \wedge R(\overrightarrow{z}, x, \overrightarrow{w})) \to R(\overrightarrow{z}, y, \overrightarrow{w})$$

- *Characteristic Functions*:

$$?_\varphi = 1 \leftrightarrow \varphi,$$

$$?_\varphi = 0 \leftrightarrow \neg\varphi,$$

- *Knowledge of Functions*:

$$K_a^{\overrightarrow{x}} f(\overrightarrow{x})$$

In fact, two of the axioms are redundant: symmetry and transitivity of = follow from the other axioms, but we chose to include them for convenience. We write $\vdash \psi$ if ψ is provable in the proof system LED. For any set of formulas Φ and any formula ψ, we write $\Phi \vdash \psi$ if there exist finitely many formulas $\phi_1, \ldots, \phi_n \in \Phi$ (for some $n \in N$) such that $\vdash (\phi_1 \wedge \ldots \phi_n) \to \psi$. We say that Φ is *logically closed* if, for every formula $\psi \in \mathcal{L}$, $\Phi \vdash \psi$ implies $\psi \in \Phi$. We say that Φ is *consistent* if $\Phi \nvdash \perp$, and that *a formula* φ *is consistent with* Φ if $\Phi \cup \{\varphi\}$ is consistent (equivalently: if $\Phi \nvdash \neg\varphi$).

Lemma 2.1 *For a set* Φ *of formulas, put* $K_a^{\overrightarrow{t}}\Phi := \{K_a^{\overrightarrow{t}}\phi : \phi \in \Phi\}$. *Then we have that:*

- *if* $\Phi \vdash \psi$ *then* $K_a^{\overrightarrow{t}}\Phi \vdash K_a^{\overrightarrow{t}}\psi$.
- $\Phi \cup \{\psi\} \vdash \theta$ *iff* $\Phi \vdash (\psi \to \theta)$.

Proposition 2.2 *Let φ be a formula and z be a variable that does* not *occur in the scope of any K_a-operator in φ. Then the following is provable in LED:*

$$\vdash (x = y \wedge \varphi[z/x]) \rightarrow \varphi[z/y].$$

NOTE The unrestricted version of the above schema is not valid! A counterexample is obtained by taking for instance φ to be the formula $K_a(z = c)$. This is related to the Phosphorus/Hesperus paradox.

Proposition 2.3 *(*Knowledge of Ground Terms and Ground Identities*). For all* ***ground terms*** *$t, t' \in \mathcal{T}^0$, all the instances of the following schema are provable in LED:*

$$\vdash K_a t;$$

$$\vdash t = t' \rightarrow K_a t = t'.$$

Proof. We prove the *first claim* by induction on t: *For the base step*, let $t := c$ be a constant. From the Knowledge De Re axiom, we get $\vdash x = c \rightarrow (K_a x \leftrightarrow K_a(x = c))$. By substituting c for x and using the first equality axiom, we get $\vdash K_a c \leftrightarrow K_a(c = c)$. But on the other hand, by applying Necessitation to the first equality axiom, we have $\vdash K_a(c = c)$, and hence we obtain $\vdash K_a c$. *For the inductive step*: consider a term of the form $f(\overrightarrow{t})$, where $\overrightarrow{t} = (t_1, \ldots, t_n)$ is a tuple of ground terms. By the induction hypothesis, we can assume that $\vdash K_a t_i$ for all $i = 1, n$. Using this and the Knowledge De Re axiom, we derive $\vdash \overrightarrow{t} = \overrightarrow{c} \rightarrow K_a \overrightarrow{t} = \overrightarrow{c}$. Combining this with $\vdash K_a \overrightarrow{t} = \overrightarrow{c} \rightarrow K_a f(\overrightarrow{t}) = f(\overrightarrow{c})$ (obtained by applying Necessitation, Kripke's axiom and Modus Ponens to the fourth equality axiom), we obtain $\vdash \overrightarrow{t} = \overrightarrow{c} \rightarrow K_a f(\overrightarrow{t}) = f(\overrightarrow{c}$. Combining this with the theorem $\vdash \left(\overrightarrow{t} = c \wedge f(\overrightarrow{c}) = d\right) \rightarrow K_a^{\overrightarrow{t}=\overrightarrow{c}} f(\overrightarrow{t}) = d$ (obtained from the axiom $\vdash K_a^{\overrightarrow{t}} f(\overrightarrow{t})$ and the Knowledge De Re axiom), we get that $\vdash \left(\overrightarrow{t} = c \wedge f(\overrightarrow{c}) = d\right) \rightarrow K_a f(\overrightarrow{t}) = d$. This, together with the obvious theorem $\vdash \overrightarrow{t} = c \rightarrow \left(f(\overrightarrow{t}) = d \rightarrow (\overrightarrow{t} = c \wedge f(\overrightarrow{c}) = d)\right)$ (an obvious consequence of the equality axioms), gives us $\vdash \overrightarrow{t} = c \rightarrow \left(f(\overrightarrow{t}) = d \rightarrow K_a f(\overrightarrow{t}) = d\right)$. Applying the the (EVR) rule, we get $\vdash f(\overrightarrow{t}) = d \rightarrow K_a f(\overrightarrow{t}) = d$, which by the Knowledge De Re axiom, yields $\vdash f(\overrightarrow{t}) = d \rightarrow K_a f(\overrightarrow{t})$. Applying again the (EVR) rule, we obtain $\vdash K_a f(\overrightarrow{t})$, as desired.

As for the second claim: given the first claim, we have $\vdash K_a t$ and $\vdash K_a t'$. This, together with (a suitable substitution instance of) the Knowledge De Re axiom and the conjunctivity of knowledge, gives us $\vdash (t = c \wedge t' = c) \rightarrow K_a(t = c \wedge t' = c)$, and hence (using equality axioms and the axioms of normal modal logic) $\vdash (t = c \wedge t' = c) \rightarrow K_a t = t'$. Together with $\vdash t = c \rightarrow (t = t' \rightarrow t' = c)$ (a consequence of the equality axioms), this yields $\vdash t = c \rightarrow (t = t' \rightarrow K_a t = t')$. By the (EVR) rule, we obtain $\vdash t = t' \rightarrow K_a t = t'$, as desired. □

Proposition 2.4 *All the following theorems are provable in LED:*

$\vdash K_a^{x_1,\ldots,x_k} y \to K_a^{x_{\pi(1)},\ldots,x_{\pi(k)}} y$, *for every permutation* $\pi : \{1,\ldots,k\} \to \{1,\ldots,k\}$

$$\vdash (K_a^{\vec{x}} \vec{y} \wedge K_a^{\vec{x},\vec{y}} \vec{z}) \to K_a^{\vec{x}} \vec{z}$$

$$\vdash K_a^{\vec{x}} \vec{y} \to K_a^{\vec{x},\vec{z}} \vec{y}$$

$$\vdash K_a^{\vec{x}} \vec{y} \to K_a^{\vec{x}} f(\vec{y})$$

$$\vdash \vec{x} = \vec{c} \to (K_a^{\vec{x}} \varphi \to K_a^{\vec{x}=\vec{c}} \varphi)$$

$$\vdash K_a^{\vec{x}} (\varphi \to \psi) \to (K_a^{\vec{x}} \varphi \to K_a^{\vec{x}} \psi)$$

$$\vdash K_a^{\vec{x}} \varphi \to \varphi$$

$$\vdash K_a^{\vec{x}} \varphi \to K_a^{\vec{x}} K_a^{\vec{x}} \varphi$$

$$\vdash \neg K_a^{\vec{x}} \varphi \to K_a^{\vec{x}} \neg K_a^{\vec{x}} \varphi$$

Proof. We only prove the first two formulas, the other proofs are similar. *For the first,* we use the obvious propositional validity $\vdash (x_1 = c_1 \wedge \ldots x_k = c_k) \to (x_{\pi(1)} = c_{\pi(1)} \wedge \ldots x_{\pi(k)} = c_{\pi(k)})$, together with two instances of Knowledge De Re axiom: $\vdash (x_1 = c_1 \wedge \ldots x_k = c_k) \to (K_a^{x_1,\ldots,x_k} y \leftrightarrow K_a^{x_1=c_1 \wedge \ldots x_k=c_k} y)$, and $\vdash (x_1 = c_1 \wedge \ldots x_k = c_k) \to (K_a^{x_{\pi(1)},\ldots,x_{\pi(k)}} y \leftrightarrow K_a^{x_{\pi(1)}=c_{\pi(1)} \wedge \ldots x_{\pi(k)}=c_{\pi(k)}} y)$. From these we derive $\vdash (x_1 = c_1 \wedge \ldots x_k = c_k) \to (K_a^{x_1,\ldots,x_k} y \to K_a^{x_{\pi(1)},\ldots,x_{\pi(k)}} y)$, then apply repeatedly the (EVR) rule to obtain the desired conclusion.

For the second, we use three instances of Knowledge De Re axiom: $\vdash (\vec{x} = \vec{c} \wedge \vec{y} = \vec{d}) \to (K_a^{\vec{x}} \vec{y} \leftrightarrow K_a^{\vec{x}=\vec{c}} \vec{y} = \vec{d})$, $\vdash (\vec{x} = \vec{c} \wedge \vec{y} = \vec{d} \wedge \vec{x} = \vec{e}) \to (K_a^{\vec{x},\vec{y}} \vec{z} \leftrightarrow K_a^{\vec{x}=\vec{c} \wedge \vec{y}=\vec{d}} \vec{z} = \vec{e})$, and $\vdash (\vec{x} = \vec{c} \wedge \vec{z} = \vec{e}) \to (K_a^{\vec{x}} \vec{z} \leftrightarrow K_a^{\vec{x}=\vec{c}} \vec{z} = \vec{e})$. From these, together with the usual properties of the normal propositional operator $K_a \phi$ (and the fact that $K^\phi \psi$ is just an abbreviation for $K_a(\phi \to \psi)$), we obtain $\vdash (\vec{x} = \vec{c} \wedge \vec{y} = \vec{d} \wedge = \vec{e}) \to ((K_a^{\vec{x}} \vec{y} \wedge K_a^{\vec{x},\vec{y}} \vec{z}) \to K_a^{\vec{x}} \vec{z})$, then we apply the (EVR) rule. □

One can also easily verify that:

Proposition 2.5 *The following Necessitation-type rule for $K_a^{\vec{t}}$ is derivable in LED: if $\vdash \varphi$ then $\vdash K_a^{\vec{t}} \varphi$.*

"Pseudo-modalities": necessitation/possibility forms. For any finite string $s \in (\mathcal{L} \cup (\mathcal{A} \times \mathcal{T}^*))^*$, consisting of formulas $\phi \in \mathcal{L}$ and/or pairs $(a, \vec{t})$ of agents $a \in \mathcal{A}$ and strings $\vec{t} \in \mathcal{T}^*$ of terms, we define "pseudo-modalities" $[s]$ and $\langle s \rangle$, mapping any formula $\phi \in \mathcal{L}$ to formulas $[s]\phi \in \mathcal{L}$ (called a "necessity form") and $\langle s \rangle \phi \in \mathcal{L}$ (called a "possibility form"). The definition is by recursion, putting for necessity forms: $[\lambda]\phi := \phi$ for the empty string λ; $[\psi, s]\phi := \psi \to [s]\phi$; and $[(a, \vec{t}), s]\phi := K_a^{\vec{t}} [s]\phi$. As for possibility forms, we put $\langle s \rangle \phi := \neg [s] \neg \phi$.

Lemma 2.6 *For every necessity form* $[s]$ *there exists some formula* $\psi \in \mathcal{L}$, *such that for all* $\theta \in \mathcal{L}$, *we have:*

$$\vdash [s]\theta \quad \textit{iff} \quad \vdash \psi \to \theta.$$

Moreover, the same constants and variables occur in ψ *as in* s.

Proof. If $s = \lambda$, then take $\psi := \top$. Otherwise, $[s]\theta$ is just a sequence of symbols of the form $\psi \to \ldots$ and $K_a^{\vec{t}} \ldots$, followed at the end by θ. Starting from the left, we can "eliminate" one by one each knowledge symbol $K_a^{\vec{t}} \ldots$ by "pushing" it into the premise, using the fact[10] that: $\vdash \psi \to K_a^{\vec{t}} \phi$ holds iff $\vdash \langle K_a^{\vec{t}} \rangle \psi \to \phi$ holds. At the end of this process, we obtain a formula of the form $\psi \to \theta$. It is easy to see that ψ depends only on s, not on θ, and that moreover ψ contains the same constants and variables as s. □

Lemma 2.7 *Given* $s \in (\mathcal{L} \cup (\mathcal{A} \times \mathcal{T}^*))^*$, $t \in \mathcal{T}$, $\varphi \in \mathcal{L}$, *let* c *be a constant that does not occur in* s, t *or* φ. *Then the following rule is admissible in LED:*

$$\textit{if} \ \vdash [s](t = c \to \varphi) \quad \textit{then} \ \vdash [s]\varphi$$

Proof. Let ψ be the formula associated to s by the previous Lemma: so for all θ, $\vdash [s]\theta$ iff $\vdash \psi \to \theta$. Suppose now that we have $\vdash [s](t = c \to \varphi)$. Then we also get $\vdash \psi \to (t = c \to \varphi)$, and hence $\vdash t = c \to (\psi \to \varphi)$. Let x be a variable not occurring in t, φ or s (and hence, by the previous Lemma, not occurring in ψ either). Using (some substitution instance of one of) the Equality Axioms, we obtain $\vdash x = c \to (t = x \to (\psi \to \varphi))$. Since c does not to occur in s, t or φ, by the previous Lemma it doesn't occur in ψ either. By the (EVR) rule, we obtain $\vdash t = x \to (\psi \to \varphi)$. Using the Variable Substitution Rule (where we substitute t for x), we get $\vdash t = t \to (\psi \to \varphi)$. But we also have $\vdash t = t$ (by another of the Equality axioms), and hence $\vdash \psi \to \varphi$. Using again the previous Lemma, we obtain $\vdash [s]\varphi$. □

Theorem 2.8 *The proof system LED is* sound *and* strongly complete *(and hence the logic LED is* compact*). Moreover, this logic has the* strong finite model property, *and hence it is* decidable.

The rest of this section is dedicating to the proof of this theorem. For any countable set of constants C, let $\mathcal{L}_C$ be the language of *LED* based only on constants in C. A *C-theory* Φ is a consistent set of formulas in $\mathcal{L}_C$; here, "consistent" means consistent with respect to the proof system *LED* formulated for the language $\mathcal{L}_C$. A *maximal C-theory* is a C-theory Φ that is maximal (w.r.t. inclusion) among all C-theories. A *C-witnessed theory* is a C-theory Φ such that, for every term $t \in \mathcal{T}_C$, string $s \in (\mathcal{L}_C \cup (\mathcal{A} \times \mathcal{T}_C^*))^*$ and formula $\varphi \in \mathcal{L}_C$, if $\Phi \vdash [s](t = c \to \varphi)$ for all $c \in C$, then $\Phi \vdash [s]\varphi$. Equivalently: if whenever $\langle s \rangle \varphi$ is consistent with Φ, then there exists some $c \in C$ s.t. $\langle s \rangle (t = c \wedge \varphi)$

[10]This is an instance of the well-known fact that in the axiomatic system $S5$, a formula $\psi \to \Box\phi$ is a theorem iff the formula $\Diamond\psi \to \phi$ is a theorem.

is consistent with Φ. A *maximal C-witnessed theory* is a C-witnessed theory which is not a proper subset of any other C-witnessed theory.

For the completeness proof, we make use of the following three easily verifiable results:

Lemma 2.9 *If Φ is a C-theory and $\Phi \nvdash \neg\phi$, then $\Phi \cup \{\phi\}$ is also a C-theory. Moreover, if Φ is C-witnessed, then $\Phi \cup \{\phi\}$ is C-witnessed.*

Lemma 2.10 *If $\Phi_0 \subseteq \Phi_1 \subseteq \ldots \subseteq \Phi_n \subseteq \ldots$ is an increasing chain of C-theories, then $\bigcup_{n\in N} \Phi_n$ is a C-theory. Moreover, if all Φ_n are C-witnessed then $\bigcup_{n\in N} \Phi_n$ is C-witnessed.*

Lemma 2.11 *A C-theory Φ is a C-witnessed maximal C-theory iff it is a maximal C-witnessed theory.*

The completeness proof goes now via the following steps:

Lemma 2.12 *(*Lindenbaum Lemma*) Every C-witnessed theory Φ can be extended to a maximal C-witnessed theory $T_\Phi \supseteq \Phi$.*

Proof. Let $\phi_0, \phi_1, \ldots, \phi_n, \ldots$ be an enumeration of formulas in $\mathcal{L}_C$. We define an increasing chain $\Phi_0 \subseteq \Phi_1 \ldots \subseteq \Phi_n \subseteq \ldots$ of C-witnessed theories: first, put $\Phi_0 := \Phi$; then, given the witnessed C-theory Φ_n, put $\Phi_{n+1} := \Phi_n$ if $\Phi \vdash \neg\phi_n$, and put $\Phi_{n+1} := \Phi_n \cup \{\phi_n\}$ otherwise (if $\Phi_n \nvdash \neg\phi_n$). Finally, we put $T_\Phi := \bigcup_{n\in N} \Phi_n$. By Lemma 2.10, this is a C-witnessed theory. Moreover, it is also a maximal C-theory (since every formula consistent with T_Φ is in T_Φ), so it is a maximal C-witnessed theory. □

Lemma 2.13 *(*Extension Lemma*) Let C be a set of constants, and let $C' = \{c_0, c_1, \ldots, c_n, \ldots\}$ be a countable set of "fresh" constants, i.e. s.t. $C \cap C' = \emptyset$. Put $\tilde{C} = C \cup C'$. Then every C-theory Φ can be extended to a $\tilde{C}$-witnessed theory $\tilde{\Phi} \supseteq \Phi$, and hence (by Lindenbaum Lemma) to a maximal $\tilde{C}$-witnessed theory $T_\Phi \supseteq \Phi$.*

Proof. Let $\gamma_1, \ldots, \gamma_n \ldots$ be an enumeration of all the triplets of the form $\gamma_n = (s_n, t_n, \phi_n)$ consisting of any necessity form $s_n \in (\mathcal{L}_{\tilde{C}} \cup (\mathcal{A} \times \mathcal{T}^*_{\tilde{C}}))^*$, any term $t_n \in \mathcal{T}_{\tilde{C}}$ and formula $\phi_n \in \mathcal{L}_{\tilde{C}}$. For every such triplet $\gamma_n = (s_n, t_n, \phi_n)$, put $C'(n) =: \{c' \in C' : c' \text{ occurs in either } s_n \text{ or } t_n \text{ or } \phi_n\}$. Note that $C'(n)$ is always finite.

We now construct an increasing chain $\Phi_0 \subseteq \Phi_1 \ldots \subseteq \Phi_n \subseteq \ldots$ of $\tilde{C}$-theories, satisfying the following three properties: (1) $\Phi_0 = \Phi$; (2) for every $n \in N$, the set $C'_n := \{c' \in C' : c' \text{ occurs in } \Phi_n\}$ is finite; (3) for every triplet $\gamma_n = (s_n, t_n, \phi_n)$ in the above enumeration, if $\Phi_n \nvdash \neg\langle s_n\rangle\phi_n$, then $\exists m \in N$ s.t. $\langle s_n\rangle(t_n = c'_m \wedge \phi_n) \in \Phi_{n+1}$. The construction is by recursion. For $n = 0$, we put $\Phi_0 := \Phi$, which takes care of condition (1) above. At step $n+1$, let Φ_n be a $\tilde{C}$-theory satisfying clause (1) above, and let $\gamma_n = (s_n, t_n, \phi_n)$ be the n-th triplet in the above enumeration. We have two cases: (a) if we have $\Phi_n \vdash [s_n]\neg\phi_n$, then we put $\Phi_{n+1} := \Phi_n$; (b) in case that we have $\Phi_n \nvdash [s_n]\neg\phi_n$, then we choose m

to be the least natural number bigger[11] than the indices of all the constants in $C'(n) \cup C'_n$, and put $\Phi_n+1 := \Phi_n \cup \{\langle s_n \rangle (t_n = c'_m \wedge \phi_n)\}$. To show that this gives us a C-theory, notice that c'_m doesn't occur in s_n, t_n, ϕ_n or Φ_n. If Φ_{n+1} were inconsistent, then we'd have $\Phi_n \vdash [s_n](t_n = c'_m \rightarrow \neg\phi_n)$, so $\exists \theta_1, \ldots, \theta_k \in \Phi_n$ s.t. $\vdash \theta_1 \rightarrow (\theta_n \rightarrow \cdots [s_n](t_n = c'_m \rightarrow \neg\phi_n))$ is a theorem in LED. But $c'_m \notin C'(n) \cup C'_n$, so c'_m doesn't occur in $s_n, t_n, \phi_n, \theta_1, \ldots, \theta_n$ (or in any other formula of Φ_n). By Lemma 2.7, we have that $\vdash \theta_1 \rightarrow (\theta_n \rightarrow \cdots [s_n]\neg\phi_n)$ is also a theorem in LED, and hence that $\Phi_n \vdash [s]\neg\phi_n$, contrary to our assumption (in case b). So in both cases Φ_{n+1} is a $\tilde{C}$-theory. It is also easy to see that it satisfies condition (2): in case (a) we have $C'_{n+1} = C'_n$ (finite by the inductive assumption); in case (b) we have $C'_{n+1} = C'_n \cup C'(n) \cup \{c'_m\}$ (still finite). Finally, it is obvious that condition (3) is satisfied.

Given now this increasing sequence $\Phi = \Phi_0 \subseteq \cdots \subseteq \Phi_n \subseteq \cdots$ of $\tilde{C}$-theories satisfying (1)-(3) above, take $\tilde{\Phi} := \bigcup_{n \in N} \Phi_n$. By Lemma 2.10, $\tilde{\Phi}$ is a $\tilde{C}$-theory, and it obviously includes $\Phi = \Phi_0$. Condition (3) above implies that $\tilde{\Phi}$ is $\tilde{C}$-witnessed. □

Together, the last three results imply that, in order to show completeness, it is enough to show that, for any countable set C of constants, *every maximal C-witnessed theory has a model.* We now proceed to prove this.

From now on, we fix the set of constants C, and we assume given a maximal C-witnessed theory T_0. For each term $t \in \mathcal{T} = \mathcal{T}_C$, we can define an equivalence relation $\sim^t$ on maximal C-witnessed theories T, T', by putting: $T \sim^t T'$ iff $\forall c \in C((t = c) \in T \Leftrightarrow (t = c) \in T')$. Put $\sim := \bigcap_{t \in \mathcal{T}^0} \sim^t$ (where recall that $\mathcal{T}^0$ is the set of ground terms). It is obvious that $\sim$ is also an equivalence relation on maximal C-witnessed theories.

In addition, we can define another equivalence relation $\equiv$ on the set of constants C by putting: $c \equiv c'$ iff $(c = c') \in T_0$. For any constant $c \in C$, let us denote by $[c] := \{c' \in C : c \equiv c'\}$ the equivalence class of c modulo $\equiv$.

Canonical Model The *canonical model for* T_0 is a model $M = (\Omega, D, [0], [1], \sim_a, \| \bullet \|, \bullet(\bullet), \mathbf{f}, \mathbf{R})_{a \in \mathcal{A}, f \in \mathcal{F}, R \in \mathcal{R}}$ for the language $\mathcal{L}_C$, defined as follows: the *state space* is $\Omega := \{T \subseteq \mathcal{L}_C :$ T maximal witnessed $\mathcal{L}_C$-theory with $T \sim T_0\}$; the set of *objects* is $D := \{[c] : c \in C\}$, where $[c]$ is the equivalence class of c modulo $\equiv$, and the equivalence classes $[0]$ and $[1]$ are the two designated objects; the *epistemic relations* are: $T \sim_a T'$ iff $\forall \varphi \in \mathcal{L}_C(K_a\varphi \in T \Rightarrow \varphi \in T')$. For $f \in \mathcal{F}$, we put $\mathbf{f}([c_1], \ldots, [c_n]) := [c]$ for $c_1, \ldots, c_n, c \in C$ with $(f(c_1, \ldots, c_n) = c) \in_0$; and for $R \in \mathcal{R}$, we put $\mathbf{R} := \{([c_1], \ldots, [c_n]) : R(c_1, \ldots, c_n) \in T_0$. The *valuation* is $\|p\| := \{T \in W : p \in T\}$. The *value* $T(\alpha)$ of $\alpha \in Var \cup C$ at world $T \in \Omega$ is given by $T(c) := [c]$ for $c \in C$, and $T(x) := [c]$, for $x \in Var$ and $c \in C$ with $(x = c) \in T$. It is easy to check that these definitions are independent of the choice of representatives, so M is indeed a *well-defined model* for $\mathcal{L}_C$.

[11] Such a number exists, due to the inductive assumption (2) above.

Lemma 2.14 *(*Intersection Lemma*) For all agents $a \in \mathcal{A}$ and finite strings of terms $\overrightarrow{t} = (t_1, \ldots, t_n)$, we have: $\sim_a^{\overrightarrow{t}} = \sim_a \cap \sim^{t_1} \cap \cdots \sim^{t_n}$.*

Proof. *The left-to-right inclusion*: the *LED*-theorem $\vdash K_a^{\overrightarrow{x}}\overrightarrow{y} \to K_a^{\overrightarrow{x},\overrightarrow{z}}\overrightarrow{y}$ (proven in Proposition 2.4) yields by substitution $\vdash K_a\varphi \to K_a^{\overrightarrow{t}}\varphi$ and $\vdash K_a^{t_i}\varphi \to K_a^{\overrightarrow{t}}\varphi$, from which we obtain $\sim_a^{\overrightarrow{t}} \subseteq \sim_a$ and $\sim_a^{\overrightarrow{t}} \subseteq \sim^{t_i}$ (for all $i = 1, n$). *For the converse inclusion*: suppose $T, S \in \Omega$ satisfy $T \sim_a S$ and $T \sim^{t_i} S$ for all $i = 1, n$. To show that $T \sim_a^{\overrightarrow{t}} S$, let $K_a^{\overrightarrow{t}}\varphi \in T$. We need to show that $\varphi \in S$: for this, notice that, since T is C-witnessed (and , for each $i = 1, n$ there must exist constants $c_i \in C$ such that T is consistent with $t_i = c_i$. Since T is maximal, it follows that $(t_i = c_i) \in T$, and moreover that $(\overrightarrow{t} = \overrightarrow{c}) \in T$. By applying the theorem $\vdash \overrightarrow{t} = \overrightarrow{c} \to (K_a^{\overrightarrow{t}}\varphi \to K_a^{\overrightarrow{t}=\overrightarrow{c}}\varphi)$, we obtain that $K_a^{\overrightarrow{t}=\overrightarrow{c}}\varphi \in T$, i.e. $K_a(\overrightarrow{t} = \overrightarrow{c} \to \varphi) \in T$. This together with $T \sim_a S$, gives us that $(\overrightarrow{t} = \overrightarrow{c} \to \varphi) \in S$. But from $(\overrightarrow{t} = \overrightarrow{c}) \in T$ and $T \sim^{t_i} S$ for all $i = 1, n$, we derive that $(\overrightarrow{t} = \overrightarrow{c}) \in T$, and hence by closure of (the maximal theory T) under modus ponens, we obtain that $\varphi \in T$. □

As a consequence of Lemma 2.14, *all $\sim_a^{\overrightarrow{t}}$ are equivalence relations.*

Lemma 2.15 *(*Diamond Lemma*) Let $T \in \Omega$, and let $a, \overrightarrow{t}, \varphi$ be such that $K_a^{\overrightarrow{t}}\varphi \notin T$. Then there exists some theory $S \in \Omega$ such that $T \sim_a^{\overrightarrow{t}} S$ but $\varphi \notin S$.*

Proof. Let $\Psi := \{\psi : K_a^{\overrightarrow{t}}\psi \in T\}$. We will show the following
Claim: The set $\Psi \cup \{\neg\varphi\}$ *is a C-witnessed theory.*

To prove this claim, we first need to show that this set is consistent. Suppose not; then there exist $\psi_1, \ldots, \psi_n \in \Psi$ (hence, $K_a^{\overrightarrow{t}}\psi_i \in T$ for all $i = 1, n$) such that $\vdash (\psi_1 \wedge \ldots \psi_n) \to \varphi$ is a theorem. But then we also have that $\vdash (K_a^{\overrightarrow{t}}\psi_1 \wedge \ldots K_a^{\overrightarrow{t}}\psi_n) \to K_a^{\overrightarrow{t}}\varphi$ and $(K_a^{\overrightarrow{t}}\psi_1 \wedge \ldots K_a^{\overrightarrow{t}}\psi_n) \in T$, hence $K_a^{\overrightarrow{t}}\varphi \in T$, in contradiction with our assumption (that $K_a^{\overrightarrow{t}}\varphi \notin T$).

Next, to show that $\Psi \cup \{\neg\varphi\}$ is C-witnessed, suppose that, for some triple (s', t', φ'), we have $\Psi \cup \{\neg\varphi\} \vdash [s'](t' = c \to \phi')$ for all $c \in C$. By a previous lemma, this gives that $\Psi \vdash (\neg\varphi \to [s'](t' = c \to \phi'))$ for all c, and by another lemma we obtain that $K_a^{\overrightarrow{t}}\Psi \vdash K_a^{\overrightarrow{t}}(\neg\varphi \to [s'](t' = c \to \phi'))$ (where recall that $K_a^{\overrightarrow{t}}\Psi = \{K_a^{\overrightarrow{t}}\psi : \psi \in \Psi\}$). But note that $K_a\Psi \subseteq T$, and hence we get $T \vdash K_a^{\overrightarrow{t}}(\neg\varphi \to [s'](t' = c \to \phi'))$ for all $c \in C$. Since $T \in \Omega$ is C-witnessed, it follows (by applying the C-witnessing condition to the necessitation form $((a, \overrightarrow{t}), \neg\varphi, s'))$ that $T \vdash K_a^{\overrightarrow{t}}(\neg\varphi \to [s']\phi')$, and hence by maximality that $K_a^{\overrightarrow{t}}(\neg\varphi \to [s']\phi') \in T$, hence $(\neg\varphi \to [s']\phi') \in \Psi$. From this we obtain that $\Psi \cup \{\neg\varphi\} \vdash [s']\phi'$, thus proving our Claim above.

Given the Claim, we can now use the Extension Lemma (in combination with Lindenbaum Lemma) to extend the set $\Psi \cup \{\neg\varphi\}$ to a maximal C-witnessed theory S. It is easy to see that we have $S \sim T \sim T_0$, hence $S \in \Omega$. We obviously have $(\neg\varphi) \in S$, so by consistency $\varphi \notin S$. Finally, $\Psi \subseteq S$ gives us that $T \sim_a^{\overrightarrow{t}} S$. □

Lemma 2.16 *("Knowledge de Re" Lemma) Let $T \in \Omega$, and let $a, \vec{t}, t'$ be such that $K_a^{\vec{t}} t' \notin T$. Then there exists some theory $S \in \Omega$ such that $T \sim_a^{\vec{t}} S$ but $T \not\sim^{t'} S$.*

Proof. Since T is a maximal C-witnessed theory, there exist $\vec{c} \in C^*$, $c \in C$ such that $\vec{t} = \vec{c}, t' = c' \in T$. By using the theorem $\vdash (\vec{t} = \vec{c} \wedge t' = c') \to (K_a^{\vec{t}} t' \leftrightarrow K_a(\vec{t} = \vec{c} \to t' = c'))$ (which is a substitution instance of the *Knowledge de Re* axiom) and the assumption that $K_a^{\vec{t}} t' \notin T$, we obtain that $K_a(\vec{t} = \vec{c} \to t' = c') \notin T$. By the Diamond Lemma, there exists some $S \in \Omega$ such that $T \sim_a^{\vec{t}} S$ but $(\vec{t} = \vec{c} \to t' = c') \notin S$. By the maximality of S, we get that $(\vec{t} = \vec{c} \wedge t' \neq c') \in S$, and hence that $T \sim_{t_i} S$ for all $i = 1, n$ but $T \not\sim_{t'} S$. Using also $T \sim_a^{\vec{t}} S$ and the Intersection Lemma above, we conclude that $T \sim_a^{\vec{t}} S$ (and $T \not\sim_{t'} S$), as desired. □

Lemma 2.17 *(Truth Lemma) Let $M = (\Omega, D, \sim_a, \| \bullet \|, \bullet(\bullet), \mathbf{f})_{a \in \mathcal{A}, f \in F}$ be the canonical model for (some theory) T_0. Then for every formula φ and every term t, we have:*

(1) $T \in \|\varphi\|_M$ iff $\varphi \in T$, and

(2) $T(t) = [c]$ iff $(t = c) \in T$.

Proof. We prove both claims by simultaneous induction on the complexity [12] of formulas φ and terms t:

To prove (1): for *atomic formulas* $p \in P$, (1) is trivial. *For relational atoms* $R(\vec{t})$, let $\vec{[c]} \in D^*$ be s.t. $T(\vec{t}) = \vec{[c]}$, so by the induction hypothesis for (2) we have $(\vec{t} = \vec{c}) \in T$. Then we have the following sequence of equivalencies: $T \in \|R(\vec{t})\|_M$ iff $T(\vec{t}) \in \mathbf{R}$ iff $\vec{[c]} \in \mathbf{R}$ iff $R(\vec{[c]}) \in T_0$ iff (using $T \sim T_0$) $R(\vec{[c]}) \in T$ iff $R(\vec{[t]}) \in T$ (where at the last step we used the fact that $(\vec{t} = \vec{c}) \in T$ and the equality axioms). For *implicational formulas* $\phi \to \psi$, this goes as usual, using the properties of maximally consistent theories. For *epistemic formulas* $K_a^{\vec{t}} t'$, with $\vec{t} = (t_1, \ldots, t_n)$: to prove the *left-to-right implication*, suppose that $T \in \|K_a^{\vec{t}} t'\|_M$ but $(K_a^{\vec{t}} t') \notin T$. By the Knowledge de Re Lemma, there exists some $S \in \Omega$ such that $T \sim_a^{\vec{t}} S$ but $T \not\sim^{t'} S$. By the Intersection Lemma, we obtain that $T \sim_a S$ and $T \sim^{t_i} S$ for all $i \in \{1, \ldots, n\}$; i.e. for every $c \in C$ and $i \in \{1, \ldots, n\}$, we have that: $(t_i = c) \in T \Leftrightarrow (t_i = c) \in S$. By the induction hypothesis for claim (2), this implies that $T(\vec{t}) = S(\vec{t})$. From this, together with $T \sim_a S$ and the fact that $T \in \|K_a^{\vec{t}} t'\|_M$ (as well as the semantic clause for knowledge de re), we obtain that $T(t') = S(t')$. Applying again the induction hypothesis for (2), we get that

---

[12] Our notion of complexity is a function $comp : \mathcal{L}_C \cup \mathcal{T}_C \to N$, defined recursively by putting: $comp(p) = comp(c) = comp(x) = 0$, $comp(R(t_1, \ldots, t_n)) = 1 + max(comp(t_1), \ldots, comp(t_n))$, $comp(\phi \to \psi) = 1 + max(comp(\phi), comp(\psi))$, $comp(K_a^{t_1, \ldots, t_n} t') = 1 + max(comp(t_1), \ldots, comp(t_n), comp(t'))$, $comp(?_\phi) = 1 + comp(\phi)$, $comp(f(t_1, \ldots, t_n)) = 1 + max(comp(t_1), \ldots, comp(t_n))$.

$(t' = c) \in T \Leftrightarrow (t' = c) \in S$ holds for all $c \in C$, i.e. $T \sim^{t'} S$ contrary to our assumption above. To show the *converse*: suppose that $(K_a^{\vec{t}} t') \in T$. Since T is C-witnessed there must exist $\vec{c} \in C^*, c' \in C$ such that $(\vec{t} = \vec{c}), (t' = c') \in T$. By the induction hypothesis for (2), we have $T(t_i) = [c_i]$ for all $i \in \{1, \ldots, n\}$, and also $T(t') = [c']$. To prove now that $T \in \|\varphi\|_M$, let $S \in \Omega$ be such that $T \sim_a S$ and $T(\vec{t}) = S(\vec{t})$. It is enough to prove that $T(t') = S(t')$. Using the Knowledge de Re axiom, and the fact that $(K_a^{\vec{t}} t') \in T$, we obtain that $(K_a^{\vec{t}=\vec{c}} t' = c') \in T$, i.e. $K_a(\vec{t} = \vec{c} \to t' = c') \in T$. Since $T \sim_a S$, we must have $(\vec{t} = \vec{c} \to t' = c') \in S$. From $T(\vec{t}) = S(\vec{t})$ and $T(t_i) = [c_i]$ for all i, we get that $S(t_i) = [c_i]$ for all i, and so by the induction hypothesis for (2) we have $(\vec{t} = \vec{c}) \in S$. This, together with $(\vec{t} = \vec{c} \to t' = c') \in S$, gives us that $(t' = c') \in S$. Applying again the induction hypothesis for the second claim, we obtain that $S(t') = [c'] = T(t')$, as desired.

To show (2): it is trivially true for variables $x \in Var$ and constants $c' \in C$. For terms of the form $?_\varphi$: we know that by definition $T(?_\varphi) = [1]$ holds iff $T \in \|\varphi\|_M$ holds, i.e. (by the induction hypothesis for (1)) iff $\varphi \in T$ iff $(?_\varphi \leftrightarrow 1) \in T$ (by the Characteristic Functions Axiom). A similar argument shows that $T(?_\varphi) = [0]$ iff $(?_\varphi \leftrightarrow 0) \in T$. Since $[0]$ and $[1]$ are the only possible values of $T(?_\varphi)$, we obtain the desired conclusion. For terms of the form $f(t_1, \ldots, t_n)$, let $c_1, \ldots, c_n \in C$ be s. t. $T(t_1) = [c_1], \ldots, T(t_n) = [c_n]$, i.e. $(t_1 = c_1), \ldots, (t_n = c_n) \in T$. (By the definition of the canonical value function, such constants must exist.) We have $T(f(t_1, \ldots, t_n) = \mathbf{f}(T(t_1), \ldots, T(t_n)) = \mathbf{f}([c_1], \ldots, [c_n]) = [f(c_1, \ldots, c_n)]$. Hence we have that: $T(f(t_1, \ldots, t_n) = [c]$ holds iff $f(c_1, \ldots, c_n)] = [c]$ holds, i.e. iff $(f(c_1, \ldots, c_n) = c) \in T$. □

In particular, T_0 is satisfied at world T_0 in M: *this finishes our proof of strong completeness.*

The *decidability proof* goes via the following two steps:

STEP 1: Reduction of *LED* validities to validities in a less expressive language. Let LED_0 be the language with the following syntax

$$\varphi ::= p \mid R(\vec{t}) \mid \varphi \to \varphi \mid K_a\varphi$$

$$t ::= x \mid \quad c \quad \mid f(\vec{t})$$

In other words: we only allow terms that do not contain characteristic functions $?_\phi$ and we only allow the usual (propositional) modalities. The semantics is the obvious one, with all constructs interpreted as in GK and with the epistemic modalities interpreted in the usual way (using the relations $\sim_a$).

In fact, for technical reasons it is convenient to also look at the *extended* language LED_1 obtained by adding the usual epistemic modalities to LED:

$$\varphi ::= p \mid R(\vec{t}) \mid \varphi \to \varphi \mid \; K_a^{\vec{t}} t \mid K_a\varphi$$

$$t ::= x \mid \quad c \quad \mid \quad ?_\varphi \quad \mid f(\vec{t})$$

(Once again, the semantics is the obvious one).

It is clear that *LED_1 and LED_0 are co-expressive*, since $K_a\varphi$ is equivalent to $\varphi \wedge K_a^{\lambda}?_{\varphi}$. In contrast, *$LED_0$ is a less expressive language than LED (and hence than LED_1)*:

Counterexample. We show that the formula $K_a x$ is not equivalent to any formula in LED_0. Suppose, towards a contradiction that $K_a x$ is equivalent to some formula ϕ_0 in LED_0. Let C_0 be the finite set of constants occurring in ϕ_0, and $C_1 := C_0 \cup \{0, 1\}$. Take a model M_1 with two distinct worlds $W_1 = \{w, w'\}$, four distinct objects $D = \{[0], [1], d, d'\}$, $f(\bullet) = [0]$ for all functions and arguments, $\sim_a = W_1 \times W_1$, $\|p\| = \emptyset$ for all p, and $w(x) = d$, $w'(x) = d'$ for all variables x. Take another model M_2 with only one world $W_2 = \{w''\}$, same D, f and $\|p\|$ as for M_1, but with $\sim_a = W_2 \times W_2$ and $w''(x) = d$ for all variables x. It is easy to see that the worlds w, w' and w'' satisfy exactly the same formulas in the language of LED_0 based only on constants in C_0. Hence, these three worlds are equivalent wrt the truth value of ϕ_0. However, $K_a x$ is true at w'', while being false at w and w'. This contradicts the equivalence between $K_a x$ and ϕ_0.

So *the modalities for knowledge of a value really increase the expressivity of our language*. Nevertheless, we can prove that *every validity of LED_1 "translates" to a validity of LED_0*:

Proposition 2.18 *("Validity Reduction") There exists a computable map τ from the language LED_1 to the language LED_0, such that, for every formula φ of LED_1, we have:*

$$\varphi \text{ is valid} \quad \text{iff } \tau(\varphi) \text{ is valid.}$$

Proof. The proof is by induction, using another notion of complexity γ that *counts only the number of nested de re modalities and nested ? symbols.*[13] Note that every term t of LED_1 can be rewritten as $t = t_0[x_1/?_{\phi_1}, \ldots, x_n/?_{\phi_n}]$, for some term t_0 of LED_0 as well as some variables $x_1, \ldots, x_n$ and formulas $\phi_1, \ldots, \phi_n$ (in LED_1), with $\gamma(\phi_i) < \gamma(t)$. For $\vec{c} \in \{0, 1\}^n$, we introduce the notations $t_0[\vec{c}] := t_0[x_1/c_1, \ldots, x_n/c_n]$, and $\vec{c}(\vec{\phi}) := \bigwedge_{i=1,n} c_i(\phi_i)$, where $c(\phi) := \phi$ for $c = 1$, and $c(\phi) = \neg\phi$ for $c = 0$. Now for any tuple of terms $\vec{t} = (t^1, \ldots, t^m)$ of LKG_1, let t_0^i (with $i \in \{1, \ldots, m\}$) be the corresponding terms in LED_0, with variables $x_1^i, \ldots, x_{n_i}^i$ and formulas $\phi_1^i, \ldots, \phi_{n_i}^i$ (for $i \in \{1, \ldots, m\}$), s.t. $t^i = t_0^i[x_1^i/?_{\phi_1^i}, \ldots, x_{n_i}^i/?_{\phi_{n_i}^i}]$ holds for all $i \in \{1, \ldots, m\}$. Then we put $(R(t^1, \ldots, t^m)_0 := \bigvee_{\vec{c}^1 \in \{0,1\}^{n_1}} \cdots \bigvee_{\vec{c}^m \in \{0,1\}^{n_m}} \left(\bigwedge_{i \in \{1,\ldots,m\}} \vec{c}^i(\vec{\phi}^i) \wedge R(t_0^1[\vec{c}^1], \ldots, t_0^m[\vec{c}^m])\right)$.

Claim A: *$R(\vec{t})$ is logically equivalent to $(R(\vec{t}))_0$.*

(The proof is an easy verification.)

[13] More precisely, we recursively put: $\gamma(p) = \gamma(x) = \gamma(c) = 0$, $\gamma(\varphi \to \psi) = max(\gamma(\varphi), \gamma(\psi))$, $\gamma(K_a\varphi) = \gamma(\neg\varphi) = \gamma(\varphi)$, $\gamma(K_a^{t_1,\ldots,t_n} t) = 1 + max(\gamma(t_1), \ldots, \gamma(t_n), \gamma(t))$, $\gamma(?_\varphi) = 1 + \gamma(\varphi)$, $\gamma(R(t_1, \ldots, t_n)) = \gamma(f(t_1, \ldots, f_n)) = max(\gamma(t_1), \ldots, \gamma(t_n))$.

Given this claim, let φ be any formula in LED_1. We can obviously bring it to conjunctive normal form, i.e. establish a validity $\models \varphi \Leftrightarrow \bigwedge_i \bigvee_j \phi^{ij}$, where the formulas ϕ^{ij} are of one the following "basic" forms p, $\neg p$, $R(\overrightarrow{t})$, $\neg R(\overrightarrow{t})$, $K_a^{\overrightarrow{t}} t'$, $\neg K_a^{\overrightarrow{t}} t'$, $K_a\psi$, or $\neg K_a\psi$. Let $\mathcal{T}_\varphi$ be the set of all terms occurring in this normal form, and let F be an *injective* map that associates to each term $t \in \mathcal{T}_\varphi$ some "fresh" constant $F(t) \in C \setminus \mathcal{T}_\varphi$ (such that $F(t) \neq F(t')$ for $t \neq t'$). For any string $\overrightarrow{t} = (t_1, \ldots, t_n)$ of terms in $\mathcal{T}_\varphi$, put $F(\overrightarrow{t}) := (F(t_1), \ldots, F(t_n))$. We now associate to each of the above "basic" formulas ϕ^{ij} some corresponding formulas ϕ_0^{ij}, as follows: $\phi_0^{ij} := \phi^{ij}$ if ϕ^{ij} is the form p or $K_a\psi$; $\phi_0^{ij} := (R(\overrightarrow{t}))_0$ (as defined above) if ϕ^{ij} is of the form $R(\overrightarrow{t})$; $\phi_0^{ij} := K_a\left(\overrightarrow{t} = F(\overrightarrow{t}) \Rightarrow t' = F(t')\right)$ if ϕ^{ij} is of the form $K_a^{\overrightarrow{t}} t'$; and finally $(\phi^{ij})_0 := \neg\psi_0$, if $\phi^{ij} = \neg\psi$ with ψ of one of the forms p, $R(\overrightarrow{t})$, $K_a\varphi$ or $K_a^{\overrightarrow{t}} t'$.

We associate now to our formula φ above a new formula φ_0 of lower γ-complexity, by putting

$$\varphi^0 := \left(\bigwedge_{t \in \mathcal{T}_\phi} t = F(t)\right) \Rightarrow \bigwedge_i \bigvee_j \phi_0^{ij}.$$

It is now easy to verify the following:

Claim B: If φ is not in LED_0, then $\gamma(\varphi^0) < \gamma(\varphi)$.

Finally, we can prove the key step of our "Validity Reduction":

Claim C: φ is valid iff φ^0 is valid.

Proof of Claim C: Note the validity $\models \left(\bigwedge_{t \in \mathcal{T}_\phi} t = F(t)\right) \Rightarrow (\phi^{ij} \Leftrightarrow \phi_0^{ij})$. (This is obvious when ϕ^{ij} is of the form p, $\neg p$, $K_a\psi$ or $\neg K_a\psi$; it follows from Claim A when ϕ^{ij} is of the form $R(\overrightarrow{t})$ or $\neg R(\overrightarrow{t})$; and it follows from the Knowledge de Re Axiom when ϕ^{ij} is of the form $K_a^{\overrightarrow{t}} t'$ or $\neg K_a^{\overrightarrow{t}} t'$.) Using the normal form of φ, we obtain the following validity:

$$(*) \models \left(\bigwedge_{t \in \mathcal{T}_\varphi} t = F(t)\right) \Rightarrow (\varphi \Leftrightarrow \bigwedge_i \bigvee_j \phi_0^{ij}).$$

To prove now *one direction of Claim C*, assume that φ is valid. Using (*), it follows that $\left(\bigwedge_{t \in \mathcal{T}_\phi} t = F(t)\right) \Rightarrow \bigwedge_i \bigvee_j \phi_0^{ij}$ is valid, i.e. φ_0 is valid. For the *other direction*: assume that φ_0 is valid, and let $M = (W, D, [0], [1], \sim_a, \| \bullet \|, \bullet(\bullet), \mathbf{f}, \mathbf{R})_{a \in \mathcal{A}, f \in \mathcal{F}, R \in \mathcal{R}}$ be a model and $w_0 \in W$ be any world. We can change this to a different model $M' = (W, D, [0], [1], \sim_a, \| \bullet \|, \bullet(\bullet)', \mathbf{f}, \mathbf{R})_{a \in \mathcal{A}, f \in \mathcal{F}, R \in \mathcal{R}}$, where we changed only the value map (and only at w_0) by putting $w_0(c)' = w_0(t)$ whenever $F(t) = c$ with $t \in \mathcal{T}_\varphi$, and $w(c)' = w(c)$ in rest. Since F is injective, this gives us a well-defined value map. The change doesn't affect the values of the terms $t \in \mathcal{T}_\varphi$ (since they don't contain any of the fresh constants whose value was changed), so we have $w_0(t)' = w_0(t) = w_0(F(t))'$ for all these terms, hence $w_0 \in |\bigwedge_{t\in\mathcal{T}_\phi} t = F(t)|_{M'}$. Using (*) and the fact that φ_0 is valid, it follows that $w_0 \in |\varphi\|_{M'}$. But φ contains none of the constants whose values were changed, so its truth value was not affected by the change, i.e. we also have $w_0 \in |\varphi\|_M$. Since M and w_0 are arbitrary, we conclude that φ is valid.

Applying repeatedly the last two Claims, we get an immediate proof of Proposition 2.18, by induction on $\gamma(\varphi)$. □

Thus, we have reduced the problem of proving FMP for LED to the corresponding problem for the simpler language LED_0.

STEP 2: Finite Model Property for LED_0

Proposition 2.19 *The logic LED_0 has* (strong) finite model property*: every satisfiable formula φ_0 is satisfiable in a finite model.*

Proof. Let φ_0 be a satisfiable formula in a language $\mathcal{L} = \mathcal{L}(P, Var, C, \mathcal{F}, \mathcal{R}, ar)$ for LED_0, and let $M = (W, D, [0], [1], \sim_a, \| \bullet \|, \bullet(\bullet), \mathbf{f}, \mathbf{R})_{a\in\mathcal{A}, f\in\mathcal{F}, R\in\mathcal{R}}$ be a model and $w_0 \in W$ be a world such that $w_0 \in \|\varphi_0\|_M$. Take $\Sigma \subseteq \mathcal{L} \cup \mathcal{T}$ be the *smallest* set of formulas and terms in LED_0 that contains φ_0, 0 and 1, and is closed under subterms and subformulas.[14] It is easy to see that Σ *is finite.*

Let us put $\mathcal{T}_\Sigma := \mathcal{T} \cap \Sigma$, $\mathcal{L}_\Sigma := \mathcal{L} \cap \Sigma$, $P_\Sigma := P \cap \Sigma$, $Var_\Sigma := Var \cap \Sigma$, $C_\Sigma := C \cap \Sigma$. We now define an equivalence relation $\cong$ on W by putting: $w \cong v$ iff $\forall\varphi \in \Sigma(w \in \|\varphi\|_M \Leftrightarrow v \in \|\varphi\|_M)$. For any $w \in W$ we denote by $|w| := \{v \in W : w \cong v\}$ the $\cong$-equivalence class of w, and we put $W_\Sigma := \{|w| : w \in W\}$ for the set of all $\cong$-equivalence classes. Note that W_Σ *is finite.* Fix now some arbitrary well-ordering $<$ of W. For every class $w \in W_\Sigma$, we denote by w_0 the first element of the class $|w|$ (wrt $<$). Let $D_0 := \{w(t) : |w| \in W_\Sigma, t \in \mathcal{T}_\Sigma\} \cup \{[0], [1]\}$. Note that D_0 is *finite.* If $D \setminus D_0 \neq \emptyset$, then choose some $d^0 \in D \setminus D_0$, and put $D_\Sigma := D_0 \cup \{d^0\}$. If however $D \setminus D_0 = \emptyset$, then put $D_\Sigma := D_0 = D$. Note that in both cases D_Σ is a *finite subset of D.*

We now define a "filtrated" model

$$M_\Sigma = (W_\Sigma, D_\Sigma, [0]_\Sigma, [1]_\Sigma, \sim_a^\Sigma, \| \bullet \|_\Sigma, \bullet(\bullet), \mathbf{f}_\Sigma, \mathbf{R}_\Sigma)_{a\in\mathcal{A}, f\in\mathcal{F}, R\in\mathcal{R}}.$$

by taking W_Σ and D_Σ as above, and putting $[0]_\Sigma := [0]$; $[1]_\Sigma := [1]$; $w \sim_a^\Sigma v$ iff $\forall K_a\phi \in \Sigma\,(w \in \|K_a\phi\|_M \Leftrightarrow v \in \|K_a\phi\|_M)$; $\|p\|_\Sigma := \{|w| : w \in \|p\|_M\}$ for $p \in P_\Sigma$, and $\|p\|_\Sigma := \emptyset$ for $p \in P \setminus P_\Sigma$; $|w|(\alpha) := w(\alpha)$ for $\alpha \in Var_\Sigma \cup C_\Sigma$, and $|w|(\alpha) := d^0$ for $\alpha \in (Var \cup \Sigma) \setminus (Var_\Sigma \cup C_\Sigma)$; $\mathbf{f}_\Sigma(d_1, \ldots, d_n) := \mathbf{f}(d_1, \ldots, d_n)$ if there exists a term $f(t_1, \ldots, t_n) \in \Sigma$ and a world $|w| \in W_\Sigma$ s.t. $|w|(t_i) = d_i$ for all $i \in \{1, \ldots, n\}$; and $\mathbf{f}_\Sigma(d_1, \ldots, d_n) := d^0$ otherwise; finally, $\mathbf{R}_\Sigma := \mathbf{R} \cap (D_\Sigma \times D_\Sigma)$. Note that $\mathbf{f}_\Sigma : D_\Sigma \to D_\Sigma$ is a *well-defined* function, $=_\Sigma$ is the diagonal $\{(d, d) : d \in D_\Sigma\}$, and M_Σ *is indeed a finite model.*

Claim D ("*Term Lemma*"): For every term $t \in \mathcal{T}_\Sigma$ and $w \in W$, we have $|w|(t)_\Sigma = w(t)$.

---

[14] More precisely, Σ has to satisfy the following closure conditions: (1) if $(\varphi \to \psi) \in \Sigma$ then $\varphi, \psi \in \Sigma$; (2) if $K_a\varphi \in \Sigma$ then $\varphi \in \Sigma$; (3) if $f(t_1, \ldots, t_n) \in \Sigma$ then $t_1, \ldots, t_n \in \Sigma$; (4) if $R(t_1, \ldots, t_n) \in \Sigma$ then $t_1, \ldots, t_n \in \Sigma$.

Proof of Claim D: Proof by induction: the base case is by definition; for the inductive step: $|w|(f(t_1,\ldots,t_n) = f_\Sigma(|w|(t_1),\ldots,|w|(t_n)) = f_\Sigma(w(t_1),\ldots,w(t_n)) = f(w(t_1),\ldots,w(t_n)) = w(f(t_1,\ldots,t_n))$, where we used the induction hypothesis, the definition of f_Σ and $f(t_1,\ldots,t_n) \in \mathcal{T}_\Sigma$.

Claim E ("*Filtration Lemma*") For all formulas $\phi \in \Sigma$, we have:

$$|w| \in \|\phi\|_{M_\Sigma} \text{ iff } w \in \|\phi\|_M, \quad \text{for every } w \in W.$$

Proof of Claim E: Proof by induction on ϕ. All steps go as in the classical proof of the Filtration Lemma (for the logic $S5$), except for the relational atoms $R(t_1,\ldots,t_n) \in \Sigma$, for which we have $t_1,\ldots,t_n \in \mathcal{T}_\Sigma$, and thus Claim D can be applied. So we have the sequence of equivalencies: $|w| \in \|R(t_1,\ldots,t_n\|_{M_\Sigma}$ iff $(|w|(t_1),\ldots,|w|(t_n)) \in \mathbf{R}_\Sigma$ iff (by definition of R_Σ) $(|w|(t_1),\ldots,|w|(t_n)) \in \mathbf{R}$ iff (by Claim D) $(w(t_1),\ldots,w(t_n)) \in \mathbf{R}$ iff $w \in \|R(t_1,\ldots,t_n)\|_M$.

This finishes our proof that LED_0 has FMP. □

Putting together Step 1 and Step 2, we conclude that *LED also has FMP*, and thus (being also axiomatizable) it is *decidable*.

3 Learning the Value of a Variable

We now extend *LED* with *public value-announcement operators* $\langle ! \overrightarrow{t} \rangle$ for every tuple $\overrightarrow{t} \in \mathcal{T}^*$. These operators act on both formulas and terms. The syntax of *Public Announcement Logic of Epistemic Dependency* (PALED) is given by:

$$\varphi ::= p \mid R(\overrightarrow{t}) \mid \varphi \to \varphi \mid K_a^{\overrightarrow{t}} t \mid \langle ! \overrightarrow{t} \rangle \varphi$$

$$t ::= x \mid \quad c \quad \mid \quad ?_\varphi \quad \mid f(\overrightarrow{t}) \mid \langle ! \overrightarrow{t} \rangle t$$

where x are variables, c are constants, t are term in $\mathcal{T}$, $\overrightarrow{t}$ are finite tuples of terms, and f and R are symbols of arity equal to the length of $\overrightarrow{t}$.

The operations of *propositional substitution* and *variable substitution* can be extended in the obvious way to the new formulas and terms. [15]

Semantics. Our notion of model is the same as for *LED*. For every model M, we define an *extended valuation* (truth map) $\|\varphi\|$, an *extended value map* $w(t)_M$, and the *update* $M^{\overrightarrow{t}}$ *of model* M *with any finite string of terms* $\overrightarrow{t} \in \mathcal{T}^*$. The truth map and the extended value map are defined as for *LED*, except that we add the clauses:

$$\| \langle ! \overrightarrow{t} \rangle \varphi \|_M \; = \; \|\varphi\|_{M^{\overrightarrow{t}}}, \quad \text{and} \quad w(\langle ! \overrightarrow{t} \rangle t')_M \; = \; w(t')_{M^{\overrightarrow{t}}}.$$

For the update $M^{\overrightarrow{t}} := (W, \sim^a_{M^{\overrightarrow{t}}}, \| \bullet \|, \bullet(\bullet), \mathbf{f}, \mathbf{R})$, we *leave all the components the same*, except for *changing the epistemic relations* as follows:

$$\sim^a_{M^{\overrightarrow{t}}} \; = \; \{(w,s) \in W \times W | w \sim^a s, w(\overrightarrow{t})_M = s(\overrightarrow{t})_M\}.$$

[15] Formally, we put: $(\langle !t_1,\ldots,t_n\rangle\varphi)[p/\theta] := \langle !t_1[p/\theta],\ldots,t_n[p/\theta]\rangle(\varphi[p/\theta])$; $(\langle !t_1,\ldots,t_n\rangle t')[p/\theta] := \langle !t_1[p/\theta],\ldots,t_n[p/\theta]\rangle(t'[p/\theta])$; $(\langle !t_1,\ldots,t_n\rangle\varphi)[x/t] := \langle !t_1[x/t],\ldots,t_n[x/t]\rangle(\varphi[x/t])$; $(\langle !t_1,\ldots,t_n\rangle t')[x/t] := \langle !t_1[x/t],\ldots,t_n[x/t]\rangle(t'[x/t])$.

So all *agents jointly learn the values of* $\overrightarrow{t}$, and nothing else changes.

Example 2 The formula $\langle !x_a \rangle (K_a x_b \wedge K_b x_b)$ is true in (all worlds of) the model from Example 1 above. This can be verified by performing an update $!x_a$, which removes epistemic arrows between worlds having different values for x_a and checking that $K_a x_b \wedge K_b x_b$ holds in the updated model. So, after the value of Alice's number is announced, everybody will know Bob's number.

Propositional Public Announcements. The standard (propositional) public announcement formulas from PAL can be defined as abbreviations in our syntax, by putting: $\langle !\phi \rangle \psi := \phi \wedge \langle !(?\phi) \rangle \psi$, and $[!\phi]\psi := \phi \rightarrow \langle !(?\phi) \rangle \psi$.

Proof system. We obtain a complete system for $PALED$ by restricting the Substitution Rules to static contexts, and adding a Necessitation Rule for announcements, as well as Reduction Axioms. More precisely:

(i) *Restricted Propositional Substitution*: From φ, infer $\varphi[p/\theta]$, *provided that* p *doesn't occur in the scope of any dynamic operator in* φ.

(ii) *Restricted Variable Substitution*: From φ, infer $\varphi[x/t]$, *provided that* x *doesn't occur in the scope of any dynamic operator in* φ.

(iii) *All the other axioms and rules of the system LED.*

(iv) *Necessitation Rule for Announcements*: From $\vdash \varphi$ infer $\vdash \langle !\overrightarrow{t} \rangle \varphi$.

(v) *Propositional Reduction Axioms*[16]:

$$\langle !\overrightarrow{t} \rangle p \leftrightarrow p$$

$$\langle !\overrightarrow{t} \rangle R(t_1, \ldots, t_n) \leftrightarrow R(\langle !\overrightarrow{t} \rangle t_1, \ldots, \langle !\overrightarrow{t} \rangle t_n)$$

$$\langle !\overrightarrow{t} \rangle (\varphi \rightarrow \psi) \leftrightarrow \left(\langle !\overrightarrow{t} \rangle \varphi \rightarrow \langle !\overrightarrow{t} \rangle \psi \right)$$

$$\langle !\overrightarrow{t} \rangle K_a^{t_1, \ldots, t_n} t' \leftrightarrow K_a^{\overrightarrow{t}, \langle !\overrightarrow{t} \rangle t_1, \ldots, \langle !\overrightarrow{t} \rangle t_n} \langle !\overrightarrow{t} \rangle t'$$

(vi) *Term Reduction Axioms*:

$$\langle !\overrightarrow{t} \rangle c = c$$

$$\langle !\overrightarrow{t} \rangle x = x$$

$$\langle !\overrightarrow{t} \rangle ?_\varphi = ?_{\langle !\overrightarrow{t} \rangle \varphi}$$

$$\langle !\overrightarrow{t} \rangle f(t_1, \ldots, t_n) = f(\langle !\overrightarrow{t} \rangle t_1, \ldots \langle !\overrightarrow{t} \rangle t_n)$$

Applying the Reduction Axiom iteratively, we can eliminate all dynamic operators in the usual way, and thus prove:

[16] Note that the third reduction axiom is the axiom K for announcements. This is usually stated for *universal* modalities $[\alpha]\psi$, but these coincide with the existential ones in our case, since value announcements are *deterministic* actions (whose transition relations are *functions*). Combining this axiom with the Necessitation Rule for announcements, one can show that formulas obtained by prefixing provably equivalent formulas with dynamic operators are provably equivalent. This is needed to apply Reduction Axioms repeatedly in order to gradually reduce (and eventually eliminate) *nested* announcement operators, e.g. $\langle \overrightarrow{t}^1 \rangle \langle \overrightarrow{t}^2 \rangle \psi$.

Theorem 3.1 The proof system *PALED* is sound and weakly complete for the logic *PALED*. Moreover, *PALED* has the same expressivity as *LED*.

4 Comparison with other work

Our epistemic dependency formulas are closely connected to Dependency Logic [18,19]. Note that $K_a^{x_1,\ldots,x_n}y$ expresses only a "local" dependency (at the actual world): this is reflected in the fact that this attitude is *not introspective* (i.e. $K_a^{x_1,\ldots,x_n}y$ does *not* imply $K_a K_a^{x_1,\ldots,x_n}y$). However, its "introspective version" $K_a K_a^{x_1,\ldots,x_n}y$ gives a more "global" dependency (across all the epistemically-possible worlds), thus capturing *knowledge of the dependency.* It is easy to see that $w \models K_a K_a^{x_1,\ldots,x_n}y$ is equivalent to the assertion that the *dependence atom* $= (x_1, \ldots, x_n, y)$ holds at the "team" $\{v : w \sim_a v\}$ comprising the set of (variable assignments associated to all) epistemic alternatives of w. But note that *LED* is *decidable*, in contrast to most variants of Dependence Logic!

Our logic has also interesting relations with the so-called 'erotetic logics' [8,12,26], including inquisitive logics [10]. First, as Hintikka [13], Schaffer [17], Aloni and others [1] argued, all types of knowledge (knowing that, knowing what, knowing who, knowing how, knowing whether, knowing which) are special cases of knowing the answer to a question: "*All knowledge involves a question. To know is to know the answer*" ([17]: 401). Second, every "variable" (mapping worlds to a set D of values) induces a partition of the state space; so variables could be used to represent any partitional question. Knowing the answer to a question is the same as knowing the value of the corresponding map. [17] Our epistemic dependency formulas capture an 'epistemic' version of *interrogative implication*, as studied in inquisitive logics. But the variable representation gives us *more information*: if we identify the values in D with abstract "answers", then we can compare answers for different questions, and thus formalize phenomena such as "knowing the answer without knowing the question", that cannot be dealt with in standard inquisitive semantics. We believe that an account of *questions as functions* from worlds to (sets of) "answers" gives a better model for interrogatives then the usual inquisitive representation.

In contrast to both Inquisitive Logic and Dependency Logic, our approach preserves the "classicality" of propositional calculus, and re-interprets the non-classical features in terms of *modal-epistemic* operators. As in Dynamic Epistemic Logic [14,7,5,22,20], and its most natural interrogative versions [3,21,6,4], our semantics is "modal" in the usual sense, with formulas evaluated at worlds, rather than at sets of worlds. We think of this as an advantage of our approach, suggesting that questions (and variable dependency) can be understood without denying the classical logical principles. To know is to know the answer. But the logic of Aristotle, Boole and Frege is not dead just yet.

[17] As for the "non-partitional" questions (having non-unique complete answers), as considered in Inquisitive Semantics, they could be represented in our framework as functions from worlds to *sets* of answers in $\mathcal{P}(D)$. It would be interesting to study the epistemic logic of "knowing *an* answer" (rather than 'the' answer) in this generalized framework.

References

[1] Aloni, M., P. Egré and T. De Jager, *Knowing whether a or b*, Synthese **190** (2013), pp. 2595–2621.

[2] Armstrong, W. J., *Dependency structures of data base relationships*, IFIP Congress **74** (1974), pp. 580–583.

[3] Baltag, A., *Logics for insecure communication*, in: *Proceedings of the 8th conference on Theoretical aspects of rationality and knowledge*, Morgan Kaufmann Publishers Inc., 2001, pp. 111–121.

[4] Baltag, A., R. Boddy and S. Smets, *Group knowledge in interrogative epistemology*, in: *Outstanding Contributions to Logic: J. Hintikka*, Springer, 2016 p. to appear.

[5] Baltag, A., L. Moss and S. Solecki, *The logic of public announcements, common knowledge, and private suspicions*, in: *Proceedings of the 7th conference on Theoretical aspects of rationality and knowledge*, Morgan Kaufmann Publishers Inc., 1998, pp. 43–56.

[6] Baltag, A. and S. Smets, *Learning by questions and answers: From belief-revision cycles to doxastic fixed points*, in: *Logic, Language, Information and Computation*, Springer, 2009 pp. 124–139.

[7] Baltag, A., H. van Ditmarsch and L. Moss, *Epistemic logic and information update*, Handbook of the Philosophy of Information, Elsevier Science Publishers, Amsterdam (2008), pp. 361–456.

[8] Belnap Jr, N. D., *Questions, answers, and presuppositions*, The Journal of Philosophy **63** (1966), pp. 609–611.

[9] Burrows, M., M. Abadi and R. Needham, *A logic of authentication*, Proceedings of the Royal Society of London **426** (1989), pp. 233–271.

[10] Ciardelli, I. and F. Roelofsen, *Inquisitive logic*, Journal of Philosophical Logic **40** (2011), pp. 55–94.

[11] Dechesne, F. and Y. Wang, *To know or not to know: epistemic approaches to security protocol verification*, Synthese **177** (2010), pp. 51–76.

[12] Groenendijk, J. and M. Stokhof, *On the semantics of questions and the pragmatics of answers*, Semantics: Critical Concepts in Linguistics (1985), p. 288.

[13] Hintikka, J., "Socratic epistemology: Explorations of knowledge-seeking by questioning," Cambridge University Press, 2007.

[14] Plaza, J., *Logics of public communication*, in: *Proceedings of the 4th International Symposium on Methodologies for Intelligent Systems* (1989), pp. 201–216.

[15] Plaza, J., *Logics of public communication*, Synthese **158** (2007), pp. 165–179.

[16] Quine, W. V., *On what there is*, Review of Metaphysics **2** (1948), pp. 21–36.

[17] Schaffer, J., *Knowing the answer*, Philosophy and Phenomenological Research **75** (2007), pp. 383–403.

[18] Vaananen, J., "Dependence logic," Cambridge University Press, 2007.

[19] Vaananen, J., *Modal dependence logic*, Texts in Logic and Games **4** (2008), pp. 237–254.

[20] van Benthem, J., "Logical dynamics of information and interaction," Cambridge University Press, 2011.

[21] van Benthem, J. and S. Minica, *Toward a dynamic logic of questions*, Journal of Philosophical Logic **41** (2012), pp. 633–669.

[22] van Ditmarsch, H., W. van Der Hoek and B. Kooi, **337**, Springer Science & Business Media, 2007.

[23] van Ditmarsch, H. P., J. Ruan and R. Verbrugge, *Sum and product in dynamic epistemic logic*, Journal of Logic and Computation **18** (2008), pp. 563–588.

[24] Wang, Y. and J. Fan, *Knowing that, knowing what, and public communication: Public announcement logic with kv operators*, in: *Proceedings of IJCAI*, 2013, pp. 1139–1146.

[25] Wang, Y. and J. Fan, *Conditionally knowing what*, in: *Advances in Modal Logic* (2014), pp. 569–587.

[26] Wisniewski, A., "The Posing of Questions, Logical Foundations of Erotetic Inferences," Synthese Library. Kluwer., 1995.

Beliefs and Evidence in Justification Models

Alexandru Baltag [1]

Institute for Logic, Language and Computation. University of Amsterdam The Netherlands

Virginie Fiutek [2]

University of Lille 3 France

Sonja Smets [3]

Institute for Logic, Language and Computation. University of Amsterdam The Netherlands

Abstract

In this paper we focus on the formal qualitative representation of an agent's evidence and justification in support of her beliefs and knowledge. Our formal setting is based on 'justification models', which we introduce as a generalization of the so-called 'evidence models' proposed by J. van Benthem and E. Pacuit in [18]. We use these structures to express how an agent's evidence supports her doxastic state, expressing as such an agent's justifiable beliefs. We study a number of specific classes of justification models as well as their relations. Overall, these structures are more general than the so-called plausibility models used to represent an agent's doxastic and epistemic states in [14,7,8]. We illustrate the models in this paper via examples and focus on the dynamics of justification models.

Keywords: Justification models, justifiable beliefs, doxastic logic, evidence logic

1 Introduction

We have gained inspiration from Keith Lehrer's informal analysis of defeasible knowledge in terms of "undefeated justified acceptance" [11]. Any formal representation of this epistemic concept will requires an intricate analysis that can link an agent's defeasible knowledge of a proposition to the sound justification that she has in support of it. We pursue this line of thought in this paper

[1] A.Baltag@uva.nl

[2] fiutek.virginie@gmail.com

[3] S.J.L.Smets@uva.nl. The contribution of S. Smets to this work is funded by The European Research Council ERC-2011-STG No. 283963, part of the European Community's FP7.

and focus on the introduction of a formal system that can express the necessary relations between an agent's epistemic or doxastic state and the evidence or justification that supports it. The models we introduce in this paper are called "justification models" and we study these structures in the framework of Dynamic Epistemic Logic and its recent extensions that can deal with belief revision theory [14,7,8].

In the tradition of logics designed to handle evidence and beliefs, we follow in this paper the semantic account that was initiated in [18] in the context of neighborhood models while the work in [2,5,3,4] explores evidence and beliefs in the context of topological models. van Benthem and Pacuit's semantic approach to evidence in terms of neighborhood models allows them to deal with possibly false and possibly mutually inconsistent evidence. If we focus on the relation between evidence and the agent's beliefs as well as her belief dynamics, we observe that the belief revision policies modelled in the context of 'evidence models' [18] do not necessarily satisfy the AGM postulates of belief revision [1]. This is due to the fact that the preorder relation that can be induced on the possible worlds in these models is not total. van Benthem and Pacuit go on to show that (uniform) evidence models can be turned into partial plausibility models and they indicate how a partial plausibility model can be extended to an evidence model. In their later work, van Benthem, Pacuit and Fernandez-Duque [16,17] study different types of evidence models as well as their relations to plausibility models. Continuing this line of work, we study in this paper different types of justification models to represent evidence and beliefs and we explore the relations between these classes of models.

In this paper we first introduce our main formal system in section 2 and we explain how to enhance our structures with a plausibility relation over possible worlds in section 3. In section 4 we study a number of different classes of justification models. An important type of justification models is given by the class of (introspective) evidence models of [18]. In our overview we include plausibility models and show how they are related to justification models. We further indicate how a justification model can be mapped into a plausibility model which allows us to define a number of epistemic and doxastic attitudes of an agent. Next we introduce counting models and weighting models, proving that they can be considered as a special kind of justification models and we show that (introspective) evidence models match to a special kind of justification models.

After introducing different types of justification models in section 4, we study their dynamics in section 5. In particular we focus on the notion of update of a justification model. We study this update operation in each of the introduced classes of justification models.

Finally, we provide a language for a logical system containing a number of sound axioms that can establish a new logic of justifiable beliefs in section 6. Given this logical language, we can express a number of interesting philosophical concepts and properties about justifiable beliefs and defeasible knowledge. We conclude this paper with some final remarks in the last section.

2 Introduction to Justification Models

We start with the following example of a specific scenario:

Example 1. Our agent Alice, a biology student, investigates an animal which is unkown to her. Alice will form a belief about the animal in front of her on the basis of the evidence that she gathers from four different sources of information (from her colleagues). Her first information source tells Alice that 'the animal can swim'. The second source states that 'it is a non-flying bird'. The third source says 'it lays eggs' and the fourth source says 'it flies'.

In this example, the collection of evidence coming from the four sources is accessbile to Alice. Yet this doesn't mean that all the evidence that is accessible to Alice forms a consistent set. Indeed, in this example we assume that an animal either does or doesn't fly but can't do both, as such the second and fourth sources are contradicting each other. Alice's evidence is not conclusive, yet Alice can reason about her evidence and she knows the evidence that is accessible to her.

To provide a formal model of this example, we start with a possible worlds model in which a piece of *evidence* is represented as a set of possible worlds. Formally, we introduce a new type of model, called *justification model*, to capture the agent's evidence or justification as well as the agent's doxastic and epistemic state.

Definition 1 *A justification model $\mathcal{M}$ is a tuple $(S, E, \preceq, \|\cdot\|)$ consisting of a finite set S of states or so-called possible worlds, a family $E \subseteq \mathcal{P}(S)$ of non-empty subsets $e \subseteq S$ ($\emptyset \notin E$), called evidence (sets) such that S is itself an evidence set ($S \in E$). We call a body of evidence (or argument) any $F \subseteq E$ such that $\bigcap F \neq \emptyset$ and we denote by $\mathcal{E} \subseteq \mathcal{P}(E)$ the family of all bodies of evidence. Any justification model comes equipped with a standard valuation map $\|\cdot\|$ and a partial preorder $\preceq$ on $\mathcal{E}$ satisfying the following constraints:*

$$F \subseteq F' \Rightarrow F \preceq F'$$

$$F \preceq F', G \preceq G' \text{ and } F' \cap G' = \emptyset \Rightarrow F \cup G \preceq F' \cup G'$$

$$F \prec F', G \preceq G' \text{ and } F' \cap G' = \emptyset \Rightarrow F \cup G \prec F' \cup G'$$

where $F, F', G, G', F \cup G, F' \cup G'$ are bodies of evidence, i.e. consistent families of evidence sets.

Note that the empty family of evidence sets $\emptyset$ is still a body of evidence, since $\bigcap \emptyset = \{s \in S \mid \forall e \in E(e \in \emptyset \Rightarrow s \in e)\}$, $\emptyset$ is a consistent family of evidence sets.

The above introduced relation $\preceq$ is a partial preorder, connecting only the consistent families of evidence sets. Here we read $F \preceq G$ as the body of evidence G is (considered) to be at least as convincing or easier to accept (by some implicit agent) as the body of evidence F. Similarly, the strict version $F \prec G$ denotes that the body of evidence G is (considered to be) more convincing, easier to accept (by some implicit agent) than the body of evidence F. The conditions in definition 1 indicate that the introduced preorder $\preceq$ does not contradict the set-theoretic inclusion order on bodies of evidence. We can impose further conditions on $\preceq$ to obtain a total preorder by requiring that either $F \preceq F'$ or $F' \preceq F$. Such a justification model with a total preorder relation is called a *total justification model.* Note that in total justification models, all evidence sets are comparable.

In this paper we assume that the agent is introspective regarding her evidence. Informally this means that the agent knows what evidence is available to her. Without this assumption, it will be natural to replace the current family of evidence sets E by a relation $E \subseteq S \times \wp(S)$, which will coincides with the definition of the evidence relation by van Benthem and Pacuit in [18].

Example 1 continued. Figure 1 illustrates the justification model for the above example. In this figure we introduce the state space S and name the possible worlds s, t, u, v, w and x, each of which satisfies a given atomic proposition coming from the set {Whale, Pigeon, Goldfish, Pengiun, Emu, Bat}. The family E consists of four evidence sets $e_1 = \{s, t, u\}$, $e_2 = \{u, v\}$, $e_3 = \{u, v, t, w\}$ and $e_4 = \{w, x\}$ and S. The agent Alice is implicitly present and we assume that she has the different arguments (or bodies of evidence) e_1, e_2, e_3, e_4 at her disposal. The collection of Alice's arguments (or bodies of evidence) $\mathcal{E}$ includes besides $\emptyset$ a number of important arguments $\{e_1\}, \{e_2\}, \{e_3\}, \{e_4\}, \{e_1, e_2\}, \{e_1, e_3\}, \{e_3, e_4\}, \{e_1, e_2, e_3\}$. In particular, her body of evidence coming from sources 3 and 4 supports the hypothesis that the animal is a typical flying bird (e.g. pigeon), while Alice's body of evidence from sources 1, 2 and 3 supports the hypothesis that the animal is a penguin (non-fying bird, lays eggs and swims). Alice's arguments are ordered, e.g. $\{e_1\} \preceq \{e_1, e_2\}$, hence the argument that the animal is a whale is less convincing to her in the light of the evidence coming from sources 1 and 2.

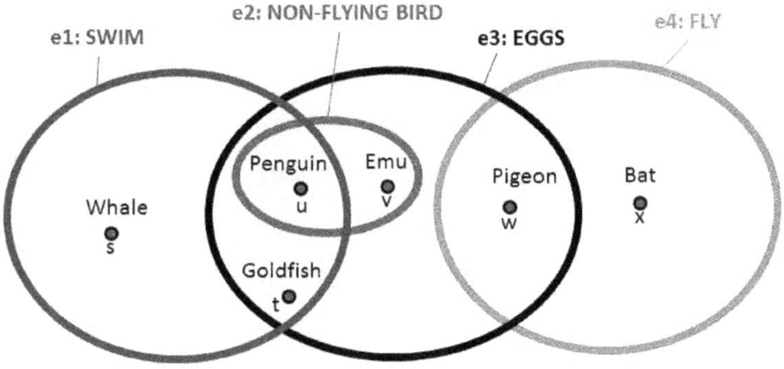

Fig. 1. Justification model for Example 1

3 Plausibility in Justification Models

In the literature on dynamic epistemic logic, a range of epistemic and doxastic attitudes of agents can be represented in the framework of so-called plausibilty models [7,8,14,19].Plausibility models are Kripke models in which the accessibility relation (or so-called plausibility relation) is given by a preorder $\leq \subseteq S \times S$ over the set of possible worlds. We will read the plausibility relation $s \leq t$ to capture that world s is at least as plausible as t. As argued for in [8,15,19], these type of models have a number of advantages over the well-known $KD45$ models in modal logic for the representation of doxastic states, especially in the context of belief dynamics and belief revision. Hence it is a natural step to investigate the options to introduce a plausibility relation over states within justification models. As we show next, this can be done in a canonical way.

We first introduce the notion of a *largest body of evidence consistent with a given state* $s \in S$ and denote it as

$$E_s := \{e \in E \mid s \in e\}$$

A plausibility relation on states can then be induced directly from the partial preorder on $\mathcal{E}$ as follows: For two states $s, t \in S$, we put

$$s \leq_E t \text{ iff } E_t \leq E_s$$

Example 1 continued. To illustrate this, we return to Figure 1 and observe that the largest body of evidence consistent with x is $E_x := \{e_4\}$ and the largest body of evidence consistent with w is $E_w := \{e_3, e_4\}$. Since $E_x \leq E_w$, we obtain $w \leq_E x$ in this example.

Epistemic and doxastic notions Given a justification model equipped with a plausibility relation, we can define all epistemic and doxastic notions usually defined on plausibility models [8,19]. This includes the notions of irrevocable knowledge (K), belief (B), conditional belief (B^-), strong belief (Sb) and defeasible knowledge (K_D), which we define below by using the plausibility order $\leq_E$ (and it's strict version $<_E$). In the following we use the notation $best_{\leq_E} P$ to denote the most plausible P-worlds in the plausiblity ordering, i.e. $best_{\leq_E} P = \text{Min}_{\leq_E} P = \{s \in P |$ there is no $t < s$ for any $t \in P\}$. We abbreviate $best_{\leq_E} S$ as *best* to denote the set of most plausible states in the given state space S.

$$\begin{aligned}
KP &:= \{s \in S : P = S\} \\
BP &:= \{s \in S : best_{\leq_E} \subseteq P\} \\
B^Q P &:= \{s \in S : best_{\leq_E} Q \subseteq P\} \\
SbP &:= \{s \in S : P \neq \emptyset \text{ and } t <_E w \text{ for all } t \in P \text{ and all } w \notin P\} \\
K_D P &= \{s \in S : t \not>_E s \text{ implies } t \in P\}
\end{aligned}$$

Note that in case the plausibility order $\leq_E$ is a total relation, the following proposition holds:

$$s \models K_D P \text{ iff } s \models B^Q P \text{ for all } Q \text{ such that } s \models Q$$

In the last section we return to these epistemic and doxastic attitudes and investigate their link to an agent's arguments or bodies of evidence.

4 Special classes of justification models

Justification models provide a very general framework, subsuming a range of different existing settings. In this section we study the relations between different classes of justification models. In particular we show that partial and total plausibility models and the evidence models of [18] are a special class of justification models. We further introduce two other special classes of justification models, called counting models and weighting models. The following overview of the relations between justification models, counting models, weighting models, plausibility models and evidence models, illustrated in the following Figure 2, will be explained below:

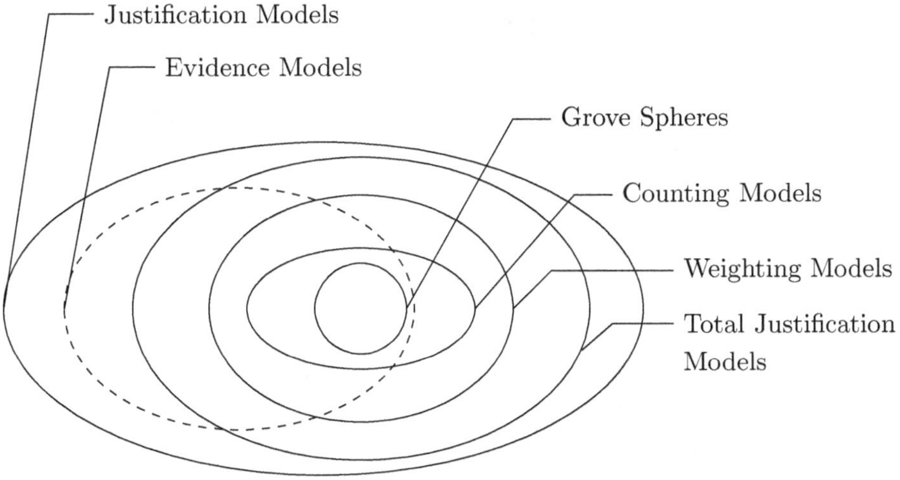

Fig. 2. Overview of Justification Models

4.1 Plausibility models

Any plausibility model $(S, \leq, \|\cdot\|)$, equipped with a preorder $\leq \subseteq S \times S$, can be viewed as a special kind of justification model $(S, E, \preceq, \|\cdot\|)$ in which S is the set of possible worlds and the set of evidence sets is given by $E = \{\downarrow w : w \in S\}$ where $\downarrow w = \{s \in S : s \leq w\}$. The preorder on bodies of evidence can be given by either one of the following options[4]:

(i) (Inclusion order) $F \preceq_1 F'$ iff $F \subseteq F'$ or

(ii) (Cardinality order) $F \preceq_2 F'$ iff $|F| \leq |F'|$.

In the second case, plausibility models are a special case of the counting models which we introduce below. In the first case, plausibility models $(S, \leq, \|\cdot\|)$ are a special kind of justification models $(S, E, \preceq_1, \|\cdot\|)$ in which the preorder on bodies of evidence is given by inclusion ($\preceq_1 = \subseteq$), the evidence sets are nested that is, $\forall e, e' \in E$ either $e \subseteq e'$ or $e' \subseteq e$ (i.e. the pre-order is a total

[4] Note that while the inclusion order is not necessarily total, the cardinality order is total.

pre-order). A body of evidence F corresponds to any family of spheres of a plausibility model[5] and $\mathcal{E}$ corresponds to all the families of spheres.

In line with the mentioned two cases, we can define two *plausibility maps* $Just_1$ and $Just_2$, mapping plausibility models to justification models:

- $(S, \leq, \|.\|) \overset{\text{Just1}}{\longmapsto} (S, E, \preceq_1, \|.\|)$
- $(S, \leq, \|.\|) \overset{\text{Just2}}{\longmapsto} (S, E, \preceq_2, \|.\|)$.

The plausibility map $Just_1$ corresponds to the case where the preorder on bodies of evidence is given by the inclusion order while the plausibility map $Just_2$ corresponds to the case where the preorder on bodies of evidence is given by the cardinality order. We call the justification models that can be obtained in one of these two ways (by applying $Just_1$ or $Just_2$ to a plausibility model), *sphere-based justification models.*

Reversly, any justification model $(S, E, \preceq, \|.\|)$ can be mapped into a type of plausibility model $(S, \leq, \|.\|)$. In order to do so, we define the *plausibility map* $Plau$, mapping justification models to plausibility models: $(S, E, \preceq, \|.\|) \overset{\text{Plau}}{\longmapsto} (S, \leq, \|.\|)$.

A justification model with a partial pre-order gives a partial plausibility model while a justification model with a total pre-order gives a total plausibility model. Following this, we observe that total justification models induce total plausibility models, i.e.

$$\forall F, F'(F \preceq F' \vee F' \preceq F) \Longleftrightarrow \forall s, s'(s \leq_E s' \vee s' \leq_E s)$$

Note that the map $Plau$ mapping justification models to plausibility models is not an injective map. So, two different justification models can give rise to the same plausibility model. If we interpret a plausibility model $\mathcal{M}$ as a justification model $\mathcal{M}'$ and then apply the map $Plau$, we obtain the initial plausibility model $\mathcal{M}$. The converse is false since if we apply the map $Plau$ on a justification model $\mathcal{M}'$ to obtain a plausibility model $\mathcal{M}$ and then interpret this plausibility model $\mathcal{M}$ as a justification model, we do not obtain the initial justification model $\mathcal{M}'$. Thus, we have both:

- $Plau(Just_1(\mathcal{M})) = \mathcal{M}$ for any plausibility model $\mathcal{M}$ and $Just_1(Plau(\mathcal{M}')) \neq \mathcal{M}'$ for any justification model $\mathcal{M}'$,
- $Plau(Just_2(\mathcal{M})) = \mathcal{M}$ for any plausibility model $\mathcal{M}$ and $Just_2(Plau(\mathcal{M}')) \neq \mathcal{M}'$ for any justification model $\mathcal{M}'$.

[5] We switch back and forth between the representation in terms of plausibility models and the equivalent setting in terms of Grove models, [10]. Semantically, the belief state of an agent can be modelled using families of sets of sets called spheres. This type of sphere model is built up from sets of possible worlds and so defines propositions as sets of possible worlds. The propositions believed by an agent (constituting her belief set) form the central sphere which is surrounded by concentric spheres, each of them representing a degree of similarity to the central sphere.

4.2 Counting models

We introduce the structure of *counting model* as follows. A counting model is a justification model $(S, E, \preceq, \|\cdot\|)$ in which the pre-order is given by the cardinality order, i.e. $F \preceq F'$ iff $| F | \leq | F' |$.[6]

In counting models, a body of evidence G is considered to be more convincing than a body of evidence F if and only if the number of evidence sets $e \in G$ is bigger than the number of evidence sets $e \in F$: $F \preceq G$ iff $| F | \leq | G |$. The intuition is that the more evidence the agent knows or has access too, the stronger is the evidence. This notion of evidence strength can be applicable to a number of scenarios, though one easily can provide examples where expressing strength of evidence in terms of counting will not be applicable.

Example 2. We represent an example of a counting model in Figure 3. In this example, we introduce four possible worlds namely s, t, v and w. We consider five evidence sets e_1, e_2, e_3, e_4 and e_5. Since the order on evidence is induced from the cardinality order, we have $F \preceq G$ iff $| F | \leq | G |$. As such we assign the following numbers to arguments: $| \{e_1\} | = 1$, $| \{e_3, e_4\} | = 2$, $| \{e_2, e_4\} | = 2$ and $| \{e_2, e_3, e_5\} | = 3$. This yields the following preorder over arguments, indicating their strenght $\{e_2, e_4\} \prec \{e_2, e_3, e_5\}$, $\{e_3, e_4\} \preceq \{e_2, e_4\}$, $\{e_1\} \prec \{e_3, e_4\}$.

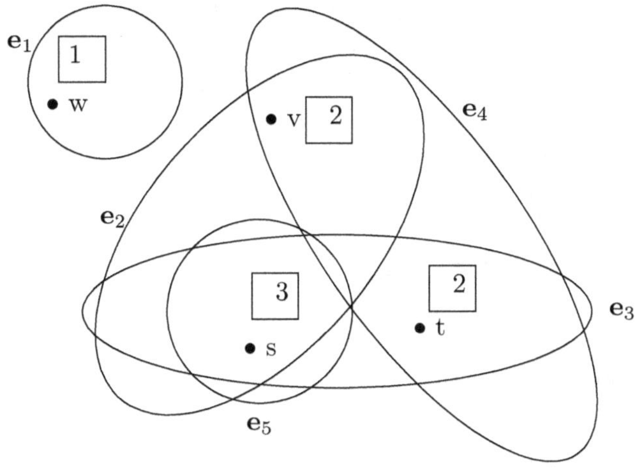

Fig. 3. Counting model

4.3 Weighting models

A special class of justification models is what we call 'Weighting models' these are structures $(S, E, f, \|\cdot\|)$ where we assign a weight (in terms of a natural number) to each piece of evidence $f : E \to \mathbb{N}$.

[6] Note that cardinality generates a total pre-order.

Let $(S, E, f, \|\cdot\|)$ be a weighting model. The function f can be extended to bodies of evidence $\mathcal{E}$ such that $f(E) = \sum_{e \in E} f(e)$. Given the weights of bodies of evidence, it is natural to introduce the preorder on bodies of evidence as follows:

$$E \preceq_f E' \text{ iff } f(E) \leq f(E')$$

As such any weighting model endowed with $\preceq_f$ is a justification model.

Example 3. We represent an example of a weighting model in Figure 4. In this example, there are four possible worlds namely s, t, v and w and five evidence sets e_1, e_2, e_3, e_4 and e_5. The model comes equipped with the function $f : E \to \mathbb{N}$ defined such that $f(e_1) = 1, f(e_2) = 2, f(e_3) = 1, f(e_4) = 3$ and $f(e_5) = 3$. We calculate that $f(e_1) = 1$, $f(\{e_3, e_4\}) = 4$, $f(\{e_2, e_4\}) = 5$ and $f(\{e_2, e_3, e_5\}) = 6$ and this then yields the following preorder over arguments $\{e_2, e_4\} \prec_f \{e_2, e_3, e_5\}$, $\{e_3, e_4\} \prec_f \{e_2, e_4\}$, $\{e_1\} \prec_f \{e_3, e_4\}$.

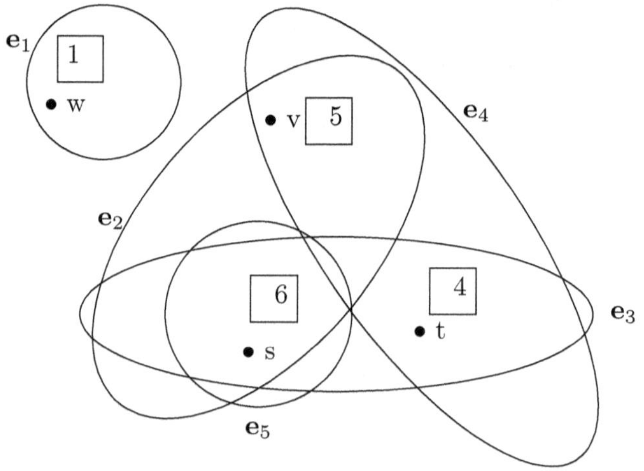

Fig. 4. Weighting model with $f(e_1) = 1$,$f(e_2) = 2$,$f(e_3) = 1$,$f(e_4) = 3$,$f(e_5) = 3$

One observes that counting models are a special case of weighting models in which $f(e) = 1$ for all $e \in E$.

Proposition 1. Every weighting model $(S, E, f, \|\cdot\|)$ is a justification model $(S, E, \preceq, \|.\|)$.

In order to prove this, we need to show that weighting models (and so counting models) satisfy the three constraints which the preorder must satisfy for it to be a justification model.

Proof. First note that since we use the order on natural numbers, transitivity follows.

- if $F \subseteq F'$ then $f(F) = \sum_{e \in F} f(e) \leq \sum_{e \in F'} f(e) = f(F')$ that is, $F \preceq F'$.

- Let $F \preceq F', G \preceq G'$ and $F' \cap G' = \emptyset$.
  $F \cup G = \sum_{e \in F \cup G} f(e) = \sum_{e \in F} f(e) + \sum_{e \in G} f(e) - \sum_{e \in F \cap G} f(e)$.
  Moreover $F' \cup G' = \sum_{e \in F' \cup G'} f(e) = \sum_{e \in F'} f(e) + \sum_{e \in G'} f(e) - \sum_{e \in F \cap G} f(e)$.
  Since $F' \cap G' = \emptyset$ then $F' \cup G' = \sum_{e \in F' \cup G'} f(e) = \sum_{e \in F'} f(e) + \sum_{e \in G'} f(e)$.
  Since $F \preceq F', G \preceq G'$ then $f(F) + f(G) - \sum_{e \in F \cap G} f(e) \leq f(F') + f(G')$.
  Hence $F \cup G \preceq F' \cup G'$.
- Let $F \prec F', G \preceq G'$ and $F' \cap G' = \emptyset$.
  $F \cup G = \sum_{e \in F \cup G} f(e) = \sum_{e \in F} f(e) + \sum_{e \in G} f(e) - \sum_{e \in F \cap G} f(e)$.
  Moreover $F' \cup G' = \sum_{e \in F' \cup G'} f(e) = \sum_{e \in F'} f(e) + \sum_{e \in G'} f(e) - \sum_{e \in F \cap G} f(e)$.
  Since $F' \cap G' = \emptyset$ then $F' \cup G' = \sum_{e \in F' \cup G'} f(e) = \sum_{e \in F'} f(e) + \sum_{e \in G'} f(e)$.
  Then $f(F) + f(G) - \sum_{e \in F \cap G} f(e) < f(F') + f(G')$.
  Hence $F \cup G \prec F' \cup G'$.

4.4 Evidence models

In [18], Johan van Benthem and Eric Pacuit introduce their so-called 'evidence models'. These models are based on the well-known neighbourhood semantics for modal logic in which the neighbourhoods are interpreted as evidence sets: pieces of evidence (possibly false, possibly mutually inconsistent) possessed by the agent. It is important to note that the plausibility relation that can be induced on possible worlds in evidence models is not a total preorder. Hence in these models, not all possible worlds are comparable.

Definition 2 *An evidence model M is a tuple $(S, E, ||.||)$ consisting of a non-empty set of worlds S. E is an evidence relation $E \subseteq S \times \mathcal{P}(S)$ and $||.||$ is a standard valuation function.*

The collection of evidence sets is defined as $E(s) = \{X | sEX, X \subseteq S\}$ we impose two constraints on the evidence function:
(Cons) For each state s, $\emptyset \notin E(s)$
(Triv) For each state s, $S \in E(s)$
These constraints ensure that no evidence set is empty and that the universe S is itself an evidence set. In this framework, the combination of different evidence sets does not necessarily yield consistent evidence. Indeed for any two evidence sets X and Y , X and Y may be disjoints sets that is, $X \cap Y = \emptyset$.

To investigate the relation between evidence models and justification models, we first introduce the notion of an *introspective evidence model*:

Definition 3 *An evidence model $\mathcal{M}$ is introspective iff we have sEX iff tEX for all $s, t \in S$ and for all $X \subseteq S$.*

Introspective evidence models are a special kind of justification models namely, they correspond exactly to those justification models in which the pre-order on bodies of evidence is given by the inclusion order.

In an introspective evidence model, the evidence relation E boils down to the concept of evidence used in justification models, that is, it becomes a family of evidence sets $E \subseteq \mathcal{P}(S)$ such that $E_s = \{e \mid s \in E\}$ for any $s \in S$. Moreover, the notions of irrevocable knowledge (K), belief (B) and conditional belief (B^-) defined in evidence models in [18] do exactly correspond to the notions we defined earlier in this paper.

4.5 Important notions in justification models

For a given argument $F \in \mathcal{E}$, an evidence set $e \in E$ and state $s \in S$ in a justification model, we can define a number of philosophical concepts indicating when an argument is sound, when an argument (conditionally) supports a proposition and when we have a (conditional) justification for a proposition:

Definition 4 *An argument F is sound at s iff $s \in \bigcap F$.*

Note that the empty argument $\emptyset$ is always sound at every state s since $s \in \bigcap \emptyset = S$.

Definition 5 *An argument F supports Q (or F is an argument for Q) iff $\bigcap F \subseteq Q$.*

Definition 6 *A justification for Q is an argument F such that all arguments at least as strong as F support Q, i.e. $\forall F'(F \preceq F' \Rightarrow \bigcap F' \subseteq Q)$.*

Definition 7 *An argument F supports Q conditional on P (or F is an argument for Q conditional on P) iff $\bigcap F \cap P \subseteq Q$.*

Definition 8 *A justification for Q given P is an argument F that is consistent with P such that all arguments at least as strong as F support Q conditional on P, i.e. $\bigcap F \cap P \neq \emptyset$ and $\forall F'(F \preceq F' \Rightarrow \bigcap F' \cap P \subseteq Q)$.*

5 Dynamics of Justification Models

In line with the work on dynamic epistemic logic, we will model the dynamics of justification models as a model transforming operation. Such a model transformation is taken to be triggered by an epistemic or doxastic event. In line with the above examples, we consider the case in which an agent is confronted with new incoming information and accommodates this new information into her epistemic or doxastic state.

While different types of events can be studied, ranging from public announcements [12] to private announcements including truthfull ones and untruthfull ones [6,15], in this paper we restrict ourselves to *updates* of justification models in which the agent receives new truthful information:

Definition 9 *Given a justification model $\mathcal{M} = (S, E, \preceq, \|\cdot\|)$ and a subset $P \subseteq S$, we define the relativization of the justification model $\mathcal{M}$ to P as $\mathcal{M}|P = (S', E', \preceq', \|\cdot\|')$ with:*

$$S' = P$$

$$E' = \{e \cap S' \mid e \in E,\ e \cap S' \neq \emptyset\}$$

$$F' \preceq' G' \text{ iff } \{e \in E \mid e \cap S' \in F'\} \preceq \{e \in E \mid e \cap S' \in G'\}$$
$$\|\cdot\|' = \|\cdot\| \cap S'$$

In the updated model, the new set of states S' is reduced to the set of states satisfying P. The new evidence set E' is taken to be the old evidence E that is consistent with the states surviving the update and the order $\preceq'$ on new bodies of evidence $F' \preceq' G'$ reflects the fact that the new evidence within G' is at least as strong as the new evidence in F'.

We define the restriction $F|\mathcal{M}'$ of the argument F to the justification model $\mathcal{M}'$ as follows: If F is an argument in a justification model $\mathcal{M}$, and if $\mathcal{M}' = \mathcal{M}|P$ is the relativization of the justification model $\mathcal{M}$ to a subset P, then $F|\mathcal{M}'$ is a body of evidence for the model $\mathcal{M}'$ (called the restriction $F|\mathcal{M}'$ of the argument F to model $\mathcal{M}'$), defined by

$$F|\mathcal{M}' = \{e \cap S' \mid e \in F\}$$

5.1 Plausibility models and Evidence Models

The dynamics of evidence models is studied in [18] and the relativization of these models coincides with the update defined by van Benthem and Pacuit. When restricting to sphere-based justification models, our update operation coincides with the usual update on plausibility models where the plausibility order $\leq'_E$ in the resulting model is given as:

$$\leq'_E = \leq_E \cap (S' \times S')$$

For $s \in S'$, we have $E_s = \{e \in E \mid s \in e\}$ and $E'_s = \{e \cap S' \mid e \in E_s\}$. Note that from $e \in E_s$ and $s \in S'$, we have $s \in e \cap S' \neq \emptyset$.

For $s, t \in S'$, we have:

$$\begin{aligned}
s \leq'_E t \iff E'_t \preceq' E'_s &\iff \{e \cap S' \mid e \in E_t\} \preceq' \{e \cap S' \mid e \in E_s\} \\
&\iff \{e \cap S' \mid e \in E, t \in e\} \preceq' \{e \cap S' \mid e \in E, s \in e\} \\
&\iff \{e \mid e \in E, t \in e\} \preceq \{e \mid e \in E, s \in e\} \\
&\iff E_t \preceq E_s \\
&\iff s \leq_E t
\end{aligned}$$

5.2 Counting models

It is interesting to study the dynamics of specific classes of justification models. In particular we observe that the class of counting models is not closed under the update operation. When a given counting model is updated, the result of updating yields a justification model but not necessarily a counting model.

The following example illustrates the problem:

Example 4. Consider the counting model depicted in Figure 5 where there are three pieces of evidence e_1, e_2 and e_3 and 4 states s, t, u and v. The most plausible states are the states s and u since $E_s := \{e_1, e_2\}$, $E_t := \{e_1\}$, $E_u := \{e_1, e_3\}$, $E_v := \{e_3\}$ and so $\mid E_t \mid < \mid E_s \mid$, $\mid E_v \mid < \mid E_s \mid$, $\mid E_t \mid < \mid E_u \mid$ and $\mid E_v \mid < \mid E_u \mid$.

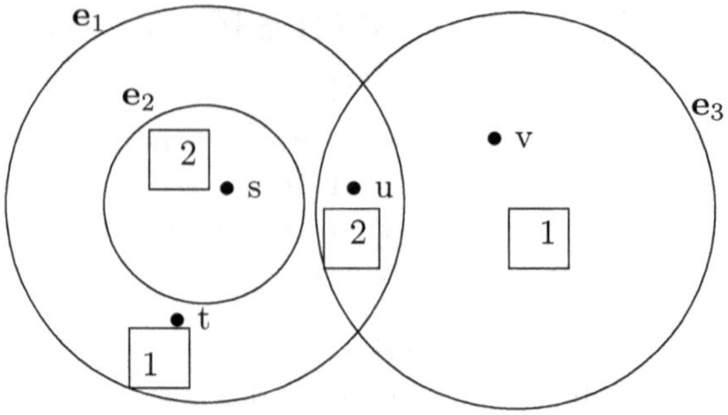

Fig. 5. Initial counting model

It is easily to see that a problem arises when dealing with certain updates of this model. Suppose that the implicit agent receives the hard information that P such that P is only true in s and v. Next, the model is updated with $!P$ and the states u and t are deleted as illustrated in Figure 6. Observe that s and v are equiplausible after the update action, since $E_v := \{e_3\}$, $E_s := \{e_4\}$ and so $\mid E_s \mid = \mid E_w \mid$.

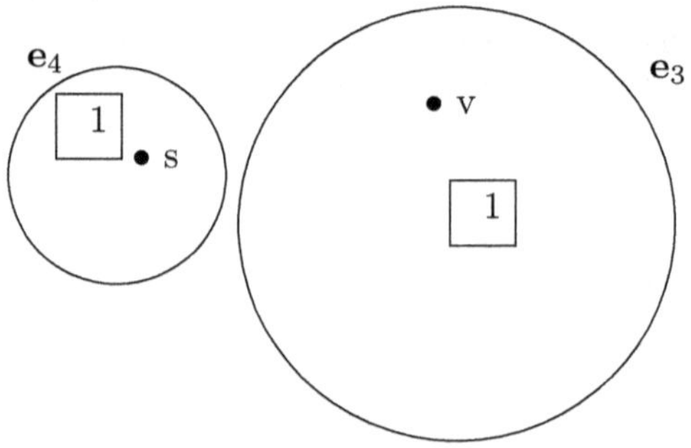

Fig. 6. Updated Counting Model

Due to the update we now lost the information that originally $\mid E_v \mid < \mid E_s \mid$, indeed one would have expected to obtain the justification model as depicted in Figure 7 yet no propositions are present to distinghuish e_1 and e_2 to make it possible that they can both be counted.

The problem is clearly visible if we work with plausibility models. We first provide the corresponding initial (total) plausibility model in Figure 8, and then represent the updated plausibility model after the update with P in Figure 9. After the update, the state s is still more plausible than the state v.

5.3 Weighting models

A solution to the problem in the previous subsection can be provided by working with weighting models instead of counting models.

We define the *map* Wei, mapping weighting models to justification models as follows: $(S, E, f, \|\cdot\|) \overset{\text{Wei}}{\longmapsto} (S, \leq, \|.\|)$.

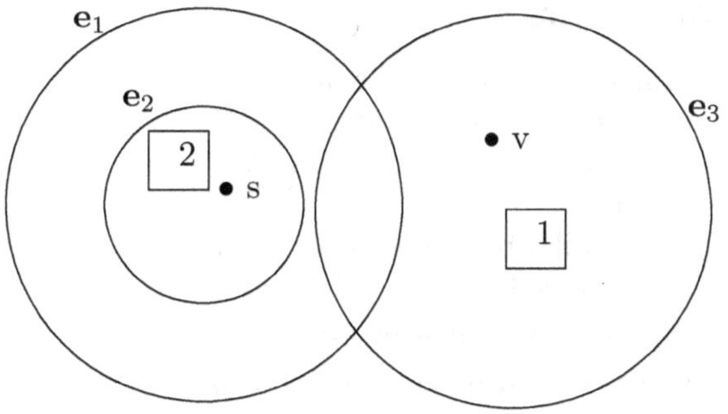

Fig. 7. Labelling solution

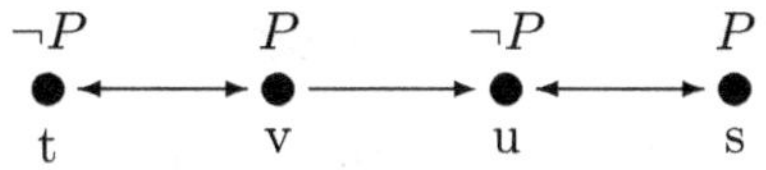

Fig. 8. Initial plausibility model

Fig. 9. Plausibility model updated

In contrast to counting models, we can see that the class of weighting models is better behaved:

Proposition 2. The class of weighting models is closed under the operation of updates

$$Wei(\mathcal{M})|\varphi = Wei(\mathcal{M}|\varphi)$$

Proof. Let $(S, E, f, \|\cdot\|)$ be a weighting model $\mathcal{M}$ where $f : E \to \mathbb{N}$. We define the result of updating this model with φ. The weighting model $\mathcal{M} = (S, E, f, \|\cdot\|)$ is changed to the weighting model $\mathcal{M}|\varphi = (S', E', f', \|\cdot\|')$ with:

$$S' = \|\varphi\|_S$$

$$E' = \{e \cap S' \mid e \in E, \ e \cap S' \neq \emptyset\}$$

$$f'(e') = \sum\{f(e) \mid e \in E \text{ such that } e \cap S' \neq \emptyset\}$$

$$\|\cdot\|' = \|\cdot\| \cap S'$$

Coming back to Figure 5, let us put $f(e_1) = f(e_2) = f(e_3) = 1$. After the update with P, we obtain $v < s$ since as depicted in Figure 7, $f(e_1) + f(e_2) > f(e_3)$.

6 Justifiable Beliefs

Now that we have introduced a number of epistemic and doxastic notions and concepts such as evidence and arguments, we will study the link between them. In particular this ties in with philosophical debates about justifiable beliefs supported by arguments.

As our language to talk about justification models we work with an extension of the setting introduced in [18].[7] This extension is necessary to capture the main interesting features of justification models.

Syntax. Formally, we build up the language $\mathcal{L}_{JB}$ as follows:

Definition 10 *Let Φ be a set of propositional atoms, we define the language in BNF format*

$$\varphi ::= p \mid \neg\varphi \mid \varphi \wedge \psi \mid K\varphi \mid K_D\varphi \mid sound \mid \forall^{ev}\varphi \mid [\preceq]\varphi$$

The language comes equipped with the standard Boolean operators of negation and conjuction and a number of modalities which come with the following intended interpretation: $K\varphi$ expresses that the agent knows that φ is the case. $K_D\varphi$ expresses that the agent defeasibly knows that φ is the case. The expression *sound* captures that the current argument F is sound (i.e. true) at the actual state s, i.e. the current pieces of evidence $e \in F$ are true. We use $\forall^{ev}\varphi$ to capture the expression that for every argument F, φ is the case. The dynamic construct $[\preceq]\varphi$ captures that for every argument F' at least as convincing as the current argument F, φ is the case.

Given the language $\mathcal{L}_{JB}$, we introduce the following abbreviations: $B\varphi := K\neg K_D\neg K_D\varphi$ is read as 'the implicit agent believes that φ'. In total justification models, this reduces to $B\varphi := \neg K_D\neg K_D\varphi$

In addition, we introduce the following abbreviations:

$$Supp\,\varphi := K(sound \rightarrow \varphi)$$

Here $Supp\,\varphi$ captures the fact that the current argument F supports φ.

$$Just\,\varphi := [\preceq]supp\,\varphi$$

The construct $Just\,\varphi$ expresses that the current argument is a justification for φ.

$$\boxplus\varphi := \exists^{ev} supp\,\varphi$$

Here $\boxplus\varphi$ captures the fact that there exists an argument in support of φ.

Semantics. The formulas of $\mathcal{L}_{JB}$ are interpreted at a state s and a body of evidence F such that F is the current argument. Given a justification model

[7] Note that the axiom system of van Benthem and Pacuit [18] holds for general justification models. In particular the reduction axioms in their dynamic account are sound in the class of all justification models.

$\mathcal{M}$, a semantics for $\mathcal{L}_{JB}$ is build up as follows:

$$\begin{array}{ll}
s, F \models p & \text{iff } s \in V(p) \\
s, F \models \neg\varphi & \text{iff } s, F \not\models \varphi \\
s, F \models \varphi \wedge \psi & \text{iff } (s, F \models \varphi) \wedge (s, F \models \psi) \\
s, F \models K\varphi & \text{iff } t, F \models \varphi \text{ for every } t \in S \\
s, F \models K_D\varphi & \text{iff } t, F \models \varphi \text{ for every } t \in S \text{ such that } t \leq s \\
s, F \models \mathit{sound} & \text{iff } s \in \bigcap F \\
s, F \models \forall^{ev}\varphi & \text{iff } s, F' \models \varphi \text{ for every } F' \in \mathcal{E} \\
s, F \models [\preceq]\varphi & \text{iff } \forall F'(F \preceq F' \Rightarrow s, F' \models \varphi)
\end{array}$$

The interpretation for the classical operators is standard. Following the above intented interpretation, note that irrevocable knowledge $K\varphi$ is expressed as truth of φ in all possible worlds while the concept of defeasible concept of knowledge $K_D\varphi$ is interpreted as truth of φ in all worlds that are at least as plausible as the point of evaluation. The interpretation of *sound*, refers to the current argument F at which the evaluation takes place. Guaranteeing that the current argument F is true at a state s, is captured by the fact that s has to be contained in $\bigcap F$. The expression $\forall^{ev}\varphi$ quantifies over all arguments in the collection $\mathcal{E}$ while the semantics of $[\preceq]\varphi$ uses the preorder over bodies of evidence to express that φ holds at every equally strong or stronger argument in the preorder.

A number of interesting axioms for the logic $\mathcal{L}_{JB}$ over the class of total justification models can be shown to be sound.[8] Besides the standard axioms of the logic for irrevocable and defeasible knowledge [8], this includes:

Necessitation Rules for both $\forall^{ev}$ and $[\preceq]$

$S5$-axioms for $\forall^{ev}$

$S4$-axioms for $[\preceq]$

$[\preceq]K\varphi \to K[\preceq]\varphi$

$\forall^{ev}K\varphi \to K\forall^{ev}\varphi$

$\forall^{ev}K_D\varphi \to K_D\forall^{ev}\varphi$

$\forall^{ev}\varphi \to [\preceq]\varphi$

In the context of total justification models, we add a *totality axiom* for bodies of evidence (arguments): $\forall^{ev}(\varphi \vee [\preceq]\psi) \wedge \forall^{ev}(\psi \vee [\preceq]\varphi) \to \forall^{ev}\varphi \vee \forall^{ev}\psi$

[8] Note that in the specific context of the logical system of [18], the axioms of van Benthem and Pacuit hold for general justification models. In particular also the reduction axioms in the dynamic approach of van Benthem and Pacuit are sound in the class of all justification models.

Justifiable beliefs. In the above logical setting we can encoded a number of important statements as follows:

Proposition 3. An agent believes Q iff every argument can be strengthened to a justification for Q, i.e.

$$\forall F \exists F' \succeq F(\forall F'' \succeq F'(\bigcap F'' \subseteq Q))$$

This fact can be captured by the following validity: $Bp \iff \forall^{ev}\langle\preceq\rangle just\, p$. Or more explicitly as: $Bp \iff \forall^{ev}\langle\preceq\rangle[\preceq] supp\, p$

To prove this proposition we first state and prove the following Lemma:

Lemma 1. An agent believes Q iff all maximal (in the sense of strength order) arguments supports Q, i.e. $\forall F \in Max_{\preceq}\mathcal{E}(\bigcap F \subseteq Q)$ where $Max_{\preceq}\mathcal{E} = \{F \in \mathcal{E} \mid F \not\prec F' \text{ for any } F' \in \mathcal{E}\}$.

Proof

- In the direction from left to right, we start from a given justification model $\mathcal{M}$ in which BQ is true at s. So we know that $bestS \subseteq Q$. Let $F \in Max_{\preceq}\mathcal{E}$ and $t \in \bigcap F$. Then $F \subseteq E_t$, so $F \preceq E_t$. Suppose $t \notin bestS$. Then $\exists w <_E t$, so $E_t \prec E_w$, so $F \prec E_w$. This contradicts $F \in Max_{\preceq}\mathcal{E}$. Then $t \in bestS$, so $\bigcap F \subseteq bestS$. Hence $\bigcap F \subseteq Q$.
- In the direction from right to left we assume as given a justification model $\mathcal{M}$ and a state s such that $\forall F \in Max_{\preceq}\mathcal{E}(\bigcap F \subseteq Q)$. Let $t \in bestS$. Suppose $E_t \notin Max_{\preceq}\mathcal{E}$. Then $\exists F' \in \mathcal{E}$ such that $E_t \prec F'$. Let $w \in \bigcap F'$, so $F' \subseteq E_w$, so $F' \preceq E_w$. Then $E_t \prec E_w$, so $w <_E t$. This contradicts $t \in bestS$. Hence, $E_t \in Max_{\preceq}\mathcal{E}$. Then, $t \in \bigcap E_t \subseteq Q$. Hence $t \in Q$, so $bestS \subseteq Q$. Hence, BQ is true at s.

Now we can prove Proposition 3.

Proof

- In the direction from left to right, we start from a given justification model $\mathcal{M}$ in which BQ is true at s. Let $F \in \mathcal{E}$. Then F can be strengthened to a maximal argument F', i.e. $\exists F' \succeq F(F' \in Max_{\preceq}\mathcal{E})$. Indeed since S is finite, so is $\mathcal{E}$. By Lemma 1, since BQ is true at s and $F' \in Max_{\preceq}\mathcal{E}$, $\bigcap F' \subseteq Q$. So F' supports Q. Let $F'' \succeq F'$. Then $F'' \in Max_{\preceq}\mathcal{E}$ and by Lemma 1, $\bigcap F'' \subseteq Q$. So F'' supports Q. Then, F' is a justification for Q. Hence, F can be strengthened to a justification for Q.
- In the direction from right to left we assume as given a justification model $\mathcal{M}$ and a state s such that $\forall F \exists F' \succeq F(\forall F'' \succeq F'(\bigcap F'' \subseteq Q))$. Let $F \in Max_{\preceq}\mathcal{E}$. Then $F \succeq F'$. Take $F'' := F$. Hence, $\bigcap F \subseteq Q$. By Lemma 1, BQ is true at s.

Proposition 4. An agent believes Q conditional on P iff every argument consistent with P can be strengthened to a justification for Q given P, i.e.

$$\forall F(\bigcap F \cap P \neq \emptyset \Rightarrow \exists F' \succeq F(\bigcap F' \cap P \notin \emptyset \wedge \forall F'' \succeq F'(\bigcap F'' \cap P \subseteq Q)))$$

To prove Proposition 4, we first state and prove the following Lemma

Lemma 2. An agent believes Q conditional on P iff all maximal (in the sense of strength order) arguments consistent with P supports Q conditional on P, i.e. $\forall F \in \mathcal{E}(F \in Max^P_{\preceq}\mathcal{E} \Rightarrow \bigcap F \cap P \subseteq Q)$ where $Max^P_{\preceq}\mathcal{E} = \{F \in \mathcal{E} \mid \bigcap F \cap P \neq \emptyset$ and $F \not\prec F'$ for any $F' \in \mathcal{E}(\bigcap F' \cap P \neq \emptyset)\}$.

Proof

- In the direction from left to right, we start from a given justification model $\mathcal{M}$ in which $B^P Q$ is true at s. So we know that $bestP \subseteq Q$. Let $F \in Max^P_{\preceq}\mathcal{E}$ and $t \in \bigcap F \cap P$. Then $F \subseteq E_t$, so $F \preceq E_t$. Suppose $t \notin bestP$. Then $\exists w <_E t$, so $E_t \prec E_w$, so $F \prec E_w$. This contradicts $F \in Max^P_{\preceq}$. Then $t \in bestP$. So we proved that $\forall t(t \in \bigcap F \cap P \Rightarrow t \in bestP)$. Hence, $\bigcap F \cap P \subseteq bestP$, i.e. $\bigcap F \cap P \subseteq Q$.
- In the direction from right to left we assume as given a justification model $\mathcal{M}$ and a state s such that $\forall F \in \mathcal{E}(F \in Max^P_{\preceq}\mathcal{E} \Rightarrow \bigcap F \cap P \subseteq Q)$. Let $t \in bestP$. Then $\bigcap E_t \cap P \neq \emptyset$. Suppose $E_t \notin Max^P_{\preceq}\mathcal{E}$. Then $\exists F' \in \mathcal{E}$ such that $(\bigcap F' \cap P \neq \emptyset)$ and $E_t \prec F'$. Let $w \in \bigcap F' \cap P$, so $F' \subseteq E_w$, so $F' \preceq E_w$. Then $E_t \prec E_w$, so $w <_E t$. This contradicts $t \in bestP$. Hence, $E_t \in Max^P_{\preceq}\mathcal{E}$. Then, $t \in \bigcap E_t \cap P \subseteq Q$. Hence $t \in Q$. So we proved that $\forall t(t \in bestP \Rightarrow t \in Q)$. Hence, $bestP \subseteq Q$, i.e. $B^P Q$ is true at s.

Now we can prove Proposition 4.

Proof.

- In the direction from left to right, we start from a given justification model $\mathcal{M}$ in which $B^P Q$ is true at s. Let $F \in \mathcal{E}$ such that F is consistent with P, i.e. $\bigcap F \cap P \neq \emptyset$. Then F can be strengthened to a maximal argument F' consistent with P, i.e. $\exists F' \succeq F(F' \in Max^P_{\preceq}\mathcal{E})$. Indeed since S is finite, so is $\mathcal{E}$. By Lemma 2, since $B^P Q$ is true at s and $F' \in Max^P_{\preceq}\mathcal{E}$, $\bigcap F' \cap P \subseteq Q$. So F' supports Q conditional on P. Let $F'' \succeq F'$. Then $F'' \in Max^P_{\preceq}\mathcal{E}$ and by Lemma 2, $\bigcap F'' \cap P \subseteq Q$. So F'' supports Q conditional on P. Then, F' is a justification for Q given P. Hence, F can be strengthened to a justification for Q given P.
- In the direction from right to left we assume as given a justification model $\mathcal{M}$ and a state s such that $\forall F(\bigcap F \cap P \neq \emptyset \Rightarrow \exists F' \succeq F(\bigcap F' \cap P \notin \emptyset \land \forall F'' \succeq F'(\bigcap F'' \cap P \subseteq Q)))$. Let $F \in Max^P_{\preceq}\mathcal{E}$. Then $F \succeq F'$. Take $F'' := F$. Hence, $\bigcap F \cap P \subseteq Q$. By Lemma 2, $B^P Q$ is true at s.

In the remainder of this section we restrict ourselves to justification models with a total preorder. In such total justification models, every belief is a justified belief.

Proposition 5. In total justification models, an agent believes Q iff there exists a justification F for Q, i.e. $\exists F \forall F' \succeq F(\bigcap F' \subseteq Q)$
This fact can be captured by the following validity:

$$Bp \iff \exists^{ev} just\, p$$

Or writing it more explicitly: $Bp \iff \exists^{ev}[\preceq] supp\, p$

Proof.

- In the direction from left to right, we start from a given total justification model $\mathcal{M}$ in which BQ is true at s. By Proposition 3, every argument can be strengthened to a justification for Q. Take any argument and strengthen it, then we have a justification for Q.
- In the direction from right to left we assume as given a total justification model $\mathcal{M}$ and a state s such that there exists a justification F for Q. We have to show that $s \models BQ$. Since F is a justification for Q, by Definition 6, $\forall F' \in \mathcal{E}(F \preceq F' \Rightarrow \bigcap F' \subseteq Q)$. Take any argument G such that $F \preceq G$ or $G \preceq F$. If $F \preceq G$, $\bigcap G \subseteq Q$. If $G \preceq F$, G can be strengthened to a justification for Q since $G \preceq F$ and $\forall F' \in \mathcal{E}$ such that $F \preceq F', \bigcap F' \subseteq Q$. By Proposition 3, BQ is true at s.

Proposition 6. In total justification models, an agent defeasibly knows Q at s iff there exists a sound (true) justification F for Q at s, i.e.

$$\exists F(s \in \bigcap F \wedge \forall F' \succeq F(\bigcap F' \subseteq Q))$$

This fact can be captured by the following validity: $K_D p \iff \exists^{ev}(sound \wedge just\, p)$

Proof.

- In the direction from left to right, we start from a given total justification model $\mathcal{M}$ in which $K_D Q$ is true at s. Then $\forall t(t \leq_E s \Rightarrow t \in Q)$. So, $\forall t(E_s \preceq E_t \Rightarrow t \in Q)$. Take $F := E_s$. Since $s \in \bigcap E_s$, $s \in \bigcap F$. Let $F' \succeq F$ and $t \in \bigcap F'$. Then $F' \subseteq E_t$, so $F' \preceq E_t$. Since $E_s \preceq F'$, $E_s \preceq E_t$. Then $t \leq_E s$, so $t \in Q$. Hence, $\exists F(s \in \bigcap F \wedge \forall F' \succeq F(\bigcap F' \subseteq Q))$.
- In the direction from right to left we assume as given a total justification model $\mathcal{M}$ and a state s such that $\exists F(s \in \bigcap F \wedge \forall F' \succeq F(\bigcap F' \subseteq Q))$. Let $t \leq_E s$. We want to show that $t \in Q$. As we know, $s \in \bigcap F$. Then $F \subseteq E_s$, so $F \preceq E_s$. Since $t \leq_E s$, $E_s \preceq E_t$. Then $F \preceq E_t$. By assumption, $\bigcap E_t \subseteq Q$. So, $t \in Q$. Hence, $K_D Q$ is true at s.

Proposition 7. An agent believes Q conditional on P iff there exists a justification F for Q given P, i.e. $\exists F(\bigcap F \cap P \neq \emptyset \wedge \forall F' \succeq F(\bigcap F' \cap P \subseteq Q))$

Proof.

- In the direction from left to right, we start from a given total justification model $\mathcal{M}$ in which $B^P Q$ is true at s. By Proposition 4, every argument consistent with P can be strengthened to a justification for Q given P. Take any argument consistent with P and strengthen it, then we have a justification for Q given P.
- In the direction from right to left we assume as given a total justification model $\mathcal{M}$ and a state s such that there exists a justification F for Q given P. We have to show that $s \models B^P Q$. Since F is a justification for Q given P, by

Definition 8, $\bigcap F \cap P \neq \emptyset$ and $\forall F' \in \mathcal{E}(F \preceq F' \Rightarrow \bigcap F' \cap P \subseteq Q)$. Take any argument G consistent with P ($\bigcap G \cap P \neq \emptyset$) such that $F \preceq G$ or $G \preceq F$. If $F \preceq G$, $\bigcap G \cap P \subseteq Q$. If $G \preceq F$, G can be strengthened to a justification for Q given P since $G \preceq F$ and $\forall F' \in \mathcal{E}$ such that $F \preceq F', \bigcap F' \cap P \subseteq Q$. By Proposition 4, $B^P Q$ is true at s.

Conclusion

We have provided a general setting of justification models which subsumes plausibility models, counting and weighting models as well as evidence models. With this line of work we enhance the investigations into different types of models that connect evidence and beliefs, relating it to the investigations in neighborhood structures in [18,16,17] and our recent work on topological models in [2,5,3,4]. From a logic-technical point of view, we have introduced a number of sound axioms for the logic of justifiable beliefs. In this context we were are able to express beliefs supported by a justification and knowledge supported by an agent's sound justification. Similar concepts are investigated in the context of topological models in [4] in which a complete axiom system is provided. An initial investigation of the relation between beliefs and arguments as studied in formal argumentation theory, ties in with the present research and is further explored in [13].

We have studied in this paper the dynamics of justification models under updates with new truthful information and indicated an interesting problem that we encountered in the context of updating counting models. There are of course a number of other ways in which a given justification model can be transformed into a new one. One can for instance add a new body of evidence to a given structure and give it a degree of plausibility in relation to the other bodies of evidence. This would correspond to the action of adding soft evidence of which the agent is not fully certain. We have left the theory of soft evidence upgrades for future work.

From a philosophical point of view, one can argue that the introduced setting in this paper is too restrictive as it can only deal with beliefs supported by genuine evidence and leaves no room for an agent's biases (or defaults). In a number of applications or scenarios, one may wantt to consider cases in which an agent may have some preferences that are not genuinely based on (external) evidence but based on the trustworthiness of her (internal) senses and reasons. To model such scenarios it would be necessary to introduce a refined type of justification model that contains evidence sets consisting of two types: genuine evidence and biases. In refined justification models, the definition of E can be given by:

Definition 11 $E = E_0 \cup B$ *such that* E_0 *is the family of evidence sets representing the genuine evidence the agent has while* B *is the family of evidence sets representing the prior biases of the agent.*

In such a refined setting it is natural to impose a condition on B that captures the fact that all biases $b \in B$ should strictly increase the strengh of a body of evidence. Formally, $F < F \cup \{b\}$ such that $F \cup \{b\} \in \mathcal{E}$. It is then interesting to note that in a refined justification model, we can consider a "softer" kind of support: an argument F *weakly supports* Q (or F is a *"soft" argument for* Q) conditional on some set $B' \subseteq B$ of biases iff $F \cup B'$ supports Q.

The (refined) justification models provide the technical underpinning that is needed to give a formal account of K. Leher's justification games, the first steps of which are provided [9].

References

[1] Alchourrón, C. E., P. Gärdenfors and D. Makinson, *On the logic of theory change: Partial meet contraction and revision functions*, J. Symb. Log. **50** (1985), pp. 510–530.

[2] Baltag, A., N. Bezhanishvili, A. Ozgun and S. Smets, *The topology of belief, belief revision and defeasible knowledge*, in: H. Huang, D. Grossi and O. Roy, editors, *Logic, Rationality and Interaction, Proceedings of the Fourth International Workshop, LORI 2013.*

[3] Baltag, A., N. Bezhanishvili, A. Ozgun and S. Smets, *The topological theory of belief*, Technical report, ILLC Pre-publication PP-2015-18. (2015).

[4] Baltag, A., N. Bezhanishvili, A. Ozgun and S. Smets, *Justified belief and the topology of evidence*, in: *WOLLIC 2016 proceedings, Springer*, 2016.

[5] Baltag, A., N. Bezhanishvili, A. Ozgun and S. Smets, *The topology of full and weak belief.*, in: *Forthcoming in the Proceedings of the 11th International Tbilisi Symposium on Language, Logic and Computation in Lecture Notes in Computer Science*, 2016.

[6] Baltag, A., L. Moss and S. Solecki, *The logic of public announcements, common knowledge, and private suspicions*, in: *Proceedings of the 7th conference on Theoretical aspects of rationality and knowledge*, Morgan Kaufmann Publishers Inc., 1998, pp. 43–56.

[7] Baltag, A. and S. Smets, *Dynamic belief revision over multi-agent plausibility models*, in: *Proceedings of the 7th Conference on Logic and the Foundations of Game and Decision (LOFT 2006), University of Liverpool.*, 2006.

[8] Baltag, A. and S. Smets, *A qualitative theory of dynamic interactive belief revision*, Logic and the Foundations of Game and Decision Theory (LOFT7) (2008), pp. 11–58.

[9] Fiutek, V., "Playing with Knowledge and Belief," ILLC Dissertation Series DS-2013-02 University of Amsterdam, 2013.

[10] Grove, A., *Two modellings for theory change*, Journal of Philosophical Logic **17** (1988), pp. 157–170.

[11] Lehrer, K., "Theory of Knowledge," Routledge, 1990.

[12] Plaza, J., *Logics of public communication*, in: *Proceedings of the 4th International Symposium on Methodologies for Intelligent Systems* (1989), pp. 201–216.

[13] Shi, C. and S. Smets, *Beliefs supported by arguments*, in: *Proceedings of the Chinese Conference on Logic and Argumentation (CLAR 2016), CEUR proceedings collection*, 2016.

[14] van Benthem, J., *Dynamic logic for belief revision*, Journal of Applied Non-Classical Logics **12(2)** (2007), pp. 129–155.

[15] van Benthem, J., "Logical dynamics of information and interaction," Cambridge University Press, 2011.

[16] van Benthem, J., D. F. Duque and E. Pacuit, *Evidence logic: A new look at neighborhood structures*, in: *Advances in Modal Logic 9, papers from the ninth conference on "Advances in Modal Logic,"*, 2012, pp. 97–118.

[17] van Benthem, J., D. Fernández-Duque and E. Pacuit, *Evidence and plausibility in neighborhood structures*, Annals of Pure and Applied Logic **165** (2014), pp. 106 – 133.

[18] van Benthem, J. and E. Pacuit, *Dynamic logics of evidence-based beliefs*, Studia Logica **99(1)** (2011), pp. 61–92.

[19] van Benthem, J. and S. Smets, *Dynamic logics of belief change.*, in: *Handbook of Logics for Knowledge and Belief*, College Publications, 2015 .

Locales, Nuclei, and Dragalin Frames

Guram Bezhanishvili [1]

Department of Mathematical Sciences
New Mexico State University

Wesley H. Holliday [2]

Department of Philosophy &
Group in Logic and the Methodology of Science
University of California, Berkeley

Abstract

It is a classic result in lattice theory that a poset is a complete lattice iff it can be realized as fixpoints of a closure operator on a powerset. Dragalin [9,10] observed that a poset is a locale (complete Heyting algebra) iff it can be realized as fixpoints of a nucleus on the locale of upsets of a poset. He also showed how to generate a nucleus on upsets by adding a structure of "paths" to a poset, forming what we call a Dragalin frame. This allowed Dragalin to introduce a semantics for intuitionistic logic that generalizes Beth and Kripke semantics. He proved that every spatial locale (locale of open sets of a topological space) can be realized as fixpoints of the nucleus generated by a Dragalin frame. In this paper, we strengthen Dragalin's result and prove that every locale—not only spatial locales—can be realized as fixpoints of the nucleus generated by a Dragalin frame. In fact, we prove the stronger result that for every nucleus on the upsets of a poset, there is a Dragalin frame based on that poset that generates the given nucleus. We then compare Dragalin's approach to generating nuclei with the relational approach of Fairtlough and Mendler [11], based on what we call FM-frames. Surprisingly, every Dragalin frame can be turned into an equivalent FM-frame, albeit on a different poset. Thus, every locale can be realized as fixpoints of the nucleus generated by an FM-frame. Finally, we consider the relational approach of Goldblatt [13] and characterize the locales that can be realized using Goldblatt frames.

Keywords: nucleus, locale, Heyting algebra, intuitionistic logic, lax logic

1 Introduction

A well-known result of Shehtman [30] (cf. [23]) shows that there are intermediate logics that cannot be characterized by Kripke frames [21]. This incompleteness result renewed interest in the earlier topological semantics for intuitionistic

[1] guram@math.nmsu.edu.

[2] wesholliday@berkeley.edu.

logic due to Tarski [33]. It remains a famous open problem of Kuznetsov [22] whether every intermediate logic is topologically complete. Dragalin [9,10] made the important but somewhat neglected observation that by generalizing Kripke frames in a way inspired by Beth semantics [4], one obtains a semantics for intuitionistic logic that is at least as general as topological semantics.

Dragalin frames are triples $(S, \leq, D)$ where $(S, \leq)$ is a poset and $D\colon S \to \wp(\wp(S))$ satisfies natural conditions stated below.[3] Dragalin called each $X \in D(x)$ a *path starting from* x. In the literature on Beth semantics, 'path' suggests a linearly ordered subset of $(S, \leq)$, so we will instead call X a *development* starting from x and elements of X *stages* of the development. For the poset $(S, \leq)$, we use the following standard notation for $Y \subseteq S$ and $y \in S$:

- $\uparrow Y = \{z \in S \mid \exists y \in Y\colon y \leq z\}$ and $\uparrow y = \uparrow\{y\}$;
- $\downarrow Y = \{z \in S \mid \exists y \in Y\colon z \leq y\}$ and $\downarrow y = \downarrow\{y\}$.

A subset U of S is an *upset* (upward closed set) if $x \in U$ implies $\uparrow x \subseteq U$. A *downset* (downward closed set) is defined dually. If $X \subseteq \downarrow Y$, so $\forall x \in X$ $\exists y \in Y\colon x \leq y$ (every stage of development in X is extended by a stage of development in Y), then we say that X is *bounded by* Y.

Definition 1.1 A Dragalin frame is a triple $\mathfrak{F} = (S, \leq, D)$ where $(S, \leq)$ is a poset and $D\colon S \to \wp(\wp(S))$ (a *Dragalin function*) is such that for all $x, y \in S$:

(1°) $\varnothing \notin D(x)$;

(2°) if $y \in X \in D(x)$, then $\exists z \in X\colon x \leq z$ and $y \leq z$;

(3°) if $x \leq y$, then $\forall Y \in D(y)\ \exists X \in D(x)\colon X \subseteq \downarrow Y$;

(4°) if $y \in X \in D(x)$, then $\exists Y \in D(y)\colon Y \subseteq \downarrow X$.

A Dragalin frame is *normal* if $D(x) \neq \varnothing$ for all $x \in S$.

Conditions (1°)–(4°) admit intuitive intepretations. (1°) says that the empty set is not a development of anything. (2°) says that every stage y in a development starting from x is at least *compatible* with x, in that x and y have a common extension z. Dragalin also mentions the stronger condition:

(2°°) if $X \in D(x)$, then $X \subseteq \uparrow x$,

so the stages in a development starting from x are extensions of x. Next, (3°) says that if at some "future" stage y a development Y will become available, then it is already possible to follow a development that is bounded by Y. Dragalin also mentions the stronger condition:

(3°°) if $x \leq y$, then $D(y) \subseteq D(x)$,

[3] Dragalin used the term 'Beth-Kripke frame'. To give due credit to Dragalin, we introduce the term 'Dragalin frame' instead. Note that Dragalin started with a preordered set $(S, \leq)$, but there is no loss of generality in starting with a poset (see Remark 2.6 and Theorem 3.5). Also note that Dragalin worked with downsets in $(S, \leq)$. Others, including Goldblatt [13,14] and Fairtlough and Mendler [11], work instead with upsets. It proves to be more convenient to flip Dragalin's approach to use upsets than to flip the other approaches to use downsets. Thus, we will work with upsets, at the expense of another flip of perspective in Theorem 2.8.

so developments available at future stages are already available. Finally, (4°) says that we "can always stay inside" a development, in the sense that for every stage y in X, we can follow a development Y from y that is bounded by X. A stronger notion of "staying inside" comes from replacing $Y \subseteq {\downarrow}X$ with $Y \subseteq X$:

(4°°) if $y \in X \in D(x)$, then $\exists Y \in D(y)$: $Y \subseteq X$.

In §3 we will see that (2°°), (3°°), and (4°°) can be assumed without loss of generality. This motivates the following definition.

Definition 1.2 A Dragalin frame is *standard* if it satisfies (2°°)–(4°°).

To use Dragalin frames to give semantics for the language of propositional logic, we use models $\mathcal{M} = (S, \leq, D, V)$ where $(S, \leq, D)$ is a Dragalin frame and V assigns to each propositional variable an upset $V(p)$ of $(S, \leq)$ with the property that $x \in V(p)$ iff $\forall X \in D(x)\ X \cap V(p) \neq \varnothing$, i.e., each development starting from x hits the interpretation of p. The forcing clauses for the connectives are the same as in intuitionistic Kripke semantics except for $\bot$ and $\vee$:

- $\mathcal{M}, x \Vdash \bot$ iff $D(x) = \varnothing$;
- $\mathcal{M}, x \Vdash \varphi \vee \psi$ iff $\forall X \in D(x)\ \exists y \in X$: $\mathcal{M}, y \Vdash \varphi$ or $\mathcal{M}, y \Vdash \psi$.

What is going on here is that we are evaluating formulas not in the full Heyting algebra $\mathsf{Up}(S, \leq)$ of upsets of $(S, \leq)$, as in Kripke semantics, but rather in the Heyting algebra of just those upsets U such that $x \in U$ iff $\forall X \in D(x)$ $X \cap U \neq \varnothing$. As Dragalin explained, and as we review in §2, the function D gives rise to a *nucleus* on the Heyting algebra $\mathsf{Up}(S, \leq)$, and we are evaluating formulas in the Heyting algebra of fixpoints of this nucleus.[4] This idea of evaluating formulas as fixpoints of a closure operator, of which a nucleus is a special case, and interpreting disjunction by taking the closure of the union appears in semantics for substructural logic [27, §12.2] and in recent philosophical discussions of the relation between intuitionistic and classical logic [28,29].

Nuclei play an important role in pointfree topology [18]. The pointfree generalization of a topological space is a complete Heyting algebra, also known as a *locale*,[5] and nuclei on a locale describe sublocales of the locale. The locale $\mathsf{Up}(S, \leq)$ is a very special locale (see Remark 2.7). Shehtman's result shows that not all intermediate logics are complete with respect to such locales. On the other hand, Dragalin proved that every locale can be represented as the algebra of fixpoints of a nucleus on some $\mathsf{Up}(S, \leq)$. This representation theorem is related to the classic result in lattice theory that every complete lattice can be represented as the lattice of fixpoints of a closure operator on a powerset.

Dragalin's representation theorem motivates the notion of a *nuclear frame* that we introduce in §2. The question is then how a nucleus on $\mathsf{Up}(S, \leq)$ can be realized more concretely. Dragalin frames do so with the function D, and

4 Dragalin [9,10] used the term 'completion operator' instead of 'nucleus'. In this paper, we follow the now standard terminology and notation from pointfree topology.

5 Locales are also known as *frames* in the pointfree topology literature, but in this paper we use the term 'frame' as it is used in the modal logic literature.

Dragalin proved that every *spatial* locale (locale of open sets of a topological space) can be represented as fixpoints of the nucleus generated by D for some Dragalin frame. In §3, we prove that every locale—not only spatial locales—can be represented as fixpoints of the nucleus generated by a Dragalin frame. In fact, we prove the stronger result that for every nucleus on the upsets of a poset, there is a Dragalin frame based on that poset that realizes the given nucleus. In §4, we compare Dragalin's approach to generating nuclei with the relational approach of Fairtlough and Mendler [11], based on what we call FM-frames. Surprisingly, every Dragalin frame can be turned into an equivalent FM-frame, albeit on a different poset. Thus, every locale can be represented as fixpoints of the nucleus generated by an FM-frame. Finally, in §5, we consider the relational approach of Goldblatt [13] and characterize the locales representable using Goldblatt frames. We conclude in §6 with directions for further research.

2 Nuclear Frames

Our basic objects of study will be nuclei on Heyting algebras and locales [24,25,12,18].

Definition 2.1 A *nucleus* on a Heyting algebra H is a function $j\colon H \to H$ such that for all $a, b \in H$: $a \leq ja$ (inflationary); $jja = ja$ (idempotent); $j(a \wedge b) = ja \wedge jb$ (multiplicative). A nucleus is *dense* if $j0 = 0$.

A *nuclear algebra* is a pair $\mathbb{H} = (H, j)$ where H is a Heyting algebra and j is a nucleus on H. It is a *localic nuclear algebra* if H is a locale.

The following result is well known (see, e.g., [10, p. 71]).

Theorem 2.2 *If* $\mathbb{H} = (H, \wedge, \vee, \to, 0, j)$ *is a nuclear algebra, then the* algebra of fixpoints $\mathbb{H}_j = (H_j, \wedge_j, \vee_j, \to_j, 0_j)$ *is a Heyting algebra where* $H_j = \{a \in H \mid a = ja\}$ *is the set of fixpoints of* j *in* H *and for* $a, b \in H_j$*:*

(i) $a \wedge_j b = a \wedge b$; (ii) $a \vee_j b = j(a \vee b)$;

(iii) $a \to_j b = a \to b$; (iv) $0_j = j0$.

If $\mathbb{H}$ *is a localic nuclear algebra, then* $\mathbb{H}_j$ *is a locale, where for* $X \subseteq H_j$*:*

(v) $\bigwedge_j X = \bigwedge X$; (vi) $\bigvee_j X = j \bigvee X$.

We will often abuse notation and conflate H_j and $\mathbb{H}_j$.

Remark 2.3 An important example of a nucleus on a Heyting algebra is the nucleus of double negation $\neg\neg$, where $\neg x = x \to 0$. It is well known that the algebra of fixpoints of double negation as in Theorem 2.2 forms a Boolean algebra, which is complete if the original Heyting algebra is complete. For example, since the Heyting algebra $\mathsf{Up}(S, \leq)$ of upsets in a poset is a locale, the algebra of fixpoints in $(\mathsf{Up}(S, \leq), \neg\neg)$ is a complete Boolean algebra. Going from a poset to a complete Boolean algebra in this way is a standard technique in set theory for relating forcing posets to Boolean-valued models [32].

The nuclei on a Heyting algebra H are naturally ordered by $j \leq k$ iff $ja \leq ka$ for all $a \in H$. It is well known that in the case of a locale L, the collection

$N(L)$ of all nuclei on L with the natural ordering is itself a locale (see, e.g., [12, Th. 2.20], [18, Prop. II.2.5]). Meets in $N(L)$ are computed pointwise, whereas joins are more difficult to describe.

There are several families of nuclei that can generate all nuclei in $N(L)$. Here we focus on the following: given $a \in L$, define the function w_a on L by

$$w_a b = (b \to a) \to a. \tag{1}$$

It can be verified that w_a is a nucleus. Note that when $a = 0$, w_0 is the nucleus of double negation.[6] One of the special roles of these nuclei is shown by the following observation of Simmons [31, p. 243], which we will utilize in Theorem 3.5. To keep the paper self-contained, we include a proof.

Lemma 2.4 (Simmons) *Given a locale L and a nucleus j on L,*

$$j = \bigwedge\{w_{ja} \mid a \in L\}.$$

Proof. First, observe that for any nucleus k on L,

$$k \leq w_a \text{ iff } ka \leq a. \tag{2}$$

From left to right, if $k \leq w_a$, then $ka \leq w_a a = (a \to a) \to a = a$. From right to left, the multiplicativity of k implies $k(b \to a) \leq kb \to ka$, and the inflationarity of k yields $(b \to a) \leq k(b \to a)$. It follows that $kb \wedge (b \to a) \leq kb \wedge k(b \to a) \leq ka$ and hence $kb \leq (b \to a) \to ka$. Thus, if $ka \leq a$, then we have $kb \leq (b \to a) \to ka \leq (b \to a) \to a = w_a b$ for every $b \in L$, so $k \leq w_a$.

It follows from (2) that j is a lower bound of $\{w_{ja} \mid a \in L\}$. To see that it is the greatest, suppose k is also a lower bound of $\{w_{ja} \mid a \in L\}$, so for every $a \in L$, we have $k \leq w_{ja}$. Then (2) implies $ka \leq ja$ for every $a \in L$, so $k \leq j$. Therefore, j is the greatest lower bound. □

In this paper, we are interested in nuclear algebras in which the underlying Heyting algebra is the locale $\mathsf{Up}(S, \leq)$ of upsets of some poset $(S, \leq)$, in which implication is defined by $U \to V = \{x \in S \mid {\uparrow}x \cap U \subseteq V\}$. The locale $\mathsf{Down}(S, \leq)$ of downsets of $(S, \leq)$ is defined dually.

Definition 2.5 A *nuclear frame* is a triple $\mathfrak{F} = (S, \leq, j)$ where $(S, \leq)$ is a poset and j is a nucleus on $\mathsf{Up}(\mathfrak{F}) := \mathsf{Up}(S, \leq)$. We say that $\mathfrak{F}$ is *dense* if j is dense. The *nuclear algebra of* $\mathfrak{F}$ is the nuclear algebra $(\mathsf{Up}(\mathfrak{F}), j)$.

Remark 2.6 One could also allow the relation $\leq$ in Definition 2.5 to be a preorder. However, as is well known, for any preordered set $(S, \leq)$, taking its quotient with respect to the equivalence relation defined by $x \sim y$ iff $x \leq y$ and $y \leq x$ produces a poset $(S', \leq')$, called the *skeleton* of $(S, \leq)$, such that $\mathsf{Up}(S', \leq')$ is isomorphic to $\mathsf{Up}(S, \leq)$. Thus, any preordered nuclear frame $(S, \leq, j)$ can be turned into a partially ordered nuclear frame $(S', \leq', j')$ such that their nuclear algebras are isomorphic.

---

[6] As in the case of double negation, for any nuclear algebra of the form (H, w_a), its algebra of fixpoints is a Boolean algebra (see [12, p. 330], [18, p. 51]).

Remark 2.7 Nuclear frames generate only a special class of localic nuclear algebras. Recall that for a locale L, an element $a \in L$ is *completely join-prime* if from $a \leq \bigvee X$ it follows that $a \leq x$ for some $x \in X$. Let $J^\infty(L)$ be the set of completely join-prime elements of L. We call a locale L *Alexandroff* if $J^\infty(L)$ is join-dense in L, i.e., each element of L is the join of completely join-prime elements below it. Then L is Alexandroff iff L is isomorphic to the locale of upsets of a poset (see, e.g., [8,5]) and hence to the locale of open sets in an Alexandroff space. Thus, nuclear frames generate exactly the localic nuclear algebras based on Alexandroff locales, which we call *Alexandroff nuclear algebras.*

Although not every localic nuclear algebra can be represented as the nuclear algebra of a nuclear frame, nonetheless every locale can be represented as the algebra of fixpoints in the nuclear algebra of a nuclear frame. To keep the paper self-contained, we include a proof of this important result from [10, p. 75].

Theorem 2.8 (Dragalin) *A poset P is a locale iff there is a dense nuclear frame $\mathfrak{F}$ such that P is isomorphic to the algebra of fixpoints in the nuclear algebra of $\mathfrak{F}$.*

Proof. From right to left, since $\mathsf{Up}(\mathfrak{F})$ is a locale, the algebra of fixpoints in the nuclear algebra of $\mathfrak{F}$ is a locale by Theorem 2.2.

From left to right, suppose P is a locale. Let $S = P \setminus \{0\}$ and $\leq$ be the restricted order. We will build a nuclear frame $\mathfrak{F}$ whose poset is $(S, \geq)$. Since $\mathsf{Up}(S, \geq) = \mathsf{Down}(S, \leq)$, we can work with the locale of *downsets* in the poset $(S, \leq)$. Define a unary function j on this locale by

$$jX = \downarrow \bigvee X, \tag{3}$$

where $\bigvee X$ is the join of X in P, which exists since P is complete, and $\downarrow$ indicates the downset in $(S, \leq)$. It is easy to see that j is inflationary, idempotent, and that $\downarrow\bigvee(X \cap Y) \subseteq (\downarrow\bigvee X) \cap (\downarrow\bigvee Y)$ for $X, Y \in \mathsf{Down}(S, \leq)$. To see that $\downarrow\bigvee(X \cap Y) \supseteq (\downarrow\bigvee X) \cap (\downarrow\bigvee Y)$, suppose that $a \in S$ is in the right hand side, so $a \leq \bigvee X$ and $a \leq \bigvee Y$, whence $a \leq (\bigvee X) \wedge (\bigvee Y)$. By the join-infinite distributive law for locales,

$$(\bigvee X) \wedge (\bigvee Y) = \bigvee\{x \wedge y \mid x \in X, y \in Y\},$$

so $a \leq \bigvee\{x \wedge y \mid x \in X, y \in Y\}$. Since X and Y are downsets, we have $\{x \wedge y \mid x \in X, y \in Y\} \subseteq (X \cap Y) \cup \{0\}$, so $\bigvee\{x \wedge y \mid x \in X, y \in Y\} \leq \bigvee((X \cap Y) \cup \{0\}) = \bigvee(X \cap Y)$. Thus, $a \leq \bigvee(X \cap Y)$ and hence $a \in \downarrow\bigvee(X \cap Y)$. Therefore, j is a nucleus. To see that j is dense, observe that $j\varnothing = \downarrow\bigvee\varnothing = \downarrow 0 = \varnothing$ since $0 \notin S$.

Finally, we must check that our original locale P is isomorphic to the algebra of fixpoints in the nuclear algebra of $\mathfrak{F}$. Observe that the fixpoints of j in the nuclear algebra of $\mathfrak{F}$ are exactly the principal downsets in $(S, \leq)$ plus $\varnothing$. Thus, the map sending each x to $\downarrow x$ is the desired isomorphism. □

Remark 2.9 The proof technique of Theorem 2.8 is related to a standard technique in set theory, whereby one goes from a complete Boolean algebra to a poset by deleting the bottom element [32]. The poset thereby obtained is a *separative* poset (if $y \not\leq x$, then $\exists y' \leq y\ \forall y'' \leq y'$: $y'' \not\leq x$). Moreover, the nucleus j defined in (3) above is the nucleus of double negation [6].

Theorem 2.8 shows that nuclear frames suffice to represent arbitrary locales. However, since nuclear frames are a mix of the concrete $(S, \leq)$ and the algebraic j, it is natural to ask if we can replace the nucleus j with more concrete data from which j can be recovered. We will answer this question in the next section.

3 Dragalin Frames

The Dragalin frames of Definition 1.1 replace the nucleus j in a nuclear frame $(S, \leq, j)$ with the function $D\colon S \to \wp(\wp(S))$. As shown by Dragalin [10, pp. 72-73], this D indeed gives rise to a nucleus, as in Proposition 3.1. Given its importance in our story, we include a proof of this result.

Proposition 3.1 (Dragalin) *Given a Dragalin frame $\mathfrak{F} = (S, \leq, D)$, define a function $[D\rangle$ on $\mathsf{Up}(\mathfrak{F})$ by*

$$[D\rangle U = \{x \in S \mid \forall X \in D(x)\colon X \cap U \neq \varnothing\}. \tag{4}$$

(i) *$[D\rangle$ is a nucleus on $\mathsf{Up}(\mathfrak{F})$;*

(ii) *$[D\rangle$ is a dense nucleus iff $\mathfrak{F}$ is normal.*

We call $(\mathsf{Up}(\mathfrak{F}), [D\rangle)$ the nuclear algebra *of $\mathfrak{F}$.*

Proof. For part (i), to see that $U \in \mathsf{Up}(\mathfrak{F})$ implies $[D\rangle U \in \mathsf{Up}(\mathfrak{F})$, suppose $x \in [D\rangle U$ and $x \leq y$. For each $Y \in D(y)$, by (3°) there is an $X \in D(x)$ with $X \subseteq {\downarrow}Y$. Since $x \in [D\rangle U$, $X \cap U \neq \varnothing$, which with $U \in \mathsf{Up}(\mathfrak{F})$ and $X \subseteq {\downarrow}Y$ implies $Y \cap U \neq \varnothing$. Since this holds for each $Y \in D(y)$, we have $y \in [D\rangle U$.

For inflationarity, for any $X \in D(x)$, there is a $z \in X$ with $x \leq z$ by (1°)–(2°). So if $x \in U \in \mathsf{Up}(\mathfrak{F})$, then $z \in U$, so $X \cap U \neq \varnothing$. Hence $x \in [D\rangle U$.

For idempotence, suppose $x \notin [D\rangle U$, so there is an $X \in D(x)$ such that $X \cap U = \varnothing$. We claim that $X \cap [D\rangle U = \varnothing$. For any $y \in X$, by (4°) we have a $Y \in D(y)$ such that $Y \subseteq {\downarrow}X$. For reductio, suppose there is a $z \in Y \cap U$. Then since $Y \subseteq {\downarrow}X$, there is a $z' \in X$ such that $z \leq z'$, which with $z \in U \in \mathsf{Up}(\mathfrak{F})$ implies $z' \in U$. But then $z' \in X \cap U$, contradicting $X \cap U = \varnothing$ from above. Hence $Y \cap U = \varnothing$, so $y \notin [D\rangle U$. Since this holds for all $y \in X$, $X \cap [D\rangle U = \varnothing$, which with $X \in D(x)$ implies $x \notin [D\rangle[D\rangle U$.

Finally, $[D\rangle$ is monotonic ($A \subseteq B$ implies $[D\rangle A \subseteq [D\rangle B$) by its definition, so $[D\rangle(U \cap U') \subseteq [D\rangle U \cap [D\rangle U'$. Conversely, if we can show that $A \cap [D\rangle A' \subseteq [D\rangle(A \cap A')$, then by two applications of this fact, plus monotonicity and idempotence, $[D\rangle U \cap [D\rangle U' \subseteq [D\rangle([D\rangle U \cap U') \subseteq [D\rangle[D\rangle(U \cap U') \subseteq [D\rangle(U \cap U')$. So suppose $x \in A \cap [D\rangle A'$. Then for $X \in D(x)$, there is a $y \in X \cap A'$. Therefore, by (2°), there is a $z \in X$ with $x \leq z$ and $y \leq z$. Since $x \in A$, $y \in A'$, and $A, A' \in \mathsf{Up}(\mathfrak{F})$, we have $z \in A \cap A'$, so $X \cap A \cap A' \neq \varnothing$. Thus, $x \in [D\rangle(A \cap A')$.

For part (ii), by (4) we have that $[D\rangle\varnothing = \{x \in S \mid D(x) = \varnothing\}$, so $[D\rangle\varnothing = \varnothing$ ($[D\rangle$ is dense) iff $D(x) \neq \varnothing$ for all $x \in S$ ($\mathfrak{F}$ is normal). □

Example 3.2 For any poset $(S, \leq)$ and $x \in S$, define $D(x) = \{\uparrow y \mid x \leq y\}$. One can easily check that $(S, \leq, D)$ is a standard Dragalin frame as in Definition 1.2. Observe that $x \in [D\rangle U$ iff $\forall y \geq x\ \exists z \geq y$: $z \in U$, so $[D\rangle$ is the nucleus of double negation (recall Remark 2.3).

The obvious next question is: which nuclear frames can be generated by Dragalin frames as in Proposition 3.1? In addition, in light of Theorem 2.8, another obvious question is: which locales can be generated as the algebra of fixpoints in the nuclear algebra of a Dragalin frame?

Dragalin gave a partial answer to the second question. We provide a sketch of his proof [10, pp. 75-76] of Theorem 3.3 to convey the main idea, but we omit the details since we will prove a more general result below. Recall that a *spatial* locale is a locale isomorphic to the locale of open sets of a topological space, and a *normal* Dragalin frame $(S, \leq, D)$ is one in which $D(x) \neq \varnothing$ for all $x \in S$.

Theorem 3.3 (Dragalin) *If L is a spatial locale, then there is a normal Dragalin frame $\mathfrak{F}$ such that L is isomorphic to the algebra of fixpoints in the nuclear algebra of $\mathfrak{F}$.*

Proof. [Sketch] Given a topological space (X, Ω) and $x \in X$, a $\mathcal{B} \subseteq \Omega$ is a *local basis of* x if (i) $x \in \bigcap \mathcal{B}$ and (ii) whenever $x \in U \in \Omega$, there is a $V \in \mathcal{B}$ with $V \subseteq U$. From (X, Ω), we define $(S, \leq, D)$ where $S = \Omega \setminus \{\varnothing\}$, $U \leq V$ iff $U \supseteq V$, and $D(U) = \{\mathcal{B} \mid \exists x \in U\colon \mathcal{B}$ is a local basis of x and $\bigcup \mathcal{B} \subseteq U\}$. Dragalin showed that $(S, \leq, D)$ is a normal Dragalin frame. In addition, for the locale $\Omega(X)$ of opens of (X, Ω), he showed that $f\colon \Omega(X) \to \mathsf{Up}(S, \leq)_{[D\rangle}$ defined by $f(U) = \{V \in S \mid U \leq V\}$ is an isomorphism. □

In the other direction, it is not the case that for every Dragalin frame $\mathfrak{F}$, the algebra of fixpoints in the nuclear algebra of $\mathfrak{F}$ is spatial.

Example 3.4 Consider the Dragalin frame $(S, \leq, D)$ where $(S, \leq)$ is the poset associated with the complete infinite binary tree and D gives the nucleus of double negation as in Example 3.2. Then the algebra of fixpoints in the nuclear algebra $(\mathsf{Up}(S, \leq), [D\rangle)$ is a complete atomless Boolean algebra [16, Ex. 2.40]. But a complete Boolean algebra is spatial iff it is atomic.

Indeed, *every* locale can be realized as the algebra of fixpoints in the nuclear algebra of a Dragalin frame. To prove this, it would suffice to show that the nuclear frame used in the proof of Theorem 2.8 can be generated by a Dragalin frame. We will prove the following stronger result.

Theorem 3.5 *Given any nuclear frame $(S, \leq, j)$, there is a standard Dragalin frame $(S, \leq, D)$ such that $j = [D\rangle$.*

To prove Theorem 3.5, we will utilize Simmons's result (Lemma 2.4) that every nucleus on a locale L is the meet, in $N(L)$, of nuclei of the w_a type. We will apply this to the locale $L = \mathsf{Up}(S, \leq)$. First, we show how nuclei of the w_a type on $\mathsf{Up}(S, \leq)$ can be realized by a Dragalin frame $(S, \leq, D)$.

Lemma 3.6 *Given a poset $(S, \leq)$ and $A \in \mathsf{Up}(S, \leq)$, let w_A be the nucleus on $\mathsf{Up}(S, \leq)$ defined as in (1) and define $D_A(x) = \{\uparrow x' \setminus A \mid x' \in \uparrow x \setminus A\}$. Then:*

(i) $(S, \leq, D_A)$ *is a standard Dragalin frame;*

(ii) $[D_A\rangle = w_A$.

Proof. For (i), D_A satisfies (1°) because if $x' \in \uparrow x \setminus A$, then $x' \in \uparrow x' \setminus A$, which shows that $\varnothing \notin D(x)$. Clearly D_A also satisfies (2°°), as well as (3°°), for if $x \leq y$ and $y' \in \uparrow y \setminus A$, then $y' \in \uparrow x \setminus A$. Finally, for (4°°), suppose $y \in X \in D(x)$, so $X = \uparrow x' \setminus A$ for some $x' \in \uparrow x \setminus A$. Let $Y = \uparrow y \setminus A$. Since $y \in X$ and hence $y \notin A$, we have $y \in Y$, which implies $Y \in D(y)$. Moreover, since $y \in X$ and hence $y \in \uparrow x'$, we have $\uparrow y \setminus A \subseteq \uparrow x' \setminus A$, so $Y \subseteq X$.

For (ii), observe that $x \in w_A U = (U \to A) \to A$ iff for all $x' \geq x$, if $x' \notin A$, then there is an $x'' \geq x'$ such that $x'' \in U \setminus A$. This is equivalent to the condition that for all $x' \in \uparrow x \setminus A$, we have $(\uparrow x' \setminus A) \cap U \neq \varnothing$. That is in turn equivalent to the condition that for all $X \in D_A(x)$, $X \cap U \neq \varnothing$, which is finally equivalent to $x \in [D_A\rangle U$. □

Next we show that we can build up meets of nuclei from Dragalin frames.

Lemma 3.7 *Given a family* $\{j_\alpha\}_{\alpha \in I}$ *of nuclei on* $\mathsf{Up}(S, \leq)$ *and a family* $\{D_\alpha\}_{\alpha \in I}$ *of Dragalin functions on* $(S, \leq)$ *such that*

$$x \in j_\alpha U \text{ iff } x \in [D_\alpha\rangle U, \tag{5}$$

the function D defined by

$$D(x) = \bigcup_{\alpha \in I} D_\alpha(x) \tag{6}$$

is a Dragalin function such that

$$x \in (\bigwedge_{\alpha \in I} j_\alpha)U \text{ iff } x \in [D\rangle U. \tag{7}$$

Moreover, if each D_α *is standard, then so is D.*

Proof. We first prove (7). Since meets of nuclei are computed pointwise and meets in $\mathsf{Up}(S, \leq)$ are intersections, we have

$$(\bigwedge_{\alpha \in I} j_\alpha)U = \bigwedge_{\alpha \in I} j_\alpha U = \bigcap_{\alpha \in I} j_\alpha U.$$

Now suppose $x \notin \bigcap_{\alpha \in I} j_\alpha U$, so there is some $\alpha \in I$ such that $x \notin j_\alpha U$. Then by (5), we have $x \notin [D_\alpha\rangle U$, so there is some $X \in D_\alpha(x)$ with $X \cap U = \varnothing$. By (6), $X \in D(x)$, which with $X \cap U = \varnothing$ implies $x \notin [D\rangle U$. Conversely, suppose $x \notin [D\rangle U$, so there is some $X \in D(x)$ with $X \cap U = \varnothing$. Then by (6), there is some $\alpha \in I$ such that $X \in D_\alpha(x)$, which with $X \cap U = \varnothing$ implies $x \notin [D_\alpha\rangle U$, which implies $x \notin j_\alpha U$ by (5), so $x \notin \bigcap_{\alpha \in I} j_\alpha U$.

Next, we show that D satisfies (1°)–(4°), assuming that each D_α does. Clearly if (1°) holds for each D_α, then it holds for D. The same is true of (2°).

For (3°), suppose $x \leq y$ and $Y \in D(y)$. Then $Y \in D_\alpha(x)$ for some $\alpha \in I$. Applying (3°) for D_α, there is an $X \subseteq {\downarrow}Y$ such that $X \in D_\alpha(x)$ and hence $X \in D(x)$, so (3°) also holds for D. Similarly, for (4°), if $y \in X \in D(x)$, then $X \in D_\alpha(x)$ for some $\alpha \in I$, in which case (4°) for D_α gives us a $Y \subseteq {\downarrow}X$ such that $Y \in D_\alpha(y)$ and hence $Y \in D(y)$, so (4°) also holds for D. It is also easy to see that if each D_α satisfies (2°°)–(4°°), then so does D. □

We can now put the pieces together to prove Theorem 3.5.

Proof. By Lemmas 2.4, 3.6, and 3.7, we have:

$$j = \bigwedge\{w_{jA} \mid A \in \mathsf{Up}(S, \leq)\} = \bigwedge\{[D_{jA}\rangle \mid A \in \mathsf{Up}(S, \leq)\} = [D\rangle,$$

where D is defined from the D_{jA}'s as in Lemma 3.7. By Lemma 3.6, each D_{jA} is standard, so D is as well by Lemma 3.7. □

Putting together Theorems 2.8 and 3.5 and Proposition 3.1(ii), we obtain the following.

Corollary 3.8 *A poset P is a locale iff there is a standard normal Dragalin frame $\mathfrak{F}$ such that P is isomorphic to the algebra of fixpoints in the nuclear algebra of $\mathfrak{F}$.*

In §4, we shall see that an analogue of Corollary 3.8 holds for frames that replace Dragalin's function D with a partial order $\preceq$. For the purposes of comparing these frames, we will use the fact that every Dragalin frame can be turned into one satisfying a property stronger than (2°°).

Definition 3.9 A Dragalin frame $\mathfrak{F} = (S, \leq, D)$ is *convex* if for all $x \in S$ and $X \in D(x)$, we have $X = {\uparrow}x \cap {\downarrow}X$.

Remark 3.10

(i) If a Dragalin frame $\mathfrak{F} = (S, \leq, D)$ is convex, then $X \in D(x)$ implies $x \in X$. For if $X \in D(x)$, then by (1°), there is a $y \in X = {\uparrow}x \cap {\downarrow}X$, so $x \leq y$ and $y \in {\downarrow}X$, which implies $x \in {\downarrow}X$ and hence $x \in {\uparrow}x \cap {\downarrow}X = X$.

(ii) A convex $\mathfrak{F}$ typically does not satisfy (3°°). Consider $x, y \in S$ such that $D(y) \neq \varnothing$. By convexity each $Y \in D(y)$ is such that $Y \subseteq {\uparrow}y$, and by (i) each $X \in D(x)$ is such that $x \in X$, so if $x < y$, then $D(x) \cap D(y) = \varnothing$.

(iii) By contrast, every convex $\mathfrak{F}$ satisfies (4°°). Suppose $y \in X \in D(x)$, so by (4°) there is a $Y \in D(y)$ such that $Y \subseteq {\downarrow}X$. By convexity, $Y \subseteq {\uparrow}y$ and $X = {\uparrow}x \cap {\downarrow}X$. Since $y \in X \in D(x)$ implies $x \leq y$ by convexity, we have ${\uparrow}y \subseteq {\uparrow}x$. Thus, $Y \subseteq {\uparrow}x$, which with $Y \subseteq {\downarrow}X$ implies $Y \subseteq {\uparrow}x \cap {\downarrow}X = X$.

Proposition 3.11 *For each Dragalin frame $\mathfrak{F}$, there is a convex Dragalin frame $\mathfrak{G}$ such that the nuclear algebras of $\mathfrak{F}$ and $\mathfrak{G}$ are isomorphic. Moreover, if $\mathfrak{F}$ is normal, then so is $\mathfrak{G}$.*

Proof. Given a Dragalin frame $\mathfrak{F} = (S, \leq, D)$, define $\mathfrak{G} = (S, \leq, D')$ by:

$$D'(x) = \{{\uparrow}x \cap {\downarrow}X \mid X \in D(x)\}.$$

We claim that for all $U \in \mathsf{Up}(S, \leq)$, $[D\rangle U = [D'\rangle U$. Suppose $x \in [D\rangle U$ and consider some $\uparrow x \cap \downarrow X \in D'(x)$, so $X \in D(x)$. Since $x \in [D\rangle U$ and $X \in D(x)$, there is a $y \in X \cap U$. Then by (2°), there is a $z \in X$ with $x \leq z$ and $y \leq z$. From $z \in X$ and $x \leq z$, we have $z \in \uparrow x \cap X \subseteq \uparrow x \cap \downarrow X$. From $y \in U$ and $y \leq z$, we have $z \in U$. Therefore, $(\uparrow x \cap \downarrow X) \cap U \neq \varnothing$ and hence $x \in [D'\rangle U$. Conversely, if $x \notin [D\rangle U$, so there is an $X \in D(x)$ such that $X \cap U = \varnothing$, then since U is an upset, $(\uparrow x \cap \downarrow X) \cap U = \varnothing$ and $\uparrow x \cap \downarrow X \in D'(x)$, so $x \notin [D'\rangle U$. It follows that the nuclear algebras of $\mathfrak{F}$ and $\mathfrak{G}$ are isomorphic.

Next we show that $\mathfrak{G}$ satisfies (1°)–(4°), so it is a Dragalin frame. Since $\mathfrak{G}$ is convex, (2°) is immediate. For (1°) for D', if $X \in D(x)$, then by (1°)–(2°) for D, $\uparrow x \cap X \neq \varnothing$, so $\uparrow x \cap \downarrow X \neq \varnothing$ and hence $\varnothing \notin D'(x)$.

For (3°) for D', suppose $x \leq y$ and $Y' \in D'(y)$, so $Y' = \uparrow y \cap \downarrow Y$ for some $Y \in D(y)$. Then by (3°) for D, there is an $X \in D(x)$ such that $X \subseteq \downarrow Y$. Setting $X' = \uparrow x \cap \downarrow X$, we have $X' \in D'(x)$, and we claim that $X' \subseteq \downarrow Y'$, which will establish (3°) for D'. If $a \in X'$, then $a \in \downarrow X$, so there is a b with $a \leq b \in X$. Then since $X \subseteq \downarrow Y$, there is a c with $b \leq c \in Y$. Given $Y \in D(y)$, it follows by (2°) for D that there is a $z \in Y$ such that $y \leq z$ and $c \leq z$. Thus, $a \leq b \leq c \leq z \in \uparrow y \cap Y \subseteq \uparrow y \cap \downarrow Y$, so $a \in \downarrow(\uparrow y \cap \downarrow Y) = \downarrow Y'$.

Next we prove (4°°). Suppose $y \in X' \in D'(x)$, so $X' = \uparrow x \cap \downarrow X$ for some $X \in D(x)$. We need a $Y' \in D'(y)$ with $Y' \subseteq X'$, so we need a $Y \in D(y)$ with $\uparrow y \cap \downarrow Y \subseteq \uparrow x \cap \downarrow X$. Since $y \in X'$, $y \in \downarrow X$, so there is a z with $y \leq z \in X$. Given $X \in D(x)$, it follows by (4°) for D that there is a $Z \in D(z)$ such that $Z \subseteq \downarrow X$. Then given $y \leq z$, it follows by (3°) for D that there is a $Y \in D(y)$ such that $Y \subseteq \downarrow Z$, which with $Z \subseteq \downarrow X$ implies $Y \subseteq \downarrow X$, which in turn implies $\downarrow Y \subseteq \downarrow X$. Since $y \in X'$, $y \in \uparrow x$, so we also have $\uparrow y \subseteq \uparrow x$. Therefore, $\uparrow y \cap \downarrow Y \subseteq \uparrow x \cap \downarrow X$.

Finally, it is obvious that if $\mathfrak{F}$ is normal, then $\mathfrak{G}$ is normal too. □

4 Fairtlough-Mendler Frames

In this section, we consider another way of replacing the nucleus j in a nuclear frame with more concrete data. Fairtlough and Mendler [11] (also see [1]) give a semantics for an intuitionistic modal logic called *propositional lax logic* with a modality $\bigcirc$ obeying the axioms of a nucleus. The frames used in their semantics therefore provide another method for representing nuclear algebras.

Definition 4.1 An *FM-frame* (Fairtlough-Mendler frame) is a tuple $\mathfrak{F} = (S, \leq, \preceq, F)$, where $\leq$ and $\preceq$ are preorders on the set S, $\preceq$ is a subrelation of $\leq$, and $F \in \mathsf{Up}(\mathfrak{F}) := \mathsf{Up}(S, \leq)$. We say that $\mathfrak{F}$ is *normal* if $F = \varnothing$, and $\mathfrak{F}$ is *partially ordered* if $\leq$ (and hence $\preceq$) is a partial order.

Each FM-frame gives rise to a nuclear algebra. To see this, for each FM-frame $\mathfrak{F} = (S, \leq, \preceq, F)$, let $\mathsf{Up}(\mathfrak{F})_F = \{U \in \mathsf{Up}(\mathfrak{F}) \mid F \subseteq U\}$ be the relativization of $\mathsf{Up}(\mathfrak{F})$ to F. Then $\mathsf{Up}(\mathfrak{F})_F$ is a locale, where the operations $\wedge, \vee, \rightarrow$ on $\mathsf{Up}(\mathfrak{F})_F$ are the restrictions of the corresponding operations on $\mathsf{Up}(\mathfrak{F})$, and F is the bottom element of $\mathsf{Up}(\mathfrak{F})_F$. When $F = \varnothing$, we obviously have $\mathsf{Up}(\mathfrak{F})_F = \mathsf{Up}(\mathfrak{F})$. In order to define a nucleus on $\mathsf{Up}(\mathfrak{F})_F$, recall that for a binary relation R on S, $x \in S$, and $U \subseteq S$, it is customary to define

$R(x) = \{y \in S \mid xRy\}$ and let

$$\Box_R(U) = \{x \in S \mid R(x) \subseteq U\} \text{ and } \Diamond_R(U) = \{x \in S \mid R(x) \cap U \neq \varnothing\}.$$

Now consider the following operator on $\mathsf{Up}(\mathfrak{F})_F$:

$$\Box_{\leq}\Diamond_{\preceq}U = \{x \in S \mid \forall y(x \leq y \Rightarrow \exists z(y \preceq z \ \&\ z \in U))\}.$$

Fairtlough and Mendler use this operator to interpret the modality $\bigcirc$ of lax logic in FM-frames. The following proposition is essentially their soundness result for lax logic [11, p. 9], which we prove for the reader's convenience.

Proposition 4.2 (Fairtlough-Mendler) *If $\mathfrak{F} = (S, \leq, \preceq, F)$ is an FM-frame, then $(\mathsf{Up}(\mathfrak{F})_F, \Box_{\leq}\Diamond_{\preceq})$ is a nuclear algebra, which we call the* nuclear algebra of $\mathfrak{F}$. *Moreover, if $F = \varnothing$, then $\Box_{\leq}\Diamond_{\preceq}$ is a dense nucleus.*

Proof. For any $U \in \mathsf{Up}(\mathfrak{F})_F$, since $\preceq$ is a reflexive subrelation of $\leq$, we have $U \subseteq \Box_{\leq}\Diamond_{\preceq}U$. Since $\leq$ is reflexive and $\preceq$ is transitive, we also have $\Box_{\leq}\Diamond_{\preceq}\Box_{\leq}\Diamond_{\preceq}U \subseteq \Box_{\leq}\Diamond_{\preceq}U$. Clearly $\Box_{\leq}\Diamond_{\preceq}(U \cap V) \subseteq \Box_{\leq}\Diamond_{\preceq}U \cap \Box_{\leq}\Diamond_{\preceq}V$. Conversely, for any $U, V \in \mathsf{Up}(\mathfrak{F})_F$, if $x \in \Box_{\leq}\Diamond_{\preceq}U \cap \Box_{\leq}\Diamond_{\preceq}V$ and $x \leq y$, then there is a z with $y \preceq z \in U$ and hence $y \leq z$. So $x \leq z$, which with $x \in \Box_{\leq}\Diamond_{\preceq}V$ implies that there is a w with $z \preceq w \in V$ and hence $z \leq w$. Then since $z \in U \in \mathsf{Up}(\mathfrak{F})_F$, we have $w \in U \cap V$. Thus, $x \in \Box_{\leq}\Diamond_{\preceq}(U \cap V)$. Finally, since $\leq$ is reflexive, $\Box_{\leq}\Diamond_{\preceq}\varnothing = \varnothing$, so $\Box_{\leq}\Diamond_{\preceq}$ is dense if $F = \varnothing$. □

Example 4.3 If $\leq = \preceq$ in an FM-frame $(S, \leq, \preceq, F)$, then $\Box_{\leq}\Diamond_{\preceq}$ is the nucleus of double negation (cf. [11, p. 23] and [3]).

Remark 4.4 Let us now consider extracting a nuclear *frame* rather than a nuclear algebra from an FM-frame. If the FM-frame $\mathfrak{F} = (S, \leq, \preceq, F)$ is normal, then $(S, \leq, \Box_{\leq}\Diamond_{\preceq})$ is a preordered nuclear frame as in Remark 2.6, which we can turn into a partially ordered nuclear frame by taking the skeleton, and the nuclear algebra of this nuclear frame is isomorphic to that of $\mathfrak{F}$. Now suppose $\mathfrak{F}$ is not normal. Define the preordered nuclear frame $(S^-, \leq^-, j)$ where $S^- = S \setminus F$, $\leq^-$ is the restriction of $\leq$ to S^-, and j is defined for $U \in \mathsf{Up}(S^-, \leq^-)$ by $jU = \Box_{\leq}\Diamond_{\preceq}(U \cup F) \setminus F$. The nuclear algebra of $(S^-, \leq^-, j)$ is then isomorphic to that of $\mathfrak{F}$, and once again we can turn $(S^-, \leq^-, j)$ into a partially ordered nuclear frame as in Remark 2.6.

Turning FM-frames into partially ordered FM-frames is more difficult. It is not clear how to define the skeleton $\mathfrak{F}' = (S', \leq', \preceq', F')$ of an FM-frame $\mathfrak{F} = (S, \leq, \preceq, F)$ such that $(\mathsf{Up}(\mathfrak{F})_F, \Box_{\leq}\Diamond_{\preceq})$ and $(\mathsf{Up}(\mathfrak{F}')_{F'}, \Box_{\leq'}\Diamond_{\preceq'})$ are isomorphic nuclear algebras. The difficulty is in defining $\preceq'$, as standard ways of defining a new binary relation on a quotient do not work in this case. Below we take a different approach by unwinding instead of collapsing clusters to produce a partially ordered FM-frame $\mathfrak{F}^\dagger$. While the nuclear algebra of $\mathfrak{F}^\dagger$ will be "larger" than that of $\mathfrak{F}$, their algebras of fixpoints will be isomorphic.

Proposition 4.5 *For any FM-frame $\mathfrak{F}$, there is a partially ordered FM-frame $\mathfrak{F}^\dagger$ such that the algebra of fixpoints in the nuclear algebra of $\mathfrak{F}$ is isomorphic to that of $\mathfrak{F}^\dagger$. Moreover, $\mathfrak{F}$ is normal iff $\mathfrak{F}^\dagger$ is normal.*

Proof. Given an FM-frame $\mathfrak{F} = (S, \sqsubseteq, \preceq, F)$, define $\mathfrak{F}^\dagger = (S^\dagger, \sqsubseteq^\dagger, \preceq^\dagger, F^\dagger)$ as follows (in this proof, $\leq$ is the usual ordering on $\mathbb{N}$):

- $S^\dagger = \{\langle x, t\rangle \mid x \in S,\, t \in \mathbb{N}\}$;
- $\langle x, t\rangle \sqsubseteq^\dagger \langle x', t'\rangle$ iff either $[x = x'$ and $t \leq t']$ or $[x \sqsubseteq x'$ and $t < t']$;
- $\langle x, t\rangle \preceq^\dagger \langle x', t'\rangle$ iff either $[x = x'$ and $t \leq t']$ or $[x \preceq x'$ and $t < t']$;
- $F^\dagger = \{\langle x, t\rangle \in S^\dagger \mid x \in F\}$.

Observe that $\sqsubseteq^\dagger$ and $\preceq^\dagger$ are partial orders. Moreover, $F = \varnothing$ iff $F^\dagger = \varnothing$, so $\mathfrak{F}$ is normal iff $\mathfrak{F}^\dagger$ is normal.

Let j be the nucleus associated with $\mathfrak{F}$ and $j^\dagger$ the nucleus associated with $\mathfrak{F}^\dagger$. Let $g\colon S^\dagger \to S$ be defined by $g(x, t) = x$. We claim that the function G that maps each fixpoint U of $j^\dagger$ to $G(U) = g[U]$ is an isomorphism between the algebras of fixpoints in the nuclear algebras of $\mathfrak{F}^\dagger$ and $\mathfrak{F}$.

First, toward showing that G sends fixpoints of $j^\dagger$ to fixpoints of j, we show that for every $\langle x, t\rangle \in S^\dagger$ and $U \in \mathsf{Up}(\mathfrak{F}^\dagger)_{F^\dagger}$:

$$\langle x, t\rangle \in j^\dagger U \text{ iff } x \in jg[U]. \tag{8}$$

From left to right, if $x' \sqsupseteq x$, then $\langle x', t+1\rangle \sqsupseteq^\dagger \langle x, t\rangle$. Since $\langle x, t\rangle \in j^\dagger U$, there is $\langle x'', t''\rangle \succeq^\dagger \langle x', t+1\rangle$ such that $\langle x'', t''\rangle \in U$. This implies $x'' \succeq x'$ and $x'' \in g[U]$. Hence we have shown that $x \in jg[U]$. From right to left, if $\langle x', t'\rangle \sqsupseteq^\dagger \langle x, t\rangle$, then $x' \sqsupseteq x$. Since $x \in jg[U]$, there is an $x'' \succeq x'$ with $x'' \in g[U]$. It follows that there is some $s \in \mathbb{N}$ such that $\langle x'', s\rangle \in U$. Since $U \in \mathsf{Up}(\mathfrak{F}^\dagger)_{F^\dagger}$, it follows that $\langle x'', t' + s + 1\rangle \in U$. Given $x'' \succeq x'$ and $t' + s + 1 > t'$, we have $\langle x'', t' + s + 1\rangle \succeq^\dagger \langle x', t'\rangle$. Therefore, $\langle x, t\rangle \in j^\dagger U$.

Now we can see that if U is a fixpoint of $j^\dagger$, then $g[U]$ is a fixpoint of j. To see that $jg[U] \subseteq g[U]$, observe that if $x \in jg[U]$, then by (8) and the assumption that U is a fixpoint of $j^\dagger$, we have $\langle x, t\rangle \in j^\dagger U \subseteq U$, whence $x \in g[U]$.

Second, we claim that G is surjective. Suppose V is a fixpoint of j. Then we claim that $g^{-1}[V]$ is a fixpoint of $j^\dagger$, which with $g[g^{-1}[V]] = V$, given by the surjectivity of g, will show that G is surjective. We begin by showing that $g^{-1}[V] \in \mathsf{Up}(\mathfrak{F}^\dagger)_{F^\dagger}$. To see that $g^{-1}[V] \in \mathsf{Up}(\mathfrak{F}^\dagger)$, observe that if $\langle x, t\rangle \in g^{-1}[V]$ and $\langle x', t'\rangle \sqsupseteq^\dagger \langle x, t\rangle$, then $x \in V$ and $x' \sqsupseteq x$, which with $V \in \mathsf{Up}(\mathfrak{F})$ implies $x' \in V$ and hence $\langle x', t'\rangle \in g^{-1}[V]$. To see that $g^{-1}[V] \in \mathsf{Up}(\mathfrak{F}^\dagger)_{F^\dagger}$, observe that since $g[F^\dagger] = F \subseteq V$, we have $F^\dagger \subseteq g^{-1}[g[F^\dagger]] \subseteq g^{-1}[V]$. Next, we must show that $j^\dagger g^{-1}[V] \subseteq g^{-1}[V]$. If $\langle x, t\rangle \in j^\dagger g^{-1}[V]$, then by (8), $x \in jg[g^{-1}[V]] = jV = V$, using that V is a fixpoint of j. Hence $\langle x, t\rangle \in g^{-1}[V]$.

Third, we show that G preserves and reflects order: for any fixpoints U and V of $j^\dagger$, we have $U \subseteq V$ iff $g[U] \subseteq g[V]$. If $U \subseteq V$ and $x \in g[U]$, then for some $t \in \mathbb{N}$, $\langle x, t\rangle \in U \subseteq V$, so $x \in g[V]$. Conversely, suppose $U \not\subseteq V$, so there is an $\langle x, t\rangle \in U \setminus V$. Then $x \in g[U]$, and since $V = j^\dagger V$, $\langle x, t\rangle \notin j^\dagger V$, which with (8) implies $x \notin jg[V]$ and hence $x \notin g[V]$. So $g[U] \not\subseteq g[V]$.

Thus, G is an isomorphism between the algebras of fixpoints of $\mathfrak{F}^\dagger$ and $\mathfrak{F}$. □

Turning preordered FM-frames into partially ordered Dragalin frames does not require the unwinding in the previous proof. In this case we may collapse clusters without difficulty, by first turning a preordered FM-frame into a partially ordered nuclear frame as in Remark 4.4 and then turning that nuclear frame into a Dragalin frame as in Theorem 3.5.

Remark 4.6 In the case of a normal FM-frame $\mathfrak{F} = (S, \leq, \preceq, F)$, there is an even more direct approach: we can define a standard Dragalin frame $\mathfrak{D} = (S', \leq', D)$ where $(S', \leq')$ is the skeleton of $(S, \leq)$ and $D([x]) = \{\Uparrow y \mid x \leq y\}$, where $[x]$ is the equivalence class of x and $\Uparrow y = \{[z] \in S' \mid y \preceq z\}$. For lack of space, we omit the proof that the nuclear algebras of $\mathfrak{F}$ and $\mathfrak{D}$ are isomorphic.

It is rather surprising that we can also go in the other direction, from Dragalin to FM-frames. This we do by "enlarging" the underlying poset.

Theorem 4.7 *For any Dragalin frame $\mathfrak{D}$, there is an FM-frame $\mathfrak{F}$ such that the nuclear algebras of $\mathfrak{D}$ and $\mathfrak{F}$ are isomorphic. Moreover, $\mathfrak{D}$ is normal iff $\mathfrak{F}$ is normal.*

Proof. By Proposition 3.11, we may assume that the Dragalin frame $\mathfrak{D}$ is convex and therefore satisfies ($4^{\circ\circ}$) by Remark 3.10(iii). First we give the proof assuming that $\mathfrak{D}$ is normal and then show how to modify the proof to lift this assumption. The construction is similar to the construction of intuitionistic relational frames from intuitionistic neighborhood frames in [19] (cf. the construction of birelational frames from monotonic neighborhood frames in [20]). Given a normal Dragalin frame $\mathfrak{D} = (S, \leq, D)$, we define an FM-frame $\mathfrak{F} = (S', \leq', \preceq', F')$ with $F' = \varnothing$ as follows:

- $S' = \{(x, X) \mid x \in S, X \in D(x)\}$;
- $(x, X) \leq' (y, Y)$ iff $x \leq y$;
- $(x, X) \preceq' (y, Y)$ iff $y \in X$ and $Y \subseteq X$.

Clearly $\leq'$ is a preorder. To see that $\preceq'$ is a preorder, $X \in D(x)$ implies $x \in X$ by convexity and Remark 3.10(i), so $\preceq'$ is reflexive. For transitivity, if $(x, X) \preceq' (y, Y) \preceq' (z, Z)$, then $Z \subseteq Y \subseteq X$, so $Z \subseteq X$, and $z \in Y \subseteq X$, so $z \in X$. Hence $(x, X) \preceq' (z, Z)$. Finally, $\preceq'$ is a subrelation of $\leq'$: if $(x, X) \preceq' (y, Y)$, then by ($2^{\circ\circ}$), $y \in X \in D(x)$ implies $x \leq y$, so $(x, X) \leq' (y, Y)$.

Define $f\colon \mathsf{Up}(\mathfrak{D}) \to \mathsf{Up}(\mathfrak{F})$ by

$$f(U) = \{(x, X) \mid x \in U, X \in D(x)\}.$$

It is routine to check that f is an isomorphism between $\mathsf{Up}(\mathfrak{D})$ and $\mathsf{Up}(\mathfrak{F})$. To show it is an isomorphism between the nuclear algebras of $\mathfrak{D}$ and $\mathfrak{F}$, we show:

$$f([D\rangle U) = \Box_{\leq'} \Diamond_{\preceq'} f(U). \tag{9}$$

Suppose $(x, X) \in f([D\rangle U)$, so $x \in [D\rangle U$ and $X \in D(x)$. Consider any (y, Y) such that $(x, X) \leq' (y, Y)$, so $x \leq y$. Then from $x \in [D\rangle U$ we have $y \in [D\rangle U$, since $[D\rangle U$ is an upset whenever U is. Given $y \in [D\rangle U$ and $Y \in D(y)$, there is a $z \in Y$ such that $z \in U$. Given $z \in Y \in D(y)$, by ($4^{\circ\circ}$) there is a $Z \in D(z)$ such that $Z \subseteq Y$. Thus, $(y, Y) \preceq' (z, Z)$. Then since $z \in U$, we have $(z, Z) \in f(U)$. Hence we have shown that $(x, X) \in \Box_{\leq'} \Diamond_{\preceq'} f(U)$.

Conversely, suppose $(x, X) \notin f([D\rangle U)$, so $x \notin [D\rangle U$. Then there is a $Y \in D(x)$ with $Y \cap U = \varnothing$ (but $Y \neq \varnothing$ by (1°)). Therefore, $(x, X) \leq' (x, Y)$, and for any (z, Z) such that $(x, Y) \preceq' (z, Z)$, we have $z \in Y$ and hence $z \notin U$, so $(z, Z) \notin f(U)$. Thus, $x \notin \Box_{\leq'}\Diamond_{\preceq'} f(U)$. This completes the proof of (9).

If $\mathfrak{D} = (S, \leq, D)$ is not normal, define the FM-frame $\mathfrak{F} = (S', \leq', \preceq', F')$ by:

- $S' = \{(x, X) \mid x \in S, X \in D(x)\} \cup \{(x, \varnothing) \mid D(x) = \varnothing\} \cup \{m\}$; $F' = \{m\}$;
- $(x, X) \leq' (y, Y)$ iff $x \leq y$; and m is the maximum of $\leq'$;
- $(x, X) \preceq' (y, Y)$ iff $y \in X$ and $\varnothing \neq Y \subseteq X$;
- for all $(x, \varnothing) \in S'$, $(x, \varnothing) \preceq' (x, \varnothing)$, $(x, \varnothing) \preceq' m$, and $m \preceq' m$.

By (1°), for all $x \in S$, $\varnothing \notin D(x)$, so adding $(x, \varnothing)$ to S' when $D(x) = \varnothing$ does not cause any ambiguity. Note that $\leq'$ and $\preceq'$ are still preorders, and $\preceq'$ is a subrelation of $\leq'$. Since $\mathfrak{F}$ is an FM-frame, its nuclear algebra is based on the locale $\mathsf{Up}(\mathfrak{F})_{F'}$. Define $g\colon \mathsf{Up}(\mathfrak{D}) \to \mathsf{Up}(\mathfrak{F})_{F'}$ by

$$g(U) = \{(x, X) \mid x \in U, X \in D(x)\} \cup \{(x, \varnothing) \mid x \in U, D(x) = \varnothing\} \cup \{m\}.$$

As in the case of f above, it is routine to check that g is an isomorphism between $\mathsf{Up}(\mathfrak{D})$ and $\mathsf{Up}(\mathfrak{F})_{F'}$. To see that it is an isomorphism between the nuclear algebras of $\mathfrak{D}$ and $\mathfrak{F}$, the proof of (9) above works for g in place of f with only small additions. In particular, we must show that $(x, \varnothing) \in g([D\rangle U)$ iff $(x, \varnothing) \in \Box_{\leq'}\Diamond_{\preceq'} g(U)$. In fact, for any $(x, \varnothing) \in S'$, both $(x, \varnothing) \in g([D\rangle U)$ and $(x, \varnothing) \in \Box_{\leq'}\Diamond_{\preceq'} g(U)$. To see the first, since $D(x) = \varnothing$, for any $U \in \mathsf{Up}(\mathfrak{D})$ we have $x \in [D\rangle U$ and hence $(x, \varnothing) \in g([D\rangle U)$. To see the second, consider any (y, Y) such that $(x, \varnothing) \leq' (y, Y)$. Then $x \leq y$, which with $D(x) = \varnothing$ and (3°) implies $D(y) = \varnothing$, so $(y, Y) = (y, \varnothing)$. By construction, $(y, \varnothing) \preceq' m$ and $m \in g(U)$. Thus, $(x, \varnothing) \in \Box_{\leq'}\Diamond_{\preceq'} g(U)$. □

From Theorems 3.5 and 4.7, we obtain the following.

Corollary 4.8 *For any nuclear frame $\mathfrak{F}$, there is an FM-frame $\mathfrak{G}$ such that the nuclear algebras of $\mathfrak{F}$ and $\mathfrak{G}$ are isomorphic.*

This is weaker than what we had for Dragalin frames in Theorem 3.5, which showed that we can always go from a nuclear frame to a Dragalin frame *based on the same poset*. The following example shows that when going from a nuclear frame to an FM-frame, changing the underlying poset may be unavoidable.

Example 4.9 If $\leq$ is identity, so $\mathsf{Up}(S, \leq) = \wp(S)$, then $\preceq = \leq$, so $\Box_{\leq}\Diamond_{\preceq}$ is the identity nucleus on $\wp(S)$ and any relativization thereof. Yet for any non-trivial Boolean algebra, there is a nucleus distinct from the identity nucleus.

Putting together Corollary 3.8, Theorem 4.7, and Proposition 4.5, we obtain the following analogue of Corollary 3.8.

Corollary 4.10 *A poset P is a locale iff there is a partially ordered normal FM-frame $\mathfrak{F}$ such that P is isomorphic to the algebra of fixpoints in the nuclear algebra of $\mathfrak{F}$.*

5 Goldblatt Frames

A nucleus on a Heyting algebra is a special case of a *dual operator* on a Heyting algebra, a unary function that preserves all finite meets (including 1). Following the tradition in modal logic, we denote such a function by $\Box$.

Definition 5.1 A *modal Heyting algebra* is a pair $(H, \Box)$ where H is a Heyting algebra and $\Box$ is a dual operator. It is a *modal locale* if H is a locale.

Typical examples of modal locales come from intuitionistic modal frames [34].

Definition 5.2 An *IM-frame* (intuitionistic modal frame) is a triple $\mathfrak{F} = (S, \leq, R)$ where $(S, \leq)$ is a poset and $R \subseteq S^2$ is such that $\leq \circ R \circ \leq = R$.

The condition that $\leq \circ R \circ \leq = R$ guarantees that for $U \in \mathsf{Up}(\mathfrak{F})$, we also have $\Box_R U \in \mathsf{Up}(\mathfrak{F})$. It is straightforward to check that $\mathfrak{F}^+ := (\mathsf{Up}(\mathfrak{F}), \Box_R)$ is a modal locale. A natural question, then, is whether there are conditions on an IM-frame $\mathfrak{F}$ that are equivalent to $\mathfrak{F}^+$ being a nuclear algebra. These conditions were identified by Goldblatt [13, pp. 500-01]. Recall that a relation $R \subseteq S^2$ is *dense* if whenever xRy, there is a $z \in S$ such that $xRzRy$.

Lemma 5.3 (Goldblatt) *Let $\mathfrak{F} = (S, \leq, R)$ be an IM-frame.*

(i) *R is a subrelation of $\leq$ iff $U \subseteq \Box_R U$ for each $U \in \mathsf{Up}(\mathfrak{F})$.*

(ii) *R is dense iff $\Box_R \Box_R U \subseteq \Box_R U$ for each $U \in \mathsf{Up}(\mathfrak{F})$.*

Remark 5.4 Goldblatt [13] did not assume the full IM-frame condition that $\leq \circ R \circ \leq = R$, but only the weaker condition that $R \circ \leq = R$, which is still sufficient for $\Box_R$ to be a function on $\mathsf{Up}(S, \leq)$.[7] Relative to frames satisfying the weaker condition, the property that $\Box_R \Box_R U \subseteq \Box_R U$ for each $U \in \mathsf{Up}(\mathfrak{F})$ corresponds to a "pseudo-density" condition on R [13, p. 501], rather than density. But any frame satisfying the weaker condition can be turned into an IM-frame such that their associated modal locales are isomorphic (by simply defining a new relation $R' = \leq \circ R \circ \leq$).

Following Goldblatt's notation, we will use $\prec$ for a dense subrelation of $\leq$.

Definition 5.5 A *Goldblatt frame* is an IM-frame $\mathfrak{F} = (S, \leq, \prec)$ such that $\prec$ is a dense subrelation of $\leq$.

For a Goldblatt frame $\mathfrak{F} = (S, \leq, \prec)$, the modal operator $\Box_\prec$ is a nucleus by Lemma 5.3, which gives us the following.

Proposition 5.6 (Goldblatt) *If $\mathfrak{F} = (S, \leq, \prec)$ is a Goldblatt frame, then $(\mathsf{Up}(\mathfrak{F}), \Box_\prec)$ is a nuclear algebra, which we call the* nuclear algebra of $\mathfrak{F}$.

Remark 5.7 A special case of the frames considered by Goldblatt [13]—with $\prec$ not only dense, but also serial—appears in the semantics for *intuitionistic epistemic logic* in [2,26], which treats $\Box_\prec$ as an intuitionistic knowledge modality. Note that in a Goldblatt frame, $\Box_\prec$ is a dense nucleus iff $\prec$ is serial. The logic IEL^+ of [2,26] is exactly the logic of a dense nucleus. This is the extension

[7] A still weaker sufficient condition is that $R \circ \leq \subseteq \leq \circ R$ [7].

of propositional lax logic with the axiom $\neg\bigcirc\bot$, which Fairtlough and Mendler [11, Thm. 4.5] prove is the propositional lax logic of normal FM-frames.

We have seen that for every nuclear frame $\mathfrak{F}$, there is a Dragalin frame and an FM-frame whose nuclear algebras are isomorphic to that of $\mathfrak{F}$ (Theorem 3.5 and Corollary 4.8). Let us now consider for which nuclear frames there is a Goldblatt frame with an isomorphic nuclear algebra.

First, we define a necessary and sufficient condition for a modal locale $(L, \Box)$ to be isomorphic to $\mathfrak{F}^+$ for some IM-frame $\mathfrak{F}$ (recall Remark 2.7).

Definition 5.8 A modal locale $(L, \Box)$ is *perfect* if L is Alexandroff and $\Box$ is completely multiplicative: for every $X \subseteq L$, $\Box \bigwedge X = \bigwedge\{\Box x \mid x \in X\}$.

As is well known, if a function on a complete lattice is completely multiplicative, then it admits an adjoint.

Lemma 5.9 *Given a modal locale $(L, \Box)$ with $\Box$ completely multiplicative, the function $\Diamond^*\colon L \to L$ defined by $\Diamond^* a = \bigwedge\{x \in L \mid a \leq \Box x\}$ is a left adjoint of $\Box$: for all $a, b \in L$, $\Diamond^* a \leq b$ iff $a \leq \Box b$.*

Lemma 5.9 is used in the proof of the following characterization.

Theorem 5.10 *Let $(L, \Box)$ be a modal locale.*

(i) *$(L, \Box)$ is isomorphic to $\mathfrak{F}^+$ for an IM-frame $\mathfrak{F}$ iff $(L, \Box)$ is perfect.*

(ii) *$(L, \Box)$ is isomorphic to $\mathfrak{F}^+$ for a Goldblatt frame $\mathfrak{F}$ iff $(L, \Box)$ is a perfect nuclear algebra.*

Proof. For part (i), for any IM-frame $\mathfrak{F}$, it is easy to see that $\mathfrak{F}^+$ is perfect. Conversely, suppose $(L, \Box)$ is perfect, so L is Alexandroff. Let $S = J^\infty(L)$ (see Remark 2.7) and $\sqsubseteq$ be the dual of the restriction of the order $\leq$ on L to S. It is well known that the function $f\colon L \to \mathsf{Up}(S, \sqsubseteq)$ defined by $f(a) = \{x \in S \mid a \sqsubseteq x\}$ is an isomorphism between L and $\mathsf{Up}(S, \sqsubseteq)$ (see, e.g., [8,5]).

If $\Box$ is a completely multiplicative operator on L, then we define R on S by xRy iff $y \leq \Diamond^* x$. To see that $\mathfrak{F} := (S, \sqsubseteq, R)$ is an IM-frame, suppose $x \sqsubseteq yRz \sqsubseteq u$. Then $u \leq z$, $z \leq \Diamond^* y$, and $y \leq x$. Therefore, $u \leq \Diamond^* y$ and $\Diamond^* y \leq \Diamond^* x$, yielding $u \leq \Diamond^* x$, so xRu. To see that f is an isomorphism between $(L, \Box)$ and $\mathfrak{F}^+$, it only remains to show that $f(\Box a) = \Box_R f(a)$. Suppose $x \in f(\Box a)$ and xRy. Then $\Box a \sqsubseteq x$, so $x \leq \Box a$, and $y \leq \Diamond^* x$, so by Lemma 5.9, $y \leq \Diamond^* x \leq a$ and hence $a \sqsubseteq y$. Thus, $y \in f(a)$, which shows $x \in \Box_R f(a)$. Conversely, if $x \notin f(\Box a)$, then $\Box a \not\sqsubseteq x$, so $x \not\leq \Box a$ and hence $\Diamond^* x \not\leq a$ by Lemma 5.9. Since L is Alexandroff, there is a $y \in S$ such that $y \leq \Diamond^* x$ and $y \not\leq a$. Hence xRy and $a \not\sqsubseteq y$, so $y \notin f(a)$, whence $x \notin \Box_R f(a)$.

For part (ii), apply part (i) and Lemma 5.3. $\Box$

Note that when L is the locale of upsets in a poset P, the poset $(J^\infty(L), \sqsubseteq)$ constructed in the proof of Theorem 5.10 is isomorphic to P. Thus, the following is an immediate corollary of Theorem 5.10.

Corollary 5.11 *Given any nuclear frame $\mathfrak{F} = (S, \leq, j)$ with j completely multiplicative, there is a Goldblatt frame $\mathfrak{G} = (S, \leq, \prec)$ such that $j = \Box_\prec$.*

Using the above results, we can also characterize the locales that can be realized as the algebra of fixpoints in the nuclear algebra of a Goldblatt frame.

Theorem 5.12 *A locale L is isomorphic to the algebra of fixpoints in the nuclear algebra of a Goldblatt frame iff L is completely distributive.*

Proof. From right to left, it suffices to observe that in the proof of Theorem 2.8, if the poset P is a completely distributive locale, then the nucleus j defined in (3) is completely multiplicative, whence Corollary 5.11 gives us the desired Goldblatt frame. The proof that j is completely multiplicative is just like the original proof that j is multiplicative, which used the join-infinite distributive law for locales, but now we use the completely distributive law.

From left to right, by Theorem 5.10, the nucleus in the nuclear algebra of any Goldblatt frame is completely multiplicative, and the underlying locale of the nuclear algebra of any Goldblatt frame is Alexandroff and hence completely distributive. Thus, it suffices to show that for any localic nuclear algebra $\mathbb{L} = (L, j)$ with L completely distributive and j completely multiplicative, its algebra of fixpoints $\mathbb{L}_j$ is completely distributive. Let $\{x_{\varphi,\psi} \mid \varphi \in \Phi,\, \psi \in \Psi_\varphi\}$ be a doubly indexed family of elements from $\mathbb{L}_j$, F the set of functions f assigning to each $\varphi \in \Phi$ some $f(\varphi) \in \Psi_\varphi$, $\bigwedge$ and $\bigvee$ the operations in L, and $\bigsqcap$ and $\bigsqcup$ the operations in $\mathbb{L}_j$. Then we have:

$$\begin{aligned}
\bigsqcap_{\varphi\in\Phi} \bigsqcup_{\psi\in\Psi_\varphi} x_{\varphi,\psi} &= \bigwedge_{\varphi\in\Phi} j \bigvee_{\psi\in\Psi_\varphi} x_{\varphi,\psi} && \text{by definition of } \mathbb{L}_j \\
&= j \bigwedge_{\varphi\in\Phi} \bigvee_{\psi\in\Psi_\varphi} x_{\varphi,\psi} && \text{by complete multiplicativity of } j \\
&= j \bigvee_{f\in F} \bigwedge_{\varphi\in\Phi} x_{\varphi,f(\varphi)} && \text{by complete distributivity of } L \\
&= \bigsqcup_{f\in F} \bigsqcap_{\varphi\in\Phi} x_{\varphi,f(\varphi)} && \text{by definition of } \mathbb{L}_j,
\end{aligned}$$

so $\mathbb{L}_j$ is completely distributive. □

Remark 5.13 While every Alexandroff locale is completely distributive, there are completely distributive non-Alexandroff locales, such as the interval $[0, 1]$.

6 Conclusion

We now have a complete picture of the ability of Dragalin frames, FM-frames, and Goldblatt frames to represent nuclear algebras and, via their algebras of fixpoints, to represent locales. This is summarized in the following table: [8]

[8] We put the dash next to FM because we must typically change the underlying poset of a nuclear frame in order to find an FM-frame with an isomorphic nuclear algebra (see Theorem 4.7 and Example 4.9). Recall that for a nuclear frame $(S, \leq, j)$, its associated locale is $\mathsf{Up}(S, \leq)$, whereas for an FM frame $(S, \leq, \preceq, F)$, its associated locale is $\mathsf{Up}(S, \leq)_F$.

Frames	Nuclear Frames	Nuclear Algebras	Locales
Dragalin	all	Alexandroff	all
FM	–	Alexandroff	all
Goldblatt	completely multiplicative j	perfect	completely distributive

The frames studied in this paper are not the only frames in the literature for representing nuclear algebras. Goldblatt [14] introduces *localic cover systems* and proves that every locale can be realized as the algebra of fixpoints in the nuclear algebra of a localic cover system, by showing that the nucleus j as in the proof of Theorem 2.8 can be generated by one of his cover systems. In future work [6], we will present a detailed comparison of this "cover" perspective and the "development" perspective of Dragalin, thereby relating Scott-Montague-style neighborhood semantics with Beth-style path semantics.

We will also explain in future work how Dragalin frames $(S, \leq, D)$ for intuitionistic propositional logic extend to *modal Dragalin frames* $(S, \leq, D, R)$ for intuitionistic modal logic, where R is a binary relation on S that interacts with $\leq$ and D in a natural way. This provides an intuitionistic generalization of the recently studied "possibility semantics" for classical modal logic [17,15,16,3].

Returning to the logical angle with which we began, it is an open problem whether every intermediate logic is the logic of some class of locales. Given the results of this paper, we can equivalently rephrase the problem as follows: is every intermediate logic the logic of some class of Dragalin frames?

Acknowledgement

We wish to thank the three anonymous AiML referees for their valuable comments, Dana Scott for helpful references, and the Department of Mathematical Sciences at NMSU and the Group in Logic at UC Berkeley for travel support.

References

[1] Alechina, N., M. Mendler, V. de Paiva and E. Ritter, *Categorical and Kripke semantics for constructive S4 modal logic*, Lecture Notes in Computer Science **2142** (2001), pp. 292–307.

[2] Artemov, S. and T. Protopopescu, *Intuitionistic epistemic logic* (2014), arXiv:1406.1582v2 [math.LO].

[3] van Benthem, J., N. Bezhanishvili and W. H. Holliday, *A bimodal perspective on possibility semantics*, Journal of Logic and Computation (Forthcoming).

[4] Beth, E., *Semantic construction of intuitionistic logic*, Mededelingen der Koninklijke Nederlandse Akademie van Wetenschappen **19** (1956), pp. 357–388.

[5] Bezhanishvili, G., *Varieties of monadic Heyting algebras. Part II: Duality theory*, Studia Logica **62** (1999), pp. 21–48.

[6] Bezhanishvili, G. and W. H. Holliday, *Development frames* (2016), in preparation.

[7] Božic, M. and K. Došen, *Models for normal intuitionistic modal logics*, Studia Logica **43** (1984), pp. 217–245.

[8] Davey, B. A., *On the lattice of subvarieties*, Houston Journal of Mathematics **5** (1979), pp. 183–192.

[9] Dragalin, A. G., "Matematicheskii Intuitsionizm: Vvedenie v Teoriyu Dokazatelstv," Matematicheskaya Logika i Osnovaniya Matematiki, "Nauka", Moscow, 1979.

[10] Dragalin, A. G., "Mathematical Intuitionism: Introduction to Proof Theory," Translations of Mathematical Monographs **67**, American Mathematical Society, Providence, RI, 1988.

[11] Fairtlough, M. and M. Mendler, *Propositional lax logic*, Information and Computation **137** (1997), pp. 1–33.

[12] Fourman, M. P. and D. S. Scott, *Sheaves and logic*, in: M. P. Fourman, C. J. Mulvey and D. S. Scott, editors, *Applications of Sheaves*, Springer, Berlin, 1979 pp. 302–401.

[13] Goldblatt, R., *Grothendieck topology as geometric modality*, Zeitschrift für Mathematische Logik und Grundlagen der Mathematik **27** (1981), pp. 495–529.

[14] Goldblatt, R., *Cover semantics for quantified lax logic*, Journal of Logic and Computation **21** (2011), pp. 1035–1063.

[15] Holliday, W. H., *Partiality and adjointness in modal logic*, in: R. Goré, B. Kooi and A. Kurucz, editors, *Advances in Modal Logic, Vol. 10*, College Publications, London, 2014 pp. 313–332.

[16] Holliday, W. H., *Possibility frames and forcing for modal logic* (2015), UC Berkeley Working Paper in Logic and the Methodology of Science.
URL `http://escholarship.org/uc/item/5462j5b6`

[17] Humberstone, L., *From worlds to possibilities*, Journal of Philosophical Logic **10** (1981), pp. 313–339.

[18] Johnstone, P. T., "Stone Spaces," Cambridge Studies in Advanced Mathematics **3**, Cambridge University Press, Cambridge, 1982.

[19] Kojima, K., *Relational and neighborhood semantics for intuitionistic modal logic*, Reports on Mathematical Logic **47** (2012), pp. 87–113.

[20] Kracht, M. and F. Wolter, *Normal monomodal logics can simulate all others*, Journal of Symbolic Logic **64** (1999), pp. 99–138.

[21] Kripke, S. A., *Semantical analysis of intuitionistic logic I*, in: J. N. Crossley and M. A. E. Dummett, editors, *Formal Systems and Recursive Functions*, North-Holland Publishing Company, Amsterdam, 1965 pp. 92–130.

[22] Kuznetsov, A. V., *On superintuitionistic logics*, in: *Proceedings of the International Congress of Mathematicians (Vancouver, B. C., 1974), Vol. 1* (1975), pp. 243–249.

[23] Litak, T., *A continuum of incomplete intermediate logics*, Reports on Mathematical Logic **36** (2002), pp. 131–141.

[24] Macnab, D. S., "An Algebraic Study of Modal Operators on Heyting Algebras with Applications to Topology and Sheafification," Ph.D. thesis, University of Aberdeen (1976).

[25] Macnab, D. S., *Modal operators on Heyting algebras*, Algebra Universalis **12** (1981), pp. 5–29.

[26] Protopopescu, T., *Intuitionistic epistemology and modal logics of verification*, in: W. van der Hoek, W. H. Holliday and W. Wang, editors, *Logic, Rationality, and Interaction*, Springer, Berlin, 2015 pp. 295–307.

[27] Restall, G., "An Introduction to Substructural Logics," Routledge, New York, 2000.

[28] Rumfitt, I., *On a neglected path to intuitionism*, Topoi **31** (2012), pp. 101–109.

[29] Rumfitt, I., "The Boundary Stones of Thought: An Essay in the Philosophy of Logic," Oxford University Press, Oxford, 2015.

[30] Shehtman, V. B., *Incomplete propositional logics*, Doklady Akademii Nauk SSSR **235** (1977), pp. 542–545 (Russian).

[31] Simmons, H., *A framework for topology*, in: A. Macintyre, L. Pacholski and J. Paris, editors, *Logic Colloquium '77*, North-Holland, Amsterdam, 1978 pp. 239–251.

[32] Takeuti, G. and W. M. Zaring, "Axiomatic Set Theory," Springer-Verlag, New York, 1973.

[33] Tarski, A., *Der Aussagenkalkuül und die Topologie*, Fundamenta Mathematicae **31** (1938), pp. 103–134.

[34] Wolter, F. and M. Zakharyaschev, *The relation between intuitionistic and classical modal logics*, Algebra and Logic **36** (1997), pp. 73–92.

Embedding formalisms: hypersequents and two-level systems of rules

Agata Ciabattoni [1]

TU Wien (Vienna University of Technology)
agata@logic.at

Francesco A. Genco [1]

TU Wien (Vienna University of Technology)
genco@logic.at

Abstract

A system of rules consists of (possibly labelled) sequent rules connected to each other by some variables and subject to the condition of appearing in a certain order in the derivation. The formalism of systems of rules is quite powerful and allows, e.g., the definition of analytic labelled sequent calculi for intermediate and modal logics characterised by frame conditions beyond the geometric fragment. Using propositional intermediate logics as a case study, we show how to use hypersequent calculus derivations to construct derivations using two-level systems of sequent rules and vice versa. Our transformations (embeddings) show that the hypersequent calculus and this proper restriction of systems of rules have the same expressive power.

Keywords: proof theory, hypersequent calculi, systems of rules, intermediate logics.

1 Introduction

Proof theory provides a constructive approach to investigating fundamental meta-logical and computational properties of a logic through the design and the study of analytic calculi. These calculi, whose proofs proceed by stepwise decomposition of the formulae to be proved, are also the base for developing computerised reasoning methods.

The difficulty in finding analytic sequent calculi for several logics of interest has lead to the introduction of many other formalisms and new ones emerge on a regular basis; prominent examples are the hypersequent calculus [1], which deals with sets of sequents rather than single sequents, and the labelled calculus [6,9], which manipulates sequents containing labelled formulae and relations on labels. The multitude and diversity of the introduced formalisms has made

[1] Supported by FWF project START Y544-N23 and W1255-N23.

it increasingly important to identify their interrelationships and relative expressive power. *Embeddings* bewteen formalisms, i.e. functions that take any calculus in some formalism and yield a calculus for the same logic in another formalism, are useful tools to prove that a formalism subsumes another one in terms of expressiveness (or, when bi-directional, that two formalisms are equi-expressive). Such embeddings can also provide useful reformulations of known calculi and allow the transfer of certain proof-theoretic results, thus alleviating the need for independent proofs in each system and avoiding duplicating work; for example the highly technical proof of cut-elimination for modal provability logic GL for tree-hypersequents [11] could have been induced from the proof for labelled sequents [9], using the subsequently discovered embedding [8] of the former formalism into the latter. Various embeddings between formalisms have appeared in the literature, see, e.g., [12,8,7,13] (and the bibliography thereof).

In this paper we focus on the hypersequent calculus and on the formalism of systems of rules which was introduced in [10] to define analytic labelled calculi for logics semantically characterised by frame conditions beyond the geometric fragment. A system[2] of rules is a set of (possibly labelled) sequent rules linked together by some variables and by the requirement for the rules of appearing in a certain order in the derivation. Systems of rules are quite powerful and enable to define analytic labelled calculi, e.g., for *all* logics characterised by frame properties that correspond to formulae in the Sahlqvist fragment. The downside of this great expressivity is the non-locality of rule applications, which appears at two levels: horizontally, because of the dependency between rules occurring in disjoint branches; and vertically, because of rules that can only be applied above other rules. A possible connection between hypersequents and systems of rules is hinted in [10]. Our paper formalises in full this intuition. Focusing on propositional logics intermediate between intuitionistic and classical logic, we define a bi-directional *embedding* between hypersequents and a subclass of systems of rules in which the vertical non-locality is restricted to at most two (non labelled) sequent rules. We call *two-level systems of rules* this proper restriction of the full formalism. Beside showing that the two seemingly different formalisms are actually a notational variant of each other, our embeddings can be used

- to recover locality in two-level systems of rules;
- to transfer analyticity from hypersequents to two-level systems of rules;
- to define new cut-free two-level systems of rules; e.g. for substructural logics or intermediate logics characterised by Hilbert axioms within the class $\mathcal{P}_3$ in the classification of [5];
- to provide a reformulation of hypersequent calculi which may be of independent interest due to its close relation to natural deduction systems.

[2] The word "system" is used in the same sense as in linear algebra, where there are systems of equations with variables in common, and each equation is meaningful and can be solved only if considered together with the other equations of the system.

2 Preliminaries

A *hypersequent* [1,2] is a |-separated multiset of ordinary sequents, called *components*. The sequents we consider in this paper have the form $\Gamma \Rightarrow \Pi$ where Γ is a (possibly empty) multiset of formulae in the language of intuitionistic logic and Π contains at most one formula.

Notation. Unless stated otherwise we use upper-case Greek letters for multisets of formulae (where Π contains at most one element), lower-case Greek letters for formulae, and G, H for (possibly empty) hypersequents.

As with sequent calculi, the inference rules of hypersequent calculi consist of initial hypersequents (i.e., axioms), the cut-rule as well as logical and structural rules. The logical and structural rules are divided into *internal* and *external rules*. The internal rules deal with formulae within one component of the conclusion. Examples of external structural rules include external weakening (EW) and external contraction (EC), see Fig. 1.

Rules are usually presented as rule schemata. Concrete instances of a rule are obtained by substituting formulae for schematic variables. Following standard practice, we do not explicitly distinguish between a rule and a rule schema.

Fig. 1 displays the hypersequent version HLJ of the propositional sequent calculus LJ for intuitionistic logic. Note that the "hyperlevel" of HLJ is in fact

$$\varphi \Rightarrow \varphi \qquad \bot \Rightarrow \Pi \qquad \frac{G \mid \Gamma, \varphi \Rightarrow \Pi \quad G \mid \Gamma, \psi \Rightarrow \Pi}{G \mid \Gamma, \varphi \vee \psi \Rightarrow \Pi}\ (\vee l) \qquad \frac{G \mid \Gamma \Rightarrow \varphi_i}{G \mid \Gamma \Rightarrow \varphi_1 \vee \varphi_2}\ (\vee r)$$

$$\frac{G \mid \Gamma, \varphi, \psi \Rightarrow \Pi}{G \mid \Gamma, \varphi \& \psi \Rightarrow \Pi}\ (\& l) \qquad \frac{G \mid \Gamma \Rightarrow \varphi \quad G \mid \Gamma \Rightarrow \psi}{G \mid \Gamma \Rightarrow \varphi \& \psi}\ (\& r) \qquad \frac{G \mid \Gamma \Rightarrow \Pi}{G \mid \varphi, \Gamma \Rightarrow \Pi}\ (w)$$

$$\frac{G \mid \Gamma \Rightarrow \varphi \quad G \mid \Gamma, \psi \Rightarrow \Pi}{G \mid \Gamma, \varphi \supset \psi \Rightarrow \Pi}\ (\supset l) \qquad \frac{G \mid \Gamma, \varphi \Rightarrow \psi}{G \mid \Gamma \Rightarrow \varphi \supset \psi}\ (\supset r) \qquad \frac{G \mid \varphi, \varphi, \Gamma \Rightarrow \Pi}{G \mid \varphi, \Gamma \Rightarrow \Pi}\ (c)$$

$$\frac{G \mid \Gamma \Rightarrow \varphi \quad G \mid \varphi, \Gamma' \Rightarrow \Pi}{G \mid \Gamma, \Gamma' \Rightarrow \Pi}\ (cut) \qquad \frac{G}{G \mid \Gamma \Rightarrow \Pi}\ (EW) \qquad \frac{G \mid \Gamma \Rightarrow \Pi \mid \Gamma \Rightarrow \Pi}{G \mid \Gamma \Rightarrow \Pi}\ (EC)$$

Fig. 1. Rules and axioms of HLJ.

redundant since a hypersequent $\Gamma_1 \Rightarrow \Pi_1 \mid \ldots \mid \Gamma_k \Rightarrow \Pi_k$ is derivable in HLJ if and only if $\Gamma_i \Rightarrow \Pi_i$ is derivable in LJ for some $i \in \{1, \ldots, k\}$. Indeed, any sequent calculus can be trivially viewed as a hypersequent calculus. The added expressive power of the latter is due to the possibility of defining new rules which act simultaneously on several components of one or more hypersequents.

Example 2.1 By adding to HLJ the structural rule introduced in [2]

$$\frac{G \mid \Phi, \Gamma_1 \Rightarrow \Pi_1 \quad G \mid \Psi, \Gamma_2 \Rightarrow \Pi_2}{G \mid \Psi, \Gamma_1 \Rightarrow \Pi_1 \mid \Phi, \Gamma_2 \Rightarrow \Pi_2}\ (com)$$

we obtain a cut-free calculus for Gödel logic, which is (axiomatised by) intuitionistic logic with the linearity axiom $(\varphi \supset \psi) \vee (\psi \supset \varphi)$.

Since the usual interpretation of the symbol "|" is disjunctive, the hypersequent calculus is suitable to capture properties (Hilbert axioms, algebraic equations...) that can be expressed in a disjunctive form. More precisely, consider the following classification of intuitionistic formulae, which adapts that in [5] for substructural logics: $\mathcal{N}_0$ and $\mathcal{P}_0$ are the set of atomic formulae

$$\begin{array}{ll} \mathcal{P}_{n+1} & ::= \bot \mid \top \mid \mathcal{N}_n \mid \mathcal{P}_{n+1} \& \mathcal{P}_{n+1} \mid \mathcal{P}_{n+1} \vee \mathcal{P}_{n+1} \\ \mathcal{N}_{n+1} & ::= \bot \mid \top \mid \mathcal{P}_n \mid \mathcal{N}_{n+1} \& \mathcal{N}_{n+1} \mid \mathcal{P}_{n+1} \supset \mathcal{N}_{n+1} \end{array}$$

As shown in [5] all axioms within the class $\mathcal{P}_3$ can be algorithmically transformed into equivalent *structural* hypersequent rules that preserve cut-elimination when added to the calculus HLJ. In particular the rule (com) in Example 2.1 can be automatedly extracted [3] from the linearity axiom.

Notation and Assumptions. Given a hypersequent rule (r) with premisses $G \mid H_1 \ldots G \mid H_n$ and conclusion $G \mid H$, we call *active* the components in the hypersequents $H_1, \ldots, H_n, H$. We call *context components* the components of G. In this paper we will only consider hypersequent rules that (i) are (external) context sharing, i.e., whose premisses all contain the same hypersequent context, and (ii) (except for (EC)) have one active component in each premiss, i.e., in which each H_i is a sequent. Note that (i) is not a restriction and, in absence of eigenvariables, neither is (ii), because we can always transform a rule into an equivalent one that satisfies these conditions.

System of rules were introduced in [10] to define analytic labelled calculi for logics semantically characterised by generalised geometric implications, a class of first-order formulae that includes the frame properties that correspond to formulae in the Sahlqvist fragment. In general, a *system of rules* consists of a set of (possibly labelled) sequent rules $\{(r_{1_1}), \ldots, (r_{1_n}), \ldots, (r_{k_1}), \ldots (r_{k_m}), (r_{end})\}$ connected to each other by (schematic) variables or labels and whose applicability conditions follow the schema

$$\frac{\begin{array}{c}\mathcal{D}_1 \\ \vdots \\ \Gamma \Rightarrow \Pi\end{array} \quad \ldots \quad \begin{array}{c}\mathcal{D}_k \\ \vdots \\ \Gamma \Rightarrow \Pi\end{array}}{\Gamma \Rightarrow \Pi}\ (r_{end}) \qquad (1)$$

where each derivation $\mathcal{D}_i$, for $1 \leq i \leq k$, may contain applications of the (r_{i_j}) rules in a specific order. Analyticity of system of rules (when added to a sequent or a labelled sequent calculus for classical or intuitionistic logic) was proved in [10] for systems acting on atomic formulae or relational atoms. We define below a proper restriction of systems of rules which manipulate LJ sequents.

Definition 2.2 A *two-level system of rules* (*2-system* for short) is a set of LJ rules $\{(r_1), \ldots, (r_k), (r_{end})\}$ with applicability condition (1) and in which each derivation $\mathcal{D}_i$, for $1 \leq i \leq k$, contain (at most) *one* application of the rule (r_i). The rule (r_{end}) is called *ending rule* while $(r_1), \ldots, (r_k)$ *non-ending rules.*

[3] Program at http://www.logic.at/people/lara/axiomcalc.html.

Example 2.3 The 2-system $Sys_{(com^*)}$ in [10] for the linearity axiom (cf. Example 2.1) is the following (φ and ψ are metavariables for formulae):

$$\cfrac{\cfrac{\psi,\varphi,\Gamma_1 \Rightarrow \Pi_1}{\psi,\Gamma_1 \Rightarrow \Pi_1}\,(com_1') \quad \vdots \quad \Gamma \Rightarrow \Pi \qquad \cfrac{\psi,\varphi,\Gamma_2 \Rightarrow \Pi_2}{\varphi,\Gamma_2 \Rightarrow \Pi_2}\,(com_2') \quad \vdots \quad \Gamma \Rightarrow \Pi}{\Gamma \Rightarrow \Pi}\,(com_{end}')$$

Given a calculus $\mathcal{C}$ and a set of rules $\mathbb{R}$, $\mathcal{C} + \mathbb{R}$ will denote the calculus obtained by adding the elements of $\mathbb{R}$ to $\mathcal{C}$, and $\vdash_{\mathcal{C}+\mathbb{R}}$ its derivability relation.

3 From 2-systems to hypersequent rules and back

Given a 2-system Sys we construct the corresponding hypersequent rule Hr_{Sys}; vice versa, from a hypersequent rule Hr we construct the corresponding 2-system Sys_{Hr}. The transformation of derivations from HLJ + Hr into LJ + Sys_{Hr} (and from LJ + Sys into HLJ + Hr_{Sys}) is shown in Section 4.

From 2-systems to hypersequent rules

Given a 2-system Sys of the form

$$\cfrac{\begin{matrix}\mathcal{D}_1 \\ \vdots \\ \Gamma \Rightarrow \Pi\end{matrix} \quad \dots \quad \begin{matrix}\mathcal{D}_k \\ \vdots \\ \Gamma \Rightarrow \Pi\end{matrix}}{\Gamma \Rightarrow \Pi}\,(r_{end})$$

where each derivation $\mathcal{D}_i$, for $1 \leq i \leq k$, may contain an application of the rule

$$\cfrac{\varphi_i^1,\dots,\varphi_i^{l_i},\Gamma_i \Rightarrow \Pi_i \quad \dots \quad \psi_i^1,\dots,\psi_i^{m_i},\Gamma_i \Rightarrow \Pi_i}{\theta_i^1,\ \dots,\ \theta_i^{n_i},\Gamma_i \Rightarrow \Pi_i}\,(r_i)$$

the corresponding hypersequent rule Hr_{Sys} is as follows:

$$\cfrac{M_1 \qquad \dots \qquad M_k}{G \mid \theta_1^1,\ \dots,\ \theta_1^{n_1},\Gamma_1 \Rightarrow \Pi_1 \mid \dots \mid \theta_k^1,\ \dots,\ \theta_k^{n_k},\Gamma_k \Rightarrow \Pi_k}$$

where M_i, for $1 \leq i \leq k$, is the multiset of premisses

$$G \mid \varphi_i^1,\dots,\varphi_i^{l_i},\Gamma_i \Rightarrow \Pi_i \quad \dots \quad G \mid \psi_i^1,\dots,\psi_i^{m_i},\Gamma_i \Rightarrow \Pi_i$$

Example 3.1 From Negri's 2-system in Example 2.3 we obtain

$$\cfrac{G \mid \varphi,\psi,\Gamma_1 \Rightarrow \Pi_1 \quad G \mid \varphi,\psi,\Gamma_2 \Rightarrow \Pi_2}{G \mid \psi,\Gamma_1 \Rightarrow \Pi_1 \mid \varphi,\Gamma_2 \Rightarrow \Pi_2}\,(com^*)$$

From hypersequent rules to 2-systems

Given any hypersequent rule Hr of the form

$$\cfrac{M_1 \qquad \dots \qquad M_k}{G \mid \Theta_1^1,\ \dots,\ \Theta_1^{n_1},\Gamma_1 \Rightarrow \Pi_1 \mid \dots \mid \Theta_k^1,\ \dots,\ \Theta_k^{n_k},\Gamma_k \Rightarrow \Pi_k}\,(r)$$

where the sets M_i, for $1 \le i \le k$, constitute a partition of the set of premisses of (r) and each M_i contains the premisses

$$G \mid C_i^1 \quad \dots \quad G \mid C_i^{m_i}$$

where $C_i^1, \dots, C_i^{m_i}$ are sequents. The corresponding 2-system Sys_{Hr} is

$$\dfrac{\begin{array}{c}\mathcal{D}_1\\ \vdots\\ \Gamma \Rightarrow \Pi\end{array} \quad \dots \quad \begin{array}{c}\mathcal{D}_k\\ \vdots\\ \Gamma \Rightarrow \Pi\end{array}}{\Gamma \Rightarrow \Pi}\ (r_{end})$$

where the derivation $\mathcal{D}_i$, for $1 \le i \le k$, may contain an instance of the rule

$$\dfrac{C_i^1 \quad \dots \quad C_i^{m_i}}{\Theta_i^1, \dots, \Theta_i^{n_i}, \Gamma_i \Rightarrow \Pi_i}\ (r_i)$$

Example 3.2 The translation $Sys_{(com)}$ of the rule (com) in Example 2.1 is

$$\dfrac{\begin{array}{c}\dfrac{\Phi, \Gamma_1 \Rightarrow \Pi_1}{\Psi, \Gamma_1 \Rightarrow \Pi_1}\ (com_1)\\ \vdots\\ \Gamma \Rightarrow \Pi\end{array} \qquad \begin{array}{c}\dfrac{\Psi, \Gamma_2 \Rightarrow \Pi_2}{\Phi, \Gamma_2 \Rightarrow \Pi_2}\ (com_2)\\ \vdots\\ \Gamma \Rightarrow \Pi\end{array}}{\Gamma \Rightarrow \Pi}\ (com_{end})$$

Definition 3.3 We say that the premisses of Hr contained in M_i, for $1 \le i < k$, are *linked* to the component $\Theta_i^1, \dots, \Theta_i^{n_1}, \Gamma_i \Rightarrow \Pi_i$ of the conclusion.

Remark 3.4 A natural deduction calculus for Gödel logic has been introduced in [3]. Defined by reformulating (a variant of) the hypersequent calculus in Example 2.1, the calculus in [3] extends Gentzen NJ calculus for intuitionistic logic with non-local rules which simulate (com) and (EC). The rule $Sys_{(com)}$ in Example 3.2 turns out to be a suitable combination of these non-local rules.

4 Embedding the two formalisms

We introduce procedures to transform 2-system derivations into hypersequent derivations and vice versa.

4.1 From 2-systems to hypersequent derivations

Given any set $\mathbb{S}$ of 2-systems and set $\mathbb{H}$ of hypersequent rules s.t. if $Sys \in \mathbb{S}$ then $Hr_{Sys} \in \mathbb{H}$. Starting from a derivation $\mathcal{D}$ in LJ + $\mathbb{S}$ we construct a derivation $\mathcal{D}'$ in HLJ + $\mathbb{H}$ of the same end-sequent. The construction proceeds by a stepwise translation of the rules in $\mathcal{D}$: non-ending rules of the systems in $\mathbb{S}$ are translated into applications of the corresponding rules in $\mathbb{H}$ (and additional (EW), if needed), ending rules are translated into applications of (EC) and rules of LJ into rules of HLJ (possibly using (EW)). To keep track of the various translation steps, we mark the derivation $\mathcal{D}$. We start by marking and translating the leaves of $\mathcal{D}$. The rules with marked premisses are then translated one by one and the marks are moved to the conclusions of the rules. The process is repeated until we reach and translate the root of $\mathcal{D}$.

Definition 4.1 A *configuration* is a pair $(\mathcal{D}, \mathfrak{M})$ where $\mathcal{D}$ is an LJ + $\mathbb{S}$ derivation, for some set of 2-systems $\mathbb{S}$, and $\mathfrak{M}$ is a set of marks s.t. (i) each mark in $\mathfrak{M}$ refers to a sequent occurrence in $\mathcal{D}$ (*marked sequent*); (ii) in each path between a leaf of $\mathcal{D}$ and its root exactly one marked sequent occurs; and (iii) if a premiss of a non-ending rule of a 2-system $S \in \mathbb{S}$ is marked, then there are no marked sequents below the premisses of any non-ending rule of such instance.

The algorithm

Input: a derivation $\mathcal{D}$ in LJ + $\mathbb{S}$. Output: a derivation $\mathcal{D}'$ of the same sequent in HLJ + $\mathbb{H}$.

Translating axioms. The leaves of $\mathcal{D}$ are marked and copied as leaves of $\mathcal{D}'$.

Translating rules. Rules are translated one by one in the following order: first the one-premiss logical and structural rules applied to marked sequents, then the two-premiss logical rules and ending rules with all premisses marked, and finally (all) non-ending rules of one 2-system instance. After having translated each rule (or all non-ending rules of an occurrence of a 2-system simultaneously), we remove the marks from the premisses of the translated rules and mark their conclusions.

Since the LJ rules are particular instances of HLJ rules, we only show how to translate 2-systems.

$(*)$ For each configuration $(\mathcal{D}, \mathfrak{M})$ we have a set of hypersequent derivations s.t. each marked sequent in $\mathcal{D}$ is translated into one component of the root of exactly one HLJ + $\mathbb{H}$ derivation, and each component of the root of a HLJ + $\mathbb{H}$ derivation translates a marked sequent. This property holds for the leaves: we show that it is preserved by each translation step.

Consider a 2-system $Sys \in \mathbb{S}$ applied in $\mathcal{D}$ with the following instances of

(i) non-ending rules:

$$\dfrac{\begin{matrix}\vdots & & \vdots\\ C_1^1 & \dots & C_1^{m_1}\end{matrix}}{\Delta_1, \Gamma_1 \Rightarrow \Pi_1}\,(r_1) \qquad \dots \qquad \dfrac{\begin{matrix}\vdots & & \vdots\\ C_k^1 & \dots & C_k^{m_k}\end{matrix}}{\Delta_k, \Gamma_k \Rightarrow \Pi_k}\,(r_k)$$

where $C_1^1, \dots, C_1^{m_1}, \dots, C_k^1, \dots, C_k^{m_k}$ are marked sequents. By $(*)$ we have hypersequent derivations of

$$G_1^1 \mid C_1^1 \quad \dots \quad G_1^{m_1} \mid C_1^{m_1} \qquad \dots \qquad G_k^1 \mid C_k^1 \quad \dots \quad G_k^{m_k} \mid C_k^{m_k}$$

Lemma .1 in the Appendix ensures that each displayed instance C_i^j (corresponding to a marked occurrence) occurs exactly once in these hypersequents. We apply Hr_{Sys} as follows

$$\dfrac{M_1 \qquad \dots \qquad M_k}{G \mid \Delta_1, \Gamma_1 \Rightarrow \Pi_1 \mid \dots \mid \Delta_k, \Gamma_k \Rightarrow \Pi_k}$$

where $G = G_1^1 \mid \dots \mid G_1^{m_1} \mid \dots \mid G_k^1 \mid \dots \mid G_k^{m_k}$, and M_i, for $1 \le i \le k$, is the set of premisses $G \mid C_i^1 \quad \dots \quad G \mid C_i^{m_i}$ obtained from

$G_i^1 \mid C_i^1 \quad \ldots \quad G_i^{m_i} \mid C_i^{m_i}$ by repeatedly applying (EW). We move the marks to the conclusions of $(r_1), \ldots, (r_k)$.

(ii) ending rule:

$$\dfrac{\begin{matrix} \vdots & \vdots \\ \Gamma \Rightarrow \Pi & \Gamma \Rightarrow \Pi \end{matrix}}{\Gamma \Rightarrow \Pi}\ (r_{end})$$

Without loss of generality we can assume that the non-ending rules of the considered 2-system have been applied above the premisses of (r_{end}) (as otherwise the application of the 2-system is redundant). Hence we have a derivation in HLJ + $\mathbb{H}$ of $G \mid \Gamma \Rightarrow \Pi \mid \ldots \mid \Gamma \Rightarrow \Pi$. The desired derivation of $G \mid \Gamma \Rightarrow \Pi$ is obtained by repeatedly applying (EC). The mark is now moved to the conclusion of (r_{end}).

It is easy to see that property $(*)$ is preserved after each step.

Theorem 4.2 *For any set $\mathbb{H}$ of hypersequent rules and set $\mathbb{S}$ of 2-systems s.t. if $Sys \in \mathbb{S}$ then $Hr_{Sys} \in \mathbb{H}$, if $\vdash_{\mathrm{LJ}+\mathbb{S}} \Gamma \Rightarrow \Pi$ then $\vdash_{\mathrm{HLJ}+\mathbb{H}} \Gamma \Rightarrow \Pi$.*

Proof. Apply the above algorithm to the LJ + $\mathbb{S}$ derivation $\mathcal{D}$ of $\Gamma \Rightarrow \Pi$ to obtain $\mathcal{D}'$. The algorithm terminates because the number of rule applications in a derivation is finite. By induction on the number u of 2-system instances whose non-ending rules are still to be translated we prove that the algorithm does not get stuck before translating the root of $\mathcal{D}$. The claim then follows by property $(*)$. If $u = 0$ all remaining rules can be translated as soon as the premisses are marked. Assume $u = n + 1$. By Lemma 4.4 below there is a 2-system instance S that still has untranslated non-ending rules and is not blocked by any other 2-system. Given that all the non-ending rules above the non-ending rules of S must be translated, the only rule applications that have to be translated before S do not belong to any 2-system. These rule applications can be translated as soon as their premisses are marked, and we obtain $u = n$.□

Definition 4.3 Let S and S' be instances of possibly different 2-systems, (r) a non-ending rule occurrence of S and (r') a non-ending rule occurrence of S'. We say that S' *blocks* S (through (r')) and S *is blocked by* S' (on (r)) if (r') occurs some steps above (r).

Lemma 4.4 *Let $\mathcal{D}$ be a derivation in* LJ + $\mathbb{S}$ *and $\overline{\mathbb{S}}$ any set of instances of 2-systems occurring in $\mathcal{D}$. There is some $S \in \overline{\mathbb{S}}$ that is not blocked by any $S' \in \overline{\mathbb{S}}$.*

Proof. First we prove that $\mathcal{D}$ cannot contain a sequence of instances of 2-systems $S_1, \ldots, S_n, S_{n+1}$ s.t. S_i blocks S_{i+1} for $1 \le i \le n$ and $S_1 = S_{n+1}$. We call such a sequence $(S_1, \ldots, S_n, S_{n+1})$ a *loop* and show that the existence of a loop leads to a contradiction. Without loss of generality we assume that $(S_1, \ldots, S_n, S_{n+1})$ is a *distributed* loop, i.e. no S_i in the loop is blocked on a rule application (r) and blocks S_{i+1} through (r). Indeed, from any non-distributed loop we can always extract a subsequence that is a distributed loop by removing some elements from the sequence.

For any element k in the distributed loop either the ending rule of S_k occurs above a premiss of the ending rule of S_{k+1}, or the ending rule of S_{k+1} occur above a premiss of the ending rule of S_k. Otherwise the subtrees of the derivation rooted in S_k and in S_{k+1} are disjoint and S_k cannot block S_{k+1}. Consider now the occurrences $(r_1), \dots, (r_n)$ of the ending rules of $S_1, \dots, S_n$. Let (r_j) be the lowermost rule of the loop, i.e. no $(r_1), \dots, (r_n)$ occurs below (r_j). We distinguish two cases: either (i) all (r_k) $(k = 1, \dots, n;\ k \neq j)$ occur above (only) one premiss of (r_j); or (ii) the (r_k)'s occur above more than one premiss of (r_j). If (i) is the case the loop is not distributed, against the assumption, as S_j is a two-level system. If (ii) holds, $S_1, \dots, S_n$ is not a loop, as systems above different premisses of (r_j) cannot block each other.

We prove now the main statement. Assume that for each $S \in \overline{\mathbb{S}}$ there is $S' \in \overline{\mathbb{S}}$ such that S' blocks S. Either there is a loop among the elements of $\overline{\mathbb{S}}$, but we proved this is impossible; or the cardinality of $\overline{\mathbb{S}}$ is infinite, but this contradicts the fact that $\mathcal{D}$ contains a finite number of rule applications. □

Example 4.5 The following derivation in the calculus LJ+$Sys_{(com^*)}$ for Gödel logic (see Example 2.3)

$$\dfrac{\dfrac{\dfrac{\dfrac{\dfrac{\alpha \Rightarrow \alpha \qquad \dfrac{\bot \Rightarrow}{\bot, \alpha \Rightarrow}\,(w)}{\alpha \supset \bot, \alpha \Rightarrow}\,(\supset l)}{\alpha \Rightarrow}\,(com'_1)}{\Rightarrow \alpha \supset \bot}\,(\supset r)}{\Rightarrow (\alpha \supset \bot) \vee ((\alpha \supset \bot) \supset \bot)}\,(\vee r) \qquad \dfrac{\dfrac{\dfrac{\dfrac{\dfrac{\bot \Rightarrow}{\alpha, \bot \Rightarrow}\,(w) \qquad \alpha \Rightarrow \alpha}{\alpha, \alpha \supset \bot \Rightarrow}\,(\supset l)}{\alpha \supset \bot \Rightarrow}\,(com'_2)}{\Rightarrow (\alpha \supset \bot) \supset \bot}\,(\supset r)}{\Rightarrow (\alpha \supset \bot) \vee ((\alpha \supset \bot) \supset \bot)}\,(\vee r)}{\Rightarrow (\alpha \supset \bot) \vee ((\alpha \supset \bot) \supset \bot)}\,(com'_{end})$$

is translated into the HLJ + (com^*) derivation (see Example 3.1)

$$\dfrac{\dfrac{\dfrac{\dfrac{\dfrac{\dfrac{\dfrac{\alpha \Rightarrow \alpha \qquad \dfrac{\bot \Rightarrow}{\bot, \alpha \Rightarrow}\,(w)}{\alpha \supset \bot, \alpha \Rightarrow}\,(\supset l) \qquad \dfrac{\dfrac{\bot \Rightarrow}{\alpha, \bot \Rightarrow}\,(w) \qquad \alpha \Rightarrow \alpha}{\alpha, \alpha \supset \bot \Rightarrow}\,(\supset l)}{\alpha \Rightarrow \mid \alpha \supset \bot \Rightarrow}\,(com^*)}{\alpha \Rightarrow \mid \Rightarrow (\alpha \supset \bot) \supset \bot}\,(\supset r)}{\Rightarrow \alpha \supset \bot \mid \Rightarrow (\alpha \supset \bot) \supset \bot}\,(\supset r)}{\Rightarrow \alpha \supset \bot \mid \Rightarrow (\alpha \supset \bot) \vee ((\alpha \supset \bot) \supset \bot)}\,(\vee r)}{\Rightarrow (\alpha \supset \bot) \vee ((\alpha \supset \bot) \supset \bot) \mid \Rightarrow (\alpha \supset \bot) \vee ((\alpha \supset \bot) \supset \bot)}\,(\vee r)}{\Rightarrow (\alpha \supset \bot) \vee ((\alpha \supset \bot) \supset \bot)}\,(EC)$$

4.2 From hypersequent to 2-system derivations

Given any set $\mathbb{H}$ of hypersequent rules and set $\mathbb{S}$ of 2-systems s.t. if $Hr \in \mathbb{H}$ then $Sys_{Hr} \in \mathbb{S}$. Starting from a derivation in HLJ + $\mathbb{H}$ we construct a derivation in LJ + $\mathbb{S}$ of the same end-sequent.

The algorithm

Input: a suitable derivation $\mathcal{D}$ of a sequent $\Gamma \Rightarrow \Pi$ in HLJ + $\mathbb{H}$. Output: a derivation $\mathcal{D}'$ of $\Gamma \Rightarrow \Pi$ in LJ + $\mathbb{S}$.

Intuitively, each application of a HLJ rule in $\mathcal{D}$ is rewritten as an application of an LJ rule in $\mathcal{D}'$. Some care is needed to handle the external structural rules in $\mathbb{H}$ as well as (EW) and (EC). To deal with the latter rules, which have no direct translation in $\mathrm{LJ} + \mathbb{S}$, we consider derivations $\mathcal{D}$ in which (i) all applications of (EC) occur immediately above the root, and (ii) all applications of (EW) occur where immediately needed, that is they introduce components of the context of rules with more than one premiss. As shown in Section 4.2.1 each hypersequent derivation (of a sequent) can be transformed into an equivalent one having this shape.

The rules in $\mathbb{H}$ are translated in two steps. First for each component of the premiss of the uppermost application of (EC) in $\mathcal{D}$ we find a *partial derivation*, that is a derivation in LJ extended by the rules of the 2-systems in $\mathbb{S}$ without any applicability condition (Lemma 4.13). The desired derivation $\mathcal{D}'$ is then obtained by suitably applying to these partial derivations the corresponding ending rules (Theorem 4.14).

Definition 4.6 A partial derivation in $\mathrm{LJ} + \mathbb{S}$ is a derivation in LJ extended with the non-ending rules of $\mathbb{S}$ (without their applicability conditions).

We show an example of translation to guide the reader's intuition through the proofs that follow.

Example 4.7 Consider the $\mathrm{HLJ} + (com)$ derivation (see Example 2.1)

$$
\begin{prooftree}
\AxiomC{$\alpha \Rightarrow \alpha$}
\RightLabel{(EW)}
\UnaryInfC{$\alpha \Rightarrow \alpha \mid \beta \Rightarrow \alpha\&\beta$}
\AxiomC{$\beta \Rightarrow \beta$}
\AxiomC{$\alpha \Rightarrow \alpha$}
\RightLabel{(com)}
\BinaryInfC{$\alpha \Rightarrow \beta \mid \beta \Rightarrow \alpha$}
\AxiomC{$\beta \Rightarrow \beta$}
\RightLabel{(EW)}
\UnaryInfC{$\alpha \Rightarrow \beta \mid \beta \Rightarrow \beta$}
\RightLabel{$(\&r)$}
\BinaryInfC{$\alpha \Rightarrow \beta \mid \beta \Rightarrow \alpha\&\beta$}
\RightLabel{$(\&r)$}
\BinaryInfC{$\alpha \Rightarrow \alpha\&\beta \mid \beta \Rightarrow \alpha\&\beta$}
\RightLabel{$(\supset r)$}
\UnaryInfC{$\alpha \Rightarrow \alpha\&\beta \mid \Rightarrow \beta \supset (\alpha\&\beta)$}
\RightLabel{$(\supset r)$}
\UnaryInfC{$\Rightarrow \alpha \supset (\alpha\&\beta) \mid \Rightarrow \beta \supset (\alpha\&\beta)$}
\RightLabel{$(\vee r)$}
\UnaryInfC{$\Rightarrow \alpha \supset (\alpha\&\beta) \mid \Rightarrow (\alpha \supset (\alpha\&\beta)) \vee (\beta \supset (\alpha\&\beta))$}
\RightLabel{$(\vee r)$}
\UnaryInfC{$\Rightarrow (\alpha \supset (\alpha\&\beta)) \vee (\beta \supset (\alpha\&\beta)) \mid \Rightarrow (\alpha \supset (\alpha\&\beta)) \vee (\beta \supset (\alpha\&\beta))$}
\RightLabel{(EC)}
\UnaryInfC{$\Rightarrow (\alpha \supset (\alpha\&\beta)) \vee (\beta \supset (\alpha\&\beta))$}
\end{prooftree}
$$

First observe that this derivation satisfies property (i) and, as (EW) cannot be moved below the two-premiss rule $(\&r)$, also (ii). The partial derivations in $\mathrm{LJ} + \{(com_1), (com_2)\}$ (see Example 3.2) for the components of the uppermost application of (EC) in the above proof are:

$$
\begin{prooftree}
\AxiomC{$\alpha \Rightarrow \alpha$}
\AxiomC{$\beta \Rightarrow \beta$}
\RightLabel{(com_1)}
\UnaryInfC{$\alpha \Rightarrow \beta$}
\RightLabel{$(\&r)$}
\BinaryInfC{$\alpha \Rightarrow \alpha\&\beta$}
\RightLabel{$(\supset r)$}
\UnaryInfC{$\Rightarrow \alpha \supset (\alpha\&\beta)$}
\RightLabel{$(\vee r)$}
\UnaryInfC{$\Rightarrow (\alpha \supset (\alpha\&\beta)) \vee (\beta \supset (\alpha\&\beta))$}
\end{prooftree}
\qquad
\begin{prooftree}
\AxiomC{$\alpha \Rightarrow \alpha$}
\RightLabel{(com_2)}
\UnaryInfC{$\beta \Rightarrow \alpha$}
\AxiomC{$\beta \Rightarrow \beta$}
\RightLabel{$(\&r)$}
\BinaryInfC{$\beta \Rightarrow \alpha\&\beta$}
\RightLabel{$(\supset r)$}
\UnaryInfC{$\Rightarrow \beta \supset (\alpha\&\beta)$}
\RightLabel{$(\vee r)$}
\UnaryInfC{$\Rightarrow (\alpha \supset (\alpha\&\beta)) \vee (\beta \supset (\alpha\&\beta))$}
\end{prooftree}
$$

These partial derivations have the same "structure" as the hypersequent derivations (see *ancestor tree* in Definition 4.11) of the corresponding components. The desired derivation of $\Rightarrow (\alpha \supset (\alpha\&\beta)) \vee (\beta \supset (\alpha\&\beta))$ in $\mathrm{LJ} + Sys_{(com)}$ is simply obtained by connecting the partial derivations via (com_{end}).

We use Definition 4.8 and 4.9 to formalise and achieve properties (i) and (ii).

Definition 4.8 For any one-premiss rule (r) we call a *queue of* (r) a sequence of consecutive applications of (r) that is neither immediately preceded nor immediately followed by applications of (r).

Definition 4.9 We say that an HLJ$+\mathbb{H}$ derivation is in *structured form iff* all (EC) applications occur in a queue immediately above the root, and all (EW) applications occur in subderivations of the form

$$\dfrac{\dfrac{\dfrac{G_1 \mid C_1}{\vdots}\,(EW)}{G \mid C_1}\,(EW) \quad \dots \quad \dfrac{\dfrac{\dfrac{G_n \mid C_n}{\vdots}\,(EW)}{G \mid C_n}\,(EW)}{G \mid C_0}\,(r)$$

where (r) is any rule with more than one premiss and each component of G is contained in at least one of the hypersequents $G_1, \dots, G_n$.

A derivation in structured form can be divided into a part containing only (EC) applications and a part containing the applications of any other rule. We introduce a notation for the hypersequent separating the two parts.

Definition 4.10 If $\mathcal{D}$ is a derivation in structured form, we denote by $\widehat{H}_{\mathcal{D}}$ the premiss of the uppermost application of (EC) in $\mathcal{D}$.

We prove below that from any HLJ $+\ \mathbb{H}$ derivation $\mathcal{D}$ of a sequent we can construct a partial derivation for each component of $\widehat{H}_{\mathcal{D}}$ having the same structure as the ancestor tree of that component (Definition 4.11). By *having the same structure* we mean that the partial derivation of a hypersequent component contains the translation of the rules in the ancestor tree of that component in $\mathcal{D}$, with the exception of (EW).

Definition 4.11 Given a HLJ$+\mathbb{H}$ derivation. A sequent (hypersequent component) C' is a *parent* of a sequent C, denoted as $p(C, C')$, if one of the following conditions holds:

- C is active in the conclusion of an application of some $Hr \in \mathbb{H}$, and C' is the active component of a premiss linked to C (see Definition 3.3);
- C is active in the conclusion of an application of a rule of HLJ, and C' is the active component of a premiss of such application;
- C is a context component in the conclusion of any rule application, and C' is the corresponding context component in a premiss of such application.

We say that a sequent C' is an *ancestor* of a sequent C, and we write $a(C, C')$, if the pair (C, C') is in the transitive closure of the relation $p(\cdot,\cdot)$. The *ancestor tree* of a sequent C is the tree whose nodes are all sequents related to C by $a(\cdot,\cdot)$ and whose edges are defined by the relation $p(\cdot,\cdot)$ between such nodes.

Remark 4.12

- In an $\text{HLJ} + \mathbb{H}$ derivation that does not use (EC), the ancestor tree of each hypersequent is a sequent derivation.
- If C is the active component of an application of (EW), then there is no C' such that $p(C, C')$.

As usual, the *length* of a derivation is the the maximal number of applications of inference rules +1 occurring on any branch.

Lemma 4.13 *Let* $\mathbb{H}$ *be a set of hypersequent rules and* $\mathbb{S}$ *of* 2*-systems s.t. if* $Hr \in \mathbb{H}$ *then* $Sys_{Hr} \in \mathbb{S}$. *Given any* $\text{HLJ} + \mathbb{H}$ *derivation* $\mathcal{D}$ *in structured form, for each component* C *of* $\widehat{H}_{\mathcal{D}}$ *we can construct a partial derivation in* $\text{LJ} + \mathbb{S}$ *having the same structure as the ancestor tree of* C *in* $\mathcal{D}$.

Proof. Let H be a hypersequent in $\mathcal{D}$ derived without using (EC). We construct a partial derivation in $\text{LJ} + \mathbb{S}$ with the required property for each of its components. The proof proceeds by induction on the length l of the derivation of H by translating each rule of $\text{HLJ} + \mathbb{H}$, with the exception of (EW), into the corresponding sequent rule in $\text{LJ} + \mathbb{S}$.
Base case. If $l = 1$ (i.e. H is an axiom) the partial derivation in $\text{LJ} + \mathbb{S}$ simply contains H.
Inductive step. We consider the last rule (r) $(\neq (EW))$ applied in the (sub)derivation $\mathcal{D}'$ of H, and we distinguish the two cases: (i) (r) is a one-premiss rule and (ii) (r) has more premisses; for the latter case, since $\mathcal{D}'$ is in structured form, we deal also with possible queues of (EW) above its premisses.

(i) Assume that the derivation ending in a one-premiss rule $(r) \in \text{HLJ}$ is

$$\dfrac{\begin{matrix}\mathcal{D}\\ \vdots\\ G \mid C\end{matrix}}{G \mid C'}\ (r)$$

By induction hypothesis there is a partial derivation of C (and of each component of G) having the same structure as the ancestor tree of C. The partial derivation of C' is simply obtained by applying (r).

The case in which (r) is a one-premiss rule belonging to $\mathbb{H}$ is a special case of (ii), for which there is no need to consider queues of (EW).

(ii) Assume that $(r) = (Hr) \in \mathbb{H}$ has more than one premiss, the remaining cases ($(r) \in \text{HLJ}$, and $(r) \in \mathbb{H}$ and has only one premiss) being simpler. Assume that the derivation $\mathcal{D}'$, of length n, ends as follows

$$\dfrac{\begin{matrix}\mathcal{D}_1^1\\ \vdots\\ G \mid C_1^1\end{matrix} \quad \dots \quad \begin{matrix}\mathcal{D}_1^{m_1}\\ \vdots\\ G \mid C_1^{m_1}\end{matrix} \quad \dots \quad \begin{matrix}\mathcal{D}_k^1\\ \vdots\\ G \mid C_k^1\end{matrix} \quad \dots \quad \begin{matrix}\mathcal{D}_k^{m_k}\\ \vdots\\ G \mid C_k^{m_k}\end{matrix}}{G \mid \Delta_1, \Gamma_1 \Rightarrow \Pi_1 \mid \dots \mid \Delta_k, \Gamma_k \Rightarrow \Pi_k}\ (Hr)$$

where the premisses $G \mid C^i_j$ of (Hr) are possibly inferred by a queue of (EW). When this is the case we consider the uppermost hypersequents in the queue. More precisely, we consider the following derivations (each of which has length strictly less than n)

$$\begin{array}{ccccccc} \mathcal{D}^1_1 & & \mathcal{D}^{m_1}_1 & & \mathcal{D}^1_k & & \mathcal{D}^{m_k}_k \\ \vdots & & \vdots & & \vdots & & \vdots \\ G^1_1 \mid C^1_1 & \dots & G^{m_1}_1 \mid C^{m_1}_1 & \dots & G^1_k \mid C^1_k & \dots & G^{m_k}_k \mid C^{m_k}_k \end{array}$$

where, for $1 \leq y \leq k$ and $1 \leq x \leq m_y$, the hypersequent G^x_y is G if there is no (EW) application immediately above $G \mid C^x_y$; otherwise, $G^x_y \mid C^x_y$ is the premiss of the uppermost (EW) application in the queue immediately above $G \mid C^x_y$.

Since $\mathcal{D}$ (and hence $\mathcal{D}'$) is in structured form each component of G must occur in at least one of the hypersequents $G^1_1, \dots, G^{m_1}_1, \dots, G^1_k, \dots, G^{m_k}_k$. Hence the induction hypothesis gives us partial derivations for each component of G[4]. We obtain partial derivations for $\Delta_1, \Gamma_1 \Rightarrow \Pi_1, \dots, \Delta_k, \Gamma_k \Rightarrow \Pi_k$ applying the non-ending rules of the 2-system Sys_{Hr} as follows

$$\frac{C^1_1 \quad \dots \quad C^{m_1}_1}{\Delta_1, \Gamma_1 \Rightarrow \Pi_1}\ (r_1) \qquad \dots \qquad \frac{C^1_k \quad \dots \quad C^{m_k}_k}{\Delta_k, \Gamma_k \Rightarrow \Pi_k}\ (r_k)$$

Indeed, by induction hypothesis we have a partial derivation for each C^x_y.

The obtained partial derivations clearly satisfy the following property: (with the exception of (EW) and of the dummy ending rules) a rule application occurs in the ancestor tree of a hypersequent component in $\mathcal{D}$ *iff* its translation occurs in the partial derivation of such component. □

Theorem 4.14 *For any set $\mathbb{H}$ of hypersequent rules and set $\mathbb{S}$ of 2-systems s.t. if $Hr \in \mathbb{H}$ then $Sys_{Hr} \in \mathbb{S}$, if $\vdash_{\mathrm{HLJ+\mathbb{H}}} \Gamma \Rightarrow \Pi$ then $\vdash_{\mathrm{LJ+\mathbb{S}}} \Gamma \Rightarrow \Pi$.*

Proof. Let $\mathcal{D}$ be a HLJ + $\mathbb{H}$ derivation of $\Gamma \Rightarrow \Pi$. By the results in Section 4.2.1 we can assume that $\mathcal{D}$ is in structured form. By applying the procedure of Lemma 4.13 to the premiss $\widehat{H}_{\mathcal{D}}$ of the uppermost application of (EC) in $\mathcal{D}$ we obtain a set of partial derivations $\{\mathcal{D}_i\}_{i \in I}$ whose rules translate those occurring in the ancestor trees of each component of $\widehat{H}_{\mathcal{D}}$. We show that we can suitably apply the ending rules of 2-systems in $\mathbb{S}$ to the roots of $\{\mathcal{D}_i\}_{i \in I}$ in order to obtain the required LJ + $\mathbb{S}$ derivation of $\Gamma \Rightarrow \Pi$.

To do that we first group all non-ending rule applications in $\{\mathcal{D}_i\}_{i \in I}$ according to the application of $Hr \in \mathbb{H}$ that these rules translate. For each such group we apply one ending rule below the partial derivations in which the non-ending

[4] In case we have different partial derivations for a component C of G we can always obtain one partial derivation by applying a "dummy" ending rule as

$$\frac{C \quad \dots \quad C}{C}$$

rules of the group occur. The needed ending rules are always applicable. This is not the case only when two non-ending rules belonging to the same 2-system instance occur in a same partial derivation $\mathcal{D}'_j$. Two cases can arise: either (i) $\mathcal{D}'_j \in \{\mathcal{D}_i\}_{i\in I}$, or (ii) $\mathcal{D}'_j$ is obtained by the application of an ending rule in $\mathrm{LJ} + \mathbb{S}$ to two partial derivations. If (i), two nodes in the ancestor tree of a component of $\widehat{H}_{\mathcal{D}}$ must be active in the conclusion of a single Hr application, which contradicts Definition 4.11 and the fact that we deal with hypersequent rules having premisses with one active component only. If (ii), assume that there are two applications $(Hr_1), (Hr_2)$ of rules in $\mathbb{H}$ s.t. both have active components in the ancestor trees of the same two components C_1 and C_2 of $\widehat{H}_{\mathcal{D}}$. If (Hr_1) occurs below (Hr_2) two active components in (Hr_1) cannot have ancestors that are active in (Hr_2) (and vice versa), against the assumptions of *(ii)*. Otherwise, (Hr_1) and (Hr_2) occur above different premisses of a rule application (r). If this holds, we show that we can modify the partial derivations in $\{\mathcal{D}_i\}_{i\in I}$ in order to apply the needed ending rule. Indeed, some nodes of the ancestor tree of one of C_1 and C_2 must be context components of (r) (otherwise the elements of the two ancestor trees never occur in the same hypersequent above (r)). Then the non-ending rules translating (Hr_1) and (Hr_2) occur in a partial derivation above different premisses of a dummy ending rule. To avoid this situation it is enough to split the two premisses in different partial derivations (removing some premisses of the dummy ending rule in each partial derivation). In general, there could be n applications of rules in $\mathbb{H}$ s.t. the 1st and the nth have active components in the ancestor tree of the same component of $\widehat{H}_{\mathcal{D}}$, and the ith application (for $1 \le i < n$) has active components in the ancestor tree of the same component of $\widehat{H}_{\mathcal{D}}$ as the $(i+1)^{\text{th}}$. We can reason in a similar way considering the whole sequence of applications instead of a pair.

Hence, we eventually obtain an $\mathrm{LJ} + \mathbb{S}$ derivation of $\Gamma \Rightarrow \Pi$. □

4.2.1 Pre-processing of hypersequent derivations

In the previous algorithm we only considered hypersequent derivations in structured form, i.e. in which (EC) applications occur immediately above the root and (EW) applications occur where needed. Here we show how to transform each hypersequent derivation into a derivation in structured form.

Definition 4.15 The *external contraction rank (ec-rank)* of an application E of (EC) in a derivation is the number of applications of rules other than (EC) between E and the root of the derivation.

Lemma 4.16 *Each* HLJ+$\mathbb{H}$ *derivation* $\mathcal{D}$ *can be transformed into a derivation of the same end-hypersequent in which all* (EC) *applications have ec-rank* 0.

Proof. Proceed by double induction on the lexicographically ordered pair $\langle \mu, \nu \rangle$, where μ is the maximum ec-rank of any (EC) application in $\mathcal{D}$, and ν is the number of (EC) applications in $\mathcal{D}$ with maximum ec-rank.
Base case. If $\mu = 0$ the claim trivially holds.
Inductive step. Assume that $\mathcal{D}$ has maximum ec-rank μ and that there are ν applications of the rule (EC) with ec-rank μ. We show how to transform $\mathcal{D}$

into a derivation $\mathcal{D}'$ having either maximum ec-rank $\mu' < \mu$ or ec-rank μ and number of (EC) applications with maximum ec-rank $\nu' < \nu$.

Consider an (EC) application with ec-rank μ in $\mathcal{D}$ and the queue of (EC) containing it. There cannot be any applications of (EC) above this queue because the ec-rank of its elements is maximal. We distinguish cases according to the rule (r) applied to the conclusion of the last element of such queue.

Assume that (r) has one premiss. If $(r) = (EW)$, we apply (EW) (with the same active component) before the queue. If $(r) \neq (EW)$, we apply (r) immediately before the queue, possibly followed by applications of (EC).

Notation. Given a hypersequent H we denote by $(H)^u$ the hypersequent $H \mid \ldots \mid H$ containing u of copies of H $(u \geq 0)$.

Let (r) be a(ny external) context-sharing rule with more than one premiss and consider any subderivation of $\mathcal{D}$ of the form

$$\dfrac{\dfrac{\dfrac{\begin{array}{c}\mathcal{D}_1\\ \vdots\\ G \mid G_1' \mid (C_1)^{m_1}\end{array}}{\vdots}\,(EC)}{G \mid C_1}\,(EC) \quad \cdots \quad \dfrac{\dfrac{\begin{array}{c}\mathcal{D}_n\\ \vdots\\ G \mid G_n' \mid (C_n)^{m_n}\end{array}}{\vdots}\,(EC)}{G \mid C_n}\,(EC)}{G \mid H}\,(r)$$

where G_i', for $1 \leq i \leq n$, only contains components in G and the derivations $\mathcal{D}_1, \ldots, \mathcal{D}_n$ contain no application of (EC). We can transform $\mathcal{D}$ into a derivation $\mathcal{D}'$ in which all applications of (EC) occurring above the hypersequent $G \mid H$ are either immediately above it or immediately above another application of (EC); their ec-rank is thus reduced by 1.

We first prove that $(\star)$ the hypersequent $G \mid G'' \mid (H)^q$, where $G'' = G_1' \mid \ldots \mid G_n'$ and $q = (\sum_{i=1}^n (m_i - 1)) + 1$ is derivable from

$$G \mid G_1' \mid (C_1)^{m_1}\ ,\ \ldots\ ,\ G \mid G_n' \mid (C_n)^{m_n}$$

using only (EW) and (r). The hypersequent $G \mid H$ then follows from $G \mid G'' \mid (H)^q$ by (EC) as all the components of G'' occur also in G. The obtained derivation $\mathcal{D}'$ has maximum ec-rank $\mu' < \mu$, or the occurrences of (EC) with ec-rank μ occurring in it are $\nu' < \nu$.

It remains to prove claim $(\star)$. For each element of the set

$$\mathbb{Q} = \{G \mid G'' \mid (H)^0 \mid (C_1)^{x_1} \mid \ldots \mid (C_n)^{x_n} \ :\ \sum_{i=1}^n x_i \ = \ (\sum_{i=1}^n (m_i - 1)) + 1\}$$

there is a derivation from $G \mid G_1' \mid (C_1)^{m_1}\ ,\ \ldots\ ,\ G \mid G_n' \mid (C_n)^{m_n}$ using only (EW). Indeed for any hypersequent in $\mathbb{Q}$ and for $1 \leq i \leq n$, there is at least one $x_i \geq m_i$, because otherwise $\sum_{i=1}^n x_i \ < \ (\sum_{i=1}^n (m_i - 1)) + 1$. The claim $(\star)$ therefore follows by Lemma 4.17 below being $G \mid G'' \mid (H)^q$ the only element of the set $(q = (\sum_{i=1}^n (m_i - 1)) + 1)$

$$\mathbb{Q}' = \{G \mid G'' \mid (H)^q \mid (C_1)^{x_1} \mid \ldots \mid (C_n)^{x_n} \ :\ \sum_{i=1}^n x_i \ = \ 0\}.$$

□

The following is the central lemma of the previous proof.

Lemma 4.17 *For any application of a hypersequent rule*

$$\frac{G \mid C_1 \qquad \dots \qquad G \mid C_n}{G \mid H}\ (r)$$

and natural number $d \geq 0$, consider the set of hypersequents

$$\mathbb{L}_d = \{G \mid (H)^c \mid (C_1)^{x_1} \mid \dots \mid (C_n)^{x_n}\ :\ \sum_{i=1}^{n} x_i = d\}$$

where G, H are hypersequents, $C_1, \dots, C_n$ sequents, and c is a natural number. For any natural number e, s.t. $0 \leq e \leq d$, each element of the set

$$\mathbb{L}_{(d-e)} = \{G \mid (H)^{c+e} \mid (C_1)^{x'_1} \mid \dots \mid (C_n)^{x'_n}\ :\ \sum_{i=1}^{n} x'_i = d - e\}$$

is derivable from hypersequents in $\mathbb{L}_d$ by repeatedly applying the rule (r).

Proof. By induction on e.
Base case: If $e = 0$, then $\mathbb{L}_d = \mathbb{L}_{d-e}$.
Inductive step: Assume that $e > 0$ and that the claim holds for all $e' < e$. By induction hypothesis there exists a derivation from the hypersequents in $\mathbb{L}_d$ for each element of the set

$$\mathbb{L}_{(d-(e-1))} = \{G \mid (H)^{c+(e-1)} \mid (C_1)^{x''_1} \mid \dots \mid (C_n)^{x''_n}\ :\ \sum_{i=1}^{n} x''_i = d - (e-1)\}$$

that only consists of applications of (r). Any hypersequent

$$G \mid (H)^{c+e} \mid (C_1)^{x'_1} \mid \dots \mid (C_n)^{x'_n}$$

in $\mathbb{L}_{(d-e)}$ can be derived from elements of $\mathbb{L}_{(d-(e-1))}$ as follows:

$$\frac{G \mid (H)^{c+(e-1)} \mid H'_1 \qquad \dots \qquad G \mid (H)^{c+(e-1)} \mid H'_n}{G \mid (H)^{c+e} \mid (C_1)^{x'_1} \mid \dots \mid (C_n)^{x'_n}}\ (r)$$

where, for $1 \leq i \leq n$, $H'_i = (C_1)^{y_1} \mid \dots \mid (C_n)^{y_n}$ is such that if $j \neq i$ then $y_j = x'_j$ and if $j = i$ then $x'_j + 1$; i.e., the components $C_1, \dots, C_n \notin G$ occur in the ith premiss as many times as in the conclusion, except for C_i which occurs one more time.

All premisses of this rule application are hypersequents in $\mathbb{L}_{(d-(e-1))}$, indeed $(x'_1 + 1) + x'_2 + \dots + x'_n = \dots = x'_1 + \dots + x'_{n-1} + (x'_n + 1) = (\sum_{i=1}^{n} x'_i) + 1$ and $(\sum_{i=1}^{n} x'_i) + 1 = (d-e) + 1 = d - (e-1)$. Given that only the rule (r) is used to derive the elements of $\mathbb{L}_{d-(e-1)}$ from the elements of $\mathbb{L}_d$, also the elements of $\mathbb{L}_{(d-e)}$ can be derived from those of $\mathbb{L}_d$ by applying only (r). □

Lemma 4.18 *Any* HLJ + $\mathbb{H}$ *derivation of a sequent can be transformed into a derivation in structured form.*

Proof. Let $\mathcal{D}$ be a derivation of a sequent S in HLJ + $\mathbb{H}$. By Lemma 4.16 we can assume that all applications of (EC) in $\mathcal{D}$ occur in a queue immediately above S. Consider an application of (EW), with premiss G and conclusion $G \mid C$, which is not as in Definition 4.9. First notice that $G \mid C$ cannot be the root of $\mathcal{D}$. We show how to shift this application of (EW) below other rule applications until the statement is satisfied for such application. Three cases can arise:

(i) C is the active component in the premiss of an application of a rule (r). The conclusion of (r) is simply obtained by applying (EW) (possibly multiple times) to G.

(ii) C is a context component in the premiss of an application of a one-premiss rule (r). The (EW) is simply shifted below (r).

(iii) C occurs actively inside the queues of (EW) above all the premisses of an application of a rule (r). We remove all the applications of (EW) with active component C in the queues and apply (r) with one context component less, followed by (EW).

The termination of the procedure follows from the fact that $\mathcal{D}$ is finite and that (i)–(iii) always reduce the number of rules different from (EW) occurring below the (EW) applications. □

5 Applications and Future Work

We provided constructive transformations (embeddings) from hypersequent derivations into derivations in 2-systems of rules and back. Defined using intermediate logics as a case study, the embeddings do not depend on the considered calculus rules and can be naturally extended to other classes of (propositional) logics, e.g., substructural or modal logics. This shows that the two seemingly different proof frameworks have the same expressive power.

For 2-systems, the benefits of the embedding include: (i) analyticity proofs, (ii) new cut-free calculi and (iii) locality of derivations using the |-notation. Ad (i): the method in [10] transforms generalised geometric formulae in the class GA_1 into analytic 2-systems. The analiticity proof in [10] relies on the fact that the obtained 2-systems manipulate atomic formulae only; this is the case for labelled 2-systems arising from frame conditions, but it does not hold anymore when translating axiom schemata, e.g. $(\varphi \supset \psi) \vee (\psi \supset \varphi)$ for Gödel logic (cf. Example 2.1). In this case analyticity for the obtained 2-systems can be recovered by (a) first translating them into hypersequent rules, (b) applying the "completion" procedure [5] in [5] to the latter, and (c) translating them back. Ad (ii): for purely propositional formulae, the class of axioms that can be automatedly

[5] This amount in transforming each structural hypersequent rule into an equivalent one (w.r.t. intuitionistic logic) that preserves cut-elimination when added to the HLJ calculus.

transformed into analytic structural hypersequent rules (i.e. the class $\mathcal{P}_3$ in [5], see the program at http://www.logic.at/people/lara/axiomcalc.html) strictly contains GA_1. E.g. $\neg\alpha \vee \neg\neg\alpha$ belongs to $\mathcal{P}_3$ (and to GA_2) but *not* to GA_1; hence when applied to $\neg\alpha \vee \neg\neg\alpha$ the method in [10] does not lead to an analytic 2-system, which can instead be defined by translating the hypersequent rule equivalent to the axiom, i.e.

$$\frac{G \mid \Gamma, \Gamma' \Rightarrow}{G \mid \Gamma \Rightarrow \mid \Gamma' \Rightarrow}\ (lq)$$

The transformation from hypersequent derivations into 2-systems allows us to reformulate the former without using |-separated components and without the need of (EC), which is "internalised" within the (ending rules of the) 2-systems. The resulting calculi can be rewritten as natural deduction systems. As future work we plan to explore their potential for extracting the computational content of the formalised logics (see, e.g. [3] for an attempt of translating the hypersequent calculus for Gödel logic – cf. Example 2.1 – into a natural deduction system to be used for establishing a Curry-Howard correspondence).

Finally, as shown in [4], all propositional axiomatisable intermediate logics are definable by adding to intuitionistic logic suitable formulae (*canonical formulae*) belonging to the class in the hierarchy of [5] immediately above the one that can be handled by hypersequents. The established connection between hypersequents and two-level systems of rules suggests the use of *three-level* systems of rules for dealing with the only class still escaping uniform analytic hypersequent calculi.

References

[1] Avron, A., *A constructive analysis of RM*, J. Symbolic Logic **52** (1987), pp. 939–951.

[2] Avron, A., *Hypersequents, logical consequence and intermediate logics for concurrency*, Ann. Math. Artif. Intell. **4** (1991), pp. 225–248.

[3] Beckmann, A. and N. Preining, *Hyper natural deduction*, in: *LICS 2015* (2015), pp. 547–558.

[4] Chagrov, A. and M. Zakharyaschev, "Modal Logic," Oxford Logic Guides **35**, Oxford University Press, 1997.

[5] Ciabattoni, A., N. Galatos and K. Terui, *From axioms to analytic rules in nonclassical logics*, in: *LICS 2008* (2008), pp. 229–240.

[6] Fitting, M., "Proof methods for modal and intuitionistic logics," Synthese Library **169**, D. Reidel Publishing Co., 1983.

[7] Fitting, M., *Prefixed tableaus and nested sequents*, Ann. Pure Appl. Logic **163** (2012), pp. 291–313.

[8] Goré, R. and R. Ramanayake, *Labelled tree sequents, tree hypersequents and nested (deep) sequents*, in: *Advances in Modal Logic* (2012), pp. 279–299.

[9] Negri, S., *Proof analysis in modal logic*, J. Philos. Logic **34** (2005), pp. 507–544.

[10] Negri, S., *Proof analysis beyond geometric theories: from rule systems to systems of rules*, J. Logic Comput. (2014).

[11] Poggiolesi, F., *A purely syntactic and cut-free sequent calculus for the modal logic of provability*, Rev. Symb. Log. **2** (2009), pp. 593–611.

[12] Ramanayake, R., *Embedding the hypersequent calculus in the display calculus*, J. Logic Comput. **3** (2015), pp. 921–942.
[13] Wansing, H., *Translation of hypersequents into display sequents*, Log. J. IGPL **6** (1998), pp. 719–733.

Appendix

Lemma .1 *Let C_1' and C_2' be two occurrences of marked sequents that are translated by the algorithm of Section 4.1 as components of the same hypersequent. Then C_1' and C_2' are neither premisses of a rule which is not the ending rule of any 2-system, nor premisses of two non-ending rules belonging to the same 2-system instance.*

Proof. Let C_1 and C_2 be the components translating C_1' and C_2', respectively. Assume that C_1 and C_2 occur in the same hypersequent.

We first specify when the algorithm of Section 4.1 translates two marked sequents into two components of the same hypersequent. In order to do this we introduce the notion of *inner path* of a 2-system instance, i.e. a path between a premiss of the ending rule of a 2-system instance and the conclusion of one of its non-ending rules. The components that occur in the same hypersequent are those that translate sequents that occur either (i) in inner paths of a single instance S of 2-system, or (ii) in inner paths of two distinct instances S_1 and S_2 of possibly different 2-systems that are related by the transitive closure of the relation of sharing part of a path. If the latter holds, we say that S_1 and S_2 are *chained*. To see that these are the only possible cases consider when the algorithm translates two conclusions of different rule applications into components of the same hypersequent. This only happens when we translate the conclusions of the non-ending rules of a 2-system instance (or the sequents occurring below these conclusions but above the ending rule of the 2-system instance), and when a single marked sequent occurs in the intersection of the inner paths of two different 2-systems instances. In this case, due to the handling of hypersequent contexts in the algorithm, all marked sequents in the inner paths of the 2-systems instances are translated into components of a single hypersequent.

Suppose that (i) holds, i.e. C_1' and C_2' occur in two different inner paths of a single instance S of 2-system. Then the sequents C_1' and C_2' cannot be the premisses of a rule which is not an ending rule because otherwise two inner paths of S would occur above the same premiss of the ending rule of S, against the definition of inner path. Moreover, the sequents C_1' and C_2' cannot be the premisses of two non-ending rules belonging to a single instance S_3 of 2-system. Otherwise, two non-ending rules of S_3 would occur along two inner paths of S, against the definition of 2-system.

Suppose now that (ii) is the case, i.e. C_1' and C_2' occur in the inner paths of two chained instances S_1 and S_2 of possibly different 2-systems. We distinguish two sub-cases: (ii.a) S_1 and S_2 share part of an inner path, (ii.b) S_1 and S_2 do not share part of an inner path.

Assume that (ii.a) holds. The shared inner path does not contain C_1' and C_2' (as otherwise the two occurrences C_1 and C_2 would coincide). If C_1' and C_2'

are the premisses of some rule (which is not an ending rule), then S_1 and S_2 also share part of the path in which C'_1 and C'_2 occur, but then the ending rule applications of S_1 and S_2 would coincide, against the definition of 2-system. If C'_1 and C'_2 are the premisses of two non-ending rules belonging to a single 2-system instance S_3, then two non-ending rules of S_3 occur above different premisses of the lower of the ending rules of S_1 and S_2, which contradicts the definition of 2-system.

If (ii.b) holds then, firstly, some instances of 2-systems are chained to both S_1 and S_2, and the ending rule of one of these instances must occur below the ending rules of S_1 and S_2. Otherwise, no 2-system instance chained to S_1 can be chained also to S_2. We call S_0 the 2-system instance with the uppermost ending rule among the instances that are chained to both S_1 and S_2. Secondly, S_1 and S_2 occur above different premisses of the ending rule of S_0. Otherwise, either C_1 and C_2 are not in the same hypersequent (because C'_1 and C'_2 occur above the rule application that joins the inner paths of the 2-systems chained to S_1 and S_2) or C'_1 and C'_2 cannot be marked (because all the inner paths of one among S_1 and S_2 occur above an inner path of the other). Finally, the non-ending rules of S_0 must have already been translated by the algorithm, otherwise C_1 and C_2 cannot occur in the same hypersequent. Assume, by contradition, that C'_1 and C'_2 are the premisses of the same rule application which is not an ending rule. This contradicts the fact that C'_1 and C'_2 occur above different premisses of the ending rule of S_0. Suppose now that C'_1 and C'_2 are the premisses of two non-ending rules belonging to a single 2-system instance S_3. Then, two non-ending rules of S_3 occur above different premisses of S_0, against the definition of 2-system (S_3 and S_0 cannot coincide as the non-ending rules of S_0 have already been translated). □

Classical and Empirical Negation in Subintuitionistic Logic

Michael De [1]

Department of Philosophy, University of Konstanz
Universitätsstraße 10, 78464 Konstanz Germany

Hitoshi Omori [2]

Department of Philosophy, Kyoto University
Yoshida Honmachi, Sakyo-ku, Kyoto, 606-8501, Japan

Abstract

Subintuitionistic (propositional) logics are those in a standard intuitionistic language that result by weakening the frame conditions of the Kripke semantics for intuitionistic logic. In this paper we consider two negation expansions of subintuitionistic logic, one by classical negation and the other by what has been dubbed "empirical" negation. We provide an axiomatization of each expansion and show them sound and strongly complete. We conclude with some final remarks, including avenues for future research.

Keywords: Subintuitionistic logic, intuitionistic logic, classical negation, empirical negation

1 Introduction

With the advent of Kripke semantics for modal logic, it became natural to investigate logics weaker than **S5** by weakening its frame properties. For intuitionistic logic things have tended in the other direction through the investigation of intermediate logics that obtain by strengthening the frame properties of intuitionistic logic. There has been less interest in subintuitionistic logics that obtain by weakening the frame properties of intuitionistic logic. To our knowledge, there has been no study of expansions of subintuitionistic logic by operators outside of the standard language. Our interest lies in expansions of subintuitionistic logic that obtain by adding classical or classical-like negations

[1] This research was funded in part by the European Research Council under the European Community's Seventh Framework Programme (FP7/2007-2013)/ERC Grant agreement nr 263227. Email: `mikejde@gmail.com`.

[2] Postdoctoral research fellow of Japan Society for the Promotion of Science (JSPS). Email: `hitoshiomori@gmail.com`

to the language. Since we are interested in logics as consequence relations, our work most closely resembles Greg Restall's [14]. [3]

In Kripke semantics, classical negation is ordinarily defined such that the negation $\neg A$ of a sentence is true at a state just in case A is not true at that state. In this sense classical negation is extensional, since it does not require looking beyond the state of evaluation. The problem with this definition, however, is that it is incompatible with Heredity, i.e., that if an atomic statement A is true at some state w, then it is true at any state accessible to w. For if A is not true at x so that $\neg A$ is, and if y is accessible from x, we cannot require that the truth of $\neg A$ be preserved up to y since A may be true there. So adding classical negation requires either eschewing Heredity, or else restricting the accessibility relation in a way that allows Heredity to stay in force. The latter option is not available in the case of intuitionistic logic since there are no restrictions that would preserve the intuitionistic fragment. For instance, one natural restriction is to let accessibility be identity but that restriction results in classical logic. [4] We must therefore eschew Heredity, which makes subintuitionistic logic ideal for the purposes of adding classical negation.

There is another way of adding a classical-like negation to intuitionistic logic. A defining characteristic of classicality is that $\neg A$ is true just in case A is not. We may then define truth in a model to be truth at a distinguished base state, and define negation so that $\neg A$ is true in a model if, and only if, A is not. This means that negation is no longer extensional in the sense that the truth of $\neg A$ at an aribtrary *state* in a model depends on whether A is true at the base state, yet it is extensional in the sense that determining whether a negation is true in a model requires going nowhere else besides the state relative to which truth-in-a-model is defined, namely, the base state. Notice that, importantly, this definition of negation is compatible with Heredity. Suppose g is the base state of a model and that A is not true there. Then $\neg A$ is true at *every* state in the model, and so its truth is trivially preserved up the accessibility relation.

In [3] and [4], Michael De and Hitoshi Omori give philosophical grounds for adding such a negation to the language of intuitionistic logic, where they there call it *empirical negation*. To put matters briefly, intuitionistic negation, standardly defined as implication to absurdity, is too strong to allow for a generalization of intuitionistic logic to empirical domains, such as the physical sciences. [5] For instance, to express that a certain proposition (such as Goldbach's conjecture) has *not* been proven, it is too strong to say that the supposition of the proposition leads to absurdity. The idea, then, is to expand the vocabulary with an empirical negation that roughly expresses that "There fails to be sufficient evidence at present to warrant the proposition that...".

---

[3] See [2], [6], [1] and the references cited therein for work on subintuitionistic logic.

[4] Compare this with classical relevance logic which results by adding classical negation to relevance logic in just this way, i.e. by letting the order relative to which truth is preserved be identity. Here, however, we do not get a collapse to classical logic. See [10] and [11] for details.

[5] This project is given its fullest defense in the works of Michael Dummett.

The negation has a number of interesting properties. One in particular is its close affinity to classical negation even though adding it to intuitionistic logic does not result in a collapse to classical logic.[6]

The paper proceeds as follows. In §2 we present the weakest subintuitionistic logic **SJ** of [14]. In §3 we expand **SJ** by empirical negation and show the axiomatization sound and strongly complete. In §4 we do the same for classical negation. In §5 we draw some comparisons to related systems, including weak relevance logics and a system of [5] in which both classical and intuitionistic negation coexist. In §6 we conclude with some final remarks.

2 Subintuitionistic logics revisited

We begin by presenting the weakest of the subintuitionistic logics, **SJ** which we then go onto expand by empirical and classical negation respectively in §3 and §4.

Definition 2.1 The language $\mathcal{L}$ consists of a finite set $\{\wedge, \vee, \rightarrow\}$ of propositional connectives and a denumerable set Prop of propositional variables which we denote by p, q, etc. Furthermore, we denote by Form the set of formulae defined as usual in $\mathcal{L}$. We denote a formula of $\mathcal{L}$ by A, B, C, etc. and a set of formulae of $\mathcal{L}$ by Γ, Δ, Σ, etc.

2.1 Semantics

Definition 2.2 A model for the language $\mathcal{L}$ is a quadruple $\langle W, g, R, V\rangle$, where W is a non-empty set (of states); $g \in W$ (the base state); R is a binary relation on W satisfying what in [14] is called *omniscience*, namely gRw for all $w \in W$; and $V : W \times \mathsf{Prop} \rightarrow \{0,1\}$ an assignment of truth values to state-variable pairs.[7] Valuations V are then extended to interpretations I to state-formula pairs by the following conditions:

- $I(w,p) = V(w,p)$
- $I(w, A \wedge B) = 1$ iff $I(w,A) = 1$ and $I(w,B) = 1$
- $I(w, A \vee B) = 1$ iff $I(w,A) = 1$ or $I(w,B) = 1$
- $I(w, A \rightarrow B) = 1$ iff for all $x \in W$, if wRx and $I(x,A) = 1$ then $I(x,B) = 1$.

Semantic consequence is now defined in terms of truth preservation at g: $\Sigma \models A$ iff for all models $\langle W, g, R, I\rangle$, $I(g,A) = 1$ if $I(g,B) = 1$ for all $B \in \Sigma$.

Remark 2.3 Note that we are not assuming

(Heredity) If $V(w,p) = 1$ and wRx then $V(x,p) = 1$,

---

[6] However, see [7] and [5] for different approaches to combining both classical and intuitionistic negation without such a collapse. It should be noted that the natural deduction system presented in [7] is trivial as stated (i.e. any formula follows from any set of formulae), an error that is to be blamed on $EFQ\neg$. To see this let Γ be empty and α a contradiction. The correct rule is obtained by deleting α from the left-hand side of the premise sequent (as Lloyd Humberstone conveyed in personal communication).

[7] Omniscient frames are sometimes referred to as *strongly generated*.

nor that R is reflexive or transitive; R is assumed only to satisfy omniscience. Typically, inuitionistic Kripke models are not pointed in the sense that they contain a base state relative to which truth in the model is defined. It plays an essential role here for logics—taken as consequence relations and not classes of theorems—properly weaker than intuitionistic logic, but also for intuitionistic logic expanded by empirical negation.

2.2 Proof Theory

Definition 2.4 The system **SJ** consists of the following axiom schemata.

(Ax1)	$A \to A$
(Ax2)	$A \to (B \to B)$
(Ax3)	$((A \to B) \wedge (B \to C)) \to (A \to C)$
(Ax4)	$(A \wedge B) \to A$
(Ax5)	$(A \wedge B) \to B$
(Ax6)	$((C \to A) \wedge (C \to B)) \to (C \to (A \wedge B)))$
(Ax7)	$A \to (A \vee B)$
(Ax8)	$B \to (A \vee B)$
(Ax9)	$((A \to C) \wedge (B \to C)) \to ((A \vee B) \to C))$
(Ax10)	$(A \wedge (B \vee C)) \to ((A \wedge B) \vee (A \wedge C))$

In addition to these axioms, we have the following rules of inference.

(MP) $$\frac{A \quad A \to B}{B}$$

(DMP) $$\frac{A \vee C \quad (A \to B) \vee C}{B \vee C}$$

(Adj) $$\frac{A \quad B}{A \wedge B}$$

(DR) $$\frac{(A \to B) \vee E \quad (C \to D) \vee E}{((B \to C) \to (A \to D)) \vee E}$$

Finally, we write $\Gamma \vdash A$ if there is a sequence of formulae $B_1, \dots, B_n, A$, $n \geq 0$, such that every formula in the sequence $B_1, \dots, B_n, A$ either (i) belongs to Γ; (ii) is an axiom of **SJ**; (iii) is obtained by one of the rules (MP)–(DR) from formulae preceding it in sequence.

Remark 2.5 Note that the following rule, included in the original formulation of **SJ**, is derivable in view of (DR), (MP), (Ax1) and (Ax9).

(R) $$\frac{A \to B \quad C \to D}{(B \to C) \to (A \to D)}$$

To see this, assume $A \to B$ and $C \to D$. Then by (Ax7) and (MP) we obtain $(A \to B) \vee E$ and $(C \to D) \vee E$ where E is $(B \to C) \to (A \to D)$. Then by applying (DR), we have $E \vee E$, and by (MP), (Ax1) and (Ax9), we obtain E, as desired.

Remark 2.6 It deserves noting that the following rules, known as Prefixing, Suffixing and Transitivity respectively, are derivable in **SJ** in view of (R), (MP) and (Ax1):

(Prefixing) $$\frac{C \to D}{(A \to C) \to (A \to D)}$$

(Suffixing) $$\frac{A \to B}{(B \to C) \to (A \to C)}$$

(Transitivity) $$\frac{A \to B \quad B \to C}{A \to C}$$

Proposition 2.7 *The following formulae and rules are provable in* **SJ**.

(1) $$\frac{A \vee B \quad A \to C}{C \vee B}$$

(2) $$((A \vee C) \wedge (B \vee C)) \to ((A \wedge B) \vee C)$$

Proof. Left as an exercise for the reader. □

2.3 Soundness and completeness

Theorem 2.8 (Soundness) *For* $\Gamma \cup \{A\} \subseteq \mathsf{Form}$, *if* $\Gamma \vdash A$ *then* $\Gamma \models A$.

Proof. The proof, by induction on the length of proof, can be found in [14]. The omniscience of g in used in showing that (DR) preserves validity. □

Following [14], the following notions will be used in the proofs of completeness.

Definition 2.9

(i) If Π is a set of sentences, let $\Pi_{\to}$ be the set of all members of Π of the form $A \to B$.
(ii) $\Sigma \vdash_{\Pi} A$ iff $\Sigma \cup \Pi_{\to} \vdash A$.
(iii) Σ is a Π-*theory* iff:
 (a) if $A, B \in \Sigma$ then $A \wedge B \in \Sigma$
 (b) if $\vdash_{\Pi} A \to B$ then (if $A \in \Sigma$ then $B \in \Sigma$).
(iv) Σ is *prime* iff (if $A \vee B \in \Sigma$ then $A \in \Sigma$ or $B \in \Sigma$).
(v) If X is any set of sets of formulae the binary relation R on X is defined thus:

$$\Sigma R \Delta \text{ iff (if } A \to B \in \Sigma \text{ then (if } A \in \Delta \text{ then } B \in \Delta)).$$

(vi) $\Sigma \vdash_{\Pi} \Delta$ iff for some $D_1, \ldots, D_n \in \Delta, \Sigma \vdash_{\Pi} D_1 \vee \cdots \vee D_n$.
(vii) $\vdash_{\Pi} \Sigma \to \Delta$ iff for some $C_1, \ldots, C_n \in \Sigma$ and $D_1, \ldots, D_m \in \Delta$:

$$\vdash_{\Pi} C_1 \wedge \cdots \wedge C_n \to D_1 \vee \cdots \vee D_m.$$

(viii) Σ is Π-*deductively closed* iff (if $\Sigma \vdash_{\Pi} A$ then $A \in \Sigma$).
(ix) $\langle \Sigma, \Delta \rangle$ is a Π-*partition* iff (i) $\Sigma \cup \Delta = \mathsf{Form}$ and (ii) $\not\vdash_{\Pi} \Sigma \to \Delta$.

In all the above, if $\Pi = \emptyset$, then the prefix 'Π-' will simply be omitted.

With these notions in mind, some lemmas are listed without their proofs. For details, see [14].

Proposition 2.10 *If $A \vdash C$ and $B \vdash C$, then $A \vee B \vdash C$.*

Lemma 2.11 *If $\langle \Sigma, \Delta \rangle$ is a Π-partition then Σ is a prime Π-theory.*

Lemma 2.12 *If $\Sigma \nvdash \Delta$ then there are $\Sigma' \supseteq \Sigma$ and $\Delta' \supseteq \Delta$ such that $\langle \Sigma', \Delta' \rangle$ is a partition, and Σ' is deductively closed.*

Corollary 2.13 *If $\Sigma \nvdash A$ then there is $\Pi \supseteq \Sigma$ such that $A \notin \Pi$, Π is a prime Π-theory and Π is Π-deductively closed.*

Lemma 2.14 *If $\nvdash_{\Pi} \Sigma \to \Delta$ then there are $\Sigma' \supseteq \Sigma$ and $\Delta' \supseteq \Delta$ such that $\langle \Sigma', \Delta' \rangle$ is a Π-partition.*

Lemma 2.15 *Let Σ be a prime Π-theory and $A \to B \notin \Sigma$. Then there is a prime Π-theory, Δ such that $\Sigma R \Delta$, $A \in \Delta$, $B \notin \Delta$.*

We are now in a position to prove completeness.

Theorem 2.16 (Completeness) *For $\Gamma \cup \{A\} \subseteq \mathsf{Form}$, if $\Gamma \models A$ then $\Gamma \vdash A$.*

Proof. The proof, due to [14], is given in the appendix. □

Remark 2.17 Restall notes the following correspondences between frame conditions and valid formulae.

Frame conditions	Characteristic formulae
Heredity	$A \to (B \to A)$
Reflexivity	$(A \wedge (A \to B)) \to B$
Transitivity	$(A \to B) \to ((B \to C) \to (A \to C))$

We will make use of this result later.

Remark 2.18 As observed by Heinirich Wansing in [15], the sets of theorems for the systems of Corsi, Došen and Restall coincide. However, Corsi and Došen show only *weak* completeness using standard techniques from modal logics, whereas Restall shows *strong* completeness using techniques from relevance logics. Restall's axiomatization is better suited for our purposes since the other axiomatizations will not work when we expand the language by empirical negation (cf. Remark 3.6).

Remark 2.19 As we noted earlier, in a semantic setting there is no straightforward way of adding classical negation to intuitionistic logic since classical negation breaks Heredity, the preservation of truth up the accessibility relation. In subintuitionistic logics where Heredity is not in force, there is no obstacle to adding classical negation. Recall that Heredity is required for validating the schema $A \to (B \to A)$, and that this schema has been criticized for introducing fallacies of relevance, since the mere truth of A does not guarantee us that an arbitrary B relevantly implies A. This suggests that **SJ** gets us at least one step closer to relevance than full intuitionistic logic. What then is the relation between subintuitionistic and relevance logic? It is, to be exact, that **SJ** is the relevance logic $\mathbf{B}^+$ of [12] extended by axioms (Ax2) and (Ax3). [8]

---

[8] We note that (Ax3) characterizes one of the many further conditions on the ternary relation. See [13, Theorem 2] for the details.

Let us now turn to the Heredity-friendly way of adding a classical-like negation to subintutionistic logic, even though we are working with the weakest subintuitionistic logic without the Heredity axiom, $A \to (B \to A)$.

3 Empirical negation

Before turning to the proof theory of subintuitionistic logic with empirical negation, we start with some preliminaries.

Definition 3.1 The language $\mathcal{L}_\sim$ consists of a finite set $\{\sim, \wedge, \vee, \to\}$ of propositional connectives and a denumerable set Prop of propositional variables which we denote by p, q, etc. Furthermore, we denote by $\mathsf{Form}_\sim$ the set of formulae defined as usual in $\mathcal{L}_\sim$. We denote a formula of $\mathcal{L}_\sim$ by A, B, C, etc. and a set of formulae of $\mathcal{L}_\sim$ by Γ, Δ, Σ, etc.

3.1 Semantics

Definition 3.2 A model for the language $\mathcal{L}_\sim$ is a quadruple $\langle W, g, R, V\rangle$, where W is a non-empty set (of states); $g \in W$ (the base state); R is a binary relation on W with g omniscient; and $V : W \times \mathsf{Prop} \to \{0,1\}$ an assignment of truth values to state-variable pairs. Valuations V are then extended to interpretations I to state-formula pairs by the following conditions:

- $I(w,p) = V(w,p)$
- $I(w, {\sim}A) = 1$ iff $I(g,A) = 0$
- $I(w, A \wedge B) = 1$ iff $I(w,A) = 1$ and $I(w,B) = 1$
- $I(w, A \vee B) = 1$ iff $I(w,A) = 1$ or $I(w,B) = 1$
- $I(w, A \to B) = 1$ iff for all $x \in W$, if wRx and $I(x,A) = 1$ then $I(x,B) = 1$.

Semantic consequence is again defined in terms of truth preservation at g: $\Sigma \models_e A$ iff for all models $\langle W, g, R, I\rangle$, $I(g,A) = 1$ if $I(g,B) = 1$ for all $B \in \Sigma$.

Remark 3.3 Note that we are neither assuming Heredity, nor that R is reflexive or transitive; R is assumed only to satisfy omniscience.

3.2 Proof Theory

Definition 3.4 The system $\mathbf{SJ}^\sim$ is obtained by adding the following axiom schemata and rule of inference to $\mathbf{SJ}$:

(Ax$_\sim$1)	$A \vee {\sim}A$	(Ax$_\sim$4)	${\sim}{\sim}{\sim}A \to {\sim}A$
(Ax$_\sim$2)	$B \to ({\sim}A \vee {\sim}{\sim}A)$	(Ax$_\sim$5)	$({\sim}A \wedge {\sim}B) \to {\sim}(A \vee B)$
(Ax$_\sim$3)	${\sim}{\sim}A \to ({\sim}A \to B)$	(DRP)	$\dfrac{(A \vee B) \vee C}{({\sim}A \to {\sim}{\sim}B) \vee C}$

Finally, we write $\Gamma \vdash_e A$ if there is a sequence of formulae $B_1, \ldots, B_n, A$, $n \geq 0$, such that every formula in the sequence $B_1, \ldots, B_n, A$ either (i) belongs to Γ; (ii) is an axiom of $\mathbf{SJ}^\sim$; (iii) is obtained by one of the rules (MP)–(DRP) from formulae preceding it in sequence.

Remark 3.5 Note that the following rule is derivable in $\mathbf{SJ}^{\sim}$:

$$\text{(RP)} \qquad \frac{A \vee B}{{\sim}A \to {\sim}{\sim}B}$$

To see this, assume $A \vee B$. Then by (Ax7) and (MP), we obtain $(A \vee B) \vee ({\sim}A \to {\sim}{\sim}B)$, and by applying (DRP) we have $({\sim}A \to {\sim}{\sim}B) \vee ({\sim}A \to {\sim}{\sim}B)$ which implies ${\sim}A \to {\sim}{\sim}B$, as desired.

Remark 3.6 Note that even though (Ax$_{\sim}$1) is valid and that $B \models_e A \vee {\sim}A$, $B \to (A \vee {\sim}A)$ is *not* valid in $\mathbf{SJ}^{\sim}$. This is the reason both (Ax$_{\sim}$1) and (Ax$_{\sim}$2) are needed. This also shows that the rule of inference from A to $B \to A$, employed in the axiomatizations of Corsi and Došen, is *not* sound in $\mathbf{SJ}^{\sim}$.

3.3 Some Basic Results

Theorem 3.7 (Classical deduction theorem) *For* $\Gamma \cup \{A, B\} \subseteq \mathsf{Form}_{\sim}$, *if* $\Gamma, A \vdash_e B$ *then* $\Gamma \vdash_e {\sim}A \vee B$.

Proof. By induction on the length n of a proof of $\Gamma, A \vdash_e B$. The details are given in the appendix. □

For the purpose of proving the other direction of the deduction theorem, we need another lemma.

Lemma 3.8 *The following are derivable in* $\mathbf{SJ}^{\sim}$*:*

$$\text{(RC}{\sim}\text{)} \quad \frac{A \to B}{{\sim}B \to {\sim}A} \qquad \text{(RIDN)} \quad \frac{A}{{\sim}{\sim}A}$$

$$\text{(3)} \quad {\sim}{\sim}(A \to A) \qquad ({\sim}\text{-DS}) \quad \frac{A \quad {\sim}A \vee B}{B}$$

Proof. For (RC$\sim$), assume $A \to B$. Then by making use of (1) and (Ax$_{\sim}$1), we have $B \vee {\sim}A$. Thus by applying (RP), we obtain ${\sim}B \to {\sim}{\sim}{\sim}A$, and finally by (Ax$_{\sim}$4) and (Transitivity), we obtain ${\sim}B \to {\sim}A$, as desired.

For (3), first apply (RC$\sim$) to ${\sim}(A \to A) \to (A \to A)$, an instance of (Ax2), and we get ${\sim}(A \to A) \to {\sim}{\sim}(A \to A)$. This together with (Ax1) and (Adj) gives us $({\sim}(A \to A) \to {\sim}{\sim}(A \to A)) \wedge ({\sim}{\sim}(A \to A) \to {\sim}{\sim}(A \to A))$, and by (Ax9) and (MP), we obtain $({\sim}(A \to A) \vee {\sim}{\sim}(A \to A)) \to {\sim}{\sim}(A \to A)$. Thus, we obtain the desired result in view of (Ax$_{\sim}$1) and (MP).

For (RIDN), assume A. Then by (Ax8), we obtain ${\sim}(A \to A) \vee A$. By applying (RP), we get ${\sim}{\sim}(A \to A) \to {\sim}{\sim}A$. The desired result follows by this, (3) and (MP).

Finally, for (${\sim}$-DS), assume A and ${\sim}A \vee B$. By the former and (RIDN), we obtain ${\sim}{\sim}A$, and by this together with (Ax$_{\sim}$3) and (MP), we obtain ${\sim}A \to B$. Therefore, by applying (1) to this and the latter assumption, we obtain $B \vee B$ which implies B, as desired. □

Remark 3.9 Note that (${\sim}$-DS) implies *ex contradictione quodlibet*:

$$({\sim}\text{-ECQ}) \qquad \frac{A \quad {\sim}A}{B}$$

Proposition 3.10 *For* $\Gamma \cup \{A, B\} \subseteq \mathsf{Form}_{\sim}$, *if* $\Gamma \vdash_e {\sim}A \vee B$ *then* $\Gamma, A \vdash_e B$.

Proof. Immediate in view of (${\sim}$-DS). □

By combining Theorem 3.7 and Proposition 3.10, we obtain the following.

Theorem 3.11 *For* $\Gamma \cup \{A, B\} \subseteq \mathsf{Form}_{\sim}$, $\Gamma, A \vdash_e B$ *iff* $\Gamma \vdash_e {\sim}A \vee B$.

Corollary 3.12 *For* $\Gamma \cup \{A, B\} \subseteq \mathsf{Form}_{\sim}$, $\Gamma, A \vdash_e B$ *iff* $\Gamma \vdash_e {\sim}B \to {\sim}A$.

We note that the following are theorems of $\mathbf{SJ}^{\sim}$.

Proposition 3.13 *The following formulae are provable in* $\mathbf{SJ}^{\sim}$.

(4)	${\sim}(A \vee B) \to ({\sim}A \wedge {\sim}B)$	(6)	${\sim}A \to {\sim}{\sim}{\sim}A$
(5)	$({\sim}A \vee {\sim}B) \to {\sim}(A \wedge B)$	(7)	${\sim}A \to ({\sim}{\sim}A \to B)$

Proof. (4) and (5) are essentially proved using (Ax4), (Ax5) and (Ax7), (Ax8) respectively together with (RC${\sim}$). (6) follows immediately by (Ax$_{\sim}$1) and (RP). For (7), apply (Transitivity) to (6) and ${\sim}{\sim}{\sim}A \to ({\sim}{\sim}A \to B)$, an instance of (Ax$_{\sim}$3). □

3.4 Soundness and completeness

We now proceed to the proof of soundness and completeness.

Theorem 3.14 *For* $\Gamma \cup \{A\} \subseteq \mathsf{Form}_{\sim}$, *if* $\Gamma \vdash_e A$ *then* $\Gamma \models_e A$.

Proof. By induction on the length of the proof, as usual. □

Proposition 3.15 *If* $A \vdash_e C$ *and* $B \vdash_e C$, *then* $A \vee B \vdash_e C$.

Proof. Assume $A \vdash_e C$ and $B \vdash_e C$. Then, by Corollary 3.12, we obtain $\vdash_e {\sim}C \to {\sim}A$ and $\vdash_e {\sim}C \to {\sim}B$ respectively. By (Ax6), we get $\vdash_e {\sim}C \to ({\sim}A \wedge {\sim}B)$, and therefore $\vdash_e {\sim}C \to {\sim}(A \vee B)$ by (Ax$_{\sim}$5) and (MP). Finally, we obtain the desired result by another application of Corollary 3.12. □

The following lemmas are useful for the completeness proof.

Lemma 3.16 *If* Σ *is prime,* Π*-deductively closed and* $A \notin \Sigma$ *then* ${\sim}A \in \Sigma$.

Proof. If Σ is Π-deductively closed, then by (Ax$_{\sim}$1) we obtain $A \vee {\sim}A \in \Sigma$. This together with $A \notin \Sigma$ and the primeness of Σ implies ${\sim}A \in \Sigma$, as desired.□

Lemma 3.17 *If* Σ *is non-trivial, prime,* Π*-deductively closed and* $A \in \Sigma$ *then* ${\sim}{\sim}A \in \Sigma$.

Proof. Assume the required assumptions and suppose for reductio that ${\sim}A \in \Sigma$. Then this implies $\Sigma \vdash_\Pi {\sim}A$. Moreover, $A \in \Sigma$ implies that $\Sigma \vdash_\Pi A$. Therefore, these together with (${\sim}$-ECQ) imply that $\Sigma \vdash_\Pi B$ for any B, and since Σ is Π-deductively closed, we obtain $B \in \Sigma$ for any B. But this contradicts the assumption that B is non-trivial. Thus the given assumptions imply ${\sim}A \notin \Sigma$, and in view of Lemma 3.16, this implies ${\sim}{\sim}A \in \Sigma$, as desired. □

Lemma 3.18 *If* Σ *is a non-empty prime* Π*-theory and* ${\sim}A \notin \Sigma$ *then* ${\sim}{\sim}A \in \Sigma$.

Proof. Since Σ is non-empty, let B be an element of Σ. By (Ax$_{\sim}$2) we have $\vdash_e B \to (\sim A \vee \sim\sim A)$ and since Σ is a Π-theory, we obtain $\sim A \vee \sim\sim A \in \Sigma$. This together with $\sim A \notin \Sigma$ and the primeness of Σ imply $\sim\sim A \in \Sigma$, as desired. □

Theorem 3.19 *For* $\Gamma \cup \{A\} \subseteq \mathsf{Form}_{\sim}$, *if* $\Gamma \models_e A$ *then* $\Gamma \vdash_e A$.

Proof. The proof is given in the appendix. □

4 Classical negation

In [9], Ryo Kashima investigates some subintuitionistic logics with classical negation, but it is important to note two points of departure from the present work. The first is that Kashima proves only weak completeness, a point that is important in the present context and the reason that our (and Restall's) frames are omniscient. Doing away with omniscience yields a class of theorems equivalent to **SJ**'s, but a different consequence relation, hence a different logic in our sense. Second, Kashima works with sequent calculi, whereas we prefer to work with Hilbert-style axioms system.

Definition 4.1 The language $\mathcal{L}_{\neg}$ consists of a finite set $\{\neg, \wedge, \vee, \to\}$ of propositional connectives and a denumerable set Prop of propositional variables which we denote by p, q, etc. Furthermore, we denote by $\mathsf{Form}_{\neg}$ the set of formulae defined as usual in $\mathcal{L}_{\neg}$. We denote a formula of $\mathcal{L}_{\neg}$ by A, B, C, etc. and a set of formulae of $\mathcal{L}_{\neg}$ by Γ, Δ, Σ, etc.

4.1 Semantics

Definition 4.2 A model for the language $\mathcal{L}_{\neg}$ is a quadruple $\langle W, g, R, V\rangle$, where W is a non-empty set (of states); $g \in W$ (the base state); R is a binary relation on W with g omniscient; and $V : W \times \mathsf{Prop} \to \{0, 1\}$ an assignment of truth values to state-variable pairs. Valuations V are then extended to interpretations I to state-formula pairs by the following conditions:

- $I(w, p) = V(w, p)$
- $I(w, \neg A) = 1$ iff $I(w, A) = 0$
- $I(w, A \wedge B) = 1$ iff $I(w, A) = 1$ and $I(w, B) = 1$
- $I(w, A \vee B) = 1$ iff $I(w, A) = 1$ or $I(w, B) = 1$
- $I(w, A \to B) = 1$ iff for all $x \in W$: if wRx and $I(x, A) = 1$ then $I(x, B) = 1$.

Semantic consequence is defined in terms of truth preservation at g: $\Sigma \models_c A$ iff for all models $\langle W, g, R, I\rangle$, $I(g, A) = 1$ if $I(g, B) = 1$ for all $B \in \Sigma$.

4.2 Proof Theory

Definition 4.3 The system $\mathbf{SJ}^{\neg}$ is obtained by adding the following axiom schemata to **SJ**.

(Ax$_{\neg}$1) $\neg\neg A \to A$ (D-Antilogism) $\dfrac{((A \wedge B) \to \neg C) \vee D}{((A \wedge C) \to \neg B) \vee D}$

Finally, we write $\Gamma \vdash_c A$ if there is a sequence of formulae $B_1, \ldots, B_n, A$, $n \geq 0$, such that every formula in the sequence $B_1, \ldots, B_n, A$ either (i) belongs to Γ; (ii) is an axiom of $\mathbf{SJ}^\neg$; (iii) is obtained by one of the rules (MP)–(D-Antilogism) from formulae preceding it in sequence.

Remark 4.4 As may be verified, the following rule is derivable in $\mathbf{SJ}^\neg$:

(Antilogism) $$\frac{(A \wedge B) \to \neg C}{(A \wedge C) \to \neg B}$$

The above axiomatization is inspired by that for classical relevant logic given by Robert Meyer and Richard Routley in [10, p.57].

4.3 Some Basic Results

We first observe two derivable rules in $\mathbf{SJ}^\neg$.

Proposition 4.5 *The following formula and rules are derivable in* $\mathbf{SJ}^\neg$.

(RC¬1) $\dfrac{A \to \neg B}{B \to \neg A}$ (8) $A \to \neg\neg A$ (RC¬2) $\dfrac{A \to B}{\neg B \to \neg A}$

Proof. For (RC¬1), assume $A \to \neg B$. By this, (Ax5) and (Transitivity), we obtain $(B \wedge A) \to \neg B$ and by (Antilogism), we obtain $(B \wedge B) \to \neg A$. Moreover, $B \to (B \wedge B)$ is provable by (Ax1), (Adj), (Ax6) and (MP). Thus we obtain the desired result by (Transitivity). (8) is provable in view of (Ax1) and (RC¬1). For (RC¬2), assume $A \to B$. Then by (8) and (Transitivity), we obtain $A \to \neg\neg B$. Thus, by (RC¬1), we have $\neg B \to \neg A$, as desired. □

Second, we observe that the complete set of de Morgan laws are provable.

Proposition 4.6 *The following formulae are provable in* $\mathbf{SJ}^\neg$.

(9)	$\neg(A \vee B) \to (\neg A \wedge \neg B)$	(11)	$\neg(A \wedge B) \to (\neg A \vee \neg B)$
(10)	$(\neg A \vee \neg B) \to \neg(A \wedge B)$	(12)	$(\neg A \wedge \neg B) \to \neg(A \vee B)$

Proof. (9) and (10) are easy in view of (RC¬2). For (11), note first that we obtain $\neg(\neg A \vee \neg B) \to (\neg\neg A \wedge \neg\neg B)$ in view of (9), and thus we obtain $\neg(\neg A \vee \neg B) \to (A \wedge B)$ in view of (Ax$_\neg$1). Now, by applying (RC¬2), we obtain $\neg(A \wedge B) \to \neg\neg(\neg A \vee \neg B)$, and this together with (Ax$_\neg$1) and (Transitivity), we obtain the desired result. For (12), note first that we have $A \to (\neg\neg A \vee \neg\neg B)$ and $B \to (\neg\neg A \vee \neg\neg B)$ in view of (8), introduction of disjunction, and (Transitivity). Thus we obtain $(A \vee B) \to (\neg\neg A \vee \neg\neg B)$ by (Adj), (Ax9) and (MP). This together with (10) and (Transitivity) implies $(A \vee B) \to \neg(\neg A \wedge \neg B)$, and finally, by applying (RC¬1), we obtain the desired result. □

Third, we observe that some basic formulae are provable in $\mathbf{SJ}^\neg$.

Proposition 4.7 *The following formulae are provable in* $\mathbf{SJ}^\neg$.

(13)	$(A \wedge \neg A) \to B$	(14)	$B \to (A \vee \neg A)$

Proof. For (13), by (Ax5), (8) and (Transitivity), we obtain $(A \wedge \neg B) \rightarrow \neg\neg A$. By applying (Antilogism), we obtain $(A \wedge \neg A) \rightarrow \neg\neg B$, and thus we obtain the desired result in view of (Ax$_\neg$1) and (Transitivity). For (14), by applying (RC$\neg$2) to $(A \wedge \neg A) \rightarrow \neg B$, an instance of (13), we obtain $\neg\neg B \rightarrow \neg(A \wedge \neg A)$. This together with (8) and (Transitivity) implies $B \rightarrow \neg(A \wedge \neg A)$. Moreover, in view of (11) and (Ax$_\neg$1), we obtain $\neg(A \wedge \neg A) \rightarrow (A \vee \neg A)$. Thus, the desired result follows by (Transitivity). □

Now we turn to prove the deduction theorem with respect to the material conditional defined in terms of classical negation.

Theorem 4.8 (Classical deduction theorem) *For* $\Gamma \cup \{A, B\} \subseteq \mathsf{Form}_\neg$, *if* $\Gamma, A \vdash_c B$ *then* $\Gamma \vdash_c \neg A \vee B$.

Proof. By the induction on the length n of the proof of $\Gamma, A \vdash_c B$. The details are given in the appendix. □

Proposition 4.9 *For* $\Gamma \cup \{A, B\} \subseteq \mathsf{Form}_\neg$, *if* $\Gamma \vdash_c \neg A \vee B$ *then* $\Gamma, A \vdash_c B$.

Proof. It suffices to prove the following:

$$(\neg\text{-DS}) \qquad \frac{A \quad \neg A \vee B}{B}.$$

Assume A and $\neg A \vee B$. Then by (Adj), we obtain $A \wedge (\neg A \vee B)$. Thus by (Ax10) and (MP), we have $(A \wedge \neg A) \vee (A \wedge B)$. Now, this together with (13) and (1) implies $B \vee (A \wedge B)$, and thus by making use of (Ax1), (Ax5), (Adj), (Ax9) and (MP), we obtain B, as desired. □

By combining Theorem 4.8 and Proposition 4.9, we obtain the following.

Theorem 4.10 *For* $\Gamma \cup \{A, B\} \subseteq \mathsf{Form}_\neg$, *we have* $\Gamma, A \vdash_c B$ *iff* $\Gamma \vdash_c \neg A \vee B$.

4.4 Soundness and completeness

We now proceed to the proof of soundness and completeness.

Theorem 4.11 *For* $\Gamma \cup \{A\} \subseteq \mathsf{Form}_\sim$, *if* $\Gamma \vdash_c A$ *then* $\Gamma \models_c A$.

Proof. By induction on the length of the proof, as usual. □

Proposition 4.12 *If* $A \vdash_c C$ *and* $B \vdash_c C$, *then* $A \vee B \vdash_c C$.

Proof. Assume $A \vdash_c C$ and $B \vdash_c C$. Then, by Theorem 3.11, we obtain $\vdash_c \neg A \vee C$ and $\vdash_c \neg B \vee C$ respectively. By (Adj), we get $\vdash_c (\neg A \vee C) \wedge (\neg B \vee C)$, and therefore $\vdash_c (\neg A \wedge \neg B) \vee C$ by (2) and (MP). Moreover, by (12) and (1), we obtain $\vdash_c \neg(A \vee B) \vee C$. Finally, we obtain the desired result by another application of Theorem 4.10. □

We are now in a position to prove completeness.

Theorem 4.13 *For* $\Gamma \cup \{A\} \subseteq \mathsf{Form}_\neg$, *if* $\Gamma \models_c A$ *then* $\Gamma \vdash_c A$.

Proof. The proof is found in the appendix. □

5 Reflections

5.1 Comparing $\mathbf{SJ}^{\sim}$ and $\mathbf{SJ}^{\neg}$

Although the logics share a lot in common, $\mathbf{SJ}^{\sim}$ and $\mathbf{SJ}^{\neg}$ are incomparable: each has a theorem the other does not, assuming that we normalize the language, assigning the same symbol for both negations, say $\neg$. For instance, we have that $B \to (A \vee {\sim}A)$ is provable in $\mathbf{SJ}^{\neg}$ but not in $\mathbf{SJ}^{\sim}$, while $\neg A \to (\neg\neg A \to B)$ is provable in $\mathbf{SJ}^{\sim}$ but not in $\mathbf{SJ}^{\neg}$.

There are further interesting points of comparison, but due to space constraints, we leave discovery of them to the interested reader.

5.2 Extensions of $\mathbf{SJ}^{\sim}$

One natural extension of $\mathbf{SJ}^{\sim}$ is to full intuitionistic logic with empirical negation, $\mathbf{IPC}^{\sim}$, a system axiomatized and shown strongly complete in [4]. Given the strength of the intuitionistic conditional, the axiomatization of $\mathbf{IPC}^{\sim}$ is much smoother than that of $\mathbf{SJ}^{\sim}$, requiring far fewer rules and a simpler and more familiar axiomatization. We need only add to **IPC** two axioms and one rule governing $\sim$: axioms (Ax$_{\sim}$1) (Law of Excluded Middle) and (Ax$_{\sim}$3) (*ex contradictione quodlibet* for empirically negated formulae) of $\mathbf{SJ}^{\sim}$, and the rule stating that from $A \vee B$, one may infer ${\sim}A \to B$ (called (RP) in [4]).

The most straightforward way to obtain $\mathbf{IPC}^{\sim}$ from $\mathbf{SJ}^{\sim}$ is by adding the axioms corresponding to Heredity and the reflexivity and transitivity of accessibilty, namely,

- $A \to (B \to A)$ (Heredity),
- $(A \to B) \to ((B \to C) \to (A \to C))$ (transitivity),
- $(A \wedge (A \to B)) \to B$ (reflexivity)

The resulting axiomatization is, however, redundant, as a number of the axioms and rules can be shown derivable from a proper subset of the others. [9]

5.3 Extensions of $\mathbf{SJ}^{\neg}$

In [5], Luis Fariñas del Cerro and Andreas Herzig provide an interesting way of adding classical negation to inuititionistic logic without a collapse to classical logic. In this section we relate their results to our own by showing, semantically, that their logic $C + J$ is an extension of the subintuitionistic logic with classical negation $\mathbf{SJ}^{\neg}$. We also strengthen their weak soundness and completeness results to their strong counterparts.

Definition 5.1 A ***CJ****-model* for the language $\mathcal{L}_{\neg}$ is a quadruple $\langle W, g, R, V\rangle$, where W is a non-empty set; $g \in W$ is omniscient; R is a reflexive and transitive relation on W; and $V : W \times \mathsf{Prop} \to \{0, 1\}$ an assignment of truth values to state-variable pairs satisfying Heredity. Valuations V are extended to interpretations I to state-formula pairs in the same way as in Definition 4.2.

[9] Note that (Ax10) in $\mathbf{IPC}^{\sim}$ can be swapped with (Ax$_{\sim}$3) of $\mathbf{SJ}^{\sim}$, though this is not the case for $\mathbf{SJ}^{\sim}$ where the order of antecedents matters.

Semantic consequence is defined in terms of truth preservation at g: $\Sigma \models_{\mathbf{CJ}} A$ iff for all models $\langle W, g, R, I\rangle$, $I(g, A) = 1$ if $I(g, B) = 1$ for all $B \in \Sigma$.

Remark 5.2 First, del Cerro and Herzig did not employ the above semantic consequence relation, but instead introduced the notion of validity defined as follows: $\models_{C+J} A$ iff for all models $\langle W, g, R, I\rangle$, $I(w, A) = 1$ for all $w \in W$. Second, while Heredity is ensured for atomic formulae, given the presence of classical negation $\neg$, it fails for arbitrary formulae, in particular, only for formulae containing classical negation.

The following essential notion is employed by del Cerro and Herzig.

Definition 5.3 A formula $A \in \mathsf{Form}_{\neg}$ is *persistent* iff $A \in \mathsf{Prop}$ or A is of the form $B \to C$ for some $B, C \in \mathsf{Form}_{\neg}$.

By making use of this notion, we introduce an extension of $\mathbf{SJ}^{\neg}$.

Definition 5.4 Let **CJ** be the extension of $\mathbf{SJ}^{\neg}$ obtained by adding the following axioms.

($\mathrm{Ax}_{\mathbf{CJ}}1$)	$A \to (B \to A)$ where A is persistent
($\mathrm{Ax}_{\mathbf{CJ}}2$)	$(A \wedge (A \to B)) \to B$
($\mathrm{Ax}_{\mathbf{CJ}}3$)	$(A \to B) \to ((B \to C) \to (A \to C))$

We denote a derivation in **CJ** of A from Γ by $\Gamma \vdash_{\mathbf{CJ}} A$.

Then, we have the following soundness and strong completeness results.

Theorem 5.5 *For* $\Gamma \cup \{A\} \subseteq \mathsf{Form}_{\neg}$, $\Gamma \models_{\mathbf{CJ}} A$ *iff* $\Gamma \vdash_{\mathbf{CJ}} A$.

Proof. Ths proof is found in the appendix. □

This leads us to the following identity between $C+J$-validity and **CJ**-validity.

Theorem 5.6 *For* $A \in \mathsf{Form}_{\neg}$, $\models_{C+J} A$ *iff* $\models_{\mathbf{CJ}} A$.

Proof. For the left-to-right direction, it is obvious in view of the definitions of $\models_{\mathbf{CJ}}$ and $\models_{C+J}$. For the other direction, it suffices to prove that if $\vdash_{\mathbf{CJ}} A$ then $\models_{C+J} A$ in light of the previous completeness theorem. This is easily shown by checking that axioms are true at an arbitrary state in the model and that rules preserves validity. We only note that the transitivity of R and Heredity for persistent formulae are used in showing that (DR) preserves validity, and the reflexivity of R is used in showing that (MP) and (DMP) preserve validity. □

6 Final remarks

In neither expansion of **SJ** is the intuitionistic conditional definable from the other connectives. However, if we add the primitive subintuitionistic negation operator $\neg\!\!\!\to$ to either, we can then define the intuitionistic conditional by $A \to B := \neg\!\!\!\to(A \wedge \otimes B)$, for $\otimes \in \{\neg, \sim\}$. Subintuitionistic logic is therefore recoverable as long as we have both negations around (recalling that the falsum is definable in the conditional-free fragments of both $\mathbf{SJ}^{\sim}$ and $\mathbf{SJ}^{\neg}$, in the former by $\sim A \wedge \sim\sim A$ and in the latter by $A \wedge \neg A$). Thus adding classical negation to the conditional-free fragment of intuitionistic logic gives a properly stronger expansion than adding the conditional. Relatedly, adding either subintuitionistic negation or the conditional to classical logic yields the same expansion of classical logic.

We wish to briefly mention some avenues for future research. One is to investigate expansions of relevant logic by empirical negation. Classical relevant logic, recall, is the expansion of the positive fragment $\mathbf{R}^+$ of $\mathbf{R}$ by classical negation, obtained by adding to $\mathbf{R}^+$ the following axioms:

- $\neg\neg A \to A$ (double negation);
- $(A \wedge B) \to \neg C \models (A \wedge C) \to \neg B$ (antilogism).

Both of these classical principles are valid in $\mathbf{SJ}^{\neg}$, but the second fails in $\mathbf{SJ}^{\sim}$, marking an important difference between classical and empirical negation. This failure to validate antilogism makes empirical negation a better fit for relevance logic while still offering a negation of a highly classical nature. The results found here concerning $\mathbf{SJ}^{\sim}$ should carry over to its relevant cousin $\mathbf{B}^+$ expanded by empirical negation, something we hope to explore in future work.

The second avenue for future research concerns the investigation of first-order subintuitionistic logics with classical and empirical negation. Some work in this direction can be found in Ryo Ishigaki and Kentaro Kikuchi's [8].[10] They prove the soundness and weak completeness for Hilbert-style axiom systems by making use of tree-sequent calculi, and the same strategy should be applicable to expansions with classical and empirical negation.

Appendix

Proof of Theorem 2.16

We prove the contrapositive. Suppose that $\Gamma \not\vdash A$. By Corollary 2.13, there is a $\Pi \supseteq \Gamma$ such that Π is a prime Π-theory and $A \notin \Pi$. Define the model $\mathfrak{A} = \langle X, \Pi, R, I \rangle$, where $X = \{\Delta : \Delta \text{ is a non-empty non-trivial prime } \Pi\text{-theory}\}$, R defined as in Definition 2.9(v) and I is defined thus. For every state Σ and propositional parameter p:

$$I(\Sigma, p) = 1 \text{ iff } p \in \Sigma$$

We show that this condition holds for an arbitrary formula B:

$$(*) \qquad I(\Sigma, B) = 1 \text{ iff } B \in \Sigma$$

It then follows by a routine induction on the complexity of B that $\mathfrak{A}$ is a counter-model for the inference, and hence that $\Gamma \not\models A$. We show only the interesting case when B has the form $C \to D$.

We have that $I(\Sigma, C \to D) = 1$ iff for all Δ s.t. $\Sigma R \Delta$, if $I(\Delta, C) = 1$ then $I(\Delta, D) = 1$; iff for all Δ s.t. $\Sigma R \Delta$, if $C \in \Delta$ then $D \in \Delta$ by IH; iff $C \to D \in \Sigma$.

[10] See also [16].

For the last equivalence, assume $C \to D \in \Sigma$ and $C \in \Delta$ for any Δ such that $\Sigma R \Delta$. Then by the definition of $\Sigma R \Delta$, we obtain $D \in \Delta$, as desired. On the other hand, suppose $C \to D \notin \Sigma$. Then by Lemma 2.15, there is a Δ such that $\Sigma R \Delta$, $C \in \Delta$, $D \notin \Delta$ and Δ is a prime Π-theory. Furthermore, the non-triviality of Δ follows from the fact that $D \notin \Delta$. Thus, we obtain the desired result.

Proof of Theorem 3.7

By the induction on the length n of the proof of $\Gamma, A \vdash_e B$. If $n = 1$, then we have the following three cases.

- If B is one of the axioms of $\mathbf{SJ}^{\sim}$, then we have $\vdash_e B$. Therefore, by (Ax8), we obtain $\vdash_e {\sim}A \vee B$ which implies the desired result.
- If $B \in \Gamma$, then we have $\Gamma \vdash_e B$, and thus we obtain the desired result by (Ax8).
- If $B = A$, then by (Ax$_{\sim}$1), we have ${\sim}A \vee B$ which implies the desired result.

For $n > 1$, then there are five additional cases to be considered.

- If B is obtained by applying (MP), then we will have $\Gamma, A \vdash_e C$ and $\Gamma, A \vdash_e C \to B$ lengths of the proof of which are less than n. Thus, by induction hypothesis, we have $\Gamma \vdash_e {\sim}A \vee C$ and $\Gamma \vdash_e {\sim}A \vee (C \to B)$, and by (DMP), we obtain $\Gamma \vdash_e {\sim}A \vee B$ as desired.
- If B is obtained by applying (DMP), then $B = D \vee E$ and we will have $\Gamma, A \vdash_e C \vee E$ and $\Gamma, A \vdash_e (C \to D) \vee E$ lengths of the proof of which are less than n. Thus, by induction hypothesis, we have $\Gamma \vdash_e {\sim}A \vee C \vee E$ and $\Gamma \vdash_e {\sim}A \vee (C \to D) \vee E$, and by (DMP), we obtain $\Gamma \vdash_e {\sim}A \vee D \vee E$ as desired.
- If B is obtained by applying (Adj), then $B = C \wedge D$ and we will have $\Gamma, A \vdash_e C$ and $\Gamma, A \vdash_e D$ lengths of the proof of which are less than n. Thus, by induction hypothesis, we have $\Gamma \vdash_e {\sim}A \vee C$ and $\Gamma \vdash_e {\sim}A \vee D$, and by (Adj), (2) and (MP), we obtain $\Gamma \vdash_e {\sim}A \vee (C \wedge D)$ as desired.
- If B is obtained by applying (DR), then $B = ((D \to E) \to (C \to F)) \vee G$ and we will have $\Gamma, A \vdash_e (C \to D) \vee G$ and $\Gamma, A \vdash_e (E \to F) \vee G$ lengths of the proof of which are less than n. Thus, by induction hypothesis, we have $\Gamma \vdash_e {\sim}A \vee (C \to D) \vee G$ and $\Gamma \vdash_e {\sim}A \vee (E \to F) \vee G$, and by (DR), we obtain $\Gamma \vdash_e {\sim}A \vee B$ as desired.
- If B is obtained by applying (DRP), then $B = ({\sim}C \to {\sim}{\sim}D) \vee E$ and we will have $\Gamma, A \vdash_e C \vee D \vee E$ length of the proof of which is less than n. Thus, by induction hypothesis, we have $\Gamma \vdash_e {\sim}A \vee (C \vee D) \vee E$. By (DRP), we obtain $\Gamma \vdash_e {\sim}A \vee ({\sim}C \to {\sim}{\sim}D) \vee E$, i.e. $\Gamma \vdash_e {\sim}A \vee B$ as desired.

This completes the proof.

Proof of Theorem 3.19

We prove the contrapositive. Suppose that $\Gamma \not\vdash_e A$. By Corollary 2.13, there is a $\Pi \supseteq \Gamma$ such that Π is a prime Π-theory and $A \notin \Pi$. Define the model $\mathfrak{A} = \langle X, \Pi, R, I\rangle$, where $X = \{\Delta : \Delta$ is a non-empty non-trivial prime Π-theory$\}$, R defined as in Definition 2.9(v) and I is defined thus. For every state Σ and propositional parameter p:

$$I(\Sigma, p) = 1 \text{ iff } p \in \Sigma.$$

We show that this condition holds for an arbitrary formula B:

$$(*) \qquad I(\Sigma, B) = 1 \text{ iff } B \in \Sigma.$$

It then follows that $\mathfrak{A}$ is a counter-model for the inference, and hence that $\Gamma \not\models_e A$. The proof of $(*)$ is by induction on the complexity of B. We show only the case for empirical negation, $\sim$.

We have that $I(\Sigma, {\sim}C) = 1$ iff $I(\Pi, C) \neq 1$; iff $C \notin \Pi$ by IH; iff ${\sim}C \in \Sigma$. For the last equivalence, suppose $C \notin \Pi$ and ${\sim}C \notin \Sigma$ for reductio. Then, by Lemmas 3.16 and 3.18, we obtain ${\sim}C \in \Pi$ and ${\sim\sim}C \in \Sigma$. The former and (7) implies that ${\sim\sim}C \to D \in \Pi$ since Π is a Π-theory. Moreover, since Σ is also a Π-theory and $\vdash_\Pi {\sim\sim}C \to D$ (since ${\sim\sim}C \to D \in \Pi_\to$), we obtain $D \in \Sigma$ in view of ${\sim\sim}C \in \Sigma$. But D is arbitrary, and this contradicts that Σ is non-trivial. For the other way around, suppose ${\sim}C \in \Sigma$ and $C \in \Pi$ for reductio. Then, by the latter and Lemma 3.17 we obtain ${\sim\sim}C \in \Pi$. The rest of the proof is similar to the proof for the other direction, but we use (Ax$_\sim$3) instead of (7).

Proof of Theorem 4.8

By the induction on the length n of the proof of $\Gamma, A \vdash_c B$. If $n = 1$, then we have the following three cases.

- If B is one of the axioms of $\mathbf{SJ}^\neg$, then we have $\vdash_c B$. Therefore, by (Ax8), we obtain $\vdash_c \neg A \vee B$ which implies the desired result.
- If $B \in \Gamma$, then we have $\Gamma \vdash_c B$, and thus we obtain the desired result by (Ax8).
- If $B = A$, then by (14), we have $\neg A \vee B$ which implies the desired result.

For $n > 1$, then there are five additional cases to be considered.

- If B is obtained by applying (MP), then we will have $\Gamma, A \vdash_c C$ and $\Gamma, A \vdash_c C \to B$ lengths of the proof of which are less than n. Thus, by induction hypothesis, we have $\Gamma \vdash_c \neg A \vee C$ and $\Gamma \vdash_c \neg A \vee (C \to B)$, and by (DMP), we obtain $\Gamma \vdash_c \neg A \vee B$ as desired.
- If B is obtained by applying (DMP), then $B = D \vee E$ and we will have $\Gamma, A \vdash_c C \vee E$ and $\Gamma, A \vdash_c (C \to D) \vee E$ lengths of the proof of which are less than n. Thus, by induction hypothesis, we have $\Gamma \vdash_c \neg A \vee C \vee E$ and $\Gamma \vdash_c \neg A \vee (C \to D) \vee E$, and by (DMP), we obtain $\Gamma \vdash_c \neg A \vee D \vee E$ as desired.

- If B is obtained by applying (Adj), then $B = C \wedge D$ and we will have $\Gamma, A \vdash_c C$ and $\Gamma, A \vdash_c D$ lengths of the proof of which are less than n. Thus, by induction hypothesis, we have $\Gamma \vdash_c \neg A \vee C$ and $\Gamma \vdash_c \neg A \vee D$, and by (Adj), (2) and (MP), we obtain $\Gamma \vdash_c \neg A \vee (C \wedge D)$ as desired.
- If B is obtained by applying (DR), then $B = ((D \to E) \to (C \to F)) \vee G$ and we will have $\Gamma, A \vdash_c (C \to D) \vee G$ and $\Gamma, A \vdash_c (E \to F) \vee G$ lengths of the proof of which are less than n. Thus, by induction hypothesis, we have $\Gamma \vdash_c \neg A \vee (C \to D) \vee G$ and $\Gamma \vdash_c \neg A \vee (E \to F) \vee G$, and by (DR), we obtain $\Gamma \vdash_c \neg A \vee B$ as desired.
- If B is obtained by applying (D-Antilogism), then $B = ((C \wedge E) \to \neg D) \vee F$ and we will have $\Gamma, A \vdash_c ((C \wedge D) \to \neg E) \vee F$ length of the proof of which is less than n. Thus, by induction hypothesis, we have $\Gamma \vdash_c \neg A \vee ((C \wedge D) \to \neg E) \vee F$. By (D-Antilogism), we obtain $\Gamma \vdash_c \neg A \vee ((C \wedge E) \to \neg D) \vee F$, i.e. $\Gamma \vdash_c \neg A \vee B$ as desired.

This completes the proof.

Proof of Theorem 4.13

We prove the contrapositive. Suppose that $\Gamma \not\vdash_c A$. By Corollary 2.13, there is a $\Pi \supseteq \Gamma$ such that Π is a prime Π-theory and $A \notin \Pi$. Define the interpretation $\mathfrak{A} = \langle X, \Pi, R, I \rangle$, where $X = \{\Delta : \Delta \text{ is a non-trivial prime } \Pi\text{-theory}\}$, R defined as in Definition 2.9(v) and I is defined thus. For every state Σ and propositional parameter p:

$$I(\Sigma, p) = 1 \text{ iff } p \in \Sigma.$$

We show that this condition holds for an arbitrary formula B:

$$(*) \qquad I(\Sigma, B) = 1 \text{ iff } B \in \Sigma.$$

It then follows that $\mathfrak{A}$ is a counter-model for the inference, and hence that $\Gamma \not\models_c A$. The proof of $(*)$ is by induction on the complexity of B. We deal only with classical negation.

We have that $I(\Sigma, \neg C) = 1$ iff $I(\Sigma, C) \neq 1$; iff $C \notin \Sigma$ by IH; iff $\neg C \in \Sigma$. For the last equivalence, assume $C \notin \Sigma$. Since Σ is non-empty, let D be an element of Σ. By (14) we have $\vdash_c D \to (C \vee \neg C)$ and since Σ is a Π-theory, we obtain $C \vee \neg C \in \Sigma$. This together with $C \notin \Sigma$ and the primeness of Σ imply $\neg C \in \Sigma$, as desired. For the other way around, suppose $\neg C \in \Sigma$ and $C \in \Sigma$ for reductio. Then, we obtain $C \wedge \neg C \in \Sigma$ since Σ is a Π-theory. And this together with (13) implies $D \in \Sigma$ for any D. But this contradicts that Σ is non-trivial.

Proof of Theorem 5.5

For soundness, we need to check that the additional axioms are valid. Since ($\mathrm{Ax}_{\mathbf{CJ}}2$) and ($\mathrm{Ax}_{\mathbf{CJ}}3$) are handled by Restall, we check ($\mathrm{Ax}_{\mathbf{CJ}}1$). We split the cases depending on the form of A. For the case when A is a propositional variable, assume, for reductio, $I(g, A \to (B \to A)) \neq 1$. Then, for some

$w_0 \in W$, we have $I(w_0, A) = 1$ and $I(w_0, B \to A) \neq 1$. The latter is equivalent to $I(w_1, B) = 1$ and $I(w_1, A) \neq 1$ for some $w_1 \in W$ such that $w_0 R w_1$. But by Heredity, we also have $I(w_1, A) = 1$ which implies a contradiction. Therefore, we obtain $I(g, A \to (B \to A)) = 1$. For the case when A is of the form $C \to D$, we make use of the transitivity of R. Details are left to the reader.

For completeness, we need to check that the binary relation is reflexive and transitive, and also the persistence condition holds for $p \in \mathsf{Prop}$. Again, we only check the last condition since the other two are shown in [14]. Assume that $p \in \Sigma$ and $\Sigma R \Delta$. Then, since Δ is non-empty, there is an element D such that $D \in \Delta$. And in view of ($\mathrm{Ax}_{\mathbf{CJ}}1$), we have $\vdash_{\Pi} p \to (D \to p)$, and since Σ is a Π-theory, we obtain $D \to p \in \Sigma$. Finally, by the definition of R and $D \in \Delta$, we obtain $p \in \Delta$, as desired.

References

[1] Celani, S. and R. Jansana, *A closer look at some subintuitionistic logics*, Notre Dame Journal of Formal Logic **42** (2001), pp. 225–255.

[2] Corsi, G., *Weak logics with strict implication*, Mathematical Logic Quarterly **33** (1987), pp. 398–406.

[3] De, M., *Empirical Negation*, Acta Analytica **28** (2013), pp. 49–69.

[4] De, M. and H. Omori, *More on empirical negation*, in: R. Goré, B. Kooi and A. Kurucz, editors, *Advances in Modal Logic* (2014), pp. 114–133.

[5] del Cerro, L. F. and A. Herzig, *Combining classical and intuitionistic logic*, in: F. Baader and K. Schulz, editors, *Frontiers of Combining Systems* (1996), pp. 93–102.

[6] Došen, K., *Modal translations in K and D*, in: M. de Rijke, editor, *Diamonds and Defaults*, Kluwer Academic Publishers, 1993 pp. 103–127.

[7] Humberstone, L., *Interval semantics for tense logic: some remarks*, Journal of Philosophical Logic **8** (1979), pp. 171–196.

[8] Ishigaki, R. and K. Kikuchi, *Tree-sequent methods for subintuitionistic predicate logics*, in: *Automated Reasoning with Analytic Tableaux and Related Methods*, Lecture Notes in Computer Science **4548** (2007), pp. 149–164.

[9] Kashima, R., *Sequent calculi of non-classical logics – Proofs of completeness theorems by sequent calculi (in Japanese)*, in: *Proceedings of Mathematical Society of Japan Annual Colloquium of Foundations of Mathematics*, 1999, pp. 49–67.

[10] Meyer, R. K. and R. Routley, *Classical Relevant Logics I*, Studia Logica **32** (1973), pp. 51–66.

[11] Meyer, R. K. and R. Routley, *Classical Relevant Logics II*, Studia Logica **33** (1974), pp. 183–194.

[12] Priest, G. and R. Sylvan, *Simplified semantics for basic relevant logic*, Journal of Philosophical Logic **21** (1992), pp. 217–232.

[13] Restall, G., *Simplified semantics for relevant logics (and some of their rivals)*, Journal of Philosophical Logic **22** (1993), pp. 481–511.

[14] Restall, G., *Subintuitionistic logics*, Notre Dame Journal of Formal Logic **35** (1994), pp. 116–126.

[15] Wansing, H., "Displaying Modal Logic," Kluwer Academic Publishers, Dordrecht, 1998.

[16] Zimmermann, E., *Predicate logical extensions of some subintuitionistic logics*, Studia Logica **91** (2009), pp. 131–138.

Axiomatizing a Real-Valued Modal Logic

Denisa Diaconescu [1]

Faculty of Mathematics and Computer Science
University of Bucharest, Romania
ddiaconescu@fmi.unibuc.ro

George Metcalfe Laura Schnüriger [2]

Mathematical Institute
University of Bern, Switzerland
{george.metcalfe,laura.schnueriger}@math.unibe.ch

Abstract

A many-valued modal logic is introduced that combines the standard (crisp) Kripke frame semantics of the modal logic K with connectives interpreted locally as abelian group operations over the real numbers. A labelled tableau system and a sequent calculus admitting cut elimination are then defined for this logic and used to establish completeness of an axiomatic extension of the multiplicative fragment of abelian logic.

Keywords: Modal Logics, Many-Valued Logics, Abelian Logic.

1 Introduction

Many-valued modal logics model modal notions such as necessity, belief, and spatio-temporal relations in the presence of multiple degrees of truth, certainty, or possibility. They are defined by extending the Kripke frames of classical modal logic with a many-valued semantics at each world, and have been used to model fuzzy belief [16,12], fuzzy similarity measures [13], many-valued tense logics [17,9], and spatial reasoning with vague predicates [28]. Such logics also provide the basis for fuzzy description logics, which, analogously to the classical case, can be viewed as many-valued multi-modal logics (see, e.g., [29,15,1]). General approaches to finite-valued modal logics are described in [10,11,3,27], while infinite-valued modal logics with propositional operations depending only on a given total order – in particular, Gödel modal logics – are investigated in [6,24,7,5,4].

Many-valued modal logics of "magnitude" typically involve reasoning about some form of addition over sets of real numbers, archetypal examples being

[1] Supported by Sciex grant 13.192 and Romanian National Authority for Scientific Research and Innovation grant, CNCS-UEFISCDI, project number PN-II-RU-TE-2014-4-0730.

[2] Supported by Swiss National Science Foundation grant 200021_146748.

Łukasiewicz modal logics, where propositional connectives are interpreted by continuous functions over the real unit interval [14,3,18,22] (see also [21,20,23] for related real-valued modal logics). Finite-valued (crisp) Łukasiewicz modal logics are axiomatized in [18], but the axiom system defined for the infinite-valued (crisp) Łukasiewicz modal logic includes a rule that has infinitely many premises. This matters because, although it is easy enough to define a many-valued modal logic semantically (simply decide on suitable sets of values and operations), studying such a logic when it lacks a finitary axiom system or algebraic semantics may be difficult; consider, for example, classical modal logic deprived of the theory of Boolean algebras with operators. Note also that, while validity in finite-valued Łukasiewicz modal logics is known to be PSPACE-complete [2], only a NEXPTIME upper bound is known for the infinite-valued case, as may be deduced from complexity results for Łukasiewicz description logics obtained in [20].

In this paper, we take a first step towards addressing these issues by defining and investigating a simple many-valued modal logic of magnitude $\mathrm{K}(\mathbb{R})$ with propositional connectives interpreted as the usual group operations over the real numbers. The next step would then be to interpret infinite-valued Łukasiewicz modal logic in an extension of $\mathrm{K}(\mathbb{R})$ with lattice connectives. The logic $\mathrm{K}(\mathbb{R})$ may be viewed as a minimal modal extension of the multiplicative fragment of abelian logic studied in [26,8,25]. We provide here a sound and complete axiom system for $\mathrm{K}(\mathbb{R})$, making use of both a labelled tableau system and a sequent calculus admitting cut elimination to establish the more difficult completeness result. We also obtain an EXPTIME upper bound for validity.

2 A Real-Valued Modal Logic

Let us fix Fm as the set of *formulas*, denoted by φ, ψ, χ, defined inductively for a language with a binary connective $\to$ and a modal connective $\Box$ over a countably infinite set Var of propositional variables, denoted by p, q. The *complexity* of a formula φ is defined as the number of occurrences of connectives in φ, and the *modal depth* of φ is defined as the deepest nesting of the modal connective $\Box$ in φ. Fixing some $p_0 \in \mathrm{Var}$, we define additional connectives

$$\overline{0} := p_0 \to p_0, \quad \neg\varphi := \varphi \to \overline{0}, \quad \varphi \& \psi := \neg\varphi \to \psi, \quad \text{and} \quad \Diamond\varphi := \neg\Box\neg\varphi.$$

We also define $0\varphi := \overline{0}$ and $(n+1)\varphi := \varphi\&(n\varphi)$ for $n \in \mathbb{N}$.

Let us remark that these (perhaps counter-intuitively) defined connectives arise as a natural feature of the multiplicative fragment of abelian logic [26,8,25] – an axiomatic extension of multiplicative linear logic that is complete with respect to the class of abelian groups with $x \to y$ interpreted as $y - x$, where a formula is valid if it is non-negative. Since the multiplicative conjunction and disjunction, and also the multiplicative constants, coincide in this logic, we can restrict to a language with just implication and define $\overline{0} := p_0 \to p_0$, where the "0" anticipates the interpretation in $\mathbb{R}$. Negation and multiplicative conjunction (equivalently, disjunction) connectives are then defined as usual.

A *frame* is a pair $\mathfrak{F} = \langle W, R \rangle$, where W is a non-empty set of *worlds* and $R \subseteq W \times W$ is an *accessibility relation.* $\mathfrak{F}$ is called *serial* if for all $x \in W$, there exists $y \in W$ such that Rxy. A K($\mathbb{R}$)-*model* $\mathfrak{M} = \langle W, R, V \rangle$ consists of a serial frame $\langle W, R \rangle$ and a map $V\colon \mathrm{Var} \times W \to [-r, r]$ for some $r \in \mathbb{R}^+$, called a *valuation.* This valuation is extended to $V\colon \mathrm{Fm} \times W \to \mathbb{R}$ by

$$V(\varphi \to \psi, x) = V(\psi, x) - V(\varphi, x)$$
$$V(\Box\varphi, x) = \bigwedge\{V(\varphi, y) : Rxy\}.$$

It follows also that

$$V(\overline{0}, x) = 0 \qquad V(\varphi \& \psi, x) = V(\varphi, x) + V(\psi, x)$$
$$V(\neg\varphi, x) = -V(\varphi, x) \qquad V(\Diamond\varphi, x) = \bigvee\{V(\varphi, y) : Rxy\}.$$

$\varphi \in \mathrm{Fm}$ is *valid* in a K($\mathbb{R}$)-model $\mathfrak{M} = \langle W, R, V \rangle$ if $V(\varphi, x) \geq 0$ for all $x \in W$. If φ is valid in all K($\mathbb{R}$)-models, then φ is K($\mathbb{R}$)-*valid*, written $\models_{\mathrm{K}(\mathbb{R})} \varphi$.

The restriction to serial frames is more or less imposed for this semantics by the fact that $\bigwedge \emptyset$ and $\bigvee \emptyset$ do not exist for $\mathbb{R}$. Note also that the seriality axiom $\Box\varphi \to \Diamond\varphi$ is derivable in any extension of the multiplicative fragment of abelian logic with the standard axiom $\Box(\varphi \to \psi) \to (\Box\varphi \to \Box\psi)$. Similarly, restricting the codomain of a valuation to a bounded subset of $\mathbb{R}$ circumvents problems with infima or suprema of unbounded sets of values and is justified to some extent by the following finite model property.

Lemma 2.1 $\models_{\mathrm{K}(\mathbb{R})} \varphi$ *if and only if* φ *is valid in all finite* K($\mathbb{R}$)*-models.*

Proof. It suffices to prove the following: for any K($\mathbb{R}$)-model $\mathfrak{M} = \langle W, R, V \rangle$, $x \in W$, finite set of formulas S, and $\varepsilon > 0$, there exists a finite K($\mathbb{R}$)-model $\mathfrak{M}' = \langle W', R', V' \rangle$ with $x \in W'$ such that $|V(\varphi, x) - V'(\varphi, x)| < \varepsilon$ for all $\varphi \in S$. We proceed by induction on the sum of the complexities of the formulas in S.

For the base case, S contains only variables and we let $\mathfrak{M}' = \langle W', R', V' \rangle$ with $W' = \{x\}$, $R' = \{(x, x)\}$, and $V'(p, x) = V(p, x)$ for each $p \in \mathrm{Var}$. For the inductive step, suppose first that $S = S' \cup \{\psi \to \chi\}$. Then we can apply the induction hypothesis with $\mathfrak{M}$, $x \in W$, $S'' = S' \cup \{\psi, \chi\}$, and $\frac{\varepsilon}{2} > 0$ to obtain a finite K($\mathbb{R}$)-model $\mathfrak{M}' = \langle W', R', V' \rangle$ with $x \in W'$ such that $|V(\varphi, x) - V'(\varphi, x)| < \frac{\varepsilon}{2}$ for all $\varphi \in S''$. It suffices then to observe that $|V(\psi \to \chi, x) - V'(\psi \to \chi, x)| = |V(\chi, x) - V(\psi, x) - V'(\chi, x) + V'(\psi, x)| \leq |V(\chi, x) - V'(\chi, x)| + |V(\psi, x) - V'(\psi, x)| < \frac{\varepsilon}{2} + \frac{\varepsilon}{2} = \varepsilon$.

Now suppose that S consists of variables and boxed formulas $\Box\psi_1, \ldots, \Box\psi_n$ ($n \geq 1$). Then for $1 \leq i \leq n$, there exists $y_i \in W$ such that Rxy_i and $|V(\Box\psi_i, x) - V(\psi_i, y_i)| < \frac{\varepsilon}{2}$. We apply the induction hypothesis to each submodel $\mathfrak{M}_i$ of $\mathfrak{M}$ generated by y_i (i.e., the restriction of $\mathfrak{M}$ to the smallest subset of W containing y_i and closed under R) with $S' = (S \setminus \{\Box\psi_1, \ldots, \Box\psi_n\}) \cup \{\psi_1, \ldots, \psi_n\}$, $y_i \in W_i$, and $\frac{\varepsilon}{2} > 0$ to obtain a finite K($\mathbb{R}$)-model $\mathfrak{M}'_i = \langle W'_i, R'_i, V'_i \rangle$ and $y_i \in W'_i$ such that $|V(\varphi, y_i) - V'(\varphi, y_i)| < \frac{\varepsilon}{2}$ for all $\varphi \in S'$. By renaming worlds, we may assume that these models are disjoint and do not include x. Now let $\mathfrak{M}' = \langle W', R', V' \rangle$ be the finite K($\mathbb{R}$)-model

$$
\begin{array}{ll}
\text{(B)} & (\varphi \to \psi) \to ((\psi \to \chi) \to (\varphi \to \chi)) \\
\text{(C)} & (\varphi \to (\psi \to \chi)) \to (\psi \to (\varphi \to \chi)) \\
\text{(I)} & \varphi \to \varphi \\
\text{(A)} & ((\varphi \to \psi) \to \psi) \to \varphi \\
\text{(K)} & \Box(\varphi \to \psi) \to (\Box\varphi \to \Box\psi) \\
(\text{D}_n) & \Box(n\varphi) \to n\Box\varphi \qquad (n \geq 2)
\end{array}
$$

$$
\frac{\varphi \quad \varphi \to \psi}{\psi}\ (\text{mp}) \qquad \frac{\varphi}{\Box\varphi}\ (\text{nec}) \qquad \frac{n\varphi}{\varphi}\ (\text{con}_n) \quad (n \geq 2)
$$

Fig. 1. The axiom system K(ℝ)

with $W' = \{x\} \cup W'_1 \cup \ldots \cup W'_n$ such that for $u, v \in W'$,

$$
R'uv = \begin{cases} R'_i uv & \text{if } u, v \in W'_i \\ 1 & \text{if } u = x,\ v \in \{y_1, \ldots, y_n\} \\ 0 & \text{otherwise,} \end{cases} \qquad V'(p, u) = \begin{cases} V'_i(p, u) & \text{if } u \in W'_i \\ V(p, x) & \text{if } u = x. \end{cases}
$$

Clearly $V'(p, x) = V(p, x)$ for each propositional variable $p \in S$. For $1 \leq i \leq n$, recall that $|V(\Box\psi_i, x) - V(\psi_i, y_i)| < \frac{\varepsilon}{2}$ and also $|V(\psi_i, y_j) - V'(\psi_i, y_j)| < \frac{\varepsilon}{2}$ for $1 \leq j \leq n$, so $|V(\Box\psi_i, x) - V'(\Box\psi_i, x)| < \varepsilon$. □

The main goal of this paper will be to prove the following soundness and completeness theorem for the axiom system K(ℝ) presented in Fig. 1.

Theorem 2.2 *For any* $\varphi \in \text{Fm}$, $\vdash_{\text{K}(\mathbb{R})} \varphi$ *if and only if* $\models_{\text{K}(\mathbb{R})} \varphi$.

Soundness (the left-to-right direction) is straightforward. It is easily checked that the axioms (B), (C), (I), (A), and (K) are valid in all K(ℝ)-models. For the less standard axioms (D_n) ($n \geq 2$), it suffices to consider a K(ℝ)-model $\mathfrak{M} = \langle W, R, V \rangle$ and $x \in W$, and to observe that

$$
\begin{aligned}
V(\Box(n\varphi), x) &= \bigwedge\{V(n\varphi, y) : Rxy\} \\
&= \bigwedge\{nV(\varphi, y) : Rxy\} \\
&= n\bigwedge\{V(\varphi, y) : Rxy\} \\
&= V(n\Box\varphi, x).
\end{aligned}
$$

Clearly, (mp) and (nec) preserve validity in K(ℝ)-models. For (con_n), note that if $V(n\varphi, x) \geq 0$ for a K(ℝ)-model $\mathfrak{M} = \langle W, R, V \rangle$ and $x \in W$, also $V(\varphi, x) \geq 0$.

Proving completeness (the right-to-left direction) will be our main aim in the remainder of this paper. First, in Section 3, we define a labelled tableau system that is sound and complete with respect to the Kripke semantics. In Section 4, we then provide a sequent calculus that proves the same formulas as the axiom system K(ℝ). In Section 5, we establish the soundness and completeness of all these systems with respect to the Kripke semantics by showing that formulas derivable in the labelled tableau calculus are derivable in the sequent calculus. Finally, in Section 6, we prove cut elimination for the sequent calculus.

3 A Labelled Tableau System

In this section we introduce a labelled tableau calculus $\mathrm{LK}(\mathbb{R})$ for checking $\mathrm{K}(\mathbb{R})$-validity, based very closely on the Kripke semantics. Intuitively, to check whether a formula φ takes a value less than 0 in a world x, we decompose the propositional structure of φ to obtain an inequation between sums of formulas at x. Box formulas on the right of inequations generate new worlds accessible to x and new inequations between sums of formulas to be processed. Box formulas on the left are decomposed by considering accessible worlds and generating new inequations for those worlds. These inequations may involve formulas evaluated at different worlds; we therefore treat inequations between formulas labelled with integers representing worlds in Kripke frames. The formula φ will be valid if and only if the generated set of inequations (suitably interpreted) is unsatisfiable over the real numbers.

More precisely, we consider *(tableau) nodes* of the following forms:

(1) $(\Gamma)^{\boldsymbol{k}} \rhd (\Delta)^{\boldsymbol{l}}$ such that $\rhd \in \{>, \geq\}$ and $(\Gamma)^{\boldsymbol{k}} = [(\varphi_1)^{k_1}, \ldots, (\varphi_n)^{k_n}]$ and $(\Delta)^{\boldsymbol{l}} = [(\psi_1)^{l_1}, \ldots, (\psi_m)^{l_m}]$ are multisets of formulas $\Gamma = [\varphi_1, \ldots, \varphi_n]$ and $\Delta = [\psi_1, \ldots, \psi_m]$ labelled by $k_1, \ldots, k_n, l_1, \ldots, l_m \in \mathbb{Z}$;

(2) rij such that $i, j \in \mathbb{N}$.

Intuitively, (1) represents an inequation between sums of values of formulas evaluated at (possibly different) worlds of a $\mathrm{K}(\mathbb{R})$-model (φ_i is evaluated at world $|k_i|$ on the left, and ψ_j at world $|k_j|$ on the right) and (2), the expression rij, denotes that world j is accessible from world i in this model.

We define the *complexity* of an inequation $(\Gamma)^{\boldsymbol{k}} \rhd (\Delta)^{\boldsymbol{l}}$ to be the sum of the complexities of the formulas in Γ and Δ, where formulas of the form $\Box\varphi$ labelled by $-i$ for $i \in \mathbb{N}$ are treated as propositional variables.

A *tableau for a formula* φ is a finite sequence of nodes starting with $[] > [(\varphi)^1]$, $r12$ generated according to the inference rules of the system presented in Fig. 2; that is, if expressions above the line in an instance of a rule occur in the sequence, then the sequence can be extended with the expressions below the line. The tableau is called *complete* if the rules have been applied exhaustively, but only once to the same set of premises, and no application of (ex) is followed by another application of (ex). As labelled inequations occurring above the line in an instance of a rule have a higher complexity than those occurring below the line, there exists a complete tableau for every formula.

We call expressions of the form $(p)^i$ or $(\Box\varphi)^{-i}$ with $i \in \mathbb{N}$ *labelled variables*. The *system of inequations* associated to a tableau consists of all inequations over the labelled variables occurring in the tableau where the comma "," is interpreted as the usual addition over the real numbers. A tableau is *closed* if its associated system of inequations is inconsistent over $\mathbb{R}$; otherwise it is *open*. A formula $\varphi \in \mathrm{Fm}$ is *derivable* in the labelled tableau calculus $\mathrm{LK}(\mathbb{R})$, written $\vdash_{\mathrm{LK}(\mathbb{R})} \varphi$, if there exists a complete closed tableau for φ.

$$\frac{(\Gamma)^{\boldsymbol{k}}, (\varphi \to \psi)^{i} \rhd (\Delta)^{\boldsymbol{l}}}{(\Gamma)^{\boldsymbol{k}}, (\psi)^{i} \rhd (\varphi)^{i}, (\Delta)^{\boldsymbol{l}}} \ (\to\rhd) \qquad \frac{(\Gamma)^{\boldsymbol{k}} \rhd (\varphi \to \psi)^{i}, (\Delta)^{\boldsymbol{l}}}{(\Gamma)^{\boldsymbol{k}}, (\varphi)^{i} \rhd (\psi)^{i}, (\Delta)^{\boldsymbol{l}}} \ (\rhd\to)$$

$$\frac{\begin{array}{c} rij \\ (\Gamma)^{\boldsymbol{k}}, (\Box\varphi)^{i} \rhd (\Delta)^{\boldsymbol{l}} \end{array}}{\begin{array}{c} (\varphi)^{j} \geq (\Box\varphi)^{-i} \\ (\Gamma)^{\boldsymbol{k}}, (\Box\varphi)^{-i} \rhd (\Delta)^{\boldsymbol{l}} \end{array}} \ (\Box\rhd) \qquad \frac{(\Gamma)^{\boldsymbol{k}} \rhd (\Box\varphi)^{i}, (\Delta)^{\boldsymbol{l}}}{\begin{array}{c} (\Box\varphi)^{-i} \geq (\varphi)^{j} \\ (\Gamma)^{\boldsymbol{k}} \rhd (\Box\varphi)^{-i}, (\Delta)^{\boldsymbol{l}} \\ rij \end{array}} \ (\rhd\Box) \quad i, j \in \mathbb{N},\ j \text{ new}$$

$$\frac{rik}{rkj} \ (\text{ex}) \quad j \in \mathbb{N} \text{ new}$$

Fig. 2. The labelled tableau calculus LK($\mathbb{R}$)

Example 3.1 The seriality axiom is derivable in LK($\mathbb{R}$) using the following complete tableau for $\Box p \to \Diamond p = \Box p \to (\Box(p \to (p \to p)) \to (p \to p))$:

$$\begin{array}{rl}
1: & [] > (\Box p \to (\Box(p \to (p \to p)) \to (p \to p)))^{1} \\
2: & r12 \\
3: & (\Box p)^{1} > (\Box(p \to (p \to p)) \to (p \to p))^{1} \\
4: & (\Box p)^{1}, (\Box(p \to (p \to p)))^{1} > (p \to p)^{1} \\
5: & (\Box p)^{1}, (\Box(p \to (p \to p)))^{1}, (p)^{1} > (p)^{1} \\
6: & (p)^{2} \geq (\Box p)^{-1} \\
7: & (\Box p)^{-1}, (\Box(p \to (p \to p)))^{1}, (p)^{1} > (p)^{1} \\
8: & (p \to (p \to p))^{2} \geq (\Box(p \to (p \to p)))^{-1} \\
9: & (\Box p)^{-1}, (\Box(p \to (p \to p)))^{-1}, (p)^{1} > (p)^{1} \\
10: & (p \to p)^{2} \geq (p)^{2}, (\Box(p \to (p \to p)))^{-1} \\
11: & (p)^{2} \geq (p)^{2}, (p)^{2}, (\Box(p \to (p \to p)))^{-1} \\
12: & r23
\end{array}$$

with the following inconsistent system of inequations over $\mathbb{R}$

$$\{y + u + v > v,\ x \geq y,\ x \geq 2x + u\}$$

where x, y, u, v stand for $(p)^2$, $(\Box p)^{-1}$, $(\Box(p \to (p \to p)))^{-1}$, $(p)^1$, respectively.

Let us call a K($\mathbb{R}$)-model $\mathfrak{M} = \langle W, R, V\rangle$ *faithful* to a tableau T if there is a map $f\colon \mathbb{N} \to W$ (said to *show* that $\mathfrak{M}$ is faithful to T) such that if rij occurs in T, then $Rf(i)f(j)$ is in $\mathfrak{M}$, and for every inequation $(\varphi_1)^{i_1}, \ldots, (\varphi_n)^{i_n} \rhd (\psi_1)^{j_1}, \ldots, (\psi_m)^{j_m}$ occurring in T,

$$V(\varphi_1, f(|i_1|)) + \ldots + V(\varphi_n, f(|i_n|)) \rhd V(\psi_1, f(|j_1|)) + \ldots + V(\psi_m, f(|j_m|)).$$

Note that whenever a K($\mathbb{R}$)-model $\mathfrak{M} = \langle W, R, V\rangle$ is faithful to a tableau T, the map defined by $e((p)^i) = V(p, i)$ and $e((\Box\varphi)^{-i}) = V(\Box\varphi, i)$ satisfies the system of inequations associated to T over $\mathbb{R}$, and hence T is open.

The following lemma establishes the soundness of the rules of LK($\mathbb{R}$).

Lemma 3.2 *Let $\mathfrak{M} = \langle W, R, V\rangle$ be a finite* K($\mathbb{R}$)*-model faithful to a tableau T. If a rule of* LK($\mathbb{R}$) *is applied to T to obtain an extension T', then $\mathfrak{M}$ is faithful to T'.*

Proof. Let f be a map showing that $\mathfrak{M} = \langle W, R, V\rangle$ is a finite K($\mathbb{R}$)-model faithful to a tableau T. The cases of $(\rightarrow\triangleright)$ and $(\triangleright\rightarrow)$ follow easily. For $(\Box\triangleright)$, suppose that $(\Gamma)^{\boldsymbol{k}}, (\Box\varphi)^i \triangleright (\Delta)^{\boldsymbol{l}}$ and rij appear in T and we obtain an extension T' of T with $(\varphi)^j \geq (\Box\varphi)^{-i}$ and $(\Gamma)^{\boldsymbol{k}}, (\Box\varphi)^{-i} \triangleright (\Delta)^{\boldsymbol{l}}$. Since $\mathfrak{M}$ is faithful to T, we have $Rf(i)f(j)$. But then $V(\varphi, f(|j|)) = V(\varphi, f(j)) \geq V(\Box\varphi, f(i)) = V(\Box\varphi, f(|-i|))$, so $\mathfrak{M}$ is faithful to T'.

For $(\triangleright\Box)$, suppose that $(\Gamma)^{\boldsymbol{k}} \triangleright (\Box\varphi)^i, (\Delta)^{\boldsymbol{l}}$ $(i \in \mathbb{N})$ appears in T and we obtain an extension T' of T with rij ($j \in \mathbb{N}$ new), $(\Box\varphi)^{-i} \geq (\varphi)^j$, and $(\Gamma)^{\boldsymbol{k}} \triangleright (\Box\varphi)^{-i}, (\Delta)^{\boldsymbol{l}}$. Because $\mathfrak{M}$ is finite and serial, there exists $v \in W$ such that $Rf(i)v$ and $V(\Box\varphi, f(i)) = V(\varphi, v)$. Hence the map f' defined to be f but with $f'(j) = v$ shows that $\mathfrak{M}$ is faithful to T'.

Finally, for (ex) suppose that rik appears in T and we obtain an extension T' of T with rkj ($j \in \mathbb{N}$ new). Since rik is in T, we have $Rf(i)f(k)$. Because $\mathfrak{M}$ is serial, there exists $v \in W$ such that $Rf(k)v$. The map f' defined to be f but with $f'(j) = v$ shows that $\mathfrak{M}$ is faithful to T'. □

To establish the completeness of LK($\mathbb{R}$), we introduce the following notion. Let T be a complete open tableau and let e be a map satisfying the system of inequations associated to T. We say that $\mathfrak{M} = \langle W, R, V\rangle$ is an *e-induced model* of T if

- $W = \{w_i : i \in \mathbb{N}$ is a label occurring in $T\}$;
- Rw_iw_j if and only if rij occurs in T or $i = j$ and rik is not in T for any k;
- $V(p, w_i) = \begin{cases} e((p)^i) & \text{if } (p)^i \text{ occurs in } T \\ 0 & \text{otherwise.} \end{cases}$

Lemma 3.3 *Let $\mathfrak{M} = \langle W, R, V\rangle$ be an e-induced model of a complete open tableau T, and extend the map e by fixing $e((\varphi)^i) = V(\varphi, w_i)$ for each $w_i \in W$. If $(\varphi_1)^{i_1}, \ldots, (\varphi_n)^{i_n} \triangleright (\psi_1)^{j_1}, \ldots, (\psi_m)^{j_m}$ appears in T, then*

$$e((\varphi_1)^{i_1}) + \ldots + e((\varphi_n)^{i_n}) \triangleright e((\psi_1)^{j_1}) + \ldots + e((\psi_m)^{j_m}).$$

Proof. We proceed by induction on the complexity of the inequation. The base case follows using the definition of $\mathfrak{M}$ and the fact that e is a map satisfying the system of inequations associated to T, while the cases where $(\varphi_1)^{i_1}, \ldots, (\varphi_n)^{i_n} \triangleright (\psi_1)^{j_1}, \ldots, (\psi_m)^{j_m}$ appears as a premise of an application of $(\rightarrow\triangleright)$ or $(\triangleright\rightarrow)$ in T follow directly using the induction hypothesis. Suppose that the inequation is of the form $(\Gamma)^{\boldsymbol{k}}, (\Box\varphi)^i \triangleright (\Delta)^{\boldsymbol{l}}$ for $i \in \mathbb{N}$. Since $\mathfrak{M}$ is finite, there is a j such that rij occurs in T and $V(\Box\varphi, w_i) = V(\varphi, w_j)$. But also $(\varphi)^j \geq (\Box\varphi)^{-i}$ occurs in T, and hence, by the induction hypothesis, $V(\varphi, w_j) = e((\varphi)^j) \geq e((\Box\varphi)^{-i})$. We also have that $(\Gamma)^{\boldsymbol{k}}, (\Box\varphi)^{-i} \triangleright (\Delta)^{\boldsymbol{l}}$ occurs in T, and the desired inequality follows by another application of the induction hypothesis. Finally, if the inequation is of the form $(\Gamma)^{\boldsymbol{k}} \triangleright (\Box\varphi)^i, (\Delta)^{\boldsymbol{l}}$ with $i \in \mathbb{N}$, then by $(\triangleright\Box)$ we must have in the tableau rij for some $j \in \mathbb{N}$, $(\Box\varphi)^{-i} \geq (\varphi)^j$, and $(\Gamma)^{\boldsymbol{k}} \rhd (\Box\varphi)^{-i}, (\Delta)^{\boldsymbol{l}}$. The desired inequality follows by applying the induction hypothesis to these two inequalities and observing that $V(\varphi, w_j) \geq V(\Box\varphi, w_i)$. □

Putting together these last two lemmas we obtain the following soundness and completeness theorem for LK($\mathbb{R}$).

Theorem 3.4 *For any $\varphi \in \mathrm{Fm}$, $\vdash_{\mathrm{LK}(\mathbb{R})} \varphi$ if and only if $\models_{\mathrm{K}(\mathbb{R})} \varphi$.*

Proof. For the left-to-right direction, assume $\not\models_{\mathrm{K}(\mathbb{R})} \varphi$. Then by Lemma 2.1, there is a finite K($\mathbb{R}$)-model $\mathfrak{M} = \langle W, R, V \rangle$ and some world $w_1 \in W$ such that $0 > V(\varphi, w_1)$. Let $f \colon \mathbb{N} \to W$ be any function such that $f(1) = w_1$ and $f(2) = w_2$, where Rw_1w_2. This function shows that $\mathfrak{M}$ is faithful to the tableau consisting just of $[] > [(\varphi)^1]$, $r12$. Suppose that by applying the decomposition rules to this tableau, we obtain a complete tableau T. Applying Lemma 3.2 inductively, $\mathfrak{M}$ is faithful to T. So the system of inequations associated with T is consistent over $\mathbb{R}$, and T is open. Hence $\not\vdash_{\mathrm{LK}(\mathbb{R})} \varphi$.

For the right-to-left direction, suppose that $\not\vdash_{\mathrm{LK}(\mathbb{R})} \varphi$. Then there is a complete open tableau T beginning with $[] > [(\varphi)^1]$, $r12$. Let e be a map satisfying the system of inequations associated to T and consider any e-induced model $\mathfrak{M} = \langle W, R, V \rangle$ of T. By Lemma 3.3, we obtain $0 > e((\varphi)^1) = V(\varphi, w_1)$. Hence $\not\models_{\mathrm{K}(\mathbb{R})} \varphi$. □

We also obtain an upper bound for the complexity of checking K($\mathbb{R}$)-validity.

Theorem 3.5 K($\mathbb{R}$)*-validity is in EXPTIME.*

Proof. Given a formula φ of modal depth d, we generate a complete tableau for φ. We do this stepwise, where after i steps, we are only left with nodes containing formulas of modal depth at most $d - i$. Step $i + 1$ is then as follows. We first apply the $(\to\rhd)$ and $(\rhd\to)$ rules exhaustively. Inequations containing implicational formulas can then be removed, since they will not belong to the set of inequations associated to the tableau. Hence we obtain nodes containing only labelled variables and modal formulas. We then apply the rule $(\rhd\Box)$ for each boxed formula occurring on the right in one of these nodes and (ex) one time for each new label. We apply $(\Box\rhd)$ exhaustively and then remove all nodes containing formulas of modal depth $d - i$. After d steps we obtain a complete tableau for φ that contains exponentially (in d) many different nodes using exponentially many (in d) labels. Hence we obtain a linear programming problem of at most exponential size in d. The result follows from the fact that the linear programming problem is in P [19]. □

4 A Sequent Calculus

We define a *sequent* to be an ordered pair of finite multisets of formulas Γ and Δ, written $\Gamma \Rightarrow \Delta$. For multisets of formulas Γ and Δ, we write Γ, Δ to denote their multiset union, $n\Gamma$ for $\Gamma, \ldots, \Gamma$ (n times), and $\Box\Gamma$ for $[\Box\varphi : \varphi \in \Gamma]$.

We define a formula translation of sequents as follows:

$$\mathcal{I}(\varphi_1, \ldots, \varphi_n \Rightarrow \psi_1, \ldots, \psi_m) \;:=\; (\varphi_1 \& \ldots \& \varphi_n) \to (\psi_1 \& \ldots \& \psi_m),$$

$$\frac{}{\Delta \Rightarrow \Delta}\ (\textsc{id}) \qquad \frac{\Gamma, \varphi \Rightarrow \Delta \quad \Pi \Rightarrow \varphi, \Sigma}{\Gamma, \Pi \Rightarrow \Sigma, \Delta}\ (\textsc{cut})$$

$$\frac{\Gamma \Rightarrow \Delta \quad \Pi \Rightarrow \Sigma}{\Gamma, \Pi \Rightarrow \Sigma, \Delta}\ (\textsc{mix}) \qquad \frac{n\Gamma \Rightarrow n\Delta}{\Gamma \Rightarrow \Delta}\ (\textsc{sc}_n) \quad (n \geq 2)$$

$$\frac{\Gamma, \psi \Rightarrow \varphi, \Delta}{\Gamma, \varphi \to \psi \Rightarrow \Delta}\ (\to\Rightarrow) \qquad \frac{\Gamma, \varphi \Rightarrow \psi, \Delta}{\Gamma \Rightarrow \varphi \to \psi, \Delta}\ (\Rightarrow\to)$$

$$\frac{\Gamma \Rightarrow n[\varphi]}{\Box\Gamma \Rightarrow n[\Box\varphi]}\ (\Box_n) \quad (n \geq 0)$$

Fig. 3. The sequent calculus GK($\mathbb{R}$)

where $\varphi_1 \& \ldots \& \varphi_n = \overline{0}$ for $n = 0$. We say that a sequent $\Gamma \Rightarrow \Delta$ is K($\mathbb{R}$)-*valid*, written $\models_{K(\mathbb{R})} \Gamma \Rightarrow \Delta$, if $\models_{K(\mathbb{R})} \mathcal{I}(\Gamma \Rightarrow \Delta)$.

A sequent calculus GK($\mathbb{R}$) is presented in Fig. 3. The following rules are derivable in this system.

$$\frac{\Gamma, \varphi, \psi \Rightarrow \Delta}{\Gamma, \varphi \& \psi \Rightarrow \Delta}\ (\&\Rightarrow) \qquad \frac{\Gamma \Rightarrow \varphi, \psi, \Delta}{\Gamma \Rightarrow \varphi \& \psi, \Delta}\ (\Rightarrow\&)$$

$$\frac{\Gamma \Rightarrow \varphi, \Delta}{\Gamma, \neg\varphi \Rightarrow \Delta}\ (\neg\Rightarrow) \qquad \frac{\Gamma, \varphi \Rightarrow \Delta}{\Gamma \Rightarrow \neg\varphi, \Delta}\ (\Rightarrow\neg)$$

$$\frac{\Gamma \Rightarrow \Delta}{\Gamma, \overline{0} \Rightarrow \Delta}\ (\overline{0}\Rightarrow) \qquad \frac{\Gamma \Rightarrow \Delta}{\Gamma \Rightarrow \overline{0}, \Delta}\ (\Rightarrow\overline{0})$$

Example 4.1 The rule $(\Box_n)$ can be used to derive instances of (D_n) as follows:

$$\cfrac{\cfrac{\cfrac{\cfrac{\cfrac{}{\varphi, \ldots, \varphi \Rightarrow \varphi, \ldots, \varphi}\ (\textsc{id})}{\vdots}\ (\&\Rightarrow)}{n\varphi \Rightarrow \varphi, \ldots, \varphi}\ (\&\Rightarrow)}{\Box(n\varphi) \Rightarrow \Box\varphi, \ldots, \Box\varphi}\ (\Box_n)}{\cfrac{\cfrac{\vdots}{\Box(n\varphi) \Rightarrow n\Box\varphi}\ (\Rightarrow\&)}{\Rightarrow \Box(n\varphi) \to n\Box\varphi}\ (\Rightarrow\to)}\ (\Rightarrow\&)$$

We note also that the "cancellation" rule

$$\frac{\Gamma, \varphi \Rightarrow \varphi, \Delta}{\Gamma \Rightarrow \Delta}\ (\textsc{can})$$

is both derivable in GK($\mathbb{R}$) and can be used, with (MIX), to derive (CUT):

$$\cfrac{\cfrac{\cfrac{}{\varphi \Rightarrow \varphi}\ (\textsc{id})}{\varphi \to \varphi \Rightarrow}\ (\to\Rightarrow) \quad \cfrac{\Gamma, \varphi \Rightarrow \varphi, \Delta}{\Gamma \Rightarrow \varphi \to \varphi, \Delta}\ (\Rightarrow\to)}{\Gamma \Rightarrow \Delta}\ (\textsc{cut}) \qquad \cfrac{\cfrac{\Gamma, \varphi \Rightarrow \Delta \quad \Pi \Rightarrow \varphi, \Sigma}{\Gamma, \Pi, \varphi \Rightarrow \varphi, \Sigma, \Delta}\ (\textsc{mix})}{\Gamma, \Pi \Rightarrow \Sigma, \Delta}\ (\textsc{can})$$

Moreover, the following rule (used in the proofs of Theorems 2.2 and 4.3)

$$\frac{\Gamma_0 \Rightarrow \quad \Gamma_1 \Rightarrow k[\varphi_1] \quad \dots \quad \Gamma_n \Rightarrow k[\varphi_n]}{\Delta, \Box\Gamma \Rightarrow \Box\varphi_1, \dots, \Box\varphi_n, \Delta}\ (\Box_{k,n})$$

$$\text{where } k \in \mathbb{N} \setminus \{0\},\ n \in \mathbb{N},\ k\Gamma = \Gamma_0 \uplus \Gamma_1 \uplus \dots \uplus \Gamma_n$$

is derivable in GK($\mathbb{R}$) as shown below:

$$\dfrac{\dfrac{}{\Delta \Rightarrow \Delta}\ (\textsc{id}) \qquad \dfrac{\dfrac{\dfrac{\Gamma_0 \Rightarrow}{\Box\Gamma_0 \Rightarrow}\ (\Box_0) \qquad \dfrac{\dfrac{\Gamma_1 \Rightarrow k[\varphi_1]}{\Box\Gamma_1 \Rightarrow k[\Box\varphi_1]}\ (\Box_k) \qquad \dfrac{\dfrac{\Gamma_n \Rightarrow k[\varphi_n]}{\Box\Gamma_n \Rightarrow k[\Box\varphi_n]}\ (\Box_k)}{\vdots}\ (\textsc{mix})}{\Box(\Gamma_1 \uplus \dots \uplus \Gamma_n) \Rightarrow k[\Box\varphi_1], \dots, k[\Box\varphi_n]}\ (\textsc{mix})}{\Box(\Gamma_0 \uplus \Gamma_1 \uplus \dots \uplus \Gamma_n) \Rightarrow k[\Box\varphi_1], \dots, k[\Box\varphi_n]}\ (\textsc{mix})}{\Box\Gamma \Rightarrow \Box\varphi_1, \dots, \Box\varphi_n}\ (\textsc{sc}_k)}{\Delta, \Box\Gamma \Rightarrow \Box\varphi_1, \dots, \Box\varphi_n, \Delta}\ (\textsc{mix})$$

We now establish the equivalence of GK($\mathbb{R}$) with the axiom system K($\mathbb{R}$):

Theorem 4.2 $\vdash_{\mathrm{GK}(\mathbb{R})} \Gamma \Rightarrow \Delta$ *if and only if* $\vdash_{\mathrm{K}(\mathbb{R})} \mathcal{I}(\Gamma \Rightarrow \Delta)$.

Proof. It suffices for the left-to-right direction to show that for any rule of GK($\mathbb{R}$) with premises $S_1, \dots, S_m$ and conclusion S, whenever $\vdash_{\mathrm{K}(\mathbb{R})} \mathcal{I}(S_i)$ for $i = 1 \dots m$, also $\vdash_{\mathrm{K}(\mathbb{R})} \mathcal{I}(S)$. For example, consider the rule $(\Box_n)$ and assume that $\vdash_{\mathrm{K}(\mathbb{R})} \mathcal{I}(\Gamma \Rightarrow n[\varphi])$. Suppose that $\Gamma = [\psi_1, \dots, \psi_m]$ and $\psi = \psi_1 \& \dots \& \psi_m$. We continue the derivation of $\mathcal{I}(\Gamma \Rightarrow n[\varphi]) = \psi \to n\varphi$ in K($\mathbb{R}$) to obtain a derivation of $\Box\psi \to n\Box\varphi$:

1.	$\psi \to n\varphi$	
2.	$\Box(\psi \to n\varphi)$	(nec)
3.	$\Box(\psi \to n\varphi) \to (\Box\psi \to \Box n\varphi)$	(K)
4.	$\Box\psi \to \Box n\varphi$	(mp) with 2,3
5.	$\Box n\varphi \to n\Box\varphi$	(D_n)
6.	$(\Box\psi \to \Box n\varphi) \to ((\Box n\varphi \to n\Box\varphi) \to (\Box\psi \to n\Box\varphi))$	(B)
7.	$(\Box n\varphi \to n\Box\varphi) \to (\Box\psi \to n\Box\varphi)$	(mp) with 4,6
8.	$\Box\psi \to n\Box\varphi$	(mp) with 5,7.

$(\Box\psi_1 \& \dots \& \Box\psi_m) \to \Box\psi$ is derivable using (B), (C), (I), and (K), so, using (B) and (mp), we obtain a derivation of $\mathcal{I}(\Box\Gamma \Rightarrow n[\Box\varphi]) = (\Box\psi_1 \& \dots \& \Box\psi_m) \to n\Box\varphi$ in GK($\mathbb{R}$).

For the right-to-left direction, it is straightforward to show that every axiom of K($\mathbb{R}$) is derivable in GK($\mathbb{R}$); see, e.g., Example 4.1 for derivations of instances of (D_n). Also, the rules of K($\mathbb{R}$) are derivable in GK($\mathbb{R}$). For example, for (con_n), starting with $\Rightarrow n\varphi$, we can apply (CUT) with the derivable sequent $n\varphi \Rightarrow n[\varphi]$ to obtain $\Rightarrow n[\varphi]$ and then, by an application of ($\textsc{sc}_n$), obtain also $\Rightarrow \varphi$. Hence, if $\vdash_{\mathrm{K}(\mathbb{R})} \mathcal{I}(\Gamma \Rightarrow \Delta)$, then $\vdash_{\mathrm{GK}(\mathbb{R})} \Rightarrow \mathcal{I}(\Gamma \Rightarrow \Delta)$ and, applying (CUT) with the derivable sequent $\Gamma, \mathcal{I}(\Gamma \Rightarrow \Delta) \Rightarrow \Delta$, also $\vdash_{\mathrm{GK}(\mathbb{R})} \Gamma \Rightarrow \Delta$. $\Box$

The rule (CUT) is not really necessary for derivations in GK($\mathbb{R}$). That is, there exists an algorithm for constructively eliminating applications of the rule (CUT) from derivations in GK($\mathbb{R}$); this may be stated as follows:

Theorem 4.3 GK($\mathbb{R}$) *admits cut elimination.*

However, as this result is not required for the proof of the completeness theorem for K($\mathbb{R}$) (Theorem 2.2, proved in Section 5), we defer its proof to Section 6.

5 Completeness

This section is devoted to proving Theorem 2.2. We begin with a simple lemma establishing a separation of propositional variables and boxed formulas.

Lemma 5.1 *If* Γ *and* Δ *are multisets of propositional variables and* $\models_{\mathrm{K}(\mathbb{R})} \Gamma, \Box\Pi \Rightarrow \Delta, \Box\Sigma$, *then* $\Gamma = \Delta$ *and* $\models_{\mathrm{K}(\mathbb{R})} \Box\Pi \Rightarrow \Box\Sigma$.

Proof. Suppose that Γ and Δ are multisets of propositional variables and $\models_{\mathrm{K}(\mathbb{R})} \Gamma, \Box\Pi \Rightarrow \Delta, \Box\Sigma$. It suffices to show that $\Gamma = \Delta$ as then clearly also $\models_{\mathrm{K}(\mathbb{R})} \Box\Pi \Rightarrow \Box\Sigma$. Suppose for a contradiction that $\Gamma \neq \Delta$. Without loss of generality, some propositional variable p occurs strictly more times in Γ than Δ. Consider a K($\mathbb{R}$)-model with worlds x, y satisfying Rxy and Ryy where $V(p,x) = 1$, $V(p,y) = 0$, and $V(q,x) = V(q,y) = 0$ for $q \neq p$. Then $V(\Box\varphi, x) = 0$ for any $\varphi \in \mathrm{Fm}$, and $\not\models_{\mathrm{K}(\mathbb{R})} \Gamma, \Box\Pi \Rightarrow \Delta, \Box\Sigma$, a contradiction. $\Box$

Theorem 2.2 is a consequence of the following result and Theorem 4.2.

Theorem 5.2 *If* $\models_{\mathrm{K}(\mathbb{R})} \Gamma \Rightarrow \Delta$, *then* $\vdash_{\mathrm{GK}(\mathbb{R})} \Gamma \Rightarrow \Delta$.

Proof. We prove the claim by induction on the lexicographically ordered pair consisting of the modal depth of $\mathcal{I}(\Gamma \Rightarrow \Delta)$ and the sum of the complexities of the formulas in $\Gamma \Rightarrow \Delta$. Assume $\models_{\mathrm{K}(\mathbb{R})} \Gamma \Rightarrow \Delta$. If $\Gamma = \Gamma' \uplus [\varphi \to \psi]$, then $\models_{\mathrm{K}(\mathbb{R})} \Gamma', \psi \Rightarrow \varphi, \Delta$ and, by the induction hypothesis, $\vdash_{\mathrm{GK}(\mathbb{R})} \Gamma', \psi \Rightarrow \varphi, \Delta$. Hence $\vdash_{\mathrm{GK}(\mathbb{R})} \Gamma', \varphi \to \psi \Rightarrow \Delta$. The case for $\Delta = \Delta' \uplus [\varphi \to \psi]$ is very similar.

If $\Gamma \Rightarrow \Delta$ has the form $\Gamma_1, \Box\Gamma_2 \Rightarrow \Delta_1, \Box\Delta_2$ where Γ_1 and Δ_1 contain only propositional variables, then, by Lemma 5.1, we obtain $\Gamma_1 = \Delta_1$ and $\models_{\mathrm{K}(\mathbb{R})} \Box\Gamma_2 \Rightarrow \Box\Delta_2$. Clearly $\vdash_{\mathrm{GK}(\mathbb{R})} \Gamma_1 \Rightarrow \Delta_1$. Hence it suffices, using (MIX), to prove that $\vdash_{\mathrm{GK}(\mathbb{R})} \Box\Gamma_2 \Rightarrow \Box\Delta_2$, where $\Box\Gamma_2 \Rightarrow \Box\Delta_2$ has the form

$$\Box\varphi_1, \ldots, \Box\varphi_n \Rightarrow \Box\psi_1, \ldots, \Box\psi_m \quad (m, n \in \mathbb{N}).$$

We know $\models_{\mathrm{K}(\mathbb{R})} \Box\varphi_1, \ldots, \Box\varphi_n \Rightarrow \Box\psi_1, \ldots, \Box\psi_m$. Hence, translating between sequents and formulas and using Theorem 3.4, there is a complete closed tableau T beginning with $r12$ and containing

$$(\Box\varphi_1)^{-1}, \ldots, (\Box\varphi_n)^{-1} > (\Box\psi_1)^{-1}, \ldots, (\Box\psi_m)^{-1}. \quad (1)$$

T must then also contain inequations for new labels $y_1, \ldots, y_m \in \mathbb{N}$

$$(\Box\psi_1)^{-1} \geq (\psi_1)^{y_1} \quad \ldots \quad (\Box\psi_m)^{-1} \geq (\psi_m)^{y_m} \quad (2)$$

and, fixing $y_0 = 2$ for convenience,

$$\begin{array}{ccc} (\varphi_1)^{y_0} \geq (\Box\varphi_1)^{-1} & \ldots & (\varphi_1)^{y_m} \geq (\Box\varphi_1)^{-1} \\ \vdots & & \vdots \\ (\varphi_n)^{y_0} \geq (\Box\varphi_n)^{-1} & \ldots & (\varphi_n)^{y_m} \geq (\Box\varphi_n)^{-1}. \end{array} \qquad (3)$$

Since T is closed, the system $\mathcal{S}$ of inequations associated to T is inconsistent over $\mathbb{R}$. Note that an inequation $\Gamma \rhd \Delta$ occurring in T may not occur in $\mathcal{S}$; however, there does always occur in $\mathcal{S}$ an inequation obtained by applying the tableau rules for $\to$ to $\Gamma \rhd \Delta$ and then the tableau rules for $\Box$ switching each $i \in \mathbb{N}$ to $-i$; we call this inequation the *derived inequation* of $\Gamma \rhd \Delta$.

Recall that a system of inequations of the form $f_i(\bar{x}) > g_i(\bar{x})$ $(1 \leq i \leq r)$ and $h_j(\bar{x}) \geq k_j(\bar{x})$ $(1 \leq j \leq s)$ where each f_i, g_i, h_j, k_j is a positive linear sum of variables in $\bar{x}$ containing no constants, is inconsistent over $\mathbb{R}$ if and only if there exist $\lambda_1, \ldots, \lambda_r \in \mathbb{N}$ (not all zero) and $\mu_1, \ldots, \mu_s \in \mathbb{N}$ such that $\lambda_1 f_1 + \ldots + \lambda_r f_r + \mu_1 h_1 + \ldots + \mu_s h_s = \lambda_1 g_1 + \ldots + \lambda_r g_r + \mu_1 k_1 + \ldots + \mu_s k_s$. For convenience, we may say that the inequation $f_i(\bar{x}) > g_i(\bar{x})$ or $h_j(\bar{x}) \geq k_j(\bar{x})$ is "used" λ_i or μ_j times, respectively, in the linear combination.

We now consider a linear combination of the inequations in $\mathcal{S}$ that witnesses inconsistency over $\mathbb{R}$ and observe:

(i) The inequation (1) is the only strict inequation occurring in $\mathcal{S}$ and hence must be used in the linear combination some fixed $k > 0$ times.

(ii) The variables $(\Box\psi_1)^{-1}, \ldots, (\Box\psi_m)^{-1}$ occur in $\mathcal{S}$ only in (1) and in the inequations derived from (2); hence, using (i), each inequation derived from (2) must be used in the linear combination k times.

(iii) The variables $(\Box\varphi_1)^{-1}, \ldots, (\Box\varphi_n)^{-1}$ occur in $\mathcal{S}$ only in (1) and in the inequations derived from (3); hence, given that the derived inequation of $(\varphi_i)^{y_j} \geq (\Box\varphi_i)^{-j}$ is used in the linear combination $\lambda_{i,j} > 0$ times, we obtain $\lambda_{i,0} + \lambda_{i,1} + \ldots + \lambda_{i,m} = k$ for $1 \leq i \leq n$.

Let $\mathcal{S}'$ be the system of inequations obtained from $\mathcal{S}$ by replacing (1) and the inequations derived from (2) and (3) with the inequations derived from

$$\begin{array}{ll} \lambda_{1,0}[(\varphi_1)^{y_0}], \ldots, \lambda_{n,0}[(\varphi_n)^{y_0}] > [] & \\ \lambda_{1,j}[(\varphi_1)^{y_j}], \ldots, \lambda_{n,j}[(\varphi_n)^{y_j}] > k[(\psi_j)^{y_j}] & 1 \leq j \leq m \end{array} \qquad (4).$$

Crucially, there is also a linear combination of the inequations in $\mathcal{S}'$ witnessing inconsistency over $\mathbb{R}$ that uses each inequation derived from one of the inequations in (4) exactly once. Moreover, all the inequations in $\mathcal{S}'$ are obtained by applying tableau rules to the inequations in (4). Observe now, however, that the different inequations in (4) contain different labels $y_0, y_1, \ldots, y_m$. Hence the inequations in $\mathcal{S}'$ obtained by applying tableau rules to different inequations in (4) will contain disjoint sets of variables. It follows that by applying tableau rules to any one particular inequation in (4) produces a subset of the inequations in $\mathcal{S}'$ that admits a linear combination witnessing inconsistency over $\mathbb{R}$.

But then, translating between sequents and formulas and using Theorem 3.4 again, we obtain

$$\models_{\mathrm{K}(\mathbb{R})} \lambda_{1,0}[\varphi_1], \ldots, \lambda_{n,0}[\varphi_n] \Rightarrow$$
$$\models_{\mathrm{K}(\mathbb{R})} \lambda_{1,j}[\varphi_1], \ldots, \lambda_{n,j}[\varphi_n] \Rightarrow k[\psi_j] \quad 1 \leq j \leq m.$$

So, by the induction hypothesis,

$$\vdash_{\mathrm{GK}(\mathbb{R})} \lambda_{1,0}[\varphi_1], \ldots, \lambda_{n,0}[\varphi_n] \Rightarrow$$
$$\vdash_{\mathrm{GK}(\mathbb{R})} \lambda_{1,j}[\varphi_1], \ldots, \lambda_{n,j}[\varphi_n] \Rightarrow k[\psi_j] \quad 1 \leq j \leq m.$$

But now we can apply the (derived) rule ($\Box_{k,n}$) to obtain the required derivation of $\Box\varphi_1, \ldots, \Box\varphi_n \Rightarrow \Box\psi_1, \ldots, \Box\psi_m$ in GK($\mathbb{R}$). □

6 Cut Elimination

This section is devoted to proving Theorem 4.3. Let $\mathrm{GK}(\mathbb{R})^{\mathrm{r}}$ be the sequent calculus consisting of the rules (ID), ($\to\Rightarrow$), ($\Rightarrow\to$), and ($\Box_{k,n}$). We show first that every cut-free derivation in GK($\mathbb{R}$) can be transformed algorithmically into a derivation in $\mathrm{GK}(\mathbb{R})^{\mathrm{r}}$. Recall that for a sequent calculus C, a sequent rule is *admissible* for C if for any instance of the rule, whenever the premises are derivable in C, the conclusion is derivable in C; the rule is *invertible* for C if for any instance of the rule, whenever the conclusion is derivable in C, the premises are derivable in C.

We begin with two preparatory lemmas:

Lemma 6.1 *The rules* ($\to\Rightarrow$) *and* ($\Rightarrow\to$) *are invertible for* $\mathrm{GK}(\mathbb{R})^{\mathrm{r}}$.

Proof. Simple (constructive) inductions on the height of a derivation of the premise in $\mathrm{GK}(\mathbb{R})^{\mathrm{r}}$ in each case. □

Lemma 6.2 *The rules* (MIX) *and* (SC_n) *are admissible in* $\mathrm{GK}(\mathbb{R})^{\mathrm{r}}$.

Proof. To show the admissibility of (MIX) in $\mathrm{GK}(\mathbb{R})^{\mathrm{r}}$, we prove that whenever $\vdash_{\mathrm{GK}(\mathbb{R})^{\mathrm{r}}} \Gamma \Rightarrow \Delta$ and $\vdash_{\mathrm{GK}(\mathbb{R})^{\mathrm{r}}} \Pi \Rightarrow \Sigma$ with $r, s \in \mathbb{N}$, then $\vdash_{\mathrm{GK}(\mathbb{R})^{\mathrm{r}}} r\Gamma, s\Pi \Rightarrow s\Sigma, r\Delta$. We proceed by induction on the sum of the heights of derivations d_1 and d_2 of $\Gamma \Rightarrow \Delta$ and $\Pi \Rightarrow \Sigma$, respectively.

For the base case, if d_1 and d_2 have height 0, then $\Gamma \Rightarrow \Delta$ and $\Pi \Rightarrow \Sigma$ are instances of (ID), i.e., $\Gamma = \Delta$ and $\Pi = \Sigma$. Hence $r\Gamma \uplus s\Pi = r\Delta \uplus s\Sigma$ and $\vdash_{\mathrm{GK}(\mathbb{R})^{\mathrm{r}}} r\Gamma, s\Pi \Rightarrow s\Sigma, r\Delta$ by (ID). If the last application in d_1 is ($\Box_{k,n}$) and d_2 has height 0, then $\Pi = \Sigma$ and the result follows by an application of ($\Box_{k,rn}$). The case where d_1 has height 0 and d_2 ends with ($\Box_{k,n}$) is symmetrical.

If the last application of a rule in d_1 or d_2 is ($\to\Rightarrow$) or ($\Rightarrow\to$), then the result follows easily by an application of the induction hypothesis and further applications of the rule. Suppose then finally that d_1 ends with

$$\frac{\Gamma_0 \Rightarrow \quad \Gamma_1 \Rightarrow k[\varphi_1] \quad \ldots \quad \Gamma_n \Rightarrow k[\varphi_n]}{\Omega, \Box\Gamma' \Rightarrow \Box\varphi_1, \ldots, \Box\varphi_n, \Omega}\ (\Box_{k,n}) \qquad \text{with } k\Gamma' = \Gamma_0 \uplus \Gamma_1 \uplus \ldots \uplus \Gamma_n$$

and that d_2 ends with an application of $(\Box_{l,m})$

$$\frac{\Pi_0 \Rightarrow \quad \Pi_1 \Rightarrow l[\psi_1] \quad \ldots \quad \Pi_m \Rightarrow l[\psi_m]}{\Theta, \Box\Pi' \Rightarrow \Box\psi_1, \ldots, \Box\psi_m, \Theta} \; (\Box_{l,m}) \text{with } l\Pi' = \Pi_0 \uplus \Pi_1 \uplus \ldots \uplus \Pi_m.$$

Then we can complete our derivation as follows

$$\frac{rl\Gamma_0, sk\Pi_0 \Rightarrow \quad \{l\Gamma_i \Rightarrow kl[\varphi_i]\}_{1 \leq i \leq n} \quad \{k\Pi_j \Rightarrow kl[\psi_j]\}_{1 \leq j \leq m}}{r\Omega, s\Theta, r\Box\Gamma', s\Box\Pi' \Rightarrow r\Box\varphi_1, \ldots, r\Box\varphi_n, s\Box\psi_1, \ldots, s\Box\psi_m, r\Omega, s\Theta} \; (\Box_{kl,rn+sm})$$

where the premises are all derivable by the induction hypothesis.

We establish the admissibility of (SC_n) by proving that whenever $\vdash_{\text{GK}(\mathbb{R})^{\text{r}}} n\Gamma \Rightarrow n\Delta$, then $\vdash_{\text{GK}(\mathbb{R})^{\text{r}}} \Gamma \Rightarrow \Delta$, proceeding by induction on the sum of the complexities of the formulas in Γ, Δ. For the base case, if $n\Gamma = n\Delta$, in particular when Γ and Δ contain only propositional variables, then $\Gamma = \Delta$ and $\vdash_{\text{GK}(\mathbb{R})^{\text{r}}} \Gamma \Rightarrow \Delta$ by (ID). If Γ contains a formula $\varphi \to \psi$, then by the invertibility of the rule ($\to\Rightarrow$) established in Lemma 6.1, $\vdash_{\text{GK}(\mathbb{R})^{\text{r}}} n(\Gamma - [\varphi \to \psi]), n\psi \Rightarrow n\varphi, n\Delta$. The induction hypothesis and an application of ($\to\Rightarrow$) gives $\vdash_{\text{GK}(\mathbb{R})^{\text{r}}} \Gamma \Rightarrow \Delta$. The case where Δ contains a formula $\varphi \to \psi$ is symmetrical. In the final case, the derivation of $n\Gamma \Rightarrow n\Delta$ must end with an application of $(\Box_{k,nl})$ where $\Gamma = \Pi \uplus [\Box\Sigma]$ and $\Delta = \Pi \uplus [\Box\varphi_1, \ldots, \Box\varphi_l]$. But then we obtain a derivation of $\Gamma \Rightarrow \Delta$ using $(\Box_{kn,l})$ and the admissibility of (MIX). □

Proof of Theorem 4.3. The rule $(\Box_n)$ is derivable in $\text{GK}(\mathbb{R})^{\text{r}}$ using $(\Box_{k,n})$ with $k = n$ and $\varphi_1 = \ldots = \varphi_n = \varphi$ and $\Gamma_1 = \ldots = \Gamma_n = \Gamma$. Hence, using the proofs of Lemma 6.2, every cut-free derivation in $\text{GK}(\mathbb{R})$ can be transformed algorithmically into a derivation in $\text{GK}(\mathbb{R})^{\text{r}}$. To establish cut-elimination for $\text{GK}(\mathbb{R})$, it suffices now to show that an uppermost application of (CUT) in a derivation in $\text{GK}(\mathbb{R})$ can be eliminated. We will prove (constructively) that

$$\vdash_{\text{GK}(\mathbb{R})^{\text{r}}} \Gamma, \varphi \Rightarrow \varphi, \Delta \quad \Longrightarrow \quad \vdash_{\text{GK}(\mathbb{R})^{\text{r}}} \Gamma \Rightarrow \Delta. \qquad (\star)$$

Suppose then that there are cut-free derivations in $\text{GK}(\mathbb{R})$ of the premises of the uppermost application of $\Gamma, \varphi \Rightarrow \Delta$ and $\Pi \Rightarrow \varphi, \Sigma$. Clearly, by (MIX), we have a cut-free derivation of $\Gamma, \Pi, \varphi \Rightarrow \varphi, \Sigma, \Delta$ in $\text{GK}(\mathbb{R})$, and hence a derivation of $\Gamma, \Pi, \varphi \Rightarrow \varphi, \Sigma, \Delta$ in $\text{GK}(\mathbb{R})^{\text{r}}$. By $(\star)$, we obtain a derivation of $\Gamma, \Pi \Rightarrow \Sigma, \Delta$ in $\text{GK}(\mathbb{R})^{\text{r}}$, which also gives the desired derivation in $\text{GK}(\mathbb{R})$.

We prove $(\star)$ by induction on the lexicographically ordered pair consisting of the modal depth of φ and the sum of the complexities of the formulas in $\Gamma, \varphi \Rightarrow \varphi, \Delta$. If $\Gamma \uplus [\varphi] = [\varphi] \uplus \Delta$, in particular if the sequent contains only propositional variables, then $\Gamma = \Delta$ and $\Gamma \Rightarrow \Delta$ is derivable using (ID). If φ has the form $\psi \to \chi$, then we use the invertibility of ($\to\Rightarrow$) and ($\Rightarrow\to$) in $\text{GK}(\mathbb{R})^{\text{r}}$ and apply the induction hypothesis twice. The cases where Γ or Δ includes a formula $\psi \to \chi$ are very similar. Lastly, suppose that $\Gamma, \varphi \Rightarrow \varphi, \Delta$ contains only propositional variables and box formulas. Then there is a derivation of the sequent ending with an application of $(\Box_{k,n})$. The case where $\Box\varphi$ does not appear in the premise is trivial, so just consider the case

$$\frac{\Pi_0, k_0[\varphi] \Rightarrow \quad \Pi_1, k_1[\varphi] \Rightarrow k[\varphi] \quad \{\Pi_i, k_i[\varphi] \Rightarrow k[\psi_i]\}_{i=2}^{n}}{\Sigma, \Box\Pi, \Box\varphi \Rightarrow \Box\varphi, \Box\psi_2, \dots, \Box\psi_n, \Sigma} \ (\Box_{k,n})$$

where $k\Pi = \Pi_0 \uplus \Pi_1 \uplus \dots \uplus \Pi_n$ and $k = k_0 + k_1 + \dots + k_n$. By the induction hypothesis, we obtain

$$\vdash_{\mathrm{GK}(\mathbb{R})^{\mathrm{r}}} \Pi_1 \Rightarrow (k - k_1)[\varphi].$$

By Lemma 6.2 (the admissibility of (MIX)), we have derivations in $\mathrm{GK}(\mathbb{R})^{\mathrm{r}}$ of

$$k_0\Pi_1, (k-k_1)\Pi_0, (k-k_1)k_0[\varphi] \Rightarrow (k-k_1)k_0[\varphi]$$
$$k_i\Pi_1, (k-k_1)\Pi_i, (k-k_1)k_i[\varphi] \Rightarrow (k-k_1)k_i[\varphi], (k-k_1)k[\psi_i] \quad 2 \le i \le n.$$

So, by the induction hypothesis, we have derivations in $\mathrm{GK}(\mathbb{R})^{\mathrm{r}}$ of

$$k_0\Pi_1, (k-k_1)\Pi_0 \Rightarrow$$
$$k_i\Pi_1, (k-k_1)\Pi_i \Rightarrow (k-k_1)k[\psi_i] \quad 2 \le i \le n.$$

Now by an application of $(\Box_{(k-k_1)k,n-1})$, we have a derivation ending with

$$\frac{k_0\Pi_1, (k-k_1)\Pi_0 \Rightarrow \quad \{k_i\Pi_1, (k-k_1)\Pi_i \Rightarrow (k-k_1)k[\psi_i]\}_{i=2}^{n}}{\Sigma, \Box\Pi \Rightarrow \Box\psi_2, \dots, \Box\psi_n, \Sigma}$$

where $(k-k_1)k\Pi = (k_0 + k_2 + \dots + k_n)(\Pi_0 \uplus \Pi_1 \uplus \dots \uplus \Pi_n)$. □

References

[1] Borgwardt, S., F. Distel and R. Peñaloza, *The limits of decidability in fuzzy description logics with general concept inclusions*, Artificial Intelligence **218** (2015), pp. 23–55.

[2] Bou, F., M. Cerami and F. Esteva, *Finite-valued Łukasiewicz modal logic is PSPACE-complete*, in: *Proceedings of IJCAI 2011*, 2011, pp. 774–779.

[3] Bou, F., F. Esteva, L. Godo and R. Rodríguez, *On the minimum many-valued logic over a finite residuated lattice*, Journal of Logic and Computation **21** (2011), pp. 739–790.

[4] Caicedo, X., G. Metcalfe, R. Rodríguez and J. Rogger, *Decidability in order-based modal logics, Journal of Computer System Sciences*, in press.

[5] Caicedo, X., G. Metcalfe, R. Rodríguez and J. Rogger, *A finite model property for Gödel modal logics*, in: *Proceedings of WoLLIC 2013, Springer LNCS 8071*, LNCS **8701** (2013), pp. 226–237.

[6] Caicedo, X. and R. Rodríguez, *Standard Gödel modal logics*, Studia Logica **94** (2010), pp. 189–214.

[7] Caicedo, X. and R. Rodríguez, *Bi-modal Gödel logic over [0,1]-valued Kripke frames*, Journal of Logic and Computation **25** (2015), pp. 37–55.

[8] Casari, E., *Comparative logics and abelian ℓ-groups*, in: C. Bonotto, R. Ferro, S. Valentini and A. Zanardo, editors, *Logic Colloquium '88*, Elsevier, 1989 pp. 161–190.

[9] Diaconescu, D. and G. Georgescu, *Tense operators on MV-algebras and Łukasiewicz-Moisil algebras*, Fundamenta Informaticae **81** (2007), pp. 379–408.

[10] Fitting, M. C., *Many-valued modal logics*, Fundamenta Informaticae **15** (1991), pp. 235–254.

[11] Fitting, M. C., *Many-valued modal logics II*, Fundamenta Informaticae **17** (1992), pp. 55–73.
[12] Godo, L., P. Hájek and F. Esteva, *A fuzzy modal logic for belief functions*, Fundamenta Informaticae **57** (2003), pp. 127–146.
[13] Godo, L. and R. Rodríguez, *A fuzzy modal logic for similarity reasoning*, in: *Fuzzy Logic and Soft Computing* (1999), pp. 33–48.
[14] Hájek, P., "Metamathematics of Fuzzy Logic," Kluwer, Dordrecht, 1998.
[15] Hájek, P., *Making fuzzy description logic more general*, Fuzzy Sets and Systems **154** (2005), pp. 1–15.
[16] Hájek, P., D. Harmancová, F. Esteva, P. Garcia and L. Godo, *On modal logics for qualitative possibility in a fuzzy setting*, in: *Proceedings of UAI 1994*, 1994, pp. 278–285.
[17] Hájek, P., D. Harmancová and R. Verbrugge, *A qualitative fuzzy possibilistic logic*, International Journal of Approximate Reasoning **12** (1995), pp. 1–19.
[18] Hansoul, G. and B. Teheux, *Extending Łukasiewicz logics with a modality: Algebraic approach to relational semantics*, Studia Logica **101** (2013), pp. 505–545.
[19] Khachiyan, L. G., *A polynomial algorithm in linear programming*, Soviet Mathematics Doklady **20** (1979), pp. 191–194.
[20] Kulacka, A., D. Pattinson and L. Schröder, *Syntactic labelled tableaux for Łukasiewicz fuzzy ALC*, in: *Proceedings of IJCAI 2013*, 2013.
[21] Kupke, C. and D. Pattinson, *On modal logics of linear inequalities*, in: *Proceedings of AiML 2010* (2010), pp. 235–255.
[22] Marti, M. and G. Metcalfe, *Hennessy-Milner properties for many-valued modal logics*, in: *Proceedings of AiML 2014* (2014), pp. 407–420.
[23] Matteo, M. and A. Simpson, *Łukasiewicz mu-calculus*, in: *Proceedings Workshop on Fixed Points in Computer Science*, EPCTS **126** (2013), pp. 87–104.
[24] Metcalfe, G. and N. Olivetti, *Towards a proof theory of Gödel modal logics*, Logical Methods in Computer Science **7** (2011), pp. 1–27.
[25] Metcalfe, G., N. Olivetti and D. Gabbay, *Sequent and hypersequent calculi for abelian and Łukasiewicz logics*, ACM Transactions on Computational Logic **6** (2005), pp. 578–613.
[26] Meyer, R. K. and J. K. Slaney, *Abelian logic from A to Z*, in: *Paraconsistent Logic: Essays on the Inconsistent*, Philosophia Verlag, 1989 pp. 245–288.
[27] Priest, G., *Many-valued modal logics: a simple approach*, Review of Symbolic Logic **1** (2008), pp. 190–203.
[28] Schockaert, S., M. D. Cock and E. Kerre, *Spatial reasoning in a fuzzy region connection calculus*, Artificial Intelligence **173** (2009), pp. 258–298.
[29] Straccia, U., *Reasoning within fuzzy description logics*, Journal of Artificial Intelligence Research **14** (2001), pp. 137–166.

Fully Arbitrary Public Announcements

Hans van Ditmarsch

LORIA - CNRS / University of Lorraine

Wiebe van der Hoek

University of Liverpool

Louwe B. Kuijer

University of Liverpool
LORIA - CNRS / University of Lorraine

Abstract

In Arbitrary Public Announcement Logic (APAL) an operator $\Box$ is used. The intended meaning of $\Box\varphi$ is "for every ψ we have $[\psi]\varphi$." However, for technical reasons the semantics of APAL do not entirely match the intended meaning: in APAL the formula $\Box\varphi$ holds if and only if for every $\Box$-free ψ we have $[\psi]\varphi$. Here we introduce Fully Arbitrary Public Announcement Logic (F-APAL), where the semantics do match the intended meaning: in F-APAL the formula $\Box\varphi$ holds if and only if for every ψ we have $[\psi]\varphi$.

Keywords: Dynamic Epistemic Logic, Public Announcements, Arbitrary Public Announcements.

1 Introduction

One line of research in dynamic epistemic logic is to add "arbitrary" versions of dynamic operators. Examples of such logics include Arbitrary Public Announcement Logic [4,5], Group Announcement Logic [1], Arbitrary Action Model Logic [14], Refinement Modal Logic [9] and Arbitrary Arrow Update Logic [11]. The intuition behind these arbitrary operators is that they represent universal quantification over their non-arbitrary counterpart.

As the title suggests, we will focus on Arbitrary Public Announcement Logic (APAL). APAL is based on Public Announcement Logic (PAL) [6], but adds an extra operator $\Box$.[1] Intuitively, $\Box\varphi$ is intended to mean "after every public

[1] The symbol $\Box_a$ is also often used for "agent a knows that..." Here we use K_a to denote the knowledge operator, and reserve $\Box$ for arbitrary public announcement operators.

announcement ψ, φ will hold." So we would like $\Box$ to satisfy the following property:

$$\mathcal{M}, w \models \Box\varphi \Leftrightarrow \forall\psi \in \mathcal{L}_{APAL} : \mathcal{M}, w \models [\psi]\varphi \tag{1}$$

where $\mathcal{L}_{APAL}$ is the language of APAL. Unfortunately, (1) is circular: if we want to use it to determine the value of $\Box\varphi$, then we have to determine whether $[\psi]\varphi$ holds for all $\psi \in \mathcal{L}_{APAL}$. So in particular, we have to determine whether $[\Box\varphi]\varphi$ holds, which in turn requires us to determine whether $\Box\varphi$ holds. But that's where we started out; if we want to find out whether $\Box\varphi$ holds, we first have to know whether $\Box\varphi$ holds. In order to avoid this circularity, [4,5] define $\Box$ not by (1) but by

$$\mathcal{M}, w \models \Box\varphi \Leftrightarrow \forall\psi \in \mathcal{L}_{PAL} : \mathcal{M}, w \models [\psi]\varphi \tag{2}$$

where $\mathcal{L}_{PAL}$ is the set of formulas that do not themselves contain the operator $\Box$. Since (2) is non-circular, it can be used as a definition of $\Box$.

If we could have used (1) as definition of $\Box$, then $\Box$ would trivially have satisfied (1). But we cannot. Still, not all hope is lost. Even though (1) is not suitable as a *definition* of $\Box$, it might be a *property* of $\Box$ under another definition. Unfortunately, as shown in [16], (2) is not such a definition: in APAL, $\Box$ does not satisfy (1).

Let us reflect briefly on what it means for $\Box$ not to satisfy (1). The operator $\Box$ is called an "arbitrary public announcement." This name is justified by the intuition that $\Box\varphi$ is supposed to hold if and only if $[\psi]\varphi$ holds for every public announcement $[\psi]$. So $\Box$ is supposed to represent any announcement $[\psi]$; an arbitrary announcement indeed. But there are φ and ψ in APAL such that $\Box\varphi$ holds but $[\psi]\varphi$ does not. In other words, $\Box$ in APAL does not represent every possible announcement $[\psi]$, so it is not a fully arbitrary public announcement.

In this paper, we introduce Fully Arbitrary Public Announcement Logic (F-APAL). The idea behind F-APAL is that our $\Box$ operator, unlike the APAL one, will represent a fully arbitrary public announcement. More precisely, we will define a logical language $\mathcal{L}$ that has $\Box$ as an operator and semantics for $\mathcal{L}$ such that

$$\mathcal{M}, w \models \Box\varphi \Leftrightarrow \forall\psi \in \mathcal{L} : \mathcal{M}, w \models [\psi]\varphi \tag{*}$$

is satisfied. The price we pay for this fully arbitrary announcement operator $\Box$ is that our language $\mathcal{L}$ contains multiple auxiliary operators. If fact, we use a proper class of indexed auxiliary operators: $\{\Box_\alpha \mid \alpha \text{ is an ordinal}\}$.

The remainder of this paper is structured as follows. First, in Section 2, we briefly discuss the problems with circularity, and show that the circularity in (*) is vicious, so (*) cannot be used as a definition of $\Box$. Then, in Section 3, we introduce the language and semantics of our logic F-APAL. In Section 4, we show that $\Box$ satisfies (*) in F-APAL. In Section 5, we prove a few other properties of F-APAL. Finally, in Section 6, we discuss the relation between F-APAL and fixed points.

2 Circularity

The easiest way to obtain semantics for $\Box$ that satisfy (*), would be to use (*) as a definition of $\Box$. Certainly, (*) is circular, and circular definitions are generally frowned upon. But not all circularity is vicious, and non-viciously circular properties can be used as definitions. For example, many fixed points can be given a circular definition. The question, then, is whether the circularity in (*) is vicious.

We are working in a two-valued modal logic. So a property can be used as a definition for an operator X of arity n if and only if, for each pointed model $\mathcal{M}, w$ and each tuple $\varphi_1, \cdots, \varphi_n$, exactly one of $\mathcal{M}, w \models X(\varphi_1, \cdots, \varphi_n)$ and $\mathcal{M}, w \not\models X(\varphi_1, \cdots, \varphi_n)$ satisfies the property. There are two ways in which a property can fail to be suitable as a definition. Firstly, the property can be *inconsistent*, and allow neither truth value for a sentence in some pointed model. A typical example of an inconsistent property is the (modal) liar: [2]

$$\text{“This sentence is false (in this pointed model).”} \tag{3}$$

If we suppose that (3) is true, then the claim made in (3) is is true, so (3) is false. In two-valued logic this is a contradiction, so (3) cannot be true. If, on the other hand, we suppose that (3) is false, then the claim made in (3) is false, so (3) is not false. This too is a contradiction, so (3) cannot be false either. It follows that (3) allows neither truth value, so it is inconsistent.

Secondly, a property can be *underdetermined*, and allow both truth values for a sentence in some pointed model. A typical example of an underdetermined sentence is the (modal) truth teller:

$$\text{“This sentence is true (in this pointed model).”} \tag{4}$$

If we suppose that (4) is true, then the claim made in (4) is true, so (4) is true. This does not lead to a contradiction, so we can consistently say that (4) is true. If, on the other hand, we suppose that (4) is false, then the claim made in (4) is false, so (4) is not true and therefore false. Again, we do not arrive at a contradiction, so we can consistently say that (4) is false. We are working in a two-valued logic, so we cannot assign (4) both truth values at the same time. We are, however, free to choose either of the truth values. So (4) is underdetermined.

If a circular property is inconsistent or underdetermined, then the circularity is vicious. But that does not mean that the two kinds of vicious circularity are equally bad. In both cases, we cannot use the circular property as a definition. But with an underdetermined property we can try to find a different definition that satisfies the property, whereas with an inconsistent property we have no choice but to give up.

Sadly, the circularity in (*) turns out to be vicious. Fortunately, however, it exhibits the less problematic kind of viciousness: (*) is underdetermined but

[2] Note that, in the terminology used above, the liar sentence is a nullary operator.

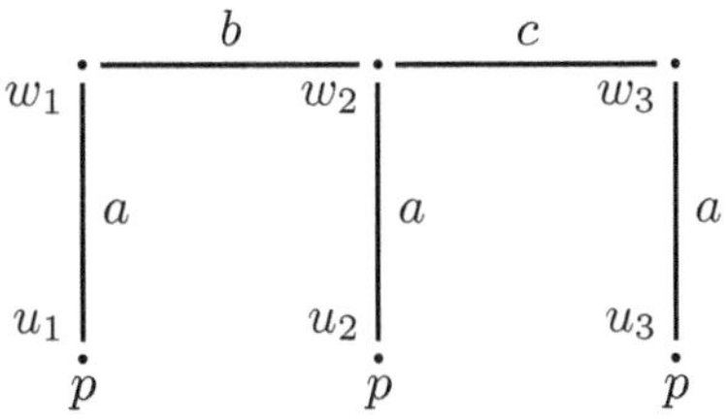

Fig. 1. The S5 model $\mathcal{M}_{Un}$. Reflexive arrows are not drawn, for reasons of clarity.

consistent. So while we cannot use (*) as a definition of $\Box$, we may be able to find semantics for $\Box$ that satisfy (*). We prove the consistency of (*) in Sections 3 and 4, by constructing semantics that satisfy it. In this section, we show that (*) is underdetermined.

A slight complication is that whether (*) is underdetermined may depend on the other connectives that are present in the language. We have not defined the language $\mathcal{L}$ yet, so we will use a smaller language in this section. We assume that our language uses only propositional atoms and the connectives $\neg, \vee, K_a, [\psi]$ and $\Box$. All except $\Box$ are given the usual semantics,[3] we make no assumptions about the semantics of $\Box$. Additionally, for ease of notation, we use $\wedge, \hat{K}_a, \langle\psi\rangle$ and $\Diamond$ as duals of $\vee, K_a, [\psi]$ and $\Box$, respectively.

In this section we do not consider the auxiliary operators $\Box_\alpha$. This is only for reasons of clarity of presentation, however. If we added the $\Box_\alpha$ operators (with the semantics as given in Section 3), then the proofs given in this section would still work with only very minor modifications.

Consider the S5 model $\mathcal{M}_{Un}$ shown in Figure 1 and let $\xi := \hat{K}_a p \wedge \hat{K}_b K_a \neg p \wedge \hat{K}_c K_a \neg p$. So ξ holds if and only if (1) there is an a-accessible p world, (2) there is a b-accessible world where there is no a-accessible p world and (3) there is a c-accessible world where there is no a-accessible p world. Regardless of which worlds of $\mathcal{M}_{Un}$ we retain or eliminate, the only world in which ξ can possibly hold is w_2. And for ξ to hold in w_2 it must be the case that exactly w_1, w_2, w_3 and u_2 are retained while u_1 and u_3 are eliminated.

In the next two lemmas, we show that $\mathcal{M}_{Un}, w_2 \not\models \Diamond\xi$ and $\mathcal{M}_{Un}, w_2 \models \Diamond\xi$ are both consistent with (*). In order to do this, we use the following observation. By the definition of public announcements, we have $\mathcal{M}, w \models [\psi]\varphi$ if and only if $\mathcal{M}, w \models \psi$ and $\mathcal{M}_\psi, w \models \varphi$, where $\mathcal{M}_\psi$ is the restriction of $\mathcal{M}$ to those worlds satisfy ψ. This implies that, for every ψ_1, ψ_2 that have the same extension, $\mathcal{M}, w \models [\psi_1]\varphi$ if and only if $\mathcal{M}, w \models [\psi_2]\varphi$. Now, consider the right hand side of (*): $\forall\psi \in \mathcal{L} : \mathcal{M}, w \models [\psi]\varphi$. Because it is only the extension of ψ that matters, that is equivalent to

$$\forall x \in \{[\![\psi]\!]_{\mathcal{M}} \mid \psi \in \mathcal{L}\} : \mathcal{M}_x, w \models \varphi,$$

where $\mathcal{M}_x$ is the restriction of $\mathcal{M}$ to x and, by convention, $\mathcal{M}_x, w \models \varphi$ for

[3] We assume that the reader is familiar with the standard semantics for these operators. If not, see Section 3.

every φ if $w \notin x$.[4] So (*) is equivalent to

$$\mathcal{M}, w \models \Box\varphi \Leftrightarrow \forall x \in \{[\![\psi]\!]_{\mathcal{M}} \mid \psi \in \mathcal{L}\} : \mathcal{M}_x, w \models \varphi. \qquad (**)$$

In the following two lemmas, we start by defining a set X. We then define the semantics for $\Box$ by

$$\mathcal{M}, w \models \Box\varphi \Leftrightarrow \forall x \in X : \mathcal{M}_x, w \models \varphi. \qquad (5)$$

We then show that, under these semantics, X is exactly the set of extensions on $\mathcal{M}$, so $X = \{[\![\psi]\!]_{\mathcal{M}} \mid \psi \in \mathcal{L}\}$. It follows immediately that (**) is satisfied, and therefore (*) is satisfied as well.

Lemma 2.1 *There is a valuation for $\Diamond\xi$ that is consistent with* (*) *such that $\mathcal{M}_{Un}, w_2 \not\models \Diamond\xi$.*

Proof. Take $X = \{\varnothing, \{w_1, w_2, w_3\}, \{u_1, u_2, u_3\}, \{w_1, w_2, w_3, u_1, u_2, u_3\}\}$, and define $\Box$ by (5). We show that X is exactly the set of extension on $\mathcal{M}_{Un}$. First, note that for every $x \in X$ there is a formula ψ such that $x = [\![\psi]\!]$: we have $\varnothing = [\![p \wedge \neg p]\!]$, $\{w_1, w_2, w_3\} = [\![\neg p]\!]$, $\{u_1, u_2, u_3\} = [\![p]\!]$ and $\{w_1, w_2, w_3, u_1, u_2, u_3\} = [\![p \vee \neg p]\!]$.

Left to show is that for every formula ψ, there is some $x \in X$ such that $x = [\![\psi]\!]$. So we need to show that ψ cannot distinguish between the three columns of the model, i.e. there is no formula that can distinguish between w_1, w_2 and w_3 or between u_1, u_2 and u_3. We do this by induction. First, as base case, note that there is no atomic formula that can distinguish between the three columns. Then, assume as induction hypothesis that ψ is not atomic and that no strict subformula of ψ can distinguish between the columns. We continue by case distinction on the main connective of ψ.

- Suppose the main connective of ψ is not $\Box$. Then ψ can only distinguish between the columns if at least one of its strict subformulas can. By the induction hypothesis this is not the case.
- Suppose $\psi = \Box\psi'$. Then ψ can distinguish between two columns only if ψ' distinguishes between the columns in $(\mathcal{M}_{Un})_x$ for some $x \in X$. For $x = \varnothing$ this is trivially not the case, by convention we have $(\mathcal{M}_{Un})_\varnothing, v \models \psi$ for every world v, since $v \notin \varnothing$.

  For $x = \{w_1, w_2, w_3\}$ and $x = \{u_1, u_2, u_3\}$ we have that all worlds in the resulting model $(\mathcal{M}_{Un})_x$ agree on all propositional variables. So no formula using atoms, $\neg, \vee$ and K_a can distinguish between any two worlds. Furthermore, this property is retained in submodels of $(\mathcal{M}_{Un})_x$, so the operators $[\chi]$ and $\Box$ cannot help distinguish between any worlds either. In particular, ψ' cannot distinguish between the columns.

  Finally, for $x = \{w_1, w_2, w_3, u_1, u_2, u_3\}$ we have $(\mathcal{M}_{Un})_x = \mathcal{M}_{Un}$. So by the induction hypothesis ψ' cannot distinguish between the columns.

[4] This convention corresponds to the convention in public announcement logic that $\mathcal{M}, w \models [\psi]\varphi$ for all φ if $\mathcal{M}, w \not\models \psi$.

For every $x \in X$, we have seen that ψ' cannot distinguish between the columns of $(\mathcal{M}_{Un})_x$. So $\psi = \Box\psi'$ cannot distinguish between the columns of $\mathcal{M}_{Un}$.

So in both cases ψ cannot distinguish between the columns of the model. This completes the induction step, thereby showing that for every ψ there is an $x \in X$ such that $[\![\psi]\!] = x$. It follows that X is indeed the set of extensions on $\mathcal{M}_{Un}$, so (*) is satisfied.

Furthermore, we have $\mathcal{M}_{Un}, w_2 \models \Diamond\xi$ if and only if $\{w_1, w_2, w_3, u_2\} \in X$. This is not the case, so $\mathcal{M}_{Un}, w_2 \not\models \Diamond\xi$ is consistent with (*).

□

Lemma 2.2 *There is a valuation for $\Diamond\xi$ that is consistent with* (*) *such that $\mathcal{M}_{Un}, w_2 \models \Diamond\xi$.*

Proof. Take $X = 2^{\{w_1,w_2,w_2,u_1,u_2,u_3\}}$, and define $\Box$ by (5). We show that X is the set of extensions on $\mathcal{M}_{Un}$. First, note that for every ψ we trivially have $[\![\psi]\!] \in X$. Left to show is that for every $x \in X$ there is some formula ψ such that $x = [\![\psi]\!]$.

We have $\mathcal{M}_{Un}, w_2 \models \Diamond\xi$, since $\{w_1, w_2, w_3, u_2\} \in X$. As discussed above, w_2 is also the only world where $\Diamond\xi$ holds. But then every world v of $\mathcal{M}_{Un}$ can be uniquely identified by some formula δ_v. Specifically, we have $\delta_{w_1} = \neg\Diamond\xi \wedge \hat{K}_b\Diamond\xi$, $\delta_{w_2} = \Diamond\xi$, $\delta_{w_3} = \neg\Diamond\xi \wedge \hat{K}_c\Diamond\xi$ and $\delta_{u_i} = p \wedge \hat{K}_a\delta_{w_i}$ for $i \in \{1, 2, 3\}$. Every $x \in X$ is then the extension of the appropriate disjunction of such δ_v. So X is indeed the set of extensions, which implies that $\mathcal{M}_{Un}, w_2 \models \Diamond\xi$ is consistent with (*). □

Corollary 2.3 *The characterization* (*) *is underdetermined.*

We should note that, although $\mathcal{M}_{Un}, w_2 \not\models \Diamond\xi$ and $\mathcal{M}_{Un}, w_2 \models \Diamond\xi$ are both consistent with (*), this does not mean that we consider both to be equally good solutions. The worlds w_1, w_2 and w_3 in $\mathcal{M}_{Un}$ are bisimilar to each other. In a well behaved modal logic we would therefore expect them to satisfy the same formulas. If we take $\mathcal{M}_{Un}, w_2 \not\models \Diamond\xi$ then they do indeed satisfy the same formulas. But if we take $\mathcal{M}_{Un}, w_2 \models \Diamond\xi$ then $\Diamond\xi$ distinguishes between w_2 on the one hand and w_1 and w_3 on the other. There seems to be no compelling reason to allow the $\Diamond$ operator to break bisimilarity, so we prefer $\mathcal{M}_{Un}, w_2 \not\models \Diamond\xi$ over $\mathcal{M}_{Un}, w_2 \models \Diamond\xi$. Fortunately, it will turn out that in our semantics for F-APAL we have $\mathcal{M}_{Un}, w_2 \not\models \Diamond\xi$.

3 Language and Semantics

In the previous section we showed that (*) is underdetermined, and therefore not suitable as a definition. If we want $\Box$ to satisfy (*), we will have to find deterministic semantics for $\Box$ that satisfy (*). In this section we introduce such semantics, in the next section we prove that the semantics satisfy (*). Our logic F-APAL uses ordinals, so before defining the language and semantics of F-APAL we give a brief reminder of the properties of ordinals that we need. A more thorough introduction to ordinals can be found in most textbooks about set theory, see for example [15].[5] The following two definitions are as usual.

Definition 3.1 A set x is *transitive* if for all $y \in x$ and $z \in y$ we have $z \in x$.

Definition 3.2 A set α is an *ordinal number* if α is transitive and, for all $\beta \in \alpha$, β is a transitive set. The class $\{\alpha \mid \alpha \text{ is an ordinal number}\}$ of all ordinal numbers is denoted Ord.

If α and β are ordinals numbers, we write $\alpha < \beta$ if $\alpha \in \beta$ and $\alpha \leq \beta$ if $\alpha < \beta$ or $\alpha = \beta$. Furthermore, we write $\alpha + 1$ for the set $\{\alpha\} \cup \alpha$.

We often omit the word "number" and speak simply of an ordinal α. We also follow the usual convention of using the natural numbers to denote the finite ordinals; 0 represents $\varnothing$, 1 represents $\varnothing + 1 = \{\varnothing\}$, and so on.

We will use a few relatively well known properties of ordinal numbers that we state here without proof.

Lemma 3.3 *The following properties hold.*

- Ord *is not a set, it is a proper class,*
- *for any* $\alpha, \beta \in \text{Ord}$, *either* $\alpha \leq \beta$ *or* $\beta \leq \alpha$,
- *for any set* X *of ordinals, the set* $\sup(X) := \bigcup_{\alpha \in X} \alpha$ *is an ordinal, and* $\alpha \leq \sup(X)$ *for all* $\alpha \in X$,
- *for any class* X *of ordinals, there is an ordinal* $\min(X) \in X$ *such that* $\min(X) \leq \alpha$ *for all* $\alpha \in X$.

This finishes our very brief discussion of ordinals. Let us continue by defining our language $\mathcal{L}$.

Definition 3.4 Let a countable set $\mathcal{P}$ of propositional variables and a finite set $\mathcal{A}$ of agents be given. The language $\mathcal{L}(\mathcal{P}, \mathcal{A})$ is given by the following normal form:

$$\varphi ::= p \mid \neg\varphi \mid \varphi \vee \varphi \mid K_a\varphi \mid [\varphi]\varphi \mid \Box_\alpha\varphi \mid \Box\varphi$$

where $p \in \mathcal{P}$, $a \in \mathcal{A}$ and $\alpha \in \text{Ord}$.

For $\alpha \in \text{Ord}$, the sub-language $\mathcal{L}_\alpha(\mathcal{P}, \mathcal{A})$ is the class of formulas that contain neither $\Box$ nor $\Box_\beta$ with $\beta \geq \alpha$.

We write $\mathcal{L}$ and $\mathcal{L}_\alpha$ for $\mathcal{L}(\mathcal{P}, \mathcal{A})$ and $\mathcal{L}_\alpha(\mathcal{P}, \mathcal{A})$ respectively where this should not cause confusion. Furthermore, we use $\wedge$, $\hat{K}_a$, $\langle\varphi\rangle$, $\Diamond_\alpha$ and $\Diamond$ in the usual way as abbreviations.

Note that $\mathcal{L}$ is a proper class. It is unusual for a logic to have a proper class of formulas, and it has certain consequences that F-APAL does. For example, the validities of F-APAL are trivially not recursively enumerable—or enumerable at all. For the results presented in this paper, however, the fact that $\mathcal{L}$ is a proper class does not give any trouble. With the exception of the proofs that depend on $\mathcal{L}$ being a proper class, all proofs proceed in the same way as they would have if $\mathcal{L}$ had been a set.

[5] Alternatively, see [7] for an introduction to ordinals and circularity.

We evaluate our language on S5-models. The choice of this class of models is more for historical reasons than practical ones: the original APAL papers [4,5] use S5 models and we follow their example, but very few of the proofs in this paper depend on the fact that we are working in S5.

Definition 3.5 A *model* $\mathcal{M}$ is a triple $\mathcal{M} = (W, R, V)$ where W is a set of worlds, $R : \mathcal{A} \to \wp(W \times W)$ assigns to each agent an equivalence relation on W and $V : \mathcal{P} \to \wp W$. A pointed model is a pair $\mathcal{M}, w$ where $\mathcal{M} = (W, R, V)$ is a model and $w \in W$. We write $R_a(w)$ for $\{w' \mid (w, w') \in R(a)\}$.

The semantics for F-APAL are as follows.

Definition 3.6 The satisfaction relation $\models$ is given recursively by

$$\begin{array}{lcl}
\mathcal{M}, w \models p & \Leftrightarrow & w \in V(p), \\
\mathcal{M}, w \models \neg\varphi & \Leftrightarrow & \mathcal{M}, w \not\models \varphi, \\
\mathcal{M}, w \models \varphi_1 \vee \varphi_2 & \Leftrightarrow & \mathcal{M}, w \models \varphi_1 \text{ or } \mathcal{M}, w \models \varphi_2, \\
\mathcal{M}, w \models K_a\varphi & \Leftrightarrow & \mathcal{M}, w' \models \varphi \text{ for all } w' \in R_a(w), \\
\mathcal{M}, w \models [\varphi_1]\varphi_2 & \Leftrightarrow & \mathcal{M}, w \not\models \varphi_1 \text{ or } \mathcal{M}_{\varphi_1}, w \models \varphi_2, \\
\mathcal{M}, w \models \Box_\alpha\varphi & \Leftrightarrow & \mathcal{M}, w \models [\psi]\varphi \text{ for all } \psi \in \mathcal{L}_\alpha, \\
\mathcal{M}, w \models \Box\varphi & \Leftrightarrow & \mathcal{M}, w \models \Box_\alpha\varphi \text{ for all } \alpha \in \text{Ord}.
\end{array}$$

Where $\mathcal{M}_\varphi$ is given by $\mathcal{M}_\varphi = (W_\varphi, R_\varphi, V_\varphi)$, $W_\varphi = \{w \in W \mid \mathcal{M}, w \models \varphi\}$, $R_\varphi(a) = R(a) \cap (W_\varphi \times W_\varphi)$ and $V_\varphi(p) = V(p) \cap W_\varphi$.

We write $\mathcal{M} \models \varphi$ if $\mathcal{M}, w \models \varphi$ for every world w of $\mathcal{M}$, and $\models \varphi$ if $\mathcal{M} \models \varphi$ for every model $\mathcal{M}$.

Note that $\mathcal{L}_0$ is the language without any $\Box_\alpha$ or $\Box$ operators, so $\mathcal{L}_0 = \mathcal{L}_{PAL}$. The operator $\Box_0$ quantifies over all announcements from $\mathcal{L}_0 = \mathcal{L}_{PAL}$, so $\Box_0$ has the same semantics as the APAL operator from the original APAL papers [4,5]. Thus, APAL is embedded in F-APAL as the fragment $\mathcal{L}_1$.

4 Fully Arbitrary Public Announcements

In this section we show that $\Box$, as defined above, is a fully arbitrary public announcement. So we show that $\Box$ satisfies (*). Before we can do so, however, we need a few auxiliary definitions and one lemma.

Definition 4.1 Let $\varphi \in \mathcal{L}$ and $\beta \in \text{Ord}$. Then $\downarrow_\beta(\varphi) \in \mathcal{L}_{\beta+1}$ is the formula obtained by replacing all occurrences of $\Box$ and $\Box_\alpha$ where $\alpha > \beta$ by $\Box_\beta$.

Definition 4.2 Let $\mathcal{M} = (W, R, V)$ be a model, $\varphi \in \mathcal{L}$ and let α be an ordinal. We say that α *approximates* Ord for φ on $\mathcal{M}$ if for every submodel $\mathcal{M}' = (W', R', V')$ of $\mathcal{M}$, every $w \in W'$, every subformula ψ of φ and every $\beta \geq \alpha$ we have

$$\mathcal{M}', w \models \psi \Leftrightarrow \mathcal{M}', w \models \downarrow_\beta(\psi).$$

We write $\text{Approx}(\mathcal{M}, \varphi)$ for the class of ordinals that approximate Ord on $\mathcal{M}$ for φ.

Lemma 4.3 (Approximation Lemma) *For every model* $\mathcal{M} = (W, R, V)$ *and every* $\varphi \in \mathcal{L}$, *the class* $\text{Approx}(\mathcal{M}, \varphi)$ *is non-empty.*

Proof. By induction on the construction of φ. If φ is atomic then it does not contain any boxes, so the lemma is trivial. Suppose therefore as induction hypothesis that φ is not atomic and that the lemma holds for all strict subformulas of φ. Since, by assumption, $\text{Approx}(\mathcal{M}, \psi)$ is nonempty for every strict subformula ψ of φ, let $\alpha_\psi \in \text{Approx}(\mathcal{M}, \psi)$.

As usual, we continue by a case distinction on the main connective of φ. Most of the cases are quite trivial, so we do not give their proofs in much detail.

- Suppose $\varphi = \neg\psi$. Then $\alpha_\psi \in \text{Approx}(\mathcal{M}, \varphi)$.
- Suppose $\varphi = \psi_1 \vee \psi_2$. Then $\max(\alpha_{\psi_1}, \alpha_{\psi_2}) \in \text{Approx}(\mathcal{M}, \varphi)$.
- Suppose $\varphi = K_a\psi$. Then $\alpha_\psi \in \text{Approx}(\mathcal{M}, \varphi)$, since α_ψ approximates Ord on every world of $\mathcal{M}$.
- Suppose $\varphi = [\psi_1]\psi_2$. Let $\alpha = \max(\alpha_{\psi_1}, \alpha_{\psi_2})$. Then, for every $\beta \geq \alpha$, $\mathcal{M}_{\psi_1} = \mathcal{M}_{\downarrow_\beta(\psi_1)}$. Furthermore, α_{ψ_2} approximates Ord not just on $\mathcal{M}$ but also on all of its submodels, so ψ_2 is equivalent to $\downarrow_\beta(\psi_2)$ on the updated model. Clearly, $\downarrow_\beta([\psi_1]\psi_2) = [\downarrow_\beta(\psi_1)]\downarrow_\beta(\psi_2)$, so it follows that $\downarrow_\beta([\psi_1]\psi_2)$ is equivalent to $[\psi_1]\psi_2$ on $\mathcal{M}$. So $\alpha \in \text{Approx}(\mathcal{M}, \varphi)$.
- Suppose $\varphi = \Box_\gamma\psi$. Then $\max(\gamma, \alpha_\psi) \in \text{Approx}(\mathcal{M}, \varphi)$.
- Suppose $\varphi = \Box\psi$. Let $\mathfrak{M}$ be the set of pointed models $\mathcal{M}', w$ such that $\mathcal{M}'$ is a submodel of $\mathcal{M}$. We can partition $\mathfrak{M}$ into $\mathfrak{M}^+ := \{\mathcal{M}', w \in \mathfrak{M} \mid \mathcal{M}', w \models \Box_\alpha\psi \text{ for all ordinals } \alpha\}$ and $\mathfrak{M}^- := \{\mathcal{M}', w \in \mathfrak{M} \mid \mathcal{M}', w \not\models \Box_\alpha\psi \text{ for some ordinal } \alpha\}$.

  For $\mathcal{M}', w \in \mathfrak{M}$, let $\alpha_{\mathcal{M}',w}$ be given by

  $$\alpha_{\mathcal{M}',w} := \begin{cases} 0 & \text{if } \mathcal{M}', w \in \mathcal{M}^+ \\ \min\{\alpha \mid \mathcal{M}', w \not\models \Box_\alpha\psi'\} & \text{if } \mathcal{M}', w \in \mathcal{M}^- \end{cases}$$

  For every $\beta \geq \alpha_{\mathcal{M}',w}$ we have $\mathcal{M}', w \models \Box\psi \Leftrightarrow \mathcal{M}', w \models \Box_\beta\psi$.

  Now, let $\alpha_\mathfrak{M} := \sup\{\alpha_{\mathcal{M}',w} \mid \mathcal{M}', w \in \mathfrak{M}\}$. This supremum exists and is itself an ordinal, because $\mathfrak{M}$ is a set. Take $\alpha = \max(\alpha_\mathfrak{M}, \alpha_\psi)$. For every $\mathcal{M}', w$ and every $\beta \geq \alpha$ we have $\mathcal{M}', w \models \Box\psi \Leftrightarrow \mathcal{M}', w \models \Box_\beta\psi$ since $\beta \geq \alpha_\mathfrak{M}$ and therefore $\beta \geq \alpha_{\mathcal{M}',w}$. Furthermore, $\mathcal{M}', w \models \Box_\beta\psi \Leftrightarrow \mathcal{M}', w \models \downarrow_\beta(\Box_\beta\psi)$ since $\beta \geq \alpha_\psi$ (and $\beta \geq \beta$, so $\downarrow_\beta(\Box_\beta\psi) = \Box_\beta\downarrow_\beta(\psi)$). It follows that $\mathcal{M}', w \models \Box\psi \Leftrightarrow \mathcal{M}', w \models \downarrow_\beta(\Box\psi)$. This holds for every $\mathcal{M}', w$ and every $\beta \geq \alpha$, so $\alpha \in \text{Approx}(\mathcal{M}, \varphi)$.

□

Remark 4.4 Alternatively, we could have proven the Approximation Lemma by using the pigeonhole principle: a model $\mathcal{M} = (W, R, V)$ has at most $|2^W|$ different extensions, so for any α with $|\alpha| > |2^W|$ and any φ we have $\mathcal{M} \models \varphi \leftrightarrow \downarrow_\alpha(\varphi)$. This alternative proof is more complicated than the one given above, however, which is why we gave this one.

Theorem 4.5 *For every pointed model* $\mathcal{M}, w$ *and every* $\varphi \in \mathcal{L}$, *we have* $\mathcal{M}, w \models \Box\varphi \Leftrightarrow \forall\psi \in \mathcal{L} : \mathcal{M}, w \models [\psi]\varphi$.

Proof. First, suppose $\forall\psi \in \mathcal{L} : \mathcal{M}, w \models [\psi]\varphi$. Then, in particular, for every ordinal α we have $\forall\psi \in \mathcal{L}_\alpha : \mathcal{M}, w \models [\psi]\varphi$ and therefore, by the semantics of $\Box_\alpha$, $\mathcal{M}, w \models \Box_\alpha\varphi$. Since this holds for every ordinal α, we have $\mathcal{M}, w \models \Box\alpha$.

We continue the proof by contraposition, so suppose there is some $\psi \in \mathcal{L}$ such that $\mathcal{M}, w \not\models [\psi]\varphi$. By the Approximation Lemma, there is an ordinal α such that ψ is equivalent to $\downarrow_\alpha(\psi)$ on all submodels of $\mathcal{M}$. In particular, ψ and $\downarrow_\alpha(\psi)$ are equivalent on $\mathcal{M}$. This implies that $\mathcal{M}, w \not\models [\downarrow_\alpha(\psi)]\varphi$. We have $\downarrow_\alpha(\psi) \in \mathcal{L}_{\alpha+1}$, so $\mathcal{M}, w \not\models \Box_{\alpha+1}\varphi$. That, finally, implies $\mathcal{M}, w \not\models \Box\varphi$. □

5 Properties of F-APAL

Here we briefly discuss a few properties of F-APAL. None of these properties are particularly surprising, this section could be seen as a "sanity check" for $\Box$, showing that it has about the properties one would expect.

Proposition 5.1 *F-APAL is invariant under bisimulation.*

Proof. First, we show that $\mathcal{L}_\alpha$ is invariant under bisimulation for all $\alpha \in$ Ord. We do this by induction on α. As base case, suppose $\alpha = 0$. Then $\mathcal{L}_\alpha$ is public announcement logic, which is known to be invariant under bisimulation. Suppose then as induction hypothesis that $\mathcal{L}_\beta$ is invariant under bisimulation for all $\beta < \alpha$.

Now, take any $\varphi \in \mathcal{L}_\alpha$. We show that φ is invariant under bisimulation by a secondary induction on the construction of φ. If φ is atomic, then it is trivially invariant under bisimulation. Suppose as secondary induction hypothesis that all strict subformulas of φ are invariant under bisimulation. Take any two pointed models $\mathcal{M}_1, w_1$ and $\mathcal{M}_2, w_2$ that are bisimilar. We proceed by case distinction on the main connective of φ, but most of the cases are trivial so we omit them. The one case that we do consider in detail is $\varphi = \Box_\beta\psi$. Since $\varphi \in \mathcal{L}_\alpha$, we have $\beta < \alpha$. By the primary induction hypothesis, this implies that, for every $\chi \in \mathcal{L}_\beta$, we have that χ is invariant under bisimulation. As a result, for every such χ, the pointed models $(\mathcal{M}_1)_\chi, w_1$ and $(\mathcal{M}_2)_\chi, w_2$ are bisimilar (if they exist). By the secondary induction hypothesis, this implies that ψ cannot distinguish between $(\mathcal{M}_1)_\chi, w_1$ and $(\mathcal{M}_2)_\chi, w_2$. It follows that $\Box_\beta\psi$ cannot distinguish between $\mathcal{M}_1, w_1$ and $\mathcal{M}_2, w_2$. this holds for any two bisimilar pointed models, so φ is invariant under bisimulation. This completes the induction step of the secondary and primary inductions, so $\mathcal{L}_\alpha$ is invariant under bisimulation for all $\alpha \in$ Ord.

Left to show is that $\mathcal{L}$ is invariant under bisimulation. Take any $\varphi \in \mathcal{L}$. Once again, we use induction on the construction of φ to show that it is invariant under bisimulation. As base case, suppose φ is atomic. Then it is trivially invariant under bisimulation. Suppose therefore as induction hypothesis that all strict subformulas of φ are invariant under bisimulation. We proceed by a case distinction on the main connective of φ. Most cases are trivial, so we omit them. The cases that we do consider are $\varphi = \Box_\alpha\psi$ and $\varphi = \Box\psi$.

Suppose $\varphi = \Box_\alpha\psi$. Then we reason as before: the formulas that $\Box_\alpha$ quantifies over are invariant under bisimulation, as is ψ. It follows that $\Box_\alpha\psi$ is also invariant under bisimulation.

Suppose $\varphi = \Box\psi$. As shown in the previous case, $\Box_\alpha\psi$ is invariant under bisimulation for all $\alpha \in \text{Ord}$. By definition, $\Box\psi$ holds if and only if $\Box_\alpha\psi$ holds for all $\alpha \in \text{Ord}$. So $\Box\psi$ is also invariant under bisimulation. This completes the induction step and thereby the proof. □

Recall that in Section 2 we showed that (*) is under-determined by showing that in the model $\mathcal{M}_{Un}$, see Figure 1, we can assign two different valuations for $\Diamond\xi$ that are both consistent with (*). In one of these valuations, $\Diamond\xi$ is false in the world w_2. In the other valuation, $\Diamond\xi$ is true in w_2 because it is self-fulfilling.

By defining our semantics for F-APAL, we made a choice between these two valuations. It follows easily from the fact that F-APAL is invariant under bisimulation that we have chosen the valuation $\mathcal{M}_{Un}, w_2 \not\models \Diamond\xi$. As discussed in Section 2, we prefer $\mathcal{M}_{Un}, w_2 \not\models \Diamond\xi$ over $\mathcal{M}_{Un}, w_2 \models \Diamond\xi$. So we are satisfied that F-APAL makes $\Diamond\xi$ false in $\mathcal{M}_{Un}, w_2$.

Proposition 5.2 *Let $\varphi, \varphi' \in \mathcal{L}$ and $\alpha \in \text{Ord}$. Then*

(i) *if* $\models \varphi$*, then* $\models \Box\varphi$ *and* $\models \Box_\alpha\varphi$*,*

(ii) $\models \Box(\varphi \to \varphi') \to (\Box\varphi \to \Box\varphi')$ *and* $\models \Box_\alpha(\varphi \to \varphi') \to (\Box_\alpha\varphi \to \Box_\alpha\varphi')$*,*

(iii) $\models \Box\varphi \to \varphi$ *and* $\models \Box_\alpha\varphi \to \varphi$*,*

(iv) $\models \varphi \to \Diamond\varphi$ *and* $\models \varphi \to \Diamond_\alpha\varphi$*,*

(v) $\models \Box\varphi \leftrightarrow \Box\Box\varphi$ *and* $\models \Box_\alpha\varphi \leftrightarrow \Box_\alpha\Box_\alpha\varphi$*.*

Proof. Let $\mathcal{M} = (W, R, V)$ be any model and let $w \in W$.

- Suppose $\models \varphi$. Then, in particular, for every ψ such that $\mathcal{M}, w \models \psi$, we have $\mathcal{M}_\psi, w \models \varphi$ and therefore $\mathcal{M}, w \models [\psi]\varphi$. For every ψ such that $\mathcal{M}, w \not\models \psi$ we trivially have $\mathcal{M}, w \models [\psi]\varphi$. So for all ψ, we have $\mathcal{M}, w \models [\psi]\varphi$. This implies that $\mathcal{M}, w \models \Box\varphi$, and therefore also the weaker statement $\mathcal{M}, w \models \Box_\alpha\varphi$.
- If $\mathcal{M}, w \models \Box(\varphi \to \varphi')$ and $\mathcal{M}, w \models \Box\varphi$, then for every ψ such that $\mathcal{M}, w \models \psi$ we have $\mathcal{M}_\psi, w \models \varphi \to \varphi'$ and $\mathcal{M}_\psi, w \models \varphi$ and therefore $\mathcal{M}_\psi, w \models \varphi'$. It follows that $\mathcal{M}, w \models [\psi]\varphi'$ and therefore $\models \Box(\varphi \to \varphi') \to (\Box\varphi \to \Box\varphi')$.

  The same holds if we restrict to $\psi \in \mathcal{L}_\alpha$ instead of all ψ, so $\models \Box_\alpha(\varphi \to \varphi') \to (\Box_\alpha\varphi \to \Box_\alpha\varphi')$.
- If $\mathcal{M}, w \models \Box\varphi$ or $\mathcal{M}, w \models \Box_\alpha\varphi$ then, in particular, $\mathcal{M}, w \models [\top]\varphi$ and therefore $\mathcal{M}, w \models \varphi$.
- If $\mathcal{M}, w \models \varphi$ then $\mathcal{M}, w \models \langle\top\rangle\varphi$ and therefore $\mathcal{M}, w \models \Diamond\varphi$ and $\mathcal{M}, w \models \Diamond_\alpha\varphi$.
- Suppose $\mathcal{M}, w \models \Box\varphi$. Then, in particular, for every ψ_1, ψ_2 we have $\mathcal{M}, w \models [\psi_1 \wedge [\psi_1]\psi_2]\varphi$. That is equivalent to $\mathcal{M}, w \models [\psi_1][\psi_2]\varphi$. This holds for every ψ_1, ψ_2, so $\mathcal{M}, w \models \Box\Box\varphi$.

  Similarly, if $\mathcal{M}, w \models \Box_\alpha\varphi$ then $\mathcal{M}, w \models [\psi_1 \wedge [\psi_1]\psi_2]\varphi$ for all $\psi_1, \psi_2 \in \mathcal{L}_\alpha$. It follows that $\mathcal{M}, w \models [\psi_1][\psi_2]\varphi$ for all $\psi_1, \psi_2 \in \mathcal{L}_\alpha$, so $\mathcal{M}, w \models \Box_\alpha\Box_\alpha\varphi$.

  The other side of the bi-implications follows from $\models \Box\varphi \to \varphi$ and $\models \Box_\alpha\varphi \to \varphi$.

□

In particular, this shows that $\Box$ and $\Box_\alpha$ are S4 operators.

Before proving the next proposition, we need an auxiliary lemma. This lemma and the next proposition are the only places in this paper where we use the fact that we use the class S5 of models, all other results apply to K as well. The lemma uses a technique very similar to the one used in [10] to show that every formula is "whether-knowable," i.e. for every formula φ, every agent a and pointed model $\mathcal{M}, w$ there is a formula ψ such that either $\mathcal{M}, w \models [\psi]K_a\varphi$ or $\mathcal{M}, w \models [\psi]K_a\neg\varphi$.

Lemma 5.3 *Let $\varphi \in \mathcal{L}$, and let P be the propositional variables that occur in φ. There is a function $f : 2^P \to \{\top, \bot\}$ such that for every model $\mathcal{M}$ and every $P' \subseteq P$: if $\mathcal{M} \models p$ for all $p \in P'$ and $\mathcal{M} \models \neg p$ for all $p \in P \setminus P'$, then $\mathcal{M} \models \varphi$ if $\mathcal{M} \models f(P')$ and $\mathcal{M} \models \neg\varphi$ if $\mathcal{M} \models \neg f(P')$.*

Proof. By induction on the construction of φ. As base case, suppose φ is atomic. Then $\varphi = p$ for some $p \in \mathcal{P}$. The function given by $f(\varnothing) = \bot$ and $f(\{p\}) = \top$ then satisfies the lemma. Suppose then as induction hypothesis that φ is not atomic and that the lemma holds for all strict subformulas of φ. Given such a strict subformula ψ, let f_ψ be the function associated with ψ. The proof continues by a case distinction on the main connective of φ.

- Suppose $\varphi = \neg\psi$. Then the function given by $f(P') = \neg f_\psi(P')$ satisfies the lemma.
- Suppose $\varphi = \psi_1 \vee \psi_2$. Then the function $f(P') = f_{\psi_1}(P') \vee f_{\psi_2}(P')$ satisfies the lemma.
- Suppose $\varphi = K_a\psi$. For every χ, we have $\mathcal{M} \models \chi \Rightarrow \mathcal{M} \models K_a\chi$. Furthermore, since we are working in S5, we have $\mathcal{M} \models \neg\chi \to \mathcal{M} \models \neg K_a\chi$. As a result, the function $f = f_\psi$ satisfies the lemma.
- Suppose $\varphi = [\psi_1]\psi_2$. Let f be given by $f(P') = f_{\psi_1}(P') \to f_{\psi_2}(P')$. Let $\mathcal{M}$ be any model such that $\mathcal{M} \models p$ for all $p \in P'$ and $\mathcal{M} \models \neg p$ for all $p \in P \setminus P'$. If $f_{\psi_1}(P') = \bot$ then $\mathcal{M} \models \neg\psi_1$, so trivially $\mathcal{M} \models [\psi_1]\psi_2$. Note that in this case $f(P') = \top$, so the lemma is satisfied. If $f_{\psi_1}(P') = \top$, then $\mathcal{M}_{\psi_1} = \mathcal{M}$, so φ is equivalent to ψ_2 on $\mathcal{M}$. Note that $f(P') = f_{\psi_2}(P')$ in this case, so the lemma is satisfied.
- Suppose $\varphi = \Box_\alpha\psi$ or $\varphi = \Box\varphi$. For every updated model $\mathcal{M}_\chi$ and every $p \in P$, we have $\mathcal{M} \models p \Rightarrow \mathcal{M}_\chi \models p$ and $\mathcal{M} \models \neg p \Rightarrow \mathcal{M}_\chi \models \neg p$. So if we take $f = f_\psi$, then the lemma is satisfied.

$\Box$

Proposition 5.4 *Let $\varphi \in \mathcal{L}$ and $\alpha \in$ Ord. Then*

(i) $\models \Box\Diamond\varphi \to \Diamond\Box\varphi$ *and* $\models \Box_\alpha\Diamond_\alpha\varphi \to \Diamond_\alpha\Box_\alpha\varphi$,

(ii) $\models \Diamond\Box\varphi \to \Box\Diamond\varphi$ *and* $\models \Diamond_\alpha\Box_\alpha\varphi \to \Box_\alpha\Diamond_\alpha\varphi$.

Proof. Let a pointed model $\mathcal{M}, w$ be given. Let $P = \{p_1, \cdots, p_n\}$ be the set of propositional variables that occur in φ. Furthermore, let $P' = \{p \in P \mid \mathcal{M}, w \models p\}$ and let $\zeta = \bigwedge_{p \in P'} p \wedge \bigwedge_{p \in P \setminus P'} \neg p$. Finally, let $f_\varphi : 2^P \to \{\top, \bot\}$ be the function from Lemma 5.3.

Now, consider the models $\mathcal{M}_\zeta$. For every $p \in P'$ we have $\mathcal{M}_\zeta \models p$ and for every $p \in P \backslash P'$ we have $\mathcal{M}_\zeta \models \neg p$. The same holds for every submodel of $\mathcal{M}_\zeta$. Lemma 5.3 therefore implies that, for all submodels $\mathcal{M}_1, \mathcal{M}_2, \mathcal{M}_3, \mathcal{M}_4$ of $\mathcal{M}_\zeta$, we have $\mathcal{M}_1, w \models \varphi \Leftrightarrow \mathcal{M}_2, w \models \Diamond\varphi \Leftrightarrow \mathcal{M}_3, w \models \Box\varphi \Leftrightarrow \mathcal{M}_4, w \models f_\varphi(P')$.

Suppose now that $\mathcal{M}, w \models \Box\Diamond\varphi$. Then, in particular, $\mathcal{M}, w \models [\zeta]\Diamond\varphi$. It follows that $\mathcal{M}_\zeta, w \models \Diamond\varphi$ and therefore $\mathcal{M}_\zeta \models \Box\varphi$. As such, we have $\mathcal{M}, w \models \langle\zeta\rangle\Box\varphi$ and therefore $\mathcal{M}, w \models \Diamond\Box\varphi$.

Suppose then that $\mathcal{M}, w \models \Diamond\Box\varphi$. Then there is some ψ such that $\mathcal{M}_\psi, w \models \Box\varphi$. In particular, $\mathcal{M}_\psi, w \models [\zeta]\varphi$, so $(\mathcal{M}_\psi)_\zeta, w \models \varphi$. But $(\mathcal{M}_\psi)_\zeta$ is a submodel of $\mathcal{M}_\zeta$, so we have $\mathcal{M}', w \models \varphi$ for every submodel of $\mathcal{M}_\zeta$ (that includes w). For every ψ, $(\mathcal{M}_\psi)_\zeta$ is a submodel of $\mathcal{M}_\zeta$, so $(\mathcal{M}_\psi)_\zeta, w \models \varphi$. It follows that, for every ψ, $\mathcal{M}, w \models [\psi]\langle\zeta\rangle\varphi$ and therefore $\mathcal{M}, w \models \Box\Diamond\varphi$.

Finally, since $\zeta \in \mathcal{L}_0$, the same reasoning holds for $\Box_\alpha$ and $\Diamond_\alpha$ instead of $\Box$ and $\Diamond$, so $\models \Box_\alpha\Diamond_\alpha\varphi \to \Diamond_\alpha\Box_\alpha\varphi$ and $\models \Diamond_\alpha\Box_\alpha\varphi \to \Box_\alpha\Diamond_\alpha\varphi$. □

Proposition 5.4.(i) is known as the Church-Rosser schema and characterizes the property known as convergence or confluence. In our terms: if in a given model $\mathcal{M}, w$ you make two different (truthful) announcements φ and ψ, you get two typically different (non-bisimilar) model restrictions $\mathcal{M}_\varphi, w$ and $\mathcal{M}_\psi, w$. Proposition 5.4.(i) then says that in such a case there are announcements φ' and ψ' such that $(\mathcal{M}_\varphi)_{\varphi'}, w$ is bisimilar to $(\mathcal{M}_\psi)_{\psi'}, w$.

Proposition 5.4.(ii) is also known as the McKinsey schema. In the presence of the schema $\Box\varphi \to \Box\Box\varphi$ (which is valid in F-APAL, see Proposition 5.2) this characterizes so-called *atomicity*. Intuitively, given any model $\mathcal{M}, w$ and any set P of propositional variables, there is a ζ such that on $\mathcal{M}_\zeta, w$ we have $\varphi \leftrightarrow \Box\varphi$ for all φ that contain only propositional variables from P. So in $\mathcal{M}_\zeta, w$ you already know all there is to know about P, any further model restriction is uninformative. Even more intuitively, Proposition 5.4.(ii) says that given any model $\mathcal{M}, w$ and any formula φ, you can make a most informative announcement with respect to the propositional variables occurring in φ; namely the ζ above.

For a more detailed description of these properties, see [8].

6 F-APAL and fixed points

The semantics of $\Box$ are reminiscent of fixed point constructions. Indeed, we can define $\Box$ as a fixed point. The relation between F-APAL and fixed points is not straightforward, however. There are also several open questions regarding the fixed points related to F-APAL. Let us therefore discuss the relation between F-APAL and fixed points in detail. The auxiliary operators $\Box_\alpha$ are not important to the fixed point behavior of $\Box$, so in this section we work with a language $\mathcal{L}'$ that includes $\neg, \vee, K_a, [\psi]$ and $\Box$ but not $\Box_\alpha$.

Let us start by recalling the definitions of (*), (**) and (5) (with $\mathcal{L}'$ substituted for $\mathcal{L}$):

$$\mathcal{M}, w \models \Box\varphi \Leftrightarrow \forall\psi \in \mathcal{L}' : \mathcal{M}, w \models [\psi]\varphi \qquad (*')$$

$$\mathcal{M}, w \models \Box\varphi \Leftrightarrow \forall x \in \{[\![\psi]\!]_{\mathcal{M}} \mid \psi \in \mathcal{L}'\} : \mathcal{M}_x, w \models \varphi. \qquad (**')$$

$$\mathcal{M}, w \models \Box\varphi \Leftrightarrow \forall x \in X : \mathcal{M}_x, w \models \varphi. \qquad (5)$$

Also, recall that (*′) and (**′) are equivalent. Now, let $\mathcal{M} = (W, R, V)$ and suppose that we would define $\Box$ not as in Definition 3.6 but by (5) for some $X \subseteq 2^W$. This alternative definition only applies to the specific model $\mathcal{M}$, so we will assume that $\Box$ is defined in some way on other models.

Because we define $\Box$ by (5), we have (**′) and (*′) if X is the set of extensions on $\mathcal{M}$. (See Section 2 for a discussion of why this is so.) In order to emphasize that our semantics, and therefore our set of extensions, depend on X let us write $[\![\mathcal{L}']\!]^X_{\mathcal{M}}$ for the set of all extensions on $\mathcal{M}$. So (*′) is satisfied if

$$X = [\![\mathcal{L}']\!]^X_{\mathcal{M}}.$$

In other words, (*′) is satisfied if X is a fixed point of $f : X \mapsto [\![\mathcal{L}']\!]^X_{\mathcal{M}}$. Unfortunately, f is not monotone,[6] nor do we have $X \subseteq f(X)$. The standard methods for proving the existence of a fixed point therefore do not apply; whether f has a fixed point (on every model) is, to the best of our knowledge, an open question.

Alternatively, we can consider the function $g : X \mapsto X \cup [\![\mathcal{L}']\!]^X_{\mathcal{M}}$. This function is also not monotone, but it does by construction satisfy $X \subseteq g(X)$. As a result, g is guaranteed to have a fixed point. For example, 2^W is a fixed point of g. More importantly, $\{[\![\psi]\!]_{\mathcal{M}} \mid \psi \in \mathcal{L}\}$ is a fixed point of g. In fact, $\{[\![\psi]\!]_{\mathcal{M}} \mid \psi \in \mathcal{L}\} = \lim_{\alpha \in \mathrm{Ord}} g^\alpha(\varnothing)$.[7]

The construction of $\{[\![\psi]\!]_{\mathcal{M}} \mid \psi \in \mathcal{L}\}$ is therefore an instance of a general kind of fixed point construction: let S be any complete lattice, and let $h : S \to S$ be any function that satisfies $s \leq h(s)$ for all $s \in S$. Then $\lim_{\alpha \in \mathrm{Ord}} h^\alpha(0)$ is guaranteed to exist and be a fixed point of h, where 0 is the least element of S and $h^\alpha(0)$ is defined as $\sup_{\beta<\alpha} h^\beta(\varnothing)$ when α is a limit ordinal.

However, unless h is monotone, it is not guaranteed that $\lim_{\alpha \in \mathrm{Ord}} h^\alpha(0)$ is the *least* fixed point of h. As such, we cannot immediately conclude that $\{[\![\psi]\!]_{\mathcal{M}} \mid \psi \in \mathcal{L}\}$ is the least fixed point of g. Whether it is in fact the least fixed point is an open question.

The fact that $\{[\![\psi]\!]_{\mathcal{M}} \mid \psi \in \mathcal{L}\} = \lim_{\alpha \in \mathrm{Ord}} g^\alpha(\varnothing)$ shows that we could have defined the semantics of $\Box$ using a fixed point, instead of using $\mathcal{M}, w \models \Box\varphi \Leftrightarrow \mathcal{M}, w \models \Box_\alpha\varphi$ for all $\alpha \in \mathrm{Ord}$. It should be noted, however, that although the auxiliary operators $\Box_\alpha$ do not occur in the fixed point definition of $\Box$, we do still need them in order to satisfy (*).

As noted before, any fixed point of f satisfies (*′). So it can be used to define a fully public arbitrary announcement for the language $\mathcal{L}'$, the language

6 i.e. it is not the case that, for all X, Y, if $X \subseteq Y$ then $f(X) \subseteq F(Y)$.

7 Where $g^\alpha(\varnothing)$ is defined as $\bigcup_{\beta<\alpha} g^\beta(\varnothing)$ if α is a limit ordinal, and we use the discrete topology (i.e. $\lim_{\alpha \in \mathrm{Ord}} Z_\alpha = l \Leftrightarrow \exists\beta\forall\alpha > \beta : Z_\alpha = l$).

without auxiliary operators. A fixed point of g is not guaranteed to share this property, however. If X is a fixed point of g, then we have $X = X \cup [\![\mathcal{L}']\!]^X_{\mathcal{M}}$. But then there might be $x \in X \setminus [\![\mathcal{L}']\!]^X_{\mathcal{M}}$, i.e. there might be some sets that are quantified over by $\Box$ that are not the extension of any formula.

For the specific fixed point $\lim_{\alpha\in\text{Ord}} f^\alpha(\varnothing)$ of g, we can solve this problem using the auxiliary operators. While some elements of $\lim_{\alpha\in\text{Ord}} f^\alpha(\varnothing)$ are not the extension of any formula of $\mathcal{L}'$, each of them is the extension of some formula of $\mathcal{L}$.

We hope that this section has clarified the relation between F-APAL and fixed points. In particular, we hope that it explains why we define the semantics of F-APAL the way we do, instead of as a fixed point.

7 Conclusion

We introduced a logic F-APAL, in which the connective $\Box$ represents a fully arbitrary public announcement, i.e. we have

$$\mathcal{M}, w \models \Box\varphi \Leftrightarrow \forall\psi \in \mathcal{L} : \mathcal{M}, w, \models [\psi]\varphi$$

for all $\varphi \in \mathcal{L}$ and every pointed model $\mathcal{M}, w$. The price we pay for this property is that we use a proper class of auxiliary operators, $\{\Box_\alpha \mid \alpha \in \text{Ord}\}$.

This suggests a few directions for further research. Firstly, we could try to use similar techniques to design semantics for other circular properties. Examples of such circular properties include "agent a knows at least as much as agent b"

$$\mathcal{M}, w \models a \succeq b \Leftrightarrow \forall\psi : \mathcal{M}, w \models K_b\psi \to K_a\psi$$

and "everything agent a believes is true" [8]

$$\mathcal{M}, w \models T(a) \Leftrightarrow \forall\psi : \mathcal{M}, w \models K_a\psi \to \psi.$$

Or consider *knowledge based programs* [12,13] (or similarly: *epistemic protocols* [2,3]). Such programs contain instructions for multiple agents to perform actions. But, importantly, every action has to come with an epistemic precondition for the agent that is supposed to carry out the action. So a knowledge based program can only contain clauses of the form "if $K_a\varphi$, then a should do x." These kinds of programs are useful when modeling distributed systems. Suppose we use $\blacksquare\psi$ to denote "there is a knowledge based program π that, if executed, guarantees outcome ψ." Then $\blacksquare\psi$ is circular, since π could contain a clause "if $K_a\blacksquare\psi$, then a should do x."

Secondly, we could attempt to reduce the conceptual cost of F-APAL by using fewer auxiliary operators. Ideally we would use no auxiliary operators at all, but barring that it would be nice to have a *set* of auxiliary operators instead of a proper class. We conjecture that it is possible to have fully arbitrary public announcements with a set of auxiliary operators, but we think it may be

[8] If we work in S5, $K_a\psi \to \psi$ is always true for every ψ. So in order to make this property interesting we would need to use a different class of models.

impossible to have fully arbitrary public announcements without any auxiliary operators.

Acknowledgments

We would like to thank the AiML reviewers for their insightful remarks. Their questions about fixed points were especially useful, and prompted the addition of Section 6. Hans van Ditmarsch and Louwe B. Kuijer acknowledge support from ERC project EPS 313360. Hans van Ditmarsch is also affiliated to IMSc, Chennai, India, and Zhejiang University, China.

References

[1] Ågotnes, T., P. Balbiani, H. van Ditmarsch and P. Seban, *Group announcement logic*, Journal of Applied Logic **8** (2010), pp. 62–81.

[2] Attamah, M., H. van Ditmarsch, D. Grossi and W. van der Hoek, *Knowledge and gossip*, in: T. Schaub, G. Friedrich and B. O'Sullivan, editors, *Proceedings of the 21st European Conference on Artificial Intelligence (ECAI 2014)*, 2014, pp. 21–26.

[3] Attamah, M., H. van Ditmarsch, D. Grossi and W. van der Hoek, *The pleasure of gossip*, in: C. Başkent, L. Moss and R. Ramanujam, editors, *Rohit Parikh on Logic, Language and Society*, To Appear.

[4] Balbiani, P., A. Baltag, H. van Ditmarsch, A. Herzig, T. Hoshi and T. de Lima, *What can we achieve by arbitrary announcements? A dynamic take on Fitch's knowability*, in: D. Samet, editor, *Proceedings of the 11th conference on Theoretical aspects of rationality and knowledge*, 2007, pp. 42–51.

[5] Balbiani, P., A. Baltag, H. van Ditmarsch, A. Herzig, T. Hoshi and T. de Lima, *'Knowable' as 'Known after an Announcement'*, The Review of Symbolic Logic **1** (2008), pp. 305–334.

[6] Baltag, A., L. Moss and S. Solecki, *The logic of public announcements, common knowledge, and private suspicions*, in: I. Gilboa, editor, *Proceedings of the 7th conference on Theoretical aspects of rationality and knowledge* (1998), pp. 43–56.

[7] Barwise, J. and L. Moss, "Vicious Circles," CSLI Publications, 1996.

[8] Blackburn, P., M. de Rijke and Y. Venema, "Modal Logic," Cambridge University Press, 2001.

[9] Bozzelli, L., H. van Ditmarsch, T. French, J. Hales and S. Pinchinat, *Refinement modal logic*, Information and Computation **239** (2014), pp. 303–339.

[10] van Ditmarsch, H., W. van der Hoek and P. Iliev, *Everything is knowable - how to get to know whether a proposition is true*, Theoria **78** (2012), pp. 93–114.

[11] van Ditmarsch, H., W. van der Hoek and B. Kooi, *Arbitrary arrow update logic*, presented at Advances in Modal Logic 2014.

[12] Fagin, R., J. Y. Halpern, Y. Moses and M. Vardi, "Reasoning About Knowledge," MIT Press, Cambridge, MA, USA, 1995.

[13] Fagin, R., J. Y. Halpern, Y. Moses and M. Y. Vardi, *Knowledge based programs*, Distributed Computing **10** (1997), pp. 199–225.

[14] Hales, J., *Arbitrary action model logic and action model synthesis*, in: *28th Annual ACM/IEEE Symposium on Logic in Computer Science*, 2013, pp. 253–262.

[15] Jech, T., "Set Theory," Springer, Berlin, 1997.

[16] Kuijer, L. B., *How arbitrary are arbitrary public announcements?*, in: M. Colinet, S. Katrenko and R. K. Rendsvig, editors, *Pristine Perspectives on Logic, Language and Computation*, 2014, pp. 109–123.

A cut-free sequent calculus for the logic of subset spaces

Birgit Elbl [1]

UniBw München
85577 Neubiberg, Germany

Abstract

Following the tradition of labelled sequent calculi for modal logics, we present a one-sided, cut-free sequent calculus for the bimodal logic of subset spaces. In labelled sequent calculi, semantical notions are internalised into the calculus, and we take care to choose them close to the original interpretation of the system. To achieve this, we introduce a variation of the standard method, considering structured labels instead of simple tokens, in our particular case pairs of labels. With this new device, we can formulate a calculus with extremely simple frame rules and good proof-theoretical properties. The logical rules are invertible, structural rules are admissible. We show the admissibility of cut and relate our system to the well-known Hilbert-style axiomatisation of the logic. Finally, we present a direct proof of completeness based on proof search.

Keywords: proof theory, cut-free sequent calculus, labelled deduction, direct completeness proof, logic of subset spaces.

1 Introduction

The logic of *subset spaces* SSL discussed here is a bimodal logic introduced in [1] for formalising reasoning about points and sets. Its extension *topologic* can be considered a refinement of Tarski's and McKinsey's topological interpretation [20,15] for the modal system **S4**. SSL is also called a logic of *knowledge* and *effort*. The relation to epistemic logic is investigated further in [17]. More recently, an interpretation of the language of public announcement logic in subset models was given [21]. Several extensions of the language of SSL have been studied, for example the addition of an *overlap operator* as a third modality [11] or announcement operators [2]. In the present work, however, we study the original language and its meaning given by subset spaces.

Subset frames consist of a set X of *points* and a collection $\mathcal{O}$ of non-empty subsets of X called *opens*. Worlds are pairs (x, u) where x is a point and u is an open containing x. The first set K, L of SSL-modalities corresponds to quantification over points in the same environment, while the second set $\Box, \Diamond$

[1] `Birgit.Elbl@unibw.de`

refers to the worlds obtained by shrinking the environment of a fixed point x. So the relation $\supseteq$ for opens determines the $\Box\Diamond$-reachability. A *sound* and *complete* Hilbert style axiomatisation is presented in [1]. It combines **S4**-axioms for $\Box, \Diamond$ with **S5**-axioms for K, L and further axioms known as *persistence* for literals and *cross axioms*. As **S5** is contained as a subsystem, a corresponding *cut-free sequent calculus* is not straightforward (see [19] for a discussion of the case of **S5**), and the combination with a second set of modalities generates further difficulties.

Labelled calculi provide not only a solution for **S5** but also a general method to construct sequent systems for modal logics, see [18,7]. In that approach, the semantics is to a certain extent internalised into the calculus. The labels denote worlds in a Kripke frame. The basic judgements of the calculus have the form $x\colon A$ or xRy which can be read as "A holds at x" or "y is reachable from x", respectively. In addition to the logical rules, one has *frame rules* that reflect the conditions for the Kripke frames of the logic.

We want to define a labelled calculus in that style based on subset frames. Corresponding to the structure of worlds in subset spaces, we use *pairs* (x, u) of simple labels x, u in judgements $(x, u) : A$ of our calculus and introduce formal judgements for "(x, u) is a world" and "u can be shrunk to obtain v".[2] The *frame rules* of the calculus reflect basic properties of these relations. From a semantic point of view, we generalise the class of models: the second components of pairs need not be sets and relations $\mathcal{W}$ and $\mathcal{R}$ are included in the frame, which have to satisfy some essential conditions but need not be identical to $\in$ and $\supseteq$. We call the elements of this more general class of models *abstract subset spaces*. As we keep the basic structure of pairs and the frame conditions for $\mathcal{W}$ and $\mathcal{R}$ are satisfied by $\in$ and $\supseteq$, subset spaces are a *special case* of abstract subset spaces, without any transformation of primitive notions.

Now the setting is different from the standard labelled systems but the general strategy can be employed to develop a cut-free calculus. In contrast to [18], we use a one-sided sequent system in the Schütte-Tait style. The logical rules correspond to the right rules of a two-sided system, the dual left rules are avoided. This cuts down the number of rules, although we retain all modalities. The interpretation of the modal operators as explained above and the conditions for abstract subset spaces determine the rules of the calculus.

As SSL does not have the finite model property w.r.t. subset spaces, the class of *cross axiom models* has also been introduced in [1] and has been used for the proof of the decidability. Alternatively to the approach presented here, we could have chosen these as the starting point and applied directly the method presented in [18], as all frame conditions satisfy the prerequisites. This leads to a cut-free system for SSL ([6]), as validity in all subset spaces and validity in all cross axiom models coincide but then the internalised semantics is significantly different from the original one. Obviously, the argument that is formalised in the deduction then uses the reachability relations instead of the notions given

[2] More precisely for their negations, see below.

by the models itself. More important, the frame rules are no longer regular rule schemes. The accessibility relations in cross axiom frames must satisfy the so-called *cross condition*. This can be rewritten to a *geometric formula* in the sense of [18] but turning it into a proof rule in the natural way yields a *frame* rule which — read from bottom to top — generates new worlds via the involved eigenvariable.

In contrast to this, the requirements for abstract subset spaces are just closure conditions. Given any $\mathcal{W}_0 \subseteq X \times \mathcal{O}$ and $\mathcal{R}_0 \subseteq \mathcal{O} \times \mathcal{O}$, there is a least extension to an abstract subset space $(X, \mathcal{O}, \mathcal{W}, \mathcal{R})$ and this can be obtained by a combination of standard relational operations (composition, inversion, reflexive-transitive closure). The corresponding frame rules are very simple. They are not subject to eigenvariable conditions. They could be readily replaced by a computation of this closure alternating with the logical rules or by complex application conditions for the logical rules that refer to that closure.

We proceed as follows: In Section 2 we present some basics concerning the logic of subset spaces and sequent calculi. Based on the model class of abstract subset spaces presented in Section 3, we develop a labelled calculus **LSSL-p** for SSL in Section 4. We show several proof-theoretic properties in Sections 5 and 6, in particular the invertibility of logical rules and the admissibility of weakening, contraction and cut. Derivations in **LSSL-p** of the Hilbert axioms from [1] are presented. Completeness, however, is proved directly in the style of [13,12]. In contrast to the proof in [1], the argument in Section 7 shows how to produce a derivation for valid formulas and yields a (in general infinite) countermodel for non-valid formulas.

2 Preliminaries

2.1 The logic of subset spaces

Following [1], a *subset frame* is a pair $\mathcal{X} = (X, \mathcal{O})$ where X is a set of *points* and $\mathcal{O}$ is a set of non-empty subsets of X called *opens*. We presuppose a fixed set PV of propositional letters. The *formulas* of the logic of subset spaces are built from the elements of PV using propositional connectives and the modalities $\Box, \Diamond, \mathrm{K}, \mathrm{L}$ where $\Box, \Diamond$ are dual to each other and so are K, L. The value of a propositional letter in a particular world is a truth value. For us, a valuation for a subset frame is a mapping $\mathcal{V} : X \to (\mathrm{PV} \to \mathbb{B})$ where $\mathbb{B}$ denotes the set of Boolean truth values, and a *subset space* $\mathcal{X} = (X, \mathcal{O}, \mathcal{V})$ consists of a subset frame $(X, \mathcal{O})$ and a valuation $\mathcal{V}$ for it. A *world* (x, u) consists of a point $x \in X$ and an open u that contains it. The *satisfaction relation* $\models_{\mathcal{X}}$ is given by the usual interpretation of the propositional connectives plus the following conditions for all $(x, u) \in X \times \mathcal{O}$ such that $x \in u$ and arbitrary formulas A:

$$
\begin{array}{ll}
x, u \models_{\mathcal{X}} \mathrm{K}A & \text{iff } y, u \models_{\mathcal{X}} A \text{ for all } y \in u \\
x, u \models_{\mathcal{X}} \Box A & \text{iff } x, v \models_{\mathcal{X}} A \text{ for all } v \in \mathcal{O} \text{ such that } x \in v \subseteq u \\
x, u \models_{\mathcal{X}} \mathrm{L}A & \text{iff there exists } y \in u \text{ such that } y, u \models_{\mathcal{X}} A \\
x, u \models_{\mathcal{X}} \Diamond A & \text{iff there exists } v \in \mathcal{O} \text{ such that } x \in v \subseteq u \text{ and } x, v \models_{\mathcal{X}} A
\end{array}
$$

Here x is a point and u an open such that $x \in u$. Hence validity in a subset frame is just validity in the corresponding Kripke frame $(\mathcal{W}, \mathcal{S}, \mathcal{R})$ with the set $\mathcal{W} := \{(x, u) \in X \times \mathcal{O} \mid x \in u\}$ of worlds and the accessibility relations

$$\begin{aligned} \mathcal{S} &:= \{((x, u), (y, u)) \mid x, y \in X \text{ and } \{x, y\} \subseteq u \in \mathcal{O}\} \\ \mathcal{R} &:= \{((x, u), (x, v)) \mid x \in v \subseteq u\} \end{aligned}$$

for $\Box, \Diamond$ and K, L respectively.

Cross axiom frames and cross axiom models are introduced in [1] in order to prove the decidability of subset space logic (also see [14] for a simplified proof). A *cross axiom frame* $(\mathcal{W}, \mathcal{S}, \mathcal{R})$ consists of a set $\mathcal{W}$, an equivalence relation $\mathcal{S}$ on $\mathcal{W}$ and a preorder $\mathcal{R}$ on $\mathcal{W}$ so that $\mathcal{R}; \mathcal{S} \subseteq \mathcal{S}; \mathcal{R}$. Here and in the sequel we use standard notation for operations on relations: ";" stands for relational composition (not $\circ$), $\cdot^+$ for the transitive closure and $\cdot^*$ for the reflexive-transitive closure. A *cross axiom model* $(\mathcal{W}, \mathcal{S}, \mathcal{R}, \mathcal{V})$ is a cross axiom frame $(\mathcal{W}, \mathcal{S}, \mathcal{R})$ together with a valuation $\mathcal{V} : \mathcal{W} \to (\mathrm{PV} \to \mathbb{B})$ so that $\mathcal{V}(w) = \mathcal{V}(w')$ whenever $(w, w') \in \mathcal{R}$. It can easily be checked that the transformation for subset frames into Kripke models described above yields a cross axiom frame. Extending this with the valuation $\mathcal{V}' : \mathcal{W} \to (\mathrm{PV} \to \mathbb{B})$ given by $\mathcal{V}'((x, u)) := \mathcal{V}(x)$, we obtain a cross axiom model, in which the same formulas are valid. [3]

The reason for introducing cross axiom models as an auxiliary concept lies in the fact that they enjoy the finite model property, in contrast to subset spaces. In particular, this tells us that there are cross axiom models which are not isomorphic to a Kripke frame induced by a subset space. A characterisation of those cross axiom frames that are isomorphic copies of transformed subset frames is presented by Heinemann in [10]. However, validity in all subset models and validity in all cross axiom models coincide, similar for satisfiability. This is a consequence of the fact that the Hilbert-system given below is sound w.r.t. cross axiom models (and hence also w.r.t. subset spaces) and complete w.r.t. subset spaces (and hence also w.r.t. cross axiom models).

The axioms of *subset space logic*, see Table 1, are given in [1]. The instances of the axiom scheme (ca) are called *cross axioms*, and (pers) is the *persistence* for literals. Furthermore, we have **S5**-axioms for K and **S4**-axioms for $\Box$. The rules of inference are *modus ponens* and the usual rules of *necessitation* for $\Box$, K. We denote the (Hilbert style) deductive system given by the axioms and rules in Table 1 by **HSS**.

In [1], the modalities $\Diamond$, L (as well as $\vee, \to$) are defined notions. There, $\Diamond A$ stands for $\neg\Box\neg A$, and LA stands for $\neg$K$\neg A$. As the focus is on the modalities, we prefer to keep all four of them as primitives, and reduce the number of logical operators in a different way: negation on non-atoms and implication are taken as defined. That means that $\wedge, \vee$ can be used freely in building a formula, while '$\to$' is excluded and '$\neg$' restricted to the case of propositional variables. This can also be understood as presupposing a second set of *negative*

[3] For the more general case of abstract subset spaces, see 3.3.

Axioms in the system **HSS**:
all substitution instances of tautologies of propositional logic
$(P \to \Box P) \wedge (\neg P \to \Box \neg P)$ (pers) for propositional letters P
$\mathrm{K}\Box A \to \Box \mathrm{K} A$ (ca)
$\mathrm{K}A \to (A \wedge \mathrm{KK}A)$ $\Box A \to (A \wedge \Box\Box A)$
$\mathrm{K}(A \to B) \to (\mathrm{K}A \to \mathrm{K}B)$ $\Box(A \to B) \to (\Box A \to \Box B)$ $\mathrm{L}A \to \mathrm{KL}A$

Rules:

$$\frac{A \to B \qquad A}{B} \qquad \frac{A}{\Box A} \qquad \frac{A}{\mathrm{K}A}$$

Table 1
The system **HSS** - axioms and rules of the logic of subset spaces

literals $\neg P, \neg Q, \neg P', \neg Q, \ldots$ equipped with a bijection '$\neg$', mapping *positive literals* (i.e. propositional variables) to negative literals. This mapping is used in the semantics of the language as well as in the logical axioms of the calculi. Negation *for non-atoms* is given by

$$\neg\Box A :\equiv \Diamond\neg A \quad \neg\Diamond A :\equiv \Box\neg A \quad \neg\mathrm{K}A :\equiv \mathrm{L}\neg A \quad \neg\mathrm{L}A :\equiv \mathrm{K}\neg A$$
$$\neg\neg P :\equiv P \quad \neg(A \wedge B) :\equiv \neg A \vee \neg B \quad \neg(A \vee B) :\equiv \neg A \wedge \neg B$$

and $A \to B$ stands for $\neg A \vee B$. For one-sided sequent systems, a significant simplification is achieved by using this defined negation for compound formulas. It is a prerequisite for the GS-calculi in [9] and part of the Schütte-Tait-style, which we will adopt for the calculus in Section 4.

2.2 One-sided labelled sequent calculi

As our starting point, we choose a propositional, one-sided sequent calculus in the Schütte-Tait style where weakening and contraction are absorbed into the logical rules, i.e. the propositional, cut-free part of the calculus **GS3** in [9]:

$$(\mathrm{ax})\ \frac{}{\Gamma, P, \neg P} \qquad (\wedge)\ \frac{\Gamma, A \qquad \Gamma, B}{\Gamma, A \wedge B} \qquad (\vee)\ \frac{\Gamma, A, B}{\Gamma, A \vee B}$$

Here and in the sequel, sequents are multisets of formulas. A two-sided sequent $A_1, \ldots, A_m \Rightarrow B_1, \ldots, B_n$ corresponds to $\neg A_1, \ldots, \neg A_m, B_1, \ldots, B_n$. Negri's system **G3K** [18] can readily be rewritten in the one-sided style. Originally, the elements of sequents are *relational atoms* xRy or *labelled formulas* $x\colon A$, where x, y are *labels* taken from a fixed set, A is a modal formula, and R is a binary relation symbol that stands for the accessibility relation. Logical axioms for the relational atoms are present but it is pointed out in [18] that they are only needed for deriving properties of the accessibility relation. Hence they can safely be removed. As a consequence, atoms tRs would be needed in the original setting on the left side of sequents only. Correspondingly, they would occur *negated* only in the one-sided system. So we can introduce relational symbols $\overline{R}$ for the complement relation right from the beginning and avoid negation. Now we obtain the system **GS3K** in Table 2. Here $!(y)$ abbreviates

$$(\text{ax})\ \frac{}{\Gamma, x\colon P, x\colon \neg P} \qquad (\wedge)\ \frac{\Gamma, x\colon A \qquad \Gamma, x\colon B}{\Gamma, x\colon A \wedge B} \qquad (\vee)\ \frac{\Gamma, x\colon A, x\colon B}{\Gamma, x\colon A \vee B}$$

$$(\Box)\ \frac{\Gamma, x\,\overline{R}\,y, y\colon A}{\Gamma, x\colon \Box A}\ !(\text{y}) \qquad (\Diamond)\ \frac{\Gamma, x\,\overline{R}\,y, x\colon \Diamond A, y\colon A}{\Gamma, x\,\overline{R}\,y, x\colon \Diamond A}$$

Table 2
The system GS3K

the usual eigenvariable condition that y does not occur in the conclusion.

In [18] a general method for generating cut-free sequent calculi for modal logics is presented. It applies to normal modal logics which are characterised by universal axioms or, more generally, geometric implications as frame conditions. The latter are formulas of the form $\forall \bar{x}\,(A \to B)$ where A, B are formulas not containing[4] $\to$ or $\forall$. Geometric frame conditions can be transformed schematically into left rules of the calculus. For example, the conditions for **S4** and **S5** are universal formulas. Using R as a symbol for the accessibility relation, they can be written as

(reflexivity)	$\forall x\,(xRx)$	(**S4**, **S5**)
(transitivity)	$\forall x \forall y \forall z\,(xRy \wedge yRz \to xRz)$	(**S4**, **S5**)
(symmetry)	$\forall x \forall y\,(xRy \to yRx)$	(**S5**)

which yield the rules:

$$\frac{xRx, \Gamma \Rightarrow \Delta}{\Gamma \Rightarrow \Delta}\ \text{Ref} \qquad \frac{xRz, xRy, yRz, \Gamma \Rightarrow \Delta}{xRy, yRz, \Gamma \Rightarrow \Delta}\ \text{Trans}$$

$$\frac{yRx, xRy, \Gamma \Rightarrow \Delta}{xRy, \Gamma \Rightarrow \Delta}\ \text{Sym}$$

By reformulating these for one-sided sequents we obtain

$$\frac{\Gamma, x\,\overline{R}\,x}{\Gamma}\ \text{Ref} \qquad \frac{\Gamma, x\,\overline{R}\,z, x\,\overline{R}\,y, y\,\overline{R}\,z}{\Gamma, x\,\overline{R}\,y, y\,\overline{R}\,z}\ \text{Trans} \qquad \frac{\Gamma, y\,\overline{R}\,x, x\,\overline{R}\,y}{\Gamma, x\,\overline{R}\,y}\ \text{Sym}$$

Adding the corresponding rules to **GS3K**, we obtain labelled systems for **S4** and **S5**. The two-sided versions of these systems, the general method, as well as systems for further modal logics are studied in detail in [18].

3 Abstract subset spaces

We introduce a class of models which is slightly more general than subset spaces. As in subset spaces, the worlds are *pairs* from a set $X \times \mathcal{O}$ but $\mathcal{O}$ need not consist of subsets of X. The relation $\mathcal{W}$ determines which pairs are indeed worlds. In subset spaces this is fixed to be $\in$. Similar to subset spaces, the accessibility

[4] In this context, negation is defined using $\to, \bot$, hence also excluded from A, B.

relation for K can be described as "equality of the second components" and the accessibility relation for $\Box$ is determined by a relation on $\mathcal{O}$ but the latter relation need no longer be $\supseteq$.

Definition 3.1 An *abstract subset frame* $(X, \mathcal{O}, \mathcal{W}, \mathcal{R})$ consists of sets $X, \mathcal{O}$, a relation $\mathcal{W} \subseteq X \times \mathcal{O}$ and a preorder $\mathcal{R} \subseteq \mathcal{O} \times \mathcal{O}$ so that $\mathcal{W}; \mathcal{R}^{-1} \subseteq \mathcal{W}$. An *abstract subset space* $(X, \mathcal{O}, \mathcal{W}, \mathcal{R}, \mathcal{V})$ consists of an abstract subset frame $(X, \mathcal{O}, \mathcal{W}, \mathcal{R})$ and a valuation $\mathcal{V} : X \to (\mathrm{PV} \to \mathbb{B})$.

Setting $\mathcal{W} = \{(x, u) \in X \times \mathcal{O} \mid x \in u\}$ and $\mathcal{R} = \{(u, v) \in \mathcal{O} \times \mathcal{O} \mid v \subseteq u\}$ turns every subset space into an abstract subset space. Choosing a set of properties for the definition of abstract subset frames can be interpreted as looking for a set of (simple and natural) axioms for $\in, \supseteq$ that is sufficient for our purpose. We use reflexivity and transitivity of $\supseteq$ as well as the obvious $\forall x \in X \forall u, v \in \mathcal{O}\, (x \in u \wedge u \subseteq v \to x \in v)$. Antisymmetry of $\supseteq$ is simply not needed.

The assignment of cross axiom models to subset spaces is generalised to the case of abstract subset spaces in the straightforward way:

Definition 3.2 Let $(X, \mathcal{O}, \mathcal{W}, \mathcal{R})$ be an abstract subset frame. The corresponding accessibility relations $\hat{\mathcal{S}}, \hat{\mathcal{R}}$ are defined as follows:

$$\begin{aligned}\hat{\mathcal{S}} &:= \{((x, u), (y, u)) \mid (x, u), (y, u) \in \mathcal{W}\} \\ \hat{\mathcal{R}} &:= \{((x, u), (x, v)) \mid (x, u), (x, v) \in \mathcal{W} \text{ and } (u, v) \in \mathcal{R}\}\end{aligned}$$

If $\mathcal{V} : X \to (\mathrm{PV} \to \mathbb{B})$ is a valuation for that frame, then the mapping $\hat{\mathcal{V}} : \mathcal{W} \to (\mathrm{PV} \to \mathbb{B})$ is given by $\hat{\mathcal{V}}(x, u) := \mathcal{V}(x)$ for all $(x, u) \in \mathcal{W}$.

Lemma 3.3 *Let $(X, \mathcal{O}, \mathcal{W}, \mathcal{R})$ be an abstract subset frame. Then $(\mathcal{W}, \hat{\mathcal{S}}, \hat{\mathcal{R}})$ is a cross axiom frame. If furthermore $\mathcal{V} : X \to (PV \to \mathbb{B})$ is a valuation for that abstract subset frame, then $(\mathcal{W}, \hat{\mathcal{S}}, \hat{\mathcal{R}}, \hat{\mathcal{V}})$ is a cross axiom model.*

Proof. Straightforward verification. We present the proof of the properties which are most characteristic for cross axiom models.
Cross property: Let $(x, u)\hat{\mathcal{R}}(y, v)$ and $(y, v)\hat{\mathcal{S}}(z, w)$. Then $x = y$, $v = w$, $(u, v) \in \mathcal{R}$, and the pairs $(x, u), (y, v), (z, w)$ are worlds in $\mathcal{W}$. As $\mathcal{W}; \mathcal{R}^{-1} \subseteq \mathcal{W}$, we can infer that $(z, u) \in \mathcal{W}$. Hence $(x, u)\hat{\mathcal{S}}(z, u)$ and $(z, u)\hat{\mathcal{R}}(z, v) = (z, w)$.
Persistence: Let $(x, u), (y, v) \in \hat{\mathcal{R}}$. Then $x = y$, and we have $\hat{\mathcal{V}}(x, u) = \mathcal{V}(x) = \mathcal{V}(y) = \hat{\mathcal{V}}(y, v)$. $\Box$

The validity of formulas in abstract subset spaces is defined as usual, using the accessibility relation $\hat{\mathcal{S}}$ for K, L and $\hat{\mathcal{R}}$ for $\Box, \Diamond$. Hence the use of K, L amounts to quantification over worlds with the same second component, and $\Box, \Diamond$ refer to all worlds with the same first and an $\mathcal{R}$-reachable second component. Furthermore, validity in an abstract subset space coincides with validity in the induced cross axiom model. The soundness of **HSS** w.r.t. abstract subset spaces is immediate from the soundness w.r.t. cross axiom models, and completeness follows from completeness w.r.t. subset spaces. So the difference lies in the class of models, not in the set of valid sentences.

Comparing abstract subset spaces with subset spaces, we find first that they are *conceptually* close. Some specific choices, however, are replaced by postulated properties, and this is part of the development of the rule set in Section 4. Second, we observe that cross axiom models induced by abstract subset spaces satisfy Heinemann's conditions in [10] for isomorphic copies of cross axiom models induced by subset spaces.

In contrast to subset spaces and cross axiom models, the requirements for abstract subset spaces are simple closure conditions. Hence, given arbitrary sets $\mathcal{W}_0 \subseteq X \times \mathcal{O}$ and $\mathcal{R}_0 \subseteq \mathcal{O} \times \mathcal{O}$, there is a unique least extension that is an abstract subset frame:

Lemma 3.4 *Let $X, \mathcal{O}$ be sets and $\mathcal{W}_0, \mathcal{R}_0$ relations so that $\mathcal{W}_0 \subseteq X \times \mathcal{O}$ and $\mathcal{R}_0 \subseteq \mathcal{O} \times \mathcal{O}$. Then $(X, \mathcal{O}, (\mathcal{W}_0; (\mathcal{R}_0^*)^{-1}), \mathcal{R}_0^*)$ is the least abstract subset frame $(X, \mathcal{O}, \mathcal{W}, \mathcal{R})$ so that $\mathcal{W}_0 \subseteq \mathcal{W}$ and $\mathcal{R}_0 \subseteq \mathcal{R}$. It is called the abstract subset frame* generated by $(X, \mathcal{O}, \mathcal{W}_0, \mathcal{R}_0)$.

Proof. The relation $\mathcal{R}_0^*$ is a preorder. Furthermore:

$$(\mathcal{W}_0; (\mathcal{R}_0^*)^{-1}); (\mathcal{R}_0^*)^{-1} = \mathcal{W}_0; (\mathcal{R}_0^*; \mathcal{R}_0^*)^{-1} \subseteq \mathcal{W}_0; (\mathcal{R}_0^*)^{-1}$$

Let $(X, \mathcal{O}, \mathcal{W}, \mathcal{R})$ be an abstract subset frame satisfying $\mathcal{W}_0 \subseteq \mathcal{W}$ and $\mathcal{R}_0 \subseteq \mathcal{R}$. Then the reflexive-transitive closure $\mathcal{R}_0^*$ is a subset of $\mathcal{R}$, and consequently $\mathcal{W}_0; (\mathcal{R}_0^*)^{-1} \subseteq \mathcal{W}; \mathcal{R}^{-1} \subseteq \mathcal{W}$. □

4 The labelled sequent calculus LSSL-p

4.1 Axioms and rules of LSSL-p

Now we define a calculus **LSSL-p**, a labelled calculus for subset space logic, following the general method of constructing labelled calculi but introducing pairs as labels.

For **LSSL-p**, we need two disjoint sets L_1, L_2 of labels. We use the symbols $x, y, z, x', x_1, \ldots$ for the elements of L_1 and $u, v, w, u', u_1, \ldots$ for the elements of L_2. Our judgements are relational atoms $x\,\overline{W}\,u$ or $u\,\overline{R}\,v$ or of the form $(x, u)\colon A$ where $(x, u) \in L_1 \times L_2$ and A is an SSL-formula as given above. Some of the rules are subject to a condition which is abbreviated to $j(\ldots)$ and will be discussed below. The $!(\ldots)$ stands for the usual *eigenvariable* condition that the label does not occur in the conclusion.

The letters $\overline{W}$, $\overline{R}$ in formulas stand for the *complement* of the corresponding relations. The judgement $(x, u)\colon A$ should be read as "*if* (x, u) is a world, then A holds at (x, u)". Note that (x, u) might be no world, in which case "A holds at (x, u)" makes no sense. This is true for abstract subset spaces — where the set of worlds is given by $\mathcal{W}$ — as well as for the original subset spaces where $x, u \models \ldots$ is defined only in case that $x \in u$. To put it differently, the "term" (x, u) is a partial term, as it may have no value in the given model. The statement *not* $x\,\overline{W}\,u$ — (x, u) is a world — then corresponds to "(x, u) denotes". From $(y, u)\colon A$ we could only deduce "if (y, u) is a world then $(x,u)\colon \mathrm{L}A$". [5] Instead of introducing $y\,\overline{W}\,u$ together with $(x,u)\colon \mathrm{L}A$, we postulate that it is already present in the context. As $(y,u)\colon B$ corresponds to "if (y,u) is a world then ...", any judgement $(y,u)\colon B$ in the context would do the job. Consequently, in the formulation of the calculus we use the condition:

$\boldsymbol{j(y,u)}$ **:** "The conclusion contains some judgement $(y,u)\colon B$ or $y\,\overline{W}\,u$."

The rule (L) could be split into

$$\frac{\Gamma, y\,\overline{W}\,u, (x,u) : \mathrm{L}A, (y,u) : A}{\Gamma, y\,\overline{W}\,u, (x,u) : \mathrm{L}A} \qquad \frac{\Gamma, (y,u) : B, (x,u) : \mathrm{L}A, (y,u) : A}{\Gamma, (y,u) : B, (x,u) : \mathrm{L}A}$$

which makes the similarity to the usual rule for 'possibly A' more explicit but the condition above abbreviated by $j(y,u)$ provides a way to combine these two possibilities. As we interpret the judgements in subset spaces where only certain pairs are worlds, it seems adequate to read the complex label (x,u) as a partial term but it comes with a straightforward totalisation: extend the domain to $X \times \mathcal{O}$ and map (x,u) to the corresponding pair, then read $(x,u) : A$ (again) as: "(x,u) is no world or it is a world, at which A holds." This extension eliminates the partiality if desired. Still, "(y,u) is no world" does not imply $(x,u) : A$. So the side condition will be kept in the calculus. Also note that in a proof search this side condition restricts the instantiation to worlds that are already present in the lower sequent. A similar remark applies to the $\Diamond$-rule, which is also subject to a condition $j(\ldots)$. Due to the $(\Diamond_{ref})$-rule, we do not need a reflexivity rule for R. The reflexivity of the accessibility relation for K/L is built into the system.

Now let us consider the necessitation rules. For soundness, the atom $x\,\overline{W}\,u$ in the premiss of the $\Box$ and K-rules would not be necessary. If preferred, we can generalise the rule so that $x\,\overline{W}\,u$ need not be present in the upper sequent. As weakening is admissible, this makes no big difference (except shortening some sequents in the derivation). We will, however, use the fact that an atom $x\,\overline{W}\,u$ can be contracted into the $(x,u)\colon \Box A$ or $(x,u)\colon \mathrm{K}A$ built in a $(\Box)/(\mathrm{K})$-inference. This improves the permutability of rules. To see this, consider a derivation ending with

$$(\Box)\,\frac{\Gamma\,,\, w\,\overline{R}\,u\,,\,(x,w)\colon \Diamond B\,,\,(x,u)\colon B\,,\,(x,v)\colon A\,,\, u\,\overline{R}\,v\,, x\,\overline{W}\,u}{(\Diamond)\,\dfrac{\Gamma\,,\, w\,\overline{R}\,u\,,\,(x,w)\colon \Diamond B\,,\,(x,u)\colon B\,,\,(x,u)\colon \Box A}{\Gamma\,,\, w\,\overline{R}\,u\,,\,(x,w)\colon \Diamond B\,,\,\boldsymbol{(x,u)}\colon \Box A}}$$

in which $(\Box)$ can be permuted downward.

The rules (R-trans) and (RW) just reflect the conditions for abstract subset frames. An example for the use of (RW) can be found in the derivation of the cross axiom in the proof of Lemma 5.5. Persistence can be combined

---

[5] Compare this with Beeson's axiom ([3], p. 98)

$$A\{t/x\} \wedge t\downarrow \rightarrow \exists x A$$

where $t\downarrow$ stands for "t denotes".

$$(\text{ax})\ \frac{}{\Gamma, (x,u)\colon P, (x,v)\colon \neg P} \qquad (\Diamond_{ref})\ \frac{\Gamma, (x,u)\colon \Diamond A, (x,u)\colon A}{\Gamma, (x,u)\colon \Diamond A}$$

$$(\Box)\ \frac{\Gamma, x\,\overline{W}\,u, u\,\overline{R}\,v, (x,v)\colon A}{\Gamma, (x,u)\colon \Box A}\ !(v) \qquad (\Diamond)\ \frac{\Gamma, u\,\overline{R}\,v, (x,u)\colon \Diamond A, (x,v)\colon A}{\Gamma, u\,\overline{R}\,v, (x,u)\colon \Diamond A}\ j(x,v)$$

$$(\mathrm{K})\ \frac{\Gamma, x\,\overline{W}\,u, (y,u)\colon A}{\Gamma, (x,u)\colon \mathrm{K}A}\ !(y) \qquad (\mathrm{L})\ \frac{\Gamma, (x,u)\colon \mathrm{L}A, (y,u)\colon A}{\Gamma, (x,u)\colon \mathrm{L}A}\ j(y,u)$$

$$(\wedge)\ \frac{\Gamma, (x,u)\colon A \qquad \Gamma, (x,u)\colon B}{\Gamma, (x,u)\colon A \wedge B} \qquad (\vee)\ \frac{\Gamma, (x,u)\colon A, (x,u)\colon B}{\Gamma, (x,u)\colon A \vee B}$$

$$(R\text{-trans})\ \frac{\Gamma, u\,\overline{R}\,v, v\,\overline{R}\,w, u\,\overline{R}\,w}{\Gamma, u\,\overline{R}\,v, v\,\overline{R}\,w} \qquad (RW)\ \frac{\Gamma, v\,\overline{R}\,u, x\,\overline{W}\,v}{\Gamma, v\,\overline{R}\,u}\ j(x,u)$$

Table 3
System **LSSL-p**

conveniently with the logical axioms. Note that this axiom is simpler than the persistence condition for cross axiom models which refers to the R-accessibility relation. The full system is given in Table 3. We use $\vdash$ for derivability and $\vdash^n$ for the existence of a derivation of height $\leq n$.

Definition 4.1 Let $\mathcal{M} = (X, \mathcal{O}, \mathcal{W}, \mathcal{R}, \mathcal{V})$ be an abstract subset spaces, and $\ell_1 : L_1 \to X$ and $\ell_2 : L_2 \to \mathcal{O}$ mappings. Then, based on the validity of formulas, we define the validity of judgements and sequents as follows:

$$\begin{array}{lll}
(\mathcal{M}, \ell_1, \ell_2) \models (x,u) : A & \iff & ((\ell_1(x), \ell_2(u)) \in \mathcal{W} \text{ implies } (\ell_1(x), \ell_2(u)) \models_{\mathcal{M}} A) \\
(\mathcal{M}, \ell_1, \ell_2) \models x\,\overline{W}\,u & \iff & (\ell_1(x), \ell_2(u)) \notin \mathcal{W} \\
(\mathcal{M}, \ell_1, \ell_2) \models u\,\overline{R}\,v & \iff & (\ell_2(u), \ell_2(v)) \notin \mathcal{R} \\
(\mathcal{M}, \ell_1, \ell_2) \models \Gamma & \iff & (\mathcal{M}, \ell_1, \ell_2) \models J \text{ for some judgement } J \text{ in } \Gamma
\end{array}$$

Lemma 4.2 (Soundness) *If* $\vdash \Gamma$ *then* $(\mathcal{M}, \ell_1, \ell_2) \models \Gamma$ *for all abstract subset spaces* $\mathcal{M} = (X, \mathcal{O}, \mathcal{W}, \mathcal{R}, \mathcal{V})$ *and mappings* $\ell_1 : L_1 \to X$ *and* $\ell_2 : L_2 \to \mathcal{O}$.

Proof. Induction on the height of a **LSSL-p** derivation. □

Corollary 4.3 ***LSSL-p*** *is sound with respect to subset spaces.*

4.2 The role of pairs

We have presented the axioms and rules of **LSSL-p** and argued that it is rather natural, as it is close to the subset space semantics. Before we demonstrate that the calculus enjoys the desired proof-theoretical properties, we want to discuss its design in relation to alternative approaches.

Labelled sequent systems have been used before in connection with non-relational semantics. Gilbert and Maffezioli [8] for example develop sequent calculi for several modal logics which are weaker than the smallest normal modal logic. Semantics for these languages is usually based on *neighbourhood* frames. The calculi utilise a translation into a multi-modal system with normal modalities. Negri and Olivetti [16] present a sequent calculus for preferential conditional logic PCL. They internalise the weak neighbourhood semantics for PCL. Similar to our approach, the set of judgements is extended in both cases. In [8] we have relational judgements that do not state accessibility in the original setting but refer to accessibility in the multi-modal system. In [16] the extension is taken even further. In order to deal with the quantifier alternation in the semantical explanation of the conditional, new primitives for certain subexpressions of that definition are introduced. Note that the shift from the topological semantics (see [4]) to the bimodal system of topologic (see [1]), also eliminates the alternation of quantifiers in the semantic definition

$$\exists u \in \mathcal{O}(x \in u \wedge \forall y \in u \text{“}A \text{ holds at } y\text{”})$$

of "necessarily A". The universal modality is mapped to $\Diamond\textsc{k}$ in topologic (see [1], p. 103). However, the worlds in topologic and in the weaker subset space logic contain a component 'point' as well as a component 'set'. Points and sets play a role also in [8] and [16]. In [8] the labels in the calculus stand for worlds in the translated system, which are points or sets of points where the points are distinguished with the help of the modality 'σ'. In [16], two types of labels are used instead. In contrast to both settings, the worlds in subset spaces are *pairs* of points and sets, and so we use pairs also in the judgements.

A noticeable feature of **LSSL-p** is the simplicity of the frame rules. They are based on universal axioms only. We could even replace them by complex application conditions for $\Diamond$ and $\textsc{l}$ that refer to the computed closure. To this end, let

$$\begin{aligned}
\mathbf{L}_1(\Gamma) &= \{x \in L_1 \mid x \text{ occurs in } \Gamma\}\\
\mathbf{L}_2(\Gamma) &= \{u \in L_2 \mid u \text{ occurs in } \Gamma\}\\
\mathbf{R}_0(\Gamma) &= \{(u,v) \mid u\,\overline{R}\,v \text{ occurs in } \Gamma\} \subseteq \mathbf{L}_2(\Gamma) \times \mathbf{L}_2(\Gamma)\\
\mathbf{W}_0(\Gamma) &= \{(x,u) \mid x\,\overline{W}\,u \text{ or some } (x,u)\colon A \text{ occurs in } \Gamma\}\\
\mathbf{R}(\Gamma) &= \mathbf{R}_0(\Gamma)^* \subseteq \mathbf{L}_2(\Gamma) \times \mathbf{L}_2(\Gamma)\\
\mathbf{W}(\Gamma) &= \mathbf{W}_0(\Gamma); \mathbf{R}(\Gamma)^{-1}
\end{aligned}$$

for multisets Γ of judgements. If we generalise ($\Diamond$) and ($\textsc{l}$) to

$$(\Diamond^*)\ \frac{\Gamma, (x,u) : \Diamond A, (x,v) : A}{\Gamma, (x,u) : \Diamond A} \quad \begin{array}{ll}\text{if} & (x,v) \in \mathbf{W}(\Gamma, (x,u) : \Diamond A)\\ \text{and} & (u,v) \in \mathbf{R}(\Gamma, (x,u) : \Diamond A)\end{array}$$

$$(\textsc{l}^*)\ \frac{\Gamma, (x,u) : \textsc{l}A, (y,u) : A}{\Gamma, (x,u) : \textsc{l}A} \text{ if } (y,u) \in \mathbf{W}(\Gamma, (x,u) : \Diamond A)$$

then we can remove the frame rules. This is also the first step in the development of the 'compressed' version of the system which is used for proof search in Section 6.

Still, the cross axiom models offer an alternative way to define a calculus without introducing pairs [6]. The only frame condition that has not already been studied is the cross condition $\mathcal{R};\mathcal{S} \subseteq \mathcal{S};\mathcal{R}$ which can be transformed to the geometric formula $\forall x, y, y(xRy \wedge ySz \rightarrow \exists y'(xSy' \wedge y'Rz))$ and yields the cross rule:

$$\frac{\Gamma, x\,\overline{R}\,y, y\,\overline{S}\,z, x\,\overline{S}\,y', y'\,\overline{R}\,z}{\Gamma, x\,\overline{R}\,y, y\,\overline{S}\,z}$$

The set of derivable formulas of these two systems coincide but neither can be translated just by replacing every step locally by a sequence of steps of the other system. Passing from the system based on cross axiom models to **LSSL-p** would require choosing an appropriate substitution of pairs for simple labels so that the R/S-relations are reduced to the special form in abstract subset spaces and, in particular, the applications of the cross rule with their eigenvariables can be removed. The other direction is a bit simpler, as we can leave the labelled formulas unchanged. The frame judgements have to be translated according to the transformation of abstract subset spaces into cross axiom models. Instead of translating the applications of frame rules one by one, the block of frame rules needed for the computation of the closure should be transformed as a whole. Further analysis of the class of derivations obtained by this translation might be useful for advanced proof-theoretic investigations of SSL but then one could use **LSSL-p** right away.

5 Basic properties of LSSL-p

We start with some simple properties of the calculus.

Lemma 5.1 *The following holds for **LSSL-p**:*

(i) (renaming) *Let d be a derivation with endsequent Γ and $x, y \in L_1$ (or $u, v \in L_2$) where y (or v) does not occur in d. Then replacing every occurrence of x in d by y (or u by v respectively) yields a derivation of the endsequent $\Gamma\{y/x\}$ (or $\Gamma\{v/u\}$ respectively).*

(ii) (label substitution) *Let $x_1, \ldots, x_n$ be pairwise distinct elements of L_1 and $u_1, \ldots, u_m$ be pairwise distinct elements of L_2. Then $\vdash^n \Gamma$ implies $\vdash^n \Gamma\{y_1/x_1, \ldots, y_n/x_n, v_1/u_1, \ldots, v_m/u_m\}$ for all $y_1, \ldots, y_n \in L_1$ and $v_1, \ldots, v_m \in L_2$.*

(iii) (weakening) $\vdash^n \Gamma \Longrightarrow \vdash^n \Gamma, J$ *for every judgement J*

(iv) (R-contraction) $\vdash^n \Gamma, u\,\overline{R}\,v, u\,\overline{R}\,v \Longrightarrow \vdash^n \Gamma, u\,\overline{R}\,v$

(v) (W-contraction)
 (a) $\vdash^n \Gamma, x\,\overline{W}\,u, x\,\overline{W}\,u \Longrightarrow \vdash^n \Gamma, x\,\overline{W}\,u$
 (b) $\vdash^n \Gamma, x\,\overline{W}\,u, (x,u)\colon A \Longrightarrow \vdash^n \Gamma, (x,u)\colon A$

(vi) (R-reflexivity) $\vdash^n \Gamma, u\,\overline{R}\,u \Longrightarrow \vdash^n \Gamma$

Proof. Straightforward induction on the height of the given derivation. The proof of Facts (ii),(iii) need Fact (i) in order to avoid a clash with eigenvariables. Intuitively, Fact (v)(b) holds because $(x,u) : A$ is treated throughout as if it were of the form $x\,\overline{W}\,u \vee \ldots$ and we can contract several occurrences of $x\,\overline{W}\,u$. Technically, we use the fact that the $j(x,u)$-condition can also be fulfilled by an occurrence of $(x,u) : A$, not only $x\,\overline{W}\,u$, and that an occurrence of $x\,\overline{W}\,u$ can be contracted into the constructed formula $\Box B$ or $\mathrm{K}B$ in $\Box/\mathrm{K}$-inferences. In the cases where A is constructed by $\Box, \mathrm{K}$ we use (v)(a) in the proof. Furthermore,

Facts (iv) and (v) are used in the proof of (vi) in the case of frame rules. □

With Fact (i), we can always rename eigenvariables in a proof so that they become different from each other and from every other variable in a given judgement, sequent or second derivation. We make use of this fact in many of the proofs below without mentioning it explicitly.

Lemma 5.2 *The* ***HSS****-rules of necessitation are admissible.*

Proof. Let $(x, u)\colon A$ be derivable. Then, by label substitution, also $(x, v)\colon A$ and $(y, u)\colon A$ are derivable for fresh $v \in L_2, y \in L_1$. By the admissibility of weakening we obtain the derivability of $x\,\overline{W}\,u\,,\ u\,\overline{R}\,v\,,\ (x, v)\colon A$ and of $x\,\overline{W}\,u\,,\ (y, u)\colon A$. Application of ($\Box$) or (K) respectively yields a derivation of $(x, u)\colon \Box A$ or $(x, u)\colon \mathrm{K}A$. □

For ($\Diamond$) and (L), inversion is just an instance of weakening. So the theorem below yields the height-preserving invertibility of all logical rules.

Theorem 5.3 (Invertibility of logical rules) *The following holds for the calculus* ***LSSL-p****:*

(i) ($\wedge$-inversion) $\vdash^n \Gamma, (x, u)\colon A \wedge B \Longrightarrow \vdash^n \Gamma, (x, u)\colon A$ *and* $\vdash^n \Gamma, (x, u)\colon B$

(ii) ($\vee$-inversion) $\vdash^n \Gamma, (x, u)\colon A \vee B \Longrightarrow \vdash^n \Gamma, (x, u)\colon A, (x, u)\colon B$

(iii) ($\Box$-inversion) $\vdash^n \Gamma, (x, u)\colon \Box A \Longrightarrow \vdash^n \Gamma, u\,\overline{R}\,v, x\,\overline{W}\,u, (x, v)\colon A$ *for every* $v \in L_2$

(iv) (K-inversion) $\vdash^n \Gamma, (x, u)\colon \mathrm{K}A \Longrightarrow \vdash^n \Gamma, x\,\overline{W}\,u, (y, u)\colon A$ *for all* $y \in L_1$

Proof. By induction on the height of the derivation. In the proof of Fact (iii) and (iv), the additional $x\,\overline{W}\,u$ is used in the case where $(x, u)\colon \Box A$ or $(x, u)\colon \mathrm{K}A$ is necessary to meet the context condition $j(x, u)$ for the last inference in the given derivation. □

Theorem 5.4 (Admissibility of contraction) *If* $\vdash^n \Gamma, (x, u)\colon A, (x, u)\colon A$ *then* $\vdash^n \Gamma, (x, u)\colon A$

Proof. By induction on the height of the derivation. In the case that one of the distinguished occurrences of $(x, u)\colon A$ is constructed by the last inference (and the principle symbol of A is not $\Diamond$ or L), we combine inversion with the induction hypothesis. In the case of ($\Box$), i.e.:

$$(\Box)\ \dfrac{\dfrac{\vdots}{\Gamma\,,\ (x, u)\colon \Box A\,,\ x\,\overline{W}\,u\,,\ u\,\overline{R}\,v\,,\ (x, v)\colon A}}{\Gamma\,,\ (x, u)\colon \Box A\,,\ (x, u)\colon \Box A}\ !(v)$$

we use height-preserving inversion of ($\Box$) first, then the IH, followed by (R-contraction) and (W-contraction), and finally an application of ($\Box$) to build the desired derivation. The case of (K) is similar, without (R-contraction). □

Next we demonstrate the strength of the calculus by presenting derivations for the **HSS** axioms. With negation defined as above, this is provided by the derivability of the sequents (ii)-(ix) in the next lemma.

Lemma 5.5 *For all SSL-formulas A, B, predicate letters P, $x \in L_1$, and $u \in L_2$, the following sequents are derivable in **LSSL-p**:*

(i) $(x,u)\colon A\,,\ (x,u)\colon \neg A$

(ii) $(x,u)\colon A$ *where A is a substitution instance of a tautology of classical propositional logic*

(iii) $(x,u)\colon (\neg P \vee \Box P) \wedge (P \vee \Box\neg P)$ *for propositional letters P*

(iv) $(x,u)\colon \mathrm{L}\neg A\,,\ (x,u)\colon A \wedge \mathrm{KK}A$

(v) $(x,u)\colon \Diamond\neg A\,,\ (x,u)\colon A \wedge \Box\Box A$

(vi) $(x,u)\colon \mathrm{L}(A \wedge \neg B)\,,\ (x,u)\colon \mathrm{L}\neg A\,,\ (x,u)\colon \mathrm{K}B$

(vii) $(x,u)\colon \Diamond(A \wedge \neg B)\,,\ (x,u)\colon \Diamond\neg A\,,\ (x,u)\colon \Box B$

(viii) $(x,u)\colon \mathrm{K}\neg A\,,\ (x,u)\colon \mathrm{KL}A$

(ix) $(x,u)\colon \mathrm{L}\Diamond\neg A\,,\ (x,u)\colon \Box\mathrm{K}A$

Proof. For Fact (i), we proceed by straightforward induction on A as usual. One can obtain a derivation as required in Fact (ii) from a cut-free derivation of the tautology in **GS3**. We present the derivation of sequent (ix), the cross axiom:

$$\dfrac{\dfrac{\dfrac{\dfrac{\dfrac{\dfrac{\text{Fact (i), weakening}}{(x,u)\colon \mathrm{L}\Diamond\neg A, (y,u)\colon \Diamond\neg A, (y,v)\colon \neg A, x\overline{W}u, x\overline{W}v, y\overline{W}u, (y,v)\colon A, u\overline{R}v}}{(x,u)\colon \mathrm{L}\Diamond\neg A\,,\ (y,u)\colon \Diamond\neg A\,,\ x\overline{W}u\,,\ x\overline{W}v\,,\ y\overline{W}u\,,\ (y,v)\colon A\,,\ u\overline{R}v}(\Diamond)}{(x,u)\colon \mathrm{L}\Diamond\neg A\,,\ x\overline{W}u\,,\ x\overline{W}v\,,\ y\overline{W}u\,,\ (y,v)\colon A\,,\ u\overline{R}v}(\mathrm{L})}{(x,u)\colon \mathrm{L}\Diamond\neg A\,,\ x\overline{W}u\,,\ x\overline{W}v\,,\ (y,v)\colon A\,,\ u\overline{R}v}(RW)}{(x,u)\colon \mathrm{L}\Diamond\neg A\,,\ x\overline{W}u\,,\ (x,v)\colon \mathrm{K}A\,,\ u\overline{R}v}(\mathrm{K})\,!(y)}{(x,u)\colon \mathrm{L}\Diamond\neg A\,,\ (x,u)\colon \Box\mathrm{K}A}(\Box)\,!(v)$$

The remaining derivations are given in the appendix. □

6 Admissibility of cut

Lemma 6.1 *If $\vdash^n \Gamma\,,\ (x,u)\colon A$ and $\vdash^m \Pi\,,\ (x,u)\colon \neg A$, then $\vdash \Gamma, \Pi\,,\ xWu$.*

Proof. By induction on A, side induction on $n+m$.
Case 1: $(x,u)\colon A$ is of no relevance for the last inference in the first of the given derivations, or $(x,u)\colon \neg A$ is of no relevance for the last inference in the second derivation. If the corresponding conclusion is an axiom then Γ, Π is an axiom. Otherwise, use the side induction hypothesis. (Rename eigenvariables first if necessary.)
Case 2: $(x,u)\colon A$ is relevant for the last inference but only to meet the context condition $j(x,u)$:

$$\frac{\Gamma'\,,\ (x,u)\colon A}{\Gamma\,,\ (x,u)\colon A}$$

By side induction hypothesis we obtain $\Gamma', \Pi, x\overline{W}u$, from which we can deduce

$\Gamma, \Pi, x\,\overline{W}\,u$.
Case 3: $(x,u)\colon \neg A$ is relevant for the last inference but only to meet the context condition $j(x,u)$: similar
Case 4: $(x,u)\colon A$ and $(x,u)\colon \neg A$ are the principal formulas in the last inference of the respective derivations. Then we distinguish cases according to A.

If A is a literal, then $\Gamma, (x,u)\colon A$ and $\Pi, (x,u)\colon \neg A$ are axioms with principal formulas $(x,u)\colon A$ and $(x,u)\colon \neg A$ respectively. In that case Γ contains $(x,v)\colon \neg A$ for some v and Π contains $(x,w)\colon A$ for some w. Hence Γ, Π is an axiom.

Furthermore, we present in detail the case of the principal symbols $\Box/\Diamond$ in $A/\neg A$. The remaining cases are similar, even a bit simpler.

W.l.o.g. $A \equiv \Box B$ and $\neg A \equiv \Diamond\neg B$. Then the derivations have the form:

$$d_1 : \quad (\Box)\,\dfrac{\dfrac{\vdots}{\Gamma\,,\; x\,\overline{W}\,u\,,\; u\,\overline{R}\,v\,,\; (x,v)\colon A}}{\Gamma\,,\; (x,u)\colon \Box A}\;!(v)$$

and

$$d_2 : \quad (\Diamond/\Diamond_{ref})\,\dfrac{\dfrac{\vdots}{\Pi'\,,\; (u\,\overline{R}\,w\,,\;)(x,u)\colon \Diamond\neg A\,,\; (x,w)\colon \neg A}}{\Pi'\,,\; (u\,\overline{R}\,w\,,\;)(x,u)\colon \Diamond A}\; j(x,w)$$

where $u\,\overline{R}\,w$ may be missing if $u = w$, and $\Pi = \Pi', u\,\overline{R}\,w$ otherwise. First we obtain $\vdash \Gamma, \Pi', (u\,\overline{R}\,w\,,\;)(x,w)\colon \neg A\,,\; x\,\overline{W}\,u$ by side induction hypothesis. Second, applying substitution to the immediate subderivation of d_1, we obtain a derivation of $\Gamma\,,\; x\,\overline{W}\,u\,,\; u\,\overline{R}\,w\,,\; (x,w)\colon A$. Combining these and using the (main) induction hypothesis, we get:

$$\vdash \Gamma\,,\; x\,\overline{W}\,u\,,\; u\,\overline{R}\,w\,,\; \Gamma\,,\; \Pi'\,,\; (u\,\overline{R}\,w\,,\;)\,,\; x\,\overline{W}\,u\,,\; x\,\overline{W}\,w$$

Subcase 1: The inference introducing $\neg A \equiv \Diamond\neg B$ was $\Diamond_{ref}$. Then the second $u\,\overline{R}\,w$ is missing, $\Pi' = \Pi$ and $u = w$. In that case, we use the admissibility of the reflexivity rule and (W-contraction) to obtain $\vdash \Gamma\,,\; x\,\overline{W}\,u\,,\; \Gamma\,,\; \Pi$.

Subcase 2: Otherwise. Then $\Pi = \Pi', u\,\overline{R}\,w$ and Π contains $x\,\overline{W}\,w$ or a judgement of the form $(x,w)\colon B$. Now we apply (R-contraction), (W-contraction) of type (a), and (W-contraction) of type (a) or (b) to obtain again $\vdash \Gamma\,,\; x\,\overline{W}\,u\,,\; \Gamma\,,\; \Pi$.

In both cases, the proof is completed by applying (contraction) for the formulas in Γ. □

Corollary 6.2 *Let Γ, Π be sequents so that Γ, Π contains $x\,\overline{W}\,u$ or some judgement of the form $(x,u)\colon B$. If $\vdash \Gamma\,,\; (x,u)\colon A$ and $\vdash \Pi\,,\; (x,u)\colon \neg A$, then $\vdash \Gamma, \Pi$.*

Proof. Use 6.1 and (W-contraction). □

Lemma 6.3 *The rule* modus ponens *is admissible.*

Proof. Let $(x,u)\colon A \to B$ and $(x,u)\colon A$ be derivable. By $(\vee)$-inversion, the sequent $(x,u)\colon \neg A$, (x,u) : B is also derivable. Using 6.2, we can conclude that $(x,u)\colon B$ is derivable. □

Lemma 6.4 *If **HSS** $\vdash A$ then **LSSL-p** $\vdash (x,u)\colon A$ for all $x \in L_1, u \in L_2$.*

Proof. By induction on the length of a derivation in **HSS**, using 5.5, 5.2 and 6.3. □

As it has been proved in [1] that **HSS** is complete w.r.t. subset spaces, we have the following theorem:

Theorem 6.5 ***LSSL-p** is complete w.r.t. abstract subset spaces.*

In the next section we present a direct proof of this result based on proof search in **LSSL-p**.

7 Direct proof of completeness

In this section, we present the definition of a *search tree* which reflects bottom-up proof search for **LSSL-p** and use it for the proof of completeness. Completeness proofs in this style for labelled calculi have been presented before e.g. in [7,5,16]. Our calculus, however, does not belong to the class considered in [7,5], and we can not expect to construct a *finite* countermodel, as we are working with (abstract) subset spaces. In particular, we will not obtain a *decision procedure*. However, we get a reasonable strategy for the construction of proofs, as evidence for validity, which produces an output for all valid formulas. To simplify a our procedure, we consider a ***compressed*** version of derivations and a corresponding system **LSSL-pc**:

- *Frame rules* are never applied explicitly. We reformulate the application conditions instead. To this end, we use the relations $\mathbf{R}, \mathbf{W}$ defined in Sec. 4.2. on page 278.
- All reductions corresponding to *disjunctive* rules, i.e. rules introducing $\vee, \Diamond, \mathrm{L}$ are performed in one step. For this, we let $\mathbf{D}(\Gamma)$ denote the least multiset extending Γ that satisfies:
  - · $(x,u)\colon A$ and $(x,u)\colon B$ are in $\mathbf{D}(\Gamma)$ if $(x,u)\colon A \vee B$ is in $\mathbf{D}(\Gamma)$.
  - · $(x,v)\colon A$ is in $\mathbf{D}(\Gamma)$ if $(x,v) \in \mathbf{W}(\Gamma)$ and $(x,u)\colon \Diamond A$ is in $\mathbf{D}(\Gamma)$ for some u so that $(u,v) \in \mathbf{R}(\Gamma)$
  - · $(y,u)\colon A$ is in $\mathbf{D}(\Gamma)$ if $(y,u) \in \mathbf{W}(\Gamma)$ and $(x,u)\colon \mathrm{L}A$ is in $\mathbf{D}(\Gamma)$ for some x

Note that in the definition of $\mathbf{R}(\Gamma)$ we build the reflexive-transitive closure *on* $\boldsymbol{L}_2(\Gamma)$. As a consequence, the sets $\mathbf{R}(\Gamma)$ and $\mathbf{W}(\Gamma)$ are finite. Obviously, $\mathbf{D}(\Gamma)$ is also finite and can be computed in a straightforward way. Applying the rule (R-trans) from bottom to top, we can turn every Γ into a sequent which contains $u\,\overline{R}\,v$ for every pair in $\mathbf{R}_0(\Gamma)^+$. With the help of (RW) we can then obtain a sequent which satisfies the $j(x,v)$-condition for all pairs (x,v) in $\mathbf{W}(\Gamma)$. With this in mind, we see that a derivation of $\mathbf{D}(\Gamma)$ can be turned into a derivation of Γ by applications of (weakening), (contraction), disjunctive rules (i.e. $(\Diamond), (\Diamond_{ref}), (\mathrm{L}), (\vee)$) and frame rules.

$$(\wedge)\ \frac{\Gamma, (x,u)\colon A \qquad \Gamma, (x,u)\colon B}{\Gamma, (x,u)\colon A \wedge B} \qquad (\mathrm{ax})\ \frac{}{\Gamma, (x,u)\colon P, (x,v)\colon \neg P} \qquad (\mathbf{D})\ \frac{\mathbf{D}(\Gamma)}{\Gamma}$$

$$(\Box)\ \frac{\Gamma, x\overline{W}\,u, u\overline{R}\,v, (x,v)\colon A}{\Gamma, (x,u)\colon \Box A}\ !(v) \qquad (\mathrm{K})\ \frac{\Gamma, x\overline{W}\,u, (y,u)\colon A}{\Gamma, (x,u)\colon \mathrm{K}A}\ !(y)$$

Table 4
System **LSSL-pc**

Now we consider trees (finite and infinite) of sequents built according to the rules in Table 4. We write seq(N) for the sequent at node N.

The rules $(\wedge), (\Box), (\mathrm{K})$ and their constructed principal formulas as well as the corresponding judgements are called *conjunctive.* An *expansion step* for such a formula consists of adding nodes with the corresponding premisses as children. The expansion step for a judgement $(x,u)\colon A \wedge B$ at leaf N will be performed only if neither $(x,u)\colon A$ nor $(x,u)\colon B$ occurs on the path $\alpha(N)$ leading from the root to N. Similarly, the step for $(x,u)\colon \Box A$ at N will be performed only if no judgement $(x,v)\colon A$ satisfying $(u,v) \in \mathbf{R}(\Gamma)$ occurs on $\alpha(N)$, and the expansion for $(x,u)\colon \mathrm{K}A$ is subject to the restriction that no judgement $(y,u)\colon A$ occurs on $\alpha(N)$. Furthermore, we apply these steps only to nodes which are no axioms.

Applying these steps successively to all conjunctive judgements at leaves that meet these conditions (but not to conjunctive formulas produced by this transformation) is called a *C-step* for the tree T. A *D-step* for it consists in adding a child N' to every leaf N and let seq$(N') := \mathbf{D}(\text{seq}(N))$.

To search for a derivation of $(x,u)\colon A$, we proceed as follows: Start with the one-node-tree with $(x,u)\colon A$ at its root, and add a single child N with seq$(N) = \mathbf{D}((x,u)\colon A)$. As long as there are still leaves N so that seq(N) is no axiom and all thoses leaves are expandable, perform a C-step followed by a D-step. Now, if this procedure terminates with a tree where all leaves contain axioms, then we have found a derivation of $(x,u)\colon A$ in **LSSL-pc** which can be uncompressed to give a derivation of $(x,u)\colon A$ in **LSSL-p**. If the procedure terminates with a leaf with no axiom that is not expandable, then we consider the path α from the root to that node, and let $(\Gamma_i)_{i \in I}$ with appropriate $I = \{0, \ldots, n\}$ be the corresponding sequence of sequents. Otherwise, the procedure generates an infinite tree which contains an infinite branch α, and we let $\alpha = (\Gamma_i)_{i \in I}$ with $I = \mathbb{N}$ denote the infinite sequence of corresponding sequents. In both cases, we let $\Gamma := \bigcup_{i \in I} \Gamma_i$.

Now it remains to be shown that we can obtain a countermodel based on a path α as described above. To this end, we extend the definitions in Sec. 4.2, page 278, to *infinite* Γ, and let $X := \mathbf{L}_1(\Gamma)$, $\mathcal{O} := \mathbf{L}_2(\Gamma)$, $\mathcal{W}(\Gamma) := \mathbf{W}(\Gamma)$, and $\mathcal{R} := \mathbf{R}(\Gamma)$. Then $(X, \mathcal{O}, \mathcal{W}, \mathcal{R})$ is the abstract subset frame generated by $(X, \mathcal{O}, \mathbf{W}_0(\Gamma), \mathbf{R}_0(\Gamma))$.

Due to the construction of the tree, judgements $(x,u)\colon P$ or $(x,u)\colon \neg P$ for $P \in PV$ occur in every Γ_j with $j \geq i$ if they occur in Γ_i, and we know that no Γ_i is an axiom. As a consequence, there is a valuation $\mathcal{V}$ for $(X, \mathcal{O}, \mathcal{W}, \mathcal{R})$ so that

$$\mathcal{V}(y)(P) = \mathbf{t} \text{ if } (y,v)\colon \neg P \text{ occurs in } \alpha \text{ for some } v \in L_2$$
$$\mathcal{V}(y)(P) = \mathbf{f} \text{ if } (y,v)\colon P \text{ occurs in } \alpha \text{ for some } v \in L_2$$

Let $\mathcal{M} = (X, \mathcal{O}, \mathcal{W}, \mathcal{R}, \mathcal{V})$, choose ℓ_1, ℓ_2 so that $\ell_1(y) = y$ for all $y \in L_1$ and $\ell_2(v) = v$ for all $v \in L_2$.

The construction of the tree also ensures that judgements $u\,\overline{R}\,v$, $x\,\overline{W}\,u$, $(x,u)\colon A \vee B$, $(x,u)\colon \Diamond A$, $(x,u)\colon \mathrm{L}A$ occur in every Γ_j with $j \geq i$ if they occur in Γ_i. Furthermore, if some Γ_i satisfies the $j(x,u)$-condition, then so does every Γ_j with $j \geq i$. This is used in proving the following facts:

(i) $\mathbf{R}(\Gamma_{i-1}) \subseteq \mathbf{R}(\Gamma_i)$ for all $i \in \mathbb{N}_+ \cap I$ and $\mathcal{R} = \bigcup_{i \in I} \mathbf{R}(\Gamma_i)$

(ii) $\mathbf{W}(\Gamma_{i-1}) \subseteq \mathbf{W}(\Gamma_i)$ for all $i \in \mathbb{N}_+ \cap I$ and $\mathcal{W} = \bigcup_{i \in I} \mathbf{W}(\Gamma_i)$

(iii) If $(y,v)\colon B \wedge C$ in Γ then $(y,v)\colon B$ in Γ or $(y,v)\colon C$ in Γ

(iv) If $(y,v)\colon B \vee C$ in Γ then $(y,v)\colon B$ in Γ and $(y,v)\colon C$ in Γ

(v) If $(y,v)\colon \Diamond B$ in Γ then $(y,w)\colon B$ in Γ for all w so that $(v,w) \in \mathcal{R}$ and $(y,w) \in \mathcal{W}$.

(vi) If $(y,v)\colon \mathrm{L}B$ in Γ then $(z,v)\colon B$ in Γ for all z so that $(z,v) \in \mathcal{W}$.

(vii) If $(y,v)\colon \Box B$ in Γ then there is w so that $(v,w) \in \mathcal{R}$ and $(y,w)\colon B$ in Γ (and hence also $(y,w) \in \mathcal{W}$).

(viii) If $(y,v)\colon \mathrm{K}B$ in Γ then there is z so that $(z,v)\colon B$ in Γ (and hence also $(z,v) \in \mathcal{W}$).

To see Fact (viii), assume that $\mathrm{K}B$ is in Γ. Then at some point of the construction, the tree had a node of α containing $\mathrm{K}B$ as a leaf and the corresponding expansion was considered. Either the expansion was performed, in which case the next node of α contains some $(z,v)\colon B$, or the expansion was rejected. If it was rejected then there is some $(z,v)\colon B$ earlier in α.

We present the proof of Fact (v) as an example for the disjunctive connectives: Let $(y,v)\colon \Diamond B$ be in Γ_i, $(v,w) \in \mathcal{R}$ and $(y,w) \in \mathcal{W}$. By Facts (i) and (ii) we get $j,k \in I$ so that $(v,w) \in \mathbf{R}(\Gamma_j)$ and $(y,w) \in \mathbf{W}(\Gamma_k)$. Now let $m \geq \max\{i,j,k\}$ so that $\Gamma_{m+1} = \mathbf{D}(\Gamma_m)$. Then $(y,v)\colon B$ is in Γ_{m+1}.

The remaining cases are similar. With these facts, an easy induction on B shows that $(\mathcal{M}, \ell_1, \ell_2) \not\models (y,v)\colon B$ for all $(y,v)\colon B$ in α, in particular $\mathcal{M} \not\models (x,u)\colon A$.

8 Conclusion and further work

The logic of subset spaces is a bimodal logic with no obvious, pure, cut-free sequent system. The labelled approach, however, offers an alternative. Introducing a new variant which makes use of compound expressions for worlds and an "is-world"-predicate, we obtain a solution which is satisfactory in several ways: it is close to the original semantics and it uses only simple frame rules. The completeness proof is based on proof search, producing a (possibly infinite) countermodel for non-valid formulas.

The cut-free sequent calculus **LSSL-p** has been developed as a contribution to the *proof theory* of subset space logic. It has the *subformula* and *separation property*, and it provides a promising starting point for further proof-theoretic investigations of this logic and its extensions. If the focus is on *automatic deduction*, other types of calculi may be preferable. The study of tableau systems for SSL based on cross axiom models is current work of others. However, the development of termination conditions for **LSSL-p** could yield an interesting candidate for comparison.

Based on subset space logic, *topologic* is also introduced in [1]. A natural continuation of our work would be an extension to that system.

The labelled approach is sometimes criticised for not being *purely proof-theoretic*, as the semantics is internalised in the system. Although it constitutes a very convincing method for constructing calculi in difficult cases as SSL, the quest for a fine, "pure" system remains interesting. The calculus **LSSL-p** presented here provides new insights which may prove helpful also for this goal. However, even if alternatives will be developed, the labelled **LSSL-p** is still a rather simple and natural solution.

Acknowledgement. I would like to thank the referees for the many helpful suggestions and comments.

Appendix

A Derivability of the HSS axioms

In most of the derivations below we use Fact 1 in 5.3 plus weakening.
Cross axiom: see Section 5; Persistence: Let $\Pi :\equiv u\,\overline{R}\,v\,,\, x\,\overline{W}\,u$ in:

$$(\wedge)\,\dfrac{(\vee)\,\dfrac{(\Box)\,\dfrac{(ax)\,\dfrac{}{(x,u)\colon \neg P\,,\,(x,v)\colon P\,,\,\Pi}}{(x,u)\colon \neg P\,,\,(x,u)\colon \Box P}\,!(v)}{(x,u)\colon \neg P \vee \Box P} \qquad \dfrac{\dfrac{\dfrac{}{(x,u)\colon P\,,\,(x,v)\colon \neg P\,,\,\Pi}\,(ax)}{(x,u)\colon P\,,\,(x,u)\colon \Box\neg P}\,!(v)}{(x,u)\colon P \vee \Box\neg P}\,(\vee)}{(x,u)\colon (\neg P \vee \Box P) \wedge (P \vee \Box\neg P)}$$

T and K4 for $\Box$:

$$(\Diamond_{ref})\,\dfrac{(x,u)\colon \Diamond\neg A\,,\,(x,u)\colon \neg A\,,\,(x,u)\colon A}{(x,u)\colon \Diamond\neg A\,,\,(x,u)\colon A}$$

$$(\Box)\,\dfrac{(\Box)\,\dfrac{(R\text{-trans})\,\dfrac{(\Diamond)\,\dfrac{(x,u)\colon \Diamond\neg A\,,\,(x,w)\colon \neg A\,,\,(x,w)\colon A\,,\,x\,\overline{W}\,v\,,\,v\,\overline{R}\,w\,,\,x\,\overline{W}\,u\,,\,u\,\overline{R}\,v\,,\,u\,\overline{R}\,w}{(x,u)\colon \Diamond\neg A\,,\,(x,w)\colon A\,,\,x\,\overline{W}\,v\,,\,v\,\overline{R}\,w\,,\,x\,\overline{W}\,u\,,\,u\,\overline{R}\,v\,,\,u\,\overline{R}\,w}}{(x,u)\colon \Diamond\neg A\,,\,(x,w)\colon A\,,\,x\,\overline{W}\,v\,,\,v\,\overline{R}\,w\,,\,x\,\overline{W}\,u\,,\,u\,\overline{R}\,v}\,!(\mathrm{w})}{(x,u)\colon \Diamond\neg A\,,\,(x,v)\colon \Box A\,,\,x\,\overline{W}\,u\,,\,u\,\overline{R}\,v}\,!(\mathrm{v})}{(x,u)\colon \Diamond\neg A\,,\,(x,u)\colon \Box\Box A}$$

Normality for $\Box$: Let $\Gamma :\equiv (x,u)\colon \Diamond(A \wedge \neg B)\,,\,(x,u)\colon \Diamond\neg A\,,\, x\,\overline{W}\,u$ in:

$$(\Box)\,\dfrac{(\Diamond)\,\dfrac{(\wedge)\,\dfrac{(\Diamond)\,\dfrac{\Gamma, (x,v)\colon A, (x,v)\colon \neg A, (x,v)\colon B, u\,\overline{R}\,v}{\Gamma, (x,v)\colon A, (x,v)\colon B, u\,\overline{R}\,v} \qquad \Gamma, (x,v)\colon \neg B, (x,v)\colon B, u\,\overline{R}\,v}{\Gamma, (x,v)\colon A \wedge \neg B, (x,v)\colon B, u\,\overline{R}\,v}}{\Gamma, (x,v)\colon B, u\,\overline{R}\,v}\,!(v)}{(x,u)\colon \Diamond(A \wedge \neg B)\,,\,(x,u)\colon \Diamond\neg A, (x,u)\colon \Box B}$$

The corresponding properties for K are proved similar.
Euclidean property:

$$\text{(L)}\ \dfrac{x\overline{W}u\,,\,(y,u)\colon \neg A\,,\,(z,u)\colon \text{L}A\,,\,(y,u)\colon A\,,\,x\overline{W}u}{\text{(K)}\ \dfrac{x\overline{W}u\,,\,(y,u)\colon \neg A\,,\,(z,u)\colon \text{L}A\,,\,x\overline{W}u}{\text{(K)}\ \dfrac{(y,u)\colon \neg A\,,\,(x,u)\colon \text{KL}A\,,\,x\overline{W}u}{(x,u)\colon \text{K}\neg A\,,\,(x,u)\colon \text{KL}A}\,!(y)}\,!(z)}$$

References

[1] A. Dabrowski, L.S. Moss and R. Parikh, *Topological reasoning and the logic of knowledge*, Annals of Pure and Applied Logic **78** (1996), pp. 73–110.

[2] Balbiani, P., H. van Ditmarsch and A. Kudinov, *Subset space logic with arbitrary announcements*, in: *ICLA 2013*, LNAI (2013), pp. 233–244.

[3] Beeson, M. J., "Foundations of Constructive Mathematics," Springer, 1985.

[4] Blackburn, P. and J. van Benthem, *Modal logic: a semantic perspective*, in: P. Blackburn et al., editors, *Handbook of modal logic*, Elsevier, 2007 pp. 1–84.

[5] D. Garg, V. Genovese and S. Negri, *Countermodels from sequent calculi in multi-modal logics*, technical report online at `http://www.mpi-sws.org/~dg` (2012).

[6] Elbl, B. and G. Krommes, *Notes an applying Negri's method to the logic of subset spaces*, unpublished notes.

[7] Garg, D., V. Genovese and S. Negri, *Countermodels from sequent calculi in multi-modal logics*, in: *Proceedings of the 2012 27th Annual IEEE/ACM Symposium on Logic in Computer Science*, LICS '12 (2012), pp. 315–324.

[8] Gilbert, D. R. and P. Maffezioli, *Modular sequent calculi for classical modal logics*, Studia Logica **103** (2015), pp. 175–217.

[9] H. Schwichtenberg and A. Troelstra, "Basic Proof Theory," Cambridge University Press, 1996.

[10] Heinemann, B., *On subset space Kripke structures*, 9th Panhellenic Logic Symposium, Athens, Greece, July 2013. (Accepted for presentation; the paper had to be withdrawn for private reasons.).

[11] Heinemann, B., *Regarding overlaps in topologic*, in: *Advances in Modal Logic*, AiML **6** (2006), pp. 259–277.

[12] J. Gallier, "Logic for Computer Science," Harper & Row, 1986.

[13] K. Schütte, "Proof Theory," Springer, 1997.

[14] Krommes, G., *A new proof of decidability for the modal logic of subset spaces*, in: *Eighth ESSLLI student session*, 2003, pp. 137–148.

[15] McKinsey, J. and A. Tarski, *The algebra of topology*, Annals of Mathematics **45** (1944), pp. 141–191.

[16] Negri, S. and N. Olivetti, *A sequent calculus for preferential conditional logic based on neighbourhood semantics*, in: *Automated Reasoning with Analytic Tableaux and Related Methods*, LNCS **9323**, 2015, pp. 115–134.

[17] Parikh, R., L. Moss and C. Steinsvold, *Topology and epistemic logic*, in: M. Aiello, I. Pratt-Hartmann and J. van Benthem, editors, *Handbook of Spatial Logics*, Springer, 2007 pp. 299–342.

[18] S. Negri, *Proof analysis in modal logic*, Journal of Philosophical Logic **34** (2005), pp. 507–544.

[19] S. Negri, *Proof theory for modal logic*, Philosophy Compass **6/8** (2011), pp. 523–538.

[20] Tarski, A., *Der Aussagenkalkül und die Topologie*, Fundamenta Mathematicae **31** (1938), pp. 103–134.

[21] Wáng, Y. N. and T. Ågotness, *Subset space public announcement logic*, in: *ICLA 2013*, LNAI (2013), pp. 245–257.

Post Completeness in Congruential Modal Logics

Peter Fritz [1]

Department of Philosophy, Classics, History of Art and Ideas, University of Oslo
Postboks 1020 Blindern
0315 Oslo, Norway

Abstract

Well-known results due to David Makinson show that there are exactly two Post complete normal modal logics, that in both of them, the modal operator is truth-functional, and that every consistent normal modal logic can be extended to at least one of them. Lloyd Humberstone has recently shown that a natural analog of this result in congruential modal logics fails, by showing that not every congruential modal logic can be extended to one in which the modal operator is truth-functional. As Humberstone notes, the issue of Post completeness in congruential modal logics is not well understood. The present article shows that in contrast to normal modal logics, the extent of the property of Post completeness among congruential modal logics depends on the background set of logics. Some basic results on the corresponding properties of Post completeness are established, in particular that although a congruential modal logic is Post complete among all modal logics if and only if its modality is truth-functional, there are continuum many modal logics Post complete among congruential modal logics.

Keywords: Propositional Modal Logic, Post Completeness, Congruential Modal Logics, Classical Modal Logics

1 Introduction

The notion of Post completeness captures the intuitive idea of a logic being maximal, in the sense of it not being possible to strengthen the logic without collapsing it into inconsistency. This can be made precise in a very abstract setting: Let L be a set, informally understood as a set of formulas, and C a subset of the power set of L containing L itself, informally understood as the set of logics under consideration. A set $\Lambda \in C$ such that $\Lambda \neq L$ can then be defined to be *Post complete in* C if there is no $\Lambda' \in C$ such that $\Lambda \subset \Lambda' \subset L$. Post completeness in C is thus simply the property of being a coatom of C,

[1] Thanks to Rohan French, Lloyd Humberstone and David Makinson for very helpful discussions and comments on drafts of this paper, and to three anonymous reviewers for their suggestions concerning presentational matters.

partially ordered by $\subseteq$. Calling a member of C *consistent* if it is distinct from L, the coatoms of C can also be described as the maximal elements of the set of consistent members of C.

This abstract account of Post completeness makes clear that logics, understood as sets of formulas, are not Post complete *simpliciter*, but only Post complete *relative to* a given set of logics. Of course, if only one such set is considered, one may naturally talk about Post completeness simpliciter, leaving the relativity to this set implicit. Most of the literature on Post completeness in propositional modal logic operates on such an assumption, considering only Post completeness relative to the set of all *modal logics*, defined as sets of formulas containing all truth-functional tautologies and closed under the rules of modus ponens and uniform substitution. This is somewhat surprising, as the vast majority of research in modal logic focuses on the much more restrictive set of normal modal logics. The matter is partly explained by results due to Makinson [11], which show that among normal modal logics, the background set of logics is irrelevant, as a normal modal logic is Post complete in the set of all logics if and only if it is Post complete in the set of normal modal logics. (This section gives an informal introduction and overview; formal definitions and results will be stated more precisely below.)

The present article explores Post completeness beyond normal modal logics, in particular in the context of congruential modal logics. As will be shown, the background set of logics matters in this context, as most logics Post complete in the set of congruential modal logics fail to be Post complete in the set of all modal logics. First, however, some remarks on why Post completeness is an interesting notion.

Much of mathematical research in modal logic it is concerned with investigating various aspects of the set of normal modal logics, which, ordered by $\subseteq$, forms a complete lattice. Indeed, Rautenberg [13, p. 219] goes so far as to suggest that any investigation of normal modal logics is in effect aimed at improving our understanding of the lattice of normal modal logics. A natural part of such an enterprise is the investigation of the lattice's coatoms – the logics Post complete in it. What about its atoms? A version of Lindenbaum's lemma guarantees that every consistent normal modal logic can be extended to (i.e., is a subset of) a maximal consistent one (i.e., a coatom). But no such result guarantees that every normal modal logic distinct from the smallest normal modal logic **K** is an extension of an atom. In fact, Kracht [10, Theorem 7.7.2] notes, drawing on results due to Blok [2], that the lattice of normal modal logics is atomless.

A further reason for studying Post completeness arises from Makinson's results mentioned above, which also show that the lattice of normal modal logics has exactly two coatoms. This fact, and various specific details concerning these two logics, have proved extremely useful in a wide variety of applications (see, e.g., the appeals to "Makinson's Theorem" at various places in [10]). The intrinsic interest and evident usefulness of investigating Post completeness among normal modal logics therefore motivate studying this notion in wider classes of modal logics, and the present article makes a start, focusing in particular on congruential modal logics.

Section 2 briefly sets out the background theory of modal logics and corresponding models. Section 3 states the fundamental facts concerning Post completeness in normal modal logics which follow from Makinson's results, and some recent results due to Humberstone on Post completeness in congruential modal logics. Section 4 determines the number of logics Post complete in the lattice of congruential modal logics to be that of the continuum, and shows that infinitely many such logics are determined by a class of neighborhood frames. Section 5 shows that there are precisely four congruential modal logics which are Post complete in the lattice of all modal logics; with the first result of section 4, this shows that Post completeness among congruential modal logics is dependent on the background set of logics. Section 5 also shows that there is a continuum of extensions of congruential modal logics Post complete in the lattice of all modal logics which are not extensions of normal modal logics. Section 6 generalizes an observation of Humberstone's that the intersection of logics Post complete in lattice of all modal logics and closed under certain rules can be axiomatized using conditionals corresponding to these rules. Section 7 concludes, highlighting a number of open questions.

2 Modal Logics and Algebraic Models

Let $\mathcal{L}$ be the set of formulas of a propositional modal language, built up as usual from a countably infinite set of proposition letters $p, q, \ldots$ using the nullary operator $\top$ (trivial truth), the unary operator $\neg$ (negation), the binary operator $\wedge$ (conjunction) and the unary operator $\Box$ (the modality). Other operators, such as $\bot$, $\vee$ and $\rightarrow$ will be used as syntactic abbreviations as usual. Let a *substitution* be a function $\sigma : \mathcal{L} \rightarrow \mathcal{L}$ such that for all $\varphi, \psi \in \mathcal{L}$, $\sigma(\top) = \top$, $\sigma(\neg\varphi) = \neg\sigma(\varphi)$, $\sigma(\varphi \wedge \psi) = \sigma(\varphi) \wedge \sigma(\psi)$ and $\sigma(\Box\varphi) = \Box\sigma(\varphi)$. Let Σ be the set of substitutions. Let a *modal logic* be a set $\Lambda \subseteq \mathcal{L}$ such that Λ contains all propositional tautologies ($\varphi \in \mathcal{L}$ not containing $\Box$ true under every classical truth-value assignment) and is closed under modus ponens (if $\varphi, \varphi \rightarrow \psi \in \Lambda$ then $\psi \in \Lambda$) and uniform substitution (if $\varphi \in \Lambda$ then $\sigma(\varphi) \in \Lambda$ for any substitution σ). Let a modal logic be *consistent* if it is distinct from $\mathcal{L}$.

Usually, restrictions on modal logics are formulated in terms of containing certain axioms and being closed under certain rules. To provide an abstract framework for such restrictions, let a *rule* be a set of finite non-empty sequences of formulas. Let any $\Gamma \subseteq \mathcal{L}$ be *closed under* a rule R just in case for all $\langle \rho_0, \ldots, \rho_n \rangle \in R$, if $\rho_i \in \Gamma$ for all $i < n$, then $\rho_n \in \Gamma$. The members of a rule will be called its *instances*. A rule R is *substitution-invariant* if for all $\langle \rho_0, \ldots, \rho_n \rangle \in R$ and substitutions σ, $\langle \sigma(\rho_0), \ldots, \sigma(\rho_n) \rangle \in R$. In this setting, the usual requirement of being closed under "necessitation" can be formulated as being closed under the substitution-invariant rule $\{\langle \sigma(p), \sigma(\Box p) \rangle : \sigma \in \Sigma\}$. The treatment of axioms can be subsumed under this account of rules, using sequences of formulas of length one. So, being a *normal* modal logic can be formulated as being a modal logic closed under the substitution-invariant rule

$N = \{\langle\sigma(K)\rangle, \langle\sigma(p), \sigma(\Box p)\rangle : \sigma \in \Sigma\}$, where K is the familiar distributivity axiom $\Box(p \to q) \to (\Box p \to \Box q)$.

For any substitution-invariant rule R, let $L(R)$ be the set of modal logics closed under R. It is routine to show that this is a complete lattice with top element $\mathcal{L}$. Let a modal logic Λ be *R-Post complete* just in case it is a coatom of $L(R)$ (a consistent member of $L(R)$ which is not a subset of any other consistent member of $L(R)$). Let R-Post be the set of $\Lambda \in L(R)$ which are R-Post complete, and, for any set $\Gamma \subseteq \mathcal{L}$, R-Post(Γ) the set of $\Lambda \in R$-Post which contain Γ. A routine version of Lindenbaum's lemma establishes that $L(R)$ is coatomic: every consistent modal logic closed under R can be extended to an R-Post complete modal logic.

Among the most important restrictions on modal logics are those of being normal, congruential, quasi-normal and quasi-congruential. (See [14] and [5] for general discussion of these classes and the models for them used below; the term "classical" is sometimes used instead of "congruential".) Normality was defined above as being closed under N; *congruentiality* can be defined analogously as being closed under the substitution-invariant rule $C = \{\langle\sigma(p \leftrightarrow q), \sigma(\Box p \leftrightarrow \Box q)\rangle : \sigma \in \Sigma\}$. A modal logic is *quasi-normal* if it is an extension of a normal modal logic, and *quasi-congruential* if it is an extension of a congruential modal logic. Note that quasi-normality and quasi-congruentiality can also be defined by appeal to substitution-invariant rules, *viz.* the rules whose instances are the singleton sequences of members of the smallest normal and congruential modal logics, respectively.

Although each of the four lattices of modal logics just defined gives rise to a distinct notion of Post completeness, not each of these notions gives rise to its own set of questions. To illustrate this, consider normal and congruential modal logics. It is routine to show that a congruential modal logic Λ (i.e., $\Lambda \in L(C)$) is normal (i.e., $\Lambda \in L(N)$) if and only if it contains both K and $\Box\top$. Thus $L(N)$ is simply the principal filter of $L(C)$ generated by the smallest congruential modal logic containing K and $\Box\top$, which is of course the familiar modal logic **K**. Thus it is clear that any $\Lambda \in L(N)$ is N-Post complete if and only if it is C-Post complete, which means that among normal modal logics, the notion of N-Post completeness coincides with that of C-Post completeness, and so the investigation of N-Post completeness is a special case of the investigation of C-Post completeness. A similar point applies to the constraints of being quasi-normal and quasi-congruential, as the notions of Post completeness to which they give rise are special cases of the notion of $\emptyset$-Post completeness.

Several of the following results will appeal to standard algebraic models for congruential modal logics. Let a *modal matrix* be a structure $\mathfrak{A} = \langle A, 1, -, \sqcap, *, D\rangle$ such that $\langle A, 1, -, \sqcap\rangle$ is a Boolean algebra, $* : A \to A$ and $D \subseteq A$ is a filter of the algebra. For such a modal matrix $\mathfrak{A}$, let an *interpretation* be a function ι mapping proposition letters to elements of A. Extend such functions implicitly to $\mathcal{L}$, by letting $\iota(\top) = 1$, $\iota(\neg\varphi) = -\iota(\varphi)$, $\iota(\varphi \wedge \psi) = \iota(\varphi) \sqcap \iota(\psi)$ and $\iota(\Box\varphi) = *\iota(\varphi)$. Define $\Lambda(\mathfrak{A})$, the *logic of* $\mathfrak{A}$, to be the set of formulas φ such that $\iota(\varphi) \in D$ for all interpretations ι. Using Lindenbaum-Tarski algebras, it is routine to show that a modal logic is quasi-congruential if and only if it is the logic of a modal matrix, and congruential if and only if it is the logic of a modal matrix with singleton filter (see, e.g., Hansson and Gärdenfors [8]).

3 Makinson's and Humberstone's Results

It follows from the results of Makinson [11] that among normal modal logics, $\emptyset$-Post completeness coincides with N-Post completeness, and that these properties are had by exactly two logics, both of which interpret $\Box$ truth-functionally. These logics are most naturally described using algebraic models:

Treating T and F as the usual truth-values, every function $* : \{T, F\} \to \{T, F\}$ gives rise to a modal matrix $\mathfrak{T}^* = \langle\{T, F\}, T, -, \sqcap, *, \{T\}\rangle$ in which $-$ and $\sqcap$ are the usual truth-functional operations of negation and conjunction. Let t and f be the constant one-place functions to T and F, respectively, i the identity function and n the function mapping T and F to each other. Each such function $*$ thus gives rise to a logic $\Lambda^* = \Lambda(\mathfrak{T}^*)$ in which the modality $\Box$ behaves according to the truth-function $*$. Call these the four *truth-functional modal logics.* It is easy to see that no two truth-functions give rise to the same truth-functional modal logic. The two normal modal logics which are $\emptyset/N$-Post complete are Λ^t and Λ^i, and thus every consistent normal modal logic can be extended to at least one of these.

Can similar results be obtained for congruential modal logics? Humberstone [9] gives a negative answer, by showing that some congruential modal logics cannot be extended to any truth-functional modal logic. Since every consistent congruential modal logic can be extended to a C-Post complete one, it follows that some C-Post complete modal logics are not truth-functional. So, what is the extent of the property of Post completeness among congruential and quasi-congruential modal logics, and how do these sets of modal logics relate to each other and to the set of truth-functional modal logics? The remainder of this paper gives some basic answers to these and closely related questions.

4 Continuum Many C-Post Complete Logics

The first result to be established shows that the number of modal logics Post complete in the lattice of congruential modal logics is $\beth_1$ ($= 2^{\aleph_0}$):

Theorem 4.1 *The number of C-Post complete modal logics is $\beth_1$.*

Proof. Let $\langle A, \omega, -, \sqcap\rangle$ be the countable Boolean algebra of finite and cofinite sets of natural numbers. Let B be the set of finite non-empty sets of natural numbers, and $\langle b_n : n \in \omega\rangle$ an enumeration of B. For each set of natural numbers $S \subseteq \omega$, define a modal matrix $\mathfrak{A}_S = \langle A, \omega, -, \sqcap, *, \{\omega\}\rangle$, where $*$ is defined as follows:

$*(\omega) = b_0$

$*(b_n) = b_{n+1}$ for all $n \in \omega$

$*(-b_n) = \omega$ for all $n \in S$

$*(-b_n) = \emptyset$ for all $n \in \omega \backslash S$

$*(\emptyset) = \omega$

Since every modal matrix with a singleton filter determines a congruential modal logic, $\Lambda(\mathfrak{A}_S)$ is a congruential modal logic. Furthermore, for all $S \subseteq \omega$ and $n \in \omega$,

$\Box\neg\Box^n\Box\top \in \Lambda(\mathfrak{A}_S)$ iff $n \in S$

$\neg\Box\neg\Box^n\Box\top \in \Lambda(\mathfrak{A}_S)$ iff $n \notin S$

where $\Box^n$ is a string of n $\Box$ operators. Thus for any distinct $S, S' \subseteq \omega$, $\Lambda(\mathfrak{A}_S)$ and $\Lambda(\mathfrak{A}_{S'})$ cannot be extended to the same consistent modal logic. As every consistent congruential modal logic can be extended to a C-Post complete modal logic, there are $\beth_1$ such logics. □

Theoremhood in $\Lambda(\mathfrak{A}_S)$ depends not only on S, but also on the choice of the enumeration of B. E.g., consider a set S containing 2 but not 3. Then $\Box(\Box\top \wedge \Box\Box\top) \in \Lambda(\mathfrak{A}_S)$ if $b_0 \cap b_1 = b_2$, but not if $b_0 \cap b_1 = b_3$.

Similar to relational frames for normal modal logics, so-called neighborhood frames are naturally used to provide possible world models for congruential modal logics. A neighborhood frame is a pair $\langle W, N\rangle$ such that W is a set (the "worlds") and $N : \mathcal{P}(W) \to \mathcal{P}(W)$, from which a model can be obtained by adding a valuation function V which maps every proposition letter to a set of worlds. Truth of a formula at a world is defined as in relational frames, with the following clause for the modal operator:

$\langle W, N, V\rangle, w \models \Box\varphi$ iff $\{v \in W : \langle W, N, V\rangle, v \models \varphi\} \in N(w)$

The logic of a class of neighborhood frames is the set of formulas true at every world in every model based on a frame in the class.

For every set W, the powerset $\mathcal{P}(W)$ forms a Boolean algebra, and it is easy to see that the modal matrices with singleton filter based on $\mathcal{P}(W)$ correspond uniquely to the neighborhood frames on W. Neighborhood frames can therefore be seen as modal matrices with singleton filters based on powerset algebras. Not every Boolean algebra is isomorphic to a powerset algebra, however, which opens up the possibility that some congruential modal logic is not the logic of any class of neighborhood frames. That there are such logics was shown by Gerson [6]. With Theorem 4.1, this raises the question how widespread such incompleteness is among logics Post complete in the set of congruential modal logics. The following result gives a partial answer, by showing that there are infinitely many C-Post complete logics which are the logic of a class of neighborhood frames. What their precise number is will be left open, as well as the question whether there are any C-Post complete logics which fail to be the logic of a class of neighborhood frames, and if so, how many such logics there are.

Theorem 4.2 *There are at least $\aleph_0$ C-Post complete modal logics each of which is the logic of a class of neighborhood frames.*

Proof. For every natural number n, let $\mathfrak{A}_n$ be a matrix based on the powerset algebra A on $n = \{0, \dots, n-1\}$ with filter n and function $*$ to be defined. Let

$B = A \backslash \{n\}$ and $\langle b_i : i < 2^n - 1 \rangle$ an enumeration of B such that $b_0 = \emptyset$. Define $*$ as follows:

$$*(n) = n$$

$$*(b_i) = b_{i+1} \text{ for all } i < 2^n - 2$$

$$*(b_{2^n-2}) = b_0$$

Let $\Lambda_n = \Lambda(\mathfrak{A}_n)$, which is congruential by construction. For any n, the smallest $l > 0$ such that $\neg\Box^l\bot \in \Lambda_n$ is $2^n - 1$. So if $n \neq n'$, $\Lambda_n \neq \Lambda_{n'}$. Since $\mathfrak{A}_n$ is based on a power set algebra, Λ_n is the logic of the corresponding neighborhood frame. It only remains to argue that Λ_n is C-Post complete.

Consider any congruential modal logic Λ properly extending Λ_n. Then there is a $\varphi \in \Lambda$ which is not in Λ_n. Since $\varphi \notin \Lambda_n$, there is an interpretation ι such that $\iota(\varphi) \neq n$. For each element $x \in A$, there is a formula $\delta(x)$ containing no proposition letters such that $\kappa(\delta(x)) = x$ for every interpretation κ: let $\delta(n) = \top$ and $\delta(b_i) = \Box^i\bot$. Let σ be the substitution mapping each proposition letter p to $\delta(\iota(p))$. A routine induction on the complexity of formulas shows that for every interpretation κ, $\kappa(\sigma(\varphi))) = \iota(\varphi)$, and so $\kappa(\sigma(\varphi))) \neq n$.

One the one hand, by construction of $\mathfrak{A}_n$, there is a number k such that $\kappa(\Box^k\sigma(\varphi)) = \emptyset$ for all interpretations κ. So $\neg\Box^k\sigma(\varphi)$ is a member of Λ_n and thus a member of Λ. On the other hand, since $\varphi \in \Lambda$, by uniform substitution, $\sigma(\varphi) \in \Lambda$. So $\top \leftrightarrow \sigma(\varphi) \in \Lambda$, and therefore by k applications of the congruentiality rule, $\Box^k\top \leftrightarrow \Box^k\sigma(\varphi)$. But $\Box^k\top$ is a member of Λ_n and so a member of Λ, hence $\Box^k\sigma(\varphi) \in \Lambda$. It follows that Λ is inconsistent. □

5 Truth-Functionality and Quasi-Congruentiality

From Theorem 4.1, it follows immediately that there are $\beth_1$ congruential modal logics which cannot be extended to a truth-functional modal logic. In this sense, C-Post completeness differs markedly from N-Post completeness. But it turns out that one consequence of Makinson's results does extend to congruential modal logics: $\emptyset$-Post completeness coincides with truth-functionality also among congruential modal logics. The proof relies on a lemma of Segerberg [15, p. 712, Lemma A]; to state it, let $\mathcal{L}_0$ be the set of formulas containing no proposition letters (i.e., built up entirely from $\top$).

Lemma 5.1 *A consistent modal logic Λ has exactly one $\emptyset$-Post complete extension if and only if for all $\varphi \in \mathcal{L}_0$, $\varphi \in \Lambda$ or $\neg\varphi \in \Lambda$.*

Theorem 5.2 *A congruential modal logic is $\emptyset$-Post complete if and only if it is truth-functional.*

Proof. The *if* direction is routine, so consider a congruential modal logic $\Lambda \in \emptyset$-Post. It follows by Lemma 5.1 that for each of $\Box\top$ and $\Box\bot$, Λ contains either it or its negation. The rest of the argument follows along the lines of [11]: Since Λ is congruential, $\Lambda = \Lambda(\mathfrak{A})$ for some modal matrix $\mathfrak{A} = \langle A, 1, -, \sqcap, \dagger, \{1\} \rangle$. Let $h : \{T, F\} \to A$ map T to 1 and F to 0 $(= -1)$. Since $\{\dagger(1), \dagger(0)\} \subseteq \{1, 0\}$, a truth-function $*$ can be defined as $h^{-1} \circ \dagger \circ h$, and it is easily seen that h is a homomorphism from $\mathfrak{T}^*$ to $\mathfrak{A}$. Consequently, Λ is a sublogic of Λ^*, and so as $\Lambda \in \emptyset$-Post, $\Lambda^* = \Lambda$. $\square$

Among congruential modal logics, C-Post completeness and $\emptyset$-Post completeness therefore come wide apart. The results established so far thus emphasize the remarkableness of the fact that these two properties coincide among normal modal logics.

Consider now quasi-congruential modal logics, which can be characterized as the modal logics closed under the substitution-invariant rule $QC = \{\langle\varphi\rangle : \varphi \in \mathbf{E}\}$, where $\mathbf{E}$ is the smallest congruential modal logic. As noted above, $\emptyset$-Post completeness and QC-Post completeness trivially coincide on $L(QC)$. Moreover, the number of such logics has already been determined to be $\beth_1$ by Segerberg [15, p. 713], who shows that there are $\beth_1$ $\emptyset$-Post complete quasi-normal modal logics. [2] Given Theorem 5.2, there must be $\beth_1$ of them which are not congruential. This observation leaves open the possibility that all of them are quasi-normal, but the proof of Theorem 4.1 can be adapted to rule this out:

Theorem 5.3 *There are $\beth_1$ $\emptyset$-Post complete quasi-congruential modal logics which are not quasi-normal.*

Proof. Since the modal matrices used in the proof of Theorem 4.1 are based on an algebra generated by the single element 1, it follows with [12, Theorem 1] that extending the filter of any such modal matrix $\mathfrak{A}_S$ to an ultrafilter produces a matrix which determines a quasi-congruential modal logic which is $\emptyset$-Post complete. Let $\mathfrak{A}_S(U)$ be such a matrix with ultrafilter U such that $-b_n \in U$ for some $n \in \omega$. Then $\neg\Box^{n+1}\top \in \Lambda(\mathfrak{A}_S(U))$, which entails that this logic is not quasi-normal. As in Theorem 4.1, $\Lambda(\mathfrak{A}_S(U)) \neq \Lambda(\mathfrak{A}_{S'}(U))$ for any distinct S and S', from which the claim to be proven follows. $\square$

6 Characterizing Intersections of Post-Complete Extensions

For any substitution-invariant rule R and set of formulas Γ, let $\Lambda_R(\Gamma)$ be the smallest modal logic closed under R which contains Γ; since $L(R)$ is a complete lattice, this is well-defined. Call $\Lambda_R(\Gamma)$ the R-logic *axiomatized by* Γ. Humberstone [9] notes that $\bigcap(\emptyset\text{-Post} \cap L(N))$, the intersection of the two $\emptyset$-Post complete normal modal logics, is the normal modal logic axiomatized by the formula $NC = p \to \Box p$, i.e., $\Lambda_N(\{NC\})$. He also notes that the intersection of the four truth-functional modal logics is the congruential modal logic axiomatized by the "extensionality conditional" $EC = (p \leftrightarrow q) \to (\Box p \leftrightarrow \Box q)$. With Theorem 5.2, it follows that this is the intersection of the $\emptyset$-Post complete congruential modal logics.

The axioms appealed to in these observations strongly suggest a general connection between $\emptyset$-Post complete logics closed under a given substitution-

[2] The claim in [7, pp. 133, 136 & 142] that there are only two such logics is therefore incorrect; this also seems to affect the discussion in [9, Coda].

invariant rule and the conditionals corresponding to the instances of this rule. This section establishes such a connection. The result to be proven shows, for any set of formulas Γ, how to characterize the intersection of the $\emptyset$-Post complete logics closed under a substitution-invariant rule R which contain Γ using the conditionals corresponding to the instances of R. The natural conjecture is that this intersection is simply the modal logic axiomatized by the union of Γ and the set of these conditionals. It turns out that this is incorrect, but that applying a natural operation to the logic so axiomatized produces the desired intersection.

To motivate the required operation, consider the case of $R = \emptyset$. The natural conjecture just mentioned says that $\bigcap\emptyset\text{-Post}(\Gamma)$ is $\Lambda_\emptyset(\Gamma)$, the modal logic axiomatized by Γ. This is not the case: As shown by Segerberg [16], $\emptyset\text{-Post}(\Gamma) = \emptyset\text{-Post}(\Delta)$ whenever $\Gamma \cap \mathcal{L}_0 = \Delta \cap \mathcal{L}_0$. Thus, as long as no $\mathcal{L}_0$ formulas are added, $\Lambda_\emptyset(\Gamma)$ can be expanded without adding formulas not in $\bigcap\emptyset\text{-Post}(\Gamma)$. (That there are logics which can be so expanded will follow from lemmas to be established presently.) This problem can be solved by expanding $\Lambda_\emptyset(\Gamma)$, adding all formulas whose substitution instances in $\mathcal{L}_0$ are already contained in $\Lambda_\emptyset(\Gamma)$. This turns out to be the required operation. It will now be defined formally, and some lemmas will be established, with which the desired result can be established.

Let a substitution σ be a 0-*substitution* if $\sigma(\varphi) \in \mathcal{L}_0$ for all formulas φ. For any set of formulas Γ, define

$$\varepsilon_0(\Gamma) = \{\varphi \in \mathcal{L} : \sigma(\varphi) \in \Gamma \text{ for every 0-substitution } \sigma\}.$$

Call this the 0-*expansion of* Γ.

Lemma 6.1 *For any modal logics Λ, Λ':*

(i) $\Lambda \subseteq \varepsilon_0(\Lambda)$

(ii) $\varepsilon_0(\Lambda)$ is a modal logic.

(iii) If Λ is closed under a given substitution-invariant rule, so is $\varepsilon_0(\Lambda)$.

(iv) If Λ is consistent, so is $\varepsilon_0(\Lambda)$.

(v) If $\Lambda \cap \mathcal{L}_0 \subseteq \Lambda'$ then $\varepsilon_0(\Lambda) \subseteq \varepsilon_0(\Lambda')$.

Proof. Routine. □

With this lemma, it is easy to see that for every substitution-invariant rule R, ε_0 is a closure operator on $L(R)\backslash\mathcal{L}$, the consistent modal logics closed under R, ordered by $\subseteq$.

Lemma 6.2 *For every substitution-invariant rule R and R-Post complete logic Λ, $\varepsilon_0(\Lambda) = \Lambda$.*

Proof. Let $\Lambda \in L(R)$ such that $\varepsilon_0(\Lambda) \neq \Lambda$. Then by Lemma 6.1 (i), $\Lambda \subset \varepsilon_0(\Lambda)$. By Lemma 6.1 (ii) and (iii), $\varepsilon_0(\Lambda) \in L(R)$. Since $\Lambda \subset \varepsilon_0(\Lambda)$, Λ is consistent, and so with Lemma 6.1 (iv), $\varepsilon_0(\Lambda)$ is consistent. So Λ is not R-Post complete.□

Before applying this lemma to establish the main theorem of this section, it is worth relating the operation of 0-expansion to the closely related notions of 0-reducibility and general Post completeness, which Chagrov and Zakharyaschev [4, chapter 13] discuss in detail. A modal logic Λ is *0-reducible* just in case for every formula $\varphi \notin \Lambda$, there is a 0-substitution σ such that $\sigma(\varphi) \notin \Lambda$. It is easy to see that Λ is 0-reducible just in case $\varepsilon_0(\Lambda) = \Lambda$, and that $\varepsilon_0(\Lambda)$ is the smallest 0-reducible modal logic containing Λ. Λ is *generally Post complete* if Λ is R-Post complete, where R is the union of substitution-invariant rules under which Λ closed. Chagrov and Zakharyaschev [4, Theorem 13.11] show that a consistent modal logic Λ is 0-reducible if and only if it is generally Post complete.[3] Since there are consistent modal logics which are not generally Post complete (see, e.g., their Theorem 13.2), it follows, as claimed above, that there are modal logics Λ such that $\varepsilon_0(\Lambda) \neq \Lambda$. The connections just drawn also show that fittingly, a modal logic is generally Post complete just in case it is R-Post complete for some substitution-invariant rule R; this observation provides an alternative route to establishing Lemma 6.2.

Returning to the theorem to be established, define, for any rule R, $\overrightarrow{R} = \{\bigwedge_{i<n} \rho_i \to \rho_n : \langle \rho_0, \ldots, \rho_n \rangle \in R\}$. A final lemma leads to the desired result:

Lemma 6.3 *For any $\emptyset$-Post complete modal logic Λ closed under a substitution-invariant rule R, $\overrightarrow{R} \subseteq \Lambda$.*

Proof. Let $\Lambda \in \emptyset\text{-Post} \cap L(R)$ and consider any $\langle \rho_0, \ldots, \rho_n \rangle \in R$. Since by Lemma 6.2, $\varepsilon_0(\Lambda) = \Lambda$, it suffices to show, for an arbitrary 0-substitution σ, that $\bigwedge_{i<n} \sigma(\rho_i) \to \sigma_n(\rho) \in \Lambda$. This can be done by a case distinction, using Lemma 5.1: If $\bigwedge_{i<n} \sigma(\rho_i) \notin \Lambda$, then $\neg \bigwedge_{i<n} \sigma(\rho_i) \in \Lambda$, and so $\bigwedge_{i<n} \sigma(\rho_i) \to \sigma(\rho_n) \in \Lambda$. If $\bigwedge_{i<n} \sigma(\rho_i) \in \Lambda$, then as Λ is closed under the substitution-invariant rule R, $\sigma(\rho_n) \in \Lambda$, and therefore $\bigwedge_{i<n} \sigma(\rho_i) \to \sigma(\rho_n) \in \Lambda$. □

Theorem 6.4 *For any set of formulas Γ and substitution-invariant rule R,*

$$\bigcap(\emptyset\text{-Post}(\Gamma) \cap L(R)) = \varepsilon_0(\Lambda_\emptyset(\Gamma \cup \overrightarrow{R})).$$

Proof. $\subseteq$: Consider any formula $\varphi \notin \varepsilon_0(\Lambda_\emptyset(\Gamma \cup \overrightarrow{R}))$. Thus there is a 0-substitution σ such that $\sigma(\varphi) \notin \Lambda_\emptyset(\Gamma \cup \overrightarrow{R})$. A routine argument shows that then, $\Lambda_\emptyset(\Gamma \cup \overrightarrow{R} \cup \{\neg\sigma(\varphi)\})$ is consistent, which can therefore be extended to a $\emptyset$-Post complete modal logic Λ. Since Λ contains the conditionals in $\overrightarrow{R}$, it is closed under R. Thus $\Lambda \in \emptyset\text{-Post}(\Gamma) \cap L(R)$. As Λ is consistent, $\varphi \notin \Lambda$, and therefore $\varphi \notin \bigcap(\emptyset\text{-Post}(\Gamma) \cap L(R))$.

$\supseteq$: Consider any $\Lambda \in \emptyset\text{-Post}(\Gamma) \cap L(R)$ (if there is no such element, this direction is trivial). It suffices to show that $\varepsilon_0(\Lambda_\emptyset(\Gamma \cup \overrightarrow{R})) \subseteq \Lambda$. Since Λ is $\emptyset$-Post complete, it follows with Lemma 6.2 that $\varepsilon_0(\Lambda) = \Lambda$. So by Lemma 6.1 (v), it suffices to show that $\Lambda_\emptyset(\Gamma \cup \overrightarrow{R}) \subseteq \Lambda$, which is immediate using Lemma 6.3. □

[3] The treatment of rules in [4] is slightly different from the present treatment, which affects the definition of general Post completeness. For present purposes, the difference is merely a matter of presentation.

The case of $R = \emptyset$ showed that the operation of 0-expansion appealed to in this result is essential, but Humberstone's observation did not appeal to it. However, the observation falls out as a corollary of Theorem 6.4 with the following lemma:

Lemma 6.5 *For any modal logic Λ containing EC, $\varepsilon_0(\Lambda) = \Lambda$.*

Proof. Assume $EC \in \Lambda$; note that this means that Λ is congruential. It follows by a routine induction that for any substitution σ and formula φ built up from proposition letters $p_0, \ldots, p_{n-1}$, $\bigwedge_{i<n}(p_i \leftrightarrow \sigma(p_i)) \to (\varphi \leftrightarrow \sigma(\varphi)) \in \Lambda$. By Lemma 6.1 (i), it suffices to show that $\varepsilon_0(\Lambda) \subseteq \Lambda$, so consider any $\varphi \in \varepsilon_0(\Lambda)$ built up from proposition letters $p_0, \ldots, p_{n-1}$. Let Σ' be a finite set of 0-substitutions such that for every truth-value assignment among $p_0, \ldots, p_{n-1}$, there is $\sigma \in \Sigma'$ mapping each proposition letter among $p_0, \ldots, p_{n-1}$ correspondingly to $\top$ or $\bot$. Then $\bigvee_{\sigma\in\Sigma'} \bigwedge_{i<n}(p_i \leftrightarrow \sigma(p_i))$ is a tautology. With the schema derived earlier, it follows that $\bigvee_{\sigma\in\Sigma'}(\varphi \leftrightarrow \sigma(\varphi)) \in \Lambda$. Since $\varphi \in \varepsilon_0(\Lambda)$, $\sigma(\varphi) \in \Lambda$ for all $\sigma \in \Sigma'$, and therefore $\varphi \in \Lambda$. □

The desired corollary follows immediately from Theorem 6.4 and Lemma 6.5:

Corollary 6.6 *For any set of formulas Γ,*

$$\bigcap(\emptyset\text{-Post}(\Gamma) \cap L(C)) = \Lambda_\emptyset(\Gamma \cup \{EC\}).$$

As a second corollary of Theorem 6.4, another characterization of general Post completeness can be obtained:[4]

Corollary 6.7 *A modal logic is generally Post complete if and only if it is the intersection of a non-empty set of $\emptyset$-Post complete modal logics.*

Proof. The claim is immediate for inconsistent modal logics, so let Λ be a consistent modal logic. As noted above, Λ is generally Post complete if and only if $\varepsilon_0(\Lambda) = \Lambda$. Letting $R = \emptyset$, it follows from Theorem 6.4 that $\bigcap \emptyset\text{-Post}(\Lambda) = \varepsilon_0(\Lambda)$. If $\varepsilon_0(\Lambda) = \Lambda$, then $\bigcap \emptyset\text{-Post}(\Lambda) = \Lambda$. If $\Lambda = \bigcap S$ for some nonempty set $S \subseteq \emptyset$-Post, then $S \subseteq \emptyset\text{-Post}(\Lambda)$, so $\Lambda = \bigcap \emptyset\text{-Post}(\Lambda)$, and thus $\varepsilon_0(\Lambda) = \Lambda$. □

Having characterized the intersection of the modal logics extending a given set which are closed under a substitution-invariant rule R and $\emptyset$-Post complete, it is natural to ask for characterizations of similar intersections. First, one might ask for a characterization of the intersection of the modal logics extending Γ which are closed under R and *generally* Post complete. Second, one might ask for a characterization of the intersection of the modal logics extending Γ which are R-Post complete (and thus closed under R). Third, given a substitution-invariant rule $R' \subseteq R$, one might ask for a characterization of the intersection of the modal logics closed under R extending Γ which are R'-Post complete.

[4] The result is due to David Makinson (p.c.), who provided a more direct proof, appealing to the observation that ε_0 commutes with intersection: $\varepsilon_0(\bigcap\{\Gamma_i : i \in I\}) = \bigcap\{\varepsilon_0(\Gamma_i) : i \in I\}$.

The first question can be answered relatively easily; the others will be left open. Writing g-Post(Γ) for the set of generally Post complete modal logics extending Γ, a natural characterization can be given as follows:

Theorem 6.8 *For any set of formulas* Γ *and substitution-invariant rule* R*,*

$$\bigcap(g\text{-Post}(\Gamma) \cap L(R)) = \varepsilon_0(\Lambda_R(\Gamma)).$$

Proof. The claim is immediate if $\Lambda_R(\Gamma)$ is inconsistent, so assume otherwise.

$\subseteq$: It suffices to show that $\varepsilon_0(\Lambda_R(\Gamma)) \in g\text{-Post}(\Gamma) \cap L(R)$. As noted earlier, $\varepsilon_0(\Lambda_R(\Gamma))$ is 0-reducible and therefore generally Post complete; by Lemma 6.1 (iii), it is a member of $L(R)$.

$\supseteq$: For any $\Lambda \in g\text{-Post}(\Gamma) \cap L(R)$, $\Lambda_R(\Gamma) \subseteq \Lambda$. As noted earlier, since Λ is generally Post complete, $\varepsilon_0(\Lambda) = \Lambda$, and therefore with Lemma 6.1 (v), $\varepsilon_0(\Lambda_R(\Gamma)) \subseteq \Lambda$. □

7 Conclusion

This paper made a start at investigating Post completeness among congruential modal logics. Both similarities and differences to the case of normal modal logics were established, most importantly that while $\emptyset$-Post completeness coincides with truth-functionality in both settings, there are $\beth_1$ modal logics Post complete in the set of congruential modal logics, in contrast to the two modal logics Post complete in the set of normal modal logics. The few elementary results established here bring out many open questions.

Two clusters of questions were already mentioned above: First, how many modal logics Post complete in the set of congruential modal logics are the logic of a class of neighborhood frames, and how many (if any) are not? An analogous question arises for quasi-congruential modal logics and neighborhood frames with distinguished elements (and quasi-normal modal logics and relational frames with distinguished elements, a question which seems not to have been considered). Second, for any substitution-invariant rules $R' \subseteq R$ and set of formulas Γ, how can one characterize the intersection of modal logics extending Γ which are R-Post complete, and the intersection of modal logics extending Γ which are R'-Post complete and closed under R?

Similar to the questions considered in Bellissima [1], one could also investigate, for each cardinal $\kappa \leq \beth_1$, the number of congruential modal logics Λ such that $|C\text{-Post}(\Lambda)| = \kappa$.[5] It would also be interesting to know whether what Segerberg [16] calls "Halldén's Theorem" holds among congruential modal logics, in the sense that for all congruential modal logics Λ and Λ', $C\text{-Post}(\Lambda \cap \Lambda') = C\text{-Post}(\Lambda) \cup C\text{-Post}(\Lambda')$. More generally, the question could be asked for any substitution-invariant rule.

Further interesting questions arise from the intersection of modal logics whose only Post complete extension is a specific logic. As Blok and Köhler [3,

[5] [1] is concerned with an analogous question for $\emptyset$-Post completeness, focusing on normal modal logics. Surprisingly, it seems to presuppose the truth of of the continuum hypothesis without mentioning it; see the four-fold case distinction in the proof of Theorem 3.1, p. 133.

p. 952–954] note, one can show with Lemma 5.1 that for any $\emptyset$-Post complete modal logic Λ, the set of modal logics whose only $\emptyset$-Post complete extension is Λ contains its intersection. This, as they note, does not carry over to normal modal logics: while the set of normal modal logics whose only N-Post complete extension is Λ^i contains its intersection, the set of normal modal logics whose only N-Post complete extension is Λ^t does not. The former intersection is **D**, the smallest normal modal logic containing the axiom $D = \Diamond\top$, and the latter intersection is **K**. Furthermore, **D** and Λ^t give rise to a so-called *splitting* of the lattice of normal modal logic, since for any normal modal logic Λ, $\mathbf{D} \subseteq \Lambda$ or $\Lambda \subseteq \Lambda^t$ but not both. (See [10, section 7.2] for more on splittings.) Since there are far more C-Post complete modal logics than N-Post complete ones, it is an interesting question to ask for which C-Post complete modal logics Λ the set of congruential modal logics whose only C-Post complete extension is Λ contains its intersection, and for cases in which the answer is affirmative, whether the relevant intersection gives rise to a splitting of the lattice of congruential modal logics. One might also ask which C-Post complete modal logics give rise to a splitting of this lattice, and investigate whether there are cases in which the intersection of congruential modal logics whose only C-Post complete extension is a given logic and another C-Post complete modal logic form a splitting pair. More generally, one could consider an arbitrary set S of C-Post complete modal logics, and ask similar questions concerning the set of congruential modal logics whose C-Post complete extensions are precisely the members of S.

Many more questions could be asked concerning Post completeness in congruential modal logics, but the ones mentioned so far should suffice to indicate that much remains to be done in this area.

References

[1] Bellissima, F., *Post complete and 0-axiomatizable modal logics*, Annals of Pure and Applied Logic **47** (1990), pp. 121–144.

[2] Blok, W. J., *The lattice of varieties of modal algebras is not strongly atomic*, Algebra Universalis **11** (1980), pp. 285–294.

[3] Blok, W. J. and P. Köhler, *Algebraic semantics for quasi-classical modal logics*, The Journal of Symbolic Logic **48** (1983), pp. 941–964.

[4] Chagrov, A. and M. Zakharyaschev, "Modal Logic," Oxford Logic Guides **35**, Oxford: Clarendon Press, 1997.

[5] Chellas, B. F., "Modal logic: an introduction," Cambridge: Cambridge University Press, 1980.

[6] Gerson, M., *The inadequacy of the neighbourhood semantics for modal logic*, The Journal of Symbolic Logic **40** (1975), pp. 141–148.

[7] Goldblatt, R. and T. Kowalski, *The power of a propositional constant*, Journal of Philosophical Logic **43** (2014), pp. 133–152.

[8] Hansson, B. and P. Gärdenfors, *A guide to intensional semantics*, in: *Modality, Morality and Other Problems of Sense and Nonsense: Essays dedicated to Sören Halldén*, Lund: CWK Gleerup Bokförlag, 1973 pp. 151–167.

[9] Humberstone, L., *Note on extending congruential modal logics*, Notre Dame Journal of Formal Logic **57** (2016), pp. 95–103.

[10] Kracht, M., "Tools and Techniques in Modal Logic," Studies in Logic and the Foundations of Mathematics **142**, Amsterdam: Elsevier, 1999.

[11] Makinson, D., *Some embedding theorems for modal logic*, Notre Dame Journal of Formal Logic **12** (1971), pp. 252–254.

[12] Makinson, D. and K. Segerberg, *Post completeness and ultrafilters*, Zeitschrift für mathematische Logik und Grundlagen der Mathematik **20** (1974), pp. 385–388.

[13] Rautenberg, W., "Klassische und nichtklassische Aussagenlogik," Braunschweig: Vieweg, 1979.

[14] Segerberg, K., "An Essay in Classical Modal Logic," Filosofiska Studier **13**, Uppsala: Uppsala Universitet, 1971.

[15] Segerberg, K., *Post completeness in modal logic*, The Journal of Symbolic Logic **37** (1972), pp. 711–715.

[16] Segerberg, K., *Halldén's theorem on Post completeness*, in: *Modality, Morality and Other Problems of Sense and Nonsense: Essays dedicated to Sören Halldén*, Lund: CWK Gleerup Bokförlag, 1973 pp. 206–209.

Synthetic Completeness Proofs for Seligman-style Tableau Systems

Klaus Frovin Jørgensen

Section of Philosophy and Science Studies, Roskilde University

Patrick Blackburn

Section of Philosophy and Science Studies, Roskilde University

Thomas Bolander

Department of Applied Mathematics and Computer Science, Technical University of Denmark

Torben Braüner

Department of People and Technology, Roskilde University

Abstract

Hybrid logic is a form of modal logic which allows reference to worlds. We can think of it as 'modal logic with labelling built into the object language' and various forms of labelled deduction have played a central role in its proof theory. Jerry Seligman's work [13,14] in which 'rules involving labels' are rejected in favour of 'rules for all' is an interesting exception to this. Seligman's approach was originally for natural deduction; the authors of the present paper recently extended it to tableau inference [1,2]. Our earlier work was syntactic: we showed completeness by translating between Seligman-style and labelled tableaus, but our results only covered the minimal hybrid logic; in the present paper we provide completeness results for a wider range of hybrid logics and languages. We do so by adapting the synthetic approach to tableau completeness (due to Smullyan, and widely applied in modal logic by Fitting) so that we can directly build maximal consistent sets of tableau blocks.

Keywords: Hybrid logic, tableaus, Seligman-style, synthetic completeness method, Bridge rule, pure axioms, tense logic, universal modality, difference operator

1 Introduction

Hybrid logic is a form of modal logic which allows us to refer to worlds: it contains nominals (special atomic formulas true at a unique world) and formulas of the form $@_i\varphi$. Here i is a nominal, and the formula $@_i\varphi$ is true iff φ is true at the unique world where i is true. So hybrid logic can be thought of as modal logic with world-labelling apparatus hard-wired into the object language, and various forms of labelled deduction have played the leading role in the development of hybrid proof theory. That is, many hybrid proof systems work by manipulating formulas of the form $@_i\varphi$ (and manipulating *only* formulas of this form) rather than arbitrary formulas. [1]

An exception to this is Jerry Seligman's work (dating back to the 1990s) in which 'rules involving labels' are rejected in favour of 'rules for all'. Seligman introduced his approach in two papers, the natural deduction based [13] and the sequent calculus based [14]; his natural deduction approach was later developed by Braüner [4]. We recently adapted Seligman's approach to tableau inference [1,2]; our key idea was to subdivide tableau branches into *blocks*, and use a rule called GoTo to navigate between them. Our investigations were syntactic: we proved completeness by explicit translation between Seligman-style and labelled tableaus, but only discussed the minimal hybrid logic.

In this paper we provide completeness results for a wider range of hybrid logics and languages. In Section 2 we introduce STB and ST, the two Seligman-style tableau systems we shall work with. [2] In Section 3 we adapt the synthetic tableau completeness method so that we can directly build maximal consistent sets of blocks, rather than maximal consistent sets of formulas or labelled formulas; [3] this yields completeness for STB. In Section 4 we eliminate a rule called Bridge to obtain completeness for ST. In Section 5 we show that this leads to completeness for richer logics and languages. Section 6 concludes.

2 Two Basic Calculi: ST and STB

We mostly work with a basic hybrid language built over a countable set of propositional symbols and a countable set of nominals. We take $\neg$, $\vee$, $\Diamond$, and for each nominal i an $@_i$-operator as primitive connectives, and build formulas as follows (here i ranges over nominals and p over propositional symbols):

$$\varphi ::= i \mid p \mid \neg\varphi \mid \varphi \vee \psi \mid \Diamond\varphi \mid @_i\varphi.$$

Other booleans are defined as usual, and $\Box\varphi$ is defined to be $\neg\Diamond\neg\varphi$. Note that nominals can occur either as subscripts to @ ("in operator position") or as formulas in their own right ("in formula position"). We typically use i, j and k for nominals and p, q and r for ordinary propositional symbols. Nominals and propositional symbols are the atomic formulas.

---

[1] Labelled deduction has a long history in modal logic, not just hybrid logic. The basic idea dates back to Fitch [6], was developed and generalized by Fitting [7,8,9], used for intuitionistic modal logic by Simpson [15], and became a proof strategy for many kinds of non-classical logics with the work of Gabbay [10]. Work by Negri has taken the approach in another direction by grounding it in the Kleene-Gentzen G3 system; see, for example, [12].

[2] Space limitations mean we cannot further discuss our reasons for finding Seligman-style tableaus so interesting, though we give motivating examples in figures 3 and 8. For a deeper discussion of Seligman's work and its links with our own, we refer the reader to [2].

[3] The synthetic method dates back to Smullyan's [16] classic work on first-order tableaus, and has been widely employed in modal logic by Fitting.

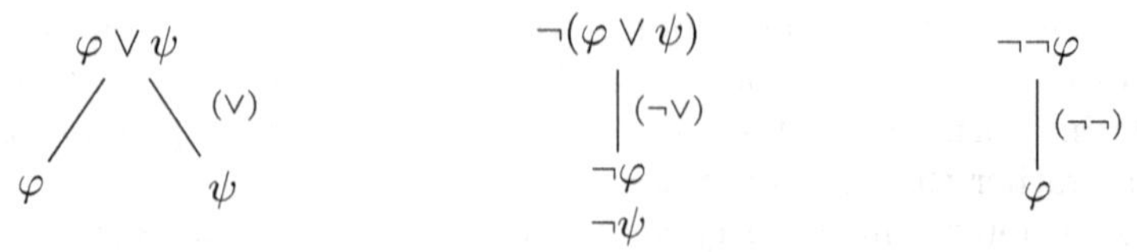

Fig. 1. Tableau rules for propositional logic.

We interpret the language in models based on frames (W, R), where W is a non-empty set (we call its elements worlds) and R is a binary relation on W (the accessibility relation). A model is a triple (W, R, V) where (W, R) is a frame and V (the valuation) maps propositional symbols p to arbitrary subsets of W, and nominals i to singleton subsets of W. A model is *named* iff for every world w there is some nominal i such that $V(i) = \{w\}$.[4]

Satisfiability in a model is defined in the usual way as a relation between a model $\mathfrak{M} = (W, R, V)$, a world $w \in W$, and a formula φ:

$$\begin{array}{ll}
\mathfrak{M}, w \models a & \text{iff } a \text{ is atomic and } w \in V(a) \\
\mathfrak{M}, w \models \neg\varphi & \text{iff } \mathfrak{M}, w \not\models \varphi \\
\mathfrak{M}, w \models \varphi \vee \psi & \text{iff } \mathfrak{M}, w \models \varphi \text{ or } \mathfrak{M}, w \models \psi \\
\mathfrak{M}, w \models \Diamond\varphi & \text{iff for some } w',\ wRw' \text{ and } \mathfrak{M}, w' \models \varphi \\
\mathfrak{M}, w \models @_i\varphi & \text{iff } \mathfrak{M}, w' \models \varphi \text{ and } w' \in V(i).
\end{array}$$

A formula φ is *valid on* $\mathfrak{M} = (W, R, V)$ when for all worlds $w \in W$ we have that $\mathfrak{M}, w \models \varphi$. A formula is *valid* if it is valid on all models.

Now for our tableau systems. For the propositional connectives we use the standard rules shown in figure 1. And we work with the usual notion of branch in a (tableau) tree. But we also need *blocks*: given a branch Θ in a tableau, we define a block to be one of the following:

- The *initial block*, consisting of all the formulas on Θ until the first horizontal line (or all formulas if there is no such line).
- The *current block*, consisting of all formulas below the last horizontal line (or all formulas if there is no such line).
- All formulas that occur between a pair of two consecutive horizontal lines.

The rule allowing us to close down one block and start up a new one is GoTo. Its precise formulation is given in figure 2. All blocks except the initial one are opened by an application of GoTo, and hence they all contain a nominal as

[4] That is, a model is named if each of its worlds is named by some nominal. Most models are not named (as our language is countable) but as we shall see in Section 5, named models have desirable properties. Our goal here is to prove completeness by building named models. Note: distinct nominals can name the same world (just as distinct first-order constants can denote the same element of a first-order model).

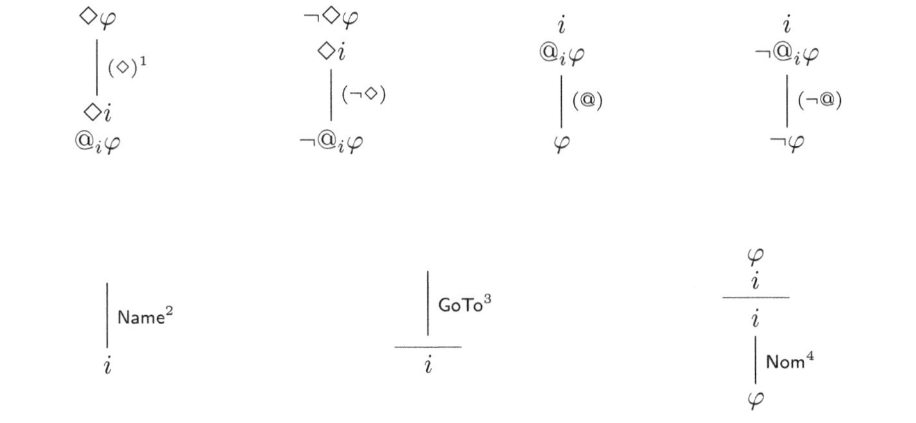

[1] The nominal i is fresh and φ is not a nominal.
[2] The nominal i is fresh.
[3] The nominal i must already occur on the branch.
[4] The horizontal line below the two uppermost premises signifies that these premises belong to a block distinct from the current one, whereas the third premise (the lowermost occurrence of i) belongs to the current block.

Fig. 2. **ST** tableau rules for basic hybrid logic.

their first formula. This nominal is called the *opening nominal* of the block. If the initial block contains one or more nominals generated by applications of the Name rule (see figure 2), then the nominal generated by the first application will be called the *opening nominal* of the initial block. Otherwise, the initial block will have no opening nominal.

The crucial rules of the Seligman-style tableau calculi are given in figure 2. The general conditions on rule applications are as follows:

- The propositional rules $(\vee)$, $(\neg\vee)$, $(\neg\neg)$ as well as $(\Diamond)$ and $(\neg\Diamond)$ can only be applied to premises that belong to the current block or a previous block with the same opening nominal.
- In the rules $(@)$ and $(\neg @)$, the first premise i must either belong to the current block or a previous block with the same opening nominal. The second premise $@_i\varphi$ $(\neg @_i\varphi)$ can appear anywhere on the branch.
- GoTo and Name can always be applied as they have no premises.
- Nom can be applied as described in the rule itself: if φ and i belong to some block distinct from the current block, and i belongs to the current block, then φ can be added to the current block.

In figure 3 we illustrate the calculus with two simple tableaus: one proves a valid formula, the other provides a counter-example for a non-valid formula.

1	$\neg(@_i j \wedge @_j \varphi \to @_i \varphi)$	
2	$@_i j \wedge @_j \varphi$	$(\neg \to)$ on 1
3	$\neg @_i \varphi$	$(\neg \to)$ on 1
4	$@_i j$	$(\wedge)$ on 2
5	$@_j \varphi$	$(\wedge)$ on 2
6	i	GoTo
7	j	$(@)$ on 4,6
8	φ	$(@)$ on 5,7
9	$\neg \varphi$	$(\neg @)$ on 3,6
	$\times$	

1	$\neg(\Diamond\Diamond i \to \Diamond i)$	
2	k	Name
3	$\Diamond\Diamond i$	$(\neg \to)$ on 1
4	$\neg \Diamond i$	$(\neg \to)$ on 1
5	$\Diamond j$	$(\Diamond)$ on 3
6	$@_j \Diamond i$	$(\Diamond)$ on 3
7	$\neg @_j i$	$(\neg \Diamond)$ on 4,5
8	j	GoTo
9	$\Diamond i$	$(@)$ on 6,8
10	$\neg i$	$(\neg @)$ on 7,8

Fig. 3. On the left the valid $@_j i \wedge @_j \varphi \to @_i \varphi$ is proved. On the right there is a non-closed tableau which can be used to construct a non-transitive model falsifying $\Diamond\Diamond i \to \Diamond i$. The model has three worlds, $\overline{k}$, $\overline{j}$ and $\overline{i}$, where only $(\overline{k}, \overline{j})$ and $(\overline{j}, \overline{i})$ are in the accessibility relation. The Name-rule gives us a name for the first world. In both tableaus we use derived rules for defined connectives.

Our first Seligman-style tableau calculus is called ST and consists of the rules given in figure 1 and 2. Tableaus are built in the expected way, but let us be explicit about our closure condition: *a branch closes either by having φ and $\neg\varphi$ inside a block, or inside two distinct blocks with the same opening nominal.* In [2] we showed that ST could prove all validities by translating between ST-tableaus and labelled tableaus. Here our goal is to prove a completeness result for ST that can be straightforwardly generalised to richer logics and languages.

To do this we will introduce a second calculus called STB. This is ST augmented by the Bridge rule, which is shown in figure 4. For the rule to be applicable three conditions have to be satisfied: (1) i and $\Diamond j$ must occur together on the same block, or on blocks with the same opening nominal; (2) j and k also have to occur on the same block, or on blocks with the same opening nominal; and (3) i must occur on the current block or on a block with the same opening nominal as the current block. Note that Bridge can be seen a restricted form of cut for nominals: reading it from bottom to top, it lets us 'cut' by introducing new symbols j and $\Diamond j$. We first prove completeness for STB, and then extend it to a proof for ST by eliminating Bridge.

3 Completeness for STB

In this section we adapt the Smullyan-Fitting synthetic completeness method so that it works with maximal consistent sets of *blocks* (rather than merely formulas or labelled formulas). This leads directly to completeness for STB.

3.1 Hintikka sets of blocks and induced models

First we generalize the notion of Hintikka set. Our tableau branches are sequences of finite blocks of formulas, each of which (with the possible exception of the first) is named by an opening nominal. So the first step is to choose a set-theoretic representation of this fundamental notion. Accordingly we define a *named block* to be a pair (A, i), where A is a set of formulas that has the

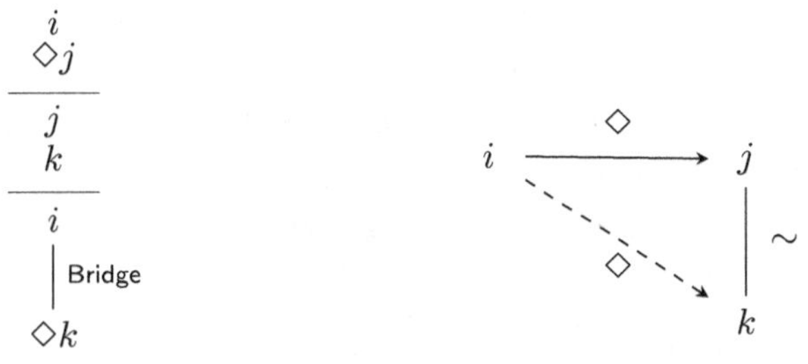

Fig. 4. The Bridge rule is shown on the left. On the right we show its semantic counterpart: if there is an edge from i to j, and if j and k are equivalent (that is, they name the same world) then there is an edge from i to k. This notion of equivalence is made precise as $\sim_H$ in the forthcoming model-constructions.

nominal i as one of its elements. So named blocks are pointed sets, and when we talk of an *i-block* we mean a named block of the form (A, i), that is, a named block with i as its distinguished element. We say that a formula φ is *on* a named block (A, i) iff $\varphi \in A$. A named block (A, i) is finite iff A is finite.

A set of finite named blocks H is called a *Hintikka set* if it satisfies the following properties:

(i) If there is an i-block in H with an atomic formula a on it, then there is no i-block in H with $\neg a$ on it.

(ii) If there is an i-block in H with $\Diamond j$ on it, then there is no i-block in H with $\neg\Diamond j$ on it.

(iii) If φ is a formula containing an occurrence of nominal i (either in formula or operator position) and φ is on some named block in H, then there is an i-block in H.

(iv) If there is an i-block in H with j on it, there is a j-block with i on it too.

(v) If there is an i-block in H with j on it, and a j-block in H with k on it, then there is an i-block in H with k on it.

(vi) If there is an i-block in H with $\Diamond j$ on it, and a j-block in H with k on it, then there is an i-block in H with $\Diamond k$ on it.

(vii) If there is an i-block in H with $\Diamond j$ on it, and an i-block in H with k on it, then there is an k-block in H with $\Diamond j$ on it.

(viii) If there is an i-block in H with $\varphi \vee \psi$ on it, then there is an i-block in H with φ or ψ on it.

(ix) If there is an i-block in H with $\neg(\varphi \vee \psi)$ on it, then there is an i-block in H with both $\neg\varphi$ and $\neg\psi$ on it.

(x) If there is an i-block in H with $\neg\neg\varphi$ on it, then there is an i-block in H with φ on it.

(xi) If there is a named block in H with $@_j\varphi$ (or $\neg @_j\varphi$) on it, then there is a j-block in H with φ (or $\neg\varphi$) on it.

(xii) If there is an i-block in H with $\Diamond\varphi$ on it, and φ is not a nominal, then there are (possibly identical) i-blocks in H with $\Diamond j$ and $@_j\varphi$ on them.

(xiii) If there are (possibly identical) i-blocks in H with $\neg\Diamond\varphi$ and $\Diamond j$ on them, then there is an i-block in H with $\neg @_j\varphi$ on it.

Think of a Hintikka set H as an (abstract versions of) an exhausted open tableau branch. We will construct a model out of equivalence classes of the nominals that name its blocks. If nominals i and j are names of blocks on H, we define $i \sim_H j$ iff there is an i-block in H with j on it; it follows from the iiird, ivth and vth Hintikka properties that $i \sim_H j$ is an equivalence relation. For a Hintikka set H of finite named blocks and a nominal i, we denote the equivalence class of i by $|i|_{\sim_H}$, suppressing $\sim_H$ when it's clear from context.

Given a Hintikka set H we construct *the named model* $\mathfrak{M}_H$ *induced by* H as follows. The set of worlds W_H of $\mathfrak{M}_H$ is the set of equivalence classes of nominals occurring in H. The relation R_H of $\mathfrak{M}_H$ is defined by:

$$|i|R_H|j| \text{ iff there is an } i\text{-block in } H \text{ with } \Diamond j \text{ on it.}$$

The well-definedness of R_H follows from the combination of the vith and viith Hintikka properties. The valuation function V_H of $\mathfrak{M}_H$ is defined by:

(i) If nominal i is in an equivalence class in W_H, set $V_H(i) = \{|i|\}$. Otherwise pick some $|j| \in W_H$ and set $V_H(i) = \{|j|\}$. Clearly V_H names every world.

(ii) If propositional variable p occurs on some block in H, then set $V_H(p)$ to be $\{|i| \ : \text{ there is an } i\text{-block in } H \text{ with } p \text{ on it}\}$. Otherwise set $V_H(p) = \emptyset$.

Lemma 3.1 (Hintikka block lemma) *Any Hintikka set H is satisfied by the named model it induces.*

Proof. Suppose H is a Hintikka set of finite named blocks. Let $\mathfrak{M} = (W, R, V)$ be the named model induced by H. Then by simultaneous induction on the complexity of φ we can show the following (stronger) fact:

A. If φ is on some i-block in H, then $\mathfrak{M}, |i| \models \varphi$,

B. If $\neg\varphi$ is is on some i-block in H, then $\mathfrak{M}, |i| \not\models \varphi$.

We need the stronger formulation to drive the inductive step for $\neg$ through. □

3.2 Maximal consistent sets of blocks

Let S be a finite set of finite named blocks. We say that T is a *named* STB-*tableau for S* when the initial part of T consists of the finite named blocks from S and the rest of T is constructed by application of the STB-rules. If T is a named STB-tableau for a set S of finite named blocks, any block in S is called a *root-block*; note that the order of the root blocks in the initial part of T doesn't matter. We then say that a (possibly infinite) set S of finite named blocks is STB-*consistent* when there doesn't exist a closed STB-tableau for any finite subset of S. We say that a set $S^\star$ of finite named blocks is *maximally* STB-*consistent* if is STB-consistent, and no proper extension of $S^\star$ is STB-consistent.

To prove completeness for STB (and later ST) we will need two technical lemmas. Let $B[i/j]$ denote the uniform substitution of nominal i for j in block

B; this means that we have uniformly substituted i for j in every formula occurring in B. We can also substitute i for j in a branch Θ if we substitute i for j in every block B occurring in Θ; this is written $\Theta[i/j]$.

Lemma 3.2 (Substitution lemma)

(i) ***We can uniformly substitute in a tableau.*** *Suppose Φ is a branch in an* STB*-tableau in which the nominals i and j occur, and Φ is extended by applying rule R to input I to obtain output O. Then $\Phi[i/j]$ can be extended by applying R to input $I[i/j]$ to obtain output $O[i/j]$.*

(ii) ***Which nominals are used as fresh nominals is irrelevant.*** *Suppose Θ is a branch in a finite tableau T and let i be a nominal which is used somewhere in Θ as a fresh nominal. Suppose, furthermore, that j is not used at all in Θ. We can then uniformly substitute j for i in Θ to obtain a branch $\Theta[j/i]$, where rule-applications in $\Theta[j/i]$ mimic rule-applications in Θ, the only change being that j is used instead of i.*

Proof. By induction on the construction of Φ and Θ respectively. □

We use the following lemma repeatedly (without comment) in what follows.

Lemma 3.3

(i) *If there exists a closed tableau for a finite set S^f of finite named blocks, then there exists a closed tableau for any finite set of finite named blocks which is a superset of S^f.*

(ii) *If S is a maximal* STB*-consistent set of finite named blocks, and if there is an i-block in S with φ on it, then there is also an i-block in S with just φ and i on it.*

Proof. Straightforward with the help of the substitution lemma. □

We are now ready for a version of the *Lindenbaum-Henkin construction* that works for named blocks. First, assume we have an enumerated infinite set of fresh nominals; these will be used to witness the diamonds.[5] Second, we will need an enumeration of the finite named blocks of this extended language; let's assume that these are given as $B_1, B_2, B_3, \ldots$, and so on.

Let S be an STB-consistent set of finite named blocks in the original (non-extended) language. We now construct, starting with S, an increasing sequence of consistent sets of finite named blocks such that the union of the whole sequence is our desired maximal STB-consistent set (in the extended language). But we need to ensure, that the diamonds are witnessed in the final union. So if some $\Diamond\varphi$ is on some block B and φ is not a nominal, we will create a block B' called the $\Diamond$-witness for B, such that for any $\Diamond\varphi$ on B, with φ not a nominal, $\Diamond i$ and $@_i\varphi$ are on B', with i being the first fresh nominal. The

[5] We say "witness" as we can view $\Diamond$ as an existential quantifier over acccessible worlds. We will witness diamond formulas with nominals—thereby mimicking Henkin's well-known first-order completeness proof strategy of witnessing existential formulas with constants.

Lindenbaum-Henkin-construction then goes like this: Let S_1 be S. Suppose S_n has been constructed. Then:

$$S_{n+1} = \begin{cases} S_n, & \text{if } S_n \cup \{B_n\} \text{ is inconsistent,} \\ S_n \cup \{B_n\}, & \text{if } S_n \cup \{B_n\} \text{ is consistent, and on } B_n \text{ there is no } \Diamond\varphi, \\ S_n \cup \{B_n\} \cup \{B'\}, & \text{if } S_n \cup \{B_n\} \text{ is consistent, and on } B_n \text{ there is at least one } \Diamond\varphi, \text{ with } \varphi \text{ not a nominal and } B' \text{ is the } \Diamond\text{-witness for } B_n. \end{cases}$$

Finally, we'll say that a set S of finite named blocks is *$\Diamond$-saturated*, if for any $\Diamond\varphi$ occurring on any i-block $B \in S$, there are (possibly identical) i-blocks B_1 and B_2 with $\Diamond j$ and $@_j\varphi$ on them.

Lemma 3.4 (Lindenbaum-Henkin) *Any* STB*-consistent set of finite named blocks can be extended into a $\Diamond$-saturated maximally* STB*-consistent set of finite named blocks.*

Proof. Let S be a consistent set of finite named blocks. Use S as input for the Lindenbaum-Henkin-construction. The claim we will prove is that $\bigcup S_n$ is a $\Diamond$-saturated maximally STB-consistent set.

We first need to prove that the sequence is a sequence of consistent sets ordered by $\subseteq$. By construction it's clear, that the sequence is ordered by $\subseteq$. It is not difficult, using lemma 3.2.ii, to show that if S_n is consistent then S_{n+1} is consistent too.

Consistency. Suppose $\bigcup S_n$ is not consistent. Then there is a finite set S^f of finite named blocks such that there is a closed STB-tableau for S^f. All the blocks in S^f is found in our enumeration of the finite named blocks. Let m be the largest index number. But then $B \in S_{m+1}$, for every $B \in S^f$, so by lemma 3.3.i we can construct a closed tableau for S_{m+1} which contradicts the consistency of S_{m+1}.

Maximality. Suppose $\bigcup S_n$ isn't maximal. Thus for some m, $B_m \notin \bigcup S_n$ and $\bigcup S_n \cup B_m$ is consistent. But $S_{m+1} \subseteq \bigcup S_n$ and $S_{m+1} = S_m \cup \{B_m\}$, (as $S_m \cup B_m$ is also consistent). But $B_m \in \bigcup S_n$, as $B_m \in S_{m+1} \subseteq \bigcup S_n$.

$\Diamond$-saturatedness. Suppose $\Diamond\varphi$ occurs on some i-block $B \in \bigcup S_n$. This block B occurs somewhere in the enumeration; suppose it's indexed by m. Then by the construction of S_{m+1} we have a witnessing i-block B' with both $\Diamond j$ and $@_j\varphi$ on B'. As $S_{m+1} \subseteq \bigcup S_n$ we have $B' \in \bigcup S_n$. □

3.3 Completeness

We have reached the heart of the synthetic completeness method:

Lemma 3.5 (Smullyan-Fitting block lemma for STB) *If S is a maximal* STB*-consistent set of finite named blocks, then S is a Hintikka set.*

Proof. Given a maximal STB-consistent set S of finite named blocks, we can show that all Hintikka properties hold for S. Here we prove four cases, including property vi, which is where Bridge is used.

Property i and ii. Let φ be either a nominal j, a propositional symbol p or the formula $\Diamond j$. Assume there is an i-block $B_1 \in S$ with φ on it. Further suppose that there is another i-block $B_2 \in S$ with $\neg\varphi$ on it. Then from the union of B_1 and B_2 we have a closed tableau, as both φ and $\neg\varphi$ occur on two (possibly identical) i-blocks, contradicting the STB-consistency of S.

Property iii. Assume that i occurs in some formula on some block $B_1 \in S$, but assume—working towards a contradiction—that there is no i-block at all in S. Let B_2 be the i-block which only has i on it. As B_2 is not in S it follows by the maximality of S that $S \cup \{B_2\}$ is STB-inconsistent. So there is a finite set S^f which is a subset of S such that there exists a closed tableau for $S^f \cup \{B_2\}$. Now, take S^f and add B_1 to it (if it isn't already there). Using GoTo we can extend $S^f \cup \{B_1\}$ by opening an i-block, and by copying the tableau we have for $S^f \cup \{B_1\}$ we can extend this to a closed tableau. But this contradicts the STB-consistency of S.

Property vi. Suppose there is an i-block $B_1 \in S$ with $\Diamond j$ on it and a j-block $B_2 \in S$ with k on it. Further suppose that there is no i-block with $\Diamond k$ on it in S. Let B_3 be the i-block consisting just of i and $\Diamond k$. By the maximality of S we have that $S \cup \{B_3\}$ is inconsistent. Thus, for a finite $S^f \subseteq S$ there is a closed tableau T for $S^f \cup \{B_3\}$. Now take $S^f \cup \{B_1\} \cup \{B_2\}$ as root-blocks. We start a tableau for these root-blocks by applying GoTo to open an i-block, and then apply Bridge to add $\Diamond k$ to it. But then we can construct a closed tableau for a finite subset of S which is a contradiction. □

We are ready to prove completeness for STB. We say that a formula φ is STB-consistent iff there is no closed STB tableau with φ as its sole root formula.

Theorem 3.6 *Every* STB*-consistent formula is satisfiable on a named model.*

Proof. Given an STB-consistent formula φ, take it as the root formula of a tableau. Immediately apply rule Name to ensure that our initial block has a name, i say. Applying Name cannot lead to a closed tableau as φ is STB-consistent. Hence $(\{\varphi, i\}, i)$ is an STB-consistent finite named block, and by the Lindenbaum-Henkin lemma it can be extended to a maximal STB-consistent set of finite named blocks S, which by the Smullyan-Fitting block lemma is a Hintikka set. We then form the named model $\mathfrak{M}$ induced by S, and by the Hintikka block lemma have $\mathfrak{M}, |i| \models \varphi$. □

4 Completeness for ST

The only place where Bridge was used in the completeness proof for STB was to show Hintikka property vi in the proof of the Smullyan-Fitting block lemma. We now show that we can prove a version of this lemma without using Bridge, that is, we can prove a Smullyan-Fitting block lemma for ST as well. We do so by showing that whatever we can do in a tableau using an i-block with $\Diamond k$ on it, we can already do using an i-block with $\Diamond j$ on it and a j-block with k on it. This is precisely what is needed for the vith property. To make this work we need to be able to keep track of formula occurrences that stem from or are influenced by $\Diamond k$ on the i-block.[6]

In a named ST-tableau T a formula occurrence of φ *descends from* $\Diamond i$ *in a root-block* B, if the occurrence of φ is:

(i) $\Diamond i$ as it occurs in the root-block B, or

(ii) the output of an application of either $(\neg\Diamond)$ or Nom which as input has a formula occurrence that descends from $\Diamond i$ in the root-block B.

Lemma 4.1 *If an occurrence of* φ *descends from* $\Diamond i$ *occurring on a root block* B*, then* φ *is either* $\Diamond i$ *or* $\neg @_i \psi$*, for some* ψ*.*

Proof. Let T be a named ST-tableau with root block B having $\Diamond i$ as an element. Suppose φ descends from $\Diamond i$ in B. The lemma is proved by induction on the construction of T.

Base case. Suppose φ is in root block and φ descends from $\Diamond i$ in B. As no rules have been applied, φ can only be $\Diamond i$ in B.

Inductive step. Suppose the lemma holds for T_0 and that T_0 is extended by applying R, thereby extending branch Θ_0 to Θ. If none of the formulas in the output of R descend from $\Diamond i$ on B, the lemma holds trivially for Θ. Suppose therefore that some of rule R's output descends from $\Diamond i$. Which rules could R be? By the definition of descendence, R can only be $(\neg\Diamond)$ or Nom.

Subcase a). Suppose R is $(\neg\Diamond)$ having input that descends from $\Diamond i$. By the induction hypothesis it can only be $\Diamond i$ of the two input-formulas that descends from $\Diamond i$ on B. The output of R, which by assumption descends from $\Diamond i$, is $\neg @_i \psi$, for some ψ. The lemma is thus proved for this case.

Subcase b). Suppose R is Nom. Its input is j and γ, for some nominal j and some formula γ. As the output of R by assumption desends from $\Diamond i$, some of the input of R has to descend (by definition of descendence). So γ, by the induction hypothesis, has to descend. Therefore γ is either $\Diamond i$ or $\neg @_i \psi$, for some ψ. Thus, the output of R which descends from $\Diamond i$ is either $\Diamond i$ or $\neg @_i \psi$, and the lemma is proved for this case. □

We next prove the crucial elimination lemma. First we need a convention. The use in a tableau proof of a $\Diamond i$ occurring in a root block has a possible trace that we need to follow and modify if we want to eliminate this occurrence of $\Diamond i$ from the root block. But recall that only two types of formulas, namely $\Diamond i$ and $\neg @_i \psi$, can descend from $\Diamond i$. This motivates the following:

- $\Diamond j$ is the j-replacement of $\Diamond i$, if the latter descends from $\Diamond i$
- $\neg @_j \psi$ is the j-replacement of $\neg @_i \psi$, if the latter descends from $\Diamond i$

In general, and to establish notation, if φ descends from $\Diamond i$ we denote its j-replacement by φ^j.

[6] Note that our elimination proof does *not* show that Bridge is derivable in ST. Rather, it shows that ST can do something equivalent to Bridge with respect to the blocks B_1, B_2 and B_3 referred to in the proof of Hintikka property vi (page 311). This is why the elimination lemma proved below is formulated in terms of blocks of this form.

Lemma 4.2 (Elimination lemma) *Suppose B_1 is the i-block consisting of i and $\Diamond j$, B_2 is the j-block consisting of j and k, and B_3 is the i-block consisting of i and $\Diamond k$. Suppose furthermore that S is any finite set of finite named blocks. Given a finite* ST*-tableau T for $S \cup \{B_3\}$ we can construct another finite* ST*-tableau T' for $S \cup \{B_1\} \cup \{B_2\}$ such that there is a correspondence between the branches of T and T' in such a way, that given any branch Θ of T, the following holds for any formula φ occurring on any l-block in Θ:*

(i) *If φ does not descend from $\Diamond k$ in B_3, then φ occurs on an l-block of the corresponding Θ' in T'.*

(ii) *If φ descends from $\Diamond k$ in B_3, then φ^j occurs on an l-block of the corresponding Θ' in T'.*

Proof. Let S, B_1, B_2 and B_3 be as supposed in the lemma, and let T be the finite ST-tableau for $S \cup \{B_3\}$. We also suppose that $j \neq k$, as otherwise there is nothing to prove. We can moreover, due to the second part of lemma 3.2, assume that j is not used as a fresh nominal in T. The proof will proceed by induction of the depth of branches of T: We systematically take branches Θ of T and produce corresponding branches Θ' of T', ensuring on the fly that the properties of the lemma hold.

Base case. Out of the root blocks of T we need to construct the root-blocks of T'. This is simple: just replace the T root-block B_3 by B_1 and B_2. Thus, the root blocks of T' are $S \cup \{B_1\} \cup \{B_2\}$. The lemma holds trivially.

Inductive step. Suppose we have gone through the initial part Θ_0 of Θ, the branch we follow in T, and that the lemma holds for Θ_0 and corresponding Θ'_0. Suppose the way Θ_0 is extended is by applying rule R. Either:

a) No formula occurrence of the input for R descends from $\Diamond k$ in B_3, or

b) Some formula occurrence of the input for R descends from $\Diamond k$ in B_3.

Case a). In this case an application of R, which extends Θ_0, will also extend Θ'_0, as all the input for R is to be found, by the induction hypothesis, on Θ'_0.

Case b). Suppose for some φ that is input for rule R which has been used to extend Θ_0, φ descends from $\Diamond k$ in B_3. The induction hypothesis will then give us φ^j on the corresponding Θ'_0. In such cases we cannot generally apply R right away to extend Θ'_0 as the input may not fit anymore; so we'll need to repair such mismatches. Now, which rules could R possibly be? If some input has to descend from $\Diamond k$ on B_3, the input must, by lemma 4.1, be either $\Diamond k$ or $\neg @_k \psi$. By inspection of the ST-rules we then see that R can only be $(\neg\Diamond)$, $(\neg @)$ or Nom. Which subcases will this give rise to? In case the input of $(\neg\Diamond)$ descends from $\Diamond k$, the input has to be $\Diamond k$ and $\neg\Diamond\psi$, for some ψ, both occuring on l-blocks of Θ_0. Analogously, in case input for $(\neg @)$ descends from $\Diamond k$, the input has to be $\neg @_k \psi$ occuring on an m-block and k occuring on an l-block. In case an input formula for Nom descends from $\Diamond k$ it has to be either $\Diamond k$ or $\neg @_k \psi$. Thus there are four subcases here:

Subcase i). The rule-application extending Θ_0 is $(\neg\Diamond)$ which has input $\Diamond k$ occurring on an l-block of Θ_0 and $\neg\Diamond\psi$ occurring on an l-block of Θ_0; the input

IH	1	j	
IH	2	k	
IH		$\vdots$	
IH	3	m	
IH	4	$\neg @_j \psi$	
IH	5	l	
IH	6	k	
IH	7	l	
	8	k	Nom on 5,6,7
	9	j	Nom on 1,2,8
	10	$\neg \psi$	$(\neg @)$ on 4,9

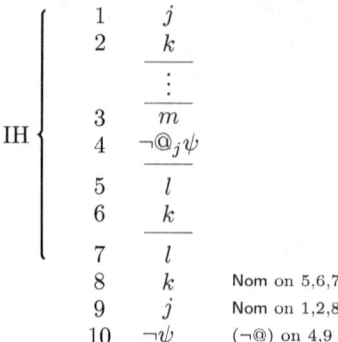

Fig. 5. The extension of Θ_0' in subcase ii.

$\Diamond k$ descends from $\Diamond k$ in B_3. The output of R is $\neg @_k \psi$ on the current l-block; this occurrence of $\neg @_k \psi$ also descends from $\Diamond k$ on B_3.

The induction hypothesis gives us the occurrence of the j-replacement $\Diamond j$ on an l-block of the corresponding Θ_0', and $\neg \Diamond \psi$ on an l-block of Θ_0'. We now apply $(\neg \Diamond)$ in order to extend Θ_0' and add $\neg @_j \psi$ on the current l-block. The output of R extending Θ_0 was $\neg @_k \psi$ descending from $\Diamond k$ in B_3. As we have created the j-replacement on the corresponding branch the lemma is proved for this subcase.

Subcase ii). Suppose the rule-application is $(\neg @)$ with the input $\neg @_k \psi$ which descends from $\Diamond k$ in B_3. Suppose furthermore that the input $\neg @_k \psi$ occurs on an m-block and the other input for $(\neg @)$ is k occurring on an l-block. The output of $(\neg @)$ is $\neg \psi$ which is added to the current l-block. Note, the output of R thus extending Θ_0 does not descend from $\Diamond k$ in B_3.

The induction hypothesis gives us a) An m-block on the corresponding Θ_0' with the j-replacement $\neg @_j \psi$, b) an l-block on Θ_0' with k on, and c) the current block is an l-block. figure 5 shows how to extend Θ_0' in this case—note, that we have suppressed quite a few dots.

Subcase iii). Suppose the rule application extending Θ_0 is Nom, where one of the input formulas is $\Diamond k$, which descends from $\Diamond k$ in B_3 and the input $\Diamond k$ occurs toegether with a nominal m on an n-block of Θ_0. The current block is an l-block on which m occurs. Θ_0 is thus extended by adding $\Diamond k$ to the current l-block. This output $\Diamond k$ descends from $\Diamond k$ in B_3.

The induction hypothesis gives us that the current block of the corresponding Θ_0' is an l-block. Moreover, on Θ_0' there is an n-block on which both m and the j-replacement $\Diamond j$ occur. We extend Θ_0' by adding $\Diamond j$ to the current l-block. This was what was required for this subcase.

Subcase iv). Suppose Θ_0 is extended by Nom having as input $\neg @_k \psi$, which descends from $\Diamond k$ on B_3. This case is precisely as the previous subcase. Here it's just $\neg @_k \psi$ instead of $\Diamond k$. Both descend from $\Diamond k$ in B_3. □

Lemma 4.3 (Smullyan-Fitting block lemma for ST) *If S is a maximal* ST*-consistent set of finite named blocks, then S is a Hintikka set.*

Proof. Given a maximal ST-consistent set S of finite named blocks, we can prove that all Hintikka properties hold for S. Here we give the crucial step for property vi, which is where Bridge was used in the proof for STB.

Property vi. Suppose there is an i-block $B'_1 \in S$ with $\Diamond j$ on it and also a j-block $B'_2 \in S$ with k on it. Suppose there is no i-block with $\Diamond k$ on it in S. Let B_3 be the i-block consisting just of i and $\Diamond k$. By maximality of S we have that $S \cup \{B_3\}$ is inconsistent. Thus, for a finite $S^f \subseteq S$ there is a closed tableau T for $S^f \cup \{B_3\}$. The procedure given in the proof of the elimination lemma converts T into a tableau T' having S^f, B_1 and B_2 as root blocks, where B_1 and B_2 are as in the elimination lemma. We now show that T' can be turned into a closed tableau (in case it isn't one already), which implies that there exists a closed tableau having S^f, B'_1 and B'_2 as root blocks.

Take any branch Θ of T. As Θ is closed there exist l-blocks B_4 and B_5 such that $\varphi \in B_4$ and $\neg\varphi \in B_5$. Either φ or $\neg\varphi$ descends from $\Diamond k$ on B_3 in T or neither of them does. If neither φ or $\neg\varphi$ descends from $\Diamond k$ on B_3, then by the elimination lemma there are l-blocks B'_4 and B'_5 occurring on Θ' such that φ and $\neg\varphi$ occur on B'_4 and B'_5, respectively. But then Θ' is closed.

Suppose therefore, that φ descends from $\Diamond k \in B_3$. Then we have two subcases: φ is either $\Diamond k$ or φ is $\neg @_k \psi$.

Subcase 1: φ is $\Diamond k$. According to the elimination lemma, we here have a branch Θ' of T' containing an l-block with $\Diamond j$ on it (as $\Diamond j$ is the j-replacement of $\Diamond k$) and an l-block with $\neg\Diamond k$ (note that $\neg\Diamond k$ cannot descend from $\Diamond k$). We extend Θ' by an application of GoTo opening an l-block. Then we apply $(\neg\Diamond)$ on input $\Diamond j$ and $\neg\Diamond k$ and retrieve $\neg @_j k$. Then open a j-block with GoTo and after that we apply $(\neg @)$ in order to add $\neg k$ to the current j-block. This extension of Θ' is closed, as k occurs on the j-block B_2.

Subcase 2: φ is $\neg @_k \psi$. In this case we have on Θ' an l-block with $\neg @_j \psi$ and an l-block with $\neg\neg @_k \psi$. We can extend Θ' by an application of GoTo, thereby opening an l-block. Then we apply $(\neg\neg)$ in order to add $@_k \psi$ to the current l-block. Then we open an j-block with GoTo, and by applying $(\neg @)$ we retrieve $\neg\psi$, by applying Nom we retrieve k since k occurs in the j-block B_2, and finally, we retrieve ψ by an application of $(@)$. This extension of Θ' is closed.

The case where $\neg\varphi$ descends from $\Diamond k$ in B_3 is similar. Then we have an l-block on Θ' with $@_k \psi$ and an l-block with $\neg @_j \psi$. We proceed as before. □

Theorem 4.4 *Every* ST*-consistent formula is satisfiable on a named model.*

Proof. Like the proof of Theorem 3.6, but making use of the Smullyan-Fitting block lemma for ST. □

5 Extensions of Basic Hybrid Logic

We now discuss Seligman-style systems for various extended hybrid logics.

5.1 Multimodal hybrid logic

Instead of working with a single $\Diamond$ we could have a collection of diamonds $\langle r \rangle$, where $r \in R$ for some (typically finite) index set R. However (assuming each modality is independent) there is little to say here: we simply add a version of the diamond (and box) rule for each additional modality. The completeness proofs generalize: the Hintikka properties are duplicated for each modality, the induced model is built as in the unimodal case, and the Lindenbaum-Henkin construction proceeds as before, but with witnessing blocks for *all* diamonds.

5.2 Pure axioms

A hybrid formula ρ is pure if it contains no ordinary propositional symbols: for example, $i \to \neg\Diamond i$, $\Diamond\Diamond i \to \Diamond i$ and $@_i\Box(\Diamond i \to i)$ are pure. Every pure formula defines a first-order definable class of frames; the examples just given define the classes of irreflexive, transitive, and antisymmetric frames respectively.[7] Moreoever, as we now show, when pure formulas are used as axioms, the resulting system is complete with respect to the class of frames the axioms define (the class of frames on which every axiom is valid).[8]

Let $i_1, \ldots, i_n$ be the nominals in ρ; then $\rho(j_1, \ldots, j_n/i_1, \ldots, i_n)$ is the formula obtained when we uniformly substitute $j_1, \ldots, j_n$ for $i_1, \ldots, i_n$ in ρ. We call $\rho(j_1, \ldots, j_n/i_1, \ldots, i_n)$ a *pure instance* of ρ. Let Axiom be some set of pure formulas. The rules of ST-Axiom are the rules of ST together with the rule that if $\rho \in \mathsf{Axiom}$, then we can add any pure instance of ρ to the current block.

Theorem 5.1 **(**Axiom**-completeness)** *Every* ST-Axiom*-consistent formula is satisfiable on a named model based on a frame that belongs to the class that* Axiom *defines.*

Proof. Let φ be the ST-Axiom-consistent formula. Take it as the root formula of a tableau, immediately apply rule Name to ensure that the initial block is named, and use the Lindenbaum-Henkin construction to form a maximal ST-Axiom-consistent S of finite named blocks. Assume for the sake of contradiction that the induced model $\mathfrak{M} = (\mathfrak{F}, V)$ is based on frame $\mathfrak{F}$ *not* belonging to the class defined by Axiom, that is, $\mathfrak{F} \not\models \rho$ for some $\rho \in \mathsf{Axiom}$. This means, that for some valuation function V' and some world w we have

$$(\mathfrak{F}, V'), w \not\models \rho.$$

Let $i_1, \ldots, i_n$ be the nominals of ρ, and suppose $V'(i_1) = \{w_1\}, \ldots, V'(i_n) = \{w_n\}$. And now the usefulness of named models becomes clear: because $\mathfrak{M}$ is named, there are nominals $i, j_1, \ldots, j_n$ such that $|i| = w$ and $V(j_1) = \{w_1\}, \ldots, V(j_n) = \{w_n\}$. Uniformly substituting $j_1 \ldots, j_n$ for $i_1, \ldots, i_n$ yields:

$$\mathfrak{M}, |i| \not\models \rho(j_1 \ldots, j_n/i_1 \ldots i_n).$$

[7] A formula φ defines a class of frames F when $\mathfrak{F} \models \varphi$ iff $\mathfrak{F}$ belongs to F. Here $\mathfrak{F} \models \varphi$ (φ is valid on $\mathfrak{F}$) means that φ is true at every world in $\mathfrak{F}$ under any valuation.

[8] Such results are familiar to hybrid logicians: see in particular [3], [5] and [11]; the first reference contains the result most closely related to the result we shall now prove.

The Hintikka block lemma implies that no i-block with $\rho(j_1, \ldots, j_n/i_1, \ldots, i_n)$ on is an element of S. Therefore, by maximal ST-Axiom-consistency of S adding the i-block B with just i and $\rho(j_1, \ldots, j_n/i_1, \ldots, i_n)$ on makes it ST-Axiom-inconsistent. So there exists a closed ST-Axiom-tableau T for $\{B\} \cup S^f$, where S^f is some finite subset of S. Now use S^f as the initial blocks of a new tableau. We assume i occurs in S^f (otherwise we just add a block from S containing i), so below S^f we open an i-block by GoTo. To this i-block we add the axiom $\rho(j_1 \ldots, j_n/i_1 \ldots i_n)$. Below this we can (modulo renaming of fresh nominals) paste our closed tableau T. But this contradicts the ST-Axiom-consistency of S as $S^f \subseteq S$, so $\mathfrak{F}$ does in fact belong to the class defined by Axiom. □

5.3 Hybrid tense logic

The best known multimodal logic with linked diamonds is probably tense logic. The diamond pair F and P both make use of the same relation: F looks forward along it (towards the future), and P looks backward (towards the past).

$$\mathfrak{M}, w \models \mathsf{F}\varphi \text{ iff for some } w',\ wRw' \text{ and } \mathfrak{M}, w' \models \varphi$$
$$\mathfrak{M}, w \models \mathsf{P}\varphi \text{ iff for some } w',\ w'Rw \text{ and } \mathfrak{M}, w' \models \varphi.$$

We call such models *bidirectional.* The box-form $\mathsf{G}\varphi$ means φ holds at all times in the future, and $\mathsf{H}\varphi$ means φ holds at all times in the past. The tableau rules for these operators are obtained by instantiating the generic diamond rules given in figure 2 by F and P. As F and P explore the same relation in opposite directions, we *also* add the transposition rules in figure 6 thereby obtaining the calculus $\mathsf{ST}_{(\mathsf{P},\mathsf{F})}$.

Theorem 5.2 ((P, F)-completeness) *Every* $\mathsf{ST}_{(\mathsf{P},\mathsf{F})}$*-consistent formula is satisfiable on a named bidirectional model.*

Proof. Let φ be the $\mathsf{ST}_{(\mathsf{P},\mathsf{F})}$-consistent formula which via the Lindenbaum-Henkin construction leads to $\mathfrak{M} = (T, R_\mathsf{F}, R_\mathsf{P}, V)$ as the induced named model. We want to show for all $t, t' \in T$ that $tR_\mathsf{F}t'$ iff $t'R_\mathsf{P}t$. Suppose for the sake of contradiction that this is not the case: either we have $tR_\mathsf{F}t'$ but not $t'R_\mathsf{P}t$, or we have $t'R_\mathsf{P}t$ but not $tR_\mathsf{F}t'$. We prove (by using F-trans) the first leads to a contradiction; the second is completely analogous (using P-trans instead) So assume $tR_\mathsf{F}t'$ but not $t'R_\mathsf{P}t$. As all elements of T are equivalence classes of nominals, this means that there are nominals i and j such that $|i|R_\mathsf{F}|j|$ but not $|j|R_\mathsf{P}|i|$. As it is not the case that $|j|R_\mathsf{P}|i|$, and as $\mathfrak{M}, |i| \models i$, it follows that $\mathfrak{M}, |j| \not\models \mathsf{P}i$. So by the construction of $\mathfrak{M}$ there cannot be any j-block in S with $\mathsf{P}i$ on it. But now we have a problem: as $|i|R_\mathsf{F}|j|$, there is there is an i-block with $\mathsf{F}j$ on it. In this block we can apply GoTo to open a j-block and we can then apply F-trans to extend it with $\mathsf{P}i$—in short, we have built a j-block in S with $\mathsf{P}i$ on it: contradiction. We conclude that the frame is bidirectional. □

5.4 The universal modality

The universal modality is a standard tool in hybrid logic. Formulas of the form $\mathsf{E}\varphi$ are satisfied at a world w in a model if there is some world in the model

Fig. 6. Transposition rules for hybrid tense logic.

that satisfies φ. That is:

$$\mathfrak{M}, w \models \mathsf{E}\varphi \text{ iff for some } w' \text{ we have } \mathfrak{M}, w' \models \varphi.$$

The dual of E is A. Formulas of the form $\mathsf{A}\varphi$ are satisfied at world w in a model if all worlds in the model satisfy φ. In short, the universal modality uses the universal relation $W \times W$ on worlds.[9] We capture its logic by adding the generic diamond rules plus the rule in figure 7, thereby obtaining the ST_E calculus. For an example of the rules at work see figure 8.

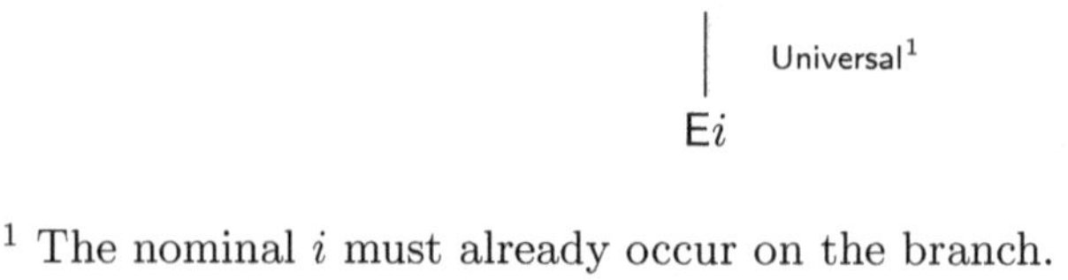
Universal¹
Ei
¹ The nominal i must already occur on the branch.

Fig. 7. The rule for the universal modality.

Theorem 5.3 (E-completeness) *Every ST_E-consistent formula is satisfiable on a named model $\mathfrak{M}$ with R_E being $W \times W$.*

Proof. Let S be the maximal ST_E-consistent set produced by the ST_E-consistent φ inducing the model $\mathfrak{M}$ which has R_E as the accessibility relation for E. Assume conversely that E is not universal. By construction of the worlds in W there exists nominals i and j, such that both i and j occur as names for blocks in S, and it is not the case that $|i|R_\mathsf{E}|j|$. By the definition of R_E this means that there are no i-blocks in S with $\mathsf{E}j$ on them. So, take the i-block with just $\mathsf{E}j$ and i on it and call it B. It follows from the maximality of S

[9] So the universal modality is just as $\mathsf{S5}$ operator. But it is usually used as a tool to express global constraints involving other modalities, for example: $A(\Box p \to \Diamond q)$. Note that both $\mathsf{E}(i \wedge \varphi)$ and $\mathsf{A}(i \to \varphi)$ are ways of expressing $@_i\varphi$.

1	$\neg(\Diamond\varphi \to \mathsf{E}\varphi)$	
2	$\Diamond\varphi$	$(\neg \to)$ on 1
3	$\neg\mathsf{E}\varphi$	$(\neg \to)$ on 1
4	$\Diamond i$	$(\Diamond)$ on 2
5	$@_i\varphi$	$(\Diamond)$ on 2
6	$\mathsf{E}i$	Universal
7	$\neg @_i\varphi$	(E) on 3 and 6
	$\times$	

1	$\neg(\mathsf{E}\varphi \to \varphi \vee \mathsf{D}\varphi)$	
2	$\mathsf{E}\varphi$	$(\neg \to)$ on 1
3	$\neg(\varphi \vee \mathsf{D}\varphi)$	$(\neg \to)$ on 1
4	$\neg\varphi$	$(\neg\vee)$ on 3
5	$\neg\mathsf{D}\varphi$	$(\neg\vee)$ on 3
6	$\mathsf{E}i$	(E) on 2
7	$@_i\varphi$	(E) on 2

8	i		9	$\neg i$	EM-nom
10	φ	(@)	11	$\mathsf{D}i$	D_1 on 9
	$\times$		12	$\neg @_i\varphi$	$(\neg\mathsf{D})$ on 5,11
				$\times$	

Fig. 8. Two examples of the rules at work. On the left we see a way of using the rules for $\Diamond$ and E to prove $\Diamond\varphi \to \mathsf{E}\varphi$: if φ is true at an *accessible* world, then it is true at *some* world in the model. The example on the right shows how to prove $\mathsf{E}\varphi \to \varphi \vee \mathsf{D}\varphi$ using the rules for D and E.

that $S \cup \{B\}$ is $\mathsf{ST_E}$-inconsistent. Therefore, there exists finite $S^f \subseteq S$, such that $S^f \cup \{B\}$ has a closed tableau. Let B_i and B_j be blocks in S containing i and j, respectively; such blocks exist by our initial assumption. Then we can construct a closed $\mathsf{ST_E}$-tableau from $S^f \cup \{B_i\} \cup \{B_j\}$, since below these initial blocks we can apply GoTo and open an i-block and then apply Universal to add $\mathsf{E}j$. Below this we simply paste (modulus renaming of fresh variables) the closed tableau we are given for $S^f \cup \{B\}$. This, however, contradicts the $\mathsf{ST_E}$-consistency of S, as $S^f \cup \{B_i\} \cup \{B_j\}$ without loss of generality can be assumed to be a subset of S □

5.5 The difference operator

The satisfaction definition for D is:

$$\mathfrak{M}, w \models \mathsf{D}\varphi \text{ iff there exists } w' \neq w \text{ such that } \mathfrak{M}, w' \models \varphi.$$

That is, D is evaluated using the difference relation ($W \times W$ minus the diagonal-elements). [10] We get a complete system $\mathsf{ST_D}$ by taking the usual diamond rules together with the rules shown figure 9. For an illustrative example we derive in the calculus $\mathsf{E}\varphi \to \varphi \vee \mathsf{D}\varphi$ using both the E- and the D-rules, see figure 8. These extra rules force R_D to be the difference relation.

Theorem 5.4 (D-completeness) *Every $\mathsf{ST_D}$-consistent formula is satisfiable on a named model $\mathfrak{M}$, where R_D is the difference relation.*

Proof. Suppose an $\mathsf{ST_D}$-consistent formula leads to the maximal $\mathsf{ST_D}$-consistent set S, which induces a model $\mathfrak{M}$ having relation R_D. Suppose R_D isn't the difference relation. Then (i) there is a world from W related by R_D to itself or (ii) there are distinct worlds in W that are not related by R_D. By construction of the set of world it follows that either:

[10] The difference operator is stronger than the universal modality. $\mathsf{E}\varphi$ can be defined to be $\varphi \vee \mathsf{D}\varphi$, but D cannot be defined in terms of E; see [11] for detailed discussion.

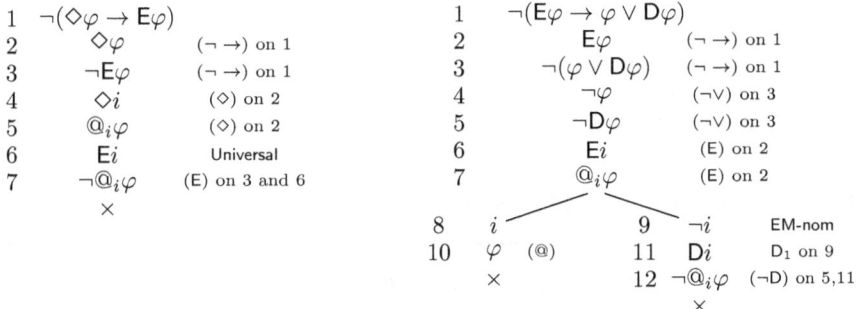

* The nominal i must already occur on the branch.

Fig. 9. Rules for the difference operator.

(i) There is a nominal i occuring in S such that $|i|R_{\mathsf{D}}|i|$, or

(ii) There are nominals i and j occuring in S such that $|i| \neq |j|$ but not $|i|R_{\mathsf{D}}|j|$.

The first case is easy: $|i|R_{\mathsf{D}}|i|$ means that there is an i-block B in S with $\mathsf{D}i$ on it. We can immediately construct a closed tableau from B by applying the rule D_2 to $\mathsf{D}i$ to get $\neg i$, which contradicts the consistency of S.

Now for the second case. As $|i| \neq |j|$ there can be no i-block with j on it in S. Let B_1 be the i-block with just i and j on it. As S is maximal ST_{D}-consistent there exists a finite $S_1^f \subseteq S$ and a closed tableau T_1 for $\{B_1\} \cup S_1^f$. Moreover, not $|i|R_{\mathsf{D}}|j|$ being the case implies (by definition of $|i|R_{\mathsf{D}}|j|$) that there cannot be any i-block with $\mathsf{D}j$ on in S. So if B_2 is the i-block with just i and $\mathsf{D}j$ on it, then (by the maximality of S) there exists a finite $S_2^f \subseteq S$ and a closed tableau T_2 for $S_2^f \cup \{B_2\}$. On the other hand, we can construct the following ST_{D}-tableau (if i or j does not occur in $S_1^f \cup S_2^f$ we just add blocks from S with these as appropriate; such blocks clearly exist):

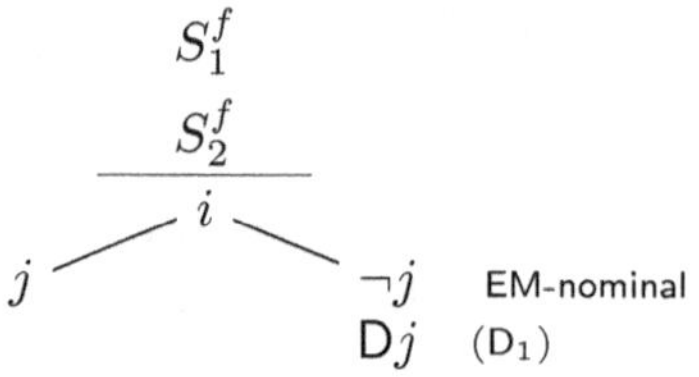

Below the left-hand branch we can paste the closed tableau T_1, and below the right-hand branch, closed tableau T_2 (modulus renaming of fresh variables). As $S_1^f \cup S_2^f$ is a subset of the ST_{D}-consistent set S, no such tableau exists. □

6 Concluding remarks

The Seligman-style approach to hybrid inference is an intriguing alternative to the better-known labelling methods, and in this paper we have presented some first results on extending Seligman-style tableau inference to a wider ranger of hybrid logics and languages. Much remains to be done. For a start, we would like to prove similar results for the terminating tableau system $\mathsf{ST}^\star$ discussed in [2], and, if possible, to try and show that Nom can be restricted in its application to nominals and near-atomic formulas of the form $\Diamond i$. In addition, we would like to develop Seligman-style tableau systems for various versions of first-order hybrid logic (with the downarrow binder $\downarrow$). This is a useful language for rethinking some traditional issues in philosophical logic (scope, equality, indexicality, actuality, existence and definite descriptions) and we believe Seligman-style tableaus are the simplest tools for working with it.

Acknowledgements

Klaus Frovin Jørgensen and Patrick Blackburn are grateful to the Spanish Ministerio de Economía y Competitividad for funding the project *Logica Intensional Hibrida* (*Hybrid Intensional Logic*), FFI2013-47126-P, hosted by the Universidad de Salamanca. Patrick Blackburn and Torben Braüner acknowledge the funding received from the VELUX FOUNDATION for the project *Hybrid-Logical Proofs at Work in Cognitive Psychology* (VELUX 33305). All authors would like to thank the three anonymous referees whose valuable comments and questions enabled us to improve the final version.

References

[1] Blackburn, P., T. Bolander, T. Braüner and K. F. Jørgensen, *A Seligman tableau system for hybrid logic*, Lecture Notes in Computer Science **8312** (2013), pp. 147–163.

[2] Blackburn, P., T. Bolander, T. Braüner and K. F. Jørgensen, *Completeness and termination for a Seligman-style tableau system*, Journal of Logic and Computation (To appear), doi:10.1093/logcom/exv052.

[3] Blackburn, P. and M. Tzakova, *Hybrid languages and temporal logic*, Logic Journal of the IGPL **7(1)** (1999), pp. 27–54.

[4] Braüner, T., "Hybrid Logic and its Proof-Theory," Applied Logic Series **37**, Springer, 2011.

[5] Bull, R., *An approach to tense logic*, Theoria **36** (1970), pp. 282–300.

[6] Fitch, F., *Tree proofs in modal logic*, Journal of Symbolic Logic **31** (1966), p. 152, abstract.

[7] Fitting, M., *Tableau methods of proof for modal logics*, Notre Dame Journal of Formal Logic **13** (1972), pp. 237–247.

[8] Fitting, M., "Proof Methods for Modal and Intuitionistic Logics," Reidel, 1983.

[9] Fitting, M., *Prefixed tableaus and nested sequents*, Annals of Pure and Applied Logic **34** (2012), pp. 291–313.

[10] Gabbay, D., "Labelled Deductive Systems," Oxford University Press, 1996.

[11] Gargov, G. and V. Goranko, *Modal logic with names*, Journal of Philosophical Logic **22** (1993), pp. 607–636.

[12] Negri, S., *Proofs and countermodels in non-classical logics*, Logica Universalis **8** (2014), pp. 25–60.

[13] Seligman, J., *The logic of correct description*, in: M. de Rijke, editor, *Advances in Intensional Logic*, Applied Logic Series **7**, Kluwer, 1997 pp. 107 – 135.

[14] Seligman, J., *Internalisation: The case of hybrid logics*, Journal of Logic and Computation **11** (2001), pp. 671–689.

[15] Simpson, A. K., "The proof theory and semantics of intuitionistic modal logic," Ph.D. thesis, University of Edinburgh (1994).

[16] Smullyan, R. M., "First-Order Logic," Springer, 1968.

The Logic of Conditional Beliefs: Neighbourhood Semantics and Sequent Calculus

M. Girlando[a], S. Negri[b], N. Olivetti[a], V. Risch[a] [1] [2]

[a] *Aix Marseille Université, CNRS, ENSAM, Université de Toulon, LSIS UMR 7296, Marseille, France*
[b] *Department of Philosophy, University of Helsinki, Finland*

Abstract

The logic of Conditional Beliefs has been introduced by Board, Baltag and Smets to reason about knowledge and revisable beliefs in a multi-agent setting. It is shown how the semantics of this logic, defined in terms of plausibility models, can be equivalently formulated in terms of neighbourhood models, a multi-agent generalisation of Lewis' spheres models. On the basis of this new semantics, a labelled sequent calculus for the logic of Conditional Beliefs is developed. The calculus has strong proof-theoretic properties, in particular admissibility of contraction and cut, and it provides a direct decision procedure for the logic. Furthermore, its semantic completeness is used to obtain a constructive proof of the finite model property of the logic.

Keywords: Epistemic logic, conditional logic, neighbourhood semantics, sequent calculus, decision procedure.

1 Introduction

Modal epistemic logic has been studied for a long time in formal epistemology, computer science, and notably in artificial intelligence. In this logic, to each agent i is associated a knowledge modality K_i, so that the formula K_iA expresses that "agent i knows A." Through agent-indexed modal operators, epistemic logic can be employed to reason about the mutual knowledge of a set of agents. The logic has been further extended by other modalities to encode various types of combined knowledge of agents (e.g., common knowledge). However, knowledge is not the only propositional attitude, and belief is equally significant to reason about epistemic interaction among agents. Board [5], and then Baltag and Smets [2], [3], [4] have proposed a logic called *CDL* (Conditional Doxastic Logic) for modelling both belief and knowledge in a multi-agent

[1] e-mail: `marianna.girlando@univ-amu.fr`, `sara.negri@helsinki.fi`, `nicola.olivetti@univ-amu.fr`, `vincent.risch@univ-amu.fr`

[2] This work was partially supported by the *LabEx Archimède*, AMU.

setting. The essential feature of beliefs is that they are *revisable* whenever the agent learns new information. To capture this revisable nature of beliefs, *CDL* contains the conditional belief operator $Bel_i(C|B)$, the meaning of which is that agent i believes C if she learnt B. Thus the conditional belief has an hypothetical meaning: if agent i learnt B, she would believe that C was true in the state of the world *before* the act of learning B. The logic captures the agent's changing beliefs in an unchanging world. For this reason Baltag and Smets [3] qualify this logic as "static" in contrast to "dynamic" epistemic logic, where the very act of learning (by some form of announcement) may change the agent's beliefs. The logic *CDL* in itself is used as the basic formalism to study further dynamic extensions of epistemic logics, determined by several kinds of epistemic/doxastic actions. Notice that both unconditional beliefs and knowledge can be defined in *CDL*: Bel_iB (agent i believe B) as $Bel_i(B|\top)$, and K_iB (agent i knows B) as $Bel(\bot|\neg B)$, the latter meaning that i considers impossible (inconsistent) to learn $\neg B$.

To exemplify the language, consider a variant of the three-wise-men puzzle, where agent a may initially believe that she has a white hat: $Bel_a W_a$. However, if a learns that agent b knows the colour of the hat b herself wears, she might change her beliefs and be convinced that she is wearing a black hat instead: $Bel_a(B_a|K_bW_b \vee K_bB_b)$. The example shows that the conditional operator is non-monotonic in the sense that $Bel_i(C|A)$ does not entail $Bel_i(C|A \wedge B)$ (here $A = \top$).

The axiomatization of the operator Bel_i in *CDL* internalises the well-known AGM postulates of belief revision[3].

The semantic interpretation of *CDL* is defined in terms of the so-called epistemic plausibility models. In these models, to each agent i is associated an equivalence relation $\sim_i$, used to interpret knowledge, and a well-founded preorder $\preceq_i$ on worlds. The relation $\preceq_i$ assesses the relative plausibility of worlds according to an agent i and is used to interpret conditional beliefs: i believes B conditionally on A in a world x if B holds in *the most plausible worlds* accessible from x in which A holds, where the "most plausible worlds" for an agent i are the $\preceq_i$-minimal ones. This semantic approach has been dominant in the studies of *CDL*; in addition to [5] and [3] we mention works by Pacuit [19], Van Ditmarsch et al. [20] and Demay [6].

In this paper we provide an alternative semantics, based on neighbourhood models, for *CDL*. Neighbourhood models are often used in the interpretation of non-normal modal logics. In the present setting they can be seen as a multi-agent generalization of Lewis' spheres models for counterfactual logics. Notice that finite sphere models have been used to define semantically (mono-agent) belief revision since Grove' seminal work [8]. In neighbourhood models to each world x and agent i is associated a set $I_i(x)$ of nested sets of worlds; each set $\alpha \in I_i(x)$ represents, so to say, a relevant piece of information that can

[3] We cannot mention here the vast literature on the relation between belief revision, conditional logics, the Ramsey Test, and Gärdenfors Triviality Result.

be used to establish the truth of an epistemic/doxastic statement. The interpretation of the conditional belief operator Bel_i then coincides with Lewis' semantics of the counterfactual operator. The equivalence between plausibility models and neighbourhood models does not come as totally unexpected: for the mono-agent case, it was suggested or stated without proof by Board [5], Pacuit [19], Marti et al [11], and it is based on an old result about the correspondence between partial orders and Alexandroff topologies [1]. We will detail the correspondence for the multi-agent case.

We believe that neighbourhood models provide by themselves a terse interpretation of the epistemic and doxastic modalities, abstracting away from the relational information specified in plausibility models. Moreover, it is worth noticing that in these models the interpretation of unconditional beliefs and knowledge results in the standard universal/existential neighbourhood modalities.

Up to this moment, the logic *CDL* has been studied only from a semantic viewpoint, and no proof-system or calculus is known for it. Our main goal is to provide one. On the basis of neighbourhood semantics we develop a labelled sequent calculus called **G3CDL**. We follow the general methodology of [13] to develop labelled calculi for modal logics. Similarly to [14], the calculus **G3CDL** makes use of world and neighbourhood labels, thereby importing the semantics, limited to the essential, into the syntax. In **G3CDL**, each connective is handled by symmetric left/right rules, whereas the properties of neighbourhood models are handled by additional rules independent of the language of *CDL*. The resulting calculus is analytical and enjoys strong proof-theoretical properties, the most important being admissibility of cut and contraction, for which we provide a syntactical proof. Through the adoption of a standard strategy, we show that the calculus **G3CDL** provides a decision procedure for *CDL*. We shall also prove the semantic completeness of the calculus: it is possible to extract from a failed derivation a finite countermodel of the initial formula. This result combined with the soundness of the calculus yields a constructive proof of the finite model property of *CDL*.

2 The logic of conditional beliefs: Axiomatization and semantics

The language of *CDL* is defined from a denumerable set of atoms Atm by means of propositional connectives and the conditional operator Bel_i, where i ranges over a set of agents $\mathcal{A}$. In the following, P denotes an atom and i an agent. The formulas of the language are generated according to the following definition:

$$A \ := \ P \mid \bot \mid \neg A \mid A \wedge A \mid A \vee A \mid A \supset A \mid Bel_i(A|A)$$

The conditional belief operator $Bel_i(C|B)$ is read "agent i believes C, given B." As mentioned in the introduction, we may define the unconditional belief and knowledge operator in terms of conditional belief:

$Bel_i A =_{def} Bel_i(A|\top)$ (belief)
$K_i A =_{def} Bel_i(\bot|\neg A)$ (knowledge)

An axiomatization of *CDL* has been discussed in [5], [19], [3]. We present below Board's axiomatization, which is formulated using only the conditional belief operator. Equivalent axiomatizations that make use of both the belief operator and the knowledge operator have been given by Baltag and Smets [2], [4], [3], and Pacuit [19]. The axiomatization of *CDL* extends the classical propositional calculus by the following axioms and rules:

(1) If $\vdash B$, then $\vdash Bel_i(B|A)$
(2) If $\vdash A \supset\subset B$, then $\vdash Bel_i(C|A) \supset\subset Bel_i(C|B)$
(3) $(Bel_i(B|A) \wedge Bel_i(B \supset C|A)) \supset Bel_i(C|A)$
(4) $Bel_i(A|A)$
(5) $Bel_i(B|A) \supset (Bel_i(C|A \wedge B) \supset\subset Bel_i(C|A))$
(6) $\neg Bel_i(\neg B|A) \supset (Bel_i(C|A \wedge B) \supset\subset Bel_i(B \supset C|A))$
(7) $Bel_i(B|A) \supset Bel_i(Bel_i(B|A)|C)$
(8) $\neg Bel_i(B|A) \supset Bel_i(\neg Bel_i(B|A)|C)$
(9)$A \supset \neg Bel_i(\bot|A)$

In terms of Belief Revision, the above axioms may be understood as an epistemic and internalized version of the AGM postulates. Some quick remarks (cf. [5] for a deeper discussion): The distribution axiom (3) and the epistemization rule (2) express deductive closure of beliefs. The success axiom (4) ensures that the learned information is included in the set of beliefs. Axioms (5) and (6) encode the *minimal change principle*, a basic assumption of belief revision (see the correspondence with AGM postulates K*7 and K*8). Axiom (9) ensures that learning a true information cannot lead to inconsistent beliefs (it roughly corresponds to AGM K*5). Axioms (7) and (8) express positive and negative introspection for belief. Observe that from the above axioms it is possible to derive the standard S5 characterization of knowledge:

$$K_iA \supset A \qquad K_iA \supset K_iK_iA \qquad \neg K_iA \supset K_i\neg K_iA$$

The semantics of *CDL* is defined in terms of *epistemic plausibility models* (*P*-models for short; they were originally called Belief Revision Structures by Board). These are Kripke structures that comprise for each agent two relations over worlds, namely an equivalence relation, which defines knowledge (as in standard epistemic models) and a plausibility relation, which is used to define beliefs. The intuition is that the beliefs of an agent are the propositions which hold in the worlds considered as the most plausible by the agent.

A *pre-order* $\preceq$ over a set W is a reflexive and transitive relation over W. Given $S \subseteq W$, $\preceq$ is *connected* over S if for all $x, y \in S$ either $x \preceq y$ or $y \preceq x$. An *infinite descending* $\preceq$*-chain* over W is a sequence of elements of W $\{x_n\}_{n\geq 0}$ such that for all n, $x_{n+1} \preceq x_n$ but $x_n \not\preceq x_{n+1}$. We say that $\preceq$ is *well-founded* over W if there are no infinite descending $\preceq$-chains over W. Given $S \subseteq W$, let $Min_{\preceq}(S) \equiv \{u \in S \,|\, \forall z \in S(z \preceq u \text{ implies } u \preceq z)\}$. Observe that whenever $\preceq$ is connected over S the definition $Min_{\preceq}(S)$ can be simplified to $Min_{\preceq}(S) = \{u \in S \,|\, \forall z \in S\ (u \preceq z)\}$. Finally, the well-foundedness property can be equivalently stated as: for each $S \subseteq W$ if $S \neq \emptyset$ then $Min_{\preceq}(S) \neq \emptyset$.

Definition 2.1 Let $\mathcal{A}$ be a set of agents; an *epistemic plausibility model*

$\mathcal{M} = \langle W, \{\sim_i\}_{i\in\mathcal{A}}, \{\preceq_i\}_{i\in\mathcal{A}}, [\![\ ]\!]\rangle$ consists of the following: a non-empty set W of elements called "worlds" or "states"; for each $i \in \mathcal{A}$, an equivalence relation $\sim_i$ over W (with $[x]_{\sim_i} \equiv \{w \mid w \sim_i x\}$); for each $i \in \mathcal{A}$, a well-founded pre-order $\preceq_i$ over W; a valuation function $[\![\ ]\!] : Atm \to \mathcal{P}(W)$. We assume $\preceq_i$ to satisfy the following properties:

- *Plausibility implies possibility*: If $w \preceq_i v$ then $w \sim_i v$;
- *Local connectedness*: If $w \sim_i v$ then $w \preceq_i v$ or $v \preceq_i w$ (in other words, $\preceq_i$ is connected over every equivalence class of $\sim_i$).

The truth conditions for formulas of the language are given by inductively extending the evaluation function $[\![\ ]\!]$ as follows:

- For the Boolean case we have the standard clauses, $[\![A \wedge B]\!] \equiv [\![A]\!] \cap [\![B]\!]$, $[\![\neg A]\!] \equiv W - [\![A]\!]$, etc.
- $[\![Bel_i(B|A)]\!] \equiv \{x \in W \mid Min_{\preceq_i}([x]_{\sim_i} \cap [\![A]\!]) \subseteq [\![B]\!]\}$.

We say that a formula A is *valid* in a model $\mathcal{M}$ if $[\![A]\!] = W$ and that A is *valid in the class of epistemic plausibility models* if A is valid in every P-model.
Notational convention: We often write $\mathcal{M}, x \Vdash A$ meaning $x \in [\![A]\!]$. The notation is further shortened to $x \Vdash A$ whenever $\mathcal{M}$ is unambiguous.

The axiomatization of *CDL* is sound and complete w.r.t. epistemic plausibility models [5].

Theorem 2.2 (Completeness of the axiomatization) *A formula A is a theorem of CDL if and only if it is valid in the class of P-models.*

The following proposition, proved by unfolding the definitions, gives an equivalent formulation of the truth condition of the conditional operator Bel_i provided in Definition 2.1. From now on, we shall use this formulation.

Proposition 2.3 *Given any P-model $\mathcal{M} = \langle W, \{\sim_i\}_{i\in\mathcal{A}}, \{\preceq_i\}_{i\in\mathcal{A}}, [\![\ ]\!]\rangle$, with $x \in W$, we have that $\mathcal{M}, x \Vdash Bel_i(B|A)$ iff: either for all y, $y \sim_i x$ implies $y \Vdash \neg A$ or there is y with $y \sim_i x$ such that $y \Vdash A$ and $\forall z, z \preceq_i y$ implies $z \Vdash A \supset B$.*

We introduce an alternative semantics for *CDL* based on neighbourhood models (N-models for short). As explained in the introduction, these are a multi-agent version of the spheres models introduced by Lewis for counterfactual logic.

Definition 2.4 Let $\mathcal{A}$ be a set of agents; a *multi-agent neighbourhood model* has the form $\mathcal{M} = \langle W, \{I\}_{i\in\mathcal{A}}, [\![\ ]\!]\rangle$ where W is a non empty set of elements; for each $i \in \mathcal{A}$, I_i is a function $I_i : W \to \mathcal{P}(\mathcal{P}(W))$, and $[\![\ ]\!] : Atm \to \mathcal{P}(W)$ is the propositional evaluation.
For $i \in \mathcal{A}$, $x \in W$, I_i satisfies the following properties:

- *Non-emptiness*: $\forall \alpha \in I_i(x), \alpha \neq \emptyset$
- *Nesting*: $\forall \alpha, \beta \in I_i(x), \alpha \subseteq \beta$ or $\beta \subseteq \alpha$
- *Total reflexivity*:[4] $\exists \alpha \in I_i(x)$ such that $x \in \alpha$

[4] Total reflexivity entails $\forall x \in W, I_i(x) \neq \emptyset$.

- *Local absoluteness*: If $\alpha \in I_i(x)$ and $y \in \alpha$ then $I_i(x) = I_i(y)$
- *Strong closure under intersection*: If $S \subseteq I_i(x)$ and $S \neq \emptyset$ then $\bigcap S \in S$.

The truth conditions for Boolean combinations of formulas are the standard ones, as in P-models; for conditional belief we have:

$x \in [\![Bel_i(B|A)]\!]$ iff $\forall \alpha \in I_i(x)$ it holds that $\alpha \cap [\![A]\!] = \emptyset$ or $\exists \beta \in I_i(x)$ such that $\beta \cap [\![A[\!\neq \emptyset$ and $\beta \subseteq [\![A \supset B]\!]$

A formula A is *valid* in $\mathcal{M}$ if $[\![A]\!] = W$. We say that A *is valid in the class of neighbourhood models* if A is valid in every N-model.

Observe that strong closure under intersection always holds in finite models, because of non-emptiness and nesting. To simplify the notation, we use the local forcing relations introduced in [12]:

$\alpha \Vdash^{\forall} A$ iff $\forall y \in \alpha, y \Vdash A$
$\alpha \Vdash^{\exists} A$ iff $\exists y \in \alpha, y \Vdash A$

With this notation, the truth condition for the conditional belief operator belief Bel_i becomes:

$x \Vdash Bel_i(B|A)$ iff $(\forall \alpha \in I_i(x),\ \alpha \Vdash^{\forall} \neg A)$ or $(\exists \beta \in I_i(x),\ \beta \Vdash^{\exists} A$ and $\beta \Vdash^{\forall} A \supset B)$

With the notation just introduced the semantic definition of unconditional belief and knowledge operators can be stated as follows:

$x \Vdash Bel_i B$ iff $\exists \beta \in I_i(x), \beta \Vdash^{\forall} B$
$x \Vdash K_i B$ iff $\forall \beta \in I_i(x), \beta \Vdash^{\forall} B$

Notice that these operators correspond to the standard modalities in neighbourhood models.

We now show the equivalence between neighbourhood models and epistemic plausibility models. The proofs make use of the basic correspondence between partial orders and topologies recalled in Marti and Pinosio [11] and Pacuit [18], and that dates back to Alexandroff [1]. However, the result must be adapted to the present setting of multi-agent epistemic neighbourhood models.

Theorem 2.5 *A formula A is valid in the class P-models if and only if it is valid in the class of multi-agent N-models.*

Proof. We first define the measure of *weight* of a *CDL* formula as follows: $\mathtt{w}(P) = \mathtt{w}(\bot) = 1$; $\mathtt{w}(\neg A) = \mathtt{w}(A) + 2$; $\mathtt{w}(A \circ B) = \mathtt{w}(A) + \mathtt{w}(B) + 1$ for $\circ = \{\wedge, \vee, \supset\}$; $\mathtt{w}(Bel_i(B|A)) = \mathtt{w}(A) + \mathtt{w}(B) + 3$ (cf. Definition 3.2).
[**only if**] Given a N-model $\mathcal{M}_N$ we build a P-model $\mathcal{M}_P$ and we show that for any formula A, if A is valid in $\mathcal{M}_P$ then A is valid in $\mathcal{M}_N$.
Let $\mathcal{M}_N = \langle W, \{I\}_{i \in \mathcal{A}}, [\![\ ]\!]\rangle$ be a multi-agent N-model. We construct a P-model $\mathcal{M}_P = \langle W, \{\sim_i\}_{i \in \mathcal{A}}, \{\preceq_i\}_{i \in \mathcal{A}}, [\![\ ]\!]\rangle$, by stipulating:

- $x \sim_i y$ iff $\exists \alpha \in I_i(x), y \in \alpha$;
- $x \preceq_i y$ iff $\forall \alpha \in I_i(y)$, if $y \in \alpha$ then $x \in \alpha$.

We can easily show that $\sim_i$ is an equivalence relation and that $\preceq_i$ satisfies the properties of reflexivity, transitivity, and plausibility implies possibility. Properties of local connectedness and well-foundedness for $\preceq_i$ require some additional work.

Local connectedness: suppose that $x \sim_i y$ holds, but neither $x \preceq_i y$ nor $y \preceq_i x$ hold. By definition of $\preceq_i$ we have for some $\beta \in I_i(y)$, $y \in \beta$ and $x \notin \beta$ and for some $\gamma \in I_i(x)$, $x \in \gamma$ and $y \notin \gamma$. Since $x \sim_i y$, by reflexivity $\exists \alpha \in I_i(x), y \in \alpha$, whence by local absoluteness $I_i(y) = I_i(x)$. Thus both $\beta, \gamma \in I_i(x)$, and by nesting either $\beta \subseteq \gamma$ or $\gamma \subseteq \beta$ holds. In case the former holds we get $y \in \gamma$, and in case the latter holds we have $x \in \beta$. In both cases we reach a contradiction.

Well-foundedness: If $\mathcal{M}_N$ is finite there is nothing to prove. Suppose then that $\mathcal{M}_N$ is *infinite*. Suppose that there is an infinite descending chain $\{z_k\}_{k \geq 0}$ w.r.t. $\preceq_i$, with all $z_k \in W$, so that for all k it holds that $z_{k+1} \preceq_i z_k$ and $z_k \not\preceq_i z_{k+1}$. Observe that by definition of $\preceq_i$, plausibility implies possibility and local absoluteness we obtain that for all $k, h \geq 0$, it holds $I_i(z_k) = I_i(z_h) = \ldots = I_i(z_0)$. Thus by definition of $\preceq_i$, for all $k \geq 0$ since $z_k \not\preceq_i z_{k+1}$, we get that for all $z_k \in \{z_k\}_{k \geq 0}$ there exists $\beta_{z_{k+1}} \in I_i(z_0)$ such that: $(*)$ $z_{k+1} \in \beta_{z_{k+1}}$ and $z_k \notin \beta_{z_{k+1}}$. Consider the set $T = \{\beta_{z_{k+1}} | z_k \in \{z_k\}_{k \geq 0}\}$. T is non-empty; thus by the strong closure under intersection it follows that $\bigcap T \in T$, and also $\bigcap T \neq \emptyset$. Obviously, we have that $(**)$ for all $\beta \in T$, $\bigcap T \subseteq \beta$. Since $\bigcap T \in T$ it must be $\bigcap T = \beta_{z_{t+1}}$ for some $z_t \in \{z_k\}_{k \geq 0}$. But by using $(*)$ *twice* (namely for z_{t+1} and for z_{t+2}) we have $z_{t+1} \in \beta_{z_{t+1}}$ and $z_{t+1} \notin \beta_{z_{t+2}}$, thus $\bigcap T = \beta_{z_{t+1}} \not\subseteq \beta_{z_{t+2}}$ against $(**)$.

We now prove that for any $x \in W$ and formula A it holds that

(a) $\mathcal{M}_N, x \Vdash A$ iff $\mathcal{M}_P, x \Vdash A$

We proceed by induction on the weight of A. The base case (A atomic) holds by definition; for the inductive cases, we consider only $A = Bel_i(C|B)$. To simplify notation we write $u \Vdash_P B$ instead of $\mathcal{M}_P, u \Vdash B$ and $u \Vdash_N B$ instead of $\mathcal{M}_N, u \Vdash B$. Direction $[\Rightarrow]$ of statement (a) easily follows from the definitions. As for the opposite direction, suppose that $x \Vdash_P Bel_i(C|B)$ holds. This means that either $\forall y\ y \sim_i x$ implies $y \Vdash_P \neg B$ or there exists w such that $w \sim_i x$ and $w \Vdash_P B$ and $\forall z, z \preceq_i w$ implies $z \Vdash_P B \supset C$. There are two cases to consider. If the first disjunct holds, by definition and by inductive hypothesis statement (a) is met. We explicitly prove the case in which the second disjunct holds. Suppose that there exists w such that $w \sim_i x$ and $w \Vdash_P B$ and $\forall z, z \preceq_i w$ implies $z \Vdash_P B \supset C$. From $w \sim_i x$ (hypothesis) it follows by definition that $\exists \alpha \in I(x), w \in \alpha$. By local absoluteness, $I(x) = I(w)$. Now consider the set $S = \{\beta \in I(x) | w \in \beta\}$. It holds that $\alpha \in S$, and that $S \neq \emptyset$. Let $\gamma = \cap S$. By strong closure under intersection, $\gamma \in S \subseteq I_i(x)$; thus $\gamma \in I_i(x)$. But $w \in \gamma$ and since we have $w \Vdash_P B$, by inductive hypothesis we also have $w \Vdash_N B$. We have obtained that $\gamma \Vdash^{\exists} B$. We still have to prove that $\gamma \Vdash^{\forall} B \supset C$. Given $u \in \gamma$, we want to prove that $u \Vdash_N B \supset C$. We first show that $u \preceq_i w$. To

this purpose (by definition of $\preceq_i$), let $\delta \in I(w)$ with $w \in \delta$ we have to show that $u \in \delta$: since $I(x) = I(w)$, also $\delta \in I(x)$, whence, $\delta \in S$, so that $\gamma \subseteq \delta$, and therefore $u \in \delta$. Since $u \preceq_i w$ by the hypothesis we have $u \Vdash_P B \supset C$ and finally by induction hypothesis $u \Vdash_N B \supset C$.
Next, we show that if A is valid in $\mathcal{M}_P$ then A is also valid in $\mathcal{M}_N$. Suppose that A is valid in $\mathcal{M}_P$. This means that for all $w \in W$, we have $w \Vdash_P A$, thus by (a) we have also $w \Vdash_N A$ for all $w \in W$, which means that A is valid in $\mathcal{M}_N$. Finally, let A be valid in the class of P-models. We want to show that A is also valid in the class of N-models. Given a N-model $\mathcal{M}_N$, we build an P-model $\mathcal{M}_P$ as above. By hypothesis A is valid in $\mathcal{M}_P$ and for what we have just shown A is valid in $\mathcal{M}_N$.

[If] Given a P-model $\mathcal{M}_P$ we build an N-model $\mathcal{M}_N$ and we show that for any A, if A is valid in $\mathcal{M}_N$ then A is valid in $\mathcal{M}_P$. Let $\mathcal{M}_P = \langle W, \{\sim_i\}_{i\in\mathcal{A}}, \{\preceq_i\}_{i\in\mathcal{A}}, [\![\ ]\!]\rangle$ be an P-model. We build an N-model $\mathcal{M}_N$ as follows. Let $u \in W$, and define its downward closed set $\downarrow^{\preceq_i} u$ w.r.t. $\preceq_i$ as $\downarrow^{\preceq_i} u = \{v \in W | v \preceq_i u\}$. We now define the model $\mathcal{M}_N = \langle W, \{I\}_{i\in\mathcal{A}}, [\![\ ]\!]\rangle$, where the neighbourhood for any $x \in W$ is $I_i(x) = \{\downarrow^{\preceq_i} u | u \sim_i x\}$.

It can be easily proved that $\mathcal{M}_N$ satisfies all the properties of an N-model; we show only the case of the *strong closure under intersection*. In the finite case, this property immediately follows from properties of non-emptiness and nesting. Let us consider the infinite case. Let $S \subseteq I_i(x)$, $S \neq \emptyset$, with S countable so that $S = \{\alpha_h | h \geqslant 0\}$ where $\alpha_h = \downarrow^{\preceq_i} x_h$ for $x_h \sim_i x$. We prove that $(*)$ $\exists \alpha_h \in S$ such that $\forall \alpha_k \in S, \alpha_h \subseteq \alpha_k$. If $(*)$ holds then $\alpha_h = \bigcap S$ and $\alpha_h \in S$ and the proof is over. Suppose by contradiction that $(*)$ does not hold. This means that 1) $\forall \alpha_h \in S\ \exists \alpha_k \in S, \alpha_h \not\subseteq \alpha_k$. Thus, by the property of spheres nesting 2) $\forall \alpha_h \in S\ \exists \alpha_k \in S, \alpha_k \subset \alpha_h$. From 2), by denumerable dependent choice we build an infinite (strictly decreasing) chain of neighbourhoods $\alpha_1 \supset \alpha_2 \supset \alpha_3 \supset \ldots$. For every $n \geq 1$ we have by definition that $\alpha_n = \downarrow^{\preceq_i} u_n$. Let $v_n \in \alpha_n - \alpha_{n+1}$, $v_{n+1} \in \alpha_{n+1} - \alpha_{n+2}$, etc. We have $v_{n+1} \preceq_i u_{n+1}$ by construction and it is enough to prove that $u_{n+1} \preceq_i v_n$ to conclude by transitivity that $v_{n+1} \preceq_i v_n$. By construction, we have $v_n \not\preceq_i u_{n+1}$ and therefore by local connectedness, $u_{n+1} \preceq_i v_n$. Moreover by $v_n \not\preceq_i u_{n+1}$ it also follows that $v_n \not\preceq_i v_{n+1}$. We have thus an infinitely descending $\preceq_i$-chain of worlds $\{v_n\}_{n\geq 1}$, against the assumption of well-foundedness of W. We reached a contradiction from the negation of $(*)$; therefore, $(*)$ holds.

We now have to prove that for any $x \in W$ and formula A, it holds that (b) $\mathcal{M}_P, x \Vdash A$ iff $\mathcal{M}_N, x \Vdash A$. The proof strategy is the same employed in the previous case. Next, as above, we show that if A is valid in $\mathcal{M}_N$ then A is also valid in $\mathcal{M}_P$. Finally, let A be valid in the class of N-models. We want to show that A is also valid in the class of P-models. Given an P-model $\mathcal{M}_P$, we build an N-model $\mathcal{M}_N$ as described. By hypothesis A is valid in $\mathcal{M}_N$ and by what we have just shown A is valid in $\mathcal{M}_P$. □

Corollary 2.6 *A formula A is a theorem of CDL if and only if it is valid in the class of neighbourhood models.*

Observe that the correspondence between plausibility and neighbourhood models holds for infinite models as well. For this reason, the correspondence can probably be used to establish *strong completeness* of *CDL*, which at present is an open issue, with respect to any of the two semantics.

Initial sequents

$$x : P, \Gamma \Rightarrow \Delta, x : P \qquad x : \bot, \Gamma \Rightarrow \Delta \qquad \Gamma \Rightarrow \Delta, x : \top$$

Rules for local forcing

$$\frac{x : A, x \in a, a \Vdash^{\forall} A, \Gamma \Rightarrow \Delta}{x \in a, a \Vdash^{\forall} A, \Gamma \Rightarrow \Delta}\ L{\Vdash^{\forall}} \qquad \frac{x \in a, \Gamma \Rightarrow \Delta, x : A}{\Gamma \Rightarrow \Delta, a \Vdash^{\forall} A}\ R{\Vdash^{\forall}}\ (x \text{ fresh})$$

$$\frac{x \in a, x : A, \Gamma \Rightarrow \Delta}{a \Vdash^{\exists} A, \Gamma \Rightarrow \Delta}\ L{\Vdash^{\exists}}\ (x \text{ fresh}) \qquad \frac{x \in a, \Gamma \Rightarrow \Delta, x : A, a \Vdash^{\exists} A}{x \in a, \Gamma \Rightarrow \Delta, a \Vdash^{\exists} A}\ R{\Vdash^{\exists}}$$

Propositional rules

$$\frac{\Gamma \Rightarrow \Delta, x : A}{x : \neg A, \Gamma \Rightarrow \Delta}\ L\neg \qquad \frac{x : A, \Gamma \Rightarrow \Delta}{\Gamma \Rightarrow \Delta, x : \neg A}\ R\neg$$

$$\frac{x : A, x : B, \Gamma \Rightarrow \Delta}{x : A \wedge B, \Gamma \Rightarrow \Delta}\ L\wedge \qquad \frac{\Gamma \Rightarrow \Delta, x : A \quad \Gamma \Rightarrow \Delta, x : B}{\Gamma \Rightarrow \Delta, x : A \wedge B}\ R\wedge$$

$$\frac{x : A, \Gamma \Rightarrow \Delta \quad x : B, \Gamma \Rightarrow \Delta}{x : A \vee B, \Gamma \Rightarrow \Delta}\ L\vee \qquad \frac{\Gamma \Rightarrow \Delta, x : A, x : B}{\Gamma \Rightarrow \Delta, x : A \vee B}\ R\vee$$

$$\frac{\Gamma \Rightarrow \Delta, x : A \quad x : B, \Gamma \Rightarrow \Delta}{x : A \supset B, \Gamma \Rightarrow \Delta}\ L\supset \qquad \frac{x : A, \Gamma \Rightarrow \Delta, x : B}{\Gamma \Rightarrow \Delta, x : A \supset B}\ R\supset$$

Rules for conditional belief

$$\frac{a \in I_i(x), a \Vdash^{\exists} A, \Gamma \Rightarrow \Delta, x \Vdash_i B|A}{\Gamma \Rightarrow \Delta, x : Bel_i(B|A)}\ RB\ (a \text{ fresh})$$

$$\frac{a \in I_i(x), x : Bel_i(B|A), \Gamma \Rightarrow \Delta, a \Vdash^{\exists} A \quad x \Vdash_i B|A, a \in I_i(x), x : Bel_i(B|A), \Gamma \Rightarrow \Delta}{a \in I_i(x), x : Bel_i(B|A), \Gamma \Rightarrow \Delta}\ LB$$

$$\frac{a \in I_i(x), \Gamma \Rightarrow \Delta, x \Vdash_i B|A, a \Vdash^{\exists} A \quad a \in I_i(x), \Gamma \Rightarrow \Delta, x \Vdash_i B|A, a \Vdash^{\forall} A \supset B}{a \in I_i(x), \Gamma \Rightarrow \Delta, x \Vdash_i B|A}\ RC$$

$$\frac{a \in I_i(x), a \Vdash^{\exists} A, a \Vdash^{\forall} A \supset B, \Gamma \Rightarrow \Delta}{x \Vdash_i B|A, \Gamma \Rightarrow \Delta}\ LC (a \text{ fresh})$$

Rules for inclusion

$$\frac{a \subseteq a, \Gamma \Rightarrow \Delta}{\Gamma \Rightarrow \Delta}\ Ref \qquad \frac{c \subseteq a, c \subseteq b, b \subseteq a, \Gamma \Rightarrow \Delta}{c \subseteq b, b \subseteq a, \Gamma \Rightarrow \Delta}\ Tr$$

$$\frac{x \in a, a \subseteq b, x \in b, \Gamma \Rightarrow \Delta}{x \in a, a \subseteq b, \Gamma \Rightarrow \Delta}\ L\subseteq$$

Rules for semantic conditions

$$\frac{a \subseteq b, a \in I_i(x), b \in I_i(x), \Gamma \Rightarrow \Delta \quad b \subseteq a, a \in I_i(x), b \in I_i(x), \Gamma \Rightarrow \Delta}{a \in I_i(x), b \in I_i(x), \Gamma \Rightarrow \Delta}\ S$$

$$\frac{x \in a, a \in I_i(x), \Gamma \Rightarrow \Delta}{\Gamma \Rightarrow \Delta}\ T\ (a \text{ fresh})$$

$$\frac{a \in I_i(x), y \in a, b \in I_i(x), b \in I_i(y), \Gamma \Rightarrow \Delta}{a \in I_i(x), y \in a, b \in I_i(x), \Gamma \Rightarrow \Delta}\ A_1 \qquad \frac{a \in I_i(x), y \in a, a \in I_i(y), \Gamma \Rightarrow \Delta}{a \in I_i(x), y \in a, \Gamma \Rightarrow \Delta}\ A_2$$

Table 1. Sequent calculus **G3CDL**

3 Sequent calculus

In this section we present a labelled sequent calculus **G3CDL** for *CDL* based on neighbourhood semantics. As shown in Table 1, the calculus **G3CDL** has two kinds of labels: labels for worlds $x, y \dots$ and labels for neighbourhoods $a, b \dots$, as in the ground calculus for neighbourhood semantics introduced in [12].

The meaning of the expressions employed in the calculus is defined as follows:

$$a \Vdash^{\exists} A \;\equiv\; \exists x(x \in a \;\&\; x \Vdash A); \quad a \Vdash^{\forall} A \;\equiv\; \forall x(x \in a \longrightarrow x \Vdash A)$$
$$x \Vdash_i B|A \;\equiv\; \exists c(c \in I_i(x) \;\&\; c \Vdash^{\exists} A \;\&\; c \Vdash^{\forall} A \supset B)$$
$$x : Bel_i(B|A) \equiv \forall a \in I_i(x)(a \Vdash^{\forall} \neg A) \text{ or } \exists b \in I_i(x)(b \Vdash^{\exists} A \;\&\; b \Vdash^{\forall} A \supset B)$$

Here $\Vdash$ denotes the forcing condition of relational semantics; to distinguish the semantic notion and its syntactic counterpart, and for the sake of a more compact notation, we employ a colon in the labelled calculus. The propositional rules of **G3CDL**, the basic labelled modal system, are given as in [13], while the rules for the local forcing relation are defined as in [12].

Furthermore, each semantic condition on neighbourhood models (Definition 2.4) is in correspondence with a rule in the calculus. Rule (S) corresponds to the property of nesting in Definition 2.4; (T) corresponds to total reflexivity, and (A_1) and (A_2) to local absoluteness. As for non-emptiness, the property is expressed by the rules for local forcing. The property of strong closure under intersection needs not be expressed, since the property holds in finite models and we shall prove that the logic has the *finite model property.*

Observe that some rules maintain their principal formula in the premisses: this is needed to ensure invertibility of the rules and admissibility of contraction.

Example 3.1 We show a derivation of the left-to-right direction of axiom (6). We omit the derivable left premisses of rule (RC) in $\mathcal{D}$ and of rule (LB) in the final derivation.

$\mathcal{D}$:

$$\begin{array}{cl}
y : A \cdots \Rightarrow \dots y : A \qquad\qquad y : B \cdots \Rightarrow \dots y : B & \\ \hline
y : A, y : B, y \in b, c \in I_i(x), c \Vdash^{\exists} A, b \in I_i(x) \cdots \Rightarrow \dots y : A \wedge B & R\wedge \\ \hline
y : A, y : B, y \in b, c \in I_i(x), c \Vdash^{\exists} A, b \in I_i(x) \cdots \Rightarrow \dots b \Vdash^{\exists} A \wedge B & R\Vdash^{\exists} \\ \hline
y \in b, c \in I_i(x), c \Vdash^{\exists} A, b \in I_i(x) \cdots \Rightarrow \dots b \Vdash^{\exists} A \wedge B, y : A \supset \neg B & R\supset, R\neg \\ \hline
c \in I_i(x), c \Vdash^{\exists} A, b \in I_i(x) \cdots \Rightarrow \dots b \Vdash^{\exists} A \wedge B, b \Vdash^{\forall} A \supset \neg B & R\Vdash^{\forall} \\ \hline
c \in I_i(x), c \Vdash^{\exists} A, b \in I_i(x) \cdots \Rightarrow \dots b \Vdash^{\exists} A \wedge B, x \Vdash_i \neg B|A & RC \\ \hline
b \in I_i(x), b \Vdash^{\exists} A, b \Vdash^{\forall} A \supset C, a \Vdash^{\exists} A \wedge B \cdots \Rightarrow \dots x : Bel_i(\neg B|A), b \Vdash^{\exists} A \wedge B & RB
\end{array}$$

$\mathcal{E}$:

$$\begin{array}{cl}
z : A \cdots \Rightarrow \dots z : A \qquad\qquad z : c \cdots \Rightarrow \dots z : C & \\ \hline
z : A \supset C, z : A, z : B, z \in b, b \in I_i(x), b \Vdash^{\exists} A, b \Vdash^{\forall} A \supset C, a \Vdash^{\exists} A \wedge B, \cdots \Rightarrow \dots z : C & L\supset \\ \hline
z : A, z : B, z \in b, b \in I_i(x), b \Vdash^{\exists} A, b \Vdash^{\forall} A \supset C, a \Vdash^{\exists} A \wedge B \cdots \Rightarrow \dots z : C & L\Vdash^{\forall} \\ \hline
z \in b, b \in I_i(x), b \Vdash^{\exists} A, b \Vdash^{\forall} A \supset C, a \Vdash^{\exists} A \wedge B \cdots \Rightarrow \dots z : (A \wedge B) \supset C & R\supset, L\wedge \\ \hline
b \in I_i(x), b \Vdash^{\exists} A, b \Vdash^{\forall} A \supset C, a \Vdash^{\exists} A \wedge B \cdots \Rightarrow \dots b \Vdash^{\forall} (A \wedge B) \supset C & R\Vdash^{\forall}
\end{array}$$

$$\dfrac{\dfrac{\dfrac{\dfrac{\dfrac{\begin{matrix}\vdots & & \vdots \\ \mathcal{D} & & \mathcal{E}\end{matrix}}{b \in I_i(x), b \Vdash^{\exists} A, b \Vdash^{\forall} A \supset C, a \in I_i(x), a \Vdash^{\exists} A \wedge B, x : Bel_i(C|A) \Rightarrow x : Bel_i(\neg B|A), x \Vdash_i C|A \wedge B}\,RC}{x \Vdash_i C|A, a \in I_i(x), a \Vdash^{\exists} A \wedge B, x : Bel_i(C|A) \Rightarrow x : Bel_i(\neg B|A), x \Vdash_i C|A \wedge B}\,LC}{a \in I_i(x), a \Vdash^{\exists} A \wedge B, x : Bel_i(C|A) \Rightarrow x : Bel_i(\neg B|A), x \Vdash_i C|A \wedge B}\,LB}{x : Bel_i(C|A) \Rightarrow x : Bel_i(\neg B|A), x : Bel_i(C|A \wedge B)}\,RB}{x : \neg(Bel_i(\neg B|A)), x : Bel_i(C|A) \Rightarrow x : Bel_i(C|A \wedge B)}\,L\neg$$

Rules for unconditional belief and knowledge

The modal operators of belief and knowledge can be defined semantically in terms of the conditional belief operator: $Bel_i A = Bel_i(A|\top)$ and $K_i A = Bel_i(\bot|\neg A)$. By adopting these definitions, we can extend **G3CDL** with the rules displayed below, which correspond to the interpretation of the two operations in the neighbourhood semantics.

$$\dfrac{a \in I_i(x), \Gamma \Rightarrow \Delta, a \Vdash^{\forall} A}{\Gamma \Rightarrow \Delta, x : K_i A}\,LK\ (a \text{ fresh}) \qquad \dfrac{a \in I_i(x), x : K_i A, a \Vdash^{\forall} A, \Gamma \Rightarrow \Delta}{a \in I_i(x), x : K_i A, \Gamma \Rightarrow \Delta}\,RK$$

$$\dfrac{a \in I_i(x), \Gamma \Rightarrow \Delta, x : Bel_i A, a \Vdash^{\forall} A}{a \in I_i(x), \Gamma \Rightarrow \Delta, x : Bel_i A}\,LUB \qquad \dfrac{a \in I_i(x), a \Vdash^{\forall} A \Rightarrow \Delta}{x : Bel_i A, \Gamma \Rightarrow \Delta}\,RUB\ (a \text{ fresh})$$

These rules are *admissible* in **G3CDL**, i.e., whenever the premiss is derivable, also the conclusion is. This can be proved employing the rules of **G3CDL** and the rules of weakening and contraction, shown admissible in next section. By means of example, we show admissibility of (LK) (the other rules can be obtained in a similar way).

$$\dfrac{\dfrac{...a \Vdash^{\exists} \neg A \Rightarrow a \Vdash^{\exists} \neg A... \qquad \dfrac{a \in I_i(x), \Gamma \Rightarrow \Delta, a \Vdash^{\forall} A}{a \in I_i(x), a \Vdash^{\exists} \neg A, \Gamma \Rightarrow \Delta, x \Vdash_i \bot|\neg A, a \Vdash^{\forall} A}\,Wk}{a \in I_i(x), a \Vdash^{\exists} \neg A, \Gamma \Rightarrow \Delta, x \Vdash_i \bot|\neg A}\,RC}{\Gamma \Rightarrow \Delta, x : Bel_i(\bot|\neg A)}\,RB$$

The left premiss of (RC), which we have not detailed, is derivable.

Structural properties

Definition 3.2 The label of formulas of the form $x : A$ is x. The label of formulas of the form $a \Vdash^{\forall} A$ and $a \Vdash^{\exists} A$ is a. The label of a formula $\mathcal{F}$ will be denoted by $l(\mathcal{F})$. The pure part of a labelled formula $\mathcal{F}$ is the part without the label and without the forcing relation, either local ($\Vdash^{\exists}$, $\Vdash^{\forall}$) or worldwide (:) and will be denoted by $p(\mathcal{F})$.
The *weight of a labelled formula* $\mathcal{F}$ is the pair $(\mathtt{w}(p(\mathcal{F})), \mathtt{w}(l(\mathcal{F})))$ where:

(i) for all world labels x and all neighbourhood labels a, $\mathtt{w}(x) = 0$, $\mathtt{w}(a) = 1$;
(ii) $\mathtt{w}(P) = \mathtt{w}(\bot) = 1$; $\mathtt{w}(\neg A) = \mathtt{w}(A) + 2$; $\mathtt{w}(A \circ B) = \mathtt{w}(A) + \mathtt{w}(B) + 1$ for $\circ$ conjunction, disjunction, or implication; $\mathtt{w}(B|A) = \mathtt{w}(A) + \mathtt{w}(B) + 2$; $\mathtt{w}(Bel_i(B|A)) = \mathtt{w}(B|A) + 1$.

Weights of labelled formulas are ordered lexicographically.

From the definition of weight it is clear that the weight gets decreased if we move from a formula labelled by a neighbourhood label to the same formula labelled by a world label, or if we move (regardless the label) to a formula with a pure part of strictly smaller weight. The following lemma is proved by induction on formula weights:

Lemma 3.3 *Sequents of the following form are derivable in* **G3CDL** *for arbitrary neighbourhoods labels a, b and formulas A and B:*
(i) $a \subseteq b, \Gamma \Rightarrow \Delta, a \subseteq b$ *(ii)* $a \Vdash^{\forall} A, \Gamma \Rightarrow \Delta, a \Vdash^{\forall} A$ *(iii)* $a \Vdash^{\exists} A, \Gamma \Rightarrow \Delta, a \Vdash^{\exists} A$
(iv) $x \Vdash_i B|A, \Gamma \Rightarrow \Delta, x \Vdash_i B|A$ *(v)* $x : A, \Gamma \Rightarrow \Delta, x : A$

The definition of substitution of labels given in [13] can be extended in an obvious way – that need not be pedantically detailed here – to all the formulas of our language and to neighbourhood labels. With this definition we have, for example, $(a \Vdash^{\exists} A)(b/a) \equiv b \Vdash^{\exists} A$, and $(x \Vdash_i B|A)(y/x) \equiv y \Vdash_i B|A$.

We denote by $\vdash_n \Gamma \Rightarrow \Delta$ a derivation whose endsequent is $\Gamma \Rightarrow \Delta$ and which has height n, where the height of a derivation is the number of nodes occurring in the longest derivation branch. The calculus is routinely shown to enjoy the property of height preserving (hp for short) substitution both of world and neighbourhood labels:

Proposition 3.4

(i) *If* $\vdash_n \Gamma \Rightarrow \Delta$, *then* $\vdash_n \Gamma(y/x) \Rightarrow \Delta(y/x)$;
(ii) *If* $\vdash_n \Gamma \Rightarrow \Delta$, *then* $\vdash_n \Gamma(b/a) \Rightarrow \Delta(b/a)$.

Hp-admissibility of weakening and contraction are then obtained by an easy induction on derivation height:

Proposition 3.5 *The rules of left and right weakening are hp-admissible in* **G3CDL**.

Theorem 3.6 *All the rules of* **G3CDL** *are* hp-invertible, *i.e. for every rule of the form* $\frac{\Gamma' \Rightarrow \Delta'}{\Gamma \Rightarrow \Delta}$, *if* $\vdash_n \Gamma \Rightarrow \Delta$ *then* $\vdash_n \Gamma' \Rightarrow \Delta'$, *and for every rule of the form* $\frac{\Gamma' \Rightarrow \Delta' \quad \Gamma'' \Rightarrow \Delta''}{\Gamma \Rightarrow \Delta}$ *if* $\vdash_n \Gamma \Rightarrow \Delta$ *then* $\vdash_n \Gamma' \Rightarrow \Delta'$ *and* $\vdash_n \Gamma'' \Rightarrow \Delta''$.

The rules of contraction of **G3CDL** have the following form, where $\mathcal{F}$ is either a "relational" atom of the form $a \in I(x)$ or $x \in a$ or a labelled formula of the form $x : A$, $a \Vdash^{\forall} A$, $a \Vdash^{\exists} A$ or a formula of the form $x \Vdash_i B|A$ or $x : Bel_i(B|A)$:

$$\frac{\mathcal{F}, \mathcal{F}, \Gamma \Rightarrow \Delta}{\mathcal{F}, \Gamma \Rightarrow \Delta}\, LCtr \qquad \frac{\Gamma \Rightarrow \Delta, \mathcal{F}, \mathcal{F}}{\Gamma \Rightarrow \Delta, \mathcal{F}}\, RCtr$$

Theorem 3.7 *The rules of left and right contraction are hp-admissible in* **G3CDL**.

Theorem 3.8 *Cut is admissible in* **G3CDL**.

Proof. By double induction, with primary induction on the weight of the cut formula and subinduction on the sum of the heights of derivations of the premisses of cut. The cases in which the premisses of cut are either initial sequents or obtained through the rules for $\wedge$, $\vee$, or $\supset$ follow the treatment of Theorem 11.9 of [16]. For the cases in which the cut formula is a side formula in at least one rule used to derive the premisses of cut, the cut reduction is dealt with in the usual way by permutation of cut, with possibly an application of hp-substitution to avoid a clash with the fresh variable in rules with variable condition. In all such cases the cut height is reduced.

For space limitations, we treat only the cases in wich the cut formula is principal in both premisses and has the form $x \Vdash_i B|A$ or $x : Bel_i(B|A)$.

(1) The cut formula is $x \Vdash_i B|A$, principal in both premisses of cut:

$$\dfrac{a \in I_i(x), \Gamma \Rightarrow \Delta, x \Vdash_i B|A, a \Vdash^{\exists} A \quad a \in I_i(x), \Gamma \Rightarrow \Delta, x \Vdash_i B|A, a \Vdash^{\forall} A \supset B}{a \in I_i(x), \Gamma \Rightarrow \Delta, x \Vdash_i B|A}\ RC$$

$$\dfrac{\overset{\mathcal{D}}{b \in I_i(x), b \Vdash^{\exists} A, b \Vdash^{\forall} A \supset B, \Gamma' \Rightarrow \Delta'}}{x \Vdash_i B|A, \Gamma' \Rightarrow \Delta'}\ LC$$

The conclusion of the cut is the sequent $a \in I_i(x), \Gamma, \Gamma' \Rightarrow \Delta, \Delta'$. The derivation is converted into the following:

$$\dfrac{\dfrac{\dfrac{a \in I_i(x), \Gamma \Rightarrow \Delta, x \Vdash_i B|A, a \Vdash^{\exists} A \quad x \Vdash_i B|A, \Gamma' \Rightarrow \Delta'}{a \in I_i(x), \Gamma, \Gamma' \Rightarrow \Delta, \Delta', a \Vdash^{\exists} A}\ Cut_1 \qquad (1)}{a \in I_i(x)^3, \Gamma^2, \Gamma'^3 \Rightarrow \Delta^2, \Delta'^3}\ Cut_4}{a \in I_i(x), \Gamma, \Gamma' \Rightarrow \Delta, \Delta'}\ Ctr^*$$

where (1) is the derivation:

$$\dfrac{a \in I_i(x), \Gamma, \Gamma' \Rightarrow \Delta, \Delta', a \Vdash^{\forall} A \supset B \quad \overset{\mathcal{D}(a/b)}{a \in I_i(x), a \Vdash^{\exists} A, a \Vdash^{\forall} A \supset B, \Gamma' \Rightarrow \Delta'}}{a \in I_i(x)^2, a \Vdash^{\exists} A, \Gamma, \Gamma'^2 \Rightarrow \Delta, \Delta'^2}\ Cut_3$$

where the left premiss is obtained by Cut_2 from the sequent $a \in I_i(x), \Gamma \Rightarrow \Delta, x \Vdash_i B|A, a \Vdash^{\forall} A \supset B$ and $x \Vdash_i B|A, \Gamma' \Rightarrow \Delta'$. Observe that all four cuts are of reduced height (Cut_1 and Cut_2) or reduced weight (Cut_3 and Cut_4) because $\mathtt{w}(a \Vdash^{\exists} A) < \mathtt{w}(a \Vdash^{\forall} A \supset B) < \mathtt{w}(x \Vdash_i B|A)$.

(2) The cut formula is $x : Bel_i(B|A)$, principal in both premisses of cut:

$$\dfrac{\overset{\mathcal{D}}{b \in I_i(x), b \Vdash^{\exists} A, \Gamma \Rightarrow \Delta, x \Vdash_i B|A}}{\Gamma \Rightarrow \Delta, x : Bel_i(B|A)}\ RB$$

$$\dfrac{a \in I_i(x), x : Bel_i(B|A), \Gamma' \Rightarrow \Delta', a \Vdash^{\exists} A \quad a \in I_i(x), x \Vdash_i B|A, x : Bel_i(B|A), \Gamma' \Rightarrow \Delta'}{a \in I_i(x), x : Bel_i(B|A), \Gamma' \Rightarrow \Delta'}\ LB$$

The conclusion is the sequent $a \in I_i(x), \Gamma, \Gamma' \Rightarrow \Delta, \Delta'$. The cut is converted to four smaller cuts as follows:

$$\dfrac{\dfrac{\dfrac{\Gamma \Rightarrow \Delta, x : Bel_i(B|A) \quad a \in I_i(x), x : Bel_i(B|A), \Gamma' \Rightarrow \Delta', a \Vdash^{\exists} A}{a \in I_i(x), \Gamma, \Gamma' \Rightarrow \Delta, \Delta', a \Vdash^{\exists} A}\ Cut_2 \qquad (2)}{a \in I_i(x)^3, \Gamma^3, \Gamma'^2 \Rightarrow \Delta^3, \Delta'^2}\ Cut_4}{a \in I_i(x), \Gamma, \Gamma' \Rightarrow \Delta, \Delta'}\ Ctr^*$$

where (2) is the derivation:

$$\dfrac{\overset{\mathcal{D}(a/b)}{a \in I_i(x), a \Vdash^{\exists} A, \Gamma \Rightarrow \Delta, x \Vdash_i B|A} \quad a \in I_i(x), x \Vdash_i B|A, \Gamma, \Gamma' \Rightarrow \Delta, \Delta'}{a \in I_i(x)^2, a \Vdash^{\exists} A, \Gamma^2, \Gamma' \Rightarrow \Delta^2, \Delta'}\ Cut_3$$

where the right premiss is derived by Cut_1 from $\Gamma \Rightarrow \Delta, x : Bel_i(B|A)$ and $a \in I_i(x), x \Vdash_i B|A, x : Bel_i(B|A), \Gamma' \Rightarrow \Delta'$. Cut_1 and Cut_2 have reduced height and the other cuts are performed on formulas of reduced weight, because $\mathtt{w}(a \Vdash^{\exists} A) < \mathtt{w}(x \Vdash_i B|A) < \mathtt{w}(x : Bel_i(B|A))$. □

4 Soundness, termination, and completeness

We first show soundness of the calculus. We need to interpret labelled sequents in neighbourhood models, and to this purpose we define the notion of realization.

Definition 4.1 Let $\mathcal{M} = \langle W, \{I\}_{i \in \mathcal{A}}, [\![\ ]\!] \rangle$ be a neighbourhood model, S a set of world labels, and N a set of neighbourhood labels. An SN-realization over $\mathcal{M}$ consists of a pair of functions (ρ, σ) such that

- $\rho : S \to W$ is a function which assigns to each $x \in S$ an element $\rho(x) = w \in W$;
- $\sigma : N \to \mathcal{P}(W)$, i.e. a function which assigns to each $a \in N$ an element $\sigma(a) \in I(w)$, for some $w \in W$.

Given a sequent $\Gamma \Rightarrow \Delta$, with S, N as above, and (ρ, σ) an SN-realization, we say that $\Gamma \Rightarrow \Delta$ is satisfied in $\mathcal{M}$ under the SN-realization (ρ, σ) if the following conditions hold:

- $\mathcal{M} \vDash_{\rho,\sigma} a \in I_i(x)$ if $\sigma(a) \in I_i(\rho(x))$ and $\mathcal{M} \vDash_{\rho,\sigma} a \subseteq b$ if $\sigma(a) \subseteq \sigma(b)$;
- $\mathcal{M} \vDash_{\rho,\sigma} x : A$ if $\rho(x) \Vdash A$;
- $\mathcal{M} \vDash_{\rho,\sigma} a \Vdash^{\exists} A$ if $\sigma(a) \Vdash^{\exists} A$ and $\mathcal{M} \vDash_{\rho,\sigma} a \Vdash^{\forall} A$ if $\sigma(a) \Vdash^{\forall} A$;
- $\mathcal{M} \vDash_{\rho,\sigma} x \Vdash_i B|A$ if for some c $\in I_i(\rho(x))$, $c \Vdash^{\exists} A$ and $c \Vdash^{\forall} A \supset B$;
- $\mathcal{M} \vDash_{\rho,\sigma} x \Vdash_i Bel_i(B|A)$ if for all $a \in I_i(\rho(x))$, $a \Vdash^{\forall} A$ or $\mathcal{M} \vDash_{\rho,\sigma} x \Vdash_i B|A$;
- $\mathcal{M} \vDash_{\rho,\sigma} \Gamma \Rightarrow \Delta$ if either $\mathcal{M} \nvDash_{\rho,\sigma} F$ for some formula $F \in \Gamma$ or $\mathcal{M} \vDash_{\rho,\sigma} G$ for some formula $G \in \Delta$.

Then, define $\mathcal{M} \vDash \Gamma \Rightarrow \Delta$ iff $\mathcal{M} \vDash_{\rho,\sigma} \Gamma \Rightarrow \Delta$ for every SN- realization (ρ, σ). A sequent $\Gamma \Rightarrow \Delta$ is said to be *valid* if $\mathcal{M} \vDash \Gamma \Rightarrow \Delta$ holds for every neighbourhood model $\mathcal{M}$, i.e. if $\Gamma \Rightarrow \Delta$ is satisfied for every model $\mathcal{M}$ and for every SN-realization (ρ, σ).

Theorem 4.2 (Soundness) *If a sequent $\Gamma \Rightarrow \Delta$ is derivable in the calculus, then it is valid in the class of multi-agent neighbourhood models.*

We now show that, by adopting a suitable proof search strategy, the calculus yields a decision procedure for *CDL*. We also prove the completeness of the calculus under the same strategy. The adoption of the strategy is not strictly necessary for completeness; however, it ensures that we can extract a finite countermodel from an open or failed derivation branch. Although the termination proof has some similarity with the one in [14], for **G3CDL** it is more difficult due to the specific semantic rules, in particular local absoluteness.

As often happens with labelled calculi, the calculus **G3CDL** in itself is non-terminating in the sense that a root-first (i.e. upwards) construction of a derivation may generate infinite branches. Here below is an example (we omit writing the derivable left premisses of *LB*):

$$\vdots$$

$$\dfrac{\dfrac{\dfrac{\dfrac{\dfrac{c \in I_i(x), c \Vdash^{\exists} A, c \Vdash^{\forall} A \supset B \ldots x : Bel_i(B|A) \Rightarrow x \Vdash_i C|A}{x \Vdash_i B|A, b \in I_i(x), b \Vdash^{\exists} A, b \Vdash^{\forall} A \supset B, a \in I_i(x), a \Vdash^{\exists} A, x : Bel_i(B|A) \Rightarrow x \Vdash_i C|A}\ LC}{b \in I_i(x), b \Vdash^{\exists} A, b \Vdash^{\forall} A \supset B, a \in I_i(x), a \Vdash^{\exists} A, x : Bel_i(B|A) \Rightarrow x \Vdash_i C|A}\ LB}{x \Vdash_i B|A, a \in I_i(x), a \Vdash^{\exists} A, x : Bel_i(B|A) \Rightarrow x \Vdash_i C|A}\ LC}{a \in I_i(x), a \Vdash^{\exists} A, x : Bel_i(B|A) \Rightarrow x \Vdash_i C|A}\ LB}{x : Bel_i(B|A) \Rightarrow x : Bel_i(C|A)}\ RB$$

The loop is generated by the application of rules (LB) and (LC). Our aim is to specify a strategy which ensures termination by preventing any kind of loop. The main point is to avoid redundant (backwards) applications of rules. To define precisely this notion we associate to each rule a saturation condition.

Definition 4.3 Given a derivation branch $\mathcal{B}$ of the form $\Gamma_0 \Rightarrow \Delta_0, ..., \Gamma_k \Rightarrow \Delta_k, \Gamma_{k+1} \Rightarrow \Delta_{k+1}, ...$ where $\Gamma_0 \Rightarrow \Delta_0$ is the sequent $\Rightarrow x_0 : A_0$, let $\downarrow \Gamma_k$ (respectively $\downarrow \Delta_k$) denote the union of the antecedents (respectively the succedents) occurring in the branch from the root $\Gamma_0 \Rightarrow \Delta_0$ up to $\Gamma_k \Rightarrow \Delta_k$.

For each rule (R), we say that a sequent $\Gamma \Rightarrow \Delta$ *satisfies the saturation condition associated to* (R) if the following hold: for rule $(L\wedge)$, if $x : A \wedge B \in \Gamma$, then $x : A \in \downarrow \Gamma$ and $x : B \in \downarrow \Gamma$. The other propositional conditions are similar, and can be found in [14]. Conditions for the other rules are the following: (Rf) If a is in Γ, Δ then $a \subseteq a$ is in Γ; (Tr) If $a \subseteq b$ and $b \subseteq c$ are in Γ, then $a \subseteq c$ is in Γ; $(L \subseteq)$ If $x \in a$ and $a \subseteq b$ are in Γ, then $x \in b$ is in Γ; $(R \Vdash^{\forall})$ If $a \Vdash^{\forall} A$ is in $\downarrow \Delta$, then for some x there is $x \in a$ in Γ and $x : A$ in $\downarrow \Delta$; $(L \Vdash^{\forall})$ If $x \in a$ and $a \Vdash^{\forall} A$ are in Γ, then $x : A$ is in Γ; $(R \Vdash^{\exists})$ If $x \in a$ is in Γ and $a \Vdash^{\exists} A$ is in Δ, then $x : A$ is in $\downarrow \Delta$; $(L \Vdash^{\exists})$ If $a \Vdash^{\exists} A$ is in $\downarrow \Gamma$, then for some x there is $x \in a$ in Γ and $x : A$ is in $\downarrow \Gamma$; (RB) If $x : Bel_i(B|A)$ is in $\downarrow \Delta$, then for some $i \in \mathcal{A}$ and for some a, $a \in I_i(x)$ is in Γ, $a \Vdash^{\exists} A$ is in $\downarrow \Gamma$ and $x \Vdash_i B|A$ is in $\downarrow \Delta$; (LB) If $a \in I_i(x)$ and $x : Bel_i(B|A)$ are in Γ, then either $a \Vdash^{\exists} A$ is in $\downarrow \Delta$ or $x \Vdash_i B|A$ is in $\downarrow \Gamma$; (RC) If $a \in I_i(x)$ is in Γ and $x \Vdash_i B|A$ is in Δ, then either $a \Vdash^{\exists} A$ or $a \Vdash^{\forall} A \supset B$ are in $\downarrow \Delta$; (LC) If $x \Vdash_i B|A$ is in $\downarrow \Gamma$, then for some $i \in \mathcal{A}$ and for some a, $a \in I_i(x)$ is in Γ, $a \Vdash^{\exists} A$ and $a \Vdash^{\forall} A \supset B$ are in $\downarrow \Gamma$; (T) For all x occurring in $\downarrow \Gamma \cup \downarrow \Delta$, for all $i \in \mathcal{A}$ there is an a such that $a \in I_i(x)$ and $x \in a$ are in Γ; (S) If $a \in I_i(x)$ and $b \in I_i(x)$ are in Γ, then $a \subseteq b$ or $b \subseteq a$ are in Γ; $(A1)$ If $a \in I_i(x)$ and $y \in a$ are in Γ, then if $b \in I_i(x)$ is in Γ also $b \in I_i(y)$ is in Γ; If $b \in I_i(y)$ is in Γ also $b \in I_i(x)$ is in Γ (for $(A2)$ is similar).

Furthermore, a sequent $\Gamma \Rightarrow \Delta$ is *saturated* if

(Init) There is no $x : P$ in $\Gamma \cap \Delta$;

$(L\bot)$ There is no $x : \bot$ in Γ;

$\Gamma \Rightarrow \Delta$ satisfies *all* saturation conditions listed above.

To analyse the interdependencies between labels in a sequent we introduce the following:

Definition 4.4 Given a branch $\mathcal{B}$ as in Definition 4.3, a neighbourhood label a and world labels x, y, all occurring in $\downarrow \Gamma_k$, we define:

- $k(x) = min\{t \mid x \text{ occurs in } \Gamma_t\}$; we similarly define $k(a)$.
- $x \rightarrow_g a$ (read "x generates a") if for some $t \leq k$ and $i \in \mathcal{A}$, $k(a) = t$ and $a \in I_i(x)$ occurs in Γ_t.
- $a \rightarrow_g x$ (read "a generates x") if for some $t \leq k$ and $i \in \mathcal{A}$, $k(x) = t$ and $x \in a$ occurs in Γ_t.
- $x \overset{w}{\rightarrow} y$ (read "x generates y") if for some a it holds that $x \rightarrow_g a$ and $a \rightarrow_g y$.

Lemma 4.5 *Given a branch $\mathcal{B}$ as in Definition 4.3, we have that (a) the relation $\overset{w}{\rightarrow}$ is acyclic and forms a tree with root x_0 and (b) all world labels occurring in $\mathcal{B}$ are nodes of the tree, that is letting $\overset{w}{\rightarrow}^*$ be the transitive closure of $\overset{w}{\rightarrow}$, if u occurs in $\downarrow \Gamma_k$, then $x_0 \overset{w}{\rightarrow}^* u$.*

Proof. (a) immediately follows from the definition of relation $\rightarrow_g$ and from the sequent calculus rules, (b) easily proven by induction on $k(u) \leqslant k$. □

We can now detail the proof-search strategy. A rule (R) is said to be *applicable* to a world label x if R is applicable to a labelled formula with label x occurring in the denominator of a rule. In case of rules $(A1), (A2)$ of local absoluteness, we say the rule is applied to x (rather than to y).

Definition 4.6 When constructing root-first a derivation tree for a sequent $\Rightarrow x_0 : A$, apply the following strategy:

(i) No rule can be applied to an initial sequent;
(ii) If $k(x) < k(y)$ all rules applicable to x are applied before any rule applicable to y.
(iii) Rule (T) is applied as the first one to each world label x.
(iv) Rules which do not introduce a new label (static rules) are applied *before* the rules which do introduce new labels (dynamic rules), with the exception of (T), as in (iii);
(v) Rule (RB) is applied *before* rule (LC);
(vi) A rule (R) cannot be applied to a sequent $\Gamma_i \Rightarrow \Delta_i$ if $\downarrow \Gamma_i$ and / or $\downarrow \Delta_i$ satisfy the saturation condition associated to (R).

It follows from the strategy that if $x \overset{w}{\rightarrow} y$, every rule applicable to x is applied before any every rule applicable to y. In the previous example, the loop would have been stopped at the second application (root-first) of (LB), because the application of (LB) would violate condition (vi): the branch already satisfies the saturation condition for (LB), because $x \Vdash_i B|A$ is already in $\downarrow \Gamma$.
As an easy consequence of conditions (ii) and (iv) of the strategy, we have:

Lemma 4.7 *Let us consider a branch $\mathcal{B}$ as in Definition 4.3 and two labels x, y such that $x \overset{w}{\rightarrow}^* y$. Then for all b, if $b \in I_i(x) \in \Gamma_k$ then also $b \in I_i(y) \in \Gamma_k$.*

As usual, the size of a formula A, denoted by $|A|$, is the number of symbols occurring in A. The size of a sequent $\Gamma \Rightarrow \Delta$ is the sum of all the sizes of the formulas occurring in it. The following Lemma and Proposition are needed to prove termination.

Lemma 4.8 *Given a branch $\mathcal{B}$ as in Definition 4.3 and a world label x, we define $N(x) = \{a \mid x \rightarrow_g a\}$ as the set of neighbourhood labels generated by x, and $W(x) = \{y \mid x \stackrel{w}{\rightarrow} y\}$ as the set of world labels generated by x. The size of $N(x)$ and $W(x)$ is finite, more precisely: $Card(N(x)) = O(|A_0|)$ and $Card(W(x)) = O(|A_0|^2)$.*

Proposition 4.9 *Any derivation branch $\mathcal{B} = \Gamma_0 \Rightarrow \Delta_0, ..., \Gamma_k \Rightarrow \Delta_k, \Gamma_{k+1} \Rightarrow \Delta_{k+1}, ...$ of a derivation which starts from $\Gamma_0 \Rightarrow \Delta_0$, where $\Gamma_0 \Rightarrow \Delta_0$ has the form $\Rightarrow x_0 : A_0$, and which is built in accordance with the Strategy, is finite.*

Proof. Let us consider a branch $\mathcal{B}$. Suppose by contradiction that $\mathcal{B}$ is not finite, let $\Gamma^* = \bigcup_k \Gamma_k$ and $\Delta^* = \bigcup_k \Delta_k$. Then Γ^* is infinite. All formulas occurring with a label in Γ^* are subformulas of A_0, but the subformulas of A_0 are finitely many (namely they are $O(|A_0|)$. Thus Γ^* must contain infinitely many labels. In the light of Lemma 4.8, we have that Γ^* must contain infinitely many *world* labels, since each world label x generates only $O(|A_0|)$ neighbourhood labels. Let us consider now the tree determined by the relation $\stackrel{w}{\rightarrow}^*$ with root x_0. By Lemma 4.5, each label in any Γ_k occurs in that tree, therefore the tree determined by $\stackrel{w}{\rightarrow}^*$ is infinite. By previous lemma, every label in the tree has $O(|A_0|^2)$ successors, thus a finite number. By König's lemma, the tree must contain an *infinite path*: $x_0 \stackrel{w}{\rightarrow} x_1 \stackrel{w}{\rightarrow} \ldots \stackrel{w}{\rightarrow} x_t \stackrel{w}{\rightarrow} x_{t+1} \ldots$, with all x_t being different. We observe that (a) infinitely many x_t must be generated by dynamic rules using some subformulas of A_0, but (b) these formulas are finitely many, thus there must be a subformula of A_0 which is used infinitely many times to "generate" world labels (or better to generate a neighbourhood label from which a further world label is generated). There are two cases: this subformula is of type $Bel_i(D|C)$ occurring in Δ^* or it is of type $\Vdash_i B|A$ occurring in Γ^* (in this latter case it is not properly a subformulas of A_0 but it comes from one of them). In the first case it must occur that for some x_t we have that $x_t : Bel_i(D|C)$ occurs in some $\Delta_{s(x_t)}$ and for some a, such that $k(a) = s(x_t)+1$, we have that $a \in I_i(x_t), a \Vdash^\exists C \in \Gamma_{s(x_t)+1}$ and $x_t \Vdash_i D|C \in \Delta_{s(x_t)+1}$. Moreover, we have $a \rightarrow_g x_{t+1}$. But at the same time there must be in the sequence an x_r with $r > t$, such that $x_r : Bel_i(D|C)$ occurs in some $\Delta_{s(x_r)}$ and for a new b, that is with $k(b) = s(x_r) + 1$, we have that $(*)$ $b \in I_i(x_r), b \Vdash^\exists C \in \Gamma_{s(x_r)+1}$ and $x_r \Vdash_i D|C \in \Delta_{s(x_r)+1}$ and $b \rightarrow_g x_{t+1}$. By Lemma 4.7, we have that $a \in I_i(x_r)$; thus a itself fulfils the saturation condition for (RB) applied to $x_r : Bel_i(D|C) \in \Delta_{s(x_r)}$, and step $(*)$ violates the strategy. We have thus reached a contradiction.

In the second case the situation is similar: for some t, $x_t \Vdash_i D|C$ occurs in some $\Gamma_{s(x_t)}$ and for a new a, with $k(a) = s(x_t)+1$, we have that $a \in I_i(x_t), a \Vdash^\exists C \in \Gamma_{s(x_t)+1}$ and $a \Vdash^\forall C \supset D \in \Gamma_{s(x_t)+1}$. Moreover, we have that $a \rightarrow_g x_{t+1}$. Similarly there must be an x_r in the sequence with $r > t$, such that $x_r \Vdash_i D|C$ occurs in some $\Gamma_{s(x_r)}$ and for a new b, with $k(b) = s(x_r) + 1$, we have that we have that (**) $b \in I_i(x_r), b \Vdash^{\exists} C \in \Gamma_{s(x_r)+1}$ and $b \Vdash^{\forall} C \supset D \in \Gamma_{s(x_r)+1}$. By Lemma 4.7, we have that $a \in I_i(x_r)$; thus a itself fulfils the saturation condition for (LC) applied to $x_r \Vdash_i D|C \in \Gamma_{s(x_r)}$, so that step (**) violates the strategy. Again, we have reached a contradiction. □

Termination of proof search under the strategy is now an obvious consequence:

Theorem 4.10 *Proof search for any sequent of the form $\Rightarrow x_0 : A_0$ always comes to an end after a finite number of steps. Furthermore, each sequent that occurs as a leaf of the derivation tree is either an initial sequent or a saturated sequent.*

The above theorem provides a decision procedure for *CDL*. Even without a precise analysis of its complexity, it is easy to see that each proof branch may have an exponential size with respect to the size of the formula A_0 at the root of the derivation. The exact complexity of logic *CDL* has not been determined. In [9] it is shown that the single-agent version of *CDL* is CoNP. However, since $S5_n$, the multi-agent version of $S5$, is embeddable in *CDL* via the definition of the knowledge operator K_i, by the results in [10] we get that PSPACE is a lower bound for the complexity of *CDL*. We strongly conjecture that this is also its upper bound; this will be the object of future research, together with a strategy to obtain from **G3CDL** an optimal decision procedure for *CDL*.
The calculus is complete under the terminating strategy.

Theorem 4.11 *Let $\Gamma \Rightarrow \Delta$ be the upper sequent of a saturated branch $\mathcal{B}$ in a derivation tree. Then there exists a finite countermodel $\mathcal{M}$ to $\Gamma \Rightarrow \Delta$ that satisfies all formulas in $\downarrow \Gamma$ and falsifies all formulas in $\downarrow \Delta$.*

Proof. Let $\Gamma \Rightarrow \Delta$ be the upper sequent of a saturated branch $\mathcal{B}$. By theorem 4.10 , $\mathcal{B}$ is finite. We construct a model $\mathcal{M}_{\mathcal{B}}$ and an $SN_{\mathcal{B}}$-realization (ρ, σ), and show that it satisfies all formulas in $\downarrow \Gamma$ and falsifies all formulas in $\downarrow \Delta$.
Let $S_{\mathcal{B}} = \{x \mid x \in (\downarrow \Gamma \cup \downarrow \Delta)\}$ and $N_{\mathcal{B}} = \{a \mid a \in (\downarrow \Gamma \cup \downarrow \Delta)\}$. Then, associate to each $a \in N_{\mathcal{B}}$ a neighbourhood α_a, such that $\alpha_a = \{y \in S_{\mathcal{B}} | y \in a$ belongs to $\Gamma\}$, thus $\alpha_a \subseteq S_{\mathcal{B}}$. We define a neighbourhood model $\mathcal{M}_{\mathcal{B}} = \langle W, I_i, [\![\ ]\!] \rangle$ as

- $W = S_{\mathcal{B}}$, i.e. the set W consists of all the labels occurring in the saturated branch $\mathcal{B}$;
- For each $x \in W$, $I_i(x) = \{\alpha_a | a \in I_i(x)$ belongs to $\downarrow \Gamma\}$;
- For P atomic, $[\![P]\!] = \{x \in W | x : P$ belongs to $\downarrow \Gamma\}$.

Employing the saturation conditions we can easily prove that if $a \subseteq b$ belongs to Γ, then $\alpha_a \subseteq \beta_b$ and that $\mathcal{M}_{\mathcal{B}}$ satisfies all properties of a multi-agent neighbourhood model, namely non-emptiness, total reflexivity, nesting, and local absoluteness (strong closure under intersection follows from finiteness). We define a realization (ρ, σ) such that $\rho(x) = x$ and $\sigma(a) = \alpha_a$. We then prove that

[Claim 1] if $\mathcal{F}$ is in $\downarrow \Gamma$, then $\mathcal{M}_{\mathcal{B}} \models \mathcal{F}$
[Claim 2] if $\mathcal{F}$ is in $\downarrow \Delta$, then $\mathcal{M}_{\mathcal{B}} \not\models \mathcal{F}$

where $\mathcal{F}$ denotes any formula of the language, i.e. $\mathcal{F} = a \in I_i(x), x \in A, a \subseteq b, x \Vdash^{\forall} A, x \Vdash^{\exists} A, x \Vdash_i B|A, x : A$. The two claim are routinely proved by induction on the weight of the formula $\mathcal{F}$ using the fact that $\Gamma \Rightarrow \Delta$ is saturated and employing, whenever needed, the induction hypothesis.

□

The completeness of the calculus is an obvious consequence:

Theorem 4.12 *If A is valid then it is provable in* **G3CDL**.

Theorem 4.11 together with soundness of **G3CDL** provide a constructive proof of *the finite model property* of the logic *CDL*: if A is satisfiable in a model, then by the soundness of **G3CDL** we have that $\neg A$ is not provable. Thus by Theorem 4.11 we can build a finite countermodel that falsifies $\neg A$, i.e. that satisfies A.

5 Conclusions, related works, and further research

We have proposed an alternative semantics, based on neighbourhood models, for the logic *CDL* of conditional beliefs. On the basis of this semantics, which is is a multi-agent version of Lewis' spheres models, we have developed the labelled sequent calculus **G3CDL**, following the methodology of [13], [12], [14]. The calculus **G3CDL** is analytical and enjoys cut elimination, admissibility of the other structural rules, and invertibility of all the rules. Moreover, on the basis of this calculus, we obtain a decision procedure for the logic *CDL* under a natural proof search strategy. The completeness of the calculus is established by means of a finite procedure which constructs a countermodel from a failed (or open) derivation branch. The finite countermodel construction provides in itself a constructive proof of the finite model property of the logic.

Although no proof-system for *CDL* was known before the calculus **G3CDL**, a few labelled calculi for conditional logics have been studied in the literature, in [15], [7], [17]. Observe however that all these calculi are based either on the relational semantics or on the selection function semantics; thus there is no direct relation between the calculus presented in this paper and these works.

A number of issues are open to further investigation. On the semantical side, other doxastic operators have been considered in the literature, such as *safe* belief and *strong* belief [3]. We conjecture that also these operators can be naturally interpreted in neighbourhood models and consequently captured by extensions of the calculus **G3CDL**. Furthermore, *CDL* is the "static" logic that underlies dynamic extensions by *doxastic actions* [3]. It should be worth studying if our calculus can be extended to deal with the dynamic systems as well.

Finally, from a computational side, to the best of our knowledge the exact complexity of *CDL* is not known. We conjecture its upper bound to be PSPACE; however, further investigations are needed to confirm this result.

References

[1] Alexandroff, P., *Diskrete Räume*, Mat.Sb. (NS) **2** (1937), pp. 501–519.

[2] Baltag, A. and S. Smets, *Conditional doxastic models: A qualitative approach to dynamic belief revision*, Electronic Notes in Theoretical Computer Science **165** (2006), pp. 5–21.

[3] Baltag, A. and S. Smets, *A qualitative theory of dynamic interactive belief revision*, Logic and the foundations of game and decision theory (LOFT 7) **3** (2008), pp. 9–58.

[4] Baltag, A., S. Smets et al., *The logic of conditional doxastic actions*, Texts in Logic and Games, Special Issue on New Perspectives on Games and Interaction **4** (2008), pp. 9–31.

[5] Board, O., *Dynamic interactive epistemology*, Games and Economic Behavior **49** (2004), pp. 49–80.

[6] Demey, L., *Some remarks on the model theory of epistemic plausibility models*, Journal of Applied Non-Classical Logics **21** (2011), pp. 375–395.

[7] Giordano, L., V. Gliozzi, N. Olivetti and C. Schwind, *Tableau calculus for preference-based conditional logics: Pcl and its extensions*, ACM Transactions on Computational Logic (TOCL) **10** (2009), p. 21.

[8] Grove, A., *Two modellings for theory change*, Journal of philosophical logic **17** (1988), pp. 157–170.

[9] Halpern, J. Y. and N. Friedman, *On the complexity of conditional logics*, in: *Principles of Knowledge Representation and Reasoning: Proceedings of the Fourth International Conference (KR'94)*, Morgan Kaufmann Pub, 1994, p. 202.

[10] Halpern, J. Y. and Y. Moses, *A guide to completeness and complexity for modal logics of knowledge and belief*, Artificial intelligence **54** (1992), pp. 319–379.

[11] Marti, J. and R. Pinosio, *Topological semantics for conditionals*, The Logica Yearbook (2013).

[12] Negri, S., *Proof theory for non-normal modal logics: The neighbourhood formalism and basic results*, IFCoLog Journal of Logic and its Applications, to appear, `http://www.helsinki.fi/~negri/negri_ifcolog.pdf`.

[13] Negri, S., *Proof analysis in modal logic*, Journal of Philosophical Logic **34** (2005), pp. 507–544.

[14] Negri, S. and N. Olivetti, *A sequent calculus for preferential conditional logic based on neighbourhood semantics*, in: *Automated Reasoning with Analytic Tableaux and Related Methods*, Springer, 2015 pp. 115–134.

[15] Negri, S. and G. Sbardolini, *Proof analysis for Lewis counterfactuals*, The Review of Symbolic Logic (2014), pp. 1–32.

[16] Negri, S. and J. von Plato, "Proof analysis: a contribution to Hilbert's last problem," Cambridge University Press, 2011.

[17] Olivetti, N., G. L. Pozzato and C. B. Schwind, *A sequent calculus and a theorem prover for standard conditional logics*, ACM Transactions on Computational Logic (TOCL) **8** (2007), p. 22.

[18] Pacuit, E., *Neighborhood semantics for modal logic*, Notes of a course on neighborhood structures for modal logic: http://ai.stanford.edu/ epacuit/classes/esslli/nbhdesslli.pdf (2007).

[19] Pacuit, E., *Dynamic epistemic logic I: Modeling knowledge and belief*, Philosophy Compass **8** (2013), pp. 798–814.

[20] van Ditmarsch, H., W. van Der Hoek and B. Kooi, *Dynamic epistemic logic*, Springer Science & Business Media (2008).

The Tangled Derivative Logic of the Real Line and Zero-Dimensional Spaces

Robert Goldblatt

School of Mathematics and Statistics, Victoria University of Wellington
sms.vuw.ac.nz/~rob

Ian Hodkinson

Department of Computing, Imperial College London
www.doc.ic.ac.uk/~imh

Abstract

In a topological setting in which the diamond modality is interpreted as the derivative (set of limit points) operator, we study a 'tangled derivative' connective that assigns to any finite set of propositions the largest set in which all those propositions are strictly dense. Building on earlier work of ourselves and others we axiomatise the resulting logic of the real line. We then show that the logic of any zero-dimensional dense-in-itself metric space is the 'tangled' extension of KD4, eliminating an assumption of separability in previous results for zero-dimensional spaces. This requires new kinds of 'dissection lemma' in the sense of McKinsey-Tarski. We extend the analysis to include the universal modality, and also show that the tangled extension of KD4 has a strong completeness result for topological models that fails for its Kripke semantics.

Keywords: derivative operator, dense-in-itself metric space, modal logic, finite model property, zero-dimensional, strong completeness

1 Introduction

The *tangle* connective applies to a finite set Γ of modal formulas to give a new formula $\langle t\rangle\Gamma$ with the following semantics in a model on Kripke frame (W, R):

> $\langle t\rangle\Gamma$ is true at w iff there is an endless R-path $wRw_1 \cdots w_nRw_{n+1} \cdots\cdots$ in W with each member of Γ being true at w_n for infinitely many n.

This pertains to arbitrary models, but in a finite transitive frame the truth condition equivalently means that w can access a cluster (maximal R-clique) in which each member of Γ is true at some point. Denoting by $[\![\varphi]\!]$ the set of points at which a formula φ is true, $[\![\langle t\rangle\Gamma]\!]$ can be shown to be equal to the union

$$\bigcup\{S \subseteq W : S \subseteq \bigcap\nolimits_{\gamma\in\Gamma} R^{-1}([\![\gamma]\!] \cap S)\}. \tag{1.1}$$

This connective was introduced by Dawar and Otto [3], who showed that over the class of finite transitive frames, the bisimulation-invariant fragment of monadic second-order logic collapses to that of first-order logic, with both fragments being expressively equivalent to the language $\mathcal{L}_{\Box}^{\langle t\rangle}$ that adds $\langle t\rangle$ to the language $\mathcal{L}_{\Box}$ of the basic modal logic of a unary modality $\Box$.

Now $\mathcal{L}_{\Box}^{\langle t\rangle}$ is translatable into the language $\mathcal{L}_{\Box}^{\mu}$ of the modal mu-calculus, since $\langle t\rangle$ has the same meaning as the $\mathcal{L}_{\Box}^{\mu}$-formula

$$\nu p \bigwedge_{\gamma\in\Gamma} \Diamond(\gamma \wedge p),$$

where ν is the greatest fixed point operator, $\Diamond$ is the dual modality to $\Box$, and p is a fresh propositional atom not occurring in Γ. But the mu-calculus is expressively equivalent to the bisimulation-invariant fragment of monadic second-order logic [7], so the upshot is that $\mathcal{L}_{\Box}^{\langle t\rangle}$ is expressively equivalent to the seemingly more powerful $\mathcal{L}_{\Box}^{\mu}$ over finite transitive frames.

The name 'tangle' was coined by Fernández-Duque [4,5], who developed the following topological interpretation of $\langle t\rangle$. A collection $\mathcal{G}$ of subsets of a topological space X is said to be *tangled in a subset* S of X if, for all $G \in \mathcal{G}$, $G \cap S$ is dense in S. In other words, each point of S is in the closure $\mathrm{cl}_X(G\cap S)$ of $G\cap S$. There is a largest subset of X in which $\mathcal{G}$ is tangled, and this is called the *tangled closure* of $\mathcal{G}$. In a model on X, $[\![\langle t\rangle\Gamma]\!]$ is defined to be the tangled closure of $\{[\![\gamma]\!] : \gamma \in \Gamma\}$, which can be described as the set

$$\bigcup\{S \subseteq X : S \subseteq \bigcap\nolimits_{\gamma\in\Gamma} \mathrm{cl}_X([\![\gamma]\!] \cap S)\}. \tag{1.2}$$

Interpreting $\Diamond\varphi$ as the closure $\mathrm{cl}_X[\![\varphi]\!]$ of $[\![\varphi]\!]$, and $\Box\varphi$ as the interior $\mathrm{int}_X[\![\varphi]\!]$, Fernańdez-Duque axiomatised the resulting $\mathcal{L}_{\Box}^{\langle t\rangle}$-logic as an extension of S4, and showed it has the finite model property. [1] In this S4 setting (1.1) is an instance of (1.2), because an S4-frame, having R reflexive and transitive, is a topological space under the Alexandroff topology generated by the sets $R(w) = \{v : wRv\}$ for all $w \in W$, and in this topology the closure $\mathrm{cl}(S)$ of S is just $R^{-1}(S)$.

The present paper is a continuation of our work in [6], where the equivalence of $\mathcal{L}_{\Box}^{\langle t\rangle}$ and $\mathcal{L}_{\Box}^{\mu}$ over finite transitive frames was lifted to the class of all topological spaces, and the finite model property over frames was established for a range of logics having the tangle connective, including some having the universal modality $\forall$ as well. We also studied the more expressive interpretation of a modal diamond as the *derivative operator* $\langle d_X\rangle$ of a space X. For that semantics the diamond and its dual box are written as $\langle d\rangle$ and $[d]$, with $[\![\langle d\rangle\varphi]\!]$ being the set $\langle d_X\rangle[\![\varphi]\!]$ of *limit points* of $[\![\varphi]\!]$. Then $\Diamond\varphi$ is spatially equivalent to $\varphi \vee \langle d\rangle\varphi$, and $\Box\varphi$ to $\varphi \wedge [d]\varphi$. Here we write $[d]^*\varphi$ for the formula $\varphi \wedge [d]\varphi$.

For this derivative interpretation we write the tangle connective as $\langle dt\rangle$. It has exactly the same 'endless R-path' meaning as $\langle t\rangle$ in frames, where $[\![\langle dt\rangle\Gamma]\!]$

[1] The notation $\langle t\rangle$ is ours. In [5], $\langle t\rangle\Gamma$ is written $\Diamond^*\Gamma$, or just $\Diamond\Gamma$, justified because in finite S4 models the $\mathcal{L}_{\Box}^{\langle t\rangle}$-formula $\Diamond^*\{\varphi\}$ has the same meaning as the $\mathcal{L}_{\Box}$-formula $\Diamond\varphi$.

continues to be the set (1.1). But what changes are the frames themselves, which no longer require all points to be reflexive. So we interpret $\langle dt \rangle$ over K4-frames rather than S4-frames. For the spatial interpretation of $\langle dt \rangle$ we replace the closure operator cl_X by $\langle d_X \rangle$, and define $[\![\langle dt \rangle \Gamma]\!]$ to be the set

$$\bigcup \{S \subseteq X : S \subseteq \bigcap\nolimits_{\gamma \in \Gamma} \langle d_X \rangle ([\![\gamma]\!] \cap S)\}. \tag{1.3}$$

The inclusion $S \subseteq \langle d_X \rangle ([\![\gamma]\!] \cap S)$ says that every point of S is a limit point of $[\![\gamma]\!] \cap S$. Since in general $\langle d_X \rangle Y \subseteq \mathrm{cl}_X Y$, and indeed $\mathrm{cl}_X Y = Y \cup \langle d_X \rangle Y$, this is a stricter form of density of $[\![\gamma]\!] \cap S$ in S. So according to (1.3), $[\![\langle dt \rangle \Gamma]\!]$ is the union of all sets S in which $\{[\![\gamma]\!] : \gamma \in \Gamma\}$ is *strictly tangled*, and may be called the *tangled derivative* of $\{[\![\gamma]\!] : \gamma \in \Gamma\}$.

In spaces in which the derivative $\langle d_X \rangle \{x\}$ of any point x is closed (so-called T_D spaces), $\langle t \rangle$ is definable from $\langle dt \rangle$, since $\langle t \rangle \Gamma$ is equivalent to the formula $\bigwedge \Gamma \vee \langle d \rangle \bigwedge \Gamma \vee \langle dt \rangle \Gamma$ (see [6, Lemma 6.5]).

Shehtman's seminal paper [13] axiomatised the logic of some classical spaces in the language $\mathcal{L}_{[d]}$ with the derivative interpretation. It proved that the $\mathcal{L}_{[d]}$-logic of any separable zero-dimensional dense-in-itself metric space is KD4, and the logic of the Euclidean space $\mathbb{R}^n$ is $\mathrm{KD4G_1}$ for all $n \geq 2$. It also conjectured that the logic of the real line $\mathbb{R}$ is $\mathrm{KD4G_2}$, which was later verified by Shehtman [12] and Lucero-Bryan [8].[2]

Our purpose in this paper is to lift these results to the language $\mathcal{L}_{[d]}^{\langle dt \rangle}$ with the tangled derivative connective. We use the name Lt for the $\mathcal{L}_{[d]}^{\langle dt \rangle}$-logic defined by adding to the axiomatisation of some $\mathcal{L}_{[d]}$-logic named L the 'fixed point' axioms

Fix: $\langle dt \rangle \Gamma \to \langle d \rangle (\gamma \wedge \langle dt \rangle \Gamma)$, all $\gamma \in \Gamma$

Ind: $[d]^*(\varphi \to \bigwedge_{\gamma \in \Gamma} \langle d \rangle (\gamma \wedge \varphi)) \to (\varphi \to \langle dt \rangle \Gamma)$.

We already dealt with $\mathbb{R}^n$ for $n \geq 2$ in [6, Theorem 9.3] which showed that if X is any dense-in-itself metric space, then the $\mathcal{L}_{[d]}^{\langle dt \rangle}$-logic of X is included in $\mathrm{KD4G_1}t$, and is exactly $\mathrm{KD4G_1}t$ if X validates $\mathrm{G_1}$.[3] In particular this holds when $X = \mathbb{R}^n$.

Here we will prove that the $\mathcal{L}_{[d]}^{\langle dt \rangle}$-logic of $\mathbb{R}$ is $\mathrm{KD4G_2}t$. The proof uses a result of [8] about the existence of *d-morphisms* from $\mathbb{R}$ onto finite $\mathrm{KD4G_2}$-frames. We show that these morphisms preserve validity of $\mathcal{L}_{[d]}^{\langle dt \rangle}$-formulas.

We then turn to zero-dimensional spaces and generalize the result of [13] by eliminating the restriction to separable spaces, and showing that the $\mathcal{L}_{[d]}^{\langle dt \rangle}$-logic of each zero-dimensional dense-in-itself metric space is KD4t. This requires certain 'dissection' lemmas about the partitioning of an open set into subsets with properties that allow them to be used to represent the structure of finite

[2] Definition of the logics KD4, $\mathrm{KD4G_1}$ and $\mathrm{KD4G_2}$ are given in Section 5 below.

[3] This result also holds when restricted to $\mathcal{L}_{[d]}$ and $\mathrm{KD4G_1}$, answering another question from [13].

frames. A variant of the dissection lemma of McKinsey and Tarski [9] allows any dense-in-itself metric space to represent finite rooted S4-frames. But KD-frames may have irreflexive points, and we need a further dissection result to handle them (Theorem 7.3), as well as one for zero-dimensional spaces about dissection into special open sets (Theorem 7.5).

We extend our results to the language with the universal modality, showing that the $\mathcal{L}^{\langle dt \rangle}_{[d]\forall}$-logic of any zero-dimensional dense-in-itself metric space is KD4t.U, and of $\mathbb{R}$ is KD4G$_2t$.UC, where C is Shehtman's axiom expressing topological connectedness.

Finally, we give a topological strong completeness result for KD4t, showing that any countable KD4t-consistent set of formulas is satisfiable at any point of any zero-dimensional dense-in-itself metric space X. By contrast, this strong completeness fails for KD4t over its Kripke semantics.

2 Formulas and Frames

We assume a set Var of *propositional variables* or *atoms*. Formulas are constructed from these variables by the standard Boolean connectives $\top$, $\neg$, $\wedge$; the unary modality $[d]$; and the *tangle* connective $\langle dt \rangle$ which assigns a formula $\langle dt \rangle\Gamma$ to each finite non-empty set Γ of formulas. The other Boolean connectives $\bot$, $\vee$, $\to$ are introduced as standard abbreviations, and the dual $\langle d \rangle$ of $[d]$ is defined to be $\neg[d]\neg$. We write $[d]^*\varphi$ for the formula $\varphi \wedge [d]\varphi$, and $\langle d \rangle^*\varphi$ for $\varphi \vee \langle d \rangle\varphi$. We denote the set of all formulas by $\mathcal{L}^{\langle dt \rangle}_{[d]}$, and the set of formulas with no occurrence of $\langle dt \rangle$ by $\mathcal{L}_{[d]}$.

A *frame* is a pair $\mathcal{F} = (W, R)$, where W is a non-empty set and R is a binary relation on W. We may write any of $R(w, v)$, Rwv, and wRv to denote that $(w, v) \in R$. We let $R(w)$ denote the set $\{v \in W : wRv\}$ An element w is called *reflexive* if wRw, and *irreflexive* otherwise.

We restrict ourselves throughout the paper to frames that have *transitive* R. Then if $R^* = R \cup id_W$, where id_W is the identity relation on W, we get that R^* is the reflexive transitive closure of R, and putting $R^*(w) = \{v \in W : wR^*v\}$ we get that $R^*(w) = \{w\} \cup R(w)$. Observe that w is reflexive iff $R^*(w) = R(w)$.

If R^{-1} is the inverse relation to R, then each subset $W' \subseteq W$ has the *R-inverse image* $R^{-1}(W') = \{w \in W : \exists v \in W'(wRv)\} = \{w : R(w) \cap W' \neq \emptyset\}$. For singleton subsets we write $R^{-1}(\{v\})$ just as $R^{-1}(v)$.

A *model* $(\mathcal{F}, h)$ on $\mathcal{F}$ is given by an *assignment* h, which is a function from Var into the powerset $\wp W$ of W. The notion $(\mathcal{F}, h), w \models \varphi$ of a formula φ being true, or satisfied, at w in model $(\mathcal{F}, h)$ is defined by induction on the formation of φ, by the following clauses.

(i) $(\mathcal{F}, h), w \models p$ iff $w \in h(p)$, for $p \in$ Var.

(ii) $(\mathcal{F}, h), w \models \top$.

(iii) $(\mathcal{F}, h), w \models \neg\varphi$ iff $(\mathcal{F}, h), w \not\models \varphi$.

(iv) $(\mathcal{F}, h), w \models \varphi \wedge \psi$ iff $(\mathcal{F}, h), w \models \varphi$ and $(\mathcal{F}, h), w \models \psi$.

(v) $(\mathcal{F}, h), w \models [d]\varphi$ iff $(\mathcal{F}, h), v \models \varphi$ for every $v \in R(w)$.

(vi) $(\mathcal{F}, h), w \models \langle dt \rangle \Gamma$ iff there is a sequence $w = w_0, w_1, \ldots$ in W with $w_n R w_{n+1}$ for each $n < \omega$ and such that for each $\gamma \in \Gamma$ there are infinitely many $n < \omega$ with $(\mathcal{F}, h), w_n \models \gamma$.

The sequence $\{w_n : n < \omega\}$ in the last clause could be described as an *endless R-path satisfying each member of* Γ *infinitely often.*

A formula φ is *satisfiable in frame* $\mathcal{F}$ if $(\mathcal{F}, h), w \models \varphi$ for some h and some w. φ is *valid in* $\mathcal{F}$ if $\neg\varphi$ is not satisfiable in $\mathcal{F}$, i.e. if φ is true at every point in every model on $\mathcal{F}$.

In any model $(\mathcal{F}, h)$ each formula defines the 'truth-set' $[\![\varphi]\!]_h = \{w \in W : (\mathcal{F}, h), w \models \varphi\}$. In particular $[\![p]\!]_h = h(p)$ for $p \in \mathsf{Var}$. The semantic clause 5 above for $[d]$ states that $[\![[d]\varphi]\!]_h = \{w : R(w) \subseteq [\![\varphi]\!]_h\}$. The truth condition for the dual modality $\langle d \rangle$ gives

$$[\![\langle d \rangle \varphi]\!]_h = \{w : R(w) \cap [\![\varphi]\!]_h \neq \emptyset\} = R^{-1}[\![\varphi]\!]_h. \tag{2.1}$$

Lemma 2.1 *In any model* $(\mathcal{F}, h)$ *on a frame, and any* $w \in W$*, we have* $w \in [\![\langle dt \rangle \Gamma]\!]_h$ *iff there is a subset* S *of* W *such that*

$$w \in S \subseteq \bigcap\nolimits_{\gamma \in \Gamma} R^{-1}([\![\gamma]\!]_h \cap S). \tag{2.2}$$

Proof. Suppose $w \in [\![\langle dt \rangle \Gamma]\!]_h$, and let $S = \{w_n : n < \omega\}$ be the resulting sequence given by clause 6 above. Then $w = w_0 \in S$, and for any $w_n \in S$ and any $\gamma \in \Gamma$ there is an $m > n$ such that $(\mathcal{F}, h), w_m \models \gamma$, so as R is transitive, $w_n R w_m \in [\![\gamma]\!]_h \cap S$, showing $w_n \in R^{-1}([\![\gamma]\!]_h \cap S)$. This proves (2.2).

Conversely, if (2.2) holds for some S, then for each $v \in S$ and each $\gamma \in \Gamma$ there is some $u \in S$ such that vRu and $(\mathcal{F}, h), u \models \gamma$. Using this, if $\Gamma = \{\gamma_1, \ldots, \gamma_m\}$ we can iteratively choose a sequence $\{w_n : n < \omega\}$ of members of S with $w_0 = w$ and $w_n R w_{n+1}$ and every index $mk + i$ with $1 \leq i \leq m$ having $(\mathcal{F}, h), w_{mk+i} \models \gamma_i$. This shows that $(\mathcal{F}, h), w \models \langle dt \rangle \Gamma$. □

Thus $[\![\langle dt \rangle \Gamma]\!]_h$ is the union of all subsets S of W such that $S \subseteq \bigcap_{\gamma \in \Gamma} R^{-1}([\![\gamma]\!]_h \cap S)$. To put this in perspective we invoke the Knaster-Tarski Theorem on fixed points of monotone functions on complete lattices [15]. Define $F(S) = \bigcap_{\gamma \in \Gamma} R^{-1}([\![\gamma]\!]_h \cap S)$. Then F is a function on the lattice of all subsets of W that is monotone for inclusion: $S \subseteq S'$ implies $F(S) \subseteq F(S')$. The Knaster-Tarski Theorem states that such a function has a *largest fixed point* S_0, namely if $S_0 = \bigcup\{S \subseteq X : S \subseteq F(S)\}$, then $S_0 = F(S_0)$ and $S \subseteq S_0$ whenever $F(S) = S$. But here $\bigcup\{S \subseteq X : S \subseteq F(S)\} = [\![\langle dt \rangle \Gamma]\!]_h$. So $[\![\langle dt \rangle \Gamma]\!]_h$ is the largest fixed point of F.

This fixed point interpretation and the 'endless R-path' interpretation of $\langle dt \rangle$ each have their uses. In Section 4 we use the fixed point approach to show that validity of formulas is preserved by certain morphisms. Here we note that the endless path approach makes it very easy to see that validity is preserved by *generated* subframes. Recall that $\mathcal{F}' = (W', R')$ is a generated subframe of $\mathcal{F} = (W, R)$ if $W' \subseteq W$, R' is the restriction of R to W', and W' is R-closed in the sense that $R(w) \subseteq W'$ for all $w \in W'$. Then given models $(\mathcal{F}, h)$ and

$(\mathcal{F}', h')$ with $h'(p) = h(p) \cap W'$ for $p \in \mathsf{Var}$, it is a standard fact that for all $\mathcal{L}_{[d]}$-formulas φ and all $w \in W'$,

$$(\mathcal{F}, h), w \models \varphi \quad \text{iff} \quad (\mathcal{F}', h'), w \models \varphi. \tag{2.3}$$

In other words, $[\![\varphi]\!]_{h'} = [\![\varphi]\!]_h \cap W'$. But this result extends readily to all $\mathcal{L}_{[d]}^{\langle dt \rangle}$-formulas, with the inductive case of a formula $\langle dt \rangle \Gamma$ holding because any endless R'-path is an R-path and, crucially, any endless R-path that starts in W' remains in W' by the R-closure of W' and so is an R'-path.

We write $\mathcal{F}^*(w)$ for the subframe of $\mathcal{F}$ based on $R^*(w)$, which is a generated subframe as R is transitive. If $W = R^*(w)$, then $\mathcal{F} = \mathcal{F}^*(w)$ and we say that w is a *root* of $\mathcal{F}$. From the result (2.3) a standard argument gives

Theorem 2.2 *A $\mathcal{L}_{[d]}^{\langle dt \rangle}$-formula is valid in a frame $\mathcal{F}$ iff it is valid in every rooted subframe $\mathcal{F}^*(w)$ of $\mathcal{F}$.* □

3 Spaces

Let X be a topological space. We do not name the topology of X, but just refer to various subsets of X as being open or closed in X. An open set O containing a point x is called an *open neighbourhood* of x. Then $O \setminus \{x\}$ is a *punctured neighbourhood* of x. We write $\mathrm{cl}_X\, S$ for the closure (smallest closed superset) of a subset $S \subseteq X$, and $\mathrm{int}_X\, S$ for the interior (largest open subset) of S. $\langle d_X \rangle S$ denotes the *derivative* or set of limit points of S. Then we have

$$\mathrm{int}_X S = \{x \in X : O \subseteq S \text{ for some open neighbourhood } O \text{ of } x\}. \tag{3.1}$$

$$\mathrm{cl}_X S = \{x \in X : S \cap O \neq \emptyset \text{ for all open neighbourhoods } O \text{ of } x\}. \tag{3.2}$$

$$\langle d_X \rangle S = \{x \in X : S \cap O \setminus \{x\} \neq \emptyset \text{ for all open neighbourhoods } O \text{ of } x\}. \tag{3.3}$$

X is called *dense-in-itself* if $\langle d_X \rangle X = X$, i.e. if every point x of X is a limit point of X and so $\{x\}$ is not open. We record some standard facts about these operators:

Lemma 3.1

(1) *$\langle d_X \rangle$ is additive: $\langle d_X \rangle (S \cup T) = \langle d_X \rangle S \cup \langle d_X \rangle T$.*

(2) *$\mathrm{cl}_X\, S = S \cup \langle d_X \rangle S$.*

(3) *S is closed iff it contains all its limit points (i.e. $\langle d_X \rangle S \subseteq S$).* □

A *model* (X, h) on X is given by an *assignment* $h : \mathsf{Var} \to \wp X$. The truth/satisfaction relation $(X, h), x \models \varphi$, and associated truth-sets $[\![\varphi]\!]_h = \{x \in X : (X, h), x \models \varphi\}$, are defined inductively as follows.

1. $(X, h), x \models p$ iff $x \in h(p)$, for $p \in \mathsf{Var}$.
2. $(X, h), x \models \top$.
3. $(X, h), x \models \neg\varphi$ iff $(X, h), x \not\models \varphi$.
4. $(X, h), x \models \varphi \wedge \psi$ iff $(X, h), x \models \varphi$ and $(X, h), x \models \psi$.

5. $(X, h), x \models [d]\varphi$ iff there is an open neighbourhood O of x with $(X, h), y \models \varphi$ for every $y \in O \setminus \{x\}$.
6. $(X, h), x \models \langle dt \rangle \Gamma$ iff there is some $S \subseteq X$ such that $x \in S \subseteq \bigcap_{\gamma \in \Gamma} \langle d_X \rangle (\llbracket \gamma \rrbracket_h \cap S)$.

By clause 5, $[d]\varphi$ is true at x iff x has a punctured neighbourhood included in $\llbracket \varphi \rrbracket_h$. This implies that

> $(X, h), x \models \langle d \rangle \varphi$ iff every open neighbourhood O of x has $(X, h), y \models \varphi$ for some $y \in O \setminus \{x\}$.

From (3.3) it then follows that $\llbracket \langle d \rangle \varphi \rrbracket_h = \langle d_X \rangle \llbracket \varphi \rrbracket_h$, the set of all limit points of $\llbracket \varphi \rrbracket_h$. Using (3.1) and Lemma 3.1(2) we get that $\llbracket [d]^* \varphi \rrbracket_h = \mathrm{int}_X \llbracket \varphi \rrbracket_h$ and $\llbracket \langle d \rangle^* \varphi \rrbracket_h = \mathrm{cl}_X \llbracket \varphi \rrbracket_h$.

A set Y is *strictly dense* in a set S containing Y if every member of S is a limit point of Y, i.e. $S \subseteq \langle d_X \rangle Y$. In a model (X, h) a finite set of formulas Γ will be called *strictly tangled in* S if $\llbracket \gamma \rrbracket_h \cap S$ is strictly dense in S for all $\gamma \in \Gamma$, i.e. if $S \subseteq \bigcap_{\gamma \in \Gamma} \langle d_X \rangle (\llbracket \gamma \rrbracket_h \cap S)$. Thus the truth condition for $\langle dt \rangle \Gamma$ gives that $\llbracket \langle dt \rangle \Gamma \rrbracket_h$ is the union of all the sets S such that Γ is strictly tangled in S.

Theorem 3.2 *Γ is tangled in $\llbracket \langle dt \rangle \Gamma \rrbracket_h$, so $\llbracket \langle dt \rangle \Gamma \rrbracket_h$ is the largest set in which Γ is strictly tangled. Moreover $\llbracket \langle dt \rangle \Gamma \rrbracket_h = \bigcap_{\gamma \in \Gamma} \langle d_X \rangle (\llbracket \gamma \rrbracket_h \cap \llbracket \langle dt \rangle \Gamma \rrbracket_h)$.*

Proof. This is another instance of the Knaster-Tarski Theorem (see the paragraph after Lemma 2.1). If $F(S) = \bigcap_{\gamma \in \Gamma} \langle d_X \rangle (\llbracket \gamma \rrbracket_h \cap S)$, then Γ is strictly tangled in S iff $S \subseteq F(S)$. Now F is a monotone function on the powerset lattice of W, so the Knaster-Tarski Theorem states that F has *largest fixed point* $\bigcup \{S \subseteq X : S \subseteq F(S)\} = \llbracket \langle dt \rangle \Gamma \rrbracket_h$. □

Given the interpretation of $\langle d \rangle$ as $\langle d \rangle_X$ in spaces and as R^{-1} in frames (2.1), we now see that the semantics of $\langle dt \rangle$ in frames and spaces is formally the same: in both cases $\llbracket \langle dt \rangle \Gamma \rrbracket_h$ is the largest solution of the equation $S = \bigcap_{\gamma \in \Gamma} \mathbf{d}(\llbracket \gamma \rrbracket_h \cap S)$, where $\mathbf{d}$ is the relevant interpretation of $\langle d \rangle$.

A formula φ is *satisfiable in space* X if $(X, h), x \models \varphi$ for some h and some x, and is *valid in* X if $\neg \varphi$ is not satisfiable in X.

Theorem 3.3 *In any topological space, all instances of the following formula schemes are valid.*

Fix: $\langle dt \rangle \Gamma \to \langle d \rangle (\gamma \wedge \langle dt \rangle \Gamma)$, *with* $\gamma \in \Gamma$.

Ind: $[d]^* (\varphi \to \bigwedge_{\gamma \in \Gamma} \langle d \rangle (\gamma \wedge \varphi)) \to (\varphi \to \langle dt \rangle \Gamma)$.

Proof. Working in any model (X, h) on X, we show that these formulas are true at all points. The result for Fix is immediate from the previous theorem, which gives $\llbracket \langle dt \rangle \Gamma \rrbracket_h \subseteq \llbracket \langle d \rangle (\gamma \wedge \langle dt \rangle \Gamma) \rrbracket_h$.

For Ind, suppose $x \models [d]^* (\varphi \to \bigwedge_{\gamma \in \Gamma} \langle d \rangle (\gamma \wedge \varphi))$ and $x \models \varphi$. Then there is an open neighbourhood O of x such that for any $\gamma \in \Gamma$, $O \subseteq \llbracket \varphi \to \langle d \rangle (\gamma \wedge \varphi) \rrbracket_h$, hence $O \cap \llbracket \varphi \rrbracket_h \subseteq \langle d_X \rangle (\llbracket \gamma \rrbracket_h \cap \llbracket \varphi \rrbracket_h)$.

But then $O \cap \llbracket \varphi \rrbracket_h \subseteq \langle d_X \rangle (\llbracket \gamma \rrbracket_h \cap O \cap \llbracket \varphi \rrbracket_h)$, because if $y \in O \cap \llbracket \varphi \rrbracket_h$, then for any open neighbourhood O' of y, $O \cap O' \setminus \{y\}$ intersects $\llbracket \gamma \rrbracket_h \cap \llbracket \varphi \rrbracket_h$, hence

$O' \setminus \{y\}$ intersects $[\![\gamma]\!]_h \cap O \cap [\![\varphi]\!]_h$.

This shows that Γ is strictly tangled in $O \cap [\![\varphi]\!]_h$, and hence that $O \cap [\![\varphi]\!]_h \subseteq [\![\langle dt\rangle\Gamma]\!]_h$. But $x \in O \cap [\![\varphi]\!]_h$, so then $x \models \langle dt\rangle\Gamma$, confirming that Ind is true at x. □

Corresponding to generated subframes we have the notion of Y being an *open subspace* of X, meaning that Y is a subspace of X that is itself open in the topology on X. The openness of Y ensures that for all $S \subseteq X$,

$$\langle d_X\rangle S \cap Y = \langle d_Y\rangle(S \cap Y). \tag{3.4}$$

Theorem 3.4 *If Y is an open subspace of X, then any $\mathcal{L}_{[d]}^{\langle dt\rangle}$-formula valid in X is valid in Y.*

Proof. This follows from the result that if models (X,h) and (Y,h') have $h'(p) = h(p) \cap Y$ for $p \in \mathsf{Var}$, then for all $\mathcal{L}_{[d]}^{\langle dt\rangle}$-formulas φ and all $y \in Y$,

$$(X,h), y \models \varphi \quad \text{iff} \quad (Y,h'), y \models \varphi.$$

This result, which gives $[\![\varphi]\!]_{h'} = [\![\varphi]\!]_h \cap Y$, is standard for $\mathcal{L}_{[d]}$-formulas, with the inductive case of a formula $\langle d\rangle\varphi$ holding because (3.4) implies that $[\![\langle d\rangle\varphi]\!]_h \cap Y = \langle d_Y\rangle([\![\varphi]\!]_h \cap Y) = [\![\langle d\rangle\varphi]\!]_{h'}$.

For the case of a formula $\langle dt\rangle\Gamma$, assuming inductively that the result holds for all members of Γ, let $y \in Y$ and $(Y,h'), y \models \langle dt\rangle\Gamma$. Then y belongs to some $S \subseteq Y$ in which Γ is strictly tangled in (Y,h'). So for any $\gamma \in \Gamma$, $S \subseteq \langle d_Y\rangle([\![\gamma]\!]_{h'} \cap S) \subseteq \langle d_X\rangle([\![\gamma]\!]_h \cap S)$. This shows that Γ is strictly tangled in S in (X,h), hence $(X,h), y \models \langle dt\rangle\Gamma$.

Conversely, if $(X,h), y \models \langle dt\rangle\Gamma$, then y belongs to some $S \subseteq X$ in which Γ is strictly tangled in (X,h). Then for any $\gamma \in \Gamma$, $S \subseteq \langle d_X\rangle([\![\gamma]\!]_h \cap S)$, and so

$$S \cap Y \subseteq \langle d_X\rangle([\![\gamma]\!]_h \cap S) \cap Y = \langle d_Y\rangle([\![\gamma]\!]_h \cap S \cap Y) = \langle d_Y\rangle([\![\gamma]\!]_{h'} \cap S \cap Y).$$

Hence Γ is strictly tangled in $S \cap Y$ in (Y,h'). Since $y \in S \cap Y$, $(Y,h'), y \models \langle dt\rangle\Gamma$ follows as required. □

4 d-Morphisms

A function $\rho : X \to W$ is called a *d-morphism* from a space X to a frame $\mathcal{F} = (W,R)$ if every subset $S \subseteq W$ has $\langle d_X\rangle\rho^{-1}(S) = \rho^{-1}(R^{-1}(S))$. When W is finite, for this to hold it is sufficient to require that it hold whenever S is a singleton, i.e. $\langle d_X\rangle\rho^{-1}(w) = \rho^{-1}(R^{-1}(w))$ for all $w \in W$. This is because the operators $\langle d_X\rangle$, ρ^{-1} and R^{-1} distribute across finite unions, as was observed in [13] where maps of this kind were first studied.[4]

It is shown in [2, Corollary 2.9] that surjective d-morphisms preserve validity of all $\mathcal{L}_{[d]}$-formulas, regardless of whether $\mathcal{F}$ is finite or infinite. We will extend that result to the present language with $\langle dt\rangle$, making use of the two results in the next Lemma, which follow from [2, Theorem 2.7].

---

[4] In [6] a d-morphism with W finite is called a *representation* of the frame $\mathcal{F}$ over the space X.

Lemma 4.1 *Let $\rho : X \to W$ be a d-morphism from space X to frame $\mathcal{F} = (W, R)$.*

(1) *For all $w \in W$, $\rho^{-1}(R^*(w))$ is an open subset of X.*

(2) *For every* ***irreflexive*** *$w \in W$, the preimage $\rho^{-1}(w)$ is a discrete subspace of X.* □

Theorem 4.2 *Let $\rho : X \to W$ be a d-morphism from space X to frame $\mathcal{F} = (W, R)$, and let (X, h) and $(\mathcal{F}, h')$ be models having $[\![p]\!]_h = \rho^{-1}[\![p]\!]_{h'}$ for all $p \in \mathsf{Var}$. Then $[\![\varphi]\!]_h = \rho^{-1}[\![\varphi]\!]_{h'}$ for all $\mathcal{L}^{\langle dt \rangle}_{[d]}$-formulas φ.*

Proof. By induction on the formation of φ, with the result holding by assumption if $\varphi \in \mathsf{Var}$; the inductive cases of the Boolean connectives being standard, and the inductive case of a formula beginning with $\langle d \rangle$ following from the definition of d-morphism.

Consider the case of a formula $\langle dt \rangle \Gamma$, on the inductive assumption that the result holds for all members of Γ. If $x \in \rho^{-1}[\![\langle dt \rangle \Gamma]\!]_{h'}$, then by Lemma 2.1 $\rho(x)$ belong to some $S \subseteq W$ such that $S \subseteq \bigcap_{\gamma \in \Gamma} R^{-1}([\![\gamma]\!]_{h'} \cap S)$. Let $Y = \rho^{-1}(S)$. If $y \in Y$, then for any $\gamma \in \Gamma$, $\rho(y) \in R^{-1}([\![\gamma]\!]_{h'} \cap S)$, so $y \in \rho^{-1}(R^{-1}([\![\gamma]\!]_{h'} \cap S))$. But as ρ is a d-morphism, this implies

$$y \in \langle d_X \rangle \rho^{-1}([\![\gamma]\!]_{h'} \cap S) = \langle d_X \rangle (\rho^{-1}[\![\gamma]\!]_{h'} \cap \rho^{-1}(S)) = \langle d_X \rangle ([\![\gamma]\!]_h \cap Y).$$

This shows that $Y \subseteq \langle d_X \rangle ([\![\gamma]\!]_h \cap Y)$ for all $\gamma \in \Gamma$, so Γ is strictly tangled in Y in (X, h). Since $x \in Y$, that gives $x \in [\![\langle dt \rangle \Gamma]\!]_h$, proving that $\rho^{-1}[\![\langle dt \rangle \Gamma]\!]_{h'} \subseteq [\![\langle dt \rangle \Gamma]\!]_h$.

For the converse inclusion, suppose $x \in [\![\langle dt \rangle \Gamma]\!]_h$. Then x belongs to some $S \subseteq X$ such that $S \subseteq \bigcap_{\gamma \in \Gamma} \langle d_X \rangle ([\![\gamma]\!]_h \cap S)$. Then $\rho(x) \in \rho(S)$. We will show that $\rho(S) \subseteq \bigcap_{\gamma \in \Gamma} R^{-1}([\![\gamma]\!]_{h'} \cap \rho(S))$. For this, take any $w \in \rho(S)$. Then $w = \rho(y)$ for some $y \in S$. Let $O = \rho^{-1}(R^*(w))$, an open neighbourhood of y by Lemma 4.1(1).

Now if w is irreflexive, then by Lemma 4.1(2), y is isolated from the rest of $\rho^{-1}(w)$, so there is an open neighbourhood O' of y such that $O' \cap \rho^{-1}(w) = \{y\}$. Put $U = O \cap O'$. If however w is reflexive, put $U = O$. Either way, U is an open neighbourhood of y. Since $y \in S$, for any $\gamma \in \Gamma$ we have $y \in \langle d_X \rangle ([\![\gamma]\!]_h \cap S)$, so there is some $y' \in U$ with $y \neq y' \in [\![\gamma]\!]_h \cap S$. Then $\rho(y') \in \rho(S)$. Also as $y' \in [\![\gamma]\!]_h$, the induction hypothesis on γ gives $\rho(y') \in [\![\gamma]\!]_{h'}$.

Since $y' \in U \subseteq O$, we get $\rho(y') \in R^*(w)$. If w is reflexive, this immediately gives $wR\rho(y')$. But if w is irreflexive, then $y' \in O'$ and $y' \neq y$, so $y' \notin \rho^{-1}(w)$, hence $\rho(y') \neq w$, and therefore again $wR\rho(y')$. Thus in any case $wR\rho(y')$. Now we have $wR\rho(y') \in [\![\gamma]\!]_{h'} \cap \rho(S)$. Hence $w \in R^{-1}([\![\gamma]\!]_{h'} \cap \rho(S))$ as required.

This complete the proof that $\rho(x) \in \rho(S) \subseteq \bigcap_{\gamma \in \Gamma} R^{-1}([\![\gamma]\!]_{h'} \cap \rho(S))$, which by Lemma 2.1 implies $\rho(x) \in [\![\langle dt \rangle \Gamma]\!]_{h'}$, hence $x \in \rho^{-1}[\![\langle dt \rangle \Gamma]\!]_{h'}$. □

Corollary 4.3 *If there exists a surjective d-morphism $\rho : X \to \mathcal{F}$, then any $\mathcal{L}^{\langle dt \rangle}_{[d]}$-formula valid in X is valid in $\mathcal{F}$.*

Proof. If φ is not valid in $\mathcal{F}$, then there is a model $(\mathcal{F}, h')$ with $w \notin [\![\varphi]\!]_{h'}$ for some w. Then $w = \rho(x)$ for some $x \in X$ as ρ is surjective. Define a model (X, h) by putting $h(p) = \rho^{-1}[\![p]\!]_{h'}$ for all $p \in \mathsf{Var}$. Then $x \notin \rho^{-1}[\![\varphi]\!]_{h'} = [\![\varphi]\!]_h$ by the Theorem, making φ not valid in X. □

5 The Logic of $\mathbb{R}$

By the *logic of a space* X, in a given language, we mean the set of all formulas of that language that are valid in X. In this section we will identify the $\mathcal{L}^{\langle dt \rangle}_{[d]}$-logic of the real line $\mathbb{R}$, considered as a space under its standard Euclidean metric topology.

Any space validates the formula-scheme $\langle d \rangle\langle d \rangle\varphi \to \langle d \rangle^*\varphi$. The stronger scheme $\langle d \rangle\langle d \rangle\varphi \to \langle d \rangle\varphi$, which corresponds to transitivity of R in frames, is valid precisely in the T_D-spaces, which are those in which the derivative $\langle d_X \rangle\{x\}$ of any point is closed. This T_D condition, introduced in [1], is implied by the T_1 separation property, which is itself equivalent to $\langle d_X \rangle\{x\} = \emptyset$. Our concern here is with the logic of metric spaces, which have the even stronger T_2 property, so this justifies our restriction to transitive frames.

By a *tangle logic* we mean any set of $\mathcal{L}^{\langle dt \rangle}_{[d]}$-formulas that includes all instances of tautologies and of the schemes

K: $[d](\varphi \to \psi) \to ([d]\varphi \to [d]\psi)$

4: $\langle d \rangle\langle d \rangle\varphi \to \langle d \rangle\varphi$

Fix: $\langle dt \rangle\Gamma \to \langle d \rangle(\gamma \wedge \langle dt \rangle\Gamma)$, all $\gamma \in \Gamma$

Ind: $[d]^*(\varphi \to \bigwedge_{\gamma \in \Gamma} \langle d \rangle(\gamma \wedge \varphi)) \to (\varphi \to \langle dt \rangle\Gamma)$

and is closed under modus ponens and the $[d]$-modality. The members of the logic will be called its *theorems*. It is readily seen that if $\mathcal{F}$ is any transitive frame, then the set of all formulas valid in $\mathcal{F}$ is a tangle logic.

The smallest tangle logic will be denoted K4t, and the smallest tangle logic containing the D-axiom $\langle d \rangle\top$ will be denoted KD4t. The frame condition for validity of $\langle d \rangle\top$ is that R be *serial*, i.e. $\forall w \exists w'(wRw')$, so $R(w) \neq \emptyset$ for all w. A serial transitive frame will be referred to as a *KD4-frame*.

Theorem 5.1 (Soundness) *The $\mathcal{L}^{\langle dt \rangle}_{[d]}$-logic of any T_D-space includes K4t and the logic of any dense-in-itself T_D-space includes KD4t.*

Proof. It is standard that the set of formulas valid in a space includes axiom K and is closed under modus ponens and $[d]$. It includes Fix and Ind by Theorem 3.3, and includes axiom 4 when the space is T_D as just noted. The D-axiom is valid in space X iff $\langle d_X \rangle X = X$, which means that X is dense-in-itself. □

There is a natural topological distinction between $\mathbb{R}$ and the higher dimensional Euclidean spaces $\mathbb{R}^n$ for $n \geq 2$. If O is an open ball in $\mathbb{R}^n$, then a punctured subspace $O \setminus \{x\}$ is connected (indeed any two points of $O \setminus \{x\}$ are joined by a continuous path lying in $O \setminus \{x\}$). But if O is an open interval (a, b) in $\mathbb{R}$, then $O \setminus \{x\}$ is disconnected, with two connected components (a, x) and (x, b). This difference is captured with the help of certain formulas G_n for $n \geq 1$. G_n has the $n+1$ variables $p_0, \ldots, p_n$. For $i \leq n$, put $Q_i = p_i \wedge \bigwedge_{i \neq j \leq n} \neg p_j$. Then G_n is

$$[d](\bigvee_{i \leq n} [d]^* Q_i) \to \bigvee_{i \leq n} [d] \neg Q_i.$$

It asserts that if some punctured neighbourhood of x can be covered by the interiors of $n+1$ disjoint sets, then there must be a punctured neighbourhood of x that is disjoint from one of those sets.

The G_n's were introduced by Shehtman in [13], where he showed that the $\mathcal{L}_{[d]}$-logic of $\mathbb{R}^n$ for $n \geq 2$ is $\mathrm{KD4G}_1$, the smallest extension of KD4 to include all substitution instances of G_1. He also conjectured that the $\mathcal{L}_{[d]}$-logic of $\mathbb{R}$ is $\mathrm{KD4G}_2$. This was later verified by himself [12] with another proof given by Lucero-Bryan [8].

Let $\mathrm{KD4G}_2t$ be the smallest tangle logic including the D-axiom and all substitution instances of G_2. In [6, Section 4.13] we proved that this logic has the finite model property for frames: if an $\mathcal{L}_{[d]}^{\langle dt \rangle}$-formula φ is not a $\mathrm{KD4G}_2t$-theorem, then it is falsified by a model on some finite frame that validates $\mathrm{KD4G}_2t$.

Theorem 5.2 *$KD4G_2t$ is the $\mathcal{L}_{[d]}^{\langle dt \rangle}$-logic of $\mathbb{R}$.*

Proof. *Soundness*: since $\mathbb{R}$ is a dense-in-itself T_D-space that validates G_2, its logic includes $\mathrm{KD4G}_2t$. *Completeness*: let φ be an $\mathcal{L}_{[d]}^{\langle dt \rangle}$-formula that is not a $\mathrm{KD4G}_2t$-theorem. Then by [6] there is a finite KD4-frame $\mathcal{F}$ falsifying φ that validates $\mathrm{KD4G}_2t$. Hence by Theorem 2.2, φ is not valid in some rooted subframe $\mathcal{F}^*(w)$ of $\mathcal{F}$ that also validates $\mathrm{K4DG}_2t$. But Lemma 4.4 of [8] showed that any finite rooted $\mathrm{KD4G}_2$-frame is a d-morphic image of any interval (x, y) of $\mathbb{R}$ with $x < y$. So there exists such a d-morphism from (x, y) onto $\mathcal{F}^*(w)$. By Corollary 4.3 it follows that φ is not valid in the open subspace (x, y) of $\mathbb{R}$. Hence φ is not valid in $\mathbb{R}$ by Theorem 3.4. □

As already noted in the Introduction, we showed in [6, Theorem 9.3] that for $n \geq 2$, the $\mathcal{L}_{[d]}^{\langle dt \rangle}$-logic of $\mathbb{R}^n$ is $\mathrm{KD4G}_1t$.

6 Universal Modality

We now extend the syntax to include the universal modality $\forall$, creating formulas $\forall\varphi$ with the spatial semantics $(X, h), x \models \forall\varphi$ iff for all $y \in X$, $(X, h), y \models \varphi$; and the frame-semantics $(\mathcal{F}, h), w \models \forall\varphi$ iff for all $v \in W$, $(\mathcal{F}, h), v \models \varphi$.

We denote the new set of formulas by $\mathcal{L}_{[d]\forall}^{\langle dt \rangle}$. It is straightforward to extend Corollary 4.3 to show that validity of these formulas is preserved by surjective d-morphisms.

A tangle logic for the new language is now required to include the S5 axioms and rules for $\forall$, and the scheme

U: $\forall\varphi \to [d]\varphi$,

all of which are valid in all frames and spaces. The smallest such logic is denoted K4t.U, and KD4t.U and KD4G$_n t$.U are defined similarly. All of these $\mathcal{L}^{\langle dt \rangle}_{[d]\forall}$-logics were shown in [6] to have the finite model property over their frames.

Let K4t.UC be the smallest tangle logic extending K4t.U in the language $\mathcal{L}^{\langle dt \rangle}_{[d]\forall}$ that includes the scheme

C: $\forall(\Box^*\varphi \vee \Box^*\neg\varphi) \to (\forall\varphi \vee \forall\neg\varphi)$,

which was introduced in [11]. The condition for validity of C in a frame is graphical connectedness: any two points are connected by an $(R \cup R^{-1})$-path. The validity condition for C in a space X is topological connectedness, i.e. that X cannot be partitioned into two non-empty open subsets.

In [6], the logics K4t.UC, KD4t.UC and KD4G$_n t$.UC were all shown to have the finite model property over their frames. Theorem 9.5 of [6] proved that if X is any dense-in-itself metric space then the $\mathcal{L}^{\langle dt \rangle}_{[d]\forall}$-logic of X is included in KD4G$_1 t$.UC, and is exactly KD4G$_1 t$.UC if X validates G$_1$. In particular the $\mathcal{L}^{\langle dt \rangle}_{[d]\forall}$-logic of $\mathbb{R}^n$ is KD4G$_1 t$.UC for all $n \geq 2$.

As to the real line $\mathbb{R}$, its $\mathcal{L}_{[d]\forall}$-logic was shown to be KD4G$_2$.UC in [8]. We extend this now to the language with $\langle dt \rangle$.

Theorem 6.1 *KD4G$_2 t$.UC is the $\mathcal{L}^{\langle dt \rangle}_{[d]\forall}$-logic of* $\mathbb{R}$.

Proof. *Soundness*: the axioms D, 4, G$_2$ and C are all valid in $\mathbb{R}$.

Completeness: let φ be an $\mathcal{L}^{\langle dt \rangle}_{[d]\forall}$-formula that is not a KD4G$_2 t$.UC-theorem. Then by [6, Section 4.13] there is a finite frame $\mathcal{F}$ that validates KD4G$_2 t$.UC and falsifies φ. But Theorem 5.25 of [8] proved that any KD4G$_2$.UC frame is a d-morphic image of $\mathbb{R}$. Since validity of $\mathcal{L}^{\langle dt \rangle}_{[d]\forall}$-formulas is preserved by surjective d-morphisms, φ is not valid in $\mathbb{R}$.

□

7 Zero-Dimensionality and Dissection

A metric space is *zero-dimensional* if its topology has a basis of clopen (closed and open) sets. Such a space is totally disconnected: distinct points can be separated by a clopen set. Examples of zero-dimensionality include the space of rationals $\mathbb{Q}$, the irrationals $\mathbb{R} - \mathbb{Q}$, the Cantor space, and the Baire space ω^ω. These are all dense-in-themselves and *separable*, i.e. they have a countable dense subset.

It was shown in [13] that the $\mathcal{L}_{[d]}$-logic of each separable zero-dimensional dense-in-itself metric space is KD4. Here we will generalise this, first by eliminating the restriction to separable spaces, and then by showing that the $\mathcal{L}^{\langle dt \rangle}_{[d]}$-logic of each zero-dimensional dense-in-itself metric space is KD4t. This section provides some prerequisite results about 'dissecting' an open set into special subsets with properties that allow us to use them to represent the structure of finite frames.

First some background. If X is a metric space, we write d for its metric, and $N_\varepsilon(x_0)$ for the open ball $\{x \in X : d(x, x_0) < \varepsilon\}$, where $x_0 \in X$, $\varepsilon \in \mathbb{R}$,

$\varepsilon > 0$. For non-empty $S \subseteq X$, define $d(x, S) = \inf\{d(x, y) : y \in S\}$. (We leave $d(x, \emptyset)$ undefined.) The following Lemma collects some standard facts.

Lemma 7.1 *Let X be a dense-in-itself metric space and $S \subseteq X$.*

(1) *If S is non-empty and open, then S is infinite.*

(2) $\langle d_X \rangle S = \{x \in X : S \cap O$ *is infinite for every open neighbourhood O of* $x\}$.

(3) $\mathrm{int}_X\, S \subseteq \langle d_X \rangle S$. *If S is open then $\langle d_X \rangle S = \mathrm{cl}_X\, S$.* □

Theorem 7.2 *Let X be a dense-in-itself metric space. If $\mathbb{G}$ is a non-empty open subset of X, then for $r, s < \omega$, $\mathbb{G}$ can be partitioned into non-empty open subsets $\mathbb{G}_1, \ldots, \mathbb{G}_r$ and other non-empty sets $\mathbb{B}_0, \ldots, \mathbb{B}_s$ such that, letting $D = \mathrm{cl}_X(\mathbb{G}) \setminus (\mathbb{G}_1 \cup \cdots \cup \mathbb{G}_r)$, we have $\mathrm{cl}_X(\mathbb{G}_i) \setminus \mathbb{G}_i = D$ for each $i = 1, \ldots, r$, and $\langle d_X \rangle \mathbb{B}_j = D$ for each $j = 0, \ldots, s$.* □

Rasiowa and Sikorski in [10, III, 7.1] proved the version of this theorem that has '$\mathrm{cl}_X\, \mathbb{B}_j = D$' in place of '$\langle d_X \rangle \mathbb{B}_j = D$'. That version follows from the above because if $\langle d_X \rangle \mathbb{B}_j = D$, then $\mathrm{cl}_X\, \mathbb{B}_j = \mathbb{B}_j \cup \langle d_X \rangle \mathbb{B}_j = \mathbb{B}_j \cup D = D$ since $\mathbb{B}_j \subseteq D$. But the two versions are equivalent, for by applying the version from [10] to r and $2s + 1$ we first obtain disjoint sets $\mathbb{B}_j^i$ with $\mathrm{cl}_X\, \mathbb{B}_j^i = D$ for $j = 0, \ldots, s$ and $i = 0, 1$, and define $\mathbb{B}_j = \mathbb{B}_j^0 \cup \mathbb{B}_j^1$ for each j. Then it can be shown that $\langle d_X \rangle \mathbb{B}_j = D$ [6, Section 7.4].

Tarski introduced the first version of the cl-formulation of this theorem in [14, satz 3.10]. It had $s = 0$ and required X to be separable. He credited the proof to Samuel Eilenberg, noting that he had originally proven the result himself for $\mathbb{R}$ and its dense-in-themselves subspaces. The restriction to $s = 0$ was removed in [9, theorem 3.5], where the property of X that the theorem asserts was called 'dissectability'. The separability restriction was removed in [10].

Theorem 7.2 allows a suitable morphism to be constructed from $\mathbb{G}$ onto any finite rooted reflexive transitive frame. But KD-frames may have irreflexive points, and we need further dissection results to handle this. First we state a result that is an instance of Theorem 7.8(1) of [6].

Theorem 7.3 *Let X be a dense-in-itself metric space and $\mathbb{U}$ be a non-empty open subset of X. Then there are disjoint non-empty subsets $\mathbb{I}_0, \mathbb{I}_1 \subseteq \mathbb{U}$ satisfying $\langle d_X \rangle \mathbb{I}_0 = \langle d_X \rangle \mathbb{I}_1 = \mathrm{cl}_X(\mathbb{U}) \setminus \mathbb{U}$.* □

The next results require X to be zero-dimensional.

Lemma 7.4 *Let X be a zero-dimensional dense-in-itself metric space and $\mathbb{G}$ be a non-empty open subset of X. Then $\mathbb{G}$ can be partitioned into non-empty open subsets $\mathbb{G}_0, \mathbb{G}_1$ such that $\mathrm{cl}_X(\mathbb{G}) \setminus \mathbb{G} = \mathrm{cl}_X(\mathbb{G}_i) \setminus \mathbb{G}_i$ for each $i = 0, 1$.*

Proof. If $\mathrm{cl}_X \mathbb{G} \subseteq \mathbb{G}$, then simply let $\mathbb{G}_0$ be any non-empty clopen proper subset of $\mathbb{G}$, and $\mathbb{G}_1 = \mathbb{G} \setminus \mathbb{G}_0$.

From now on, assume that $\mathrm{cl}_X \mathbb{G} \setminus \mathbb{G} \neq \emptyset$. By Theorem 7.3, there are disjoint non-empty $\mathbb{I}_0, \mathbb{I}_1 \subseteq \mathbb{G}$ such that $\langle d_X \rangle \mathbb{I}_i = \mathrm{cl}_X(\mathbb{G}) \setminus \mathbb{G}$ for each $i = 0, 1$. Both $\mathbb{I}_i$ are infinite (otherwise, $\langle d_X \rangle \mathbb{I}_i = \emptyset \neq \mathrm{cl}_X \mathbb{G} \setminus \mathbb{G}$).

For each $x \in \mathbb{I}_0$, let $m(x) = \min(d(x, \mathbb{I}_0 \setminus \{x\}), d(x, \mathbb{I}_1))$. By assumption, $\mathbb{G} \cap \langle d_X \rangle \mathbb{I}_i = \emptyset$ for each i. So $m(x) > 0$. For each such x, *using zero-dimensionality*, choose a clopen neighbourhood $B(x)$ of x with

$$B(x) \subseteq N_{m(x)/6}(x). \tag{7.1}$$

Note that $B(x) \cap \mathbb{I}_1 = \emptyset$. Now define $\mathbb{G}_0 = \bigcup_{x \in \mathbb{I}_0} B(x)$ and $\mathbb{G}_1 = \mathbb{G} \setminus \mathbb{G}_0$. Plainly, $\mathbb{G}_0$ is open and $\mathbb{I}_0 \subseteq \mathbb{G}_0$, and $\mathbb{G}_0 \cap \mathbb{I}_1 = \emptyset$ so $\mathbb{I}_1 \subseteq \mathbb{G}_1$.

Claim. $\mathbb{G}_1$ is open (in X).
Proof of claim. Let $a \in \mathbb{G}_1$ be arbitrary, and let $s = d(a, \mathbb{I}_0)$. Again, $s > 0$. Fix $x \in \mathbb{I}_0$ with $d(a, x) < 2s$. Since $a \notin \mathbb{G}_0 \supseteq B(x)$, and $B(x)$ is clopen, we can choose an open neighbourhood N of a disjoint from $B(x)$. We can further suppose that $N \subseteq \mathbb{G}$ and $N \subseteq N_{s/2}(a)$.

The claim will be proved if we show that $N \subseteq \mathbb{G}_1$, which we do by showing that $N \cap B(y) = \emptyset$ for each $y \in \mathbb{I}_0 \setminus \{x\}$. Take such a y, and let $d(a, y) = t$, say (we have $s \leq t$). Then

$$m(y) \leq d(y, \mathbb{I}_0 \setminus \{y\}) \leq d(y, x) \leq d(y, a) + d(a, x) < t + 2s \leq 3t.$$

So $m(y)/6 \leq t/2$. By (7.1), $B(y) \subseteq N_{t/2}(y)$. So $B(y) \cap N_{t/2}(a) = \emptyset$. But $N \subseteq N_{s/2}(a) \subseteq N_{t/2}(a)$, so $B(y) \cap N = \emptyset$ as required. This proves the claim.

So we have partitioned $\mathbb{G}$ into two non-empty open sets $\mathbb{G}_i$ with $\mathbb{I}_i \subseteq \mathbb{G}_i$ $(i = 0, 1)$. It remains to check that $\mathrm{cl}_X(\mathbb{G}) \setminus \mathbb{G} = \mathrm{cl}_X(\mathbb{G}_i) \setminus \mathbb{G}_i$ for each i:

$$\begin{aligned}
\mathrm{cl}_X(\mathbb{G}) \setminus \mathbb{G} &= \langle d_X \rangle \mathbb{I}_i && \text{by choice of } \mathbb{I}_i \\
&= \langle d_X \rangle \mathbb{I}_i \setminus \mathbb{G} && \text{since } \langle d_X \rangle \mathbb{I}_i \text{ is disjoint from } \mathbb{G} \\
&\subseteq \mathrm{cl}_X(\mathbb{G}_i) \setminus \mathbb{G} && \text{since } \mathbb{I}_i \subseteq \mathbb{G}_i \text{ so } \langle d_X \rangle \mathbb{I}_i \subseteq \langle d_X \rangle \mathbb{G}_i \subseteq \mathrm{cl}_X \mathbb{G}_i \\
&\subseteq \mathrm{cl}_X(\mathbb{G}_i) \setminus \mathbb{G}_i && \text{since } \mathbb{G}_i \subseteq \mathbb{G} \\
&= \mathrm{cl}_X(\mathbb{G}_i) \setminus (\mathbb{G}_i \cup \mathbb{G}_{1-i}) && \text{since } \mathbb{G}_{1-i} \text{ is open and disjoint from } \mathbb{G}_i, \\
& && \quad \text{so also disjoint from } \mathrm{cl}_X \mathbb{G}_i \\
&= \mathrm{cl}_X(\mathbb{G}_i) \setminus \mathbb{G} && \text{since } \mathbb{G}_0 \cup \mathbb{G}_1 = \mathbb{G} \\
&\subseteq \mathrm{cl}_X(\mathbb{G}) \setminus \mathbb{G} && \text{since } \mathbb{G}_i \subseteq \mathbb{G}.
\end{aligned}$$

This proves the Lemma. □

The partitioning of $\mathbb{G}$ can now be extended to any finite number of cells.

Theorem 7.5 (dissection) *Let $\mathbb{G}$ be a non-empty open subset of a zero-dimensional dense-in-itself metric space X, and let $n < \omega$. Then $\mathbb{G}$ can be partitioned into non-empty open subsets $\mathbb{G}_0, \ldots, \mathbb{G}_n$ such that $\mathrm{cl}_X(\mathbb{G}) \setminus \mathbb{G} = \mathrm{cl}_X(\mathbb{G}_i) \setminus \mathbb{G}_i$ for each $i \leq n$.*

Proof. By induction on n. If $n = 0$ we let $\mathbb{G}_0 = \mathbb{G}$. Now we assume the result for n and prove it for $n + 1$. By the inductive hypothesis, there is a partition $\mathbb{G}_0, \ldots, \mathbb{G}_n$ of $\mathbb{G}$ into non-empty open sets with $\mathrm{cl}_X(\mathbb{G}) \setminus \mathbb{G} = \mathrm{cl}_X(\mathbb{G}_i) \setminus \mathbb{G}_i$ for each $i \leq n$. By the preceding lemma, $\mathbb{G}_n$ can be partitioned into non-empty open sets $\mathbb{G}_n^0, \mathbb{G}_n^1$ with $\mathrm{cl}_X(\mathbb{G}_n) \setminus \mathbb{G}_n = \mathrm{cl}_X(\mathbb{G}_n^i) \setminus \mathbb{G}_n^i$ for each $i = 0, 1$. So $\mathrm{cl}_X(\mathbb{G}) \setminus \mathbb{G} = \mathrm{cl}_X(\mathbb{G}_n^i) \setminus \mathbb{G}_n^i$ for each $i = 0, 1$. The required partition is now $\mathbb{G}_0, \ldots, \mathbb{G}_{n-1}, \mathbb{G}_n^0, \mathbb{G}_n^1$.

□

8 d-Morphisms on Open Subspaces

Let X be a topological space, $\mathbb{G}$ a non-empty open subset of X, $\mathcal{F} = (W, R)$ a *finite* frame, and $\rho : \mathbb{G} \to W$ a map. Recall from Section 4 that ρ is a *d-morphism* from the space $\mathbb{G}$ to $\mathcal{F}$ when

$$\langle d_{\mathbb{G}} \rangle \rho^{-1}(w) = \rho^{-1}(R^{-1}(w)) \text{ for all } w \in W. \tag{8.1}$$

But the openness of $\mathbb{G}$ ensures (3.4) that $\langle d_{\mathbb{G}} \rangle \rho^{-1}(w) = \mathbb{G} \cap \langle d_X \rangle \rho^{-1}(w)$, so ρ is a d-morphism from $\mathbb{G}$ to $\mathcal{F}$ iff $\mathbb{G} \cap \langle d_X \rangle \rho^{-1}(w) = \rho^{-1}(R^{-1}(w))$ for all $w \in W$.

The following are some useful facts about the relations between d-morphisms on open subspaces. The proofs are left to the reader.

Lemma 8.1 *Let $\mathcal{F}' = (W', R')$ be a generated subframe of $\mathcal{F} = (W, R)$, let $\mathbb{T}$ and $\mathbb{G}$ be open subsets of a space X with $\emptyset \neq \mathbb{T} \subseteq \mathbb{G}$, and let $\rho : \mathbb{G} \to W'$ be a map.*

(1) *ρ is a d-morphism from $\mathbb{G}$ to $\mathcal{F}$ iff it is a d-morphism from $\mathbb{G}$ to $\mathcal{F}'$.*

(2) *$\rho \restriction \mathbb{T}$ is a d-morphism from $\mathbb{T}$ to $\mathcal{F}$ iff $\mathbb{T} \cap \langle d_X \rangle \rho^{-1}(w) = \mathbb{T} \cap \rho^{-1}(R^{-1}(w))$ for every $w \in W$.*

(3) *If $\mathbb{T} = \bigcup_{i \in I} \mathbb{G}_i$, where for each i, $\mathbb{G}_i$ is non-empty and open and $\rho \restriction \mathbb{G}_i$ is a d-morphism from $\mathbb{G}_i$ to $\mathcal{F}$, then $\rho \restriction \mathbb{T}$ is a d-morphism from $\mathbb{T}$ to $\mathcal{F}$.* □

A map $\rho : \mathbb{G} \to W$ is said to be *full* if $\mathrm{cl}_X(\mathbb{G}) \setminus \mathbb{G} \subseteq \langle d_X \rangle \rho^{-1}(w)$ for all $w \in W$. The following is the key result on the existence of full d-morphisms.

Theorem 8.2 *Let X be a zero-dimensional dense-in-itself metric space, and $\mathcal{F} = (W, R)$ be a finite, rooted, KD4-frame (i.e. R is transitive and serial). If $\mathbb{G} \subseteq X$ is a non-empty open set, then there is a surjective full d-morphism $\rho : \mathbb{G} \to W$ from $\mathbb{G}$ to $\mathcal{F}$.*

Proof. We construct a full d-morphism from $\mathbb{G}$ onto the rooted frame $\mathcal{F} = (W, R)$. To describe the structure of $\mathcal{F}$, define $R^\circ = R \cap R^{-1}$ and $R^\bullet = R \setminus R^{-1}$. Then $R^\circ(w) = \{v \in W : wRvRw\}$ and $R^\bullet(w) = \{v \in W : wRv \text{ and not } vRw\}$. The set $\{w\} \cup R^\circ(w)$ is the *cluster* of w. These clusters partition W. If w_0 is a root of $\mathcal{F}$, then $W = \{w_0\} \cup R^\circ(w_0) \cup R^\bullet(w_0)$, and there are two kinds of structure that $\mathcal{F}$ could have:

1. **w_0 is reflexive.** Then $w_0 \in R^\circ(w_0)$ and W is the disjoint union of $R^\circ(w_0)$ and $R^\bullet(w_0)$, with $R^\circ(w_0) \times W \subseteq R$. In this case it is possible that $R^\bullet(w_0) = \emptyset$, so then $\mathcal{F}$ consists of the single cluster $R^\circ(w_0)$. Either way, the relation R is universal on $R^\circ(w_0)$, whch has at least one element.
2. **w_0 is irreflexive.** Then $R^\circ(w_0) = \emptyset$ and W is the disjoint union of $\{w_0\}$ and $R^\bullet(w_0)$, with $\{w_0\} \times R^\bullet(w_0) \subseteq R$. As R is serial, $R^\bullet(w_0) \neq \emptyset$.

The proof proceeds by induction on the size of W. We make the induction hypothesis that the result holds for all frames smaller than $\mathcal{F}$, and consider the two cases for w_0 just described.

Case 1: w_0 is reflexive. Then $W = R^\circ(w_0) \cup R^\bullet(w_0)$ as above. By Theorem 7.2, $\mathbb{G}$ can be partitioned into non-empty sets $\mathbb{B}_v$ ($v \in R^\circ(w_0)$) and non-empty open sets $\mathbb{G}_u$ ($u \in R^\bullet(w_0)$) such that, for each $u \in R^\bullet(w_0)$ and $v \in R^\circ(w_0)$, we have

$$\mathrm{cl}_X(\mathbb{G}_u) \setminus \mathbb{G}_u = \langle d_X \rangle \mathbb{B}_v = \mathrm{cl}_X(\mathbb{G}) \setminus \bigcup_{w \in R^\bullet(w_0)} \mathbb{G}_w = D, \text{ say.} \tag{8.2}$$

Each $\mathcal{F}^*(u)$ for $u \in R^\bullet(w_0)$ is a finite rooted KD4-frame smaller than $\mathcal{F}$, since it does not contain w_0, so by the inductive hypothesis, there is a surjective full d-morphism ρ_u from $\mathbb{G}_u$ to $\mathcal{F}^*(u)$. Define $\rho : \mathbb{G} \to W$ by

$$\rho(x) = \begin{cases} \rho_u(x), & \text{if } x \in \mathbb{G}_u \text{ for some (unique) } u \in R^\bullet(w_0), \\ v, & \text{if } x \in \mathbb{B}_v \text{ for some (unique) } v \in R^\circ(w_0). \end{cases}$$

That is, $\rho = \bigcup_{u \in R^\bullet(w_0)} \rho_u \cup \bigcup_{v \in R^\circ(w_0)} \mathbb{B}_v \times \{v\}$. We will show that ρ is a surjective full d-morphism from $\mathbb{G}$ to $\mathcal{F}$. The following claim will help.

Claim. $D \subseteq \langle d_X \rangle \rho^{-1}(w)$ for every $w \in W$, where D is defined in (8.2).
Proof of claim. There are two cases. The first is when $w \in R^\bullet(w_0)$. Now (8.2) gives $D = \mathrm{cl}_X\, \mathbb{G}_w \setminus \mathbb{G}_w$. As $\rho_w : \mathbb{G}_w \to \mathcal{F}^*(w)$ is full, $\mathrm{cl}_X\, \mathbb{G}_w \setminus \mathbb{G}_w \subseteq \langle d_X \rangle \rho_w^{-1}(w) \subseteq \langle d_X \rangle \rho^{-1}(w)$ (since $\rho_w \subseteq \rho$).
The second case is when $w \notin R^\bullet(w_0)$. Since $w \in W = R^\circ(w_0) \cup R^\bullet(w_0)$, we have $w \in R^\circ(w_0)$. Hence $\rho^{-1}(w) = \mathbb{B}_w$. By (8.2), $D = \langle d_X \rangle \mathbb{B}_w = \langle d_X \rangle \rho^{-1}(w)$. This proves the claim.

We now check that ρ is a d-morphism from $\mathbb{G}$ to $\mathcal{F}$. So let $w \in W$. We require

$$\mathbb{G} \cap \langle d_X \rangle \rho^{-1}(w) = \rho^{-1}(R^{-1}(w)). \tag{8.3}$$

Since $\mathbb{G}$ is partitioned into the $\mathbb{G}_u$'s and $\mathbb{B}_v$'s, it is enough to prove, for all $u \in R^\bullet(w_0)$ and $v \in R^\circ(w_0)$, that the following equations hold.

$$\mathbb{G}_u \cap \langle d_X \rangle \rho^{-1}(w) = \mathbb{G}_u \cap \rho^{-1}(R^{-1}(w)), \tag{8.4}$$

$$\mathbb{B}_v \cap \langle d_X \rangle \rho^{-1}(w) = \mathbb{B}_v \cap \rho^{-1}(R^{-1}(w)). \tag{8.5}$$

(i) **Proof of** (8.4). Since $\mathbb{G}_u$ is open and $\rho \restriction \mathbb{G}_u = \rho_u$, a d-morphism from $\mathbb{G}_u$ to the generated subframe $\mathcal{F}^*(u)$ of $\mathcal{F}$, Lemma 8.1(1) implies that $\rho \restriction \mathbb{G}_u$ is a d-morphism from $\mathbb{G}_u$ to $\mathcal{F}$. Then Lemma 8.1(2) yields (8.4).
(ii) **Proof of** (8.5). $\mathbb{B}_v \subseteq D$ by definition of D (8.2), so $\mathbb{B}_v \subseteq \langle d_X \rangle \rho^{-1}(w)$ by the claim. Since $v \in R^\circ(w_0)$, we have vRw_0Rw (as w_0 is a reflexive root), hence vRw and $\mathbb{B}_v = \rho^{-1}(v) \subseteq \rho^{-1}(R^{-1}(w))$. Thus $\mathbb{B}_v \cap \langle d_X \rangle \rho^{-1}(w) = \mathbb{B}_v = \mathbb{B}_v \cap \rho^{-1}(R^{-1}(w))$.

So ρ is indeed a d-morphism from $\mathbb{G}$ to $\mathcal{F}$. Now $\mathrm{cl}_X\,\mathbb{G} \setminus \mathbb{G} \subseteq D$ by (8.2), so by the claim, $\mathrm{cl}_X\,\mathbb{G} \setminus \mathbb{G} \subseteq \langle d_X \rangle \rho^{-1}(w)$ for every $w \in W$, showing that ρ is full.

We also need that ρ is surjective. But if $u \in R^\bullet(w_0)$ then $u = \rho(x)$ for some $x \in \mathbb{G}_u$ as $\rho_u : \mathbb{G}_u \to R^*(u)$ is surjective, and if $v \in R^\circ(w_0)$ then $v = \rho(x)$ for all $x \in \mathbb{B}_v$ by definition.

Case 2: w_0 is irreflexive. Then $W = \{w_0\} \cup R^\bullet(w_0)$. By Theorem 7.3 with $\mathbb{U} = \mathbb{G}$, there are disjoint non-empty $\mathbb{I}, \mathbb{I}' \subseteq \mathbb{G}$ with $\langle d_X \rangle \mathbb{I} = \langle d_X \rangle \mathbb{I}' = \mathrm{cl}_X(\mathbb{G}) \setminus \mathbb{G}$.

Let $\mathbb{G}' = \mathbb{G} \setminus \mathbb{I}$. Plainly, $\mathbb{G}'$ is non-empty (it contains $\mathbb{I}'$) and open (since $\langle d_X \rangle \mathbb{I}$ is disjoint from $\mathbb{G}$, we have $\mathbb{G}' = \mathbb{G} \setminus \mathbb{I} = \mathbb{G} \setminus (\mathbb{I} \cup \langle d_X \rangle \mathbb{I}) = \mathbb{G} \setminus \mathrm{cl}_X\,\mathbb{I}$).

Claim. $\mathrm{cl}_X(\mathbb{G}') \setminus \mathbb{G}' = (\mathrm{cl}_X(\mathbb{G}) \setminus \mathbb{G}) \cup \mathbb{I}$.

Proof of claim. Plainly $\mathrm{cl}_X(\mathbb{G}') \setminus \mathbb{G}' = \mathrm{cl}_X(\mathbb{G}') \setminus (\mathbb{G} \setminus \mathbb{I}) \subseteq (\mathrm{cl}_X(\mathbb{G}') \setminus \mathbb{G}) \cup \mathbb{I} \subseteq (\mathrm{cl}_X(\mathbb{G}) \setminus \mathbb{G}) \cup \mathbb{I}$. Conversely, $\mathrm{cl}_X(\mathbb{G}) \setminus \mathbb{G} = \langle d_X \rangle \mathbb{I}' \subseteq \langle d_X \rangle \mathbb{G}' \subseteq \mathrm{cl}_X\,\mathbb{G}'$, and

$\mathbb{I} \subseteq \mathbb{G} \cap \mathrm{cl}_X\,\mathbb{G}$	clear
$= \mathbb{G} \cap \langle d_X \rangle \mathbb{G}$	since $\mathrm{cl}_X\,\mathbb{G} = \langle d_X \rangle \mathbb{G}$ by Lemma 7.1(3)
$= \mathbb{G} \cap (\langle d_X \rangle \mathbb{I} \cup \langle d_X \rangle \mathbb{G}')$	since $\mathbb{G} = \mathbb{I} \cup \mathbb{G}'$ and $\langle d_X \rangle$ is additive
$= \mathbb{G} \cap \langle d_X \rangle \mathbb{G}'$	since $\mathbb{G} \cap \langle d_X \rangle \mathbb{I} = \emptyset$
$\subseteq \langle d_X \rangle \mathbb{G}' \subseteq \mathrm{cl}_X\,\mathbb{G}'$	clear.

So $(\mathrm{cl}_X(\mathbb{G}) \setminus \mathbb{G}) \cup \mathbb{I} \subseteq \mathrm{cl}_X\,\mathbb{G}'$. The converse inclusion $(\mathrm{cl}_X(\mathbb{G}) \setminus \mathbb{G}) \cup \mathbb{I} \subseteq \mathrm{cl}_X(\mathbb{G}') \setminus \mathbb{G}'$ now follows, since both $\mathrm{cl}_X(\mathbb{G}) \setminus \mathbb{G}$ and $\mathbb{I}$ are disjoint from $\mathbb{G}'$. This proves the claim.

By Theorem 7.5, $\mathbb{G}'$ can be partitioned into non-empty open sets $\mathbb{G}_u$ ($u \in R^\bullet(w_0)$) with $\mathrm{cl}_X(\mathbb{G}') \setminus \mathbb{G}' = \mathrm{cl}_X(\mathbb{G}_u) \setminus \mathbb{G}_u$ for each such u. Here we need $R^\bullet(w_0) \neq \emptyset$, which holds by seriality of R and irreflexivity of w_0.

For each $u \in R^\bullet(w_0)$, the frame $\mathcal{F}^*(u)$ is a finite rooted KD4-frame smaller than $\mathcal{F}$. Inductively, there is a surjective full d-morphism ρ_u from $\mathbb{G}_u$ to $\mathcal{F}^*(u)$. Define $\rho : \mathbb{G} \to W$ by

$$\rho(x) = \begin{cases} \rho_u(x), & \text{if } x \in \mathbb{G}_u \text{ for some (unique) } u \in R^\bullet(w_0), \\ w_0, & \text{if } x \in \mathbb{I}. \end{cases}$$

Then ρ is surjective, similarly to Case 1. It helps below to note that

$$(\mathrm{cl}_X(\mathbb{G}) \setminus \mathbb{G}) \cup \mathbb{I} \subseteq \langle d_X \rangle \rho^{-1}(w) \quad \text{for all } w \in R^\bullet(w_0). \tag{8.6}$$

For, by the claim, for each $w \in R^\bullet(w_0)$ we have $(\mathrm{cl}_X(\mathbb{G}) \setminus \mathbb{G}) \cup \mathbb{I} = \mathrm{cl}_X(\mathbb{G}') \setminus \mathbb{G}' = \mathrm{cl}_X(\mathbb{G}_w) \setminus \mathbb{G}_w$. But $\mathrm{cl}_X(\mathbb{G}_w) \setminus \mathbb{G}_w \subseteq \langle d_X \rangle \rho^{-1}(w)$, by fullness of ρ_w and the fact that w is in $\mathcal{F}^*(w)$.

We now check that ρ is a d-morphism from $\mathbb{G}$ to $\mathcal{F}$. So let $w \in W$. We require (8.3) to hold. Since $\mathbb{G}$ is partitioned as $\mathbb{I} \cup \mathbb{G}'$, it is enough to prove the two equations

$$\mathbb{I} \cap \langle d_X \rangle \rho^{-1}(w) = \mathbb{I} \cap \rho^{-1}(R^{-1}(w)), \tag{8.7}$$

$$\mathbb{G}' \cap \langle d_X \rangle \rho^{-1}(w) = \mathbb{G}' \cap \rho^{-1}(R^{-1}(w)). \tag{8.8}$$

(i) **Proof of** (8.7).

(a) If $w \in R^\bullet(w_0)$ then $\mathbb{I} \cap \langle d_X \rangle \rho^{-1}(w) = \mathbb{I}$ by (8.6), and $\mathbb{I} \cap \rho^{-1}(R^{-1}(w)) = \mathbb{I}$ since $\mathbb{I} \subseteq \rho^{-1}(w_0)$ and $w_0 \in R^{-1}(w)$.

(b) Suppose instead that $w \notin R^\bullet(w_0)$. As R is transitive, $W = R^\bullet(w_0) \cup \{w_0\}$, so $w = w_0$ and $\rho^{-1}(w) = \mathbb{I}$. Thus $\mathbb{I} \cap \langle d_X \rangle \rho^{-1}(w) = \mathbb{I} \cap \langle d_X \rangle \mathbb{I} = \emptyset$. Also, $R^{-1}(w) = \emptyset$, so $\mathbb{I} \cap \rho^{-1}(R^{-1}(w)) = \emptyset$ as well.

(ii) **Proof of** (8.8). For each $w \in R^\bullet(w_0)$, $\mathbb{G}_w$ is non-empty and open, and $\rho \restriction \mathbb{G}_w = \rho_w$ is a d-morphism from $\mathbb{G}_w$ to $\mathcal{F}$. Since $\mathbb{G}' = \bigcup_{w \in R^\bullet(w_0)} \mathbb{G}_w$, Lemma 8.1(3) tells us that $\rho \restriction \mathbb{G}'$ is a d-morphism from $\mathbb{G}'$ to $\mathcal{F}$. Now (8.8) follows by Lemma 8.1(2).

So ρ is a d-morphism from $\mathbb{G}$ onto $\mathcal{F}$. It remains to show that it is full, i.e. that $\mathrm{cl}_X(\mathbb{G}) \setminus \mathbb{G} \subseteq \langle d_X \rangle \rho^{-1}(w)$ for all $w \in W$. For $w \in R^\bullet(w_0)$, the result follows from (8.6). But for $w = w_0$, we have $\mathrm{cl}_X(\mathbb{G}) \setminus \mathbb{G} = \langle d_X \rangle \mathbb{I} = \langle d_X \rangle \rho^{-1}(w_0)$.

This completes the induction and proves the Theorem. □

Corollary 8.3 *Let X be a zero-dimensional dense-in-itself metric space. Then for every finite KD4-frame $\mathcal{F} = (W, R)$, there is a surjective d-morphism from any non-empty open subset $\mathbb{G}$ of X to $\mathcal{F}$.*

Proof. Let $W = \{w_0, \ldots, w_n\}$. Then $W = R^*(w_0) \cup \cdots \cup R^*(w_n)$ and so $\mathcal{F}$ is the union of its rooted subframes $\mathcal{F}^*(w_0), \ldots, \mathcal{F}^*(w_n)$, which are also KD4-frames. As an open subspace of X, $\mathbb{G}$ is infinite (Lemma 7.1(1)) and zero-dimensional, so can be partitioned into non-empty open subsets $\mathbb{G}_0, \ldots, \mathbb{G}_n$. This follows from Theorem 7.5, but does not depend on it and is a standard fact that holds for any infinite zero-dimensional metric space.[5] For each $i \leq n$, by Theorem 8.2 there is a surjective d-morphism $\rho_i : \mathbb{G}_i \to R^*(w_i)$ from $\mathbb{G}_i$ to $\mathcal{F}^*(w_i)$. By Lemma 8.1(1), ρ_i is a d-morphism from $\mathbb{G}_i$ to $\mathcal{F}$. Put $\rho = \bigcup_{i \leq n} \rho_i$. Then ρ is a map $\mathbb{G} \to W$ that is surjective as the $R^*(w_i)$'s cover W and each ρ_i is surjective. For each i, $\rho \restriction \mathbb{G}_i$ is the d-morphism ρ_i from $\mathbb{G}_i$ to $\mathcal{F}$, so by Lemma 8.1(3), ρ is a d-morphism from $\mathbb{G}$ to $\mathcal{F}$. □

In [6, Section 4.8] we showed that KD4t has the finite model property over serial transitive frames. We can now apply that to zero-dimensional metric spaces.

Theorem 8.4 *If X is any zero-dimensional dense-in-itself metric space, then the $\mathcal{L}^{\langle dt \rangle}_{[d]}$-logic of X is KD4t.*

Proof. (A similar argument to Theorem 5.2.) *Soundness*: By Theorem 5.1, the $\mathcal{L}^{\langle dt \rangle}_{[d]}$-logic of any dense-in-itself metric space includes KD4t.

Completeness: if φ is an $\mathcal{L}^{\langle dt \rangle}_{[d]}$-formula that is not a KD4t-theorem, then by [6, Section 4.8] there is a finite frame $\mathcal{F}$ validating KD4t in which φ is not

---

[5] Actually it holds for any infinite totally disconnected space, a weaker property than zero-dimensionality.

valid. Let $\mathbb{G} = X$. Then by Corollary 8.3 there is a surjective d-morphism $\rho : X \to W$ from X to $\mathcal{F}$, so by Corollary 4.3, φ is not valid in X. □

The same argument shows that KD4t.U is the $\mathcal{L}^{\langle dt \rangle}_{[d]\forall}$-logic of X, because KD4t.U has the finite model property over KD4-frames [6, Section 4.9]. Of course the connectedness axiom C is not relevant here, as X is totally disconnected.

In conclusion we extend Theorem 8.4 to a *strong completeness* result.

Theorem 8.5 *If Γ is any countable KD4t-consistent set of formulas, and x_0 is any point of any zero-dimensional dense-in-itself metric space X, then Γ is satisfiable at x_0 in X.* □

Proof. Space restrictions prevent us from giving full details of the proof, but we can outline the main construction involved. First, a refinement of the proof of Theorem 8.4 shows that any non-theorem of KD4t is falsifiable at *any* x_0 in X. Hence any finite subset of Γ, being KD4t-consistent, is satisfiable at x_0 in some model.

We can assume without loss of generality that Γ is maximally KD4t-consistent. As Γ is countable, we can express it as $\bigcup_{n<\omega} \Gamma_n$, where each Γ_n is finite and $\Gamma_0 \subseteq \Gamma_1 \subseteq \cdots$. Then for each $n < \omega$, Γ_n is satisfiable at x_0, so there is an assignment $g_n : \mathsf{Var} \to \wp X$ such that $(X, g_n), x_0 \models \bigwedge \Gamma_n$. Because X is zero-dimensional and the Γ_n are finite, we can choose, by induction on n, a sequence $(C_n : n < \omega)$ of clopen sets with $X = C_0 \supseteq C_1 \supseteq \cdots$ and with the following properties holding for each $n > 0$, where we write $D_n = C_n \setminus C_{n+1}$:

C1 $x_0 \in C_n \subseteq N_{1/n}(x_0)$,

C2 for each formula $\langle d \rangle \varphi \in \Gamma_n$, there is $x \in D_n$ with $(X, g_n), x \models \varphi$,

C3 for each $\neg\langle d \rangle \varphi \in \Gamma_n$, we have $(X, g_n), x \not\models \varphi$ for each $x \in C_n \setminus \{x_0\}$.

Then by C1, $\bigcap_{n<\omega} C_n = \{x_0\}$. The D_n are clopen and pairwise disjoint and partition $X \setminus \{x_0\}$, and $C_m \setminus \{x_0\} = \bigcup_{m \leq n < \omega} D_n$ for each $m < \omega$.

Now an assignment $g : \mathsf{Var} \to \wp X$ can be defined by declaring $x_0 \in g(p)$ iff $p \in \Gamma$, and if $x \neq x_0$, then $x \in g(p)$ iff $x \in g_n(p)$ for the unique n such that $x \in D_n$. Since each D_n is open, an induction on φ (cf. Theorem 3.4) shows that for each formula φ and each $x \in D_n$ we have

$$(X, g), x \models \varphi \text{ iff } (X, g_n), x \models \varphi.$$

Using this, a further inductive argument involving C2 and C3 shows that for each formula φ we have $\varphi \in \Gamma$ iff $(X, g), x_0 \models \varphi$. Hence Γ is satisfiable in X at x_0 as required. □

By contrast, strong completeness fails for the Kripke frame semantics. If

$$\Sigma = \{p_0, q, [d]^*(p_{2n} \to \langle d \rangle (p_{2n+1} \wedge \neg q)), [d]^*(p_{2n+1} \to \langle d \rangle (p_{2n+2} \wedge q)) : n < \omega\},$$

then the set $\Sigma \cup \{\neg\langle dt \rangle \{q, \neg q\}\}$, discussed in [6, Section 4.4], is KD4t-consistent because each of its finite subsets is satisfiable in a KD4-frame. But

$\Sigma \cup \{\neg\langle dt\rangle\{q, \neg q\}\}$ is not itself satisfiable at any point of a transitive frame. For if Σ is satisfied at some such point w, then there is an endless R-path in the frame starting from w along which q and $\neg q$ are each satisfied infinitely often in the model in question, ensuring that $\neg\langle dt\rangle\{q, \neg q\}$ is false at w.

Strong completeness for topological semantics does not hold in general if the language is enriched by the universal modality. In [6, Section 10.5] it is shown that there exists a countable set of $\mathcal{L}_{[d]\forall}$-formulas that is finitely satisfiable, but not satisfiable, in any dense-in-itself metric space that is compact and locally connected.

References

[1] Aull, C. E. and W. J. Thron, *Separation axioms between T_0 and T_1*, Indagationes Mathematicae (Proceedings) **65** (1962), pp. 26–37.

[2] Bezhanishvili, G., L. Esakia and D. Gabelaia, *Some results on modal axiomatization and definability for topological spaces*, Studia Logica **81** (2005), pp. 325–355.

[3] Dawar, A. and M. Otto, *Modal characterisation theorems over special classes of frames*, Annals of Pure and Applied Logic **161** (2009), pp. 1–42.

[4] Fernández-Duque, D., *On the modal definability of simulability by finite transitive models*, Studia Logica **98** (2011), pp. 347–373.

[5] Fernández-Duque, D., *Tangled modal logic for spatial reasoning*, in: T. Walsh, editor, *Proceedings of the Twenty-Second International Joint Conference on Artificial Intelligence (IJCAI)* (2011), pp. 857–862.

[6] Goldblatt, R. and I. Hodkinson, *Spatial logic of modal mu-calculus and tangled closure operators* (2014), preprint at `http://arxiv.org/abs/1603.01766`.

[7] Janin, D. and I. Walukiewicz, *On the expressive completeness of the propositional mu-calculus with respect to monadic second order logic*, in: U. Montanari and V. Sassone, editors, *CONCUR '96: Concurrency Theory*, Lecture Notes in Computer Science **1119** (1996), pp. 263–277.

[8] Lucero-Bryan, J. G., *The d-logic of the real line*, Journal of Logic and Computation **23** (2013), pp. 121–156, `doi:10.1093/logcom/exr054`.

[9] McKinsey, J. C. C. and A. Tarski, *The algebra of topology*, Annals of Mathematics **45** (1944), pp. 141–191.

[10] Rasiowa, H. and R. Sikorski, "The Mathematics of Metamathematics," PWN–Polish Scientific Publishers, Warsaw, 1963.

[11] Shehtman, V., *«Everywhere» and «Here»*, Journal of Applied Non-Classical Logics **9** (1999), pp. 369–379.

[12] Shehtman, V., "Modal logic of Topological Spaces," Habilitation thesis, Moscow (2000), in Russian.

[13] Shehtman, V., *Derived sets in Euclidean spaces and modal logic*, Technical Report X-1990-05, University of Amsterdam (1990), `http://www.illc.uva.nl/Research/Publications/Reports/X-1990-05.text.pdf`.

[14] Tarski, A., *Der Aussagenkalkül und die Topologie*, Fundamenta Mathematicae **31** (1938), pp. 103–134, English translation by J. H. Woodger as *Sentential Calculus and Topology* in [16], 421–454.

[15] Tarski, A., *A lattice-theoretical fixpoint theorem and its applications*, Pacific Journal of Mathematics **5** (1955), pp. 285–309.

[16] Tarski, A., "Logic, Semantics, Metamathematics: Papers from 1923 to 1938," Oxford University Press, 1956, translated into English and edited by J. H. Woodger.

“Knowing value” logic as a normal modal logic

Tao Gu and Yanjing Wang

Department of Philosophy, Peking University

Abstract

Recent years witness a growing interest in nonstandard epistemic logics of “knowing whether”, “knowing what”, “knowing how”, and so on. These logics are usually not normal, i.e., the standard axioms and reasoning rules for modal logic may be invalid. In this paper, we show that the conditional “knowing value” logic proposed by Wang and Fan [12] can be viewed as a disguised normal modal logic by treating the negation of the Kv operator as a special diamond. Under this perspective, it turns out that the original first-order Kripke semantics can be greatly simplified by introducing a ternary relation R_i^c in standard Kripke models, which associates one world with two i-accessible worlds that do not agree on the value of constant c. Under intuitive constraints, the modal logic based on such Kripke models is exactly the one studied by Wang and Fan [12,13]. Moreover, there is a very natural binary generalization of the “knowing value” diamond, which, surprisingly, does not increase the expressive power of the logic. The resulting logic with the binary diamond has a transparent normal modal system, which sharpens our understanding of the “knowing value” logic and simplifies some previously hard problems.

Keywords: knowing value, normal modal logic, ternary relation, binary modality, first-order modal logic

1 Introduction

Classic epistemic logic à la von Wright and Hintikka mainly studies the inference patterns about propositional knowledge by using a modal operator K_i to express that agent i *knows that* a proposition is true. Epistemic logic has been successfully applied to various fields to capture knowledge and its change in multi-agent settings, such as distributed systems and imperfect information games (cf. e.g. [3,9]). However, in everyday life, knowledge is often expressed in terms of knowing the answer to an embedding question, such as “I know *whether* the claim is true”, “I know *what* your password is”, “I know *how* to prove the theorem” and so on. Recent years witness a growing interest in the logics of such knowledge expressions [7,8,12,13,4,5,10]. The fundamental idea is to simply treat “knowing

* Yanjing Wang acknowledges the support from the National Program for Special Support of Eminent Professionals and NSSF key projects 12&ZD119. The authors thank the anonymous reviewers for their helpful comments on an earlier version of this paper.

whether", "knowing what", "knowing how" as new modalities, just as "knowing that" in standard epistemic logic (cf. the survey [11]).

The resulting logics are usually not *normal* in the technical sense that usual modal axioms and rules may be invalid. For example, the K axiom for normal modal logic is not valid for the knowing whether operator, i.e., $Kw(p \to q) \land Kwp \to Kwq$ does not hold, e.g., knowing that p is false makes sure that you know whether p and also whether $p \to q$, but it does not tell you anything about the truth value of q. Similarly, knowing how to swim and knowing how to cook does not mean knowing how to swim and cook at the same time, thus invalidating $Khp \land Khq \to Kh(p \land q)$, a theorem in normal modal logic when taking Kh as a box modality.

On the other hand, the non-normality does not necessarily mean that we have to abandon Kripke models for more general models. As demonstrated in [5], we can still use Kripke models to accommodate those non-normal modal logics by using nonstandard yet intuitive truth conditions for the new modalities. However, there is usually a clear asymmetry between the relatively simple modal language and the "rich" model which may cause troubles in axiomatizing the logic. For example, the conditional knowing value logic proposed in [12] has the following language $\mathbf{ELKv}^r$ (where $i \in \mathbf{I}, p \in \mathbf{P}, c \in \mathbf{C}$ and $\mathbf{I}, \mathbf{P}, \mathbf{C}$ are countably infinite):

$$\phi ::= \top \mid p \mid \neg\phi \mid (\phi \land \phi) \mid K_i\phi \mid Kv_i(\phi, c)$$

where $Kv_i(\phi, c)$ says that i knows [what] the value of c [is], given ϕ, e.g., I know the password of this website given it is 4-digit, since I may have only one 4-digit password ever, although I am not sure which password I used for this website without the information on the digits. The language is interpreted on first-order Kripke models $\mathcal{M} = \langle S, D, \{\to_i : i \in \mathbf{I}\}, V, V_C \rangle$ where $\langle S, \{\to_i : i \in \mathbf{I}\}, V \rangle$ is a standard Kripke model, and D is a *constant* domain, and V_C assigns to each (non-rigid) $c \in \mathbf{C}$ an element in D on each $s \in S$. The semantics for the new Kv_i operator is as follows:

$$\mathcal{M}, s \vDash Kv_i(\phi, c) \iff \text{for any } t_1, t_2 : \text{ if } s \to_i t_1, s \to_i t_2, \mathcal{M}, t_1 \vDash \phi \text{ and } \mathcal{M}, t_2 \vDash \phi, \text{ then } V_C(c, t_1) = V_C(c, t_2).$$

According to this semantics, the formula $Kv_i(\phi, c)$ can also be understood as a first-order modal formula: $\exists x K_i(\phi \to c = x)$.[1] Thus $\mathbf{ELKv}^r$ can be viewed as a (small) fragment of first-order modal logic where a quantifier is packed with a modality. It is shown in [12] that $\mathbf{ELKv}^r$ is equally expressive as public announcement logic extended with unconditional Kv_i operators proposed in [7] (i.e., only $Kv_i(\top, c)$ are allowed). Satisfiability of $\mathbf{ELKv}^r$ over arbitrary models is PSPACE-complete, as proved in [2]. Note that although values are assigned to the constants in the model, we cannot talk about them explicitly in the language. In fact, we only care about whether on some worlds a given constant has exactly the *same* value. The contrast between the rich model and the simple language made the completeness proof of the following axiomatization $\mathbb{SELKV}^r\mathbb{S}5$ quite involved over multi-agent

[1] Note that there is a constant domain D and each c has a unique value on each state.

S5 models (cf. [13]).[2]

System $\mathbb{SELKV}^r\mathbb{S}5$

Axiom Schemas	
TAUT	all the instances of tautologies
DISTK	$K_i(p \to q) \to (K_i p \to K_i q)$
T	$K_i p \to p$
4	$K_i p \to K_i K_i p$
5	$\neg K_i p \to K_i \neg K_i p$
DISTKv^r	$K_i(p \to q) \to (Kv_i(q,c) \to Kv_i(p,c))$
Kv^{r}4	$Kv_i(p,c) \to K_i Kv_i(p,c)$
Kv$^r\bot$	$Kv_i(\bot,c)$
Kv$^r\vee$	$\hat{K}_i(p \wedge q) \wedge Kv_i(p,c) \wedge Kv_i(q,c) \to Kv_i(p \vee q, c)$

Rules	
MP	$\dfrac{p, p \to q}{q}$
NECK	$\dfrac{\phi}{K_i \phi}$
SUB	$\dfrac{\phi}{\phi[p/\psi]}$
RE	$\dfrac{\psi \leftrightarrow \chi}{\phi \leftrightarrow \phi[\psi/\chi]}$

Since the Kv_i operator is not a modality taking only propositions as arguments, it is hard to say whether the above logic is normal or not. DISTKv^r looks a little bit like the K axiom but it is in fact about the interaction between K_i and Kv_i. Kv^{r}4 is a variation of the positive introspection axiom, and the corresponding negative introspection is derivable. Kv$^r\bot$ says that the Kv_i operator is essentially a conditional. Axiom Kv$^r\vee$ handles the composition of the conditions, where $\hat{K}_i$ is the dual of K_i.

In this paper, we look at **ELKv^r** from a new yet "normal modal logic" perspective in order to answer the following questions:

(i) Since we do not talk about values in the language, is there a simpler value-free Kripke-model based semantics for **ELKv^r** that can keep the logic (valid formulas) the same? If so, we can restore the symmetry between the language and the model and understand the essence of our logic.

(ii) Can **ELKv^r** be linked to a normal modal logic (modulo some syntactic transformation)? If so, we can apply many standard modal logic techniques to simplify previously complicated discussions.

We give positive answers to both questions, inspired by a crucial observation:

Observation *$\neg Kv_i(\phi, c)$ can be viewed as a diamond operator $\Diamond_i^c \phi$ which says that there are two i-accessible ϕ-worlds, which do not agree on the value of c.*

Note that to simplify the technical discussion in order to reveal the crucial points, in this paper we focus on the logic over arbitrary models. Our techniques can be applied to the S5 setting.

The contributions of this paper are summarized as below:

- We give a simple alternative Kripke semantics to **ELKv^r** without value assignments, which does not change the set of valid formulas. The completeness proof is much simpler compared to the one in [13].

[2] $\phi[\psi/\chi]$ in the rule RE (replacement of equivalents) denotes any formula obtained by replacing *some* occurrences of ψ by χ.

- We generalize $\Diamond_i^c \phi$ in a natural way to a binary diamond operator $\Diamond_i^c(\phi, \psi)$. It turns out the generalization does not increase the expressive power of the logic but it can give us a transparent normal modal logic proof system.
- The normal modal logic perspective helps us to discover a bisimulation notion for **ELKv**^{r} and obtain a proof system for a weaker language proposed by [7].

Our findings show that **ELKv**^{r} is essentially a "disguised" normal logic, and this may help us to understand such nonstandard epistemic operators better.

The rest of the paper is organized as follows: we first introduce in Section 2 the language with the unary diamond $\Diamond_i^c$ and a semantics based on Kripke model with both binary and ternary relations under three intuitive constraints. We show that this semantics is equivalent to the original FO Kripke semantics of **ELKv**^{r} modulo validity (under a straightforward syntactic translation). In Section 3 we prove the completeness of the translated $\mathbb{SELKV}^r$ system w.r.t. the new semantics directly. This demonstrates the advantages of using this simplified semantics. In Section 4 we generalize the $\Diamond_i^c$ naturally to a binary one and show that the extended language is in fact equally expressive as **ELKv**^{r}. On the other hand, the extended language facilitates a transparent normal logic proof system. It then helps us in Section 5 to come up with a notion of bisimulation for **ELKv**^{r} and obtain a proof system for a weaker language proposed earlier. We conclude in the end with future directions.

2 Negation of Kv_i as a diamond

As we mentioned in the introduction, $\neg Kv_i(\phi, c)$ can be viewed as a diamond formula $\Diamond_i^c \phi$: there are two i-accessible ϕ-worlds which do not agree on the value of c. Then $\Box_i^c \phi := \neg \Diamond_i^c \neg \phi$ means that all the i-accessible $\neg\phi$-worlds agree on the value of c (and it is $Kv_i(\neg\phi, c)$ essentially.). For uniformity of the language, we take $\Box_i^c$ as the primitive symbol and introduce the following language (**MLKv**^{r}):

$$\phi ::= \top \mid p \mid \neg\phi \mid (\phi \wedge \phi) \mid \Box_i \phi \mid \Box_i^c \phi$$

where $p \in \mathbf{P}, c \in \mathbf{C}, i \in \mathbf{I}$. We can, without difficulty, inductively define a translation function T from the original **ELKv**^{r} to this modal language (the other way is also straightforward):

Definition 2.1 *A translation function T from **ELKv**^{r} to **MLKv**^{r} formulas is defined as follows:*

$$\begin{aligned} T(p) &= p \\ T(\neg\phi) &= \neg T(\phi) \\ T(\phi \wedge \psi) &= T(\phi) \wedge T(\psi) \\ T(K_i \phi) &= \Box_i T(\phi) \\ T(Kv_i(\phi, c)) &= \Box_i^c \neg T(\phi) \end{aligned}$$

Now we have the following translated axioms (the names are kept):

$$\begin{aligned} T(\texttt{DISTKv}^r) &= \Box_i(p \to q) \to (\Box_i^c \neg q \to \Box_i^c \neg p) \\ T(\texttt{Kv}^r\vee) &= \Diamond_i(p \wedge q) \wedge \Box_i^c \neg p \wedge \Box_i^c \neg q \to \Box_i^c \neg(p \vee q) \\ T(\texttt{Kv}^r\bot) &= \Box_i^c(\neg\bot) \end{aligned}$$

We can massage the axioms,[3] and obtain the following equivalent system (modulo translation T) from $\mathbb{SELKV}^r$ (i.e., $\mathbb{SELKV}^r$-$\mathbb{S}5$ without the S5 related axioms $\mathtt{T}, 4, 5, \mathtt{Kv}^r4$):

System $\mathbb{SMLKV}^r$

Axiom Schemas	
TAUT	all the instances of tautologies
DISTK	$\Box_i(p \to q) \to (\Box_i p \to \Box_i q)$
DISTKv^r	$\Box_i(p \to q) \to (\Box_i^c p \to \Box_i^c q)$
Kv^r∨	$\Diamond_i(p \wedge q) \wedge \Diamond_i^c(p \vee q) \to (\Diamond_i^c p \vee \Diamond_i^c q)$

Rules	
MP	$\dfrac{\phi, \phi \to \psi}{\psi}$
NECK	$\dfrac{\phi}{\Box_i \phi}$
NECKv^r	$\dfrac{\phi}{\Box_i^c \phi}$
SUB	$\dfrac{\phi}{\phi[p/\psi]}$
RE	$\dfrac{\psi \leftrightarrow \chi}{\phi \leftrightarrow \phi[\psi/\chi]}$

Note that instead of $\mathtt{Kv}^r\bot$, we have a more classic-looking rule NECKv^r. In fact, it is equivalent to have either $\mathtt{Kv}^r\bot$ or NECKv^r in the system: from NECKv^r, it is trivial to derive $\mathtt{Kv}^r\bot$, and from $\mathtt{Kv}^r\bot$, DISTKv^r and NECK it is also straightforward to derive each instance of NECKv^r by taking p in DISTKv^r as $\top$.

To a modal logician, $\mathbb{SMLKV}^r$ may look much more friendly compared to $\mathbb{SELKV}^r$. In particular, Kv^r∨ is simply a conditional distribution axiom for $\Diamond_i^c$ over disjunction. Note that $\Diamond_i(p \vee q) \to (\Diamond_i p \vee \Diamond_i q)$ is valid but $\Diamond_i^c(p \vee q) \to (\Diamond_i^c p \vee \Diamond_i^c q)$ is not, e.g., all the p worlds agree on the value of c and all the q worlds agree on the value of c but they just cannot agree with each other. This demonstrates that $\Diamond_i^c$ is apparently not a normal modality. However, as we will discover later that this apparent non-normality is a bit misleading and we will restore the normality in the next section by considering a natural binary generalization of the $\Diamond_i^c$ operator.

Now we are going to give a simplified but equivalent semantics to **MLKv**^{r} such that the system $\mathbb{SMLKV}^r$ is sound and complete. The idea is to abandon the first-order Kripke model and use a rather standard Kripke model for propositional modal logics since much of the information in the FO Kripke model is not relevant for the language **MLKv**^{r}.

Definition 2.2 *A model for **MLKv**^{r} is a tuple $\langle S, \{\to_i : i \in \mathbf{I}\}, \{R_i^c : i \in \mathbf{I}, c \in \mathbf{C}\}, V\rangle$, where*

- *$\langle S, \{\to_i : i \in \mathbf{I}\}, V\rangle$ is a standard Kripke model with binary relations.*
- *For each $c \in \mathbf{C}$, R_i^c is a triple relation over S satisfying for any $s, t, u, v \in S$:*

(i) *SYM: $sR_i^c tu \iff sR_i^c ut$*
(ii) *INCL: $sR_i^c tu$ only if $s \to_i t$ and $s \to_i u$*
(iii) *ATEUC: $sR_i^c tu$ and $s \to_i v$ imply that at least one of $sR_i^c tv$ and $sR_i^c uv$ holds*

Intuitively, $sR_i^c tu$ roughly means that s can see two i-accessible worlds t, u which do not agree on the value of c, although we do not have value assignments

[3] For example, T(DISTKv^r) is equivalent to $\Box_i(\neg q \to \neg p) \to (\Box_i^c \neg q \to \Box_i^c \neg p)$ under RE, which is equivalent to $\Box_i(p \to q) \to (\Box_i p \to \Box_i q)$ under SUB.

for c in the model. Further conditions are imposed to let the ternary relation really capture what we want. (i) is a symmetry condition on the later two arguments of R_i^c. Condition (ii) establishes the connection between the ternary and binary relations. The most crucial condition is (iii), an anti-euclidean property[4] that says if two i-accessible worlds do not agree on the value of c then for any third i-accessible world it must disagree with one of the two worlds on c.[5]

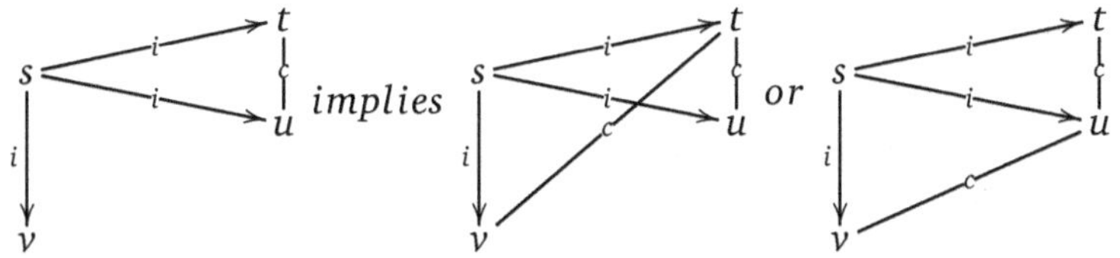

The new semantics is defined as follows which reflects the intuition behind R_i^c:

$\mathcal{M}, s \Vdash \top$		always
$\mathcal{M}, s \Vdash p$	$\Leftrightarrow$	$s \in V(p)$
$\mathcal{M}, s \Vdash \neg\phi$	$\Leftrightarrow$	$\mathcal{M}, s \nVdash \phi$
$\mathcal{M}, s \Vdash \phi \wedge \psi$	$\Leftrightarrow$	$\mathcal{M}, s \Vdash \phi$ and $\mathcal{M}, s \Vdash \psi$
$\mathcal{M}, s \Vdash \Diamond_i \phi$	$\Leftrightarrow$	there exists t such that $s \to_i t$ and $\mathcal{M}, t \Vdash \phi$
$\mathcal{M}, s \Vdash \Diamond_i^c \phi$	$\Leftrightarrow$	there exist u, v such that $sR_i^c uv$, $\mathcal{M}, u \Vdash \phi$ and $\mathcal{M}, v \Vdash \phi$

In the rest of this section, we show that the above semantics of **MLKv**^{r} is equivalent to the semantics for **ELKv**^{r} modulo the syntactic translation T. To show this, the difficult part is to saturate an **MLKv**^{r} model with value assignments while keeping the truth values of formulas modulo translation T. Note that this is not straightforward, as it is possible in an **MLKv**^{r} model that $sR_i^c tt$ and there is no way to assign to c two different values on the same world t. Moreover, it can happen that $sR_i^c uv$, $s \to_j u$ and $s \to_j v$ while it is not the case that $sR_j^c uv$. However, we can avoid such problem by preprocessing the **MLKv**^{r} model before assigning valuations.

Lemma 2.3 *For any set of* **ELKv**^{r} *formula* $\Sigma \cup \{\phi\}$, $\Sigma \vDash \phi$ *iff* $T(\Sigma) \Vdash T(\phi)$.

Proof It suffices to prove that for any set of **ELKv**^{r} formula Σ, Σ is $\vDash$-satisfiable iff $T(\Sigma)$ is $\Vdash$-satisfiable. We say that an **ELKv**^{r} model $\mathcal{M}, s$ is *equivalent* to an **MLKv**^{r} model $\mathcal{N}, t$ if for all $\phi \in$ **ELKv**^{r}: $\mathcal{M}, s \vDash \phi \iff \mathcal{N}, t \Vdash T(\phi)$. In the following we show that for any pointed FO Kripke model $\mathcal{M}, s$ for **ELKv**^{r}, there is an equivalent **MLKv**^{r} model $\mathcal{N}, t$, and vice versa.

$\Rightarrow$: For any pointed FO Kripke model $\mathcal{M}, s$, we can naturally define the ternary relation R_i^c as follows: $sR_i^c tu$ iff $s \to_i t$, $s \to_i u$ and $V_C(c, t) \neq V_C(c, u)$. It's straight-

[4] Euclidean property says that $\forall x, y, z : xRy \wedge xRz \to yRz$. Taking $R' = \bar{R}$ we have $\forall x, y, z : \neg xR'y \wedge \neg xR'z \to \neg yR'z$, i.e., $\forall x, y, z : yR'z \to xR'y \vee xR'z$. Our condition is inspired by this observation.

[5] Careful readers may wonder about whether we can break the the ternary relation into two: the i-relation and an anti-equivalence relation. We will come back to this point at the end of the paper.

forward to check that the resulting model $\mathcal{N}$ is an $\mathbf{MLKv}^r$ model (satisfying the three conditions) and $\mathcal{N}, s$ is equivalent to $\mathcal{M}, s$.

$\Leftarrow$: Recall that what we need to show is the following: given an $\mathbf{MLKv}^r$ model $\mathcal{N} = \langle S, \{\rightarrow_i : i \in \mathbf{I}\}, \{R_i^c : i \in \mathbf{I}, c \in \mathbf{C}\}, V\rangle$ and $t \in S$, find an $\mathbf{ELKv}^r$ model $\mathcal{M}, s$ such that $\mathcal{M}, s \vDash \phi \iff \mathcal{N}, t \Vdash T(\phi)$. As mentioned, we need to preprocess $\mathcal{N}$ before assigning values.

The preprocessing consists of two steps: splitting and unraveling. We first split the states in $\mathcal{N}$ into two copies in order to handle the $uR_i^c vv$ problem mentioned before. Let $\mathcal{N}'$ be $\langle S \times \{0,1\}, \{\rightarrow'_i : i \in \mathbf{I}\}, \{P_i^c : i \in \mathbf{I}, c \in \mathbf{C}\}, V'\rangle$, where:

- $(u,x) \rightarrow'_i (v,y) \iff u \rightarrow_i v$
- $(u,x)P_i^c(v,y)(w,z) \iff uR_i^c vw$ and $(v,y) \neq (w,z)$
- $V'((u,x)) = V(u)$

It can be verified that $\mathcal{N}'$ has the three properties of $\mathbf{MLKv}^r$ models,[6] and there is no state v such that $uP_i^c vv$ for any u, v. We can prove the following claim by a simple induction on the structure of $\mathbf{MLKv}^r$ formulas:

$$\mathcal{N}, u \equiv_{\mathbf{MLKv}^r} \mathcal{N}', (u,x) \text{ where } x \in \{0,1\}$$

The only non-trivial case is when $\phi = \Diamond_i^c \psi$. Suppose $\mathcal{N}, u \Vdash \Diamond_i^c \psi$, then there exist $v, v' \in S$ (v and v' are not necessarily different) such that $uR_i^c vv'$, $\mathcal{N}, v \Vdash \psi$ and $\mathcal{N}, v' \Vdash \psi$. By the definition of P_i^c and the induction hypothesis, $(u,x)P_i^c(v,0)(v',1)$, $\mathcal{N}', (v,0) \Vdash \psi$ and $\mathcal{N}', (v',1) \Vdash \psi$, so $\mathcal{N}', (u,x) \Vdash \Diamond_i^c \psi$. Suppose $\mathcal{N}', (u,x) \Vdash \Diamond_i^c \psi$, then there exist $(v,y)(v',z)$ such that $(u,x)P_i^c(v,y)(v',z)$, $\mathcal{N}, (v,y) \Vdash \psi$ and $\mathcal{N}, (v',z) \Vdash \psi$. According to the definition of P_i^c, this entails $(v,y) \neq (v',z)$ and $uR_i^c vv'$. By induction hypothesis, $\mathcal{N}, (v,y) \Vdash \psi$ and $\mathcal{N}, (v',z) \Vdash \psi$, so $\mathcal{N}, u \Vdash \Diamond_i^c \psi$. As a simple consequence,

$$\mathcal{N}, s \equiv_{\mathbf{MLKv}^r} \mathcal{N}', (s,0) \tag{1}$$

To simplify notation, we shall write $(s,0)$ as s' in the rest of this proof.

Now we unravel $\mathcal{N}'$ at s' into $\mathcal{M}' = \langle W, \{\hookrightarrow_i : i \in I\}, \{Q_i^c : i \in \mathbf{I}, c \in \mathbf{C}\}, U'\rangle$:

- $W = \{\langle s', i_1, v_1, \ldots, i_k, v_k\rangle$: there is a path $s' \xrightarrow{i_1} v_1 \ldots \xrightarrow{i_k} v_k$ in $\mathcal{N}'\}$. Note that the trivial path $\langle s'\rangle \in W$,
- $\langle s', i_1, \ldots, v_k\rangle \hookrightarrow_i \langle s', j_1, \ldots, u_m\rangle$ iff $m = k+1$, $\langle s', i_1, \ldots, v_k\rangle = \langle s', j_1, \ldots, u_k\rangle$, $j_m = i$ and $v_k \rightarrow'_i u_m$ in $\mathcal{N}'$,
- $\langle s', i_1, \ldots, v_k\rangle Q_i^c \langle s', j_1, \ldots, u_m\rangle\langle s', l_1, \ldots, l_n\rangle$ iff $v_k P_i^c u_m l_n$, $\langle s', i_1, \ldots, v_k\rangle \hookrightarrow_i \langle s', j_1, \ldots, u_m\rangle$ and $\langle s', i_1, \ldots, v_k\rangle \hookrightarrow_i \langle s', l_1, \ldots, l_n\rangle$,
- $U'(\langle s', i_1, \ldots, u\rangle) = V'(u)$.

[6] Take the anti-euclidean property as an example. Suppose $(u,x)P_i^c(v,y)(v',y')$ and $(u,x) \rightarrow'_i (w,z)$. If (w,z) is one of (v,y) and (v',y'), done. If not, then $(u,x)P_i^c(v,y)(v',y')$ implies $uR_i^c vw$, and $(u,x) \rightarrow'_i (w,z)$ implies $u \rightarrow_i w$. By the anti-euclidean property of the original model $\mathcal{N}$, we have either $uR_i^c vw$ or $uR_i^c v'w$. So either $(u,x)R_i^c(v,y)(w,z)$ or $(u,x)R_i^c(v',y')(w,z)$.

Intuitively, the new model $\mathcal{M}'$ starts from $\langle s'\rangle$, and each state corresponds to a path which is accessible from s' in $\mathcal{N}'$. It is not hard to verify the three properties of **MLKv**^{r} models.[7] By definition, the $\hookrightarrow$ skeleton of $\mathcal{M}'$ is a tree-like structure: acyclic, every state except the root $\langle s'\rangle$ can be reached eventually by $\langle s'\rangle$ and has one and only one predecessor. It follows that for any u,v in $\mathcal{M}'$, it is not the case that $u \hookrightarrow_i v$ and $u \hookrightarrow_j v$ for any $i \neq j$.

Now we can prove the following by induction on the structure of $\phi \in \mathbf{MLKv}^r$: $\mathcal{M}', \langle s', i_1, v_1, \ldots, i_k, v_k\rangle \equiv_{\mathbf{MLKv}^r} \mathcal{N}', v_k$ In particular,

$$\mathcal{N}', s' \equiv_{\mathbf{MLKv}^r} \mathcal{M}', \langle s'\rangle \tag{2}$$

Now the only thing left is to transform the **MLKv**^{r} model $\mathcal{M}'$ into an equivalent **ELKv**^{r} model $\mathcal{M}$. Basically, we just need to give values to $c \in \mathbf{C}$ on each state according to the ternary relations Q^c_j. Let $\mathcal{M} = \langle W, D, \{\hookrightarrow_i : i \in \mathbf{I}\}, U, V_{\mathbf{C}}\rangle$ where:

- W and $\{\hookrightarrow_i : i \in \mathbf{I}\}$ are exactly the same as in $\mathcal{M}'$;
- $U = U'$;
- $V_{\mathbf{C}}(c,w) = |(c,w)|_\sim$. That is, $V_{\mathbf{C}}(c,w)$ is the equivalence class under the equivalence relation $\sim$ over $\mathbf{C} \times W$ defined as:
  $\sim\ = \{\langle (c,u),(e,v)\rangle : c = e, \exists s \exists j : s \hookrightarrow_j u, s \hookrightarrow_j v, \forall w \in W : \neg w Q^c_j uv\} \cup \{\langle (c,u),(c,u)\rangle \mid (c,u) \in \mathbf{C} \times W\}$
- $D = \{|(c,w)|_\sim \mid (c,w) \in \mathbf{C} \times W\}$;

To make sure $\mathcal{M}$ is well-defined, we need to show that $\sim$ is an equivalence relation. Reflexivity and symmetry are obvious, and for transitivity: If $(c,w) \sim (d,u)$, $(d,u) \sim (e,v)$, then $c = d = e$, there exist s, i such that $s \hookrightarrow_i w$, $s \hookrightarrow_i u$ while for any t not $t Q^c_i wu$, and there exist s', j such that $s' \hookrightarrow_j u$, $s' \hookrightarrow_j v$ while for any t not $t Q^c_j uv$. Since every state in W has at most one predecessor, $s = s'$. Since there is at most one relation between two different states, $i = j$. Therefore $s \hookrightarrow_i w$, $s \hookrightarrow_i v$ and $s \hookrightarrow_i u$. Suppose towards contradiction that there exists $o \in W$ such that $o Q^c_i wv$, then $o = s$. Thus $s Q^c_i wu$ or $s Q^c_i uv$ by anti-euclidean property, contradiction. Therefore $(c,w) \sim (e,v)$.

We still need to verify that this assignment is good, in the sense that: for any **ELKv**^{r} formula ϕ, $\mathcal{M}', w \Vdash T(\phi) \iff \mathcal{M}, w \vDash \phi$ for any $w \in W$. We prove this by induction on ϕ and only show the non-trivial case:

If $\phi = Kv_i(\psi, c)$, then $T(\phi) = \Box^c_i \neg T(\psi)$.

$\Rightarrow$: Suppose $\mathcal{M}, w \nvDash Kv_i(\psi, c)$ then there exist t, t' such that $w \hookrightarrow_i t$, $w \hookrightarrow_i t$, $\mathcal{M}, t \vDash \psi$, $\mathcal{M}, t' \vDash \psi$ and $(c,t) \nsim (c,t')$. According to the definition of $\sim$, this implies $\exists u$ such that $u Q^c_i tt'$. But we have shown that every state has exactly one predecessor, so $u = w$, and $w Q^c_i tt'$. By induction hypothesis, $\mathcal{M}', t \Vdash T(\psi)$ and $\mathcal{M}', t' \Vdash T(\psi)$. Therefore, $\mathcal{M}', w \Vdash \Diamond^c_i T(\psi)$, i.e., $\mathcal{M}', w \nVdash \Box^c_i \neg T(\psi)$.

---

[7] Again, take the anti-euclidean property as an example. Suppose $u Q^c_i vv'$, $u \hookrightarrow_i t$. Suppose $u = \langle s', \ldots, u_k\rangle$, $v = \langle s', \ldots, v_m\rangle$, $v' = \langle s', \ldots, v'_n\rangle$, $w = \langle s', \ldots, w_l\rangle$. Then $u_k P^c_i v_m v'_n$ and $u_k \to'_i w_l$, which implies at least one of $u_k P^c_i v_m w_l$ and $u_k P^c_i v'_n w_l$ holds. This together with $u \hookrightarrow_i v$, $u \hookrightarrow_i v'$ and $u \hookrightarrow_i w$ imply either $u Q^c_i vw$ or $u Q^c_i v'w$.

$\Leftarrow$: Suppose $\mathcal{M}', w \nVdash \Box_i^c \neg T(\psi)$, i.e. $\mathcal{M}', w \Vdash \Diamond_i^c T(\psi)$. Then there exist $t, t' \in W$ such that $wQ_i^c tt'$, $\mathcal{M}', t \Vdash T(\psi)$ and $\mathcal{M}', t' \Vdash T(\psi)$. So $w \hookrightarrow_i t, w \hookrightarrow_i t'$ but $(c, t) \not\sim (c, t')$, i.e. $V_\mathbf{C}(c, t) \neq V_\mathbf{C}(c, t')$. By induction hypothesis, $\mathcal{M}, t \vDash \psi$ and $\mathcal{M}, t' \vDash \psi$. Therefore, $\mathcal{M}, w \nvDash Kv_i(\psi, c)$.

It follows that for any **ELKv**^{r} formula ϕ,

$$\mathcal{M}', \langle s' \rangle \Vdash T(\phi) \iff \mathcal{M}, \langle s' \rangle \vDash \phi \tag{3}$$

With (1), (2) and (3), we can now conclude that for any **MLKv**^{r} model $\mathcal{N}, t$ there is always an equivalent **ELKv**^{r} model $\mathcal{M}, s$ and this concludes the proof. □

Remark 2.4 The above lemma implies that for any **ELKv**^{r} formula ϕ:

$$\vDash \phi \iff \Vdash T(\phi)$$

which asserts the validities are the same modulo the translation. We need the stronger version to handle strong completeness later.

3 Completeness of $\mathbb{SMLKV}^r$

In this section, we show a direct proof of the strong completeness of $\mathbb{SMLKV}^r$ proposed in the previous section. As we will see, this proof is much simpler compared to the original completeness proof of $\mathbb{SELKV}^r$ in [13] due to the fact that we do not need to construct a FO canonical Kripke model with value assignments anymore.

Definition 3.1 *The canonical model of* $\mathbb{SMLKV}^r$ *is a tuple*

$$\mathcal{M} = \langle S, \{\to_i : i \in \mathbf{I}\}, \{R_i^c : i \in \mathbf{I}, c \in \mathbf{C}\}, V \rangle$$

where:

- *S is the set of all maximal $\mathbb{SMLKV}^r$-consistent sets of **MLKv**^{r} formulas,*
- $s \to_i t \iff \{\phi : \Box_i \phi \in s\} \subseteq t$,
- $sR_i^c tu \iff$ *(1)*$\{\phi : \Box_i \phi \in s\} \subseteq t \cap u$ *and (2)*$\{\psi : \Box_i^c \psi \in s\} \subseteq t \cup u$,
- $V(s) = \{p : p \in s\}$.

Note that condition (2) for R_i^c says that if s can see two i-accessible worlds which do not agree on c then at least one should satisfy ψ for each $Kv_i(\neg\psi, c) \in s$.

Proposition 3.2 *The canonical model $\mathcal{M}$ is an **MLKv**^{r} model.*

Proof We only need to check the three conditions of R_i^c.

(i) $sR_i^c uv \Rightarrow sR_i^c vu$: Obvious.

(ii) $sR_i^c uv \Rightarrow s \to_i u$: By condition (1) in the definition of R_i^c.

(iii) $sR_i^c uv$ and $s \to_i t \Rightarrow$ either $sR_i^c ut$ or $sR_i^c tv$: Suppose not. Then according to the definition of R_i^c, we have $\{\psi : \Box_i^c \psi \in s\} \nsubseteq u \cup t$ and $\{\psi : \Box_i^c \psi \in s\} \nsubseteq v \cup t$. So there exist $\psi_1, \psi_2 \in \{\psi : \Box_i^c \psi \in s\}$ such that $\psi_1 \notin u \cup t$ and

$\psi_2 \notin v \cup t$. According to the property of maximal consistent sets, this entails $\psi_1 \wedge \psi_2 \notin u \cup t$ and $\psi_1 \wedge \psi_2 \notin v \cup t$. Now, we distinguish two situations: $\Diamond_i(\neg\psi_1 \wedge \neg\psi_2) \in s$ and $\Diamond_i(\neg\psi_1 \wedge \neg\psi_2) \notin s$, and go on to show that in both cases we would arrive at contradiction.

Suppose $\Diamond_i(\neg\psi_1 \wedge \neg\psi_2) \in s$. Note that since $\psi_1, \psi_2 \in \{\psi : \Box_i^c\psi \in s\}$, $\Box_i^c\psi_1 \in s$ and $\Box_i^c\psi_2 \in s$. Then according to $\mathtt{Kv}^r\vee$, we have $\Box_i^c(\psi_1 \wedge \psi_2) \in s$. So $\psi_1 \wedge \psi_2 \in \{\psi : \Box_i^c\psi \in s\}$. Since $\{\psi : \Box_i^c\psi \in s\} \subseteq u \cup v$, $\psi_1 \wedge \psi_2 \in u \cup v$. But this means that $\psi_1 \wedge \psi_2 \in u \cup t$ or $v \cup t$, contradiction.

Suppose $\Diamond_i(\neg\psi_1 \wedge \neg\psi_2) \notin s$, then $\Box_i(\psi_1 \vee \psi_2) \in s$. According to the definition of R_i^c, we have $\psi_1 \vee \psi_2 \in t$. By the property of MCS, at least one of ψ_1 and ψ_2 is in t. However, since $\psi_1 \notin u \cup t$ and $\psi_2 \notin v \cup t$, we have $\psi_1, \psi_2 \notin t$, contradiction.

Therefore, the canonical model $\mathcal{M}$ is indeed an $\mathbf{MLKv}^r$ model. □

By a Lindenbaum-like argument, every consistent set of $\mathbf{MLKv}^r$ formulas can be extended to a maximal consistent set (of $\mathbf{MLKv}^r$ formulas). In the following we (as routine) prove the existence lemma for both modalities $\Diamond_i$ and $\Diamond_i^c$ in order to obtain the truth lemma. The proof is the $\mathbb{SMLKV}^r$ adaption of the proof of $\mathbb{SELKV}^r$ in [13].

Given a state $s \in S$ such that $\Diamond_i^c\phi \in s$. We let $Z = \{\psi \mid \Box_i\psi \in s\} \cup \{\phi\}$ and $X = \{\chi \mid \Box_i^c\chi \in s\}$. Since X is countable, we list the elements in X as χ_i for $i \in \mathbb{N}$. Note that since $\vdash \Box_i^c\top$, $\top \in X$, namely X is non-empty.

Fact 3.3 *For any $\chi \in X$, $\{\chi\} \cup Z$ is consistent. Therefore Z and every χ are also consistent.*

Proof Suppose not, then there exists $\chi \in X, \psi_1, \ldots, \psi_n \in Z$ such that $\vdash \psi_1 \wedge \cdots \wedge \psi_n \wedge \phi \to \neg\chi$. By NECK and DISTK, we have $\vdash \Box_i(\psi_1 \wedge \cdots \wedge \psi_n) \to \Box_i(\phi \to \neg\chi)$. Since $\Box_i\psi_1, \ldots, \Box_i\psi_n \in s$, $\Box_i(\phi \to \neg\chi) \in s$. Note that $\mathtt{Kv}^r\vee$ is equivalent to $\Box_i(p \to q) \to (\Diamond_i^c p \to \Diamond_i^c q)$. By SUB, we have $\vdash \Box_i(\phi \to \neg\chi) \wedge \Diamond_i^c\phi \to \Diamond_i^c\neg\chi$. This together with the fact that $\Box_i(\phi \to \neg\chi) \in s$ and $\Diamond_i^c\phi \in s$ (assumption), we have $\Diamond_i^c\neg\chi \in s$, contradiction. Since $\top \in X$, $\{\top\} \cup Z$ is consistent thus Z is consistent. □

Let $B_0 = Z \cup \{\chi_0\}$, $C_0 = Z$. We inductively construct B_n and C_n as following:

- If $B_n \cup \{\chi_{n+1}\}$ is consistent, then $B_{n+1} = B_n \cup \{\chi_{n+1}\}$, $C_{n+1} = C_n$.
- Else, $B_{n+1} = B_n$, $C_{n+1} = C_n \cup \{\chi_{n+1}\}$.
- Finally, let $B = \bigcup_{n<\omega} B_n$, $C = \bigcup_{n<\omega} C_n$.

In order to show that B and C are consistent we first show that B_n and C_n are consistent for each $n < \omega$.

Proposition 3.4 *For any $k \geq 0$, if B_k is consistent and χ_{k+1} is not consistent with B_k, then χ_{k+1} is consistent with C_k. Therefore B_k and C_k are consistent for $k \in \mathbb{N}$.*

Proof Suppose not, i.e., χ_{k+1} is not consistent with both B_k and C_k. Let $U = B_k \setminus Z$, $V = C_k \setminus Z$, $\overline{U} = \{\neg\psi : \psi \in U\}$, and $\overline{V} = \{\neg\psi : \psi \in V\}$. Then there exist $\alpha_1, \ldots, \alpha_l, \beta_1, \ldots, \beta_m, \gamma_1, \ldots, \gamma_n \in Z$ such that:

- $\vdash \alpha_1 \wedge \cdots \wedge \alpha_l \wedge \bigwedge U \wedge \phi \to \neg\chi_{k+1}$

- $\vdash \beta_1 \wedge \cdots \wedge \beta_m \wedge \bigwedge V \wedge \phi \to \neg\chi_{k+1}$
- $\vdash \gamma_1 \wedge \cdots \wedge \gamma_n \wedge \bigwedge U \wedge \phi \to \bigwedge \overline{V}$

The last one is due to the fact that any formula in $C_k \setminus Z$ is inconsistent with B_k by construction. By NECK, DISTK and the definition of Z and X, we have

- $\Box_i(\bigwedge U \wedge \phi \to \neg\chi_{k+1}) \in s$
- $\Box_i(\bigwedge V \wedge \phi \to \neg\chi_{k+1}) \in s$
- $\Box_i(\bigwedge U \wedge \phi \to \bigwedge \overline{V}) \in s$

First, we claim that $\Diamond_i(\bigwedge U \wedge \phi) \in s$. If not, then $\Box_i\neg(\bigwedge U \wedge \phi) \in s$, which means that $\neg(\bigwedge U \wedge \phi) \in Z \subseteq B_k$. But as $U \subseteq B_k$, $\phi \in B_k$, this implies that B_k is inconsistent, contradiction.

Then, we claim that $\Diamond_i(\neg\chi_{k+1} \wedge \bigwedge \overline{V}) \in s$. Since $\Box_i(\bigwedge U \wedge \phi \to \neg\chi_{k+1}) \in s$ and $\Box_i(\bigwedge U \wedge \phi \to \bigwedge \overline{V}) \in s$ then $\Box_i(\bigwedge U \wedge \phi \to \neg\chi_{k+1} \wedge \bigwedge \overline{V})$, we immediately get $\Diamond_i(\neg\chi_{k+1} \wedge \bigwedge \overline{V}) \in s$ due to the fact $\Diamond_i(\bigwedge U \wedge \phi) \in s$ that we just showed.

Finally, since $\Box_i^c \chi_{k+1} \in s$ and $\Box_i^c \psi \in s$ for all $\psi \in V$, $\Box_i^c(\chi_{k+1} \wedge \bigwedge V) \in s$. Therefore $\Box_i^c \neg\phi \in s$, contradiction to $\Diamond_i^c \phi \in s$.

Now, we can prove that B_k and C_k are consistent for any $k \in \mathbb{N}$. We do induction on k. For $k = 0$, then $B_0 = C_0 = Z$, whose consistency is shown in *Fact 3.3*. For $k = i+1$, consider whether χ_{i+1} is consistent with B_i. If χ_{i+1} is consistent with B_i, then $B_k = B_i \cup \chi_{i+1}$ and $C_k = C_i$ (by induction hypothesis) are consistent. If χ_{i+1} is inconsistent with B_i, then by induction hypothesis, $B_i = B_{i+1}$ and C_i are consistent. So according to the above conclusion C_{i+1} is also consistent. □

Proposition 3.5 *$B = \bigcup_{n<\omega} B_n$ and $C = \bigcup_{n<\omega} C_n$ are both consistent.*

Proof Suppose B is not consistent. That is, there exist $\phi_1, \ldots, \phi_n \in B$ such that $\vdash \phi_1 \wedge \cdots \wedge \phi_n \to \bot$. Therefore, there must be a finite m such that $\phi_1, \ldots, \phi_n \in B_m$. But this means that B_m is already inconsistent, contradictory to the construction of B_k. The case for C is similar. □

It is routine to prove the following:

Lemma 3.6 *(Existence Lemma for $\Diamond_i$) Given a state $s \in S$. If $\Diamond_i \phi \in s$, then there exists $t \in S$ such that $s \to_i t$ and $\phi \in t$;*

Also we have the existence lemma for $\Diamond_i^c$:

Lemma 3.7 *(Existence Lemma for $\Diamond_i^c$) Given a state $s \in S$. If $\Diamond_i^c \psi \in s$, then there exist $t, u \in S$ such that $sR_i^c tu$ and $\psi \in t \cap u$.*

Proof Let Z, B and C be defined as above. Due to Proposition 4.5 B and C are both consistent. Therefore, both can be extended into maximal consistent sets, say t and u. Now, the construction of B and C itself guarantee that $sR_i^c tu$ and $\phi \in t, u$. □

Lemma 3.8 *(Truth Lemma) For any state $s \in \mathcal{M}$ and ϕ, $\mathcal{M}, s \Vdash \phi \iff \phi \in s$.*

Proof Prove by induction. We only give the $\Diamond_i^c \psi$ case; the others are routine.

$\Rightarrow$: Suppose $\mathcal{M}, s \Vdash \Diamond_i^c \psi$. Then there exist t, u such that $sR_i^c tu$, $\mathcal{M}, t \Vdash \psi$ and $\mathcal{M}, u \Vdash \psi$. By induction hypothesis, $\psi \in t \cap u$. If $\Diamond_i^c \psi \notin s$, then $\Box_i^c \neg\psi \in s$, which implies $\neg\psi \in t \cup u$ by the construction of R_i^c, contradiction. Therefore, $\Diamond_i^c \psi \in s$.

$\Leftarrow$:Suppose $\Diamond_i^c \psi \in s$. Then according to the existence lemma for $\Diamond_i^c$, there exist t, u such that $sR_i^c tu$ and $\psi \in t \cap u$. By induction hypothesis, $\mathcal{M}, t \Vdash \psi$, $\mathcal{M}, u \Vdash \psi$. Therefore $\mathcal{M}, s \Vdash \Diamond_i^c \psi$. □

The completeness result then follows immediately:

Theorem 3.9 *(Completeness)* $\mathbb{SMLKV}^r$ *is strongly complete over arbitrary models.*

Remark 3.10 At this point, it is interesting to compare our canonical model with the canonical model used in [13]. A complication in [13] is that merely maximal consistent sets are not enough to build a FO canonical Kripke model. However, as we have seen, we only use the maximal consistent sets in our canonical **MLKv**^{r} model: it does not involve value assignments. Thus we have restored the symmetry between the logical language and the model to some extent: there is no longer too much information in the model, which cannot be talked about by the language. Note that we allow $sR_i tt$, which also helps to have compact models.

4 Extended language with binary modalities

In the previous sections, we treat $\Diamond_i^c$ as a unary modality interpreted by a ternary relation. Essentially, $\Diamond_i^c$ can be viewed as a binary modality where the two arguments are the same. In this section, we restore the symmetry between the semantics and the syntax one step further by having the *binary* $\Diamond_i^c(\cdot,\cdot)$ in the language. Surprisingly, this extension does *not* increase the expressive power of **MLKv**^{r}. What is more, the new logic is normal. Consequently, the extension will help us to understand **MLKv**^{r} more deeply from a normal modal logic point of view.

The extended language **MLKv**^{b} is given by the following BNF (^{b} for *binary*):

$$\phi ::= \top \mid p \mid \neg\phi \mid (\phi \wedge \phi) \mid \Box_i \phi \mid \Box_i^c(\phi, \phi)$$

We define $\Diamond_i^c(\psi, \phi)$ as $\neg\Box_i^c(\neg\psi, \neg\phi)$. And $\Diamond_i^c \phi$ is now equivalent to the **MLKv**^{b} formula $\Diamond_i^c(\phi, \phi)$. To see the intuition, for example, $\Diamond_i^c(p, \neg p)$ says that i can see a p world and a $\neg p$ world which do not agree on the value of c. Formally, the semantics is defined on the same **MLKv**^{r} models $\mathcal{M} = \langle S, \{\rightarrow_i : i \in \mathbf{I}\}, \{R_i^c : i \in \mathbf{I}, c \in \mathbf{C}\}, V, V_\mathbf{C}\rangle$:

$\mathcal{M}, s \Vdash \Diamond_i^c(\phi, \psi)$	$\Leftrightarrow$	there exist $t, u \in S$ such that $sR_i^c tu$, $\mathcal{M}, t \Vdash \phi$ and $\mathcal{M}, u \Vdash \psi$.

The above semantics coincides with the standard semantics for binary diamond modalities [1].[8] Note that $\Diamond_i^c(\phi, \psi)$ is essentially different from $\Diamond_i^c(\phi \vee \psi)$: the latter only says that there are two $\phi \vee \psi$-successors that have different values of c,

[8] Binary modalities appear in many modal logics, such as the until operator in temporal logic, and the relevant implication in relevance logic interpreted on Kripke models with a ternary relation.

but not necessarily one ϕ world and one ψ world. So, on first sight, **MLKv**^{r} seems to be weaker than **MLKv**^{b}.

However, we will show by the following lemma that **MLKv**^{r} and **MLKv**^{b} are equally expressive, by reducing the binary $\Diamond_i^c$ to the unary $\Diamond_i^c$ in presence of the diamond $\Diamond_i$.

Lemma 4.1 *$\Diamond_i^c(\phi,\psi)$ is equivalent to the disjunction of the following three formulas:*

(i) $\Diamond_i^c\phi \wedge \Diamond_i\psi$

(ii) $\Diamond_i^c\psi \wedge \Diamond_i\phi$

(iii) $\Diamond_i\phi \wedge \Diamond_i\psi \wedge \neg\Diamond_i^c\phi \wedge \neg\Diamond_i^c\psi \wedge \Diamond_i^c(\phi \vee \psi)$

Proof The proof consists of two directions.

First, we show that each of the three disjuncts entails $\Diamond_i^c(\phi,\psi)$.

(i) For any model $\mathcal{M},s$ that satisfies $\Diamond_i^c\phi \wedge \Diamond_i\psi$, there exists $t,u \in S$ such that $sR_i^c tu$, $t \Vdash \phi$ and $u \Vdash \phi$, and exists v such that $s \to_i v$ and $v \Vdash \psi$. According to the property of R_i^c, at least one of $sR_i^c tv$ and $sR_i^c uv$ holds. W.l.o.g. suppose $sR_i^c tv$. Then according to the semantics of $\Diamond_i^c$, we have $s \Vdash \Diamond_i^c(\phi,\psi)$.

(ii) For $\Diamond_i^c\psi \wedge \Diamond_i\phi$, the proof is similar to (i).

(iii) If $\mathcal{M},s \Vdash \Diamond_i\phi \wedge \Diamond_i\psi \wedge \neg\Diamond_i^c\phi \wedge \neg\Diamond_i^c\psi \wedge \Diamond_i^c(\phi \vee \psi)$, then: s has ϕ-successors and ψ-successors; all ϕ-successors have the same value of c, all ϕ-successors have the same value of c, but the two values are different due to $\mathcal{M},s \Vdash \Diamond_i^c(\phi \vee \psi)$. So we can easily guarantee that there are two states, one ϕ-successor and one ψ-successor of s such that they have different values with regard to c. This means $\mathcal{M},s \Vdash \Diamond_i^c(\phi,\psi)$.

Second, we prove that if $\mathcal{M},s \Vdash \Diamond_i^c(\phi,\psi)$, then at least one of (i), (ii) and (iii) holds.

Suppose $\mathcal{M},s \Vdash \Diamond_i^c(\phi,\psi)$, namely there exist $t,u \in \mathcal{M}$ such that $sR_i^c tu$, $\mathcal{M},t \Vdash \phi$ and $\mathcal{M},u \Vdash \psi$. We immediately have $\mathcal{M},s \Vdash \Diamond_i\phi \wedge \Diamond_i\psi \wedge \Diamond_i^c(\phi \vee \psi)$. If neither $\Diamond_i^c\phi$ nor $\Diamond_i^c\psi$ holds on s, then $\mathcal{M},s \Vdash \Diamond_i\phi \wedge \Diamond_i\psi \wedge \neg\Diamond_i^c\phi \wedge \neg\Diamond_i^c\psi \wedge \Diamond_i^c(\phi \vee \psi)$. Therefore, $\mathcal{M},s \Vdash (\Diamond_i^c\phi \wedge \Diamond_i\psi) \vee (\Diamond_i^c\psi \wedge \Diamond_i\phi) \vee (\Diamond_i\phi \wedge \Diamond_i\psi \wedge \neg\Diamond_i^c\phi \wedge \neg\Diamond_i^c\psi \wedge \Diamond_i^c(\phi \vee \psi))$

In sum, we can now conclude the equivalence. □

With this lemma in hand, the reduction theorem is straightforward:

Theorem 4.2 *(Reduction) For any* **MLKv**^{b} *formula ϕ, there exists an* **MLKv**^{r} *formula ψ such that for any pointed model $\mathcal{M},s$: $\mathcal{M},s \Vdash \phi \iff \mathcal{M},s \Vdash \psi$.*

Proof We define a reduction function r inductively:

- $r(p) = p$; $r(\neg\phi) = \neg r(\phi)$; $r(\phi \wedge \psi) = r(\phi) \wedge r(\psi)$; $r(\Diamond_i\phi) = \Diamond_i r(\phi)$;
- $r(\Diamond_i^c(\phi,\psi)) = (\Diamond_i^c r(\phi) \wedge \Diamond_i r(\psi)) \vee (\Diamond_i^c r(\psi) \wedge \Diamond_i r(\phi)) \vee (\Diamond_i r(\phi) \wedge \Diamond_i r(\psi) \wedge \neg\Diamond_i^c r(\phi) \wedge \neg\Diamond_i^c r(\psi) \wedge \Diamond_i^c(r(\phi) \vee r(\psi)))$.

The correctness of the reduction is guaranteed based on Lemma 4.1. It is not hard (but important) to see that the rewriting always terminates. □

Remark 4.3 Although $\mathbf{MLKv}^r$ is equally expressive as $\mathbf{MLKv}^b$ in presence of $\Diamond_i$, it is not the case if $\Diamond_i$ is absent. To see this, consider the following two pointed models $\mathcal{M}, s$ and $\mathcal{N}, x$ where $sR_i^c tu, sR_i^c uv$ in the left model, and $xR_i^c yz$ in the other model:

We can use $\Diamond_i^c(p,q)$ to distinguish the two pointed models; however, they are indistinguishable by using any formula with the unary $\Diamond_i^c$ but no $\Diamond_i$, which can be proved by a simple induction.

Now observe that $\mathbf{MLKv}^b$ is a standard modal language defined on standard Kripke models with standard semantics. It is a relatively routine exercise to propose a normal modal logic system with the following axioms SYM, INCL and ATEUC to capture the corresponding special properties of the models:

System $\mathbb{SMLKV}^b$

Axiom Schemas

TAUT	all the instances of tautologies
DISTK	$\Box_i(p \to q) \to (\Box_i p \to \Box_i q)$
DISTKv^b	$\Box_i^c(p \to q, r) \to (\Box_i^c(p,r) \to \Box_i^c(q,r))$
SYM	$\Box_i^c(p,q) \to \Box_i^c(q,p)$
INCL	$\Diamond_i^c(p,q) \to \Diamond_i p$
ATEUC	$\Diamond_i^c(p,q) \wedge \Diamond_i r \to \Diamond_i^c(p,r) \vee \Diamond_i^c(q,r)$

Rules

MP	$\dfrac{\phi, \phi \to \psi}{\psi}$
NECK	$\dfrac{\phi}{\Box_i \phi}$
NECKv^b	$\dfrac{\phi}{\Box_i^c(\phi, \psi)}$
SUB	$\dfrac{\phi}{\phi[p/\psi]}$
RE	$\dfrac{\psi \leftrightarrow \chi}{\phi \leftrightarrow \phi[\psi/\chi]}$

Note that due to SYM, we do not need to include the variations of DISTKv^b and NECKv^b w.r.t. the second argument in the binary $\Box_i^c$ (cf. [1] for the standard proof systems of polyadic normal modal logics.)

In this system $\mathbb{SMLKV}^b$ we can derive all the axioms in $\mathbb{SMLKV}^r$. Before proving it, we first show the following handy propositions.

Proposition 4.4 $\vdash_{\mathbb{SMLKV}^b} \Diamond_i^c(p \vee q, r) \to \Diamond_i^c(p,r) \vee \Diamond_i^c(q,r)$.

Proof This proposition captures the interaction between boolean operator $\vee$ and $\Diamond_i^c$ [9]. So we can only start from the axiom DISTKv^b. Note that RE is used frequently.

(1) $\Box_i^c(p \to q, r) \to (\Box_i^c(p,r) \to \Box_i^c(q,r))$ (DISTKv^b)
(2) $\neg(\Box_i^c(p,r) \to \Box_i^c(q,r)) \to \neg\Box_i^c(p \to q, r)$
(3) $\Box_i^c(p,r) \wedge \Diamond_i^c(\neg q, \neg r) \to \Diamond_i^c(p \wedge \neg q, \neg r)$
(4) $\Diamond_i^c(\neg q, \neg r) \to (\neg\Box_i^c(p,r) \vee \Diamond_i^c(p \wedge \neg q, \neg r))$
(5) $\Diamond_i^c(\neg q, \neg r) \to (\Diamond_i^c(\neg p, \neg r) \vee \Diamond_i^c(p \wedge \neg q, \neg r))$

[9] Actually this is a standard axiom for normal modal logic. In case the binary case might not be that familiar, we give the proof here.

(6) $\Diamond_i^c(p \vee q, r) \to (\Diamond_i^c(p, r) \vee \Diamond_i^c(\neg p \wedge (p \vee q), r))$ ((5) & SUB)
(7) $\Diamond_i^c(p \vee q, r) \to (\Diamond_i^c(p, r) \vee \Diamond_i^c(q, r))$ □

Proposition 4.5 $\vdash_{\mathbb{SMLKV}^b} \Diamond_i^c(p \wedge q, r) \to \Diamond_i^c(p, r) \wedge \Diamond_i^c(q, r)$.

Proof This is similar to the above proof. □

Proposition 4.6 $\vdash_{\mathbb{SMLKV}^b} \Box_i^c(p, r) \wedge \Box_i^c(q, r) \wedge \Diamond_i \neg r \to \Box_i^c(p, q)$.

Proof Easily derived from ATEUC: $\Diamond_i^c(p, q) \wedge \Diamond_i r \to \Diamond_i^c(p, r) \vee \Diamond_i^c(q, r)$. □

Proposition 4.7 *All the* $\mathbb{SMLKV}^r$ *axioms are provable in* $\mathbb{SMLKV}^b$ *and the rules of* $\mathbb{SMLKV}^r$ *are admissible in* $\mathbb{SMLKV}^b$ *(viewing* $\Diamond_i^c \phi$ *as* $\Diamond_i^c(\phi, \phi)$*).*

Proof We need to check $\mathtt{DISTKv}^r$, $\mathtt{Kv}^r\vee$ and $\mathtt{NECKv}^r$ in $\mathbb{SMLKV}^r$.

(i) $\mathtt{DISTKv}^r$:

(1) $\Diamond_i^c(p, q) \to \Diamond_i p$ (INCL)
(2) $\Box_i \neg p \to \Box_i^c(\neg p, \neg q)$
(3) $\Box_i(p \to q) \to \Box_i^c(p \to q, p)$ ((2) SUB)
(4) $\Box_i(p \to q) \to \Box_i^c(p \to q, q)$ ((2) SUB)
(5) $\Box_i(p \to q) \to (\Box_i^c(p \to q, p) \wedge \Box_i^c(p \to q, q))$ ((3) (4))
(6) $\Box_i(p \to q) \to (\Box_i^c(p, p) \to \Box_i^c(q, p)) \wedge (\Box_i^c(p, q) \to \Box_i^c(q, q))$ ($\mathtt{DISTKv}^b$)
(7) $\Box_i(p \to q) \to (\Box_i^c p \to \Box_i^c q)$ ((6) SYM)

(ii) $\mathtt{Kv}^r\vee$:

(1) $\Diamond_i^c(p, q) \wedge \Diamond_i r \to \Diamond_i^c(q, r) \vee \Diamond_i^c(p, r)$ (ATEUC)
(2) $\Diamond_i^c(p \vee q, p \vee q) \wedge \Diamond_i(p \wedge q) \to \Diamond_i^c(p \vee q, p \wedge q)$ (SUB)
(3) $\Diamond_i^c(p \vee q) \wedge \Diamond_i(p \wedge q) \to (\Diamond_i^c(p, p \wedge q) \vee \Diamond_i^c(q, p \wedge q))$ (Prop. 4.4)
(4) $\Diamond_i^c(p \vee q) \wedge \Diamond_i(p \wedge q) \to (\Diamond_i^c(p, p) \vee \Diamond_i^c(q, q))$ (Prop. 4.5)
(5) $\Diamond_i^c(p \vee q) \wedge \Diamond_i(p \wedge q) \to (\Diamond_i^c p \vee \Diamond_i^c q)$

(iii) $\mathtt{NECKv}^r$: It is a special case of $\mathtt{NECKv}^b$ in $\mathbb{SMLKV}^b$ where the two arguments are the same.

Other axioms and rules in $\mathbb{SMLKV}^r$ are exactly the same as in $\mathbb{SMLKV}^b$. □

Now as we can see below, the standard technique suffices to prove the completeness of $\mathbb{SMLKV}^b$. The only tricky point is the ternary canonical relation.

Theorem 4.8 $\mathbb{SMLKV}^b$ *is sound and strongly complete w.r.t.* $\mathbf{MLKv}^r$ *models.*

Proof The soundness is straightforward to check. For the completeness we build a canonical model:

$$\mathcal{M} = \langle S, \{\to_i : i \in \mathbf{I}\}, \{R_i^c : i \in \mathbf{I}, c \in \mathbf{C}\}, V_{\mathbf{C}} \rangle$$

- S is the set of all maximal $\mathbb{SMLKV}^b$-consistent sets of $\mathbf{MLKv}^b$ formulas,
- $s \to_i t \iff \{\phi : \Box_i \phi \in s\} \subseteq t$,
- $sR_i^c tu \iff$ (1) $\{\phi : \Box_i \phi \in s\} \subseteq t \cap u$ and (2) for any $\Box_i^c(\phi, \psi) \in s$, $\phi \in t$ or $\psi \in u$.
- $V_{\mathbf{C}}(s) = \{p : p \in s\}$.

Note that the existence lemma for $\Diamond_i^c$ is quite routine for normal polyadic modal logic, cf. [1]. The idea is to build two i-successors t,u of s if $\Diamond_i^c(\phi,\psi) \in s$, such that $\phi \in t$, $\psi \in u$ and $sR_i^c tu$. According to the method in [1, pp. 200], we can build two maximal consistent sets t,u such that $\phi \in t$ and $\psi \in u$, and for all $\Box_i^c(\chi_1,\chi_2) \in s$ we have $\chi_1 \in t$ or $\chi_2 \in u$. To make sure $sR_i^c tu$ we just need to check condition (1). To see this, note that by INCL we have $\Box_i\chi \to \Box_i^c(\chi,\theta) \in s$ and by SYM we have $\Box_i\chi \to \Box_i^c(\theta,\chi) \in s$. Therefore for each $\Box_i\chi \in s$ we have $\Box_i^c(\chi,\bot) \in s$ and $\Box_i^c(\bot,\chi) \in s$. Due to the construction of maximal consistent sets t,u, we have $\chi \in t$ or $\bot \in u$, and $\bot \in t$ or $\chi \in u$, which implies $\chi \in t \cap u$. Thus $\{\chi : \Box_i\chi \in s\} \subseteq t \cap u$. This concludes the proof that $sR_i^c tu$. Based on the existence lemmas for both $\Diamond_i$ and $\Diamond_i^c$ we can prove the truth lemma $\phi \in s \iff s \Vdash \phi$ using standard techniques.

In the rest of this proof we verify that the canonical model satisfies the three properties of **MLKv**^{r} models. Note that condition (1) in the definition of R_i^c is symmetric, and condition (2) is also implicitly symmetric due to axiom SYM. It is also obvious that $sR_i^c tu$ implies $sR_i t$ and $sR_i u$ by definition. We only need to verify the anti-euclidean property.

Towards contradiction suppose $sR_i^c tu$, $s \to_i v$ but neither $sR_i^c tv$ nor $sR_i^c uv$. Then according to the definition of R_i^c, there exist $\Box_i^c(\phi_1,\psi_1), \Box_i^c(\phi_2,\psi_2) \in s$ such that $\neg\phi_1 \in t, \neg\psi_1 \in v$, $\neg\phi_2 \in u$ and $\neg\psi_2 \in v$. Therefore $\neg\psi_1 \wedge \neg\psi_2 \in v$. Since $s \to_i v$, $\Diamond_i(\neg\psi_1 \wedge \neg\psi_2) \in s$. By DISTKv^b, SYM, and NECKv^b, it is not hard to show $\vdash_{\mathbb{SMLKV}^b} \Box_i^c(\phi_1,\psi_1) \to \Box_i^c(\phi_1,\psi_1 \vee \psi_2)$ and $\vdash_{\mathbb{SMLKV}^b} \Box_i^c(\phi_2,\psi_2) \to \Box_i^c(\phi_2,\psi_1 \vee \psi_2)$. So $\Box_i^c(\phi_1,\psi_1 \vee \psi_2), \Box_i^c(\phi_2,\psi_1 \vee \psi_2) \in s$. By Proposition 4.6 and SUB, $\vdash_{\mathbb{SMLKV}^b}$ $\Box_i^c(\phi_1,\psi_1 \vee \psi_2) \wedge \Box_i^c(\phi_2,\psi_1 \vee \psi_2) \wedge \Diamond_i(\psi_1 \vee \psi_2) \to \Box_i^c(\phi_1,\phi_2)$. Since $\Box_i^c(\phi_1,\psi_1 \vee \psi_2)$, $\Box_i^c(\phi_2,\psi_1 \vee \psi_2)$ and $\Diamond_i(\psi_1 \vee \psi_2)$ are all in s, $\Box_i^c(\phi_1,\phi_2) \in s$. This together with $sR_i^c tu$ imply that $\phi_1 \in t$ or $\phi_2 \in u$, contradictory to the assumption that $\neg\phi_1 \in t$ and $\neg\phi_2 \in u$. □

5 Applications

5.1 Bisimulation

In the field of modal logic, various bisimulation notions help to characterize the expressive power of the new semantics-driven logics. As a normal modal logic, **MLKv**^{b} has a natural notion of bisimulation (cf. [1]), and it will in turn help us to find a notion of bisimulation over FO Kripke models for the original **ELKv**^{r}.

Definition 5.1 (C-Bisimulation) *Let $\mathcal{M}_1 = \langle S_1, \{\to_i^1 : i \in I\}, \{R_i^c : i \in I, c \in \mathbf{C}\}, V_1\rangle$, $\mathcal{M}_2 = \langle S_2, \{\to_i^2 : i \in I, c \in \mathbf{C}\}, \{Q_i^c : i \in I\}, V_2\rangle$ be two models for **MLKv**^{b} (also for **MLKv**^{r}). A* **C***-bisimulation between $\mathcal{M}_1$ and $\mathcal{M}_2$ is a non-empty binary relation $Z \subseteq S_1 \times S_2$ such that for all $s_1 Z s_2$, the following conditions are satisfied:*

Inv : $V_1(s_1) = V_2(s_2)$;

Zig : $s_1 \to_i^1 t_1 \Rightarrow \exists t_2$ *such that* $s_2 \to_i^2 t_2$ *and* $t_1 Z t_2$;

Zag : $s_2 \to_i^2 t_2 \Rightarrow \exists t_1$ *such that* $s_1 \to_i^1 t_1$ *and* $t_1 Z t_2$;

Kvb-Zig : $s_1 R_i^c t_1 u_1 \Rightarrow \exists t_2, u_2 \in S_2$ *such that* $t_1 Z t_2$, $u_1 Z u_2$ *and* $s_2 Q_i^c t_2 u_2$;

Kvb-Zag : $s_2 Q_i^c t_2 u_2 \Rightarrow \exists t_1, u_1 \in S_1$ *such that* $t_1 Z t_2$, $u_1 Z u_2$ *and* $s_1 R_i^c t_1 u_1$.

We say $\mathcal{M},s$ and $\mathcal{N},t$ are **C**-*bisimilar* ($\mathcal{M},s \leftrightarroweq_{C} \mathcal{N},t$) *if there is a* **C**-*bisimulation* Z *between* $\mathcal{M}$ *and* $\mathcal{N}$ *and* $(s,t) \in Z$.

Theorem 5.2 *If* $\mathcal{M}_1,s_1 \leftrightarroweq_{C} \mathcal{M}_2,s_2$, *then* $\mathcal{M}_1,s_1 \equiv_{MLKv^b} \mathcal{M}_2,s_2$.

Proof Suppose $\mathcal{M}_1,s_1 \leftrightarroweq_{C} \mathcal{M}_2,s_2$. We prove by induction on the structure of **MLKv**^{b} formulas, and the only non-trivial case is when $\phi = \Diamond_i^c(\psi,\chi)$.

Suppose $\mathcal{M}_1,s_1 \Vdash \Diamond_i^c(\psi,\chi)$, then $\exists t_1,u_1 \in S_1$ such that $s_1 R_i^c t_1 u_1$ with $\mathcal{M}_1,t_1 \Vdash \psi$ and $\mathcal{M}_1,u_1 \Vdash \chi$. By **Kvb-Zig**, there exist $t_2,u_2 \in S_2$ such that $t_1 Z t_2$, $u_1 Z u_2$ and $s_2 Q_i^c t_2 u_2$. By induction hypothesis, $\mathcal{M}_2,t_2 \Vdash \psi$ and $\mathcal{M}_2,u_2 \Vdash \chi$. Therefore, $\mathcal{M}_2,s_2 \Vdash \Diamond_i^c(\psi,\chi)$. The other side is similar by **Kvb-Zag**. □

As in normal modal logic, we have the following theorem for **MLKv**^{b} (we omit the rather standard proof, but one can try to see how the binary modality $\Diamond_i^c$ facilitates the proof):

Theorem 5.3 *Suppose* $\mathcal{M}$, $\mathcal{N}$ *are finite models. Then* $\mathcal{M},s \leftrightarroweq_{C} \mathcal{N},t \iff \mathcal{M},s \equiv_{MLKv^b} \mathcal{N},t$.

Since **MLKv**^{r} and **MLKv**^{b} have the same expressive power we immediately have:

Corollary 5.4 *Suppose* $\mathcal{M}$ *and* $\mathcal{N}$ *are finite models. Then* $\mathcal{M},s \leftrightarroweq_{C} \mathcal{N},t \iff \mathcal{M},s \equiv_{MLKv^r} \mathcal{N},t$.

In [12], a notion of bisimulation has been offered for **ELKv**, the epistemic logic with unconditional Kv_i operators. However, for **ELKv**^{r} it was not that clear about the suitable bisimulation notion. Now we can recast **C**-bisimulation back to the setting of **ELKv**^{r} over FO Kripke models since **ELKv**^{r} and **MLKv**^{r} are essentially the same language.

Definition 5.5 (C-bisimulation over FO Kripke models) *Given two pointed FO Kripke models* $\mathcal{M} = \langle S_1, D_1, \{\to_i^1 : i \in \mathbf{I}\}, V_1, V_C^1\rangle$, *and* $\mathcal{N} = \langle S_2, D_2, \{\to_i^2 : i \in \mathbf{I}\}, V_2, V_C^2\rangle$, *a relation* $Z \subseteq S_1 \times S_2$ *is a* **C**-*bisimulation between the two models* $\mathcal{M},\mathcal{N}$ *if whenever* $s_1 Z s_2$ *we have:*

Inv $V_1(s_1) = V_2(s_2)$;

Zig : $s_1 \to_i^1 t_1 \Rightarrow \exists t_2$ *such that* $s_2 \to_i^2 t_2$ *and* $t_1 Z t_2$;

Zag : $s_2 \to_i^2 t_2 \Rightarrow \exists t_1$ *such that* $s_1 \to_i^1 t_1$ *and* $t_1 Z t_2$;

Kvr-Zig *If* $s_1 \to_i^1 t_1$ *and* $s_1 \to_i^1 u_1$ *and* $V_C^1(c,t_1) \neq V_C^1(c,u_1)$ *then there are* t_2 *and* u_2 *in* $\mathcal{N}$ *such that* $s_2 \to_i^2 t_2$, $s_2 \to_i^2 u_2$, $t_1 Z t_2$, $u_1 Z u_2$, *and* $V_C^2(c,t_2) \neq V_C^2(c,u_2)$ *in* $\mathcal{N}$.

Kvr-Zag *If* $s_2 \to_i^2 t_2$ *and* $s \to_i^2 u_2$ *and* $V_C^2(c,t_2) \neq V_C^2(c,u_2)$ *then there are* t_1 *and* u_1 *in* $\mathcal{M}$ *such that* $s_1 \to_i^1 t_1$, $s \to_i^1 u_1$, $t_1 Z t_2$, $u_1 Z u_2$ *and* $V_C^1(c,t_1) \neq V_C^1(c,u_1)$ *in* $\mathcal{M}$.

Abusing the notation, FO Kripke models $\mathcal{M},s$ *and* $\mathcal{N},t$ *are* **C**-*bisimilar* ($\mathcal{M},s \leftrightarroweq_{C} \mathcal{N},t$) *iff there exists a* **C**-*bisimulation* Z *between* $\mathcal{M}$ *and* $\mathcal{N}$ *such that* $s,t \in Z$.

Now since **MLKv**^{b} and **MLKv**^{r} have exactly the same expressive power, and **MLKv**^{r} is equivalent to **ELKv**^{r} modulo translation. The above **C**-bisimilation works for **ELKv**^{r}, as proved in detailed in [6]:

Theorem 5.6 *For finite FO Kripke models $\mathcal{M}_1, \mathcal{M}_2$: $\mathcal{M}_1, s_1 \leftrightarrows_C \mathcal{M}_2, s_2$ iff $\mathcal{M}_1, s_1 \equiv_{\mathbf{ELKv}^r} \mathcal{M}_2, s_2$.*

5.2 Completeness of $\mathbb{SMLKV}$

The unconditional *Kv* operator was introduced in [7] in the context of epistemic logic (call the language **ELKv**):

$$\phi ::= \top \mid p \mid \neg\phi \mid (\phi \wedge \phi) \mid K_i\phi \mid Kv_i c$$

Essentially, $Kv_i c$ is $Kv_i(\top, c)$ in $\mathbf{ELKv}^r$. The semantics is as in the case of $\mathbf{ELKv}^r$, which is based on FO Kripke models. Plaza gave two axioms on top of $\mathbb{S}5$ which are the counterparts of the introspection axioms in standard epistemic logic (over FO epistemic models):

$$Kv_i c \to K_i Kv_i c \qquad\qquad \neg Kv_i c \to K_i \neg Kv_i c$$

However, neither [7] nor [12,13] gave a complete proof of this simple logic. Here we look at this language from our $\Diamond_i^c$ perspective, and consider the corresponding simple language (**MLKv**) over the class of all the models:

$$\phi ::= \top \mid p \mid \neg\phi \mid (\phi \wedge \phi) \mid \Box_i\phi \mid \Box_i^c\bot$$

Note that $\neg Kv_i c$ can be viewed as $\Diamond_i^c\top$. Thus $Kv_i c$ is indeed $\Box_i^c\bot$.

The semantics is just as in $\mathbf{MLKv}^r$ but we only allow $\top$ as the argument for $\Diamond_i^c$. As in the case of $\mathbf{ELKv}^r$ and $\mathbf{MLKv}^r$ we can simply show that:

Proposition 5.7 *For any set of **ELKv** formula $\Sigma \cup \{\phi\}$, $\Sigma \vDash \phi$ iff $T(\Sigma) \Vdash T(\phi)$.*

Now based on this view, a natural system $\mathbb{SMLKV}$ is obtained by simplifying the system $\mathbb{SMLKV}^r$:

System $\mathbb{SMLKV}$

Axiom Schemas		Rules	
TAUT	all the instances of tautologies	MP	$\dfrac{\phi, \phi \to \psi}{\psi}$
DISTK	$\Box_i(p \to q) \to (\Box_i p \to \Box_i q)$	NECK	$\dfrac{\phi}{\Box_i\phi}$
INCLT	$\Diamond_i^c\top \to \Diamond_i\top$	SUB	$\dfrac{\phi}{\phi[p/\psi]}$
		RE	$\dfrac{\psi \leftrightarrow \chi}{\phi \leftrightarrow \phi[\psi/\chi]}$

Note that due to the fact that the only $\Diamond_i^c$ formula is $\Diamond_i^c\top$ (and $\Box_i^c\bot$), most of the previous axioms and rules do not apply. We only need to add one axiom INCLT on top of the usual normal modal logic, inspired by the INCL axioms of $\mathbb{SMLKV}^b$.

We go on to prove the completeness of $\mathbb{SMLKV}$.

Definition 5.8 *The canonical model $\mathcal{M}$ is a tuple $\langle S, \{\to_i : i \in \mathbf{I}\}, \{R_i^c : i \in \mathbf{I}, c \in \mathbf{C}\}, V\rangle$ where:*

- *S is the set of maximal $\mathbb{SMLKV}$-consistent sets,*
- *$s \to_i t \iff \{\phi : \Box_i\phi \in s\} \subseteq t$,*

- $sR_i^c tt' \iff s \to_i t, s \to_i t'$ *and* $\Diamond_i^c \top \in s$,
- $V(s) = \{p : p \in s\}$

The only tricky point is the definition of canonical relations R_i^c. The intuition is that as long as a state can see two states having different values, then we can safely assume all the states that it can see have different values. Note that in **MLKv**^{r} models we also allow $sR_i^c tt$. We first need to verify that $\mathcal{M}$ is an **MLKv**^{r} model:

Proposition 5.9 *The canonical model $\mathcal{M}$ is an **MLKv**^{r} model.*

Proof We only need to verify the three conditions. The first two are again obvious by definition, so we only prove the anti-euclidean property. Suppose $sR_i^c tt'$ and $s \to_i u$. Then $s \to_i t$, $s \to_i t'$ and $\Diamond_i^c \top \in s$. So both $sR_i^c tu$ and $sR_i^c t'u$ by the definition of R_i^c. □

The existence lemma for $\Diamond_i$ is routine. As for the case of $\Diamond_i^c$:

Lemma 5.10 *(Existence Lemma for $\Diamond_i^c$) If $\Diamond_i^c \top \in s$, then there exist t, u such that $sR_i^c tu$.*

Proof Suppose $\Diamond_i^c \top \in s$ then due to INCLT $\Diamond_i \top \in s$. By the existence lemma for $\Diamond_i$, it follows that there exists t such that $s \to_i t$. Therefore by definition $sR_i^c tt$. □

Lemma 5.11 *For any s in $\mathcal{M}$ and **MLKv** formula ϕ, $\mathcal{M}, s \Vdash \phi \iff \phi \in s$.*

Proof The only interesting case is when $\phi = \Diamond_i^c \top$.

$\Rightarrow$: Suppose $\mathcal{M}, s \Vdash \Diamond_i^c \top$. Then there exist t, t' such that $sR_i^c tt'$. By the definition of R_i^c, $\Diamond_i^c \top \in s$.

$\Leftarrow$: Suppose $\Diamond_i^c \top \in s$. By the existence lemma for $\Diamond_i^c$, there exist t, t' such that $sR_i^c tt'$. Therefore $\mathcal{M}, s \Vdash \Diamond_i^c \top$. □

Theorem 5.12 $\mathbb{SMLKV}$ *is strongly complete w.r.t. **MLKv** models.*

A corollary follows immediately based on Proposition 5.7:

Corollary 5.13 $\mathbb{SMLKV}$ *(viewing* INCLT *as $\neg Kv_i \top \to \hat{K}_i \top$) is strongly complete w.r.t. **ELKv**^{r} models.*

6 Discussion and future work

In this paper, we introduce a ternary relation based simple semantics to the "knowing value" logic without explicit value assignments. Under this semantics, the logic can be viewed as a disguised normal modal logic with both standard unary and binary modalities. The use of this perspective is demonstrated by various applications.

Another intuitive way to simplify the original FO-Kripke semantics is to introduce a binary relation $\asymp_c$ for each c representing the inequality of the value of c. Correspondingly, in the language, besides $\Diamond_i \phi$ we may introduce $\Diamond_c \phi$ formulas saying that there is a different world where ϕ holds but c has a different value compared to the current world. However, it is not straightforward to express $\neg Kv(\psi, c)$ in this language. The closest counterpart $\Diamond_i(\psi \wedge \Diamond_c \psi)$ will not do the job alone. We probably need to add a further condition: $\Diamond_i \Diamond_c p \to \Diamond_i p$ which says the $\asymp_c$ successors of an i-reachable world are again i-reachable. Actually it means that we should combine $\to_i$ and $\asymp_c$ which is almost our ternary R_i^c. Moreover, to axiomatize this $\asymp_c$ we need the axioms of anti-equivalence (irreflexivity, symmetry, and anti-euclidean property[10]). However, irreflexivity and anti-euclidean property for the binary $\asymp_c$ are not definable in modal logic. We probably need to do the same as in the $\Diamond_i^c$ case: use $\mathrm{Kv}\vee$ to capture the i-accessible anti-euclidean property to some extent. Having said the above, it is clear that our approach in this paper is more intuitive and technically natural.

To close, we list a few directions which we leave for future occasions:

- The corresponding results in the setting of epistemic (S5) models.
- Characterization theorem of $\mathbf{ELKv}^r$ ($\mathbf{MLKv}^r$) within first-order modal logic via **C**-bisimulation.
- A decision procedure for $\mathbf{ELKv}^r$ ($\mathbf{MLKv}^r$) based on the simplified models.
- In similar ways, we can try to simplify the semantics for other "knowing-X" logics, such as knowing whether, knowing how, and so on.

References

[1] Blackburn, P., M. de Rijke and Y. Venema, "Modal Logic," Cambridge University Press, 2002.

[2] Ding, Y., *Axiomatization and complexity of modal logic with knowing-what operator on model class K* (2015), unpublished manuscript.
URL `http://www.voidprove.com/research.html`

[3] Fagin, R., J. Halpern, Y. Moses and M. Vardi, "Reasoning about knowledge," MIT Press, Cambridge, MA, USA, 1995.

[4] Fan, J., Y. Wang and H. van Ditmarsch, *Almost necessary*, in: *Advances in Modal Logic Vol. 10*, 2014, pp. 178–196.

[5] Fan, J., Y. Wang and H. van Ditmarsch, *Contingency and knowing whether*, The Review of Symbolic Logic **8** (2015), pp. 75–107.

[6] Gattinger, M., J. van Eijck and Y. Wang, *Knowing value and public inspection* (2016), manuscript.

[7] Plaza, J. A., *Logics of public communications*, in: M. L. Emrich, M. S. Pfeifer, M. Hadzikadic and Z. W. Ras, editors, *Proceedings of the 4th International Symposium on Methodologies for Intelligent Systems*, 1989, pp. 201–216.

[8] van der Hoek, W. and A. Lomuscio, *A logic for ignorance*, Electronic Notes in Theoretical Computer Science **85** (2004), pp. 117–133.

[9] van Ditmarsch, H., J. Halpern, W. van der Hoek and B. Kooi, editors, "Handbook of Epistemic Logic," College Publications, 2015.

[10] Wang, Y., *A logic of knowing how*, in: *Proceedings of LORI 2015*, 2015, pp. 392–405.

[11] Wang, Y., *Beyond knowing that: a new generation of epistemic logics*, in: *Hintikka's volume in outstanding contributions to logic*, Springer, 2016 Forthcoming.
URL `http://arxiv.org/abs/1605.01995`

[12] Wang, Y. and J. Fan, *Knowing that, knowing what, and public communication: Public announcement logic with* Kv *operators*, in: *Proceedings of IJCAI 13*, 2013, pp. 1139–1146.

[13] Wang, Y. and J. Fan, *Conditionally knowing what*, in: *Advances in Modal Logic Vol. 10*, 2014, pp. 569–587.

---

[10] Here anti-euclidean property means if $x \asymp_c y$ and there is another world z then $x \asymp_c z$ or $y \asymp_c z$. A simple disjoint union argument can show it is not modally definable. Thanks to Zhiguang Zhao for pointing it out.

Decidable first-order modal logics with counting quantifiers

Christopher Hampson

King's College London
Strand, London
WC2R 2LS

Abstract

In this paper, we examine the computational complexity of various natural one-variable fragments of first-order modal logics with the addition of arbitrary counting quantifiers. The addition of counting quantifiers provides us a rich language with which to succinctly express statements about the quantity of objects satisfying a given property, using only a single variable. We provide optimal upper bounds on the complexity of the decision problem for several one-variable fragments, by establishing the finite model property. In particular, we show that the decision (validity) problem for the one-variable fragment of the minimal first-order modal logic **QK** with counting quantifiers is coNExpTime-complete. In the propositional setting, these results also provide optimal upper bounds for many two-dimensional modal logics in which one component is von Wright's logic of 'elsewhere'.

Keywords: first-order modal logic, quantified modal logic, two-dimensional modal logic, counting quantifiers, decidable fragment, finite model property, quasimodel

1 Introduction

First-order modal logics are notorious for their poor computational behaviour, and even the modal versions of many decidable fragments of classical first-order logic are undecidable. For example, even the two-variable, monadic fragment of many first-order modal logics is already undecidable [8]. However, more restrictive decidable fragments are known to exist, such as the *monodic* fragment, in which modalities *de re* are restricted to formulas containing at most one free variable, with no restrictions are placed on modalities *de dicto* [22].

In classical first-order logic, counting quantifiers allow us to succinctly express statements about the quantity of objects satisfying a given property, without requiring many auxiliary variables to address each object. It is well-known that counting quantifiers can be safely added to the two-variable fragment of classical first-order logic without affecting the computational complexity [13].

They, therefore, provide an attractive addition in the quest to gain greater expressive power from finite variable fragments of first-order modal logics, without jeopardizing their decidability. Some examples of first-order formulas with counting quantifiers include:

- *"It is possible that there are more than eight planets"*: $\Diamond\exists_{>8}x\ \mathsf{Planet}(x)$
- *"Two components are believed to be faulty"*: $\exists_{=2}x(\mathsf{Component}(x) \wedge \Box\mathsf{Faulty}(x))$
- The generalised Barcan formula: $\exists_{=c}x\Diamond P(x) \leftrightarrow \Diamond\exists_{=c}x\ P(x)$, for $c < \omega$,

Unfortunately, many decidable first-order *temporal* logics become undecidable with the addition of counting quantifiers; even if we restrict the quantifiers to those of the form $\exists_{\leq c}x$, for $c = 0, 1$ [7]. On the other hand, less expressive first-order modal logics, such as the one-variable fragment of quantified **S5** with counting quantifiers can be easily embedded into the two-variable fragment of classical first-order logic with counting quantifiers, and are therefore no more complex than their counting-free counterparts. It is, therefore, interesting to establish where the boundary lies between decidable and undecidable fragments of first-order modal logics with counting quantifiers.

In this paper, we provide optimal upper bounds for various one-variable fragments of the quantified versions **K**, **KT**, **KB**, **S5**, and **Alt**, with arbitrary counting quantifiers whose subscripts are encoded as binary strings.

In Section 2, we introduce the definitions for the fragments of first-order modal logics that will be working with, and in Section 3 we prove the main results of this paper. Section 4 describes the connection between certain fragments of first-order modal logics with counting quantifiers and two-dimensional propositional modal logics in which one component is von Wright's logic of 'elsewhere'. We conclude with a discussion of some open problems in Section 5. Supplementary proofs and polynomial reductions between several first-order modal logics are provided in the Appendices.

2 First-order Modal Logics

Given a countably infinite set of predicate symbols $\mathsf{Pred} = \{P_0, P_1, \dots\}$, each with an associated arity, and a countable set of first-order variables Var. Let $\mathcal{Q}^{\#}\mathcal{ML}$ denote the set of all first-order modal formulas with *counting quantifiers* defined by the following grammar:

$$\varphi ::= P_i(x_1, \dots, x_n) \mid \neg\varphi \mid (\varphi_1 \wedge \varphi_2) \mid \Diamond\varphi \mid (\exists_{\leq c}x\ \varphi)$$

where $P_i \in \mathsf{Pred}$ is an n-ary predicate symbol, $x, x_1, \dots, x_n \in \mathsf{Var}$ are first-order variables, and $c < \omega$ is a natural number quantifier subscript. Other boolean connectives are defined in the usual way, with the addition of $\forall x\ \varphi := \exists_{\leq 0}x\ \neg\varphi$, $\exists_{\geq c}x\ \varphi := \neg\exists_{\leq(c-1)}x\ \varphi$, and $\exists_{=c}x\ \varphi := \exists_{\geq c}x\ \varphi \wedge \exists_{\leq c}x\ \varphi$, for $c > 0$. The results contained herein can be easily modified to accommodate taking either $\exists_{=c}$ or $\exists_{\geq c}$ as primitive.

Throughout this paper, we will assume that formulas of $\mathcal{Q}^{\#}\mathcal{ML}$ can be encoded as strings over some finite alphabet, and define the *size* of a formula $\varphi \in \mathcal{Q}^{\#}\mathcal{ML}$, denoted $||\varphi||$, to be the length of the encoding string, with quantifier subscripts encoded in *binary*. We define $\mathsf{sub}(\varphi) \subseteq \mathcal{Q}^{\#}\mathcal{ML}$ to be the set of all *subformulas* of φ, $\mathsf{md}(\varphi) < \omega$ to be the *modal depth* of φ, taken to be the maximum nesting depth of modal operators, and $\mathsf{count}(\varphi) < \omega$ to be the value of the largest quantifier subscript occurring in φ.

For each $\ell < \omega$, let $\mathcal{Q}^{\#}\mathcal{ML}^{\ell}$ denote the ℓ-variable fragment comprising only those formulas containing the variables $x_1, \ldots, x_\ell$, and denote by $\mathcal{Q}^{\#}\mathcal{ML}_k$ the set of all $\mathcal{Q}^{\#}\mathcal{ML}$ formulas that do not contain quantifiers with subscripts larger than $k < \omega$. In particular, we identify $\mathcal{Q}^{\#}\mathcal{ML}_0$ with the language of regular (counting-free) first-order modal logic. We write $\mathcal{Q}^{\#}\mathcal{ML}_k^{\ell} = \mathcal{Q}^{\#}\mathcal{ML}^{\ell} \cap \mathcal{Q}^{\#}\mathcal{ML}_k$ for the ℓ-variable fragment with quantifiers subscripts not exceeding k.

Formulas of $\mathcal{Q}^{\#}\mathcal{ML}$ are interpreted in *first-order Kripke models* of the form $\mathfrak{M} = (\mathfrak{F}, D, I)$, where $\mathfrak{F} = (W, R)$ is a *Kripke frame*, D is a non-empty *domain*, and I is a function associating each $w \in W$ with a first-order structure

$$I(w) = \left\langle D, P_0^{I(w)}, P_1^{I(w)}, \ldots \right\rangle$$

where $P_i^{I(w)} \subseteq D^n$ is an n-ary relation on the domain D, for every n-ary predicate symbol $P_i \in \mathsf{Pred}$. The *size of* $\mathfrak{M}$ is taken to be $|W| \cdot |D|$.

In this paper we consider only logics that can be characterised by models with constant domains. However, all the results proved here can be extended to those cases characterised by models with *expanding* or *decreasing* domains via a standard reduction [22].

A *variable assignment* on $\mathfrak{M}$ is a function $\mathfrak{a} : \mathsf{Var} \to D$ mapping variables to elements of the domain. Given a model $\mathfrak{M} = (\mathfrak{F}, D, I)$ and a variable assignment $\mathfrak{a}$, we define satisfiability in $\mathfrak{M}$ by taking, for all $w \in W$:

$$\begin{aligned}
\mathfrak{M}, w \models^{\mathfrak{a}} P_i(x_1, \ldots, x_n) \quad &\iff \quad (\mathfrak{a}(x_1), \ldots, \mathfrak{a}(x_n)) \in P_i^{I(w)}, \\
\mathfrak{M}, w \models^{\mathfrak{a}} \neg\varphi \quad &\iff \quad \mathfrak{M}, w \not\models^{\mathfrak{a}} \varphi, \\
\mathfrak{M}, w \models^{\mathfrak{a}} (\varphi_1 \wedge \varphi_2) \quad &\iff \quad \mathfrak{M}, w \models^{\mathfrak{a}} \varphi_1 \text{ and } \mathfrak{M}, w \models^{\mathfrak{a}} \varphi_2, \\
\mathfrak{M}, w \models^{\mathfrak{a}} \Diamond\varphi \quad &\iff \quad wRv \text{ and } \mathfrak{M}, v \models^{\mathfrak{a}} \varphi, \text{ for some } v \in W,
\end{aligned}$$

where $P_i \in \mathsf{Pred}$ is an n-ary predicate symbol, and

$$\mathfrak{M}, w \models^{\mathfrak{a}} (\exists_{\leq c} x\, \varphi) \quad \iff \quad \left|\{a \in D \;:\; \mathfrak{M}, w \models^{\mathfrak{a}(x/a)} \varphi\}\right| \;\leq\; c,$$

for $c < \omega$, where $|X|$ denotes the cardinality of X, and $\mathfrak{a}(x/a) : \mathsf{Var} \to D$ is the variable assignment that agrees with $\mathfrak{a}$ on all variables except x, for which it assigns the value $a \in D$.

A formula φ is said to be *satisfiable* in a model $\mathfrak{M} = (\mathfrak{F}, D, I)$, based on $\mathfrak{F} = (W, R)$, if there is some $w \in W$ and some variable assignment $\mathfrak{a} : \mathsf{Var} \to D$ such that $\mathfrak{M}, w \models^{\mathfrak{a}} \varphi$. We say that φ is *valid* in $\mathfrak{F}$ if its negation $\neg\varphi$ cannot be satisfied in any model based on $\mathfrak{F}$; in which case we write that $\mathfrak{F} \models \varphi$.

Given a non-empty class $\mathcal{C}$ of Kripke frames, we define the first-order modal logic of $\mathcal{C}$ to be the set

$$\mathsf{Q^{\#}Log}(\mathcal{C}) = \{\varphi \in \mathcal{Q}^{\#}\mathcal{ML} \;:\; \mathfrak{F} \models \varphi \text{ for all } \mathfrak{F} \in \mathcal{C}\}$$

of all formulas that are valid in all first-order Kripke models based on frames belonging to $\mathcal{C}$. For a propositional modal logic L, we define $\mathbf{Q}^{\#}L = \mathsf{Q}^{\#}\mathsf{Log}(\mathsf{Fr}\ L)$, where $\mathsf{Fr}\ L$ denotes the class of all frames for L; *i.e.* those frames validating every formula of L.

We will be interested in the *decision (validity) problem* for $\mathbf{Q}^{\#}L$ which asks whether a given formula belongs to $\mathbf{Q}^{\#}L$, and is complementary to the *satisfiability problem* which asks whether a given formula is *satisfiable with respect to* $\mathbf{Q}^{\#}L$; *i.e.* satisfiable in a model based on some frame $\mathfrak{F} \in \mathsf{Fr}\ L$.

Definition 2.1 A first-order modal logic $\mathbf{Q}^{\#}L$ is said to have the *poly-size (resp. exponential) finite model property (fmp)* if every φ that is satisfiable with respect to $\mathbf{Q}^{\#}L$ can be satisfied in a finite model $\mathfrak{M} = (\mathfrak{F}, D, I)$ whose size is at most polynomial (resp. exponential) in the size of φ, with $\mathfrak{F} \in \mathsf{Fr}\ L$.

In what follows, we will be interested in the quantified versions of the propositional modal logics **K**, **KT**, **KB**, **S5**, and **Alt**, whose formulas are validated by the class of all frames, all reflexive frames, all symmetric frames, all equivalence relations, and all partial functions, respectively.

3 Main results

The main result of this section will be to show that the one-variable fragment of $\mathbf{Q}^{\#}\mathbf{K}$, characterised by the class of all frames, enjoys the exponential finite model property. Following this, the analogous results for the one-variable fragments of each of the logics $\mathbf{Q}^{\#}\mathbf{KT}$, $\mathbf{Q}^{\#}\mathbf{KB}$, $\mathbf{Q}^{\#}\mathbf{S5}$, and $\mathbf{Q}^{\#}\mathbf{Alt}$ can be obtained by reducing them to $\mathbf{Q}^{\#}\mathbf{K}$ (see Appendix A).

Theorem 3.1 *The fragment* $\mathbf{Q}^{\#}\mathbf{K} \cap \mathcal{Q}^{\#}\mathcal{ML}^{1}$ *has the exponential fmp.*

To prove this, we employ a version of the method of *quasimodels* [20,2]. Our quasimodels closely resemble full Kripke models, however, each first-order structure is replaced with a *quasistate*, each of which may be finitely represented. The basic structure of our quasimodels can still be infinite, and may require additional non-trivial 'pruning' techniques to ensure that large quasimodels can be reduced to smaller finite quasimodels without affecting satisfiability. Therein lies the crux of the problem we must solve.

First, let us fix some arbitrary first-order modal formula $\varphi \in \mathcal{Q}^{\#}\mathcal{ML}$, and throughout what follows let $n = |\mathrm{sub}(\varphi)|$ denote the number of subformulas of φ, $m = \mathsf{md}(\varphi)$ denote the modal depth of φ, and $C = \mathsf{count}(\varphi)$ denote the value of the largest quantifier subscript occurring in φ. In particular we note that $n, m < ||\varphi||$, while $C < 2^{||\varphi||}$, owing to the binary encoding of subscripts.

We define a *type* for φ to be any subset $t \subseteq \mathrm{sub}(\varphi)$ that is *Boolean-saturated* in the sense that:

(tp1) $\neg\psi \in t$ if and only if $\psi \notin t$, for all $\neg\psi \in \mathrm{sub}(\varphi)$, and

(tp2) $\psi_1 \wedge \psi_2 \in t$ if and only if $\psi_1 \in t$ and $\psi_2 \in t$, for all $\psi_1 \wedge \psi_2 \in \mathrm{sub}(\varphi)$.

Definition 3.2 We define a *quasistate* for φ to be a pair (T, μ) such that:

(qs1) T is a non-empty set of *types* for φ,

(qs2) $\mu : T \to \{1, \dots, C, C+1\}$ is a *'multiplicity'* function,

(qs3) *($\exists_{\leq c}$-saturation)* For all $t \in T$ and $(\exists_{\leq c} x\ \xi) \in \mathrm{sub}(\varphi)$,

$$(\exists_{\leq c} x\ \xi) \in t \qquad \iff \qquad \sum_{t' \in T(\xi)} \mu(t') \ \leq \ c,$$

where $T(\xi) = \{t \in T \ : \ \xi \in t\}$ denotes the set of types belonging to T that contain ξ.

Note that the size of each quasistate cannot exceed the number of distinct types for φ, which is to say that $|T| \leq 2^n$. The multiplicity function indicates how many 'duplicates' of each type are required in order to transform the quasistate into an appropriate first-order structure. Note that φ is indifferent to any duplicates in excess of the value of its largest quantifier subscript.

A *basic structure* for φ is a triple $(W, \prec, \boldsymbol{q})$, where $(W, \prec)$ is an intransitive, irreflexive tree of depth $\leq m$, and $\boldsymbol{q}$ is a function associating each $w \in W$ with a quasistate $\boldsymbol{q}(w) = (T_w, \mu_w)$. An *(indexed) run through* $(W, \prec, \boldsymbol{q})$ is a pair $r = (f_r, i_r)$, where f_r is a function associating each $w \in W$ with a type $f_r(w) \in T_w$, and i_r is an index used to distinguish otherwise identical runs. For convenience, we do not distinguish between the run and the function described by its first argument, writing $r(w)$ in place of $f_r(w)$, for $w \in W$.

Definition 3.3 A *quasimodel for* φ is a tuple $\mathfrak{Q} = (W, \prec, \boldsymbol{q}, \mathfrak{R})$ such that:

(qm1) $(W, \prec, \boldsymbol{q})$ is a basic structure for φ, and $\mathfrak{R}$ is an set of indexed runs through $(W, \prec, \boldsymbol{q})$,

(qm2) There is some $w_0 \in W$ and $t_0 \in T_{w_0}$ such that $\varphi \in t_0$,

(qm3) *(coherence)* For all $r \in \mathfrak{R}$, $w \in W$ and $\Diamond\xi \in \mathrm{sub}(\varphi)$,

$$\exists v \in W;\ w \prec v \text{ and } \xi \in r(v) \quad \Longrightarrow \quad \Diamond\xi \in r(w),$$

(qm4) *(saturation)* For all $r \in \mathfrak{R}$, $w \in W$ and $\Diamond\xi \in \mathrm{sub}(\varphi)$,

$$\Diamond\xi \in r(w) \quad \Longrightarrow \quad \exists v \in W;\ w \prec v \text{ and } \xi \in r(v),$$

(qm5) For all $w \in W$ and $t \in T_w$,

$$\mu_w(t) = \min\big(\,|\{r \in \mathfrak{R} : r(w) = t\}|\,,\ C+1\big).$$

The *size* of $\mathfrak{Q}$ is taken to be $|W| \cdot |\mathfrak{R}|$.

The following lemma establishes that our quasimodels precisely capture the notion of satisfiability with respect to $\mathbf{Q^{\#}K}$, and that every quasimodel for φ can be transformed into model for φ of proportional size.

Lemma 3.4 *Let $\varphi \in \mathcal{Q}^{\#}\mathcal{ML}^1$ be an arbitrary formula in one-variable. Then φ is satisfiable with respect to $\mathbf{Q}^{\#}\mathbf{K}$ iff there is a quasimodel for φ.*

Proof. Suppose that φ is satisfiable with respect to $\mathbf{Q}^{\#}\mathbf{K}$. Then $\mathfrak{M}, w_0 \models^{\mathfrak{a}} \varphi$ for some first-order Kripke model $\mathfrak{M} = (\mathfrak{F}, D, I)$, where $\mathfrak{F} = (W, \prec) \in \mathsf{Fr}\ \mathbf{K}$ is a frame for $\mathbf{K}$, with $w_0 \in W$. By a standard unravelling argument [1], we may assume without any loss of generality that $\mathfrak{F}$ is an intransitive, irreflexive tree of depth at most $m = \mathsf{md}(\varphi)$.

With each $w \in W$ and $a \in D$, we associate the type

$$\mathrm{tp}_w^{\mathfrak{M}}[a] = \{\xi \in \mathrm{sub}(\varphi) \ : \ \mathfrak{M}, w \models^{\mathfrak{a}(x/a)} \xi\},$$

and define a basic structure $(W, \prec, \boldsymbol{q})$, by taking $\boldsymbol{q}(w) = (T_w, \mu_w)$, for all $w \in W$, where

$$T_w = \{\mathrm{tp}_w^{\mathfrak{M}}[a] \ : \ a \in D\} \quad \text{and} \quad \mu_w(t) = \min\left(\left|\{a \in D : \mathrm{tp}_w^{\mathfrak{M}}[a] = t\}\right|, C+1\right)$$

for all $t \in T_w$. It is straightforward to check that $\boldsymbol{q}(w)$ is a quasistate, for each $w \in W$. Indeed, suppose that $\mathrm{tp}_w^{\mathfrak{M}}[a] \in T_w$ and that $(\exists_{\leq c} x\ \xi) \in \mathrm{sub}(\varphi)$, for some $c \leq C$, then we have that:

$$\begin{aligned}
(\exists_{\leq c} x\ \xi) \in \mathrm{tp}_w^{\mathfrak{M}}[a] \quad &\iff \quad \mathfrak{M}, w \models^{\mathfrak{a}(x/a)} (\exists_{\leq c} x\ \xi) \quad \text{by definition,} \\
&\iff \quad \left|\{b \in D \ : \ \mathfrak{M}, w \models^{\mathfrak{a}(x/b)} \xi\}\right| \ \leq \ c, \\
&\iff \quad \sum_{t' \in T_w(\xi)} \left|\{b \in D \ : \ \mathrm{tp}_w^{\mathfrak{M}}[b] = t'\}\right| \ \leq \ c, \\
&\iff \quad \sum_{t' \in T_w(\xi)} \mu_w(t') \ \leq \ c.
\end{aligned}$$

The final equivalence follows from the fact that each summand strictly less than $(C+1)$, since $c \leq C$. Hence, it follows from the definition that $\mu_w(t) = \left|\{b \in D \ : \ \mathrm{tp}_w^{\mathfrak{M}}[b] = t'\}\right|$, for all $t' \in T_w(\xi)$.
Furthermore, for each $a \in D$ we define an indexed run $r_a = (f_a, a)$, where $f_a : W \to 2^{\mathrm{sub}(\varphi)}$ is the function defined such that

$$f_a(w) = \mathrm{tp}_w^{\mathfrak{M}}[a],$$

for all $w \in W$. We then take $\mathfrak{R} = \{r_a \ : \ a \in D\}$ to be the set of all such indexed runs through $(W, \prec, \boldsymbol{q})$. Note that there may be many runs in $\mathfrak{R}$ that differ only in their index. It is straightforward to check that $(W, \prec, \boldsymbol{q}, \mathfrak{R})$ is a quasimodel for φ.

Conversely, suppose that $\mathfrak{Q} = (W, \prec, \boldsymbol{q}, \mathfrak{R})$ is a quasimodel for φ. We define a first-order Kripke model $\mathfrak{M} = (\mathfrak{F}, D, I)$, by taking

$$\mathfrak{F} = (W, \prec) \in \mathsf{Fr}\ \mathbf{K}, \qquad D = \mathfrak{R}, \qquad \text{and} \qquad P_i^{I(w)} = \{r \in \mathfrak{R} : P_i(x) \in r(w)\},$$

for all predicate symbols $P_i \in \mathsf{Pred}$ and $w \in W$. It remains to check that $\mathfrak{M}$ is a model for φ.

Claim 3.5 *We claim that, for all* $w \in W$, $r \in \mathfrak{R}$, *and* $\psi \in \mathrm{sub}(\varphi)$,

$$\mathfrak{M}, w \models^{\mathfrak{a}(x/r)} \psi \qquad \Longleftrightarrow \qquad \psi \in r(w).$$

This can be established by induction on the construction of ψ, the details for which can be found in Appendix B.

By **(qm2)**, there is some $w_0 \in W$ and $t_0 \in T_{w_0}$ such that $\varphi \in t_0$, while by **(qm5)** we have that there is some $r_0 \in \mathfrak{R}$ such that $r_0(w_0) = t_0$. Hence, it follows from (I.H.) that $\mathfrak{M}, w_0 \models^{\mathfrak{a}(x/r_0)} \varphi$, which is to say that φ is satisfiable with respect to $\mathbf{Q^{\#}K}$, as required. □

Hence, to show that the one-variable fragment $\mathbf{Q^{\#}K} \cap \mathcal{Q}^{\#}\mathcal{ML}^1$ has the exponential fmp, it is enough to show that every quasimodel for φ can be transformed into a finite quasimodel that is at most exponential in the size of φ.

Lemma 3.6 *If φ has a quasimodel, then φ has a quasimodel that is at most exponential in the size of φ.*

Proof. Suppose that $\mathfrak{Q} = (W, \prec, \boldsymbol{q}, \mathfrak{R})$ is a quasimodel for φ. The proof follows two stages: the first involves pruning both the basic structure and the set of runs so that they are both at most exponential in the size of φ. During this stage we inadvertently destroy some of the defining properties of our quasimodel; in particular the saturation condition **(qm4)**. In the second stage we remedy this deficiency by adding multiple 'copies' of each quasistate and performing 'surgery' on a finite set of runs to repair saturation.

Step 1) Firstly, it follows from **(qm2)** that there is some $w_0 \in W$ and $t_0 \in T_{w_0}$ such that $\varphi \in t_0$. By **(qm5)**, for each $w \in W$ and each $t \in T_w$ we may fix some run $s_{(w,t)} \in \mathfrak{R}$ such that $s_{(w,t)}(w) = t$. Take $\mathfrak{S}(w) = \{s_{(w,t)} : t \in T_w\}$ be to the set comprising all such runs, for each $w \in W$. In particular, we note that $|\mathfrak{S}(w)| = |T_w| \leq 2^n$. Furthermore, by **(qm4)**, for each $\Diamond\alpha \in t$ we may fix some $v = v_{(w,t,\alpha)} \in W$ such that $w \prec v$ and $\alpha \in s_{(w,t)}(v)$.

We define inductively a sequence of subsets $W_i \subseteq W$, for $i = 0, \dots, m$, by taking $W_0 = \{w_0\}$, and

$$W_{k+1} = \big\{v_{(w,t,\alpha)} \in W \ : \ w \in W_k,\ t \in T_w,\ \text{and } \Diamond\alpha \in t\big\},$$

for $k < m$. We then define a new basic structure $(W', \prec', \boldsymbol{q}')$, by taking

$$W' = \bigcup_{k=0}^{m} W_k \qquad u \prec' v \iff u \prec v, \qquad \text{and} \qquad \boldsymbol{q}'(u) = \boldsymbol{q}(u),$$

for all $u, v \in W'$.

Let $\mathfrak{S} = \bigcup\{\mathfrak{S}(w) : w \in W'\}$, and note that $\mathfrak{S}$ is finite since it is a finite union of finite sets of runs. However, $\mathfrak{S}$ need not be plentiful enough to accommodate **(qm5)**. Hence we must extend $\mathfrak{S}$ to a 'small' subset $\mathfrak{R}'$ of $\mathfrak{R}$ by choosing sufficiently many runs so as to satisfy **(qm5)**.

More precisely, for each $w \in W'$, $t \in T_w$ and $i < \mu_w(t)$ we can fix some $r_{(w,t,i)} \in \mathfrak{R}$ such that $r_{(w,t,i)}(w) = t$, and $r_{(w,t,i)} \neq r_{(w,t,j)}$ for $i \neq j$, which we are able to do since $\mathfrak{Q}$ satisfies **(qm5)**. Furthermore, we may assume without any loss of generality that $r_{(w,t,0)} = s_{(w,t)} \in \mathfrak{S}(w)$, as defined above. We take

$$\mathfrak{R}' = \{r_{(w,t,i)} \in \mathfrak{R} \ : \ w \in W',\ t \in T_w \text{ and } i < \mu_w(t)\}$$

to be the set of all such runs, and define $\mathfrak{Q}' = (W', \prec', \boldsymbol{q}', \mathfrak{R}')$. We note that

$$|W'| \leq |W_0| + \cdots + |W_m| \leq (m+1) \cdot |W_m| \leq (m+1) \cdot (n \cdot 2^n)^m \tag{1}$$

$$|\mathfrak{R}'| \leq |W'| \cdot \max_{w \in W} |T_w| \cdot (C+1) \leq (m+1) \cdot (n \cdot 2^n)^m \cdot 2^n \cdot (C+1) \tag{2}$$

Furthermore, by construction, $\mathfrak{Q}'$ satisfies each of the conditions **(qm1)**, **(qm2)**, **(qm3)**, and **(qm5)**, as can be easily verified. However $\mathfrak{Q}'$ fails to satisfy the saturation condition **(qm4)**. To remedy this, we diverge from the techniques of [20,2] by extending our basic structure with not one but *multiple* 'copies' of each quasistate; each associated with a given transposition of runs.

Step 2) Let $\mathsf{Sym}(\mathfrak{R}')$ denote the set of all permutations $\sigma : \mathfrak{R}' \to \mathfrak{R}'$ on the set of runs $\mathfrak{R}'$, with $\mathsf{id} \in \mathsf{Sym}(\mathfrak{R}')$ denoting the identity function. For each $w \in W'$ and each $r \in \mathfrak{R}'$, let $\tau_{(w,r)} \in \mathsf{Sym}(\mathfrak{R}')$ denote the permutation that transposes r and $s_{(w,t)} \in \mathfrak{S}(w)$, where $t = r(w)$, and let $\mathsf{Trans}(w) = \{\tau_{(w,r)} : r \in \mathfrak{R}'\}$ denote the set of all such transpositions. In particular, we have that $|\mathsf{Trans}(w)| \leq |\mathfrak{R}'|$ is at most exponential in the size of φ.

In what follows, we construct a new basic structure based on some 'small' subset of $W' \times \mathsf{Sym}(\mathfrak{R}')$. Naturally, we cannot construct a basic structure out of the set of all pairs from $W' \times \mathsf{Sym}(\mathfrak{R}')$ if we are to insist on an exponential upper bound on the size of the quasimodel. Instead, for each $(w, \sigma) \in W' \times \mathsf{Sym}(\mathfrak{R}')$, we may define a small set of *successors* $S(w, \sigma) \subseteq W' \times \mathsf{Sym}(\mathfrak{R}')$, by taking

$$S(w, \sigma) = \{(v, \sigma') \ : \ w \prec v \text{ and } \sigma' = (\tau \circ \sigma) \text{ for some } \tau \in \mathsf{Trans}(w)\}.$$

In particular, we note that $|S(w, \sigma)| \leq |W'| \cdot |\mathsf{Trans}(w)| \leq |W'| \cdot |\mathfrak{R}'|$ is at most exponential in the size of φ. We construct a new sequence of sets $W'_i \subseteq W' \times \mathsf{Sym}(\mathfrak{R}')$, for $i = 0, \ldots, m$, by taking

$$W'_0 = \{(w_0, \mathsf{id})\} \qquad \text{and} \qquad W'_{k+1} = \bigcup \{S(w, \sigma) \ : \ (w, \sigma) \in W'_k\}.$$

for $k < m$. The new basic structure is defined to be the triple $(W'', \prec'', \boldsymbol{q}'')$, where

$$W'' = \bigcup_{k=0}^{m} W'_k, \qquad (u, \sigma) \prec'' (v, \rho) \iff (v, \rho) \in S(u, \sigma)$$

and $\boldsymbol{q}''(u, \sigma) = \boldsymbol{q}'(u)$, for all $(u, \sigma), (v, \rho) \in W''$.

Finally, for each run $r \in \mathfrak{R}'$ we define a new run $\widehat{r}$ through $(W'', \prec'', \boldsymbol{q}'')$, by taking

$$\widehat{r}(w, \sigma) = \sigma(r)(w)$$

for all $(w, \sigma) \in W''$. That is to say that the new run $\widehat{r}$ behaves at $(w, \sigma) \in W''$ as $\sigma(r) \in \mathfrak{R}'$ does at $w \in W'$. Take $\mathfrak{R}'' = \{\widehat{r} : r \in \mathfrak{R}'\}$ to be the set of all such runs, and define $\mathfrak{Q}'' = (W'', \prec'', \boldsymbol{q}'', \mathfrak{R}'')$, where

$$|W''| \leq (m+1) \cdot (|W'| \cdot |\mathfrak{R}'|)^m \qquad \text{and} \qquad |\mathfrak{R}''| = |\mathfrak{R}'| \tag{3}$$

are both at most exponential in the size of φ. All that remains is to show that $\mathfrak{Q}''$ is a quasimodel for φ.

- It follows from the construction that $\varphi \in t_0$ for some $t_0 \in T_{w_0} = T_{(w_0, \mathsf{id})}$, where $(w_0, \mathsf{id}) \in W'$, as required for **(qm2)**.

- For **(qm3)**, suppose that $\widehat{r} \in \mathfrak{R}''$, $(w, \sigma), (v, \rho) \in W''$ and $\Diamond\alpha \in \mathrm{sub}(\varphi)$ are such that $(w, \sigma) \prec'' (v, \rho)$ and $\alpha \in \widehat{r}(v, \rho)$.

  By definition we have that $(v, \rho) \in S(w, \sigma)$, which is to say that $w \prec v$ and $\rho = \tau \circ \sigma$ for some transposition $\tau \in \mathsf{Trans}(w)$. Hence we have that

  $$\alpha \in \widehat{r}(v, \rho) \;=\; \rho(r)(v) \;=\; (\tau \circ \sigma)(r)(v) \;=\; \tau(\sigma(r))(v),$$

  where $\tau(\sigma(r)) \in \mathfrak{R}'$. Since $\mathfrak{Q}'$ is coherent and $w \prec' v$, we have that $\Diamond\alpha \in \tau(\sigma(r))(w)$. However, we have that $\tau \in \mathsf{Trans}(w)$ and hence by definition $\tau(\sigma(r))(w) = \sigma(r)(w)$, since τ transposes only runs that coincide at w. In particular, we have that $\Diamond\alpha \in \sigma(r)(w)$, which is to say that $\Diamond\alpha \in \widehat{r}(w, \sigma)$, as required.

- For **(qm4)**, suppose that $\widehat{r} \in \mathfrak{R}'$, $(w, \sigma) \in W''$ and $\Diamond\alpha \in \mathrm{sub}(\varphi)$ are such that $\Diamond\alpha \in \widehat{r}(w, \sigma)$. This is to say that $\Diamond\alpha \in \sigma(r)(w)$. Let $t = \sigma(r)(w)$ and let $s_{(w,t)} \in \mathfrak{S}(w)$ be such that $s_{(w,t)}(w) = t$. By construction there is some $v = v_{(w,t,\alpha)} \in W'$ such that $w \prec' v$ and $\alpha \in s_{(w,t)}(v)$.

  Let $\tau = \tau_{(w, \sigma(r))} \in \mathsf{Trans}(w)$ be the transposition that swaps $\sigma(r) \in \mathfrak{R}'$ and $s_{(w,t)} \in \mathfrak{S}(w)$. It follows from the construction that there is some $(v, \tau \circ \sigma) \in S(w, \sigma) \subseteq W''$ such that

  $$\alpha \;\in\; s_{(w,t)}(v) \;=\; \tau(\sigma(r))(v) \;=\; (\tau \circ \sigma)(r)(v) \;=\; \widehat{r}(v, \tau \circ \sigma),$$

  and $(w, \sigma) \prec'' (v, \tau \circ \sigma)$, as required.

- For **(qm5)**, suppose that $(w, \sigma) \in W''$ and $t \in T_{(w,\sigma)} = T_w$ and consider the following sets:

  $$X = \{\widehat{r} \in \mathfrak{R}'' : \widehat{r}(w, \sigma) = t\} \qquad \text{and} \qquad Y = \{r \in \mathfrak{R}' : r(w) = t\}.$$

  We define a bijection $f : X \to Y$ by taking $f(\widehat{r}) = \sigma(r)$, for all $\widehat{r} \in X$, since by definition $\widehat{r}(w, \sigma) = \sigma(r)(w) = f(\widehat{r})(w)$. Hence $\widehat{r} \in X$ if and only if $f(\widehat{r}) \in Y$, and thus $|X| = |Y|$. That is to say that the number of runs passing through each type remains unaffected by Step 2 of our construction. It then follows from the definitions that

$$\mu_{(w,\sigma)}(t) = \mu_w(t) = \min(|Y|, C+1) = \min(|X|, C+1)$$

as required.

Hence we have established that $\mathfrak{Q}''$ is a quasimodel for φ, whose size is at most exponential in the size of φ, as can be deduced from (1)–(3), as required. □

Theorem 3.1 now follows from Lemmas 3.4–3.6, and hence the one-variable fragment $\mathbf{Q^{\#}K} \cap \mathcal{Q}^{\#}\mathcal{ML}^1$ has the exponential fmp. This provides us with an optimal upper bound on the complexity of the decision problem for this fragment; the lower bound being provided by the coNExpTime-hardness of the the regular (counting-free) one-variable fragment $\mathbf{Q^{\#}K} \cap \mathcal{Q}^{\#}\mathcal{ML}_0^1$ [10].

Corollary 3.7 *The one-variable fragment of* $\mathbf{Q^{\#}K}$ *is* coNExpTime*-complete.*

We note that the decision problems for each of the logics $\mathbf{Q^{\#}KT}$, $\mathbf{Q^{\#}KB}$, $\mathbf{Q^{\#}S5}$ and $\mathbf{Q^{\#}Alt}$ can be polynomially reduced to that of $\mathbf{Q^{\#}K}$, such that the exponential fmp is preserved. Hence, we are able to deduce that the one-variable fragments of each of these logics also enjoys the exponential finite model property.

Note that the existence of a polynomial reduction between two *propositional* modal logics does not guarantee an analogous reduction between their first-order counterparts, without a complementary model transformation. (see, for example, [2, Remark 6.19]). However, in the aforementioned cases, such 'model-level' reductions are not hard to construct; examples of which are sketched in Appendix A.

Corollary 3.8 *Each of the fragments* $L \cap \mathcal{Q}^{\#}\mathcal{ML}^1$ *has the exponential fmp, for* $L \in \{\mathbf{Q^{\#}KT}, \mathbf{Q^{\#}KB}, \mathbf{Q^{\#}S5}, \mathbf{Q^{\#}Alt}\}$.

It is clear that we can do no better than this bound, since even the classical one-variable fragment with counting quantifiers contains 'small' formulas such as $\varphi_k = \exists_{>2^k} x\ \top$, whose size is linear in k but can only be satisfied in models containing at least 2^k elements, where $\top := P_0(x) \vee \neg P_0(x)$. Indeed, even without counting quantifiers, each of the logics $L \cap \mathcal{Q}^{\#}\mathcal{ML}_0^1$, for $L \in \{\mathbf{Q^{\#}K}, \mathbf{Q^{\#}KT}, \mathbf{Q^{\#}KB}, \mathbf{Q^{\#}S5}\}$ admit formulas that cannot be satisfied in any 'small' model.

On the other hand, it is known that $\mathbf{Q^{\#}Alt} \cap \mathcal{Q}^{\#}\mathcal{ML}_0^1$ enjoys the poly-size model property, and is therefore coNP-complete [10]. In fact, it can be shown that the decision problem for $\mathbf{Q^{\#}Alt} \cap \mathcal{Q}^{\#}\mathcal{ML}^1$ is also coNP-complete, by a reduction to the one-variable fragment of classical first-order logic with counting quantifiers, despite lacking the poly-size model property. The following result follows the approach taken in [2, Proposition 5.35].

Theorem 3.9 *The one-variable fragment of* $\mathbf{Q^{\#}Alt}$ *is* coNP*-complete.*

Proof. Let $\varphi \in \mathcal{Q}^{\#}\mathcal{ML}^{1}$ be an arbitrary formula in one variable with quantifier. For each $\psi \in \mathrm{sub}(\varphi)$, let $Q^0_\psi, \ldots, Q^{m+1}_\psi \in \mathsf{Pred}$ be fresh monadic predicate symbols not occurring in φ, where $m = \mathsf{md}(\varphi)$. For each $i \leq m$, let ζ_i denote the conjunction of the following formulas:

$$\bigwedge_{(\psi_1 \wedge \psi_2) \in \mathrm{sub}(\varphi)} Q^i_{\psi_1 \wedge \psi_2}(x) \leftrightarrow (Q^i_{\psi_1}(x) \wedge Q^i_{\psi_2}(x)),$$

$$\bigwedge_{\neg\psi \in \mathrm{sub}(\varphi)} Q^i_{\neg\psi}(x) \leftrightarrow \neg Q^i_{\psi}(x),$$

$$\bigwedge_{(\exists_{=c} x\ \psi) \in \mathrm{sub}(\varphi)} Q^i_{(\exists_{=c} x\ \psi)}(x) \leftrightarrow \exists_{=c} x\ Q^i_{\psi}(x),$$

$$\bigwedge_{\Diamond\psi \in \mathrm{sub}(\varphi)} Q^i_{\Diamond\psi}(x) \leftrightarrow Q^{i+1}_{\psi}(x)$$

and let $\mathsf{t}_\ell(\varphi) = \bigwedge_{i \leq \ell} \forall x \zeta_i \to Q^0_\varphi$. In particular, we note that the size of each $\mathsf{t}_\ell(\varphi)$ is at most polynomial in the size of φ. We claim that $\varphi \in \mathbf{Q^{\#}Alt}$ if and only if $\mathsf{t}_\ell(\varphi)$ is valid with respect to (classical) first-order logic, with counting quantifiers, for all $\ell \leq m$.

($\Rightarrow$) Suppose that there is some $\ell \leq m$ such that $\mathsf{t}_\ell(\varphi)$ is not valid. Then $\mathfrak{A} \models^{\mathfrak{a}} \bigwedge_{i \leq \ell} \forall x \zeta_i$ and $\mathfrak{A} \not\models^{\mathfrak{a}} Q^0_\varphi$ for some classical first-order model $\mathfrak{A} = (D, J)$. We define a first-order Kripke model $\mathfrak{M} = (\mathfrak{F}, D, I)$, where $\mathfrak{F} = (W, R)$, by taking $W = \{w_i : i \leq \ell\}$, $R = \{(w_i, w_{i+1}) : i < \ell\}$, and $P^{I(w_i)} = (Q^i_{P(x)})^J$, for all predicate symbols $P \in \mathsf{Pred}$ occurring in φ, and all $w_i \in W$. We claim that, for all $w_i \in W$, $b \in D$ and $\psi \in \mathrm{sub}(\varphi)$,

$$\mathfrak{M}, w_i \models^{\mathfrak{a}(x/b)} \psi \quad \Longleftrightarrow \quad \mathfrak{A} \models^{\mathfrak{a}(x/b)} Q^i_\psi$$

as can be verified by induction on the construction of ψ. Hence it follows that $\mathfrak{M}, w_0 \not\models^{\mathfrak{a}} \varphi$, which is to say that $\varphi \notin \mathbf{Q^{\#}Alt}$, as required.

($\Leftarrow$) Suppose that $\varphi \notin \mathbf{Q^{\#}Alt}$. Then $\mathfrak{M}, w \models^{\mathfrak{a}} \varphi$ for some first-order Kripke model $\mathfrak{M} = (\mathfrak{F}, D, I)$, where $\mathfrak{F} = (W, R) \in \mathsf{Fr}\ \mathbf{Alt}$. Without loss of generality we may suppose that $W = \{w_i : i \leq \ell\}$ and $R = \{(w_i, w_{i+1}) : i < \ell\}$, for some $\ell \leq m$. We define a classical first-order model $\mathfrak{A} = (D, J)$ by taking

$$(Q^i_\psi)^J = \{b \in D \ : \ \mathfrak{M}, w_i \models^{\mathfrak{a}(x/b)} \psi\}$$

for $\psi \in \mathrm{sub}(\varphi)$ and $i \leq \ell$. It then follows from this definition that $\mathfrak{A} \not\models^{\mathfrak{a}} \mathsf{t}_\ell(\varphi)$, which is to say that there is some $\mathsf{t}_\ell(\varphi)$ is not valid, for $\ell \leq m$.

Since the validity problem for the one-variable fragment of (classical) first-order logic, with counting quantifiers, is decidable in coNP [14], so too must be that of the one-variable fragment of $\mathbf{Q^{\#}Alt}$, as required. □

Finally, we note that Theorem 3.9 holds if we replace $\mathbf{Q^{\#}Alt}$ with its *serial* [1] extension $\mathbf{Q^{\#}AltD}$. We need only add reflexive loop to the last element of $\mathfrak{F}$.

[1] A frame (W, R) is said to be serial if for every $w \in W$ there is some $v \in W$ such that wRv.

4 Two-dimensional Modal Logics

First-order modal logics are intimately related to another extensively studied formalism; that of many-dimensional modal logics [17,15,2,9,11]. Given a countably infinite set of propositional variables $\mathsf{Prop} = \{p_0, p_1, \dots\}$, let $\mathcal{ML}_2$ denote the set of bimodal formulas defined by the following grammar:

$$\varphi \ ::= \ p_i \mid \neg\varphi \mid (\varphi_1 \wedge \varphi_2) \mid \Diamond_h\varphi \mid \Diamond_v\varphi$$

where $p_i \in \mathsf{Prop}$.

Formulas of $\mathcal{ML}_2$ are interpreted over Kripke models $\mathfrak{M} = (\mathfrak{F}, \mathfrak{V})$, where $\mathfrak{F} = (W, R_h, R_v)$ is a *bimodal Kripke frame*, with $R_h, R_v \subseteq W \times W$, and $\mathfrak{V} : \mathsf{Prop} \to 2^W$ is a propositional valuation. Satisfiability is defined in the usual way with $\Diamond_j\varphi$ being interpreted by the relation R_j, for $j = h, v$.

Of particular interest are product models in which the two modal operators act orthogonally. We define the *product* of two unimodal frames $\mathfrak{F}_h = (W_h, R_h)$ and $\mathfrak{F}_v = (W_v, R_v)$ to be the bimodal frame $\mathfrak{F}_h \times \mathfrak{F}_v = (W_h \times W_v, \overline{R}_h, \overline{R}_v)$, where

$$(u,v)\overline{R}_h(u',v') \iff uR_hu' \text{ and } v = v',$$

$$(u,v)\overline{R}_v(u',v') \iff u = u' \text{ and } vR_vv',$$

for all $u, u' \in W_h$ and $v, v' \in W_v$. The *product* of two unimodal logics L_h and L_v is defined to be the bimodal logic

$$L_h \times L_v = \{\varphi \in \mathcal{ML}_2 \ : \ \mathfrak{F}_h \times \mathfrak{F}_v \models \varphi \text{ where } \mathfrak{F}_j \in \mathsf{Fr}\ L_j \text{ for } j = h, v\}$$

Although product logics are characterised by their product frames, in general there may be frames for product logics that are not product frames. Moreover, it is not always obvious whether an arbitrary bimodal frame is a frame for a given product logic, if the logic happens to be non-finitely axiomatisable. For this reason, it is often more convenient to work with the following, more restrictive, variant of the general ('abstract') fmp.

Definition 4.1 A product logic $L_h \times L_v$ is said to have the *poly-size (resp. exponential) product fmp* if every $\varphi \notin L_h \times L_v$ can be refuted in a model based on a *product frame* for $L_h \times L_v$ that is at most polynomial (resp. exponential) in the size of φ.

Clearly, any logic possessing the product fmp also enjoys the more general 'abstract' fmp.

It is well established that products of the form $L \times \mathbf{S5}$ can be interpreted as syntactic variants of the one-variable fragment of first-order modal logic $\mathbf{Q}^{\#}L \cap \mathcal{Q}^{\#}\mathcal{ML}^1_0$, in which only quantifiers with zero subscripts are permitted [3]. A further connection can be established between products of the form $L \times \mathbf{Diff}$ and the one-variable fragment $\mathbf{Q}^{\#}L \cap \mathcal{Q}^{\#}\mathcal{ML}^1_1$, where $\mathbf{Diff}$ denotes von Wright's logic of 'elsewhere' [19,16], characterised by all those unimodal formulas that are valid in all *difference frames* of the form $(W, \neq)$.

We define the translation $(\cdot)^\dagger : \mathcal{ML}_2 \to \mathcal{Q}^\#\mathcal{ML}_1^1$ by taking

$$p_i^\dagger = P_i(x), \qquad (\neg\psi)^\dagger = \neg\psi^\dagger, \qquad (\psi_1 \wedge \psi_2)^\dagger = \psi_1^\dagger \wedge \psi_2^\dagger,$$
$$(\Diamond_h\psi)^\dagger = \Diamond\psi^\dagger, \qquad (\Diamond_v\psi)^\dagger = \exists^{\neq} x\, \psi^\dagger,$$

where $P_i \in \mathsf{Pred}$ is a unique monadic predicate associated with each propositional variable $p_i \in \mathsf{Prop}$, and $\exists^{\neq} x\, \varphi := \big(\neg\varphi \wedge \exists_{>0} x\, \varphi\big) \vee \exists_{>1} x\, \varphi$. Furthermore, with each first-order Kripke model $\mathfrak{M} = (\mathfrak{F}, D, I)$, we associate a propositional product model $\mathfrak{M}^\star = (\mathfrak{F}_h \times \mathfrak{F}_v, \mathfrak{V})$, by taking

$$\mathfrak{F}_h = \mathfrak{F} \in \mathsf{Fr}\ L, \qquad \mathfrak{F}_v = (D, \neq) \in \mathsf{Fr}\ \mathbf{Diff}, \qquad \mathfrak{V}(p_i) = \big\{(w, v) : v \in P_i^{I(w)}\big\}$$

for all $p_i \in \mathsf{Prop}$. It is then straightforward to check that φ is satisfiable in $\mathfrak{M}^\star$ if and only if $\varphi^\dagger$ is satisfiable in $\mathfrak{M}$. Furthermore, since the model transformation $(\cdot)^\star$ maps bijectively onto the class of product frames characterising $L \times \mathbf{Diff}$, we have the following polynomial reduction from $L \times \mathbf{Diff}$ to $\mathbf{Q}^\# L \cap \mathcal{Q}^\#\mathcal{ML}_1^1$.

Proposition 4.2 $\varphi \in L \times \mathbf{Diff}$ *if and only if* $\varphi^\dagger \in \mathbf{Q}^\# L \cap \mathcal{Q}^\#\mathcal{ML}_1^1$.

That is to say that products of the form $L \times \mathbf{Diff}$ can be embedded within the one-variable fragment $\mathbf{Q}^\# L \cap \mathcal{Q}^\#\mathcal{ML}_1^1$, with quantifier subscripts not exceeding one. Moreover, it follows that $\mathbf{Q}^\# L \cap \mathcal{Q}^\#\mathcal{ML}_1^1$ has the poly-size (resp. exponential) fmp if and only if $L \times \mathbf{Diff}$ has the poly-size (resp. exponential) *product* fmp. Consequently, we have the following corollary of Theorem 3.1.

Corollary 4.3 *Let* $L \in \{\mathbf{K}, \mathbf{KT}, \mathbf{KB}, \mathbf{S5}\}$. *Then*

(i) $L \times \mathbf{Diff}$ *has the exponential product fmp,*

(ii) The decision problem for $L \times \mathbf{Diff}$ *is* coNExpTime*-complete.*

The lower bounds follow from [10, Theorem 3.2], in which every bimodal logic between $\mathbf{K} \times \mathbf{K}$ and $\mathbf{S5} \times \mathbf{S5}$ is shown to be coNExpTime-hard.

Previously, only non-elementary upper bounds on the complexity of $\mathbf{K} \times \mathbf{Diff}$ were known. In particular, $\mathbf{K} \times \mathbf{Diff}$ can be embedded into the product $\mathbf{K} \times \mathbf{Lin}$, whose decidability can be established by a variant on the 'mosaic' approach [20,2]. More recently, the *abstract* fmp of $\mathbf{K} \times \mathbf{Diff}$ had been established via a method of 'canonical filtrations' [18]. However, this yields only a non-elementary bound on the size of the filtrated models. Furthermore, it is known that $\mathbf{K} \times \mathbf{Diff}$ is non-finitely axiomatisable [5], and there is currently no known procedure for deciding whether an arbitrary bimodal frame is a frame for $\mathbf{K} \times \mathbf{Diff}$. Consequently, we cannot construct a feasible decision procedure from the abstract fmp alone.

It should be noted that the upper bound on the decision problem for $\mathbf{S5} \times \mathbf{Diff}$ follows from the coNExpTime-completeness of the two-variable fragment $\mathcal{C}_1^2$ with quantifier subscripts not exceeding one [12]. For instance, take $(\cdot)^\ddagger : \mathcal{ML}_2 \to \mathcal{C}^2$ to be the translation given:

$$p_i^\ddagger = P_i(x,y), \qquad (\neg\psi)^\ddagger = \neg\psi^\ddagger, \qquad (\psi_1 \wedge \psi_2)^\ddagger = \psi_1^\ddagger \wedge \psi_2^\ddagger,$$
$$(\Diamond_h\psi)^\ddagger = \exists_{>0}x\ \psi^\ddagger, \qquad (\Diamond_v\psi)^\ddagger = \exists^{\neq}y\ \big(D(y) \wedge \psi^\ddagger\big),$$

where $P_i \in \mathsf{Pred}$ is a unique binary predicate symbol associated with each propositional variable $p_i \in \mathsf{Prop}$ and $D \in \mathsf{Pred}$ is an auxiliary monadic predicate symbol, providing a guard on the $\exists^{\neq}y$ quantifier, defined above. We have that $\varphi \in \mathbf{S5} \times \mathbf{Diff}$ if and only if $\varphi^\ddagger$ is valid with respect to classical first-order logic.

Note, however, that the full two-variable fragment $\mathcal{C}^2$, with counting quantifiers, does not enjoy the finite model property, even if we restrict the quantifier subscripts to ≤ 1; as evidenced by the following formulas of [4]:

$$\forall x \exists y\ P(x,y)\ \wedge\ \forall y \exists_{\leq 1} x\ P(x,y)\ \wedge\ \exists y \forall x\ \neg P(x,y).$$

Hence, we cannot infer the exponential fmp of $\mathbf{S5} \times \mathbf{Diff}$ by appealing to the classical two-variable fragment, and that $(\cdot)^\ddagger$ maps onto a proper fragment of $\mathcal{C}^2$ possessing the exponential fmp.

Finally, since $\mathbf{Alt} \times \mathbf{Diff}$ is reducible to the fragment $\mathbf{Q^\#Alt} \cap \mathcal{Q}^\#\mathcal{ML}_1^1$, we note the following corollary of Theorem 3.9.

Corollary 4.4 *The decision problem for* $\mathbf{Alt} \times \mathbf{Diff}$ *is* coNP*-complete.*

5 Discussion

We conclude with a discussion of several related open problems:

- It remains open whether these results can be extended to the *monodic* fragment, appropriately extended to counting quantifiers. However, it should be noted that there is no immediate application of the techniques developed in [21], which are able to prove that the (counting-free) monodic fragment of $\mathbf{Q^\#K^*}$ is decidable, since even the one-variable fragment $\mathbf{Q^\#K^*} \cap \mathcal{Q}^\#\mathcal{ML}_1^1$ is known to be non-recursively enumerable [6]. Here, $\mathbf{K^*}$ denotes the bimodal logic of all frames whose second relation is the transitive closure of the first.
- It is known that the one-variable fragment $\mathbf{Q^\#K4} \cap \mathcal{Q}^\#\mathcal{ML}_0^1$, with the sole quantifier $\exists_{\leq 0}x$, has the fmp and is decidable in coN2ExpTime [3,2], where $\mathbf{K4}$ denotes the logic characterised by the class of all transitive frames. However, it remains open whether the full one-variable fragment $\mathbf{Q^\#K4} \cap \mathcal{Q}^\#\mathcal{ML}^1$ is decidable or has even the 'abstract' fmp.
- For many first-order modal logics characterised by *linear orders*, the addition of counting quantifiers causes a jump from decidability to undecidability [7]. In particular the fragment $\mathbf{Q^\#K4.3} \cap \mathcal{Q}^\#\mathcal{ML}_1^1$ is known to be undecidable, where $\mathbf{K4.3}$ denotes the logic of all linear orders. Moreover, the same is true if we consider the sub-logic characterised by models with *decreasing* domains. However, it remains open whether the same result holds in the case of *expanding* domains.

Acknowledgements I would like to thank Agi Kurucz for her comments on an early version of this manuscript, as well as the anonymous reviewers whose comments have improved the quality of this paper.

References

[1] Blackburn, P., M. de Rijke and Y. Venema, "Modal Logic," Cambridge University Press, New York, NY, USA, 2001.

[2] Gabbay, D., A. Kurucz, F. Wolter and M. Zakharyaschev, "Many-Dimensional Modal Logics: Theory and Applications," Studies in Logic and the Foundations of Mathematics **148**, Elsevier, 2003.

[3] Gabbay, D. M. and V. Shehtman, *Products of modal logics. Part I*, Logic Journal of the IGPL **6** (1998), pp. 73–146.

[4] Grädel, E., M. Otto and E. Rosen, *Two-variable logic with counting is decidable*, in: *Logic in Computer Science, 1997. LICS'97. Proceedings., 12th Annual IEEE Symposium on*, IEEE, 1997, pp. 306–317.

[5] Hampson, C. and A. Kurucz, *Axiomatisation and decision problems of modal product logics with the difference operator*, (manuscript).

[6] Hampson, C. and A. Kurucz, *On modal products with the logic of 'elsewhere'*, in: T. Bolander, T. Braüner, S. Ghilardi and L. Moss, editors, *Advances in Modal Logic*, Advances in Modal Logic **9** (2012), pp. 339–347.

[7] Hampson, C. and A. Kurucz, *Undecidable propositional bimodal logics and one-variable first-order linear temporal logics with counting*, ACM Transactions on Computational Logic (TOCL) **16** (2015).

[8] Kripke, S. A., *The undecidability of monadic modal quantification theory*, Mathematical Logic Quarterly **8** (1962), pp. 113–116.

[9] Kurucz, A., *Combining modal logics*, in: P. Blackburn, J. van Benthem and F. Wolter, editors, *Handbook of Modal Logic*, Studies in Logic and Practical Reasoning **3**, Elsevier, 2007 pp. 869–924.

[10] Marx, M., *Complexity of products of modal logics*, Journal of Logic and Computation **9** (1999), pp. 197–214.

[11] Marx, M. and Y. Venema, "Multi-dimensional modal logic," Springer, 1997.

[12] Pacholski, L., W. Szwast and L. Tendera, *Complexity of two-variable logic with counting*, in: *Proceedings of the 12th Annual IEEE Symposium on Logic in Computer Science (LICS'97)*, IEEE, 1997, pp. 318–327.

[13] Pratt-Hartmann, I., *Complexity of the two-variable fragment with counting quantifiers*, Journal of Logic, Language and Information **14** (2005), pp. 369–395.

[14] Pratt-Hartmann, I., *On the computational complexity of the numerically definite syllogistic and related logics*, Bulletin of Symbolic Logic **14** (2008), pp. 1–28.

[15] Segerberg, K., *Two-dimensional modal logic*, Journal of Philosophical Logic **2** (1973), pp. 77–96.

[16] Segerberg, K., *A note on the logic of elsewhere*, Theoria **46** (1980), pp. 183–187.

[17] Shehtman, V., *Two-dimensional modal logics*, Mathematical Notices of the USSR Academy of Sciences **23** (1978), pp. 417–424, (Translated from Russian).

[18] Shehtman, V., *Canonical filtrations and local tabularity*, in: R. Goré, B. Kooi and A. Kurucz, editors, *Advances in Modal Logic*, Advances in Modal Logic **10** (2014), pp. 498–512.

[19] von Wright, G. H., *A modal logic of place*, in: E. Sosa, editor, *The philosophy of Nicolas Rescher*, Dordrecht, 1979 pp. 65–73.

[20] Wolter, F., *The product of converse PDL and polymodal K*, Journal of Logic and Computation **10** (2000), pp. 223–251.

[21] Wolter, F. and M. Zakharyaschev, *Temporalizing description logics*, in: D. M. Gabbay and M. de Rijke, editors, *Frontiers of Combining Systems 2*, 1998, pp. 104–109.

[22] Wolter, F. and M. Zakharyaschev, *Decidable fragments of first-order modal logics*, The Journal of Symbolic Logic **66** (2001), pp. 1415–1438.

A Modal Reductions

In this appendix we outline the reductions between the first-order modal logic $\mathbf{Q^{\#}K}$ and the logics $\mathbf{Q^{\#}KT}$, $\mathbf{Q^{\#}KB}$, $\mathbf{Q^{\#}S5}$ and $\mathbf{Q^{\#}Alt}$, discussed above in Section 3. Each of the reductions respects the number of first-order variables, thereby completing the proof of Corollary 3.8.

Let $\varphi \in \mathcal{Q}^{\#}\mathcal{ML}^1$ be an arbitrary first-order modal formula, and for each $\alpha \in \mathrm{sub}(\varphi)$, let Q_α be a fresh propositional variable not occurring in φ. We define a translation $(\cdot)^\dagger : \mathrm{sub}(\varphi) \to \mathcal{Q}^{\#}\mathcal{ML}^1$ by taking

$$P_i(x)^\dagger = P_i(x), \qquad (\neg\psi)^\dagger = \neg\psi^\dagger, \qquad (\psi_1 \wedge \psi_2)^\dagger = \psi_1^\dagger \wedge \psi_2^\dagger,$$
$$(\Diamond\psi)^\dagger = Q_\psi(x), \qquad (\exists_{\leq c}x\ \psi)^\dagger = \exists_{\leq c}x\ \psi^\dagger.$$

That is to say that we replace each subformula of the form $\Diamond\psi$ with a monadic predicate $Q_\psi(x)$. We then define the following formulas:

$$\zeta_1 := \bigwedge_{\xi\in\mathrm{sub}(\varphi)} \Box^{(m)}\big(Q_\xi(x) \to \Box\xi^\dagger\big) \ \wedge\ \Box^{(m)}\big(\Diamond\xi^\dagger \to Q_\xi(x)\big) \tag{A.1}$$

$$\zeta_2 := \bigwedge_{\xi\in\mathrm{sub}(\varphi)} \Box^{(m)}\big(Q_\xi(x) \leftrightarrow (\xi^\dagger \vee \Diamond\xi^\dagger)\big) \tag{A.2}$$

$$\zeta_3 := \bigwedge_{\xi\in\mathrm{sub}(\varphi)} \big(Q_\xi(x) \to \Box Q_\xi(x)\big) \ \wedge\ \big(\Diamond Q_\xi(x) \to Q_\xi(x)\big) \\ \wedge\ \big(Q_\xi(x) \leftrightarrow (\xi^\dagger \vee \Diamond\xi^\dagger)\big) \tag{A.3}$$

$$\zeta_4 := \bigwedge_{\xi\in\mathrm{sub}(\varphi)} \Box^{(m)}\big(\Diamond\xi^\dagger \to Q_\xi(x)\big) \ \wedge\ \Box^{(m)}\big(\xi^\dagger \to \Box Q_\xi(x)\big) \\ \wedge\ \Box^{(m)}\big(\Diamond(Q_\xi(x) \wedge \neg\Diamond\xi^\dagger) \to \xi^\dagger\big) \ \wedge\ \big(Q_\xi(x) \to \Diamond\xi^\dagger\big) \tag{A.4}$$

where $\Box^{(0)}\varphi := \varphi$ and $\Box^{(k)}\varphi := \varphi \wedge \Box\Box^{(k-1)}\varphi$, for $k > 0$.

Note that each of the formulas $\zeta_1, \zeta_2, \zeta_3$ and ζ_4, are at most linear in the size of φ, since $||\xi^\dagger|| \leq ||\xi||$, for all $\xi \in \mathrm{sub}(\varphi)$.

Proposition A.1 *Let $\varphi \in \mathcal{Q}^{\#}\mathcal{ML}^1$ and let $\zeta_1, \zeta_2, \zeta_3$, and ζ_4 be as above, then:*

(i) $\varphi \in \mathbf{Q^{\#}Alt}$ *if and only if* $(\forall x\zeta_1 \to \varphi^\dagger) \in \mathbf{Q^{\#}K}$,

(ii) $\varphi \in \mathbf{Q^{\#}KT}$ *if and only if* $(\forall x\zeta_2 \to \varphi^\dagger) \in \mathbf{Q^{\#}K}$,

(iii) $\varphi \in \mathbf{Q^{\#}S5}$ *if and only if* $(\forall x\zeta_3 \to \varphi^\dagger) \in \mathbf{Q^{\#}K}$,

(iv) $\varphi \in \mathbf{Q^{\#}KB}$ *if and only if* $(\forall x\zeta_4 \to \varphi^\dagger) \in \mathbf{Q^{\#}K}$.

Proof.

(i) ($\Rightarrow$) Suppose that $(\forall x\zeta_1 \to \varphi^\dagger) \notin \mathbf{Q^{\#}K}$. Then $\mathfrak{M}, r \models^{\mathfrak{a}} \forall x\zeta_1$ and $\mathfrak{M}, r \not\models^{\mathfrak{a}} \varphi$ for some model $\mathfrak{M} = (\mathfrak{F}, D, I)$, where $\mathfrak{F} = (W, R) \in \mathsf{Fr}\ \mathbf{K}$. Without loss of generality we may suppose that $\mathfrak{F}$ is an irreflexive, intransitive tree, rooted at $r \in W$. Let $w_0Rw_1R\ldots Rw_\ell$ denote the longest R-chain in $\mathfrak{F}$ such that

$w_0 = r$ and $\ell \leq m$, and define a new model $\mathfrak{M}' = (\mathfrak{F}', D, I)$ where $\mathfrak{F}' = (W', R') \in \mathsf{Fr}\ \mathbf{Alt}$, by taking $W' = \{w_i : i \leq \ell\}$ and $R' = R \cap (W' \times W')$. We claim that, for all $w \in W'$, $b \in D$, and $\psi \in \mathrm{sub}(\varphi)$,

$$\mathfrak{M}', w \models^{\mathfrak{a}(x/b)} \psi \quad \Longleftrightarrow \quad \mathfrak{M}, w \models^{\mathfrak{a}(x/b)} \psi^\dagger,$$

whenever $\mathsf{md}(\psi) + d(r, w) \leq m$, as can be verified by induction on the construction of ψ, where $d(r, w)$ denotes the distance[2] between r and w.

Hence it follows that $\mathfrak{M}', r \not\models^{\mathfrak{a}} \varphi$, which is to say that $\varphi \notin \mathbf{Q^{\#}Alt}$, as required.

($\Leftarrow$) Conversely, suppose that $\varphi \notin \mathbf{Q^{\#}Alt}$. Then $\mathfrak{M}, r \not\models^{\mathfrak{a}} \varphi$ for some model $\mathfrak{M} = (\mathfrak{F}, D, I)$, where $\mathfrak{F} = (W, R) \in \mathsf{Fr}\ \mathbf{Alt} \subseteq \mathsf{Fr}\ \mathbf{K}$ describes a partial function. We define a new model $\mathfrak{M}' = (\mathfrak{F}, D, I')$ by taking $P_i^{I'(w)} = P_i^{I(w)}$ for all $P_i(x) \in \mathrm{sub}(\varphi)$ and $b \in Q_\xi^{I'(w)}$ iff $\mathfrak{M}, w \models^{\mathfrak{a}(x/b)} \Diamond\xi$. We claim that, for all $w \in W$, $b \in D$, and $\psi \in \mathrm{sub}(\varphi)$,

$$\mathfrak{M}', w \models^{\mathfrak{a}(x/b)} \psi^\dagger \quad \Longleftrightarrow \quad \mathfrak{M}, w \models^{\mathfrak{a}(x/b)} \psi,$$

as can be verified by induction on the construction of ψ.

Furthermore, since $\mathfrak{F}$ is a frame for **Alt**, we have that $\mathfrak{M}, r \models^{\mathfrak{a}} \forall x \zeta_1$, as can be easily verified. Hence it follows that $\mathfrak{M}', r \not\models^{\mathfrak{a}} (\forall x \zeta_1 \to \varphi)$, which is to say that $(\forall x \zeta_1 \to \varphi) \notin \mathbf{Q^{\#}K}$, as required.

(ii) ($\Rightarrow$) Suppose that $(\forall x \zeta_2 \to \varphi^\dagger) \notin \mathbf{Q^{\#}K}$. Then $\mathfrak{M}, r \models^{\mathfrak{a}} \forall x \zeta_2$ and $\mathfrak{M}, r \not\models^{\mathfrak{a}} \varphi$ for some model $\mathfrak{M} = (\mathfrak{F}, D, I)$, where $\mathfrak{F} = (W, R) \in \mathsf{Fr}\ \mathbf{K}$. Without loss of generality we may suppose that $\mathfrak{F}$ is an irreflexive, intransitive tree, rooted at $r \in W$. Let $\mathfrak{F}^+ = (W, R^+) \in \mathsf{Fr}\ \mathbf{KT}$ denote the reflexive closure of $\mathfrak{F}$, where $R^+ = R \cup \{(w, w) : w \in W\}$, and let $\mathfrak{M}' = (\mathfrak{F}^+, D, I)$ be a new model. We claim that, for all $w \in W'$, $b \in D$, and $\psi \in \mathrm{sub}(\varphi)$,

$$\mathfrak{M}', w \models^{\mathfrak{a}(x/b)} \psi \quad \Longleftrightarrow \quad \mathfrak{M}, w \models^{\mathfrak{a}(x/b)} \psi^\dagger,$$

whenever $\mathsf{md}(\psi) + d(r, w) \leq m$, as can be verified by induction on the construction of ψ, where $d(r, w)$ denotes the distance between r and w.

Hence it follows that $\mathfrak{M}', r \not\models^{\mathfrak{a}} \varphi$, which is to say that $\varphi \notin \mathbf{Q^{\#}KT}$, as required.

($\Leftarrow$) The converse direction is similar to that of (i), using the fact that that $\mathsf{Fr}\ \mathbf{KT} \subseteq \mathsf{Fr}\ \mathbf{K}$, and verifying that the new model satisfies $\forall x \zeta_2$, which follows from the structure of the frames for **KT**.

(iii) ($\Rightarrow$) Suppose that $(\forall x \zeta_3 \to \varphi^\dagger) \notin \mathbf{Q^{\#}K}$. Then $\mathfrak{M}, w \models^{\mathfrak{a}} \forall x \zeta_3$ and $\mathfrak{M}, w \not\models^{\mathfrak{a}} \varphi$ for some model $\mathfrak{M} = (\mathfrak{F}, D, I)$, where $\mathfrak{F} = (W, R) \in \mathsf{Fr}\ \mathbf{K}$. Without loss of

[2] The (geodesic) distance $d(w, v)$ is the smallest integer $n < \omega$ such that $w_0 R w_1 R \ldots R w_n$ is an R-chain, with $w_0 = w$ and $w_n = v$.

generality we may suppose that $\mathfrak{F}$ is an irreflexive, intransitive tree of depth ≤ 1, since both $\mathsf{md}(\zeta_3), \mathsf{md}(\varphi^\dagger) \leq 1$, rooted at $r \in W$. Let $\mathfrak{F}^\circ = (W, W \times W)$ denote the universal closure of $\mathfrak{F}$, and let $\mathfrak{M}' = (\mathfrak{F}^\circ, D, I)$ be a new model. We claim that, for all $w \in W$, $b \in D$, and $\psi \in \mathrm{sub}(\varphi)$,

$$\mathfrak{M}', w \models^{\mathfrak{a}(x/b)} \psi \quad \Longleftrightarrow \quad \mathfrak{M}, w \models^{\mathfrak{a}(x/b)} \psi^\dagger,$$

as can be verified by induction on the construction of ψ.

Hence it follows that $\mathfrak{M}', r \not\models^{\mathfrak{a}} \varphi$, which is to say that $\varphi \notin \mathbf{Q^\#S5}$, as required.

($\Leftarrow$) Again, the converse direction is similar to that of (i), using the fact that that $\mathsf{Fr}\ \mathbf{S5} \subseteq \mathsf{Fr}\ \mathbf{K}$, and verifying that the new model satisfies $\forall x \zeta_3$, which follows from the structure of the frames for **S5**.

(iv) ($\Rightarrow$) Suppose that $(\forall x \zeta_4 \to \varphi^\dagger) \notin \mathbf{Q^\#K}$. Then $\mathfrak{M}, w \models^{\mathfrak{a}} \forall x \zeta_4$ and $\mathfrak{M}, w \not\models^{\mathfrak{a}} \varphi$ for some model $\mathfrak{M} = (\mathfrak{F}, D, I)$, where $\mathfrak{F} = (W, R) \in \mathsf{Fr}\ \mathbf{K}$. Without loss of generality we may suppose that $\mathfrak{F}$ is an irreflexive, intransitive tree of depth, rooted at $r \in W$. Let $\mathfrak{F}^\smile = (W, R \cup R^\smile) \in \mathsf{Fr}\ \mathbf{KB}$ denote the symmetric closure of $\mathfrak{F}$, where $R^\smile = \{(v, w) : (w, v) \in R\}$, and let $\mathfrak{M}' = (\mathfrak{F}^\smile, D, I)$ be a new model. We claim that, for $w \in W$, $b \in D$, and $\psi \in \mathrm{sub}(\varphi)$,

$$\mathfrak{M}', w \models^{\mathfrak{a}(x/b)} \psi \quad \Longleftrightarrow \quad \mathfrak{M}, w \models^{\mathfrak{a}(x/b)} \psi^\dagger,$$

whenever $\mathsf{md}(\psi) + d(r, w) \leq m$, as can be verified by induction on the construction of ψ, where $d(r, w)$ denotes the distance between r and w.

Hence it follows that $\mathfrak{M}', r \not\models^{\mathfrak{a}} \varphi$, which is to say that $\varphi \notin \mathbf{Q^\#KB}$, as required.

($\Leftarrow$) The converse direction is similar to that of (i), using the fact that that $\mathsf{Fr}\ \mathbf{KB} \subseteq \mathsf{Fr}\ \mathbf{K}$, and verifying that the new model satisfies $\forall x \zeta_4$, which follows from the structure of the frames for **KB**.

□

B Supplementary Proof

In this appendix we explore the details of Claim 3.5 from the proof of Lemma 3.4, wherein we established the correspondence between satisfiability with respect to $\mathbf{Q^{\#}K}$ and the existence of quasimodels.

Claim 3.5 *For all* $w \in W$, $r \in \mathfrak{R}$, *and* $\psi \in \mathrm{sub}(\varphi)$,

$$\mathfrak{M}, w \models^{\mathfrak{a}(x/r)} \psi \qquad \Longleftrightarrow \qquad \psi \in r(w). \tag{I.H.}$$

Proof. The claim follows by induction on the construction of ψ, with each of the five inductive cases detailed below:

- *Case* $\psi = P_i(x)$: This follows immediately from the definition of $P_i^{I(w)}$, since
$$\mathfrak{M}, w \models^{\mathfrak{a}(x/r)} P_i(x) \iff r \in P_i^{I(w)} \iff P_i(x) \in r(w).$$
- *Case* $\psi = \neg\xi$: We have that
$$\mathfrak{M}, w \models^{\mathfrak{a}(x/r)} \neg\xi \iff \mathfrak{M}, w \not\models^{\mathfrak{a}(x/r)} \xi \overset{\text{(I.H.)}}{\iff} \xi \notin r(w) \overset{\textbf{(tp1)}}{\iff} \neg\xi \in r(w).$$
- *Case* $\psi = (\xi_1 \wedge \xi_2)$: We have that
$$\begin{aligned}\mathfrak{M}, w \models^{\mathfrak{a}(x/r)} (\xi_1 \wedge \xi_2) &\iff \mathfrak{M}, w \not\models^{\mathfrak{a}(x/r)} \xi_1 \text{ and } \mathfrak{M}, w \not\models^{\mathfrak{a}(x/r)} \xi_2, \\ &\overset{\text{(I.H.)}}{\iff} \xi_1 \in r(w) \text{ and } \xi_2 \in r(w), \\ &\overset{\textbf{(tp2)}}{\iff} (\xi_1 \wedge \xi_2) \in r(w).\end{aligned}$$
- *Case* $\psi = \Diamond\xi$: We have that
$$\begin{aligned}\mathfrak{M}, w \models^{\mathfrak{a}(x/r)} \Diamond\xi &\iff w \prec v \text{ and } \mathfrak{M}, v \models^{\mathfrak{a}(x/r)} \xi, \quad \text{for some } v \in W, \\ &\overset{\text{(I.H.)}}{\iff} w \prec v \text{ and } \xi \in r(v), \quad \text{for some } v \in W, \\ &\iff \Diamond\alpha \in r(w) \quad \text{by } \textbf{(qm3)} \text{ and } \textbf{(qm4)}.\end{aligned}$$
- *Case* $\psi = \exists_{\leq c} x\ \xi$: We have that
$$\begin{aligned}\mathfrak{M}, w \models^{\mathfrak{a}(x/r)} (\exists_{\leq c} x\ \xi) &\overset{\text{(def)}}{\iff} \left|\{s \in \mathfrak{R} : \mathfrak{M}, w \models^{\mathfrak{a}(x/s)} \xi\}\right| \leq c \\ &\overset{\text{(I.H.)}}{\iff} |\{s \in \mathfrak{R} : \xi \in s(w)\}| \leq c \\ &\iff \sum_{t \in T_w(\xi)} |\{s \in \mathfrak{R} : s(w) = t\}| \leq c \\ &\overset{\textbf{(qm5)}}{\iff} \sum_{t \in T_w(\xi)} \mu_w(t) \leq c \\ &\overset{\textbf{(qs3)}}{\iff} (\exists_{\leq c} x\ \xi) \in r(w).\end{aligned}$$

  The penultimate equivalence follows from the fact that each summand strictly less than $(C+1)$, since $c \leq C$. Hence, it follows from **(qm5)** that $\mu_w(t) = |\{s \in \mathfrak{R} : s(w) = t\}|$, for all $t \in T_w(\xi)$.

Hence, it follow that $\mathfrak{M}, w \models^{\mathfrak{a}(x/r)} \psi$ if and only if $\psi \in r(w)$, for all $\psi \in \mathrm{sub}(\varphi)$, as required. □

The Succinctness of First-order Logic over Modal Logic via a Formula Size Game

Lauri Hella Miikka Vilander

School of Information Sciences
University of Tampere

Abstract

We propose a new version of formula size game for modal logic. The game characterizes the equivalence of pointed Kripke-models up to formulas of given numbers of modal operators and binary connectives. Our game is similar to the well-known Adler-Immerman game. However, due to a crucial difference in the definition of positions of the game, its winning condition is simpler, and the second player (duplicator) does not have a trivial optimal strategy. Thus, unlike the Adler-Immerman game, our game is a genuine two-person game. We illustrate the use of the game by proving a nonelementary succinctness gap between bisimulation invariant first-order logic FO and (basic) modal logic ML.

Keywords: Succinctness, formula size game, bisimulation invariant first-order logic, n-bisimulation.

1 Introduction

Succinctness is an important research topic that has been quite active in modal logic for the last couple of decades; see, e.g., [3,13,11,14,1,12,5] for earlier work on this topic and [6,20,7,10,19,21] for recent research. If two logics $\mathcal{L}$ and $\mathcal{L}'$ have equal expressive power, it is natural to ask, whether there are properties that can be expressed in $\mathcal{L}$ by a substantially shorter formula than in $\mathcal{L}'$ (or vice versa). For example, $\mathcal{L}$ is *exponentially more succinct* than $\mathcal{L}'$, if for every integer n there is an $\mathcal{L}$-formula φ_n of length $\mathcal{O}(n)$ such that any equivalent $\mathcal{L}'$-formula ψ_n is of length at least 2^n.

Often such a gap in succinctness comes together with a similar gap in the complexity of the logics. For example, Etessami, Vardi and Wilke [3] proved that, over ω-words, the two-variable fragment FO^2 of first-order logic has the same expressive power as unary-TL (a weak version of temporal logic), but FO^2 is exponentially more succinct than unary-TL, and furthermore, the complexity of satisfiability for FO^2 is NEXPTIME-complete, while the complexity of unary-TL is in NP [17]. However, succinctness does not always lead to a penalty in terms of complexity: an example is public announcement logic PAL which is exponentially more succinct than epistemic logic EL, but both have the same complexity, as proved by Lutz in [12].

In order to prove succinctness results we need a method for proving lower bounds for the length of formulas expressing given properties. The two most common methods used in the recent literature are the *formula size game* introduced by Adler and Immerman [1], and *extended syntax trees* due to Grohe and Schweikardt [8]. The latter was inspired by the former, and in fact, an extended syntax tree is essentially a witness for the existence of a winning strategy in the Adler-Immerman game. Thus, these two methods are equivalent, and the choice between them is often a matter of convenience.

Originally, Adler and Immerman [1] formulated their game for the branching-time temporal logic CTL. They used it for proving an $n!$ lower bound on the size of CTL-formulas for expressing that there is a path on which each of the propositions $p_1, \ldots, p_n$ is true. As it is straightforward to express this property by a formula of CTL^+ of size linear in n, their result established that CTL^+ is $n!$ times more succinct than CTL, thus improving an earlier exponential succinctness result of Wilke [22].

After its introduction in [1], the Adler-Immerman game, as well as the method of extended syntax trees, has been adapted to a host of modal languages. These include epistemic logic [6], multimodal logics with union and intersection operators on modalities [20] and modal logic with contingency operator [21], among others.

The Adler-Immerman game can be seen as a variation of the Ehrenfeucht-Fraïssé game, or, in the case of modal logics, the bisimulation game. In the Adler-Immerman game, quantifier rank (or modal depth) is replaced by a parameter, usually called formula size, that is closely related to the length of the formula. Moreover, in order to use the game for proving that a property is not definable by a formula of a given size, it is necessary to play the game on a pair $(\mathbb{A}, \mathbb{B})$ of sets of structures instead of just a pair of single structures.

The basic idea of the Adler-Immerman game is that one of the players, S (spoiler), tries to show that the sets $\mathbb{A}$ and $\mathbb{B}$ can be separated by a formula of size n, while the other player, D (duplicator), aims to show that no formula of size at most n suffices for this. The moves that S makes in the game reflect directly the logical operators in a formula that is supposed to separate the sets $\mathbb{A}$ and $\mathbb{B}$. Any pair (σ, δ) of strategies for the players S and D produces a finite game tree $T_{\sigma,\delta}$, and S wins this play if the size of $T_{\sigma,\delta}$ is at most n. The strategy σ is a winning strategy for S if using it, S wins every play of the game. If this is the case, then there is a formula of size at most n that separates the sets, and this formula can actually be read from the strategy σ.

A peculiar feature of the Adler-Immerman game is that the second player, duplicator, can be completely eliminated from it. This is because D has an optimal strategy $\delta_{\max}$, which is to always choose the maximal allowed answer; this strategy guarantees that the size of the tree $T_{\sigma,\delta}$ is as large as possible. Thus, in this sense the Adler-Immerman game is not a genuine two-person game, but rather a one-person game.

In the present paper, we propose another type of formula size game for modal logic. Our game is a natural adaptation of the game introduced by

Hella and Väänänen [9] for propositional logic and first-order logic. The basic setting in our game is the same as in the Adler-Immerman game: there are two players, S and D, and two sets of structures that S claims can be separated by a formula of some given size. The crucial difference is that in our game we define positions to be tuples $(m, k, \mathbb{A}, \mathbb{B})$ instead of just pairs $(\mathbb{A}, \mathbb{B})$ of sets of structures, where m and k are parameters referring to the number of modal operators and binary connectives in a formula. In each move S has to decrease at least one of the parameters m or k. The game ends when the players reach a position $(m^*, k^*, \mathbb{A}^*, \mathbb{B}^*)$ such that either there is a literal separating $\mathbb{A}^*$ and $\mathbb{B}^*$, or S cannot make any moves, usually because $m^* = k^* = 0$. In the former case, S wins the play; otherwise D wins.

Thus, in contrast to the Adler-Immerman game, to determine the winner in our game it suffices to consider a single "leaf-node" $(m^*, k^*, \mathbb{A}^*, \mathbb{B}^*)$ of the game tree. This also means that our game is a real two-person game: the final position $(m^*, k^*, \mathbb{A}^*, \mathbb{B}^*)$ of a play depends on the moves of D, and there is no simple optimal strategy for D that could be used for eliminating the role of D in the game.

We believe that our game is more intuitive and thus, in some cases it may be easier to use than the Adler-Immerman game. On the other hand, it should be remarked that the two games are essentially equivalent: The moves corresponding to connectives and modal operators are the same in both games (when restricting to the sets $\mathbb{A}$ and $\mathbb{B}$ in a position $(m, k, \mathbb{A}, \mathbb{B})$). Hence, in principle, it is possible to translate a winning strategy in one of the games to a corresponding winning strategy in the other.

We illustrate the use of our game by proving a nonelementary succinctness gap between first-order logic FO and (basic) modal logic ML. More precisely, we define a bisimulation invariant property of pointed Kripke-models by a first-order formula of size $\mathcal{O}(2^n)$, and show that this property cannot be defined by any ML-formula of size less than the exponential tower of height $n-1$. Furthermore, we show that the same property of pointed Kripke-models is already definable by a formula of size $\mathcal{O}(2^n)$ in ML^2, which is a version of 2-dimensional modal logic defined by Otto in [16]. Hence the same nonelementary succinctness result holds for ML^2 over ML.

A similar gap between FO and temporal logic follows from a construction in the PhD thesis [18] of Stockmeyer. He proved that the satisfiability problem of FO over words is of nonelementary complexity. Etessami and Wilke [4] observed that from Stockmeyer's proof it is possible to extract FO-formulas of size $\mathcal{O}(n)$ whose smallest models are words of length nonelementary in n. On the other hand, it is well known that any satisfiable formula of temporal logic has a model of size $\mathcal{O}(2^n)$, where n is the size of the formula.

2 Preliminaries

In this section we fix notation, define the syntax and semantics of basic modal logic and define our notions of formula size. For more on the notions used in the paper, we refer to the textbook [2] of Blackburn, de Rijke and Venema.

Basic modal logic and first-order logic

Let Φ be a set of proposition symbols, and let $\mathcal{M} = (W, R, V)$, where W is a set, $R \subseteq W \times W$ and $V : \Phi \to \mathcal{P}(W)$, and let $w \in W$. The structure $(\mathcal{M}, w)$ is called a *pointed Kripke-model for* Φ.

Let $(\mathcal{M}, w)$ be a pointed Kripke-model. We use the notation

$$\Box(\mathcal{M}, w) := \{(\mathcal{M}, v) \mid v \in W, wR^{\mathcal{M}}v\}.$$

If $\mathbb{A}$ is a set of pointed Kripke-models, we use the notation

$$\Box\mathbb{A} := \bigcup_{(\mathcal{M},w)\in\mathbb{A}} \Box(\mathcal{M}, w).$$

Furthermore, if f is a function $f : \mathbb{A} \to \Box\mathbb{A}$ such that $f(\mathcal{M}, w) \in \Box(\mathcal{M}, w)$ for every $(\mathcal{M}, w) \in \mathbb{A}$, then we use the notation

$$\Diamond_f\mathbb{A} := f(\mathbb{A}).$$

Now we define the syntax and semantics of basic modal logic for pointed models.

Definition 2.1 Let Φ be a set of proposition symbols. The set of formulas of $\mathrm{ML}(\Phi)$ is generated by the following grammar

$$\varphi := p \mid \neg p \mid (\varphi \wedge \varphi) \mid (\varphi \vee \varphi) \mid \Diamond\varphi \mid \Box\varphi,$$

where $p \in \Phi$.

As is apparent from the definition of the syntax, we assume that all ML-formulas are in negation normal form. This is useful for the formula size game that we introduce in the next section.

Definition 2.2 The satisfaction relation $(\mathcal{M}, w) \vDash \varphi$ between pointed Kripke-models $(\mathcal{M}, w)$ and $\mathrm{ML}(\Phi)$-formulas φ is defined as follows:

(1) $(\mathcal{M}, w) \vDash p \Leftrightarrow w \in V(p)$,

(2) $(\mathcal{M}, w) \vDash \neg p \Leftrightarrow w \notin V(p)$,

(3) $(\mathcal{M}, w) \vDash (\varphi \wedge \psi) \Leftrightarrow (\mathcal{M}, w) \vDash \varphi$ and $(\mathcal{M}, w) \vDash \psi$,

(4) $(\mathcal{M}, w) \vDash (\varphi \vee \psi) \Leftrightarrow (\mathcal{M}, w) \vDash \varphi$ or $(\mathcal{M}, w) \vDash \psi$,

(5) $(\mathcal{M}, w) \vDash \Diamond\varphi \Leftrightarrow$ there is $(\mathcal{M}, v) \in \Box(\mathcal{M}, w)$ such that $(\mathcal{M}, v) \vDash \varphi$,

(6) $(\mathcal{M}, w) \vDash \Box\varphi \Leftrightarrow$ for every $(\mathcal{M}, v) \in \Box(\mathcal{M}, w)$ it holds that $(\mathcal{M}, v) \vDash \varphi$.

Furthermore, if $\mathbb{A}$ is a class of pointed Kripke-models, then

$$\mathbb{A} \vDash \varphi \Leftrightarrow (\mathcal{A}, w) \vDash \varphi \text{ for every } (\mathcal{A}, w) \in \mathbb{A}.$$

For the sake of convenience we also use the notation

$$\mathbb{A} \vDash \neg\varphi \Leftrightarrow (\mathcal{A}, w) \nvDash \varphi \text{ for every } (\mathcal{A}, w) \in \mathbb{A}.$$

In Section 4, we also consider the case $\Phi = \emptyset$. For this purpose, we add the atomic constants $\top$ and $\bot$ to ML, where $(\mathcal{M}, w) \vDash \top$ and $(\mathcal{M}, w) \nvDash \bot$ for all pointed Kripke-models $(\mathcal{M}, w)$.

The syntax and semantics for first-order logic are defined in the standard way. Each ML-formula φ defines a class $\mathrm{Mod}(\varphi)$ of pointed Kripke-models:

$$\mathrm{Mod}(\varphi) := \{(\mathcal{M}, w) \mid (\mathcal{M}, w) \vDash \varphi\}.$$

In the same way, any FO-formula $\psi(x)$ in the vocabulary consisting of the accessibility relation symbol R and unary relation symbols U_p for $p \in \Phi$ defines a class $\mathrm{Mod}(\psi)$ of pointed Kripke-models:

$$\mathrm{Mod}(\psi) := \{(\mathcal{M}, w) \mid \mathcal{M} \vDash \psi[w/x]\}.$$

The formulas $\varphi \in$ ML and $\psi(x) \in$ FO are *equivalent* if $\mathrm{Mod}(\varphi) = \mathrm{Mod}(\psi)$.

The well-known link between ML and FO is the following theorem.

Theorem 2.3 (van Benthem Characterization Theorem) *A first-order formula $\psi(x)$ is equivalent to some formula in* ML *if and only if* $\mathrm{Mod}(\psi)$ *is bisimulation invariant.*

If a property of pointed Kripke-models is n-bisimulation invariant for some $n \in \mathbb{N}$, then it is also bisimulation invariant. Thus, FO-definability and n-bisimulation invariance imply ML-definability for any property of pointed Kripke-models. We will use this version of van Benthem's characterization in Section 4.1 for showing that certain property is ML-definable. For the sake of easier reading, we give here the definition of n-bisimulation.

Definition 2.4 Let $(\mathcal{M}, w)$ and $(\mathcal{M}', w')$ be pointed Φ-models. We say that $(\mathcal{M}, w)$ and $(\mathcal{M}', w')$ are *n-bisimilar*, $(\mathcal{M}, w) \leftrightarroweq_n (\mathcal{M}', w')$, if there are binary relations $Z_n \subseteq \cdots \subseteq Z_0$ such that for every $0 \leq i \leq n-1$ we have

(1) $(\mathcal{M}, w) Z_n (\mathcal{M}', w')$,

(2) if $(\mathcal{M}, v) Z_0 (\mathcal{M}', v')$, then $(\mathcal{M}, v) \vDash p \Leftrightarrow (\mathcal{M}', v') \vDash p$ for each $p \in \Phi$,

(3) if $(\mathcal{M}, v) Z_{i+1} (\mathcal{M}', v')$ and $(\mathcal{M}, u) \in \Box(\mathcal{M}, v)$ then there is $(\mathcal{M}', u') \in \Box(\mathcal{M}', v')$ such that $(\mathcal{M}, u) Z_i (\mathcal{M}', u')$,

(4) if $(\mathcal{M}, v) Z_{i+1} (\mathcal{M}', v')$ and $(\mathcal{M}', u') \in \Box(\mathcal{M}', v')$ then there is $(\mathcal{M}, u) \in \Box(\mathcal{M}, v)$ such that $(\mathcal{M}, u) Z_i (\mathcal{M}', u')$.

It is well known that if Φ is finite, two pointed Φ-models are n-bisimilar if and only if they are equivalent with respect to ML(Φ)-formulas of modal depth at most n.

Formula size

We define notions of formula size for ML and FO. These notions are related to the length of the formula as a string rather than the DAG-size[1] of it. For ML

[1] The DAG-size of a formula φ is the number of edges of the syntactic structure of φ in the form of a DAG. Thus since the fan-out in the DAG is at most two, the DAG-size is at most two times the number of subformulas of φ.

we define separately the number of modal operators and the number of binary connectives in the formula.

Definition 2.5 The *modal size* of a formula $\varphi \in \mathrm{ML}$, denoted $\mathrm{ms}(\varphi)$, is defined recursively as follows:

(1) If φ is a literal, then $\mathrm{ms}(\varphi) = 0$.

(2) If $\varphi = \psi \vee \vartheta$ or $\varphi = \psi \wedge \vartheta$, then $\mathrm{ms}(\varphi) = \mathrm{ms}(\psi) + \mathrm{ms}(\vartheta)$.

(3) If $\varphi = \Diamond\psi$ or $\varphi = \Box\psi$, then $\mathrm{ms}(\varphi) = \mathrm{ms}(\psi) + 1$.

Definition 2.6 The *binary connective size* of a formula $\varphi \in \mathrm{ML}$, denoted by $\mathrm{cs}(\varphi)$, is defined recursively as follows:

(1) If φ is a literal, then $\mathrm{cs}(\varphi) = 0$.

(2) If $\varphi = \psi \vee \vartheta$ or $\varphi = \psi \wedge \vartheta$, then $\mathrm{cs}(\varphi) = \mathrm{cs}(\psi) + \mathrm{cs}(\vartheta) + 1$.

(3) If $\varphi = \Diamond\psi$ or $\varphi = \Box\psi$, then $\mathrm{cs}(\varphi) = \mathrm{cs}(\psi)$.

The size of an ML formula is defined as the sum of modal size and connective size. We do not count literals or parentheses since their number can be derived from the number of binary connectives.

Definition 2.7 The *size* of a formula $\varphi \in \mathrm{ML}$ is $\mathrm{s}(\varphi) = \mathrm{ms}(\varphi) + \mathrm{cs}(\varphi)$.

Similarly we define formula size for FO to be the number of binary connectives and quantifiers in the formula. In general this could lead to an arbitrarily large difference between formula size and actual string length. For an example if f is a unary function symbol, then atomic formulas of the form $f(x) = x$, $f(f(x)) = x$ and so on, all have size 0. In this paper however, we only consider formulas with one binary relation so this is not an issue.

Definition 2.8 The *size* of a formula $\varphi \in \mathrm{FO}$, denoted by $\mathrm{s}(\varphi)$, is defined recursively as follows:

(1) If φ is a literal, then $\mathrm{s}(\varphi) = 0$.

(2) If $\varphi = \neg\psi$, then $\mathrm{s}(\varphi) = \mathrm{s}(\psi)$.

(3) If $\varphi = \psi \vee \vartheta$ or $\varphi = \psi \wedge \vartheta$, then $\mathrm{s}(\varphi) = \mathrm{s}(\psi) + \mathrm{s}(\vartheta) + 1$.

(4) If $\varphi = \exists x\psi$ or $\varphi = \forall x\psi$, then $\mathrm{s}(\varphi) = \mathrm{s}(\psi) + 1$.

To refer to some rather large formula sizes we need the exponential tower function.

Definition 2.9 We define the function $\mathrm{twr} : \mathbb{N} \to \mathbb{N}$ recursively as follows:

$$\mathrm{twr}(0) = 1$$
$$\mathrm{twr}(n+1) = 2^{\mathrm{twr}(n)}.$$

We will also use in the sequel the binary logarithm function, denoted by log.

Separating classes by formulas

The definition of the formula size game in the next section is based on the notion of separating classes of pointed Kripke-models by formulas.

Definition 2.10 Let $\mathbb{A}$ and $\mathbb{B}$ be classes of pointed Kripke-models.
(a) We say that a formula $\varphi \in \mathrm{ML}$ *separates the classes* $\mathbb{A}$ *and* $\mathbb{B}$ if $\mathbb{A} \models \varphi$ and $\mathbb{B} \models \neg\varphi$.
(b) Similarly, a formula $\psi(x) \in \mathrm{FO}$ separates the classes $\mathbb{A}$ and $\mathbb{B}$ if for all $(\mathcal{M}, w) \in \mathbb{A}$, $\mathcal{M} \models \psi[w/x]$ and for all $(\mathcal{M}, w) \in \mathbb{B}$, $\mathcal{M} \models \neg\psi[w/x]$.

In other words, a formula $\varphi \in \mathrm{ML}$ separates the classes $\mathbb{A}$ and $\mathbb{B}$ if $\mathbb{A} \subseteq \mathrm{Mod}(\varphi)$ and $\mathbb{B} \subseteq \overline{\mathrm{Mod}(\varphi)}$, where $\overline{\mathrm{Mod}(\varphi)}$ is the complement of $\mathrm{Mod}(\varphi)$.

3 The formula size game

As in the Adler-Immerman game, the basic idea in our formula size game is that there are two players, S (spoiler) and D (duplicator), who play on a pair $(\mathbb{A}, \mathbb{B})$ of two sets of pointed Kripke-models. The aim of S is to show that $\mathbb{A}$ and $\mathbb{B}$ can be separated by a formula with modal size at most m and connective size at most k, while D tries to refute this. The moves of S reflect the connectives and modal operators of a formula that is supposed to separate the sets. The parameters m and k decrease with every move and act as resources indicating how many connectives and modal operators S has left to spend.

The crucial difference between our game and the Adler-Immerman game is that we define positions in the game to be tuples $(m, k, \mathbb{A}, \mathbb{B})$ instead of just pairs $(\mathbb{A}, \mathbb{B})$. This means that in the connective moves, D has a genuine choice to make. Furthermore, the winning condition of the game is based on a natural property of single positions instead of the size of the entire game tree.

We give now the precise definition of our game.

Definition 3.1 Let $\mathbb{A}_0$ and $\mathbb{B}_0$ be sets of pointed Φ-Kripke-models and let $m_0, k_0 \in \mathbb{N}$. *The formula size game between the sets* $\mathbb{A}_0$ *and* $\mathbb{B}_0$, denoted $\mathrm{FS}_{m_0,k_0}(\mathbb{A}_0, \mathbb{B}_0)$, has two players, S and D. The number m_0 is the *modal parameter* and k_0 is the *connective parameter* of the game. The starting position of the game is $(m_0, k_0, \mathbb{A}_0, \mathbb{B}_0)$. Let the position after n moves be $(m, k, \mathbb{A}, \mathbb{B})$. To continue the game, S has the following four moves to choose from:

- *Left splitting move*: First, S chooses natural numbers m_1, m_2, k_1 and k_2 and sets $\mathbb{A}_1$ and $\mathbb{A}_2$ such that $m_1 + m_2 = m$, $k_1 + k_2 + 1 = k$ and $\mathbb{A}_1 \cup \mathbb{A}_2 = \mathbb{A}$. Then D decides whether the game continues from the position $(m_1, k_1, \mathbb{A}_1, \mathbb{B})$ or the position $(m_2, k_2, \mathbb{A}_2, \mathbb{B})$.
- *Right splitting move*: First, S chooses natural numbers m_1, m_2, k_1 and k_2 and sets $\mathbb{B}_1$ and $\mathbb{B}_2$ such that $m_1+m_2 = m$, $k_1+k_2+1 = k$ and $\mathbb{B}_1 \cup \mathbb{B}_2 = \mathbb{B}$. Then D decides whether the game continues from the position $(m_1, k_1, \mathbb{A}, \mathbb{B}_1)$ or the position $(m_2, k_2, \mathbb{A}, \mathbb{B}_2)$.
- *Left successor move*: S chooses a function $f : \mathbb{A} \to \Box\mathbb{A}$ such that $f(\mathcal{A}, w) \in \Box(\mathcal{A}, w)$ for all $(\mathcal{A}, w) \in \mathbb{A}$ and the game continues from the position $(m - 1, k, \Diamond_f\mathbb{A}, \Box\mathbb{B})$.
- *Right successor move*: S chooses a function $g : \mathbb{B} \to \Box\mathbb{B}$ such that $g(\mathcal{B}, w) \in \Box(\mathcal{B}, w)$ for all $(\mathcal{B}, w) \in \mathbb{B}$ and the game continues from the position $(m - 1, k, \Box\mathbb{A}, \Diamond_g\mathbb{B})$.

The game ends and S wins in a position $(m, k, \mathbb{A}, \mathbb{B})$ if there is a Φ-literal φ which separates the sets $\mathbb{A}$ and $\mathbb{B}$. The game ends and D wins in a position $(m, k, \mathbb{A}, \mathbb{B})$ if S cannot move and S does not win in this position.

The modal and connective parameters m and k can be thought of as resources for S, since in a position $(m, k, \mathbb{A}, \mathbb{B})$ S cannot make a successor move if $m = 0$ or a splitting move if $k = 0$. Note also that if $\Box(\mathcal{M}, w) = \emptyset$ for some $(\mathcal{M}, w) \in \mathbb{A}$ ($\in \mathbb{B}$) then S cannot make a left (right) successor move.

We prove now that the formula size game indeed characterizes the separation of two sets of pointed Kripke-models by a formula of a given size.

Theorem 3.2 *Let $\mathbb{A}$ and $\mathbb{B}$ be sets of pointed Φ-models and let m and k be natural numbers. Then the following conditions are equivalent:*

$(\mathrm{win})_{m,k}$ *S has a winning strategy in the game* $\mathrm{FS}_{m,k}(\mathbb{A}, \mathbb{B})$.

$(\mathrm{sep})_{m,k}$ *There is a formula $\varphi \in \mathrm{ML}(\Phi)$ such that* $\mathrm{ms}(\varphi) \leq m$, $\mathrm{cs}(\varphi) \leq k$ *and the formula φ separates the sets $\mathbb{A}$ and $\mathbb{B}$.*

Proof. The proof proceeds by induction on the number $m+k$. If $m+k = 0$, no moves can be made. Thus if S wins, then there is a literal φ that separates the sets $\mathbb{A}$ and $\mathbb{B}$. In this case $\mathrm{s}(\varphi) = 0$ so $(\mathrm{win})_{0,0} \Rightarrow (\mathrm{sep})_{0,0}$. On the other hand, if there is a formula φ such that $\mathrm{s}(\varphi) \leq 0$ and φ separates the sets $\mathbb{A}$ and $\mathbb{B}$, then φ is a literal. Thus S wins the game, and we see that $(\mathrm{sep})_{0,0} \Rightarrow (\mathrm{win})_{0,0}$.

Suppose then that $m + k > 0$ and $(\mathrm{win})_{n,l} \Leftrightarrow (\mathrm{sep})_{n,l}$ for all $n, l \in \mathbb{Z}_+$ such that $n + l < m + k$. Assume first that $(\mathrm{win})_{m,k}$ holds. Consider the following cases according to the first move in the winning strategy of S.

(a) Assume that the first move of the winning strategy of S is a left splitting move choosing numbers $m_1, m_2, k_1, k_2 \in \mathbb{N}$ such that $m_1 + m_2 = m$ and $k_1 + k_2 + 1 = k$, and sets $\mathbb{A}_1, \mathbb{A}_2 \subseteq \mathbb{A}$ such that $\mathbb{A}_1 \cup \mathbb{A}_2 = \mathbb{A}$. Since this move is given by a winning strategy, S has a winning strategy for both possible continuations of the game, $(m_1, k_1, \mathbb{A}_1, \mathbb{B})$ and $(m_2, k_2, \mathbb{A}_2, \mathbb{B})$. Since $m_i + k_i < m_i + k_i + 1 \leq m + k$ for $i \in \{1, 2\}$, by induction hypothesis there is a formula ψ such that $\mathrm{ms}(\psi) \leq m_1$, $\mathrm{cs}(\psi) \leq k_1$ and ψ separates the sets $\mathbb{A}_1$ and $\mathbb{B}$ and a formula ϑ such that $\mathrm{ms}(\vartheta) \leq m_2$, $\mathrm{cs}(\vartheta) \leq k_2$ and ϑ separates the sets $\mathbb{A}_2$ and $\mathbb{B}$. Thus $\mathbb{A}_1 \vDash \psi$ and $\mathbb{A}_2 \vDash \vartheta$ so $\mathbb{A} \vDash \psi \vee \vartheta$. On the other hand $\mathbb{B} \vDash \neg\psi$ and $\mathbb{B} \vDash \neg\vartheta$ so $\mathbb{B} \vDash \neg(\psi \vee \vartheta)$. Therefore the formula $\psi \vee \vartheta$ separates the sets $\mathbb{A}$ and $\mathbb{B}$. In addition $\mathrm{ms}(\psi \vee \vartheta) = \mathrm{ms}(\psi) + \mathrm{ms}(\vartheta) \leq m_1 + m_2 = m$ and $\mathrm{cs}(\psi \vee \vartheta) = \mathrm{cs}(\psi) + \mathrm{cs}(\vartheta) + 1 \leq k_1 + k_2 + 1 = k$ so $(\mathrm{sep})_{m,k}$ holds.

(b) Assume that the first move of the winning strategy of S is a right splitting move choosing numbers $m_1, m_2, k_1, k_2 \in \mathbb{N}$ such that $m_1 + m_2 = m$ and $k_1 + k_2 + 1 = k$, and sets $\mathbb{B}_1, \mathbb{B}_2 \subseteq \mathbb{B}$ such that $\mathbb{B}_1 \cup \mathbb{B}_2 = \mathbb{B}$. Since this move is given by a winning strategy, player I has a winning strategy for both possible continuations of the game, $(m_1, k_1, \mathbb{A}, \mathbb{B}_1)$ and $(m_2, k_2, \mathbb{A}, \mathbb{B}_2)$. By induction hypothesis there is a formula ψ such that $\mathrm{ms}(\psi) \leq m_1$, $\mathrm{cs}(\psi) \leq k_1$ and ψ separates the sets $\mathbb{A}$ and $\mathbb{B}_1$ and a formula ϑ such that $\mathrm{ms}(\vartheta) \leq m_2$, $\mathrm{cs}(\vartheta) \leq k_2$ and ϑ separates the sets $\mathbb{A}$ and $\mathbb{B}_2$. Thus $\mathbb{A} \vDash \psi$ and

$\mathbb{A} \models \vartheta$ so $\mathbb{A} \models \psi \wedge \vartheta$. On the other hand $\mathbb{B}_1 \models \neg\psi$ and $\mathbb{B}_2 \models \neg\vartheta$ so $\mathbb{B} \models \neg(\psi \wedge \vartheta)$. Therefore the formula $\psi \wedge \vartheta$ separates the sets $\mathbb{A}$ and $\mathbb{B}$. In addition $\mathrm{ms}(\psi \wedge \vartheta) = \mathrm{ms}(\psi) + \mathrm{ms}(\vartheta) \leq m_1 + m_2 = m$ and $\mathrm{cs}(\psi \wedge \vartheta) = \mathrm{cs}(\psi) + \mathrm{cs}(\vartheta) + 1 \leq k_1 + k_2 + 1 = k$ so $(\mathrm{sep})_{m,k}$ holds.

(c) Assume that the first move of the winning strategy of S is a left successor move choosing a function $f : \mathbb{A} \to \Box\mathbb{A}$ such that $f(\mathcal{A}, w) \in \Box(\mathcal{A}, w)$ for all $(\mathcal{A}, w) \in \mathbb{A}$. The game continues from the position $(m-1, k, \Diamond_f\mathbb{A}, \Box\mathbb{B})$ and S has a winning strategy from this position. By induction hypothesis there is a formula ψ such that $\mathrm{ms}(\psi) \leq m-1$, $\mathrm{cs}(\psi) \leq k$ and ψ separates the sets $\Diamond_f\mathbb{A}$ and $\Box\mathbb{B}$. Now for every $(\mathcal{A}, w) \in \mathbb{A}$ we have $f(\mathcal{A}, w) \in \Box(\mathcal{A}, w)$ and $f(\mathcal{A}, w) \models \psi$. Therefore $\mathbb{A} \models \Diamond\psi$. On the other hand $\Box\mathbb{B} \models \neg\psi$ so for every $(\mathcal{B}, w) \in \mathbb{B}$ and every $(\mathcal{B}, v) \in \Box(\mathcal{B}, w)$ we have $(\mathcal{B}, v) \not\models \psi$. Thus $\mathbb{B} \models \neg\Diamond\psi$. So the formula $\Diamond\psi$ separates the sets $\mathbb{A}$ and $\mathbb{B}$ and since $\mathrm{ms}(\Diamond\psi) = \mathrm{ms}(\psi) + 1 \leq m$ and $\mathrm{cs}(\Diamond\psi) = \mathrm{cs}(\psi) \leq k$, $(\mathrm{sep})_{m,k}$ holds.

(d) Assume that the first move of the winning strategy of player I is a right successor move choosing a function $g : \mathbb{B} \to \Box\mathbb{B}$ such that $g(\mathcal{B}, w) \in \Box(\mathcal{B}, w)$ for every $(\mathcal{B}, w) \in \mathbb{B}$. The game continues from the position $(m-1, k, \Box\mathbb{A}, \Diamond_g\mathbb{B})$ and player I has a winning strategy from this position. By induction hypothesis there is a formula ψ such that $\mathrm{ms}(\psi) \leq m-1$, $\mathrm{cs}(\psi) \leq k$ and ψ separates the sets $\Box\mathbb{A}$ and $\Diamond_g\mathbb{B}$. Thus $\Box\mathbb{A} \models \psi$ so for every $(\mathcal{A}, w) \in \mathbb{A}$ and every $(\mathcal{A}, v) \in \Box(\mathcal{A}, w)$ we have $(\mathcal{A}, v) \models \psi$ so $\mathbb{A} \models \Box\psi$. On the other hand $\Diamond_g\mathbb{B} \models \neg\psi$ so for every $(\mathcal{B}, w) \in \mathbb{B}$ we have $g(\mathcal{B}, w) \in \Box(\mathcal{B}, w)$ and $g(\mathcal{B}, w) \not\models \psi$. Thus $\mathbb{B} \models \neg\Box\psi$. Therefore the formula $\Box\psi$ separates the sets $\mathbb{A}$ and $\mathbb{B}$ and since $\mathrm{ms}(\Box\psi) = \mathrm{ms}(\psi) + 1 \leq m$ and $\mathrm{cs}(\Box\psi) = \mathrm{cs}(\psi) \leq k$, $(\mathrm{sep})_{m,k}$ holds.

Now assume $(\mathrm{sep})_{m,k}$ holds, and φ is the formula separating $\mathbb{A}$ and $\mathbb{B}$. We obtain a winning strategy of S for the game $\mathrm{FS}_{m,k}(\mathbb{A}, \mathbb{B})$ using φ as follows:

(a) If φ is a literal, S wins the game with no moves.

(b) Assume that $\varphi = \psi \vee \vartheta$. Let $\mathbb{A}_1 := \{(\mathcal{A}, w) \in \mathbb{A} \mid (\mathcal{A}, w) \models \psi\}$ and $\mathbb{A}_2 := \{(\mathcal{A}, w) \in \mathbb{A} \mid (\mathcal{A}, w) \models \vartheta\}$. Since $\mathbb{A} \models \varphi$ we have $\mathbb{A}_1 \cup \mathbb{A}_2 = \mathbb{A}$. In addition, since $\mathbb{B} \models \neg\varphi$, we have $\mathbb{B} \models \neg\psi$ and $\mathbb{B} \models \neg\vartheta$. Thus ψ separates the sets $\mathbb{A}_1$ and $\mathbb{B}$ and ϑ separates the sets $\mathbb{A}_2$ and $\mathbb{B}$. Since $\mathrm{ms}(\psi) + \mathrm{ms}(\vartheta) = \mathrm{ms}(\varphi) \leq m$, there are $m_1, m_2 \in \mathbb{N}$ such that $m_1 + m_2 = m$, $\mathrm{ms}(\psi) \leq m_1$ and $\mathrm{ms}(\vartheta) \leq m_2$. Similarly since $\mathrm{cs}(\psi) + \mathrm{cs}(\vartheta) + 1 = \mathrm{cs}(\varphi) \leq k$, there are $k_1, k_2 \in \mathbb{N}$ such that $k_1 + k_2 + 1 = k$, $\mathrm{cs}(\psi) \leq k_1$ and $\mathrm{cs}(\vartheta) \leq k_2$. By induction hypothesis S has winning strategies for the games $\mathrm{FS}_{m_1,k_1}(\mathbb{A}_1, \mathbb{B})$ and $\mathrm{FS}_{m_2,k_2}(\mathbb{A}_2, \mathbb{B})$. Since $k \geq \mathrm{cs}(\varphi) \geq 1$, S can start the game $\mathrm{FS}_{m,k}(\mathbb{A}, \mathbb{B})$ with a left splitting move choosing the numbers m_1, m_2, k_1 and k_2 and the sets $\mathbb{A}_1$ and $\mathbb{A}_2$. Then S wins the game by following the winning strategy for whichever position D chooses.

(c) Assume that $\varphi = \psi \wedge \vartheta$. Let $\mathbb{B}_1 := \{(\mathcal{B}, w) \in \mathbb{B} \mid (\mathcal{B}, w) \not\models \psi\}$ and $\mathbb{B}_2 := \{(\mathcal{B}, w) \in \mathbb{B} \mid (\mathcal{B}, w) \not\models \vartheta\}$. Since $\mathbb{B} \models \neg\varphi$, we have $\mathbb{B}_1 \cup \mathbb{B}_2 = \mathbb{B}$. In addition, since $\mathbb{A} \models \varphi$, we have $\mathbb{A} \models \psi$ and $\mathbb{A} \models \vartheta$. Thus ψ separates the sets $\mathbb{A}$ and $\mathbb{B}_1$ while ϑ separates the sets $\mathbb{A}$ and $\mathbb{B}_2$. As in the previous case, there are $m_1, m_2, k_1, k_2 \in \mathbb{N}$ such that $m_1 + m_2 = m$, $\mathrm{ms}(\psi) \leq m_1$, $\mathrm{ms}(\vartheta) \leq m_2$, $k_1 + k_2 = k$, $\mathrm{cs}(\psi) \leq k_1$ and $\mathrm{cs}(\vartheta) \leq k_2$. By induction hypothesis player I has a winning strategy for the games $\mathrm{FS}_{m,k}(\mathbb{A}, \mathbb{B}_1)$ and $\mathrm{FS}_{m,k}(\mathbb{A}, \mathbb{B}_2)$. Player I wins the game $\mathrm{FS}_{m,k}(\mathbb{A}, \mathbb{B})$ by starting with a right splitting move choosing the numbers m_1, m_2, k_1, and k_2 and the sets $\mathbb{B}_1$ and $\mathbb{B}_2$ and proceeding according to the winning strategies for the games $\mathrm{FS}_{m,k}(\mathbb{A}, \mathbb{B}_1)$ and $\mathrm{FS}_{m,k}(\mathbb{A}, \mathbb{B}_2)$.

(d) Assume that $\varphi = \Diamond\psi$. Since $\mathbb{A} \models \varphi$, for every $(\mathcal{A}, w) \in \mathbb{A}$ there is $(\mathcal{A}, v_w) \in \Box(\mathcal{A}, w)$ such that $(\mathcal{A}, v_w) \models \psi$. We define the function $f : \mathbb{A} \to \Box\mathbb{A}$ by $f(\mathcal{A}, w) = (\mathcal{A}, v_w)$. Clearly $\Diamond_f\mathbb{A} \models \psi$. On the other hand $\mathbb{B} \models \neg\varphi$ so for each $(\mathcal{B}, w) \in \mathbb{B}$ and each $(\mathcal{B}, v) \in \Box(\mathcal{B}, w)$ we have $(\mathcal{B}, v) \not\models \psi$. Therefore $\Box\mathbb{B} \models \neg\psi$ and the formula ψ separates the sets $\Diamond_f\mathbb{A}$ and $\Box\mathbb{B}$. Moreover, $\mathrm{ms}(\psi) = \mathrm{ms}(\varphi) - 1 \leq m - 1$ and $\mathrm{cs}(\psi) = \mathrm{cs}(\varphi) \leq k$ so by induction hypothesis S has a winning strategy for the game $\mathrm{FS}_{m-1,k}(\Diamond_f\mathbb{A}, \Box\mathbb{B})$. Since $m \geq \mathrm{ms}(\varphi) \geq 1$, S can start the game $\mathrm{FS}_{m,k}(\mathbb{A}, \mathbb{B})$ with a left successor move choosing the function f. Then S wins the game by following the winning strategy for the game $\mathrm{FS}_{m-1,k}(\Diamond_f\mathbb{A}, \Box\mathbb{B})$.

(e) Assume finally that $\varphi = \Box\psi$. Since $\mathbb{A} \models \varphi$, as in the previous case we obtain $\Box\mathbb{A} \models \psi$. On the other hand, since $\mathbb{B} \models \neg\varphi$, for every $(\mathcal{B}, w) \in \mathbb{B}$ there is $(\mathcal{B}, v_w) \in \Box(\mathcal{B}, w)$ such that $(\mathcal{B}, v_w) \not\models \psi$. We define the function $g : \mathbb{B} \to \Box\mathbb{B}$ by $g(\mathcal{B}, w) = (\mathcal{B}, v_w)$. Clearly $\Diamond_g\mathbb{B} \models \neg\psi$ so the formula ψ separates the sets $\Box\mathbb{A}$ and $\Diamond_g\mathbb{B}$. By induction hypothesis player I has a winning strategy for the game $\mathrm{FS}_{m-1,k}(\Box\mathbb{A}, \Diamond_g\mathbb{B})$. Player wins the game $\mathrm{FS}_{m,k}(\mathbb{A}, \mathbb{B})$ by starting with a right successor move choosing the function g and proceeding according to the winning strategy of the game $\mathrm{FS}_{m-1,k}(\Box\mathbb{A}, \Diamond_g\mathbb{B})$.

□

Note that in Theorem 3.2 we allow the set of proposition symbols Φ to be infinite. This is in contrast with other similar games, such as the bisimulation game and the n-bisimulation game. For an example let $\Phi = \{p_i \mid i \in \mathbb{N}\}$ and $W = \{w\} \cup \{w_i \mid i \in \mathbb{N}\}$. Furthermore let $(\mathcal{A}, w)$ be a pointed model, where $\mathrm{dom}(\mathcal{A}) = W$, $R^{\mathcal{A}} = \{(w, w_i) \mid i \in \mathbb{N}\}$ and $V^{\mathcal{A}}(p_i) = \{w_j \mid j \geq i\}$ for each $i \in \mathbb{N}$. Let $(\mathcal{B}, w)$ be the same model with the addition of a point $w_{\mathbb{N}}$ in which all propositions are true. In other words $\mathrm{dom}(\mathcal{B}) = W \cup \{w_{\mathbb{N}}\}$, $R^{\mathcal{B}} = R^{\mathcal{A}} \cup \{(b, w_{\mathbb{N}})\}$ and $V^{\mathcal{B}}(p_i) = V^{\mathcal{A}}(p_i) \cup \{w_{\mathbb{N}}\}$ for each $i \in \mathbb{N}$.

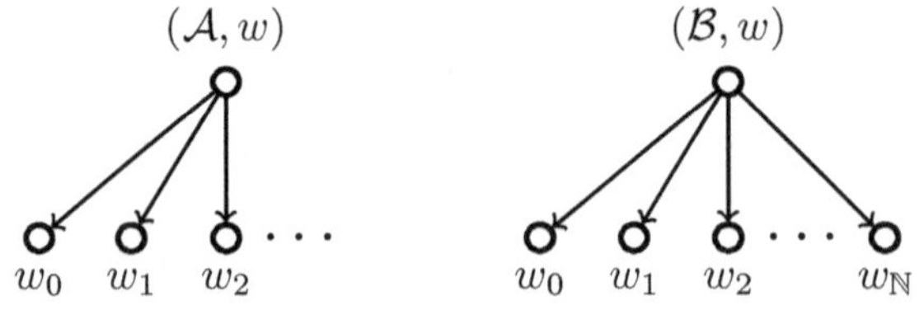

Fig. 1. The pointed models $(\mathcal{A}, w)$ and $(\mathcal{B}, w)$.

We see that by moving to $w_{\mathbb{N}}$, S wins the (n-)bisimulation game between the models $(\mathcal{A}, w)$ and $(\mathcal{B}, w)$, even though the models satisfy exactly the same ML-formulas.

We prove next that m-bisimilarity implies that D has winning strategy in the formula size game with modal parameter m. This simple observation is used in the next section, when we apply the game $\mathrm{FS}_{m,k}$ for proving a succinctness result for FO over ML.

Theorem 3.3 *Let $\mathbb{A}$ and $\mathbb{B}$ be sets of pointed models and let $m, k \in \mathbb{N}$. If there are m-bisimilar pointed models $(\mathcal{A}, w) \in \mathbb{A}$ and $(\mathcal{B}, v) \in \mathbb{B}$, then D has a winning strategy for the game $\mathrm{FS}_{m,k}(\mathbb{A}, \mathbb{B})$.*

Proof. The proof proceeds by induction on the number $m+k \in \mathbb{N}$. If $m+k = 0$ and $(\mathcal{A}, w) \in \mathbb{A}$ and $(\mathcal{B}, v) \in \mathbb{B}$ are m-bisimilar, then they are 0-bisimilar and thus satisfy the same literals. Thus there is no literal $\varphi \in \mathrm{ML}$ that separates the sets $\mathbb{A}$ and $\mathbb{B}$. Since S cannot make any moves and S does not win the game in this position, D wins the game $\mathrm{FS}_{0,0}(\mathbb{A}, \mathbb{B})$.

Assume that $m + k > 0$ and $(\mathcal{A}, w) \in \mathbb{A}$ and $(\mathcal{B}, v) \in \mathbb{B}$ are m-bisimilar. As in the basic step, S does not win the game in this position. We consider the cases of the first move of S in the game $\mathrm{FS}_{m,k}(\mathbb{A}, \mathbb{B})$.

If S starts with a left splitting move choosing the numbers m_1, m_2, k_1 and k_2 and the sets $\mathbb{A}_1$ and $\mathbb{A}_2$, then since $\mathbb{A}_1 \cup \mathbb{A}_2 = \mathbb{A}$, D can choose the next position $(m_i, k_i, \mathbb{A}_i, \mathbb{B})$, $i \in \{1, 2\}$ in such a way that $(\mathcal{A}, w) \in \mathbb{A}_i$. Then we have $m_i \leq m$ and $m_i + k_i < m + k$ so by induction hypothesis D has a winning strategy for the game $\mathrm{FS}_{m_i,k_i}(\mathbb{A}_i, \mathbb{B})$. The case of a right splitting move is similar.

If S starts with a left successor move choosing a function $f : \mathbb{A} \to \Box\mathbb{A}$, then since $(\mathcal{A}, w)$ and $(\mathcal{B}, v)$ are m-bisimilar, there is a pointed model $(\mathcal{B}, v') \in \Box(\mathcal{B}, v)$ that is $m-1$-bisimilar with the pointed model $f(\mathcal{A}, w)$. Since $m-1+k < m+k$, by induction hypothesis D has a winning strategy in $\mathrm{FS}_{m-1,k}(\Diamond_f\mathbb{A}, \Box\mathbb{B})$. The case of a right successor move is similar. □

4 Succinctness of FO over ML

In this section, we illustrate the use of the formula size game $\mathrm{FS}_{m,k}$ by proving a nonelementary succinctness gap between bisimulation invariant first-order logic and modal logic. We also show that this gap is already present between the 2-dimensional modal logic ML^2 introduced in [16] and basic modal logic.

4.1 A property of pointed frames

For the remainder of this paper we consider only the case where the set Φ of propositional symbols is empty. This makes all points in Kripke-models propositionally equivalent so we call pointed models in this section pointed frames. The only formulas available for the win condition of S in the game $\mathrm{FS}_{m,k}$ are $\bot$ and $\top$. Thus S only wins the game from the position $(m, k, \mathbb{A}, \mathbb{B})$ if either $\mathbb{A} = \emptyset$ and $\mathbb{B} \neq \emptyset$, or $\mathbb{A} \neq \emptyset$ and $\mathbb{B} = \emptyset$.

We will use the following two classes in our application of the formula size game $\mathrm{FS}_{m,k}$:

- $\mathbb{A}_n$ is the class of all pointed frames $(\mathcal{A}, w)$ such that for all $(\mathcal{A}, u), (\mathcal{A}, v) \in \Box(\mathcal{A}, w)$, the frames $(\mathcal{A}, u)$ and $(\mathcal{A}, v)$ are n-bisimilar.
- $\mathbb{B}_n$ is the complement of $\mathbb{A}_n$.

Lemma 4.1 *For each $n \in \mathbb{N}$ there is a formula $\varphi_n(x) \in \mathrm{FO}$ that separates the classes $\mathbb{A}_n$ and $\mathbb{B}_n$ such that the size of $\varphi_n(x)$ is exponential with respect to n, i.e., $\mathrm{s}(\varphi_n) = \mathcal{O}(2^n)$.*

Proof. We first define formulas $\psi_n(x, y) \in \mathrm{FO}$ such that $(\mathcal{M}, u) \leftrightarroows_n (\mathcal{M}, v)$ if and only if $\mathcal{M} \models \psi_n[u/x, v/y]$. The formulas $\psi_n(x, y)$ are defined recursively as follows:

$$\begin{aligned}\psi_1(x,y) &:= \exists s R(x,s) \leftrightarrow \exists t R(y,t) \\ \psi_{n+1}(x,y) &:= \forall s(R(x,s) \rightarrow \exists t(R(y,t) \wedge \psi_n(s,t)) \\ &\quad \wedge \forall t(R(y,t) \rightarrow \exists s(R(x,s) \wedge \psi_n(s,t)).\end{aligned}$$

Clearly these formulas express n-bisimilarity as intended. When we interpret the equivalences and implications as shorthand in the standard way, we get the sizes $\mathrm{s}(\psi_1) = 11$ and $\mathrm{s}(\psi_{n+1}) = 2 \cdot \mathrm{s}(\psi_n) + 13$. Thus $\mathrm{s}(\psi_n) = 3 \cdot 2^{n+2} - 13$.

Now we can define the formulas φ_n:

$$\varphi_n(x) := \forall y \forall z (R(x,y) \wedge R(x,z) \rightarrow \psi_n(y,z)).$$

Clearly for every $(\mathcal{A}, w) \in \mathbb{A}_n$ we have $\mathcal{A} \models \varphi_n[w/x]$ and for every $(\mathcal{B}, v) \in \mathbb{B}_n$ we have $\mathcal{B} \models \neg\varphi_n[w/x]$ so the formula φ_n separates the classes $\mathbb{A}_n$ and $\mathbb{B}_n$. Furthermore, $\mathrm{s}(\varphi_n) = \mathrm{s}(\psi_n) + 6 = 3 \cdot 2^{n+2} - 7$ so the size of φ_n is exponential with respect to n. □

Lemma 4.2 *For each $n \in \mathbb{N}$, the formula φ_n is $n+1$-bisimulation invariant.*

Proof. Let $(\mathcal{A}, w)$ and $(\mathcal{B}, v)$ be $n+1$-bisimilar pointed models. Assume that $\mathcal{A} \models \varphi_n[w/x]$. If $(\mathcal{B}, v_1), (\mathcal{B}, v_2) \in \Box(\mathcal{B}, v)$, by $n+1$-bisimilarity there are $(\mathcal{A}, w_1), (\mathcal{A}, w_2) \in \Box(\mathcal{A}, w)$ such that $(\mathcal{A}, w_1) \leftrightarroows_n (\mathcal{B}, v_1)$ and $(\mathcal{A}, w_2) \leftrightarroows_n (\mathcal{B}, v_2)$. Since $\mathcal{A} \models \varphi_n[w/x]$, we have $(\mathcal{B}, v_1) \leftrightarroows_n (\mathcal{A}, w_1) \leftrightarroows_n (\mathcal{A}, w_2) \leftrightarroows_n (\mathcal{B}, v_2)$ so $\mathcal{B} \models \psi_n[v_1/x, v_2/y]$. Thus, we see that $\mathcal{B} \models \varphi_n[v/x]$. □

It follows now from van Benthem's characterization theorem that each φ_n is equivalent to some ML-formula. Thus, we get the following corollary.

Corollary 4.3 *For each $n \in \mathbb{N}$, there is a formula $\vartheta_n \in \mathrm{ML}$ that separates the classes $\mathbb{A}_n$ and $\mathbb{B}_n$.*

4.2 Set theoretic construction of pointed frames

We have shown that the classes $\mathbb{A}_n$ and $\mathbb{B}_n$ can be separated both in ML and in FO. Furthermore the size of the FO-formula is exponential with respect to n. It only remains to ask: what is the size of the smallest ML-formula that separates the classes $\mathbb{A}_n$ and $\mathbb{B}_n$? To answer this we will need suitable subsets of $\mathbb{A}_n$ and $\mathbb{B}_n$ to play the formula size game on.

Definition 4.4 Let $n \in \mathbb{N}$. *The finite levels of the cumulative hierarchy* are defined recursively as follows:

$$\mathsf{V}_0 = \emptyset$$
$$\mathsf{V}_{n+1} = \mathcal{P}(\mathsf{V}_n)$$

For every $n \in \mathbb{N}$, V_n is a *transitive set*, i.e., for every $a \in \mathsf{V}_n$ and every $b \in a$ it holds that $b \in \mathsf{V}_n$. Thus it is reasonable to define a frame $\mathcal{F}_n = (\mathsf{V}_n, R_n)$, where for all $a, b \in \mathsf{V}_n$ it holds that $(a,b) \in R_n \Leftrightarrow b \in a$.

For every point $a \in \mathsf{V}_n$ we denote by $(\mathcal{M}_a, a)$ the pointed frame, where $\mathcal{M}_a$ is the subframe of $\mathcal{F}_n$ generated by the point a.

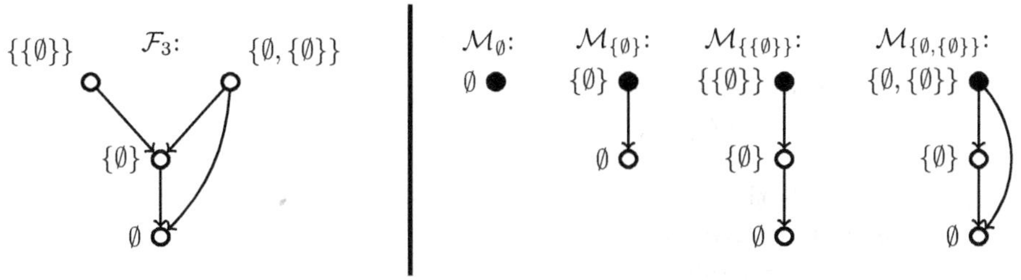

Fig. 2. The frame $\mathcal{F}_3$ and its generated subframes

Lemma 4.5 *Let $n \in \mathbb{N}$ and $a, b \in \mathsf{V}_{n+1}$. If $a \neq b$, then $(\mathcal{M}_a, a) \not\leftrightarrows_n (\mathcal{M}_b, b)$.*

Proof. We prove the claim by induction on n. The basic step $n = 0$ is trivial since V_1 only has one element. For the induction step, assume that $a, b \in \mathsf{V}_{n+1}$ and $a \neq b$. Assume further for contradiction that $(\mathcal{M}_a, a) \leftrightarrows_n (\mathcal{M}_b, b)$. Since $a \neq b$, by symmetry we can assume that there is $x \in a$ such that $x \notin b$. By n-bisimilarity there is $y \in b$ such that $(\mathcal{M}_x, x)$ and $(\mathcal{M}_y, y)$ are $n-1$-bisimilar. Since $x \in a \in \mathsf{V}_{n+1}$ and $y \in b \in \mathsf{V}_{n+1}$, we have $x, y \in \mathsf{V}_n$. By induction hypothesis we obtain $x = y$. This is a contradiction, since $x \notin b$ and $y \in b$. □

If $\mathbb{A}$ is a set of pointed frames we use the notation $\triangle\mathbb{A}$ for the pointed frame which is formed by taking all the pointed frames of $\mathbb{A}$ and connecting a new root to their distinguished points as illustrated in Figure 3. To make sure that $(\triangle\mathbb{A}, v)$ is bisimilar with $(\mathcal{A}, v)$ for any $(\mathcal{A}, v) \in \triangle\mathbb{A}$, we require that the frames in $\mathbb{A}$ are compatible in possible intersections. The precise definition is the following.

Let $\mathbb{A}$ be a set of pointed frames such that for all $(\mathcal{A}, v), (\mathcal{A}', v') \in \mathbb{A}$ it holds that $R^{\mathcal{A}} \upharpoonright (\mathrm{dom}(\mathcal{A}) \cap \mathrm{dom}(\mathcal{A}')) = R^{\mathcal{A}'} \upharpoonright (\mathrm{dom}(\mathcal{A}) \cap \mathrm{dom}(\mathcal{A}'))$ and let $w \notin \mathrm{dom}(\mathcal{A})$ for all $(\mathcal{A}, v) \in \mathbb{A}$. We use the notation $\triangle\mathbb{A} := (\mathcal{M}, w)$, where

$$\mathrm{dom}(\mathcal{M}) = \{w\} \cup \bigcup\{\mathrm{dom}(\mathcal{A}) \mid (\mathcal{A}, v) \in \mathbb{A}\}, \text{ and}$$
$$R^{\mathcal{M}} = \{(w, v) \mid (\mathcal{A}, v) \in \mathbb{A}\} \cup \bigcup\{R^{\mathcal{A}} \mid (\mathcal{A}, v) \in \mathbb{A}\}.$$

For each $n \in \mathbb{N}$ we define the following sets of pointed frames:

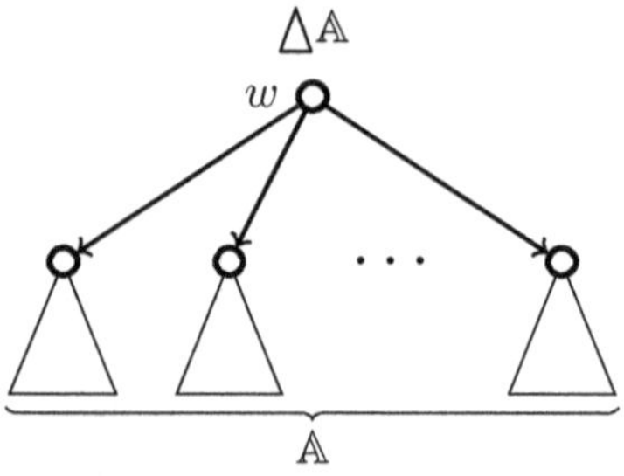

Fig. 3. The pointed frame $\Delta\mathbb{A}$

$$\mathbb{C}_n := \{\Delta\{(\mathcal{M}_a, a)\} \mid a \in \mathsf{V}_{n+1}\}$$
$$\mathbb{D}_n := \{\Delta\{(\mathcal{M}_a, a), (\mathcal{M}_b, b)\} \mid a, b \in \mathsf{V}_{n+1}, a \neq b\}.$$

In other words the pointed frames in $\mathbb{C}_n$ have a single successor from level $n+1$ of the cumulative hierarchy, whereas the pointed frames in $\mathbb{D}_n$ have two different successors from the same set. Therefore clearly $\mathbb{C}_n \subseteq \mathbb{A}_n$ and by Lemma 4.5 also $\mathbb{D}_n \subseteq \mathbb{B}_n$. In the next subsection we will use these sets in the formula size game.

It is well known that the cardinality of V_n is the exponential tower of $n-1$. Thus, the cardinality of $\mathbb{C}_n$ is $\text{twr}(n)$.

Lemma 4.6 *If* $n \in \mathbb{N}$, *we have* $|\mathbb{C}_n| = |\mathsf{V}_{n+1}| = \text{twr}(n)$. □

4.3 Graph colorings and winning strategies in $\text{FS}_{m,k}$

Our aim is to prove that any ML-formula ϑ_n separating the sets $\mathbb{C}_n$ and $\mathbb{D}_n$ is of size at least $\text{twr}(n-1)$. To do this, we make use of a surprising connection between the chromatic numbers of certain graphs related to pairs of the form $(\mathbb{V}, \mathbb{E})$, where $\mathbb{V} \subseteq \mathbb{C}_n$ and $\mathbb{E} \subseteq \mathbb{D}_n$, and existence of a winning strategy for D in the game $\text{FS}_{m,k}(\mathbb{V}, \mathbb{E})$.

Let $n \in \mathbb{N}$, $\emptyset \neq \mathbb{V} \subseteq \mathbb{C}_n$ and $\mathbb{E} \subseteq \mathbb{D}_n$. Then $\mathcal{G}(\mathbb{V}, \mathbb{E})$ denotes the graph (V, E), where

$$V = \{(\mathcal{M}, w) \mid \Delta\{(\mathcal{M}, w)\} \in \mathbb{V}\}, \text{ and}$$
$$E = \{((\mathcal{M}, w), (\mathcal{M}', w')) \in V \times V \mid \Delta\{(\mathcal{M}, w), (\mathcal{M}', w')\} \in \mathbb{E}\}.$$

Definition 4.7 Let $\mathcal{G} = (V, E)$ be a graph and let C be a set. A function $\chi : V \to C$ is a *coloring* of the graph $\mathcal{G}$ if for all $u, v \in V$ it holds that if $(u, v) \in E$, then $\chi(u) \neq \chi(v)$. If the set C has k elements, then χ is called a k-*coloring* of $\mathcal{G}$.

The *chromatic number* of $\mathcal{G}$, denoted by $\chi(\mathcal{G})$, is the smallest number $k \in \mathbb{N}$ for which there is a k-coloring of $\mathcal{G}$.

When playing the formula size game $\text{FS}_{m,k}(\mathbb{V}, \mathbb{E})$, splitting moves correspond with dividing either the vertex set or the edge set of the graph $\mathcal{G}(\mathbb{V}, \mathbb{E})$

into two parts, forming two new graphs. In the next lemma we get simple arithmetic estimates for the behaviour of chromatic numbers in such divisions.

Lemma 4.8 *Let $\mathcal{G} = (V, E)$ be a graph.*

(i) *Let $V_1, V_2 \subseteq V$ be nonempty such that $V_1 \cup V_2 = V$ and let $\mathcal{G}_1 = (V_1, E \restriction V_1)$ and $\mathcal{G}_2 = (V_2, E \restriction V_2)$. Then we have $\chi(\mathcal{G}) \leq \chi(\mathcal{G}_1) + \chi(\mathcal{G}_2)$.*

(ii) *Let $E_1, E_2 \subseteq E$ such that $E_1 \cup E_2 = E$ and let $\mathcal{G}_1 = (V, E_1)$ and $\mathcal{G}_2 = (V, E_2)$. Then $\chi(\mathcal{G}) \leq \chi(\mathcal{G}_1)\chi(\mathcal{G}_2)$.*

Proof.

(i) Let V_1, V_2, $\mathcal{G}_1$ and $\mathcal{G}_2$ be as in the claim and let $k_1 = \chi(\mathcal{G}_1)$ and $k_2 = \chi(\mathcal{G}_2)$. Let $\chi_1 : V_1 \to \{1, \ldots, k_1\}$ be a k_1-coloring of the graph $\mathcal{G}_1$ and let $\chi_2 : V_2 \to \{k_1+1, \ldots, k_1+k_2\}$ be a k_2-coloring of the graph $\mathcal{G}_2$. Then it is straightforward to show that $\chi = \chi_1 \cup (\chi_2 \restriction (V_2 \setminus V_1))$ is a $k_1 + k_2$-coloring of the graph $\mathcal{G}$, whence $\chi(\mathcal{G}) \leq k_1 + k_2 = \chi(\mathcal{G}_1) + \chi(\mathcal{G}_2)$.

(ii) Let $\chi_1 : V \to \{1, \ldots, k_1\}$ and $\chi_2 : V \to \{1, \ldots, k_2\}$ be colorings of the graphs $\mathcal{G}_1$ and $\mathcal{G}_2$, respectively. Then it is easy to verify that the map $\chi : V \to \{1, \ldots, k_1\} \times \{1, \ldots, k_2\}$ defined by $\chi(v) = (\chi_1(v), \chi_2(v))$ is a coloring of $\mathcal{G}$. Thus we obtain $\chi(\mathcal{G}) \leq |\{1, \ldots, k_1\} \times \{1, \ldots, k_2\}| = \chi(\mathcal{G}_1)\chi(\mathcal{G}_2)$.

□

Lemma 4.9 *Assume $\emptyset \neq \mathbb{V} \subseteq \mathbb{C}_n$ and $\mathbb{E} \subseteq \mathbb{D}_n$ for some $n \in \mathbb{N}$ and let $m, k \in \mathbb{N}$. If $\chi(\mathcal{G}(\mathbb{V}, \mathbb{E})) \geq 2$ and $k < \log(\chi(\mathcal{G}(\mathbb{V}, \mathbb{E})))$, then D has a winning strategy in the game $\mathrm{FS}_{m,k}(\mathbb{V}, \mathbb{E})$.*

Proof. Let $n, m, k \in \mathbb{N}$ and assume that $\emptyset \neq \mathbb{V} \subseteq \mathbb{C}_n$, $\mathbb{E} \subseteq \mathbb{D}_n$, $\chi(\mathcal{G}(\mathbb{V}, \mathbb{E})) \geq 2$ and $k < \log(\chi(\mathcal{G}(\mathbb{V}, \mathbb{E})))$. We prove the claim by induction on k.

If $k = 0$, S can only make successor moves. Since $\chi(\mathcal{G}(\mathbb{V}, \mathbb{E})) \geq 2$, there are $(\mathcal{M}, w), (\mathcal{M}', w') \in V$ such that $((\mathcal{M}, w), (\mathcal{M}', w')) \in E$. Thus $\Delta\{(\mathcal{M}, w)\}$, $\Delta\{(\mathcal{M}', w')\} \in \mathbb{V}$ and $\Delta\{(\mathcal{M}, w), (\mathcal{M}', w')\} \in \mathbb{E}$. If S makes a left or right successor move, then in the resulting position $(m - 1, 0, \mathbb{V}', \mathbb{E}')$ it holds that $(\mathcal{M}, w) \in \mathbb{V}' \cap \mathbb{E}'$ or $(\mathcal{M}', w') \in \mathbb{V}' \cap \mathbb{E}'$. Thus the same pointed model is present on both sides of the game and by Theorem 3.3, D has a winning strategy for the game $\mathrm{FS}_{m,k}(\mathbb{V}', \mathbb{E}')$.

Assume then that $k > 0$. If S starts the game with a successor move, then D wins as described above.

Assume that S begins the game with a left splitting move choosing the numbers $m_1, m_2, k_1, k_2 \in \mathbb{N}$ and the sets $\mathbb{V}_1, \mathbb{V}_2 \subseteq \mathbb{V}$. Consider the graphs $\mathcal{G}(\mathbb{V}, \mathbb{E}) = (V, E)$, $\mathcal{G}(\mathbb{V}_1, \mathbb{E}) = (V_1, E_1)$ and $\mathcal{G}(\mathbb{V}_2, \mathbb{E}) = (V_2, E_2)$. Since $\mathbb{V}_1 \cup \mathbb{V}_2 = \mathbb{V}$, we have $V_1 \cup V_2 = V$. In addition, by the definition of the graphs $\mathcal{G}(\mathbb{V}, \mathbb{E})$, $\mathcal{G}(\mathbb{V}_1, \mathbb{E})$ and $\mathcal{G}(\mathbb{V}_2, \mathbb{E})$ we see that $E_1 = E \restriction V_1$ and $E_2 = E \restriction V_2$. Thus by Lemma 4.8, we obtain $\chi(\mathcal{G}(\mathbb{V}, \mathbb{E})) \leq \chi(\mathcal{G}(\mathbb{V}_1, \mathbb{E})) + \chi(\mathcal{G}(\mathbb{V}_2, \mathbb{E}))$. It must hold that $k_1 < \log(\chi(\mathcal{G}(\mathbb{V}_1, \mathbb{E})))$ or $k_2 < \log(\chi(\mathcal{G}(\mathbb{V}_2, \mathbb{E})))$, since otherwise we would have

$$\begin{aligned} k &< \log(\chi(\mathcal{G}(\mathbb{V}, \mathbb{E}))) \leq \log(\chi(\mathcal{G}(\mathbb{V}_1, \mathbb{E})) + \chi(\mathcal{G}(\mathbb{V}_2, \mathbb{E}))) \\ &\leq \log(\chi(\mathcal{G}(\mathbb{V}_1, \mathbb{E}))) + \log(\chi(\mathcal{G}(\mathbb{V}_2, \mathbb{E}))) + 1 \leq k_1 + k_2 + 1 = k. \end{aligned}$$

Thus D can choose the next position of the game, $(m_i, k_i, \mathbb{V}_i, \mathbb{E})$, in such a way that $k_i < \log(\chi(\mathcal{G}(\mathbb{V}_i, \mathbb{E})))$. By induction hypothesis D has a winning strategy in the game $\mathrm{FS}_{m_i,k_i}(\mathbb{V}_i, \mathbb{E})$.

Assume then that S begins the game with a right splitting move choosing the numbers $m_1, m_2, k_1, k_2 \in \mathbb{N}$ and the sets $\mathbb{E}_1, \mathbb{E}_2 \subseteq \mathbb{E}$. Consider now the graphs $\mathcal{G}(\mathbb{V}, \mathbb{E}) = (V, E)$, $\mathcal{G}(\mathbb{V}, \mathbb{E}_1) = (V_1, E_1)$ and $\mathcal{G}(\mathbb{V}, \mathbb{E}_2) = (V_2, E_2)$. Clearly $V_1 = V_2 = V$ and since $\mathbb{E}_1 \cup \mathbb{E}_2 = \mathbb{E}$, we have $E_1 \cup E_2 = E$. Thus by Lemma 4.8, we obtain $\chi(\mathcal{G}(\mathbb{V}, \mathbb{E})) \leq \chi(\mathcal{G}(\mathbb{V}, \mathbb{E}_1))\chi(\mathcal{G}(\mathbb{V}, \mathbb{E}_2))$. It must hold that $k_1 < \log(\chi(\mathcal{G}(\mathbb{V}, \mathbb{E}_1)))$ or $k_2 < \log(\chi(\mathcal{G}(\mathbb{V}, \mathbb{E}_2)))$, since otherwise we would have

$$\begin{aligned} k &< \log(\chi(\mathcal{G}(\mathbb{V}, \mathbb{E}))) \leq \log(\chi(\mathcal{G}(\mathbb{V}, \mathbb{E}_1))\chi(\mathcal{G}(\mathbb{V}, \mathbb{E}_2))) \\ &= \log(\chi(\mathcal{G}(\mathbb{V}, \mathbb{E}_1))) + \log(\chi(\mathcal{G}(\mathbb{V}, \mathbb{E}_2))) \leq k_1 + k_2 + 1 = k. \end{aligned}$$

Thus D can again choose the next position of the game, $(m_i, k_i, \mathbb{V}, \mathbb{E}_i)$, in such a way that $k_i < \log(\chi(\mathcal{G}(\mathbb{V}, \mathbb{E}_i)))$. By induction hypothesis D has a winning strategy in the game $\mathrm{FS}_{m_i,k_i}(\mathbb{V}, \mathbb{E}_i)$. □

Lemma 4.10 *If* $k < \mathrm{twr}(n-1)$ *and* $m \in \mathbb{N}$, *then D has a winning strategy in the game* $\mathrm{FS}_{m,k}(\mathbb{C}_n, \mathbb{D}_n)$.

Proof. By Lemma 4.6, we have $|\mathbb{C}_n| = \mathrm{twr}(n)$ and the set $\mathbb{D}_n$ consists of all the pointed frames $\Delta\{(\mathcal{M}, w), (\mathcal{M}', w')\}$, where $(\mathcal{M}, w), (\mathcal{M}', w') \in \mathbb{C}_n$, $(\mathcal{M}, w) \neq (\mathcal{M}', w')$. Thus the graph $\mathcal{G}(\mathbb{C}_n, \mathbb{D}_n)$ is isomorphic with the complete graph $K_{\mathrm{twr}(n)}$. Therefore we obtain

$$\chi(\mathcal{G}(\mathbb{C}_n, \mathbb{D}_n)) = \chi(K_{\mathrm{twr}(n)}) = \mathrm{twr}(n).$$

By the assumption, $k < \mathrm{twr}(n-1) = \log(\mathrm{twr}(n)) = \log(\chi(\mathcal{G}(\mathbb{C}_n, \mathbb{D}_n)))$, so by Lemma 4.9, D has a winning strategy in the game $\mathrm{FS}_{m,k}(\mathbb{C}_n, \mathbb{D}_n)$. □

Theorem 4.11 *Let* $n \in \mathbb{N}$. *If a formula* $\vartheta_n \in \mathrm{ML}$ *separates the classes* $\mathbb{A}_n$ *and* $\mathbb{B}_n$, *then* $\mathrm{s}(\vartheta_n) \geq \mathrm{twr}(n-1)$.

Proof. Assume that a formula $\vartheta_n \in \mathrm{ML}$ separates the classes $\mathbb{A}_n$ and $\mathbb{B}_n$. As observed in the end of Subsection 4.2, it holds that $\mathbb{C}_n \subseteq \mathbb{A}_n$ and $\mathbb{D}_n \subseteq \mathbb{B}_n$. Therefore ϑ_n also separates the sets $\mathbb{C}_n$ and $\mathbb{D}_n$.

Assume for contradiction that $\mathrm{s}(\vartheta_n) < \mathrm{twr}(n-1)$. By Theorem 3.2, S has a winning strategy in the game $\mathrm{FS}_{m,k}(\mathbb{C}_n, \mathbb{D}_n)$ for $m = \mathrm{ms}(\vartheta_n)$ and $k = \mathrm{cs}(\vartheta_n)$. On the other hand, $k < \mathrm{twr}(n-1)$, whence by Lemma 4.10, D has a winning strategy in the same game. □

We now have everything we need for proving the nonelementary succinctness of FO over ML. By Lemma 4.1, for each $n \in \mathbb{N}$ there is a formula $\varphi_n(x) \in \mathrm{FO}$ such that φ_n separates the classes $\mathbb{A}_n$ and $\mathbb{B}_n$ with $s(\varphi) = \mathcal{O}(2^n)$. On the other hand by Corollary 4.3, there is an equivalent formula $\vartheta_n \in \mathrm{ML}$, but by Theorem 4.11 the size of ϑ_n must be at least $\mathrm{twr}(n-1)$. So the property of a pointed models all successors being n-bisimilar with each other can be expressed in FO with a formula of exponential size, but in ML expressing it requires a formula of non-elementary size.

Corollary 4.12 *Bisimulation invariant* FO *is nonelementarily more succinct than* ML.

Remark 4.13 It is well known that the DAG-size of any formula φ is greater than or equal to the logarithm of the size of φ. Thus if ϑ_n is a formula as in Theorem 4.11, the DAG-size of ϑ_n must be at least $\mathrm{twr}(n-2)$. Consequently the result of Corollary 4.12 also holds for DAG-size.

4.4 Succinctness of 2-dimensional modal logic

Our proof for the nonelementary succinctness gap between bisimulation invariant FO and ML is based on the fact that n-bisimilarity of two points $u, v \in W$ of a Kripke-frame $\mathcal{M} = (W, R)$ is definable by an FO-formula $\psi_n(x, y)$ (see the proof of Lemma 4.1). However, it is not difficult to see that the property $(\mathcal{M}, u) \leftrightarroweq_n (\mathcal{M}, v)$ is already expressible in 2-*dimensional modal logic*.

The idea in 2-dimensional modal logic is that the truth of formulas is evaluated on pairs (u, v) of elements of Kripke-models instead of single points. We refer to the book [15] of Marx and Venema for a detailed exposition on 2-dimensional and multi-dimensional modal logics. For our purposes it suffices to consider the modal fragment ML^2 of the 2-dimensional modal μ-calculus L^2_μ, introduced by Otto [16].

A Kripke-model $\mathcal{T}$ for ML^2 consists of a set W of points, a binary accessibility relation R, and a valuation V. Note that proposition symbols are interpreted as sets of pairs, whence V is a function $\Phi \to \mathcal{P}(W^2)$. Since accessibility is defined separately for the two components of pairs $(u, v) \in W^2$, there are two modal operators $\Diamond_1$ and $\Diamond_2$ in ML^2. The semantics of these operators and their duals are defined as follows:

- $(\mathcal{T}, (u, v)) \vDash \Diamond_1 \varphi \Leftrightarrow$ there is $u' \in W$ such that uRu' and $(\mathcal{T}, (u', v)) \vDash \varphi$,
- $(\mathcal{T}, (u, v)) \vDash \Diamond_2 \varphi \Leftrightarrow$ there is $v' \in W$ such that vRv' and $(\mathcal{T}, (u, v')) \vDash \varphi$,
- $(\mathcal{T}, (u, v)) \vDash \Box_1 \varphi \Leftrightarrow$ for all $u' \in W$, if uRu', then $(\mathcal{T}, (u', v)) \vDash \varphi$,
- $(\mathcal{T}, (u, v)) \vDash \Box_2 \varphi \Leftrightarrow$ for all $v' \in W$, if vRv', then $(\mathcal{T}, (u, v')) \vDash \varphi$.

In addition to proposition symbols, connectives and modal operators, the logic ML^2 has variable substitution operators (see [16], p. 242–43), but we will not need them here.

Any pointed Kripke-model $(\mathcal{M}, w) = ((W, R, V), w)$ can be interpreted as the 2-dimensional pointed model $(\mathcal{M}_2, (w, w))$, where $\mathcal{M}_2 = (W, R, V_2)$ and $V_2(p) = \{(w, w) \mid w \in V(p)\}$ for each $p \in \Phi$. This gives us a meaningful way of defining properties of pointed models $(\mathcal{M}, w)$ by formulas of ML^2. In particular, we say that a formula $\varphi \in \mathrm{ML}^2$ separates two classes $\mathbb{A}$ and $\mathbb{B}$ of pointed models if for all $(\mathcal{M}, w) \in \mathbb{A}$, $(\mathcal{M}_2, (w, w)) \vDash \varphi$ and for all $(\mathcal{M}, w) \in \mathbb{B}$, $(\mathcal{M}_2, (w, w)) \nvDash \varphi$.

The *size* $\mathrm{s}(\varphi)$ of a formula $\varphi \in \mathrm{ML}^2$ is defined in the same way as for formulas of ML; see Definitions 2.5, 2.6 and 2.7. In other words, $\mathrm{s}(\varphi)$ is the total number of modal operators and binary connectives occurring in φ.

Observe now that two pointed frames $(\mathcal{M}, u)$ and $(\mathcal{M}, v)$ are 1-bisimilar if and only if $(\mathcal{M}_2,(u,v)) \models \rho_1$, where $\rho_1 := \Diamond_1\top \leftrightarrow \Diamond_2\top$. Furthermore if $\rho_n \in \mathrm{ML}^2$ defines the class of all 2-dimensional pointed frames $(\mathcal{M}_2,(u,v))$ such that $(\mathcal{M},u) \leftrightarrows_n (\mathcal{M},v)$, then $\rho_{n+1} := \Box_1\Diamond_2\rho_n \wedge \Box_2\Diamond_1\rho_n$ defines the class of all $(\mathcal{M}_2,(u,v))$ such that $(\mathcal{M},u) \leftrightarrows_{n+1} (\mathcal{M},v)$.

Lemma 4.14 *For each $n \in \mathbb{N}$ there is a formula $\zeta_n \in \mathrm{ML}^2$ that separates the classes $\mathbb{A}_n$ and $\mathbb{B}_n$ such that the size of ζ_n is exponential with respect to n, i.e.,* $\mathrm{s}(\zeta_n) = \mathcal{O}(2^n)$.

Proof. Let ζ_n be the formula $\Box_1\Box_2\rho_n$. Then $(\mathcal{M}_2,(w,w)) \models \zeta_n$ if and only if $(\mathcal{M},u)$ and $(\mathcal{M},v)$ are n-bisimilar for all $(\mathcal{M},u),(\mathcal{M},v) \in \Box(\mathcal{M},w)$, whence ζ_n separates $\mathbb{A}_n$ from its complement $\mathbb{B}_n$. An easy calculation shows that the size of ζ_n is $3 \cdot 2^{n+1} - 5$. □

By Theorem 4.3, for each $n \in \mathbb{N}$ there is a formula $\vartheta_n \in \mathrm{ML}$ that is equivalent with ζ_n. On the other hand, by Theorem 4.11 the size of ϑ_n is at least $\mathrm{twr}(n-1)$. Thus, we obtain the nonelementary succinctness gap already between ML^2 and ML.

Corollary 4.15 *The 2-dimensional modal logic* ML^2 *is nonelementarily more succinct than* ML.

Acknowledgement. We are grateful to an anonymous referee for pointing out the possibility of using 2-dimensional modal logic for separating the classes $\mathbb{A}_n$ and $\mathbb{B}_n$.

References

[1] Adler, M. and N. Immerman, *An* n! *lower bound on formula size*, ACM Trans. Comput. Log. **4** (2003), pp. 296–314.
URL http://doi.acm.org/10.1145/772062.772064

[2] Blackburn, P., M. de Rijke and Y. Venema, "Modal Logic," Cambridge University Press, New York, NY, USA, 2001.

[3] Etessami, K., M. Y. Vardi and T. Wilke, *First-order logic with two variables and unary temporal logic*, Inf. Comput. **179** (2002), pp. 279–295.
URL http://dx.doi.org/10.1006/inco.2001.2953

[4] Etessami, K. and T. Wilke, *An until hierarchy and other applications of an ehrenfeucht-fraïssé game for temporal logic*, Inf. Comput. **160** (2000), pp. 88–108.
URL http://dx.doi.org/10.1006/inco.1999.2846

[5] Figueira, S. and D. Gorín, *On the size of shortest modal descriptions*, in: *Advances in Modal Logic 8, papers from the eighth conference on "Advances in Modal Logic," held in Moscow, Russia, 24-27 August 2010*, 2010, pp. 120–139.
URL http://www.aiml.net/volumes/volume8/Figueira-Gorin.pdf

[6] French, T., W. van der Hoek, P. Iliev and B. P. Kooi, *Succinctness of epistemic languages*, in: *IJCAI 2011, Proceedings of the 22nd International Joint Conference on Artificial Intelligence, Barcelona, Catalonia, Spain, July 16-22, 2011*, 2011, pp. 881–886.
URL http://ijcai.org/papers11/Papers/IJCAI11-153.pdf

[7] French, T., W. van der Hoek, P. Iliev and B. P. Kooi, *On the succinctness of some modal logics*, Artif. Intell. **197** (2013), pp. 56–85.
URL http://dx.doi.org/10.1016/j.artint.2013.02.003

[8] Grohe, M. and N. Schweikardt, *The succinctness of first-order logic on linear orders*, Logical Methods in Computer Science **1** (2005).
URL http://dx.doi.org/10.2168/LMCS-1(1:6)2005

[9] Hella, L. and J. Väänänen, *The size of a formula as a measure of complexity*, in: *Logic Without Borders - Essays on Set Theory, Model Theory, Philosophical Logic and Philosophy of Mathematics*, 2015 pp. 193–214.
URL `http://dx.doi.org/10.1515/9781614516873.193`

[10] Iliev, P., "On the Relative Succinctness of Some Modal Logics," Ph.D. thesis, University of Liverpool (2013).

[11] Laroussinie, F., N. Markey and P. Schnoebelen, *Temporal logic with forgettable past*, in: *17th IEEE Symposium on Logic in Computer Science (LICS 2002), 22-25 July 2002, Copenhagen, Denmark, Proceedings*, 2002, pp. 383–392.
URL `http://dx.doi.org/10.1109/LICS.2002.1029846`

[12] Lutz, C., *Complexity and succinctness of public announcement logic*, in: *5th International Joint Conference on Autonomous Agents and Multiagent Systems (AAMAS 2006), Hakodate, Japan, May 8-12, 2006*, 2006, pp. 137–143.
URL `http://doi.acm.org/10.1145/1160633.1160657`

[13] Lutz, C., U. Sattler and F. Wolter, *Modal logic and the two-variable fragment*, in: *Computer Science Logic, 15th International Workshop, CSL 2001. 10th Annual Conference of the EACSL, Paris, France, September 10-13, 2001, Proceedings*, 2001, pp. 247–261.
URL `http://dx.doi.org/10.1007/3-540-44802-0_18`

[14] Markey, N., *Temporal logic with past is exponentially more succinct, concurrency column*, Bulletin of the EATCS **79** (2003), pp. 122–128.

[15] Marx, M. and Y. Venema, "Multi-dimensional modal logic," Applied Logic Series **4**, Kluwer Academic Publishers, Dordrecht, 1997, xiv+239 pp.
URL `http://dx.doi.org/10.1007/978-94-011-5694-3`

[16] Otto, M., *Bisimulation-invariant PTIME and higher-dimensional μ-calculus*, Theor. Comput. Sci. **224** (1999), pp. 237–265.
URL `http://dx.doi.org/10.1016/S0304-3975(98)00314-4`

[17] Sistla, A. P. and E. M. Clarke, *The complexity of propositional linear temporal logics*, J. ACM **32** (1985), pp. 733–749.
URL `http://doi.acm.org/10.1145/3828.3837`

[18] Stockmeyer, L. J., "The Complexity of Decision Problems in Automata Theory and Logic," Ph.D. thesis, Massachusetts Institute of Technology (1974).

[19] van der Hoek, W. and P. Iliev, *On the relative succinctness of modal logics with union, intersection and quantification*, in: *International conference on Autonomous Agents and Multi-Agent Systems, AAMAS '14, Paris, France, May 5-9, 2014*, 2014, pp. 341–348.
URL `http://dl.acm.org/citation.cfm?id=2615788`

[20] van der Hoek, W., P. Iliev and B. P. Kooi, *On the relative succinctness of two extensions by definitions of multimodal logic*, in: *How the World Computes - Turing Centenary Conference and 8th Conference on Computability in Europe, CiE 2012, Cambridge, UK, June 18-23, 2012. Proceedings*, 2012, pp. 323–333.
URL `http://dx.doi.org/10.1007/978-3-642-30870-3_33`

[21] van Ditmarsch, H., J. Fan, W. van der Hoek and P. Iliev, *Some exponential lower bounds on formula-size in modal logic*, in: *Advances in Modal Logic 10, invited and contributed papers from the tenth conference on "Advances in Modal Logic," held in Groningen, The Netherlands, August 5-8, 2014*, 2014, pp. 139–157.
URL `http://www.aiml.net/volumes/volume10/Ditmarsch-Fan-Hoek-Iliev.pdf`

[22] Wilke, T., *Ctl^+ is exponentially more succinct than CTL*, in: *Foundations of Software Technology and Theoretical Computer Science, 19th Conference, Chennai, India, December 13-15, 1999, Proceedings*, 1999, pp. 110–121.
URL `http://dx.doi.org/10.1007/3-540-46691-6_9`

A canonical model construction for intuitionistic distributed knowledge

Gerhard Jäger

Institute of Computer Science, University of Bern
Neubrückstrasse 10, CH-3012 Bern, Switzerland
jaeger@inf.unibe.ch

Michel Marti [1]

Institute of Computer Science, University of Bern
Neubrückstrasse 10, CH-3012 Bern, Switzerland
mmarti@inf.unibe.ch

Abstract

Intuitionistic epistemic logic is an active research field. However, so far no consensus has been reached what the correct form of intuitionistic epistemic logic is and more technical and conceptual work is needed to obtain a better understanding. This article tries to make a small technical contribution to this enterprise.
Roughly speaking, a proposition is distributed knowledge among a group of agents if it follows from their combined knowledge. We are interested in formalizing intuitionistic distributed knowledge. Our focus is on two theories IDK and IDT, presented as Hilbert-style systems, and the proof of the completeness of these theories; their correctness is obvious.
Intuitionistic distributed knowledge is semantically treated following the standard lines of intuitionistic modal logic. Motivated by an approach due to Fagin, Halpern, and Vardi, though significantly simplified for the treatment of IDK and IDT, we show completeness of these systems via a canonical model construction.

Keywords: Distributed knowledge, intuitionistic modal logic, canonical models.

1 Introduction

Intuitionistic epistemic logic is an active research field; see, for example, Artemov and Protopopescu [1], Hirai [6], Jäger and Marti [7], Krupski and Yatmanov [8], Proietti [11], Suzuki [14] and also the somewhat older Williamson [16]. The two main pillars of most present approaches are:

- Epistemic logic based on classical modal logic. There exists a huge amount of work making case for classical multi-modal systems providing an adequate

[1] Research partly supported by the Swiss National Science Foundation.

and useful framework for reasoning about knowledge and belief. The textbooks Fagin, Halpern, Moses, and Vardi [2] and Meyer and van der Hoek [9] provide a solid introduction into this area.

- Systems of intuitionistic modal logic. There is also the interesting – though not so popular – world of intuitionistic modal logic. A fundamental result is the completeness proof for the logic **IK** in Fischer Servi [12], and Simpson [13] provides an excellent survey of intuitionistic modal logics, contains some further results and leads to present research in this area.

However, so far no consensus has been reached what the "correct" form of intuitionistic epistemic logic is. Different approaches have been proposed, varying in their philosophical justifications and taking into account various fields of possible applications. We believe that intuitionistic epistemic logic may provide an approach to dealing with knowledge that is more "constructive" than the treatment of epistemic logic based on classical logic. In particular, it has to be seen whether the process of building up (or acquiring) knowledge by an agent can be more naturally formalized in an intuitionistic environment. It is clear that more technical and conceptual work is needed and that we have to develop a better understanding of the general methodology behind intuitionistic epistemic reasoning.

This article tries to make a small technical contribution to this enterprise. It can be considered as a twin of Jäger and Marti [7] which deals with intuitionistic common knowledge. Now we are interested in formalizing intuitionistic distributed knowledge. Our focus is on two theories **IDK** and **IDT** and the proof of the completeness of these theories. This is achieved by adapting a canonical model construction for our framework.

2 The language $\mathcal{L}_{DK}$ and its semantics

Our general scenario is that we want to deal with ℓ agents $ag_1, \ldots, ag_\ell$, the individual knowledge/belief of these agents and knowledge/belief distributed among them. In order to avoid a trivial situation, $\ell \geq 2$ is a general assumption. We begin with introducing a language $\mathcal{L}_{DK}$ tailored for this purpose and interpret its formulas over so-called epistemic Kripke structures, thus providing a semantic approach to intuitionistic distributed knowledge/belief. To formally express that agent ag_i knows or believes α, we will write $\mathsf{K}_i(\alpha)$, whereas $\mathsf{D}(\alpha)$ says that α is knowledge distributed among $ag_1, \ldots, ag_\ell$. Hence the language $\mathcal{L}_{DK}$ comprises the following primitive symbols:

PS.1 Countably many atomic propositions p, q, r (possibly with subscripts); the collection of all atomic propositions is called *PROP*.

PS.2 The logical constant $\bot$ and the logical connectives $\vee$, $\wedge$, and $\rightarrow$.

PS.3 The modal operators $\mathsf{K}_1, \ldots, \mathsf{K}_\ell, \mathsf{D}$.

The *formulas* $\alpha, \beta, \gamma, \delta$ (possibly with subscripts) of $\mathcal{L}_{DK}$ are generated by the following BNF:

$$\alpha ::\equiv p \mid \bot \mid (\alpha \vee \alpha) \mid (\alpha \wedge \alpha) \mid (\alpha \rightarrow \alpha) \mid \mathsf{K}_i(\alpha) \mid \mathsf{D}(\alpha).$$

It is common in intuitionistic logic to define negation $\neg\alpha$ by $(\alpha \rightarrow \bot)$ and equivalence $(\alpha \leftrightarrow \beta)$ by $((\alpha \rightarrow \beta) \wedge (\beta \rightarrow \alpha))$. We often omit parentheses and brackets if there is no danger of confusion.

As in the classical setting all operators K_i will have the normality axiom

$$\mathsf{K}_i(\alpha \rightarrow \beta) \rightarrow (\mathsf{K}_i(\alpha) \rightarrow \mathsf{K}_i(\beta)).$$

Sometimes it is argued that interpreting $\mathsf{K}_i(\alpha)$ as "agent ag_i knows α" requires the presence of the truth property

$$\mathsf{K}_i(\alpha) \rightarrow \alpha$$

and possibly positive as well as negative introspection; otherwise $\mathsf{K}_i(\alpha)$ should be seen as stating that agent ag_i only believes α. However, since we are primarily interested in technical questions, we do not make this distinction and speak of knowledge and distributed knowledge to simplify matters.

In this paper we do not enter into a discussion of the modal logic approach to knowledge and distributed knowledge. As mentioned above, this is done in great detail in the textbooks Fagin, Halpern, Moses, and Vardi [2] and Meyer and van der Hoek [9] as well as in many research articles; see, e.g., Fagin, Halpern, and Vardi [3], Gerbrandy [4], Hakli and Negri [5], and Wang and Ågotnes [15]. However, all these texts are about distributed knowledge based on classical logic, whereas here we work in the context of intuitionistic logic. As mentioned above, intuitionistic epistemic logic is interesting by its own. Here we add the facet of intuitionistic distributed knowledge/belief to the general discussion.

First we have to fix the adequate structures over which the formulas of $\mathcal{L}_{DK}$ will be interpreted. First some notation: Given a non-empty set W and a binary relation R on W, we often write aRb for $(a, b) \in R$ and set $R[a] := \{b \in W : aRb\}$. We say that $R[a]$ is the collection of all elements of W that are accessible from a via R.

Definition 2.1 An *epistemic Kripke structure (EK-structure for short) of order* ℓ is an $(\ell + 3)$-tuple $\mathfrak{M} = (W, \preceq, R_1, \ldots, R_\ell, V)$ with the following properties:

(EK.1) W is a nonempty set (the set of the so-called worlds of $\mathfrak{M}$) and $\preceq$ is a preorder on W.

(EK.2) Every R_i for $1 \leq i \leq \ell$ is a binary relation on W such that for any $a, b \in W$,

$$a \preceq b \quad \Longrightarrow \quad R_i[b] \subseteq R_i[a].$$

(EK.3) V is a function from W to the power set of $PROP$ such that for any $a, b \in W$,

$$a \preceq b \quad \Longrightarrow \quad V(a) \subseteq V(b).$$

$\mathfrak{M}$ is called *a reflexive EK-structure* iff all relations $R_1, \ldots, R_\ell$ are reflexive.

(EK.1) and (EK.3) are the usual properties of a Kripke structure for intuitionistic propositional logic. Given the EK-structure of order ℓ

$$\mathfrak{M} = (W, \preceq, R_1, \ldots, R_\ell, V),$$

the relation R_i is the accessibility relation associated with agent ag_i that tells which worlds b are accessible for ag_i from world a; agent ag_i "knows" α in world a iff α holds in all worlds that are accessible for ag_i from world a. On the other hand, α is considered to be distributed knowledge in world a iff α holds in those worlds that are accessible for all agents $ag_1, \ldots, ag_\ell$ from a. The condition (EK.2) ensures monotonicity for formulas of the form $\mathsf{K}_i(\alpha)$. Whenever agent ag_i progresses along $\preceq$, the collection of worlds that are accessible for ag_i can go down, reflecting the fact that some worlds are ruled out as being accessible due to new information. If R_i is reflexive then all worlds b such that $a \preceq b$ are accessible for agent ag_i from a.

Definition 2.2 [Value] Given an EK-structure $\mathfrak{M} = (W, \preceq, R_1, \ldots, R_\ell, V)$ of order ℓ, the set $\|\alpha\|_{\mathfrak{M}}$ of worlds satisfying α is inductively defined as follows:

(1) $\|\bot\|_{\mathfrak{M}} := \emptyset$,

(2) $\|p\|_{\mathfrak{M}} := \{a \in W : p \in V(a)\}$ for any $p \in PROP$,

(3) $\|\alpha \vee \beta\|_{\mathfrak{M}} := \|\alpha\|_{\mathfrak{M}} \cup \|\beta\|_{\mathfrak{M}}$,

(4) $\|\alpha \wedge \beta\|_{\mathfrak{M}} := \|\alpha\|_{\mathfrak{M}} \cap \|\beta\|_{\mathfrak{M}}$,

(5) $\|\alpha \to \beta\|_{\mathfrak{M}} := \{a \in W : \{b \in W : a \preceq b\} \cap \|\alpha\|_{\mathfrak{M}} \subseteq \|\beta\|_{\mathfrak{M}}\}$,

(6) $\|\mathsf{K}_i(\alpha)\|_{\mathfrak{M}} := \{a \in W : R_i[a] \subseteq \|\alpha\|_{\mathfrak{M}}\}$,

(7) $\|\mathsf{D}(\alpha)\|_{\mathfrak{M}} := \{a \in W : \bigcap_{i=1}^{\ell} R_i[a] \subseteq \|\alpha\|_{\mathfrak{M}}\}$.

A simple proof by induction on the structure of α shows that the sets $\|\alpha\|_{\mathfrak{M}}$ satisfy the usual monotonicity condition of intuitionistic logic.

Lemma 2.3 *For all EK-structures $\mathfrak{M} = (W, \preceq, R_1, \ldots, R_\ell, V)$ of order ℓ, all elements $a, b \in W$, and all α we have that*

$$a \preceq b \text{ and } a \in \|\alpha\|_{\mathfrak{M}} \quad \Longrightarrow \quad b \in \|\alpha\|_{\mathfrak{M}}.$$

We call $\|\alpha\|_{\mathfrak{M}}$ the *value* of α in $\mathfrak{M}$ and often write $(\mathfrak{M}, a) \models \alpha$ instead of $a \in \|\alpha\|_{\mathfrak{M}}$. In addition, α is *valid in the EK-structure* $\mathfrak{M}$, written $\mathfrak{M} \models \alpha$, iff $(\mathfrak{M}, a) \models \alpha$ for all worlds a of $\mathfrak{M}$. Finally, α is called *EK-valid*, written $\models \alpha$, iff α is valid in every EK-structure. Analogously, α is called *reflexive EK-valid*, written $\models_{ref} \alpha$, iff α is valid in every reflexive EK-structure.

We end this section with comparing our semantics to some common approaches in the literature, in particular that of Fischer Servi, Plotkin and Sterling, and Simpson. Since distributed knowledge is not treated there, we confine ourselves to D-free $\mathcal{L}_{DK}$ formulas for this comparison. The mentioned authors impose certain restrictions on their frames to deal with the interplay between $\Box$- and $\Diamond$-formulas. Our modal operators K_i are boxes, as is the operator D for distributed knowledge. This means that we work in multi-agent versions of the $\Box$-fragment, and therefore do not need these frame conditions.

Intuitionistic logic requires monotonicity, and in Fischer Servi [12], Plotkin and Sterling [10], and Simpson [13] this is done by building it into the truth definition. As shown in Jäger and Marti [7], both approaches lead to equivalent notions of validity. So our semantics for intuitionistic distributed knowledge builds on established semantic concepts.

The question now is whether there exist deductive systems that prove exactly the EK-valid and reflexive EK-valid formulas, respectively.

3 The Hilbert systems IDK and IDT

In the following we present a Hilbert-style axiomatization **IDK** of intuitionistic distributed knowledge and the system **IDT** for intuitionistic distributed knowledge with the truth property. There are also natural sequent calculi that prove the same formulas, but for the model construction and completeness proofs below it is irrelevant what kind of deductive system we use.

The **axioms of IDK** comprise the usual axioms of intuitionistic propositional logic, the normality axioms, sometimes also called K-axioms,

$$\mathsf{K}_i(\alpha \to \beta) \to (\mathsf{K}_i(\alpha) \to \mathsf{K}_i(\beta)) \qquad \textbf{(K)}$$

plus the D-axioms

$$\mathsf{D}(\alpha \to \beta) \to (\mathsf{D}(\alpha) \to \mathsf{D}(\beta)), \qquad \textbf{(D1)}$$

$$\mathsf{K}_i(\alpha) \to \mathsf{D}(\alpha), \qquad \textbf{(D2)}$$

always for all i with $1 \leq i \leq \ell$ and all α, β. Because of (**D1**) the operator D is normal, and in view of (**D2**) anything known by any agent is distributed knowledge.

The rules of inference of **IDK** are modus ponens and necessitation for the operators $\mathsf{K}_1, \ldots, \mathsf{K}_\ell$ and all α, β:

$$\frac{\alpha \qquad \alpha \to \beta}{\beta}\ \textbf{(MP)} \qquad \text{and} \qquad \frac{\alpha}{\mathsf{K}_i(\alpha)}\ \textbf{(NEC)}.$$

Because of (**D2**) and (**NEC**) the necessitation rule for D

$$\frac{\alpha}{\mathsf{D}(\alpha)}$$

is derivable in **IDK**. It is easy to see that all axioms of **IDK** are EK-valid; (**D1**) and (**D2**) follow directly from the intersection-interpretation of D. Furthermore, all EK-structures are clearly closed under the rules of inference of **IDK**.

The theory **IDT** is obtained from **IDK** by adding the claim that the distributed knowledge of α implies α, i.e.,

$$\mathsf{D}(\alpha) \to \alpha \tag{\textbf{T}}$$

for all α. In view of (**D2**) this implies the truth property $\mathsf{K}_i(\alpha) \to \alpha$ for all operators K_i.

Those EK-structures of order ℓ in which all (**T**)-axioms are valid are called *(**T**)-models*. The intended structures for **IDT** are reflexive EK-structures, and (**T**) is obviously valid in those. However, there are non-reflexive EK-structures of order ℓ in which all (**T**)-axioms are valid.

Let **ID•** be one of the theories **IDK** or **IDT**. We write $\mathbf{ID\bullet} \vdash \alpha$ to state that α is provable in the theory **ID•** in the usual sense. The following soundness theorem is then straightforwardly proved by induction on the length of the derivations.

Theorem 3.1 (Soundness) *For all α we have:*

(i) $\mathbf{IDK} \vdash \alpha \quad \Longrightarrow \quad \models \alpha.$

(ii) $\mathbf{IDT} \vdash \alpha \quad \Longrightarrow \quad \models_{ref} \alpha.$

As mentioned above, there exist non-reflexive (**T**)-models. Nevertheless, validity of (**T**) in EK-structures is closely related to reflexivity, as show in the following lemma. First a definition.

Definition 3.2 Let $\mathfrak{M} = (W, \preceq, R_1, \ldots, R_\ell, V)$ be an EK-structure of order ℓ. The *reflexive extension* $\overline{\mathfrak{M}}$ of $\mathfrak{M}$ is defined to be the structure

$$(W, \preceq \overline{R}_1, \ldots, \overline{R}_\ell, V),$$

where (for $1 \leq i \leq \ell$) the relation $\overline{R}_i$ is defined to be the reflexive closure of R_i, i.e., $\overline{R}_i := R_i \cup \{(a, a) : a \in W\}$.

It is an easy observation that any (**T**)-model can be extendend to a reflexive EK-structure of the same order.

Lemma 3.3 *If $\mathfrak{M} = (W, \preceq, R_1, \ldots, R_\ell, V)$ is a (**T**)-model, then $\overline{\mathfrak{M}}$ is a reflexive EK-structure; for all $a \in W$ and all α we have that*

$$(\overline{\mathfrak{M}}, a) \models \alpha \quad \Longleftrightarrow \quad (\mathfrak{M}, a) \models \alpha.$$

Proof. The reflexivity of $\overline{\mathfrak{M}}$ is clear. The second part is proved by induction on α. If α is the logical constant $\bot$ or an atomic proposition, the assertion is obvious; if α is a disjunction, a conjunction, or an implication it follows directly from the induction hypothesis. Hence we can concentrate on the cases that α is of the form $\mathsf{K}_i(\beta)$ or $\mathsf{D}(\beta)$.

(i) Let α be the formula $\mathsf{K}_i(\beta)$. The direction from left to right is evident. So asssume $(\mathfrak{M}, a) \models \mathsf{K}_i(\beta)$, from which we obtain

$$(\mathfrak{M}, b) \models \beta \quad \text{for all } b \in R_i[a].$$

$\mathfrak{M}$ is a (**T**)-model, thus $\mathsf{K}_i(\beta) \to \beta$ is valid in $\mathfrak{M}$ and our assumption also yields $(\mathfrak{M}, a) \models \beta$. From the induction hypothesis we obtain $(\overline{\mathfrak{M}}, c) \models \beta$ for all $c \in R_i[a] \cup \{a\} = \overline{R}_i[a]$. Therefore, $(\overline{\mathfrak{M}}, a) \models \mathsf{K}_i(\beta)$.

(ii) Let α be the formula $\mathsf{D}(\beta)$. The direction from left to right is evident again. To show the converse direction, let $(\mathfrak{M}, a) \models \mathsf{D}(\beta)$. Hence we have

$$(\mathfrak{M}, b) \models \beta \quad \text{for all } b \in \bigcap_{i=1}^{\ell} R_i[a].$$

Since $\mathfrak{M}$ is a (**T**)-model, we also have $(\mathfrak{M}, a) \models \beta$. Hence the induction hypothesis implies $(\overline{\mathfrak{M}}, c) \models \beta$ for all $c \in \bigcap_{i=1}^{\ell} R_i[a] \cup \{a\} = \bigcap_{i=1}^{\ell} \overline{R}_i[a]$. This is what we had to show. □

4 Pseudo-validity

Now we build up some machinery that will lead to the canonical models and the completeness proofs for the systems **IDK** and **IDT** in the next section. What we do here is motivated by the approach presented in Fagin, Halpern, and Vardi [3] and Wang and Ågotnes [15]. However, our version is a significant simplification, tailored for the treatment of **IDK** and **IDT**.

The idea is to introduce the notion of pseudo-validity. In doing that, we interpret the formulas in EK-structures of order $(\ell+1)$ where the operator D is interpreted by the additional binary accessibility relation $R_{\ell+1}$. Afterwards we will extend these EK-structures of order $(\ell+1)$ to strict EK-structures of order $(\ell+1)$ and then collapse these strict EK-structures of order $(\ell+1)$ to EK-structures of order ℓ, suitable for our purpose.

Definition 4.1 Given an EK-structure $\mathfrak{M} = (W, \preceq, R_1, \dots, R_{\ell+1}, V)$ of order $(\ell+1)$, the set $\|\alpha\|^{ps}_{\mathfrak{M}}$ of worlds pseudo-satisfying α is inductively defined as follows: Clauses (1) to (6) are as in Definition 2.2, but clause (7) is replaced by

(7') $\|\mathsf{D}(\alpha)\|^{ps}_{\mathfrak{M}} := \{a \in W : R_{\ell+1}[a] \subseteq \|\alpha\|^{ps}_{\mathfrak{M}}\}$.

As can be seen by a trivial induction on α also this assignment of sets of worlds to formulas is monotone.

Lemma 4.2 *For every EK-structure $\mathfrak{M} = (W, \preceq, R_1, \dots, R_{\ell+1}, V)$ of order $(\ell+1)$, all elements $a, b \in W$, and all α we have that*

$$a \preceq b \text{ and } a \in \|\alpha\|^{ps}_{\mathfrak{M}} \quad \Longrightarrow \quad b \in \|\alpha\|^{ps}_{\mathfrak{M}}.$$

We call $\|\alpha\|^{ps}_{\mathfrak{M}}$ the *pseudo-value* of α in $\mathfrak{M}$ and often write $(\mathfrak{M}, a) \models^{ps} \alpha$ instead of $a \in \|\alpha\|^{ps}_{\mathfrak{M}}$. In addition, α is *pseudo-valid in the EK-structure* $\mathfrak{M}$ of order $(\ell+1)$, written $\mathfrak{M} \models^{ps} \alpha$, iff $(\mathfrak{M}, a) \models^{ps} \alpha$ for all worlds a of $\mathfrak{M}$.

Since the operator D is interpreted by the relation $R_{\ell+1}$ in an EK-structure $\mathfrak{M} = (W, \preceq, R_1, \dots, R_{\ell+1}, V)$ of order $(\ell+1)$, the axioms (**D2**) are not necessarily pseudo-valid in $\mathfrak{M}$. Those EK-structures of order $(\ell+1)$ in which all (**D2**)-axioms are pseudo-valid are called *(**D2**)-pseudo-models*. An EK-structure $\mathfrak{M}$ of order $(\ell+1)$ is a *(**D2T**)-pseudo-model* iff all (**D2**)-axioms and all (**T**)-axioms are pseudo-valid in $\mathfrak{M}$.

Of course, for every EK-structure $\mathfrak{M}$ of order ℓ there is an EK-structure $\mathfrak{M}'$ of order $(\ell+1)$ such that validity in $\mathfrak{M}$ is equivalent to pseudo-validity in $\mathfrak{M}'$. The following lemma is an immediate consequence of Definition 2.2 and Definition 4.1.

Lemma 4.3 *Let $\mathfrak{M} = (W, \preceq, R_1, \ldots, R_\ell, V)$ be an EK-structure of order ℓ and define*

$$\mathfrak{M}' \ := \ (W, \preceq, R_1, \ldots, R_\ell, \bigcap_{i=1}^{\ell} R_i, V).$$

Then $\mathfrak{M}'$ is an EK-structure of order $(\ell+1)$ and for all $a \in W$ and all α we have that

$$(\mathfrak{M}, a) \models \alpha \quad \Longleftrightarrow \quad (\mathfrak{M}', a) \models^{ps} \alpha.$$

*In particular, $\mathfrak{M}'$ is a (**D2**)-pseudo-model and if $\mathfrak{M}$ is a (**T**)-model, then $\mathfrak{M}'$ is a (**D2T**)-pseudo-model.*

Our next step is to transform a given EK-structure of order $(\ell+1)$ into what we call its strict extension. The purpose of this extension is to enforce a well-controlled behavior of the intersection of the accessibility relations. From now on we write I for the set $\{1, \ldots, (\ell+1)\}$.

Definition 4.4 [Strict extension] Given an EK-structure

$$\mathfrak{M} = (W, \preceq, R_1, \ldots, R_{\ell+1}, V)$$

of order $(\ell+1)$, its *strict extension* is defined to be the structure

$$\mathfrak{M}^\sharp \ = \ (W^\sharp, \preceq^\sharp, R_1^\sharp, \ldots, R_{\ell+1}^\sharp, V^\sharp),$$

where we set:

($\sharp$1) $W^\sharp \ := \ W \times I$,

($\sharp$2) $\preceq^\sharp \ := \ \{((a,i),(b,j)) : a \preceq b \text{ and } i,j \in I\}$,

($\sharp$3) $R_i^\sharp \ := \ \{((a,j),(b,i)) : (a,b) \in R_i \text{ and } j \in I\}$ for any $i \in I$,

($\sharp$4) $V^\sharp((a,i)) \ := \ V(a)$ for any $(a,i) \in W^\sharp$.

It is obvious that $\mathfrak{M}^\sharp$ is an EK-structure of order $(\ell+1)$. Further properties of strict extensions are summarized in the following lemma whose proof is obvious.

Lemma 4.5 *Let $\mathfrak{M} = (W, \preceq, R_1, \ldots, R_{\ell+1}, V)$ be an EK-structure of order $(\ell+1)$. Then we have:*

(i) *If i and j are different elements of I, then $R_i^\sharp[(a,k)] \cap R_j^\sharp[(a,k)] = \emptyset$ for any $(a,k) \in W^\sharp$.*

(ii) *$\bigcap_{i=1}^{\ell} R_i^\sharp[(a,k)] = \emptyset$ for any $(a,k) \in W^\sharp$.*

Proof. The second assertion is an immediate consequence of the first since we deal with at least two agents. The first assertion follows from (♯3), which claims that all elements of $W^\sharp$ accessible from (a, k) via $R_i^\sharp$ are of the form (b, i) and those accessible from (a, k) via $R_j^\sharp$ are of the form (c, j). □

The following lemma is important and shows that the strict extension of an EK-structure of order $(\ell+1)$ does not affect the class of pseudo-valid formulas.

Lemma 4.6 *Let* $\mathfrak{M} = (W, \preceq, R_1, \ldots, R_{\ell+1}, V)$ *be an EK-structure of order* $(\ell+1)$. *Then we have for all* $(a, i) \in W^\sharp$ *and all* α *that*

$$(\mathfrak{M}, a) \models^{ps} \alpha \quad \Longleftrightarrow \quad (\mathfrak{M}^\sharp, (a, i)) \models^{ps} \alpha.$$

Proof. We show this claim by induction on the structure of α and distinguish the following cases.

(i) α is the logical constant $\bot$ or an atomic proposition. Then the situation is clear.

(ii) α is a disjunction or a conjunction. Then we simply have to apply the induction hypothesis.

(iii) α is of the form $\beta \to \gamma$. To show the direction from left to right we assume $(\mathfrak{M}, a) \models^{ps} \beta \to \gamma$ and thus have

$$(\mathfrak{M}, b) \models^{ps} \beta \quad \Longrightarrow \quad (\mathfrak{M}, b) \models^{ps} \gamma \quad \text{for all } b \text{ such that } a \preceq b. \tag{1}$$

In order to prove $(\mathfrak{M}^\sharp, (a, i)) \models^{ps} \beta \to \gamma$, we pick an arbitrary $(c, j) \in W^\sharp$ for which $(a, i) \preceq^\sharp (c, j)$ and $(\mathfrak{M}^\sharp, (c, j)) \models^{ps} \beta$. By the induction hypothesis we obtain $(\mathfrak{M}, c) \models^{ps} \beta$, and in view of the definition of $\preceq^\sharp$ we also have $a \preceq c$. Hence (1) gives us $(\mathfrak{M}, c) \models^{ps} \gamma$, and a further application of the induction hypothesis yields $(\mathfrak{M}^\sharp, (c, j)) \models^{ps} \gamma$, as we had to show. The proof of the converse directions follows exactly the same pattern.

(iv) α is of the form $\mathsf{K}_j(\beta)$. For establishing the direction from left to right assume $(\mathfrak{M}, a) \models^{ps} \mathsf{K}_j(\beta)$, yielding that

$$(\mathfrak{M}, b) \models^{ps} \beta \quad \text{for all } b \in R_j[a]. \tag{2}$$

Now we pick an arbitrary element (c, k) of $R_j^\sharp[(a, i)]$. According to the definition of $R_j^\sharp$ this implies that $c \in R_j[a]$, and in view of (2), we thus obtain $(\mathfrak{M}, c) \models^{ps} \beta$. Now we can apply the induction hypothesis and have $(\mathfrak{M}^\sharp, (c, k)) \models^{ps} \beta$. Therefore, $(\mathfrak{M}^\sharp, (a, i)) \models^{ps} \mathsf{K}_j(\beta)$.

For the converse direction we proceed from $(\mathfrak{M}^\sharp, (a, i)) \models^{ps} \mathsf{K}_j(\beta)$, i.e. from

$$(\mathfrak{M}^\sharp, (b, j)) \models^{ps} \beta \quad \text{for all } (b, j) \in R_j^\sharp[(a, i)]. \tag{3}$$

Given any element c of $R_j[a]$, we obtain $(c, j) \in R_j^\sharp[(a, i)]$, thus that (3) implies $(\mathfrak{M}^\sharp, (c, j)) \models^{ps} \beta$. Applying the induction hypothesis then immediately leads to $(\mathfrak{M}, c) \models^{ps} \beta$. Hence we have $(\mathfrak{M}, a) \models^{ps} \mathsf{K}_j(\beta)$.

(v) α is of the form $\mathsf{D}(\beta)$. This case can be handled as the previous case since D is interpreted by the relations $R_{\ell+1}$ and $R^{\sharp}_{\ell+1}$, respectively. □

An immediate consequence of this lemma is that the property of being a (**D2**)-pseudo-model or a (**D2T**)-pseudo-model is inherited from an EK-structure $\mathfrak{M}$ to its strict extension $\mathfrak{M}^{\sharp}$.

Corollary 4.7 *If $\mathfrak{M}$ is a (**D2**)-pseudo-model, then $\mathfrak{M}^{\sharp}$ is a (**D2**)-pseudo-model as well; if $\mathfrak{M}$ is a (**D2T**)-pseudo-model, then also $\mathfrak{M}^{\sharp}$ is a (**D2T**)-pseudo-model.*

The strict extensions of (**D2**)-pseudo-models have a further property that will be needed in the proof of Lemma 4.10.

Lemma 4.8 *Let $\mathfrak{M} = (W, \preceq, R_1, \ldots, R_{\ell+1}, V)$ be a (**D2**)-pseudo-model and j one of the numbers $1, \ldots, \ell$. Then we have for all $(a, i), (b, k) \in W^{\sharp}$ and all α that*

$$\left.\begin{array}{l} (\mathfrak{M}^{\sharp}, (a,i)) \models^{ps} \mathsf{K}_j(\alpha) \ \textit{and} \\ (b,k) \in (R^{\sharp}_j \cup R^{\sharp}_{\ell+1})[(a,i)] \end{array}\right\} \implies (\mathfrak{M}^{\sharp}, (b,k)) \models^{ps} \alpha.$$

Proof. Since $\mathfrak{M}^{\sharp}$ is a (**D2**)-pseudo-model, $(\mathfrak{M}^{\sharp}, (a,i)) \models^{ps} \mathsf{D}(\alpha)$ follows from the assumption $(\mathfrak{M}^{\sharp}, (a,i)) \models^{ps} \mathsf{K}_j(\alpha)$. In this pseudo-model the operator D is interpreted by means of the accessibility relation $R^{\sharp}_{\ell+1}$, hence the conclusion is an immediate consequence. □

EK-structures of order $(\ell+1)$ provide only intermediate tools for the canonical model construction. In the end we are interested in EK-stuctures of order ℓ, and in order to build those, we now collapse EK-structures $\mathfrak{M}$ of order $(\ell+1)$ to so-called associated structures $\mathfrak{M}^{\star}$ of order ℓ, via their strict extensions $\mathfrak{M}^{\sharp}$.

Definition 4.9 [Associated structure] Given an EK-structure

$$\mathfrak{M} = (W, \preceq, R_1, \ldots, R_{\ell+1}, V)$$

of order $(\ell+1)$, the structure *associated with* $\mathfrak{M}$ is defined to be the structure

$$\mathfrak{M}^{\star} \ = \ (W^{\star}, \preceq^{\star}, R^{\star}_1, \ldots, R^{\star}_{\ell}, V^{\star}),$$

where we set:

($\star$1) $W^{\star} \ := \ W^{\sharp}, \quad \preceq^{\star} := \preceq^{\sharp}, \quad V^{\star} \ := \ V^{\sharp}$,

($\star$2) $R^{\star}_i \ := \ R^{\sharp}_i \cup R^{\sharp}_{\ell+1}$ for $i = 1, \ldots, \ell$.

It is clear that $\mathfrak{M}^{\star}$ is an EK-structure of order ℓ. The decisive property of this construction is that validity with respect to the structure associated with an EK-structure $\mathfrak{M}$ of order $(\ell+1)$ coincides with pseudo-validity with respect to its strict extension $\mathfrak{M}^{\sharp}$.

Lemma 4.10 *Given a (**D2**)-pseudo-model $\mathfrak{M} = (W, \preceq, R_1, \ldots, R_{\ell+1}, V)$, we have for all $(a,i) \in W^{\sharp}$ and all α that*

$$(\mathfrak{M}^{\star}, (a,i)) \models \alpha \iff (\mathfrak{M}^{\sharp}, (a,i)) \models^{ps} \alpha.$$

Proof. The proof of this equivalence is by induction on the structure of α. We distinguish the following cases:

(i) α is the logical constant $\bot$ or an atomic proposition. Then the claim follows immediately.

(ii) α is a disjunction, a conjunction, or an implication. Then we simply have to apply the induction hypothesis.

(iii) α is of the form $\mathsf{K}_j(\beta)$. In view of the definition of $R_j^\star$, the direction from left to right is obtained by a straightforward application of the induction hypothesis. For proving the converse direction, assume $(\mathfrak{M}^\sharp, (a,i)) \models^{ps} \mathsf{K}_j(\beta)$. However, then Lemma 4.8 implies $(\mathfrak{M}^\sharp, (b,k)) \models^{ps} \beta$ for all elements (b,k) of $(R^\sharp \cup R^\sharp_{\ell+1})[(a,i)] = R_j^\star[(a,i)]$. For all those (b,k) the induction hypothesis yields $(\mathfrak{M}^\star, (b,k)) \models \beta$, and thus we have $(\mathfrak{M}^\star, (a,i)) \models \mathsf{K}_j(\beta)$.

(iv) α is of the form $\mathsf{D}(\beta)$. Now we observe that

$$\begin{aligned}\textstyle\bigcap_{j=1}^{\ell} R_j^\star[(a,i)] &= \textstyle\bigcap_{j=1}^{\ell}(R_j^\sharp \cup R^\sharp_{\ell+1})[(a,i)] \\ &= (\textstyle\bigcap_{j=1}^{\ell} R_j^\sharp[(a,i)]) \cup R^\sharp_{\ell+1}[(a,i)] \;=\; R^\sharp_{\ell+1}[(a,i)],\end{aligned}$$

where the last equality follows from Lemma 4.5. Hence the operator D is interpreted in $\mathfrak{M}^\star$ as in $\mathfrak{M}^\sharp$, and our assertion is immediate from the induction hypothesis. □

We now come to the main theorem of this section. It is an immediate consequence of Lemma 4.6 and the previous lemma.

Theorem 4.11 *If $\mathfrak{M} = (W, \preceq, R_1, \ldots, R_{\ell+1}, V)$ is a **(D2)**-pseudo-model, then we have for all $(a,i) \in W^\sharp$ and all α that*

$$(\mathfrak{M}, a) \models^{ps} \alpha \quad \Longleftrightarrow \quad (\mathfrak{M}^\star, (a,i)) \models \alpha.$$

*In particular, if $\mathfrak{M}$ is a **(D2T)**-pseudo-model, then $\mathfrak{M}^\star$ is a **(T)**-model.*

5 Prime sets and completeness

Now we introduce syntactic EK-structures that are based on so-called prime sets. This is a standard approach to proving completeness of intuitionistic modal systems also used in, for example, Fischer Servi [12] and Simpson [13].

Recall that **ID•** stands for one of the theories **IDK** or **IDT**. If P is a set of formulas, then we write $P \vdash_{\mathbf{ID\bullet}} \beta$ iff there exist finitely many formulas $\gamma_1, \ldots, \gamma_n \in P$ such that $\mathbf{ID\bullet} \vdash (\gamma_1 \wedge \ldots \wedge \gamma_n) \to \beta$.

Definition 5.1 [Prime set] A set P of formulas is called **ID•**-*prime* iff it satisfies the following conditions:

(P.1) $P \vdash_{\mathbf{ID\bullet}} \beta \quad \Longrightarrow \quad \beta \in P$,

(P.2) $\beta \vee \gamma \in P \quad \Longrightarrow \quad \beta \in P$ or $\gamma \in P$,

(P.3) $\bot \notin P$.

The following prime lemma describes a crucial property of prime sets. Its proof is standard and similar to that in Jäger and Marti [7].

Lemma 5.2 (Prime lemma) *Suppose that $P \nvdash_{\mathbf{ID}\bullet} \alpha$ for some set of formulas P and some α. Then there exists an* **ID•***-prime set Q such that $P \subseteq Q$ and $Q \nvdash_{\mathbf{ID}\bullet} \alpha$.*

Relative to the theory **ID•** we now introduce the canonical EK-structure $\mathfrak{C}$. In order to keep the notation readable, we refrain from explicitly mentioning **ID•** (for example as sub- or superscript), but it should always be clear from the context to which theory we refer. If N is a set of formulas and H one of the modal operators $\mathsf{K}_1, \ldots, \mathsf{K}_\ell$, or D, we write $\mathsf{H}^{-1}(N)$ for $\{\gamma : \mathsf{H}(\gamma) \in N\}$.

Definition 5.3 [Canonical structure] The *canonical structure for* **ID•** is the $(\ell + 4)$ tuple

$$\mathfrak{C} \; = \; (\mathcal{W}, \subseteq, \mathcal{R}_1, \ldots, \mathcal{R}_{\ell+1}, \mathcal{V}),$$

where we define:

(Can1) $\mathcal{W} \; := \; \{P : P \text{ is an } \mathbf{ID}\bullet\text{-prime set of formulas}\}$,

(Can2) For any $i = 1, \ldots, \ell$: $\mathcal{R}_i \; := \; \{(P, Q) \in \mathcal{W} \times \mathcal{W} : \mathsf{K}_i^{-1}(P) \subseteq Q\}$,

(Can3) $\mathcal{R}_{\ell+1} \; := \; \{(P, Q) \in \mathcal{W} \times \mathcal{W} : \mathsf{D}^{-1}(P) \subseteq Q\}$,

(Can4) $\mathcal{V}$ is the function from $\mathcal{W}$ to the power set of *PROP* given by

$$\mathcal{V}(P) \; := \; \{p : p \in P\}.$$

It is evident that $\mathfrak{C}$ is an EK-structure of order $(\ell + 1)$. All further relevant properties follow more or less directly from the following truth property.

Lemma 5.4 (Truth lemma) *Let $\mathfrak{C} = (\mathcal{W}, \subseteq, \mathcal{R}_1, \ldots, \mathcal{R}_{\ell+1}, \mathcal{V})$ be the canonical structure for* **ID•***. Then we have for all α and all $P \in \mathcal{W}$ that*

$$\alpha \in P \quad \Longleftrightarrow \quad (\mathfrak{C}, P) \models^{ps} \alpha.$$

Proof. We establish this equivalence by induction on the structure of α and distinguish the following cases.

(i) It trivially holds in case that α is the logical constant $\bot$ or an atomic proposition.

(ii) If α is a disjunction or a conjunction it follows from the induction hypothesis and the properties of **ID•**-prime sets.

(iii) α is of the form $\beta_1 \to \beta_2$. We first assume that

$$\beta_1 \to \beta_2 \in P, \qquad P \subseteq Q \in \mathcal{W}, \qquad \text{and} \qquad (\mathfrak{C}, Q) \models^{ps} \beta_1.$$

Then we have $\beta_1 \to \beta_2 \in Q$ and (by the induction hypothesis) $\beta_1 \in Q$. Since Q is deductively closed, this yields $\beta_2 \in Q$ and thus again by the induction hypothesis that $(\mathfrak{C}, Q) \models^{ps} \beta_2$. Q has been an arbitrary superset of P within $\mathcal{W}$, and thus we conclude $(\mathfrak{C}, P) \models^{ps} \beta_1 \to \beta_2$.

Now assume $(\mathfrak{C}, P) \models^{ps} \beta_1 \rightarrow \beta_2$ and $\beta_1 \rightarrow \beta_2 \notin P$. Since P is deductively closed, we have $P \cup \{\beta_1\} \nvdash_{\mathbf{ID\bullet}} \beta_2$. By the prime lemma there exists a $Q \in \mathcal{W}$ such that

$$P \cup \{\beta_1\} \subseteq Q \qquad \text{and} \qquad Q \nvdash_{\mathbf{ID\bullet}} \beta_2, \quad \text{hence} \quad \beta_2 \notin Q.$$

Together with the induction hypothesis we thus obtain

$$(\mathfrak{C}, Q) \models^{ps} \beta_1 \qquad \text{and} \qquad (\mathfrak{C}, Q) \not\models^{ps} \beta_2.$$

Since $P \subseteq Q$, this contradicts $(\mathfrak{C}, P) \models^{ps} \beta_1 \rightarrow \beta_2$.

(iv) α is of the form $\mathsf{K}_i(\beta)$. For the direction from left to right assume

$$\mathsf{K}_i(\beta) \in P \qquad \text{and} \qquad \mathsf{K}_i^{-1}(P) \subseteq Q$$

for an arbitrary $Q \in \mathcal{W}$. This implies $\beta \in Q$, and in view of the induction hypothesis we thus have $(\mathfrak{C}, Q) \models^{ps} \beta$. Therefore, $(\mathfrak{C}, P) \models^{ps} \mathsf{K}_i(\beta)$.

For the converse direction we assume $(\mathfrak{C}, P) \models^{ps} \mathsf{K}_i(\beta)$. We first claim that

$$\mathsf{K}_i^{-1}(P) \vdash_{\mathbf{ID\bullet}} \beta. \qquad (*)$$

To establish this claim, assume for contradiction that $\mathsf{K}_i^{-1}(P) \nvdash_{\mathbf{ID\bullet}} \beta$. According to the prime lemma we thus have a $Q \in \mathcal{W}$ such that $\mathsf{K}_i^{-1}(P) \subseteq Q$ and $Q \nvdash_{\mathbf{ID\bullet}} \beta$. In particular, $\beta \notin Q$. By the induction hypothesis, this yields $(\mathfrak{C}, Q) \not\models^{ps} \beta$; a contradiction to $(\mathfrak{C}, P) \models^{ps} \mathsf{K}_i(\beta)$ and $\mathsf{K}_i^{-1}(P) \subseteq Q$.

From (*) we conclude that there are $\gamma_1, \ldots, \gamma_n \in \mathsf{K}_i^{-1}(P)$ such that

$$\mathbf{ID\bullet} \vdash (\gamma_1 \wedge \ldots \wedge \gamma_n) \rightarrow \beta.$$

Thus we also have

$$\mathbf{ID\bullet} \vdash (\mathsf{K}_i(\gamma_1) \wedge \ldots \wedge \mathsf{K}_i(\gamma_n)) \rightarrow \mathsf{K}_i(\beta),$$

with $\mathsf{K}_i(\gamma_1), \ldots, \mathsf{K}_i(\gamma_n) \in P$, implying that $P \vdash_{\mathbf{ID\bullet}} \mathsf{K}_i(\beta)$. Hence $\mathsf{K}_i(\beta) \in P$ since P is deductively closed.

(v) α is of the form $\mathsf{D}(\beta)$. Because of the pseudo-validity interpretation of D, this case is treated exactly as the previous cases. □

Corollary 5.5

(i) *If $\mathfrak{C}$ is the canonical structure for* **IDK**, *then $\mathfrak{C}$ is a (**D2**)-pseudo-model.*

(ii) *If $\mathfrak{C}$ is the canonical structure for* **IDT**, *then $\mathfrak{C}$ is a (**D2T**)-pseudo-model.*

Proof. We only have to remember that an **IDK**-prime set of formulas P is deductively closed with respect to derivability in **IDK** and, therefore, contains $\mathsf{K}_i(\alpha) \rightarrow \mathsf{D}(\alpha)$ for all $i = 1, \ldots, \ell$ and all α. Analogously, any **IDT**-prime set of formulas Q contains, in addition, the formulas $\mathsf{D}(\alpha) \rightarrow \alpha$ for any α. Thus the truth lemma implies our assertions. □

Now the stage is set, and combining what we have obtained so far, we can state the following first main result.

Theorem 5.6 *Let* $\mathfrak{C} = (\mathcal{W}, \subseteq, \mathcal{R}_1, \ldots, \mathcal{R}_{\ell+1}, \mathcal{V})$ *be the canonical structure for* **ID•** *and* $\mathfrak{C}^\star$ *the EK-structure of order* ℓ *associated with* $\mathfrak{C}$*. Then we have for all* **ID•***-prime sets of formulas* P*, all* α*, and all* $i = 1, \ldots, \ell$ *that*

$$\alpha \in P \iff (\mathfrak{C}^\star, (P, i)) \models \alpha.$$

Proof. In view of the truth lemma and Lemma 4.6 we have

$$\alpha \in P \iff (\mathfrak{C}, P) \models^{ps} \alpha \iff (\mathfrak{C}^\sharp, (P, i)) \models^{ps} \alpha$$

for the strict extension $\mathfrak{C}^\sharp$ of $\mathfrak{C}$. Furthermore, $\mathfrak{C}^\sharp$ is a (**D2**)-pseudo-model according to the previous corollary. Hence we can apply Lemma 4.10 and see that

$$(\mathfrak{C}^\star, (P, i)) \models \alpha \iff (\mathfrak{C}^\sharp, (P, i)) \models^{ps} \alpha.$$

Therefore, we have what we want. □

Theorem 5.7 (Completeness) *For all* α *we have:*

(i) $\models \alpha \implies$ **IDK** $\vdash \alpha$.

(ii) $\models_{ref} \alpha \implies$ **IDT** $\vdash \alpha$.

Proof. For the first assertion, assume $\models \alpha$ and **IDK** $\nvdash \alpha$. Note that then the prime lemma thus tells us that there exists an **IDK**-prime set P for which $P \nvdash_{\mathbf{IDK}} \alpha$. Hence $\alpha \notin P$. Consider the canonical structure $\mathfrak{C}$ for **IDK** and the EK-structure $\mathfrak{C}^\star$ associated with $\mathfrak{C}$. According to Theorem 5.6 we have $(\mathfrak{C}^\star, (P, i)) \not\models \alpha$ for $i = 1, \ldots, \ell$. This is a contradiction to $\models \alpha$.

We come to the second assertion. Now we assume $\models_{ref} \alpha$ and **IDT** $\nvdash \alpha$. In this case the prime lemma gives us an **IDT**-prime set Q for which $Q \nvdash_{\mathbf{IDK}} \alpha$ and, consequently, $\alpha \notin Q$. Now we work with the canonical structure $\mathfrak{C}$ for **IDT** and the EK-structure $\mathfrak{C}^\star$ associated with $\mathfrak{C}$. We see that $\mathfrak{C}$ is a (**D2T**)-pseudomodel by Corollary 5.5 and, consequently, $\mathfrak{C}^\star$ is a (**T**)-model by Theorem 4.11. In view of Theorem 5.6 we also have $(\mathfrak{C}^\star, (Q, i)) \not\models \alpha$ for any $i = 1, \ldots, \ell$. It only remains to move to the reflexive extension $\overline{\mathfrak{C}^\star}$ of $\mathfrak{C}^\star$ and to apply Lemma 3.3. It follows that $(\overline{\mathfrak{C}^\star}, (Q, i)) \not\models \alpha$. Since $\overline{\mathfrak{C}^\star}$ is reflexive, this is a contradiction to $\models_{ref} \alpha$. □

Together with Theorem 3.1 we thus have that **IDK** and **IDT** are sound and complete formalizations of intuitionistic distributed knowledge. In work in progress extensions of these results to systems including positive introspection and common knowledge as well as the question of the finite model property will be considered.

As in classical epistemic logic, positive introspection can be formalized by the axioms $\mathsf{D}(\alpha) \to \mathsf{D}(\mathsf{D}(\alpha))$ and $\mathsf{K}_i(\alpha) \to \mathsf{K}_i(\mathsf{K}_i(\alpha))$, where the latter semantically corresponds to the transitivity of the relations R_i. We think that an approach to completeness of intuitionistic **S4** with distributed knowledge can be done along the lines of the constructions in Fagin, Halpern, and Vardi [3] and Wang and Ågotnes [15]. Negative introspection is less straightforward, as intuitionistic **S5** is typically formulated by making use of the box and the diamond operator; see, e.g., Fischer Servi [12] and Simpson [13], and in intuitionistic modal logic $\Diamond(\alpha)$ is not equivalent to $\neg\Box(\neg\alpha)$. In our present framework the operators $\mathsf{K}_1, \ldots, \mathsf{K}_\ell$ correspond to boxes. But the corresponding diamonds are not available.

References

[1] Artemov, S. and T. Protopopescu, *Intuitionistic Epistemic Logic*, arXiv:1406.1582v4 [math.LO] (2014).

[2] Fagin, R., J. Halpern, Y. Moses and M. Vardi, "Reasoning about Knowledge," MIT Press, 1995.

[3] Fagin, R., J. Halpern and M. Vardi, *What can machines know? On the properties of knowledge in distributed systems*, Journal of the Association for Computing Machinery **39** (1992), pp. 328–376.

[4] Gerbrandy, J., *Distributed knowledge*, in: J. Hulstijn and A. Nijholt, editors, *Formal Semantics and Pragmatics of Dialogue, Twendia '98* (1998), pp. 111–123.

[5] Hakli, R. and S. Negri, *Proof theory for distributed knowledge*, in: F. Sadri and K. Satoh, editors, *Computational Logic in Multi-Agent Systems*, Lecture Notes in Artificial Intelligence **5056** (2008), pp. 100–116.

[6] Hirai, Y., *An intuitionistic epistemic logic for sequential consistency on shared memory*, in: E. Clarke and A. Voronkov, editors, *Logic for Programming, Artificial Intelligence, and Reasoning*, Lecture Notes in Computer Science **6355** (2010), pp. 272–289.

[7] Jäger, G. and M. Marti, *Intuitionistic common knowledge or belief*, to appear in Journal of Applied Logic.

[8] Krupski, V. and A. Yatmanov, *Sequent Calculus for Intuitionistic Epistemic Logic*, arXiv:1508.07851v1 [cs.LO] (2015).

[9] Meyer, J.-J. C. and W. van der Hoek, "Epistemic Logic for AI and Computer Science," Cambridge Tracts in Theoretical Computer Science **41**, Cambridge University Press, 2004.

[10] Plotkin, G. and C. Sterling, *A framework for intuitionistic modal logic,*, Theoretical Aspects of Reasoning About Knowledge (J.Y.Halpern, ed.), Morgan Kaufmann Publishers (1986), pp. 399–406.

[11] Proietti, C., *Intuitionistic epistemic logic, Kripke models and Fitch's paradox*, Journal of Philosophical Logic **41** (2012), pp. 877–900.

[12] Servi, G. F., *Axiomatizations for some intuitionistic modal logics*, Rendiconti del Seminario Matematico Università Politecnico di Torino **42** (1984), pp. 179–194.

[13] Simpson, A., "The Proof Theory and Semantics of Intuitionistic Modal Logic (revised version)," Ph.D. thesis, University of Edinburgh (1994).

[14] Suzuki, N., *Semantics for intuitionistic epistemic logics of shallow depths for game theory*, Economic Theory **53** (2012), pp. 85–110.

[15] Wáng, Y. and T. Ågotnes, *Public announcement logic with distributed knowledge*, in: H. van Ditmarsch, J. Lang and S. Ju, editors, *LORI 2011*, Lecture Notes in Artificial Intelligence **6953** (2011), pp. 328–341.

[16] Williamson, T., *On intuitionistic modal epistemic logic*, Journal of Philosophy **21** (1992), pp. 63–89.

Logics of Infinite Depth

Marcus Kracht

Fakultät Linguistik und Literaturwissenschaft
Universität Bielefeld
Postfach 10 01 31
33501 Bielefeld
Germany
marcus.kracht@uni-bielefeld.de

Abstract

Consider a definition of depth of a logic as the supremum of ordinal types of well-ordered descending chains. This extends the usual definition of codimension to infinite depths. Logics may either have no depth, or have countable depth in case a maximal well-ordered chain exists, or be of depth ω_1. We shall exhibit logics of all three types. We show in particular that many well-known systems, among them K, $\mathsf{K4}$, G, Grz and $\mathsf{S4}$, have depth ω_1. Basically, if a logic is the intersection of its splitting logics and has finite model property, then either the splitting logics have an infinite antichain (and the depth is therefore ω_1) or the splitting logics form a well-partial order whose supremum type is realised and therefore countable, though it may be different from the supremum type of the splitting logics alone.

Keywords: Lattices of Modal Logics, Well Partial Orders, Splittings

1 Introduction

The lattice Ext K is very complex. It contains continuously many complete logics. Moreover, completeness is a rather rare property. Early results by Blok revealed that complete logics above K are either intrinsically complete, that is, they are the only logics having the same class of Kripke-frames; or there are $2^{\aleph_0}$ logics with the same class of Kripke-frames. Most known systems are of the latter kind. The so-called *degree of incompleteness* is thus either 1 or of size continuum. The largest of these (which is the only complete logic in the spectrum) has $2^{\aleph_0}$ cocovers. So, order and chaos are quite close together.

In order to understand the structure of the lattice Ext K one may either study some special systems, or obtain some rough outline of the global structure. There are many ways to go. From the abovementioned results it emerged that the notion of a splittings is important for studying the structure of lattices of logics. However, even this structure is sometimes very complex and therefore some reduction in complexity may be useful. In [5] we have for example looked at groups of automorphisms of the lattices Ext L for various L. Obviously, the larger the group the more homogeneous and simpler the lattice. Alternatively, one may study some numerical invariants of logics such as depth. These topics are linked. Clearly, if Ext L admits an automorphism then that automorphism must leave the depth of logics invariant. However, there is a problem in that depth is typically a cardinal number and therefore of limited usefulness. Most interesting logics have infinite depth, and that depth is countable.

Thus we have decided to look at a definition of depth that yields an ordinal number instead. We define the (*ordinal*) *depth* of a logic L as the supremum of all κ such that there exists a downgoing chain of order type $\kappa + 1$ of logics above L such that $L = L_\kappa$. Here, a sequence $\langle L_\lambda : \lambda < \mu \rangle$ is a *downgoing chain* if for all $\lambda < \mu$ either (i) $\lambda = \lambda' + 1$ and L_λ is immediately below $L_{\lambda'}$, or (ii) λ is a limit ordinal and $L_\lambda = \bigsqcap_{\lambda' < \lambda} L_{\lambda'}$. In the finite case this equals the standard definition of depth or codimension. (L has codimension n in lattice theoretic terms iff there is a downgoing chain $L_0 \succ L_1 \succ \cdots \succ L_n = L$, starting at the top, that is, $L_0 = \mathsf{K} \oplus \bot$. This chain has order type $n+1$. By definition, L thus has depth n.) We shall see that there exist logics of varying infinite depth. For a start we note that a logic has depth 0 iff it is the inconsistent logic $\mathsf{K} \oplus \bot$. Obviously, the ordinal depth is invariant under automorphisms.

If the extension lattice of a logic has an antichain of size $\aleph_0$ then the order type can grow up to ω_1, the first uncountable ordinal. We establish this for a number of logics, including S4, G, K4 and K. If however the lattice has only finite antichains, chances are that it is a continuous lattice. Its depth is then a countable ordinal, and in that case the results by [3] can be applied to give exact bounds for the depth of the logic.

Notice that the order type of upgoing chains is generally different. For example, Ext K has no atoms, hence there is no well-ordered upgoing chain starting at K. The same applies to many other logics. [1]

2 Preliminaries

The lattice operations are denoted by $\sqcap$ and $\sqcup$ (and their infinitary versions by $\bigsqcap$ and $\bigsqcup$). x is a *lower cover* or *cocover* of y, in symbols $x \prec y$, if there is no z such that $x < z < y$. x has *dimension* n *over* y (and y has *codimension* n *under* x) if there is a finite chain $y = y_0 \prec y_1 \prec y_2 \prec \cdots \prec y_n = x$. In a modular lattice, any such chain has the same length, so the number does not depend on the choice of the sequence.

The lattice Ext K—in general any lattice Ext L of normal extensions of a modal logic L—is a locale, that is, a complete and distributive lattice that enjoys the following infinitary distributive law (see [6]).

$$L \sqcap \bigsqcup_{i \in I} M_i = \bigsqcup_{i \in I} (L \sqcap M_i) \tag{1}$$

[1] This paper is dedicated to the memory of Alexander Chagrov. I also wish to thank Stefan Geschke for his help.

L is called *continuous* if in addition

$$L \sqcup \bigsqcap_{i \in I} M_i = \bigsqcap_{i \in I} (L \sqcup M_i) \tag{2}$$

The law (2) is *not* valid in Ext K. However, for certain logics L the lattive Ext L is in fact continuous. An example is $\mathsf{S4.3}$. A logic L is $\bigsqcap$-*prime* if for every family of logics such that $\bigsqcap_{i \in I} M_i \leqslant L$ there is an $i \in I$ such that $M_i \leqslant L$. Dually for $\bigsqcup$-*prime*. L is $\bigsqcap$-*irreducible*, if for every family of logics such that $\bigsqcap_{i \in I} M_i = L$ there is an $i \in I$ such that $M_i = L$. Dually for $\bigsqcup$-irreducible. A logic is $\bigsqcup$-irreducible iff it is $\bigsqcup$-prime. Indeed, suppose that L is $\bigsqcup$-irreducible and let M_i, $i \in I$, be logics such that $\bigsqcup_{i \in I} M_i \geqslant L$. Then $L = L \sqcap \bigsqcup_{i \in I} M_i = \bigsqcup_{i \in I} L \sqcap M_i$, so there is an $i \in I$ such that $L = L \sqcap M_i$, which implies that $L \leqslant M_i$. Every $\bigsqcap$-prime logic is also $\bigsqcap$-irreducible, but the converse does not hold. This is because (2) fails to hold. (The logic of the one-point reflexive frame is a case in point.)

Prime logics are related to *splittings*. A *splitting* of a lattice $\mathcal{L} = \langle L, \sqcap, \sqcup \rangle$ is a disjoint sum $L = F + I$, where F is a principal filter and I a principal ideal of $\mathcal{L}$. Thus, $I = \downarrow L_1$ and $F = \uparrow L_2$ for some elements L_1 and L_2. It can be shown that L_1 is $\bigsqcap$-prime and L_2 is $\bigsqcup$-prime. L_2 is called the *splitting companion* of L_1 and is denoted by $\mathcal{L}/L_1$. Every $\bigsqcap$-prime logic has a unique splitting companion. Moreover, every $\bigsqcup$-prime logic is the splitting companion of some $\bigsqcap$-prime logic. This induces an order preserving bijection between $\bigsqcap$-prime and $\bigsqcup$-prime logics. In a continuous lattice this is automatically also an order preserving bijection between the $\bigsqcap$-irredicible and the $\bigsqcap$-irreducible elements.

If a logic is $\bigsqcap$-irreducible it is the theory of a subdirectly irreducible algebra; if L is in addition complete, it is the theory of a one-generated frame. (Recall that a frame $\langle W, R \rangle$ is one-generated iff there is a single point x such that $x\, R^* \, y$ for all $y \in W$, where R^* is the reflexive transitive closure of R.) Furthermore, [1] has established that L is $\bigsqcap$-prime in Ext K iff it is the logic of a one-generated finite, cycle-free frame. Thus, not every one-generated frame determines a splitting logic. This shows that Ext K is not continuous: for the logic of a one-generated finite frame is $\bigsqcap$-irreducible. However, K is the intersection of all its splitting logics. This is because if a formula has a model, it has a model on a finite cycle-free frame, by unravelling.

From this it follows that Ext K can have no atoms. For these must be $\bigsqcup$-irreducible, hence $\bigsqcup$-prime. And so they have the form Ext K/L' for some L'. L' must be a $\bigsqcap$-prime logic, and a minimal one. But it is the logic of a finite frame, and so there is a $\bigsqcap$-prime $L'' < L'$. A similar argument applies to Ext $\mathsf{K4}$, Ext $\mathsf{S4}$ and many other lattices. This explains why we focus here on downgoing chains rather than upgoing chains.

3 WPOs and their height

Take a poset $\mathcal{P} := \langle P, \leqslant \rangle$ that has no infinite descending chains. $\mathcal{P}$ is called a *well-partial order* (WPO) if it has no infinite antichains. A poset is a WPO iff every linear order extending it is a well-order. Denote by $o(\mathcal{P})$ the supremum of all well-orders on P that extend $\leqslant$. Further, by a result of [3], in a WPO, the supremum is actually realised, that is, there is an actual chain of order type $o(\mathcal{P})$ extending $\leqslant$.

A set $A \subseteq P$ is a *lower set* or *ideal* of $\mathcal{P}$ if for all $x \in A$ and $y \leqslant x$ also $y \in A$. The ideals form a distributive lattice. It is known that $\langle P, \leqslant\rangle$ is a WPO iff $\langle \mathcal{I}(\mathcal{P}), \subseteq\rangle$ is well-founded. And in that case, the height of the set $\mathcal{P}$ in this space of ideals is nothing but $o(\mathcal{P})$ ([7]).

We shall use this theory to establish some bounds on chains in lattices of modal logics. In a locale $\mathcal{L}$ (which includes all lattices of the form Ext L for some modal logic L), every logic is the intersection of $\sqcap$-irreducible elements. So, given a logic L, let Irr_L be the set of $\sqcap$-irreducible logics $\geqslant L$. Irr_L forms a poset. However, note that we are looking here at the reverse inclusion of logics. Thus, to apply the theory of WPOs notice that $L \leqslant L'$ in the notation of WPOs is the same as $L \supseteq L'$. In most cases of interest this order has no infinite descending chains, which translates into an absence of ascending chains of irreducible logics. However, at the end of this paper we shall exhibit a logic with a linear lattice of extensions with an ascending chain of irreducible (even splitting) extensions.

Now, for $L' \supseteq L$ let $h(L') \subseteq \mathrm{Irr}_L$ be the set of all $\sqcap$-irreducible logics containing L'; by definition, this set does not depend on the particular base logic L. h is a map from Ext L into $\wp(\mathrm{Irr}_L)$. Its image are lower closed sets in the poset order; equivantly, they are upper closed in the lattice or containment order. However, not all lower closed sets are of the form $h(L')$. This is the case only if Ext L is continuous. (See Chapter 7 of [6] for background. I shall make this paper self-contained, not assuming the heavy structure theory of locales.) In that case, a downgoing chain of logics $\langle L_\lambda : \lambda < \mu\rangle$ thanslates into a well-ordered sequence $\langle h(L_\lambda) : \lambda < \mu\rangle$ of closed sets of Irr_L (ordered by set inclusion). Furthermore, $h(L_{\lambda+1}) - h(L_\lambda) = \{M_\lambda\}$ for a single $\sqcap$-irreducible logic M_λ as well as

$$h(L_\lambda) = \bigcup_{\mu<\lambda} h(L_\mu) \tag{3}$$

for a limit ordinal λ. Put $L_\kappa \leqslant L_\mu$ iff $\kappa \leqslant \mu$. This order is clearly compatible with the poset order. Moreover, as all elements of Irr_L are $\sqcap$-prime, every element is of the form M_λ for some λ. Thus, a downgoing chain of logics translates into a well-ordering of Irr_L that is consistent with the partial ordering on Irr_L inherited from the lattice. This explains the connection with the theory of [7]. The depth of L is then nothing but $o(\mathrm{Irr}_L)$.

One can establish more than that, however. It is not always necessary to assume that every $\sqcap$-irreducible logic is also $\sqcap$-prime. However, if that is so, not every linear order on the irreducibles defines a downgoing chain of logics. We can conclude the following general fact.

Theorem 3.1 *Let L be a logic and $\langle \mathrm{Irr}_L, \leqslant\rangle$ be the poset of its $\sqcap$-irreducible logics. Then if $\langle \mathrm{Irr}_L, \leqslant\rangle$ is a WPO, the depth of L is $\leqslant o(\langle \mathrm{Irr}_L, \leqslant\rangle)$. Moreover, if every element of Irr_L is also $\sqcap$-prime, equality holds.*

Proof. The second claim has been proved already. So we concentrate on the latter. First of all, L has a depth, and so there is a downgoing chain $\langle L_\alpha : \alpha < \kappa\rangle$. This chain translates into an increasing sequence of subsets of Irr_L. As before, $h(L_{\lambda+1}) = h(L_\lambda) \cup \{M_\lambda\}$ for some $\bigsqcap$-irreducible M_λ. However, it is not necessarily the case that

$$h(L_\lambda) = \bigcup_{\mu<\lambda} h(L_\mu) \tag{4}$$

For if L' is $\bigsqcap$-irreducible but not $\bigsqcap$-prime it may happen that $L_\mu \nleqslant L'$ for all L_μ with $\mu < \lambda$ but $L_\lambda \leqslant L'$. The increasing sequence does therefore not define a linear well-order on the Irr_L; it can be extended to such a sequence. This latter sequence has length $\leqslant o(\langle \mathrm{Irr}_L, \leqslant\rangle)$.

A particular example may suffice to establish the point of divergence. Take the locale $G = 1 + \omega^{op} \times \omega^{op}$, consisting of a bottom element, denoted $\perp$, and all pairs (i, j) of natural numbers. ($\omega^{op} := \langle \omega, \geqslant\rangle$.) Moreover, $(i, j) \leqslant (i', j')$ iff $i \geqslant i'$ and $j \geqslant j'$. Then the $\bigsqcap$-irreducibles are the elements of the form $(i, 0)$ or $(0, j)$. The order type of the space of $\bigsqcap$-irreducibles is $\omega + \omega$. It is worth explaining why. Let $\alpha\#\beta$ denote the Hessenberg-sum of α and β. This is defined on the basis of the so-called Cantor normal form of ordinals. Every ordinal has a representation as a finite sum $\sum_{i<m} \omega^{\gamma_i}$ where the γ_i form a finite (not necessarily strictly) descending sequence. If $\alpha = \sum_{i<m} \omega^{\zeta_i}$ and $\beta = \sum_{j<m} \omega^{\theta_j}$ are in Cantor normal form with descending sequences ζ_i and θ_j, form the sequence η_k, $k < m + n$, by interleaving the ζ_i and θ_j to form a descending sequence again. Then $\alpha\#\beta := \sum_{k<m+n} \omega^{\eta_k}$. Then a result of [3] states that $o(\mathcal{P} + \mathcal{Q}) = o(\mathcal{P})\#o(\mathcal{Q})$, $\mathcal{P} + \mathcal{Q}$ being the disjoint sum of the posets $\mathcal{P}$ and $\mathcal{Q}$. In the case at hand, $o(\omega + \omega) = \omega\#\omega = \omega + \omega$.

The depth of $\perp$ however is only ω. For consider a downgoing chain of elements of order type ω in G. It consists in a sequence of elements (i_α, j_α) where $i_\alpha \leqslant i_{\alpha'}$ and $j_\alpha \leqslant j_{\alpha'}$ for all $\alpha < \alpha'$. If both the sequence of i_α, $\alpha < \omega$, and j_α, $\alpha < \omega$, are unbounded, the intersection $\bigsqcap_{\alpha<\omega}(i_\alpha, j_\alpha) = \perp$. However, if either sequence is bounded while the other is unbounded, the intersection also equals $\perp$. Hence, all downgoing chains of length ω end in $\perp$. Notice that in this particular locale there are no $\bigsqcap$-prime elements. This situation is not so uncommon in tense logic, see Section 7.9 of [6].

A downgoing chain of logics must be countable because the language is countable. Now, given a logic has a depth at all, either a largest ordinal chain exists or it does not. In the second case the upper limit of these ordinals is ω_1.

Proposition 3.2 *If a modal logic has a depth, the depth is at most ω_1.*

We start with some easy examples. A logic is *pretabular* if it is not tabular but all of its proper extensions are. Tabular logics have finite codimension, but the converse need not hold (take the logic of the so-called veiled recession frame).

Proposition 3.3 *Pretabular logics have depth $\leqslant \omega$.*

Proof. Let L be pretabular. Consider a downgoing chain $\langle L_m : m < \alpha \rangle$ of logics. This chain may be finite. If it is infinite, the logic L_ω does not have finite depth, hence is not tabular. So, L_ω is pretabular and therefore $L = L_\omega$ and $\alpha = \omega$.

In particular, $\mathsf{S5}$ and $\mathsf{Grz.3}$ have depth ω.

Consider next the logic $\mathsf{K.Alt}_1 = \mathsf{K} \oplus \Diamond p \to \Box p$. This is the least normal extension of K containing $\Diamond p \to \Box p$. Every extension of this logic has the finite model property and is finitely axiomatisable ([8]). The generated finite frames are either of the form $\mathfrak{Ch}_n := \langle \{0, 1, \cdots, n-1\}, I_n \rangle$, $i \, I_n \, j$ iff $j = i + 1$, or of the form $\mathfrak{Ch}_n^\bullet := \langle \{0, 1, \cdots, n-1\}, I_n \cup \{\langle n-1, n-1 \rangle\} \rangle$. In the lattice Ext $\mathsf{K.Alt}_1$ only the logics $\mathsf{Th}\,\mathfrak{Ch}_n$ are $\bigsqcap$-prime. (As usual, $\mathsf{Th}, \mathfrak{F}$ denotes the logic of the frame $\mathfrak{F}$.) Moreover, their intersection is $\mathsf{K.Alt}_1$. This shows that there is a downgoing chain of ordinal type ω ending at $\mathsf{K.Alt}_1$. However, the ordinal depth of that logic is larger still. The logics $\mathsf{Th}\,\mathfrak{Ch}_n^\bullet$ form a downgoing chain of type ω whose intersection is $\mathsf{K.Alt}_1.\mathsf{D}$. The logics $\mathsf{K.Alt}_1.\mathsf{D} \sqcap \mathsf{Th}\,\mathfrak{Ch}_n$ form a downgoing chain of order type ω from $\mathsf{K.Alt}_1.\mathsf{D}$ ending at $\mathsf{K.Alt}_1$. Concatenating these sequences yields a chain of type $\omega + \omega$. Since we have used up all $\bigsqcap$-irreducible elements, this cannot be improved upon.

Theorem 3.4 *The logic* $\mathsf{K.Alt}_1$ *has depth* $\omega + \omega$.

The logic $\mathsf{K45}$ is another logic with depth $\omega + \omega$. Its space of irreducible elements is different, though. It consists of (i) clusters of size n or (ii) clusters of size n preceded by an irreflexive point. Frames of the type (ii) contain frames of type (i), unlike in the previous case. Nevertheless, there is a chain of type $\omega + \omega$ which first passes through $\mathsf{S5}$, by picking up first the clusters with increasing n and then the frames of type (ii) with increasing n.

It is essential to consider maximal sequences, otherwise the results are trivial. For example, notice that there is another chain in Ext $\mathsf{K.Alt}_1$, formed by the logics $\mathsf{Th}\,\mathfrak{Ch}_n$, which has order type ω, which starts at the top and goes down to $\mathsf{K.Alt}_1$. Thus not all downgoing chains have the same order type, quite unlike the case of logics of finite depth. Moreover, a chain of type ω almost always exists.

Theorem 3.5 *Let L be a logic without finite depth which has finite model property. Then there exists a downgoing chain of order type ω starting at $\mathsf{K} \oplus \bot$ whose limit is L. In particular, L has a depth.*

Proof. Let $\{\mathfrak{F}_n : n \in \omega\}$ be an enumeration of the finite one-generated frames for L. Then form the following sequence $S := \langle L_m : m \in \omega + 1 \rangle$:

$$\begin{aligned} L_0 &:= \mathsf{K} \oplus \bot \\ L_n &:= \textstyle\bigcap_{m<n} \mathsf{Th}\,\mathfrak{F}_m \\ L_\omega &:= L \end{aligned} \tag{5}$$

This is a descending sequence of logics, that is $L_{n+1} \subseteq L_n$. Since $\mathsf{Th}\,\mathfrak{F}_n$ has finite codimension, $L_{n+1} = L_n \sqcap \mathsf{Th}\,\mathfrak{F}_n$ has finite codimension under L_n. If $L_{n+1} = L_n$, we drop L_{n+1}. We obtain a strictly descending sequence of logics of type $\leqslant \omega$ where each member has finite dimension over the next. To make this a chain, we fill in some finite number of logics at each step. This chain cannot be finite. Hence it is of order type ω. The intersection of its members is L. Hence, there is a descending chain to L of order type $\omega + 1$, making L of depth ω.

A familiar lattice theoretic argument has been used in this proof. Consider elements x, y, z in a distributive lattice such that $y = x \sqcup z$. There are maps $\psi : [z, y] \to [z \sqcap x, x] : u \mapsto u \sqcap x$ and $\chi : [z \sqcap x, x] \to [z, y] : v \mapsto v \sqcup z$. These maps are order preserving; moreover, they are inverses of each other. Assume $z \leqslant u \leqslant y$. Then

$$\chi(\psi(u)) = (u \sqcap x) \sqcup z = (u \sqcup z) \sqcap (x \sqcup z) = u \sqcap y = u \tag{6}$$

Now assume $x \sqcap z \leqslant v \leqslant x$:

$$\psi(\chi(v)) = (v \sqcup z) \sqcap x = (v \sqcap x) \sqcup (z \sqcap x) = v \sqcup (x \sqcap x) = v \tag{7}$$

It follows that the two intervals are isomorphic. A special case worth noting is the case of a cocover. u is a cocover of v iff the interval $[u, v]$ contains exactly two elements. Hence we have the following

Lemma 3.6 *Assume that x, y, z are elements in a distributive lattice such that $x \sqcup z = y$. Then z is a cocover of y iff $x \sqcap z$ is a cocover of x.*

We call the pair (y, z) a *prime quotient*, and write $y \succ z$, to say that z is a cocover of y. One says that intersecting this quotient with x *projects* it onto a prime quotient $x = x \sqcap z \succ x \sqcap y$. In general, if $x \prec y$ then either $x \sqcap z = y \sqcap z$ or $x \sqcap z \prec y \sqcap z$. As a consequence, if y has finite dimension over x then $y \sqcap z$ has finite dimension over $x \sqcap z$ as well. We shall use this type of argument quite frequently.

Notice even though in a modular lattice if x has finite dimension over y then every downward chain from x to y has the same length, this does not carry over to infinite dimensions even if the lattice is continuous (as is the case with Ext K45). The result above suggests that dimension is a cardinal invariant (thus asking about the cardinality of the chain), while depth is construed here as an ordinal number. The price to pay is to define it as the supremum of order types, since they are not unique.

Consider next the logic S4.3.1. Its frames can be represented by sequences $\gamma = \langle c_i : i < m \rangle$ of nonzero numbers, where it is understood that c_i is the size of the cluster of depth $i + 1$. The final cluster (which we define to be of depth 0) has size 1, and is *not* represented in the sequences, which may therefore be empty (in that case $m = 0$). Given a sequence, the frame is constructed over the set $\{(\delta, i) : \delta = i = 0 \text{ or } \delta - 1 < m \text{ and } i < c_{\delta-1}\}$, and $(\delta, i)\, R\, (\delta', i')$ iff $\delta \geqslant \delta'$. Now, γ can also be understood as a finite word over $\omega - \{0\}$. Given another such word, $\eta = \langle d_i : i < n \rangle$, the frame of γ is a p-morphic image of η iff there is a strictly ascending map $v : \{0, \cdots, m-1\} \to \{0, \cdots, n-1\}$ such that $d_{v(i)} \geqslant c_i$ for all $i < m$. Also, if γ is a generated subframe of η then η is contractible to γ. Now, $\mathsf{Th}\,\gamma \supseteq \mathsf{Th}\,\eta$ iff γ is a p-morphic image of a generated subframe of η iff γ is a p-morphic image of η iff $\gamma \preccurlyeq \eta$, where $\preccurlyeq$ is the Higman ordering on $(\omega - \{0\})^*$, which in turn is isomorphic to the Higman ordering on ω^*. Here, P^* for a set P denotes the set of finite sequences of elements of U. (The asterisk is thus the so-called Kleene-star.) Given a WPO P, the Higman ordering on P^* is as follows. If $x = x_0x_1 \cdots x_{m-1}$ for certain $x_i \in P$, then $x \preccurlyeq z$ iff $z = z_0z_1 \cdots z_{n-1}$, $z_i \in P$, and there is a strictly ascending map $v : \{0, \cdots, m-1\} \to \{0, \cdots, n-1\}$ such that $x_i \preccurlyeq z_{v(i)}$ for all $i < m$.

Lemma 3.7 *Let γ and η represent two finite S4.3.1-frames. Then η is contractible to γ iff $\gamma \preccurlyeq \eta$.*

Proof. Basically, it is known, e. g. [2], that for S4.3-frames $\mathfrak{F}$ and $\mathfrak{G}$, $\mathfrak{F}$ is contractible to $\mathfrak{G}$ iff $\mathfrak{G}$ is a cofinal subframe of $\mathfrak{F}$. Final clusters are of size 1, so the final cluster of $\mathfrak{G}$ is always embeddable in the final cluster of $\mathfrak{F}$. The remainder is the condition that $\gamma \preccurlyeq \eta$, where γ represents $\mathfrak{G}$ and η represents $\mathfrak{F}$.

It turns out that $o(\omega^*) = \omega^{\omega^\omega}$ (Theorem 16 of [7], based on results of [3]).

Theorem 3.8 *The depth of S4.3.1 is ω^{ω^ω}.*

We can extend this result to S4.3. A frame for S4.3 can be represented as a pair (γ, p), where $p > 0$ represents the size of the final cluster, and γ is a sequence representing the sizes of nonfinal clusters as above. Now (γ, p) is subreducible to (γ', p') iff $p \geqslant p'$ and $\gamma \preccurlyeq \gamma'$. Thus, we have a WPO of order type $\omega^* \times \omega$. The following is an application of Theorem 3.5 in [3] stating that the order type of a product of WPOs is the so-called Hessenberg product of the order types of the individual WPOs. Namely, if $\alpha = \sum_{i<m} \omega^{\zeta_i}$ and $\beta = \sum_{j<n} \omega^{\theta_j}$ are in Cantor normal form, put $\alpha \circledast \beta = \sum_{i<mn} \omega^{\eta_i}$, where $\langle \eta_i : i \leqslant mn \rangle$ is a suitable rearrangement of the sequence $\langle \zeta_i \# \theta_j : i \leqslant m, j \leqslant n \rangle$ so as to make the sequence nonincreasing. Applying this to the case at hand, we get $\omega^{\omega^\omega} \circledast \omega = \omega^{\omega^\omega \# 1} = \omega^{\omega^\omega + 1}$.

Theorem 3.9 *The depth of S4.3 is $\omega^{\omega^\omega + 1}$.*

4 Logics with Uncountable Depth

Let us now consider the case when the space of irreducibles is not a WPO. This may essentially have two reasons: the poset of irreducible logics possesses infinite upgoing chains, or it has an infinite antichain. We simplify the matter by looking at splitting elements. We first establish a result that deals with the case where the set of $\bigsqcap$-prime logics contains an infinite antichain.

Let $\mathfrak{D} := \langle \{v_0\}, \varnothing \rangle$, $\mathfrak{C}_\varepsilon := \langle \{v_1, v_0\}, \{\langle v_1, v_0 \rangle\} \rangle$. Now define frames $\mathfrak{C}_{\boldsymbol{x}} = \langle W_{\boldsymbol{x}}, R_{\boldsymbol{x}} \rangle$, $\boldsymbol{x} \in \{0,1\}^*$, inductively as follows. Assume that $\boldsymbol{x}$ has length n.

$$\begin{aligned} \mathfrak{C}_{\boldsymbol{x}0} &:= \{W_{\boldsymbol{x}} \cup \{v_{n+2}\}, R_{\boldsymbol{x}} \cup \{\langle v_{n+2}, v_{n+1} \rangle\}\} \\ \mathfrak{C}_{\boldsymbol{x}1} &:= \{W_{\boldsymbol{x}} \cup \{v_{n+2}\}, R_{\boldsymbol{x}} \cup \{\langle v_{n+2}, v_{n+1} \rangle, \langle v_{n+2}, v_0 \rangle\}\} \end{aligned} \tag{8}$$

See Figure 1 for an example.

Fig. 1. The frame $\mathfrak{C}_{11010}$

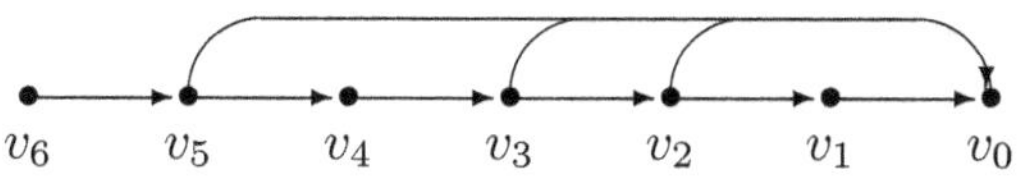

Proposition 4.1 *For all $\boldsymbol{x}, \boldsymbol{y} \in \{0,1\}^*$ the following holds.*

(i) $\mathsf{Th}\,\mathfrak{C}_{\boldsymbol{x}} \subseteq \mathsf{Th}\,\mathfrak{D}$.

(ii) $\mathsf{Th}\,\mathfrak{C}_{\boldsymbol{x}} \subseteq \mathsf{Th}\,\mathfrak{C}_{\boldsymbol{y}}$ *iff $\boldsymbol{y}$ is a prefix of $\boldsymbol{x}$.*

(iii) $\mathsf{Th}\,\mathfrak{C}_{\boldsymbol{x}}$ *has codimension $n+2$, where n is the length of $\boldsymbol{x}$.*

Proof. The first claim is obvious. For the second notice that the frames are noncontractible, one-generated and finite. Hence $\mathsf{Th}\,\mathfrak{C}_{\boldsymbol{x}} \subseteq \mathsf{Th}\,\mathfrak{C}_{\boldsymbol{y}}$ if and only if $\mathfrak{C}_{\boldsymbol{y}}$ is a generated subframe of $\mathfrak{C}_{\boldsymbol{x}}$, which by notation is the case if and only if $\boldsymbol{y}$ is a prefix of $\boldsymbol{x}$.

The logic L_P of these frames can be axiomatized. It is an extension of $\mathsf{K.Alt}_2$, where $\mathsf{Alt}_2 = \bigwedge_{i<3} \Diamond p_i \rightarrow \bigvee_{i<j<3} \Diamond(p_i \wedge p_j)$. The additional axiom characterising these frames is

$$\Diamond(p \wedge \Diamond\top) \rightarrow \Box(\neg p \rightarrow \Box\bot) \tag{9}$$

Indeed, the axiom is first-order stating that for every v such that there are w, w' and w'' with $v\,R\,w$, $v\,R\,w'$, $w\,R\,w''$ and $w \neq w'$ the world w' has no successor.

Let us note that from the results of [1] it follows that

Proposition 4.2 (Blok) $\mathsf{Th}\,\mathfrak{D}$, $\mathsf{Th}\,\mathfrak{C}_{\boldsymbol{x}}$ *are $\sqcap$-prime or all $\boldsymbol{x} \in \{0,1\}^*$.*

Define $V \subseteq \{0,1\}^*$. Put $L(\varnothing) := \mathsf{Th}\,\mathfrak{D}$, and for $V \neq \varnothing$ put $L(V) := \bigsqcap_{\boldsymbol{x}\in V} \mathsf{Th}\,\mathfrak{C}_{\boldsymbol{x}}$.

Proposition 4.3 *The following holds for all $V \subseteq \{0,1\}^*$.*

(i) *$\mathfrak{C}_{\boldsymbol{y}}$ is a frame for $L(V)$ iff there is $\boldsymbol{x} \in V$ such that $\boldsymbol{y}$ is a prefix of $\boldsymbol{x}$.*

(ii) *If in addition V is prefix closed then $L(V) \subseteq \mathsf{Th}\,\mathfrak{C}_{\boldsymbol{y}}$ iff $\boldsymbol{y} \in V$.*

Proof. (i) The logic $\mathsf{Th}\,\mathfrak{C}_{\boldsymbol{y}}$ is $\sqcap$-prime. Hence, if $L(V) \subseteq \mathsf{Th}\,\mathfrak{C}_{\boldsymbol{y}}$ there is an $\boldsymbol{x} \in V$ such that $\mathsf{Th}\,\mathfrak{C}_{\boldsymbol{x}} \subseteq \mathsf{Th}\,\mathfrak{C}_{\boldsymbol{y}}$, and conversely. The latter is the case if and only if $\boldsymbol{y}$ is a prefix of $\boldsymbol{x}$, by Proposition 4.1(ii). (ii) If $L(V) \subseteq \mathsf{Th}\,\mathfrak{C}_{\boldsymbol{y}}$ then by (i), $\boldsymbol{y}$ is a prefix of some $\boldsymbol{x} \in V$. Since V is prefix closed, $\boldsymbol{y} \in V$. The converse is clear.

Evidently, if $L = L(V)$, then for the prefix closure V° of V, $L = L(V^\circ)$. So, without loss of generality we may assume a representation of $L(V)$ with a prefix closed set V.

Proposition 4.4 *Let V, W be prefix closed.*

(i) $L(V) \subseteq L(W)$ *iff* $V \supseteq W$.

(ii) $L(W) \prec L(V)$ *iff* $W = V \cup \{\boldsymbol{x}i\}$ *where* $i \in \{0,1\}$, $\boldsymbol{x} \in V$ *but* $\boldsymbol{x}i \notin V$.

Proof. (i) follows from Proposition 4.2. (ii) Suppose that $L(W) \prec L(V)$. Then $W \supsetneq V$ and so there is a $\boldsymbol{y} \in W - V$. Every proper prefix of $\boldsymbol{y}$ must be in V otherwise $L(V \cup \{\boldsymbol{z}\})$ for such a prefix is strictly between $L(W)$ and $L(V)$. So for the largest prefix $\boldsymbol{y}$ of $\boldsymbol{x}$ we have $\boldsymbol{y} \in V$, while $\boldsymbol{x} = \boldsymbol{y}i$ for some $i \in \{0,1\}$. And conversely.

Alternatively, one may use Lemma 5.5 below to prove this proposition. The set $\{0,1\}^*$ forms a complete binary branching tree T under the prefix ordering. Let $U(T)$ be the set of prefix closed subsets of T.

Proposition 4.5 *The map $V \mapsto L(V)$ is an order preserving injection from $\langle U(T), \supseteq\rangle$ into* Ext L_P.

The mapping is not onto, however. The logic K.D is an extension of L_P that is not of that form.

The tree T has $2^{\aleph_0}$ branches. Each branch defines a logic.

Proposition 4.6 Ext L_P *has an antichain of length* $2^{\aleph_0}$.

Proof. Each branch consists of an infinite linear sequence S of words $\boldsymbol{x}_i$ such that $\boldsymbol{x}_i$ is a prefix of $\boldsymbol{x}_{i+1}$. Obviously, S is prefix closed. If S' is another such set, $L(S) \neq L(S')$. However, if $L(S) \subsetneq L(S')$, then all $\boldsymbol{x}_i$ would be in S', which cannot be the case if S' is a chain different from S. Likewise, $L(S') \subsetneq L(S)$. So the logics form an antichain.

Now let $J := \{0^n 1 : n \in \omega\}$. This is a countably infinite antichain in T. Choose a well-order κ on J, so $J = \{\boldsymbol{x}_\lambda : \lambda < \kappa\}$. κ is countable. Define now the following chain of logics. $L(V_\lambda)$, $\lambda < \omega + \kappa$, with

$$\begin{aligned} V_n &:= \{0^m : m < n\} \\ V_\omega &:= \textstyle\bigcap_{n \in \omega} V_n \\ V_{\omega+\beta} &:= V_\omega \cup \{\boldsymbol{x}_\lambda : \lambda < \beta\} \\ V_\kappa &:= L_P \end{aligned} \tag{10}$$

Proposition 4.7 *(1) $L(V_{\alpha+1})$ is a lower cover of $L(V_\alpha)$ for all $\alpha < \kappa$. (2) If α is a limit ordinal, $L(V_\alpha) = \bigsqcap_{\gamma<\alpha} L(V_\gamma)$.*

Proof. (1) By Proposition 4.4, $L(V_{\alpha+1})$ is a lower cover iff $V_{\alpha+1} - V_\alpha$ contains a single element $\boldsymbol{x}$ such that all its proper prefixes are in V_α. Two cases arise. (Case 1) $\alpha < \omega$. Then $\boldsymbol{x} = 0^\alpha$, and all proper prefixes are indeed in V_α. (Case 2) $\alpha = \omega + \beta$, and $\boldsymbol{x} = 0^n 1$ for some n. All of the proper prefixes are in V_α. (2) is clear.

The following is now evident.

Proposition 4.8 *The $\langle L(V_\alpha) : \alpha < \kappa + 1\rangle$ form a downgoing chain of logics of order type $\kappa + 1$.*

It can easily be shown that for every countable limit ordinal κ there is a downgoing chain from the top of the lattice to L_P.

Proposition 4.9 *L_P has depth ω_1.*

Let us now attack the problem of K. We shall use an abstract argument, which can be applied in a number of cases. It is based on the insight that the intersection of all $\sqcap$-prime logics is K, and that this set contains an infinite antichain.

Proposition 4.10 *Let L be a complete bounded lattice such that the bottom element 0 is the intersection of countably many elements of finite depth. Suppose there exists a countably infinite antichain $\{x_n : n \in \omega\}$ of splitting elements of finite depth. Then the depth of 0 exists and is ω_1.*

Proof. Let S be a countable set such that $\bigsqcap S = 0$. Let $X := \{x_n : n \in \omega\}$ be an antichain of $\sqcap$-prime logics. We can assume that $X \subseteq S$. Let $A := (\uparrow X) - X$, $B := S - \uparrow X$. We choose an enumeration $\langle s_\alpha : \alpha < \eta \rangle$ of S that starts by enumerating the elements from A first, then proceeds with an enumeration of X, and ends in an enumeration of B. The enumerations of A and B can be arbitrary. So, for simplicity we assume A to be of order type ω (if infinite). Let η by an arbitrary countable well-order on X. Then S has order type $\omega + \eta + \omega$, which is at least η. Now define a sequence by

$$y_\alpha := \bigsqcap_{\gamma < \alpha} s_\gamma \tag{11}$$

This is a downgoing sequence of elements with limit $0 = \bigsqcap_{\alpha<\eta} s_\alpha$ (which can be thought of as y_η). Notice also the following. By choice of A and the fact that the elements of X are prime, $X \cap \uparrow y_\omega = \varnothing$. Thus, for $\omega \leqslant \alpha < \omega + \eta$, $y_{\alpha+1} \neq y_\alpha$, so the sequence is actually properly descending at least for $\omega \leqslant \alpha < \omega + \eta$. Now, $y_{\alpha+1}$ is either a lower cover of y_α or it has finite codimension > 1 under y_α.

Hence by inserting finitely many elements between successive elements of the sequence the sequence can be extended to a downgoing chain of order type at least η. η can be choosen arbitrarily. So we can exceed any given countable ordinal. Thus the supremum of these order types is ω_1.

We conclude the following.

Theorem 4.11 *The depth of* K *is ω_1.*

From [4] we know that the logic Grz_3 of posets of depth at most 3 has an infinite antichain of splitting logics. Thus, any logic contained in Grz_3 with the finite model property has depth ω_1. Similarly for logics contained in G_3 (just take the irreflexive counterpart of Fine's frames).

Theorem 4.12 *K4, S4, G and Grz have depth ω_1.*

5 Logics with and without depth

Consider next the second type of failure, when the poset of irreducibles has infinite ascending chains. In this case not all logics have a depth. Here is an example taken from [6], Page 360. The logic called $\mathsf{G}.\Omega_2$ has an extension lattice isomorphic to $\omega + 2 + \omega^{op}$, by Theorem 7.5.14. ($\omega^{op} := \langle \omega, \geqslant \rangle$. The plus here denotes the ordered sum, making all elements of the first set lower to all elements of the second. It is thus distinct from the independent sum of posets.) Hence, it lacks well-ordered downgoing chains ending in $\mathsf{G}.\Omega_2$. The following summarizes the facts.

Theorem 5.1 Ext $\mathsf{G}.\Omega_2$ *is continuous. It contains an infinite ascending chain of splitting logics.* $\mathsf{G}.\Omega_2$ *has no depth.*

Note that the existence of infinite ascending chains of prime logics is instrumental in establishing lack of depth, as we shall see below.

On the other hand, such logics are hard to construct, as the following series of observations shows.

Lemma 5.2 *If every logic in* Ext L *that properly includes* L *has a cocover,* L *has a depth.*

Proof. For a proof, observe that a descending chain can be constructed inductively as follows. Start with $L_0 := L \oplus \bot$ and let $L_{\kappa+1}$ be a cocover of L_κ in Ext L if $L_\kappa \neq L$; and let $L_\lambda := \bigsqcap_{\mu<\lambda} L_\mu$ for a limit ordinal λ such that $L_\mu \neq L$ for all $\mu < \lambda$.

So on what conditions do cocovers exist? Here are a few cases.

Lemma 5.3 *Let* L' *be finitely axiomatizable over* L*. Then* L' *has a cocover in* Ext L.

Proof. This follows from Tukey's Lemma. Define a property $\mathcal{P}$ of sets of formulae by $\mathcal{P}(\Delta)$ iff $L \oplus \Delta \subsetneq L'$. This is finitely based: it is true of an infinite set if and only if it is true of all its finite subsets. By Tukey's Lemma there is a maximal set Δ^* having $\mathcal{P}$. Then $L \oplus \Delta^*$ is a cocover of L'.

Another example concerns splittings. Notice the following.

Lemma 5.4 *Let* L' *be a splitting logic of* Ext L *with splitting companion* L''*. Then* L'' *has a unique cocover,* $L' \sqcap L''$*. Dually,* L' *has a unique cover,* $L' \sqcup L''$*.*

Proof. Since Ext L is complete, we can form $L^\circ := \bigsqcup_{M<L''} M$. As L'' is $\bigsqcup$-prime, $L^\circ < L''$. Thus, L° is the cocover of L''. Since L'' is not below L', $L'' \sqcap L' < L''$ and hence $L'' \sqcap L' \leqslant L^\circ$. On the other hand, by definition of a splitting, since $L^\circ < L''$, we must have $L^\circ \leqslant L'$, from which it follows that $L^\circ \leqslant L'' \sqcap L'$. The two logics are thus equal. Hence, $L'' \sqcap L'$ is the (unique) cocover of L''. The other claim is dual.

Lemma 5.5 *Let* L' *be a splitting logic of* L*. If* $M \supseteq (\text{Ext } L)/L'$ *then* M *has a cocover, namely* $M \sqcap L'$*.*

Proof. By the previous lemma, we have a prime quotient $(\text{Ext } L)/L' \succ L' \sqcap (\text{Ext } L)/L'$. Now let $M \supseteq (\text{Ext } L)/L'$. Consider the quotient $M \succ M \sqcap L'$. Intersecting this quotient with $(\text{Ext } L)/L'$ we get the quotient $M = M \sqcap (\text{Ext } L)/L' \succ (M \sqcap L') \sqcap (\text{Ext } L)/L' = L' \sqcap (M \sqcap (\text{Ext } L)/L') =$

$L' \sqcap (\text{Ext } L)/L'$. This is a prime quotient. Hence the original quotient is also prime, by Lemma 3.6.

Proposition 5.6 *Assume that the set of splitting logics of* Ext L *contains no infinite ascending chains. Assume further that L is the intersection of its splitting logics. Then L has a depth.*

Proof. Construct a sequence as follows. L_0 is the inconsistent logic. For a limit ordinal λ, put $L_\lambda = \bigsqcap_{\mu<\lambda} L_\mu$. Assume that $L_\lambda \neq L$. Consider the set $U_\lambda := \{M : M \text{ splits Ext } L, L_\lambda \not\subseteq M\}$. U_λ is not empty, otherwise all splitting logics are already above L_λ, whence $L_\lambda = L$. U_λ contains no infinite ascending chains, hence it has a maximal element L^*. Now put $L_{\lambda+1} := L_\lambda \sqcap L^*$. This is a cocover of L_λ, by Lemma 5.5.

Compare this last result with Theorem 5.1. If the lattice Ext L contains an infinite ascending chain of splitting logics, then the intersection of all splitting logics of Ext L lacks depth.

6 Open Problems

Although K has depth ω_1 we cannot conclude that for any given countable ordinal κ there is a specific logic with depth κ. This remains a problem to be solved.

Problem. Construct for given countable ordinal β a logic of depth β, if such a logic exists.

If Ext L is continuous, then any $\bigsqcap$-irreducible is also $\bigsqcap$-prime. If furthermore Irr_L is a WPO then there is a chain of order type $o(\text{Irr}_L) + 1$. It turns out that this chain is optimal also for any extension of L. That is, if $L' \supseteq L$ then the depth of L' is given as λ where $L' = L_\lambda$ in this well order. This is a consequence of the fact that the space of upper closed sets is order perfect (Theorem 9 in [7]). As a consequence the problem is solved for ordinals below $\omega^{\omega^\omega+1}$.

I close with the following conjecture. Recall that ε_0 is the limit of the sequence $1, \omega, \omega^\omega, \omega^{\omega^\omega}, \cdots$. Equivalently, it is the least fixed point of the ordinal exponentiation function $\beta \mapsto \omega^\beta$. ε_0 is countable.

Conjecture. There are no logics of countable depth larger than ε_0.

References

[1] Blok, W. J., *On the degree of incompleteness in modal logic and the covering relation in the lattice of modal logics*, Technical report (1978), report 78-07, Dept. of Math., University of Amsterdam.

[2] Chagrov, A. and M. Zakharyaschev, "Modal Logic," Oxford University Press, Oxford, 1997.

[3] de Jongh, D. and R. Parikh, *Well-partial orderings and hierarchies*, Indagationes Mathematicae **39** (1977), pp. 195–207.

[4] Fine, K., *An ascending chain of S4 logics*, Theoria **40** (1974), pp. 110–116.

[5] Kracht, M., *Lattices of Modal Logics and Their Groups of Automorphisms*, Journal of Pure and Applied Logic **100** (1999), pp. 99–139.
[6] Kracht, M., "Tools and Techniques in Modal Logic," Number 142 in Studies in Logic, Elsevier, Amsterdam, 1999.
[7] Pouzet, M. and M. Sobrani, *The order type of the collection of finite series-parallel posets*, Discrete Mathematics **265** (2003), pp. 189–211.
[8] Segerberg, K., *Modal logics with functional alternative relations*, Notre Dame Journal of Formal Logic **27** (1986), pp. 504–522.

It ain't necessarily so: Basic sequent systems for negative modalities

Ori Lahav

Max Planck Institute for Software Systems (MPI-SWS), Germany

João Marcos

Federal University of Rio Grande do Norte, Brazil

Yoni Zohar

Tel Aviv University, Israel

Abstract

We look at non-classical negations and their corresponding adjustment connectives from a modal viewpoint, over complete distributive lattices, and apply a very general mechanism in order to offer adequate analytic proof systems to logics that are based on them. Defining non-classical negations within usual modal semantics automatically allows one to treat equivalent formulas as synonymous, and to have a natural justification for a global version of the contraposition rule. From that perspective, our study offers a particularly useful environment in which negative modalities and their companions may be used for dealing with inconsistency and indeterminacy. After investigating modal logics based on arbitrary frames, we extend the results to serial frames, reflexive frames, functional frames, and symmetric frames. In each case we also investigate when and how classical negation may thereby be defined.

Keywords: negative modalities, sequent systems, cut-admissibility, analyticity.

1 Capturing the impossible, and its dual

Many well-known subclassical logics —including intuitionistic logic and several many-valued logics— share the conjunction-disjunction fragment of classical logic, but disagree about the exact notion of opposition and the specific logical features to be embodied in *negation*. In contrast, modal logics are often thought of as superclassical, and are obtained by the addition of *identity*-like 'positive modalities' $\Box$ and $\Diamond$. For various well-known cases, such modalities fail to have a finite-valued characterization. Notwithstanding, each m-ary connective $\pentagon$ of a modal logic is typically *congruential* (with respect to the underlying consequence relation $\vdash$), in treating equivalent formulas as synonymous: if $\alpha_i \vdash \beta_i$ and $\beta_i \vdash \alpha_i$, for every $1 \leq i \leq m$, then $\pentagon(\alpha_1, \ldots, \alpha_m) \vdash \pentagon(\beta_1, \ldots, \beta_m)$. To logical systems containing only such sort of connectives one might associate semantics in terms of neighborhood frames (see ch.5 of [23]), and the same applies if one uses 1-ary 'negative modalities' instead, as in [19]. But normal modal logics make their 1-ary positive modalities respect indeed a stronger property: if $\alpha \vdash \beta$ then $\pentagon(\alpha) \vdash \pentagon(\beta)$. Such monotone behavior may be captured by semantics based on Kripke frames, and the same applies to the antitone behavior that characterize negative modalities, namely: if $\alpha \vdash \beta$ then $\pentagon(\beta) \vdash \pentagon(\alpha)$.

In [7] an investigation of negative modalities is accomplished on top of the $\wedge\vee\top\bot$-fragment of classical logic, and the same base language had already been considered in [18] for the combination of positive and negative modalities. Typically, in studies of positive and negative modalities the so-called compatibility (bi-relational) frames are used, and certain appropriate conditions upon the commutativity of diagrams involving their two relations are imposed, having as effect the heredity of truth (i.e., its persistence towards the future) with respect to one of the mentioned relations (assumed to be a partial order).

There are a number of studies (e.g. [21,6]) in which the above mentioned languages for dealing with negative modalities are upgraded in order to count on an (intuitionistic or classical) implication, and sometimes also its dual, co-implication (cf. [17]). If one may count on classical implication, however, it suffices to add to it the modal paraconsistent negation given by 'unnecessity' (cf. [13]), and all other connectives of normal modal logics turn out to be definable from such impoverished basis (indeed, where $\smallsmile$ is a primitive symbol for unnecessity and $\to$ represents classical implication, we have that ${\sim}\alpha := \alpha \to {\smallsmile}(\alpha \to \alpha)$ behaves as the classical negation of α, and $\Box\alpha := {\sim}{\smallsmile}\alpha$ behaves as the usual positive modality box).

Our intuition about the relation between a paracomplete (a.k.a. 'intuitionistic-like') negation and a paraconsistent negation is that the former would be expected to be more demanding than the latter, while classical negation should sit between the two (whenever it also turns out to be expressible). It takes indeed more effort to assert a negated statement constructively, while such statements are more readily asserted should some contradictions be allowed to subsist; in other words, negations in a paracomplete logic come at a greater cost than classical negations, while paraconsistent logics indulge on negations in which classical logic would show greater restraint. The presence of a classical negation, however, often makes it too easy to forget that there are two distinct kinds of deviations equally worth studying, concerning non-classical negation, as one of these deviations may then be recovered in the standard way as the dual of the other. In order to get a better grasp of the duality between paraconsistent and paracomplete modal negations (namely, unnecessity vs. impossibility), we purposefully make an effort to prevent the underlying language from being sufficiently expressive so as to allow for the definition of a classical negation (or a classical implication) — whenever that goal lies within reach. Here we do however in all cases enrich our object language with certain 'adjustment connectives' expressing negation-consistency and negation-determinacy, allowing for the simulation of usual features of classical negation and for the (partial) recovery of classical reasoning. It should be noted, however, that as a byproduct of the presence of such adjustment connectives truth will no longer be hereditary in our Kripke models, that is, it will not in general be preserved for all compound formulas towards the future, in contrast with what happens with models of compatibility frames.

In what follows, first and foremost we will concentrate on the logic PK, determined by the class of all Kripke frames, which has been introduced and received a presentation as a sequent system in [5]. We show here that it can be reintroduced in terms of a so-called 'basic sequent system', which allows one to take advantage of general techniques developed in [10], including a method for obtaining sound and complete Kripke semantics and a uniform recipe for semantic proofs of cut-admissibility or analyticity. The next section adopts a semantical perspective to explain why and how our study is done.

2 On negative modalities

We briefly recall the now familiar elements of a Kripke semantics. A *frame* is a structure consisting of a nonempty set W (of 'worlds') and a binary ('accessibility') relation R on W. A *model* $\mathcal{M} = \langle \mathcal{F}, V \rangle$ is based on a frame $\mathcal{F} = \langle W, R \rangle$ and on a *valuation* $V : W \times \mathcal{L} \to \{f, t\}$ that assigns truth-values to worlds $w \in W$ and sentences φ of a propositional language $\mathcal{L}$ generated over a denumerable set of propositional variables $\mathcal{P}$. The valuations must satisfy certain conditions that are induced by the fixed interpretation of the connectives of the language. When $V(w, \varphi) = t$ we say that V *satisfies* φ *at* w, and denote this by $\mathcal{M}, w \Vdash \varphi$; otherwise we write $\mathcal{M}, w \nVdash \varphi$ and say that V *leaves* φ *unsatisfied at* w. The connectives from the positive fragment of classical logic receive their standard boolean interpretations locally, world-wise, by recursively setting:

[S$\top$]	$\mathcal{M}, w \Vdash \top$		
[S$\wedge$]	$\mathcal{M}, w \Vdash \varphi \wedge \psi$	iff	$\mathcal{M}, w \Vdash \varphi$ and $\mathcal{M}, w \Vdash \psi$
[S$\vee$]	$\mathcal{M}, w \nVdash \varphi \vee \psi$	iff	$\mathcal{M}, w \nVdash \varphi$ and $\mathcal{M}, w \nVdash \psi$

Given formulas $\Gamma \cup \Delta$ of $\mathcal{L}$, and given a class of frames $\mathcal{E}$, we say that Γ *entails* Δ *in* $\mathcal{E}$, and denote this by $\Gamma \models_{\mathcal{E}} \Delta$, if for each model $\mathcal{M}$ based on a frame $\mathcal{F} \in \mathcal{E}$ and each world w of $\mathcal{M}$ we have either $\mathcal{M}, w \nVdash \gamma$ for some $\gamma \in \Gamma$ or $\mathcal{M}, w \Vdash \delta$ for some $\delta \in \Delta$. The assertion $\Gamma \models_{\mathcal{E}} \Delta$ will be called a *consecution*. As usual, in what follows we will focus most of the time on consecutions $\Gamma \models_{\mathcal{E}} \Delta$ involving a singleton Δ, and in the next section we will extend the notion of entailment so as to cover sequents instead of formulas. The subscript $\mathcal{E}$ shall be omitted in what follows whenever there is no risk of ambiguity.

In the following subsections we extend the above language with connectives whose modal interpretations will be useful for the investigation of non-classical negations.

2.1 Adding negations

Our first extension of the above language proceeds by the addition of a 1-ary connective $\smallsmile$, to be interpreted non-locally as follows:

$[\mathrm{S}\smallsmile]$ $\mathcal{M}, w \Vdash \smallsmile\varphi$ iff $\mathcal{M}, v \nVdash \varphi$ for <u>some</u> $v \in W$ such that wRv

Accordingly, a formula $\smallsmile\varphi$ is said to be satisfied at a given world of a model precisely when the formula φ fails to be satisfied at some world accessible from this given world. In the following paragraph we will show that $\smallsmile$ respects some minimal conditions to deserve being called a 'negation', namely, we will demonstrate its ability to invert truth-values assigned to certain formulas (at certain worlds).

Let $\#$ represent an arbitrary 1-ary connective, and let $\#^j$ abbreviate a j-long sequence of $\#$'s. The least we will demand from $\#$ to call it a *negation* is that, for every $p \in \mathcal{P}$ and every $k \in \mathbb{N}$:

$[\![\textit{falsificatio}]\!]$ $\#^k p \not\models \#^{k+1} p$ $\qquad$ $[\![\textit{verificatio}]\!]$ $\#^{k+1} p \not\models \#^k p$

To witness $[\![\textit{falsificatio}]\!]$, some sentence φ is to be satisfied while the sentence $\#\varphi$ is not simultaneously satisfied; for $[\![\textit{verificatio}]\!]$ some sentence φ is left unsatisfied while at the same time $\#\varphi$ is satisfied. To check that the connective $\smallsmile$ fulfills such requisites, it suffices for instance to build a frame in which $W = \{w_n : n \in \mathbb{N}\}$ and wRv iff $v = w^{++}$ (namely, v is the successor of w), and consider a valuation V such that $V(w_n, p) = t$ iff n is odd.

It is very easy to see that our connective $\smallsmile$ satisfies *global contraposition* in the sense that $\alpha \models \beta$ implies $\smallsmile\beta \models \smallsmile\alpha$. Indeed, assume $\alpha \models \beta$ and suppose that $\mathcal{M}, w \Vdash \smallsmile\beta$ for some world w of an arbitrary model $\mathcal{M}$. Then, $[\mathrm{S}\smallsmile]$ informs us that there must be some world v in $\mathcal{M}$ such that wRv and $\mathcal{M}, v \nVdash \beta$. By the definition of entailment, the initial assumption gives us $\mathcal{M}, v \nVdash \alpha$. Using again $[\mathrm{S}\smallsmile]$ we conclude that $\mathcal{M}, w \Vdash \smallsmile\alpha$. As a byproduct of this, if one defines an equivalence relation $\equiv$ on $\mathcal{L}$ by setting $\alpha \equiv \beta$ whenever both $\alpha \models \beta$ and $\beta \models \alpha$, then an easy structural induction on $\mathcal{L}$ establishes that $\equiv$ is not only compatible with $\smallsmile$ but also with the other connectives that are used in constructing the algebra of formulas; in other words, $\equiv$ constitutes a congruence relation on $\mathcal{L}$.

It is straightforward to see that any 1-ary connective $\#$ satisfying global contraposition is such that, given $p, q \in \mathcal{P}$:

(DM1.1#) $\#(p \vee q) \models \#p \wedge \#q$ $\qquad$ (DM2.1#) $\#p \vee \#q \models \#(p \wedge q)$

If $\#$ also respects the following consecutions, then it is said to be a *full type diamond-minus connective*:

(DM2.2#) $\#(p \wedge q) \models \#p \vee \#q$ $\qquad$ (DT#) $\#\top \models p$

Note that $\smallsmile$ is a full type diamond-minus connective. To check that $\smallsmile$ satisfies (DM2.2#), indeed, suppose that $\mathcal{M}, w \Vdash \smallsmile(p \wedge q)$ for some arbitrary world w of an arbitrary model $\mathcal{M}$. By $[\mathrm{S}\smallsmile]$ we know that there is some world v such that wRv and $\mathcal{M}, v \nVdash p \wedge q$. It follows by $[\mathrm{S}\wedge]$ that $\mathcal{M}, v \nVdash p$ or $\mathcal{M}, v \nVdash q$. Using $[\mathrm{S}\smallsmile]$ again we conclude that $\mathcal{M}, w \Vdash \smallsmile p$ or $\mathcal{M}, w \Vdash \smallsmile q$ and $[\mathrm{S}\vee]$ gives us $\mathcal{M}, w \Vdash \smallsmile p \vee \smallsmile q$. In addition, to check that $\smallsmile$ satisfies (DT#) one may invoke $[\mathrm{S}\smallsmile]$ and $[\mathrm{S}\top]$. Note that satisfying (DT#) means that the nullary connective $\bot$ taken as an abbreviation of $\#\top$ is interpretable by setting, for every world w of every model $\mathcal{M}$:

$[\mathrm{S}\bot]$ $\mathcal{M}, w \nVdash \bot$

Given a negation $\#$, we call the logic containing it *#-paraconsistent* if the following consecution fails, for $p, q \in \mathcal{P}$:

$[\![\#\text{-explosion}]\!]$ $\quad p, \#p \models q$

This means that there must be valuations that satisfy both some sentence φ and the sentence $\#\varphi$ while not satisfying every other sentence. It is worth noticing that $[\![\smile\text{-explosion}]\!]$ holds good in frames containing exclusively worlds that are accessible to themselves, and themselves only (call such worlds 'narcissistic') and worlds that do not access any other world (call them 'dead ends'): in the former case, it is impossible to simultaneously satisfy both φ and $\smile\varphi$; in the latter case, the sentence $\smile\varphi$ is never satisfied. Note moreover that in the class of all narcissistic frames the connective $\smile$ happens to behave like classical negation, i.e., it behaves like the symbol $\sim$ in the following semantic clause:

[S$\sim$] $\quad \mathcal{M}, w \Vdash \sim\varphi \quad$ iff $\quad \mathcal{M}, w \nVdash \varphi$

In contrast, in the class of all frames whose worlds are all dead ends the connective $\smile$ does not respect [verificatio], and cannot be said thus to be a negation.

We now make a further extension of the above language by adding a 1-ary connective $\frown$, non-locally interpreted as follows:

[S$\frown$] $\quad \mathcal{M}, w \Vdash \frown\varphi \quad$ iff $\quad \mathcal{M}, v \nVdash \varphi$ for <u>every</u> $v \in W$ such that wRv

It is not difficult to check that again we have a connective that qualifies as a negation, and satisfies global contraposition. To reinforce the meta-theoretical duality between the latter negation and the negation introduced above through [S$\smile$], we will henceforth refer to the previous interpretation clause in the following equivalent form:

[S$\frown$] $\quad \mathcal{M}, w \nVdash \frown\varphi \quad$ iff $\quad \mathcal{M}, v \Vdash \varphi$ for <u>some</u> $v \in W$ such that wRv

A *full type box-minus connective* is a 1-ary connective $\#$ that respects:

(DM1.2#) $\quad \#p \wedge \#q \models \#(p \vee q) \qquad$ (DF#) $\quad p \models \#\bot$

One may easily check that $\frown$ is indeed a full type box-minus connective.

Given a negation $\#$, we call the logic contaning it *#-paracomplete* if it fails the following consecution, for $p, q \in \mathcal{P}$:

$[\![\#\text{-implosion}]\!]$ $\quad q \models \#p, p$

Such failure will clearly be the case for $\# = \frown$ as soon as we entertain frames that contain worlds that are neither dead ends nor narcissistic. Otherwise, we see that $\frown$ will behave either like classical negation (if all worlds are narcissistic) or like $\top$ (if all worlds are dead ends).

In the following sections, unless noted otherwise, we will no longer consider classes of frames containing only frames with worlds that are either dead ends or narcissistic — so we will only consider entailment relations that are $\smile$-paraconsistent and $\frown$-paracomplete, for the negative modalities $\smile$ (assumed to be full-type diamond-minus) and $\frown$ (assumed to be full-type box-minus).

2.2 Recovering negation-consistency and negation-determinacy

In what follows we will call a model *dadaistic* when it contains some world in which all formulas are satisfied, and call it *nihilistic* if it leaves all formulas unsatisfied at some world. It is straightforward to see that the language based on $\wedge\vee\top\smile$, with the above interpretations, admits dadaistic models, while the language based on $\wedge\vee\bot\frown$ admits nihilistic models.

Recall that a #-paraconsistent logic allows for valuations that satisfy certain formulas φ and $\#\varphi$ while leaving some other formula ψ unsatisfied (at some fixed world). There might be reasons for disallowing this phenomenon to occur with an arbitrary φ, or for restricting to certain formulas ψ but not others. A particularly useful way of keeping a finer control over which 'inconsistencies' of the form φ and $\#\varphi$ are to be acceptable within non-dadaistic models is to mark down the formula thereby involved so as to recover a 'gentle' version of $[\![\#\text{-explosion}]\!]$. Concretely, for us here, a 1-ary connective $\circledast$ that strongly internalizes the meta-theoretic 'consistency assumption' at the object language level will be such that:

$$[\mathrm{SC}\#]\quad \mathcal{M},w\Vdash\circledast\varphi \quad\text{iff}\quad \mathcal{M},w\nVdash\varphi \text{ or } \mathcal{M},w\nVdash\#\varphi$$

It is easy to check that any connective $\circledast$ respecting [SC#] is such that:

$$(\mathrm{C1}\#)\quad \circledast p, p, \#p\vDash \qquad (\mathrm{C2}\#)\quad \vDash p,\circledast p \qquad (\mathrm{C3}\#)\quad \vDash\#p,\circledast p$$

Note in particular that (C1#) guarantees that there are no valuations that satisfy (at a fixed world) both p and $\#p$ if these are put in the presence of $\circledast p$. Thus, in case # fails $[\![\#\text{-explosion}]\!]$ we may look at the latter formula involving $\circledast$ as guaranteeing that a weaker form of explosion is available. On these grounds we shall call the connective $\circledast$ an *adjustment* companion to #: it allows one to recover explosion from within a non-#-explosive (i.e., paraconsistent) logical context, and adjust the consecutions of the underlying logic so as to allow for the simulation of the consecutions that would otherwise be justified by reference to $[\![\#\text{-explosion}]\!]$. Semantically, the presence of such connective also guarantees that dadaistic models are not admissible over the language based on $\wedge\vee\top\smile\ominus$, with the above interpretations. This is because a formula of the form $\ominus\varphi\wedge(\varphi\wedge\smile\varphi)$ is equivalent to a formula $\bot$ respecting [S$\bot$].

Dually, a #-paracomplete logic allows for valuations that leave the formulas φ and $\#\varphi$ both unsatisfied (at some fixed world), while satisfying some other formula ψ. A particular way of keeping a finer control over which 'indeterminacies' of the form φ and $\#\varphi$ are to be acceptable within non-nihilistic models is to allow for a 'gentle' version of $[\![\#\text{-implosion}]\!]$, where a 1-ary connective $\circledast$ internalizes the meta-theoretic 'determinacy assumption' at the object language level, in such a way that:

$$[\mathrm{SD}\#]\quad \mathcal{M},w\nVdash\circledast\varphi \quad\text{iff}\quad \mathcal{M},w\Vdash\varphi \text{ or } \mathcal{M},w\Vdash\#\varphi$$

Clearly, any connective $\circledast$ respecting [SD#] is such that:

$$(\mathrm{D1}\#)\quad \vDash\#p, p,\circledast p \qquad (\mathrm{D2}\#)\quad \circledast p, p\vDash \qquad (\mathrm{D3}\#)\quad \circledast p,\#p\vDash$$

Note that a formula of the form $(\varphi\vee\frown\varphi)\vee\ominus\varphi$ is equivalent to a formula $\top$ respecting [S$\top$]. Note, moreover, that whenever it turns out that a connective #

respects $[\![\#\text{-explosion}]\!]$ and at the same time its adjustment companion $\circledast$ respects [SC#], then the formula $\circledast\varphi$ is equivalent to $\top$. In an analogous way, whenever a connective # respects $[\![\#\text{-implosion}]\!]$ and at the same time its adjustment companion $\circledast$ respects [SD#], the formula $\circledast\varphi$ is equivalent to $\bot$. This stresses the fact that the adjustment connectives with which we deal in this subsection are more interesting when they accompany the respective non-classical negations to whose meaning they contribute.

At this point we have finally finished constructing the richest language that will be used throughout the rest of the paper: It will contain the connectives $\wedge\vee\top\bot\smile\ominus\frown\circleddash$, disciplined by the [S#] conditions above. In the following subsection we will explain precisely when a classical negation, that is a 1-ary connective $\sim$ subject to condition [S$\sim$], is definable with the use of our language. Fixed such language, the logic characterized over it by the class $\mathcal{E}$ of all frames will be called PK; the logic characterized by the class $\mathcal{E}_{\mathbf{D}}$ of all frames with serial accessibility relations will be called PKD; the logic characterized by the class $\mathcal{E}_{\mathbf{T}}$ of all frames with reflexive accessibility relations will be called PKT; the logic characterized by the class $\mathcal{E}_{\mathbf{Fun}}$ of all frames whose accessibility relations are total functions will be called PKF; the logic characterized by the class $\mathcal{E}_{\mathbf{B}}$ of all symmetric frames (those with symmetric accessibility relations) will be called PKB.

2.3 Around classical negation

According to the intuitions laid down at Section 1, one could expect that in general (a) $\frown\alpha \vdash \sim\alpha$ and (b) $\sim\alpha \vdash \smile\alpha$. It is easy to see that these consecutions are sanctioned by PKT, for the classical negation $\sim$ that may be defined by setting $\sim\varphi := \smile\varphi \wedge \ominus\varphi$ (alternatively, one may set $\sim\varphi := \frown\varphi \vee \circleddash\varphi$).

Meanwhile, in the deductively weaker logic PKD one cannot in general prove (a) nor (b), even though a classical negation may be defined in this logic by setting $\sim\varphi := (\frown\varphi \wedge \ominus\varphi) \vee \circleddash\varphi$. However, one can still easily prove in PKD that (c) $\frown\alpha \vdash \smile\alpha$. In the logic PKF, deductively stronger than PKD (but neither stronger nor weaker than PKT) one may also prove the converse consecution, (d) $\smile\alpha \vdash \frown\alpha$. Indeed, suppose $\mathcal{M}, w \Vdash \smile\alpha$. There is, by the fact that the accessibility relation is a total function, a single world v such that wRv. Then $\mathcal{M}, v \nVdash \alpha$, by [S$\smile$]. For a similar reason, invoking now [S$\frown$] we conclude that $\mathcal{M}, w \Vdash \frown\alpha$. Note that (c) and (d) together make our two modal non-classical negations indistinguishable from the viewpoint of PKF, yet there would still be no reason for them to collapse into classical negation.

The situation concerning classical negation and its relation to its non-classical neighbours gets even more interesting if one acknowledges that *no* classical negation is definable in PK, the weakest of our logics, but also no classical negation is definable in the fragment of PKT without neither of the adjustment connectives, or in the fragment of PKF (or PKD) without either one of the adjustment connectives, or in PKB. Detailed proofs concerning the mentioned results about (non)definability of classical negation in the weak modal logics that constitute our present object of study may be found in Section 6.

Notice that in PKD and its extensions there are no negated formulas that happen to be true or false at a given world just because there are no worlds accessible from it. Note also that the logic PKT is: paraconsistent but not paracomplete with respect to the connective $\smallsmile$; paracomplete but not paraconsistent with respect to $\smallfrown$ (even though we will not prove it here, this logic is indeed the least extension of the positive implicationless fragment of classical logic with the latter mentioned properties). The logic PKT will have its word in the following sections, for it also allows for the straightforward application of the techniques that will be hereby illustrated. In the other four mentioned logics, in contrast, both non-classical negations behave at once as paracomplete and paraconsistent negations (recall, though, that each is associated to a different adjustment connective). We take the cases among these in which no classical negation is available to be particularly attractive for the task of revealing the 'uncontamined' nature of non-classical negation. Establishing well-behaved proof theoretical counterparts for such logics, as we shall do in what follows, is meant to allow for them to be better understood and dealt with.

3 A proof system for PK

A sequent calculus for PK, that we denote by PK, was introduced in [5], and consists of the following rules:

$$[id]\ \frac{}{\Gamma,\varphi\Rightarrow\varphi,\Delta} \qquad [cut]\ \frac{\Gamma,\varphi\Rightarrow\Delta \quad \Gamma\Rightarrow\varphi,\Delta}{\Gamma\Rightarrow\Delta}$$

$$[W{\Rightarrow}]\ \frac{\Gamma\Rightarrow\Delta}{\Gamma,\varphi\Rightarrow\Delta} \qquad [{\Rightarrow}W]\ \frac{\Gamma\Rightarrow\Delta}{\Gamma\Rightarrow\varphi,\Delta}$$

$$[\bot{\Rightarrow}]\ \frac{}{\Gamma,\bot\Rightarrow\Delta} \qquad [{\Rightarrow}\top]\ \frac{}{\Gamma\Rightarrow\top,\Delta}$$

$$[\wedge{\Rightarrow}]\ \frac{\Gamma,\varphi,\psi\Rightarrow\Delta}{\Gamma,\varphi\wedge\psi\Rightarrow\Delta} \qquad [{\Rightarrow}\wedge]\ \frac{\Gamma\Rightarrow\varphi,\Delta \quad \Gamma\Rightarrow\psi,\Delta}{\Gamma\Rightarrow\varphi\wedge\psi,\Delta}$$

$$[\vee{\Rightarrow}]\ \frac{\Gamma,\varphi\Rightarrow\Delta \quad \Gamma,\psi\Rightarrow\Delta}{\Gamma,\varphi\vee\psi\Rightarrow\Delta} \qquad [{\Rightarrow}\vee]\ \frac{\Gamma\Rightarrow\varphi,\psi,\Delta}{\Gamma\Rightarrow\varphi\vee\psi,\Delta}$$

$$[\smallsmile{\Rightarrow}]\ \frac{\Gamma\Rightarrow\varphi,\Delta}{\smallfrown\Delta,\smallsmile\varphi\Rightarrow\smallsmile\Gamma} \qquad [{\Rightarrow}\smallfrown]\ \frac{\Gamma,\varphi\Rightarrow\Delta}{\smallfrown\Delta\Rightarrow\smallfrown\varphi,\smallsmile\Gamma}$$

$$[\ominus{\Rightarrow}]\ \frac{\Gamma\Rightarrow\varphi,\Delta \quad \Gamma\Rightarrow\smallsmile\varphi,\Delta}{\Gamma,\ominus\varphi\Rightarrow\Delta} \qquad [{\Rightarrow}\ominus]\ \frac{\Gamma,\varphi,\smallsmile\varphi\Rightarrow\Delta}{\Gamma\Rightarrow\ominus\varphi,\Delta}$$

$$[\circleddash{\Rightarrow}]\ \frac{\Gamma\Rightarrow\varphi,\smallfrown\varphi,\Delta}{\Gamma,\circleddash\varphi\Rightarrow\Delta} \qquad [{\Rightarrow}\circleddash]\ \frac{\Gamma,\varphi\Rightarrow\Delta \quad \Gamma,\smallfrown\varphi\Rightarrow\Delta}{\Gamma\Rightarrow\circleddash\varphi,\Delta}$$

Above, sequents are taken to have the form $\Sigma\Rightarrow\Pi$ where Σ and Π are finite sets of formulas, and given a unary connective $\#$ and $\Psi\subseteq\mathcal{L}$, by $\#\Psi$ we denote the set $\{\#\psi \mid \psi\in\Psi\}$. We write $S\vdash_{\mathrm{PK}} s$ to say that there is a derivation in PK of a sequent s from a set S of sequents. That establishes a consequence relation between sequents. A consequence relation between formulas is defined by setting $\Gamma\vdash_{\mathrm{PK}}\varphi$ if $\vdash_{\mathrm{PK}}\Gamma'\Rightarrow\varphi$ for some finite subset Γ' of Γ. The overloaded notation $\vdash_{\text{PK}}$ will always be resolved by the pertinent context.

Next, we utilize in what follows the general mechanisms and methods applicable to the so-called 'basic systems' of [10] in order to prove soundness, completeness and cut-admissibility. From the viewpoint of basic systems, each sequent is seen as a union of a 'main sequent' and a 'context sequent'. For example, in $[\Rightarrow\vee]$, the main sequent of the premise is $\Rightarrow \varphi, \psi$; the main sequent of the conclusion is $\Rightarrow \varphi \vee \psi$; and the context sequent of both is $\Gamma \Rightarrow \Delta$. Note that in the rules for $\smile$ and $\frown$, the context sequent of the premise is different from the one of the conclusion. Accordingly, [10] introduces the notion of a *basic rule*, whose premises take the form $\langle s, \pi\rangle$, where s is a sequent that corresponds to the main sequent of the premise, and π is a relation between singleton-sequents (that is, sequents of the form $\varphi \Rightarrow$ or $\Rightarrow \varphi$) called a *context relation* that determines the behavior of the context sequents. The sequent calculus PK may be naturally regarded as a basic system that employs two context relations, namely: $\pi_0 = \{\langle p_1 \Rightarrow\ ;\ p_1 \Rightarrow\rangle, \langle \Rightarrow p_1\ ;\ \Rightarrow p_1\rangle\}$, and $\pi_1 = \{\langle p_1 \Rightarrow\ ;\ \Rightarrow {\smile} p_1\rangle, \langle \Rightarrow p_1\ ;\ {\frown} p_1 \Rightarrow\rangle\}$. The rules of PK may then be presented as particular instances of basic rules. For example, the following are the basic rules for $\wedge, \smile, \frown$ and $\ominus$:

$[\Rightarrow\wedge]$ $\langle \Rightarrow p_1; \pi_0\rangle, \langle \Rightarrow p_2; \pi_0\rangle \;/\; \Rightarrow p_1 \wedge p_2$ $\quad$ $[\wedge\Rightarrow]$ $\langle p_1, p_2 \Rightarrow; \pi_0\rangle \;/\; p_1 \wedge p_2 \Rightarrow$

$[\smile\Rightarrow]$ $\langle \Rightarrow p_1; \pi_1\rangle \;/\; {\smile} p_1 \Rightarrow$ $\quad$ $[\Rightarrow\frown]$ $\langle p_1 \Rightarrow; \pi_1\rangle \;/\; \Rightarrow {\frown} p_1$

$[\ominus\Rightarrow]$ $\langle \Rightarrow p_1; \pi_0\rangle, \langle \Rightarrow {\smile} p_1; \pi_0\rangle \;/\; \ominus p_1 \Rightarrow$ $\quad$ $[\Rightarrow\ominus]$ $\langle p_1, {\smile} p_1 \Rightarrow; \pi_0\rangle \;/\; \Rightarrow \ominus p_1$

In applications of $[\Rightarrow\ominus]$, the context sequent is left unchanged, as two singleton-sequents relate to each other (with respect to π_0) iff they are the same. In contrast, applications of $[\smile\Rightarrow]$ are based on π_1. A sequent $\Gamma_1 \Rightarrow \Delta_1$ relates (with respect to π_1) to a sequent $\Gamma_2 \Rightarrow \Delta_2$ iff $\Gamma_2 = {\frown}\Delta_1$ and $\Delta_2 = {\smile}\Gamma_1$.

We extend the notion of satisfaction from Section 2 to sequents by setting $\mathcal{M}, w \Vdash \Gamma \Rightarrow \Delta$ if $\mathcal{M}, w \nVdash \gamma$ for some $\gamma \in \Gamma$ or $\mathcal{M}, w \Vdash \delta$ for some $\delta \in \Delta$. Semantics for PK may then be obtained using the general method introduced in [10], by having each derivation rule and each context relation match a semantic condition, and the semantics of the system is obtained by conjoining all these semantic conditions. For example, the basic rule $[\smile\Rightarrow]$ induces the condition: "if $\mathcal{M}, v \Vdash\ \Rightarrow \varphi$ for every world v such that wRv, then $\mathcal{M}, w \Vdash {\smile}\varphi \Rightarrow$", which is equivalent to: "If $\mathcal{M}, w \Vdash {\smile}\varphi$ then $\mathcal{M}, v \nVdash \varphi$ for some $v \in W$ such that wRv". This is half of clause $[\text{S}{\smile}]$, from Section 2. Furthermore, the context relation π_1 induces an additional semantic condition: "if wRv then $\mathcal{M}, w \Vdash\ \Rightarrow {\smile}\varphi$ whenever $\mathcal{M}, v \Vdash \varphi \Rightarrow$". This amounts to the other half of clause $[\text{S}{\smile}]$, namely: "$\mathcal{M}, w \Vdash {\smile}\varphi$ whenever $\mathcal{M}, v \nVdash \varphi$ for some $v \in W$ such that wRv". Systematically applying this semantic reading to all rules and all context relations of PK (according to Definitions 4.5 and 4.12 of [10]), one obtains soundness and completeness with respect to the class of all Kripke models $\langle \mathcal{F}, V\rangle$, where $\mathcal{F}$ is an arbitrary frame and each valuation $V : W \times \mathcal{L} \to \{f, t\}$ respects the following conditions, for every $w \in W$ and $\varphi, \psi \in \mathcal{L}$:

$[\mathbf{T}\top]$ $\quad \mathbf{T}_w(\top)$

$[\mathbf{F}\bot]$ $\quad \mathbf{F}_w(\bot)$

[$\mathbf{T}\wedge$] if $\mathbf{T}_w(\varphi)$ and $\mathbf{T}_w(\psi)$, then $\mathbf{T}_w(\varphi \wedge \psi)$
[$\mathbf{F}\wedge$] if $\mathbf{F}_w(\varphi)$ or $\mathbf{F}_w(\psi)$, then $\mathbf{F}_w(\varphi \wedge \psi)$

[$\mathbf{T}\vee$] if $\mathbf{T}_w(\varphi)$ or $\mathbf{T}_w(\psi)$, then $\mathbf{T}_w(\varphi \vee \psi)$
[$\mathbf{F}\vee$] if $\mathbf{F}_w(\varphi)$ and $\mathbf{F}_w(\psi)$, then $\mathbf{F}_w(\varphi \vee \psi)$

[$\mathbf{T}\smallsmile$] if $\mathbf{F}_v(\varphi)$ for some $v \in W$ such that wRv, then $\mathbf{T}_w(\smallsmile\varphi)$
[$\mathbf{F}\smallsmile$] if $\mathbf{T}_v(\varphi)$ for every $v \in W$ such that wRv, then $\mathbf{F}_w(\smallsmile\varphi)$

[$\mathbf{T}\smallfrown$] if $\mathbf{F}_v(\varphi)$ for every $v \in W$ such that wRv, then $\mathbf{T}_w(\smallfrown\varphi)$
[$\mathbf{F}\smallfrown$] if $\mathbf{T}_v(\varphi)$ for some $v \in W$ such that wRv, then $\mathbf{F}_w(\smallfrown\varphi)$

[$\mathbf{T}\ominus$] if $\mathbf{F}_w(\varphi)$ or $\mathbf{F}_w(\smallsmile\varphi)$, then $\mathbf{T}_w(\ominus\varphi)$
[$\mathbf{F}\ominus$] if $\mathbf{T}_w(\varphi)$ and $\mathbf{T}_w(\smallsmile\varphi)$, then $\mathbf{F}_w(\ominus\varphi)$

[$\mathbf{T}\circleddash$] if $\mathbf{F}_w(\varphi)$ and $\mathbf{F}_w(\smallfrown\varphi)$, then $\mathbf{T}_w(\circleddash\varphi)$
[$\mathbf{F}\circleddash$] if $\mathbf{T}_w(\varphi)$ or $\mathbf{T}_w(\smallfrown\varphi)$, then $\mathbf{F}_w(\circleddash\varphi)$

where we take '$\mathbf{T}_u(\alpha)$' as abbreviating '$V(u,\alpha) = t$', and '$\mathbf{F}_u(\alpha)$' as abbreviating '$V(u,\alpha) = f$'. If alternatively one just *rewrites* $V(v,\alpha) = t$ as $\mathcal{M}, v \Vdash \alpha$ and rewrites $V(v,\alpha) = f$ as $\mathcal{M}, v \nVdash \alpha$, where $\mathcal{M} = \langle\langle W, R\rangle, V\rangle$, what results thereby is a collection of conditions that are essentially identical to the [S#] clauses introduced in our Section 2.

Two brief comments are in order here. First, our valuation functions assign truth-values to *every* formula in every world. However, as the values of compound formulas are uniquely determined by the values of their subformulas, we could have rested content above with assigning truth-values to propositional variables. Second, given that for the above valuations $\mathbf{T}_u(\alpha)$ is the case iff $\mathbf{F}_u(\alpha)$ fails to be the case, the semantic conditions [**T**#] and [**F**#], for each connective #, are clearly the converse of each other. In setting the two conditions apart, we have just given them directionality, pointing from less complex to more complex formulas, and have separated between conditions induced by rules from those induced by context relations. While neither of these manoeuvres are very useful here, they will allow us to more easily relate, in Section 4, valuations to 'quasi valuations' that have non truth-functional semantics.

Fix in what follows a Kripke model $\mathcal{M} = \langle\langle W, R\rangle, V\rangle$. We say that $w, v \in W$ *agree with respect to the formula* α, *according to* V, if either ($\mathbf{T}_w(\alpha)$ and $\mathbf{T}_v(\alpha)$) or ($\mathbf{F}_w(\alpha)$ and $\mathbf{F}_v(\alpha)$). We say that $\mathcal{M}$ is *differentiated* if we have $w = v$ whenever w and v agree with respect to every $\alpha \in \mathcal{L}$, according to V. We call $\mathcal{M}$ a *strengthened* model if wRv iff ($\mathbf{T}_v(\alpha)$ implies $\mathbf{F}_w(\smallfrown\alpha)$) and ($\mathbf{F}_v(\alpha)$ implies $\mathbf{T}_w(\smallsmile\alpha)$), for every $\alpha \in \mathcal{L}$. It is worth stressing that the accessibility relation of a strengthened model is uniquely determined by the underlying collection of worlds and valuation. The following result follows directly from Corollary 4.26 in [10], thus there is no need to prove it again here:

Theorem 3.1 PK *is sound and complete with respect to any class of Kripke models that: (i) contains only models that satisfy all the above* [**T**#] *and* [**F**#] *conditions; and (ii) contains all strengthened differentiated models that satisfy all the above* [**T**#] *and* [**F**#] *conditions.*

This theorem provides a mechanism that will be recycled in the subsequent sections, when we consider extensions of PK. The following result from [5] comes as a byproduct of it:

Corollary 3.2 $\Gamma \models_{\mathcal{E}} \varphi$ *iff* $\Gamma \vdash_{\mathrm{PK}} \varphi$ *for every* $\Gamma \cup \{\varphi\} \subseteq \mathcal{L}$*, where* $\mathcal{E}$ *denotes the class of all frames.*

4 (Almost) Free Lunch: cut-elimination and analyticity

In this section we make further use of the powerful machinery introduced in [10] to prove that PK enjoys strong cut-admissibility, in other words, we show that $S \vdash_{\mathrm{PK}} s$ implies that there is a derivation in PK of the sequent s from the set of sequents S such that in every application of the cut rule the cut formula φ appears in S. In particular, $\vdash_{\mathrm{PK}} s$ implies that s is derivable in PK without any use of the cut rule. The proof is done in two steps. *First*, we present an adequate semantics for the cut-free fragment of PK. *Second*, we show that a countermodel in this new semantics entails the existence of a countermodel in the form of a Kripke model as defined in the previous section. This, together with Corollary 3.2, entails that PK is equivalent to its cut-free fragment.

Step 1. Semantics for cut-free PK

Semantics for cut-free basic systems may be obtained through the use of 'quasi valuations'. Models based on quasi valuations differ from usual Kripke models in two main aspects: (*a*) the underlying interpretation is three-valued; (*b*) the underlying interpretation is non-deterministic — the truth-value of a compound formula in a given world is *not* always uniquely determined by the truth values of its subformulas in the collection of worlds of the underlying frame.

To obtain such semantics for PK, as before, one reads off a semantic condition on quasi valuations from each derivation rule and from each context relation. As per Theorems 5.24 and 5.31 of [10], we know that the class of models based on quasi valuations that respect all these conditions is sound and complete for the cut-free fragment of PK. Concretely, given a frame $\mathcal{F} = \langle W, R\rangle$, a *quasi valuation* over it is a function $QV : W \times \mathcal{L} \to \{\{f\}, \{t\}, \{f,t\}\}$ satisfying precisely the same semantic conditions laid down in Section 3, where we now take '$\mathbf{T}_u(\alpha)$' as abbreviating '$t \in V(u,\alpha)$', and '$\mathbf{F}_u(\alpha)$' as abbreviating '$f \in V(u,\alpha)$'. Whenever we need to distinguish between a semantic condition on a tuple $\langle w, \varphi\rangle$ as constraining a valuation V or a quasi valuation QV, we will use $\mathbf{X}_w(\varphi)$ for the former and $\mathbf{X}^Q_w(\varphi)$ for the latter, where $\mathbf{X} \in \{\mathbf{T}, \mathbf{F}\}$. A *quasi model* is a structure $\mathcal{QM} = \langle \mathcal{F}, QV\rangle$, where QV is a quasi valuation over $\mathcal{F}$. The notions of a differentiated quasi model and of a strengthened quasi model are defined as before, assuming the same abbreviations.

Step 2. Semantic cut-admissibility

The next step is to show that the existence of a countermodel in the form of a strengthened differentiated quasi model implies the existence a countermodel in the form of an ordinary Kripke model (following Corollary 5.48 of [10]). For this purpose we define an *instance* of a quasi model $\mathcal{QM} = \langle\langle W, R\rangle, QV\rangle$ as any model of the form $\mathcal{M} = \langle\langle W, R'\rangle, V\rangle$ such that $\mathbf{X}^Q_w(\varphi)$ whenever $\mathbf{X}_w(\varphi)$, for every $\mathbf{X} \in \{\mathbf{T}, \mathbf{F}\}$, every $w \in W$ and every $\varphi \in \mathcal{L}$. Note that a quasi model and its instances may have different accessibility relations.

In what follows, the construction of appropriate instances is done by a recursive definition over the following well-founded relation $\prec$ on the set of formulas: $\alpha \prec \beta$ if either (i) α is a proper subformula of β; (ii) $\alpha = \smile\gamma$ and $\beta = \ominus\gamma$ for some $\gamma \in \mathcal{L}$; or (iii) $\alpha = \frown\gamma$ and $\beta = \circleddash\gamma$ for some $\gamma \in \mathcal{L}$.

Lemma 4.1 *Every quasi model has an instance.*

Proof. Let $\mathcal{QM} = \langle \mathcal{F}, QV \rangle$ be a quasi model based on a frame $\mathcal{F} = \langle W, R \rangle$. We set us now an appropriate valuation $V : W \times \mathcal{L} \to \{f, t\}$. For every world w and formula φ, the valuation V is inductively defined (with respect to $\prec$) on φ as follows: (R1) if $\mathbf{T}^Q_w(\varphi)$ fails for QV, we postulate $\mathbf{F}_w(\varphi)$ to be the case for V; (R2) if $\mathbf{F}^Q_w(\varphi)$ fails for QV, we postulate $\mathbf{T}_w(\varphi)$ to be the case for V; (R3) otherwise both $\mathbf{T}^Q_w(\varphi)$ and $\mathbf{F}^Q_w(\varphi)$ hold good for QV, and in this case we postulate $\mathbf{T}_w(\varphi)$ to be the case for V if one of the following holds:

(M1)	φ is a propositional variable or φ is $\top$
(M2)	$\varphi = \varphi_1 \wedge \varphi_2$, and both $\mathbf{T}_w(\varphi_1)$ and $\mathbf{T}_w(\varphi_2)$
(M3)	$\varphi = \varphi_1 \vee \varphi_2$, and either $\mathbf{T}_w(\varphi_1)$ or $\mathbf{T}_w(\varphi_2)$
(M4)	$\varphi = \smile\psi$, and $\mathbf{F}_v(\psi)$ for some $v \in W$ such that wRv
(M5)	$\varphi = \frown\psi$, and $\mathbf{F}_v(\psi)$ for every $v \in W$ such that wRv
(M6)	$\varphi = \ominus\psi$, and either $\mathbf{F}_w(\psi)$ or $\mathbf{F}_w(\smile\psi)$
(M7)	$\varphi = \circleddash\psi$, and both $\mathbf{F}_w(\psi)$ and $\mathbf{F}_w(\frown\psi)$

Otherwise, we postulate $\mathbf{F}_w(\varphi)$ to be the case for V. Obviously, $\mathbf{X}_w(\varphi)$ implies $\mathbf{X}^Q_w(\varphi)$ for every $w \in W$, every $\varphi \in \mathcal{L}$ and every $\mathbf{X} \in \{\mathbf{T}, \mathbf{F}\}$. It is routine to verify that $\langle \mathcal{F}, V \rangle$ is a model. We show here that the semantic conditions for $\smile$ and $\ominus$ hold:

[**Case of** $\smile$] Let $\psi \in \mathcal{L}$. Suppose first that $\mathbf{F}_v(\psi)$ is the case for some $v \in W$ such that wRv. Then $\mathbf{F}^Q_v(\psi)$. Since $\mathcal{QM}$ is a quasi model, then $\mathbf{T}^Q_w(\smile\psi)$ is the case. If, on the one hand, $\mathbf{F}^Q_w(\smile\psi)$ fails, then we must have $\mathbf{T}_w(\smile\psi)$, by (R2). If, on the other hand, neither $\mathbf{T}^Q_w(\smile\psi)$ nor $\mathbf{F}^Q_w(\smile\psi)$ fail, we are in case (R3). Since we have $\mathbf{F}_v(\psi)$ and wRv we conclude by (M4) that $\mathbf{T}_w(\smile\psi)$ must be the case. Suppose now that $\mathbf{T}_v(\psi)$ is the case for every world v such that wRv. Then we have $\mathbf{T}^Q_v(\psi)$ for every such world. Since $\mathcal{QM}$ is a quasi model, it follows that $\mathbf{F}^Q_w(\smile\psi)$ is the case. If, on the one hand, $\mathbf{T}^Q_w(\smile\psi)$ fails, then we must have $\mathbf{F}_w(\smile\psi)$, by (R1). If, on the other hand, neither $\mathbf{T}^Q_w(\smile\psi)$ nor $\mathbf{F}^Q_w(\smile\psi)$ fail, we are in case (R3). Since we have $\mathbf{T}_v(\psi)$ for every world v such that wRv we conclude that none of (M1)–(M7) applies, thus $\mathbf{F}_w(\smile\psi)$ must be the case.

[**Case of** $\ominus$] Let $\psi \in \mathcal{L}$. Suppose first that either $\mathbf{F}_w(\psi)$ or $\mathbf{F}_w(\smile\psi)$ are the case for some $w \in W$. Then either $\mathbf{F}^Q_w(\psi)$ or $\mathbf{F}^Q_w(\smile\psi)$. Since $\mathcal{QM}$ is a quasi model, it follows that $\mathbf{T}^Q_w(\ominus\psi)$. If, on the one hand, $\mathbf{F}^Q_w(\ominus\psi)$ fails, then we must have $\mathbf{T}_w(\ominus\psi)$, by (R2). If, on the other hand, neither $\mathbf{T}^Q_w(\ominus\psi)$ nor $\mathbf{F}^Q_w(\ominus\psi)$ fail, we are in case (R3) and we conclude by (M6) that $\mathbf{T}_w(\ominus\psi)$ must be the case. Suppose now that both $\mathbf{T}_w(\psi)$ and $\mathbf{T}_w(\smile\psi)$ are the case for some $w \in W$. Then $\mathbf{T}^Q_w(\psi)$ and $\mathbf{T}^Q_w(\smile\psi)$. Since $\mathcal{QM}$ is a quasi model, then $\mathbf{F}^Q_w(\ominus\psi)$. If, on the one hand, $\mathbf{T}^Q_w(\ominus\psi)$ fails, then we must have $\mathbf{F}_w(\ominus\psi)$, by (R1). If, on the other hand, neither $\mathbf{T}^Q_w(\ominus\psi)$ nor $\mathbf{F}^Q_w(\ominus\psi)$ fail, we are in case (R3) and $\mathbf{F}_w(\ominus\psi)$ must be the case because none of (M1)–(M7) applies. □

Since the class of all quasi models contains the strengthened differentiated quasi models, it follows that:

Corollary 4.2 PK *enjoys strong cut-admissibility.*

Corollary 4.3 PK *is $\prec$-analytic: If a sequent s is derivable from a set S of sequents in* PK, *then there is a derivation of s from S such that every formula φ that occurs in the derivation satisfies $\varphi \prec \psi$ for some ψ in $S \cup s$.*

Proof. By induction on the length of the derivation of s from S in PK: In all rules except for (*cut*), the premises include only formulas φ that satisfy $\varphi \prec \psi$ for some formula ψ in the conclusion. □

5 Some special classes of frames

In this section we present three very natural deductive extensions of PK. Given a property X of binary relations, we call a frame $\langle W, R\rangle$ an X *frame* if R enjoys X. A (quasi) model $\langle \mathcal{F}, V\rangle$ is called an X (*quasi*) *model* if $\mathcal{F}$ is an X frame. In addition, and similarly to what we did in the case of PK, for every proof system Y we write $S \vdash_Y s$ if there is a derivation of s from S in Y.

5.1 Seriality

Let PKD be the system obtained by augmenting PK with the following rule:

$$[\mathbf{D}] \quad \frac{\Gamma \Rightarrow \Delta}{\frown\Delta \Rightarrow \smile\Gamma}$$

This rule may be formulated as the basic rule: $\langle \Rightarrow\ ;\ \pi_1\rangle\ /\ \Rightarrow$. Since its premise is the empty sequent, the semantic condition it imposes (following [10]) is seriality: indeed, respecting $[\mathbf{D}]$ in a world w of a model $\mathcal{M}$ based on a frame $\langle W, R\rangle$ means that if $\mathcal{M}, v \Vdash\ \Rightarrow$ for every world v such that wRv, then also $\mathcal{M}, w \Vdash\ \Rightarrow$. Since the empty sequent is not satisfied at any world, this condition would hold iff for every world w there exists a world v such that wRv. A similar argument shows that every serial frame satisfies this semantic condition.

As in Corollary 3.2, we obtain a completeness theorem for PKD with respect to serial models:

Corollary 5.1 $\Gamma \models_{\mathcal{E}_\mathbf{D}} \varphi$ *iff* $\Gamma \vdash_{\mathrm{PKD}} \varphi$ *for every* $\Gamma \cup \{\varphi\} \subseteq \mathcal{L}$, *where* $\mathcal{E}_\mathbf{D}$ *is the class of serial models.*

Additionally, we may prove cut-admissibility also for PKD, going through serial quasi models.

Lemma 5.2 *Every serial quasi model has a serial instance.*

Proof. The proof is the same as the proof of Lemma 4.1. Note indeed that no property of the accessibility relation was assumed, and the constructed instance has the same accessibility relation as the original quasi model. □

Corollary 5.3 PKD *enjoys cut-admissibility and is $\prec$-analytic.*

5.2 Reflexivity

Let PKT be the system obtained by augmenting PK with the following rules:

$$[\Rightarrow\smile]\ \frac{\Gamma, \varphi \Rightarrow \Delta}{\Gamma \Rightarrow \smile\varphi, \Delta} \qquad [\frown\Rightarrow]\ \frac{\Gamma \Rightarrow \varphi, \Delta}{\Gamma, \frown\varphi \Rightarrow \Delta}$$

These rules may be formulated as the basic rules: $\langle p_1 \Rightarrow ;\ \pi_0\rangle\ /\ \Rightarrow \smile p_1$ and $\langle\Rightarrow p_1 ;\ \pi_0\rangle\ /\ \frown p_1 \Rightarrow$. It should be clear that PKT allows thus for the derivation of the consecutions representing $[\![\smile$-implosion$]\!]$ and $[\![\frown$-explosion$]\!]$.

Semantically, they impose reflexivity not on all models, but only on *strengthened* models. Indeed, since the underlying context relation is π_0, for every model $\mathcal{M} = \langle\mathcal{F}, V\rangle$ based on a frame $\mathcal{F} = \langle W, R\rangle$ that respects $[\Rightarrow\smile]$ and $[\frown\Rightarrow]$, and every world w, if $\mathcal{M}, w \models \varphi \Rightarrow$ then $\mathcal{M}, w \models \Rightarrow \smile\varphi$ and if $\mathcal{M}, w \models \Rightarrow \varphi$ then $\mathcal{M}, w \models \frown\varphi \Rightarrow$. To put it otherwise, if $\mathbf{F}_w(\varphi)$ then $\mathbf{T}_w(\smile\varphi)$, and if $\mathbf{T}_w(\varphi)$ then $\mathbf{F}_w(\frown\varphi)$. Clearly, every reflexive model satisfies these conditions. To show that every strengthened model that satisfies them is reflexive, consider an arbitrary strengthened model $\mathcal{M} = \langle\langle W, R\rangle, V\rangle$. Then for every world $w \in W$ we have that for every formula φ, ($\mathbf{T}_w(\varphi)$ implies $\mathbf{F}_w(\frown\varphi)$) and ($\mathbf{F}_w(\varphi)$ implies $\mathbf{T}_w(\smile\varphi)$), which in strengthened models means precisely that wRw. We obtain thus a completeness theorem for PKT with respect to reflexive models:

Corollary 5.4 $\Gamma \models_{\mathcal{E}_\mathbf{T}} \varphi$ *iff* $\Gamma \vdash_{\text{PKT}} \varphi$ *for every* $\Gamma \cup \{\varphi\} \subseteq \mathcal{L}$*, where* $\mathcal{E}_\mathbf{T}$ *is the class of reflexive models.*

Such semantics for PKT allows one to easily confirm that the full type diamond-minus connective $\smile$ fails (DM1.2#), and that the full type box-minus connective $\frown$ fails (DM2.2#). These properties transfer to the weaker logics PKD and PK, of course.

Cut-admissibility for PKT may be obtained using arguments similar to those used in proving Lemma 4.1. It follows thus that:

Lemma 5.5 *Every reflexive strengthened quasi model has a reflexive instance.*

Corollary 5.6 PKT *enjoys cut-admissibility and is* $\prec$*-analytic.*

5.3 Functionality

In this section we address functional frames, that is, frames whose accessibility relations are *total functions*. In every model $\langle\langle W, R\rangle, V\rangle$ of a functional frame and world $w \in W$, we have $\mathbf{T}_w(\smile\varphi)$ iff $\mathbf{T}_w(\frown\varphi)$. Hence $\smile$ and $\frown$ are indistinguishable. Accordingly, here we consider a restricted language, without $\frown$.

Let PKF be the system obtained from PK by substituting $\smile$ for $\frown$ in rules $[\Rightarrow\ominus]$ and $[\ominus\Rightarrow]$, and replacing both rules $[\Rightarrow\frown]$ and $[\smile\Rightarrow]$ with the single rule:

$$[\mathbf{Fun}]\ \frac{\Gamma \Rightarrow \Delta}{\smile\Delta \Rightarrow \smile\Gamma}$$

It is straightforward to see that rule [**Fun**] may be formulated as the following basic rule: $\langle\Rightarrow ;\ \pi_2\rangle\ /\ \Rightarrow$, for $\pi_2 = \{\langle\varphi \Rightarrow ;\ \Rightarrow \smile\varphi\rangle, \langle\Rightarrow \varphi ;\ \smile\varphi \Rightarrow\rangle\}$.

The latter rule and context relation impose functionality on *differentiated* models. Indeed, respecting the basic rule [**Fun**] corresponds to seriality, similarly to the case of the rule [**D**]. Additionally, the context relation π_2 forces the accessibility relation to be a partial function: respecting π_2 in a world w of a model $\mathcal{M} = \langle\langle W, R\rangle, V\rangle$ means that for every $v_1, v_2 \in W$ such that wRv_1 and wRv_2 and for every formula φ we have that $\mathbf{T}_{v_1}(\varphi)$ iff $\mathbf{F}_w(\smile\varphi)$ iff $\mathbf{T}_{v_2}(\varphi)$. When $\mathcal{M}$ is differentiated, this implies that $v_1 = v_2$. Now, every functional model satisfies these semantic conditions and every differentiated model that satisfies them is functional. We thus obtain a completeness result for PKF with respect to functional models:

Corollary 5.7 $\Gamma \models_{\mathcal{E}_{\mathbf{Fun}}} \varphi$ *iff* $\Gamma \vdash_{\mathrm{PKF}} \varphi$ *for every* $\Gamma \cup \{\varphi\} \subseteq \mathcal{L}$, *where* $\mathcal{E}_{\mathbf{Fun}}$ *is the class of functional models.*

In contrast with what was the case for PKT, within such semantics for PKF there are no longer countermodels for (DM1.2$\smile$) or for (DM2.2$\frown$). At any rate, it should be clear that PKF extends PKD, but does not extend PKT.

Going through quasi models we may prove cut-admissibility also for PKF. However, unlike in previous cases, considering functional quasi models will not suffice. Indeed, there exist differentiated strengthened quasi models that respect [**Fun**] whose accessibility relation is not a total function. Let a **Fun** quasi model $\mathcal{QM} = \langle\mathcal{F}, QV\rangle$ based on a frame $\mathcal{F} = \langle W, R\rangle$ be a serial quasi model in which for every $w, v \in W$ such that wRv we have, for every $\varphi \in \mathcal{L}$, both ($\mathbf{F}^Q_v(\varphi)$ implies $\mathbf{T}^Q_w(\smile\varphi)$) and ($\mathbf{T}^Q_v(\varphi)$ implies $\mathbf{F}^Q_w(\smile\varphi)$). We note that, although the accessibility relation in **Fun** quasi models may not be a total function, we are still able to extract a functional model from it:

Lemma 5.8 *Every* **Fun** *quasi model has a functional instance.*

Proof. Let $\mathcal{QM} = \langle\mathcal{F}, QV\rangle$ be an **Fun** quasi model based on a frame $\langle W, R\rangle$. Since $\mathcal{QM}$ is an **Fun** quasi model, we have in particular that R is serial. Therefore, there exists some total function $R' : W \to W$ such that $R' \subseteq R$. Let $\mathcal{F}' = \langle W, R'\rangle$. We define an appropriate valuation $V : W \times \mathcal{L} \to \{f, t\}$ as in Lemma 4.1, while disregarding (M5), and using the following instead of (M4) and (M7):

(M4') $\varphi = \smile\psi$, and $\mathbf{F}_{R'(w)}(\psi)$
(M7') $\varphi = \ominus\psi$, and $\mathbf{F}_w(\varphi)$ and $\mathbf{F}_w(\smile\varphi)$

The proof then carries on in a similar fashion to the proof of Lemma 4.1. □

Corollary 5.9 PKF *enjoys cut-admissibility and is* $\prec'$*-analytic, where* $\prec'$ *is the restriction of* $\prec$ *to the* $\frown$*-free fragment of* $\mathcal{L}$, *with an additional clause according to which* $\smile\varphi \prec \ominus\varphi$.

We include a word about further developments which could not be included here for reasons of space. It is easy to see that $\smile$ and $\frown$ may be defined using the customary presentation of the modal logic **K** by $\smile\varphi := \sim\Box\varphi$ and $\frown\varphi := \Box\sim\varphi$. When considering only functional frames (like in PKF), we get a translation to KF — the ordinary modal logic of functional Kripke models. For the $\ominus\circleddash$-free fragment of this logic, we may apply the general reduction to SAT proposed in [11], which in particular means that the derivability problem for it is in co-NP. We further note that if one dismisses [$\vee\Rightarrow$] from the proof system, derivability can be decided in linear time, by producing SAT-instances that consist solely of Horn clauses. Such 'half-disjunction' was also suggested in the context of primal infon logic [2], to obtain a linear time decision procedure.

5.4 Symmetry

Let PKB be the system obtained from PK by replacing [$\smile\Rightarrow$] and [$\Rightarrow\smile$] with the following rules:

$$[\mathbf{B}_1]\ \frac{\Gamma, \smile\Gamma', \varphi \Rightarrow \Delta, \frown\Delta'}{\frown\Delta, \Delta' \Rightarrow \frown\varphi, \smile\Gamma, \Gamma'} \qquad [\mathbf{B}_2]\ \frac{\Gamma, \smile\Gamma' \Rightarrow \varphi, \Delta, \frown\Delta'}{\frown\Delta, \Delta', \smile\varphi \Rightarrow \smile\Gamma, \Gamma'}$$

These correspond to the following basic rules: $\langle p_1 \Rightarrow\ ;\ \pi_3\rangle\ /\ \Rightarrow \frown p_1$ and $\langle \Rightarrow p_1\ ;\ \pi_3\rangle\ /\smile p_1\ \Rightarrow$, for the context relation $\pi_3 = \{\langle p_1 \Rightarrow\ ;\ \Rightarrow \smile p_1\rangle, \langle \smile p_1 \Rightarrow\ ;\ \Rightarrow p_1\rangle, \langle \Rightarrow p_1\ ;\ \frown p_1 \Rightarrow\rangle, \langle \Rightarrow \frown p_1\ ;\ p_1 \Rightarrow\rangle\}$. This relation satisfies the following property: $s\,\pi_3\,q$ iff $\overline{q}\,\pi_3\,\overline{s}$, where $(\overline{\Rightarrow \varphi})$ denotes $(\varphi \Rightarrow)$ and $(\overline{\varphi \Rightarrow})$ denotes $(\Rightarrow \varphi)$. By Proposition 4.28 of [10], the semantic condition these rules impose on strengthened models is symmetry of the accessibility relation. In addition, every symmetric model respects these rules, as well as the context relation π_3. It follows that:

Corollary 5.10 *$\Gamma \models_{\mathcal{E}_\mathbf{B}} \varphi$ iff $\Gamma \vdash_{\text{PKB}} \varphi$ for every $\Gamma \cup \{\varphi\} \subseteq \mathcal{L}$, where $\mathcal{E}_\mathbf{B}$ is the class of symmetric models.*

Symmetric frames are relevant from the viewpoint of sub-classical properties of negation. They validate, for instance, the consecutions $\smile\smile p \models p$ and $p \models \frown\frown p$. Paraconsistent logics based on symmetric (and reflexive) frames are also studied in [1], a paper that investigates in detail a conservative extension of the corresponding logic, obtained by the addition of a classical implication (but without primitive $\frown$ and $\ominus$), and offers for this logic a sequent system for which cut is not eliminable.

Quasi models for PKB are not necessarily symmetric, making it harder to convert them into instances in the form of symmetric models. This is why cut-admissibility for our system PKB is here left open as a matter for further research. However, using a similar technique of basic systems, it can be straightforwardly shown that PKB is $\prec$-analytic. This does not require quasi models at all: one only has to show that every *partial* model, whose valuation's domain is closed under $\prec$-subformulas, may be extended to a full model (see Corollary 5.44 in [10]).

6 Definability of classical negation

In this section we investigate definability of classical negation in the modal logics studied in this paper. Given a set C of connectives and a logic $\mathbf{L}$, we denote by $\mathbf{L}$-C the C-free fragment of $\mathbf{L}$, that is, the restriction of $\mathbf{L}$ to the language without the connectives in C.

Theorem 6.1

(i) *Classical negation is definable in the logics: PKT-$\{\smile, \ominus\}$, PKT-$\{\frown, \ominus\}$, PKD, and PKF.*

(ii) *Classical negation is not definable in the logics: PK, PKB, PKT-$\{\ominus, \circleddash\}$, PKD-$\{\ominus\}$, PKD-$\{\circleddash\}$, PKF-$\{\ominus\}$, and PKF-$\{\circleddash\}$.*

Proof.
(i) For PKT-$\{\smile, \ominus\}$ we set $\sim\varphi := \frown\varphi \vee \circleddash\varphi$, for PKT-$\{\frown, \circleddash\}$ we set $\sim\varphi := \smile\varphi \wedge \ominus\varphi$, and for PKD and PKF we set $\sim\varphi := (\frown\varphi \wedge \ominus\varphi) \vee \circleddash\varphi$. It is easy to see that $\Rightarrow \varphi, \sim\varphi$ and $\varphi, \sim\varphi \Rightarrow$ are derivable in each system for the defined connective $\sim$. Using cut, one obtains the usual sequent rules for classical negation. Table 1 provides the derivations for PKD. (Given that PKF is a deductive extension of PKD, the derivation in Table 1 is also good for PKF.)

$$\dfrac{\dfrac{\dfrac{\varphi \Rightarrow \varphi \qquad \dfrac{\varphi \Rightarrow \varphi}{\frown\varphi \Rightarrow \smile\varphi}[\mathbf{D}]}{\varphi, \frown\varphi, \ominus\varphi \Rightarrow}[\ominus\Rightarrow]}{\varphi, \frown\varphi \wedge \ominus\varphi \Rightarrow}[\wedge\Rightarrow] \qquad \dfrac{\varphi \Rightarrow \varphi, \frown\varphi}{\varphi, \circleddash\varphi \Rightarrow}[\circleddash\Rightarrow]}{\varphi, (\frown\varphi \wedge \ominus\varphi) \vee \circleddash\varphi \Rightarrow}[\vee\Rightarrow]$$

$$\dfrac{\dfrac{\dfrac{\varphi \Rightarrow \varphi \qquad \frown\varphi \Rightarrow \frown\varphi}{\Rightarrow \varphi, \frown\varphi, \circleddash\varphi}[\Rightarrow\circleddash] \qquad \dfrac{\varphi, \smile\varphi \Rightarrow \varphi, \circleddash\varphi}{\Rightarrow \varphi, \ominus\varphi, \circleddash\varphi}[\Rightarrow\ominus]}{\Rightarrow \varphi, (\frown\varphi \wedge \ominus\varphi), \circleddash\varphi}[\Rightarrow\wedge]}{\Rightarrow \varphi, (\frown\varphi \wedge \ominus\varphi) \vee \circleddash\varphi}[\Rightarrow\vee]$$

Table 1

(ii) Let $X \in \{PK, PKB, PKT\text{-}\{\ominus, \circleddash\}, PKD\text{-}\{\ominus\}, PKD\text{-}\{\circleddash\}, PKF\text{-}\{\ominus\}, PKF\text{-}\{\circleddash\}\}$. Suppose for the sake of contradiction that classical negation $\sim$ is definable in X. Let $p \in \mathcal{P}$ and let φ be $\sim(p)$. Then both $\Rightarrow \varphi, p$ and $p, \varphi \Rightarrow$ are valid in X. Consider a set W that consists of two worlds, w and v, and a valuation V such that $V(w, q) = 1$ and $V(v, q) = 0$ for every atomic formula q (including p). Now, for each relation R_X on W, consider the model $\mathcal{M}_X = \langle\langle W, R_X\rangle, V\rangle$. If $\mathcal{M}_X$ belongs to the class of models that semantically characterize X, then we must have that $\mathcal{M}_X, w \Vdash \varphi, p \Rightarrow$ and $\mathcal{M}_X, v \Vdash\ \Rightarrow p, \varphi$. Since in $\mathcal{M}_X$ we have $\mathbf{T}_w(p)$ and $\mathbf{F}_v(p)$, we must then have $\mathbf{F}_w(\varphi)$ and $\mathbf{T}_v(\varphi)$. We show that this is impossible, by structural induction on φ. More precisely, we claim that if $\mathbf{F}_w(\varphi)$ then $\mathbf{F}_v(\varphi)$. To show this, we consider the possible values for X, and define the accessibility relation R_X in each case. For $X \in \{PK, PKB\}$ define $R_X = \varnothing$, for $X = PKT$-$\{\ominus, \circleddash\}$ define $R_X = W \times W$, for $X \in \{PKD\text{-}\{\ominus\}, PKF\text{-}\{\ominus\}\}$ define $R_X = \{\langle w, v\rangle, \langle v, v\rangle\}$, and for $X \in \{PKD\text{-}\{\circleddash\}, PKF\text{-}\{\circleddash\}\}$ define $R_X = \{\langle w, w\rangle, \langle v, w\rangle\}$. We describe in detail only the third case. For this case, note that since R_X is a total function, $\mathcal{M}_X$ belongs to the appropriate class of models, and $\smile$ and $\frown$ are indistinguishable, hence we may choose to consider $\frown$ instead of $\smile$. The cases where φ is atomic, a conjunction, or a disjunction are trivial. If $\varphi = \frown\psi$ for some ψ and $\mathbf{F}_w(\varphi)$, then we must have $\mathbf{T}_v(\psi)$ by $[\mathbf{T}\frown]$, which implies by $[\mathbf{F}\frown]$ that $\mathbf{F}_v(\varphi)$. If $\varphi = \circleddash\psi$ for some ψ, then $\mathbf{F}_v(\varphi)$ must hold good: indeed, if on the one hand $\mathbf{T}_w(\frown\psi)$ then $\mathbf{F}_v(\psi)$ by [$\mathbf{F}\frown$], and hence $\mathbf{T}_v(\frown\psi)$ by [$\mathbf{T}\frown$], which implies by [$\mathbf{F}\ominus$] that $\mathbf{F}_v(\varphi)$; if on the other hand $\mathbf{F}_w(\frown\psi)$ then $\mathbf{T}_v(\psi)$ by [$\mathbf{T}\frown$], and hence again $\mathbf{F}_v(\varphi)$ follows by [$\mathbf{F}\ominus$]. □

7 This is possibly not the end

In contrast to the usual 'positive modalities' of normal modal logics, which are monotone with respect to the underlying notion of consequence, we have devoted this paper to antitone connectives known as 'negative modalities' — specifically, to full type box-minus and full-type diamond-minus connectives.

Be they monotone or antitone on each of their arguments, the connectives of normal modal logics are always congruential: they treat equivalent formulas as synonymous. The phenomenon seems to be an exception rather than the rule if many-valued logics with non-classical negations are involved. For instance, Kleene's 3-valued logic fails to be congruential, as $p\wedge\neg p$ is equivalent to $q\wedge\neg q$, but their respective negations, $\neg(p\wedge\neg p)$ and $\neg(q\wedge\neg q)$, are not equivalent. Also, the earliest paraconsistent logic in the literature (cf. [8]) fails to be congruential, in spite of having been defined in terms of a translation into a fragment of the modal logic $S5$, and this failure remained unknown for decades (cf. [12]). The same holds for the other early paraconsistent logics developed later on, containing extra 'strong negations' that live in the vicinity of classical negation (cf. [15,4]). Of course, there are important 'non-exceptions': intuitionistic logic and other intermediate logics constitute congruential paracomplete logics. For another example perhaps more to the point, consider the four-valued logic of FDE, whose semantics may be formulated having as truth-values $\{\mathbf{t},\mathbf{b},\mathbf{n},\mathbf{f}\}$, where $\{\mathbf{t},\mathbf{b}\}$ are designated, the transitive reflexive closure of the order $\leq$ such that $\mathbf{f}\leq\mathbf{b},\mathbf{n}\leq\mathbf{t}$ may be used to define $\wedge$ and $\vee$, respectively, as its meet and its join, while $\neg\langle\mathbf{t},\mathbf{b},\mathbf{n},\mathbf{f}\rangle := \langle\mathbf{f},\mathbf{b},\mathbf{n},\mathbf{t}\rangle$. It is not hard to see that this logic is congruential and by defining the operators $\ominus\langle\mathbf{t},\mathbf{b},\mathbf{n},\mathbf{f}\rangle := \langle\mathbf{t},\mathbf{n},\mathbf{b},\mathbf{t}\rangle$ and $\circleddash\langle\mathbf{t},\mathbf{b},\mathbf{n},\mathbf{f}\rangle := \langle\mathbf{f},\mathbf{n},\mathbf{b},\mathbf{f}\rangle$ it gets conservatively extended into another congruential logic that deductively extends our logic PKF (but does not deductively extend PKT), if we interpret $\smile$ as $\neg$. It is worth noting that the latter logic is equivalent to the expansion of FDE by the addition of a classical negation.

Some terminological conventions and some concepts used in the present paper were borrowed or adapted from other fonts, sometimes without explicit reference. For instance, in Section 2, dadaistic and nihilistic models come from [13], and that paper also introduces the connectives $\ominus$ and $\circleddash$ of the so-called Logics of Formal Inconsistency (cf. [3]) and the dual Logics of Formal Undeterminedness (cf. [13], where the adjustment connectives are called connectives 'of perfection'). The minimal conditions on negation, called ⟦*falsificatio*⟧ and ⟦*verificatio*⟧, come from [14]. What we in the present paper call 'determinacy' has in [5] been called 'determinedness'. The 'strengthened models' from Section 3 correspond to models with strongly-legal valuations in the terminology of [10]. In Section 5, Rule [$\mathbf{D}$] may be thought of as a variation on the following well-known sequent rule for the modal logic KD: $\Gamma\Rightarrow\ /\ \Box\Gamma\Rightarrow$, and rules for PKT are variations on the usual sequent rule for the modal logic

KT: $\Gamma, \varphi \Rightarrow \Delta \;/\; \Gamma, \Box\varphi \Rightarrow \Delta$ (cf. [22]). Also, the rule for PKF is a variation on the sequent rule from [9] for the 'Next' operator in the temporal logic LTL, namely: $\Gamma \Rightarrow \Delta \;/\; \Box\Gamma \Rightarrow \Box\Delta$. We have not been able to find in the literature the obvious rules $\Gamma, \varphi \Rightarrow \Delta \;/\; \Box\Gamma, \Diamond\varphi \Rightarrow \Diamond\Delta$ and $\Gamma \Rightarrow \varphi, \Delta \;/\; \Box\Gamma \Rightarrow \Box\varphi, \Diamond\Delta$ for the modal logic K of which our rules $[\smallsmile\Rightarrow]$ and $[\Rightarrow\smallfrown]$ from Section 3 would be variations on. In Section 4, the trick behind using three-valued models for addressing the admissibility of the cut rule goes at least as far back as [20].

The main feature of our approach here has been to rely on theoretical technology built elsewhere and show how it may be adapted to the present case. Our hope is that this should prove a beneficial methodology, and that the idea of obtaining completeness and cut admissibility as particular applications of more general results will become more common, rather than proceeding always through *ad hoc* completeness and cut elimination theorems.

While we have directed our attention, in this paper, to classes of frames that turned out to be particularly significative from the viewpoint of the relation between negative modalities of different types, we see two very natural ways of extending such study. The *first* natural extension would be to look at other classes of frames that prove to be relevant from the viewpoint of subclassical properties of negation. For instance, it is easy to see that the class of frames with the Church-Rosser property validates $\smallsmile\smallsmile p \models \smallfrown\smallfrown p$, pinpointing an interesting consecution involving the interaction between negations of different types. Some other classes of frames deserving study do not seem to show the same amount of promise, from the viewpoint of paraconsistency or paracompleteness. For instance, euclidean frames validate $[\![\smallsmile$-explosion$]\!]$ if in the set of formulas $\{\smallsmile p, p\}$ one replaces p by $\smallsmile r$, and validate $[\![\smallfrown$-implosion$]\!]$ if in $\{\smallfrown p, p\}$ one replaces p by $\smallfrown r$; also, transitive frames cause a similar behavior, but now swapping the roles of $\smallsmile r$ and $\smallfrown r$ in replacing p. Alternatively, a *second* avenue worth exploring would lead us into logics containing more than one negative modality of the same type (as it has been done for logics with multiple paracomplete negations in [18]). One could for instance consider not only the 'forward-looking' negative modalities defined by the semantic clauses [S$\smallsmile$] and [S$\smallfrown$], but also 'backward-looking' negative modalities $\smallsmile^{-1}$ and $\smallfrown^{-1}$ defined by the clauses obtained from the latter ones by replacing wRv by vRw (such 'converse modalities' have been studied in the context of temporal logic [16], as well as in the context of the so-called Heyting-Brouwer logic [17]). The interaction between the various negations would then be witnessed, in such extended language, by the validity over arbitrary frames of 'pure' consecutions such as $\smallsmile^{-1}\smallsmile p \models p$ and $\smallsmile\smallsmile^{-1} p \models p$ (as well as $p \models \smallfrown^{-1}\smallfrown p$ and $p \models \smallfrown\smallfrown^{-1} p$), and the validity over symmetric frames of 'mixed' consecutions such as $\smallfrown^{-1}\smallsmile p \models p$ and $\smallfrown\smallsmile^{-1} p \models p$ (as well as $p \models \smallsmile^{-1}\smallfrown p$ and $p \models \smallsmile\smallfrown^{-1} p$). In our view, it seems worth the effort applying the machinery employed in the present paper to the above mentioned systems, and still others, in order to investigate results analogous to the ones we have here looked at.[1]

1 The authors acknowledge partial support by the Marie Curie project GeTFun (PIRSES-

References

[1] Avron, A. and A. Zamansky, *A paraconsistent view on B and S5*, this volume.

[2] Beklemishev, L. and Y. Gurevich, *Propositional primal logic with disjunction*, Journal of Logic and Computation **24** (2012), pp. 257–282.

[3] Carnielli, W. A. and J. Marcos, *A taxonomy of C-systems*, in: W. A. Carnielli, M. E. Coniglio and I. M. L. D'Ottaviano, editors, *Paraconsistency: The logical way to the inconsistent*, Lecture Notes in Pure and Applied Mathematics **228**, Marcel Dekker, 2002 pp. 1–94.

[4] da Costa, N. C. A., *Calculs propositionnels pour le systèmes formels inconsistants*, Comptes Rendus Hebdomadaires des Séances de l'Académie des Sciences, Séries A–B **257** (1963), pp. 3790–3793.

[5] Dodó, A. and J. Marcos, *Negative modalities, consistency and determinedness*, Electronic Notes in Theoretical Computer Science **300** (2014), pp. 21–45.

[6] Došen, K., *Negative modal operators in intuitionistic logic*, Publications de L'Institut Mathématique (Beograd) (N.S.) **35(49)** (1984), pp. 3–14.

[7] Dunn, J. M. and C. Zhou, *Negation in the context of Gaggle Theory*, Studia Logica **80** (2005), pp. 235–264.

[8] Jaśkowski, S., *A propositional calculus for inconsistent deductive systems (in Polish)*, Studia Societatis Scientiarum Torunensis, Sectio A **5** (1948), pp. 57–77, translated into English in *Studia Logica*, 24:143–157, 1967, and in *Logic and Logical Philosophy*, 7:35–56, 1999.

[9] Kawai, H., *Sequential calculus for a first order infinitary temporal logic*, Mathematical Logic Quarterly **33** (1987), pp. 423–432.

[10] Lahav, O. and A. Avron, *A unified semantic framework for fully structural propositional sequent systems*, ACM Transactions on Computational Logic **14** (2013), pp. 27:1–27:33.

[11] Lahav, O. and Y. Zohar, *SAT-based decision procedure for analytic pure sequent calculi*, in: S. Demri, D. Kapur and C. Weidenbach, editors, *Automated Reasoning*, Lecture Notes in Computer Science **8562**, Springer International Publishing, 2014 pp. 76–90.

[12] Marcos, J., *Modality and paraconsistency*, in: M. Bilkova and L. Behounek, editors, *The Logica Yearbook 2004*, Filosofia, 2005 pp. 213–222.

[13] Marcos, J., *Nearly every normal modal logic is paranormal*, Logique et Analyse (N.S.) **48** (2005), pp. 279–300.

[14] Marcos, J., *On negation: Pure local rules*, Journal of Applied Logic **3** (2005), pp. 185–219.

[15] Nelson, D., *Negation and separation of concepts in constructive systems*, in: A. Heyting, editor, *Constructivity in Mathematics*, Studies in Logic and the Foundations of Mathematics, North-Holland, Amsterdam, 1959 pp. 208–225.

[16] Prior, A., "Past, Present and Future," Oxford University Press, 1967.

[17] Rauszer, C., *An algebraic and Kripke-style approach to a certain extension of Intuitionistic Logic*, Dissertationes Mathematicae **167** (1980).

[18] Restall, G., *Combining possibilities and negations*, Studia Logica **59** (1997), pp. 121–141.

[19] Ripley, D. W., "Negation in Natural Language," Ph.D. thesis, University of North Carolina at Chapel Hill (2009).

[20] Schütte, K., "Beweistheorie," Springer-Verlag, Berlin, 1960.

[21] Vakarelov, D., *Consistency, completeness and negation*, in: G. Priest, R. Sylvan and J. Norman, editors, *Paraconsistent Logic: Essays on the inconsistent*, Philosophia Verlag, 1989 pp. 328–363.

[22] Wansing, H., *Sequent systems for modal logics*, in: D. M. Gabbay and F. Guenthner, editors, *Handbook of Philosophical Logic*, Springer, 2002, 2nd edition pp. 61–145, vol. 8.

[23] Wójcicki, R., "Theory of Logical Calculi," Kluwer, Dordrecht, 1988.

GA-2012-318986) funded by EU-FP7, by CNPq and by The Israel Science Foundation (grant no. 817-15). They also take the chance to thank Hudson Benevides and three anonymous referees for the careful reading of an earlier version of this manuscript.

A focused framework for emulating modal proof systems

Sonia Marin, Dale Miller and Marco Volpe

Inria and LIX, École Polytechnique, France

Abstract

Several deductive formalisms (e.g., sequent, nested sequent, labeled sequent, hypersequent calculi) have been used in the literature for the treatment of modal logics, and some connections between these formalisms are already known. Here we propose a general framework, which is based on a focused version of the labeled sequent calculus by Negri, augmented with some parametric devices allowing to restrict the set of proofs. By properly defining such restrictions and by choosing an appropriate polarization of formulas, one can obtain different, concrete proof systems for the modal logic K and for its extensions by means of geometric axioms. In particular, we show how to use the expressiveness of the labeled approach and the control mechanisms of focusing in order to emulate in our framework the behavior of a range of existing formalisms and proof systems for modal logic.

Keywords: Modal logic, sequent calculi, labeled proof systems, focusing.

1 Introduction

Modal proof theory is a notoriously difficult subject and several proposals for it have been given in the literature (a general account is in [6]). Such proposals range over a set of different proof formalisms (e.g., sequent, nested sequent, labeled sequent, hypersequent calculi), each of them presenting its own features and drawbacks. For instance, proof systems based on ordinary sequents present a good behavior in terms of proof search, but they are typically designed for a specific modal logic and lack modularity when one tries to capture modal logics with particular frame conditions. Moreover, cut-elimination for an important modal logic like S5 is problematic. For this reason, more sophisticated formalisms have been adapted or introduced, e.g., several hypersequent cut-free formulations have been given for S5, while nested and labeled sequents have been used for giving modular presentations of large classes of modal logics. Several results concerning correspondences and connections between the different formalisms are known [7,10,13].

We propose a general framework for emulating and comparing existing modal proof systems as well as for generating new proof systems. We shall do this in the familiar setting of *labeled deduction systems* [8] in which the axiomatization of a particular Kripke semantics is given. The resulting encoding of a modal logic formula has its logical connectives and propositional atoms "cluttered" with additional (relational) atoms and assumptions that describe the reachability relationship of a class of Kripke models. While such clutter has been described as both "impure" and "semantic pollution" we provide here an additional defense of this approach to complement the defenses found in [19,21]. In particular, we introduce *focused* variants of sequent calculi: in such systems, we can build *synthetic* inference rules from Gentzen-style introduction and structural rules. Such synthetic rules are built from the clutter in controllable and rigid shapes. For example, when geometric formulas are used to axiomatize Kripke frames, the role of those formulas in proofs can be restricted to uses that correspond to synthetic inference rules [17,18]. By adding elements of polarization to the labeled sequent setting and by defining a few other parameters of the general framework, we are able to exploit the control mechanisms provided by focusing to reproduce proofs of the original calculi with precision.

The emulation of modal logic proof systems described in this paper can also be used to build proof checkers for modal logic formulas given checkers for first-order logic (such as those described in [4]) that do not have any special knowledge of modal operators and Kripke frames. In other words, the emulation results described here make it possible to build a modal logic proof checker using familiar proof search techniques such as backtracking search and (first-order) unification. A particularly challenging aspect of such emulation is, predictably, the promotion rule, such as the one for K (in a one-sided sequent formulation):

$$\frac{\vdash \Gamma, B}{\vdash \Diamond\Gamma, \Box B}.$$

Since many introductions must be performed at once, this inference rule corresponds to more than one synthetic inference rule in our emulation. If Γ contains n occurrences of formulas, then we could emulate this one inference rules using $n+1$ synthetic rules as follows (reading proof rules from conclusion to premises): one of these rules performs the $\Box$-introduction (which corresponds to creating a new world that is assumed to be reachable) and n of these rules perform the $\Diamond$-introduction rules (which correspond to moving all the assumptions of the form $\Diamond A$ to that new world). Notice that all of these n inference rules can, in fact, be performed in parallel. We capture such parallel application of inference rules using synthetic inference rules built with *multifocusing*. As a result, we will capture this promotion rule via two synthetic inference rules: one for the $\Box$-introduction and one capturing all $\Diamond$-introductions.

We proceed as follows. After providing background notions concerning modal logic and focusing (Section 2), we present the general framework LMF_*^X (Section 3) and prove some results about the emulation of existing modal proof systems (Section 4). In this paper, we restrict our attention to the emulation of ordinary and nested sequent systems. We remark, however, that the framework has been designed with the goal of capturing more modal calculi in a wider range of formalisms, as we discuss in the concluding remarks (Section 5), where we also sum up our contributions and propose some directions for future work.

Axiom	Condition	First-Order Formula
T: $\Box A \supset A$	Reflexivity	$\forall x.R(x,x)$
4: $\Box A \supset \Box\Box A$	Transitivity	$\forall x,y,z.(R(x,y) \wedge R(y,z)) \supset R(x,z)$
5: $\Box A \supset \Box\Diamond A$	Euclideaness	$\forall x,y,z.(R(x,y) \wedge R(x,z)) \supset R(y,z)$
B: $A \supset \Box\Diamond A$	Symmetry	$\forall x,y.R(x,y) \supset R(y,x)$
D: $\Box A \supset \Diamond A$	Seriality	$\forall x \exists y.R(x,y)$

Table 1
Axioms and corresponding first-order conditions on the accessibility relation R.

2 Background : Focusing and modal logic

2.1 Modal logic

We will consider *(propositional) modal formulas* in negation normal form based on a functionally complete set of classical connectives, a *modal operator* $\Box$, together with its dual $\Diamond$, and a denumerable set $\mathcal{P}$ of *propositional symbols*, according to the following grammar (where $P \in \mathcal{P}$):

$$A ::= P \mid \neg P \mid A \vee A \mid \top \mid A \wedge A \mid \bot \mid \Box A \mid \Diamond A,$$

The negation $\neg A$ of a formula A is defined via the De Morgan laws (so the only formally negated formulas are the atoms), and $A \supset B$ is defined as usual as $\neg A \vee B$. A is a *$\Box$-formula* (*$\Diamond$-formula*) if the main connective of A is $\Box$ ($\Diamond$).

The semantics is defined by means of *Kripke frames*, *i.e.*, pairs $\mathcal{F} = (W, R)$ where W is a non empty set of *worlds* and R is a binary relation on W. A *Kripke model* is a triple $\mathfrak{M} = (W, R, V)$ where (W, R) is a Kripke frame and $V : W \to 2^{\mathcal{P}}$ is a function that assigns to each world in W a (possibly empty) set of propositional symbols.

Truth of a modal formula at a point w in a Kripke structure $\mathfrak{M} = (W, R, V)$ is the smallest relation $\models$ satisfying:

$$\begin{array}{rcl}
\mathfrak{M}, w \models P & \text{iff} & P \in V(w) \\
\mathfrak{M}, w \models A \vee B & \text{iff} & \mathfrak{M}, w \models A \text{ or } \mathfrak{M}, w \models B \\
\mathfrak{M}, w \models A \wedge B & \text{iff} & \mathfrak{M}, w \models A \text{ and } \mathfrak{M}, w \models B \\
\mathfrak{M}, w \models \Box A & \text{iff} & \text{for all } w' \text{ s.t. } R(w, w')\ \mathfrak{M}, w' \models A \\
\mathfrak{M}, w \models \Diamond A & \text{iff} & \text{there exists } w' \text{ s.t. } R(w, w') \text{ and } \mathfrak{M}, w' \models A.
\end{array}$$

By extension, we write $\mathfrak{M} \models A$ when $\mathfrak{M}, w \models A$ for all $w \in W$ and we write $\models A$ when $\mathfrak{M} \models A$ for every Kripke structure $\mathfrak{M}$.

The former definition characterizes the basic modal logic K. Several further modal logics can be defined as extensions of K by simply restricting the class of frames we consider. Many of the restrictions we are interested in are definable as formulas of first-order logic where the binary predicate $R(x, y)$ refers to the corresponding accessibility relation. Table 1 summarizes some of the most common frame logics, describing the corresponding frame property, together with the modal axiom capturing it [22]. We will refer to the logic satisfying a set of axioms $\{F_1, \ldots, F_n\}$ as $K\{F_1, \ldots, F_n\}$.

ASYNCHRONOUS INTRODUCTION RULES

$$\frac{}{\mathcal{G} \vdash \Theta \Uparrow x : t^-, \Gamma}\ t^- \qquad \frac{\mathcal{G} \vdash \Theta \Uparrow x : A, \Gamma \quad \mathcal{G} \vdash \Theta \Uparrow x : B, \Gamma}{\mathcal{G} \vdash \Theta \Uparrow x : A \wedge^- B, \Gamma}\ \wedge^-$$

$$\frac{\mathcal{G} \vdash \Theta \Uparrow \Gamma}{\mathcal{G} \vdash \Theta \Uparrow x : f^-, \Gamma}\ f^- \quad \frac{\mathcal{G} \vdash \Theta \Uparrow x : A, x : B, \Gamma}{\mathcal{G} \vdash \Theta \Uparrow x : A \vee^- B, \Gamma}\ \vee^- \quad \frac{\mathcal{G} \cup \{xRy\} \vdash \Theta \Uparrow y : B, \Gamma}{\mathcal{G} \vdash \Theta \Uparrow x : \Box B, \Gamma}\ \Box$$

SYNCHRONOUS INTRODUCTION RULES

$$\frac{}{\mathcal{G} \vdash \Theta \Downarrow x : t^+}\ t^+ \qquad \frac{\mathcal{G} \vdash \Theta \Downarrow x : A \quad \mathcal{G} \vdash \Theta \Downarrow x : B}{\mathcal{G} \vdash \Theta \Downarrow x : A \wedge^+ B}\ \wedge^+$$

$$\frac{\mathcal{G} \vdash \Theta \Downarrow x : A}{\mathcal{G} \vdash \Theta \Downarrow x : A \vee^+ B}\ \vee_1^+ \quad \frac{\mathcal{G} \vdash \Theta \Downarrow x : B}{\mathcal{G} \vdash \Theta \Downarrow x : A \vee^+ B}\ \vee_2^+ \quad \frac{\mathcal{G} \cup \{xRy\} \vdash \Theta \Downarrow y : B}{\mathcal{G} \cup \{xRy\} \vdash \Theta \Downarrow x : \Diamond B}\ \Diamond$$

IDENTITY RULES

$$\frac{}{\mathcal{G} \vdash x : \neg B, \Theta \Downarrow x : B}\ init \qquad \frac{\mathcal{G} \vdash \Theta \Uparrow x : B \quad \mathcal{G} \vdash \Theta \Uparrow x : \neg B}{\mathcal{G} \vdash \Theta \Uparrow \cdot}\ cut$$

STRUCTURAL RULES

$$\frac{\vdash \Theta, x : B \Uparrow \Gamma}{\vdash \Theta \Uparrow x : B, \Gamma}\ store \quad \frac{\vdash \Theta \Uparrow x : B}{\vdash \Theta \Downarrow x : B}\ release \quad \frac{\vdash x : B, \Theta \Downarrow x : B}{\vdash x : B, \Theta \Uparrow \cdot}\ decide$$

In *decide*, B is a positive formula; in *release*, B is a negative formula; in *store*, B is a positive formula or a negative literal; in *init*, B is a positive literal. In $\Box$, y does not occur in Θ nor in Γ.

Fig. 1. *LMF*: a focused labeled proof system for the modal logic K

2.2 Focused labeled proof system for modal logic

This work takes up the labeled approach to the proof theory of modal logics which internalizes the Kripke semantics into the syntax to give sequent calculi for numerous modal logics [19,24]. Traditionally, labeled sequents are composed by both *labeled formulas* of the form $x : A$ and *relational atoms* of the form xRy, where x, y range over a set of variables (called *labels*) and A is a modal formula. A (one-sided) labeled sequent will therefore be of the form $\mathcal{G} \vdash \Gamma$ where $\mathcal{G}$ denotes a set of relational atoms, and Γ a multiset of labeled formulas.

Here we present a variant of the focused labeled system that was introduced in [17]. In general, a *focused sequent calculus* is one where introduction rules are placed into one of two *phases*. The *asynchronous* phase contains all invertible introduction rules: during this phase, non-atomic formulas are decomposed without external information being supplied to them (that is, they decompose asynchronously). The *synchronous* phase contains introduction rules in which decomposition may require additional information to be supplied: for example, the $\Diamond$-introduction rule needs to check the context for the suitable relational atom. Thus, these inference rules need to synchronize with some source of information such as an oracle or a proof certificate [4].

Figure 1 contains the (subset of rules for the logic K of the) focused proof system *LMF* from [17] that was designed for a range of modal logics. The key features of this proof system, which follow the design of focused proof systems for classical and intuitionistic logic given in [15], are the following.

Polarized formula *LMF* is a proof system of *polarized formulas* built using atomic formulas, the usual modalities $\Box$ and $\Diamond$, and polarized versions of the logical connectives $\vee^-$, $\vee^+$, $\wedge^-$, $\wedge^+$, and constants t^-, t^+ for $\top$, and f^-, f^+ for $\bot$. The positive and negative versions of connectives and constants have identical truth conditions but different inference rules. All polarized formulas are either positive or negative: if a formula's top-level connective is t^+, f^+, $\vee^+$, $\wedge^+$, or $\Diamond$, then that formula is positive. Dually, if a formula's top-level connective is t^-, f^-, $\vee^-$, $\wedge^-$, or $\Box$, then it is negative. In this way, every polarized formula is classified except for literals: to polarize them, we are allowed to fix the polarity of atomic formulas in any way we see fit. We may ask that all atomic formulas are positive, that they are all negative, or we can mix polarity assignments. In any case, if P is a positive atomic formula, then it is a positive formula and $\neg P$ is a negative formula: conversely, if P is a negative atomic formula, then it is a negative formula and $\neg P$ is a positive formula.

Two sequent judgments Sequents in *LMF* are of the form $\mathcal{G} \vdash \Theta \Uparrow \Delta$ or $\mathcal{G} \vdash \Theta \Downarrow x : A$, where $\mathcal{G}$ is a set of relational atoms, x is a label, A is a polarized modal formula, Θ is a multiset of labeled polarized formulas (called the *storage*), and Δ is a list of labeled polarized formulas. The formula $x : A$ in $\Downarrow$ sequents is called the *focus* of that sequent. If Γ is a multiset of formulas then $\Diamond\Gamma$ denotes the multiset $\{\Diamond B \mid B \in \Gamma\}$ and $x : \Gamma$ denotes the multiset of labeled formulas $\{x : B \mid B \in \Gamma\}$.

Two phases of inference rules All the asynchronous inference rules of *LMF* have $\Uparrow$-sequents in their premises and conclusion while all the synchronous inference rules have $\Downarrow$-sequents in their premises and conclusion. The only rules that mix these sequents are the *release* and *decide* rules. A maximal sequence of asynchronous or synchronous inferences form *phases* with interfaces between phases given by instances of the *release* and *decide* rules. These phases form, in fact, macro-level (synthetic) inference rules constructed from collections of the smaller rules of the focused sequent calculus.

A polarized formula B is a *bipolar formula* if B is a positive formula and no positive subformula occurrence of B is in the scope of a negative connective in B. A *bipole* is a pair of a synchronous phase below an asynchronous phase within *LMF*: thus, bipoles are macro inference rules in which the conclusion and the premises are $\Uparrow$-sequents with no formulas to the right of the up-arrow.

Delays We shall find it important to break a sequence of negative or positive connectives by inserting *delays*: if B is a polarized formula then we define $\partial^-(B)$ to be (always negative) $B \wedge^- t^-$ and $\partial^+(B)$ to be (always positive) $B \wedge^+ t^+$. From such a definition, the following rules can be derived:

$$\frac{\mathcal{G} \vdash \Theta \Uparrow x : B, \Delta}{\mathcal{G} \vdash \Theta \Uparrow x : \partial^-(B), \Delta}\ \partial^- \qquad \frac{\mathcal{G} \vdash \Theta \Downarrow x : B}{\mathcal{G} \vdash \Theta \Downarrow x : \partial^+(B)}\ \partial^+$$

To illustrate the use of delays, note that the sequent $xRy, yRz \vdash \cdot \Downarrow x : \Diamond\Diamond B$ must be the result of applying two $\Diamond$-introduction rules in a synchronous phase before further processing the formula B. In contrast, the sequent $xRy, yRz \vdash \cdot \Downarrow x : \Diamond\partial^-(\Diamond B)$ must be the conclusion of only one $\Diamond$-introduction rule and allows one to store an instance of $\Diamond B$ such that a separate occurrence of $\Diamond$ can take place elsewhere in the proof.

$$\dfrac{\dfrac{xRy, yRz \vdash \cdot \Downarrow z : B}{xRy, yRz \vdash \cdot \Downarrow y : \Diamond B}\ \Diamond}{xRy, yRz \vdash \cdot \Downarrow x : \Diamond\Diamond B}\ \Diamond \qquad \dfrac{\dfrac{\dfrac{\dfrac{xRy, yRz \vdash y : \Diamond B \Uparrow \cdot}{xRy, yRz \vdash \cdot \Uparrow y : \Diamond B}\ store}{xRy, yRz \vdash \cdot \Uparrow y : \partial^-(\Diamond B)}\ \partial^-}{xRy, yRz \vdash \cdot \Downarrow y : \partial^-(\Diamond B)}\ release}{xRy, yRz \vdash \cdot \Downarrow x : \Diamond\partial^-(\Diamond B)}\ \Diamond$$

The completeness of *LMF* is stated as follows [17]. We say that $\hat{B}$ is a *polarization* of the (unpolarized) B if it results from placing superscripts $+$ and $-$ on the propositional connectives, assigning atomic formulas any mix of positive or negative polarization, and inserting any number of delays. Completeness is now the statement that if B is an (unpolarized) modal logic theorem and $\hat{B}$ is *any* polarization of B, then $\vdash \cdot \Uparrow x : \hat{B}$ is provable in *LMF*. That is, the choice of polarization does not affect provability but it can have a big impact on the structure of proofs.

3 A focused labeled framework for modal logic

In this section, we present a multifocused version of *LMF* further augmented with some devices aimed at enabling the emulation of different modal proof systems. In order to motivate the need for such devices, consider the following typical sequent calculus rule for modal logic:

$$\frac{\vdash \Gamma, A}{\vdash \Diamond\Gamma, \Box A}\ .$$

Augmentation of the system *LMF* is driven by the following considerations of inference rules of this kind.

(i) As already noticed in Section 1, this rule works at the same time on one $\Box$-formula and on n $\Diamond$-formulas. In order to process such $\Diamond$-formulas, in our labeled deduction setting, it is necessary to apply the $\Diamond$-introduction rule n times. Since these applications do not interfere with each other, they can, in fact, be applied in parallel. For this reason, we move to a multifocused version of *LMF*, i.e., a variant where we can focus on several positive formulas at the same time. In this way, we can group all the $\Diamond$-introductions inside a single phase (in the following, we will sometimes call it a $\Diamond$-phase).

(ii) Intuitively, one can read this inference rule (reading from conclusion to premise) as moving from one world to another (reachable) world in a suitable Kripke structure. Such a change of world becomes apparent when we consider the corresponding deduction steps in a labeled system, as, in this case, modal introduction rules will explicitly change the label of the formulas under consideration. In order to properly mimic the behavior of the original rule, in the labeled system we need to be able to force all the formulas involved in the rule to move to the same new world. We therefore modify the notion of a labeled formula to have the form $x\sigma : A$, for σ a sequence of labels. Here x indicates in which world such a formula holds, while the sequence σ gets initialized when one multifocuses on the multiset of $\Diamond$-formulas and is used to drive future applications of $\Diamond$-rules. E.g., if $x : \Diamond\Gamma$ is on the left of $\Uparrow$, then we can multifocus on $xy : \Diamond\Gamma$ for a given y reachable from x. This y will be used as a witness in the application of a (properly modified) $\Diamond$-introduction rule, in such a way that at the end of the bipole, we will have the multiset $y : \Gamma$ on the left of $\Uparrow$. [1]

(iii) Finally, we observe that in *LMF*, when constructing a proof tree (going from the root towards the leaves), formulas we decide on are duplicated and stay in the storage (that is, on the left of $\Uparrow$ or $\Downarrow$). It follows that all along a proof, it is possible to switch freely from one label to another in the deduction process. On the contrary, in a sequent calculus rule like the one given above, only formulas having a modal operator as the main connective can be "promoted" to a different world. According to the Kripke-style interpretation presented, this amounts to considering a single world at a time, in such a way that when moving to a new one, formulas standing at previously encountered worlds are not accessible anymore. In order to emulate this aspect, labelled sequents are further decorated with a set $\mathcal{H}$ of labels, specifying which worlds are currently enabled, with the intended meaning that we can decide on a formula only if its label belongs to $\mathcal{H}$ [2].

In the following, we will formalize the intuitions given above, introduce some terminology and present the general framework LMF_*^X (Figure 2).

In the rest of this paper, a *labeled formula* will have the form $\varphi \equiv x\sigma : A$, where σ is a (possibly empty) sequence of labels. We say that x is the *present* of φ and σ is the *future* of φ. An LMF_*^X *sequent* has the form $\mathcal{G} \vdash_{\mathcal{H}} \Theta \Uparrow \Omega$ or $\mathcal{G} \vdash_{\mathcal{H}} \Theta \Downarrow \Omega$, where the *relational set (of the sequent)* $\mathcal{G}$ is a set of relational atoms, the *present (of the sequent)* $\mathcal{H}$ is a non-empty multiset of pairs $(x, \mathcal{F})$, where x is a label and $\mathcal{F}$ is a set of labels, and Θ and Ω are multisets of labeled formulas. Intuitively, a pair $(x, \mathcal{F})$ specifies that x is among the worlds we are currently working on and $\mathcal{F}$ indicates which worlds, among the reachable ones, are not accessible from x. E.g., if we are in the position of applying a $decide_*$

[1] We note that for the emulation of the calculi presented in this paper, a future consisting of a single label is enough. We prefer, however, to present the framework in this more general version that allows for capturing also other behaviors.

[2] In fact, this representation of $\mathcal{H}$ as a set of labels is good for giving an insight of the technique, but it is slightly simpler than the one concretely used in the framework, which is formalized in the next paragraphs.

Asynchronous introduction rules

$$\frac{}{\mathcal{G} \vdash_{\mathcal{H}} \Theta \Uparrow x : t^-, \Omega} t_*^- \qquad \frac{\mathcal{G} \vdash_{\mathcal{H}} \Theta \Uparrow \Omega}{\mathcal{G} \vdash_{\mathcal{H}} \Theta \Uparrow x : f^-, \Omega} f_*^-$$

$$\frac{\mathcal{G} \vdash_{\mathcal{H}} \Theta \Uparrow x : A, \Omega \quad \mathcal{G} \vdash_{\mathcal{H}} \Theta \Uparrow x : B, \Omega}{\mathcal{G} \vdash_{\mathcal{H}} \Theta \Uparrow x : A \wedge^- B, \Omega} \wedge_*^- \qquad \frac{\mathcal{G} \vdash_{\mathcal{H}} \Theta \Uparrow x : A, x : B, \Omega}{\mathcal{G} \vdash_{\mathcal{H}} \Theta \Uparrow x : A \vee^- B, \Omega} \vee_*^-$$

$$\frac{\mathcal{G} \cup \{xRy\} \vdash_{\mathcal{H}} \Theta \Uparrow y : B, \Omega}{\mathcal{G} \vdash_{\mathcal{H}} \Theta \Uparrow x : \Box B, \Omega} \Box_*$$

Synchronous introduction rules

$$\frac{}{\mathcal{G} \vdash_{\mathcal{H}} \Theta \Downarrow x\sigma : t^+} t_*^+ \qquad \frac{\mathcal{G} \vdash_{\mathcal{H}} \Theta \Downarrow x\sigma : B_1, \Omega_1 \quad \mathcal{G} \vdash_{\mathcal{H}} \Theta \Downarrow x\sigma : B_2, \Omega_2}{\mathcal{G} \vdash_{\mathcal{H}} \Theta \Downarrow x\sigma : B_1 \wedge^+ B_2, \Omega_1, \Omega_2} \wedge_*^+$$

$$\frac{\mathcal{G} \vdash_{\mathcal{H}} \Theta \Downarrow x\sigma : B_i, \Omega}{\mathcal{G} \vdash_{\mathcal{H}} \Theta \Downarrow x\sigma : B_1 \vee^+ B_2, \Omega} \vee_*^+, i \in \{1,2\} \qquad \frac{\mathcal{G} \cup \{xRy\} \vdash_{\mathcal{H}} \Theta \Downarrow y\sigma : B, \Omega}{\mathcal{G} \cup \{xRy\} \vdash_{\mathcal{H}} \Theta \Downarrow xy\sigma : \Diamond B, \Omega} \Diamond_*$$

Identity rules

$$\frac{}{\mathcal{G} \vdash_{\mathcal{H}} x : \neg B, \Theta \Downarrow x : B} init_* \qquad \frac{\mathcal{G} \vdash_{\mathcal{H}} \Theta \Uparrow x : B \quad \mathcal{G} \vdash_{\mathcal{H}} \Theta \Uparrow x : \neg B}{\mathcal{G} \vdash_{\mathcal{H}} \Theta \Uparrow \cdot} cut_*$$

Structural rules

$$\frac{\mathcal{G} \vdash_{\mathcal{H}} \Theta, x : B \Uparrow \Omega}{\mathcal{G} \vdash_{\mathcal{H}} \Theta \Uparrow x : B, \Omega} store_* \qquad \frac{\mathcal{G} \vdash_{\mathcal{H}} \Theta \Uparrow \Omega'}{\mathcal{G} \vdash_{\mathcal{H}} \Theta \Downarrow \Omega} release_* \qquad \frac{\mathcal{G} \vdash_{\mathcal{H}'} \Theta \Downarrow \Omega}{\mathcal{G} \vdash_{\mathcal{H}} \Theta \Uparrow \cdot} decide_*$$

Relational rules

$$\frac{\mathcal{G} \cup \{yRy\} \vdash_{\mathcal{H}} \Theta \Uparrow \cdot}{\mathcal{G} \vdash_{\mathcal{H}} \Theta \Uparrow \cdot} T_* \qquad \frac{\mathcal{G} \cup \{yRz\} \vdash_{\mathcal{H}} \Theta \Uparrow \cdot}{\mathcal{G} \vdash_{\mathcal{H}} \Theta \Uparrow \cdot} D_* \qquad \frac{\mathcal{G} \cup \{xRy, yRx\} \vdash_{\mathcal{H}} \Theta \Uparrow \cdot}{\mathcal{G} \cup \{xRy\} \vdash_{\mathcal{H}} \Theta \Uparrow \cdot} B_*$$

$$\frac{\mathcal{G} \cup \{xRy, yRz, xRz\} \vdash_{\mathcal{H}} \Theta \Uparrow \cdot}{\mathcal{G} \cup \{xRy, yRz\} \vdash_{\mathcal{H}} \Theta \Uparrow \cdot} 4_* \qquad \frac{\mathcal{G} \cup \{xRy, xRz, yRz\} \vdash_{\mathcal{H}} \Theta \Uparrow \cdot}{\mathcal{G} \cup \{xRy, xRz\} \vdash_{\mathcal{H}} \Theta \Uparrow \cdot} 5_*$$

In $store_*$, B is a positive formula or a negative literal.
In $init_*$, B is a positive literal.
In $\Box_*$, y is different from x and does not occur in $\mathcal{G}$ nor in Θ.
In $decide_*$, if $x\sigma : A \in \Omega$ then $x : A \in \Theta$. Moreover, Ω contains only positive formulas of the form: (i) $x\sigma : A$, where A is not a $\Diamond$-formula and $(x, \mathcal{F}) \in \mathcal{H}$ for some $\mathcal{F}$; or (ii) $zy\sigma : A$ where A is a $\Diamond$-formula, $xRy, zRy \in \mathcal{G}$, $(x, \mathcal{F}) \in \mathcal{H}$ for some $\mathcal{F}$ and $y \notin \mathcal{F}$.
In $release_*$, Ω contains no positive formulas and $\Omega' = \{x : A \mid x\sigma : A \in \Omega\}$.
In D_*, z is different from y and does not occur in $\mathcal{G}$ and Θ.

Fig. 2. LMF_*^X: a focused labeled framework for modal logic.

(by proceeding bottom-up), then a pair $(x, \mathcal{F})$ contained in $\mathcal{H}$ says that: (*i*) we can (multi)focus on non-$\Diamond$-formulas labeled with x; or (*ii*) we can "move" to a y reachable from x (by (multi)focusing on $\Diamond$-formulas) if y is not in the set $\mathcal{F}$ of *forbidden futures* for x. In this general formulation, the set $\mathcal{H}'$ that we get in the premise of the rule can be defined in an arbitrary way; specific ways of defining it will be proposed in next sections in order to obtain particular behaviors. In $decide_*$, we are also allowed to assign a future σ to a formula $x : A$ in the storage, so that we actually focus on $x\sigma : A$. Such futures are relevant in the treatment of $\Diamond$-formulas. In fact, when we apply $\Diamond_*$ with respect to a formula $xy\tau : \Diamond A$, we are forced to "move" to the world y thus getting $y\tau : A$ in the premise. Since futures of formulas are only relevant during the synchronous phase, applications of $release_*$ remove all such futures. The other rules of the system are simple adaptations of the ones in *LMF*.

We say that an LMF_*^X sequent is a *synchronized sequent* if it has the form $\mathcal{G} \vdash_{\mathcal{H}} \Theta \Uparrow \cdot$. In fact, a sequent that contains no formulas on the right of the arrow is what we get at the end of a bipole. This is the moment when we can more easily compare the status of an LMF_*^X proof with the status of the proof to be emulated, i.e., in a sense, synchronize the two proofs.

The parameter X is a subset of $\{T, 4, 5, B, D\}$, specifying which modal logic we are considering. The system $LMF_*^\emptyset$ is a system for the logic K and is obtained by including only the first four classes of rules (i.e., no relational rules). Any other system LMF_*^X is obtained by adding to $LMF_*^\emptyset$ the set of relational rules $\{C_F \mid C \in X\}$.

Theorem 3.1 *The system LMF_*^X is sound and complete with respect to the logic KX, for any polarization of formulas.*

Proof. The system LMF_*^X is a multifocused version of the system *LMF* presented in [17] and recalled in Section 2.2, augmented with some devices for controlling the application of rules. Soundness follows from the fact that such devices can only introduce restrictions to the application of rules and multifocusing can be simulated in *LMF* by several rule applications. Completeness is also a direct consequence of completeness of *LMF*, since in the liberal version presented in this section all new devices (including multifocusing) can just be ignored, or used in a trivial way, so that each proof in the previous system is also a valid proof in LMF_*^X. □

In addition to being modular with respect to the relational properties considered, we can (and in the following will) obtain different concrete proof systems by properly specifying the behavior of the new devices introduced in LMF_*^X. These will be defined by specializing the rule $decide_*$, i.e., in particular, by playing with the following parameters:

- restrictions on the class of formulas on which multifocusing can be applied;
- restrictions on the definition of the future σ of formulas in Ω;
- restriction of the multiset $\mathcal{H}'$ (in the premise of $decide_*$).

IDENTITY AND STRUCTURAL RULES

$$\frac{}{\vdash \Gamma, P, \neg P}\ init \qquad \frac{\vdash \Gamma, A \quad \vdash \Delta, \neg A}{\vdash \Gamma, \Delta}\ cut \qquad \frac{\vdash \Gamma, A, A}{\vdash \Gamma, A}\ contr$$

CLASSICAL CONNECTIVES RULES

$$\frac{\vdash \Gamma, A \quad \vdash \Gamma, B}{\vdash \Gamma, A \wedge B}\ \wedge \qquad \frac{\vdash \Gamma, A, B}{\vdash \Gamma, A \vee B}\ \vee$$

$\Box$-RULES

$$\frac{\vdash \Gamma, A}{\vdash \Diamond\Gamma, \Box A, \Delta}\ \Box_K \qquad \frac{\vdash \Diamond\Gamma, \Gamma', A}{\vdash \Diamond\Gamma, \Box A, \Delta}\ \Box_{K4} \qquad \frac{\vdash \Diamond\Gamma, \Gamma', \Box\Sigma, A}{\vdash \Diamond\Gamma, \Box\Sigma, \Box A, \Delta}\ \Box_{K45}$$

$\Diamond$-RULES

$$\frac{\vdash \Diamond A, A, \Sigma}{\vdash \Diamond A, \Sigma}\ \Diamond_T \qquad \frac{\vdash \Gamma}{\vdash \Diamond\Gamma, \Delta}\ \Diamond_D$$

In $\Box_{K4}$, Δ does not contain any formula whose main connective is $\Diamond$.
In $\Box_{K45}$, Δ does not contain any formula whose main connective is $\Diamond$ or $\Box$.
$\Gamma' \subseteq \Gamma$. $\neg A$ is the negation normal form of the negation of A.

Fig. 3. OS^X: a family of ordinary sequent proof systems for modal logic.

4 Emulation of other modal proof systems

In order to emulate proofs given in other proof calculi by means of the focused framework LMF_*^X, we need to: (i) define a proper polarization of modal formulas; and (ii) give a specialized version of the rule $decide_*$. As an illustration of this potentiality, we consider in this section some standard sequent and nested sequent calculi.

4.1 Ordinary sequent calculi

Several "ordinary" sequent systems have been proposed in the literature for different modal logics (a general account is, e.g., in [11,20]). In our treatment, we will use a formalization of a class of modal sequent systems, presented in Figure 3, which is adapted mainly from the presentations in [6,23]. It can be seen as a family of proof systems, where the system of a specific logic is obtained by adding to the base classical system (consisting of *identity*, *structural* and *classical connective* rules) one of the $\Box$-rules and any (possibly empty) set of $\Diamond$-rules. As the name of the rule suggests, the rule $\Box_K$ alone gives a system for the logic K. We replace it with $\Box_{K4}$ or $\Box_{K45}$ in case we want to capture logics characterized by transitive or both transitive and euclidean frames, respectively. The rules $\Diamond_T$ and $\Diamond_D$ can be further added, modularly, in order to get systems for those logics enjoying reflexivity and seriality, respectively. For instance, by adding $\Box_{K4}$ and $\Diamond_T$ to the base system, we get a system for the logic $S4$, while by adding $\Box_{K45}$ and $\Diamond_T$, we get a system for $S5$. Formulas are assumed to be in negation normal form.

First, we present a polarization that allows us to enforce the behavior of these rules in our framework. When translating a modal formula into a polarized one, we are often in a situation where we are interested in putting a delay in front of the formula only in the case when it is negative and not a literal. For that purpose, we define A^{∂^+}, where A is a modal formula in negation normal form, to be A if A is a literal or a positive formula and $\partial^+(A)$ otherwise. We extend such a notion to a multiset Γ of formulas by defining $\Gamma^{\partial^+} = \{A^{\partial^+} | A \in \Gamma\}$. Then we define the translation $\lfloor \cdot \rfloor$ from modal formulas in negation normal form into polarized modal formulas as follows:

$$\begin{array}{rclrcl} \lfloor P \rfloor & = & P & \lfloor A \wedge B \rfloor & = & \lfloor A \rfloor^{\partial^+} \wedge^- \lfloor B \rfloor^{\partial^+} \\ \lfloor \neg P \rfloor & = & \neg P & \lfloor A \vee B \rfloor & = & \lfloor A \rfloor^{\partial^+} \vee^- \lfloor B \rfloor^{\partial^+} \\ \lfloor \Box A \rfloor & = & \Box(\lfloor A \rfloor^{\partial^+}) & \lfloor \Diamond A \rfloor & = & \Diamond(\partial^-(\lfloor A \rfloor^{\partial^+})) \end{array}$$

In the following, we will sometimes use the natural extension of this translation to multisets of modal formulas, i.e., $\lfloor \Gamma \rfloor = \{\lfloor A \rfloor \mid A \in \Gamma\}$.

We note that the use of delays in this translation is motivated by the desire of keeping the correspondence between rule inferences in the emulated calculus and bipoles in our framework as strict as possible. In a sense, delays ensure that in a single phase we do not emulate more than one rule application of the original proof system.

Furthermore, we specialize the rule $decide_*$ as follows:

$$\frac{\mathcal{G} \vdash_{\mathcal{H}'} \Theta \Downarrow \Omega}{\mathcal{G} \vdash_{\{(x,\mathcal{F})\}} \Theta \Uparrow \cdot}\ decide_{OS}$$

where (in addition to the general conditions of Figure 2) we have that either:

(i) there exists y s.t.:
- $xRy \in \mathcal{G}$;
- if $x \neq y$, then $\mathcal{H}' = \{(y, \mathcal{F} \cup \{x\})\}$ and Ω is a multiset of formulas of the form $zy : \Diamond A$, s.t. $zRy \in \mathcal{G}$, $z \in \mathcal{F}$;
- if $x = y$ then $\mathcal{H}' = \{(x, \mathcal{F})\}$ and $\Omega = \{xx : \Diamond A\}$ for some A; or

(ii) $\Omega = \{x : A\}$ for some A and $\mathcal{H}' = \{(x, \mathcal{F})\}$.

Intuitively, the specialization with respect to the general framework consists in: (*i*) restricting the use of multifocusing to $\Diamond$-formulas; (*ii*) forcing such $\Diamond$-formulas to be labeled with the same future; (*iii*) when moving to a new label, adding the current label to the set of forbidden futures.

The structure of the proofs obtained by using these restrictions can be described as a sequence of blocks, each of which is related to a specific world (label). For each such a block, we first apply a number of classical and $\Box$-introductions on the given world (and some relational rules, if we are beyond K) and then move to a new one by means of a $\Diamond$-phase. The mechanism that we use, in $decide_{OS}$, for updating the present $\mathcal{H}$ of the sequent ensures that we never go back to an already encountered world.

We call LMF^X_{OS} the system obtained from LMF^X_* by replacing the rule $decide_*$ with the rule $decide_{OS}$. It is easy to notice that, given the polarization above and this new rule, we can in fact restrict LMF^X_{OS} to deal with sequents whose present is always a singleton and such that the future of each labeled formula has length at most 1. In the rest of this section, for simplicity, we will then write sequents using the following notation: $\mathcal{G} \vdash_{x,\mathcal{F}} \Theta \Updownarrow \Omega$.

As remarked in the discussion at the beginning of Section 3, even for a simple modal rule like $\Box_K$, at least two corresponding bipoles (one involving the $\Box$-formula and one involving $\Diamond$-formulas) are necessary in our framework. This means that in an LMF^X_{OS} proof, we can encounter synchronized sequents that do not correspond "precisely" to any sequent in the original proof, e.g., because (by reading the LMF^X_{OS} proof bottom-up) we are at a stage where a $\Box$-rule has been applied but the corresponding $\Diamond$-phase has not started yet. We will thus base our correspondence results on an interpretation that takes this fact into account. In the case of logics whose frames enjoy transitivity, such an interpretation will also have to consider that in the rule $\Box_{K4}$, $\Diamond$-formulas stay in the sequent when going from conclusion to premise, and such a behavior can only be captured in LMF^X_{OS} by applying more than one step. Formally, we define the interpretation $\mathcal{I}^X_{OS}(\cdot)$ of synchronized sequents as multisets of modal formulas (where the X denotes the fact that the interpretation is also parametric in the logic considered) as follows:

$$\mathcal{I}^X_{OS}(\mathcal{G} \vdash_{x,\mathcal{F}} \Theta \Uparrow \cdot) = \begin{cases} \bullet\ \{A \mid x : \lfloor A \rfloor^{\partial^+} \in \Theta\} \cup \{\Box B \mid y : \partial^+(\lfloor B \rfloor) \in \Theta, xRy \in \mathcal{G}^*, y \notin \mathcal{F}\}, \\ \quad \text{if } 4 \notin X \\ \bullet\ \{A \mid x : \lfloor A \rfloor^{\partial^+} \in \Theta\} \cup \{\Box B \mid y : \partial^+(\lfloor B \rfloor) \in \Theta, xRy \in \mathcal{G}^*, y \notin \mathcal{F}\} \cup \\ \quad \{\Diamond C \mid z : \lfloor \Diamond C \rfloor \in \Theta, zRx \in \mathcal{G}^*, z \in \mathcal{F}\}, \qquad \text{otherwise} \end{cases}$$

where $\mathcal{G}^*$ denotes the closure of $\mathcal{G}$ with respect to those properties among reflexivity, transitivity and euclideaness contained in X.

We notice that in an LMF^X_{OS} derivation (reading from the end-sequent upwards), when we decide on a formula, we keep a copy of it in the storage, i.e., we implicitly apply a contraction. For this reason, we have that an LMF^X_{OS} derivation tends to keep some information that is lost in the corresponding OS^X derivation (again, reading bottom-up). We define a notion of extension of a sequent that will help compare the two systems. Given a synchronized sequent $S \equiv \mathcal{G} \vdash_{\mathcal{H}} \Theta \Uparrow \cdot$ and an OS sequent $\vdash \Gamma$, we say that S *extends* $\vdash \Gamma$ if there exists $S' \equiv \mathcal{G} \vdash_{\mathcal{H}} \Theta' \Uparrow \cdot$ such that $\Theta \supseteq \Theta'$ and $\mathcal{I}^X_{OS}(S') = \Gamma$. Furthermore, we say that a synchronized sequent $\mathcal{G} \vdash_{\mathcal{H}} \Theta \Uparrow \cdot$ is in *OS form* if for all $x : A \in \Theta$, $A = \lfloor B \rfloor$ for some modal formula B.

Lemma 4.1 *Let* $\dfrac{S_1}{S}\ r$ $\left(\dfrac{S_1 \quad S_2}{S}\ r\right)$ *be an application of a non-structural rule in* OS^X. *Then for any synchronized sequent* S' *that is in OS form and extends* S, *there exists a derivation* $\begin{matrix} S'_1 \\ \vdots \\ S' \end{matrix}$ $\left(\begin{matrix} S'_1 \quad S'_2 \\ \vdots \\ S' \end{matrix}\right)$ *in* LMF^X_{OS}, *such that* S'_1 *is in OS form and extends* S_1 (S'_1, S'_2 *are in OS form and extend* S_1, S_2, *respectively*). *Furthermore, if* $\overline{S}\ ^r$ *is a rule application in* OS^X, *then for any synchronized sequent* S' *in OS form extending* S, *there exists a proof of* S' *in* LMF^X_{OS}.

Proof. The proof proceeds by considering all the non-structural rules of OS^X.

The cases for initial and the introduction of classical connectives are trivial and we omit them. We consider some key cases. We will also use the rules ∂_*^- and ∂_*^+, which are trivial adaptations of ∂^- and ∂^+ to the case of LMF_*^X sequents.

(1) Let us take an application of the rule $\Box_K$:

$$\frac{\vdash \Gamma, A}{\vdash \Diamond\Gamma, \Box A, \Delta}\ \Box_K$$

where Δ does not contain any formula whose main connective is $\Diamond$. Now assume that $S' \equiv \mathcal{G} \vdash_{x,\mathcal{F}} \Theta \Uparrow \cdot$ is in OS form and extends $\vdash \Diamond\Gamma, \Box A, \Delta$. Notice that we are in the case when 4 does not occur in X. It follows that $x : \lfloor\Diamond\Gamma\rfloor \subseteq \Theta$. We have two cases: either (a) $x : \partial^+(\lfloor\Box A\rfloor) \in \Theta$ or (b) $y : \lfloor A\rfloor^{\partial^+} \in \Theta$ and $xRy \in \mathcal{G}$. Then the LMF_*^X derivation corresponding to this rule application consists in the following steps (reading the derivation bottom-up):

(i) decide on $x : \partial^+(\lfloor\Box A\rfloor)$, ending up by adding xRy to $\mathcal{G}$ (note that this step is only required if we are in case (a));

$$\cfrac{\cfrac{\cfrac{\cfrac{\mathcal{G}\cup\{xRy\} \vdash_{x,\mathcal{F}} \Theta, y : \lfloor A\rfloor^{\partial^+} \Uparrow \cdot}{\mathcal{G}\cup\{xRy\} \vdash_{x,\mathcal{F}} \Theta \Uparrow y : \lfloor A\rfloor^{\partial^+}}\ store_*}{\mathcal{G} \vdash_{x,\mathcal{F}} \Theta \Uparrow x : \Box\lfloor A\rfloor^{\partial^+}}\ \Box_*}{\mathcal{G} \vdash_{x,\mathcal{F}} \Theta \Downarrow x : \partial^+(\Box\lfloor A\rfloor^{\partial^+})}\ \partial_*^+, release_*}{\mathcal{G} \vdash_{x,\mathcal{F}} \Theta \Uparrow \cdot}\ decide_{OS}$$

(ii) multifocus on $x : \lfloor\Diamond\Gamma\rfloor$ choosing y as the future.

$$\cfrac{\cfrac{\cfrac{\cfrac{\mathcal{G}\cup\{xRy\} \vdash_{y,\mathcal{F}\cup\{x\}} \Theta, y : \lfloor A\rfloor^{\partial^+}, y : \lfloor\Gamma\rfloor^{\partial^+} \Uparrow \cdot}{\mathcal{G}\cup\{xRy\} \vdash_{y,\mathcal{F}\cup\{x\}} \Theta, y : \lfloor A\rfloor^{\partial^+} \Uparrow y : \partial^-(\lfloor\Gamma\rfloor^{\partial^+})}\ \partial_*^-, store_*}{\mathcal{G}\cup\{xRy\} \vdash_{y,\mathcal{F}\cup\{x\}} \Theta, y : \lfloor A\rfloor^{\partial^+} \Downarrow y : \partial^-(\lfloor\Gamma\rfloor^{\partial^+})}\ release_*}{\mathcal{G}\cup\{xRy\} \vdash_{y,\mathcal{F}\cup\{x\}} \Theta, y : \lfloor A\rfloor^{\partial^+} \Downarrow xy : \Diamond\partial^-(\lfloor\Gamma\rfloor^{\partial^+})}\ \Diamond_*}{\mathcal{G}\cup\{xRy\} \vdash_{x,\mathcal{F}} \Theta, y : \lfloor A\rfloor^{\partial^+} \Uparrow \cdot}\ decide_{OS}$$

(2) Let us consider an application of the rule $\Box_{K4}$:

$$\frac{\vdash \Diamond\Gamma, \Gamma, A}{\vdash \Diamond\Gamma, \Box A, \Delta}\ \Box_{K4}$$

where Δ does not contain any formula whose main connective is $\Diamond$. Assume that $S' \equiv \mathcal{G} \vdash_{x,\mathcal{F}} \Theta \Uparrow \cdot$ is in OS form and extends $\vdash \Diamond\Gamma, \Box A, \Delta$. As in (i), we can have two cases: either (a) $x : \partial^+(\lfloor\Box A\rfloor) \in \Theta$ or (b) $y : \lfloor A\rfloor^{\partial^+} \in \Theta$ and $xRy \in \mathcal{G}^*$. Moreover, for each $B \in \Gamma$, one of the following two cases holds: either (c) $x : \lfloor\Diamond B\rfloor \in \Theta$ or (d) $z : \lfloor\Diamond B\rfloor \in \Theta$ and $zRx \in \mathcal{G}^*$ for some z. After possible applications of relational rules that lead to a sequent containing xRy (if we are in case (b)) and zRx (if we are in case (d)), the LMF_*^X derivation corresponding to this rule application consists in the following bipoles:

(i) decide on $x : \partial^+(\lfloor\Box A\rfloor)$, ending up by adding xRy to $\mathcal{G}$ (note that this step is only required if we are in case (a));

$$\dfrac{\dfrac{\dfrac{\mathcal{G}\cup\{xRy\}\vdash_{x,\mathcal{F}}\Theta,y:\lfloor A\rfloor^{\partial+}\Uparrow\cdot}{\mathcal{G}\vdash_{x,\mathcal{F}}\Theta\Uparrow x:\Box\lfloor A\rfloor^{\partial+}}\ \Box_*,store_*}{\mathcal{G}\vdash_{x,\mathcal{F}}\Theta\Downarrow x:\partial^+(\Box\lfloor A\rfloor^{\partial+})}\ \partial^+_*,release_*}{\mathcal{G}\vdash_{x,\mathcal{F}}\Theta\Uparrow\cdot}\ decide_{OS}$$

(ii) for those $B\in\Gamma$ such that case (d) holds, we apply the rule 4_* to zRx and xRy (ending up by adding zRy to the relation set);

$$\dfrac{\mathcal{G}\cup\{zRx,xRy,zRy\}\vdash_{x,\mathcal{F}}\Theta\Uparrow\cdot}{\mathcal{G}\cup\{zRx,xRy\}\vdash_{x,\mathcal{F}}\Theta\Uparrow\cdot}\ 4_*$$

(iii) multifocus on all the $w:\lfloor\Diamond B\rfloor$ such that wRy is in the relation set and $B\in\Gamma$, choosing y as the future.

$$\dfrac{\dfrac{\dfrac{\dfrac{\mathcal{G}\cup\{zRx,xRy,zRy\}\vdash_{y,\mathcal{F}\cup\{x\}}\Theta,y:\lfloor B\rfloor^{\partial+},\Omega''\Uparrow\cdot}{\mathcal{G}\cup\{zRx,xRy,zRy\}\vdash_{y,\mathcal{F}\cup\{x\}}\Theta\Uparrow y:\partial^-(\lfloor B\rfloor^{\partial+}),\Omega'}\ \partial^-_*,store_*}{\mathcal{G}\cup\{zRx,xRy,zRy\}\vdash_{y,\mathcal{F}\cup\{x\}}\Theta\Downarrow y:\partial^-(\lfloor B\rfloor^{\partial+}),\Omega'}\ release_*}{\mathcal{G}\cup\{zRx,xRy,zRy\}\vdash_{y,\mathcal{F}\cup\{x\}}\Theta\Downarrow wy:\Diamond\partial^-(\lfloor B\rfloor^{\partial+}),\Omega}\ \Diamond_*}{\mathcal{G}\cup\{zRx,xRy,zRy\}\vdash_{x,\mathcal{F}}\Theta\Uparrow\cdot}\ decide_{OS}$$

(3) Let us consider an application of the rule $\Box_{K45}$:

$$\dfrac{\vdash\Diamond\Gamma,\Gamma,\Box\Sigma,A}{\vdash\Diamond\Gamma,\Box\Sigma,\Box A,\Delta}\ \Box_{K45}$$

where Δ does not contain any formula whose main connective is $\Diamond$ or $\Box$. Assume that $S'\equiv\mathcal{G}\vdash_{x,\mathcal{F}}\Theta\Uparrow\cdot$ is in OS form and extends $\vdash\Diamond\Gamma,\Box\Sigma,\Box A,\Delta$. We focus on the treatment of the formulas in Σ, which is the difference with respect to case (ii). Let $B\in\Sigma$. By hypothesis, either (a) $x:\partial^+(\lfloor\Box B\rfloor)\in\Theta$ or $(b) y:\lfloor B\rfloor^{\partial+}\in\Theta$ and $xRy\in\mathcal{G}^*$. If we are in case (a), then an application of $\Box_*$ followed by an application of 5_* will eventually lead to a synchronized sequent S'_1 such that $\Box B\in\mathcal{I}^X_{OS}(S'_1)$. If we are in case (b), then an application of 5_*, plus possible relational rules to get xRy in the relational set, will suffice.

$$\dfrac{\dfrac{\dfrac{\dfrac{\mathcal{G}\cup\{xRy,xRu,yRu\}\vdash_{x,\mathcal{F}}\Theta,u:\lfloor B\rfloor^{\partial+}\Uparrow\cdot}{\mathcal{G}\cup\{xRy,xRu\}\vdash_{x,\mathcal{F}}\Theta,u:\lfloor B\rfloor^{\partial+}\Uparrow\cdot}\ 5_*}{\mathcal{G}\cup\{xRy,xRu\}\vdash_{x,\mathcal{F}}\Theta\Uparrow x:\Box\lfloor B\rfloor^{\partial+}}\ \Box_*,store_*}{\mathcal{G}\cup\{xRy\}\vdash_{x,\mathcal{F}}\Theta\Downarrow x:\partial^+(\Box\lfloor B\rfloor^{\partial+})}\ \partial^+_*,release_*}{\mathcal{G}\cup\{xRy\}\vdash_{x,\mathcal{F}}\Theta\Uparrow\cdot}\ decide_{OS}$$

(4) Let us consider an application of the rule $\Diamond_T$:

$$\dfrac{\vdash\Diamond A,A,\Sigma}{\vdash\Diamond A,\Sigma}\ \Diamond_T$$

and assume that $S'\equiv\mathcal{G}\vdash_{x,\mathcal{F}}\Theta\Uparrow\cdot$ is in OS form and extends $\vdash\Diamond A,\Sigma$. We have that either (a) $x:\lfloor\Diamond A\rfloor\in\Theta$ or (b) we are in a case where X contains

4 and $z : \lfloor \Diamond A \rfloor \in \Theta$ and $zRx \in \mathcal{G}^*$. After possible applications of relational rules that lead to a sequent whose relational set contains zRx (if we are in case (*b*)), the LMF_*^X derivation corresponding to this rule application consists in the following bipoles (reading the derivation bottom-up):

(i) if we are in case (*a*), apply the rule T_* in order to add xRx to $\mathcal{G}$; then decide on $x : \lfloor \Diamond A \rfloor$;

$$\dfrac{\dfrac{\dfrac{\dfrac{\mathcal{G} \cup \{xRx\} \vdash_{x,\mathcal{F}} \Theta, x : \lfloor A \rfloor^{\partial^+} \Uparrow \cdot}{\mathcal{G} \cup \{xRx\} \vdash_{x,\mathcal{F}} \Theta \Downarrow x : \partial^-(\lfloor A \rfloor^{\partial^+})}\ release_*, \partial_*^-, store_*}{\mathcal{G} \cup \{xRx\} \vdash_{x,\mathcal{F}} \Theta \Downarrow xx : \Diamond\partial^-(\lfloor A \rfloor^{\partial^+})}\ \Diamond_*}{\mathcal{G} \cup \{xRx\} \vdash_{x,\mathcal{F}} \Theta \Uparrow \cdot}\ decide_{OS}}{\mathcal{G} \vdash_{x,\mathcal{F}} \Theta \Uparrow \cdot}\ T_*$$

(ii) if we are in case (*b*), then decide on $z : \lfloor \Diamond A \rfloor$ and choose x as the future.

$$\dfrac{\dfrac{\dfrac{\mathcal{G} \cup \{zRx\} \vdash_{x,\mathcal{F}} \Theta, x : \lfloor A \rfloor^{\partial^+} \Uparrow \cdot}{\mathcal{G} \cup \{zRx\} \vdash_{x,\mathcal{F}} \Theta \Downarrow x : \partial^-(\lfloor A \rfloor^{\partial^+})}\ release_*, \partial_*^-, store_*}{\mathcal{G} \cup \{zRx\} \vdash_{x,\mathcal{F}} \Theta \Downarrow zx : \Diamond\partial^-(\lfloor A \rfloor^{\partial^+})}\ \Diamond_*}{\mathcal{G} \cup \{zRx\} \vdash_{x,\mathcal{F}} \Theta \Uparrow \cdot}\ decide_{OS}$$

(5) Let us consider an application of the rule $\Diamond_D$: $\dfrac{\vdash \Gamma}{\vdash \Diamond\Gamma, \Delta}\ \Diamond_D$ where Δ does not contain any formula whose main connective is $\Diamond$. Assume that $S' \equiv \mathcal{G} \vdash_{x,\mathcal{F}} \Theta \Uparrow \cdot$ is in OS form and extends $\vdash \Diamond\Gamma, \Delta$. For each $B \in \Gamma$, one of the following two cases holds: either (*a*) $x : \lfloor \Diamond B \rfloor \in \Theta$ or (*b*) $z : \lfloor \Diamond B \rfloor \in \Theta$ and $zRx \in \mathcal{G}^*$ (note that this is only possible if X contains 4). After possible applications of relational rules that lead to a sequent whose relational set contains zRx (if we are in case (*b*)), the LMF_*^X derivation corresponding to this application consists in the following bipoles (reading the derivation bottom-up): (i) apply the rule D_* in order to add xRy to $\mathcal{G}$ for some "fresh" y; (ii) for those $B \in \Gamma$ such that case *c* holds, apply the rule 4_* to zRx and xRy (ending up by adding zRy to the relation set); (iii) multifocus on all the $w : \lfloor \Diamond B \rfloor$ such that wRy is in the relation set and $B \in \Gamma$, choosing y as the future. □

Theorem 4.2 *Let Π be an OS^X derivation of a sequent $S \equiv\vdash A$ from the sequents $S_1, \ldots, S_n$ and let $S' \equiv \emptyset \vdash_{\{x,\emptyset\}} x : (\lfloor A \rfloor)^{\partial^+} \Uparrow \cdot$ for some x. Then there exists an LMF_{OS}^X derivation Π' of S' from $S'_1, \ldots, S'_n$, where $S'_1, \ldots, S'_n$ extend $S_1, \ldots, S_n$, respectively. Moreover, Π' is such that each rule application in Π, deriving a sequent $\hat{S}$, corresponds to a sequence s of bipoles in Π' such that s ends with a synchronized sequent $\hat{S}'$ extending $\hat{S}$.*

Proof. For simplicity, we assume that in Π the rule *contr* is only applied to a given formula immediately below a rule that introduces an occurrence of such a formula. We proceed bottom-up by starting from the root of Π and build Π' by repeatedly applying Lemma 4.1. At each step, we get as leaves sequents that are extensions of the ones in Π, so that Lemma 4.1 can be applied again. □

We say that a synchronized sequent $S \equiv \mathcal{G} \vdash_{\mathcal{H}} \Theta \Uparrow \cdot$ is a *contraction of an OS sequent* $\vdash \Gamma$ if S is in OS form, Γ contains $\mathcal{I}^X_{OS}(S)$ and for each formula A in Γ there is at least one occurrence of A in $\mathcal{I}^X_{OS}(S)$.

Lemma 4.3 *Let* $S' \equiv \mathcal{G} \vdash_{\{x,\mathcal{F}\}} \Theta \Uparrow \cdot$ *be a synchronized sequent in OS form. For each derivation of the form* $\begin{matrix} S_1' \\ \vdots \\ S' \end{matrix} \left(\begin{matrix} S_1' \quad S_2' \\ \vdots \\ S' \end{matrix}\right)$ *in* LMF^X_{OS} *that is a bipole, there exists an OS sequent S such that: (i) S′ is a contraction of S; and (ii) if* $\mathcal{I}^X_{OS}(S_1') \neq \mathcal{I}^X_{OS}(S')$ $(\mathcal{I}^X_{OS}(S_1') \neq \mathcal{I}^X_{OS}(S')$ *and* $\mathcal{I}^X_{OS}(S_2') \neq \mathcal{I}^X_{OS}(S'))$*, then there exists a rule application* $\frac{S_1}{S}$ $\left(\frac{S_1 \quad S_2}{S}\right)$ *in* OS^X *such that* $\mathcal{I}^X_{OS}(S_1') = S_1$ $(\mathcal{I}^X_{OS}(S_1') = S_1$ *and* $\mathcal{I}^X_{OS}(S_2') = S_2)$*. Furthermore, for each proof of S′ that is a bipole, there exist: (i) an OS sequent S such that S′ is a contraction of S; and (ii) a rule application* $\overline{S}$ *in* OS^X*.*

Proof. We can distinguish cases according to the main connective of the formula(s) on which we decide. The case of classical connectives is trivial, since we have that there is an exact correspondence between a bipole in LMF^X_{OS} and a rule application in OS^X. The case of a formula with $\Box$ as the main connective is also simple, because we have that $\mathcal{I}^X_{OS}(S') = \mathcal{I}^X_{OS}(S_1')$. Relational rules do not change interpretation of the sequent either. If we consider a decide on a multiset of formulas, whose main operator is $\Diamond$, we have that, by inspecting the cases arising from condition (i) in the definition of the rule $decide_{OS}$, one can see that this corresponds to an application of $\Box_K$, $\Box_{K4}$, $\Box_{K45}$, $\Diamond_T$ or $\Diamond_D$ according to the logic considered and the label chosen as the next one. □

Theorem 4.4 *Let* Π' *be a proof of a sequent* $S' \equiv \emptyset \vdash_{\{x,\emptyset\}} x : (\lfloor A \rfloor)^{\partial^+} \Uparrow \cdot$ *for some x. Then there exists a proof* Π *of a sequent S in* OS^X*, where S′ is a contraction of S, such that each bipole in* Π' *corresponds to one rule application in* Π*, plus possible applications of contr.*

Proof. We proceed top-down starting from the leaves of Π' and build Π by repeatedly applying Lemma 4.3. At each step, we get as the conclusion of an OS^X rule a sequent S^* such that the one obtained in the corresponding step of Π' is a contraction of S^*. By applying *contr*, we transform the OS^X derivation built so far and remove possible undesired multiple occurrences of a formula.□

Theorem 4.5 *The system* LMF^X_{OS} *is sound and complete for the logic KX.*

Proof. Soundness is obvious, since LMF^X_{OS} is just a restriction of LMF^X_*. Completeness follows from Theorem 4.2 and completeness of the system OS^X.□

4.2 A different formulation for ordinary sequent systems

The system LMF^X_{OS} is designed with the aim of emulating the behavior of OS^X as much as possible in a rule-by-rule way. However, we can also give a different polarization (obtained by introducing delays less intensively) that makes the role of focusing even more significant, i.e., such that a bipole in the focused system corresponds now to a larger, but well identified, block of an

OS^X derivation. In fact, as observed in the previous section, we can read an LMF^X_{OS} derivation (from the root upwards) as composed of blocks, where at each block we first apply all the classical reasoning and the $\Box$-rules relative to a given world and then execute a $\Diamond$-phase (thus moving to a different world). With the polarization given below, we can group all such classical+$\Box$ reasoning in a single asynchronous phase, so that (at least for the logic K) each block will correspond to exactly two phases. We define $\lfloor \cdot \rfloor_{OS'}$ as follows:

$$\begin{array}{rclrcl} \lfloor P \rfloor_{OS'} & = & P & \lfloor A \wedge B \rfloor_{OS'} & = & \lfloor A \rfloor_{OS'} \wedge^- \lfloor B \rfloor_{OS'} \\ \lfloor \neg P \rfloor_{OS'} & = & \neg P & \lfloor A \vee B \rfloor_{OS'} & = & \lfloor A \rfloor_{OS'} \vee^- \lfloor B \rfloor_{OS'} \\ \lfloor \Box A \rfloor_{OS'} & = & \Box(\lfloor A \rfloor_{OS'}{}^{\partial^+}) & \lfloor \Diamond A \rfloor_{OS'} & = & \Diamond(\partial^-(\lfloor A \rfloor_{OS'})) \end{array}$$

In this setting, each new bipole is started by choosing a successor y of the current world and by multifocusing on $\Diamond$-formulas labeled with a world from which y is reachable and non-$\Diamond$-formulas labeled with y, i.e., we define:

$$\frac{\mathcal{G} \vdash_{\mathcal{H}'} \Theta \Downarrow \Omega}{\mathcal{G} \vdash_{\{(x,\mathcal{F})\}} \Theta \Uparrow \cdot}\ decide_{OS'}$$

where (in addition to the general conditions of Figure 2) we have that either (i) $\Omega = \{x : B\}$ for B a positive literal and $\mathcal{H}' = \{(x, \mathcal{F})\}$ (special case for closing branches); or (ii) there exists $xRy \in \mathcal{G}$ such that:

- if $x \neq y$, then $\mathcal{H}' = \{(y, \mathcal{F} \cup \{x\})\}$ and Ω is a multiset of formulas of the form
  - $zy : \Diamond A$, s.t. $zRy \in \mathcal{G}$, $z \in \mathcal{F}$; or
  - $y : B$ for B positive but not a $\Diamond$-formula;
- if $x = y$ then $\mathcal{H}' = \{(x, \mathcal{F})\}$ and Ω is a multiset of formulas of the form $xx : \Diamond A$ or $x : B$ for B positive but not a $\Diamond$-formula.

We remark that with such a polarization and such a version of the decide rule, we obtain an instantiation of the framework which behaves very similarly to the focused system of [14].

4.3 Nested sequent calculi

Nested sequents (first introduced by Kashima [12], and then independently rediscovered by Poggiolesi [20], as *tree-hypersequents*, and by Brünnler [2]) are an extension of ordinary sequents to a structure of tree, where each [ ]-node represents the scope of a modal $\Box$. We write a nested sequent as a multiset of formulas and *boxed sequents*, according to the following grammar, where A can be any modal formula in negative normal form: $\mathcal{N} ::= \emptyset \mid A, \mathcal{N} \mid [\mathcal{N}], \mathcal{N}$

In a nested sequent calculus, a rule can be applied at any depth in this tree structure, that is, inside a certain nested sequent context. A *context* written as $\mathcal{N}\{\ \}\cdots\{\ \}$ is a nested sequent with a number of holes occurring in place of formulas (and never inside a formula). Given a context $\mathcal{N}\{\ \}\cdots\{\ \}$ with n holes, and n nested sequents $\mathcal{M}_1, \ldots, \mathcal{M}_n$, we write $\mathcal{N}\{\mathcal{M}_1\}\cdots\{\mathcal{M}_n\}$ to denote the nested sequent where the i-th hole in the context has been replaced by $\mathcal{M}_i$, with the understanding that if $\mathcal{M}_i = \emptyset$ then the hole is simply removed.

We are going to consider the nested sequent system (on Figure 4) introduced by Brünnler in [2], that we call here NS^X. The first two categories of rules

Identity and structural rules

$$\frac{}{\mathcal{N}\{P, \neg P\}}\ init \qquad \frac{\mathcal{N}\{A\} \quad \mathcal{N}\{\neg A\}}{\mathcal{N}\{\emptyset\}}\ cut$$

Connectives rules

$$\frac{\mathcal{N}\{A\} \quad \mathcal{N}\{B\}}{\mathcal{N}\{A \wedge B\}}\ \wedge \qquad \frac{\mathcal{N}\{A, B\}}{\mathcal{N}\{A \vee B\}}\ \vee \qquad \frac{\mathcal{N}\{A\}}{\mathcal{N}\{\Box A\}}\ \Box \qquad \frac{\mathcal{N}\{\Diamond A, [A, \mathcal{M}]\}}{\mathcal{N}\{\Diamond A, [\mathcal{M}]\}}\ \Diamond$$

$\Diamond$-rules

$$\frac{\mathcal{N}\{\Diamond A, A\}}{\mathcal{N}\{\Diamond A\}}\ T^{\Diamond} \qquad \frac{\mathcal{N}\{\Diamond A, [A]\}}{\mathcal{N}\{\Diamond A\}}\ D^{\Diamond} \qquad \frac{\mathcal{N}\{A, [\Diamond A, \mathcal{M}]\}}{\mathcal{N}\{[\Diamond A, \mathcal{M}]\}}\ B^{\Diamond}$$

$$\frac{\mathcal{N}\{\Diamond A, [\Diamond A, \mathcal{M}]\}}{\mathcal{N}\{\Diamond A, [\mathcal{M}]\}}\ 4^{\Diamond} \qquad \frac{\mathcal{N}\{\Diamond A\}\{\Diamond A\}}{\mathcal{N}\{\Diamond A\}\{\emptyset\}}\ 5^{\Diamond}$$

In $5^{\Diamond}$, the first hole in $\mathcal{N}\{\ \}\{\ \}$ can not occur at the root of the sequent tree.

Fig. 4. NS^X: a family of nested sequent proof systems for modal logic.

constitute a complete system for the modal logic K. It can then be extended modularly by a subset $X^{\Diamond}$ of the $\Diamond$-rules to give a complete system for any logic built from *45-closed*[3] set of axioms X among D, T, B, 4 and 5.

We want to specify the general framework LMF_*^X in order to emulate the proofs produced by NS^X. We can use here the same polarization $\lfloor \cdot \rfloor$ presented for ordinary sequents in Section 4.1 and specialize the rule $decide_*$ as follows:

$$\frac{\mathcal{G} \vdash_{\{(x,\emptyset) \mid x \in \mathcal{L}\}} \Theta \Downarrow x : A}{\mathcal{G} \vdash_{\{(x,\emptyset) \mid x \in \mathcal{L}\}} \Theta \Uparrow \cdot}\ decide_{NS}$$

where, as defined in Section 3, $\mathcal{L}$ denotes the set of all labels.

We can also use LMF_*^X in order to emulate the behavior of focused nested sequent calculi, like the one in [3]. Such a system can be captured by defining a polarization that does not apply delays intensively, such as the one given in Section 4.2 for an ordinary sequent focused system.

5 Concluding remarks

We have presented LMF_*^X as a general framework for emulating the behavior of several known modal proof systems based on ordinary sequents and on nested sequents. Our framework relies on using both labeled sequents (which injects semantics-related items into sequents) as well as focused proof rules (which

[3] X is said to be 45-closed: if whenever 4 is derivable in $K + X$, $4 \in X$, and whenever 5 is derivable in $K + X$, $5 \in X$. This condition is not restrictive as any logic obtained from a combination of axioms among D, T, B, 4 and 5, i.e. any logic in the so called S_5-cube, always has an equivalent 45-closed axiomatization.

can organize those injected items to support more high-level proof rules). The emulation of ordinary sequents is interesting because such calculi are proved to be optimal from the point of view of the efficiency of proof search. By decorating the sequents used in our framework with information (the *present* of the sequent) that specifies which world we are currently working on and which worlds are not reachable anymore, we are able to reproduce the mechanism that constrains (and improves) proof search in such calculi. Lellmann and Pimentel in [14] also use focusing but employ a notion of decorated sequent in order to constrain the construction of proofs in a fashion that only captures ordinary sequents (also, without invoking multifocusing). Section 4.2 shows how to instantiate our parametric framework so as to emulate the proofs of [14].

By analyzing the case of ordinary sequents, we conclude that modal rules in such a setting correspond to the application of two bipoles in our (1-sided) focused framework: the first bipole concerns a formula whose main connective is a $\Box$, while the second phase multifocuses on formulas with $\Diamond$ as the main connective. In the case of logics extending K, additional bipoles capturing the application of relational rules may also be required. The case of nested sequents illustrates the use of sequents decorated with a present that can contain more than just one world.

We believe that our framework is general enough to capture modal proof systems defined in other formalisms, such as prefixed tableaux systems [5,9] and 2-sequents [16] and their generalization to linear nested sequents [13]. In particular, we are currently working on formalizing a parametrization of LMF_*^X that can capture the modal hypersequent systems of, e.g., [1]. The basic idea consists in (1) using a present which is a multiset, (2) representing external structural rules as operations on such a present, and (3) viewing modal communication rules as a combination of relational and modal rules.

We have shown that LMF_*^X, when properly instantiated, can emulate several modal proof systems with high precision: individual modal inference rules correspond to certain chains of bipoles in the encoded LMF_*^X system *and vice versa.* Thus implementations of the LMF_*^X proof system can be seen as providing a theorem prover and a proof checker for the emulated proof systems. Although the LMF_*^X proof system imposes a lot of structure on the search for proofs, several important details are free to be implemented in differing ways. For example, one is free to implement the closure of the underlying world structure $\mathcal{G}^*$ via saturated bottom-up or top-down proof search.

While we have concentrated here on emulating existing calculi, we believe that LMF_*^X can be used to develop new and original (focused) proof systems for modal logics, all achieved by properly tuning the parametrical aspects of the framework: this is the object of ongoing research.

Acknowledgment. This work was funded by the ERC Advanced Grant ProofCert.

References

[1] Avron, A., *The method of hypersequents in the proof theory of propositional non-classical logics*, in: *Logic: From Foundations to Applications, European Logic Colloquium* (1994), pp. 1–32.

[2] Brünnler, K., *Deep sequent systems for modal logic*, Archive for Mathematical Logic **48** (2009), pp. 551–577.

[3] Chaudhuri, K., S. Marin and L. Straßburger, *Focused and synthetic nested sequents*, in: B. Jacobs and C. Löding, editors, *FoSSaCS*, 2016, pp. 390–407.

[4] Chihani, Z., D. Miller and F. Renaud, *Foundational proof certificates in first-order logic*, in: M. P. Bonacina, editor, *CADE 24: Conference on Automated Deduction 2013*, LNAI **7898**, 2013, pp. 162–177.

[5] Fitting, M., "Proof methods for modal and intuitionistic logics," Synthese Library **169**, D. Reidel Publishing Co., Dordrecht, 1983.

[6] Fitting, M., *Modal proof theory*, in: P. Blackburn, J. van Benthem and F. Wolter, editors, *Handbook of Modal Logic*, Elsevier, 2007 pp. 85–138.

[7] Fitting, M., *Prefixed tableaus and nested sequents*, Ann. Pure Appl. Logic **163** (2012), pp. 291–313.

[8] Gabbay, D. M., "Labelled Deductive Systems," Clarendon Press, 1996.

[9] Goré, R., *Tableau methods for modal and temporal logics*, in: M. D'Agostino, D. Gabbay, R. Hahnle and J.Posegga, editors, *Handbook of Tableau Methods*, Kluwer Academic Publishers, 1999 pp. 297–396.

[10] Goré, R. and R. Ramanayake, *Labelled tree sequents, tree hypersequents and nested (deep) sequents*, in: *Advances in Modal Logic 9*, Copenhagen, Denmark, 2012, pp. 279–299.

[11] Indrzejczak, A., "Natural Deduction, Hybrid Systems and Modal Logics," Springer, 2010.

[12] Kashima, R., *Cut-free sequent calculi for some tense logics*, Studia Logica **53** (1994), pp. 119–136.

[13] Lellmann, B., *Linear nested sequents, 2-sequents and hypersequents*, in: *TABLEAUX: Automated Reasoning with Analytic Tableaux and Related Methods*, Wrocław, Poland, September 21-24, 2015, pp. 135–150.

[14] Lellmann, B. and E. Pimentel, *Proof search in nested sequent calculi*, in: M. Davis, A. Fehnker, A. McIver and A. Voronkov, editors, *LPAR: Logic for Programming, Artificial Intelligence, and Reasoning*, Suva, Fiji, 2015, pp. 558–574.

[15] Liang, C. and D. Miller, *Focusing and polarization in linear, intuitionistic, and classical logics*, Theor. Comput. Sci. **410** (2009), pp. 4747–4768.

[16] Masini, A., *2-sequent calculus: A proof theory of modalities*, Ann. Pure Appl. Logic **58** (1992), pp. 229–246.

[17] Miller, D. and M. Volpe, *Focused labeled proof systems for modal logic*, in: M. Davis, A. Fehnker, A. McIver and A. Voronkov, editors, *LPAR: Logic for Programming, Artificial Intelligence, and Reasoning*, Suva, Fiji, 2015, pp. 266–280.

[18] Negri, S., *Proof analysis in modal logic*, J. Philosophical Logic **34** (2005), pp. 507–544.

[19] Negri, S., *Proof analysis in non-classical logics*, in: *Logic Colloquium*, Lecture Notes in Logic **28**, 2005, pp. 107–128.

[20] Poggiolesi, F., "Gentzen Calculi for Modal Propositional Logic," Springer, 2011.

[21] Read, S., *Semantic pollution and syntactic purity*, The Review of Symbolic Logic **8** (2015), pp. 649–661.

[22] Sahlqvist, H., *Completeness and correspondence in first and second order semantics for modal logic*, in: N. H. S. Kanger, editor, *Proceedings of the Third Scandinavian Logic Symposium*, 1975, pp. 110–143.

[23] Stewart, C. and P. Stouppa, *A systematic proof theory for several modal logics*, in: R. A. Schmidt, I. Pratt-Hartmann, M. Reynolds and H. Wansing, editors, *5th Conference on "Advances in Modal logic," Manchester (UK), September 2004* (2004), pp. 309–333.

[24] Viganò, L., "Labelled Non-Classical Logics," Kluwer Academic Publishers, 2000.

The structure of the lattice of normal extensions of modal logics with cyclic axioms

Yutaka Miyazaki [1]

Faculty of Liberal Arts and Sciences
Osaka University of Economics and Law
6-10, Gakuonji, Yao, Osaka
581-8511, JAPAN

Abstract

Irreflexive frames sometimes play a crucial role in the theory of modal logics, although the class of all such frames that consist of only irreflexive points can not be determined by any set of modal formulas. For instance, the modal logic determined by the frame of one irreflexive point is one of the two coatoms of the lattice of all normal modal logics. Another important result is that every rooted cycle-free frame, that consists of irreflexive points only, splits the lattice of all normal modal logics.

In this paper, we consider a family of axioms $\mathbf{Cycl}(n)$ (for $n \geq 0$), which forces frames to be *n-cyclic.* Seeking out the distribution of modal logics of irreflexive frames in the lattice of normal extensions of the modal logic with a cyclic axiom gives us information about the structure of this lattice.

We mainly discuss the case $n = 1$ (the structure of the lattice of normal extensions of $\mathbf{K} \oplus \mathbf{Cycl}(1)$) and the case $n = 2$ (that of normal extensions of $\mathbf{K} \oplus \mathbf{Cycl}(2)$). Finally we discuss the possibility that a similar or a refined argument may bring us information on the structure of the lattice of normal extensions of the logic $\mathbf{K} \oplus \mathbf{Cycl}(n)$ for every $n \geq 1$.

Keywords: irreflexive frame, cyclic axiom, splitting

1 Introduction

In frames for propositional modal logics, each point can be either *reflexive* or *irreflexive* in general. Whereas the class of all reflexive frames can be determined by simple one axiom $\mathbf{T}$, the class of all irreflexive frames cannot be characterized by any set of formulas in a usual modal language. However, irreflexive frames sometimes play a crucial role in the theory of modal logics. For instance, the modal logic determined by the frame of one irreflexive point is one of the two coatoms of the lattice of all normal modal logics [8]. Another important example is that every rooted cycle-free frame, that consists of irreflexive points only, splits the lattice of all normal modal logics [1].

[1] e-mail: y-miya@keiho-u.ac.jp

On the other hand, on *cyclicity* of modal algebras, the following result is established.

Theorem 1.1 ([7]) *Let $\mathcal{V}$ be a variety of modal algebras whose signature is finite. The following are equivalent.*

(1) *$\mathcal{V}$ is semisimple.*

(2) *$\mathcal{V}$ is a discriminator variety.*

(3) *$\mathcal{V}$ is weakly-transitive and cyclic.*

This theorem tells us that the cyclicity of modal algebras is as significant as the weak-transitivity of algebras is, although the definition of cyclicity in the above theorem is slightly different from ours. There have been accumulated an enormous amount of works on weakly-transitive modal logics. But as far as we know, there are very few results on cyclic modal logics.

In this paper, we focus on irreflexive frames that also validate the following cyclic axioms. We consider a family of axioms $\mathbf{Cycl}(n)$ (for $n \geq 0$), which forces frames to be *n-cyclic*. Seeking out the distribution of modal logics of irreflexive frames in the lattice of normal extensions of modal logics with a cyclic axiom gives us information about the structure of this lattice.

We mainly discuss the case $n = 1$ (the structure of the lattice of normal extensions of $\mathbf{K} \oplus \mathbf{Cycl}(1)$) and the case $n = 2$ (that of normal extensions of $\mathbf{K} \oplus \mathbf{Cycl}(2)$). Finally we discuss the possibility that a similar or a refined argument may bring us information on the structure of the lattice of normal extensions of the logic $\mathbf{K} \oplus \mathbf{Cycl}(n)$ for every $n \geq 1$.

2 Preliminaries

The propositional modal language is defined in a usual way, where a countably infinite set $Var := \{p_0, p_1, \ldots, p_k, \ldots\}$ of variables is used, and a nullary connective is $\bot$ (falsum), unary connectives are $\neg$ (negation) and $\Box$ (necessity), and a binary connective is $\wedge$ (conjunction). Several others are only abbreviations as, $\top := \neg\bot$, $\varphi \vee \psi := \neg(\neg\varphi \wedge \neg\psi)$, $\varphi \to \psi := \neg\varphi \vee \psi$, and $\Diamond\varphi := \neg\Box\neg\varphi$. The set of all modal formulas is denoted by Φ, that is also used for the *inconsistent logic.*

A set $\mathbf{L} \subseteq \Phi$ of formulas is called a *normal modal logic*, if it contains: (1) all classical tautologies, and (2) the formula of the form $\Box(p_0 \to p_1) \to (\Box p_0 \to \Box p_1)$, and is closed under the following rules: (3) Modus Ponens $(\varphi, \varphi \to \psi/\psi)$, (4) Uniform Substitution $(\varphi/\varphi[p_i/\psi])$, and (5) Necessitation $(\varphi/\Box\varphi)$. The smallest normal modal logic on Φ is denoted by $\mathbf{K}$. We call $\mathbf{L}$ simply a *logic* if it is a normal modal logic, since we deal with only propositional normal modal logics.

For a normal modal logic $\mathbf{L}_0$ and a set Γ of formulas, the *smallest normal extension* of $\mathbf{L}_0$ by Γ, or the smallest normal modal logic that contains both $\mathbf{L}_0$ and Γ, is denoted by $\mathbf{L}_0 \oplus \Gamma$. If Γ is a finite set $\{\varphi_1, \varphi_2, \ldots, \varphi_n\}$ of formulas, then the logic $\mathbf{L}_0 \oplus \Gamma$ is simply denoted by $\mathbf{L}_0\varphi_1\varphi_2\cdots\varphi_n$. The class of all *normal extensions* of $\mathbf{L}_0$, that is, the class $\{\mathbf{L} \subseteq \Phi \mid \mathbf{L}_0 \subseteq \mathbf{L} \text{ and } \mathbf{L} \text{ is normal}\}$

is denoted by $\mathrm{NExt}(\mathbf{L}_0)$. This forms a complete lattice, and also satisfies the (finite) distributive law [2].

Two types of mathematical structure are used to interpret modal formulas semantically: one is modal algebras and the other frames.

A *modal algebra* is a structure $\mathfrak{A} := \langle A, \cap, \cup, -, \Box, 1, 0\rangle$, where $\langle A, \cap, \cup, -, 1, 0\rangle$ is a Boolean algebra and $\Box$ is a unary modal operator (we use the same symbol as in formulas) that satisfies (a) $\Box 1 = 1$ and (b) $\Box(x \cap y) = \Box x \cap \Box y$.

To interpret a formula φ in a modal algebra $\mathfrak{A}$, we use a *valuation* $v : \Phi \to \mathfrak{A}$ which satisfies the following: (1) $v(\bot) = 0$, (2) $v(\neg\varphi) = -v(\varphi)$, (3) $v(\varphi \wedge \psi) = v(\varphi) \cap v(\psi)$, and $v(\Box\varphi) = \Box(v(\varphi))$. A formula φ is *valid* in an algebra $\mathfrak{A}$, if $v(\varphi) = 1$ holds for any valuation v on $\mathfrak{A}$. For any class $\mathcal{C}$ of modal algebras, the set $\mathbf{L}(\mathcal{C})$ of formulas that are valid in all members of the class $\mathcal{C}$ defines a normal modal logic.

On the other hand, for every modal logic $\mathbf{L}$, a particular class $\mathcal{V} = \mathcal{V}(\mathbf{L})$ of modal algebras corresponds to it. This class $\mathcal{V}$ is called an equationally definable class of modal algebras for $\mathbf{L}$, or the *variety* for $\mathbf{L}$. All subvarieties of $\mathcal{V}(\mathbf{L})$ form a complete lattice, which is dual isomorphic to $\mathrm{NExt}(\mathbf{L})$. Therefore, an investigation of modal logics can also be seen as an investigation of the varieties of modal algebras which correspond to the modal logics considered.

The smallest variety $\mathcal{V}$ that contains a class $\mathcal{C}$ of algebras can be *generated* by: $\mathcal{V} = HSP(\mathcal{C})$, where H, S, P are the following class operators of algebras of a same type. $H(\mathcal{C}) := \{\mathfrak{B} \mid \mathfrak{B}$ is a homomorphic image of some $\mathfrak{A} \in \mathcal{C}\}$, $S(\mathcal{C}) := \{\mathfrak{B} \mid \mathfrak{B}$ is a subalgebra of some $\mathfrak{A} \in \mathcal{C}\}$, and $P(\mathcal{C}) := \{\mathfrak{B} \mid \mathfrak{B}$ is a direct product of some members $\{\mathfrak{A}_i\}_{i\in I} \subseteq \mathcal{C}\}$.

Another way of generating the smallest variety $\mathcal{V}$ which contains a class $\mathcal{C}$ of algebras is: $\mathcal{V} = HP_S(\mathcal{C}_{s.i.})$, where $P_S(\mathcal{D}) := \{\mathfrak{B} \mid \mathfrak{B}$ is a subdirect product of some members $\{\mathfrak{A}_i\}_{i\in I} \subseteq \mathcal{D}\}$, and $\mathcal{C}_{s.i.}$ is the class of all *subdirectly irreducible* members in $\mathcal{C}$. Here we do not describe the detail of what the subdirect product is, and what the subdirectly irreducible (s.i. for short) members are. But by this fact, we see that it is the s.i. members in $\mathcal{C}$ that determine the variety $\mathcal{V}$. All s.i. members in $\mathcal{C}$ behave as *building blocks* of the variety $\mathcal{V}$, or dually, the corresponding modal logic. The following algebraic characterization of subdirectly irreducible modal algebras is a key for our analysis.

Theorem 2.1 ([9]) *A non-trivial modal algebra $\mathfrak{A} = \langle A, \cap, \cup, -, \Box, 1, 0\rangle$ is subdirectly irreducible if and only if there exists an element $d(\neq 1) \in A$, and for any element $x(\neq 1) \in A$, $x \cap \Box x \cap \Box^2 x \cap \cdots \cap \Box^n x \leq d$ holds for some $n \in \omega$.*

The other type of semantics, a (*general*) *frame* is a structure $\mathcal{F} := \langle W, R, P\rangle$, where W is a set of points, R a binary relation on W, and P is a subset of $\mathcal{P}(W)$ which contains W and $\emptyset$, and is closed under $\cap$ (the set-theoretic intersection, $-$ (the set-theoretic complement) and a unary operator $\Box_R$, that is defined as: $\Box_R(X) := \{y \in W \mid \forall x \in W, (xRy \text{ implies } y \in X)\}$ for $X \in \mathcal{P}(W)$.

In order to interpret a formula in a frame, we use a *valuation* $V : \Phi \to P$. For a frame $\mathcal{F} := \langle W, R, P\rangle$, a valuation V, and a point $a \in W$, we define the truth condition of a formula φ at a in a *model* $\langle \mathcal{F}, V\rangle$ ($\langle \mathcal{F}, V\rangle \models_a \varphi$ in symbol) as usual. In particular, $\langle \mathcal{F}, V\rangle \models_a \Box\varphi$ if and only if for any $b \in W$, aRb implies $\langle \mathcal{F}, V\rangle \models_b \varphi$. A formula φ is *valid* in a frame $\mathcal{F}$ ($\mathcal{F} \models \varphi$ in symbol), if $\langle \mathcal{F}, V\rangle \models_a \varphi$ holds for any valuation V on $\mathcal{F}$ and for any point $a \in \mathcal{F}$. For any class $\mathcal{D}$ of frames, the set $\mathbf{L}(\mathcal{D})$ of formulas that are valid in all members in $\mathcal{D}$ defines a normal modal logic. On the other hand, for a normal modal logic $\mathbf{L}$, a frame $\mathcal{F}$ is a *frame for* $\mathbf{L}$ ($\mathcal{F} \models \mathbf{L}$) if $\mathcal{F} \models \varphi$ holds for every $\varphi \in \mathbf{L}$.

There are some formulas (axioms) in normal modal logics that can characterize some classes of frames whose members satisfy a first order condition written in a language with predicate symbols R and $=$. We denote $\mathcal{F} \models \Xi$ to mean that the frame $\mathcal{F}$ satisfies a condition Ξ. For example, the famous axioms $\mathbf{T} := p \to \Diamond p, \mathbf{B} := p \to \Box\Diamond p$ and $\mathbf{D} := \Diamond\top$ characterize the classes of frames with the following conditions respectively.

Fact 2.2 *For a frame* $\mathcal{F} = \langle W, R, P\rangle$,

(1) $\mathcal{F} \models \mathbf{T}$ *if and only if* $\mathcal{F} \models \forall x(xRx)$.

(2) $\mathcal{F} \models \mathbf{B}$ *if and only if* $\mathcal{F} \models \forall x, y(xRy \text{ implies } yRx)$.

(3) $\mathcal{F} \models \mathbf{D}$ *if and only if* $\mathcal{F} \models \forall x \exists y(xRy)$.

According to the famous Jónsson-Tarski representation [6], for a given frame $\mathcal{F} = \langle W, R, P\rangle$, the modal algebra $\mathcal{F}^*$ which corresponds to the frame $\mathcal{F}$ is constructed as: $\mathcal{F}^* = \langle P, \cap, \cup, -, \Box_R, W, \emptyset\rangle$. The algebra $\mathcal{F}^*$ is indeed a modal algebra, and there is a following correspondence between these two semantics: for any formula $\varphi \in \Phi$, $\mathcal{F} \models \varphi$ if and only if $\mathcal{F}^* \models \varphi$. Conversely, for a given modal algebra $\mathfrak{A}$, we can construct a frame, which is denoted by $\mathfrak{A}_*$ such that both $\mathfrak{A}$ and $\mathfrak{A}_*$ validate the same set of formulas.

We have explained only a minimal set of knowledge and definitions of some technical terms. We will follow notions and nomenclature of modal logics from [4] and those of universal algebras from [3].

3 Irreflexive frames for modal logics with cyclic axioms

3.1 Irreflexive frames of a particular form

Let $\mathcal{F} = \langle W, R, P\rangle$ be a frame. A point $a \in W$ is *irreflexive* if aRa does not hold. We draw an irreflexive point in a frame by a circle ($\circ$). A frame $\mathcal{F}$ is *irreflexive* if every point in $\mathcal{F}$ is irreflexive. In this paper, we employ a family $\{\mathcal{I}_n\}_{n\in\omega}$ and $\mathcal{I}_\infty$ of irreflexive frames of the following form for $n \geq 0$. $\mathcal{I}_n$ is a finite Kripke frame $\langle W, R\rangle$, where $W := \{a_0, a_1, a_2, \cdots, a_n\}$ and $R := \{(a_0, a_k) \mid 1 \leq k \leq n\}$. $\mathcal{I}_\infty := \langle W, R\rangle$, where W is a countably infinite set $\{a_i \mid i \geq 0\}$ and $R := \{(a_0, a_i) \mid i \geq 1\}$. Figures of these frames are below. On the modal logics determined by $\mathcal{I}_n$'s and $\mathcal{I}_\infty$, the following holds.

Proposition 3.1

(1) $\mathbf{L}(\mathcal{I}_0) \supsetneq \mathbf{L}(\mathcal{I}_1) \supsetneq \mathbf{L}(\mathcal{I}_2) \supsetneq \cdots \supsetneq \mathbf{L}(\mathcal{I}_\infty)$.

(2) $\mathbf{L}(\mathcal{I}_\infty) = \bigcap_{i\in\omega} \mathbf{L}(\mathcal{I}_i)$.

Proof. (1) For $k \leq \ell$, there is a p-morphism from $\mathcal{I}_\ell$ to $\mathcal{I}_k$, and so, $\mathbf{L}(\mathcal{I}_\ell) \subseteq \mathbf{L}(\mathcal{I}_k)$. By using the axiom $\mathbf{alt}_n := \Box p_0 \vee \Box(p_0 \to p_1) \vee \cdots \vee \Box((p_0 \wedge p_1 \wedge \cdots \wedge p_{n-1}) \to p_n)$, it is easy to see that $\mathcal{I}_k \models \mathbf{alt}_k$ but that $\mathcal{I}_{k+1} \not\models \mathbf{alt}_k$. Therefore $\mathbf{L}(\mathcal{I}_k) \not\subseteq \mathbf{L}(\mathcal{I}_{k+1})$.
(2) Suppose that $\varphi \notin \mathbf{L}(\mathcal{I}_\infty)$. Then there exists a valuation V and a point a in $\mathcal{I}_\infty$ such that $\langle \mathcal{I}_\infty, V\rangle \not\models_a \varphi$. Since the length of the formula φ is finite, there is $k \in \omega$, a valuation V' on $\mathcal{I}_k$ and a point b in $\mathcal{I}_k$ such that $\langle \mathcal{I}_k, V'\rangle \not\models_b \varphi$. Therefore we have $\mathbf{L}(\mathcal{I}_\infty) \supseteq \bigcap_{i\in\omega} \mathbf{L}(\mathcal{I}_i)$. □

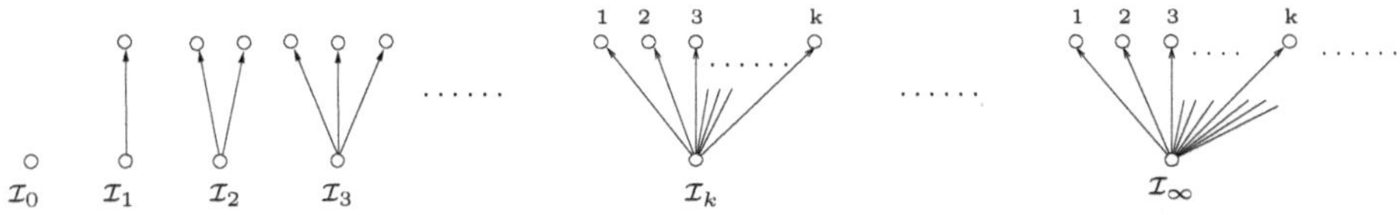

Fig. 1. Frames $\mathcal{I}_k$

3.2 Cyclic axioms and serial axioms

The *cyclic axiom* is defined as: $\mathbf{Cycl}(n) := p \to \Box^n \Diamond p$ for $n \geq 0$. This axiom characterizes the class of frames with the following property.

Fact 3.2 *For a frame* $\mathcal{F} = \langle W, R, P\rangle$ *and for* $n \geq 0$, $\mathcal{F} \models \mathbf{Cycl}(n)$ *if and only if* $\mathcal{F} \models \forall x_0, x_1, x_2, \ldots, x_n \in W, (x_0 R x_1 R x_2 R \cdots R x_n$ *implies* $x_n R x_0)$.

For $n \geq 0$, a frame $\mathcal{F}$ is *n-cyclic* if $\mathcal{F} \models \mathbf{Cycl}(n)$ holds. Note that this $\mathbf{Cycl}(n)$ is a generalization of the well known axioms $\mathbf{T} := \mathbf{Cycl}(0)$ and $\mathbf{B} := \mathbf{Cycl}(1)$.

On the other hand, the (generalized) *serial* axiom is defined as: $\mathbf{D}_n := \Box^n \Diamond \top$ for $n \geq 0$. This axiom is a generalization of the serial axiom $\mathbf{D}$, and it characterizes the following property of frames.

Fact 3.3 *For a frame* $\mathcal{F} = \langle W, R, P\rangle$ *and for* $n \geq 0$, $\mathcal{F} \models \mathbf{D}_n$ *if and only if* $\mathcal{F} \models \forall x_0, x_1, \ldots, x_n \in W, (x_0 R x_1 R \cdots R x_n$ *implies* $\exists y \in W(x_n R y))$.

For $n \geq 0$, a frame $\mathcal{F}$ is *n-serial* if $\mathcal{F} \models \mathbf{D}_n$ holds.

3.3 Levels of points in a frame

Let $\mathcal{F} = \langle W, R, P\rangle$ be a frame. Subsets $W^{(k)}$ for $k = 0, 1, 2, \cdots$ and W^∞ of W is defined as: $W^{(0)} := \{x \in W \mid \mathrm{NOT}(\exists y \in W(xRy))\}$, $W^{(n+1)} := \{x \in W \mid \exists y \in W^{(n)}(xRy)\}$, and $W^\infty := \{x \in W \mid \mathrm{NOT}(\exists n \in \omega, \exists y \in W^{(0)}(xR^n y))\}$.

In general, these $W^{(k)}$'s and W^∞ are not disjoint. But in a frame for $\mathbf{Cycl}(0)$, $\mathbf{Cycl}(1)$, or $\mathbf{Cycl}(2)$, they are disjoint. And moreover, in a frame for $\mathbf{Cycl}(0)$, $W = W^\infty$, in a frame for $\mathbf{Cycl}(1)$, $W = W^{(0)} \cup W^\infty$, and in a frame for $\mathbf{Cycl}(2)$, $W = W^{(0)} \cup W^{(1)} \cup W^\infty$. This observation is another key fact for our analysis. In these frames, a point $x \in W$ is called a point of *level n* if and only if $x \in W^{(n)}$ for $n = 0, 1$ and x is called a point of *level* ∞ if and only if $x \in W^\infty$.

4 A splitting of NEXT(**KCycl**(n)) for $n \geq 1$

Definition 4.1 [Splitting] Let $\mathcal{L} = \langle L, \wedge, \vee, 0, 1 \rangle$ be a complete lattice and $a \in L$. Then a *splits* $\mathcal{L}$ if there exists $b \in L$ such that for any $x \in L$, either $x \leq a$ or $b \leq x$, but not both. Such a pair (a, b) is called a *splitting pair* of the lattice $\mathcal{L}$. In this case the element b is a *splitting* of $\mathcal{L}$.

When we think about splittings of a lattice of logics, if a logic $\mathbf{L}(\mathfrak{A})$ (or $\mathbf{L}(\mathcal{F})$) splits the lattice, then we say that the algebra $\mathfrak{A}$ (or the frame $\mathcal{F}$) splits it.

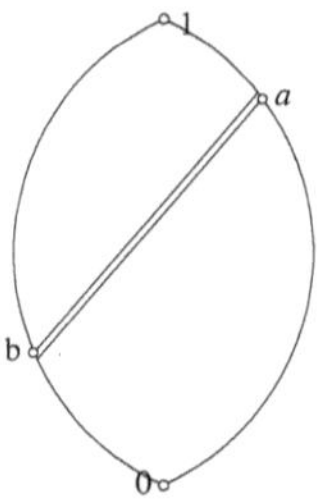

Fig. 2. A splitting of a complete lattice $\mathcal{L}$

A simple calculation for an n-cyclic s.i. modal algebra proves the following lemma first, which leads us to a splitting of the complete lattice NEXT(**KCycl**(n)).

Lemma 4.2 *Let $\mathfrak{A}$ be a non-trivial, subdirectly irreducible modal algebra for* **Cycl**(n) ($n \geq 1$). *Suppose $\Box^{n-1}\Diamond 1 \neq 1$ in $\mathfrak{A}$. Then $\Box^n 0 = 1$ holds in $\mathfrak{A}$.*

Proof. Since $\mathfrak{A}$ is s.i., there is some element $d(\neq 1) \in A$ and for the element $\Box^{n-1}\Diamond 1$, there is a number $m \in \omega$ such that $\Box^{n-1}\Diamond 1 \cap \Box^n \Diamond 1 \cap \cdots \cap \Box^m \Diamond 1 \leq d$ holds. Due to the n-cyclic axiom, we have that $\Box^{n-1}\Diamond 1 \leq d$. Now suppose that $\Box^n 0 \neq 1$ in $\mathfrak{A}$. Then similarly, there is a number $\ell \in \omega$ such that $\Box^n 0 \cap \Box^{n+1} 0 \cap \cdots \cap \Box^\ell 0 \leq d$ holds. Among the conjuncts in the left hand side, $\Box^n 0$ is the smallest, and so, we have $\Box^n 0 \leq d$. By the former, $-d \leq \Diamond^{n-1}\Box 0$, and so, $\Diamond - d \leq \Diamond^n \Box 0 \leq 0$ because of (the dual) of n-cyclic axiom. Therefore we have $\Diamond - d = 0$. Finally we have $-d \leq \Box^n \Diamond - d = \Box^n 0 \leq d$, which implies that $d = 1$. This is a contradiction. Hence we have $\Box^n 0 = 1$. □

Let Ch_n be the irreflexive frame of n point directed chain ($n \geq 1$). That is, $Ch_n := \langle W, R \rangle$, where $W := \{b_i \mid 0 \leq i \leq n-1\}$ (All b_i's are distinct.) and $R := \{(b_i, b_{i+1}) \mid 0 \leq i \leq n-2\}$. The figure of Ch_n is below.

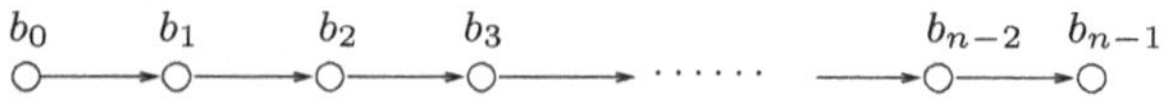

Fig. 3. Frames Ch_n

Theorem 4.3 *For any $n \geq 1$, $(\mathbf{KD}_{n-1}\mathbf{Cycl}(n), \mathbf{L}(Ch_n))$ is a splitting pair of the lattice* NEXT(**KCycl**(n)).

Proof. Suppose that $\mathbf{KD}_{n-1}\mathbf{Cycl}(n) \nsubseteq \mathbf{L}$ for a logic $\mathbf{L} \in \mathrm{NExt}(\mathbf{KCycl}(n))$. Then, there exists an s.i. algebra $\mathfrak{A}$ for $\mathbf{L}$, such that $\mathfrak{A} \not\models \mathbf{D}_{n-1}(= \Box^{n-1}\Diamond\top)$. This means that $\Box^{n-1}\Diamond 1 \neq 1$ in $\mathfrak{A}$. Therefore, by the previous lemma, $\Box^n 0 = 1$ holds in $\mathfrak{A}$. Here we claim that in $\mathfrak{A}$, $0 < \Box 0 < \Box^2 0 < \cdots < \Box^n 0 = 1$. So, suppose $\Box^k 0 = 1$ for some k $(0 \leq k < n)$. Then, since $0 \leq \Diamond 1$, we have $1 = \Box^k 0 \leq \Box^k \Diamond 1 = 1$, which implies that $\Box^{n-1}\Diamond 1 = 1$. This is a contradiction. Thus $\Box^k 0 \neq 1$ for all k $(0 \leq k < n)$. Suppose that $\Box^k 0 = \Box^{k+1} 0$ for some k $(0 \leq k < n)$. But this leads to a conclusion that $\Box^k 0 = \Box^{k+1} 0 = \Box^{k+2} 0 = \cdots = \Box^n 0 = 1$, which implies that $\Box^k 0 = 1$. This is also a contradiction.

Let $\mathfrak{A}'$ be a subalgebra of $\mathfrak{A}$ generated by the subset $B = \{0, \Box 0, \Box^2 0, \ldots, \Box^{n-1} 0, 1\}$ of A. A map $f : B \to \mathcal{C}h_n{}^*$ is defined as follows: $f(0) := \emptyset, f(\Box 0) := \{b_{n-1}\}, f(\Box^2 0) := \{b_{n-1}, b_{n-2}\}, \ldots, f(\Box^k 0) := \{b_{n-1}, b_{n-2}, \ldots, b_{n-k}\}, \ldots, f(\Box^{n-1} 0) := \{b_{n-1}, b_{n-2}, \ldots, b_1\}$, and $f(1) := \{b_{n-1}, b_{n-2}, \ldots, b_1, b_0\} = W$. Then this f can be extended to a map $F : \mathfrak{A}' \to \mathcal{C}h_n{}^*$ as follows: $F(x) := f(x)$ for $x \in B$ and suppose for any $x, y \in A'$, $F(x)$ and $F(y)$ are already defined. Then we define $F(x \cap y) := F(x) \cap F(y)$, $F(-x) := -F(x)$, and $F(\Box x) := \Box_R(F(x))$. Then, it is easy to see that this F is an embedding, and since the subset $f(B) = \{f(0), f(\Box 0), \ldots, f(\Box^{n-1} 0), f(1)\}$ of the universe of $\mathcal{C}h_n{}^*$ generates the whole $\mathcal{C}h_n{}^*$, and so, F is an isomorphism from $\mathfrak{A}'$ to $\mathcal{C}h_n{}^*$. Therefore $\mathfrak{A}' \cong \mathcal{C}h_n{}^*$, which implies that $\mathcal{C}h_n{}^* \in S(\mathfrak{A})$. Hence we have $\mathbf{L} \subseteq \mathbf{L}(\mathfrak{A}) \subseteq \mathbf{L}(\mathcal{C}h_n{}^*) = \mathbf{L}(\mathcal{C}h_n)$. □

This splitting theorem holds for any $n \geq 1$. We take this theorem as a clue to investigate the structure of the lattice $\mathrm{NExt}(\mathbf{KCycl}(n))$. In what follows, we mainly consider the case $n = 1$ and the case $n = 2$.

5 The structure of $\mathrm{NExt}(\mathbf{KCycl}(1))$

Among all connected frames for $\mathbf{Cycl}(1)$, it is only the frame $\mathcal{I}_0$ that contains a point of level 0. By Theorem 4.3, we can see that this frame splits the lattice $\mathrm{NExt}(\mathbf{KCycl}(1))$.

Theorem 5.1 $(\mathbf{KD}_0\mathbf{Cycl}(1), \mathbf{L}(\mathcal{I}_0))$ *is a splitting pair of the lattice* $\mathrm{NExt}(\mathbf{KCycl}(1))$. □

Now a question arises: what sort of modal logics are located under the logic $\mathbf{L}(\mathcal{I}_0)$ in $\mathrm{NExt}(\mathbf{KCycl}(1))$? Since $\mathcal{I}_0$ is a frame of only one irreflexive point, it is impossible that this frame is a p-morphic image of some connected frames. It is the case that some suitable generated subframes of $\mathcal{I}_n$ for $n \geq 1$ or $\mathcal{C}h_k$ for $k \geq 1$ are isomorphic to $\mathcal{I}_0$, but they are not frames for $\mathbf{Cycl}(1)$. Our argument in this section will proceed to give an answer to this question.

First of all, the following equality holds.

Proposition 5.2 $\mathbf{KCycl}(1) = \mathbf{KD}_0\mathbf{Cycl}(1) \cap \mathbf{L}(\mathcal{I}_0)$.

Proof. It is obvious that $\mathbf{KCycl}(1) \subseteq \mathbf{KD}_0\mathbf{Cycl}(1) \cap \mathbf{L}(\mathcal{I}_0)$. Conversely suppose $\varphi \notin \mathbf{KCycl}(1)$ for some $\varphi \in \Phi$. Then, there exists a frame $\mathcal{F} = \langle W, R, P\rangle$ for $\mathbf{Cycl}(1)$, a valuation V on $\mathcal{F}$, and a point $a \in W$ such that $\langle \mathcal{F}, V\rangle \not\models_a \varphi$. Now since this $\mathcal{F}$ is 1-cyclic, $W = W^{(0)} \cup W^{\infty}$. If $\mathcal{F}$ is also 0-serial, then $W^{(0)} = \emptyset$ and it is a frame for $\mathbf{Cycl}(1)$ and $\mathbf{D}_0$. Thus we have $\varphi \notin \mathbf{KD}_0\mathbf{Cycl}(1)$. Otherwise, $W^{(0)} \neq \emptyset$. However, since $\mathcal{F}$ is 1-cyclic, any point in $W^{(0)}$ is isolated from other part of the frame. So, if $a \in W^{\infty}$, then the subframe $\mathcal{F}'$ generated by the singleton $\{a\}$ is both 1-cyclic and 0-serial. With the valuation V' which is a restriction of V to $\mathcal{F}'$, we have $\langle \mathcal{F}', V'\rangle \not\models_a \varphi$. This means that $\varphi \notin \mathbf{KD}_0\mathbf{Cycl}(1)$. If $a \in W^{(0)}$, then the subframe $\mathcal{F}''$ generated by the singleton $\{a\}$ is just the frame $\mathcal{I}_0$. With the valuation V'' which is a restriction of V to $\mathcal{F}''$, $\langle \mathcal{F}'', V''\rangle \not\models_a \varphi$ holds. This means that $\varphi \notin \mathbf{L}(\mathcal{I}_0)$. Therefore we have $\varphi \notin \mathbf{KD}_0\mathbf{Cycl}(1) \cap \mathbf{L}(\mathcal{I}_0)$. Hence $\mathbf{KCycl}(1) \supseteq \mathbf{KD}_0\mathbf{Cycl}(1) \cap \mathbf{L}(\mathcal{I}_0)$.□

For normal modal logics $\mathbf{L}_1, \mathbf{L}_2$ such that $\mathbf{L}_1 \subseteq \mathbf{L}_2$, the *interval* between these two logics is denoted by $[\mathbf{L}_1, \mathbf{L}_2]$, that is, $[\mathbf{L}_1, \mathbf{L}_2] := \{\mathbf{L} \in \mathrm{NExt}(\mathbf{K}) \mid \mathbf{L}_1 \subseteq \mathbf{L} \subseteq \mathbf{L}_2\}$.

Define maps σ and τ as follows: $\sigma : \mathrm{NExt}(\mathbf{KD}_0\mathbf{Cycl}(1)) \to [\mathbf{KCycl}(1), \mathbf{L}(\mathcal{I}_0)]$ is defined as: $\sigma(\mathbf{L}) := \mathbf{L} \cap \mathbf{L}(\mathcal{I}_0)$. $\tau : [\mathbf{KCycl}(1), \mathbf{L}(\mathcal{I}_0)] \to \mathrm{NExt}(\mathbf{KD}_0\mathbf{Cycl}(1))$ is defined as: $\tau(\mathbf{M}) := \mathbf{M} \oplus \mathbf{KD}_0\mathbf{Cycl}(1)$. Then on the map σ we can show the following facts.

Lemma 5.3 *σ is a lattice-homomorphism.*

Proof. For logics $\mathbf{L}_1, \mathbf{L}_2 \in \mathrm{NExt}(\mathbf{KD}_0\mathbf{Cycl}(1))$, $\sigma(\mathbf{L}_1 \cap \mathbf{L}_2) = \mathbf{L}_1 \cap \mathbf{L}_2 \cap \mathbf{L}(\mathcal{I}_0) = \mathbf{L}_1 \cap \mathbf{L}(\mathcal{I}_0) \cap \mathbf{L}_2 \cap \mathbf{L}(\mathcal{I}_0) = \sigma(\mathbf{L}_1) \cap \sigma(\mathbf{L}_2)$. Since the lattice of normal modal logics is distributive, $\sigma(\mathbf{L}_1 \oplus \mathbf{L}_2) = (\mathbf{L}_1 \oplus \mathbf{L}_2) \cap \mathbf{L}(\mathcal{I}_0) = (\mathbf{L}_1 \cap \mathbf{L}(\mathcal{I}_0)) \oplus (\mathbf{L}_2 \cap \mathbf{L}(\mathcal{I}_0)) = \sigma(\mathbf{L}_1) \oplus \sigma(\mathbf{L}_2)$. □

Lemma 5.4 *σ is one to one.*

Proof. Suppose $\mathbf{L}_1 \not\subseteq \mathbf{L}_2$ for logics $\mathbf{L}_1, \mathbf{L}_2 \in \mathrm{NExt}(\mathbf{KD}_0\mathbf{Cycl}(1))$. Then there exists a formula φ such that $\varphi \in \mathbf{L}_1$ and $\varphi \notin \mathbf{L}_2$. By the latter, there is a frame $\mathcal{F} = \langle W, R, P\rangle$ for $\mathbf{D}_0, \mathbf{Cycl}(1)$, a valuation V on $\mathcal{F}$ and a point $a \in W$ such that $\langle \mathcal{F}, V\rangle \not\models_a \varphi$. Here, because the frame $\mathcal{F}$ is 0-serial and 1-cyclic, there exists a point $b \in W$ such that $aRbRa$. Since we have $a \not\models \varphi$, $a \not\models \Box^2\varphi$ is also the case. This means that $\langle \mathcal{F}, V\rangle \not\models_a \varphi \vee \Box^2\varphi$, and so, $\varphi \vee \Box^2\varphi \notin \mathbf{L}_2 \cap \mathbf{L}(\mathcal{I}_0)$.

On the other hand, since $\Box\bot \to \Box^2\varphi \in \mathbf{K} \subseteq \mathbf{L}(\mathcal{I}_0)$, $\Box^2\varphi \in \mathbf{L}(\mathcal{I}_0)$. Therefore, we have $\varphi \vee \Box^2\varphi \in \mathbf{L}_1 \cap \mathbf{L}(\mathcal{I}_0)$. Thus, $\sigma(\mathbf{L}_1) = \mathbf{L}_1 \cap \mathbf{L}(\mathcal{I}_0) \not\subseteq \mathbf{L}_2 \cap \mathbf{L}(\mathcal{I}_0) = \sigma(\mathbf{L}_2)$. □

Lemma 5.5 *σ is onto.*

Proof. Due to the distributivity, and by Proposition 5.2, for any $\mathbf{M} \in [\mathbf{KCycl}(1), \mathbf{L}(\mathcal{I}_0)]$, $\sigma \circ \tau(\mathbf{M}) = (\mathbf{M} \oplus \mathbf{KD}_0\mathbf{Cycl}(1)) \cap \mathbf{L}(\mathcal{I}_0) = (\mathbf{M} \cap \mathbf{L}(\mathcal{I}_0)) \oplus (\mathbf{KD}_0\mathbf{Cycl}(1) \cap \mathbf{L}(\mathcal{I}_0)) = \mathbf{M} \oplus \mathbf{KCycl}(1) = \mathbf{M}$. Hence the map σ is onto. □

So far, we have established the following theorem.

Theorem 5.6 *$[\mathbf{KCycl}(1), \mathbf{L}(\mathcal{I}_0)]$ is isomorphic to $\mathrm{NExt}(\mathbf{KD}_0\mathbf{Cycl}(1))$.*

By Lemma 5.5, we see that for any $\mathbf{M} \in [\mathbf{KCycl}(1), \mathbf{L}(\mathcal{I}_0)]$, there exists a logic $\mathbf{L} \in \mathrm{NExt}(\mathbf{KD}_0\mathbf{Cycl}(1))$ such that $\mathbf{M} = \mathbf{L} \cap \mathbf{L}(\mathcal{I}_0)$. This answers our question in the front of this section.

NEXT(**KCycl**(1)) looks like a *two-story building* as drawn below.

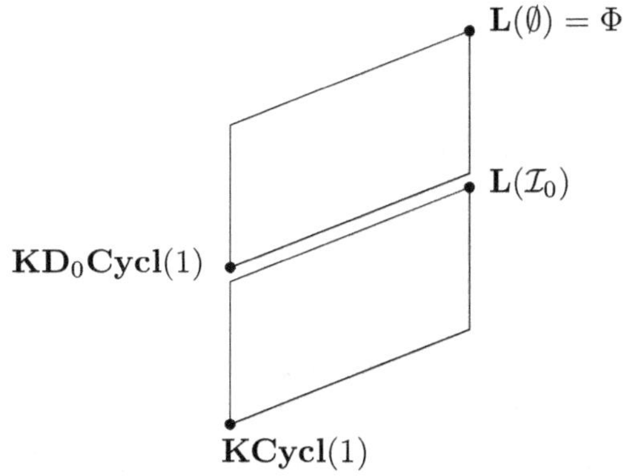

Fig. 4. The structure of NEXT(**KCycl**(1))

6 The structure of NEXT(KCycl(2))

In the case of $n = 2$, the situation is a little different. A frame for the axiom **Cycl**(2) consists of points of level 0, level 1, and level ∞ only. Examples of connected irreflexive frames for this axiom are the family $\{\mathcal{I}_i\}_{i\in\omega}$ and $\mathcal{I}_\infty$, whose logics form an infinite descending chain as shown in Proposition 3.1.

First of all, by Theorem 4.3 we can find one splitting pair.

Theorem 6.1 $(\mathbf{KD}_1\mathbf{Cycl}(2), \mathbf{L}(\mathcal{I}_1))$ *is a splitting pair of the lattice* NEXT(**KCycl**(2)).

Note that the frame $\mathcal{I}_0$ is a frame for $\mathbf{D}_1$ and **Cycl**(2), and so, $\mathbf{KD}_1\mathbf{Cycl}(2) \subseteq \mathbf{L}(\mathcal{I}_0)$, whereas all the members in $\{\mathcal{I}_k\}_{k\geq 1}$ and $\mathcal{I}_\infty$ are not 1-serial, and so, $\mathbf{L}(\mathcal{I}_k), \mathbf{L}(\mathcal{I}_\infty) \subseteq \mathbf{L}(\mathcal{I}_1)$. We can find another spitting pair in NEXT(**KCycl**(2)).

Lemma 6.2 *Let* $\mathfrak{A}$ *be a non-trivial s.i. modal algebra for* $\mathbf{D}_1$ *and* **Cycl**(2). *Suppose* $\Diamond 1 \neq 1$ *in* $\mathfrak{A}$. *Then* $\Box 0 = 1$ *in* $\mathfrak{A}$.

Proof. Since $\mathfrak{A}$ is s.i., there is some element $d(\neq 1) \in A$ and for the element $\Diamond 1$, there is a number $m \in \omega$ such that $\Diamond 1 \cap \Box\Diamond 1 \cap \Box^2\Diamond 1 \cap \cdots \cap \Box^m\Diamond 1 \leq d$ holds. Due to the 1-serial axiom and the 2-cyclic axiom, we have that $\Diamond 1 \leq d$. Now suppose that $\Box 0 \neq 1$ in $\mathfrak{A}$. Then similarly, there is a number $\ell \in \omega$ such that $\Box 0 \cap \Box^1 0 \cap \cdots \cap \Box^\ell 0 \leq d$ holds, and so, we have $\Box 0 \leq d$. By these two inequalities, $-d \leq -\Diamond 1 = \Box 0 \leq d$, which leads to a contradiction. Hence we have $\Box 0 = 1$ in $\mathfrak{A}$. □

Theorem 6.3 $(\mathbf{KD}_0\mathbf{Cycl}(2), \mathbf{L}(\mathcal{I}_0))$ *is a splitting pair of the lattice* NEXT($\mathbf{KD}_1$**Cycl**(2)).

Proof. Suppose $\mathbf{KD}_0\mathbf{Cycl}(2) \not\subseteq \mathbf{L}$ for some logic $\mathbf{L} \in$ NEXT($\mathbf{KD}_1$**Cycl**(2)). Then there exists an s.i. algebra for $\mathbf{L}, \mathbf{D}_1$ and **Cycl**(2) such that $\mathfrak{A} \not\models \mathbf{D}_0$.

This means that $\Diamond 1 \neq 1$ in $\mathfrak{A}$. By the previous lemma, $\Box 0 = 1$ holds in $\mathfrak{A}$. Consider the subalgebra $\mathfrak{A}'$ of $\mathfrak{A}$ generated by the set of elements $\{0, 1\}$, then it is just the modal algebra $\mathcal{I}_0{}^*$. Therefore we have $\mathcal{I}_0{}^* \in S(\mathfrak{A})$, and so, $\mathbf{L} \subseteq \mathbf{L}(\mathfrak{A}) \subseteq \mathbf{L}(\mathcal{I}_0{}^*) = \mathbf{L}(\mathcal{I}_0)$. □

By Theorem 6.1 and Theorem 6.3, the following is proved.

Corollary 6.4 $(\mathbf{KD_0Cycl}(2), \mathbf{L}(\mathcal{I}_0))$ *is also a splitting pair of the lattice* NEXT($\mathbf{KCycl}(2)$).

Now another question arises: what sort of modal logics are located under the logic $\mathbf{L}(\mathcal{I}_0)$, or under the logic $\mathbf{L}(\mathcal{I}_k)$ for $k \geq 1$ in NEXT($\mathbf{KCycl}(2)$)? Does there exist any intriguing structure as in NEXT($\mathbf{KCycl}(1)$)?

We will compare the top-most part NEXT($\mathbf{KD_0Cycl}(2)$) with the bottom part $[\mathbf{KCycl}(2), \mathbf{L}(\mathcal{I}_\infty)]$ in NEXT($\mathbf{KCycl}(2)$) first.

Proposition 6.5 $\mathbf{KCycl}(2) = \mathbf{KD_0Cycl}(2) \cap \mathbf{L}(\mathcal{I}_\infty)$.

Proof. It is trivial that $\mathbf{KCycl}(2) \subseteq \mathbf{KD_0Cycl}(2) \cap \mathbf{L}(\mathcal{I}_\infty)$. Conversely suppose $\varphi \notin \mathbf{KCycl}(2)$ for some $\varphi \in \Phi$. Then, there exists a frame $\mathcal{F} = \langle W, R, P\rangle$ for $\mathbf{Cycl}(2)$, a valuation V on $\mathcal{F}$, and a point $a \in W$ such that $\langle \mathcal{F}, V\rangle \not\models_a \varphi$. Now since this $\mathcal{F}$ is 2-cyclic, $W = W^{(0)} \cup W^{(1)} \cup W^{\infty}$ and points in $W^{(0)} \cup W^{(1)}$ are isolated from the rest part in $\mathcal{F}$. So, if $a \in W^{(0)}$, then the subframe generated by the singleton $\{a\}$ is just $\mathcal{I}_0$. Therefore $\varphi \notin \mathbf{L}(\mathcal{I}_0) \supseteq \mathbf{L}(\mathcal{I}_\infty)$. If $a \in W^{(1)}$, the subframe generated by $\{a\}$ is $\mathcal{I}_k$ for some $k \in \omega$ or $\mathcal{I}_\infty$. Therefore $\varphi \notin \mathbf{L}(\mathcal{I}_\infty)$. If $a \in W^{\infty}$, then the subframe $\mathcal{F}'$ generated by $\{a\}$ contains no element in $W^{(0)} \cup W^{(1)}$, which means that the frame $\mathcal{F}'$ is also a frame for $\mathbf{D}_0$. Therefore $\varphi \notin \mathbf{KD_0Cycl}(2)$. Hence we proved that $\mathbf{KCycl}(2) \supseteq \mathbf{KD_0Cycl}(2) \cap \mathbf{L}(\mathcal{I}_\infty)$. □

Similarly in the case $n = 1$, we define maps σ_∞ and τ_∞ as follows: σ_∞ : NEXT($\mathbf{KD_0Cycl}(2)$) $\to [\mathbf{KCycl}(2), \mathbf{L}(\mathcal{I}_\infty)]$ is defined as: $\sigma_\infty(\mathbf{L}) := \mathbf{L} \cap \mathbf{L}(\mathcal{I}_\infty)$. $\tau_\infty : [\mathbf{KCycl}(2), \mathbf{L}(\mathcal{I}_\infty)] \to$ NEXT($\mathbf{KD_0Cycl}(2)$) is defined as: $\tau_\infty(\mathbf{M}) := \mathbf{M} \oplus \mathbf{KD_0Cycl}(1)$. Then on the map σ_∞ we can show the following facts also in this case.

Lemma 6.6

(1) σ_∞ *is a lattice-homomorphism.*

(2) σ_∞ *is onto.*

(3) σ_∞ *is one to one.*

Proof. The fact (1) can be obtained immediately by an easy calculation. To show the fact (2), we use Proposition 6.5. To show the fact (3), suppose $\mathbf{L}_1 \not\subseteq \mathbf{L}_2$ for logics $\mathbf{L}_1, \mathbf{L}_2 \in$ NEXT($\mathbf{KD_0Cycl}(2)$). Then there exists a formula φ such that $\varphi \in \mathbf{L}_1$ and $\varphi \notin \mathbf{L}_2$. By the latter, there is a frame $\mathcal{F} = \langle W, R, P\rangle$ for $\mathbf{D}_0, \mathbf{Cycl}(2)$, a valuation V on $\mathcal{F}$ and a point $a \in W$ such that $\langle \mathcal{F}, V\rangle \not\models_a \varphi$. Now the frame $\mathcal{F}$ is 0-serial and 2-cyclic, $a \in W = W^{\infty}$ and there exist points $b, c \in W$ such that $aRbRcRa$. Since we have $a \not\models \varphi$, $a \not\models \Box^3\varphi$ is also the case.

This means that $\langle \mathcal{F}, V\rangle \not\models_a \varphi \vee \Box^3\varphi$, and so, $\varphi \vee \Box^3\varphi \notin \mathbf{L}_2 \cap \mathbf{L}(\mathcal{I}_\infty)$.

On the other hand, since $\Box^2\bot \to \Box^3\varphi \in \mathbf{K} \subseteq \mathbf{L}(\mathcal{I}_\infty)$, $\Box^3\varphi \in \mathbf{L}(\mathcal{I}_\infty)$. Therefore, we have $\varphi \vee \Box^3\varphi \in \mathbf{L}_1 \cap \mathbf{L}(\mathcal{I}_\infty)$. Thus, $\sigma_\infty(\mathbf{L}_1) = \mathbf{L}_1 \cap \mathbf{L}(\mathcal{I}_\infty) \not\subseteq \mathbf{L}_2 \cap \mathbf{L}(\mathcal{I}_\infty) = \sigma_\infty(\mathbf{L}_2)$. □

This lemma states that the map σ_∞ is an isomorphism. So far, we have shown the following.

Theorem 6.7 $[\mathbf{KCycl}(2), \mathbf{L}(\mathcal{I}_\infty)]$ *is isomorphic to* $\mathrm{NE{\scriptstyle XT}}(\mathbf{KD}_0\mathbf{Cycl}(2))$.

Next, we will compare the top-most part $\mathrm{NE{\scriptstyle XT}}(\mathbf{KD}_0\mathbf{Cycl}(2))$ with an interval $[\mathbf{KD}_0\mathbf{Cycl}(2) \cap \mathbf{L}(\mathcal{I}_k), \mathbf{L}(\mathcal{I}_k)]$ for $k \geq 0$ in the lattice $\mathrm{NE{\scriptstyle XT}}(\mathbf{KCycl}(2))$.

As in the above analysis, we define maps σ_k and τ_k as follows: $\sigma_k : \mathrm{NE{\scriptstyle XT}}(\mathbf{KD}_0\mathbf{Cycl}(2)) \to [\mathbf{KD}_0\mathbf{Cycl}(2) \cap \mathbf{L}(\mathcal{I}_k), \mathbf{L}(\mathcal{I}_k)]$ is defined as: $\sigma_k(\mathbf{L}) := \mathbf{L} \cap \mathbf{L}(\mathcal{I}_k)$. $\tau_k : [\mathbf{KD}_0\mathbf{Cycl}(2) \cap \mathbf{L}(\mathcal{I}_k), \mathbf{L}(\mathcal{I}_k)] \to \mathrm{NE{\scriptstyle XT}}(\mathbf{KD}_0\mathbf{Cycl}(2))$ is defined as: $\tau_k(\mathbf{M}) := \mathbf{M} \oplus \mathbf{KD}_0\mathbf{Cycl}(2)$. Then on the map σ_k we can also show the following facts in this case.

Lemma 6.8

(1) σ_k *is a lattice-homomorphism.*

(2) σ_k *is onto.*

(3) σ_k *is one to one.*

Proof. Proofs of the fact (1) and the fact (2) are just similar for the proofs of Lemma 6.6 To show the fact (3), suppose $\mathbf{L}_1 \not\subseteq \mathbf{L}_2$ for logics $\mathbf{L}_1, \mathbf{L}_2 \in \mathrm{NE{\scriptstyle XT}}(\mathbf{KD}_0\mathbf{Cycl}(2))$. Then there exists a formula φ such that $\varphi \in \mathbf{L}_1$ and $\varphi \notin \mathbf{L}_2$. By the latter, there is a frame $\mathcal{F} = \langle W, R, P\rangle$ for $\mathbf{D}_0, \mathbf{Cycl}(2)$, a valuation V on $\mathcal{F}$ and a point $a \in W$ such that $\langle \mathcal{F}, V\rangle \not\models_a \varphi$. Now we see that $a \in W = W^\infty$ and there exist points $b, c \in W$ such that $aRbRcRa$. Since we have $a \not\models \varphi$, $a \not\models \Box^3\varphi$ is also the case. This means that $\langle \mathcal{F}, V\rangle \not\models_a \varphi \vee \Box^3\varphi$, and so, $\varphi \vee \Box^3\varphi \notin \mathbf{L}_2 \cap \mathbf{L}(\mathcal{I}_k)$.

On the other hand, for the case $k = 0$, $\Box\bot \to \Box^3\varphi \in \mathbf{K} \subseteq \mathbf{L}(\mathcal{I}_0)$ and for the case $k \geq 1$, $\Box^2\bot \to \Box^3\varphi \in \mathbf{K} \subseteq \mathbf{L}(\mathcal{I}_k)$. Therefore for any $k \geq 0$, $\Box^3\varphi \in \mathbf{L}(\mathcal{I}_k)$. Thus we have $\varphi \vee \Box^3\varphi \in \mathbf{L}_1 \cap \mathbf{L}(\mathcal{I}_k)$. Hence, $\sigma_k(\mathbf{L}_1) = \mathbf{L}_1 \cap \mathbf{L}(\mathcal{I}_k) \not\subseteq \mathbf{L}_2 \cap \mathbf{L}(\mathcal{I}_k) = \sigma_k(\mathbf{L}_2)$. □

We have just established the following theorem.

Theorem 6.9 *For any* $k \geq 0$, $[\mathbf{KD}_0\mathbf{Cycl}(2) \cap \mathbf{L}(\mathcal{I}_k), \mathbf{L}(\mathcal{I}_k)]$ *is isomorphic to* $\mathrm{NE{\scriptstyle XT}}(\mathbf{KD}_0\mathbf{Cycl}(2))$.

Finally we show that there are countably infinite splitting pairs in the lattice $\mathrm{NE{\scriptstyle XT}}(\mathbf{KCycl}(2))$ by using these isomorphisms σ_k's. The following fact must be checked.

Proposition 6.10 $\mathbf{KD}_1\mathbf{Cycl}(2) = \mathbf{KD}_0\mathbf{Cycl}(2) \cap \mathbf{L}(\mathcal{I}_0)$.

Proof. It is trivial that $\mathbf{KD}_1\mathbf{Cycl}(2) \subseteq \mathbf{KD}_0\mathbf{Cycl}(2) \cap \mathbf{L}(\mathcal{I}_0)$. Conversely suppose $\varphi \notin \mathbf{KD}_1\mathbf{Cycl}(2)$ for some $\varphi \in \Phi$. Then, there exists a frame $\mathcal{F} = \langle w, R, P\rangle$ for $\mathbf{D}_1, \mathbf{Cycl}(2)$, a valuation V on $\mathcal{F}$, and a point $a \in W$ such that

$\langle \mathcal{F}, V \rangle \not\models_a \varphi$. Now since this $\mathcal{F}$ is 1-serial and 2-cyclic, $W = W^{(0)} \cup W^{\infty}$ and points in $W^{(0)}$ are isolated from the rest part in $\mathcal{F}$. So, if $a \in W^{(0)}$, then the subframe generated by the singleton $\{a\}$ is just $\mathcal{I}_0$. Therefore $\varphi \notin \mathbf{L}(\mathcal{I}_0)$. If $a \in W^{\infty}$, then the subframe $\mathcal{F}'$ generated by $\{a\}$ contains no element in $W^{(0)}$ at all, which means that $\mathcal{F}'$ is also a frame for $\mathbf{D}_0$. Therefore $\varphi \notin \mathbf{KD}_0\mathbf{Cycl}(2)$. Hence we proved that $\mathbf{KD}_1\mathbf{Cycl}(2) \supseteq \mathbf{KD}_0\mathbf{Cycl}(2) \cap \mathbf{L}(\mathcal{I}_0)$. □

Theorem 6.11 *For any $k \geq 1$, the pair $\big(\mathbf{KD}_0\mathbf{Cycl}(2) \cap \mathbf{L}(\mathcal{I}_{k-1}), \mathbf{L}(\mathcal{I}_k)\big)$ is a splitting pair in* NEXT($\mathbf{KCycl}(2)$).

Proof. By Theorem 6.9 and Proposition 6.10, the interval $[\mathbf{KD}_0\mathbf{Cycl}(2) \cap \mathbf{L}(\mathcal{I}_0), \Phi]$ is isomorphic to $[\mathbf{KD}_0\mathbf{Cycl}(2) \cap \mathbf{L}(\mathcal{I}_k), \mathbf{L}(\mathcal{I}_{k-1})]$. By Theorem 6.3, $\big(\mathbf{KD}_0\mathbf{Cycl}(2), \mathbf{L}(\mathcal{I}_0)\big)$ is a splitting pair in NEXT($\mathbf{KD}_1\mathbf{Cycl}(2)$) $= [\mathbf{KD}_0\mathbf{Cycl}(2) \cap \mathbf{L}(\mathcal{I}_0), \Phi]$. By the above isomorphisms, $\mathbf{KD}_0\mathbf{Cycl}(2)$ is mapped to $\mathbf{KD}_0\mathbf{Cycl}(2) \cap \mathbf{L}(\mathcal{I}_{k-1})$, and $\mathbf{L}(\mathcal{I}_0)$ to $\mathbf{L}(\mathcal{I}_k)$. Therefore, the pair $\big(\mathbf{KD}_0\mathbf{Cycl}(2) \cap \mathbf{L}(\mathcal{I}_{k-1}), \mathbf{L}(\mathcal{I}_k)\big)$ is a splitting pair in the interval $[\mathbf{KD}_0\mathbf{Cycl}(2) \cap \mathbf{L}(\mathcal{I}_k), \mathbf{L}(\mathcal{I}_{k-1})]$. Hence this pair is also a splitting pair in NEXT($\mathbf{KCycl}(2)$). □

Corollary 6.12 *There exist at least countably infinite splitting pairs in* NEXT($\mathbf{KCycl}(2)$).

In this case, NEXT($\mathbf{KCycl}(2)$) looks like a *ω-story building* as drawn below.

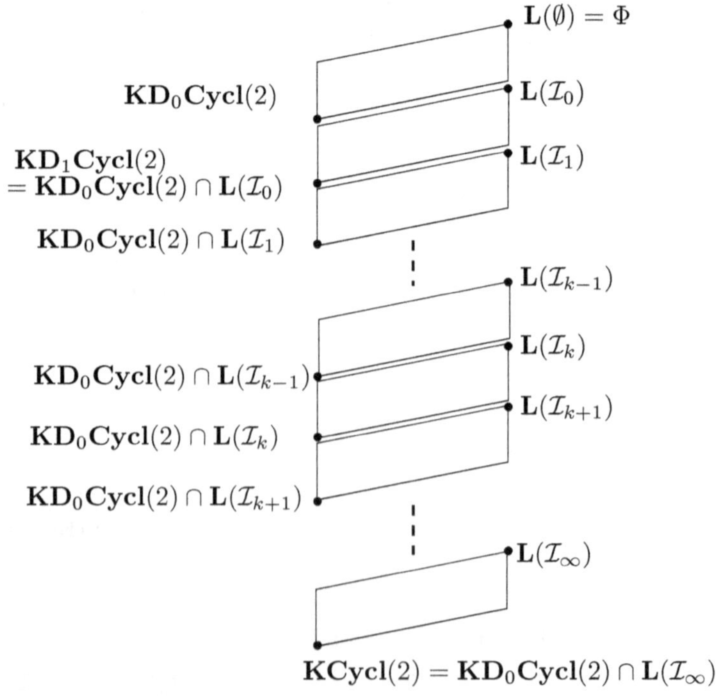

Fig. 5. The structure of NEXT($\mathbf{KCycl}(2)$)

7 Outlook

For $n \geq 1$, let $\mathcal{I}rr_n$ be the class of normal modal logics above $\mathbf{KCycl}(n)$ that are determined by a class of *irreflexive* frames. That is, $\mathcal{I}rr_n := \{\mathbf{L}(\mathcal{C}) \in \text{NExt}(\mathbf{KCycl}(n)) \mid \mathcal{C}$ is a class of some irreflexive frames$\}$. Then this $\mathcal{I}rr_n$ forms at least a meet-semilattice. In particular, $\mathcal{I}rr_1$ forms a two element chain ($\Phi = \mathbf{L}(\emptyset)$ and $\mathbf{L}(\mathcal{I}_0)$), and $\mathcal{I}rr_2$ forms an infinite descending chain $(\Phi, \mathbf{L}(\mathcal{I}_0), \mathbf{L}(\mathcal{I}_1), \mathbf{L}(\mathcal{I}_2), \cdots, \mathbf{L}(\mathcal{I}_k), \cdots, \mathbf{L}(\mathcal{I}_\infty))$. In this terminology, the results presented in this paper are summed up in the following way. In both cases there exists a essential lattice structure $\mathcal{B}_n$ at the top of the lattice $\text{NExt}(\mathbf{KCycl}(n))$ and the whole lattice can be expressed in:

$$\mathbf{KCycl}(n) \cong \mathcal{B}_n \times \mathcal{I}rr_n$$

for $n = 1, 2$. Does this beautiful expression also hold for the cases $n \geq 3$? If it is solved in the affirmative, then each lattice $\text{NExt}(\mathbf{KCycl}(n))$ has its own essential part $\mathcal{B}_n$ (it may be at around the top-most region) for any $n \geq 1$, and so, the investigation of such lattice of logics can be concentrated only in this $\mathcal{B}_n$.

In this paper, we discover a splitting pair in $\text{NExt}(\mathbf{KCycl}(1))$, and countably infinite splitting pairs in $\text{NExt}(\mathbf{KCycl}(2))$. Is there any other splitting pair in $\text{NExt}(\mathbf{KCycl}(1))$ or in $\text{NExt}(\mathbf{KCycl}(2))$?

In [5], K. Fine introduced a notion of the degree of Kripke incompleteness of a modal logic as follows: For modal logics $\mathbf{L}_1, \mathbf{L}_2 \in \text{NExt}(\mathbf{L}_0)$, $\mathbf{L}_1$ and $\mathbf{L}_2$ are *Kripke equivalent* ($\mathbf{L}_1 \equiv_K \mathbf{L}_2$ in symbol), if for any Kripke frame $\mathcal{F}$, $\mathcal{F} \models \mathbf{L}_1$ if and only if $\mathcal{F} \models \mathbf{L}_2$. *The degree of Kripke incompleteness* of $\mathbf{L}$ over $\mathbf{L}_0$ ($\delta_{\mathbf{L}_0}(\mathbf{L})$) is defined as: $card\{\mathbf{M} \in \text{NExt}(\mathbf{L}_0) \mid \mathbf{L} \equiv_K \mathbf{M}\}$. $\mathbf{L}$ is called *intrinsically complete* over $\mathbf{L}_0$ if $\delta_{\mathbf{L}_0}(\mathbf{L}) = 1$.

On the maps appeared in this paper, σ's (σ, σ_k's, and σ_∞) preserve the Kripke completeness and the finite model property. For example, σ_k maps $\mathbf{L}$ to $\mathbf{L} \cap \mathbf{L}(\mathcal{I}_k)$. Since the frame $\mathcal{I}_k$ is a finite Kripke frame, if $\mathcal{L}$ is Kripke complete (and also has the f.m.p), then so is $\mathbf{L} \cap \mathbf{L}(\mathcal{I}_k)$. (and also with f.m.p). Although $\mathcal{I}_\infty$ is not a finite Kripke frame, it can be said that the logic $\mathbf{L}(\mathcal{I}_\infty)$ has the finite model property. Thus the same argument goes through in the case of σ_∞. Then, to the converse, is it the case that τ's (τ, τ_k's and τ_∞) preserve the intrinsically completeness?

References

[1] Blok, W., *On the degree of incompleteness in modal logics and the covering relation in the lattice of modal logics*, Technical Report 78-07, Department of Mathematics, University of Amsterdam (1978).

[2] Blok, W., *The lattice of modal logics: an algebraic investigation*, Journal of Symbolic Logic **45** (1980), pp. 221–236.

[3] Blyth, S. and Sankappanavar, H.P., "A Course in Universal Algebra," Splinger Verlag, 1981.

[4] Chagrov, A. and Zakharyaschev, M., "Modal Logic," Oxford University Press, 1997.

[5] Fine, K., *An incomplete logic containing S4*, Theoria **40** (1974), pp. 23–29.
[6] Jónsson, B. and Tarski, A., *Boolean algebras with operators part i*, American Journal of Mathematics **73** (1951), pp. 891 – 939.
[7] Kowalski, T. and Kracht, M., *Semisimple varieties of modal algebras*, Studia Logica **83** (2006), pp. 351–363.
[8] Makinson, D., *Some embedding theorems for modal logic*, Notre Dame Journal of Formal Logic **12** (1971), pp. 252–254.
[9] Rautenberg, W., *Splitting lattices of logics*, Archiv für Mathematische Logik **20** (1980), pp. 155–159.

Propositional dynamic logic with Belnapian truth values

Igor Sedlár [1]

Institute of Computer Science, The Czech Academy of Sciences
Pod Vodárenskou věží 271/2
182 07 Prague 8, Czech Republic

Abstract

We introduce **BPDL**, a combination of propositional dynamic logic **PDL** with the basic four-valued modal logic **BK** studied by Odintsov and Wansing ('Modal logics with Belnapian truth values', J. Appl. Non-Class. Log. 20, 279–301 (2010)). We modify the standard arguments based on canonical models and filtration to suit the four-valued context and prove weak completeness and decidability of **BPDL**.

Keywords: Belnap–Dunn logic, Four-valued logic, Propositional dynamic logic.

1 Introduction

Propositional dynamic logic **PDL** is a well-known logical framework that allows to express properties of regular programs and formalises reasoning about these properties [8,15]. The framework sees programs as state transitions, or binary relations on states, where states of the computer are viewed as complete and consistent possible worlds. A more general notion of computer state has been put forward by Belnap and Dunn [3,2,6]. In a possible world, every formula is either true or false. In a Belnap–Dunn state, formulas can be (only) true, (only) false, both true and false, or neither true nor false. Informally, Belnap–Dunn states are seen as bodies of information about some domain and the four truth values correspond to presence or absence of information about the domain. More precisely, the four possible truth values of a formula ϕ express

[1] E-mail: sedlar@cs.cas.cz. This work was supported by the long-term strategic development financing of the Institute of Computer Science (RVO:67985807) and the Czech Science Foundation, grant GBP202/12/G061. Previous versions of this article were presented at the Knowledge Representation Seminar of the Department of Applied Informatics, Faculty of Mathematics, Physics and Informatics, Comenius University in Bratislava, and the Applied Mathematical Logic Seminar of the Department of Theoretical Computer Science, Institute of Computer Science, The Czech Academy of Sciences. I am grateful to the audiences at these seminars for useful feedback. Martin Baláž, Petr Cintula, Rostislav Horčík and Ján Šefránek have provided especially helpful comments. Thanks are also due to the AiML reviewers for suggesting a number of improvements.

four possible answers to the query 'What is the available information about ϕ?', namely:

- there is information that ϕ is true and no information that ϕ is false ('true');
- there is information that ϕ is false and no information that ϕ is true ('false');
- there is information that ϕ is true but also information that ϕ is false ('both');
- there is no information about ϕ ('neither').

Belnap and Dunn stress the importance of this generalisation to computer science, pointing mainly to databases as a potential area of application. Later work on bilattices, a generalisation of the Belnap–Dunn notion of state, has confirmed their assessment and extended the applications to other areas [12,11,1,9,10].

Putting things together, a version of **PDL** using Belnap–Dunn states would formalise reasoning about regular programs that modify (possibly incomplete and inconsistent) database-like structures. Such structures abound and a logical formalisation of reasoning about their algorithmic transformations could be of vital importance to AI and related areas. In addition to practical applications, theoretical questions pertaining to the properties of such generalised versions of **PDL** are interesting in their own right. However, Belnap–Dunn versions of **PDL** are yet to be investigated.

This article fills the gap. We discuss **BPDL**, a logic that adds program modalities to Odintsov and Wansing's [16] basic modal logic with Belnapian truth values **BK** (see also [17,18]). Our main technical results concerning **BPDL** (introduced in Section 3 of the article) are a decidability proof using a variation of the standard argument based on filtration (Section 4) and a sound and weakly complete axiomatisation (Section 5). We assume familiarity with **PDL**, but a short overview of **BK** is provided in Section 2.

We note that there are other well-known four-valued modal logics, but there are reasons to favour **BK** when it comes to combinations with **PDL**. Priest's basic modal First-Degree-Entailment $\mathbf{K_{FDE}}$ [19] lacks a sensible implication connective (e.g., Modus ponens fails), which is a problem given the importance of implication in stating properties of programs such as partial correctness. Goble's **KN4** [13] corresponds to a fragment of **BK**. The framework of Rivieccio, Jung and Jansana [20] is more complicated than **BK** in that it treats the modal accessibility relation itself as many-valued. As a result, for instance, the familiar '**K** axiom' $\Box(\phi \to \psi) \to (\Box\phi \to \Box\psi)$ is not valid. This is problematic from the viewpoint of **PDL** which is a normal modal logic. (However, a non-normal version of **PDL** built on this framework might still be interesting to look at in the future.) Another approach is to add to **PDL** a modal DeMorgan negation in the style of [7]. However, the modal negation in this framework does not fit in with implication as nicely as the negation in **BK** (for instance, $\sim(\phi \to \psi)$ does not entail ϕ, where '$\sim$' is the DeMorgan negation). Nevertheless, this approach is pursued by the present author in [21].

The general idea of providing many-valued versions of **PDL** is not new. Teheux [22] formulates **PDL** over finitely-valued Łukasiewicz logics to model the Rényi–Ulam searching game with errors. However, the non-modal fragments of his logics are non-classical, as opposed to **BK** which can be seen as an extension of the classically-based logic **K** with a strong negation. Běhounek [4,5] suggests that **PDL** with fuzzy accessibility relations is suitable for reasoning about costs of program executions, but the states in his models remain classical.

2 Modal logic with Belnapian truth values

This section provides background on **BK** and motivates our extension of the logic with program modalities. The language $\mathcal{L}_{mod}$ consists of AF, a countable set of atomic formulas, a nullary connective $\bot$, unary connectives $\sim, \Box, \Diamond$ and binary connectives $\wedge, \vee, \rightarrow$. $\neg\phi$ is defined as $\phi \rightarrow \bot$, $\top$ is defined as $\neg\bot$ and $\phi \leftrightarrow \psi$ is defined as $(\phi \rightarrow \psi) \wedge (\psi \rightarrow \phi)$. F_{mod} is the set of formulas of $\mathcal{L}_{mod}$.

Definition 2.1 [16, 285–286] An *Odintsov–Wansing model* is a tuple $M = \langle S, R, V^+, V^- \rangle$ where $S \neq \emptyset$, $R \subseteq (S \times S)$ and $V^\circ : AF \mapsto 2^S$, $\circ = \{+, -\}$. Every M induces a pair of relations $\models^+_M, \models^-_M \subseteq (S \times F_{mod})$ such that (we usually drop the subscript 'M'):

(i) $x \models^+ p$ iff $x \in V^+(p)$; $x \models^- p$ iff $x \in V^-(p)$

(ii) $x \models^+ \bot$ for no x; $x \models^- \bot$ for all x

(iii) $x \models^+ {\sim}\phi$ iff $x \models^- \phi$; $x \models^- {\sim}\phi$ iff $x \models^+ \phi$

(iv) $x \models^+ \phi \wedge \psi$ iff $x \models^+ \phi$ and $x \models^+ \psi$; $x \models^- \phi \wedge \psi$ iff $x \models^- \phi$ or $x \models^- \psi$

(v) $x \models^+ \phi \vee \psi$ iff $x \models^+ \phi$ or $x \models^+ \psi$; $x \models^- \phi \vee \psi$ iff $x \models^- \phi$ and $x \models^- \psi$

(vi) $x \models^+ \phi \rightarrow \psi$ iff $x \not\models^+ \phi$ or $x \models^+ \psi$; $x \models^- \phi \rightarrow \psi$ iff $x \models^+ \phi$ and $x \models^- \psi$

(vii) $x \models^+ \Box\phi$ iff for all y, if Rxy, then $y \models^+ \phi$
$x \models^- \Box\phi$ iff there is y such that Rxy and $y \models^- \phi$

(viii) $x \models^+ \Diamond\phi$ iff there is y such that Rxy and $y \models^+ \phi$
$x \models^- \Diamond\phi$ iff for all y, if Rxy, then $y \models^- \phi$

$|\phi|^+_M = \{x \mid x \models^+_M \phi\}$ and $|\phi|^-_M = \{x \mid x \models^-_M \phi\}$. Entailment in the resulting logic, **BK**, is defined as $\models^+$-preservation in every state of every model ($X \models_{\mathbf{BK}} \phi$ iff, for all M, $\bigcap_{\psi \in X} |\psi|^+_M \subseteq |\phi|^+_M$). Validity is defined as usual (ϕ is valid in **BK** iff $\emptyset \models_{\mathbf{BK}} \phi$).

States $x \in S$ can be seen as database-like bodies of information. The fact that $x \models^+ \phi$ can then be read as 'x provides information that ϕ is true' (or 'x supports ϕ', 'x verifies ϕ') and $x \models^- \phi$ as 'x provides information that ϕ is false' ('x falsifies ϕ'). Consequently, $|\phi|^+$ is seen as the set of states in which ϕ is true (the truth set of ϕ) and $|\phi|^-$ as the set of states in which ϕ is false (falsity set). Entailment then boils down to the usual notion of truth-preservation. The distinguishing feature of the Belnap–Dunn picture is that some bodies of information x may support conflicting information about some ϕ (if $x \models^+ \phi$ and

$x \models^- \phi$) and some bodies of information x may not provide any information about some ϕ at all (if $x \not\models^+ \phi$ and $x \not\models^- \phi$). In other words, $|\phi|^+$ and $|\phi|^-$ may have a non-empty intersection and their union is not necessarily identical to S.

The two negations '$\sim$' and '$\neg$' can be explained as follows. The formula $\sim\phi$ may be read as 'ϕ is false' (as $x \models^+ \sim\phi$ iff $x \models^- \phi$). On the other hand, the formula $\neg\phi$ is read as 'ϕ is not true' (note that $x \models^+ \phi \to \bot$ iff $x \not\models^+ \phi$). In general, neither $\sim\phi \to \neg\phi$ nor $\neg\phi \to \sim\phi$ are valid. In other words, the present framework treats 'false' and 'not true' as two independent notions. The presence of '$\sim$' and '$\neg$' in our language allows to express the four possible Belnapian truth values of a formula ϕ:

- $\phi \land \neg\sim\phi$ (ϕ is only true, i.e., true and not false);
- $\neg\phi \land \sim\phi$ (ϕ is only false, i.e., not true and false);
- $\phi \land \sim\phi$ (ϕ is both true and false);
- $\neg\phi \land \neg\sim\phi$ (ϕ is neither true nor false).

Theorem 2.2 *The following axiom system, $H(\mathbf{BK})$, is a sound and strongly complete axiomatisation of* **BK***:*

(i) *Axioms of classical propositional logic in the language $\{AF, \bot, \to, \land, \lor\}$ and Modus ponens;*

(ii) *Strong negation axioms:*

$$\begin{aligned}
\sim\sim\phi &\leftrightarrow \phi,\\
\sim(\phi \land \psi) &\leftrightarrow (\sim\phi \lor \sim\psi),\\
\sim(\phi \lor \psi) &\leftrightarrow (\sim\phi \land \sim\psi),\\
\sim(\phi \to \psi) &\leftrightarrow (\phi \land \sim\psi),\\
\top &\leftrightarrow \sim\bot;
\end{aligned}$$

(iii) *The* **K** *axiom $\Box(\phi \to \psi) \to (\Box\phi \to \Box\psi)$ and the Necessitation rule $\phi/\Box\phi$;*

(iv) *Modal interaction principles:*

$$\begin{aligned}
\neg\Box\phi &\leftrightarrow \Diamond\neg\phi,\\
\neg\Diamond\phi &\leftrightarrow \Box\neg\phi,\\
\sim\Box\phi &\leftrightarrow \Diamond\sim\phi,\\
\Box\phi &\leftrightarrow \sim\Diamond\sim\phi,\\
\sim\Diamond\phi &\leftrightarrow \Box\sim\phi,\\
\Diamond\phi &\leftrightarrow \sim\Box\sim\phi.
\end{aligned}$$

Proof. See [16]. □

The logic **BK** enjoys the deduction theorem in the sense that $\phi \models \psi$ iff $\models \phi \to \psi$.[2] An interesting feature of **BK** is that the set of valid formulas is

[2] Proof: $\not\models \phi \to \psi$ iff, for some x, $x \not\models^+ \phi \to \psi$ iff, for some x, $x \models^+ \phi$ and $x \not\models^+ \psi$ iff $\phi \not\models \psi$.

not closed under the Replacement rule $\phi \leftrightarrow \psi / \chi(\phi) \leftrightarrow \chi(\psi)$.[3] However, it is closed under the Positive replacement rule $\phi \leftrightarrow \psi / \gamma(\phi) \leftrightarrow \gamma(\psi)$ for $\sim$-free γ and the Weak replacement rule $(\phi \leftrightarrow \psi) \wedge (\sim\phi \leftrightarrow \sim\psi) / \chi(\phi) \leftrightarrow \chi(\psi)$. (See [16] for details.) Schemas $(\phi \wedge \sim\phi) \to \bot$ and $\phi \vee \sim\phi$ are not valid (but, of course, $(\phi \wedge \neg\phi) \to \bot$ and $\phi \vee \neg\phi$ both are).

Languages interpreted over bilattices often contain two additional binary connectives '$\otimes$' and '$\oplus$'. Their meaning can be outlined by the following example using the reading of the four Belnapian truth values as subsets of the set of 'classical' values {true, false}. If ϕ is only true and ψ is only false (the value of ϕ is {true} and the value of ψ is {false}), then $\phi \otimes \psi$ is neither true nor false ({true} $\cap$ {false} $= \emptyset$) whereas $\phi \oplus \psi$ is both true and false ({true} $\cup$ {false} $=$ {true, false}).[4] Odintsov and Wansing [16] do not use these connectives in the modal setting and, for the sake of simplicity, we omit them as well. We note, however, that there is no technical obstacle in introducing them to the framework and, speaking in terms of informal interpretation, they fit in nicely also to our combination of **BK** with **PDL**.

Let us now return to the informal interpretation of **BK**. If states in the model are seen as database-like bodies of information, then the accessibility relation can be construed as any binary relation between such bodies of information. Interpretations related to transformations of such bodies (adding or removing information, for example) are a natural choice. For instance, with a set of available transformations in mind, we may read Rxy as 'y is the result of transforming x in some available way'. $\Diamond\phi$ then means that there is an available transformation of the present body of information that leads to ϕ being supported and $\Box\phi$ means that all available transformations lead to ϕ being supported. Hence, **BK** can be seen as a general formalism for reasoning about such transformations.

This reading of R invites us to generalise the framework to a multi-modal setting. We may want to distinguish between different types of transformation and so we may need R_i for each type i instead of a single relation R. The corresponding formulas of a multi-modal extension of $\mathcal{L}_{mod}$, $\Box_i\phi$ ($\Diamond_i\phi$), would then express that ϕ is supported after every (some) transformation of type i. With a number of basic types at hand, the natural next step is to introduce complex transformations consisting of transformations of the basic types. This brings us to extending **BK** with program operators provided by **PDL**, i.e., choice, composition, iteration and test. Additional motivation for considering a combination of **PDL** with **BK** is given by the following examples.

---

[3] Note, for example, that $\sim(\phi \to \psi) \leftrightarrow (\phi \wedge \sim\psi)$ is valid by the completeness theorem but $\sim\sim(\phi \to \psi) \leftrightarrow \sim(\phi \wedge \sim\psi)$ is not. The latter is provably equivalent to $(\phi \to \psi) \leftrightarrow (\sim\phi \vee \psi)$. Now consider a model where $x \not\models^+ p$, $x \not\models^- p$ and $x \not\models^+ q$. Then $x \models^+ p \to q$ but $x \not\models^+ \sim p \vee q$. By the deduction theorem, $(p \to q) \to (\sim p \vee q)$ is not valid. It is easily shown that the converse implication is not valid either.

[4] Closer to the present setting, $\phi \otimes \psi$ is taken to be verified (falsified) iff both ϕ and ψ are verified (falsified); and $\phi \oplus \psi$ is verified (falsified) iff at least one of ϕ, ψ is verified (falsified). (Hence, for example, extending **BK** with these connectives would result in $(\phi \otimes \psi) \leftrightarrow (\phi \wedge \psi)$ and $(\sim\phi \otimes \sim\psi) \leftrightarrow (\sim\phi \wedge \sim\psi)$ being both valid, and similarly for $\oplus$ and $\vee$.)

Example 2.3 If Belnap–Dunn states are seen as bodies of information, then state transitions (programs) may be seen as general inference rules. Formulas of the combined language may express the nature and properties of these rules. Introducing a Belnapian negation $\sim$ into the language of **PDL** opens the possibility of expressing inferences beyond the scope of classical logic. Take, for example, *default rules* of the form

$$\frac{\psi : \phi}{\chi} \tag{1}$$

read 'If ψ is true and there is no information that ϕ is false, then infer that χ is true'. Such a default rule may be expressed by

$$(\psi \wedge \neg{\sim}\phi) \to [\alpha]\chi, \tag{2}$$

a formula that reads 'If ψ is true and ϕ is not false, then every terminating execution of α leads to a state where χ true'. If (2) holds in a state, then executing the program α in the state is equivalent to using (1) in the state. Hence, (1) and α are 'locally equivalent' in the given state. Moreover, if

$$[\beta^*]\,((\psi \wedge \neg{\sim}\phi) \to [\alpha]\chi)$$

holds in a state, then (1) and α are 'β-equivalent', or locally equivalent in every state reachable by a finite iteration of β.

Formulas of the form (2) may even be seen as *defining* α to be a counterpart of a specific default rule. On this view, it is natural to focus only on models where (2) holds in every state (is valid). This motivates a notion of global consequence to be introduced below.

Example 2.4 A special case of (1) is the *closed-world assumption* rule

$$\frac{\top : \neg\phi}{\sim\phi}, \tag{3}$$

inferring that ϕ is false from the assumption that ϕ is not known to be true. Applications of (3) correspond to executions of α in states where it is the case that

$$\neg\phi \to [\alpha]{\sim}\phi$$

Example 2.5 More generally, state transitions (programs) on Belnap–Dunn states may be seen as arbitrary modifications of states. Program α_1 is locally equivalent to 'marking ϕ as true' and α_2 to 'marking ψ as false' if

$$[\alpha_1]\phi \wedge [\alpha_2]{\sim}\psi$$

holds in the given state and similarly for β-equivalence. More interestingly, the formula

$$(\phi \wedge {\sim}\phi) \wedge \langle\alpha^*\rangle\neg(\phi \wedge {\sim}\phi)$$

says that there is inconsistent information about ϕ in the present state, but the inconsistency is removed after some finite number of executions of α. In other words, α is a ϕ-inconsistency-removing modification.

Again, we may see the above formulas as defining the respective programs to be counterparts of specific modifications of states.

3 BPDL

The language $\mathcal{L}_{dyn}$ is a variant of the language of **PDL**, containing two kinds of expressions, namely, programs P and formulas F:

$$
\begin{array}{ll}
P & \alpha ::= a \mid \alpha;\alpha \mid \alpha \cup \alpha \mid \alpha^* \mid \phi? \\
F & \phi ::= p \mid \bot \mid {\sim}\phi \mid \phi \wedge \phi \mid \phi \vee \phi \mid \phi \to \phi \mid [\alpha]\phi \mid \langle\alpha\rangle\phi
\end{array}
$$

($a \in AP$, a countable set of atomic programs, and $p \in AF$) $\neg\phi$, $\top$ and $\phi \leftrightarrow \psi$ are defined as in $\mathcal{L}_{mod}$.

Definition 3.1 A *standard dynamic Odintsov–Wansing model* is a tuple $\mathcal{M} = \langle S, R, V^+, V^- \rangle$, where S, V^+ and V^- are as in Odintsov–Wansing models. $\models^+_{\mathcal{M}}$ and $\models^-_{\mathcal{M}}$ are defined as before for $\{AF, \bot, \sim, \wedge, \vee, \to\}$. R is a function from P to binary relations on S such that $R(\alpha;\beta)$ ($R(\alpha \cup \beta)$) is the composition (union) of $R(\alpha)$ and $R(\beta)$; $R(\alpha^*)$ is the reflexive transitive closure $R(\alpha)^*$ of $R(\alpha)$; and $R(\phi?)$ is the identity relation on $|\phi|^+$. Moreover (R_α is short for $R(\alpha)$):

(i) $x \models^+ [\alpha]\phi$ iff for all y, if $R_\alpha xy$, then $y \models^+ \phi$

(ii) $x \models^- [\alpha]\phi$ iff there is y such that $R_\alpha xy$ and $y \models^- \phi$

(iii) $x \models^+ \langle\alpha\rangle\phi$ iff there is y such that $R_\alpha xy$ and $y \models^+ \phi$

(iv) $x \models^- \langle\alpha\rangle\phi$ iff for all y, if $R_\alpha xy$, then $y \models^- \phi$

Entailment in **BPDL** is defined as $\models^+$-preservation in every state of every standard dynamic Odintsov–Wansing model. Validity in $\mathcal{M}$ and (logical) validity $\models \phi$ are defined as usual. ($\mathcal{M} \models \phi$ iff $x \models^+_{\mathcal{M}} \phi$ for all states $x \in S$ of $\mathcal{M}$; $\models \phi$ iff $\mathcal{M} \models \phi$ for every standard dynamic Odintsov–Wansing model $\mathcal{M}$.) In addition to 'local' entailment, we define the *global consequence* relation as follows: $X \models^g \phi$ iff, for all $\mathcal{M}$, if every $\psi \in X$ is valid in $\mathcal{M}$, then so is ϕ.

A *non-standard* dynamic Odintsov–Wansing model is defined exactly as a standard model, with one exception: $R(\alpha^*)$ is required to be a superset of $R(\alpha)^*$ (the converse inclusion is not assumed) such that

$$|[\alpha^*]\phi|^+ = |\phi \wedge [\alpha][\alpha^*]\phi|^+ \tag{4}$$

$$|[\alpha^*]\phi|^+ \supseteq |\phi \wedge [\alpha^*](\phi \to [\alpha]\phi)|^+ \tag{5}$$

and

$$|\langle\alpha^*\rangle\phi|^+ = |\phi \vee \langle\alpha\rangle\langle\alpha^*\rangle\phi|^+ \tag{6}$$

$$|\langle\alpha^*\rangle\phi|^+ \subseteq |\phi \vee \langle\alpha^*\rangle(\neg\phi \wedge \langle\alpha\rangle\phi)|^+ \tag{7}$$

In dynamic Odintsov–Wansing models, $\phi?$ tests whether ϕ is true. Hence, test $\phi?$ executes successfully in two cases: if ϕ is only true and if ϕ is both true and false. However, if a more precise assessment of ϕ is needed, one can use $(\phi \wedge \neg{\sim}\phi)?$ and $(\phi \wedge {\sim}\phi)?$.

Lemma 3.2 *All the $H(\mathbf{BK})$ axiom schemata of Theorem 2.2, with all $\Box$ replaced by $[\alpha]$ and all $\Diamond$ replaced by $\langle\alpha\rangle$, are valid in* **BPDL**. *Moreover, the set of formulas valid in any (standard or non-standard) model is closed under Modus ponens and the Necessitation rule $\phi/[\alpha]\phi$.*

Lemma 3.3 *The following schemata are valid in every (standard or non-standard) model:*

(i) $[\alpha \cup \beta]\phi \leftrightarrow ([\alpha]\phi \wedge [\beta]\phi)$ *and* $\langle\alpha \cup \beta\rangle\phi \leftrightarrow (\langle\alpha\rangle\phi \vee \langle\beta\rangle\phi)$

(ii) $[\alpha;\beta]\phi \leftrightarrow [\alpha][\beta]\phi$ *and* $\langle\alpha;\beta\rangle\phi \leftrightarrow \langle\alpha\rangle\langle\beta\rangle\phi$

(iii) $[\psi?]\phi \leftrightarrow (\psi \to \phi)$ *and* $\langle\psi?\rangle\phi \leftrightarrow (\psi \wedge \phi)$

(iv) $[\alpha^*]\phi \leftrightarrow (\phi \wedge [\alpha][\alpha^*]\phi)$ *and* $\langle\alpha^*\rangle\phi \leftrightarrow (\phi \vee \langle\alpha\rangle\langle\alpha^*\rangle\phi)$

(v) $(\phi \wedge [\alpha^*](\phi \to [\alpha]\phi)) \to [\alpha^*]\phi$ *and* $\langle\alpha^*\rangle\phi \to (\phi \vee \langle\alpha^*\rangle(\neg\phi \wedge \langle\alpha\rangle\phi))$

Proof. The proofs are virtually identical to arguments used in the context of standard **PDL** [15]. As an example, we show that $[\alpha^*]\phi \to (\phi \wedge [\alpha][\alpha^*]\phi)$ is valid. The validity of $[\alpha^*]\phi \to \phi$ follows from the fact that $R(\alpha^*)$ is reflexive. Now if $x \not\models^+ [\alpha][\alpha^*]\phi$, then there are y, z such that $R(\alpha)xy$, $R(\alpha^*)yz$ and $z \not\models^+ \phi$. But obviously $R(\alpha^*)xz$, so $x \not\models^+ [\alpha^*]\phi$. $\Box$

It is plain that compactness fails for **BPDL** for the same reason as for **PDL** [15, 181]. Every finite subset of

$$M = \{\langle\alpha^*\rangle\phi\} \cup \{\neg\phi\} \cup \{\neg\langle\alpha^n\rangle\phi \mid n \in \omega\}$$

is satisfiable, but M itself is not ($\alpha^n = \underbrace{\alpha; \ldots ; \alpha}_{n \text{ times}}$).

Examples 2.3 – 2.5 suggest that some $\mathcal{L}_{dyn}$-formulas can be seen as definitions of specific features of programs (α represents a default rule, α removes inconsistency in the information about a specific formula, etc.). Global consequence is a natural notion here. If X is a set of such definitions, then $X \models^g \phi$ iff ϕ is valid in every model that respects the definitions 'globally'. In other words, ϕ is a consequence of the assumption that the definitions in X are satisfied in every possible state. Similarly as in the case of **PDL** (see [15, 209], global consequence for finite X corresponds to validity of specific formulas.

Proposition 3.4 *Let $\{a_1, \ldots, a_n\}$ be the set of all atomic programs appearing in some formula in (finite) X or in ϕ. Then*

$$X \models^g \phi \iff \models [(a_1 \cup \ldots \cup a_n)^*] \bigwedge X \to \phi$$

Proof. The right-to-left implication is trivial. The converse implication is established as follows. If $[(a_1 \cup \ldots \cup a_n)^*] \bigwedge X \to \phi$ is not valid (the antecedent of this implication is abbreviated as X^*), then there is a state x of a model $\mathcal{M}$ such that $x \models^+ X^* \wedge \neg\phi$. Define $\mathcal{M}_x$ by setting $S_x = \{y \mid \langle x, y\rangle \in R((a_1 \cup \ldots \cup a_n)^*)\}$ and taking R_x, V_x^+ and V_x^- to be restrictions of the original R, V^+, V^- to S_x. It is plain that $\bigwedge X$ is valid in $\mathcal{M}_x$, but ϕ is not (the key fact, easily established by induction on the complexity of α, is that if every atomic program appearing in α is in $\{a_1 \cup \ldots \cup a_n\}$, then $R(\alpha)zz'$ only if $R_x(\alpha)zz'$, for all $z, z' \in S_x$). Hence, $X \not\models^g \phi$. □

4 Decidability

In this section we establish decidability of the satisfiability problem of $\mathcal{L}_{dyn}$ formulas in (standard and non-standard) dynamic Odintsov–Wansing models. We modify the standard technique using filtration trough the Fischer–Ladner closure of a formula. Our definition of the Fisher–Ladner closure is a simplified version of the definition used in [15].

Definition 4.1 The *Fisher-Ladner closure* of ϕ, $FL(\phi)$, is the smallest set of formulas such that

- $\phi \in FL(\phi)$ and $FL(\phi)$ is closed under subformulas;
- if $[\psi?]\chi \in FL(\phi)$, then $\psi \in FL(\phi)$;
- if $[\alpha \cup \beta]\chi \in FL(\phi)$, then $[\alpha]\chi \in FL(\phi)$ and $[\beta]\chi \in FL(\phi)$;
- if $[\alpha; \beta]\chi \in FL(\phi)$, then $[\alpha][\beta]\chi \in FL(\phi)$;
- if $[\alpha^*]\chi \in FL(\phi)$, then $[\alpha][\alpha^*]\chi \in FL(\phi)$;
- variants of the above conditions with all '$[\cdot]$' replaced by '$\langle\cdot\rangle$'.

Lemma 4.2 *For all ϕ, $FL(\phi)$ is finite.*

Proof. Standard argument, see [14]. □

Definition 4.3 Let T be a set of formulas and $\mathcal{M}$ a (standard or non-standard) model with $x, y \in S$. Let $x \equiv_T y$ iff, for all $\phi \in T$,

$$x \models^+_{\mathcal{M}} \phi \iff y \models^+_{\mathcal{M}} \phi$$
$$x \models^-_{\mathcal{M}} \phi \iff y \models^-_{\mathcal{M}} \phi.$$

Let $[x]_T = \{y \mid x \equiv_T y\}$. The *filtration of $\mathcal{M}$ trough T* is $\mathcal{M}_T = \langle S_T, R_T, V_T^+, V_T^-\rangle$, where

(i) $S_T = \{[x]_T \mid x \in S\}$;

(ii) $R_T(a) = \{\langle [x]_T, [y]_T\rangle \mid R_a xy\}$ for all $a \in AP$;

(iii) $V_T^+(p) = \{[x]_T \mid x \in V^+(p)\}$;

(iv) $V_T^-(p) = \{[x]_T \mid x \in V^-(p)\}$.

Relations $\models^+_{\mathcal{M}_T}$, $\models^-_{\mathcal{M}_T}$ and $R_T(\alpha)$ for complex α are defined as in standard models.

It is plain that $\mathcal{M}_T$ is a standard model. We write $[x]$ instead of $[x]_T$ if T is clear from the context.

Lemma 4.4 *For all $\mathcal{M}_T$, $|S_T| \leq 4^{|T|}$.*

Proof. There are four possible truth values of each member of T. □

Lemma 4.5 (Filtration Lemma) *Let $\mathcal{M}$ be a (standard or non-standard) model and ϕ a formula.*

(i) *If $[\alpha]\psi \in FL(\phi)$ or $\langle\alpha\rangle\psi \in FL(\phi)$, then $R(\alpha)xy$ only if $R_{FL(\phi)}(\alpha)[x][y]$;*

(ii) *If $[\alpha]\psi \in FL(\phi)$, then $R_{FL(\phi)}(\alpha)[x][y]$ and $x \models^+ [\alpha]\psi$ only if $y \models^+ \psi$;*

(iii) *If $\langle\alpha\rangle\psi \in FL(\phi)$, then $R_{FL(\phi)}(\alpha)[x][y]$ and $y \models^+ \psi$ only if $x \models^+ \langle\alpha\rangle\psi$;*

(iv) *If $\langle\alpha\rangle\psi \in FL(\phi)$, then $R_{FL(\phi)}(\alpha)[x][y]$ and $x \models^- \langle\alpha\rangle\psi$ only if $y \models^- \psi$;*

(v) *If $[\alpha]\psi \in FL(\phi)$, then $R_{FL(\phi)}(\alpha)[x][y]$ and $y \models^- \psi$ only if $x \models^- [\alpha]\psi$;*

(vi) *If $\psi \in FL(\phi)$, then $x \models^+ \psi$ iff $[x] \models^+ \psi$;*

(vii) *If $\psi \in FL(\phi)$, then $x \models^- \psi$ iff $[x] \models^- \psi$.*

Proof. A simple but tedious variation of the standard proof using simultaneous induction on the subexpression relation [14,15]. Details of some of the steps are given in Appendix A. □

Theorem 4.6 *The satisfiability problem for* **BPDL** *is decidable.*

Proof. Standard argument. If ϕ is satisfiable in some $\mathcal{M}$, then, by Lemmas 4.4 and 4.5(vi), ϕ is satisfiable in a standard model of size at most 4^k where $k = |FL(\phi)|$. There is a finite number of such models, so a naive satisfiability algorithm is to determine $k = |FL(\phi)|$ and check all models of size 4^k. □

5 Completeness

The axiom system H(**BPDL**) results from H(**BK**) by replacing all '$\Box$' by '$[\alpha]$' and all '$\Diamond$' by '$\langle\alpha\rangle$' and adding the schemata explicitly stated in Lemma 3.3. (See Appendix B.) The notion of a maximal H(**BPDL**)-consistent set (m.c. set) of formulas is defined as usual (X is consistent iff $\neg \bigwedge X'$ is not provable for all finite $X' \subseteq X$; X is m.c. iff X is consistent all $X' \supset X$ are inconsistent). Hence, m.c. sets have all the usual properties.

Definition 5.1 The canonical model $\mathcal{M}_c = \langle S_c, R_c, V_c^+, V_c^- \rangle$ is a quadruple such that

(i) S_c is the set of all m.c. sets;

(ii) $R_c(\alpha)XY$ iff for all $[\alpha]\phi \in X$, $\phi \in Y$ (iff for all $\phi \in Y$, $\langle\alpha\rangle\phi \in X$);

(iii) $V_c^+(p) = \{X \mid p \in X\}$;

(iv) $V_c^-(p) = \{X \mid {\sim}p \in X\}$;

$|\phi|_c^+ = \{X \mid \phi \in X\}$ and $|\phi|_c^- = \{X \mid {\sim}\phi \in X\}$.

Lemma 5.2 *$|\phi|_c^+$ and $|\phi|_c^-$ behave like $|\phi|^+$ and $|\phi|^-$ (in standard and non-standard models), respectively:*

- *$X \in |p|_c^+$ iff $X \in V_c^+(p)$; $X \in |p|_c^-$ iff $X \in V_c^-(p)$;*
- *$|\bot|_c^+ = \emptyset$; $|\bot|_c^- = S_c$;*

- $|{\sim}\phi|_c^+ = |\phi|_c^-$; $|{\sim}\phi|_c^- = |\phi|_c^+$;
- $|\phi \wedge \psi|_c^+ = |\phi|_c^+ \cap |\psi|_c^+$; $|\phi \wedge \psi|_c^- = |\phi|_c^- \cup |\psi|_c^-$;
- $|\phi \vee \psi|_c^+ = |\phi|_c^+ \cup |\psi|_c^+$; $|\phi \wedge \psi|_c^- = |\phi|_c^- \cap |\psi|_c^-$;
- $|\phi \to \psi|_c^+ = (S_c - |\phi|_c^+) \cup |\psi|_c^+$; $|\phi \to \psi|_c^- = |\phi|_c^+ \cap |\psi|_c^-$;
- $|[\alpha]\phi|_c^+ = \{X \mid (\forall Y)(\textit{if } R_c(\alpha)XY, \textit{then } Y \in |\phi|_c^+)\}$;
  $|[\alpha]\phi|_c^- = \{X \mid (\exists Y)(R_c(\alpha)XY \textit{ and } Y \in |\phi|_c^-)\}$;
- $|\langle\alpha\rangle\phi|_c^+ = \{X \mid (\exists Y)(R_c(\alpha)XY \textit{ and } Y \in |\phi|_c^+)\}$;
  $|\langle\alpha\rangle\phi|_c^- = \{X \mid (\forall Y)(\textit{if } R_c(\alpha)XY, \textit{then } Y \in |\phi|_c^-)\}$.

Proof. Standard inductive argument, we state only three cases explicitly. Firstly, $|{\sim}\phi|_c^- = \{X \mid {\sim}{\sim}\phi \in X\}$ and, as $\phi \leftrightarrow {\sim}{\sim}\phi$ is an axiom, this set is identical to $\{X \mid \phi \in X\}$, i.e., to $|\phi|_c^+$.

Secondly, $|[\alpha]\phi|^+ = \{X \mid [\alpha]\phi \in X\}$. We have to show that $[\alpha]\phi \in X$ iff for all Y, $R_c(\alpha)XY$ only if $Y \in |\phi|^+$. The left-to-right implication is trivial. The right-to-left implication is established by the following standard argument. Assume that $\neg[\alpha]\phi \in X$. We want to show that there is Y such that $R_c(\alpha)XY$ and $\neg\phi \in Y$. We claim that the set

$$M = \{\neg\phi\} \cup \{\psi \mid [\alpha]\psi \in X\} \tag{8}$$

is consistent. (We denote $\{\psi \mid [\alpha]\psi \in X\}$ as $X^{-\alpha}$.) To see this, take an arbitrary finite $\Psi = \{\psi_1, \ldots, \psi_m\} \subseteq X^{-\alpha}$. It is plain that $\langle\alpha\rangle\neg\phi \wedge [\alpha]\psi_1 \wedge \ldots \wedge [\alpha]\psi_m \in X$. Hence, by the **K**-style properties of $[\alpha]$ and $\langle\alpha\rangle$, $\langle\alpha\rangle(\neg\phi \wedge \psi_1 \wedge \ldots \wedge \psi_m) \in X$. By the Necessitation rule, $\neg(\neg\phi \wedge \psi_1 \wedge \ldots \wedge \psi_m)$ is not provable, so $\{\neg\phi, \psi_1, \ldots, \psi_m\}$ is consistent. But Ψ was chosen as an arbitrary finite subset of $X^{-\alpha}$. Consequently, $\{\neg\phi\} \cup \Psi'$ for every finite $\Psi' \subseteq X^{-\alpha}$ can be shown to be consistent in this way. Hence, M itself is consistent. By the Lindenbaum Lemma, M can be extended to a m.c. Y and it is plain that $R_c(\alpha)XY$ and $\neg\phi \in Y$.

Thirdly, $|[\alpha]\phi|_c^- = \{X \mid {\sim}[\alpha]\phi \in X\}$, a set identical to $\{X \mid \langle\alpha\rangle{\sim}\phi \in X\}$ as ${\sim}[\alpha]\phi \leftrightarrow \langle\alpha\rangle{\sim}\phi$ is an axiom. A straightforward adaptation of the argument given by [15, p. 206] shows that this set is identical to $\{X \mid (\exists Y)(R_c(\alpha)XY \text{and } Y \in |\phi|_c^-)\}$. □

Lemma 5.3 *$\mathcal{M}_c$ is a non-standard model.*

Proof. We need to be establish that R satisfies the conditions required by the definition of a non-standard model. The argument for $\cup$, ;, ? and the iteration equations (4) – (7) is virtually identical to that given by [15, p.206–8]. To show that $R_c(\alpha)^* \subseteq R_c(\alpha^*)$, assume that $\langle X, Y\rangle \in R_c(\alpha)^*$ but $\langle X, Y\rangle \notin R_c(\alpha^*)$. Hence, there is ϕ such that $[\alpha^*]\phi \in X$ but $\phi \notin Y$. However, $\langle X, Y\rangle \in R_c(\alpha)^*$ implies that either $X = Y$ or else there are $Z_0, \ldots, Z_m$ such that $Z_0 = X$, $Z_m = Y$ and $\langle Z_k, Z_{k+1}\rangle \in R_c(\alpha)$ for $1 \leq k < m$. In the former case, $\phi \in Y$ by the axiom $[\alpha^*]\phi \leftrightarrow (\phi \wedge [\alpha][\alpha^*]\phi)$, a contradiction. In the latter case, $[\alpha^*]\phi \in Z_k$ entails $[\alpha][\alpha^*]\phi \in Z_{k+1}$ by the same axiom for all $1 \leq k < n$ and, hence, $[\alpha^*]\phi \in Y$. Hence, $\phi \in Y$, a contradiction. □

Since **BPDL** is not compact, it cannot enjoy a strongly complete axiomatisation (as **BK** does). However, weak completeness is another story.

Theorem 5.4 *ϕ is provable in H(**BPDL**) iff ϕ is valid in* **BPDL**.

Proof. Soundness follows from Lemmas 3.2 and 3.3. Completeness follows from Lemmas 4.5 and 5.3. If ϕ is not provable, then $X \in |\neg\phi|_c^+$ for some m.c. set X. By the Filtration Lemma, $(\mathcal{M}_c)_{FL(\neg\phi)}$ is a standard model such that $[X] \in |\neg\phi|^+_{FL(\neg\phi)}$. □

6 Conclusion

This article introduced **BPDL**, a combination of propositional dynamic logic **PDL** with the four-valued Belnapian modal logic **BK**. The logic is expected to be useful in formalising reasoning about the properties of algorithmic transformations of possibly incomplete and inconsistent database-like bodies of information. We modified the standard proofs based on filtration and the canonical-model technique and, as the main technical results of the article, established decidability of **BPDL** and provided it with a sound and weakly complete axiomatisation. The main message here is that the standard techniques are easily adapted to the four-valued setting.

The number one topic for future research is the complexity of the satisfiability problem for **BPDL**. The problem is *EXPTIME*-complete for **PDL** and it will be interesting to see whether the situation gets worse in the case of **BPDL**. Our strategy of tackling the problem will be, as for the results already achieved, to try to adapt the proof technique used in the case of **PDL** to the four-valued setting. We shall also investigate Belnapian versions of some extensions of **PDL**. The obvious choice is the first-order dynamic logic **DL**, but also concurrent **PDL** modelling parallel execution of programs. Last but not least, a more thorough examination of possible applications of **BPDL** will be an interesting enterprise.

A Proof of the Filtration Lemma

The proof is a variation of the standard proof using simultaneous induction on the subexpression (subformula or subprogram) relation [14,15]. In proving the claim of any item (i)–(vii) for any special case of α or ψ, we assume that all the items hold for all subexpressions of α and ψ. Only some steps of the proof are explicitly stated here (and, perhaps, in more detail than an expert reader needs).

A.1 $\mathcal{M}$ is a standard model

(i), $\alpha = \beta^*$. If $R(\beta^*)xy$, then, since $R(\beta^*)$ is the reflexive transitive closure of $R(\beta)$, there are $z_0, \ldots, z_n$ such that $z_0 = x$, $z_n = y$ and either $n = 0$ or else $R(\beta)z_i z_{i+1}$ for $0 \leq i < n$. If $n = 0$, then $R_{FL(\phi)}(\beta^*)[z_0][z_n]$ by the definition of $R_{FL(\phi)}(\beta^*)$. Assume $n > 0$. If $[\beta^*]\psi$ ($\langle\beta^*\rangle$) is in $FL(\phi)$, then so is $[\beta][\beta^*]\psi$ ($\langle\beta\rangle\langle\beta^*\rangle\psi$). β is a subexpression of β^*, so, in both cases, we may apply the induction hypothesis (IH): $R(\beta)z_i z_{i+1}$ implies $R_{FL(\phi)}(\beta)[z_i][z_{i+1}]$ for $0 \leq i < n$.

Hence, $R_{FL(\phi)}(\beta^*)[z_0][z_n]$ by the definition of $R_{FL(\phi)}$.

(i), $\alpha = \chi?$. If $R(\chi?)xy$, then $x = y$ and $x \models^+ \chi$. If $[\chi?]\psi$ ($\langle\chi?\rangle\psi$) is in $FL(\phi)$, then so is χ. χ is a subexpression of $[\chi?]\psi$ ($\langle\chi?\rangle\psi$) and a formula, so we may apply the IH of (vi): $x \models^+ \chi$ entails $[x] \models^+ \chi$. But $[x] = [y]$ and, hence, $R_{FL(\phi)}(\chi?)[x][y]$ by the definition of $R_{FL(\phi)}$.

(ii), $\alpha = \beta^*$. If $R_{FL(\phi)}(\beta^*)[x][y]$, then there are $z_0, \ldots, z_n$ such that $[z_0] = [x]$, $[z_n] = [y]$ and either $n = 0$ or else $R_{FL(\phi)}(\beta)[z_i][z_{i+1}]$ for $0 \leq i < n$. If $n = 0$, then $x \models^+ [\beta^*]\psi$ entails $y \models^+ \psi$ by the assumption $[\beta^*]\psi \in FL(\phi)$ and Lemma 3.3(iv). Assume $n > 0$. We prove that

$$x \models^+ [\beta^*]\psi \Longrightarrow z_k \models^+ [\beta^*]\psi \qquad (0 \leq k \leq n) \tag{A.1}$$

by induction on k. If $k = 0$, then the claim follows from the assumption $[\beta^*]\psi \in FL(\phi)$. Assume that the claim holds for $k = l$. We prove that it holds for $k = l + 1$ as well. The assumption is that $x \models^+ [\beta^*]\psi$ entails $z_l \models^+ [\beta^*]\psi$. By Lemma 3.3(iv), $z_l \models [\beta][\beta^*]\psi$. β is a subexpression of β^* and $[\beta][\beta^*]\psi \in FL(\phi)$, so we may use IH of item (ii) of the Filtration Lemma: $R_{FL(\phi)}(\beta)[z_k][z_{l+1}]$ entails $z_{l+1} \models^+ [\beta^*]\psi$. This proves (A.1). Now $x \models^+ [\beta^*]\psi$ entails $z_n \models [\beta^*]\psi$ by (A.1) and $z_n \models [\beta^*]\psi$ entails $z_n \models^+ \psi$ by Lemma 3.3(iv). But $[z_n] = [y]$ and $\psi \in FL(\phi)$, so $y \models^+ \psi$.

(iv), $\alpha = \beta^*$. If $R_{FL(\phi)}(\beta^*)[x][y]$, then there are $z_0, \ldots, z_n$ such that $[z_0] = [x]$, $[z_n] = [y]$ and either $n = 0$ or else $R_{FL(\phi)}(\beta)[z_i][z_{i+1}]$ for $0 \leq i < n$. If $n = 0$, then $x \models^- \langle\beta^*\rangle\psi$ entails $y \models^- \langle\beta^*\rangle\psi$ by the assumption $\langle\beta^*\rangle\psi \in FL(\phi)$. Hence, $y \models^+ [\beta^*]{\sim}\psi$ by Lemma 3.2. By Lemma 3.3(iv), $y \models^+ {\sim}\psi$. Hence, $y \models^- \psi$. Next, assume that $n > 0$. We prove that

$$x \models^- \langle\beta^*\rangle\psi \Longrightarrow z_k \models^- \langle\beta^*\rangle\psi \qquad (0 \leq k \leq n) \tag{A.2}$$

by induction on k. If $k = 0$, then the claim follows from the assumption $\langle\beta^*\rangle\psi \in FL(\phi)$. Assume that the claim holds for $k = l$. We prove that it holds for $k = l + 1$ as well. The assumption is that $x \models^- \langle\beta^*\rangle\psi$ entails $z_l \models^- \langle\beta^*\rangle\psi$. By Lemmas 3.2 and 3.3(iv), $z_l \models^- \langle\beta\rangle\langle\beta^*\rangle\psi$. β is a subexpression of β^*, so we may use the IH to infer $z_{l+1} \models^- \langle\beta^*\rangle\psi$. This proves (A.2). Assume that $x \models^- \langle\beta^*\rangle\psi$. By (A.2), $z_n \models^- \beta^*\rangle\psi$. By the assumption that $\langle\beta^*\rangle\psi \in FL(\phi)$, $y \models^- \beta^*\rangle\psi$. By Lemmas 3.2 and 3.3(iv), $y \models^- \psi$.

(v), $\alpha = \beta^*$. If $R_{FL(\phi)}(\beta^*)[x][y]$, then there are $z_0, \ldots, z_n$ such that $[z_0] = [x]$, $[z_n] = [y]$ and either $n = 0$ or else $R_{FL(\phi)}(\beta)[z_i][z_{i+1}]$ for $0 \leq i < n$. If $n = 0$, then the reasoning is similar as in the above cases. Hence, assume that $n > 0$. We prove that

$$z_n \models^- \psi \Longrightarrow z_{n-k} \models^- [\beta^*]\psi \qquad (0 \leq k \leq n) \tag{A.3}$$

by induction on k. The case $k = 0$is trivial. Assume that the claim holds for $k = l$. We prove that it holds for $k = l + 1$ as well. The assumption is that $z_n \models^- \psi$ entails $z_{n-l} \models^- [\beta^*]\psi$. There are two possibilities. Either (a) $z_{n-l} \models^- \psi$ or (b) $z_{n-l} \models^- [\beta][\beta^*]\psi$. If (a), then $z_{n-(l+1)} \models^- [\beta]\psi$ by IH and $z_{n-(l+1)} \models^- [\beta^*]\psi$ by $R_\beta \subseteq (R_\beta)^*$. If (b), then IH entails that $z_{n-(l+1)} \models^-$

$[\beta][\beta][\beta^*]\psi$. By $R_\beta \subseteq (R_\beta)^*$, $z_{n-(l+1)} \models^- [\beta^*]\psi$. This proves (A.3). Now $y \models^- \psi$ only if $z_n \models^- \psi$ ($\psi \in FL(\phi)$) only if $z_0 \models^- [\beta^*]\psi$ (A.3) only if $x \models^- [\beta^*]\psi$ ($[\beta^*]\psi \in FL(\phi)$).

(vi), $\psi = \sim\chi$. $x \models^+ \sim\chi$ iff $x \models^- \chi$. χ is a subexpression of $\sim\chi$, so we may use IH of (vii): and infer $x \models^- \chi$ iff $[x] \models^- \chi$ iff $[x] \models^+ \sim\chi$ (by the definition of $\models^+_{\mathcal{M}_{FL(\phi)}}$. In fact, this case requires to introduce item (vii) into the Filtration Lemma.

(vi), $\psi = [\alpha]\chi$. Assume $[\alpha]\chi \in FL(\phi)$. Then $\chi \in FL(\phi)$. To prove the left-to-right implication, assume that $x \models^+ [\alpha]\chi$ and $R_{FL(\phi)}(\alpha)[x][y]$. α is a subexpression of $[\alpha]\chi$, so we may use IH of (ii) to infer $y \models^+ \chi$. By IH, $[y] \models^+ \chi$. To prove the right-to-left implication, assume that $[x] \models^+ [\alpha]\chi$ and $R_\alpha xy$. IH of (i) implies that $R_{FL(\phi)}(\alpha)[x][y]$. Consequently, $[y] \models^+ \chi$ and, by IH, $y \models^+ \chi$.

(vi), $\psi = \langle\alpha\rangle\chi$. We prove only the right-to-left implication. If $[x] \models^+ \langle\alpha\rangle\chi$, then there is y such that $R_{FL(\phi)}(\alpha)[x][y]$ and $[y] \models^+ \chi$. By IH, $y \models^+ \chi$. By IH of (iii), $x \models^+ \langle\alpha\rangle\chi$. This was the reason we had to include item (iii) of the Lemma.

(vii), $\psi = p$. Let $p \in FL(\phi)$. The left-to-right implication is trivial. To prove the converse, assume that $[x] \models^- p$. This means that there is $x' \equiv x$ such that $x' \models^- p$. By the definition of filtration, $x \models^- p$ as well. Note that to prove this implication it was necessary to define $\equiv$ in terms of both $\models^+$ and $\models^-$.

(vii), $\psi = [\alpha]\chi$. We prove only the right-to-left implication. If $[x] \models^- [\alpha]\chi$, then there is y such that $R_{FL(\phi)}(\alpha)[x][y]$ and $[y] \models^- \chi$. By IH of (v), $y \models^- [\alpha]\chi$. This case required to introduce item (v).

(vii), $\psi = \langle\alpha\rangle\chi$. We prove only the left-to-right implication. Assume $x \models^- \langle\alpha\rangle\chi$ and $R_{FL(\phi)}(\alpha)[x][y]$. By IH of (iv), $y \models^- \chi$. By IH, $[y] \models^- \chi$. Hence, $[x] \models^- \langle\alpha\rangle\chi$. This case required to introduce item (iv) of the Lemma.

Proofs of other cases are similar or standard. □

A.2 $\mathcal{M}$ is a non-standard model

As in the standard proof for this case, the only claim where the assumption $R_{\alpha^*} = (R_\alpha)^*$ was used is (i), $\alpha = \beta^*$. Hence, we have to prove that if $[\beta^*]\psi \in FL(\phi)$ or $\langle\beta^*\rangle\psi \in FL(\phi)$, then $R(\beta^*)xy$ only if $R_{FL(\phi)}(\beta^*)[x][y]$. Our argument is very close to the one given in [15, sec. 6.3].

Assume that $\langle x, y\rangle \in R(\beta^*)$. We want to show that $\langle [x], [y]\rangle \in R_{FL(\phi)}(\beta^*)$, or equivalently that $y \in E$, where

$$E = \{z \mid \langle [x], [z]\rangle \in R_{FL(\phi)}(\beta^*)\}$$

Recall that $[\cdot]$ is given by some specific finite set$FL(\phi)$ of formulas. For any $[z]_{FL(\phi)}$, define $X_{[z]}$ to be the smallest set of formulas such that, for all $\chi \in FL(\phi)$:

- If $z \models^+ \chi$, then $\chi \in X_{[z]}$;
- If $z \not\models^+ \chi$, then $\neg\chi \in X_{[z]}$.

(Note that we are using '$\neg$' not '$\sim$'.) Obviously, $X_{[z]}$ is finite for all z. Define

$$\psi_{[z]} = \bigwedge X_{[z]}$$

It is not hard to show that, for all $w \in S$,

$$w \models^{+} \psi_{[z]} \iff w \equiv z \tag{A.4}$$

(For instance, assume that $w \not\equiv z$ because there is $\theta \in FL(\phi)$ such that $w \models^{-} \theta$ and $z \not\models^{-} \theta$. But then $z \not\models^{+} {\sim}\theta$ and, consequently, $\neg{\sim}\theta \in X_{[z]}$. But then $w \not\models^{+} \psi_{[z]}$ because $w \models^{+} {\sim}\theta$.) Now define

$$\psi_E = \bigvee_{z \in E} \psi_{[z]}$$

It is not hard to show that ψ_E defines E, i.e., for all $w \in S$

$$w \in E \iff w \models^{+} \psi_E \tag{A.5}$$

(For instance, assume that $w \in E$ but $w \not\models^{+} \psi_E$. Then $w \not\models^{+} \psi_{[z]}$ for all $z \in E$. In particular, $w \not\models^{+} \psi_{[w]}$. (A.4) entails that this is impossible.)

It is easy to show that E is closed under R_β, i.e., for all z, z',

$$z \in E \;\&\; R_\beta z z' \implies z' \in E \tag{A.6}$$

(β is a subexpression of β^*, so $R_\beta z z'$ entails $R_{FL(\phi)}(\beta)[z][z']$ by IH. By the definition of E, $z \in E$ means that $R_{FL(\phi)}(\beta^*)[x][z]$. Consequently, $R_{FL(\phi)}(\beta^*)[x][z']$. In other words, $z' \in E$.) (A.6) means that $\psi_E \to [\beta]\psi_E$ is valid in $\mathcal{M}$. By Lemma 3.2(Nec. rule), so is $[\beta^*](\psi_E \to [\beta]\psi_E)$. By Lemma 3.3(v), the induction axiom $(\phi \wedge [\alpha^*](\phi \to [\alpha]\phi)) \to [\alpha^*]\phi$ is also valid in $\mathcal{M}$.

It is also easy to show that $x \in E$ ($R(\beta^*)$ is a superset of the reflexive transitive closure of $R(\beta)$, so it contains the identity relation of $S_{FL(\phi)}$.) Hence, $x \models^{+} \psi_E \wedge [\beta^*](\psi_E \to [\beta]\psi_E)$. By the validity of the induction axiom in $\mathcal{M}$, $x \models^{+} [\beta^*]\psi_E$. Hence, if $R(\beta^*)xy$, then $y \models^{+} \psi_E$. By (A.5), $y \in E$. □

B The axiom system H(BPDL)

(i) Axioms of classical propositional logic in the language $\{AF, \bot, \to, \wedge, \vee\}$ and Modus ponens;

(ii) Strong negation axioms:

$$\begin{gathered}
{\sim}{\sim}\phi \leftrightarrow \phi, \\
{\sim}(\phi \wedge \psi) \leftrightarrow ({\sim}\phi \vee {\sim}\psi), \\
{\sim}(\phi \vee \psi) \leftrightarrow ({\sim}\phi \wedge {\sim}\psi), \\
{\sim}(\phi \to \psi) \leftrightarrow (\phi \wedge {\sim}\psi), \\
\top \leftrightarrow {\sim}\bot;
\end{gathered}$$

(iii) Modal axiom $[\alpha](\phi \to \psi) \to ([\alpha]\phi \to [\alpha]\psi)$ and the Necessitation rule $\phi/[\alpha]\phi$;

(iv) **PDL** axiom schemata

$$[\alpha \cup \beta]\phi \leftrightarrow ([\alpha]\phi \wedge [\beta]\phi) \text{ and } \langle\alpha \cup \beta\rangle\phi \leftrightarrow (\langle\alpha\rangle\phi \vee \langle\beta\rangle\phi)\,,$$
$$[\alpha;\beta]\phi \leftrightarrow [\alpha][\beta]\phi \text{ and } \langle\alpha;\beta\rangle\phi \leftrightarrow \langle\alpha\rangle\langle\beta\rangle\phi,$$
$$[\psi?]\phi \leftrightarrow (\psi \to \phi) \text{ and } \langle\psi?\rangle\phi \leftrightarrow (\psi \wedge \phi)\,,$$
$$[\alpha^*]\phi \leftrightarrow (\phi \wedge [\alpha][\alpha^*]\phi) \text{ and } \langle\alpha^*\rangle\phi \leftrightarrow (\phi \vee \langle\alpha\rangle\langle\alpha^*\rangle\phi)\,,$$
$$(\phi \wedge [\alpha^*](\phi \to [\alpha]\phi)) \to [\alpha^*]\phi \text{ and } \langle\alpha^*\rangle\phi \to (\phi \vee \langle\alpha^*\rangle(\neg\phi \wedge \langle\alpha\rangle\phi))\,;$$

(v) Modal interaction principles:

$$\neg[\alpha]\phi \leftrightarrow \langle\alpha\rangle\neg\phi,$$
$$\neg\langle\alpha\rangle\phi \leftrightarrow [\alpha]\neg\phi,$$
$$\sim[\alpha]\phi \leftrightarrow \langle\alpha\rangle\sim\phi,$$
$$[\alpha]\phi \leftrightarrow \sim\langle\alpha\rangle\sim\phi,$$
$$\sim\langle\alpha\rangle\phi \leftrightarrow [\alpha]\sim\phi,$$
$$\langle\alpha\rangle\phi \leftrightarrow \sim[\alpha]\sim\phi.$$

References

[1] Arieli, O. and A. Avron, *Reasoning with logical bilattices*, Journal of Logic, Language and Information **5** (1996), pp. 25–63.

[2] Belnap, N., *A useful four-valued logic*, in: J. M. Dunn and G. Epstein, editors, *Modern Uses of Multiple-Valued Logic*, Springer Netherlands, Dordrecht, 1977 pp. 5–37.

[3] Belnap, N., *How a computer should think*, in: G. Ryle, editor, *Contemporary Aspects of Philosophy*, Oriel Press Ltd., 1977 .

[4] Běhounek, L., *Modeling costs of program runs in fuzzified propositional dynamic logic*, in: F. Hakl, editor, *Doktorandské dny '08* (2008), pp. 6 – 14.

[5] Běhounek, L., M. Bílková and P. Cintula, *Modeling the costs of programs by fuzzy dynamic logic* (2008), talk at ECAP 2008 in Krakow.

[6] Dunn, J. M., *Intuitive semantics for first-degree entailments and "coupled trees"*, Philosophical Studies **29** (1976), pp. 149–168.

[7] Fagin, R., J. Y. Halpern and M. Vardi, *A nonstandard approach to the logical omniscience problem*, Artificial Intelligence **79** (1995), pp. 203–240.

[8] Fischer, M. J. and R. E. Ladner, *Propositional dynamic logic of regular programs*, Journal of Computer and System Sciences **18** (1979), pp. 194–211.

[9] Fitting, M., *Bilattices and the semantics of logic programming*, The Journal of Logic Programming **11** (1991), pp. 91–116.

[10] Fitting, M., *Bilattices are nice things*, in: T. Bolander, V. Hendricks and S. A. Pedersen, editors, *Self-Reference*, CSLI Press, 2006 pp. 53–77.

[11] Gargov, G., *Knowledge, uncertainty and ignorance in logic: Bilattices and beyond*, Journal of Applied Non-Classical Logics **9** (1999), pp. 195–283.

[12] Ginsberg, M. L., *Multivalued logics: a uniform approach to reasoning in artificial intelligence*, Computational Intelligence **4** (1988), pp. 265–316.

[13] Goble, L., *Paraconsistent modal logic*, Logique et Analyse **193** (2006), pp. 3–29.

[14] Goldblatt, R., "Logics of Time and Computation," CSLI Publications, 1992.

[15] Harel, D., D. Kozen and J. Tiuryn, "Dynamic Logic," MIT Press, 2000.

[16] Odintsov, S. and H. Wansing, *Modal logics with Belnapian truth values*, Journal of Applied Non-Classical Logics **20** (2010), pp. 279–301.

[17] Odintsov, S. P. and E. I. Latkin, *BK-lattices. Algebraic semantics for Belnapian modal logics*, Studia Logica **100** (2012), pp. 319–338.

[18] Odintsov, S. P. and S. O. Speranski, *The lattice of Belnapian modal logics: Special extensions and counterparts*, Logic and Logical Philosophy **25** (2016), pp. 3–33.

[19] Priest, G., *Many-valued modal logics: A simple approach*, The Review of Symbolic Logic **1** (2008), pp. 190–203.

[20] Rivieccio, U., A. Jung and R. Jansana, *Four-valued modal logic: Kripke semantics and duality*, Journal of Logic and Computation (2015), forthcoming.

[21] Sedlár, I., *Non-classical PDL on the cheap* (2016), manuscript.

[22] Teheux, B., *Propositional dynamic logic for searching games with errors*, Journal of Applied Logic **12** (2014), pp. 377–394.

Local tabularity without transitivity

Ilya Shapirovsky [1]

Institute for Information Transmission Problems

Valentin Shehtman [2]

Institute for Information Transmission Problems,
Moscow State University,
National Research University Higher School of Economics

Abstract

According to the classical result by Segerberg and Maksimova, a modal logic containing **K4** is locally tabular iff it is of finite height. The notion of finite height can also be defined for logics, in which the master modality is expressible ('pretransitive' logics). We observe that every locally tabular logic is a pretransitive logic of finite height. Then we prove some semantic criteria of local tabularly. By applying them we extend the Segerberg – Maksimova theorem to a certain larger family of pretransitive logics.

Keywords: local tabularity, pretransitive logic, logic of finite height

1 Introduction

Recall that a propositional logic is called tabular (or finite) if it is characterized by a finite logical matrix (by a finite Kripke frame in the case of normal modal or intermediate logics). A weaker property is local tabularity (or local finiteness): a logic is locally tabular if each of its finite-variable fragments is tabular. Both these properties have been studied since the 1920s. They also appeared in universal algebra as properties of varieties [14], and they were investigated for different kinds of algebras: lattices, groups, rings, Lie algebras, etc. The most famous example is the Burnside problem on local tabularity of varieties of groups with the identity $x^n = 1$, cf. [1].

In this paper we are interested in modal logics, so let us mention the corresponding results in this area. Tabularity for modal and intuitionistic logics has been studied quite well, cf. [6], chapter 12. As for local tabularity, our knowledge is still very incomplete. The first well-known examples of logics with this

---

[1] ilya.shapirovsky@gmail.com. The research was supported be the Russian Sciences Foundation (project no. 16-11-10252)

[2] vshehtman@gmail.com. The research was supported be the Russian Sciences Foundation (project no. 16-11-10252)

property are the modal logic **S5** and the intermediate logic **LC**. Beginning from the 1970s, the studies of local tabularity became more systematic. First of all, a nice theorem was found by Krister Segerberg [15] and Larisa Maksimova [13], characterizing locally tabular extensions of **K4** as logics of finite height (in another terminology, logics of finite depth, or logics of finite slices).

Around the same time an analogue to Segerbergs theorem was established for intermediate logics: Alexander Kuznetsov [12], and Yuichi Komori [9] proved local tabularity of intermediate logics of finite height. However, unlike the modal case, for intermediate logics the converse theorem does not hold, and **LC** is the simplest counterexample. Characterization of locally tabular intermediate logics remains a challenging open problem.

Research of local tabularity for intuitionistic modal logics and the corresponding class of monadic Heyting algebras was started by Guram Bezhanishvili in [2] and continued in his joint work with Revaz Grigolia [4]. These papers considered extensions of the well-known logic **MIPC**.

Another interesting result on local tabularity was obtained by Nick Bezhanishvili for bimodal logics [5]: he proved that the logic $\mathbf{S5}^2$ is pre-locally tabular. G. Bezhanishvili [3] considered some other bimodal logics, e.g. extensions of the fusions $\mathbf{Grz} * \mathbf{S5}$ and $\mathbf{GL} * \mathbf{S5}$ with (half)-commutation axioms. In these cases local tabularity can be reduced to local tabularity of intuitionistic modal logics studied in [2], [4].

For non-transitive normal extensions of **K** the first family of locally tabular logics was probably the logics $\mathbf{K} + \Box^n\bot$ considered in [7] (also, cf. [17], where local tabularity is extended to $(\mathbf{K} + \Box^n\bot) \times \mathbf{S5}$).

Characterization of locally tabular modal logics above **K** is an open problem. This paper makes a step towards its solution. The notion of finite height can also be defined for logics, in which the master modality is expressible ('pretransitive' logics). We observe that every locally tabular logic is a pretransitive logic of finite height (Theorem 3.7). Then we prove some semantic criteria of local tabularity (Theorems 4.3 and 5.2). By applying them we extend the Segerberg – Maksimova theorem to a certain family of pretransitive logics (Theorem 6.2).

2 Preliminaries

By a *logic* we mean a normal propositional monomodal logic, cf. [6]. ML denotes the set of all modal formulas; they are built from a countable set $\mathrm{PL} = \{p_1, p_2, \ldots\}$ of proposition letters, the classical connectives $\to$, $\bot$, and the modal connectives $\Diamond$, $\Box$.

An *n-formula* is a formula in proposition letters $p_1, p_2, \ldots, p_n$. $\mathrm{ML}\lceil n$ denotes the set of all n-formulas. For a set $\Gamma \subseteq \mathrm{ML}$, put $\Gamma\lceil n = \Gamma \cap (\mathrm{ML}\lceil n)$.

For a logic L, formulas φ, ψ are called L-*equivalent* if $(\varphi \leftrightarrow \psi) \in \mathrm{L}$.

L is called *locally tabular* if for any finite n there exist finitely many n-formulas up to L-equivalence. Equivalently, a logic is locally tabular if the variety of L-algebras is *locally finite*, i.e., every finitely generated L-algebra is finite.

By *a frame* we mean a Kripke frame (W, R), $W \neq \varnothing, R \subseteq W \times W$. For a frame $\mathrm{F} = (W, R)$, $\varnothing \neq V \subseteq W$, put $\mathrm{F}{\restriction}V = (V, R{\restriction}V)$, where $R{\restriction}V = R \cap (V \times V)$; $\mathrm{F}{\restriction}V$ is called *a generated subframe* of F if $R(V) \subseteq V$.

A *model* (W, R, θ) over a frame (W, R) is a tuple (W, R, θ), where $\theta : \mathrm{PL} \to 2^W$ is a *valuation.* An *n-weak model* is (W, R, θ), where $\theta : \mathrm{PL}\lceil n \to 2^W$ is an *n-valuation*; in this case we can only evaluate n-formulas.

$\mathrm{M_L} = (W_\mathrm{L}, R_\mathrm{L}, \theta_\mathrm{L})$ denotes the canonical model of L. $\mathrm{M}_{\mathrm{L}\lceil n} = (W_{\mathrm{L}\lceil n}, R_{\mathrm{L}\lceil n}, \theta_{\mathrm{L}\lceil n})$ denotes the canonical model of $\mathrm{L}\lceil n$ [17]. $\mathrm{M}_{\mathrm{L}\lceil n}$ is also called a *weak canonical model of* L.

Note that if $\mathrm{M}_{\mathrm{L}\lceil n}$ is finite, its points bijectively correspond to atoms in the Lindenbaum algebra of $\mathrm{L}\lceil n$.

Proposition 2.1 *A modal logic is locally tabular iff all its weak canonical models are finite.*

Proposition 2.2 *Every extension of a locally tabular logic is locally tabular.*

As usual, *a partition* $\mathcal{A}$ of a non-empty set W is a set of non-empty pairwise disjoint sets such that $W = \cup\mathcal{A}$. The corresponding equivalence relation is denoted by $\sim_\mathcal{A}$, so $\mathcal{A} = W/{\sim_\mathcal{A}}$. By a partition of a Kripke model or a frame we mean a partition of its set of worlds.

For a partition $\mathcal{A}$ of W and nonempty $V \subseteq W$, put

$$\mathcal{A}{\restriction}V = \{V \cap X \mid X \in \mathcal{A} \;\&\; V \cap X \neq \varnothing\}.$$

Let $\mathrm{M} = (W, R, \theta)$ be a model, Γ a set of formulas. Put

$$x \equiv_\Gamma y \text{ iff } \forall\varphi \in \Gamma(\mathrm{M}, x \vDash \varphi \iff \mathrm{M}, y \vDash \varphi).$$

A partition $\mathcal{A}$ of M *respects* Γ if $\sim_\mathcal{A} \subseteq \equiv_\Gamma$; $\mathcal{A}$ is *induced by* Γ if $\sim_\mathcal{A} = \equiv_\Gamma$.

If M is n-weak, the partition of M induced by $\mathrm{ML}\lceil n$ is called *canonical* and denoted by $\mathcal{A}(\mathrm{M})$. For $V \in \mathcal{A}(\mathrm{M})$ put

$$\mathfrak{t}_n(V) = \{\varphi \in \mathrm{ML}\lceil n \mid \mathrm{M}, x \vDash \varphi \text{ for some (for all) } x \in V\}.$$

Proposition 2.3 *If* M *is n-weak and* $\mathrm{M} \vDash \mathrm{L}\lceil n$*, then* $\mathfrak{t}_n$ *is an injection* $\mathcal{A}(\mathrm{M}) \to W_{\mathrm{L}\lceil n}$.

Proof. Note that membership in $\mathfrak{t}_n(V)$ respects all Boolean connectives. Since $\mathrm{M} \vDash \mathrm{L}\lceil n$, we have $\mathfrak{t}_n(V) \in W_{\mathrm{L}\lceil n}$ for any $V \in \mathcal{A}(\mathrm{M})$. By definition, if $V_1 \neq V_2$ for some $V_1, V_2 \in \mathcal{A}(\mathrm{M})$, then $\mathfrak{t}_n(V_1) \neq \mathfrak{t}_n(V_2)$. □

Proposition 2.4 *If* $\mathcal{A}(\mathrm{M})$ *is finite, then every* $U \in \mathcal{A}(\mathrm{M})$ *is definable in* M*, i.e.,* $\mathrm{M}, x \vDash \varphi_U \iff x \in U$ *for some n-formula* φ_U.

Proof. For any distinct $U, V \in \mathcal{A}(\mathrm{M})$, there is an n-formula φ_{UV} such that $\varphi_{UV} \in \mathfrak{t}_n(U)$ and $\varphi_{UV} \notin \mathfrak{t}_n(V)$; then put

$$\varphi_U = \bigwedge_{V \neq U} \varphi_{UV}.$$

□

Proposition 2.5 *Suppose* M *is* n*-weak,* $\mathcal{A}(\mathrm{M})$ *is finite and for any* n*-formula* φ*,*

$$\mathrm{M} \models \varphi \iff \varphi \in \mathrm{L}\lceil n.$$

Then $\mathsf{t}_n : \mathcal{A}(\mathrm{M}) \to W_{\mathrm{L}\lceil n}$ *is a bijection.*

Proof. Since $\mathrm{M} \models \mathrm{L}\lceil n$, t_n is an injection (Proposition 2.3). To check the surjectivity, suppose $\mathcal{A}(\mathrm{M}) = \{V_0, \ldots, V_l\}$, $x_i = \mathsf{t}_n(V_i)$ and there exists $x \in W_{\mathrm{L}\lceil n}$ such that $x \neq x_i$ for each $i \leq l$. For each $i \leq l$ choose a formula φ_i such that $\varphi_i \in x$ and $\varphi_i \notin x_i$; put $\varphi = \wedge_{i \leq l}\varphi_i$. Then $\varphi \in x$, so $\neg\varphi \notin \mathrm{L}$, thus for some w in M we have $\mathrm{M}, w \models \varphi$. On the other hand, for each $i \leq l$ we have $\mathrm{M}, w \models \varphi_i$ and $\varphi_i \notin \mathsf{t}_n(V_i)$. Hence $w \notin V_i$ for each $i \leq l$, which is a contradiction. □

3 All locally tabular logics are pretransitive logics of finite height

In this section we formulate a necessary condition for local tabularity.

For a binary relation R on W, R^* denotes its transitive reflexive closure: $R^* = \bigcup_{i \geq 0} R^i$, where $R^0 = Id(W) = \{(x,x) \mid x \in W\}$, $R^{i+1} = R^i \circ R$. $\sim_R$ denotes the equivalence relation $R^* \cap R^{*-1}$. A *cluster* in a frame (W, R) is an equivalence class modulo $\sim_R$; for a cluster C, the frame $(W, R)\lceil C$ is also called a cluster.

For clusters C, D, put $C \leq_R D$ iff xR^*y for some $x \in C, y \in D$. The frame $(W/{\sim_R}, \leq_R)$ is a poset; it is called the *skeleton of* F.

A poset F is of *finite height* $\leq n$ (in symbols, $ht(\mathrm{F}) \leq n$) if every of its chains contains at most n elements. F is *of height* n (in symbols, $ht(\mathrm{F}) = n$) if $ht(\mathrm{F}) \leq n$ and $ht(\mathrm{F}) \not\leq n-1$.

More generally, the *height of a frame* F is the height of its skeleton; it is also denoted by $ht(\mathrm{F})$. A class $\mathcal{F}$ of frames is of *(uniformly) finite height* if there exists $h \in \mathbb{N}$ such that for every $\mathrm{F} \in \mathcal{F}$ we have $ht(\mathrm{F}) \leq h$.

For any transitive F, $\mathrm{F} \models B_h \iff ht(\mathrm{F}) \leq h$, where

$$B_1 = p_1 \to \Box\Diamond p_1, \quad B_{i+1} = p_{i+1} \to \Box(\Diamond p_{i+1} \vee B_i)$$

(see e.g. [6, Proposition 3.44]).[3]

Theorem 3.1 *[15,13] A logic* $\mathrm{L} \supseteq \mathbf{K4}$ *is locally tabular iff it contains* B_h *for some* $h \geq 0$.

Let $R^{\leq m} = \bigcup_{0 \leq i \leq m} R^i$. A binary relation R is called *m-transitive* if $R^{\leq m} = R^*$, or equivalently, $R^{m+1} \subseteq R^{\leq m}$. R is called *pretransitive* if it is m-transitive for some $m \geq 0$.

Put

$$\Diamond^0\varphi = \varphi, \ \Diamond^{i+1}\varphi = \Diamond\Diamond^i\varphi, \ \Diamond^{\leq m}\varphi = \bigvee_{i=0}^{m} \Diamond^i\varphi, \ \Box^{\leq m}\varphi = \neg\Diamond^{\leq m}\neg\varphi.$$

[3] Is another terminology, a poset of height n is called a poset of depth $(n-1)$.

Proposition 3.2 *R is m-transitive iff $(W,R) \vDash \Diamond^{m+1}p \to \Diamond^{\leq m}p$.*

$\varphi^{[m]}$ denotes the formula obtained from φ by replacing $\Diamond$ with $\Diamond^{\leq m}$ and $\Box$ with $\Box^{\leq m}$. In particular,

$$B_1^{[m]} = p_1 \to \Box^{\leq m}\Diamond^{\leq m}p_1, \quad B_{i+1}^{[m]} = p_{i+1} \to \Box^{\leq m}(\Diamond^{\leq m}p_{i+1} \vee B_i^{[m]}).$$

Since in the case of m-transitive relation $\Diamond^{\leq m}$ corresponds to the master modality, we have

Proposition 3.3 *[11] For an m-transitive frame* F,

$$\mathrm{F} \vDash B_h^{[m]} \iff ht(\mathrm{F}) \leq h.$$

A logic L is called *m-transitive* if $\mathrm{L} \vdash \Diamond^{m+1}p \to \Diamond^{\leq m}p$. In this case, it is easy to see that the formula $\Diamond^{\leq m}\varphi$ is equivalent to every $\Diamond^{\leq n}\varphi$ for $n \geq m$. L is *pretransitive* if it is m-transitive for some $m \geq 0$. Note that every m-transitive logic is n-transitive for $n > m$.

Definition 3.4 An m-transitive logic L is of *finite height* $\leq h$ if $\mathrm{L} \vdash B_h^{[m]}$.

If L is pretransitive, there exists the least m such that L is m-transitive; then put $\mathrm{L}[h] = \mathrm{L} + B_h^{[m]}$. Note that $\mathrm{L}[h] \vdash B_h^{[n]}$ for $n > m$.

Theorem 3.5 *[11] Let* L *be a pretransitive logic. Then*

(i) $\mathrm{L}[1] \supseteq \mathrm{L}[2] \supseteq \mathrm{L}[3] \supseteq \ldots \supseteq \mathrm{L}$.

(ii) *If* L *is consistent, then* L[1] *(and, consequently, every* L[h]*) is consistent.*

(iii) *If* L *is canonical, then every* L[h] *is canonical.*

Theorem 3.6 *For an m-transitive logic* L*, the set $\{\varphi \mid \mathrm{L} \vdash \varphi^{[m]}\}$ is a logic containing* **S4**.

This is an easy consequence of Lemma 1.3.45 from [8].

Theorem 3.7 *Every locally tabular logic is a pretransitive logic of finite height.*

Proof. If L is locally tabular, $W_{\mathrm{L}\lceil 1}$ is finite. It follows that $R_{\mathrm{L}\lceil 1}^{m+1} \subseteq R_{\mathrm{L}\lceil 1}^{\leq m}$ for some m. So $\Diamond^{m+1}p \to \Diamond^{\leq m}p \in \mathrm{L}\lceil 1 \subset \mathrm{L}$, hence, L is m-transitive.

Then ${}^{[m]}\mathrm{L} = \{\varphi \mid \varphi^{[m]} \in \mathrm{L}\}$ is a logic containing **S4** (Theorem 3.6). It follows that ${}^{[m]}\mathrm{L}$ is locally tabular. So by Theorem 3.1 ${}^{[m]}\mathrm{L} \vdash B_h$ for some $h > 0$; hence $\mathrm{L} \vdash B_h^{[m]}$. □

4 Ripe frames

Now let us show that local tabularity of a Kripke complete logic is equivalent to a quite simple semantic condition on its frames.

Definition 4.1 Let $\mathrm{F} = (W,R)$ be a frame. A partition $\mathcal{A}$ of W *is R-tuned* if for any $U, V \in \mathcal{A}$

$$\exists u \in U\ \exists v \in V\ uRv \ \Rightarrow\ \forall u \in U\ \exists v \in V\ uRv. \quad (1)$$

A frame F is called *f-partitionable* for a function $f : \mathbb{N} \to \mathbb{N}$ if for any $n \in \mathbb{N}$, for any finite partition $\mathcal{A}$ of W with $|\mathcal{A}| \leq n$ there is an R-tuned refinement $\mathcal{B}$ such that $|\mathcal{B}| \leq f(n)$.

A class of frames $\mathcal{F}$ is called *ripe* if there exists $f : \mathbb{N} \to \mathbb{N}$ such that every $\mathrm{F} \in \mathcal{F}$ is f-partitionable.

Remark 4.2 If F is *f-partitionable*, then it is f'-partitionable for some monotonic f': put $f'(n) = \max\{f(n_0) \mid n_0 \leq n\}$. On the other hand, if f is monotonic, then F is f-partitionable iff for every finite partition $\mathcal{A}$ of W there exists an R-tuned refinement $\mathcal{B}$ of $\mathcal{A}$ such that $|\mathcal{B}| \leq f(|\mathcal{A}|)$.

Theorem 4.3 *For a class $\mathcal{F}$ of Kripke frames, $Log(\mathcal{F})$ is locally tabular iff $\mathcal{F}$ is ripe.*

This theorem can be proved by algebraic methods from [14],[3]. Our proof is based on the subsequent simple propositions and (more technical) Lemma 4.8 giving an upper bound for f.

Proposition 4.4 *[16] Let $\mathrm{F} = (W, R)$ be a frame, $\mathcal{A}$ an R-tuned partition of* F. *Then there is a p-morphism from* F *onto $(\mathcal{A}, R_{\mathcal{A}})$, where*

$$UR_{\mathcal{A}}V \iff \exists u \in U\ \exists v \in V\ uRv.$$

Proof. The required p-morphism is $x \mapsto V$ for $x \in V \in \mathcal{A}$. □

Proposition 4.5 *Let $\mathrm{M} = (W, R, \theta)$ be an n-weak model, $\mathcal{A}$ an R-tuned partition of* M. *If $\mathcal{A}$ respects $\{p_1, \dots, p_n\}$ then $\mathcal{A}$ refines $\mathcal{A}(\mathrm{M})$.*

Proof. By induction on the length of an n-formula φ, let us show that $x \sim_{\mathcal{A}} y$ implies $\mathrm{M}, x \vDash \varphi \iff \mathrm{M}, y \vDash \varphi$. The base and the Boolean cases are trivial. Suppose $\mathrm{M}, x \vDash \Diamond\varphi$, so $\mathrm{M}, z \vDash \varphi$ and xRz for some z. Since $x \sim_{\mathcal{A}} y$, we have yRz' and $z' \sim_{\mathcal{A}} z$ for some z'. By the induction hypothesis, $\mathrm{M}, z' \vDash \varphi$, so $\mathrm{M}, y \vDash \Diamond\varphi$. □

Proposition 4.6 *Let $\mathrm{M} = (W, R, \theta)$ be an n-weak model. If $\mathcal{A}(\mathrm{M})$ is finite, then $\mathcal{A}(\mathrm{M})$ is R-tuned.*

Proof. Suppose $U, V \in \mathcal{A}(\mathrm{M})$, $u, u' \in U$, $v \in V$, uRv. By Proposition 2.4 we have $\mathrm{M}, v' \vDash \varphi_V \iff v' \in V$. Then $\mathrm{M}, u \vDash \Diamond\varphi_V$, so $\mathrm{M}, u' \vDash \Diamond\varphi_V$, thus $u'Rv'$ for some $v' \in V$. □

Proposition 4.7 *Suppose $f : \mathbb{N} \to \mathbb{N}$,* M *is an n-weak model over an f-partitionable frame. Then $\mathcal{A}(\mathrm{M})$ is finite and $|\mathcal{A}(\mathrm{M})| \leq f(2^n)$.*

Proof. Let $\mathcal{A}$ be the partition of M induced by $\{p_1, \dots, p_n\}$. Clearly, $|\mathcal{A}| \leq 2^n$. There exists a refinement $\mathcal{B}$ of $\mathcal{A}$ such that $\mathcal{B}$ is R-tuned (where R is the accessability relation in M) and $|\mathcal{B}| \leq f(2^n)$. By Proposition 4.5, $\mathcal{B}$ refines $\mathcal{A}(\mathrm{M})$. □

Lemma 4.8 *Suppose $f : \mathbb{N} \to \mathbb{N}$ is monotonic, $(\mathrm{F}_i)_{i \in I}$ is a non-empty family of f-partitionable frames. Then the disjoint sum $\sum\limits_{i \in I} \mathrm{F}_i$ is g-partitionable for the function $g(n) = 2^{f(n)^2} \cdot n^{f(n)} \cdot f(n)^2$.*

Proof. Let $\mathrm{F}_i = (W_i, R_i)$, $i \in I$. Let $\sum_{i \in I} \mathrm{F}_i = (W, R)$, i.e.,

$$W = \{(w,i) \mid i \in I,\ w \in W_i\},\quad (w,i)R(v,j) \iff i = j \& wR_iv.$$

Fix $n \in \mathbb{N}$. Suppose $\mathcal{A}$ is a partition of W, $|\mathcal{A}| = n$, $\mathcal{A} = \{A_1, \ldots, A_n\}$.

For $V \subseteq W$, $i \in I$, put $pr_i(V) = \{w \mid (w,i) \in V\}$. For $i \in I$, put $\mathcal{A}_i = \{pr_i(A) \mid A \in \mathcal{A},\ pr_i(A) \neq \varnothing\}$; so $\mathcal{A}_i$ is a partition of W_i and $|\mathcal{A}_i| \leq n$. For every $i \in I$ there exists an R_i-tuned partition $\mathcal{B}_i$ of W_i such that $\mathcal{B}_i$ refines $\mathcal{A}_i$, and $|\mathcal{B}_i| \leq f(n)$.

Consider the signature $\Omega_n = (S^{(2)}, P_1^{(1)}, \ldots, P_n^{(1)}, c)$, where $S^{(2)}$ is a binary relation symbol, $P_1^{(1)}, \ldots, P_n^{(1)}$ are unary relation symbols, c is a constant.

For every $(w,i) \in W$ we define an Ω_n-structure $\mathrm{S}(w,i)$ as follows.

For $i \in I$, let S_i be the binary relation on $\mathcal{B}_i$ such that

$$US_iV \iff \exists u \in U\ \exists v \in V\ uR_iv.$$

For $i \in I$, $1 \leq l \leq n$ let

$$P_{il} = \{B \in \mathcal{B}_i \mid B \subseteq pr_i(A_l)\}.$$

For $w \in W_i$, $[w]_i$ denotes the element of $\mathcal{B}_i$ containing w. For $(w,i) \in W$ put

$$\mathrm{S}(w,i) = (\mathcal{B}_i; S_i, P_{i1}, \ldots, P_{in}, [w]_i).$$

For $(w,i), (u,j) \in W$ put $(w,i) \sim (v,j)$ iff the structures $\mathrm{S}(w,i)$ and $\mathrm{S}(v,j)$ are isomorphic. Clearly, $\sim$ is an equivalence on W.

We claim that $W/{\sim}$ is the required partition of F.

Note that for any $(w,i) \in W$, for any l,

$$(w,i) \in A_l \iff [w]_i \in P_{il}. \tag{2}$$

Indeed, $(w,i) \in A_l$ iff $w \in pr_i(A_l)$ iff $[w]_i \subseteq pr_i(A_l)$ (since $\mathcal{B}_i$ is a refinement of $\mathcal{A}_i$) iff $[w]_i \in P_{il}$.

Let us show that $W/{\sim}$ is a refinement of $\mathcal{A}$. Suppose $(w,i) \sim (v,j)$ and $(w,i) \in A_l$ for some l. We have to check that $(w,i) \sim_{\mathcal{A}} (v,j)$, i.e., $(v,j) \in A_l$. Since $(w,i) \in A_l$, by (2) we have $[w]_i \in P_{il}$. By the definition of $\sim$, there exists an isomorphism $h : \mathrm{S}(w,i) \to \mathrm{S}(v,j)$; then $h([w]_i) \in P_{jl}$ and $h([w]_i) = [v]_j$. It follows that $[v]_j \in P_{jl}$. By (2), $(v,j) \in A_l$.

Let us show that $W/{\sim}$ is R-tuned. To this end, suppose $(w,i) \sim (v,j)$ and $(w,i)R(w',i')$. So $i' = i$ and wR_iw'. Then we have $[w]_iS_i[w']_i$. Now, there exists an isomorphism $h : \mathrm{S}(w,i) \to \mathrm{S}(v,j)$. Then $h([w]_i)S_jh([w']_i)$ and $h([w]_i) = [v]_j$. So $[v]_jS_jh([w']_i)$. Since $\mathcal{B}_j$ is R_j-tuned, we have vR_jv' for some $v' \in h([w']_i)$. Hence $(v,j)R(v',j)$. h is also an isomorphism between $\mathrm{S}(w',i)$ and $\mathrm{S}(v',j)$, since $[v']_j = h([w']_i)$. Thus $(w',i) \sim (v',j)$.

Finally, let us check that $|W/{\sim}| \leq 2^{f(n)^2} \cdot n^{f(n)} \cdot f(n)^2$. Indeed, up to isomorphism, every $\mathrm{S}(w,i)$ is a structure with the carrier $\{1, \ldots, k\}$ for some $k > 0$.

In this case there are 2^{k^2} binary relations. An interpretation of $P_1^{(1)}, \ldots, P_n^{(1)}$ is a sequence of pairwise disjoint sets covering the carrier, so it corresponds to a function $\{1, \ldots, k\} \to \{1, \ldots, n\}$; thus there are n^k possible interpretations of these predicates. There are also k interpretations of the constant. Since $|\mathcal{B}_i| \leq f(n)$ for every i, we have

$$|W/{\sim}| \leq \sum_{k=1}^{f(n)} \left(2^{k^2} \cdot n^k \cdot k\right) \leq f(n) \cdot 2^{f(n)^2} \cdot n^{f(n)} \cdot f(n).$$

□

Proof of Theorem 4.3 Suppose $\mathrm{L} = Log(\mathcal{F})$ is locally tabular. Put $f(n) = |W_{\mathrm{L}\lceil n}|$. Suppose $\mathrm{F} = (W, R) \in \mathcal{F}$, $\mathcal{A} = \{V_1, \ldots, V_n\}$ is a finite partition of F. Put $\theta(p_i) = V_i$, $1 \leq i \leq n$ and consider the model $\mathrm{M} = (\mathrm{F}, \theta)$. The partition $\mathcal{A}(\mathrm{M})$ refines $\mathcal{A}$. Since $\mathrm{M} \vDash \mathrm{L}\lceil n$, by Proposition 2.3 we have $|\mathcal{A}(\mathrm{M})| \leq f(n)$. By Proposition 4.6, $\mathcal{A}(\mathrm{M})$ is R-tuned.

Suppose there is $f : \mathbb{N} \to \mathbb{N}$ such that every frame from $\mathcal{F}$ is f-partitionable. Since $\mathrm{L} = Log(\mathcal{F})$, then there exists a model $\mathrm{M} = (\mathrm{G}, \eta)$ such that G is a disjoint sum of some frames from $\mathcal{F}$, and $\mathrm{M} \vDash \varphi$ iff $\varphi \in \mathrm{L}$. By Lemma 4.8, G is g-partitionable for $g(n) = 2^{f(n)^2} \cdot n^{f(n)} \cdot f(n)^2$. By Propositions 2.5 and 4.7, $|W_{\mathrm{L}\lceil n}| \leq g(2^n)$ for every n. □

Since every locally tabular logic is Kripke complete, in view of Theorem 4.3 we readily obtain

Corollary 4.9 *The following conditions are equivalent:*

- *a logic* L *is locally tabular;*
- L *is the logic of a ripe class of frames;*
- L *is Kripke complete and the class of all its frames is ripe.*

The following fact from [11] gives many examples of ripe classes: if $\mathcal{F}$ is of uniformly finite height, and there exists a common finite upper bound for cardinalities of clusters in frames from $\mathcal{F}$, then $\mathcal{F}$ is ripe. By Theorem 4.3, logics of these classes are locally tabular. In the next section we will prove a stronger result: instead of boundedness of clusters, we only need them to be ripe.

5 Ripe cluster property

Definition 5.1 For a class $\mathcal{F}$ of frames let $cl\mathcal{F}$ be the class of all clusters occurring in frames from $\mathcal{F}$.

$\mathcal{F}$ has the *ripe cluster property* if $cl\mathcal{F}$ is ripe. A logic L has the *ripe cluster property* if the class of all its frames has.

Theorem 5.2 *A logic $Log(\mathcal{F})$ is locally tabular iff $\mathcal{F}$ is of uniformly finite height and has the ripe cluster property.*

Before proving this theorem let us discuss some examples.

Example 5.3 **S4** has the ripe cluster property.

Indeed, every cluster in a preorder is a frame with the universal relation $(C, C \times C)$. Every partition of C is $(C \times C)$-tuned. So every cluster in a preorder is f-partitionable, where f is the identity function on $\mathbb{N}$.

Example 5.4 $\mathbf{wK4} = \mathbf{K} + \Diamond\Diamond p \to \Diamond p \vee p$ has the ripe cluster property.

To show this, consider a cluster (C, R) in a **wK4**-frame. Then $R \cup R^0 = C \times C$. Every partition $\mathcal{A}$ of C is R-tuned, due to the following proposition.

Proposition 5.5 *If $R \cup R^0 = W \times W$, then every partition $\mathcal{A}$ of W is R-tuned.*

Proof. Let $x, y \in U$, $z \in V$, xRz for some $U, V \in \mathcal{A}$. Suppose $z \neq y$; then yRz. Suppose $z = y$; in this case $U = V$, so $x \in V$; since R is symmetric, we have yRx. □

Example 5.6 $\mathbf{K} + \Diamond\Diamond\Diamond p \to \Diamond p$ has the ripe cluster property.

Clusters in frames validating $\Diamond\Diamond\Diamond p \to \Diamond p$ are frames (C, R) such that R^* is universal and $R^3 \subseteq R$. Let us show that these clusters are f-partitionable for $f(n) = 2n$. Suppose $\mathcal{A}$ is a finite partition of C. Put $x \sim y$ iff there exists $l \geq 0$ such that $xR^{2l}y$. It is not difficult to see that $\sim$ is an equivalence relation on C and $|C/{\sim}| \leq 2$ (more details will be given in Lemma 6.4). Let $\mathcal{B} = C/({\sim} \cap {\sim_{\mathcal{A}}})$. Then $|\mathcal{B}| \leq 2|\mathcal{A}|$. To show that $\mathcal{B}$ is R-tuned, suppose $x, y \in U$, $z \in V$, xRz for some $U, V \in \mathcal{B}$. Then $yR^{2l}x$ for some $l \geq 0$, so $yR^{2l+1}z$. Since $R^3 \subseteq R$, we have $R^{2l+1} \subseteq R$, thus yRz.

Proposition 5.7 *If $\mathcal{B}$ is R-tuned, then $\mathcal{B}$ is R^l-tuned for any $l \geq 0$.*

Proof. By an easy indiction on l. □

Lemma 5.8 *Suppose* $\mathrm{F} = (W, R)$ *is a frame,* $\{X_1, X_2\}$ *is a partition of* W, $\mathrm{F}{\upharpoonright}X_1$ *is a generated subframe of* F, $\mathrm{F}{\upharpoonright}X_1$ *is* f_1*-partitionable,* $\mathrm{F}{\upharpoonright}X_2$ *is* f_2*-partitionable for some monotonic* $f_1, f_2 : \mathbb{N} \to \mathbb{N}$. *Then* F *is* g*-partitionable for*

$$g(n) = f_2\left(n \cdot 2^{f_1(n)}\right) + f_1(n).$$

Proof. Let $\mathcal{A}$ be a partition of W, $|\mathcal{A}| = n \in \mathbb{N}$. Put $\mathcal{A}_i = \mathcal{A}{\upharpoonright}X_i$, $i = 1, 2$.

There exists a partition $\mathcal{B}_1$ of X_1 such that $\mathcal{B}_1$ is $(R{\upharpoonright}X_1)$-tuned, $\mathcal{B}_1$ is a refinement of $\mathcal{A}_1$, and $|\mathcal{B}_1| \leq f_1(n)$. Let ${\sim_1} = {\sim_{\mathcal{B}_1}}$.

For $x \in X_2$ put $\mathcal{V}(x) = \{B \in \mathcal{B}_1 \mid \exists y \in B\ xRy\}$. For $x, y \in X_2$ put $x \equiv y$ iff x and y belong to the same element of $\mathcal{A}_2$ and $\mathcal{V}(x) = \mathcal{V}(y)$. Clearly, $\equiv$ is an equivalence on X_2. Since $\mathcal{V}(x) \subseteq \mathcal{B}_1$ for every $x \in X_2$, we have

$$|X_2/{\equiv}| \leq |\mathcal{A}_2| \cdot 2^{|\mathcal{B}_1|} \leq n \cdot 2^{f_1(n)}.$$

Then there exists a partition $\mathcal{B}_2$ of X_2 such that $\mathcal{B}_2$ is $(R{\upharpoonright}X_2)$-tuned, $\mathcal{B}_2$ is a refinement of $X_2/{\equiv}$, and

$$|\mathcal{B}_2| \leq f_2\left(n \cdot 2^{f_1(n)}\right).$$

Let ${\sim_2} = {\sim_{\mathcal{B}_2}}$.

Let $\sim$ be the union of $\sim_1$ and $\sim_2$; clearly, $\sim$ is an equivalence on W.

We claim that $\mathcal{C} = W/{\sim}$ is the required partition of W. By the construction, $\mathcal{C}$ refines $\mathcal{A}$ and

$$|\mathcal{C}| \leq f_2\left(n \cdot 2^{f_1(n)}\right) + f_1(n).$$

Let us check that $\mathcal{C}$ is R-tuned. Suppose $x \sim y$ and xRx' for some $x, y, x' \in W$.

First assume that $x \in X_1$. Then $x \sim_1 y$. Since $R \cap (X_1 \times X_2) = \varnothing$, we have $x' \in X_1$. Since $\mathcal{B}_1$ is $(R{\restriction}X_1)$-tuned, for some $y' \in X_1$ we have yRy' and $x' \sim_1 y'$; the latter implies $x' \sim y'$.

Assume now that $x \in X_2$ and $x' \in X_1$. Let B be the element of $\mathcal{B}_1$ containing x'. Then $B \in \mathcal{V}(x)$. Since $x \equiv y$, $\mathcal{V}(x) = \mathcal{V}(y)$. So $B \in \mathcal{V}(y)$. It follows that for some y' we have yRy' and $y' \in B$; the latter implies $x' \sim y'$.

Finally, assume that $x, x' \in X_2$. Since $x \sim_2 y$ and $\mathcal{B}_2$ is $(R_2{\restriction}X_2)$-tuned, for some y' we have yRy' and $x' \sim_2 y'$; the latter implies $x' \sim y'$. □

Lemma 5.9 *Suppose $f : \mathbb{N} \to \mathbb{N}$, (W, R) is an f-partitionable frame, $V \subset W$, $V \neq \varnothing$. Then $(V, R{\restriction}V)$ is g-partitionable for $g(n) = f(n+1) - 1$.*

Proof. Let $\mathcal{A}$ be a partition of V, $|\mathcal{A}| \leq n \in \mathbb{N}$. Put $\mathcal{A}' = \mathcal{A} \cup \{W \setminus V\}$. So $\mathcal{A}'$ is a partition of W and $|\mathcal{A}'| \leq n+1$. There exists an R-tuned refinement $\mathcal{B}$ of $\mathcal{A}'$ such that $|\mathcal{B}| \leq f(n+1)$. Then $\mathcal{C} = \{B \in \mathcal{B} \mid B \subseteq V\}$ is a partition of V, $\mathcal{C}$ refines $\mathcal{A}$ and is $(R{\restriction}V)$-tuned. There exists $B_0 \in \mathcal{B}$ such that $B_0 \subseteq W \setminus V$. Then $B_0 \notin \mathcal{C}$, so $|\mathcal{C}| < |\mathcal{B}|$. □

Lemma 5.10 *For $h \in \mathbb{N}$, $f : \mathbb{N} \to \mathbb{N}$, let $\mathcal{F}_{h,f}$ be the class*

$$\{\mathrm{F} \mid ht(\mathrm{F}) \leq h \textit{ and every cluster in } \mathrm{F} \textit{ is } f\textit{-partitionable}\}$$

Then $\mathcal{F}_{h,f}$ is ripe.

Proof. By induction on h, let us show that there exists a monotonic $g_h : \mathbb{N} \to \mathbb{N}$ such that every $\mathrm{F} \in \mathcal{F}_{h,f}$ is g_h-partitionable.

If $\mathrm{F} \in \mathcal{F}_{1,f}$, then $ht(\mathrm{F}) = 1$, so F is a disjoint sum of clusters. So F is g_1-partitionable for $g_1(n) = 2^{f(n)^2} \cdot n^{f(n)} \cdot f(n)^2$ by Lemma 4.8 (without any loss of generality, we may assume that f is monotonic).

For the induction step, suppose $\mathrm{F} = (W, R) \in \mathcal{F}_{h+1,f}$.

If $ht(\mathrm{F}) \leq h$, then F is g_h-partitionable by the induction hypothesis.

Suppose $ht(\mathrm{F}) = h+1$. Let X_1 be the union of all maximal clusters in F, $X_2 = W \setminus X_1$. Then $\mathrm{F}{\restriction}X_1$ is a disjoint sum of some clusters in F; this frame is g_1-partitionable by Lemma 4.8. The height of the frame $\mathrm{F}{\restriction}X_2$ is h, so this frame is g_h-partitionable by the induction hypothesis. Since $\mathrm{F}{\restriction}X_1$ is a generated subframe of F, by Lemma 5.8, F is g-partitionable for $g(n) = g_h\left(n \cdot 2^{g_1(n)}\right) + g_1(n)$.

Since $g_h(n) \leq g(n)$, if follows that F is g-partitionable. □

Lemma 5.11 *A class $\mathcal{F}$ of Kripke frames is ripe iff $\mathcal{F}$ is of uniformly finite height and has the ripe cluster property.*

Proof. Suppose $\mathcal{F}$ is ripe. Then its logic is locally tabular by Theorem 4.3. It follows that all frames in $\mathcal{F}$ are of height $\leq h$ for some $h \in \mathbb{N}$ (Theorem 3.7). Since $\mathcal{F}$ is ripe, the class $cl\mathcal{F}$ is also ripe by Lemma 5.9.

The right-to-left direction: for some f, h, the class $\mathcal{F}$ is a subclass of the class $\mathcal{F}_{h,f}$ described in Lemma 5.10; thus $\mathcal{F}$ is a subclass of a ripe class. □

Proof of Theorem 5.2 Follows from Lemma 5.11 and Theorem 4.3. □

Corollary 5.12 *A logic* L *is locally tabular iff* L *is a Kripke complete pretransitive logic of finite height with the ripe cluster property.*

Theorem 5.13 *Suppose* L_0 *is a canonical pretransitive logic with the ripe cluster property. Then for any logic* $\mathrm{L} \supseteq \mathrm{L}_0$*:*

L *is locally tabular iff* L *is of finite height.*

Proof. If an extension L of L_0 is locally tabular, then L is of finite height by Theorem 3.7.

To show that every extension of L_0 of finite height is locally tabular, it is sufficient to show that $\mathrm{L}_0[h]$ is locally tabular for every $h > 0$ (Proposition 2.2). By Theorem 3.5, $\mathrm{L}_0[h]$ is canonical, so it is Kripke complete. By Theorem 5.2, $\mathrm{L}_0[h]$ is locally tabular. □

6 Criterion for logics containing $\Diamond^{m+1}p \to \Diamond p \vee p$

Proposition 6.1 *Let* $m \geq 1$.

(i) $\mathbf{K} + \Diamond^{m+1}p \to \Diamond p \vee p$ *is* m*-transitive.*

(ii) $\Diamond^{m+1}p \to \Diamond p \vee p$ *is a Sahlqvist formula, the logic* $\mathbf{K} + \Diamond^{m+1}p \to \Diamond p \vee p$ *is canonical and Kripke complete, and*

$$(W, R) \models \Diamond^{m+1}p \to \Diamond p \vee p \text{ iff } R^{m+1} \subseteq R \cup R^0.$$

(iii) (W, R) *is a cluster in a frame validating* $\Diamond^{m+1}p \to \Diamond p \vee p$ *iff* $R^{\leq m} = W \times W$ *and* $R^{m+1} \subseteq R \cup R^0$.

Theorem 6.2 *Let* $m \geq 1$, $\Diamond^{m+1}p \to \Diamond p \vee p \in \mathrm{L}$. *Then* L *is locally tabular iff* $\mathrm{L} \vdash B_h^{[m]}$ *for some* $h > 0$.

We will prove this theorem later on.

Consider a frame (W, R). For $d > 0$ put $x \preceq_d y$ iff $wR^{ld}y$ for some $l \geq 0$. Clearly, $\preceq_d$ is a preorder.

Proposition 6.3 *Let* $R^* = W \times W$. *Then for every* $k > 0$ *there exists* d *such that* $0 < d \leq k$, $\preceq_d$ *is an equivalence on* W, $|W/{\preceq_d}| \leq d$, *and* $\preceq_d \subseteq \preceq_k$.

For the proof cf. [11]. The idea of is that d is a common divisor of k and the lengths of all cycles in (W, R).

Lemma 6.4 *Let* $m \geq 1$, $R^* = W \times W$, $R^{m+1} \subseteq R$. *Then* (W, R) *is* f*-partitionable for* $f(n) = mn$.

Proof. Consider a finite partition $\mathcal{A}$ of W. Let $\preceq_d$ be the equivalence described in Proposition 6.3 for $k = m$. Put $\mathcal{B} = W/(\preceq_d \cap \sim_{\mathcal{A}})$. Since $|W/\preceq_d| \leq d \leq m$, we have $|\mathcal{B}| \leq m|\mathcal{A}|$.

Suppose $x \sim_{\mathcal{B}} y$, xRz. We have $y \preceq_d x$, and then $y \preceq_m x$ (Proposition 6.3), so for some l we have $yR^{ml}x$, and thus $yR^{ml+1}z$. Since $R^{m+1} \subseteq R$, we have $R^{ml+1} \subseteq R$ (by an easy induction on l), thus yRz. It follows that $\mathcal{B}$ is R-tuned. □

Lemma 6.5 *Let $m \geq 1$, $R^* = W \times W$, and suppose $R^{m+1} \subseteq R \cup R^0$ and $R^{m+1} \not\subseteq R$. Then:*

- *there exists a cyclic path $x_0Rx_1R \dots Rx_mRx_0$;*
- *if $W \neq \{x_0, \dots, x_m\}$, then $R \cup R^0 = W \times W$.*

Proof. There exist x_0, y_0 such that $x_0R^{m+1}y_0$ and not x_0Ry_0. Since $R^{m+1} \subseteq R \cup R^0$, we have $x_0 = y_0$. So $x_0R^{m+1}x_0$, i.e., there exists a cyclic path $x_0Rx_1R \dots Rx_mRx_0$. Put $U = \{x_0, x_1, \dots, x_m\}$.

If $m = 1$, then (W, R) is a **wK4**-cluster, so $R \cup R^0 = C \times C$ (see Example 5.4).

Suppose $m \geq 2$.

Let $\overrightarrow{x_i}$ and $\overleftarrow{x_i}$ be the successor and the predecessor of x_i in our cycle, i.e.: $\overrightarrow{x_i} = x_{i+1}$ for $i < m$, $\overrightarrow{x_m} = x_0$; $\overleftarrow{x_i} = x_{i-1}$ for $i > 0$, $\overleftarrow{x_0} = x_m$.

Suppose $z \notin U$. Then for $0 \leq i \leq m$

- zRx_i implies $zR\overleftarrow{x_i}$;
- x_iRz implies $\overrightarrow{x_i}Rz$.

Indeed, $x_iR^m\overleftarrow{x_i}$, so zRx_i implies $zR^{m+1}\overleftarrow{x_i}$; since $z \neq \overleftarrow{x_i}$ we have $zR\overleftarrow{x_i}$. Similarly, $\overrightarrow{x_i}R^mx_i$, so x_iRz implies $\overrightarrow{x_i}R^{m+1}z$; since $\overrightarrow{x_i} \neq z$ we have $\overrightarrow{x_i}Rz$. It follows that

(1) $\exists x \in U\ zRx \iff \forall x \in U\ zRx$,

(2) $\exists x \in U\ xRz \iff \forall x \in U\ xRz$.

Let us check that

(3) $\forall z \notin U\ \forall x \in U\ (zRx\ \&\ xRz)$.

Indeed, since $R^{\leq m}$ is universal on W and $z \notin U$, for some l, $0 < l \leq m$, we have $zR^lx_0R^{m+1-l}x_{m+1-l}$, so $zR^{m+1}x_{m+1-l}$, and since $z \neq x_{m+1-l}$ we obtain zRx_{m+1-l}. By (1), it follows that $\forall x \in U\ zRx$.

Similarly, we have x_mR^lz for some l, $0 < l \leq m$, so $x_{l-1}R^{m+1}z$, and since $x_{l-1} \neq z$ we obtain $x_{l-1}Rz$. By (2), it follows that $\forall x \in U\ xRz$.

Now let us show that

(4) $\forall y \forall y'(y \neq y' \Rightarrow yR^{m+1}y')$.

Consider the following cases.

(a) $y, y' \notin U$. By (3) we have $yRx_0R^{m-1}x_{m-1}Ry'$, so $yR^{m+1}y'$.

(b) $y \notin U$, $y' \in U$. Then by (3) $yR\overrightarrow{y'}R^my'$, and again $yR^{m+1}y'$.

(c) $y \in U$, $y' \notin U$. Then by (3) $yR^m\overleftarrow{y'}Ry'$, and thus $yR^{m+1}y'$.

(d) $y, y' \in U$. By (3), for $z \notin U$ we have $yR^{m-1}\overleftarrow{y}RzRy'$, so $yR^{m+1}y'$ again.

If $y \neq y'$ and $yR^{m+1}y'$, then we have yRy'. Thus $R \cup R^0 = W \times W$. □

Lemma 6.6 $\mathbf{K} + \Diamond^{m+1}p \to \Diamond p \vee p$ *has the ripe cluster property for every* $m \geq 1$.

Proof. Let(C, R) be a cluster in a frame validating $\Diamond^{m+1}p \to \Diamond p \vee p$, $\mathcal{A}$ a finite partition of C. We have $R^{m+1} \subseteq R \cup R^0$.

If $R^{m+1} \subseteq R$, then by Lemma 6.4 there exists an R-tuned refinement $\mathcal{B}$ of $\mathcal{A}$ such that $|\mathcal{B}| \leq m|\mathcal{A}|$.

Otherwise, by Lemma 6.5, we have two cases: $|C| \leq m+1$, or $R \cup R^0 = C \times C$. In the first case, put $\mathcal{B} = \{\{x\} \mid x \in C\}$; trivially, $\mathcal{B}$ is R-tuned and $|\mathcal{B}| \leq m+1$. In the second case, $\mathcal{A}$ is R-tuned by Proposition 5.5.

It follows that every cluster in any frame validating $\Diamond^{m+1}p \to \Diamond p \vee p$ is f-partitionable for $f(n) = \max\{mn, m+1\}$. □

Proof of Theorem 6.2 If L is locally tabular, then $B_h^{[m]} \in \mathrm{L}$ for some $h > 0$ by Theorem 3.7.

Suppose $B_h^{[m]} \in \mathrm{L}$ for some $h > 0$. L has the ripe cluster property by Lemma 6.6. Since L is of finite height, by Theorem 5.13, L is locally tabular.□

7 Properties of partitionable frames

Theorem 3.7 implies that every f-partitionable frame is pretransitive. The next two propositions give more details.

Proposition 7.1 *If* F *is* f*-partitionable then* F *is* $(f(2)-1)$*-transitive.*

Proof. Let $\mathrm{F} = (W, R)$. Put $m = f(2) - 1$. Suppose that $xR^{m+1}y$, so there is a path $x = x_0Rx_1 \ldots Rx_{m+1} = y$.

If $x_i = x_j$ for some $i < j$, then $xR^{m+1-(j-i)}y$, so $xR^{\leq m}y$.

Suppose all the x_i are different. Consider the two-element partition $\{\{x_{m+1}\}, W \setminus \{x_{m+1}\}\}$. There exists its refinement $\mathcal{B}$ such that $\mathcal{B}$ is R-tuned and $|\mathcal{B}| \leq f(2)$. Since $\{x_{m+1}\} \in \mathcal{B}$, $\mathcal{B}$ splits $W \setminus \{x_{m+1}\}$ into $|\mathcal{B}| - 1$ parts. Since $|\mathcal{B}| - 1 \leq f(2) - 1 = m$, there are at least two $\sim_{\mathcal{B}}$-equivalent points among $x_0, \ldots, x_m$. Suppose $x_i \sim_{\mathcal{B}} x_j$ for $i < j$.

By Proposition 5.7, $\mathcal{B}$ is R^{m+1-j}-tuned, and $x_i \sim_{\mathcal{B}} x_j$, so there exists y such that $x_iR^{m+1-j}y$ and $y \sim_{\mathcal{B}} x_{m+1}$. But the element of $\mathcal{B}$ containing x_{m+1} is the singleton $\{x_{m+1}\}$, thus $y = x_{m+1}$, and therefore $xR^{\leq m}y$. □

Proposition 7.2 *If all clusters in* F *are* f*-partitionable and* $ht(\mathrm{F}) = h \in \mathbb{N}$*, then* F *is* m*-transitive for* $m = h \cdot f(2) - 1$.

Proof. Put $m_0 = f(2) - 1$. By Proposition 7.1, all clusters in F are m_0-transitive. Suppose xR^*y. Let C and D be the clusters containing x and y, respectively. In the skeleton of F, consider a maximal chain $C = C_1 \leq_R \cdots \leq_R C_l = D$ from C to D. Then for each $i < l$ we have y_iRx_{i+1} for some $y_i \in C_i$ and $x_{i+1} \in C_{i+1}$ (indeed, there are no clusters between C_i and C_{i+1}). Since C_i is m_0-transitive, we have $x_iR^{\leq m_0}y_i$; we also have y_iRx_{i+1}. Then $xR^{\leq l \cdot m_0 + l - 1}y$.

To finish the proof, note that $l \leq h$, so $l \cdot m_0 + l - 1 \leq h \cdot f(2) - 1$. □

For $m \in \mathbb{N}$, consider the following first-order property P_m:

$$\forall x_0, \ldots, x_{m+1} \left(x_0 R x_1 \ldots R x_{m+1} \to \bigvee_{i<j} x_i = x_j \vee \bigvee_{i+1<j} x_i R x_j \right).$$

Note that this property implies m-transitivity. The next theorem gives another necessary condition for local tabularity.

Theorem 7.3 *If $\mathcal{F}$ is a ripe class, then $\mathcal{F}$ satisfies P_m for some m.*

Thus theorem follows from

Proposition 7.4 *If (W, R) is f-partitionable, then $\mathrm{P}_{f(3)-1}$ holds in (W, R).*

Proof. Put $m = f(3) - 2$. Let $X = \{x_0, \ldots, x_{m+1}\}$ and suppose $x_0 R x_1 \ldots R x_{m+1}$.

First, suppose $X = W$. Then $\mathrm{P}_{f(3)-1}$ holds in (W, R), since $|W| \leq f(3)$ in this case.

Next, suppose $X \neq W$. By Lemma 5.9, $(X, R\restriction X)$ is g-partitionable for $g(n) = f(n+1) - 1$. By Proposition 7.1, $(X, R\restriction X)$ is $(g(2) - 1)$-transitive and $g(2) - 1 = f(3) - 2 = m$. Thus for some $l \leq m$ we have $x_0 R^l x_{m+1}$. Thus there is a path $x_{i_0} R x_{i_1} R \ldots R x_{i_l}$, $i_0 = 0$, $i_l = m + 1$.

Now suppose that $|X| = m + 2$ and $i_{k+1} \leq i_k + 1$ for each $k \leq m$. In this case $i_k \leq k$ for each $k \leq l$ (by trivial induction on k), so $i_l \neq m + 1$, which is a contradiction.

It follows that $|X| < m + 2$ or $i_{k+1} > i_k + 1$ for some $k \leq l$.

Thus we have $\mathrm{P}_{f(3)-2}$ in (W, R).

Therefore, $\mathrm{P}_{f(3)-1}$ holds in (W, R) anyway. □

8 Conclusion

Let us summarize our main results.

- The necessary syntactic condition for local tabularity: pretransitivity and finite height.
- Two semantic criteria of local tabularity: (i) a logic is locally tabular iff it is complete with respect to a ripe class of frames; (ii) a logic is locally tabular iff it is complete with respect to a class of frames of uniformly finite height with the ripe cluster property.
- A syntactic criterion of local tabularity above $\mathbf{K} + \Diamond^{m+1} p \to \Diamond p \vee p$.

A syntactic criterion of local tabularity above $\mathbf{K}$ remains an open problem. The necessary syntactic condition from Theorem 3.7 is not sufficient: for example, the logic $\mathbf{K}_3^2 = \mathbf{K} + \Diamond\Diamond\Diamond p \to \Diamond\Diamond p$ is 2-transitive, but even $\mathbf{K}_3^2[1]$ has Kripke incomplete extensions [10], so it is not locally tabular. However, there is a hope for a criterion. In fact, every ripe class has the property P_m for some m (Theorem 7.3). So one may ask if the following holds:

Problem 8.1 *Suppose that $\mathcal{F}$ is a class of clusters satisfying P_m for some m. Is $\mathcal{F}$ ripe?*

In view of Theorem 7.3, the positive solution of the above problem would provide us with a syntactic criterion of local tabularity over **K**.

Another important field is local tabularity of polymodal logics. The notion of a ripe class and the ripe cluster property can be reformulated in a straightforward way, and the analogues of Theorems 3.7, 4.3 and 5.2 can be proved for logics with finitely many modalities.

The authors would like to thank the anonymous reviewers for their comments.

References

[1] Adian, S. I., *The burnside problem on periodic groups and related questions*, Proceedings of the Steklov Institute of Mathematics **272** (2011), pp. 2–12.
URL `http://dx.doi.org/10.1134/S0081543811030023`

[2] Bezhanishvili, G., *Varieties of monadic Heyting algebras. Part I*, Studia Logica **61** (1998), pp. 367–402.

[3] Bezhanishvili, G., *Locally finite varieties*, algebra universalis **46** (2001), pp. 531–548.

[4] Bezhanishvili, G. and R. Grigolia, *Locally tabular extensions of MIPC*, in: *Advances in Modal Logic*, 1998, pp. 101–120.

[5] Bezhanishvili, N., *Varieties of two-dimensional cylindric algebras. Part I: Diagonal-free case*, Algebra Universalis **48** (2002), pp. 11–42.

[6] Chagrov, A. and M. Zakharyaschev, "Modal Logic," Oxford Logic Guides **35**, Oxford University Press, 1997.

[7] Gabbay, D. and V. Shehtman, *Products of modal logics, part 1*, Logic Journal of the IGPL **6** (1998), pp. 73–146.

[8] Gabbay, D., V. Shehtman and D. Skvortsov, "Quantification in Nonclassical Logic," Elsevier, 2009.

[9] Komori, Y., *The finite model property of the intermediate propositional logics on finite slices*, Journal of the Faculty of Science, the University of Tokyo **22** (1975), pp. 117–120.

[10] Kostrzycka, Z., *On non-compact logics in NEXT(KTB)*, Mathematical Logic Quarterly **54** (2008), pp. 617–624.

[11] Kudinov, A. and I. Shapirovsky, *On partitioning Kripke frames of finite height*, Izvestiya: Mathematics (2016), submitted. In Russian.
URL `http://arxiv.org/pdf/1511.09092v1`

[12] Kuznetsov, A., *Some properties of the structure of varieties of pseudo-boolean algebras*, in: *Proceedings of the XIth USSR Algebraic Colloquium*, 1971, pp. 255–256.

[13] Maksimova, L., *Modal logics of finite slices*, Algebra and Logic **14** (1975), pp. 304–319.

[14] Malcev, A., "Algebraic systems," Springer, 1973.

[15] Segerberg, K., "An Essay in Classical Modal Logic," Filosofska Studier, vol.13, Uppsala Universitet, 1971.

[16] Segerberg, K., *Franzen's proof of Bull's theorem*, Ajatus **35** (1973), pp. 216–221.

[17] Shehtman, V. B., *Canonical filtrations and local tabularity*, in: R. Goré, B. P. Kooi and A. Kurucz, editors, *Advances in Modal Logic 10*, 2014, pp. 498–512.
URL `http://www.aiml.net/volumes/volume10/Shehtman.pdf`

The Logic of Where and While in the 13th and 14th Centuries

Sara L. Uckelman

Department of Philosophy, Durham University

Abstract

Medieval analyses of molecular propositions include many non-truthfunctional connectives in addition to the standard modern binary connectives (conjunction, disjunction, and conditional). Two types of non-truthfunctional molecular propositions considered by a number of 13th- and 14th-century authors are temporal and local propositions, which combine atomic propositions with 'while' and 'where'. Despite modern interest in the historical roots of temporal and tense logic, medieval analyses of 'while' propositions are rarely discussed in modern literature, and analyses of 'where' propositions are almost completely overlooked. In this paper we introduce 13th- and 14th-century views on temporal and local propositions, and connect the medieval theories with modern temporal and spatial counterparts.

Keywords: Jean Buridan, Lambert of Auxerre, local propositions, Roger Bacon, temporal propositions, Walter Burley, William of Ockham

1 Introduction

Modern propositional logicians are familiar with three kinds of compound propositions: Conjunctions, disjunctions, and conditionals. Thirteenth-century logicians were more liberal, admitting variously five, six, seven, or more types of compound propositions. In the middle of the 13th century, Lambert of Auxerre [1] in his *Logic* identified six types of 'hypothetical' (i.e., compound, as opposed to atomic 'categorical', i.e., subject-predicate, propositions) propositions: the three familiar ones, plus causal, local, and temporal propositions [17, ¶99]. Another mid-13th century treatise, Roger Bacon's *Art and Science of Logic*, lists these six and adds expletive propositions (those which use the connective 'however'), and other ones not explicitly classified such as "Socrates

[1] The identity of the author of this text is not known for certain. Only one manuscript identifies him further than simply *Lambertus*, and there his place of origin is given as Ligny-le-Châtel. Two candidates for who Lambert is have been advanced in the literature: Lambert of Auxerre, a Dominican and canon of the cathedral of Auxerre in the late 1230s or early 1240s, and Lambert of Lagny, a secular cleric and teacher of Theobald II who later became a Dominican and a papal penitentiary. The introduction of [17] lays out the arguments for and against both positions, and comes down in the favor of an attribution of Auxerre. We follow this status quo here with the caveat that nothing we say turns on his precise identification.

is such as Plato is", "Socrates runs as often as Plato argues", etc. [2, ¶170]). Numerous other anonymous treatises from the first half of the 13th century include similar divisions. The *Ars Burana* (AB) [9, pp. 175–213], dating around 1200 [8, pp. 42, 397], and the *Ars Emmerana* (AE) [9, pp. 143–174], from the first half of the 13th century [8, p. 43] both identify the same six types of hypothetical sentences as Lambert, and add a seventh type, the 'adjunctive' [9, pp. 158, 190]. The *Dialectica Monacensis* (DM) [9, pp. 453–638], composed between the 1160s and the first decade of the 13th century [8, pp. 410–414] gives the modern logician's three types as the primary types, but subdivides conditionals into a further four categories: temporal, local, causal, and subconditional [9, p. 484].[2]

From this we can see that the other two of the four main textbooks from the middle of the century, Peter of Spain's *Summaries of Logic* and William of Sherwood's *Introduction to Logic*, are unusual, as neither of them mention these types of sentences [6, p. 115]; [19, p. 34]. In his early 14th century *Summary of Dialectic*, Jean Buridan explains that "some texts do not provide the species 'temporal' and 'local', because they can be reduced to conjunctive propositions, for saying 'Socrates is where Plato is' amounts to the same as saying 'Socrates is somewhere and Plato is there', and in the same way, saying 'Socrates lectured when Plato disputed' is equivalent to saying 'Socrates lectured sometime and then Plato disputed'" [3, p. 60]. (Note that this 'then' must be interpreted logically, not temporally.) Despite giving this reduction from temporal or local sentences to conjunctions, Buridan himself discusses these two types separately from conjunctive propositions, as do Walter Burley (c. 1275–1344) and William of Ockham (c. 1287–1347) in their textbooks, ensuring that temporal propositions remained entrenched in logical discourse throughout the rest of the 14th century and beyond.[3]

Much attention has been devoted to medieval temporal logic since the birth of modern temporal logic in the works of Arthur Prior, whose debt to medieval logicians is explicit [22, Ch. 1, §7]. However, the focus has tended to be on the semantics of tensed categorical propositions (cf. [26,27]), as opposed to the use of temporal connectives. (Two exception are [20] and [21, ch. 1.8].) When it comes to local propositions, involving the connective 'where', the case is even worse: We know of no modern investigation of medieval theories of local propositions (perhaps because so few of the medieval authors discussed these themselves!).

Thus, our first goal in this paper is to collect and present 13th- and early 14th-century views on the logic of temporal (§2) and local propositions (§4). Our second aim is to identify and model the formal properties of the medieval doctrines, so as to compare them to modern approaches. In §3 we consider two modern approaches to the logic of 'while', and show that both of them fail to

[2] References to AB, AE, and DM are to the Latin editions; all translations are my own.

[3] For example, Richard Lavenham (fl. 1380), Paul of Venice (c. 1369–1429), and John Dorp (fl. 1499) also discuss temporal propositions [20, pp. 167, 169].

capture the medieval ideas. We introduce a new definition of 'while' that does. Next, in §5 we provide a spatial analogue of the medieval 'while', as well as look at modern approaches to logics of space and spatial reasoning, including the logic of elsewhere. We conclude in §6.

2 Temporal propositions

2.1 Temporal propositions in the 13th century

Lambert defines temporal propositions as follows [17, ¶105]:

Definition 2.1 *A* temporal *proposition is one whose parts are joined by the adverb 'while', as in 'Socrates runs while Plato argues'.*

Similar definitions are given in AE [9, pp. 158–159], AB [9, pp. 190–191], and DM [9, pp. 484–485]. All four definitions crucially include reference to the use of an adverb; as Bacon notes, local and temporal propositions differ from the other type of compound propositions because they are complex 'in virtue of a relation' rather than a connective [2, ¶170]. Bacon himself doesn't mention the presence of an adverb, but that is because he defines local and temporal propositions by ostension. His example of a temporal proposition is the sentence 'Socrates hauls [the boat] in when Plato runs' [2, ¶170], similar to the examples given in the other texts, such as AE's 'While Socrates runs, Plato moves' and AB's 'Socrates reads while Plato disputes'.

None of Bacon, AE, or DM give truth conditions for temporal propositions. AB gives the same truth conditions for both temporal and causal propositions:

Definition 2.2 *If the antecedent is false and the consequent true, the proposition is worthless (*nugatoria*) [9, p. 191].*

The most explicit condition is given by Lambert:

Definition 2.3 *A temporal proposition is true if the two actions stated in the temporal proposition are carried out at the same time; it is false otherwise [17, ¶105].*[4]

The same basic idea is expressed in late-12th-century *Tractatus Anagnini*, when its author notes that "Generally, every temporal proposition is true of which each part is true" [9, p. 252]—a definition which supports Buridan's contention that some authors treated temporal propositions as if they were conjunctions.

2.2 Temporal propositions in the 14th century

The 14th-century views are distinguished from the 13th-century accounts by their sophistication; they are more nuanced, and have greater scope.

Ockham, Buridan, and Burley define the syntax of temporal propositions and their truth conditions almost identically to Lambert; however, all three include a constraint, namely that a temporal proposition be composed "out of categoricals mediated by a temporal adverb" [4, p. 127], [3, p. 65], [18, p. 191]. This has a consequence of disallowing embedded temporal propositions.

[4] More accurately, it is "in the same time"; the Latin phrase is *in eodem tempore* [7, p. 17].

Additionally, Ockham allows temporal propositions to be composed out of more than two propositions [18, p. 191] (though he gives no explanation of how this would be done, and all his examples involve only two).

Of all the accounts we consider, Ockham's and Burley's are the most detailed. They both allow for temporal adverbs beyond those signifying simultaneity, such as those indicating priority or posteriority in the temporal order. Examples of temporal adverbs of the first type include *dum* 'while, as long as, until' and *quando* 'when, at which time', while examples of the second type include *ante* 'before', *post* and *postquam* 'after', and *priusquam* 'before, until' [4, p. 128]. For temporal adverbs indicating simultaneity, Burley gives the following truth conditions:

Definition 2.4 *For the truth of a temporal [proposition], in which categorical propositions are conjoined by means of an adverb conveying simultaneity of time, it is required that both parts be true for the same time [4, p. 128].*

This condition is further specified depending on whether the time signified is present, past, or future:

> For if the parts of such a temporal [proposition] are propositions of the present, then it is required that both parts be now true for this present time, and if it is of the past, it is required that both parts were true for some past time, this is, because they themselves were true in the present tense for some past time. And if they are propositions of the future, then it is required that both parts be true for some future time, that is, because they themselves will be true in the present tense for some future time [4, p. 128].

Thus, the truth conditions of temporal propositions will depend on the tenses of the sentences being conjoined with the temporal adverb.

There is an important point in which the 14th-century truth conditions offered by Ockham and Burley differ from the 13th-century ones as typified by Lambert, a point which Buridan makes explicit. Lambert requires that the two actions indicated by the temporal propositions are carried out "*at* the same time", and his view is typical of the 13th century. However, the 14th-century conditions change *at* to *for*. Buridan makes this point explicitly, grounded in his token-based semantics which allows for a distinction between a proposition being 'true of' vs. being 'true at' something [23,25]:

> It does not suffice for its categoricals to be true at the same time; for the propositions 'Aristotle existed' and 'The Antichrist will exist' are true at the same time, namely now, but it is required and sufficient that the copulas of the categoricals consignify the same time and that they be true *for* the same time, although not at that time [3, p. 65].

That is, given that the propositions which go into a temporal proposition can themselves be tensed, there is a difference between 'while' statements where the time of reference is the same and 'while' statements where the time of evaluation is the same. Which is taken as primary is one of the ways in which the 14th-century views differ from the 13th-century views.

For adverbs indicating something other than simultaneity, Ockham notes that it is necessary that "the propositions [be] true for different times" [18, p. 191].[5] He also notes that these two conditions do not exclude each other: 'The apostles preached while Christ preached' and 'The apostles preached after Christ preached' are consistent with each other [18, p. 191]. Both Ockham and Burley are also the only ones to consider when a temporal proposition is necessary, impossible, or contingent: "In order for a temporal proposition to be necessary it is required that each part be necessary" [18, p. 191], [4, p. 129]. Given the definitions of impossibility and contingency in terms of necessity and negation, the conditions under which a temporal proposition is impossible or contingent follow naturally [18, p. 192]. A consequence of this definition is that many statements which seem to be expressing necessities, such as 'Wood becomes warm when fire is brought near it', or 'A donkey is risible when it is a man', are not necessary, because neither of their parts is necessary (and in the case of the second example, both parts are necessarily false) [18, p. 191]. These only look like they are necessary because if we interpreted them as conditionals rather than temporal statements, they would be always true, and hence necessary [18, p. 192].

From these definitions, Burley offers a few corollaries concerning inferences involving temporal propositions:

Corollary 2.5 *The negation (*oppositum*) of a temporal [proposition] is a disjunction composed from the opposites of those which were required for the truth of the temporal [4, p. 131].*

However, this is merely a sufficient condition for the falsity of the temporal proposition; it is not a necessary condition.

Corollary 2.6 *A temporal [proposition] implies both of its parts, and not conversely [4, p. 131].*

Corollary 2.7 *A temporal [proposition] implies a conjunction made of the temporal parts, but not conversely [4, p. 131].*

The second two can also be found in Ockham [18, p. 192].

3 The logic of while

Let P and F be the usual backward- and forward-facing unary temporal possibility operators and $\Box$ be universal necessity, both past and future. Our temporal models $\mathfrak{T} = \langle W, \leq, V \rangle$ are linear and continuous, reflecting the fact that for the medievals, 'time' and 'place' are two of the five continuous quantities. These two are differentiated from the other three—line, surface, and body—in that time and place both have corresponding categories, the categories of 'when' and 'where' (cf. [17, ¶561]). We define the truth conditions of these operators in the usual fashion:

[5] The same condition is given for both 'after' and 'before', which is clearly incomplete.

Definition 3.1 (Unary temporal operators) *For $w \in W$:*

$$\begin{array}{lll} w \vDash Pp & \textit{iff} & \textit{there is } w' \leq w, w' \vDash p \\ w \vDash Fp & \textit{iff} & \textit{there is } w' \geq w, w' \vDash p \\ w \vDash \Box p & \textit{iff} & \textit{for all } w', w' \vDash p \end{array}$$

That some medieval authors argue that temporal propositions are reducible to conjunctions, while others argue that they are a type of conditional is grounded in the intuition that, in principle, there are two ways in which 'the same time' can be construed, either existentially or universally. In the first, temporal statements are equivalent to conjunctions, while in the second, temporal statements are equivalent to strict implications. We introduce Q (*quando* 'while', 'at every time') as a binary connective 'while'[6]. The two cases then are:

(i) $w \vDash pQq$ iff there exists w', $w' \vDash p$ and $w' \vDash q$.

(ii) $w \vDash pQq$ iff for all w', if $w' \vDash q$ then $w' \vDash p$.

In order to see which of these is correct, we must make more precise the way in which temporal compounds interact with tensed propositions.

Case (i) can be divided into three possibilities: (a) $w = w'$, (b) $w < w'$, and (c) $w' < w$. Case (a) corresponds to the case when p and q are both present-tensed statements referring to now, in which case, if time is not extended (that is, propositions are true or false at single points, or instants, of time), then for any two present-tensed propositions p and q, pQq is equivalent to $p \wedge q$: For the only way in which two present-tensed propositions can be true at the same time is if they are true now, and if now is a single instant, this means they must both be true at this instant, which is equivalent to their conjunction being true. If two present-tensed propositions are both true now, then they are true at the same instant, and hence each is true while the other is true. Cases (b) and (c) correspond to when the point of reference is either in the past or in the future, that is, statements such as 'Socrates lectured while Plato disputed' or—forgive the somewhat awkward English grammar—'Socrates will lecture while Plato will dispute'. Let us focus on the past-tensed case: Given that the components of 'Socrates lectured while Plato disputed' are 'Socrates lectured' and 'Plato disputed', themselves past-tensed, it would be natural to formalize this as a 'while' connective between two past-tensed statements, e.g., $PpQPq$, in analogy to when the atomic statements are present-tensed. However, it is not clear that this gets the truth conditions right. On an Ockham-Buridan-Burley account, 'Socrates lectured while Plato disputed' being true implies that there was a time for which both propositions were true, i.e., $P(p \wedge q)$. By ordinary temporal reasoning this implies $Pp \wedge Pq$, but from $Pp \wedge Pq$, one cannot make the reverse inference, because there is no guarantee that the time for which p is true and the time for which q is true is the same, a fact which both Ockham and Burley point out. Ockham says:

[6] We use Q rather than W to avoid confusion with both W a set of worlds and with 'where'.

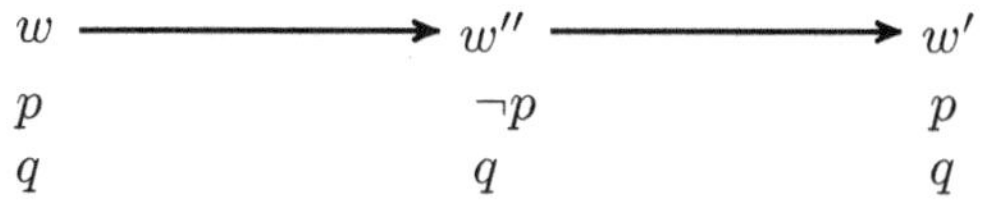

Fig. 1. $w \models pQq$

> ...a conjunctive proposition follows from a temporal proposition—but not conversely. For this does not follow: 'Adam existed and Noah existed, therefore Adam existed when Noah existed'. Nor does this follow: 'Jacob existed and Esau existed, therefore Jacob existed when Esau existed' [18, p. 192].[7]

Whether $PpQPq$ implies $P(p \wedge q)$ depends on how, precisely, we interpret Q.

In modern temporal logic, the 'while' operator is most commonly found in the context of dynamic temporal logic, where 'while φ, do α' constructions are commonly used. It is clear that this imperative, dynamic conception of 'while' is *not* what the medieval logicians had in mind. Instead, their notion is static. Modern static temporal logic tends to omit discussion of 'while', taking as basic instead the forward-looking $\mathcal{U}$ 'until' and the backward-looking $\mathcal{S}$ 'since':

Definition 3.2 (Weak until) *For $w \in W$:*

$$\begin{aligned} w \models p\mathcal{U}q \text{ iff } & \text{if there is a } w' \geq w \text{ s.t. } w' \models q \\ & \text{then for every } w'',\ w \leq w'' < w', w'' \models p \end{aligned}$$

$\mathcal{S}$ is defined symmetrically.

Malachi and Owicki use this weak notion of 'until' to define a correspondingly weak notion of 'while' [14, p. 206]:[8]

Definition 3.3 (Malachi & Owicki 'while') *For $w \in W$:*

$$\begin{aligned} w \models pQq \quad & \text{iff} \quad w \models p\mathcal{U}(\neg q) \\ & \text{iff} \quad \text{if there is a } w' \geq w \text{ s.t. } w' \models \neg q \\ & \qquad\ \text{then for every } w'',\ w \leq w'' < w', w'' \models p \end{aligned}$$

Intuitively, defining 'while' in terms of 'until' makes sense, for "p is true while q is true" seems to mean nothing more than "p is true until q is false". However, the 'until' used in the English sentence here is not the same as the weak until defined in Definition 3.2, since if q is always true, then the antecedent will be false and p can be either true or false, as illustrated in Figure 1. As a result, we must reject Malachi and Owicki's definition as unsuitable. An alternative analysis of 'while' is given in [15, p. 260]:

---

[7] Burley's example is similar: "Adam was when Noah was, therefore Adam was and Noah was" follows, but "Adam was and Noah was, therefore Adam was when Noah was" does not [4, p. 131].

[8] They read pQq as "q is true as long as p is true", but this cannot be a correct interpretation of $p\mathcal{U}(\neg q)$.

$$\begin{array}{ccccccccc} t_1 & \longrightarrow & t_2 & \longrightarrow & t_3 & \longrightarrow & t_4 & \longrightarrow & t_5 \\ p & & \neg p & & Pp & & Pp & & Pp \\ \neg q & & q & & Pq & & Pq & & Pq \end{array}$$

Fig. 2. $t_3 \models PpQPq \wedge \neg P(p \wedge q)$

Definition 3.4 (Manna & Pnueli 'while') *For* $w \in W$*:*

$$w \models pQq \text{ iff } w' \models p \text{ for every } w' \geq w \text{ such that } w'' \models q \text{ for all } w'', w \leq w'' \leq w'$$

But this also fails to capture the medieval account, in two ways. First, on this definition, $PpQPq$ does not imply $P(p \wedge q)$. When p and q are past-tensed statements, it is possible for them to both be true at the same time without there being any time for which the present-tense conjunction is true (see Figure 2), which is contrary to what Ockham and Burley argue for above. Thus, if we were to adopt these truth conditions for Q, 'Socrates lectured while Plato disputed' could not be formalized as a temporal compound of two past-tensed sentences. This raises two questions: (1) How should it be formalized?, and (2) What, if anything, does $PpQPq$ represent, given that it appears to be well-formed?

We answer the second question first, by returning to Buridan's distinction of a proposition being *true for* a time vs. being *true at* a time; this is a distinction between the time of evaluation and the time of reference. In Figure 2, Pp and Pq are *true at* the same time, namely t_3, but they are not *true for* the same time; Pp is true for t_1, because t_1 is the witness for the truth of Pp at t_3, while Pq is true for t_2, because t_2 is the witness for the truth of Pq at t_3. Thus, Buridan would reject $PpQPq$ as an appropriate analysis of temporal compounds of past-tensed sentences. However, one needn't deny the acceptability of the 'true at' analysis, and indeed, inspection of the 13th-century views show that this is precisely how they differ from the 14th-century ones: Recall that Lambert says that a temporal proposition is true "if the two actions stated in the temporal proposition are carried out at the same time", rather than that the two actions described are *true for* the same time. In taking this account, he shows similarities with Peter Abelard's views in the middle of the 12th century. As Martin describes Abelard's views:

> The compound temporal proposition formed from the two propositions 'Socrates was a youth' and 'Socrates was an infant' is true at a given time, [Abelard] tells us, just in case each component is true at that time, and so true now of the old Socrates [16, p. 165].

Martin doesn't say how the compound proposition is formed from these two statements, and if it is merely 'Socrates was an infant while Socrates was a youth', then this statement should never be true, since being an infant excludes being a youth. However, if the compound proposition expresses that these two past-tensed statements can be true *at* (as opposed to *for*) the same time, e.g.,

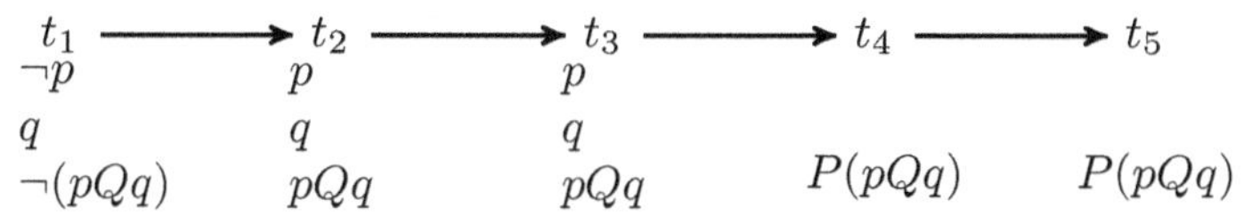

Fig. 3. $t_4 \models P(pQq)$

'Socrates was a youth is true at the same time as Socrates was an infant is true', then this is true any time Socrates is alive and was both previously an infant and previously a youth.

Ideally, we would like an account of 'while' which doesn't require us to exclude either the *true for* or the *true at* analysis, and which treats both of these in a uniform fashion. If we adopt $PpQPq$, i.e., applying the tenses directly to the atomic propositions, to indicate that the two tensed propositions are true at the same time, then the natural alternative to represent the 'true for' case is to make the entire temporal compound past-tensed, e.g., $P(pQq)$. Figure 3 gives a model where $P(pQq)$ is true. From this figure, it should be clear that $P(pQq)$ will always imply $PpQPq$, but not conversely. It should also be clear that $\varphi Q \psi$ implies $\varphi \wedge \psi$, regardless of the tenses of the individual components φ and ψ, and regardless of the tense of the entire compound, from which it follows that $P(pQq)$ implies $Pp \wedge Pq$, but also not conversely.

A problem still remains with analysing Q via the conditions given in Definition 3.4. When p and q are both present-tensed, if q is always false, pQq will always be true.[9] For Definition 3.4 can be rewritten, informally, as "for every $w' \geq w$, if w'''s being between w and w' implies that $w'' \models q$, then $w' \models p$. When q is always false, w'''s being between w and w' does not imply that $w'' \models q$, and hence the antecedent of the conditional is falsified, making the entire condition satisfied. But this goes against the medieval requirement that pQq imply $p \wedge q$. If we add this requirement to Definition 3.4, we obtain a characterization of Q that captures the medieval account, wherein pQq implies $p \wedge q$ but is not implied by it:

Definition 3.5 (Medieval 'while') *For* $w \in W$*:*

$$\begin{aligned} w \models pQq \quad \textit{iff} \quad & w \models p \wedge q \textit{ and for all } w' \geq w \\ & \textit{if for all } w'', w \leq w'' < w', w'' \models q \textit{ then } w' \models p \end{aligned}$$

An advantage of this account is that it helps understand why some medieval authors try to reduce temporal propositions to conjunctions and others to implications, because the truth conditions have both conjunctive and implicative

[9] A similar objection can be levied against Kröger's definition of Q [13, p. 22]: For $w \in W$:

$$\begin{aligned} w \models pQq \text{ iff } & (1) \text{ there is } w' > w \text{ s.t. } w' \models \neg q \text{ and} \\ & \text{for all } w'', w < w'' < w', w'' \models p \text{ or} \\ & (2) \text{ for all } w' > w, w' \models p \end{aligned}$$

Because the relations are strict, there is no requirement for $p \wedge q$ to be true at w.

conditions. A further advantage is that on this definition we are able to prove that Ockham's account of the necessity of temporal propositions is correct:

Lemma 3.6 (Ockham) $\Box(pQq)$ *iff* $\Box p \wedge \Box q$.

Proof. Fix $w \in W$.

($\Rightarrow$) If $w \vDash \Box(pQq)$, then for all w', $w' \vDash pQq$. By Definition 3.5, $w' \vDash p \wedge q$, so $w \vDash \Box(p \wedge q)$. Since necessity distributes, $w \vDash \Box p \wedge \Box q$.

($\Leftarrow$) If $w \vDash \Box p \wedge \Box q$, then $w \vDash \Box(p \wedge q)$. By Definition 3.1, for all $w' \in W$, $w' \vDash p \wedge q$, and in particular, $w' \vDash p$. From this it follows that for all $w'' < w' \in W$, $w'' \vDash q$ implies $w' \vDash p$, and for any $w \leq w''$, $w \vDash pQq$. But since w' was arbitrary, this holds of all $w \in W$, so $w \vDash \Box(pQq)$. □

4 Local propositions

The 13th-century discussions of local propositions are not as extensive, and resemble each other quite a bit. AE [9, p. 159], AB [9, p. 191], and DM [9, p. 485] all give similar accounts which closely resemble Lambert's definition:

Definition 4.1 (Lambert) *A* local *proposition is one whose parts are joined by the adverb 'where', as in 'Socrates runs where Plato argues' [17, ¶104].*

Bacon says that 'Socrates is where Plato is' is called local, and this is a complex proposition in virtue of a relation (namely 'being in the same place as') instead of a connective [2, ¶170], treating the syntactic construction of local propositions analogously to temporal ones. Of the 13th-century authors, only Lambert provides any truth conditions:

Definition 4.2 (Lambert) *A local proposition is true if the two actions stated in the local proposition are carried out in the same place; it is false otherwise [2, ¶104].*

The 14th-century discussions of local propositions are more detailed, but still relatively circumscribed compared to the temporal analyses. Buridan's definition of local propositions mirrors his definition of temporal ones, with the difference that 'temporal adverb' is replaced with 'local adverb' and 'when' with 'where' (though he notes it is possible to replace the local adverb with an equivalent phrase, such as 'Socrates is in the place in which Plato is' [3, p. 66]). For their truth, he provides a necessary but not sufficient condition:

Definition 4.3 (Buridan) *A local proposition is true if "both categoricals [are] true for the place designated by the word 'where', and it is not sufficient that they are true for the same place" [3, p. 66].*

Thus, local propositions will entail conjunctions but not vice versa, as with temporal propositions.

Ockham allows local propositions to be composed of more than two categoricals [18, p. 192], as he does with temporal propositions and with the same caveat noted earlier. An interesting point of textual interpretation arises in the statement of the truth conditions. Some manuscripts say that "for the truth of [a local] proposition, it is required that each part be true for the same place or for different places" while others give the last clause as "*and not for* different places" (emphasis added) [18, pp. 193, 205]. There is clear evidence that the latter reading should be preferred: After giving this condition, Ockham points it out as the very characteristic by which local propositions differ from temporal ones:

> In this regard it [a local proposition] differs from a temporal one. For in order for a temporal proposition to be true it is required that both parts be true for the same place or for different places, while in order for a local proposition to be true it is required that both parts be true for the same place *and not for* different places [18, p. 193].[10]

For local propositions to be distinguished from temporal ones on precisely these grounds, it must be the case that one of the two causes of truth is excluded. Burley does not discuss local propositions as a separate type. Instead, he says that local propositions such 'Socrates is moved where he runs' are reducible to one of the five basic types of hypothetical propositions: conditional, causal, temporal, conjunctive, and disjunctive [4, p. 107]. He does not say which, but from other accounts, it is clear that local propositions are analogous to temporal ones.

5 The logic of where

It is important to note that we have the same "true at (or in)" vs. "true for" distinction for local propositions that we had for temporal ones. Therefore, we should expect our analysis of the logic of where to account for this distinction, as we required of our logic of while.

The first, and simplest, modern attempt to capture a local notion in modal logic is the logic of elsewhere introduced by von Wright [29] and completely axiomatized by Segerberg [24]. In this system, $\Box$ is interpreted as 'everywhere else' and $\Diamond$ is read as 'somewhere else'; the frames for this logic are ones where xRy iff $x \neq y$, that is, R is the relation of nonidentity. The logic of elsewhere is the smallest normal modal logic containing all instances of the schemata:[11]

A: $p \rightarrow (\Box p \rightarrow \Box\Box p)$

B: $p \rightarrow \Box\Diamond p$

This logic, of course, describes exactly the opposite of what we need to capture the medieval analyses, because what we want is not the logic of elsewhere but the logic of, so to say, 'here'. However, the natural approach, taking R as identity, is clearly not correct: Such a system would collapse into the trivial system Triv. Adopting this would make local propositions equivalent to conjunctions and destroy the analogy with temporal propositions, something which all medieval authors we consider maintain.

---

[10] The italicized portion is our correction of the reading in the translation "or for", following the other manuscript tradition.

[11] In [10,11], axiom A is replaced by the axiom $\Diamond\Diamond p \rightarrow (p \vee \Diamond p)$.

Since von Wright, substantially more complex spatial logics, such as ones characterizing topological notions such as 'nearness' and 'distance', have been introduced [1,28]. These logics are aimed at capturing what Aiello and van Bentham call "the ontological structure [of space]: What are the primitive objects and their relations?" [1, p. 320]. This approach can be contrasted with one that looks at "some existing human practice, e.g., a language with spatial expressions (say locative propositions) or a diagrammatic way of visualizing things" [1, p. 320], and identifies and classifies the types of spatial structures that 'fit' these linguistic expressions. Many modern modal logicians prefer the former approach, whereas clearly here it is the latter that will be most beneficial for understanding the medieval practices: It is how we use local propositions, not the nature of the underlying structure of space, that guides the correct analysis of 'where' compounds.

All of the authors we have looked at agree that temporal propositions and local propositions should be treated analogously (though not necessarily identically: As noted in the previous section Ockham requires that the component propositions not be true for different places): 'where' is substituted for 'while', 'same place' is substituted for 'same time'. We begin with looking at how far the analogy between local propositions and temporal propositions can be pushed, specifically at our assumptions concerning the nature of time and whether they are appropriate to transfer to space. We first point out a clear disanalogy between local propositions and temporal ones. The truth of 'p where q' does not depend on the place of evaluation, whereas the truth of pQq depends on the time of evaluation. This reflects the fact that if no explicit time is specified, a temporal proposition must have both of its temporal parts true now; but if no explicit place is specified, a local proposition can be true without either of its parts being true here. A consequence of this is that we can ignore location when considering temporal propositions, but we cannot ignore time or tense when considering local propositions, for it is perfectly natural to say such things as "I will walk where Socrates disputed" or "That church stands where a Roman temple used to stand". In order for sentences like these to be sensible, we need to evaluate propositions not merely at places, but at place/time pairs. This will be reflected below when we evaluate local propositions in $\mathbb{R}^3$.

One assumption about time that we did not address explicitly in the previous section is whether time is point-based or extended (i.e., interval-based). We adopted a point-based approach without comment, for two reasons: First, it is within the bounds of medieval approaches to temporal reasoning to consider at least some statements as being true or false at instants [12], even though many sentences express propositions that are extended in duration. [13] Second,

[12] 13th- and 14th-century literature on *incipit* 'begins' and *desinit* 'ceases' define these in terms of instants.

[13] For example, DM's example is "When the sun is above the earth, it is day" [9, p. 485] (rewritable into the equivalent "it is day while the sun is above the earth"), or the other examples involving activities of extended duration such as walking, reading, and disputing, all of which are more naturally suited to evaluation at intervals than at points.

actions with extended duration are taken into account by the second conjunct of the truth conditions in Definition 3.5; while intervals are not used in the definition under that name, it is clear that the truth of every temporal proposition depends upon intervals bounded below by some point where $p \wedge q$ is true and extending until q is false.[14] If we adopt the same approach for space, we are committed to the view that an infinite number of actions can occur at a single spatial point; while this is unproblematic for temporal points, it may be less palatable for spatial points. As an alternative, we could adopt an approach where regions rather than points are taken as basic, such that truth in a region does not necessarily propagate down to subregions. For example, taking an entire football pitch as a region, if Socrates is disputing on the left half while Plato is running on the right half, "Socrates is disputing where Plato is running" is true for the entire pitch; but it is false for either of the individual halves. However, on an alternative view of events, they "unlike material objects, do not *occupy* the space at which they are located" [5, p. 17], meaning we can, in principle, allow for an infinite number of events happening at a single point, even if the material objects acting in those events do not co-exist at that point.

A further assumption that we made about the nature of time—that it is linear—is not plausible for space. Dropping linearity means that we are no longer looking at points and intervals on a line but rather points in $\mathbb{R}^2$ and 2D regions around these points, regions which will fill the same role as intervals play in the temporal case. These regions need to be 'well-behaved' or 'normal' in some intuitive sense: They shouldn't be disjoint, they shouldn't contain holes,[15] they should be extended in the spatial dimensions only, etc. How precisely this notion of 'region' is to be defined is ultimately immaterial—many different possibilities are acceptable, and we do not wish to discriminately unduly—but since we need a definition, we take the following:

Definition 5.1 (Neighborhood) *A set $A \subset \mathbb{R}^2$ is a* neighborhood *of a point $(x, y) \in \mathbb{R}^2$ if A contains an open set containing (x, y).*[16]

Definition 5.2 (Region) *Let $(x, y) \in \mathbb{R}^2$. We say that $\mathcal{R}(x, y) = A$ is a* region *of (x, y) if A is a simply connected neighborhood of (x, y).*[17]

This captures the definition of 'where' given by Lambert, quoting the anony-

---

[14] It is for this reason that hybrid approaches are not immediately adaptable to our present needs, because in standard hybrid logics, nominals range over time points, not time intervals.

[15] The requirement that regions not have holes may be too strong. For suppose that Socrates is a moderately good hammer thrower, such that no hammer he throws ever lands less than 1 meter from him. However, he's not *very* good, so they could land in any direction. Then, one might want the region referred to in the sentence "Plato walks where Socrates's hammers fall" to be a torus, rather than a circle. If this example, suggested to me by Charles Walker, is motivating enough, the requirement of simple connectedness in Definition 5.1 can be dropped.

[16] An alternative definition of 'neighborhood' requires that A itself be the open set containing (x, y), rather than merely containing an open set, but we do not need this constraint.

[17] Note that this differs from the definition of 'regions' as 'regular closed sets' in [12, p. 514].

mous author of the 12th-century *Book of Six Principles*: [18]

> "[W]here is the circumscription of a body proceeding from the circumscription of a place." For example, water collected in a container adopts the figure of the container and is transfigured in accord with the figure of the interior surface of the container, and the configuration that it has from the interior surface of the container is named 'where' [17, ¶567]. [19]

We are now in a position to define the binary local connective 'U' (*ubi* 'where'), analogous to 'Q but replacing intervals with regions, representing 'wheres':

Definition 5.3 (Medieval 'where') *For* $(t, x, y) \in (\mathbb{R}^3, \leq)$:

$$
\begin{array}{lll}
(t,x,y) \models pUq & \textit{iff} & \textit{there is } x', y' \textit{ s.t. } (t,x',y') \models p \wedge q \textit{ and for all } \mathcal{R}(x',y'),\\
& & \textit{if for all } (x'',y'') \in \mathcal{R}(x',y'), (t,x'',y'') \models q\\
& & \textit{then for all } (x'',y'') \in \mathcal{R}(x',y'), (t,x'',y'') \models p
\end{array}
$$

We then redefine the truth conditions for the relevant unary temporal operators with a specification of place:

Definition 5.4 *For* $(t, x, y) \in (\mathbb{R}^3, \leq)$:

$$
\begin{array}{lll}
(t,x,y) \models Pp & \textit{iff} & \textit{there is } t' \leq t, (t',x,y) \models p\\
(t,x,y) \models Fp & \textit{iff} & \textit{there is } t' \geq t, (t',x,y) \models p\\
(t,x,y) \models \Box p & \textit{iff} & \textit{for all } t', (t',x,y) \models p
\end{array}
$$

Because the place is kept fixed, this is equivalent to Definition 3.1 when the place is unspecified. Updating Definition 3.5 to include reference to place is also straightforward.

6 Conclusion

One of the most interesting scientific results of the 20th century was the production of a model under which time and place can be considered in symmetric fashion: One can simply treat the dimension of time as another dimension on a par with the three dimensions of space, length, breadth, and width. This parity is arguably reflected in the medieval theories we've discussed: We have seen a variety of 13th- and 14th-century views of temporal and local propositions, on which the analyses of being true at the same time and being true at the same place are (almost) symmetric, and both arise from natural uses of everyday language. We gave truth conditions for the operators Q and U which both capture the medieval views concerning the analogous structure and truth conditions of these operators as well as provide interesting alternatives to modern approaches. Our definition of Q differs from two modern definitions by requiring the implication from pQq to $p \wedge q$, while our definition of

[18] And that author in turn takes the definition from Aristotle's *Physics*.

[19] Note that this example is three-dimensional. In our discussion, we are restricting ourselves to the two-dimensional case, but only for simplicity's sake. Our definitions should be generalizable.

U is unique. This paper is only a first step to a complete analysis: The next step after defining the semantics would be to identify the characteristic axioms governing these new operators. We will pursue this in future work.

Acknowledgments

This paper could not have been completed without two crucial conversations, one on topology face-to-face with David Sheard, and the other on how 'while' interacts with tensed statements conducted virtually with an incredibly diverse cross-section of my Facebook friends list.[20] Social media as research scaffolding: It's a grand thing!

References

[1] Aiello, M. and J. van Benthem, *A modalwalk through space*, Journal of Applied Non-Classical Logics **12** (2002), pp. 319–363.

[2] Bacon, R., "The Art and Science of Logic," Pontifical Institute of Medieval Studies, 2009, T. S. Maloney, trans.

[3] Buridan, J., "John Buridan: Summulae de Dialectica," Yale University Press, 2001, G. Klima, trans.

[4] Burleigh, W., "*De Puritate Artis Logicae* With a Revised Edition of the *Tractatus Brevior*," Franciscan Institute, 1955, Philotheus Boehner, ed.

[5] Casati, R. and A. C. Varzi, "Parts and Places: The Structures of Spatial Representation," MIT Press, 1999.

[6] Copenhaver, B. P., editor, "Peter of Spain: Summaries of Logic, Text, Translation, Introduction, and Notes," Oxford University Press, 2014, with Calvin Normore and Terence Parsons.

[7] d'Auxerre, L., "Logica (Summa Lamberti)," La Nuova Italia, 1971, F. Alessio, editor.

[8] de Rijk, L. M., "Logica Modernorum, vol. II, part I," Van Gorcum & Comp., 1967.

[9] de Rijk, L. M., "Logica Modernorum, vol. II, part II," Van Gorcum & Comp., 1967.

[10] Demri, S., *A simple tableau system for the logic of elsewhere*, in: *Proc. 5th Int. Workshop on Theorem Proving with Analytical Tableaux and Related Methods (TABLEAUX96)*, Lecture Notes in Artifical Intelligence **1071** (1996), pp. 177–192.

[11] Hampson, C. and A. Kurucz, *On modal products with the logic of 'elsewhere'*, in: T. Bolander, T. Braüner, S. Ghilardi and L. Moss, editors, *Advances in Modal Logic 9* (2012), pp. 339–347.

[20] Thanks to Nicholas Adams, Thomas Ball, Hadley Foster Barth, Melissa Barton, Kate Bell, Malin Berglund, Wendel Bordelon, Edward Boreham, Liam Kofi Bright, Edward Buckner, Don Campbell, Karen Carlisle, Erin Childs, Riia M. Chmielowski, Kay Ellis, Katherine Gensler, Andrew Grosser, Robyn Hodgkin, Justine Jacot, Esther Johnston, Earl P. Jones, Heather Rose Jones, Susanne Kalejaiye, Linse Rose Kelbe, Marleen de Kramer, Jennifer Knox, Barteld Kooi, Jean Kveberg, Christer Romson Lande, Lee Large, Dan Long, Christy Mackenzie, Dave Majors, Alex Malpass, Jennifer McGowan, Lesley McIntee, Liz McKinnell, Tom McKinnell, Sonia Murphy, Katherine Napolitano, Gabriela Aslı Rino Nesin, Paddy Neumann, Lynette Nusbacher, Peryn Westerhof Nyman, Caroline Orr, Sy Delta Parker, Susanne de Paulis, Judith Marie Phillips, Mike Prendergast, Daria Rakowski, Stephanie Rebours-Smith, Kevin Rhodes, Sarah Rossiter, Angela Sanders, Fiona Scerri, Amy Selman, Phil Selman, Jennifer Smith, Lena Thane-Clarke, Petra Träm, Joel Uckelman, Nicole Uhl, Rineke Verbrugge, Miesje de Vogel, Elmar Vogt, Ursula Whitcher, Brooke White, Nik Whitehead, and Anna Wilson.

[12] Kontchakov, R., A. Kurucz, F. Wolter and M. Zakharyaschev, *Spatial logic + temporal logic = ?*, in: M. Aiello, I. Pratt-Hartmann and J. van Benthem, editors, *Handbook of Spatial Logics*, Springer, 2007 pp. 497–564.
[13] Kröger, F., "Temporal Logic of Programs," EATCS Monographs on Theoretical Computer Science **8**, Springer, 1987.
[14] Malachi, Y. and S. S. Owicki, *Temporal specifications of self-time systems*, in: H. T. Kung, B. Sproull and G. Steele, editors, *VLSI Systems and Computations*, Springer-Verlag, 1981 pp. 203–212.
[15] Manna, Z. and A. Pnueli, "The Temporal Logic of Reactive and Concurrent Systems: Specification," Springer Verlag, 1992.
[16] Martin, C. J., *Denying conditionals: Abaelard and the failure of Boethius' account of the hypothetical syllogism*, Vivarium **45** (2007), pp. 153–168.
[17] of Auxerre, L., "Logica or Summa Lamberti," University of Notre Dame, 2015, T. S. Maloney, trans.
[18] of Ockham, W., "Ockham's Theory of Propositions: Part II of the *Summa Logicae*," St. Augustine's Press, 1998, A. J. Freddoso and H. Schuurman, trans.
[19] of Sherwood, W., "William of Sherwood's *Introduction to Logic*," University of Minnesota Press, 1966, N. Kretzmann, trans.
[20] Øhrstrøm, P., *"Temporalis" in medieval logic*, Franciscan Studies **42** (1982), pp. 166–179.
[21] Øhrstrøm, P. and P. Hasle, "Temporal Logic: From Ancient Ideas to Artificial Intelligence," Kluwer Academic Publishers, 1995.
[22] Prior, A. N., "Past, Present, and Future," Clarendon Press, 1967.
[23] Prior, A. N., *The possibly-true and the possible*, Mind **78** (1969), pp. 481–492.
[24] Segerberg, K., *A note on the logic of elsewhere*, Theoria **46** (1980), pp. 183–187.
[25] Uckelman, S. L., *Prior on an* insolubilium *of Jean Buridan*, Synthese **188** (2012), pp. 487–498.
[26] Uckelman, S. L., *A quantified temporal logic for ampliation and restriction*, Vivarium **51** (2013), pp. 485–510.
[27] Uckelman, S. L. and S. Johnston, *John Buridan's* Sophismata *and interval temporal semantics*, Logical Analysis and History of Philosophy **13** (2010), pp. 133–147.
[28] van Benthem, J. and G. Bezhanishvili, *Modal logics of space*, in: M. Aiello, I. Pratt-Hartmann and J. van Benthem, editors, *Handbook of Spatial Logics*, Springer, 2007 pp. 217–298.
[29] von Wright, G. H., *A modal logic of place*, in: E. Sosa, editor, *The Philosophy of Nicholas Rescher*, Reidel, 1979 pp. 65–73.

Update, Probability, Knowledge and Belief

Jan van Eijck [1]

CWI and ILLC
Science Park 123
1098 XG Amsterdam, Netherlands

Bryan Renne [2]

University of British Columbia, Faculty of Medicine
1320-2350 Health Sciences Mall
Vancouver BC V6T 1Z3, Canada

Abstract

The paper compares two kinds of models for logics of knowledge and belief, neighbourhood models and epistemic weight models. We give sound and complete calculi for both, and we show that our calculus for neighbourhood models is sound but not complete for epistemic weight models. Epistemic weight models combine knowledge and probability by using epistemic accessibility relations and weights to define subjective probabilities. Our Probability Comparison Calculus for this class of models is a further simplification of the calculus that was presented in AIML 2014.

Keywords: Probability, epistemic modal logic, dynamic epistemic logic, probabilistic update, Bayesian learning.

---

1 Probability and Information

A Bayesian learner is an agent who uses new information to update a subjective probability distribution that somehow captures what she knows or believes about the world. In a multi-agent setting, various learners could receive differents pieces of information, and if multi-agent logics of knowledge of belief are extended with update procedures that implement the processing of new information, then this also gives a perspective on learning in a multi-agent setting. So it is natural to combine probability theory and the update perspective on knowledge and belief from dynamic modal logic.

Indeed, there exists already a modest tradition in combining DEL (Dynamic Epistemic Logic) and probabibility theory. The first combinations are in Kooi's thesis [23], in Van Benthem's [3], and in the combined effort of Van Benthem,

[1] jve@cwi.nl

[2] bryan@renne.org

Gerbrandy and Kooi in [5]; [10] gives an overview. Inspiration for this goes back to work of Fagin and Halpern in the 1990s [13]. A simplified combined system of DEL and probability theory is presented in [12]. Further simplication of the base logic is provided in [9]. The present paper extends work that was presented at AIML 2014 in [12].

A natural notion of belief that turns up in a setting of probabilistic dynamic epistemic logic is betting belief (or: Bayesian belief) in ϕ: $P(\phi) > P(\neg\phi)$. Van Eijck & Renne [11] give a logical calculus for this, combined with an S5 operator for knowledge, that we will review and extend below. This is in fact an extension of a calculus proposed by Burgess in [7]. A variation on this is threshold belief in ϕ: $P(\phi) > t$, for some specific t with $\frac{1}{2} \leq t < 1$. This is also known as Lockean belief. John Locke suggests in his work that a person's belief in a proposition ϕ is somehow connected to that person's confidence in ϕ. This confidence should then be connected in turn to the evidence that the person has for ϕ. If it makes sense to talk about degree of belief at all, then subjective probability is one way of making this precise [16]. A logic with KD45 belief and an explicit belief comparison operator is presented in [21]. See [28] for an overview of the extensive literature on belief comparison operators.

In this paper we will use a neighbourhood semantics ([8, Ch. 8] and [19]) for 'belief as willingness to bet'. Related to neighbourhood models for belief are the evidence models proposed in [4].

Attempts to develop a qualitative notion of probability date back to De Finetti [14,15]. He proposed the following requirements for a binary relation $\succeq$ on a (finite and non-empty) set W:

$$\begin{aligned} \text{nonnegativity } & A \succeq \emptyset \\ \text{nontriviality } & \emptyset \not\succeq W \\ \text{totality } & A \succeq B \text{ or } B \succeq A \\ \text{transitivity } & \text{if } A \succeq B \text{ and } B \succeq C \text{ then } A \succeq C \\ \text{quasi-additivity } & \text{if } (A \cup B) \cap C = \emptyset \text{ then } A \succeq B \text{ iff } A \cup C \succeq B \cup C \end{aligned}$$

A probability measure on W is a function $\mu : \mathcal{P}(W) \to \mathbb{R}$ satisfying $\mu(\emptyset) = 0$, $\mu(W) = 1$ and $\mu(A \cup B) = \mu(A) + \mu(B)$ for $A, B \subseteq W$ with $A \cap B = \emptyset$ (additivity). The conjecture that the five requirements completely determine a probability measure on W was refuted in [24].

Theorem 1.1 *There is a relation satisfying De Finetti's axioms that does not agree with any probability measure [24].*

Proof. Consider $W = \{p, q, r, s, t\}$ with a weight map $\nu : W \to \mathbb{N}$ given by $\nu(p) = 4, \nu(q) = 1, \nu(r) = 3, \nu(s) = 2, \nu(t) = 6$. Extend ν to subsets of W by putting $\nu(A) = \sum_{a \in A} \nu(a)$. Let the relation $\succeq_\nu$ on W be given by $A \succeq_\nu B$ iff $\nu(A) \geq \nu(B)$. Next, writing pq for $\{p, q\}$, define $\succeq$ as

$$\succeq \; := \; \succeq_\nu - \{(st, pqr)\}.$$

This yields (writing $A \approx B$ for $A \succeq B \wedge B \succeq A$ and $A \succ B$ for $A \succeq B \wedge B \not\succeq A$): $p \approx qr, rs \approx pq, qt \approx pr, pqr \succ st$. Check that this relation satisfies the De

Fenetti axioms: Transitivity still holds, because pqr and st are the only two sets that have ν-weights adding up to 8, so there can be no set A different from pqr and st with $st \succeq A$ and $A \succeq pqr$. Quasi-additivity still holds since the only C with $C \cap pqrst = \emptyset$ is $C = \emptyset$. The other three properties are obvious. But the relation does not agree with any probability measure μ. For it follows from $\mu(p) = \mu(qr)$, $\mu(rs) = \mu(pq)$, $\mu(qt) = \mu(pr)$ that $\mu(st) = \mu(pqr)$. Thus, μ cannot agree with $\{p, q, r\} \succ \{s, t\}$. □

Scott [29] gave an algebraic reformulation of the new axioms proposed in [24]. Formulated in terms of subsets of a universe W, a pair of k-length sequences of sets $(A_1, \dots, A_k)$ and $(B_1, \dots, B_k)$ is *balanced* if for each $w \in W$ it holds that $|\{i \mid w \in A_i\}| = |\{i \mid w \in B_i\}|$. The Scott axiom for $\succeq$ for length k says:

k-cancellation if $(A_1, \dots, A_k, X)$ and $(B_1, \dots, B_k, Y)$ are balanced, and $A_i \succeq B_i$ for each i with $1 \leq i \leq k$, then $Y \succeq X$.

It is not hard to see that if a relation $\succeq$ is representable by a probability measure, then $\succeq$ must satisfy cancellation for any k. Scott showed that any $\succeq$ relation satisfying nonnegativity, nontriviality, totality and cancellation for any $k \in \mathbb{N}$ determines a probability measure, thus replacing the set of conditions proposed by De Finetti by a complete set.

Notice that the example in the proof of Theorem 1.1 does not satisfy 3-cancellation, for the pair (p, rs, qt, pqr) and (qr, pq, pr, st) is balanced and satisfies $p \succeq qr$, $rs \succeq pq$, $qt \succeq pr$, but $st \succeq pqr$ does not hold.

Segerberg [30] showed how the balancedness requirements could be translated into modal logic (be it at formidable coding cost), and Segerberg [30] and Gärdenfors [17] proposed axiomatisations for a logic with $\succeq$, using an infinite number of modal schemes (one for each sequence length k) to cover the balancedness conditions identified by Scott. This approach was later adopted in Lenzen [26] for a logic of conviction (German: *Überzeugung*) and belief where an agent's conviction of ϕ is identified with assigning probability 1 to ϕ, and belief in ϕ with assignming a probability greater than $\frac{1}{2}$. Later, Herzig also incorporated action, in [20]. A difference between Lenzen's approach and ours are that Lenzen's conviction does not imply truth. Another difference is that we will not use the Scott axioms. The main difference between the approach of Herzig and ours is that he relates betting belief to a KD45 modality, while we relate betting belief to S5 knowledge.

2 Epistemic Neighbourhood Models

We now introduce epistemic neighbourhood models, where belief is represented as truth in a neighbourhood.

Definition 2.1 [Epistemic Neighbourhood Models] An **Epistemic Neighbourhood Model** $\mathcal{M}$ is a tuple $(W, \sim, N, V)$ where

- W is a non-empty set of worlds.

- $\sim$ is a function that assigns to every agent $i \in Ag$ an equivalence relation $\sim_i$ on W. We use $[w]_i$ for the $\sim_i$ class of w, i.e., for the set $\{v \in W \mid w \sim_i v\}$.
- N is a function that assigns to every agent $i \in Ag$ and world $w \in W$ a collection $N_i(w)$ of sets of worlds—each such set called a *neighbourhood* of w—subject to the following conditions.
  (c) $\forall X \in N_i(w) : X \subseteq [w]_i$.
  (n) $[w]_i \in N_i(w)$.
  (a) $\forall v \in [w]_i : N_i(v) = N_i(w)$.
  (m) $\forall X \subseteq Y \subseteq [w]_i :$ if $X \in N_i(w)$, then $Y \in N_i(w)$.
  (d) $\forall X \in N_i(w), [w]_i - X \notin N_i(w)$.
  (sc) $\forall X, Y \subseteq [w]_i$: if $[w]_i - X \notin N_i(w)$ and $X \subsetneq Y$, then $Y \in N_i(w)$.
- V is a valuation function that assigns to every $w \in W$ a subset of *Prop*.

Epistemic neighbourhood models are a variation on the well-known neighbourhood models from modal logic. The difference is that they include an epistemic component $\sim_i$ for each agent i. Since $[w]_i$ is the set of worlds agent i knows to be possible at w, each $X \in N_i(w)$ represents a proposition that the agent believes at w. Notice that it follows from these semantics that knowledge implies belief.

Property (c) This ensures that what is known is also believed.

Property (n) This ensures that what is logically true is believed.

Property (a) If X is believed, then it is known that X is believed.

Property (m) Belief is monotonic: if an agent believes X, then she believes all propositions $Y \supseteq X$ that logically follow from X.

Property (d) If i believes a proposition then i does not also believe the complement of that proposition.

Property (sc) This is a form of "strong commitment": if the agent does not believe the complement $\overline{X}$, then she must believe any strictly weaker Y implied by X.

If follows from (n) and (d) that $\emptyset \notin N_i(w)$. Indeed, by (n) $[w]_i \in N_i(w)$, hence by (d), $\emptyset = [w]_i - [w]_i \notin N_i(w)$. It follows from (d) and (m) that the intersection of two neighbourhoods of w is non-empty. Indeed, suppose $X, Y \in N_i(w)$. Then $[w]_i - X \notin N_i(w)$, by (d). Therefore, $Y \nsubseteq [w]_i - X$, by (m). Thus, $X \cap Y \neq \emptyset$. Incidentally, (m) follows from (d) and (sc). For let $X \in N_i(w)$. Then $[w]_i - X \notin N_i(w)$ by (d). Let $[w]_i \supseteq Y \supseteq X$. If $Y = X$ then $Y \in N_i(w)$ by what is given about X. If $Y \supsetneq X$ then $Y \in N_i(w)$ by (sc).

Definition 2.2 [ED Language] Let p range over a set of basic propositions P and i over a finite set of agents A.

$$\phi ::= \top \mid p \mid \neg\phi \mid (\phi \wedge \phi) \mid K_i\phi \mid B_i\phi.$$

We will employ the usual abbreviations for $\bot, \vee, \rightarrow$ and $\leftrightarrow$, and we use $\check{K}_i\phi$ for $\neg K_i \neg\phi$ and $\check{B}_i\phi$ for $\neg B_i \neg \phi$. This makes $\check{K}_i, \check{B}_i$ behave as duals (diamonds) to the boxes K_i, B_i.

Definition 2.3 [Truth in Neighbourhood Models] Key clauses are

$$\mathcal{M}, w \models K_i\phi \text{ iff for all } v \in [w]_i : \mathcal{M}, v \models \phi.$$
$$\mathcal{M}, w \models B_i\phi \text{ iff for some } X \in N_i(w)$$
$$\text{it holds that } X = \{v \in [w]_i \mid \mathcal{M}, v \models \phi\}.$$

Our first example illustrates that neighbourhood belief is not closed under conjunction.

Example 2.4

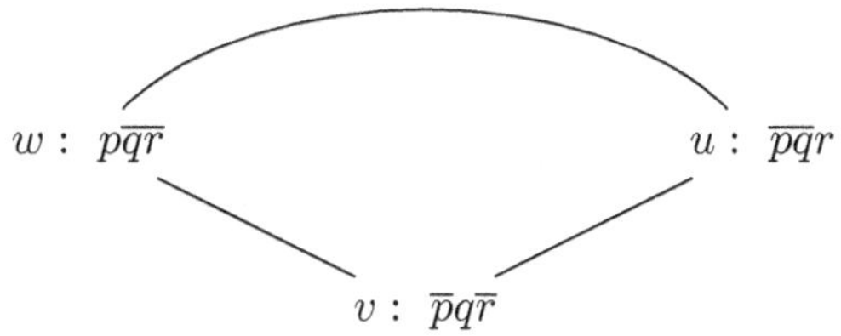

$$N(w) = N(v) = N(u) = \{\{w, v\}, \{v, u\}, \{w, u\}, \{w, v, u\}\}$$

- In all worlds, $K(p \vee q \vee r)$ is true.
- In all worlds $B\neg p$, $B\neg q$, $B\neg r$ are true.
- In all worlds $B(\neg p \wedge \neg q)$, $B(\neg p \wedge \neg r)$, $B(\neg q \wedge \neg r)$ are false.

Example 2.4 illustrates that the lottery puzzle [25] is "solved" in neighbourhood models for belief by non-closure of belief under conjunction. A calculus for ED logic is given in Figure 1.

Theorem 2.5 *The schema $K_i\phi \to B_i\phi$ is derivable in the ED calculus.*

Proof. Write $\vdash \phi$ for derivability in the ED calculus. By propositional logic, $\vdash K_i\phi \to K_i(\top \to \phi)$. From this and (M), by propositional reasoning: $\vdash K_i\phi \to (B_i\top \to B_i\phi)$. By propositional reasoning: $\vdash B_i\top \to (K_i\phi \to B_i\phi)$. From this and (N), by MP: $\vdash K_i\phi \to B_i\phi$. □

Henceforth we use $K_i\phi \to B_i\phi$ as an extra axiom scheme, and call it (KB).

Theorem 2.6 *The rule*

$$\frac{B_i\phi \qquad \phi \to \psi}{B_i\psi}$$

is derivable in the calculus of epistemic-doxastic neighbourhood logic.

Proof. Here is the derivation:

$$\dfrac{B_i\phi \qquad \dfrac{\dfrac{\phi \to \psi}{K_i(\phi \to \psi)}\,\text{Nec-K} \qquad K_i(\phi \to \psi) \to B_i\phi \to B_i\psi \;\; \text{M}}{B_i\phi \to B_i\psi}\,\text{MP}}{B_i\psi}\,\text{MP}$$

□

Theorem 2.7 (Soundness) *The calculus for ED logic is sound for epistemic neighbourhood models.*

AXIOMS

(Taut)	All instances of propositional tautologies
(Dist-K)	$K_i(\phi \to \psi) \to K_i\phi \to K_i\psi$
(T)	$K_i\phi \to \phi$
(PI-K)	$K_i\phi \to K_iK_i\phi$
(NI-K)	$\neg K_i\phi \to K_i\neg K_i\phi$
(N)	$B_i\top$.
(PI-KB)	$B_i\phi \to K_iB_i\phi$
(NI-KB)	$\neg B_i\phi \to K_i\neg B_i\phi$
(M)	$K_i(\phi \to \psi) \to B_i\phi \to B_i\psi$
(D)	$B_i\phi \to \check{B}_i\phi$.
(SC)	$\check{B}_i\phi \wedge \check{K}_i(\neg\phi \wedge \psi) \to B_i(\phi \vee \psi)$

RULES

$$\frac{\phi \to \psi \quad \phi}{\psi}\ (\text{MP}) \qquad \frac{\phi}{K_i\phi}\ (\text{Nec-K})$$

Fig. 1. ED Calculus

Proof. The soundness of Dist-K follows from the fact that the K_i are modal operators. The soundness of T, PI-K and NI-K follows from the fact that the K_i are interpreted as equivalences. The soundness of N follows from property (n). The soundness of PI-KB follows from property (a). The soundness of NI-KB also follows from property (a). The soundness of M follows from property (m). The soundness of D follows from property (d). The soundness of SC follows from property (sc). The two rules preserve soundness, so all theorems of the calculus are sound. □

The following theorem was stated and proved for a weaker notion of neighbourhood model and a weaker calculus, in the unpublished [11].

Theorem 2.8 *Every consistent formula ϕ determines a canonical epistemic neighbourhood model $\mathcal{M}_\phi$.*

Proof. Suppose ϕ is consistent, i.e., $\nvdash \neg\phi$. We construct a canonical epistemic neighbourhood model for ϕ.

Let Φ be the closure of ϕ, that is, Φ is the minimal set such that (i) $\phi \in \Phi$, $\top \in \Phi$, (ii) Φ is closed under taking subformulas (i.e., if χ is a subformula of a formula ψ in Φ, then χ is in Φ), (iii) if $\psi \in \Phi$, and ψ is not a negation, then $\neg\psi \in \Phi$.

We define the canonical model $\mathcal{M}_\phi = (W, \sim, N, V)$. W is the set of all maximal consistent subsets of Φ. W is non-empty because ϕ is supposed to be consistent. Valuations are defined as follows: $V(\mathbf{w}) = Prop \cap \mathbf{w}$.

Since each $\mathbf{w} \in W$ is a set of formulas, we can talk about what is derivable from this set: $\mathbf{w} \vdash \psi$ means that ψ is derivable in the calculus from $\mathbf{w}$.

Relations are defined as follows:

$$\mathbf{w} \sim_i \mathbf{u} \text{ iff for all } \psi \in \Phi : \mathbf{w} \vdash K_i\psi \text{ iff } \mathbf{u} \vdash K_i\psi \text{ and } \mathbf{w} \vdash K_iB_i\psi \text{ iff } \mathbf{u} \vdash K_iB_i\psi.$$

Clearly, all $\sim_i$ are equivalence relations.

Next, we define N_i by means of:

$$N_i(\mathbf{w}) := \{\{\mathbf{v} \in [\mathbf{w}]_i \mid \psi \in \mathbf{v}\} \mid \psi \in \Phi, \mathbf{w} \vdash B_i\psi\}.$$

We check that the properties of neighbourhood functions hold. (c) holds by definition of N_i. (n) holds by the fact that $\mathbf{w} \vdash B_i\top$ (by (N)), and $\{\mathbf{v} \in [\mathbf{w}]_i \mid \top \in \mathbf{v}\} = [\mathbf{w}]_i$.

To show that (a) holds assume $\mathbf{w} \sim_i \mathbf{u}$. Let $X \in N_i(\mathbf{w})$. Then for some $\psi \in \Phi$, $X = \{\mathbf{v} \in [\mathbf{w}]_i \mid \psi \in \mathbf{v}\}$ and $\mathbf{w} \vdash B_i\psi$. Therefore, by (KB), $\mathbf{w} \vdash K_iB_i\psi$. By the definition of $\sim_i$ it follows that $\mathbf{u} \vdash K_iB_i\psi$, and therefore by (T), $\mathbf{u} \vdash B_i\psi$, and we have that $X \in N_i(\mathbf{u})$.

To see that (m) holds, let $\psi \in \Phi$ and assume $\{\mathbf{v} \in [\mathbf{w}]_i \mid \psi \in \mathbf{v}\} \in N_i(\mathbf{w})$. Then $\mathbf{w} \vdash B_i\psi$. Let $\psi' \in \Phi$ and assume $\vdash \psi \to \psi'$. Then by Theorem 2.6, $\mathbf{w} \vdash B_i\psi'$. Therefore $\{\mathbf{v} \in [\mathbf{w}]_i \mid \psi' \in \mathbf{v}\} \in N_i(\mathbf{w})$.

To see that (d) holds, assume for some $\psi \in \Phi$, $\{\mathbf{v} \in [\mathbf{w}]_i \mid \psi \in \mathbf{v}\} \in N_i(\mathbf{w})$. Then $\mathbf{w} \vdash B_i\psi$. By (D), $\mathbf{w} \vdash \neg B_i\neg\psi$. Thus, $\{\mathbf{v} \in [\mathbf{w}]_i \mid \neg\psi \in \mathbf{v}\} \notin N_i(\mathbf{w})$.

Finally, we show that (sc) holds. Let $\psi \in \Phi$ and let $X = \{\mathbf{v} \in [\mathbf{w}]_i \mid \psi \in \mathbf{v}\}$. Then $[\mathbf{w}]_i - X = \{\mathbf{v} \in [\mathbf{w}]_i \mid \neg\psi \in \mathbf{v}\}$. Assume $[\mathbf{w}]_i - X \notin N_i(\mathbf{w})$. Then $\mathbf{w} \vdash \neg B_i\neg\psi$. If $X = [\mathbf{w}]_i$ there is nothing to prove. So assume there is some $\mathbf{v} \in [\mathbf{w}]_i - X$. Then there is $\psi' \in \Phi$ with $\psi' \in \mathbf{v}$ and for all $\mathbf{u} \in X : \psi' \notin \mathbf{u}$. Then $\mathbf{v} \vdash (\psi' \wedge \neg\psi)$, and therefore by (T), $\mathbf{v} \vdash \check{K}(\psi' \wedge \neg\psi)$. Thus by (SC), $\mathbf{v} \vdash B_i(\psi \vee \psi')$, and by (KB), $\mathbf{v} \vdash K_iB_i(\psi \vee \psi')$. From this, by the definition of $\sim_i$, $\mathbf{w} \vdash K_iB_i(\psi \vee \psi')$, and by (T), $\mathbf{w} \vdash B_i(\psi \vee \psi')$. It follows that $X \cup \{\mathbf{v}\} \in N_i(\mathbf{w})$. So we have indeed defined an epistemic neighbourhood model. □

Lemma 2.9 (Truth Lemma) *For all formulas $\psi \in \Phi$, we have $\mathcal{M}_\phi, \mathbf{w} \models \psi$ iff $\psi \in \mathbf{w}$.*

Proof. Induction on ψ. The cases of $\top$, p and the Boolean combinations are straightforward. For the case of $K_i\chi$. We switch for convenience to the dual, and prove $\mathcal{M}_\phi, \mathbf{w} \models \check{K}_i\chi$ iff $\check{K}_i\chi \in \mathbf{w}$.

From left to right: $\mathcal{M}_\phi, \mathbf{w} \models \check{K}_i\chi$ iff there is a $\mathbf{v} \in [\mathbf{w}]_i$ with $\mathcal{M}_\phi, \mathbf{v} \models \chi$ iff there is a $\mathbf{v} \in [\mathbf{w}]_i$ with $\chi \in \mathbf{v}$ by induction hypothesis only if $\check{K}_i\chi \in \mathbf{v}$ by (T) and maximal consistency of $\mathbf{v}$ iff $\check{K}_i\chi \in \mathbf{w}$ by definition of $[\mathbf{w}]_i$ and maximal consistency of $\mathbf{w}$.

From right to left: Suppose $\check{K}_i\chi \in \mathbf{w}$. We have to show $\mathcal{M}_\phi, \mathbf{w} \models \check{K}_i\chi$. For that we have to construct $\mathbf{v} \in [\mathbf{w}]_i$ with $\chi \in \mathbf{v}$. So let

$$\mathbf{v}^- := \{\chi\} \cup \{K_i\psi \in \Phi \mid K_i\psi \in \mathbf{w}\} \cup \{\neg K_i\psi \in \Phi \mid \neg K_i\psi \in \mathbf{w}\}.$$

Then $\mathbf{v}^-$ is consistent. For suppose not. Then there are

$$K_i\psi_1, \ldots, K_i\psi_n, \neg K_i\sigma_1, \ldots, \neg K_i\sigma_m \in \mathbf{v}^-$$

with

$$\vdash K_i\psi_1 \to \cdots \to K_i\psi_n \to \neg K_i\sigma_1 \to \cdots \to \neg K_i\sigma_m \to \neg\chi.$$

With the Nec rule:

$$\vdash K_i(K_i\psi_1 \to \cdots \to K_i\psi_n \to \neg K_i\sigma_1 \to \cdots \to \neg K_i\sigma_m \to \neg\chi).$$

From this, with $n+m$ applications of (Dist-K) and propositional reasoning:

$$\vdash K_iK_i\psi_1 \to \ldots \to K_iK_i\psi_n \to K_i\neg K_i\sigma_1 \to \cdots \to K_i\neg K_i\sigma_m \to K_i\neg\chi.$$

By construction of $\mathbf{v}^-$, $K_i\psi_1, \ldots, K_n\psi_n, \neg K_i\sigma_1, \ldots, \neg K_i\sigma_m$ are in $\mathbf{w}$. Therefore, by (PI-K) and (NI-K),

$$K_iK_i\psi_1, \ldots, K_iK_i\psi_n, K_i\neg K_i\sigma_1, \ldots, K_i\neg K_i\sigma_m$$

follow from $\mathbf{w}$. But this means $K_i\neg\chi \in \mathbf{w}$, by maximal consistency of $\mathbf{w}$, and contradiction with $\check{K}_i\chi \in \mathbf{w}$. It follows that $\mathbf{v}^-$ is consistent. Therefore an extension $\mathbf{v}$ that is Φ maximal consistent exists. By construction, $\mathbf{v} \in [\mathbf{w}]_i$, and $\chi \in \mathbf{v}$. Therefore, by the induction hypothesis, $\mathcal{M}, \mathbf{v} \models \chi$, and it follows that $\mathcal{M}, \mathbf{w} \models \check{K}_i\chi$.

This leaves the case of $B_i\chi$. $\mathcal{M}_\phi, \mathbf{w} \models B_i\chi$ iff (truth definition for $B_i\chi$) $\{\mathbf{v} \in [\mathbf{w}]_i \mid \mathcal{M}_\phi, \mathbf{v} \models \chi\} \in N_i(\mathbf{w})$ iff (induction hypothesis) $\{\mathbf{v} \in [\mathbf{w}]_i \mid \chi \in \mathbf{v}\} \in N_i(\mathbf{w})$ iff ($B_i\chi$ and χ are in Φ) $\{\mathbf{v} \in [\mathbf{w}]_i \mid \chi \in \mathbf{v}\} \in N_i(\mathbf{w})$ iff (definition of N_i, plus the fact that $B_i\chi \in \Phi$) $B_i\chi \in \mathbf{w}$. □

Theorem 2.10 (Completeness of ED Logic) *The calculus of epistemic-doxastic neighbourhood logic is complete for epistemic neighbourhood models:*

$$\text{If } \models \phi \text{ then } \vdash \phi.$$

Proof. Let $\not\vdash \phi$. Then $\neg\phi$ is consistent, and one can find a maximal consistent set $\mathbf{w}$ in the closure of $\neg\phi$ with $\neg\phi \in \mathbf{w}$. By the Truth Lemma, $\mathcal{M}_{\neg\phi}, \mathbf{w} \models \neg\phi$, i.e., $\mathcal{M}_{\neg\phi}, \mathbf{w} \not\models \phi$. Therefore, $\not\models \phi$. □

3 Epistemic Weight Models and Incompleteness

An alternative semantics for the language of epistemic doxastic logic can be given with respect to *Epistemic Weight Models.* As it turns out, the calculus given above is *sound* but *incomplete* for this alternative semantics.

Definition 3.1 [Epistemic Weight Models] An **epistemic weight model** for agents I and basic propositions P is a tuple $\mathcal{M} = (W, R, L, V)$ where

- W is a non-empty countable set of worlds,
- R assigns to every agent $i \in I$ an equivalence relation $\sim_i$ on W,
- L assigns to every $i \in I$ a function $\mathbb{L}_i$ from W to $\mathbb{Q}^+$ (the positive rationals), subject to the following boundedness condition (*).

$$\forall i \in I \forall w \in W \sum_{u \in [w]_i} \mathbb{L}_i(u) < \infty. \qquad (*)$$

where $[w]_i$ is the cell of w in the partition induced by $\sim_i$.

- V assigns to every $w \in W$ a subset of P,

Use $\mathbb{L}_i(X)$ for $\sum_{x \in X} \mathbb{L}_i(x)$.

Definition 3.2 [Truth of ED formulas in Epistemic Weight Models] Key clauses are:

$$\mathcal{M}, w \models K_i\phi \text{ iff } \quad \text{for all } v \in [w]_i : \mathcal{M}, v \models \phi.$$

$$\mathcal{M}, w \models B_i\phi \text{ iff}$$

$$\mathbb{L}_i(\{v \in [w]_i \mid \mathcal{M}, v \models \phi\}) > \mathbb{L}_i(\{v \in [w]_i \mid \mathcal{M}, v \models \neg\phi\}).$$

If we interpret ED formulas in epistemic weight models, we get:

Theorem 3.3 *The ED calculus is sound for epistemic weight models.*

Proof. The axioms are sound, and the rules preserve soundness. □

Definition 3.4 [Agreement] Let $\mathcal{M} = (W, R, N, V)$ be a neighbourhood model and let L be a weight function for $\mathcal{M}$. Then L *agrees with* $\mathcal{M}$ if it holds for all agents i and all $w \in W$ that

$$X \in N_i(w) \text{ iff } \mathbb{L}_i(X) > \mathbb{L}_i([w]_i - X).$$

The following theorem shows that the ED calculus is incomplete for epistemic weight models. A version of this theorem for a weaker notion of neighbourhood model was proved in the unpublished [11], as an adaptation of example 2 from [31, pp. 344-345].

Theorem 3.5 *There exists an epistemic neighbourhood model $\mathcal{M}$ that has no agreeing weight function.*

Proof. Consider the Fano plane from finite geometry (see, e.g., [27, Chapter 9]).

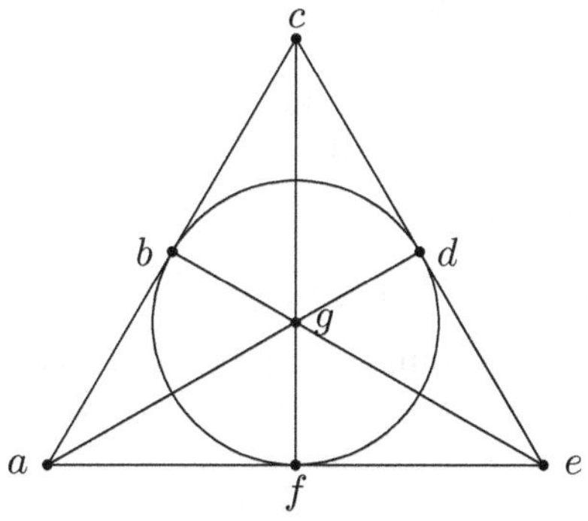

Every pair of distinct points determines a line, every line has exactly three points on it. Let $Prop := \{a, b, c, d, e, f, g\}$. Assume a single agent 0. Define $\mathcal{X}$ as the set of lines in the Fano plane (notation: xyz for $\{x, y, z\}$):

$$\mathcal{X} := \{abc, cde, afe, agd, cgf, egb, bdf\}.$$

No complement of a line contains a line (check this in the figure). Moreover, if one extends the complement of a line with another point, the result will contain a line. This is because any element of a line lies on two points of the complement (check this in the figure). Thus, the members of $\mathcal{X}'$ are the maximal sets that do not contain a line:

$$\begin{aligned}\mathcal{X}' &:= \{\overline{abc}, \overline{cde}, \overline{afe}, \overline{agd}, \overline{cgf}, \overline{egb}, \overline{bdf}\}\\ &= \{defg, abfg, bcdg, bcef, abde, acdf, aceg\}.\end{aligned}$$

Now define the neighbourhoods $\mathcal{Y}$ as the sets that contain (the points of) at least one line:

$$\mathcal{Y} := \{Y \mid \exists X \in \mathcal{X} : X \subseteq Y \subseteq W\}.$$

Let $\mathcal{M} := (W, R, N, V)$ be defined by $W := Prop$, $R_0 = W \times W$, $V(w) = \{w\}$, and for all $w \in W$, $N_0(w) = \mathcal{Y}$. So the neighbourhoods are the sets that contain (the points of) at least one line from the Fano plane.

Check that $\mathcal{X}' \cap \mathcal{Y} = \emptyset$. This shows that condition (d) holds. Condition (sc) holds because adding a point to any member of $\mathcal{X}'$ yields a neighbourhood. The other conditions for neighbourhood models are also easily checked. So $\mathcal{M}$ is a neighbourhood model.

Toward a contradiction, suppose there exists a weight function L that agrees with $\mathcal{M}$. Since each letter $p \in W$ occurs in exactly three of the seven members of $\mathcal{X}$, we have:

$$\sum_{X \in \mathcal{X}} \mathbb{L}_0(X) = \sum_{p \in W} 3 \cdot \mathbb{L}_0(\{p\}).$$

Since each letter $p \in W$ occurs in exactly four of the seven members of $\mathcal{X}'$, we have:

$$\sum_{X \in \mathcal{X}'} \mathbb{L}_0(X) = \sum_{p \in W} 4 \cdot \mathbb{L}_0(\{p\}).$$

On the other hand, from the fact that $\mathbb{L}_0(X) > \mathbb{L}_o(\overline{X})$ for all members X of $\mathcal{X}$ we get:

$$\sum_{X \in \mathcal{X}} \mathbb{L}_0(X) > \sum_{X \in \mathcal{X}} \mathbb{L}_0(\overline{X}) = \sum_{X \in \mathcal{X}'} \mathbb{L}_0(X).$$

Contradiction. So no such $\mathbb{L}_0$ exists. □

4 Epistemic Weight Models and Completeness

We will now rephrase the system of simplified probabilistic epistemic logic of [12], using an alternative syntax, following [9]. We call the result *Epistemic Comparison Logic*. In the next definition, we use $\oplus$ as a list-forming operation for formulas, so $\phi_1 \oplus \cdots \oplus \phi_n$ should be thought of as the n-element formula list $(\phi_1, \ldots, \phi_n)$. We will use Φ to range over formula lists, and $\phi \oplus \Phi$ for the extension of the formula list Φ at the front with the formula ϕ.

Definition 4.1 [EC Language]

$$\phi ::= \top \mid p \mid \neg\phi \mid \phi \wedge \phi \mid \Phi \leq_i \Phi \qquad \Phi ::= \phi \mid \phi \oplus \Phi$$

Abbreviations: As usual for $\bot, \vee, \rightarrow, \leftrightarrow$. $\Phi <_i \Psi$ for $\Phi \leq_i \Psi \wedge \neg\Psi \leq_i \Phi$. $\Phi =_i \Psi$ for $\Phi \leq_i \Psi \wedge \Psi \leq_i \Phi$. $B_i\phi$ for $(\neg\phi) <_i \phi$, $\breve{B}_i\phi$ for $(\neg\phi) \leq_i \phi$ ("Belief as willingness to bet"), $K_i\phi$ for $\top \leq_i \phi$, $\breve{K}_i\phi$ for $\bot <_i \phi$ ("Knowledge as certainty").

Definition 4.2 [Truth for EC Logic] Let $\mathcal{M} = (W, R, L, V)$ be an epistemic weight model, let $w \in W$.

$$
\begin{array}{lll}
[\![\phi]\!]_{\mathcal{M}} & := & \{w \in W \mid \mathcal{M}, w \models \phi\} \\
[\![\phi]\!]_{\mathcal{M}}^{w,i} & := & [\![\phi]\!]_{\mathcal{M}} \cap [w]_i \\
\mathbb{L}_{w,i}\phi & := & \mathbb{L}_i([\![\phi]\!]_{\mathcal{M}}^{w,i}) \\
\mathcal{M}, w \models \top & & \text{always} \\
\mathcal{M}, w \models \neg\phi & \text{iff} & \text{not } \mathcal{M}, w \models \phi \\
\mathcal{M}, w \models \phi_1 \wedge \phi_2 & \text{iff} & \mathcal{M}, w \models \phi_1 \text{ and } \mathcal{M}, w \models \phi_2 \\
\mathcal{M}, w \models \Phi \leq_i \Psi & \text{iff} & \sum_{\phi \in \Phi} \mathbb{L}_{w,i}\phi \leq \sum_{\psi \in \Psi} \mathbb{L}_{w,i}\psi
\end{array}
$$

Note that in $\sum_{\phi \in \Phi}$, we sum over *occurrences* of ϕ in the list Φ.

Weight function and epistemic accessibility relation together determine probability:

$$P_{w,i}^{\mathcal{M}}\phi := \frac{\mathbb{L}_{w,i}\phi}{\mathbb{L}_{w,i}\top} \left(= \frac{\mathbb{L}_i([\![\phi]\!]_{\mathcal{M}} \cap [w]_i)}{\mathbb{L}_i([w]_i)} \right)$$

In a slogan: "Probabilities are weights normalized for epistemic partition cells."

Example 4.3 Two bankers i, j consider buying stocks in three firms a, b, c that are involved in a takeover bid. There are three possible outcomes: a for "a wins", b for "b wins", and c for "c wins." i takes the winning chances to be $3:2:1$, j takes them to be $1:2:1$. i: solid lines, j: dashed lines.

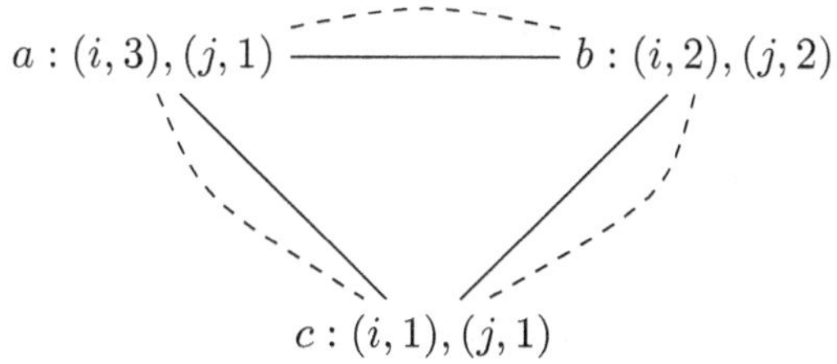

We see that i is willing to bet $1:1$ on a, while j is willing to bet $3:1$ against a. It follows that in this model i and j have an opportunity to gamble, for, to put it in Bayesian jargon, they do not have a common prior.

Suppose j has foreknowledge about what firm c will do.

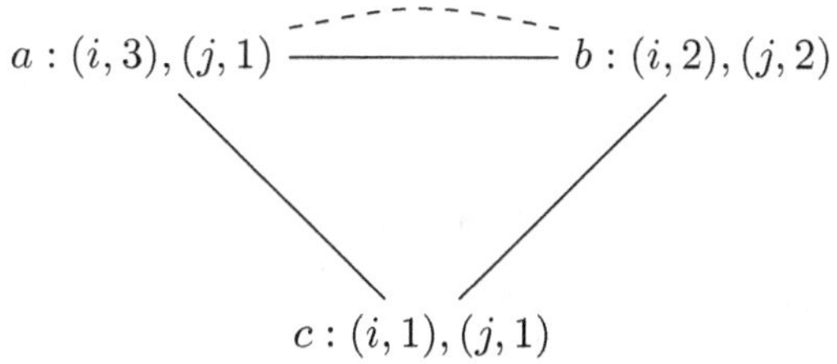

Taut instances of propositional tautologies

ProbT $(\top \leq_i \phi) \to \phi$

ProbImpl $\top \leq_i (\phi \to \psi) \to (\phi \leq_i \psi)$

PropPos $(\Phi \leq_i \Psi) \to \top \leq_i (\Phi \leq_i \Psi)$

PropNeg $(\Phi >_i \Psi) \to \top \leq_i (\Phi >_i \Psi)$

PropAdd $(\phi \wedge \psi) \oplus (\phi \wedge \neg\psi) =_i \phi$

Tran $(\Phi \leq_i \Psi) \wedge (\Psi \leq_i \Xi) \to (\Phi \leq_i \Xi)$

Tot $(\Phi \leq_i \Psi) \vee (\Psi \leq_i \Phi)$

ComL $(\Phi_1 \oplus \Phi_2 \leq_i \Psi) \leftrightarrow (\Phi_2 \oplus \Phi_1 \leq_i \Psi)$

ComR $(\Phi \leq_i \Psi_1 \oplus \Psi_2) \leftrightarrow (\Phi \leq_i \Psi_2 \oplus \Psi_1)$

Add $(\Phi_1 \leq_i \Psi_1) \wedge (\Phi_2 \leq_i \Psi_2) \to (\Phi_1 \oplus \Phi_2 \leq_i \Psi_1 \oplus \Psi_2)$

Succ $(\Phi \oplus \top \leq_i \Psi \oplus \top) \to (\Phi \leq_i \Psi)$

MP From $\vdash \phi$ and $\vdash \phi \to \psi$ derive $\vdash \psi$

NEC From $\vdash \phi$ derive $\vdash 1 \leq_i \phi$

Fig. 2. EC Calculus

The probabilities assigned by i remain as before. The probabilities assigned by j have changed, as follows. In worlds a and b, j assigns probability $\frac{1}{3}$ to a and $\frac{2}{3}$ to b. In world c, j is sure of c.

We may suppose that this new model results from j being informed about the truth value of c, while i is aware that j received this information, but without i getting the information herself. So i is aware that j's subjective probabilities have changed, and it would be unwise for i to put her beliefs to the betting test. For although i cannot distinguish the three situations, she knows that j can distinguish the c situation from the other two. Willingness of j to bet against a at any odds can be interpreted by i as an indication that c is true, thus forging an intimate link between action and information update.

The probability comparison calculus is given in Figure 2.

Definition 4.4 [EC Derivability] $\Gamma \vdash \phi$ holds if either $\phi \in \Gamma$, or ϕ is an axiom, or ϕ follows by means of the rules of the calculus from axioms or members of Γ, while taking care that application of NEC only is allowed when the set of premisses Γ is empty.

Let $\vdash_{EC}$ denote derivability in the probability comparison calculus. The deduction theorem holds for this calculus:

Theorem 4.5 $\Gamma \cup \{\phi\} \vdash \psi$ *iff* $\Gamma \vdash \phi \to \psi$.

Proof. The proof for this is folklore, but see [18] for an explanation of why the restriction on the use of NEC is crucial for this. □

The deduction theorem will help us to state a number of useful derivable principles.

(i) From ProbImpl, with propositional reasoning:

$$\vdash \top \leq_i (\phi \leftrightarrow \psi) \rightarrow (\phi =_i \psi).$$

(ii) Consider the following instance of ProbT: $\vdash (\top \leq_i \bot) \rightarrow \bot$. By propositional logic, $\vdash \neg(\top \leq_i \bot)$, i.e., $\vdash \bot <_i \top$.

(iii) $\vdash \Phi =_i \Phi$ by Tot.

(iv) From PropAdd, $\vdash (\phi \wedge \bot) \oplus (\phi \wedge \top) =_i \phi$. By propositional reasoning, $\vdash \bot \oplus \phi =_i \phi$.

(v) $\vdash \bot \oplus \Phi =_i \Phi$ by an easy induction using the previous two items.

(vi) Plugging in $\top$ as a special case in (3), we get $\vdash \bot \oplus \top =_i \top$.

(vii) Plugging in $\bot$ as a special case in (3), we get $\vdash \bot \oplus \bot =_i \bot$.

(viii) From $\vdash \phi \rightarrow \psi$, derive $\vdash \top \leq_i (\phi \rightarrow \psi)$, by NEC. Combine this with $\vdash \top \leq_i (\phi \rightarrow \psi) \rightarrow (\phi \leq_i \psi)$ (ProbImpl) to get $\vdash \phi \leq_i \psi$. Thus we have derived the rule PR:

$$\frac{\vdash \phi \rightarrow \psi}{\vdash \phi \leq_i \psi}$$

(ix) Assume $\vdash \phi \leftrightarrow \psi$. Then also $\vdash \phi \rightarrow \psi$ and $\vdash \psi \rightarrow \phi$. From the former, with PR (previous item), $\vdash \phi \leq_i \psi$, and from the latter $\vdash \psi \leq_i \phi$. Combining these, we get $\vdash \phi =_i \psi$. This gives the derived inference rule:

$$\frac{\vdash \phi \leftrightarrow \psi}{\vdash \phi =_i \psi}$$

(x) From PropAdd, $\vdash (\top \wedge \phi) \oplus (\top \wedge \neg\phi) =_i \top$, so by propositional reasoning, $\vdash \phi \oplus \neg\phi =_i \top$.

(xi) Assume $\vdash \phi =_i \top$. Then $\vdash \neg\phi \oplus \phi =_i \neg\phi \oplus \top$, and hence $\vdash \neg\phi \oplus \top =_i \top$, by the previous item. Since $\vdash \bot \oplus \top =_i \top$ we get $\vdash \neg\phi \oplus \top =_i \bot \oplus \top$, and therefore, by Succ, $\vdash \neg\phi =_i \bot$. By the deduction theorem, we have derived $\vdash \phi =_i \top \rightarrow \neg\phi =_i \bot$.

(xii) In a similar way we can derive: $\vdash \phi =_i \bot \rightarrow \neg\phi =_i \top$.

(xiii) Assume $\vdash \top \leq_i \phi$ and $\vdash \top \leq_i \phi \rightarrow \psi$. From this: $\vdash \phi =_i \top$ and $\vdash \phi \rightarrow \psi =_i \top$. From $\vdash \phi =_i \top, \vdash \neg\phi \wedge \neg\psi =_i \bot$ and from $\vdash \phi \rightarrow \psi =_i \top$ we get that $\vdash \phi \wedge \neg\psi =_i \bot$. Since $\vdash \neg\psi =_i (\phi \wedge \neg\psi) \oplus (\neg\phi \wedge \neg\psi)$, by PropAdd, we derive that $\vdash \neg\psi =_i \bot$, and it follows that $\vdash \psi =_i \top$, hence $\vdash \top \leq_i \psi$. By the deduction theorem, we have derived the distribution principle for certainty: $\vdash (\top \leq_i \phi \wedge \top \leq_i (\phi \rightarrow \psi) \rightarrow \top \leq_i \psi$.

(xiv) Using the previous item, we get:

$$\cfrac{\cfrac{\vdash \phi \rightarrow \psi}{\vdash \top \leq_i (\phi \rightarrow \psi)}\ \text{NEC}}{\vdash \top \leq_i \phi \rightarrow \top \leq_i \psi}\ \text{Distribution}$$

(xv) From Tot, by the definitions of $<_i$, $>_i$ and $=_i$:

$$\vdash \Phi <_i \Psi \vee \Phi =_i \Psi \vee \Phi >_i \Psi.$$

(xvi) By Tran and the definition of $<_i$:
$\vdash \Phi <_i \Psi \wedge \Psi \leq_i \Xi \rightarrow \Phi <_i \Xi,$
$\vdash \Phi <_i \Psi \wedge \Psi \leq_i \Xi \rightarrow \Phi <_i \Xi,$
$\vdash \Phi <_i \Psi \wedge \Psi <_i \Xi \rightarrow \Phi <_i \Xi.$

(xvii) By Add, and the definition of $=_i$:

$$\vdash \Phi_1 =_i \Phi_2 \wedge \Psi_1 =_i \Psi_2 \rightarrow \Phi_1 \oplus \Psi_1 =_i \Phi_2 \oplus \Psi_2.$$

(xviii) Assume $\vdash \phi \leq_i \psi$. Then by Add, $\vdash \neg\phi \oplus \neg\psi \oplus \phi \leq_i \neg\phi \oplus \neg\psi \oplus \psi$. By rearranging and applying (x), $\vdash \neg\psi \oplus 1 \leq_i \neg\phi \oplus 1$. By Succ, $\vdash \neg\psi \leq_i \neg\phi$. By the deduction theorem we have shown $\vdash \phi \leq_i \psi \rightarrow \neg\psi \leq_i \neg\phi$.

Given the abbreviations for $K_i\phi$ and $B_i\phi$ given after Definition 4.1, we see that we can take the ED language of Definition 2.2 to be a fragment of the EC language of Definition 4.1. It then turns out that everything that is derived in the ED calculus is also derived in the EC calculus.

Theorem 4.6 *For all ϕ in the ED language and Γ of ED-formula sets: if $\Gamma \vdash_{ED} \phi$ then $\Gamma \vdash_{EC} \phi$.*

Proof. We show that the (translations of the) axioms and rules of the ED calculus are all derivable in the EC calculus.

K_i necessitation. This is NEC.

Dist-K $K_i(\phi \rightarrow \psi) \rightarrow K_i\phi \rightarrow K_i\psi$. This was proved in (xiii) above.

T: Follows from PropT.

PI-K: Follows from ProbPos

NI-K: Follows from ProbNeg

N: This is $\bot <_i \top$. Proved in (ii) above.

PI-KB: Follows from ProbNeg.

NI-KB: Follows from ProbPos.

M: Assume $\vdash \top \leq_i (\phi \rightarrow \psi)$ and $\vdash \neg\phi <_i \phi$. We have to show that $\vdash \neg\psi <_i \psi$. From $\vdash \top \leq_i (\phi \rightarrow \psi)$ get $\vdash \phi \leq_i \psi$ by ProbImpl. From $\vdash \phi \leq_i \psi$ we get $\vdash \neg\psi \leq_i \neg\phi$ by (xviii) above. From $\vdash \neg\phi <_i \phi$ and $\vdash \phi \leq_i \psi$, by Trans, $\vdash \neg\phi < \psi$. From this and $\vdash \neg\psi \leq_i \neg\phi$ by Trans $\vdash \neg\psi < \psi$.

D: From $\vdash \neg\phi <_i \phi$ by Tot $\vdash \neg\phi \leq_i \phi$.

SC: Assume $\vdash (\neg\phi <_i \phi)$ and $\vdash \bot <_i (\neg\phi \wedge \psi)$. We have to show $\vdash (\neg\phi \wedge \neg\psi) <_i (\phi \vee \psi)$. For this we can use the equivalence of $\phi \vee \psi$ and $(\phi \wedge \neg\psi) \vee (\phi \wedge \psi) \vee (\neg\phi \wedge \psi)$. □

Theorem 4.7 *Every consistent EC formula ϕ determines a canonical epistemic weight model $\mathcal{M}_\phi$.*

Proof. Suppose ϕ is consistent, i.e., $\nvdash \neg\phi$. We construct a canonical epistemic weight model for ϕ.

Let $\mathbf{\Phi}$ be the set of all subformulas of ϕ, closed under single negations. The subformulas of $\Psi \leq_i \Xi$ are all subformulas of formulas ψ that occur as $\oplus$ terms in Ψ or Ξ, plus the results $\Psi' \leq_i \Xi'$ of leaving out $\oplus$ terms in Ψ or Ξ while taking care that Ψ' and Ξ' are not empty.

We define the canonical model $\mathcal{M}_\phi = (W, R, L, V)$. W is the set of all maximal consistent subsets of $\mathbf{\Phi}$. W is non-empty because ϕ is supposed to be consistent.

Valuations are defined as follows: $V(\mathbf{w}) = \mathit{Prop} \cap \mathbf{w}$.

Let $\text{sat}(\mathbf{w}) = \{\psi \in \mathbf{\Phi} \mid \mathbf{w} \vdash \psi\}$, that is, $\text{sat}(\mathbf{w})$ is the set of $\mathbf{\Phi}$-formulas that are provable from $\mathbf{w}$.

Notice that it follows by the soundness of the probability comparison calculus that all members of $\text{sat}(\mathbf{w})$ are true in $\mathbf{w}$.

Relations $\sim_i$ are defined as follows: $\mathbf{w} \sim_i \mathbf{u}$ iff $\text{sat}(\mathbf{w})$ and $\text{sat}(\mathbf{u})$ contain the same i-comparison formulas. Clearly, all $\sim_i$ are equivalence relations.

Now it remains to define L. Consider an agent i and an equivalence class $[\mathbf{w}]_i$ in the canonical model $\mathcal{M}_\phi$. All worlds $\mathbf{u}$ of $[\mathbf{w}]_i$ contain the same i-comparison formulas.

We show how to transform all these i-comparison formulas in a system of linear inequalities that is consistent.

For all $\mathbf{u} \in W$, we write $\phi_{\mathbf{u}}$ for the conjunction of all formulas in $\mathbf{u}$. We have:

- $\vdash \phi_{\mathbf{u}} \rightarrow \neg\phi_{\mathbf{v}}$ if $\mathbf{u} \neq \mathbf{v}$ by propositional logic.

Given any formula ψ of $\mathbf{\Phi}$, we have

- $\vdash \psi \leftrightarrow \bigvee_{\{\mathbf{u} \in W \mid \psi \in \mathbf{u}\}} \phi_{\mathbf{u}}$ by propositional logic.

Since the $\phi_{\mathbf{u}}$ are all mutually inconsistent we can prove in the calculus:

- $\vdash \psi =_i \bigoplus\{\phi_{\mathbf{u}} \mid \mathbf{u} \in W \text{ and } \psi \in \mathbf{u}\}$.

Now, when we i-compare $\phi_{\mathbf{u}}$ to $\bot$ in $\mathbf{w}$, we should obtain $>_i$ iff $\mathbf{u} \in [\mathbf{w}]_i$. Let us prove this fact.

- If $\mathbf{u} \in [\mathbf{w}]_i$, we have:
  (i) $\vdash \phi_{\mathbf{u}} \rightarrow \bot <_i \phi_{\mathbf{u}}$ by (ProbT) and (PropAdd).
  (ii) $\bot <_i \phi_{\mathbf{u}} \in \text{sat}(\mathbf{u})$;
  (iii) $\bot <_i \phi_{\mathbf{u}} \in \text{sat}(\mathbf{w})$ because $\mathbf{u} \in [\mathbf{w}]_i$.
  Therefore, $\bot <_i \phi_{\mathbf{u}}$ follows from $\mathbf{w}$.
- Suppose $\mathbf{u} \notin [\mathbf{w}]_i$. Then $\mathbf{u}$ and $\mathbf{w}$ differ by at least one i-comparison formula $\Psi \leq_i \Xi \in \mathbf{\Phi}$. Without loss of generality, assume $\Psi \leq_i \Xi \in \mathbf{w}$ and $\Psi \leq_i \Xi \notin \mathbf{u}$. Then we have:
  (i) $\vdash \phi_{\mathbf{w}} \rightarrow \Psi \leq_i \Xi$ by propositional logic;
  (ii) $\vdash \Psi \leq_i \Xi \rightarrow \neg\phi_{\mathbf{u}}$ by propositional logic;
  (iii) $\vdash \Psi \leq_i \Xi \rightarrow \top \leq_i (\Psi \leq_i \Xi)$ by axiom (PropT);
  (iv) $\vdash \phi_{\mathbf{w}} \rightarrow \top \leq_i (\Psi \leq_i \Xi)$ by propositional logic;
  (v) $\vdash \phi_{\mathbf{w}} \rightarrow \top \leq_i (\neg\phi_{\mathbf{u}})$ by ii and iv.
  (vi) $\vdash \phi_{\mathbf{w}} \rightarrow (\phi_{\mathbf{u}} \leq_i \bot)$ from the above by PropAdd and propositional reasoning.
  Therefore $\phi_{\mathbf{u}} \leq_i \bot$ follows from $\mathbf{w}$.

Thus, ψ has the same i-weight as $\{\phi_{\mathbf{u}} \mid \mathbf{u} \in [\mathbf{w}]_i \text{ and } \psi \in \mathbf{u}\}$. We can prove in the calculus:

- $\vdash \psi =_i \bigoplus\{\phi_{\mathbf{u}} \mid \mathbf{u} \in [\mathbf{w}]_i \text{ and } \psi \in \mathbf{u}\}$.

Now let $\Psi \leq_i \Xi$ be any i-comparison formula of $\mathbf{w}$. Then we can replace any $\oplus$ term ψ occurring in either Ψ or Ξ by a list of terms $\bigoplus\{\phi_{\mathbf{u}} \mid \mathbf{u} \in [\mathbf{w}]_i \text{ and } \psi \in \mathbf{u}\}$ with the same i-weight. Let the result of this be $\Psi' \leq_i \Xi'$. Regrouping the $\oplus$ terms in Ψ' and Ξ', using the abbreviation $n\chi$ for $\underbrace{\chi \oplus \cdots \oplus \chi}_{n \text{ times}}$, 0 for $\bot \oplus \cdots \oplus \bot$, m for $\underbrace{\top \oplus \cdots \oplus \top}_{m \text{ times}}$, and replacing $\oplus$ by $+$ and $\leq_i$ by $\leq$ gives a linear inequality

$$a_1\phi_{\mathbf{u}_1} + \cdots + a_n\phi_{\mathbf{u}_n} + k \leq b_1\phi_{\mathbf{v}_1} + \cdots + b_m\phi_{\mathbf{v}_m} + l$$

where a_i, b_j, k, l are non-negative integers, and the $\phi_{\mathbf{u}}$ and $\phi_{\mathbf{v}}$ figure as variables. Applying this recipe to each i-comparison formula in $\mathbf{w}$, we get a system of linear inequalities made up of i-inequalities in $\mathbf{w}$.

The set sat($\mathbf{w}$) is consistent so the above system, which is a rephrasing of inequations that are in sat($\mathbf{w}$), is also consistent and therefore satisfiable [13, Theorem 2.2]. Let $(x_u^*)_{\mathbf{u}\in[\mathbf{w}]_i}$ be a solution, and define $\mathbb{L}_i(\mathbf{u}) = x_u^*$. □

Lemma 4.8 (Truth Lemma) *Let ϕ be a consistent EC formula and $\mathbf{\Phi}$ the set of all its subformulas closed under single negations. Then for all formulas $\psi \in \mathbf{\Phi}$, we have $\mathcal{M}_\phi, \mathbf{w} \models \psi$ iff $\psi \in \mathbf{w}$.*

Proof. Induction on ψ. □

Theorem 4.9 (Completeness of Epistemic Comparison Logic) *The calculus of epistemic comparison logic given in Figure 2 is complete for epistemic weight models:*

$$\text{If } \models \phi \text{ then } \vdash_{EC} \phi.$$

Proof. Let $\not\vdash \phi$. Then $\neg\phi$ is consistent, and one can find a maximal consistent set $\mathbf{w}$ in the closure of $\neg\phi$ with $\neg\phi \in \mathbf{w}$. By the Truth Lemma, $\mathcal{M}_{\neg\phi}, \mathbf{w} \models \neg\phi$, i.e., $\mathcal{M}_{\neg\phi}, \mathbf{w} \not\models \phi$. Therefore, $\not\models \phi$. □

From Epistemic Probability Models to Epistemic Neighbourhood Models: If $\mathcal{M} = (W, R, L, V)$ is an epistemic weight model, then $\mathcal{M}^\bullet$ is the tuple (W, R, N, V) given by replacing the weight function by a function N, where N is defined as follows, for $i \in Ag$, $w \in W$.

$$N_i(w) = \{X \subseteq [w]_i \mid \mathbb{L}_i(X) > \mathbb{L}_i([w]_i - X)\}.$$

Theorem 4.10 *For any epistemic weight model $\mathcal{M}$ it holds that $\mathcal{M}^\bullet$ is a neighbourhood model.*

Proof. Check that all neighbourhood conditions hold for $\mathcal{M}^\bullet$. In particular, (sc) holds, because for every agent every world has strictly positive weight. □

Finally, it follows from the fact that the interpretation of $B_i\phi$ formulas only uses neighbourhoods that it makes no difference whether we interpret ED formulas in epistemic weight models or in their corresponding neighbourhood models.

Theorem 4.11 *For all ED formulas* ϕ*, for all epistemic probability models* $\mathcal{M}$*, for all worlds* w *of* $\mathcal{M}$*:* $\mathcal{M}^\bullet, w \models \phi$ *iff* $\mathcal{M}, w \models \phi$.

5 Updates

You are from a population with a statistical chance of 1 in 100 of having disease D. The initial screening test for this has a false positive rate of 0.2 and a false negative rate of 0.1. You tested positive (T). Should you believe you have disease D? We can model this with public announcement update.

Example 5.1 [Disease and Test]

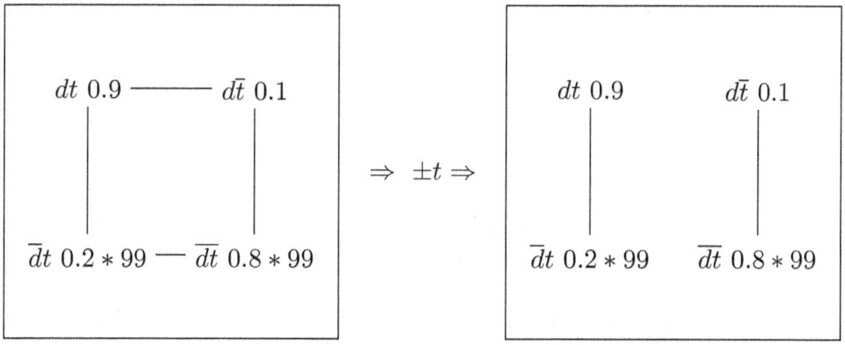

Extend the EC language with an operator $[\pm\phi]$, for publicly announcing the *value* of ϕ. This operator is defined as a map on epistemic weight models that maps $\mathcal{M}$ to $\mathcal{M}^{\pm\phi}$, where $\mathcal{M}^{\pm\phi}$ is given by the next definition.

Definition 5.2 If $\mathcal{M} = (W, \sim, L, V)$ is an epistemic weight model and ϕ is a formula of the EC language, then $\mathcal{M}^{\pm\phi} = (W^{\pm\phi}, \sim^{\pm\phi}, L^{\pm\phi}, V^{\pm\phi})$ is given by:

- $W^{\pm\phi} = W$,
- $\sim_i^{\pm\phi} = \{(w, v) \in W^2 \mid w \sim_i v \text{ and } \mathcal{M}, w \models \phi \text{ iff } \mathcal{M}, v \models \phi\}$.
- $L^{\pm\phi} = L$,
- $V^{\pm\phi} = V$.

Intuitively, what the operation $[\pm\phi]$ does is cut the i-accessibility links between ϕ and $\neg\phi$ worlds, for all agents i. Everything else remains the same.

The model shows that after the update with t the probability of d equals $\frac{0.9}{0.9+0.2*99} = \frac{9}{207} = \frac{1}{23}$, and after the update with $\neg t$ the probability of d equals $\frac{0.1}{0.1+0.8*88} = \frac{1}{704}$. A huge difference, but with both outcomes it is vastly more probable that you do not have the disease than that you have it.

Compare this with applying Bayes' Rule:

Example 5.3 [Applying Bayes' Rule]

$$P(D|T) = \frac{P(T|D)P(D)}{P(T)} = \frac{P(T|D)P(D)}{P(T|D)P(D) + P(T|\neg D)P(\neg D)}$$

Filling in $P(T|D) = 0.9, P(D) = 0.01, P(\neg D) = 0.99, P(T|\neg D) = 0.2$ gives $P(D|T) = \frac{1}{23}$.

Public announcement update of an epistemic weight model and application of Bayes' rule give the same result, in the precise sense that the probability in the input model gives the prior and the probability in the updated model the posterior.

Public announcement can also be defined for epistemic neighbourhood models, which gives us a kind of poor man's Bayesian update. To interpret $[\pm\phi]$ in epistemic neighbourhood models, define:

Definition 5.4 If $\mathcal{M} = (W, \sim, N, V)$ is an epistemic neighbourhood model and ϕ is a formula of the ED language, then $\mathcal{M}^{\pm\phi} = (W^{\pm\phi}, \sim^{\pm\phi}, N^{\pm\phi}, V^{\pm\phi})$ is given by:

- $W^{\pm\phi} = W$,
- $\sim_i^{\pm\phi} = \{(w, v) \in W^2 \mid w \sim_i v \text{ and } \mathcal{M}, w \models \phi \text{ iff } \mathcal{M}, v \models \phi\}$.
- $N_i^{\pm\phi}(w) = \{X \cap [w]_i^{\pm\phi} \mid X \in N_i(w)\}$,
- $V^{\pm\phi} = V$.

Thus, in $\mathcal{M}^{\pm\phi}$ the epistemic accessibilities for all agents are adjusted by cutting the links between ϕ and $\neg\phi$ worlds, and all neighbourhoods are restricted to the appropriate new epistemic accessibility cells. It is not hard to check that the definition is correct: if $\mathcal{M}$ is an epistemic neighbourhood model, then $\mathcal{M}^{\pm\phi}$ is an epistemic neighbourhood model. Axiomatizing the logic of poor man's Bayesian belief (the logic of ED plus the $[\pm\phi]$ operation on epistemic neighbourhood models) is future work.

6 Conclusion and Further Work

We have compared two classes of models for knowledge and belief, one based on neighbourhood semantics, one based on adding weights to worlds in S5 Kripke models, and we have given sound and complete calculi for both model classes. We have also shown that the calculus for neighbourhood models is sound but incomplete for weight models. One can view this as a point in favour of weight models, but one might just as well say that neighbourhood models allow us to make distinctions between shades of belief that are lost in weight models.

In [5], update models for probabilistic epistemic logic are built from sets of formulas that are mutually exclusive. In the present set-up, it is possible to stay a bit closer to the original update model from [2]. A weighted update model is like a weighted epistemic model, but with the valuation function replaced by a function that assigns preconditions and actions (substitutions) to events. The actions take care of factual changes: changes in the values of proposition letters. Update is a product operation, as in [2]. The new i-weight for (w, e) is computed as the product the weights of w and of e. Update products satisfy the requirements for epistemic weight models. Example 5.1 is a special case of this definition. A full blown probabilistic logic of communication and change, in the spirit of [6], is given in [1]. It is future work to give a similar axiomatic treatment to extend the EC calculus.

Alternatively, one might wish to carve out a 'natural' subspace of the set of all possible operations on epistemic weight models, by focussing on a limited set of operations, such as $[\pm\phi_Q]$, for the operation that reveals the value of ϕ to a subset Q of the set of all agents, and $[(\phi, N, M)_Q]$ (with N, M strictly positive natural numbers), for the operation that multiplies the i-weights of the ϕ worlds by N/M (and that of the other worlds by $1 - N/M$) for agents $i \in Q$, $[\sigma_Q]$, for the operation that applies the factual change operation σ, and makes the result visible to agents in Q, while other agents confuse this with 'no change', and finally, $[(p :=?, N, M)]$ for nondetermined change, an update action that makes p true with probability N/M. Each of these operators presents an axiomatisation challenge of its own. Next, one would like to find an illuminating description of the class of updates that the combination of $[\pm\phi_Q]$, $[(\phi, N, M)_Q]$, $[\sigma_Q]$ and $[(p :=?, N, M)]$ allows.

In future, we also would like to extend our treatment to capture the important distinction made in [22] between **risk** and **uncertainty**.

Acknowledgement

The paper has benefited from the insightful remarks of three anonymous AiML referees. During the *Trimestre Thématique CIPPMI* in Toulouse, Spring 2016, Professor Jaruslav Nešetřil drew the attention of the first author to the fact that the example used in the proof of Theorem 3.5 can be interpreted in the Fano plane in finite geometry This allowed us to give a much neater presentation of the incompleteness result.

References

[1] Achimescu, A. C., "Games and Logics for Informational Cascades," Master's thesis, ILLC, Amsterdam (2014).

[2] Baltag, A., L. Moss and S. Solecki, *The logic of public announcements, common knowledge, and private suspicions*, in: I. Bilboa, editor, *Proceedings of TARK'98*, 1998, pp. 43–56.

[3] Benthem, J. v., *Conditional probability meets update logic*, Journal of Logic, Language and Information **12** (2003), pp. 409–421.

[4] Benthem, J. v., D. Fernández-Duque and E. Pacuit, *Evidence logic: A new look at neighborhood structures*, Annals of Pure and Applied Logic **165** (2014), pp. 106–133.

[5] Benthem, J. v., J. Gerbrandy and B. Kooi, *Dynamic update with probabilities*, Studia Logica **93** (2009), pp. 67–96.

[6] Benthem, J. v., J. van Eijck and B. Kooi, *Logics of communication and change*, Information and Computation **204** (2006), pp. 1620–1662.

[7] Burgess, J., *Probability logic*, Journal of Symbolic Logic **34** (1969), pp. 264–274.

[8] Chellas, B., "Modal Logic: An Introduction," Cambridge University Press, 1980.

[9] Delgrande, J. P. and B. Renne, *The logic of qualitative probability*, in: *Proceedings of the Twenty-Fourth International Conference on Artificial Intelligence (IJCAI 2015)*, Buenos Aires, 2015, pp. 2904–2910.

[10] Demey, L. and J. Sack, *Epistemic probabilistic logic*, in: H. v. Ditmarsch, J. Halpern, W. van der Hoek and B. Kooi, editors, *Handbook of Epistemic Logic*, College Publications, London, 2015 pp. 147–202.

[11] Eijck, J. v. and B. Renne, *Belief as willingness to bet*, E-print, arXiv.org (2014), arXiv:1412.5090v1 [cs.LO].
URL `http://arxiv.org/abs/1412.5090v1`

[12] Eijck, J. v. and F. Schwarzentruber, *Epistemic probability logic simplified*, in: R. Goré, B. Kooi and A. Kurucz, editors, *Advances in Modal Logic, Volume 10*, 2014, pp. 158–177.

[13] Fagin, R., J. Y. Halpern and N. Megiddo, *A logic for reasoning about probabilities*, Information and computation **87** (1990), pp. 78–128.

[14] Finetti, B. d., *La prevision: ses lois logiques, se sources subjectives*, Annales de l'Institut Henri Poincaré **7** (1937), pp. 1–68, translated into English and reprinted in Kyburg and Smokler, Studies in Subjective Probability (Huntington, NY: Krieger; 1980).

[15] Finetti, B. d., *La logica del plausibile secondo la concezione di polya*, in: *Atti della XLII Riunione, Societa Italiana per il Progresso delle Scienze*, 1951, pp. 227–236.

[16] Foley, R., *Beliefs, degrees of belief, and the Lockean thesis*, in: F. Huber and C. Schmidt-Petri, editors, *Degrees of Belief*, Synthese Library **342**, Springer, 2009 pp. 37–47.

[17] Gärdenfors, P., *Qualitative probability as an intensional logic*, Journal of Philosophical Logic **4** (1975), pp. 171–185.

[18] Hakli, R. and S. Negri, *Does the deduction theorem fail for modal logic?*, Synthese **187** (2012), pp. 849–867.

[19] Hansen, H. H., C. Kupke and E. Pacuit, *Neighbourhood structures: Bisimilarity and basic model theory*, Logical Methods in Computer Science **5** (2009), pp. 1–38.

[20] Herzig, A., *Modal probability, belief, and actions*, Fundamenta Informaticae **27** (2003), pp. 323–344.

[21] Jongh, D. d. and S. Ghosh, *Comparing strengths of belief explicitly*, Logic Journal of the IGPL **21** (2013), pp. 488–514.

[22] Knight, F. H., "Risk, Uncertainty, and Profit," Hart, Schaffner & Marx; Houghton Mifflin Company, Boston, MA, 1921.

[23] Kooi, B. P., "Knowledge, Chance, and Change," Ph.D. thesis, Groningen University (2003).

[24] Kraft, C. H., J. W. Pratt and A. Seidenberg, *Intuitive probability on finite sets*, The Annals of Mathematical Statistics **30** (1959), pp. 408–419.

[25] Kyburg, H., "Probability and the Logic of Rational Belief," Wesleyan University Press, Middletown, CT, 1961.

[26] Lenzen, W., "Glauben, Wissen und Wahrscheinlichkeit — Systeme der epistemischen Logik," Springer Verlag, Wien & New York, 1980.

[27] Matoušek, J. and J. Nešetřil, "An Invitation to Discrete Mathematics," Oxford University Press, 2008, second edition.

[28] Narens, L., "Theories of Probability," World Scientific, 2007.

[29] Scott, D., *Measurement structures and linear equalities*, Journal of Mathematical Psychology **1** (1964), pp. 233–247.

[30] Segerberg, K., *Qualitative probability in a modal setting*, in: J. Fenstad, editor, *Proceedings of the 2nd Scandinavian Logic Symposium* (1971), pp. 341–352.

[31] Walley, P. and T. Fine, *Varieties of modal (classificatory) and comparative probability*, Synthese **41** (1979), pp. 321–374.

www.ingramcontent.com/pod-product-compliance
Lightning Source LLC
Chambersburg PA
CBHW071308150426
43191CB00007B/542
* 9 7 8 1 8 4 8 9 0 2 0 1 5 *